MW00838213

Translational Control
in Biology and Medicine

COLD SPRING HARBOR MONOGRAPH SERIES

Translational Control
in Biology and Medicine

EDITED BY

Michael B. Mathews

New Jersey Medical School
University of Medicine and Dentistry of New Jersey, Newark

Nahum Sonenberg

McGill University, Montréal

John W.B. Hershey

University of California, Davis

COLD SPRING HARBOR LABORATORY PRESS
Cold Spring Harbor, New York

Translational Control in Biology and Medicine

Monograph 48
© 2007 by Cold Spring Harbor Laboratory Press

Printed in the United States

Publisher	John Inglis	**Production Editor**	Patricia Barker
Acquisitions Editors	John Inglis and	**Compositor**	Compset, Inc.
	Alexander Gann	**Desktop Editor**	Lauren Heller
Production Manager	Denise Weiss	**Cover Designer**	Ed Atkeson
Project Coordinator	Joan Ebert		

Front Cover Artwork: The cover art evokes the breadth of the translation control field with the structure of a bacterial ribosome superimposed on a network of rat hippocampal neurons. The neurons were grown in tissue culture and immunostained for a somatodendritic marker, MAP2. In the cross section of the 70S ribosome, the small subunit is below (16S rRNA in *blue*, 30S proteins in *purple*) and the large subunit above (23S rRNA in *salmon*, 5S rRNA in *gray-blue*, 50S proteins in *magenta*). The peptidyl-tRNA (*orange*) is at the subunit interface, with the nascent polypeptide chain (*light blue*) modeled as an α-helix threaded through the polypeptide exit tunnel in the large subunit. The mRNA (*red*, *yellow*, and *green* segments) can be seen wrapped around the neck of the small subunit (see Yusupov et al. [2001] *Science* **292:** 883–896). Images were kindly provided by Paul De Koninck (Université Laval, Québec) and Harry Noller (University of California, Santa Cruz).

Library of Congress Cataloging-in-Publication Data

Translational control in biology and medicine / edited by Michael B.
   Mathews, Nahum Sonenberg, John W.B. Hershey.
     p. ; cm. -- (Cold Spring Harbor monograph series ; monograph 48)
   Rev. ed. of: Translational control of gene expression / edited by
Nahum Sonenberg, John W.B. Hershey, Michael B. Mathews. 2nd ed.
c2000.
   Includes bibliographical references and index.
   ISBN-13: 978-0-87969-767-9 (hardcover : alk. paper)
   1. Genetic translation. 2. Genetic regulation.   I. Mathews, Michael B.
II. Sonenberg, Nahum. III. Hershey, John W. B. IV. Translational
control of gene expression. V. Series: Cold Spring Harbor monograph
series ; 48.
   [DNLM: 1. Protein Biosynthesis.   QU 475 T772 2007]
QH450.5.T725 2007
572'.645--dc22

                                                                     2006030240

10   9   8   7   6   5   4   3   2   1

All Cold Spring Harbor Laboratory Press publications may be ordered directly from Cold Spring Harbor Laboratory Press, 500 Sunnyside Boulevard, Woodbury, New York 11797-2924. Phone: 1-800-843-4388 in Continental U.S. and Canada. All other locations: (516) 422-4100. FAX: (516) 422-4097. E-mail: cshpress@cshl.edu. For a complete catalog of Cold Spring Harbor Laboratory Press publications, visit our World Wide Web Site http://www.cshlpress.com/

Contents

Preface

THE OBJECTIVE OF THIS BOOK, like its precursors *Translational Control* (1996) and *Translational Control of Gene Expression* (2000), is to provide a comprehensive, up-to-date, and readable survey of the translational control field. The field is broad and expanding rapidly in all directions. Yet the publishers were at pains to impress upon us the need to keep the book within reasonable bounds—a precept that was easy to accept in principle but much thornier in practice. How to resolve the quandary? We decided that the book would embrace three themes.

As unabashedly conveyed by the title *Translational Control in Biology and Medicine*, the first theme emphasizes the engagement of this discipline in systems and processes at the cutting edge of biomedical research. Several chapters in the book discuss the impact that has been made on long-standing problems in diverse areas. These include learning and memory, embryonic development, and human diseases such as cancer, diabetes, and obesity, and disorders due to mitochondrial dysfunction or viral infection. Numerous antibiotics and toxins are known to target the translation system, and research and development efforts are under way to discover new drugs for further therapeutic uses. In one sense, this theme represents the fruits of efforts to apply the understanding of basic scientific principles to practical matters (aptly named translational research!); in another, it reflects a natural maturation of research into the mechanism and control of protein synthesis from some of its historical roots.

The second theme explores fundamental mechanisms and processes related to protein synthesis. Central to all progress in the field is our deepening appreciation of the translation system itself, as manifested vividly on the front cover of the book. Ribosome structures are revealing its actions and interactions in ever-increasing detail. Chapters in the book review this area, the mechanisms of protein synthesis, the structure and functions of translation factors, and the dynamic interplay among components of the translation system. Other chapters are devoted to cellular mechanisms such as apoptosis, signal transduction, and mRNA turnover, which have intimate relationships with the translation system. The field of translational control both benefits from and furthers our understanding of these critical physiological processes.

Last, but not least, our third theme encompasses some new, underrepresented and fast-growing areas. Topics accommodated in this motley category include cellular IRESs, microRNAs, mRNA localization, plants, and riboswitches. We anticipate that many of the most exciting advances will emerge from research in these areas, illuminating the regulation of gene expression across the biosphere. Of course, readers with access to previous monographs in this series can have some fun checking our success in spotting trends in the field.

While it is aimed to meet the expectations of the discriminating specialist, the book is intended to provide the newcomer with a readily intelligible entrée to the field. All of the chapters are completely new or have been extensively rewritten and updated by their authors, many of whom are themselves contributing for the first time. Most chapters do not have direct counterparts in previous monographs; those that do are radically changed in scope and emphasis, and even the introductory chapter on the origins and principles of translational control has been revised and expanded. Given the eclectic nature of the topics covered, we doubt that many will choose to read this volume from cover to cover, and certainly not in sequence, but we are confident that there is much to be gained from at least dipping into each chapter.

We are indebted to all of the authors, who have striven to write chapters that are stimulating, edifying, and authoritative. Without their expertise, there would be no book. At the Cold Spring Harbor Laboratory Press, thanks go to John Inglis who conceived the idea for the original monograph, to Alex Gann, Denise Weiss, Pat Barker, and Jan Argentine for their help and encouragement with this volume, and especially to Joan Ebert, who kept the project moving with skill and good humor. Finally, we are grateful to Marvin Wickens (University of Wisconsin) for cover art suggestions, to Harry Noller (University of California, Santa Cruz) and Paul De Koninck (Université Laval) for images, and to Cheyenne Moorman (New Jersey Medical School) for compiling the montage.

MICHAEL B. MATHEWS
NAHUM SONENBERG
JOHN W. B. HERSHEY
October, 2006

1

Origins and Principles of Translational Control

Michael B. Mathews
Department of Biochemistry and Molecular Biology
New Jersey Medical School
University of Medicine and Dentistry of New Jersey
Newark, New Jersey 07103

Nahum Sonenberg
Department of Biochemistry
McGill University, Montreal
Quebec H3G 1Y6, Canada

John W.B. Hershey
Department of Biochemistry and Molecular Medicine
University of California, School of Medicine
Davis, California 95616

PROTEINS OCCUPY A POSITION HIGH ON THE LIST of molecules important for life processes. They account for a large fraction of biological macromolecules—about 44% of the human body's dry weight, for example (Davidson et al. 1973)—they catalyze most of the reactions on which life depends, and they serve numerous structural, transport, regulatory, and other roles in all organisms. Accordingly, a large proportion of the cell's resources is devoted to translation. The magnitude of this commitment can be appreciated in genetic, biochemical, and cell biological terms.

Translation is a sophisticated process requiring extensive biological machinery. One way to gauge the amount of genetic information needed to assemble the protein synthetic machinery is to compile a "parts list" of essential proteins and RNAs. Analyses of the genomes of several microorganisms have converged on similar estimates (Hutchison et al.

1999; Tamas et al. 2002; Kobayashi et al. 2003; Waters et al. 2003). These organisms get by with about 130 genes for components of the translation machinery, including about 90 protein-coding genes (specifying 50–60 ribosomal proteins, about 20 aminoacyl-tRNA synthetases, and 10–15 translation factors) and about 40 genes for ribosomal and transfer RNAs (rRNA and tRNAs). A somewhat larger number of genes are involved in eukaryotes, which have more ribosomal proteins and initiation factors, for example. Discounting genes that are dispensable for growth in the laboratory, it can be calculated that approximately 40% of the genes in a theoretical minimal cellular genome are devoted to the translation apparatus.

This heavy genomic commitment is matched by the high proportion of a cell's energy budget and components that are devoted to translation. Protein synthesis consumes 5% of the human caloric intake but as much as 30–50% of the energy generated by rapidly growing *Escherichia coli* (Meisenberg and Simmons 1998). A portion of this is accounted for by the substantial input of energy required during translation itself (4 high-energy bonds per peptide bond or ~28 kcal/mole, plus additional consumption for initiation and termination). Extensive resources are invested in the ribosomes, tRNAs, and enzymes required for making proteins. A rapidly growing yeast cell, for example, contains nearly 200,000 ribosomes occupying as much as 30–40% of its cytoplasmic volume (Warner 1999). Growth alone demands that the yeast cell produce 2000 ribosomes/min, an operation which absorbs about 60% of its transcriptional activity in manufacturing rRNA, as well as a large fraction of its translational capacity, since ribosomal protein messenger RNAs (mRNAs) account for almost one-third of the cell's mRNA population (Warner 1999).

It would be surprising if a process of such importance were not closely monitored and regulated. In this chapter, we review the origins, mechanisms, and targets of translational control, a topic that impinges on biological fields as varied as medicine, agriculture, and biotechnology.

ORIGINS OF TRANSLATIONAL CONTROL

The central idea of translational control is that gene expression is regulated by the efficiency of utilization of mRNA in specifying protein synthesis. This notion emerged only a few years after the articulation of the central dogma of molecular biology (Crick 1958) and very soon after the formulation of the messenger hypothesis. In 1961, Jacob and Monod perceived that "the synthesis of individual proteins may be provoked or suppressed within a cell, under the influence of specific external agents, and . . . the relative rates at which different proteins are synthesized may be profoundly altered, depending on external conditions." They pointed out that such regulation "is absolutely essential to the survival of the cell," and went on to advance the concept of an unstable RNA intermediary between gene and protein as a key feature of their elegant model for transcriptional control (Jacob and Monod 1961). The idea that this mRNA could be subject to differential utilization depending on the circumstances was accorded scant attention at the time, but it was taken up enthusiastically by workers in related fields, to the extent that 10 years later, one writer could allude to the "now classical conclusion" that eggs contain translationally silent mRNA that is activated upon fertilization (Humphreys 1971).

The term Translational Control was certainly in use as early as 1968, by which date at least four clearly distinct exemplars had been recognized and were already coming under mechanistic scrutiny. The groundwork for these four paradigms—developing embryos, reticulocytes, virus- and phage-infected cells, and higher cells responding to stimuli ranging from heat to hormones and starvation to mitosis—had all been laid by the middle of the 1960s. They founded a thriving and expanding field of study that has advanced from its largely eukaryotic origins to embrace bacteria (although not yet the archaea, as far as we are aware).

Early History of Translation

The genesis of the translational control field took place at a time when studies of the translation system itself were in their infancy; many (although not all) of the reactions had been observed, but most of the components were not yet characterized and mechanistic details were essentially unknown. To place the origins of translational control in context, we briefly outline the development of protein synthesis.

Biochemical investigations of the process began in the 1950s, at the same time as the concept of proteins as unique, nonrandom linear arrays of just 20 amino acid residues was solidifying (Sanger and Tuppy published the first protein sequence, that of the insulin B chain, in 1951; Sanger and Tuppy 1951). Radioactive isotopes had begun to revolutionize many areas of biomedical science in the late 1940s, and labeled amino acids came into use as tracers around 1950. Initially, the radiolabeled amino acids had to be synthesized from simple labeled compounds such as formaldehyde or cyanide by the researchers themselves as a first step in their experiments (see, e.g., Borsook et al. 1950; Levine and Tarver 1950), but they became commercially available in the latter part of the decade. Enabled by this profound technical advance, biochemistry ran ahead of genetics, as it continued to do in this field until the advent of cloning and the systematic exploitation of the yeast system, which began to make their mark in the 1980s.

Siekevitz and Zamecnik (1951) produced a cell-free preparation from rat liver that incorporated amino acids into protein, showing that energy was required in the form of ATP and GTP. The system was refined by stages and resolved into subfractions including a microsomal fraction that contained ribosomes attached to fragments of intracellular membrane (for review, see Zamecnik 1960). Elegant pulse-chase experiments demonstrated that the ribosomes are the site of protein synthesis, not an easy task in bacterial cells where protein synthesis was found to be very rapid: The assembly of a protein chain on a ribosome was estimated to take only about 5 seconds (McQuillen et al. 1959). It is salutary to recall that this was accomplished in advance of an understanding of the central role of RNA in the flow of genetic information to protein, well before the first RNA sequence was completed (Holley et al. 1965), and in an era when theories of protein synthesis via enzyme assembly and peptide intermediates were entertained along with template theories (Campbell and Work 1953). However, it was not until the early 1960s that polysomes were observed and their function appreciated in light of the messenger hypothesis (Marks et al. 1962; Warner et al. 1963). Technical advances in electron microscopy and high-speed centrifugation made indispensable contributions during this phase of the field's development.

At much the same time, the role of aminoacyl-tRNA was being established. The existence of an intermediate, activated amino acid state was detected (Hultin and Beskow 1956) and characterized (Hoagland et al. 1958), then understood as the physical manifestation of the adaptor RNA predicted on theoretical grounds (Crick 1958). Once its function had been realized, the name transfer RNA rapidly displaced the original term, "soluble" RNA (sRNA). Later, chemical modification of the amino acid moiety of a charged tRNA confirmed that it is the RNA component that decodes the template (Chapeville et al. 1962). Thus, responsibility for the fidelity of information transfer from nucleic acid to protein rests in part on the aminoacyl-tRNA synthetases, which became the first macromolecular component of the protein synthetic apparatus to be purified (Berg and Ofengand 1958). These, together with the other enzymes, or protein "factors" as they became known, were steadily characterized and purified such that nearly all of the protein components have been known for more than 20 years. Yet, the activities of some factors remain obscure (e.g., EFP and its homolog eIF5A; Kang and Hershey 1994; Aoki et al. 1997) while others are still emerging (e.g., eIF2A; Komar et al. 2005; Ventoso et al. 2006). Even today there is no certainty that the full complement of protein factors involved in translation has been identified.

It was genetics rather than biochemistry that supplied the missing cornerstone of the protein synthetic system, mRNA. According to the messenger hypothesis, the ribosomes and other components of the protein synthesis machinery constitute a relatively stable decoding and synthetic apparatus that is programmed by an unstable template (Jacob and Monod 1961). This insight soon received confirmation in bacteria (Brenner et al. 1961; Gros et al. 1961) and in bacterial cell-free systems. The discovery that poly(U) can direct the synthesis of polyphenylalanine (Nirenberg and Matthaei 1961) was particularly fruitful, greatly speeding the elucidation of the genetic code by the mid-1960s. Because of the greater stability of most eukaryotic mRNAs, the applicability of the messenger hypothesis to higher cells was less readily apparent. Nonetheless, the existence of a class of rapidly labeled RNA, heterogeneous in size and with distinct chromatographic properties, was recognized. Its essential features as informational intermediary were confirmed and it was universally accepted several years before the discovery in the early 1970s of 5' caps and 3' poly(A) tails, the modern hallmarks of eukaryotic mRNAs (apart from those histone mRNAs that lack poly(A) and some viral mRNAs that lack one or even both of these modifications). The mRNA concept immediately revolutionized thinking about gene expression in all cells.

To appreciate the pace at which protein synthesis advanced during the decade of the 1960s, it is instructive to compare the Cold Spring Harbor Symposium volume of 1962 (on Cellular Regulatory Mechanisms) with that of 1970, a much thicker book devoted to a narrower topic (the Mechanism of Protein Synthesis). By the end of the decade, much of the translational apparatus had been characterized (although much remained to be done), many problems of regulation had been laid out, and translational control came to receive increasing attention.

General Features of Translational Control

In a multistep, multifactorial pathway like that of protein synthesis, regulation can be exerted at many levels. Examples of translational control are indeed found at different levels, but the overwhelming preponderance of known instances—including all of the earliest cases recognized—is at the level of initiation. This empirical observation conforms to the biological (and logical) principle that it is more efficient to govern a pathway at its outset than to interrupt it in midstream and have to deal with the resultant logjam of recyclable components and the accumulation of intermediates as by-products. Nevertheless, well-characterized cases do occur at later steps in the translational pathway, especially at the elongation level, where it seems that a translational block may be imposed as a safety measure to halt further peptide bond formation.

One of the chief virtues of translation as a site of regulation is that it offers the possibility of rapid response to external stimuli without invoking nuclear pathways for mRNA synthesis, processing, and transport. Predictably, the first cases to be recognized were mostly in eukaryotes and were those in which it was either self-evident or relatively simple to establish that transcription and other nuclear events were not responsible. By the same token, the relative scarcity of prokaryotic examples and their generally later recognition can be largely attributed to the lack of a nuclear barrier between the sites of mRNA synthesis and translation. The greater speed of macromolecular synthesis in bacteria and their lesser dependence on mRNA processing are other factors. These circumstances allow a coupling of transcription and translation that all but obviates the need for translational control. That it occurs at all in bacteria is due to the exigencies of particular circumstances and to the potency of translational control mechanisms.

The earliest cases of translational control to be explored in depth, in fertilized invertebrate eggs and mammalian reticulocytes, were those in which the departure from the transcription-based regulatory model was the most obvious and extreme. Protein synthesis is abruptly turned on (in fertilized eggs) and off (in iron-starved reticulocytes) in the absence of ongoing transcription. A further distinction which made it easier to define and study these two particular cases is that their regulation is apparently indiscriminate in that it affects protein synthesis generically, rather than the synthesis of specific proteins. Not all translational controls are of this type, however, as evidenced early on during studies of phage-infected bacteria. A distinction is often drawn between global and selective controls, sometimes referred to, rather misleadingly, as quantitative and qualitative controls.

Global controls, such as those operating in eggs and reticulocytes, affect the entire complement of mRNAs within a cell, switching their translation on or off or modulating it by degrees in unison. This kind of regulation is usually implemented by substantial alteration in the activity of general components of the protein synthesis machinery that act in a nonspecific manner. Selective controls, on the other hand, affect a subset of the mRNAs within a cell, in the extreme case a single species only. This can be accomplished through mechanisms that target ligands to individual mRNAs or classes of mRNAs, or by exploiting the differential sensitivity of mRNAs to more subtle changes in the activity of general components of the translation system, e.g., eIF4E (Chapters 14, 15, 16, and 20) or eIF2 (Chapters 9, 13, 16, and 20). Although examples of all these exist and are discussed at length in this volume, in the context of the historical origins of translational control, it should come as no surprise that the earliest examples were mainly of the global variety and that (with notable exceptions) definitive evidence in favor of selective translational control accumulated more slowly.

Early Paradigms of Translational Control

In large part, the origins of translational control can be traced to studies of four early examples. These are described below, followed by an example involving elongation control.

Sea Urchin Eggs

The eggs of sea urchins and other invertebrates synthesize protein at a very low rate but are triggered to incorporate amino acids within a few minutes of fertilization with little or no concomitant RNA synthesis (Hultin 1961; Nemer 1962; Gross et al. 1964). The first wave of increased translation, which lasts for several hours, is not blocked by actinomycin D (Gross et al. 1964) because the eggs contain preexisting mRNA in a masked form that is not translated until a stimulus dependent on fertilization is received. In principle, the limitation could be due to a deficiency in the translational machinery, but there is little evidence to support this possibility (Humphreys 1969). On the other hand, egg ribosomes are able to translate added poly(U) even though they display little intrinsic protein synthetic activity (Nemer 1962; Wilt and Hultin 1962). The deproteinized egg RNA can be translated in a cell-free system (Maggio et al. 1964; Monroy et al. 1965), and cytoplasmic messenger ribonucleoprotein (mRNP) particles have been observed (Spirin and Nemer 1965). Because the assembly of masked mRNP complexes must take place during oogenesis, the sea urchin system exemplifies a reversible process of mRNA repression and activation. Recent developments in this active area of research are discussed in Chapter 19.

Mammalian Reticulocytes

Because they were enucleate, it was taken for granted that protein synthesis in these immature red cells—mainly hemoglobin—would be regulated at the translational level. In the intact rabbit reticulocyte, the synthesis of heme parallels that of globin (Kruh and Borsook 1956), and globin synthesis is controlled by the availability of heme or of ferrous ions (Bruns and London 1965). Regulation by heme occurs in the highly active unfractionated reticulocyte lysate translation system (Lamfrom and Knopf 1964), the forerunner of the widely used messenger-dependent system (Pelham and Jackson 1976) and of commercially available coupled transcription–translation systems. When globin synthesis is inhibited in cells or extracts, the polysomes dissociate to monosomes (Hardesty et al. 1963; Waxman and Rabinowitz 1966), arguing that regulation affects translation initiation. The effects of heme deprivation are mediated by the protein kinase HRI (heme-regulated inhibitor) and are mimicked by unrelated stimuli such as double-stranded RNA (dsRNA) and oxidized glutathione (Ehrenfeld and Hunt 1971; Kosower et al. 1971). They extend to all mRNAs in the reticulocyte lysate (Mathews et al. 1973), implying that a general mechanism of translational control is being invoked. This mechanism centers on the phosphorylation of the α-subunit of initiation factor eIF2, which results in reduced levels of ternary complex (eIF2:GTP:Met-tRNA$_i$) and impaired loading of the 40S ribosomal subunit with Met-tRNA$_i$ (Farrell et al. 1977). Considerable attention has been given to the family of eIF2 kinases, which confer sensitivity to a wide range of stimuli. HRI, PKR, GCN2, and PERK are activated by heme deprivation, structured RNA, uncharged tRNA, and endoplasmic reticulum stress, respectively, inter alia, whereas PKZ, recently found in fish, is potentially regulated by Z-DNA (for review, see Chapter 12).

Virus-infected Cells

Translation of cellular mRNAs is suppressed during infection with many viruses (Chapter 20). This inhibition may begin before the onset of viral protein synthesis and without any apparent interference with cellular mRNA production or stability. In polioviral infection, for example, the shutoff of host-cell translation can be complete within 2 hours after infection and is followed by a wave of viral protein synthesis (Summers et al. 1965). In the first phase, polysomes break down without any effect on translation elongation or termination (Penman and Summers 1965; Summers and Maizel 1967). In the second phase, virus-specific polysomes form (Penman et al. 1963). Cellular mRNA remains intact and translatable in a cell-free system (Leibowitz and Penman 1971), evidence that initiation has become selective for viral mRNA. Translational inhibition extends to mRNAs produced by several other viruses introduced together with poliovirus in a double infection (Ehrenfeld and Lund 1977), indicative of a general effect that later work ascribed to modification of the cap-binding complex, eIF4F. Cleavage of the eIF4G subunit of this complex prevents cap-dependent initiation on cellular mRNAs but does not interfere with initiation on the viral mRNA which occurs by internal ribosome entry (Chapter 20).

Bacteriophage f2 provided the first evidence for prokaryotic translational control, as well as the first clear case of mechanisms specific for the synthesis of individual protein species. The phage RNA genome encodes four polypeptides, the maturation protein, coat protein, lysis protein, and replicase, that are initiated individually but produced at dissimilar rates. Several regulatory interactions among them are now known. One was revealed by the observation that a nonsense mutation early in the cistron coding for phage coat protein down-regulates replicase synthesis (Lodish and Zinder 1966). Apparently, passage of ribosomes through a critical region of the coat protein cistron is required to melt long-range RNA structure and allow replicase translation. In contrast, a second nonsense mutation leads to overproduction of the replicase because the coat protein acts as a repressor of replicase translation. The binding of phage coat protein to the hairpin structure containing the replicase AUG is now one of the best-characterized RNA–protein interactions (Witherell et al. 1991). Subsequent studies have disclosed translational control mechanisms in the DNA phages as well as in bacterial genes themselves (Chapter 28), but it was eukaryotic systems that made most of the early running.

Physiological Stimuli

The cells and tissues of higher organisms regulate the expression of individual genes or of whole classes of genes at the translational level in response to a wide variety of stimuli or conditions. Early examples include cell state changes, such as mitosis (Steward et al. 1968; Hodge et al. 1969; Fan and Penman 1970) and differentiation (Heywood 1970); stress resulting from heat shock (McCormick and Penman 1969), treatment with noxious substances, or the incorporation of amino acid analogs (Thomas and Mathews 1984); and normal cellular responses to ions (Drysdale and Munro 1965) and hormones (Eboué-Bonis et al. 1963; Garren et al. 1964; Martin and Young 1965; Tomkins et al. 1965). These reports strengthened the view that translational control is widespread and important even though in some cases the trail has gone cold or been erased upon further investigation. Proving that control is being exerted at the translational level can be a challenging task in nucleated cells, let alone in a tissue or whole organism, and this constituted one of the chief stumbling blocks. Although several methods are available that can give rigorous evidence (described below), simpler approaches can be misleading. One popular approach took advantage of selective inhibitors of transcription or translation, such as actinomycin D and cycloheximide, but the results were liable to be complicated (if not confounded) by the drugs' side effects or indirect sequelae in complex systems. Another argument that could be made for an effect at the translational level, although not without some reservations, came from its rapidity (see below). Timing alone cannot provide definite evidence, however, and the most convincing proofs often came from investigations of the underlying biochemical processes for example, by demonstrating changes in polysome profiles or initiation factor phosphorylation states as discussed later in this chapter and in a number of chapters in this volume (see, e.g., Chapters 13, 14, 17, and 20). The goal is to achieve an understanding of the regulatory mechanisms set in train by the stimuli applied, and within this wide array of phenomena lie many of the challenges for the future.

Secretory Pathway

One of the best-studied examples of regulation during the elongation phase is found in the synthesis of proteins that are destined for secretion or for a life within a cellular membrane (for review, see Chapter 21). Most such proteins are made on polysomes that are attached to the endoplasmic reticulum (ER), isolated from cellular homogenates in the form of microsomes. In the early 1970s, it began to seem likely that ribosomes become associated with cell membranes only after protein synthesis has been initiated (Lisowska-Bernstein et al. 1970; Rosbash 1972), and the existence of what came to be called a signal peptide was reported on secreted proteins (Milstein et al. 1972; Devilliers-Thiery et al. 1975). These findings lent substance to the signal hypothesis which proposed that an amino-terminal sequence might ensure secretion (Blobel and Sabatini 1971). The development of cell-free systems enabled the biochemical dissection of the secretory pathway (Blobel and Dobberstein 1975) and led to the discovery of the signal recognition particle (SRP). This RNP particle interacts with the signal peptide, the ribosome, and the ER. Binding of the SRP to a nascent signal peptide protruding from the ribosome causes translational arrest in the absence of cell membranes (Walter and Blobel 1981). The elongation block is relieved when the ribosome docks with its receptor on the ER, allowing the protein chain to be completed and simultaneously translocated into the lumen of the ER. It appears that this mechanism serves both to ensure cotranslational protein export and to prevent the accumulation of secretory proteins in an improper subcellular compartment (the cytosol). A similar rationale may account for control at the elongation level during heat shock to prevent the synthesis of improperly folded proteins (Theodorakis et al. 1988). Thus, elongation blocks might be used under exceptional circumstances to preserve cellular integrity when it is threatened by the production of protein at the wrong time or in the wrong place, or perhaps in the event of a sudden shortage of energy or an essential metabolite.

WHAT LIMITS PROTEIN SYNTHESIS IN PRINCIPLE?

A major goal of the regulation of gene expression is to determine the levels of proteins in the cell. A protein's level is proportional to its rate of synthesis and inversely proportional to its rate of degradation (ignoring dilution through cell division). This volume is concerned with how changes in the rate of protein synthesis affect the cellular level of proteins. For example, when the rate of protein synthesis is stimulated twofold, the protein level will increase twofold if the degradation rate constant is unaffected. A critical concern is how much time is required to reach the higher protein level. The rate of approach to the new equilibrium level is equal to the first-order rate of degradation, as shown in Figure 1. Therefore, in cells where the regulation of a protein's level must be accomplished rapidly, such proteins have a high turnover rate. Regulation of protein levels by affecting protein degradation, although frequently encountered in eukaryotic cells, is outside the scope of this volume.

Given that translational controls are so widespread in eukaryotic cells, it is appropriate to examine the fundamental principles on which these controls are based. Translational control is defined as a change in the rate (efficiency) of translation of one or more mRNAs; i.e., in the number of completed protein products per mRNA per unit time. It is generally believed that during protein synthesis, the number of protein chains initiated is about the same as the number of proteins completed; in other words, few nascent polypeptides abort and fall off the ribosome (Tsung et al. 1989), even when elongation is stalled (Fang et al. 2004). However, it was also reported that a significant fraction of nascent misfolded or degradation-targeted proteins are degraded cotranslationally (Turner and Varshavsky 2000), which could lead to an underestimation

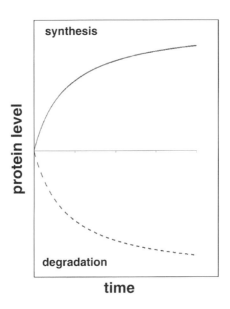

Figure 1. Rate of reaching a new protein level. The lower half of the figure shows the rate of first-order decay of a hypothetical protein. The upper half shows the rate that the protein's level increases following a doubling of its synthesis rate while the first-order rate constant for degradation is unchanged.

of the number of initiation events. Despite these caveats, it is reasonable to conclude that under steady-state conditions, the number of protein products produced approximates the number of initiation events. It follows logically that the rate of protein synthesis is determined by the number of initiation events per unit of time; i.e., the rate of initiation. What determines the number of initiation events per unit time? Four major parameters influence or define the rate of global protein synthesis. Each is considered briefly below.

Amount or Efficiency of mRNAs

The level of mRNA in the cytoplasm is determined by the rate of transcription, the proportion of primary transcripts that are processed and transported into the cytoplasm as mature mRNA, and the degradation rate of mature mRNAs in the cytoplasm. In actively translating mammalian cells, mRNAs often are found entirely in polysomes, as shown for actin (Endo and Nadal-Ginard 1987); thus, the rates of synthesis of such specific proteins are mRNA-limited. However, total mRNA in the cytoplasm frequently appears to be present in excess, with about 30% of the mRNA in cultured cells present as free mRNP particles (Geoghegan et al. 1979; Kinniburgh et al. 1979; Ouellette et al. 1982). Therefore, the level of mRNA appears not to limit the overall number of translational initiation events in these cells. In cells exhibiting low translational activity, many mRNAs are repressed and apparently unavailable to the translational apparatus (masked), as seen most dramatically in oocytes and unfertilized eggs as described above, but also in somatic cells in culture (Lee and Engelhardt 1979). Such repression sometimes appears to be "all or none," as some mRNAs are distributed bimodally in polysome profiles; a fraction of the specific mRNA is completely repressed (nontranslating mRNP particles), whereas a portion is actively translated as large polysomes (Yenofsky et al. 1982; Agrawal and Bowman 1987). The mRNAs encoding ribosomal proteins are prime examples that exhibit these characteristics (Meyuhas and Hornstein 2000). In instances of specific regulation of protein synthesis, mRNA repression and availability to the translational apparatus likely have a dominant role, for example, in the translation of ferritin mRNA (Rouault and Harford 2000; Chapter 10) and ribosomal protein mRNAs (Meyuhas and Hornstein 2000). Furthermore, individual activated mRNAs differ greatly in their efficiencies of translation as deduced from polysome sizes, thereby contributing to regulation of gene expression. These innate efficiencies are determined in large part by the primary and higher-order structures of the mRNAs (Chapter 4).

Abundance of Ribosomes

The cellular levels of ribosomes may be rate-limiting under some circumstances. Cells active in protein synthesis, for example, liver cells in newly fed rats, engage 90–95% of their ribosomes in protein synthesis (Henshaw et al. 1971), suggesting that still higher rates of protein synthesis might have been possible were there a greater number of ribosomes. On the other hand, in translationally repressed cells, such as liver cells from fasted rats (Henshaw et al. 1971) or in quiescent cells in culture (Duncan and McConkey 1982; Meyuhas et al. 1987), fewer than half of the ribosomes may be actively translating mRNAs. The level of ribosomes surely is not limiting in these cells, since a rapid increase in the rate of protein synthesis can be induced within 20 minutes, before the assembly of more ribosomes is possible (Duncan and McConkey 1982). In many mammalian tissues and in some rapidly proliferating cells in culture, nontranslating ribosomes and their subunits constitute greater than 20% of the total ribosome population, as seen by analysis of cell or tissue lysates by sucrose gradient centrifugation (Penman et al. 1963). This suggests that, in general, ribosome levels are not limiting for protein synthesis, thereby enabling the cells to rapidly activate translation in response to external signals.

Activity of the Protein Synthesis Machinery

In cells not limited by ribosome and mRNA levels, the rate of protein synthesis may nevertheless be reduced when a different translational component (e.g., a soluble factor) is limiting in amount or if one or more components have reduced specific activities. Such regulation frequently involves the phosphorylation status of translational components, as detailed in numerous chapters in this book. A priori, regulation of the overall activity of the translational apparatus is expected to affect the translation of essentially all mRNAs. However, as argued earlier by Lodish (1976), down-regulation of the initiation steps is expected to lead to greater inhibition of those mRNAs whose initiation rate constants are relatively low ("weak" mRNAs), as compared to "strong" mRNAs. Reciprocally, activation of initiation may stimulate more greatly the translation of weak mRNAs. Alteration of the activities of components that interact with mRNAs and affect their binding to ribosomes would be expected to generate differential effects on the translation of the mRNA population (Godefroy-Colburn and Thach 1981). The mechanisms affecting mRNA binding and differences in the translational efficiency of specific mRNAs are reviewed in many chapters in this volume.

Rate of Elongation

The initiation rate on an mRNA can be inhibited if a ribosome, having already initiated, vacates the initiation region too slowly. A ribosome bound at the AUG initiation codon occupies about 12–15 nucleotides (4–5 codons) downstream from the AUG and about 20 nucleotides upstream. Another ribosome can occupy the initiation site only after the first ribosome has moved about seven codons down the mRNA. When the time needed to vacate the initiation region approaches or exceeds the time required for initiation, the elongation rate becomes limiting. In general, it is believed that the elongation rate is similar for most mRNAs, with 3–8 amino acids incorporated per second per ribosome in eukaryotes, and even faster in bacteria (e.g., 15 residues for β-galactosidase), because measurements of a few specific examples gave similar results in this range (Lodish and Jacobsen 1972; Palmiter 1974).

Nevertheless, the rate of elongation may be much slower than normal, as is seen with the mRNAs encoding eIF2β (Chiorini et al. 1993) and reovirus σ1 (Fajardo and Shatkin 1990). Furthermore, the elongation rate may not be uniform throughout the coding region of an mRNA, as pausing may occur at specific locations, possibly due to the occurrence of rare codons or RNA secondary structure (Wolin and Walter 1988). If ribosome pausing occurs such that it impedes initiation, mRNA efficiency is decreased. The question of which translation phase is rate-limiting, initiation or elongation/termination, is addressed in greater detail below.

WHICH PHASE OF PROTEIN SYNTHESIS IS RATE-LIMITING AND REGULATED?

The analysis above identifies three ways in which the rate of protein synthesis may be limited and thus regulated over a relatively short time period (on the order of minutes): the rate of initiation, the rate of elongation/termination, and the repression/activation of mRNAs/mRNPs. How is the rate of protein synthesis measured and how is the rate-limiting step identified? The overall rate of protein synthesis can be measured by assaying the time course of incorporation into protein of radiolabeled amino acid precursors added to the culture medium. The method is complicated mainly by the uncertainty of the specific radioactivities of the precursors within cells, as uptake rates, intracellular de novo synthesis of amino acids, and degradation of proteins may influence these values. By assaying protein synthesis rates with different amounts of exogenous radiolabeled amino acid precursors, it is possible to estimate the nonradioactive intracellular pool size and thereby correct specific activities in different cell populations (Damm 1966).

A second approach to measuring the rate of global protein synthesis in cells is to determine the elongation/termination rate and the absolute number of active ribosomes per cell. The elongation/termination rate is equal to the average size (number of amino acid residues) of synthesized proteins divided by the ribosome transit time (Fan and Penman 1970). The average size of proteins being synthesized is determined by radiolabeling and analysis of synthesized proteins by SDS–polyacrylamide gel electrophoresis (SDS-PAGE), followed by quantitation of the density of an autoradiograph as a function of molecular weight. Ribosome transit times (RTTs) reflect the time required to elongate and terminate a nascent protein following initiation. RTTs are measured by treating cells with a radiolabeled amino acid as described above, then at various times removing two aliquots for analysis. Total radiolabeled amino acid incorporation is determined in one aliquot by filtration of trichloroacetic-acid-precipitated protein (after hydrolyzing aminoacyl-tRNA by heating or treatment with NaOH). The other aliquot is subjected to centrifugation to pellet ribosomes, allowing measurement of radiolabeled amino acid incorporation into "released" proteins (supernatant fraction) and "nascent" proteins (pellet fraction). When these three values are plotted versus time, one obtains two parallel ascending lines (total and released) (Fig. 2) which are separated on the time axis by a value that corresponds to half of the ribosome transit time (Fan and Penman 1970).

The absolute number of active ribosomes per cell is determined first by measuring the total number of ribosomes, using centrifugation to pellet the ribosomes from a known number of cells, and measuring their optical density. The ribosomes in the total population that are actively engaged in protein synthesis are measured by subjecting the cell lysate to sucrose gradient centrifugation (Merrick and Hensold 2000) to yield a polysome profile (Fig. 3, left panel). Active ribosomes are those present in polysomes (mRNAs carrying more than one ribosome) plus a minor amount as monosomes (the 80S peak is a mixture of translating and nontranslating ribosomes). The optical density in the polysome region, divided by the total optical density from 40S riboso-

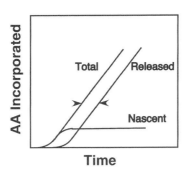

Time

Figure 2. Determination of the ribosome transit time. Following labeling of cells with [^{35}S]methionine, cells are incubated and lysed at different times. An aliquot of the lysate is subjected to hot TCA precipitation to measure incorporation of [^{35}S]methionine into protein (labeled total). A second aliquot is centrifuged to pellet ribosomes, and [^{35}S]methionine incorporation into the supernatant (released) and pellet (nascent) fractions is measured. The separation on the time axis between the total and released lines (indicated by the arrowheads) equals half of the average ribosome transit time. The method is explained in detail in Fan and Penman (1970).

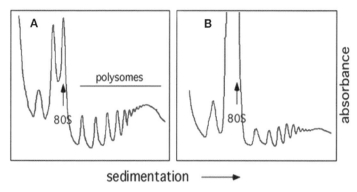

Figure 3. Analysis of translation by polysome profiles. Cell or tissue lysates are analyzed by sucrose gradient centrifugation (Merrick and Hensold 2000) followed by scanning by absorbance at 254 nm. (*A*) Profile from cells relatively active in protein synthesis; (*B*) profile from cells that have a reduced rate of initiation, indicated by the reduced number of ribosomes in the fast-sedimenting ("heavy") polysomes.

mal subunits to the bottom of the gradient, equals the percent of ribosomes that are actively engaged in translating mRNAs. The number of amino acids (aa) incorporated per unit time is calculated (Palmiter 1975) from the values above as follows:

$$\text{Rate of global protein synthesis} = \text{number of aa incorporated/time/cell}$$
$$= \text{number of active ribosomes} \times \frac{\text{number of aa in avg. protein}}{\text{ribosome transit time}}$$

This method, although more laborious than the measurement of the incorporation of labeled amino acids, is not complicated by uncertainties of amino acid transport and intracellular specific radioactivities. Both methods serve to analyze global rates of protein synthesis. In addition, the synthesis rate of a specific protein can be measured by radiolabeling proteins, followed by immunoprecipitation or high-resolution two-dimensional gel electrophoresis, to enable measurement of radioactivity in the protein of interest.

We next turn our attention to determining which phase of protein synthesis is rate-limiting, initiation or elongation/termination. Most mRNAs are thought to be limited by their initiation rate. Therefore, the question is best addressed to specific mRNAs, rather than to the whole population. One way to determine whether elongation/termination is rate-limiting for an mRNA is to treat cells with low concentrations of an elongation inhibitor such as cycloheximide or sparsomycin. If translation of the specific mRNA is limited by the elongation rate, its synthesis will be sensitive to the inhibitors of elongation. For example, a level of cycloheximide is chosen that inhibits total protein synthesis by about 5%. If an mRNA is elongation-rate-limited, it will be strongly inhibited by this level of cycloheximide, whereas if initiation is rate-limiting, it will be rather insensitive to the cycloheximide, exhibiting inhibition by only a few percent. When mRNAs encoding β-globin (Lodish and Jacobsen 1972) or reovirus proteins (Walden et al. 1981) are analyzed in this way, it is clear that initiation is the sensitive step. Because the majority of mRNAs in cells are resistant to low concentrations of cycloheximide, it is thought that the translation of most cellular mRNAs is limited at the initiation phase.

Insight into which phase of protein synthesis is rate-limiting is gained by an examination of polysome profiles (described above). The time-honored technique of polysome profile analysis remains one of the most flexible and powerful tools for diagnosing and investigating translational control. The size of a polysome (number of ribosomes per mRNA) is determined by three parameters: It is proportional to the coding length of the mRNA and to the rate of initiation and is inversely proportional to the rate of elongation/termination. On the average, ribosomes in native polysomes occur once every 80–100 nucleotides. For example, the average polysome size for globins is about five ribosomes per mRNA, or one ribosome per 90 nucleotides over the coding region. When protein synthesis is inhibited by cycloheximide such that elongation becomes rate-limiting, polysomes increase in size (e.g., to more than 12 ribosomes per globin mRNA). Therefore, polysome densities of one ribosome per 30–40 nucleotides are possible. This approaches the limit for close packing, since a ribosome occupies about 30 nucleotides of mRNA. That average polysome densities in cells are much less is due to the relatively low rate of initiation versus elongation/termination. Under most physiological circumstances when the rate of global protein synthesis is reduced, polysomes become smaller if the inhibition affects the initiation rate (Fig. 3, right panel), whereas polysomes remain large or become even larger if the elongation rate is affected.

Polysome analysis is widely used to evaluate changes in the translation of specific mRNAs. In this case, the amounts of the specific mRNA in the gradient fractions of the polysome profile are determined by hybridization techniques. Changes in the position (number of ribosomes per mRNA) or amounts (amplitude) of the mRNA in the various gradient fractions may be diagnostic of the phase of global protein synthesis that is being modulated (Fig. 4). If the size of polysomes decreases (Fig. 4, middle), either initiation is inhibited or elongation/termination is stimulated, or a combination of both occurs. In the vast majority of cases, translation initiation is the phase affected. Conversely, an increase in polysome size can be caused by an increased rate of initiation and/or a decreased rate of elongation/termination. To interpret polysome profiles unambiguously, it is advisable to measure the elongation rate by determining the ribosome transit time of the mRNA being translated, although unfortunately this is seldom done. In cases where the rate of specific protein synthesis is repressed and polysomes are smaller, initiation clearly has been inhibited. Repression or activation of the synthesis of a protein need not always affect polysome size, however. Instead, the number of translating mRNAs may be affected by masking or mobilizing the mRNA into polysomes, thereby changing the proportion of mRNAs in nonactive mRNPs versus polysomes (Fig. 4, bottom). In this case, there is a change in the *amount* (i.e., amplitude) of the mRNA in polysomes, but the average *size* of the polysomes remains the same.

The methods above also can be used to measure the absolute rate of translation initiation for a specific mRNA. The rate of initiation, i.e., the number of initiation events per minute, can be calculated from the number of ribosomes translating an mRNA (polysome size) and the ribosome transit time (the time required for the ribosome to traverse the mRNA).

$$\text{Number of initiation events/time} = \frac{\text{polysome size}}{\text{ribosome transit time}}$$

As elegantly determined for ovalbumin mRNA in chick oviducts (Palmiter 1975), ovalbumin polysomes average 12 ribosomes, and the ribosome transit time is 1.3 minutes, giving a rate of initiation of 9.2 events per minute (or one initiation every 6.5 seconds). Since the elongating ribosome requires only about 2 seconds to vacate the initiation site, it is clear that the initiation rate is slower than potentially possible and thus is rate-limiting.

Are there cases where the elongation rate is regulated? Examination of a number of specific mRNAs shows that little or a modest change in the rate of elongation is found for most mRNAs following treatment of cells with hormones and other agents (Proud 2000). In contrast, a five-fold stimulation of the rate of elongation of tyrosine aminotransferase is seen when rat hepatoma cells are treated with dibutyryl-cAMP (Roper and Wicks 1978). Similarly, the elongation rate on vitellogenin mRNA drops about fourfold when cockerel liver explants are treated

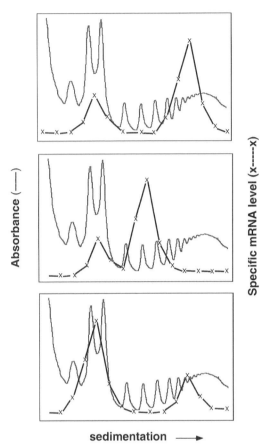

Figure 4. Analysis of the polysome profiles for a specific mRNA. The lysates from three hypothetical cell populations are analyzed by sucrose gradient sedimentation, followed by scanning of the gradients to measure absorbance (*solid lines*) and hybridization analysis of gradient fractions for the level of a specific mRNA (×———×). (*Top*) Cells active in the translation of the mRNA, with most of the mRNA present in polysomes of about 8–9 ribosomes per mRNA and a smaller fraction in nonactive mRNP particles that sediment somewhat more slowly than 80S ribosomes. (*Middle*) Cells in which the apparent initiation rate is reduced about threefold, generating smaller polysomes of about three ribosomes per mRNA but no change in the proportion of polysomes to mRNP particles. (*Bottom*) Cells in which the translation of the mRNA is reduced about threefold by shifting mRNAs from active polysomes to nonactive mRNP particles, but where the mRNAs remaining in polysomes contain 8–9 ribosomes, indicative of no apparent change in the initiation rate on the active mRNAs.

with 17β-estradiol (Gehrke and Ilan 1987). Even small changes in the elongation rate will affect the efficiencies of those mRNAs that are elongation-limited; whether or not moderate inhibition of elongation affects initiation-limited mRNA expression depends on the degree to which initiation is limiting.

TARGETS AND MECHANISMS OF TRANSLATIONAL CONTROL

Cellular protein levels in general do not correlate with their corresponding mRNA levels. In many instances, the innate efficiency and temporal regulation of mRNA translation contribute importantly in establishing the cellular levels of proteins (protein turnover being the other determining factor). Cells respond rapidly to external stimuli by modulating global rates of protein synthesis and/or the translational efficiencies of specific mRNAs. Because initiation is usually rate-limiting, most, but not all, translational controls operate at this phase of protein synthesis. Such regulation is realized through multiple mechanisms that target structural features of the mRNA, the translational apparatus, and *trans*-acting protein or RNA factors. The survey that follows identifies principal targets and mechanisms of translational control, citing chapters in this monograph where these topics are considered in greater detail.

mRNA

The intrinsic translational efficiency of an mRNA is dependent on several *cis*-acting structural elements that also have critical roles in the regulation of mRNA utilization. It is convenient to divide these elements into two categories: those that act through the general translational apparatus and those whose effects are mediated by specific *trans*-acting factors.

In prokaryotes, the first category is of overriding importance; translational efficiency is heavily influenced by the primary and secondary structures of the mRNA. The innate efficiency of a mRNA is determined by the initiation codon (e.g., AUG or GUG), by the location and strength of the Shine-Dalgarno sequence (i.e., its distance upstream of the initiation codon and its extent of complementarity with the 3′ end of the 16S rRNA), and perhaps most importantly, by the strength of secondary structure in the initiation region. Efficient mRNAs lack secondary structure around the AUG initiation codon and possess a strong Shine-Dalgarno region, enabling the mRNA to bind stably to the 30S ribosomal subunit. Temporal regulation can be achieved by stabilizing inhibitory secondary structures through the binding of *trans*-acting factors or, conversely, by preventing formation of the structure by such factors, leading to stimulation of protein synthesis. Recent discoveries of microRNAs and riboswitches that sense small metabolites demonstrate the vast potential for modulating the structure/function of bacterial mRNAs (Chapter 28).

In eukaryotes, where initiation by the 5′-cap-dependent scanning mechanism predominates, structural elements distributed along the entire length of an mRNA determine and modulate translational efficiency (Chapter 4). Efficiently translated mRNAs possess a 5′-cap structure that is readily accessible to initiation factors, a 5′-proximal initiation codon flanked by a purine at −3 and a G at +4, and a poly(A) tail. mRNA efficiency is reduced by secondary structure in the 5′-untranslated region (UTR) and by the presence of upstream AUGs or upstream open reading frames (uORFs). These structural elements interact with the translational apparatus to generate a vast range of mRNA efficiencies. In addition to the innate properties of the mRNA, *cis*-acting elements may be incorporated in the mRNA sequence that enable a temporal regulation of the mRNA's efficiency by *trans*-acting factors (Chapters 10 and 19). mRNAs that initiate internally by ribosome binding to an internal ribosomal entry site (IRES) may require only a subset of the canonical initiation factors—or even none at all in a few cases—and are stimulated by additional proteins (IRES *trans*-acting factors, or ITAFs) (Chapters 5 and 6). These differences enable some IRES-promoted mRNAs to evade regulatory mechanisms that affect 5′-cap-dependent translation, thereby allowing translation to proceed when global protein synthesis is severely inhibited. Some mRNAs can be translated by both the 5′-cap-dependent and IRES-dependent mechanisms.

cis-Acting elements that are recognized by *trans*-acting factors are sometimes found in the 5′UTRs of mRNAs (Chapter 10). One of the best-characterized elements is the iron-responsive element (IRE), a sequence- and structure-specific negative regulatory element found in the 5′UTR of ferritin mRNA (and in other mRNAs), which modulates ferritin synthesis in accordance with the level of cellular iron. Regulation is mediated by a *trans*-acting iron repressor protein (IRP) that binds to the IRE and inhibits translation (Rouault and Harford 2000; Chapter 10). *Positive* mRNA-specific regulators of 5′-cap-dependent translation have not yet been identified in eukaryotes, but the ITAFs that stimulate IRES function fall into this category. uORFs can participate in translational control in yeast and higher eukaryotes (Chapters 9 and 10). Regulation by an uORF is dependent on many factors, including the amino acid sequence encoded by the uORF, the distance between cistrons, and whether or not reinitiation occurs after uORF translation (Geballe and Sachs 2000; Chapter 8).

The 3'UTR is a rich repository of *cis*-acting elements that determine mRNA stability (Chapter 25) and localization in the cytoplasm (Chapter 24) and also serve to regulate translation initiation. These controls are often mediated by *trans*-acting proteins, and we are beginning to appreciate the regulatory roles played by microRNAs (Chapter 11). Many examples of translational control operating through the 3'UTR occur during early development, and some cases have been described in somatic cells (Chapters 10 and 19). For example, a number of mRNAs contain a cytoplasmic polyadenylation element (CPE) that binds a bifunctional protein, CPEB, which stimulates polyadenylation of the mRNA and represses translation. The repression is caused in part by recruitment by CPEB of still another protein, maskin, which binds to eIF4E and prevents its association with eIF4G. The caudal mRNA binds bicoid to a site in its 3'UTR; bicoid in turn binds d4EHP, a repressor that mimics eIF4E by binding to the m^7G-cap but cannot interact with eIF4G. At the 3' end of eukaryotic mRNAs, the poly(A) tail also has an important role as an enhancer of translation. Intriguingly, the poly(A) tail acts in synergy with the mRNA 5'-cap structure, and the translational activity of the poly(A) tail may be mediated by the poly(A)-binding protein (PABP) (Chapters 4 and 10).

mRNA stability is an important determinant of cytoplasmic mRNA levels and therefore of protein synthesis. In many instances, translation has a direct role in determining mRNA stability, because mRNA degradation may be coupled to translation (Chapter 25). Most, but not all, of the *cis*-acting elements that trigger mRNA degradation are localized to the 3'UTR. The poly(A) tail influences the degradation of mRNAs via the PABP, and short-lived mRNAs possess sequence-specific elements that mediate mRNA degradation. A separate pathway exists to degrade mRNAs that contain premature termination codons (nonsense-mediated decay) (Chapter 23). This pathway has most probably evolved to prevent the synthesis of truncated proteins that might function in a dominant-negative manner. The nonsense-mediated decay pathway poses intriguing questions concerning the possible coordination between translation and nuclear-cytoplasmic mRNA transport (Moore 2005). Similarly, mRNA degradation coupled to translation may occur when ribosomes stall ("no-go" decay; Chapter 25) or fail to terminate (nonstop decay; Chapter 29).

Within the coding sequence of some mRNAs are elements that signal ribosome frameshifting, hopping, and termination codon readthrough, all often used by viruses (Chapter 20), and the incorporation of seleno-cysteine and pyrrolysine (Chapter 22). Some of these processes are known to be regulated. For example, ribosomal frameshifting is regulated in both eukaryotes (in antizyme) and prokaryotes (in RF2 and tryptophanase).

Initiation Factors

The effects of the various *cis*-acting elements in the 5′ UTR of an mRNA are modulated through the activity of initiation factors and other *trans*-acting factors. Phosphorylation of initiation factors provides the chief means to control the rate of mRNA binding. Several factors that promote mRNA binding to ribosomes (eIF4E, eIF4G, eIF4B, and eIF3 in mammalian cells; also eIF4A in plants) are phosphorylated, and the phosphorylation status of these proteins correlates positively with both translation and growth rates of the cell (Chapters 4, 14, and 26). The phosphorylation state of these initiation factors affects translation and is modulated in a wide variety of circumstances: following cell stress (Chapter 13), during apoptosis (Chapter 16) and viral infection (Chapter 20), or in response to growth factors and hormones (Chapters 14, 15, and 17). A key target of regulation is the recognition and binding of mRNA, allowing modulation not only of translation rates, but also of mRNA selectivity. Especially important in this regard is the regulation of eIF4E activity by the 4E-BP family of proteins. The 4E-BPs in the hypophosphorylated state bind to eIF4E, thereby preventing eIF4E association with eIF4G. Phosphorylation of 4E-BPs reduces their affinity to eIF4E and reverses the inhibition. The phosphorylation of eIF4E, eIF4G, eIF4B, and eIF3 correlates with activation of protein synthesis, but how phosphorylation affects their activities is just beginning to be elucidated.

A second key target is the binding of Met-tRNA$_i$, which can be regulated through phosphorylation of eIF2. In contrast to the eIF4 group, phosphorylation of eIF2 inhibits protein synthesis by preventing the eIF2B-catalyzed exchange of GDP for GTP (Chapter 9). Extensive analyses of the mechanism of eIF2 phosphorylation led to the identification and characterization of four mammalian protein kinases: PKR, HRI, PERK, and GCN2 (Chapter 12), the first having a key role in the antiviral host defense mechanism that is mediated by interferons (Chapter 20). Phosphorylation of eIF2 has a dominant role in the inhibition of protein synthesis that accompanies the integrated stress response (ISR) caused by heat shock, hypoxia, and amino acid deprivation (Chapter 13). It also occurs following viral infection (Chapter 20) and serum deprivation (Chapter 14) and is involved in IRES function (Chapter 6), reinitiation following translation of uORFs (Chapters 9 and 13), and neuronal function (Chapter 18).

eIF2B, eIF5, and PABP also are phosphoproteins, but less is known about how posttranslational modification affects their activities. Translation initiation factor activity also can be modulated in principle by other reversible or irreversible modifications. One important example is the cleavage of eIF4G that occurs as a result of infection with certain picornaviruses. This cleavage event separates the eIF4E and eIF3 binding sites and is responsible in part for the shutoff of host protein synthesis after viral infection (Chapter 20). A different pattern of eIF4G cleavage occurs in cells undergoing apoptosis, during which a number of other initiation factors also are degraded (Chapter 16). eIF5A (originally known as eIF-4D) contains a unique modified lysine residue called hypusine that appears to be required for its activity (Park 1989). Parenthetically, it is still unknown why the eukaryotic translation system has two factors with apparently unique posttranslational modifications, diphthamide (eEF2) and hypusine (eIF5A), entailing dedicated modifying enzymes, when their homologs in the bacterial translation system lack these modifications. Mass spectroscopic analyses of initiation factors likely will lead to the identification of still other posttranslational modifications that may have roles in translational control.

Initiation factor activity can be modulated by other proteins that interact with initiation factors. Besides the eIF4E/4E-BP interaction discussed above, a number of other proteins bind to eIF4E and regulate its activity (Chapters 10, 18, and 19). Proteins exhibiting homology with eIF4G (p97/DAP5/NAT1 and Paip1) have been described (Chapter 4); they modulate translation most likely via their interaction with eIF4G-binding proteins.

Elongation Factors

Elongation rates also are modulated by phosphorylation, particularly through the activity of the translation elongation factor eEF2 (Chapter 21). This factor undergoes phosphorylation in response to growth-promoting stimuli, calcium ion fluxes, and other agents to negatively affect translation. eEF2 and the other elongation factors are also altered posttranslationally by other modifications. For example, eEF2 is a substrate for ADP-ribosylation by diphtheria toxin on the unique diphthamide residue (a uniquely modified histidine). There is evidence that diphthamide has a role in polypeptide chain elongation; deletion of one of the enzymes that participates in diphthamide synthesis causes embryonic lethality in mice (Liu et al. 2006), although the modified residue is not absolutely required for cell viability in yeast (Chapter 3). Both bacterial EF-Tu and eukaryotic eEF1A also contain other types of modifications such as methylation, and the latter factor, phosphorylation, but the effects on their functions are not yet clear.

Ribosomes

Phosphorylation of ribosomal proteins may also affect translation initiation. Of these, ribosomal protein S6 provides the best-studied example. Although its phosphorylation by S6 kinases correlates with stimulation of initiation, especially of mRNAs possessing a 5′-terminal oligopyrimidine sequence (5′TOP mRNAs), S6 phosphorylation in fact is not required in this case (Chapter 17). A number of other ribosomal proteins are phosphorylated under various conditions, but convincing proof that this results in altered ribosome activity is lacking. Ribosomal proteins also are modified by methylation (Chapter 3) and ubiquitination (Chapter 14), suggesting some heterogeneity in the ribosome population that may be important for the translation of specific classes of mRNAs.

WHY CONTROL TRANSLATION?

Thus far, we have considered the basis and principles of translational control. As mentioned above, there is a clear-cut rationale for regulating a biochemical pathway at its first step; this principle holds true, by and large, for protein synthesis, in that regulation is most often exercised at the initiation phase. From a broader perspective, however, matters become less clear-cut. Viewing gene expression in totality, translation occupies a position somewhere in the middle of an extended pathway that begins with transcription, continues with RNA processing and transport, and ends with protein translocation, modification, folding, assembly, and degradation. Each of these steps is known to be regulated in one or another biological system. Yet, two of the steps in this grand scheme, transcription and translation, are especially critical for the cell. Both are biosynthetic steps in which the cell makes large investments of energy. Consequently, both are steps at which the cell's expenditure of resources is checked. Indeed, transcription is subject to a multitude of controls. So, why control translation, too? And where and when is this option exercised?

To these frequently asked questions, there is no single answer. Rather, there are several compelling reasons for cells to deploy translational control in their arsenal of regulatory mechanisms. Some of the advantages offered by translational control are considered briefly below. Evidently, the benefits more than compensate for the energetic and other penalties paid for the privilege of exerting regulation over a midstream reaction in a long pathway.

Directness and Rapidity

Immediacy is the most conspicuous advantage of translational control over transcriptional and other nuclear control mechanisms. Whereas transcriptional control affects one of the first steps in the flow of genetic information, translational control affects one of the last steps. When control is applied at a step prior to translation, the cell has to confront subsequent biochemical reactions (splicing, nuclear transport, etc.) that might be rate-limiting and inevitably entail a delay in implementing changes in protein synthesis. No such time lag applies in the case of translational control.

Reversibility

Translational control can be either positive or negative. Most translational controls are effected by reversible modifications of translation factors, chiefly through phosphorylation. The readily reversible nature of translational control mechanisms is economical in energetic terms, a feature that is of particular biological significance in energy-deprived cells.

Fine Control

There are numerous examples of genes that are under both transcriptional and translational control (e.g., TNF-α, C/EBPβ, VEGF, ornithine decarboxylase, and *myc*). In most instances, but not all, the changes in transcription rates are considerably greater in magnitude than the changes in translation rates. Thus, regulation of gene expression at the translational level provides a means for fine control.

Regulation of Large Genes

Some genes are extremely long (e.g., dystrophin, >2000 kb), and their transcription is estimated to take an extended period of time (>24 hours for dystrophin). It is reasonable to assume that if their expression must be regulated in a relatively short period, it is likely to be accomplished at the level of translation.

Systems That Lack Transcriptional Control

In some systems (e.g., reticulocytes, oocytes, and RNA viruses), there is little or no opportunity for transcriptional control, and gene expression is modulated mostly at the translational level. The widespread use of translational controls to regulate gene expression during development suggests that this mode of control preceded transcriptional control in evolution. Such a hypothesis is consistent with the notion of the existence of an RNA World prior to the emergence of DNA. Is it therefore possible that translational control was more prevalent early in evolution and that we are now witnessing only the relics of such control mechanisms?

Spatial Control

Regulation of the site of protein synthesis within the cell can generate concentration gradients of proteins. Such gradients are known to affect the translational efficiency of other specific mRNAs that determine patterning in early development (Chapters 19 and 24). Similar mechanisms are likely to explain synaptic plasticity (see section below).

Flexibility

Because of the wide variety of mechanisms for translational control, it can be focused by specific effector mechanisms on a single or a few gene(s) or cistrons, such as the coat protein and replicase of RNA phages, antizyme, and ferritin (Farabaugh et al. 2000; Rouault and Harford 2000; Chapter 28); alternatively, by influencing general factors, it can encompass whole classes of mRNAs, as in heat shock and virus-induced host-cell shutoff (Chapter 20). Such flexibility affords the cell a powerful and adaptable means to regulate gene expression.

TRANSLATION IN THE MODERN ERA: CURRENT THEMES AND FUTURE TRENDS

The translation field, like many others, has exploited the suite of techniques that owe their existence to recombinant DNA technology (mRNA cloning, protein tagging and overexpression, reverse genetics, transgenic and knockout mice, RNA interference, etc.). The application of such methods, together with genetics (notably in the yeast *Saccharomyces cerevisiae*, but also in other organisms) and visualization techniques operating over a range of resolution (X-ray crystallography, NMR analysis, confocal microscopy, in situ hybridization, in vivo imaging, etc.), has ushered in an era where research is proceeding apace over a broad front.

As a result, many details have been filled in and an essentially complete picture of the translational apparatus is now available. Most of the translational components have been cloned and purified, and the three-dimensional structures are known for a steadily increasing fraction of them. A good deal of the mechanism of protein synthesis is understood, although much still remains to be learned. Systems at all levels of complexity are amenable to study, from the brain and memory to plants (for review, see Chapters 18 and 26), and investigators are now addressing the translational aspects of a wide spectrum of fundamental biological and cellular processes. These extend from apoptosis to development, and from signal transduction to nonsense-mediated mRNA decay (Chapters 14, 16, 19, and 23) to name just a few topics. The established systems that comprised the paradigms of translational control are still under study, of course, albeit with tools of increasing sophistication, and inroads are being made into disease processes and therapy (Chapters 15, 17, 29, and 30). Fresh impetus comes from conceptual advances in several areas, some of which are summarized below.

Multiplicity of RNA Roles

The distinct roles of mRNA (informational), tRNA (adaptor), and rRNA (structural) were identified early on in the dissection of the translation system. These RNAs are indispensable parts of the general translation apparatus. Subsequently, other RNA molecules were found to participate in specific aspects of the translation process: For example, 7SL RNA is a major component of the SRP involved in protein secretion (Chapter 21), and VA RNA$_I$ is a viral regulator of the eIF2 kinase PKR (Chapter 20). It is now realized that RNA has even more extensive roles in translation. The 23S rRNA serves as more than a passive skeleton for the 50S ribosomal subunit: It almost certainly carries out the key catalytic function of protein synthesis, namely, peptide bond formation, and conformational changes in rRNAs are probably involved in the ribosome elongation cycle (Chapter 2). In addition, the discovery of two further regulatory roles for RNA, microRNAs (miRNAs) and "riboswitches," has generated considerable excitement.

Perhaps the most dramatic and unexpected finding is that miRNAs are very abundant in most organisms and that they control the expression of as much as a third of the genome. The miRNAs are small RNAs that act *in trans* to control protein synthesis at the level of mRNA stability (via RNA silencing) and translation in eukaryotes and bacteria (Chapters 11 and 28). Current estimates are that there are approximately 1000 miRNAs in mammals and that each regulates approximately 10 mRNAs. Much of the evidence points to translational control by miRNAs, but the exact mechanism is controversial. The first report on miRNA identified the small, noncoding *Caenorhabditis elegans* RNA, *lin-4*, as a negative regulator of *lin-14*-encoded protein expression, suggesting that it functions via an antisense RNA–RNA interaction with the *lin-14* 3'UTR (Lee et al. 1993). Subsequently, Ambros's laboratory showed that the polysome distribution of *lin-14* mRNA was not changed, suggesting that inhibition of translation takes place at a postinitiation step (Olsen and Ambros 1999). Similar conclusions have been drawn in human cells, where premature termination was observed (Petersen et al. 2006). Surprisingly, however, two other groups have reported that miRNA inhibits translation at the initiation step in human cells with eIF4E as a target. Preiss and colleagues also implicated the poly(A) tail and, by inference, PABP, as an additional target for miRNA (Humphreys et al. 2005), whereas Filipowicz and coworkers provided evidence that P-bodies, a site for mRNA decapping and degradation as discussed below, are the depot where translationally inactive mRNAs are stored for later use (Pillai et al. 2005). It is clear that an in vitro system recapitulating the in vivo miRNA activity will be necessary to elucidate the molecular mechanism of miRNA function.

"Riboswitches," regulatory sequences present in the 5'UTR of prokaryotic mRNAs, constitute another remarkable discovery. These *cis*-acting structured RNA elements act as sensors of metabolite levels that control the translation of the downstream open reading frame (Chapter 28). The pervasive nature of RNA involvement in translation seems like an echo of the primordial RNA World.

P-bodies and Stress Granules

An important advance in the understanding of mRNA metabolism that has come to be appreciated in the past 5 years is the identification of cytoplasmic bodies where mRNAs can be stored or degraded (Chapter 25). The first such structure to be described, in tomato cells subjected to heat shock (Nover et al. 1989), is the stress granule where mRNAs, 40S ribosomal subunits, and most of the initiation factors are stored in response to translation inhibition as a consequence of eIF2α phosphorylation (Anderson and Kedersha 2006). Subsequently, another body, termed the P (for processing) or GW182 body, was discovered in yeast (Sheth and Parker 2003) and in mammals (van Dijk et al. 2002). These structures contain decapping enzymes, the 5′ exonuclease XRN, and the exosome, thus identifying them as the prime sites for mRNA degradation. Consistent with this, nonsense-mediated decay and siRNA-mediated degradation of mRNA occur in P-bodies. Surprisingly, recent results in yeast (Brengues et al. 2005) demonstrated that inhibition of translation resulted in mRNA *accumulation* in P-bodies, without degradation, and that these mRNAs can be recycled for translation. As noted above, miRNA-mediated inhibition of translation can also result in mRNA accumulation in P-bodies with only minor degradation. In this regard, it is of interest that mRNAs can be transported from P-bodies to stress granules in mammalian cells (Kedersha et al. 2005). This process may not occur in yeast, however, since yeast appear not to have stress granules. Future studies are bound to elucidate the mechanisms by which mRNAs are sorted into different cytoplasmic compartments, which function in mRNA metabolism.

Synaptic Plasticity and Memory

Modifications of synaptic strength are thought to serve as the cellular basis that underlies learning and memory. The best experimental model of synaptic plasticity is long-term potentiation (LTP), which is induced by stimulation of neural pathways in the hippocampus. Early LTP, which is induced by a single train of impulses, is labile and is independent of translation, whereas late LTP, which is the cellular model of long-term memory, is dependent on translation. Significantly, translation is increased in stimulated synapses by recruiting mRNAs and translational components, including ribosomes, which are stored at the base of the dendritic spines (Chapter 18). Signaling pathways that are known to stimulate translation, including the MAPK and PI3K signaling pathways, were demonstrated to activate synaptic local translation. Evidence for the important role of translational control in synaptic plasticity was obtained in the past 2 years through the use of "knockout" mice with targeted deletions of regulators of the translation machinery. For example, deletion of cytoplasmic polyadenylation element-binding protein (CPEB), which regulates translation via PABP, results in deficits in learning and memory. Similarly, two other established translational regulators, 4E-BP2 and GCN2, have also been shown to have important roles in synaptic plasticity, learning, and memory. Interestingly, simultaneous deletion of both 4E-BP2 and GCN2 lowers the threshold of synaptic stimulation needed to achieve sustained late LTP, further supporting the idea that translational control has a key role in synaptic plasticity.

Formation of Supramolecular Assemblies

It has long been known that translation involves complexes containing a number of protein subunits (e.g., eIF4F and eIF3), several enzymes (e.g., the aminoacyl-tRNA synthetase multienzyme and GAIT complexes; Chapter 29), or protein and RNA molecules (e.g., ribosomes and mRNP complexes). Increasingly, the existence and significance of larger and more elaborate arrays are being recognized, as exemplified by the multifactor complex containing several initiation factors (Chapter 9) and by the protein–RNA interactions that mediate circularization of mRNA (Chapters 4 and 9). The latter form the basis for a growing series of regulatory interactions that participate in processes such as development (Chapter 19). With the ribosome itself as the leading example, the detailed structures of these complexes are now coming to light (Chapters 2 and 3). These studies are also yielding insights into the functions, dynamics, and control of translation, which is steadily coming to be seen as a process mediated less by individual components than by large assemblies with multiple activities.

Translation Initiation Site Selection Mechanisms

The scanning model for initiation site selection on eukaryotic mRNAs posits that ribosomal subunits enter the mRNA at its 5′ end and scan along the RNA strand until they encounter an initiation codon in a favorable context, which is where initiation takes place (Chapters 4 and 9). Founded on the idea that eukaryotic mRNAs are characteristically monocistronic (Jacobson and Baltimore 1968), scanning was originally proposed on the basis of mRNA structural comparisons as well as experimental data (Kozak and Shatkin 1978). The model has been largely sustained and enjoys considerable predictive value, but it cannot account for initiation events that occur on a substantial number of viral mRNAs and a smaller proportion of cellular mRNAs (Chapters 5 and 6). For most of these mRNAs, access to the initiation codon is provided by an IRES, which causes the ribosomes or their subunits to be loaded on the mRNA directly in the vicinity of the appropriate site. The IRES thereby achieves a result similar to that afforded by the Shine-Dalgarno element in prokaryotic mRNAs, albeit probably via different interactions. Intriguingly, however, the involvement of 18S rRNA in initiation on a mouse mRNA has recently been documented (Dresios et al. 2006).

Additional mechanisms of initiation have also been uncovered. The best-known of these is ribosome shunting, or discontinuous scanning, in which ribosomal subunits traversing the 5'UTR ostensibly suspend scanning through downstream RNA sequence and structures to reach the initiation site. This behavior has been observed in mRNAs of diverse human and higher plant viruses, and some mechanistic and regulatory aspects have been worked out (Chapter 20), although much remains to be learned.

Localization, Localization, Localization: The Role of Cell Architecture

The first indication that intracellular position can have both selective and regulatory roles in mRNA translation came from the study of protein secretion, as discussed above. In this case, a protein signal on the nascent chain is recognized by the SRP, an RNP particle that controls elongation and targets synthesis to the ER. More recently, mRNA relocalization into P-bodies and stress granules has attracted considerable attention. Another mechanism—or family of mechanisms—that couples mRNA selectivity and translational control to intracellular localization is specified by RNA signals residing in the mRNA 3'UTR. These "zip codes" are recognized by proteins that nucleate formation of "locasomes" which keep the mRNAs translationally inactive until they have been transported to the appropriate destination within the cell (Chapter 24). Such mechanisms are instrumental in achieving the site-specific synthesis of a wide range of proteins that are essential for general cell activities (e.g., motility and cell division) as well as specialized functions such as developmental polarity and synaptic plasticity (Chapters 18 and 19), and we are only beginning to appreciate their full scope.

CONCLUDING REMARKS

The recognition of translational control formally requires the measurement of two parameters—the rate of protein synthesis and the concentration of the corresponding mRNA—so its rigorous demonstration can be demanding. Nevertheless, appreciation of the range of biological processes that entail translational control is expanding rapidly. At the same time, our understanding of the underlying protein synthetic apparatus is well advanced and provides a solid platform to address the mechanisms exploited by cells to control gene expression at this level. Goals for the future lie in many directions: to identify and characterize the *cis-* and *trans-*acting elements that mediate translational control, to visualize the interactions at the atomic level and characterize their kinetics, and to integrate this information within the framework of the physiology and evolution of intact cells and organisms. It is also of great significance that studies on the process of translation have advanced to the translational phase between basic research and clinical practice. For example, new antibiotics are being synthesized based on the bacterial ribosome structure, and new drugs that target initiation factors are entering cancer clinical trials (Chapters 15 and 30). On the basis of our expanding knowledge of the role of translational control in metabolism and in learning and memory, it is expected that drugs targeting the translation apparatus will be developed to treat metabolic diseases and memory loss. Yet there are undoubtedly surprises in store—in sperm translation, for example (Gur and Breitbart 2006)—and open questions remain at many levels, from the fundamental to the sophisticated.

ACKNOWLEDGMENTS

The authors' work has been supported by grants from the National Institutes of Health (M.B.M., J.W.B.H., and N.S.) and from the Canadian Institute of Health Research, National Cancer Institute of Canada, and Howard Hughes Medical Institute (N.S.).

REFERENCES

Agrawal M.G. and Bowman L.H. 1987. Transcriptional and translational regulation of ribosome protein formation during mouse myoblast differentiation. *J. Biol. Chem.* **262:** 4868–4875.

Anderson P. and Kedersha N. 2006. RNA granules. *J. Cell Biol.* **172:** 803–808.

Aoki H., Dekany K., Adams S.L., and Ganoza M.C. 1997. The gene encoding the elongation factor P protein is essential for viability and is required for protein synthesis. *J. Biol. Chem.* **272:** 32254–32259.

Berg P. and Ofengand E.J. 1958. An enzymatic mechanism for linking amino acids to RNA. *Proc. Natl. Acad. Sci.* **44:** 78–86.

Blobel G. and Dobberstein B. 1975. Transfer of proteins across membranes. II. Reconstitution of functional rough microsomes from heterologous components. *J. Cell Biol.* **67:** 852–862.

Blobel G. and Sabatini D.D. 1971. Ribosome-membrane interaction in eukaryotic cells. In *Biomembranes* (ed. L.A. Manson), pp. 193–195. Plenum Press, New York.

Borsook H., Deasy C.L., Haagensmit A.J., Keighley G., and Lowy P.H. 1950. Incorporation in vitro of labeled amino acids into bone marrow cell proteins. *J. Biol. Chem.* **186:** 297–307.

Brengues M., Teixeira D., and Parker R. 2005. Movement of eukaryotic mRNAs between polysomes and cytoplasmic processing bodies. *Science* **310:** 486–489.

Brenner S., Jacob F., and Meselson M. 1961. An unstable intermediate carrying information from genes to ribosomes for protein synthesis. *Nature* **190:** 576–581.

Bruns G.P. and London I.M. 1965. The effect of hemin on the synthesis of globin. *Biochem. Biophys. Res. Commun.* **18:** 236–242.

Campbell P.N. and Work T.S. 1953. Biosynthesis of proteins. *Nature* **171:** 997–1001.

Chapeville F., Lipmann F., von Ehrenstein G., Weisblum B., Ray W.J., and Benzer S. 1962. On the role of soluble ribonucleic acid in coding for amino acids. *Proc. Natl. Acad. Sci.* **48:** 1086–1092.

Chiorini J.A., Boal T.R., Miyamoto S., and Safer B. 1993. A difference in the rate of ribosomal elongation balances the synthesis of eukaryotic translation initiation factor (eIF)-2 alpha and eIF-2 beta. *J. Biol. Chem.* **268:** 13748–13755.

Crick F.H.C. 1958. On protein synthesis. *Symp. Soc. Exp. Biol.* **12:** 138–163.

Damm H.C., Ed. 1966. *Methods and references in biochemistry and biophysics.* World Publishing, Cleveland, Ohio.

Davidson S.D., Passmore R., and Brock J.F. 1973. *Human nutrition and dietetics.* Churchill Livingstone, London, United Kingdom.

Devilliers-Thiery A., Kindt T., Scheele G., and Blobel G. 1975. Homology in amino-terminal sequence of precursors to pancreatic secretory proteins. *Proc. Natl. Acad. Sci.* **72:** 5016–5020.

Dresios J., Chappell S.A., Zhou W., and Mauro V.P. 2006. An mRNA-rRNA base-pairing mechanism for translation initiation in eukaryotes. *Nat. Struct. Mol. Biol.* **13:** 30–34.

Drysdale J.W. and Munro H.N. 1965. Failure of actinomycin D to prevent induction of liver apofentin after iron administration. *Biochim. Biophys. Acta* **103:** 185–188.

Duncan R. and McConkey E.H. 1982. Rapid alterations in initiation rate and recruitment of inactive RNA are temporally correlated with S6 phosphorylation. *Eur. J. Biochem.* **123:** 539–544.

Eboué-Bonis D., Chambaut A.M., Volfin P., and Clauser H. 1963. Action of insulin on the isolated rat diaphragm in the presence of actinomycin D and puromycin. *Nature* **199:** 1183–1184.

Ehrenfeld E. and Hunt T. 1971. Double-stranded poliovirus RNA inhibits initiation of protein synthesis by reticulocyte lysates. *Proc. Natl. Acad. Sci.* **68:** 1075–1078.

Ehrenfeld E. and Lund H. 1977. Untranslated vesicular stomatitis virus messenger RNA after poliovirus infection. *Virology* **80:** 297–308.

Endo T. and Nadal-Ginard B. 1987. Three types of muscle-specific gene expression in fusion-blocked rat skeletal muscle cells: Translational control in EGTA-treated cells. *Cell* **49:** 515–526.

Fajardo J.E. and Shatkin A.J. 1990. Translation of bicistronic viral mRNA in transfected cells: Regulation at the level of elongation. *Proc. Natl. Acad. Sci.* **87:** 328–332.

Fan H. and Penman S. 1970. Regulation of protein synthesis in mammalian cells. II. Inhibition of protein synthesis at the level of initiation durig mitosis. *J. Mol. Biol.* **50:** 655–670.

Fang P., Spevak C.C., Wu C., and Sachs M.S. 2004. A nascent polypeptide domain that can regulate translation elongation. *Proc. Natl. Acad. Sci.* **101:** 4059–4064.

Farabaugh P.J., Qian Q., and Stahl G. 2000. Programmed translational frameshifting, hopping and readthrough of termination codons. In *Translational control of gene expression* (ed. N. Sonenberg et al.), pp. 741–761. Cold Spring Harbor Laboratory Press, Cold Spring Harbor, New York.

Farrell P.J., Balkow K., Hunt T., Jackson R.J., and Trachsel H. 1977. Phosphorylation of initiation factor eIF-2 and the control of reticulocyte protein synthesis. *Cell* **11:** 187–200.

Garren L.D., Howell R.R., Tomkins G.M., and Crocco R.M. 1964. A paradoxical effect of actinomycin D: The mechanism of regulation of enzyme synthesis by hydrocortisone. *Proc. Natl. Acad. Sci.* **52:** 1121–1129.

Geballe A.P. and Sachs M.S. 2000. Translational control by upstream open reading frames. In *Translational control of gene expression* (ed. N. Sonenberg et al.), pp. 595–614. Cold Spring Harbor Laboratory Press, Cold Spring Harbor, New York.

Gehrke L. and Ilan J. 1987. Regulation of messenger RNA translation at the elongation step during estradiol-induced vitellogenin synthesis in avian liver. In *Translational regulation of gene expression* (ed. J. Ilan), pp. 165–186. Plenum Press, New York.

Geoghegan T., Cereghini S., and Brawerman G. 1979. Inactive mRNA-protein complexes from mouse sarcoma-180 ascites cells. *Proc. Natl. Acad. Sci.* **76:** 5587–5591.

Godefroy-Colburn T. and Thach R.E. 1981. The role of mRNA competition in regulating translation. IV. Kinetic model. *J. Biol. Chem.* **256:** 11762–11773.

Gros F., Hiatt H., Gilbert W., Kurland G.G., Risebrough R.W, and Watson J.D. 1961. Unstable ribonucleic acid revealed by pulse labelling of *Escherichia coli. Nature* **190:** 581–585.

Gross P.R., Malkin L.I., and Moyer W.A. 1964. Templates for the first proteins of embryonic development. *Proc. Natl. Acad. Sci.* **51:** 407–414.

Gur Y. and Breitbart H. 2006. Mammalian sperm translate nuclear-encoded proteins by mitochondrial-type ribosomes. *Genes Dev.* **20:** 411–416.

Hardesty B., Miller R., and Schweet R. 1963. Polyribosome breakdown and hemoglobin synthesis. *Proc. Natl. Acad. Sci.* **50:** 924–931.

Henshaw E.C., Hirsch C.A., Morton B.E., and Hiatt H.H. 1971. Control of protein synthesis in mammalian tissues through changes in ribosome activity. *J. Biol. Chem.* **246:** 436–446.

Heywood S.M. 1970. Specificity of mRNA binding factor in eukaryotes. *Proc. Natl. Acad. Sci.* **67:** 1782–1788.

Hoagland M.B., Stephenson M.L., Scott J.F., Hecht L.I., and Zamecnik P.C. 1958. A soluble ribonucleic acid intermediate in protein synthesis. *J. Biol. Chem.* **231:** 241–257.

Hodge L.D., Robbins E., and Scharff M.D. 1969. Persistence of messenger RNA through mitosis in HeLa cells. *J. Cell Biol.* **40:** 497–507.

Holley R.W., Apgar J., Everett G.A., Madison J.T., Marquisee M., Merrill S.H., Penswick J.R., and Zamir A. 1965. Structure of a ribonucleic acid. *Science* **147:** 1462–1465.

Hultin T. 1961. Activation of ribosomes in sea urchin eggs in response to fertilization. *Exp. Cell Res.* **25:** 405–417.

Hultin T. and Beskow G. 1956. The incorporation of ^{14}C-L-leucine into rat liver proteins *in vitro* visualized as a two-step reaction. *Exp. Cell Res.* **11:** 664–666.

Humphreys D.T., Westman B.J., Martin D.I., and Preiss T. 2005. MicroRNAs control translation initiation by inhibiting eukaryotic initiation factor 4E/cap and poly(A) tail function. *Proc. Natl. Acad. Sci.* **102:** 16961–16966.

Humphreys T. 1969. Efficiency of translation of messenger-RNA before and after fertilization in sea urchins. *Dev. Biol.* **20:** 435–458.

———. 1971. Measurements of messenger RNA entering polysomes upon fertilization of sea urchin eggs. *Dev. Biol.* **26:** 201–208.

Hutchison C.A., Peterson S.N., Gill S.R., Cline R.T., White O., Fraser C.M., Smith H.O., and Venter J.C. 1999. Global transposon mutagenesis and a minimal *Mycoplasma* genome (comments). *Science* **286:** 2165–2169.

Jacob F.C. and Monod J. 1961. Genetic regulatory mechanisms in the synthesis of proteins. *J. Mol. Biol.* **3:** 318–356.

Jacobson M.F. and Baltimore D. 1968. Polypeptide cleavages in the formation of poliovirus proteins. *Proc. Natl. Acad. Sci.* **61:** 77–84.

Kang H.A. and Hershey J.W. 1994. Effect of initiation factor eIF-5A depletion on protein synthesis and proliferation of *Saccharomyces cerevisiae. J. Biol. Chem.* **269:** 3934–3940.

Kedersha N., Stoecklin G., Ayodele M., Yacono P., Lykke-Andersen J., Fitzler M.J., Scheuner D., Kaufman R.J., Golan D.E., and Anderson P. 2005. Stress granules and processing bodies are dynamically linked sites of mRNP remodeling. *J. Cell Biol.* **169:** 871–884.

Kinniburgh A.J., McMullen M.D., and Martin T.E. 1979. Distribution of cytoplasmic poly(A^+) RNA sequences in free messenger ribonucleoprotein and polysomes of mouse ascites cells. *J. Mol. Biol.* **132:** 695–708.

Kobayashi K., Ehrlich S.D., Albertini A., Amati G., Andersen K.K., Arnaud M., Asai K., Ashikaga S., Aymerich S., Bessieres P., et al. 2003. Essential *Bacillus subtilis* genes. *Proc. Natl. Acad. Sci.* **100:** 4678–4683.

Komar A.A., Gross S.R., Barth-Baus D., Strachan R., Hensold J.O., Goss Kinzy T., and Merrick W.C. 2005. Novel characteristics of the biological properties of the yeast *Saccharomyces cerevisiae* eukaryotic initiation factor 2A. *J. Biol. Chem.* **280:** 15601–15611.

Kosower N.S., Vanderhoff G.A., Benerofe B., Hunt T., and Kosower E.M. 1971. Inhibition of protein synthesis by glutathione disulfide in the presence of glutathione. *Biochem. Biophys. Res. Commun.* **45:** 816–821.

Kozak M. and Shatkin A.J. 1978. Migration of 40 S ribosomal subunits on messenger RNA in the presence of edeine. *J. Biol. Chem.* **253:** 6568–6577.

Kruh J. and Borsook H. 1956. Hemoglobin synthesis in rabbit reticulocytes *in vitro. J. Biol. Chem.* **220:** 905–915.

Lamfrom H. and Knopf P.M. 1964. Initiation of haemoglobin synthesis in cell-free systems. *J. Mol. Biol.* **9:** 558–575.

Lee G.T.-Y. and Engelhardt D.L. 1979. Peptide coding capacity of polysomal and non-polysomal messenger RNA during growth of animal cells. *J. Mol. Biol.* **129:** 221–233.

Lee R.C., Feinbaum R.L., and Ambros V. 1993. The *C. elegans* heterochronic gene lin-4 encodes small RNAs with antisense complementarity to lin-14. *Cell* **75:** 843–854.

Leibowitz R. and Penman S. 1971. Regulation of protein synthesis in HeLa cells. III. Inhibition during poliovirus infection. *J. Virol.* **8:** 661–668.

Levine M. and Tarver H. 1950. On the synthesis and some applications of serine-beta-C14. *J. Biol. Chem.* **184:** 427–436.

Lisowska-Bernstein, B., M.E. Lamm, and P. Vassalli. 1970. Synthesis of immunoglobulin heavy and light chains by the free ribosomes of a mouse plasma cell tumor. *Proc. Natl. Acad. Sci.* **66:** 425–432.

Liu S., Wiggins J.F., Sreenath T., Kulkarni A.B., Ward J.M., and Leppla S.H. 2006. Dph3, a small protein required for diphthamide biosynthesis, is essential in mouse development. *Mol. Cell. Biol.* **26:** 3835–3841.

Lodish H.F. 1976. Translational control of protein synthesis. *Annu. Rev. Biochem.* **45:** 39–72.

Lodish H.F. and Jacobsen M. 1972. Regulation of hemoglobin synthesis. Equal rates of translation and termination of α- and β-globin chains. *J. Biol. Chem.* **247:** 3622–3629.

Lodish H.F. and Zinder N.D. 1966. Mutants of the bacteriophage f2. VIII. Control mechanisms for phage-specific syntheses. *J. Mol. Biol.* **19:** 333–348.

Maggio R., Vittorelli M.L., Rinaldi A.M., and Monroy A. 1964. *In vitro* incorporation of amino acids into proteins stimulated by RNA from unfertilized sea urchin eggs. *Biochem. Biophys. Res. Commun.* **15:** 436–441.

Marks P.A., Burka E.R., and Schlessinger D. 1962. Protein synthesis in erythroid cells. I. Reticulocyte ribosomes active in stimulating amino acid incorporation. *Proc. Natl. Acad. Sci.* **48:** 2163–2171.

Martin T.E. and Young F.G. 1965. An *in vitro* action of human growth hormone in the presence of actinomycin D. *Nature* **208:** 684–685.

Mathews M.B., Hunt T., and Brayley A. 1973. Specificity of the control of protein synthesis by Haemin. *Nature New Biol.* **243:** 230–233.

McCormick W. and Penman S. 1969. Regulation of protein synthesis in HeLa cells: Translation at elevated temperatures. *J. Mol. Biol.* **39:** 315–333.

McQuillen K., Roberts R.B., and Britten R.J. 1959. Synthesis of nascent protein by ribosomes in *Escherichia coli. Proc. Natl. Acad. Sci.* **45:** 1437–1447.

Meisenberg G. and Simmons W.H. 1998. *Principles of medical biochemistry.* Mosby, St. Louis, Missouri.

Merrick W.C. and Hensold J.O. 2000. The use of sucrose gradients in studies on eukaryotic translation. In *Current protocols in cell biology,* pp. 11.9.1–11.9.26. John Wiley & Sons, New York.

Meyuhas O. and Hornstein E. 2000. Translational control of TOP mRNAs. In *Translational control of gene expression* (ed. N. Sonenberg et al.), pp. 671–693. Cold Spring Harbor Laboratory Press, Cold Spring Harbor, New York.

Meyuhas O., Thompson J.E.A., and Perry R.P. 1987. Glucocorticoids selectively inhibit translation of ribosomal protein mRNAs in P1798 lymphosarcoma cells. *Mol. Cell. Biol.* **7:** 2691–2699.

Milstein C., Brownlee G.G., Harrison T.M., and Mathews M.B. 1972. A possible precursor of immunoglobulin light chains. *Nat. New Biol.* **239:** 117–120.

Monroy A., Maggio R., and Rinaldi A.M. 1965. Experimentally induced activation of the ribosomes of the unfertilized sea urchin egg. *Proc. Natl. Acad. Sci.* **54:** 107–111.

Moore M.J. 2005. From birth to death: The complex lives of eukaryotic mRNAs. *Science* **309:** 1514–1518.

Nemer M. 1962. Interrelation of messenger polyribonucleotides and ribosomes in the sea urchin egg during embryonic development. *Biochem. Biophys. Res. Commun.* **8:** 511–515.

Nirenberg M.W. and Matthaei J.H. 1961. The dependence of cell-free protein synthesis in *E. coli* upon naturally occurring or synthetic polyribonucleotides. *Proc. Natl. Acad. Sci.* **47:** 1588–1602.

Nover L., Scharf K.D., and Neumann D. 1989. Cytoplasmic heat shock granules are formed from precursor particles and are associated with a specific set of mRNAs. *Mol. Cell. Biol.* **9:** 1298–1308.

Olsen P.H. and Ambros V. 1999. The lin-4 regulatory RNA controls developmental timing in *Caenorhabditis elegans* by blocking LIN-14 protein synthesis after the initiation of translation. *Dev. Biol.* **216:** 671–680.

Ouellette A.J., Ordahl C.P., Van Ness J., and Malt R.A. 1982. Mouse kidney nonpolysomal messenger ribonucleic acid: Metabolism, coding function, and translational activity. *Biochemistry* **21:** 1169–1177.

Palmiter R.D. 1974. Differential rates of initiation on conalbumin and ovalbumin messenger ribonucleic acid in reticulocyte lysates. *J. Biol. Chem.* **249:** 6779–6787.

———. 1975. Quantitation of parameters that determine the rate of ovalbumin synthesis. *Cell* **4:** 189.

Park M.H. 1989. The essential role of hypusine in eukaryotic translation initiation factor 4D (eIF-4D). Purification of eIF-4D and its precursors and comparison of their activities. *J. Biol. Chem.* **264:** 18531–18535.

Pelham H.R.B. and Jackson R.J. 1976. An efficient mRNA-dependent translation system from reticulocyte lysates. *Eur. J. Biochem.* **67**: 247–256.

Penman S. and Summers D. 1965. Effects on host cell metabolism following synchronous infection with poliovirus. *Virology* **27**: 614–620.

Penman S., Scherrer K., Becker Y., and Darnell J.E. 1963. Polyribosomes in normal and poliovirus-infected HeLa cells and their relationship to messenger RNA. *Proc. Natl. Acad. Sci.* **49**: 654–662.

Petersen C.P., Bordeleau M.E., Pelletier J., and Sharp P.A. 2006. Short RNAs repress translation after initiation in mammalian cells. *Mol. Cell* **21**: 533–542.

Pillai R.S., Bhattacharyya S.N., Artus C.G., Zoller T., Cougot N., Basyuk E., Bertrand E., and Filipowicz W. 2005. Inhibition of translational initiation by Let-7 MicroRNA in human cells. *Science* **309**: 1573–1576.

Proud C.G. 2000. Control of the elongation phase of protein synthesis. In *Translational control of gene expression* (ed. N. Sonenberg et al.), pp. 719–739. Cold Spring Harbor Laboratory Press, Cold Spring Harbor, New York.

Roper M.D. and Wicks W.D. 1978. Evidence for acceleration of the rate of elongation of tyrosine aminotransferase nascent chains by dibutyrl cyclic AMP. *Proc. Natl. Acad. Sci.* **75**: 140–144.

Rosbash M. 1972. Formation of membrane-bound polyribosomes. *J. Mol. Biol.* **65**: 413–422.

Rouault T.A. and Harford J.B. 2000. Translational control of ferritin synthesis. In *Translational control of gene expression* (ed. N. Sonenberg et al.), pp. 655–670. Cold Spring Harbor Laboratory Press, Cold Spring Harbor, New York.

Ruvinsky I. and Meyuhas O. 2006. Ribosomal protein S6 phosphorylation: From protein synthesis to cell size. *Trends Biochem Sci.* **31**: 342–348.

Sanger F. and Tuppy H. 1951. The amino-acid sequence in the phenylalanyl chain of insulin. 2. The investigation of peptides from enzymic hydrolysates. *Biochem. J.* **49**: 481–490.

Sheth U. and Parker R. 2003. Decapping and decay of messenger RNA occur in cytoplasmic processing bodies. *Science* **300**: 805–808.

Siekevitz P. and Zamecnik P.C. 1951. *In vitro* incorporation of 1-^{14}C-DL-alanine into proteins of rat-liver granular fractions. *Fed. Proc.* **10**: 246–247.

Spirin A.S. and Nemer M. 1965. Messenger RNA in early sea-urchin embryos: Cytoplasmic particles. *Science* **150**: 214–217.

Steward D.L., Shaeffer J.R., and Humphrey R.M. 1968. Breakdown and assembly of polyribosomes in synchronized Chinese hamster cells. *Science* **161**: 791–793.

Summers D.F. and Maizel J.V. 1967. Disaggregation of HeLa cell polysomes after infection with poliovirus. *Virology* **31**: 550–552.

Summers D.F., Maizel J.V., and Darnell J.E. 1965. Evidence for virus-specific noncapsid proteins in poliovirus-infected HeLa cells. *Proc. Natl. Acad. Sci.* **54**: 505–513.

Tamas I., Klasson L., Canback B., Naslund A.K., Eriksson A.S., Wernegreen J.J., Sandstrom J.P., Moran N.A., and Andersson S.G. 2002. 50 million years of genomic stasis in endosymbiotic bacteria. *Science* **296**: 2376–2379.

Theodorakis N.G., Banerji S.S., and Morimoto R.I. 1988. HSP70 mRNA translation in chicken reticulocytes is regulated at the level of elongation. *J. Biol. Chem.* **263**: 14579–14585.

Thomas G.P. and Mathews M.B. 1984. Alterations of transcription and translation in HeLa cells exposed to amino acid analogues. *Mol. Cell. Biol.* **4**: 1063–1072.

Tomkins G.M., Garren L.D., Howell R.R., and Peterkofsky B. 1965. The regulation of enzyme synthesis by steroid hormones: The role of translation. *J. Cell. Comp. Physiol.* **66**: 137–151.

Tsung K., Inouye S., and Inouye M. 1989. Factors affecting the efficiency of protein synthesis in *Escherichia coli:* Production of a polypeptide of more than 6000 amino acid residues. *J. Biol. Chem.* **264:** 4428–4433.

Turner G.C. and Varshavsky A. 2000. Detecting and measuring cotranslational protein degradation in vivo. *Science* **289:** 2117–2120.

van Dijk E., Cougot N., Meyer S., Babajko S., Wahle E., and Seraphin B. 2002. Human Dcp2: A catalytically active mRNA decapping enzyme located in specific cytoplasmic structures. *EMBO J.* **21:** 6915–6924.

Ventoso I., Sanz M.A., Molina S., Berlanga J.J., Carrasco L., and Esteban M. 2006. Translational resistance of late alphavirus mRNA to eIF2alpha phosphorylation: A strategy to overcome the antiviral effect of protein kinase PKR. *Genes Dev.* **20:** 87–100.

Walden W.E., Godefroy-Colburn T., and Thach R.E. 1981. The role of mRNA competition in regulating translation. I. Demonstration of competition *in vivo*. *J. Biol. Chem.* **256:** 11739–11746.

Walter P. and Blobel G. 1981. Translocation of proteins across the endoplasmic reticulum. III. Signal recognition protein (SRP) causes signal sequence-dependent and site-specific arrest of chain elongation that is released by microsomal membranes. *J. Cell Biol.* **91:** 557–561.

Warner J.R. 1999. The economics of ribosome biosynthesis in yeast. *Trends Biochem. Sci.* **24:** 437–440.

Warner J.R., Knopf P.M., and Rich A. 1963. A multiple ribosomal structure in protein synthesis. *Proc. Natl. Acad. Sci.* **49:** 122–129.

Waters E., Hohn M.J., Ahel I., Graham D.E., Adams M.D., Barnstead M., Beeson K.Y., Bibbs L., Bolanos R., Keller M., et al. 2003. The genome of *Nanoarchaeum equitans:* Insights into early archaeal evolution and derived parasitism. *Proc. Natl. Acad. Sci.* **100:** 12984–12988.

Waxman H.S. and Rabinowitz M. 1966. Control of reticulocyte polyribosome content and hemoglobin synthesis by heme. *Biochim. Biophys. Acta* **129:** 369–379.

Wilt F.H. and Hultin T. 1962. Stimulation of phenylalanine incorporation by polyuridylic acid in homogenates of sea urchin eggs. *Biochem. Biophys. Res. Commun.* **9:** 313–317.

Witherell G.W., Gott J.M., and Uhlenbeck O.C. 1991. Specific interaction between RNA phage coat proteins and RNA. *Prog. Nucleic Acid Res. Mol. Biol.* **40:** 185–220.

Wolin S.L. and Walter P. 1988. Ribosome pausing and stacking during translation of a eukaryotic mRNA. *EMBO J.* **7:** 3559–3569.

Yenofsky R., Bergmann I., and Brawermann G. 1982. Messenger RNA species partially in a repressed state in mouse sarcoma ascites cells. *Proc. Natl. Acad. Sci.* **79:** 5876–5880.

Zamecnik P.C. 1960. Historical and current aspects of the problem of protein synthesis. *Harvey Lect.* **54:** 256–281.

2

Structure of the Bacterial Ribosome and Some Implications for Translational Regulation

Harry F. Noller

Center for Molecular Biology of RNA and
Department of Molecular, Cell, and Developmental Biology
University of California, Santa Cruz
Santa Cruz, California 95064

TRANSLATIONAL REGULATION IS BASED ON MODULATION of translational function, most often involving the initiation phase. Not surprisingly, regulation of protein synthesis differs markedly between bacteria and eukarya, reflecting the many differences between their respective mechanisms of initiation. Although the structures of all ribosomes share commonly conserved cores, which are responsible for the main processes of translational elongation, many of the molecular components involved in translational initiation are specific to the different phylogenetic domains. These include the initiation factors, the Shine-Dalgarno sequence, formylation of the methionyl initiator tRNA, the ability to reinitiate on polycistronic mRNAs, and so on. Thus, it is not at all clear how far our knowledge of 70S (prokaryotic) ribosome structure will go toward providing insight into the mechanisms of eukaryotic translational regulation. Nevertheless, this information will help to understand prokaryotic initiation, and at least provide a starting point for interpreting the emerging structures of eukaryotic ribosomes.

Most of the steps of protein synthesis appear to be based on RNA, including the many interactions between mRNA, tRNA, and rRNA that occur during the elongation phase. The roles of the proteins, such as the elongation factors and ribosomal proteins, may be to refine underlying RNA-based mechanisms, optimizing the speed and accuracy of translation. Translational initiation, at least in part, is therefore *likely* to involve

Translational Control in Biology and Medicine ©2007 Cold Spring Harbor Laboratory Press 978-087969767-9
41

modulation of RNA-based processes by proteins such as the initiation factors. We are beginning to understand how some of these processes work, from several decades of biochemical and genetic studies combined with the more recent X-ray and cryo-electron microscopic (cryo-EM) structures of ribosomes and their functional complexes. This chapter reviews the structure of the bacterial ribosome and its relevance to translational initiation and elongation.

STRUCTURE OF THE 70S RIBOSOME

Bacterial 70S ribosomes have a molecular weight of about 2.5 MD and are formed from a small 30S (0.8 MD) and a large 50S (1.7 MD) subunit. They are composed of 60% RNA and 40% protein. The 30S subunit contains a 16S rRNA of about 1540 nucleotides and a single copy of each of about 20 different proteins, which in *Escherichia coli* are named S1 through S21. The 50S subunit contains a 23S rRNA (~2900 nucleotides), a 5S rRNA (~120 nucleotides), and more than 30 different proteins, named L1 through L34 in *E. coli*. The 50S proteins are also present as single copies, except for protein L7/L12, for which there are four copies.

Many crystal structures of ribosomes and their subunits have been obtained during the past few years (Ban et al. 2000; Wimberly et al. 2000; Schluenzen et al. 2001; Yusupov et al. 2001; Schuwirth et al. 2005), including a 5.5 Å structure of a functional complex of the 70S ribosome from *Thermus thermophilus* containing mRNA and tRNAs bound to the P and E sites (Yusupov et al. 2001). The path of the mRNA and the position of an A-site tRNA have also been determined at about 7 Å resolution (Yusupova et al. 2001). More recently, the structure of the *T. thermophilus* 70S ribosome bound with a threonyl tRNA synthetase mRNA operator and two tRNAs has been solved (Jenner et al. 2005). Fitting of the RNA and protein components to the 5.5 Å electron density map was greatly facilitated by higher-resolution structures of the 30S subunit from *T. thermophilus* (Wimberly et al. 2000) and the 50S subunits from the archaeon *Haloarcula marismortui* (Ban et al. 2000). Very recently, 3.5 Å structures of two different conformational states of the vacant *E. coli* 70S ribosome have appeared, providing the most detailed structural description so far of the complete ribosome (Schuwirth et al. 2005). In addition, structures of many functional complexes have been obtained by cryo-EM reconstruction (Frank 2003; Valle et al. 2003; Allen et al. 2005). Many of the molecular features of these low-resolution structures can be fitted by reference to the X-ray structures, providing important informa-

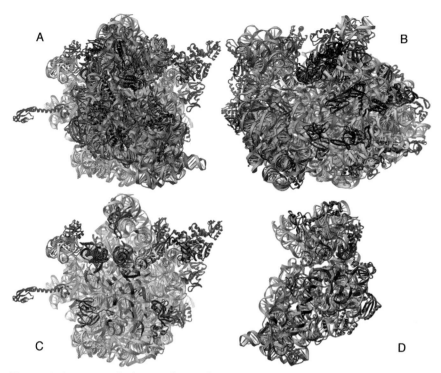

Figure 1. Structure of a bacterial 70S ribosome from *T. thermophilus* (Yusupov et al. 2001). (*A*) A view from the 30S subunit solvent face; (*B*) a view looking down the subunit interface, with the 30S subunit on the left and the 50S on the right. (*C*) Interface view of the 50S subunit, showing the three tRNAs bound to the E, P, and A sites. (*D*) Interface view of the 30S subunit, showing the binding locations for the mRNA and the anticodon stem-loops of the three tRNAs. (*Cyan*) 16S rRNA; (*dark blue*) 30S proteins; (*gray*) 23S rRNA; (*gray-blue*) 5S rRNA; (*magenta*) 50S proteins; (*green*) mRNA; (*yellow*) A-site tRNA; (*orange*) P-site tRNA; (*red*) E-site tRNA. In addition, the intersubunit contacts are colored *magenta* in *C* and *D*.

tion about functional states of the ribosome that have not yielded to crystallization attempts.

Figure 1 shows the structure of the *T. thermophilus* 70S ribosome and its subunits, as well as the locations of the mRNA and tRNAs bound to the A, P, and E sites (Yusupov et al. 2001). The 70S ribosome is approximately spherical, with a diameter of about 250 Å. The mRNA wraps around the neck of the 30S subunit, in a groove between the head and the body of the subunit (Fig. 1D) (Yusupova et al. 2001). The three tRNAs bind inside the 70S ribosome, at the interface between the 30S and 50S subunits. The anticodon stem-loop regions of the tRNAs bind to the 30S subunit, and the rest of their structures (D stem, elbow, and acceptor arm) bind to the 50S subunit. The anticodons of the A and P tRNAs are base-paired to the A and P codons of the mRNA, which are exposed on the subunit interface side of the subunit. The acceptor ends of the A and P tRNAs converge deep inside the peptidyl transferase cavity of the 50S subunit, and the acceptor end of the deacylated tRNA is located in a separate E-site cavity, to the left of the peptidyl transferase region (Fig. 1C).

The ribosomal proteins are mainly distributed over the outer surface of the ribosome, away from the RNA-rich functional interface. In fact, most of the interactions between tRNA and the ribosome involve contacts with 16S or 23S rRNA. Nevertheless, several ribosomal proteins are found in the interface cavity, and some of them contact the tRNAs. In the 30S subunit, the carboxy-terminal tails of proteins S9 and S13 contact the anticodon stem-loop of the P-site tRNA, although it has been shown that these tails can be deleted without loss of viability (Hoang et al. 2004). Protein S7 contacts the anticodon loop of the E-site tRNA, apparently helping to displace the mRNA. Protein S12 is close to the site of codon–anticodon interaction in the 30S A site, but it mainly serves to bolster interactions between 16S rRNA and the A-site codon and anticodon (Ogle et al. 2001). The elbows of all three tRNAs contact 50S proteins: L16 with the T stem-loop of A-tRNA, L5 with the T loop of P-tRNA, and L1 with the T stem-loop of the E-tRNA. In addition, the 3′-terminal tail of E-site tRNA interacts with protein L31. Many lines of evidence support the conclusion that tRNA binding and probably all ribosomal functions are based fundamentally on rRNA and that the roles of the ribosomal proteins are to enhance the functional capabilities of the RNA, as well as to assist its folding and assembly.

Subunit interaction is mediated by a dozen intersubunit bridges. The core interactions, comprising a triangular patch of contacts at the center of the interface surfaces below the tRNA-binding sites (Fig. 1C,D), are formed by contacts between 16S and 23S rRNA. The peripheral bridges include protein–RNA as well as RNA–RNA interactions; the single protein–protein bridge (bridge B1b, between the head of the 30S subunit and the central protuberance of the 50S subunit) is formed between proteins S13 and L5. The intersubunit bridges do not simply serve to hold the two subunits together; some of them rearrange during relative movement of the subunits in the elongation phase of protein synthesis (Frank and Agrawal 2000; Valle et al. 2003) and must be disrupted after each round of translation to allow formation of the 30S initiation complex.

INITIATION FACTORS

Following termination of the polypeptide chain, 70S ribosomes are recycled by removal of the remaining mRNA and deacylated tRNA, catalyzed by ribosome recycling factor (RRF), elongation factor EF-G, and initiation factor IF3 (Janosi et al. 1996; Gao et al. 2005; for review, see Chapter 7). IF3 binds to the dissociated 30S subunit, preventing its reassociation with the 50S subunit and, in addition, helps to discriminate initiator tRNA from elongator tRNAs (Hartz et al. 1989; Gualerzi et al. 2000). IF3 has been localized on the 30S subunit by chemical probing (Muralikrishna and Wickstrom 1989; Dallas and Noller 2001), cryo-EM reconstruction (McCutcheon et al. 1999), and crystallography (Pioletti et al. 2001), with sometimes conflicting results. The chemical probing studies used both chemical footprinting (with kethoxal, DMS, CMCT, and hydroxyl radicals) and directed hydroxyl radical probing, using Fe(II)-EDTA probes tethered to 15 different positions on the surface of IF3 (Muralikrishna and Wickstrom 1989; Moazed et al. 1995; Dallas and Noller 2001). Chemical probing and cryo-EM studies place the carboxy-terminal domain of IF3 at the 50S contact surface of the 30S subunits, at the front of the platform of the subunit, covering the interaction sites for bridges B2b, B2c, and B7a, consistent with its ability to prevent subunit reassociation (Subramanian and Davis 1970; Grunberg-Manago et al. 1975). An X-ray cocrystal structure of the carboxy-terminal domain of IF3 and 30S subunits, in contrast, places the factor on the solvent surface of the 30S subunit, on the side opposite to that of the subunit interface (Pioletti et al. 2001). However, the position for the IF3 C domain deduced from chemical probing is located at a lattice contact in the crystal form used in the X-ray studies, which would prevent any potential cocrystallization of the 30S subunit with the factor bound at this site. Furthermore, it would place the C domain of IF3 in the same location occupied by the Shine-Dalgarno helix during initiation (see below), suggesting that binding of the IF3 C domain to the back of the subunit is either an artifact of crystallization or a secondary binding site. Localization of the N domain of IF3 by chemical probing was less well determined than for the C domain, but all data constrain its placement in the vicinity of the 30S E site, consistent with its cross-linking to proteins S7 and S11 and with recent cryo-EM findings (MacKeen et al. 1980; Boileau et al. 1983; Allen et al. 2005). The modeled location and orientation of the two domains of IF3, based on the chemical probing data, are shown in Figure 2.

IF1 was first localized by chemical footprinting to the vicinity of the 30S A site, actually protecting some of the same bases in 16S rRNA that

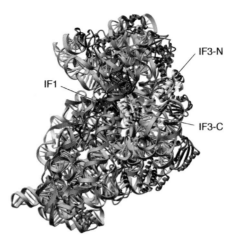

Figure 2. Structure of the *T. thermophilus* 30S ribosomal subunit (Wimberly et al. 2000; Yusupov et al. 2001) showing the crystallographically determined location of initiation factor IF1 (Carter et al. 2001) and the positions of the amino- and carboxy-terminal domains of IF3 modeled from data obtained from chemical footprinting and directed hydroxyl radical probing (Dallas and Noller 2001). The position of initiator tRNA^{fMet} as seen in the structure of the 70S complex is shown in *orange* (Yusupov et al. 2001) and the mRNA in *green* (Yusupova et al. 2001).

are protected by A-site tRNA (Moazed et al. 1995). This localization was demonstrated directly by a cocrystal structure of IF1 bound to the 30S subunit (Carter et al. 2001). IF1 binds to the bottom of the 30S A site (Fig. 2), overlapping with the position of the A-tRNA, and flipping the essential bases A1492 and A1493 out of their positions in the penultimate stem (helix 44) and into a pocket in IF1, preventing any possible binding of tRNA to the A site. At the same time, it distorts the conformation of the penultimate stem and alters the relative orientations of the three domains of the subunit. Finally, its binding to the top of the penultimate stem interferes with formation of intersubunit bridge B2a, helping to prevent subunit reassociation.

IF1 thus blocks the A site, and the N domain of IF3 most likely blocks the E site of the 30S subunit during translational initiation, helping to direct tRNA to the 30S P site, where the initiator tRNA is bound. Interactions between initiator tRNA and the 30S subunit in an initiation complex have not been determined directly by crystallography, but they are likely to share many similarities with those seen for a 70S functional complex bound with an initiator tRNA in the P site (Yusupov et al. 2001), judging from the very similar 16S rRNA footprints observed for initiator tRNA bound in a 30S initiation complex and in 70S ribosomes (Moazed and Noller 1986, 1990; Moazed et al. 1995). In addition to the afore-mentioned interactions involving the carboxy-terminal tails of proteins S9 and S13, the tRNA and mRNA interact with several conserved nucleotide bases of 16S rRNA. G1338 and A1339 contact the minor groove of the anticodon stem, a special interaction that is discussed in more detail below; m^2G966 packs against the ribose of the wobble nucleotide; C1400 stacks on the wobble base of the tRNA anticodon; and G926 contacts the phosphate of nucleotide 1 of the P-site codon. In addition, there are contacts between tRNA and the backbone moieties of 16S rRNA nucleotides 790, 1229, 1339, and 1498 (A. Korostelev et al., unpubl.).

Recently, Frank, Ehrenberg, and coworkers have reported a 13.8 Å cryo-EM structure of a 70S initiation complex formed with IF2·GDPNP along with IF1, IF3, and initiator tRNA (Allen et al. 2005). Apparent low occupancy of IF2 in the complexes required "supervised classification" of the images in order to obtain maps showing solid density for the ligand portion. The ligand density and its structural interpretation (Fig. 3) provide the first view of IF2 bound to the ribosome. It was assumed that the G domain of IF2 binds in a similar position and orientation as observed previously for the G domains of EF-Tu and EF-G (bacterial factors also named EF1A and EF2, respectively), and this provided the starting point for fitting its structure. From the shape of the density, it was inferred that

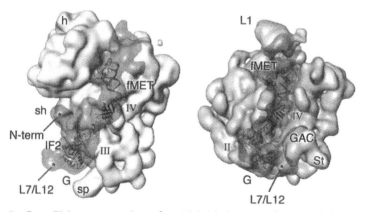

Figure 3. Cryo-EM reconstruction of a 70S initiation complex containing IF1, IF2, IF3, fMet-tRNA, and mRNA (Allen et al. 2005). Difference electron density at 13.8 Å is shown in semitransparent *red*, with the modeled positions of IF2 (its amino terminus, G domain, and domains II, III, and IV are indicated), fMet-tRNA, and an L7/L12 dimer. (GAC) GTPase-associated center; (St), L7/L12 stalk; (L1), the L1 stalk; (sh), shoulder; (h), head; (sp), spur. (Reprinted, with permission, from Allen et al. 2005 [© Elsevier].)

domains III and IV of IF2 extend in an elongated conformation into the subunit interface cavity, differing substantially from that observed for the crystal structure of IF2 from the archaeon *Methanobacterium thermoautotrophicum* (Roll-Mecak et al. 2000). The modeled position of the initiator tRNA differs from that seen for the classic P/P state either in the 70S crystal structure or in cryo-EM reconstructions of the 70S ribosome. In the 70S initiation complex, the initiator tRNA is rotated about its anticodon by about 20°, moving the acceptor end of the tRNA by 28 Å into a novel "I" position on the 50S subunit; accordingly, Allen et al. term this mode of tRNA binding the P/I state. Density features were also found attributable to IF1, corresponding to its crystallographically observed location (Carter et al. 2001) and for the carboxy- and amino-terminal domains of IF3 similar to those predicted from chemical probing studies (Dallas and Noller 2001). Finally, there is a 4° rotation of the 30S subunit relative to the 50S subunit, recalling that seen in cryo-EM reconstructions of the pretranslocation state of the 70S ribosome (Frank and Agrawal 2000; Valle et al. 2003). It was proposed that GTP hydrolysis is required to allow movement of the initiator tRNA from the P/I state into the P/P state, possibly as a proofreading step to help exclude binding of an elongator tRNA.

INITIATION MECHANISMS

One of the first structurally interpretable mechanisms of the translational initiation process was the discovery of the Shine-Dalgarno interaction (Shine and Dalgarno 1974; Steitz and Jakes 1975). In this mechanism, the authentic mRNA start codon is positioned in or very near the 30S P site by base-pairing of an upstream mRNA sequence with the CCUCC sequence found near the 3' end of 16S rRNA in all bacterial and archaeal ribosomes. An intramolecular mRNA–rRNA double helix, whose length depends on the extent of complementarity of the particular mRNA sequence with 16S rRNA, is formed and bound by the 30S subunit. A 6.5 Å difference map calculated from 70S ribosomes with and without mRNA bound revealed electron density corresponding to the predicted 8-bp helix in a pocket formed between the head, platform, and body of the 30S subunit (Fig. 4) (Yusupova et al. 2001). The pocket is formed by portions of proteins S7, S11, and S18, as well as parts of helices 20, 28, and 37 of 16S rRNA. The mRNA then passes through a short tunnel formed between protein S7 and 16S rRNA (the "upstream tunnel") to the subunit interface, where the P site is located about ten nucleotides downstream from the Shine-Dalgarno pocket. The close fit of the pocket

Figure 4. Binding of the Shine-Dalgarno helix to the ribosome, viewed from the solvent side of the 30S subunit. The helix, shown modeled into a 6.5 Å mRNA difference map (Yusupova et al. 2001), is lodged in a pocket formed by elements of the head and platform of the 30S subunit. The CCUCC sequence of 16S rRNA and complementary GGAGG sequence of mRNA are shown in *magenta*. Elsewhere, the molecular components are colored as depicted in Fig. 1. Ribosomal proteins S2, S7, S11, and S18 and helices 20, 28, and 37 of 16S rRNA are indicated.

to the Shine-Dalgarno helix, which precisely positions the upstream region of the mRNA relative to the P site, explains the accuracy of start codon selection.

It has been shown that IF3, and specifically its C domain, has a role in discrimination of initiator from elongator tRNA during formation of the 30S initiation complex (Pon and Gualerzi 1974; Hartz et al. 1989). Furthermore, only the anticodon stem-loop (ASL) region of the tRNA interacts with the 30S subunit, and in fact, the ASLs of the initiator and elongator tRNAs can be discriminated efficiently (Hartz et al. 1989). Because the initiator and elongator methionine tRNAs have identical codons, the discriminatory mechanism must recognize some other structural feature of their ASLs. RajBhandary and coworkers observed that initiator tRNAs contain a unique set of three G-C base pairs in their anticodon stems adjacent to the anticodon loop (RajBhandary and Chow 1995). In the structure of the 70S ribosome with an initiator tRNA bound to the P site, there is contact between the minor groove of the anticodon stem at this very position with bases G1338 and A1339 (Yusupov et al. 2001). Although it is not possible to make detailed structural interpretations from the 5.5 Å X-ray structure, it was proposed that these two bases might make a steric check of the identity of the ASL by recognition of the G-C base pairs in the minor groove of the helix (Dallas and Noller 2001). Recently, this model has been tested by mutation of G1338 and A1339, making use of an affinity purification scheme to obtain pure populations of mutant ribosomes for in vitro analysis (Lancaster and Noller 2005). A detailed analysis of ribosomes carrying seven different mutations at these two positions supports the conclusion that A1339 forms a type I A-minor interaction (Nissen et al. 2001) with the G30-C40 base pair and that G1338 forms a type II "G-minor" interaction with the G29-C41 base pair. It was suggested that the third tRNA base pair, G31-C29, could have a role in a proofreading step, in which the "reading frame" of G1338 and A1339 is shifted to check the lower two G-C pairs instead of the upper two. This conformational shift might be induced by IF3, possibly by movement of the head of the 30S subunit, in which G1338 and A1339 are located. Indeed, the position of IF3 inferred from chemical probing and cryo-EM studies is remote from the conserved G-C pairs of the initiator tRNA, and so must in some way affect discrimination indirectly. Most likely, discrimination is based on tRNA-30S subunit interactions that are modulated by the binding of IF3. This interpretation is supported by the finding that deletion of the carboxy-terminal tail of ribosomal protein S9, which contacts the anticodon loop of P-site tRNA, causes the mutant ribosomes to favor binding to tRNAs that have initiator-like G-C pairs in their anticodon stems (Hoang et al. 2004). Thus, the recognition of the initiator stem may be an inherent specificity of the 30S subunit that is masked by protein S9 and unmasked by IF3. This specificity is suppressed by binding of the 50S subunit (Hoang et al. 2004; Lancaster and Noller 2005), suggesting that primitive RNA-based ribosomes could have initiated on the small subunit without the help of initiation factors. It also suggests that the main task of the founding initiation factor was to promote subunit dissociation (Lancaster and Noller 2005).

An example of how mRNA structure and ribosome structure interact to regulate gene expression has come recently from the work of Yusupov and coworkers, who solved a cocrystal structure of the *T. thermophilus* 70S ribosome bound with an mRNA containing the threonyl-tRNA synthetase translational operator at 5.5 Å resolution (Jenner et al. 2005). The operator contains a modified stem-loop structure, located just upstream of the Shine-Dalgarno sequence of the mRNA, that binds the synthetase (Romby and Springer 2003). The structural results reveal electron density for both the Shine-Dalgarno helix in its binding pocket and double-helical density corresponding to the stem-loop operator structure, oriented at a 110° angle to the Shine-Dalgarno helix and perpendicular to the surface of the 30S subunit. Modeling studies show that binding of the synthetase to the operator helix would be prevented by clash between the synthetase and the ribosome when the mRNA is bound in a 30S initiation complex, explaining how the synthetase acts as a translational repressor. Curiously, when the operator-containing mRNA was bound, the normally observed E-site tRNA was absent, and a tRNA was found bound to the A site. This unexpected observation has yet to be explained, but it may be related to a previously proposed negative cooperativity between A- and E-site binding (Blaha and Nierhaus 2001).

THE ELONGATION CYCLE

Following initiation, protein synthesis proceeds by stepwise attachment of amino acids to the peptidyl chain, a process that we are now beginning to understand in structural terms (Fig. 5). The elongation phase begins with an initiator tRNA bound to the 70S ribosome in the P/P state (the classic "P-site" binding state), in which its anticodon pairs with the codon in the

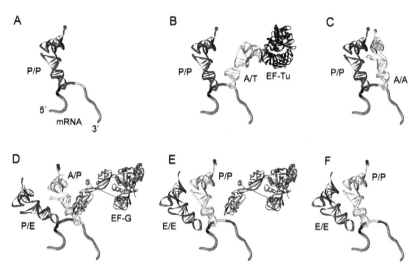

Figure 5. Structural representation of the binding states of tRNA in the ribosome during the translational elongation cycle from crystallographic (Yusupov et al. 2001) and cryo-EM (Agrawal et al. 1999; Valle et al. 2002; Frank 2003) studies. (A) Initiator fMet-tRNA bound to mRNA in the P/P state; (B) initial recognition of the aminoacyl-tRNA·EF-Tu·GTP ternary complex (A/T hybrid state); (C) release of EF-Tu·GDP and accommodation of aminoacyl-tRNA in the A/A state, leading to peptide bond formation; (D) binding of elongation factor EF-G and movement of peptidyl- and deacylated tRNAs into the A/P and P/E hybrid states, in the first step of translocation; (E) EF-G-catalyzed translocation of mRNA and tRNAs into the P/P and E/E states in the second step of translocation; (F) release of EF-G·GDP, vacating the A site for the next incoming aminoacyl-tRNA.

30S subunit P site and its CCA-(fMet) acceptor end is bound to the pep-tidyl transferase center of the 50S P site (Fig. 5A). Elongation consists of repeated cycles of three processes: (1) aminoacyl-tRNA selection, (2) pep-tide bond formation, and (3) translocation. Aminoacyl-tRNA selection is the process in which the tRNA specified by the codon exposed in the 30S A site is selected. The aminoacyl-tRNA is delivered to the ribosome as an aminoacyl-tRNA·EF-Tu·GTP ternary complex. The EF-Tu moiety of the ternary complex first binds nonspecifically to the ribosome, followed by insertion of the anticodon stem-loop of the tRNA into the 30S A site; the aminoacyl end of the tRNA is bound by EF-Tu, which prevents incorrect aminoacyl-tRNAs from entering the peptidyl transferase center; this hy-brid binding state is called the A/T state (Fig. 5B) (Moazed and Noller 1989a,b). When a cognate tRNA is bound, recognized, and stabilized by RNA–RNA interactions between bases G530, A1492, and A1493 of 16S rRNA and the codon–anticodon duplex in the 30S subunit decoding site (Ogle et al. 2001; Ogle and Ramakrishnan 2005), the GTPase activity of EF-Tu is stimulated dramatically, by an as yet unknown mechanism (Rod-nina et al. 1995). Upon hydrolysis of GTP, EF-Tu undergoes a conforma-tional change to its GDP form, which has a greatly decreased affinity for aminoacyl-tRNA and for the ribosome; consequently, EF-Tu·GDP is re-leased, leaving behind the aminoacyl-tRNA. Coupled to GTP hydrolysis and EF-Tu release, the aminoacyl end of the tRNA moves into the peptidyl transferase center of the 50S subunit, i.e., from the A/T state to the A/A state, corresponding to classic "A-site" binding (Fig. 5C). This step is called accommodation (Pape et al. 1998). Accommodation proceeds sur-prisingly slowly in the absence of EF-Tu·GTP, revealing the existence of a poorly understood kinetic barrier that suggests a requirement for some structural rearrangement within the ribosome.

When a peptidyl- (or initiator fMet-) tRNA is bound in the P/P state, and an aminoacyl-tRNA is bound in the A/A state (Fig. 5C), peptide bond formation proceeds spontaneously. The catalytic activity is peptidyl trans-ferase, a function of the 50S subunit itself. The reaction involves nucleophilic attack of the carbonyl group of the peptidyl-tRNA ester linkage by the α-amino group of the aminoacyl-tRNA, resulting in transfer of the pep-tidyl moiety from the peptidyl-tRNA to the aminoacyl-tRNA. The imme-diate result is a deacylated tRNA bound in the P/P state and an elongated peptidyl group attached to the tRNA bound in the A/A state. It is virtually certain that the catalytic center is composed exclusively of RNA, showing that the ribosome is a ribozyme (Noller et al. 1992; Nissen et al. 2000). The detailed mechanism of action of peptidyl transferase has been the topic of some controversy and is an intensive area of ongoing investigation (Muth et al. 2000; Nissen et al. 2000; Barta et al. 2001; Polacek et al. 2001; Thompson et al. 2001; Schmeing et al. 2005). Since hydrolysis of the peptidyl-tRNA linkage is also catalyzed by peptidyl transferase during the termination phase, it is remarkable that spontaneous peptide release is virtually undetectable, given the ready access of water to the catalytic site.

Finally, the process of translocation consists of coupled movement of the mRNA and tRNAs to vacate the A site for the next incoming aminoacyl-tRNA. Translocation is catalyzed by EF-G·GTP. EF-G shows a striking overall structural similarity to the aminoacyl-tRNA·EF-Tu·GTP ternary complex (Fig. 5D); its GTP-binding domain is structurally related to those of EF-Tu and IF2 and binds to the ribosome at the same place (Moazed and Noller 1987; Frank 2003). Translocation occurs in two steps. In the first step, the acceptor ends of the tRNAs move with respect to the 50S subunit, while maintaining their positions in the 30S subunit. Thus, the acceptor end of the deacylated tRNA moves into the 50S E site, and the acceptor end of the peptidyl-tRNA moves into the 50S P site, leaving the two tRNAs in the P/E and A/P hybrid states, respectively (Fig. 5D). This step may be catalyzed by EF-G prior to GTP hydrolysis (Zavialov and Ehrenberg 2003), but it was originally shown to occur spontaneously under in vitro conditions (Moazed and Noller 1989a). Recent single-molecule FRET (fluorescence resonance energy transfer) studies suggest that the pretranslocation tRNAs are in dynamic equilibrium with the hybrid state (Blanchard et al. 2004). The second step of translocation (Fig. 5E) consists of coupled movement of the mRNA and its associated tRNA anticodon ends, moving the tRNAs from their P/E and A/P states into the E/E and P/P states, corresponding to the classic E and P sites (Moazed and Noller 1989a). This step requires EF-G-catalyzed hydrolysis of GTP and is followed by release of EF-G (Fig. 5F). Although it was previously believed that GTP hydrolysis was required only for release of EF-G from the ribosome, more recent pre-steady-state kinetic studies showed that GTP hydrolysis precedes (the second step of) translocation (Rodnina et al. 1997). The final result is an empty A site in which the next codon is positioned in the 30S decoding site.

The molecular mechanism of translocation is not well understood, but some basic properties have been established. The minimal ribosomal complex needed for in vitro translocation includes mRNA, a full-length P-site tRNA, and an anticodon stem-loop bound to the A site (Joseph and Noller 1998). A major question is how such large-scale molecular movements, on the order of 20–50 Å, are driven and calibrated. A likely explanation is that they are accompanied by structural rearrangements in the ribosome of a comparable magnitude (Bretscher 1968;

Spirin 1970). The two-step hybrid states model (Moazed and Noller 1989a) implied relative movement of the two ribosomal subunits, which has now been directly observed. Cryo-EM studies of pre- and post-translocation complexes have revealed differences in the relative orientation of the 30S and 50S subunits between the two states, amounting to a 6° rotation of the 30S subunit relative to the 50S (Frank and Agrawal 2000; Valle et al. 2003). On the basis of these findings, Frank and coworkers proposed a ratcheting model for the mechanics of translocation in which back and forth rotational movement between the subunits, coupled with sequential binding and release of the two ends of the tRNAs, drives the stepwise movement of mRNA and tRNA through the ribosome (Frank and Agrawal 2000; Valle et al. 2003). A further question is whether mRNA and tRNA are both actively moved, or whether movement of one or the other is passive, following movement of its associated RNA. Studies by Spirin and coworkers showed that, for certain in vitro systems, tRNAs bound to the ribosome in the absence of mRNA can be translocated in an EF-G-dependent manner (Belitsina et al. 1981), showing that the translocation mechanism acts directly on tRNA; accordingly, it is possible that mRNA translocation may be passive. Another central question is whether translocation is a property of EF-G, which has often been called a "translocase," or whether it is a more fundamental property of the tRNA–ribosome structural complex. Under some in vitro conditions, EF-G-independent translocation has been observed (Gavrilova and Spirin 1971; Southworth et al. 2002). Indeed, the peptidyl transferase inhibitor sparsomycin is able to trigger accurate and efficient translocation in the absence of EF-G and GTP, albeit at reduced rates (Fredrick and Noller 2003). These findings show that translocation is a property of the ribosome itself and that EF-G catalyzes the process with the expenditure of GTP. Moreover, they show that the energy stored in the tRNA–ribosome complex is sufficient to drive translocation. This energy must ultimately come from the free energy of peptide bond formation. The direction of movement is determined by the changing chemical identity of the acceptor end of the tRNA, from aminoacyl to peptidyl to deacyl and the corresponding binding specificities of the A, P, and E sites. It may well turn out to be the case that translocation, like most other ribosomal processes, is a mechanism based on RNA, reflecting the molecular origins of this ancient machine.

We have now entered the era of explaining mechanisms of translational regulation in stereochemical terms. In the coming years, our knowledge of ribosome and factor structure will expand to include eukaryotic systems, which promise to be even more complex than the prokaryotic examples now in hand (Spahn et al. 2001). And because regulatory mechanisms are in principle infinite, research in this challenging field is not likely to slow down anytime soon.

ACKNOWLEDGMENTS

I thank the members of my laboratory for stimulating discussions, Phyllis Tveit for preparation of the manuscript, and Carl Gorringe for preparing the figures. Work in my laboratory is supported by grants from the National Institutes of Health, National Science Foundation, and the Agouron Institute.

REFERENCES

Agrawal R.K., Heagle A.B., Penczek P., Grassucci R.A., and Frank J. 1999. EF-G-dependent GTP hydrolysis induces translocation accompanied by large conformational changes in the 70S ribosome. *Nat. Struct. Biol.* **6:** 643–647.

Allen G.S., Zavialov A., Gursky R., Ehrenberg M., and Frank J. 2005. The cryo-EM structure of a translation initiation complex from *Escherichia coli. Cell* **121:** 703–712.

Ban N., Nissen P., Hansen J., Moore P.B., and Steitz T.A. 2000. The complete atomic structure of the large ribosomal subunit at 2.4 Å resolution. *Science* **289:** 905–920.

Barta A., Dorner S., and Polacek N. 2001. Mechanism of ribosomal peptide bond formation. *Science* **291:** 203.

Belitsina N.V., Tnalina G.Z., and Spirin A.S. 1981. Template-free ribosomal synthesis of polylysine from lysyl-tRNA. *FEBS Lett.* **131:** 289–292.

Blaha G. and Nierhaus K.H. 2001. Features and functions of the ribosomal E site. *Cold Spring Harbor Symp. Quant. Biol.* **66:** 135–146.

Blanchard S.C., Kim H.D., Gonzalez R.L., Jr., Puglisi J.D., and Chu S. 2004. tRNA dynamics on the ribosome during translation. *Proc. Natl. Acad. Sci.* **101:** 12893–12898.

Boileau G., Butler P., Hershey J.W., and Traut R.R. 1983. Direct cross-links between initiation factors 1, 2, and 3 and ribosomal proteins promoted by 2-iminothiolane. *Biochemistry* **22:** 3162–3170.

Bretscher M.S. 1968. Translocation in protein synthesis: A hybrid structure model. *Nature* **218:** 675–677.

Carter A.P., Clemons W.M., Jr., Brodersen D.E., Morgan-Warren R.J., Hartsch T., Wimberly B.T., and Ramakrishnan V. 2001. Crystal structure of an initiation factor bound to the 30S ribosomal subunit. *Science* **291:** 498–501.

Dallas A. and Noller H.F. 2001. Interaction of translation initiation factor 3 with the 30S ribosomal subunit. *Mol. Cell* **8:** 855–864.

Frank J. 2003. Electron microscopy of functional ribosome complexes. *Biopolymers* **68:** 223–233.

Frank J. and Agrawal R.K. 2000. A ratchet-like inter-subunit reorganization of the ribosome during translocation. *Nature* **406:** 318–322.

Fredrick K. and Noller H.F. 2003. Catalysis of ribosomal translocation by sparsomycin. *Science* **300:** 1159–1162.

Gao N., Zavialov A.V., Li W., Sengupta J., Valle M., Gursky R.P., Ehrenberg M., and Frank J. 2005. Mechanism for the disassembly of the posttermination complex inferred from cryo-EM studies. *Mol. Cell* **18:** 663–674.

Gavrilova L.P. and Spirin A.S. 1971. Stimulation of "non-enzymic" translocation in ribosomes by p-chloromercuribenzoate. *FEBS Lett.* **17:** 324–326.

Grunberg-Manago M., Dessen P., Pantaloni D., Godefroy-Colburn T., Wolfe A.D., and Dondon J. 1975. Light-scattering studies showing the effect of initiation factors on the reversible dissociation of *Escherichia coli* ribosomes. *J. Mol. Biol.* **94:** 461–478.

Gualerzi C., Caserta E., La Teana A., Spurio R., Tomsic J., and Pon C.L. 2000. Translation initiation in bacteria. In *The ribosome: Structure, function, antibiotics and cellular iterations* (ed. R. Garrett et al.), pp. 477–494. ASM Press, Washington, D.C.

Hartz D., McPheeters D.S., and Gold L. 1989. Selection of the initiator tRNA by *Escherichia coli* initiation factors. *Genes Dev.* **3:** 1899–1912.

Hoang L., Fredrick K., and Noller H.F. 2004. Creating ribosomes with an all-RNA 30S subunit P site. *Proc. Natl. Acad. Sci.* **101:** 12439–12443.

Janosi L., Hara H., Zhang S., and Kaji A. 1996. Ribosome recycling by ribosome recycling factor (RRF)—An important but overlooked step of protein biosynthesis. *Adv. Biophys.* **32:** 121–201.

Jenner L., Romby P., Rees B., Schulze-Briese C., Springer M., Ehresmann C., Ehresmann B., Moras D., Yusupova G., and Yusupov M. 2005. Translational operator of mRNA on the ribosome: How repressor proteins exclude ribosome binding. *Science* **308:** 120–123.

Joseph S. and Noller H.F. 1998. EF-G-catalyzed translocation of anticodon stem-loop analogs of transfer RNA in the ribosome. *EMBO J.* **17:** 3478–3483.

Lancaster L. and Noller H.F. 2005. Involvement of 16S rRNA nucleotides G1338 and A1339 in discrimination of initiator tRNA. *Mol. Cell* **20:** 623–632.

MacKeen L.A., Kahan L., Wahba A.J., and Schwartz I. 1980. Photochemical cross-linking of initiation factor-3 to *Escherichia coli* 30 S ribosomal subunits. *J. Biol. Chem.* **255:** 10526–10531.

McCutcheon J.P., Agrawal R.K., Philips S.M., Grassucci R.A., Gerchman S.E., Clemons W.M., Jr., Ramakrishnan V., and Frank J. 1999. Location of translational initiation factor IF3 on the small ribosomal subunit. *Proc. Natl. Acad. Sci.* **96:** 4301–4306.

Moazed D. and Noller H.F. 1986. Transfer RNA shields specific nucleotides in 16S ribosomal RNA from attack by chemical probes. *Cell* **47:** 985–994.

———. 1987. Interaction of antibiotics with functional sites in 16S ribosomal RNA. *Nature* **327:** 389–394.

———. 1989a. Interaction of tRNA with 23S rRNA in the ribosomal A, P, and E sites. *Cell* **57:** 585–597.

———. 1989b. Intermediate states in the movement of transfer RNA in the ribosome. *Nature* **342:** 142–148.

———. 1990. Binding of tRNA to the ribosomal A and P sites protects two distinct sets of nucleotides in 16S rRNA. *J. Mol. Biol.* **211:** 135–145.

Moazed D., Samaha R.R., Gualerzi C., and Noller H.F. 1995. Specific protection of 16S rRNA by translational initiation factors. *J. Mol. Biol.* **248:** 207–210.

Muralikrishna P. and Wickstrom E. 1989. *Escherichia coli* initiation factor 3 protein binding to 30S ribosomal subunits alters the accessibility of nucleotides within the conserved central region of 16S rRNA. *Biochemistry* **28:** 7505–7510.

Muth G.W., Ortoleva-Donnelly L., and Strobel S.A. 2000. A single adenosine with a neutral pKa in the ribosomal peptidyl transferase center. *Science* **289:** 947–950.

Nissen P., Hansen J., Ban N., Moore P.B., and Steitz T.A. 2000. The structural basis of ribosome activity in peptide bond synthesis. *Science* **289:** 920–930.

Nissen P., Ippolito J.A., Ban N., Moore P.B., and Steitz T.A. 2001. RNA tertiary interaction in the large ribosomal subunit: The A-minor motif. *Proc. Natl. Acad. Sci.* **98:** 4899–4903.

Noller H.F., Hoffarth V., and Zimniak L. 1992. Unusual resistance of peptidyl transferase to protein extraction procedures. *Science* **256:** 1416–1419.

Ogle J.M. and Ramakrishnan V. 2005. Structural insights into translational fidelity. *Annu. Rev. Biochem.* **74:** 129–177.

Ogle J.M., Brodersen D.E., Clemons W.M., Tarry M.J., Carter A.P., and Ramakrishnan V. 2001. Recognition of cognate transfer RNA by the 30S ribosomal subunit. *Science* **292:** 897–902.

Pape T., Wintermeyer W., and Rodnina M. 1998. Complete kinetic mechanism of elongation factor Tu-dependent binding of aminoacyl-tRNA to the A site of the *E. coli* ribosome. *EMBO J.* **17:** 7490–7497.

Pioletti M., Schlunzen F., Harms J., Zarivach R., Gluhmann M., Avila H., Bashan A., Bartels H., Auerbach T., Jacobi C., et al. 2001. Crystal structures of complexes of the small ribosomal subunit with tetracycline, edeine and IF3. *EMBO J.* **20:** 1829–1839.

Polacek N., Gaynor M., Yassin A., and Mankin A.S. 2001. Ribosomal peptidyl transferase can withstand mutations at the putative catalytic nucleotide. *Nature* **411:** 498–501.

Pon C.L. and Gualerzi C. 1974. Effect of initiation factor 3 binding on the 30S ribosomal subunits of *Escherichia coli. Proc. Natl. Acad. Sci.* **71:** 4950–4954.

RajBhandary U. and Chow C.M. 1995. Initiator tRNAs and initiation of protein synthesis. In *tRNA: Structure, biosynthesis and function* (ed. D. Soll and U. RajBhandary), pp. 511–528. ASM Press, Washington, D.C.

Rodnina M.V., Pape T., Fricke R., and Wintermeyer W. 1995. Elongation factor Tu, a GTPase triggered by codon recognition on the ribosome: Mechanism and GTP consumption. *Biochem. Cell Biol.* **73:** 1221–1227.

Rodnina M.V., Savelsbergh A., Katunin V.I., and Wintermeyer W. 1997. Hydrolysis of GTP by elongation factor G drives tRNA movement on the ribosome. *Nature* **385:** 37–41.

Roll-Mecak A., Cao C., Dever T.E., and Burley S.K. 2000. X-Ray structures of the universal translation initiation factor IF2/eIF5B: Conformational changes on GDP and GTP binding. *Cell* **103:** 781–792.

Romby P. and Springer M. 2003. Bacterial translational control at atomic resolution. *Trends Genet.* **19:** 155–161.

Schlunzen F., Zarivach R., Harms J., Bashan A., Tocilj A., Albrecht R., Yonath A., and Franceschi F. 2001. Structural basis for the interaction of antibiotics with the peptidyl transferase centre in eubacteria. *Nature* **413:** 814–821.

Schmeing T.M., Huang K.S., Strobel S.A., and Steitz T.A. 2005. An induced-fit mechanism to promote peptide bond formation and exclude hydrolysis of peptidyl-tRNA. *Nature* **438:** 520–524.

Schuwirth B.S., Borovinskaya M.A., Hau C.W., Zhang W., Vila-Sanjurjo A., Holton J.M., and Cate J.H. 2005. Structures of the bacterial ribosome at 3.5 Å resolution. *Science* **310:** 827–834.

Shine J. and Dalgarno L. 1974. The 3′-terminal sequence of *E. coli* 16S ribosomal RNA complementarity to nonsense triplets and ribosome binding sites. *Proc. Natl. Acad. Sci.* **71:** 1342–1346.

Southworth D.R., Brunelle J.L., and Green R. 2002. EFG-independent translocation of the mRNA:tRNA complex is promoted by modification of the ribosome with thiol-specific reagents. *J. Mol. Biol.* **324:** 611–623.

Spahn C.M., Beckmann R., Eswar N., Penczek P.A., Sali A., Blobel G., and Frank J. 2001. Structure of the 80S ribosome from *Saccharomyces cerevisiae*—tRNA-ribosome and subunit-subunit interactions. *Cell* **107:** 373–386.

Spirin A.S. 1970. A model of the functioning ribosome: Locking and unlocking of the ribosome subparticles. *Cold Spring Harbor Symp. Quant. Biol.* **34:** 197–207.

Steitz J.A. and Jakes K. 1975. How ribosomes select initiator regions in mRNA: Base pair formation between the 3′ terminus of 16S rRNA and the mRNA during initiation of protein synthesis in *Escherichia coli*. *Proc. Natl. Acad. Sci.* **72:** 4734–4738.

Subramanian A.R. and Davis B.D. 1970. Activity of initiation factor F3 in dissociating *Escherichia coli* ribosomes. *Nature* **228:** 1273–1275.

Thompson J., Kim D.F., O'Connor M., Lieberman K.R., Bayfield M.A., Gregory S.T., Green R., Noller H.F., and Dahlberg A.E. 2001. Analysis of mutations at residues A2451 and G2447 of 23S rRNA in the peptidyltransferase active site of the 50S ribosomal subunit. *Proc. Natl. Acad. Sci.* **98:** 9002–9007.

Valle M., Zavialov A., Sengupta J., Rawat U., Ehrenberg M., and Frank J. 2003. Locking and unlocking of ribosomal motions. *Cell* **114:** 123–134.

Valle M., Sengupta J., Swami N., Grassucci R., Burkhardt N., Nierhaus K., Agrawal R., and Frank J. 2002. Cryo-EM reveals an active role for aminoacyl-tRNA in the accommodation process. *EMBO J.* **21:** 3557–3567.

Wimberly B.T., Brodersen D.E., Clemons W.M., Jr., Morgan-Warren R.J., Carter A.P., Vonrhein C., Hartsch T., and Ramakrishnan V. 2000. Structure of the 30S ribosomal subunit. *Nature* **407:** 327–339.

Yusupov M.M., Yusupova G.Z., Baucom A., Lieberman K., Earnest T.N., Cate J.H., and Noller H.F. 2001. Crystal structure of the ribosome at 5.5 Å resolution. *Science* **292:** 883–896.

Yusupova G.Z., Yusupov M.M., Cate J.H., and Noller H.F. 2001. The path of messenger RNA through the ribosome. *Cell* **106:** 233–241.

Zavialov A.V. and Ehrenberg M. 2003. Peptidyl-tRNA regulates the GTPase activity of translation factors. *Cell* **114:** 113–122.

3

Structure and Function of the Eukaryotic Ribosome and Elongation Factors

Derek J. Taylor and Joachim Frank
Howard Hughes Medical Institute, HRI, Wadsworth Center
Empire State Plaza, Albany, New York 12201-0509

Terri Goss Kinzy
Department of Molecular Genetics, Microbiology & Immunology
UMDNJ Robert Wood Johnson Medical School
Piscataway, New Jersey 08854-5635

THE TRANSLATION OF THE GENETIC MESSAGE IS PERFORMED by the ribosome, a large macromolecular assembly made of ribosomal RNA (rRNA) and proteins (rps). In eukaryotes, it is composed of a small 40S subunit (SSU) and a large 60S subunit (LSU), which perform decoding and peptidyl transfer, respectively. In bacteria, the ribosome has a molecular mass of 2.5 MD and is composed of three rRNA molecules and 51 proteins (Chapter 2). Due to additional rps and rRNA expansion segments, the complete eukaryotic 80S ribosome is significantly larger ($>$ 3 MD) than its bacterial counterpart. Functionally, however, the ribosome is relatively conserved throughout all domains of life. The initiation of translation places an mRNA, with a Met-tRNA$_i^{Met}$, at the start codon in the 80S ribosome (Chapter 4). The subsequent elongation phase of protein synthesis is a cycle of aminoacyl-tRNA (aa-tRNA) delivery and peptide bond formation repeated hundreds of times during the synthesis of an average protein. This process is facilitated by the soluble eukaryotic elongation factors (eEFs), which have an important role in ensuring the accuracy of gene expression. The eEFs differ in many ways from their bacterial homologs, and they are posttranslationally modified and regulated (Chapter 21). The higher-order cellular structure of eukaryotes also necessitates highly specific interactions of the ribosome with membranes and its active role in protein transport. In the past decade, significant advances in the structural analysis of the eukaryotic ribosomes, eEFs, and their complexes have allowed new insights into the dynamic processes of translation elongation.

EVOLUTION OF STRUCTURAL STUDIES OF PROTEIN SYNTHESIS

A significant advance in our understanding of how proteins are synthesized can be attributed to the emergence of atomic structures over the past decade (Chapter 2). The crystal structures of the large 50S ribosomal subunit from halophilic archaebacterium and mesophilic bacterium, the small 30S ribosomal subunit from thermophilic bacterium, and the entire bacterial ribosome from *Thermus thermophilus* and *Escherichia coli* have been solved. One of the remarkable findings of these studies was that the catalytic core of the ribosome is composed entirely of RNA.

It has been more challenging to obtain crystals of ribosomes actively engaged in translation with functional ligands bound. For such complexes, cryo-electron microscopy (cryo-EM) has been used successfully to establish the binding positions of tRNA (Agrawal et al. 1996; Stark et al. 1997) and factors involved in bacterial initiation (Allen et al. 2005), elongation (Stark et al. 1997; Agrawal et al. 1998), termination (Klaholz et al. 2003; Rawat et al. 2003), and recycling (Agrawal et al. 2004; N. Gao et al. 2005). The fitting of atomic models from X-ray crystallography and nuclear magnetic resonance (NMR) spectroscopy into density envelopes obtained from cryo-EM maps allows the generation of quasi-atomic models with a precision that exceeds the resolution of the cryo-EM maps severalfold (Rossmann 2000). By the use of these "hybrid methods," it is possible to infer conformational changes and sites of binding interactions in the ribosome.

STRUCTURE OF THE EUKARYOTIC RIBOSOME

Compared to the eubacterial system, structural data and thus functional knowledge about eukaryotic ribosomes and elongation factors were quite sparse until recently. Although no crystal structure of the eukaryotic ribosome or its individual subunits has been solved to date, a quasi-atomic structure of the 80S ribosome from the yeast *Saccharomyces cerevisiae* (Spahn et al. 2001, 2004b) can be used for guidance in the

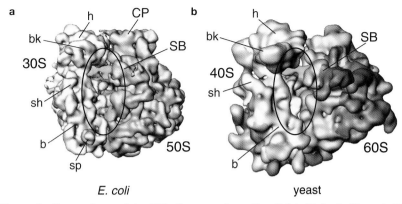

Figure 1. Comparison of the 70S ribosome from *E. coli* (*a*) (Gabashvili et al. 2000) and the 80S ribosome from *S. cerevisiae* (*b*) (Spahn et al. 2001). Extra density in the 80S ribosome attributed to additional rps and rRNA expansion segments are shown in *orange* on the SSU and *purple* on the LSU in (*b*). The basic "core" of both ribosomes is conserved at the intersubunit space (inside the oval), suggesting that critical processes of the elongation cycle, such as peptide bond formation, translocation, and decoding, are also conserved. Landmarks in the cryo-EM reconstructions are: (h) head; (bk) beak; (sh) shoulder; (b) body; (sp) spur; (SB) stalk base; (CP) central protuberance. P-site tRNA (*green*).

interpretation of experimental findings. This structure was obtained by docking homologous regions of the bacterial rRNA crystal structure into the cryo-EM density map. Most of the bacterial ribosomal proteins have homologous counterparts in eukaryotes, whose proteins generally are much larger, but clearly related based on a comparison of their amino acid sequences. Homology models of these proteins were built and docked into the cryo-EM density maps, yielding a satisfactory explanation of the observed densities.

Several rRNA expansion segments (ESs) and 20–30 extra rps (Gerbi 1996) make the eukaryotic ribosome much larger than its bacterial counterpart (Fig. 1). The eukaryotic ribosomal subunits, however, share a common rRNA core with those from prokaryotes (Gutell et al. 1985), indicating that the fundamental mechanism of peptide bond formation is likely conserved in these systems. A comparison of cryo-EM reconstructions between the 70S ribosome from *E. coli* and the 80S ribosome from *S. cerevisiae* indicates that the intersubunit space, where the key events of decoding, peptide bond formation, and binding of mRNA and tRNA occur, is very similar in the two species (Fig. 1, oval region). Cryo-EM structures of eukaryotic ribosomes from human (Spahn et al. 2004a), trypanosome (H. Gao et al. 2005), rabbit reticulocytes (Dube et al. 1998a), rat liver (Dube et al. 1998b), *Thermomyces lanuginosus* (Sengupta et al. 2004), and yeast (Spahn et al. 2001, 2004b) have been obtained, showing the extent of the conservation (Fig. 2). Because the most extensive structural and genetic information is available from the yeast *S. cerevisiae*, it is used as the basis for the discussions below.

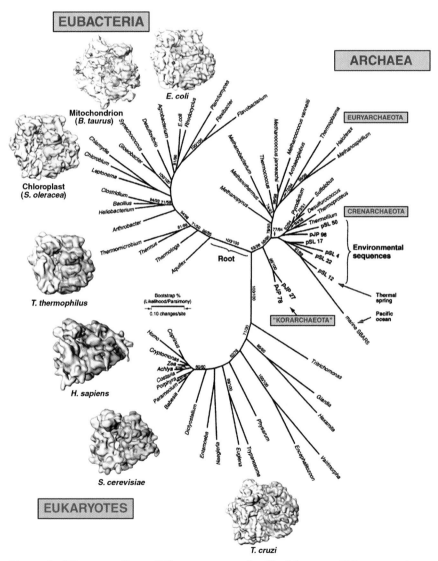

Figure 2. Ribosomes from different sources visualized by cryo-EM, arranged according to a phylogenetic tree of the three domains of life. The tree is based on a sequence comparison of 5S rRNA (Barns et al. 1996). In all reconstructions, the SSU is *yellow*, the LSU *blue*, and the P-site tRNA *green* if bound.

The 40S SSU of S. cerevisiae

The eukaryotic SSU, termed 40S based on its sedimentation value, contains the 18S rRNA responsible for decoding the mRNA and selecting cognate tRNA (Ofengand et al. 1982; Carter et al. 2000). The 40S subunit of S. cerevisiae consists of a 1798-nucleotide 18S rRNA and 32 ribosomal proteins. Compared to the 16S rRNA of the bacterial SSU, the S. cerevisiae 18S rRNA contains four significant expansion segments (ESs, Fig. 3a) (Gutell et al. 1985). The relative locations of these ESs were identified in the cryo-EM map (Fig. 3b) (Spahn et al. 2001). If we disregard the expansion segments, the remaining core of the SSU is structurally conserved, as evidenced by the fact that the X-ray structure of the SSU from T. thermophilus (Wimberly et al. 2000) fits reasonably well into the cryo-EM map of the SSU obtained from S. cerevisiae (Spahn et al. 2001). The overall shape of the yeast SSU is similar to that of bacteria, with distinguishing landmarks such as head, beak, shoulder, spur, and platform (Verschoor et al. 1996; Dube et al. 1998a,b; Spahn et al. 2001). The eukaryotic SSU has additional landmarks, such as the left foot and the right foot, which can be attributed to rRNA ES and extra rps (Fig. 3b).

Unlike the basic core of rRNA, ESs vary in length and sequence among organisms. ES7 is located at the tip of helix 27 (Fig. 3a,b). ES12 is an expansion to helix 44 that forms the "right foot" of the SSU (Dube et al. 1998a). The ES6 insertion into helix 21 averages 250 nucleotides (Neefs and De Wachter 1990), making it the largest ES of 18S rRNA. The 3′ region of ES6 is complemented by a conserved 3′ region of ES3 located near the "left foot" and juxtaposed to ES6 in the cryo-EM map (Fig. 3b). Chemical and enzymatic modifications support the existence of a direct helical interaction between the bases of ES6 and ES3 in yeast, wheat, and mouse ribosomes (Alkemar and Nygard 2004).

The 60S LSU of S. cerevisiae

Landmarks identified in the bacterial LSU, such as the L1 stalk, central protuberance, stalk base, and sarcin-ricin loop are conserved in the eukaryotic LSU (Fig. 3d). The yeast LSU is composed of three rRNA molecules, 25S (3392 nucleotides), 5.8S (158 nucleotides), and 5S (121 nucleotides), and 46 rps. The rRNA of the LSU catalyzes the peptidyl transferase activity without requiring any soluble nonribosomal translation factors or nucleotides (for review, see Moldave 1985). The structural organization, substrate specificity, and other properties of the peptidyl transferase in prokaryotes have been thoroughly reviewed (Chapter 2). Studies on eukaryotic peptide bond formation have been less extensive, but,

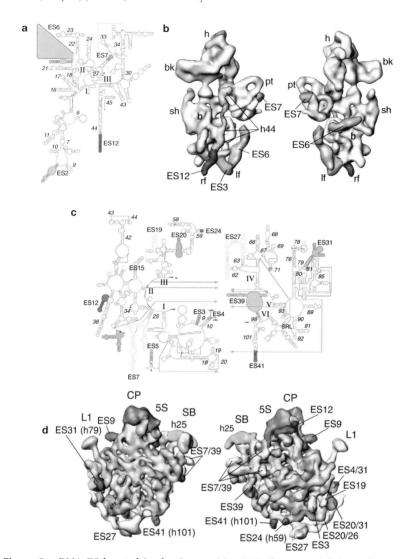

Figure 3. rRNA ES located in the *S. cerevisiae* 80S ribosome. (*a*) Secondary structure diagram of 18S rRNA; (*b*) cryo-EM reconstruction of the computationally
separated rRNA of the SSU; (*c*) secondary structure of 5.8/25S rRNA; (*d*) computationally separated rRNA of the LSU (Spahn et al. 2001). ES in the secondary sequences (*a,c*) are color-coded to map their physical locations in their respective
cryo-EM reconstructions (*b,d*). The catalytic core of rRNA is colored *yellow* in the
SSU (*b*) and *blue* in the LSU (*d*). The intersubunit surface of both ribosomal subunits is shown on the left in *b* and *d*, and the right is a 180° rotation. Landmarks
are as depicted in Fig. 1: (pt) platform; (lf) left foot; (rf) right foot; (5S rRNA)
(*purple, d*).

based on the observed conservation of rRNA that composes the peptidyl transferase center, the mechanism of action is likely similar to that of prokaryotes.

Several ESs have been identified in the secondary sequence (Gerbi 1996; Gutell et al. 2001) and the three-dimensional structure (Spahn et al. 2001) of the LSU from yeast (Fig. 3c,d). Most of these ESs are located near the periphery of the structure at two opposite ends of the LSU: one behind the central protuberance and stalk region and the other extending from the base of the L1 protuberance to the bottom of the subunit (Fig. 3d). Several of the ESs interact via tertiary and quaternary interactions, and two (ES31 and ES41) form additional bridges to the SSU.

STRUCTURES AND FUNCTIONS OF EUKARYOTIC RIBOSOMAL PROTEINS

Comparative analysis of rat, human, yeast, archaeal, and *E. coli* rps in 1995 (Wool et al. 1995), and more recently, the complete genome sequences of 45 bacterial, 14 archaeal, and 7 eukaryotic species (Lecompte et al. 2002), led to the conclusion that there are 32 (15 SSU, 17 LSU) rps that are strictly conserved in all three domains of life (Table 1). Two more LSU rps (rpL7 and rpL8) are missing in some bacteria, but are otherwise conserved throughout the three domains. Twenty-three rps (8 SSU, 15 LSU) are specific to bacteria; one rp in the LSU (rpLXa) is specific to only archaea, and 11 rps (4 SSU, 7 LSU) are found exclusively in eukaryotes.

Although the overall size of the archaeal ribosome and the length of its rRNA are closer to those of bacteria, the archaeal ribosome is essentially a small-scale model of the eukaryotic ribosome in terms of rps. There are 33 proteins (13 SSU, 20 LSU) common to archaeal and eukaryotic ribosomes, but not bacterial ribosomes. With the exception of rpLXa, all

Table 1. The number of ribosomal proteins that are conserved between archaea, eukaryotes, and bacteria

Domain	Total	SSU	LSU
All three	32	15	17
All except some bacteria	2	–	2
Bacteria only	23	8	15
Archaea only	1	–	1
Eukaryotes only	11	4	7
Archaea and eukaryotes	33	13	20

of the rps from archaea are also found in eukaryotes (Lecompte et al. 2002). This observation provides additional structures of archaeal rps solved by crystallography and NMR for use in homology modeling of eukaryotic rps.

Because the basic mechanism of protein synthesis is relatively conserved between bacteria and eukaryotes, the purpose of rRNA ES and extra rps in higher organisms is still unclear. Extra rps may be necessary to fold the rRNA ES during ribosome assembly and maturation. Other eukaryotic rps that are absent in bacterial ribosomes are known to bind rRNA, tRNA, mRNA, or translation factors. The increased need for regulation and the more complex initiation pathway in eukaryotes may require additional sites of interaction provided by these expansions. This includes the increased importance of cotranslational protein export through membranes, which would lead to properties conducive to the association of the ribosome with membranes (Dube et al. 1998b). Another potential function is to assist in nuclear transport of other molecules, as evidenced by a nuclear localization signal in several eukaryotic rps (Herve du Penhoat et al. 2004; Lipsius et al. 2005). Extraribosomal functions have also been attributed to eukaryotic rps (for review, see Kinzy and Goldman 2000; Wool 1996).

The structures of 43 (15 SSU, 28 LSU) eukaryotic rps are inferred from the crystal structures of the LSU from *Haloarcula marismortui* and the SSU from *T. thermophilus*, as detailed elsewhere (Fig. 4) (see Spahn et al. 2001; Chapter 2). Structural and biochemical data for the remaining ribosomal proteins in eukaryotes remain limited. The three-dimensional structures of three such ribosomal proteins, rpS27, rpS28, and rpL30, have recently been solved, as detailed in the following.

Ribosomal Proteins S27 and S28

The archaeal SSU protein rpS27, from *Archaeoglobus fulgidis,* was recently solved by NMR spectroscopy (Herve du Penhoat et al. 2004). Eukaryotic rpS27 has been found to be involved in a series of functions such as processing rRNA, directly binding mRNA, and degrading damaged mRNA. The solution structure of rpS27 is a β-sandwich consisting of two three-stranded sheets with a unique topology that presents a previously uncharacterized protein fold. Two more features of the *A. fulgidis* rpS27 structure conserved throughout 11 archaeal and 19 eukaryotic species correspond to a surface patch of hydrophobic residues and a C4 zinc finger motif. This suggests that rpS27 may mediate RNA–protein interactions, as has been shown for other rps that contain zinc fingers (Frankel

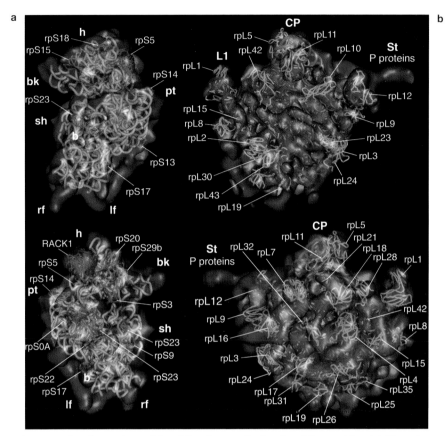

Figure 4. Quasi-atomic models of the ribosomal subunits of *S. cerevisiae*. (*a*) Homology modeling of 18S rRNA (*yellow*) and several SSU rps (*orange*) docked into the computationally separated cryo-EM density of the SSU. (*b*) Homology modeling of 5.8S/25S rRNA (*blue*), 5S rRNA (*purple*), and several LSU rps (*cyan*) docked into the cryo-EM density of the computationally separated LSU. (*Top*) Intersubunit surface of the individual ribosomal subunits; (*bottom*) 180° rotated view.

2000); however, the zinc finger motif is not universally conserved (Herve du Penhoat et al. 2004).

The solution structure of rpS28 from *Methanobacterium thermoautotrophicum* (Wu et al. 2003) and *Pyrococcus horikoshii* (Aramini et al. 2003) presents a variation of the oligonucleotide/oligosaccharide β-barrel (OB) fold with a positively charged surface on one side of the β-barrel, implying an ability to bind RNA. The carboxy-terminal segments from both organisms are unstructured in solution, but they possess a conserved signature sequence motif that is predicted to form an α-helix upon interaction with RNA. These features are consistent with the assumption that one of the primary functions of rps is to assure correct rRNA folding by stabilizing specific RNA structures.

Ribosomal Protein L30

The crystal (Chen et al. 2003) and NMR solution structures (Wong et al. 2003) of the rpL30 protein from the thermophilic archaeon, *Thermococcus celer*, have been solved, as well as the mRNA·rpL30 complex from *S. cerevisiae* (Mao et al. 1999). The overall fold and secondary structural elements are relatively conserved between the archaeal and eukaryotic rpL30. The *S. cerevisiae* rpL30 protein demonstrates feedback inhibition by binding to a helix-loop-helix structure of its own pre-mRNA and mRNA. This results in disruption of the splicing of the pre-mRNA molecule in the nucleus and reduction of the translation efficiency of the mRNA in the cytoplasm (Eng and Warner 1991; Li et al. 1996). The location of the rpL30 protein on the 80S ribosome from wheat germ has been identified using cryo-EM (Halic et al. 2005). Density attributed to rpL30 is involved in the formation of two dynamic bridges between the LSU and the SSU, suggesting that the function of rpL30, once bound to the ribosome, may be to facilitate conformational changes in the ribosomal subunits that occur during translocation (Halic et al. 2005).

P PROTEINS

In bacteria, the L7/L12 stalk is formed by the combination of an L7/L12 complex bound to the GTPase region of the 23S rRNA of the LSU through an L10 attachment (Wahl and Moller 2002; Diaconu et al. 2005). In archaea and eukaryotes, the P (P0, P1, P2) proteins constitute the flexible stalk of the LSU subunit that is connected to a small portion of the rRNA. P0, the homolog of bacterial L10, is bound to the GTPase region and to rpL12 (L11 in bacteria). Two copies each of the acidic proteins, P1 and P2 (L7 and L12 homologs, respectively), complete the assembly of the stalk. The flexible stalk of the L7/L12 proteins is disordered in the crystal structure of the 50S subunit from *H. marismortui*, but was modeled into the crystal structures of the 50S subunit from *Deinococcus radiodurans* and the 70S ribosome from *T. thermophilus* (Chapter 2). Because of the extreme flexibility of the L7/L12 stalk, its orientation is entirely different in the two crystal structures, and a cryo-EM reconstruction of the 70S ribosome from *E. coli* revealed that the L7/L12 stalk is bifurcated (Agrawal et al. 1999). The P proteins in eukaryotes have been visualized as the stalk in the cryo-EM reconstructions, but atomic models are currently limited to an amino-terminal fragment of the P2 protein from rat liver (Mandelman et al. 2002).

Although the P proteins possess bacterial homologs, they are quite different functionally and probably structurally as well. Although L7/L12 and P1/P2 proteins are of similar lengths (~110 residues), they share no sequence homology except for a flexible hinge region between the amino- and carboxy-terminal domains (Szick et al. 1998). This suggests that, like the bacterial L7/L12 stalk, the P1/P2 stalk is highly dynamic. The amino-terminal domain of the P1/P2 proteins is significantly larger than in L7/L12, whereas the carboxy-terminal domain is significantly shorter and charged, containing two serine phosphorylation sites (Hasler et al. 1991). P1 and P2 are extremely acidic proteins that exchange between ribosomal and cytoplasmic pools, implicating their involvement in factor recruitment to the ribosome (Zinker and Warner 1976).

RACK1

The RACK1 (receptor for activated C kinase) protein is a highly con-served integral eukaryotic ribosomal protein that has been identified as a component of the SSU (for review, see Nilsson et al. 2004). Its position on the SSU has recently been localized and modeled into its attributed cryo-EM density in *S. cerevisiae* and *T. lanuginosus* (Sengupta et al. 2004). The structure of RACK1 is a seven-bladed β-propeller located on the solvent-accessible back of the head of the SSU (Fig. 4a, bottom). Inter-acting with helices 39 and 40 of the 18S rRNA, RACK1 lies in close vicinity to the mRNA exit channel and serves as a platform for several kinases such as protein kinase C (PKC) and Src kinase, and the integrin β and NMDA (*N*-methyl-D-aspartate) receptors (for review, see Nilsson et al. 2004). The function and location of RACK1 suggest its involvement in eukaryotic signal transduction pathways; it has been linked to trans-lation initiation in budding yeast and stimulates ribosomal subunit asso-ciation in humans by the PKC-induced phosphorylation of eIF6.

INTERSUBUNIT BRIDGES IN EUKARYOTES

Seven bridges, designated B1–B7, were identified in the cryo-EM recon-structions of the 70S ribosome from *E. coli* (Frank et al. 1995; Gabashvili et al. 2000). These were later expanded to 30 contacts spread over 12 bridges based on the X-ray crystal structure of the *T. thermophilus* 70S ribosome

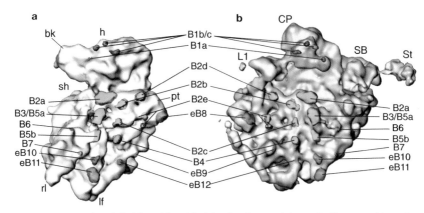

Figure 5. Intersubunit bridges identified in the *S. cerevisiae* 80S ribosome. Locations of bridges are colored in the computationally separated SSU (*a*) and LSU (*b*). Bridges B1–B7 have been identified in *E. coli*, whereas bridges eB8–eB12 have only been visualized in yeast. The yeast 80S ribosome adopts slightly different conformations depending on the presence of P-site tRNA or binding of eEF2, which affect the location of some intersubunit bridges (Spahn et al. 2004b). (*Green*) Bridges visualized in both ribosomal conformations; (*orange*) bridges that move depending on the ribosomal conformation; (*pink*) bridges that have been visualized in only one ribosomal conformation and not the other. Landmarks are as depicted in Figs. 1 and 2; (st) stalk.

(Cate et al. 1999; Yusupov et al. 2001). A comparison between *E. coli* and *T. thermophilus* revealed that the rRNA and rps involved in forming the intersubunit bridges were relatively conserved (Gabashvili et al. 2000; Gao et al. 2003). All intersubunit bridges in the bacterial 70S ribosome have a corresponding bridge in the yeast 80S ribosome (Spahn et al. 2001).

Five additional bridges were identified in the yeast 80S ribosome, eB8–eB12 (Fig. 5) (Spahn et al. 2001, 2004b). eB8 is formed by an unknown rp of the SSU and a portion of ES31 in helix 39 of the LSU. eB9 is formed by a small ES of helix 21 in the SSU and unknown rps from both ribosomal subunits. eB10 and eB11 form the strongest intersubunit connections in the 80S yeast ribosome. Bridge eB10 is composed of an unknown SSU rp and an RNA–RNA contact between helix 11 of the 18S rRNA and helices 63 and 101 of the 25S rRNA. eB11 is formed by RNA–RNA interactions between helix 9 of the 18S rRNA and ES41 in helix 101 of the 25S rRNA. An unknown SSU rp also contributes to eB11, and rpL19 contributes to the formation of eB12.

The binding of eukaryotic elongation factor 2 (eEF2), the EF-G homolog, induces rearrangements of the ribosomal subunits and results in changes of the intersubunit bridges (Spahn et al. 2004b) (see below, Dynamics of Ribosome Function in Elongation). eEF2 binding results in a shift of the components that form some bridges (Fig. 5, orange) or in the formation of bridges that are unique to only one of the ribosome conformations (Fig. 5, pink). For example, binding of eEF2 alters the contact of the unknown SSU rp that forms eB8, but eB9 is no longer discernible in the cryo-EM reconstruction (Spahn et al. 2004b). Upon eEF2 binding, the constituents of eB10 remain the same, but the location of the entire bridge changes slightly. eB12 forms after the binding of eEF2 and subsequent conformational changes in the ribosome.

CYCLIC FORMATION OF PEPTIDE BONDS

In contrast to initiation, elongation is more highly conserved throughout the three domains of life, although some key differences are evident. After initiation, the 80S ribosome consists of the LSU, SSU, mRNA, and Met-tRNA$_i^{Met}$ in the P (or peptidyl) site. The next codon to be translated is in an open ribosomal position called the A (or acceptor) site. A number of soluble protein synthesis factors engage the ribosome during the eukaryotic translation elongation cycle: eEF1, eEF2, and eEF3. eEF1 consists of three or four subunits: eEF1A (formerly EF-1α), eEF1Bα (formerly EF-1β), eEF1Bγ (formerly EF-1γ) and, in metazoans, eEF1Bβ (formerly EF-1δ). eEF2 is a GTPase that catalyzes the translocation of tRNA by one codon, and fungi uniquely require a third factor, eEF3.

DECODING AND PEPTIDYL TRANSFER

eEF1A·GTP (EF-Tu·GTP in bacteria) binds and recruits aa-tRNAs to the A site of the ribosome (Carvalho et al. 1984). The SSU of the ribosome decodes the incoming anticodon of the eEF1A·GTP·aa-tRNA ternary complex and ensures the formation of a proper codon–anticodon match. The crystal structures of the bacterial 30S subunit in complex with stem-loops of cognate and near-cognate tRNA demonstrate that only the correct codon–anticodon match results in a conformational change in the head of the SSU (Ogle et al. 2003), leading to a "closed" conformation. This conformational change in the SSU, induced by the delivery of the cognate aa-tRNA, leads to GTP hydrolysis of the ternary complex. Hydrolysis is stimulated by a region of the LSU named the GTP-associated center (GAC). The eEF1A·GDP complex is released from the ribosome, leaving the aa-tRNA in the A site. Spontaneous GDP dissociation from eEF1A is slow, and the eEF1Bαβγ complex stimulates the exchange of GDP for GTP, maintaining the level of active eEF1A·GTP (Slobin and Möller 1978).

The LSU catalyzes peptide bond formation between the P-site tRNA and the incoming aminoacyl moiety of the A-site tRNA. The E (or exit)-site tRNA leaves the ribosome, with the assistance of the L1 ribosomal protein, during each round of elongation. In fungi, eEF3 is proposed to aid in the release of deacylated tRNA from the E site (Triana-Alonso et al. 1995) and interacts with eEF1A (Anand et al. 2003).

TRANSLOCATION

Following peptide bond formation, binding of eEF2 (EF-G in bacteria) and subsequent GTP hydrolysis catalyze the translocation of the peptidyl-tRNA on the ribosome. Cryo-EM studies have shown that the factor binds in the same position as EF-G binds the bacterial ribosome (Gomez-Lorenzo et al. 2000) and that it induces a similar ratchet-like relative rotation of the two subunits (Spahn et al. 2004b) as observed for *E. coli* (Frank and Agrawal 2000). Although most work on translocation has been performed in the bacterial system (Chapter 2), the questions as to whether eEF2/EF-G acts as a motor or via a switch mechanism and which nucleotide-bound form of eEF2/EF-G binds the ribosome remain controversial (Bretscher 1968; Rodnina et al. 1997; Zavialov et al. 2005).

STRUCTURES OF EUKARYOTIC ELONGATION FACTORS

eEF1ABαγ

Structural information has recently become available for the elongation factors, almost exclusively from yeast. The crystal structure of yeast eEF1A in complex with the carboxy-terminal catalytic fragment of eEF1Bα demonstrates that the complex is fundamentally different from the prokaryotic EF-Tu·EF-Ts complex (Andersen et al. 2000). One end of eEF1Bα is buried between the switch-1 and switch-2 regions of eEF1A, thereby destroying the binding site for the Mg^{2+} ion associated with the nucleotide. The second end interacts with domain II of eEF1A in the region hypothesized to bind the CCA-aminoacyl end of the tRNA. The competition between eEF1Bα and aa-tRNA may be a central element in channeling the reactants in eukaryotic protein synthesis (Andersen et al. 2000). The averaged NMR structure of the free carboxyl terminus of human eEF1Bα places the domain-II-binding loop in an alternative position (Perez et al. 1999); thus, the conformation may change upon eEF1A binding. This hypothesis is supported by experiments showing that the *Artemia salina* eEF1A·eEF1Bα complex bound to the nonhydrolyzable GTP analog, GDPCP, dissociates upon the addition of aa-tRNA (Janssen and Moller 1988).

Additional structures of this complex with bound GDP·Mg^{2+}, GDP, and GDPNP show minor changes compared to the apo structure (Andersen et al. 2001). These changes are consistent with the positions of mutations in eEF1A that affect the dependence on the exchange factor (Kinzy and Woolford 1995; Carr-Schmid et al. 1999a) and in eEF1Bα that affect protein synthesis (Carr-Schmid et al. 1999b). The base, sugar, and α-phosphate bind similarly to known nucleotide·G-protein complexes, whereas the β- and γ-phosphates are disordered. Comparison of these structures led to the conclusion that nucleotide exchange relies on the eEF1Bα-induced disruption of the switch-2 region of eEF1A that forms part of the Mg^{2+}-binding pocket (Andersen et al. 2001). Such a disruption also occurs in EF-Tu (Kawashima et al. 1996; Wang et al. 1997) and is probably a conserved mechanism of nucleotide exchange in all G proteins that bind Mg^{2+} in the GDP state (Sprang and Coleman 1998).

The X-ray structure of the eEF1A·GDP complex from the archaeon *Sulfolobus solfataricus*, in comparison with eEF1A from the *S. cerevisiae* eEF1A·eEF1Bα complex, revealed that domain I and domains II and III can be superimposed independently with an RMSD of less than 2 Å (Vitagliano et al. 2001). Comparing all three domains together reveals that binding of the nucleotide exchange factor induces a rotation of domain I by approximately 73° with respect to domains II and III. This domain movement is comparable to that seen in bacterial EF-Tu (Berchtold et al. 1993; Kjeldgaard et al. 1993). Thus, the dynamics of EF-Tu/eEF1A are likely conserved and have a critical role during nucleotide exchange.

On the basis of these structures, new models have been developed for understanding the mechanisms of translational fidelity (Valente and Kinzy 2003). To understand the functional interaction of the eEF1A·aa-tRNA·GTP ternary complex with the ribosome, a model was prepared by overlaying the domains of the crystal structure of eEF1A with its homologous regions of the bacterial EF-Tu ternary complex (Fig. 6a,b) (Nissen et al. 1995). This model was used to understand the effect of the ribosome·eEF1A interaction in maintaining translational fidelity. Several mutant forms of eEF1A function as dominant or recessive frameshift and nonsense suppressors in vivo and missense suppressor in vitro. These mutations are all located in domain I or II, highlighting the key role of GTP binding and hydrolysis and aa-tRNA binding, respectively.

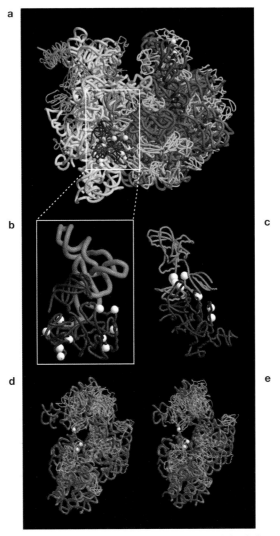

Figure 6. (*a*) Ribbon diagram of the quasi-atomic model of the translating 80S ribosome. The predicted three-dimensional structure of the eEF1A·tRNA·GTP ternary complex is modeled showing eEF1A (*red*) and aa-tRNA (*pink*). (*Green*) P-site tRNA; (*beige*) rRNA from the SSU; (*orange*) rps from the SSU; (*blue*) rRNA from the LSU (5S is *purple*); (*cyan*) proteins from the LSU. Mutations that affect translation fidelity are shown as *white spheres* for eEF1A or rps and *red spheres* for rRNA. (*b*) eEF1A ternary complex model alone; (*c*) quasi-atomic model of eEF2 as it is bound to the 80S ribosome. eEF2 is color-coded with domains I, G′ and II in *red* and III–V in *pink* to demonstrate its similarity to the eEF1A ternary complex. (*White spheres*) Point mutations of eEF2 that affect translation. The SSU is shown in two conformations: *d* is the SSU quasi-atomic model when P-site tRNA is bound, and *e* is the quasi-atomic model of the SSU when eEF2 is bound without P-site tRNA. Point mutations that affect translation are shown as in *a*.

As shown in Figure 6b, these mutations map on the perimeter of eEF1A where the factor likely interacts with the ribosomal subunits. Structures have been determined independently for the two domains of eEF1Bγ. The crystal structure of the amino-terminal 219 residues of the *TEF3*-encoded form of eEF1Bγ from *S. cerevisiae* shows structural similarity to glutathione *S*-transferase (GST) proteins (Jeppesen et al. 2003). Interestingly, the carboxy-terminal helices of GST cover the active site, and the carboxyl terminus of the eEF1Bγ domain is also near this site. Thus, the carboxyl terminus of eEF1Bγ may also cover this pocket in a manner similar to that of the GSTs and affect the structure and function of the full complex. The NMR structure of the 19-kD carboxyl terminus of human eEF1Bγ revealed it has a lens-like shape, with the concave surface containing most of the highly conserved amino acids (Vanwetswinkel et al. 2003). Overall, these structural and other biochemical data support the model that yeast eEF1 is organized as an (eEF1A·eEF1Bα·eEF1Bγ)$_2$ complex.

eEF2

Crystal structures of *S. cerevisiae* eEF2 were determined with and without the inhibitor sordarin (Jørgensen et al. 2003). The overall conformation of apo-eEF2 is similar to that of the bacterial EF-G·GDP complex. Upon drug binding, the three carboxy-terminal domains (III–V) undergo a hinge-like conformational change, with respect to the amino-terminal domains (I, G′ and II). The conformation of eEF2 in complex with the 80S ribosome as observed by cryo-EM is between those of the two crystal structures of eEF2, suggesting that the hinge movement of eEF2 upon binding to the ribosome may facilitate translocation (Spahn et al. 2004b). A study of sordarin-resistant eEF2 mutants in the hinge region indicates that many also show altered +1 frameshift effects (Fig. 5) (for review, see Valente and Kinzy 2003).

The cryo-EM structure of the yeast eEF2–ribosome complex led to predictions of a role of the tip of domain IV in reading frame maintenance (Spahn et al. 2004b). This region contains H699, which is converted to diphthamide as the result of a multistep reaction (Herve du Penhoat et al. 2004). This highly conserved modification is not essential for viability in yeast but is required for ADP-ribosylation by several toxins (Perentesis et al. 1992). Multiple substitutions of H699 in yeast are inviable or show growth defects (Kimata and Kohno 1994). The H699K eEF2 mutant is nonviable and shows a dominant-negative growth phenotype, paromomycin sensitivity, and translation effects (Ortiz and Kinzy 2005), which support the proposed role of the tip of domain IV, and likely the diphthamide modification, in translocation and translational fidelity. The structure of the ADP-ribosylated form of yeast eEF2 shows that the modified region interacts with several conserved residues in the tip of domain IV, which may indicate how it affects the ability of eEF2 to function in translocation (Jørgensen et al. 2004).

eEF3

eEF3 possesses distinct motifs including HEAT repeats, tandem ATP-binding cassettes, and a highly basic but not essential carboxyl terminus (Andersen et al. 2004). The mechanism by which eEF3 utilizes these different domains to catalyze the proposed functions at different sites is still enigmatic. eEF3 is well characterized to bind ribosomes, and the requirement for the protein in vitro is only seen with fungal ribosomes (Kambampati and Chakraburtty 1997; Gontarek et al. 1998). eEF3 also binds eEF1A, although mutants unable to bind eEF1A are viable (Anand et al. 2003). Thus, the lack of eEF3 in bacteria and metazoans most likely relates to differences in the ribosome.

Dynamics of Ribosome Function in Elongation

The binding of EF-G to the 70S ribosome induces a ratchet-like reorganization of the ribosomal subunits (Frank and Agrawal 2000). Docking of the *T. thermophilus* crystal structure into the cryo-EM density of the *E. coli* ribosome in an initiation-like state (Gao et al. 2003) identified several new intersubunit bridges, when compared to the translating ribosome with EF-G·GTP bound (Gao et al. 2003). This included the involvement of rpS19 in the B1a bridge that resulted from conformational changes in the ribosome when compared to that of the EF-G·GTP-bound state. An analogous ratchet-like rearrangement of the ribosomal subunits has been described for the 80S yeast ribosome, following the comparison of the posttranslocational ribosome carrying tRNA in the P site with the 80S yeast ribosome complexed with eEF2 (Spahn et al. 2004b). This also results in changes of intersubunit bridges, as discussed earlier. It has been suggested that the presence of P-site tRNA "locks" the ribosome in a posttranslational state (Valle et al. 2003). The ratchet-like movement results in a general rotation of the SSU with respect to the LSU that may assist in the translocation of tRNA from the A and P sites to the P and E sites, respectively (Frank and Agrawal 2000). In yeast, binding of eEF2 induces an SSU head movement by approximately 15 Å (measured at the tip), as well as significant movements of several rps (Spahn et al. 2004b). Such movements can be outlined by comparing quasi-atomic models of the yeast SSU in the P-site tRNA-bound form (Fig. 6d) with the eEF2-bound form (Fig. 6e).

Genetic Analysis of Ribosome Function

The ribosome has a key role in translational accuracy (for review, see Valente and Kinzy 2003). Since much of the rRNA is highly conserved in sequence and structure throughout the three domains of life, work from bacteria has frequently directed the studies of the yeast ribosome (for review, see Liebman et al. 1995). Yeast strains lacking the rRNA locus on the chromosome, which permits introduction of plasmid-borne rRNA mutations, allows the analysis of mutant ribosomes in a homogeneous population (Chernoff et al. 1994; Wai et al. 2000). On the basis of the role of the SSU in decoding, many 18S rRNA mutations in helix 18, helix 27, helix 34, and helix 44 have been isolated or studied to address fidelity. The 25S rRNA, in particular the sarcin/ricin domain, is also important for the accuracy of translation. Finally, saturation mutagenesis has demonstrated effects of mutations throughout the 5S rRNA. In yeast, as in *E. coli*, many ribosomal proteins are involved in maintaining quality control during translation, including rpS2, rpS3, rpS5, rpS9, and rpS28 of the SSU, as well as rpL3, rpL5, and rpL39 of the LSU. To understand these mutants in the context of the growing structural information on the ribosome, these point mutations were mapped onto the quasi-atomic model of the translating yeast 80S ribosome (Fig. 6a). Several conclusions can be drawn from the location of the mutants. First, many are far removed from the factor-binding site. Second, many rRNA point mutations are clustered near rp point mutations. Last and most dramatically, the SSU models with and without bound P-site tRNA show that the sites defined by point mutations that affect translation move significantly from one SSU conformation to the other (Fig. 6d,e).

Polypeptide Export

The nascent polypeptide chain follows a path from the peptidyl transferase center through a tunnel to the back of the LSU. Many eukaryotic mRNAs code for secretory and membrane proteins that are translocated across, or into, the membrane of the endoplasmic reticulum. These proteins contain a signal sequence that is recognized by a ribonucleo-protein (RNP) complex, the signal recognition particle (SRP), which is associated with the translating ribosome (Chapter 21). Upon signal sequence recognition, the SRP–ribosome–nascent chain complex docks onto the protein-conducting channel (PCC) in the membrane via the membrane-bound SRP receptor. The growing polypeptide chain is translocated vectorially across the membrane or partitioned laterally into the membrane, usually with the help of additional membrane protein complexes. The ribosome–PCC complex from yeast has been visualized using cryo-EM in several studies (Beckmann et al. 1997, 2001; Menetret et al. 2000). These low-resolution studies show the PCC as a drum-shaped mass with a diameter of approximately 100 Å that sits directly on the exit of the polypeptide tunnel. A recent cryo-EM structure of a translating ribosome–PCC complex from *E. coli* (Mitra et al. 2005) revealed more-detailed features of the PCC, such that the X-ray crystal structure of the heterotrimer building block of the PCC from the archaeon, *Methanococcus jannaschii* (Van den Berg et al. 2004), could be docked into the cryo-EM density with relative precision. Structural analysis reveals the most likely dimeric arrangement of the heterotrimers in forming the PCC and suggests an intricate mechanism for its double role in export and membrane integration. To date, there has been no similar study for eukaryotes, but conservation of the PCC mechanism is highly likely, based on the high sequence homology of its components.

SUMMARY AND FUTURE DIRECTIONS

Given the lack of any X-ray structures for the 80S ribosome of eukaryotes or its subunits, it is not surprising that the knowledge and insights in eukaryotic translation lag behind what has been accomplished for bacteria. Nevertheless, significant progress has been made by a combination of mutation studies, cryo-EM visualizations, docking of prokaryotic core structures, and homology modeling. Integrating the results of genetic and structural studies of the elongation factors has further advanced our understanding of the translation elongation process. Although the much higher complexity of eukaryotic translation can be inferred from biochemistry and the acquisition, by the ribosome, of rRNA expansions and additional proteins, the present data suggest that all the basic mechanisms of translation, including the conformational dynamics, are preserved. Further implications of the regulation of eukaryotic elongation, as well as the variation among bacterial, organellar, and eukaryotic ribosomes, have arisen from the increased number of cryo-EM structures from diverse species (Fig. 2). Thus, structural studies of eukaryotic ribosomes remain a rich area for further insight into the unique aspect of eukaryotic translation. The exploration of the eukaryotic ribosome to a similar level of resolution and detail as has now been achieved for bacteria is the challenge of the next decade, if not decades.

ACKNOWLEDGMENTS

We thank the members of the Kinzy and Frank labs for helpful comments. T.G.K. is supported by National Institutes of Health grants GM62789 and GM57483, and J.F. by the Howard Hughes Medical Institute and National Institutes of Health grant GM29169.

REFERENCES

Agrawal R.K., Penczek P., Grassucci R.A., and Frank J. 1998. Visualization of elongation factor G on the *Escherichia coli* 70S ribosome: The mechanism of translocation. *Proc. Natl. Acad. Sci.* **95:** 6134–6138.

Agrawal R.K., Heagle A.B., Penczek P., Grassucci R.A., and Frank J. 1999. EF-G-dependent GTP hydrolysis induces translocation accompanied by large conformational changes in the 70S ribosome. *Nat. Struct. Biol.* **6:** 643–647.

Agrawal R.K., Penczek P., Grassucci R.A., Li Y., Leith A., Nierhaus K.H., and Frank J. 1996. Direct visualization of A-, P-, and E-site transfer RNAs in the *Escherichia coli* ribosome. *Science* **271:** 1000–1002.

Agrawal R.K., Sharma M.R., Kiel M.C., Hirokawa G., Booth T.M., Spahn C.M., Grassucci R.A., Kaji A., and Frank J. 2004. Visualization of ribosome-recycling factor on the *Escherichia coli* 70S ribosome: Functional implications. *Proc. Natl. Acad. Sci.* **101:** 8900–8905.

Alkemar G. and Nygard O. 2004. Secondary structure of two regions in expansion segments ES3 and ES6 with the potential of forming a tertiary interaction in eukaryotic 40S ribosomal subunits. *RNA* **10:** 403–411.

Allen G.S., Zavialov A., Gursky R., Ehrenberg M., and Frank J. 2005. The cryo-EM structure of a translation initiation complex from *Escherichia coli*. *Cell* **121:** 703–712.

Anand M., Chakraburtty K., Marton M.J., Hinnebusch A.G., and Kinzy T.G. 2003. Functional interactions between yeast translation eukaryotic elongation factor (eEF) 1A and eEF3. *J. Biol. Chem.* **278:** 6985–6991.

Andersen C.F., Anand M., Boesen T., Van L.B., Kinzy T.G., and Andersen G.R. 2004. Purification and crystallization of the yeast translation elongation factor eEF3. *Acta Crystallogr. D Biol. Crystallogr.* **60:** 1304–1307.

Andersen G.R., Valente L., Pedersen L., Kinzy T.G., and Nyborg J. 2001. Crystal structures of nucleotide exchange intermediates in the eEF1A-eEF1Balpha complex. *Nat. Struct. Biol.* **8:** 531–534.

Andersen G.R., Pedersen L., Valente L., Chatterjee I., Kinzy T.G., Kjeldgaard M., and Nyborg J. 2000. Structural basis for nucleotide exchange and competition with tRNA in the yeast elongation factor complex eEF1A:eEF1Bα. *Mol. Cell* **6:** 1261–1266.

Aramini J.M., Huang Y.J., Cort J.R., Goldsmith-Fischman S., Xiao R., Shih L.Y., Ho C.K., Liu J., Rost B., Honig B., et al. 2003. Solution NMR structure of the 30S ribosomal protein S28E from *Pyrococcus horikoshii*. *Protein Sci.* **12:** 2823–2830.

Barns S.M., Delwiche C.F., Palmer J.D., and Pace N.R. 1996. Perspectives on archaeal diversity, thermophily and monophyly from environmental rRNA sequences. *Proc. Natl. Acad. Sci.* **93:** 9188–9193.

Beckmann R., Bubeck D., Grassucci R., Penczek P., Verschoor A., Blobel G., and Frank J. 1997. Alignment of conduits for the nascent polypeptide chain in the ribosome-Sec61 complex. *Science* **278:** 2123–2126.

Beckmann R., Spahn C.M., Eswar N., Helmers J., Penczek P.A., Sali A., Frank J., and Blobel G. 2001. Architecture of the protein-conducting channel associated with the translating 80S ribosome. *Cell* **107:** 361–372.

Berchtold H., Reshetnikova L., Reiser C.O.A., Schirmer N.K., Sprinzl M., and Hilgenfeld R. 1993. Crystal structure of active elongation factor Tu reveals major domain rearrangements. *Nature* **365:** 126–132.

Bretscher M.S. 1968. Translocation in protein synthesis: A hybrid structure model. *Nature* **218:** 675–677.

Carr-Schmid A., Durko N., Cavallius J., Merrick W.C., and Kinzy T.G. 1999a. Mutations in a GTP-binding motif of eEF1A reduce both translational fidelity and the requirement for nucleotide exchange. *J. Biol. Chem.* **274:** 30297–30302.

Carr-Schmid A., Valente L., Loik V.I., Williams T., Starita L.M., and Kinzy T.G. 1999b. Mutations in elongation factor 1β, a guanine nucleotide exchange factor, enhance translational fidelity. *Mol. Cell. Biol.* **19:** 5257–5266.

Carter A.P., Clemons W.M., Brodersen D.E., Morgan-Warren R.J., Wimberly B.T., and Ramakrishnan V. 2000. Functional insights from the structure of the 30S ribosomal subunit and its interactions with antibiotics. *Nature* **407:** 340–348.

Carvalho M.G., Carvalho J.F., and Merrick W.C. 1984. Biological characterization of various forms of elongation factor 1 from rabbit reticulocytes. *Arch. Biochem. Biophys.* **234:** 603–611.

Cate J.H., Yusupov M.M., Yusupova G.Z., Earnest T.N., and Noller H.F. 1999. X-ray crystal structures of 70S ribosome functional complexes. *Science* **285:** 2095–2133.

Chen Y.W., Bycroft M., and Wong K.B. 2003. Crystal structure of ribosomal protein L30e from the extreme thermophile *Thermococcus celer:* Thermal stability and RNA binding. *Biochemistry* **42:** 2857–2865.

Chernoff Y.O., Vincent A., and Liebman S.W. 1994. Mutations in eukaryotic 18S ribosomal RNA affect translational fidelity and resistance to aminoglycoside antibiotics. *EMBO J.* **13:** 906–913.

Diaconu M., Kothe U., Schlunzen F., Fischer N., Harms J.M., Tonevitsky A.G., Stark H., Rodnina M.V., and Wahl M.C. 2005. Structural basis for the function of the ribosomal L7/12 stalk in factor binding and GTPase activation. *Cell* **121:** 991–1004.

Dube P., Bacher G., Stark H., Mueller F., Zemlin F., van Heel M., and Brimacombe R. 1998a. Correlation of the expansion segments in mammalian rRNA with the fine structure of the 80S ribosome; a cryoelectron microscopic reconstruction of the rabbit reticulocyte ribosome at 21 Å resolution. *J. Mol. Biol.* **279:** 403–421.

Dube P., Wieske M., Stark H., Schatz M., Stahl J., Zemlin F., Lutsch G., and van Heel M. 1998b. The 80S rat liver ribosome at 25 Å resolution by electron cryomicroscopy and angular reconstitution. *Structure* **6:** 389–399.

Eng F.J. and Warner J.R. 1991. Structural basis for the regulation of splicing of a yeast messenger RNA. *Cell* **65:** 797–804.

Frank J. and Agrawal R.K. 2000. A ratchet-like inter-subunit reorganization of the ribosome during translocation. *Nature* **406:** 318–322.

Frank J., Zhu J., Penczek P., Li Y., Srivastava S., Verschoor A., Radermacher M., Grassucci R., Lata R.K., and Agrawal R.K. 1995. A model of protein synthesis based on cryo-electron microscopy of the *E. coli* ribosome. *Nature* **376:** 441–444.

Frankel A.D. 2000. Fitting peptides into the RNA world. *Curr. Opin. Struct. Biol.* **10:** 332–340.

Gabashvili I.S., Agrawal R.K., Spahn C.M., Grassucci R.A., Svergun D.I., Frank J., and Penczek P. 2000. Solution structure of the *E. coli* 70S ribosome at 11.5 Å resolution. *Cell* **100:** 537–549.

Gao H., Ayub M.J., Levin M.J., and Frank J. 2005. The structure of the 80S ribosome from *Trypanosoma cruzi* reveals unique rRNA components. *Proc. Natl. Acad. Sci.* **102:** 10206–10211.

Gao H., Sengupta J., Valle M., Korostelev A., Eswar N., Stagg S.M., Van Roey P., Agrawal R.K., Harvey S.C., Sali A., et al. 2003. Study of the structural dynamics of the *E. coli* 70S ribosome using real-space refinement. *Cell* **113:** 789–801.

Gao N., Zavialov A.V., Li W., Sengupta J., Valle M., Gursky R.P., Ehrenberg M., and Frank J. 2005. Mechanism for the disassembly of the posttermination complex inferred from cryo-EM studies. *Mol. Cell* **18:** 663–674.

Gerbi S.A. 1996. Expansion segments: Regions of variable size that interrupt the universal core secondary structure of ribosomal RNA. In *Ribosomal RNA: Structure, evolution, processing, and function in protein biosynthesis* (ed. R.A. Zimmermann and A.E. Dahlberg), pp. 71–87. CRC Press, Boca Raton, Florida.

Gomez-Lorenzo M.G., Spahn C.M., Agrawal R.K., Grassucci R.A., Penczek P., Chakraburtty K., Ballesta J.P., Lavandera J.L., Garcia-Bustos J.F., and Frank J. 2000. Three-dimensional cryo-electron microscopy localization of EF2 in the *Saccharomyces cerevisiae* 80S ribosome at 17.5 Å resolution. *EMBO J.* **19:** 2710–2718.

Gontarek R.R., Li H., Nurse K., and Prescott C.D. 1998. The N terminus of eukaryotic translation elongation factor 3 interacts with 18S rRNA and 80S ribosomes. *J. Biol. Chem.* **273:** 10249–10252.

Gutell R.R., Weiser B., Woese C.R., and Noller H.F. 1985. Comparative anatomy of 16-S-like ribosomal RNA. *Prog. Nucleic Acid Res. Mol. Biol.* **32:** 155–216.

Gutell R.R., Subashchandran S., Schnare M., Du Y., Lin N., Madabusi L., Muller K., Pande N., Shang Z., and Date S., et al. 2001. Comparative Sequence Analysis and the Prediction of RNA Structure (http://www.rna.icmb.utexas.edu).

Halic M., Becker T., Frank J., Spahn C.M., and Beckmann R. 2005. Localization and dynamic behavior of ribosomal protein L30e. *Nat. Struct. Mol. Biol.* **12:** 467–468.

Hasler P., Brot N., Weissbach H., Parnassa A.P., and Elkon K.B. 1991. Ribosomal proteins P0, P1, and P2 are phosphorylated by casein kinase II at their conserved carboxyl termini. *J. Biol. Chem.* **266:** 13815–13820.

Herve du Penhoat C., Atreya H.S., Shen Y., Liu G., Acton T.B., Xiao R., Li Z., Murray D., Montelione G.T., and Szyperski T. 2004. The NMR solution structure of the 30S ribosomal protein S27e encoded in gene RS27_ARCFU of *Archaeoglobus fulgidis* reveals a novel protein fold. *Protein Sci.* **13:** 1407–1416.

Janssen G.M. and Moller W. 1988. Kinetic studies on the role of elongation factors 1 beta and 1 gamma in protein synthesis. *J. Biol. Chem.* **263:** 1773–1778.

Jeppesen M.G., Ortiz P., Shepard W., Kinzy T.G., Nyborg J., and Andersen G.R. 2003. The crystal structure of the glutathione S-transferase-like domain of elongation factor 1Bgamma from *Saccharomyces cerevisiae*. *J. Biol. Chem.* **278:** 47190–47198.

Jørgensen R., Ortiz P.A., Carr-Schmid A., Nissen P., Kinzy T.G., and Andersen G.R. 2003. Two crystal structures demonstrate large conformational changes in the eukaryotic ribosomal translocase. *Nat. Struct. Biol.* **10:** 379–385.

Jørgensen R., Yates S.P., Teal D.J., Nilsson J., Prentice G.A., Merrill A.R., and Andersen G.R. 2004. Crystal structure of ADP-ribosylated ribosomal translocase from *Saccharomyces cerevisiae*. *J. Biol. Chem.* **279:** 45919–45925.

Kambampati R. and Chakraburtty K. 1997. Functional subdomains of yeast elongation factor 3. *J. Biol. Chem.* **272:** 6377–6381.

Kawashima T., Berthet-Colominas C., Wulff M., Cusack S., and Leberman R. 1996. The structure of the *Escherichia coli* EF-Tu·EF-Ts complex at 2.5 Å resolution. *Nature* **379:** 511–518.

Kimata Y. and Kohno K. 1994. Elongation factor 2 mutants deficient in diphthamide formation show temperature-sensitive cell growth. *J. Biol. Chem.* **269:** 13497–13501.

Kinzy T.G. and Goldman E. 2000. Non-translational functions of the translational apparatus. In *Translational control of gene expression* (ed. J.W.B. Hershey et al.), pp. 973–997. Cold Spring Harbor Laboratory Press, Cold Spring Harbor, New York.

Kinzy T.G. and Woolford J.L., Jr. 1995. Increased expression of *Saccharomyces cerevisiae* translation elongation factor EF-1α bypasses the lethality of a *TEF5* null allele encoding EF-1β. *Genetics* **141:** 481–489.

Kjeldgaard M., Nissen P., Thirup S., and Nyborg J. 1993. The crystal structure of elongation factor EF-Tu from *Thermus aquaticus* in the GTP conformation. *Structure* **1:** 35–50.

Klaholz B.P., Pape T., Zavialov A.V., Myasnikov A.G., Orlova E.V., Vestergaard B., Ehrenberg M., and van Heel M. 2003. Structure of the *Escherichia coli* ribosomal termination complex with release factor 2. *Nature* **421:** 90–94.

Lecompte O., Ripp R., Thierry J.C., Moras D., and Poch O. 2002. Comparative analysis of ribosomal proteins in complete genomes: An example of reductive evolution at the domain scale. *Nucleic Acids Res.* **30:** 5382–5390.

Li B., Vilardell J., and Warner J.R. 1996. An RNA structure involved in feedback regulation of splicing and of translation is critical for biological fitness. *Proc. Natl. Acad. Sci.* **93:** 1596–1600.

Liebman S.W., Chernoff Y.O., and Liu R. 1995. The accuracy center of a eukaryotic ribosome. *Biochem. Cell Biol.* **73:** 1141–1149.

Lipsius E., Walter K., Leicher T., Phlippen W., Bisotti M.A., and Kruppa J. 2005. Evolutionary conservation of nuclear and nucleolar targeting sequences in yeast ribosomal protein S6A. *Biochem. Biophys. Res. Commun.* **333:** 1353–1360.

Mandelman D., Gonzalo P., Lavergne J.P., Corbier C., Reboud J.P., and Haser R. 2002. Crystallization and preliminary X-ray study of an N-terminal fragment of rat liver ribosomal P2 protein. *Acta Crystallogr. D Biol. Crystallogr.* **58:** 668–671.

Mao H., White S.A., and Williamson J.R. 1999. A novel loop-loop recognition motif in the yeast ribosomal protein L30 autoregulatory RNA complex. *Nat. Struct. Biol.* **6:** 1139–1147.

Menetret J.F., Neuhof A., Morgan D.G., Plath K., Radermacher M., Rapoport T.A., and Akey C.W. 2000. The structure of ribosome-channel complexes engaged in protein translocation. *Mol. Cell* **6:** 1219–1232.

Mitra K., Schaffitzel C., Shaikh T., Tama F., Jenni S., Brooks C.L., III, Ban N., and Frank J. 2005. Structure of the *E. coli* protein-conducting channel bound to a translating ribosome. *Nature* **438**: 318–324.

Moldave K. 1985. Eukaryotic protein synthesis. *Annu. Rev. Biochem.* **54**: 1109–1149.

Neefs J.M. and De Wachter R. 1990. A proposal for the secondary structure of a variable area of eukaryotic small ribosomal subunit RNA involving the existence of a pseudo-knot. *Nucleic Acids Res.* **18**: 5695–5704.

Nilsson J., Sengupta J., Frank J., and Nissen P. 2004. Regulation of eukaryotic translation by the RACK1 protein: A platform for signalling molecules on the ribosome. *EMBO Rep.* **5**: 1137–1141.

Nissen P., Kjeldgaard M., Thirup S., Polekhina G., Reshetnikova L., Clark B.F.C., and Nyborg J. 1995. Crystal structure of the ternary complex of Phe-tRNA^{Phe}, EF-Tu, and a GTP analog. *Science* **270**: 1464–1472.

Ofengand J., Gornicki P., Chakraburtty K., and Nurse K. 1982. Functional conservation near the $3'$ end of eukaryotic small subunit RNA: Photochemical crosslinking of P site-bound acetylvalyl-tRNA to 18S RNA of yeast ribosomes. *Proc. Natl. Acad. Sci.* **79**: 2817–2821.

Ogle J.M., Carter A.P., and Ramakrishnan V. 2003. Insights into the decoding mechanism from recent ribosome structures. *Trends Biochem. Sci.* **28**: 259–266.

Ortiz P.A. and Kinzy T.G. 2005. Dominant-negative mutant phenotypes and the regulation of translation elongation factor 2 levels in yeast. *Nucleic Acids Res.* **33**: 5740–5748.

Perentesis J.P., Miller S.P., and Bodley J.W. 1992. Protein toxin inhibitors of protein synthesis. *Biofactors* **3**: 173–184.

Perez J.M., Siegal G., Kriek J., Hard K., Dijk J., Canters G.W., and Moller W. 1999. The solution structure of the guanine nucleotide exchange domain of human elongation factor 1beta reveals a striking resemblance to that of EF-Ts from *Escherichia coli*. *Structure* **7**: 217–226.

Rawat U.B., Zavialov A.V., Sengupta J., Valle M., Grassucci R.A., Linde J., Vestergaard B., Ehrenberg M., and Frank J. 2003. A cryo-electron microscopic study of ribosome-bound termination factor RF2. *Nature* **421**: 87–90.

Rodnina M.V., Savelsbergh A., Katunin V.I., and Wintermeyer W. 1997. Hydrolysis of GTP by elongation factor G drives tRNA movement on the ribosome. *Nature* **385**: 37–41.

Rossmann M.G. 2000. Fitting atomic models into electron-microscopy maps. *Acta Crystallogr. D Biol. Crystallogr.* **56**: 1341–1349.

Sengupta J., Nilsson J., Gursky R., Spahn C.M., Nissen P., and Frank J. 2004. Identification of the versatile scaffold protein RACK1 on the eukaryotic ribosome by cryo-EM. *Nat. Struct. Mol. Biol.* **11**: 957–962.

Slobin L.I. and Möller W. 1978. Purification and properties of an elongation factor functionally analogous to bacterial elongation factor Ts from embryos of *Artemia salina*. *Eur. J. Biochem.* **84**: 69–77.

Spahn C.M., Jan E., Mulder A., Grassucci R.A., Sarnow P., and Frank J. 2004a. Cryo-EM visualization of a viral internal ribosome entry site bound to human ribosomes: The IRES functions as an RNA-based translation factor. *Cell* **118**: 465–475.

Spahn C.M., Beckmann R., Eswar N., Penczek P.A., Sali A., Blobel G., and Frank J. 2001. Structure of the 80S ribosome from *Saccharomyces cerevisiae*-tRNA-ribosome and subunit-subunit interactions. *Cell* **107**: 373–386.

Spahn C.M., Gomez-Lorenzo M.G., Grassucci R.A., Jorgensen R., Andersen G.R., Beckmann R., Penczek P.A., Ballesta J.P., and Frank J. 2004b. Domain movements of elongation factor eEF2 and the eukaryotic 80S ribosome facilitate tRNA translocation. *EMBO J.* **23:** 1008–1019.

Sprang S.R. and Coleman D.E. 1998. Invasion of the nucleotide snatchers: Structural insights into the mechanism of G protein GEFs. *Cell* **95:** 155–158.

Stark H., Rodnina M.V., Rinke-Appel J., Brimacombe R., Wintermeyer W., and van Heel M. 1997. Visualization of elongation factor Tu on the *Escherichia coli* ribosome. *Nature* **389:** 403–406.

Szick K., Springer M., and Bailey-Serres J. 1998. Evolutionary analyses of the 12-kDa acidic ribosomal P-proteins reveal a distinct protein of higher plant ribosomes. *Proc. Natl. Acad. Sci.* **95:** 2378–2383.

Triana-Alonso F.J., Chakraburtty K., and Nierhaus K.H. 1995. The elongation factor 3 unique in higher fungi and essential for protein biosynthesis is an E site factor. *J. Biol. Chem.* **270:** 20473–20478.

Valente L. and Kinzy T.G. 2003. Yeast as a sensor of factors affecting the accuracy of protein synthesis. *Cell. Mol. Life Sci.* **60:** 2115–2130.

Valle M., Zavialov A., Sengupta J., Rawat U., Ehrenberg M., and Frank J. 2003. Locking and unlocking of ribosomal motions. *Cell* **114:** 123–134.

Van den Berg B., Clemons W.M., Jr., Collinson I., Modis Y., Hartmann E., Harrison S.C., and Rapoport T.A. 2004. X-ray structure of a protein-conducting channel. *Nature* **427:** 36–44.

Vanwetswinkel S., Kriek J., Andersen G.R., Guntert P., Dijk J., Canters G.W., and Siegal G. 2003. Solution structure of the 162 residue C-terminal domain of human elongation factor 1Bgamma. *J. Biol. Chem.* **278:** 43443–43451.

Verschoor A., Srivastava S., Grassucci R., and Frank J. 1996. Native 3D structure of eukaryotic 80S ribosome: Morphological homology with the *E. coli* 70S ribosome. *J. Cell Biol.* **133:** 495–505.

Vitagliano L., Masullo M., Sica F., Zagari A., and Bocchini V. 2001. The crystal structure of *Sulfolobus solfataricus* elongation factor 1alpha in complex with GDP reveals novel features in nucleotide binding and exchange. *EMBO J.* **20:** 5305–5311.

Wahl M.C. and Moller W. 2002. Structure and function of the acidic ribosomal stalk proteins. *Curr. Protein Pept. Sci.* **3:** 93–106.

Wai H.H., Vu L., Oakes M., and Nomura M. 2000. Complete deletion of yeast chromosomal rDNA repeats and integration of a new rDNA repeat: Use of rDNA deletion strains for functional analysis of rDNA promoter elements *in vivo*. *Nucleic Acids Res.* **28:** 3524–3534.

Wang Y., Jiang Y., Meyering-Voss M., Sprinzl M., and Sigler P.B. 1997. Crystal structure of the EF-Tu·EF-Ts complex from *Thermus thermophilus*. *Nat. Struct. Biol.* **4:** 650–656.

Wimberly B.T., Brodersen D.E., Clemons W.M., Jr., Morgan-Warren R.J., Carter A.P., Vonrhein C., Hartsch T., and Ramakrishnan V. 2000. Structure of the 30S ribosomal subunit. *Nature* **407:** 327–339.

Wong K.B., Lee C.F., Chan S.H., Leung T.Y., Chen Y.W., and Bycroft M. 2003. Solution structure and thermal stability of ribosomal protein L30e from hyperthermophilic archaeon *Thermococcus celer*. *Protein Sci.* **12:** 1483–1495.

Wool I.G. 1996. Extraribosomal functions of ribosomal proteins. *Trends Biochem. Sci.* **21:** 164–165.

Wool I.G., Chan Y.L., and Gluck A. 1995. Structure and evolution of mammalian ribosomal proteins. *Biochem. Cell Biol.* **73:** 933–947.

Wu B., Yee A., Pineda-Lucena A., Semesi A., Ramelot T.A., Cort J.R., Jung J., Edwards A., Lee W., Kennedy M., and Arrowsmith C.H. 2003. Solution structure of ribosomal protein S28E from *Methanobacterium thermoautotrophicum*. *Protein Sci.* **12:** 2831–2837.

Yusupov M.M., Yusupova G.Z., Baucom A., Lieberman K., Earnest T.N., Cate J.H., and Noller H.F. 2001. Crystal structure of the ribosome at 5.5 Å resolution. *Science* **292:** 883–896.

Zavialov A.V., Hauryliuk V.V., and Ehrenberg M. 2005. Guanine-nucleotide exchange on ribosome-bound elongation factor G initiates the translocation of tRNAs. *J. Biol.* **4: 9**.

Zinker S. and Warner J.R. 1976. The ribosomal proteins of *Saccharomyces cerevisiae*. Phosphorylated and exchangeable proteins. *J. Biol. Chem.* **251:** 1799–1807.

4

The Mechanism of Translation Initiation in Eukaryotes

Tatyana V. Pestova
Department of Microbiology and Immunology
State University of New York Downstate Medical Center
Brooklyn, New York 11203-2098; and
A.N. Belozersky Institute of Physicochemical Biology
Moscow State University, 119899 Moscow, Russia

Jon R. Lorsch
Department of Biophysics and Biophysical Chemistry
Johns Hopkins University School of Medicine
Baltimore, Maryland 21205

Christopher U.T. Hellen
Department of Microbiology and Immunology
State University of New York Downstate Medical Center
Brooklyn, New York 11203-2098

STANDARD TRANSLATION INITIATION IN EUKARYOTES is the process that leads to assembly of an 80S ribosome on an mRNA in which the initiation codon is base-paired to the CAU anticodon of aminoacylated initiator methionyl-transfer RNA (Met-tRNA$_i^{Met}$) in the ribosomal peptidyl (P) site. The process requires separated small (40S) and large (60S) ribosomal subunits and involves at least 12 eukaryotic initiation factors (eIFs) and the binding and hydrolysis of ATP and GTP. The resulting 80S initiation complex is competent to enter the elongation phase of translation. This chapter describes the canonical mechanism of 5'-end-dependent initiation, with a bias toward the initiation process in higher eukaryotes. This process differs in detail from that in plants and yeast, in which the subunit structure and composition of some factors differ substantially. For a more detailed review of initiation in yeast and in plants, see Chapters 9 and 26, respectively. For a review of mechanisms dependent on internal ribosome entry, see Chapters 5 and 6.

STRUCTURE OF EUKARYOTIC CYTOPLASMIC mRNAs

The translational efficiency of eukaryotic mRNAs is limited by the rate of initiation (see, e.g., Palmiter 1972), which is in turn determined by structural features of mRNAs that influence ribosomal recruitment, scanning to the initiation codon, and initiation codon recognition. Eukaryotic mRNAs associate dynamically with proteins that mediate nuclear export, subcellular localization, stability, and translational repression, and therefore exist in cells as messenger ribonucleoproteins (mRNPs) rather than as free polynucleotides. The influence of mRNP proteins on initiation is outside the scope of this review.

Almost all nucleus-transcribed mRNAs have a 5′-terminal $m^7G[5′]ppp[5′]N$ cap (where N is any nucleotide) and a 50–300-nucleotide-long 3′ poly(A) tail that synergistically enhance initiation on the mRNA. The cap is important but not essential for initiation; in its absence, initiation is 5′-end-dependent and occurs by a scanning mechanism just as on capped mRNAs (Gunnery et al. 1997). The mRNA 3′ poly(A) tail alone stimulates initiation modestly, but under normal cellular conditions, it enhances initiation strongly and synergistically with the cap (for review, see Jacobson 2000). The synergism of these two RNA elements in initiation is mediated by interactions between the proteins that bind to them: the eIF4F cap-binding complex and the poly(A)-binding protein (PABP), respectively. Disruption of the PABP–eIF4F interaction dramatically reduces cap–poly(A) synergy (Michel et al. 2000; Kahvejian et al. 2005). The activity of the 5′ terminus correlates with its accessibility, so that initiation on native capped and synthetic uncapped mRNAs is substantially diminished by even a modest secondary structure adjacent to the 5′ terminus (Lawson et al. 1988; Kozak 1991; Pestova and Kolupaeva 2002). Hairpins located distally in the 5′ leader permit attachment of the 43S complex but impair translation beyond a threshold of stability (ΔG about -60 kcal/mole for complete inhibition) by obstructing the progress of scanning ribosomal complexes to the initiation codon (Pelletier and Sonenberg 1985; Kozak 1991; Pestova and Kolupaeva 2002). It is not known whether secondary structures of intermediate stability slow ribosomal scanning or increase the off-rate of 43S complexes. Initiation efficiency is also influenced by the length and, particularly, by the degree of secondary structure in the 5′ leader. Shortening the 5′ leader below approximately 12 nucleotides leads to leaky scanning past the first initiation codon, whereas increasing its length progressively enhances initiation (Kozak 1991; Pestova and Kolupaeva 2002). The mRNA-binding cleft of the 40S subunit covers about 12 nucleotides upstream of and about 15 nucleotides downstream from the initiation codon (see, e.g., Lazarowitz and Robertson 1977), so that initiation complexes assembled on mRNAs with short 5′ leaders may lack stabilizing interactions upstream of the initiation codon.

Initiation occurs predominantly at the first AUG triplet; an AUG triplet inserted upstream of the normal start site supplants it as the initiation codon (Kozak 1983). The principal determinants of initiation codon recognition are its position and its potential for base-pairing with the anticodon of initiator methionyl-tRNA$_i$ (Met-tRNA$_i^{Met}$) (Cigan et al. 1988). Nevertheless, initiation can occur with lower efficiency at non-AUG codons (most frequently at CUG, ACG, and GUG triplets) and is modulated by flanking sequences and structures just as at AUG triplets (Peabody 1989; Kozak 1991). A significant proportion of mammalian mRNAs contain AUG triplets upstream of the principal open reading frame (ORF), often in good context and in evolutionarily conserved locations (Churbanov et al. 2005). Such triplets are frequently initiation codons for short upstream ORFs that may attenuate and potentially regulate translation of the principal ORF, which occurs by subsequent reinitiation (Geballe and Sachs 2000; Chapter 8).

Other upstream AUG triplets may be bypassed if they have weak context. Initiation codon selection (at least in higher eukaryotes) is determined by context: the sequence GCC(**A/G**)CC**AUGG** (in which the most important positions are bold) is optimal in mammalian cells, and deviations, particularly at −3 and +4 positions, lead to initiation at the next downstream AUG triplet by leaky scanning (Kozak 1991). The influence of context nucleotides is synergistic, which is suggestive of multiple interactions of mRNA with components of the arrested 43S complex. The experimentally determined optimum context occurs in a minority of vertebrate mRNAs, even when only the −3 and +4 positions are taken into account, and only these two nucleotides constitute a consensus sequence in higher eukaryotes (Pesole et al. 2000). A "suboptimal" context that leads to inefficient translation may even be a desirable property of mRNAs that encode regulatory proteins. Initiation at non-AUG triplets or at AUG triplets with poor context can be enhanced by a hairpin 12–15 nucleotides downstream from that codon (Kozak 1991), which is thought to stall the scanning 43S complex so that a weak initiation codon has

Table 1. Mammalian initiation factors

Name	Mass (kD)	Accession no.[a]	Protein DataBase no.[b]
eIF1	12.7	NM_005801	2IF1
eIF1A	16.4	L18960	1D7Q, 1JT8
eIF2α	36.1	NM_004094	1Q8K, 1Q46, 1KL9, 2A1A, 2A19,1YZ6, 1YZ7, 2AH0
eIF2β	38.4	NM_003908	1VB5, 1NEE, 1K8B, 1K81
eIF2γ	51.1	NM_001415	1S0U, 1KK0, 1KK1, 1KK2, 1KK3, 1KJZ, 2AH0
eIF2A	65.0	NM_032025	
eIF2Bα	33.7	NM_001414	**1T9K**, 1VB5
eIF2Bβ	39.0	NM_014239	2A0U
eIF2Bγ	50.2	AK024006	
eIF2Bδ	59.7	NM_172195	1T5O
eIF2Bε	80.3	NM_003907	1PAQ
eIF3a (p170)	166.5	NM_003750	
eIF3b (p116)	92.5	U78525	
eIF3c (p110)	105.3	U46025	
eIF3d (p66)	64.1	NM_003753	
eIF3e (p48)	52.2	NM_001568	
eIF3f (p47)	37.5	NM_003754	
eIF3g (p44)	35.7	U96074	
eIF3h (p40)	40.0	NM_003756	CQ0
eIF3i (p36)	36.5	U39067	
eIF3j (p35)	29.0	NM_003758	
eIF3k	25.0	AY245432	1RZ4
eIF3l (HSPC021)	66.7	AF077207	
eIF3m (GA17)	42.6	NM_006360	
eIF4AI	46.1	D13748	1FUU, 1FUK, 1QDE, 1QVA
eIF4AII	46.4	BC013708	
eIF4B	69.3	BC098437	1WI8
eIF4E	24.7	NM_001968	1RF8, 1L8B, 1EJH, 1EJ1, 1EJ4, 1WKW, 1AP8, 1IPB, 1IPC
eIF4GI	175.6	AY082886	1UG3, 1LJ2
eIF4GII	176.7	NM_003760	1HU3
eIF4H (WBCSR5)	27.4	NM_022170	
eIF5	49.2	NM_001969	2G2K, 2FUL, 2IU1
eIF5A	16.8	NM_001970	1XTD, 1EIF, 2EIF, 1BKB, 1IZ6,
eIF5B	138.9	NM_015904	1G7R, 1G7S, 1G7T
eIF6	26.6	AF022229	*1G61*, 1G62
PABP	70.7	NM_002568	1CVJ, 1IFW, 1NMR, 1JGN, 1JH4

All sequences are human; structures are eukaryotic (light-face text), archaeal (underlined), or eubacterial (boldface text).

[a]Accession numbers are from the National Center for Biotechnology Information database (http://www.ncbi.nlm.nih.gov).

[b]PDB (Protein DataBase) numbers are from the Research Collaboratory for Structural Bioinformatics (RCSB) database (http://www.rcsb.org/pdb).

longer to base-pair with the anticodon of Met-tRNA$_i^{Met}$ and to respond to establishment of this interaction. The statistically significant sequence conservation in the 30 nucleotides upstream of the initiation codon in orthologous sets of mammalian mRNAs suggests that they may contain as yet undiscovered determinants that influence initiation codon selection (Shabalina et al. 2004).

THE INITIATION PATHWAY

The initiation process consists of a series of steps, each of which is promoted by one or more eIFs. There are at least 12 well-characterized eIFs comprising 30 or more polypeptides (Table 1), several isoforms of canonical initiation factors, and numerous factors whose role and importance in initiation are incompletely characterized. Canonical factors are grouped according to the stage at which they first act (Fig. 1); however, many factors are multifunctional and also act at subsequent stages. Moreover, many eIFs bind stably to one another and may therefore be recruited to ribosomal preinitiation complexes as larger complexes, such as eIF4F (which consists of the eIF4A, eIF4E, and eIF4G subunits).

DISSOCIATION OF POSTTERMINATION RIBOSOMES

Posttermination ribosomes are associated with mRNA and deacylated tRNA, so that their dissociation may differ mechanistically from dissociation of free 80S monosomes. Ribosomal dissociation is essential because initiation can only proceed with free subunits. Dissociation of prokaryotic posttermination ribosomes requires a dedicated ribosome-recycling factor, but eukaryotes do not encode a cytoplasmic equivalent, and the factors that mediate this step are not known (Kisselev and Buckingham 2000). Ribosomes are dissociated and released from mRNA as 60S and native (factor-associated) 40S subunits after termination of translation; the latter are stably associated with eIF3 (Freienstein and Blobel 1975). Experiments done using ribosomes reassembled from purified subunits suggested that eIF3 had RNA-dependent dissociation activity (Kolupaeva et al. 2005). This observation prompted the hypothesis that eIF3 may have a major role in posttermination disassembly. eIF3's RNA-dependent dissociation activity was enhanced by eIF1A and particularly by eIF1, which binds to the interface surface of the platform and blocks access of the 60S subunit to 18S rRNA elements of the intersubunit bridges B2b and B2d (Fig. 2B) (Spahn et al.

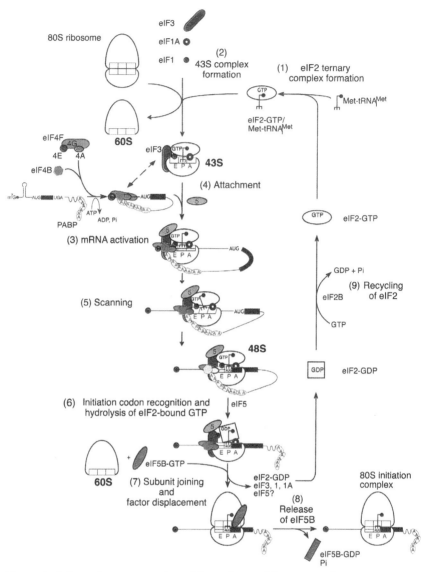

Figure 1. Schematic representation of the pathway of formation of an 80S ribosomal translation initiation complex on a capped and 3′-polyadenylated mRNA. This working model divides this process into nine different stages, showing initiation factors at the stage at which they first participate in this process. After completion of initiation, Met-tRNA$_i^{Met}$ is base-paired with the initiation codon of mRNA in the P site of the 80S ribosome, and the vacant A site is able to accept delivery by eEF1A of an elongator tRNA.

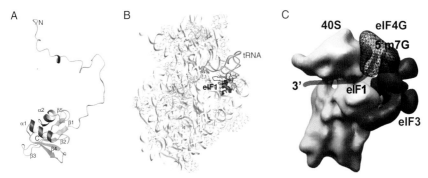

Figure 2. Positions of eIF1 and eIF3 on the small ribosomal subunit. (*A*) Solution structure of human eIF1, showing the single α/β domain and the unstructured amino-terminal tail (Fletcher et al. 1999). (*B*) Position of eIF1 (*magenta* ribbon) modeled on the crystal structure of the 30S ribosomal subunit (Yusupov et al. 2001), based on hydroxyl radical cleavage data (Lomakin et al. 2003) mapped onto 16S rRNA (*gray*) and tRNA (*light coral*). (*Green*) mRNA in the A and P sites. (*C*) Model of the 40S ribosomal subunit with bound mRNA, eIF3, and eIF4G, based on cryo-EM reconstructions (Siridechadilok et al. 2005), and eIF1 (based on Lomakin et al. 2003). (*C*, Image generously provided by B. Siridechadilok and J.A. Doudna.)

2001; Lomakin et al. 2003). eIF1 binds directly to the 40S subunit and to yeast and mammalian eIF3c subunits; it is the smallest eIF and forms a tightly packed α/β domain (Fig. 2A) (Fletcher et al. 1999). Termination of translation is linked to reinitiation by the interaction of PABP with eRF3, which may promote recycling of ribosomes to the 5′end of the same mRNA (Uchida et al. 2002).

Recent data (Chaudhuri et al. 1999; Kolupaeva et al. 2005) suggest that contrary to previous reports (for review, see Moldave 1985), eIF3 alone or together with eIF1 and/or eIF1A is insufficient to dissociate translationally inactive 80S monosomes or to prevent their formation and that either RNA or the eIF2-GTP/Met-tRNA$_i^{Met}$ complex is required as a cofactor. Dissociation of empty 80S ribosomes is therefore directly linked to 43S complex formation (Fig. 1, step 2).

By analogy with its prokaryotic homolog IF1, eIF1A may occlude the ribosomal A site and induce changes in the relative positions of domains of the 40S subunit (Battiste et al. 2000; Carter et al. 2001). eIF1A consists of a β-barrel oligonucleotide/oligosaccharide (OB) fold, a downstream α-helical domain, and random coil amino- and carboxy-terminal extensions that contain clusters of basic and acidic regions (see Fig. 8C); the OB fold is highly homologous to an analogous domain in IF1 (Battiste et al. 2000). Mammalian eIF3 contains up to 13 subunits and has a mass of >650 kD (Mayeur et al. 2003; Unbehaun et al. 2004). eIF3 in other eukaryotes contains the same core units and differing subsets of noncore subunits (see, e.g., Zhou et al. 2005). For a review of their interactions and proposed structures, see Marintchev and Wagner (2004). Modeling based on cryo-electron microscopy (cryo-EM) data (Siridechadilok et al. 2005) suggests that eIF3 binds primarily to the solvent side of the 40S subunit, behind the platform (Fig. 2C), but that one domain of eIF3 may interact with rpS13, a component of intersubunit bridge B4 (Spahn et al. 2001). Physical blockage of the interface surface of the 40S subunit could therefore contribute to the mechanism by which eIF3 prevents subunit association. Little is known concerning the basis for the ribosome-binding activity of mammalian eIF3 except that its eIF3d subunit can be cross-linked to 18S rRNA (Nygard and Nilsson 1990) and that binding of eIF3 to 40S subunits independently of other translation components is enhanced by the weakly associated eIF3j subunit (Fraser et al. 2004), which is displaced from the resulting eIF3/40S subunit complex (Unbehaun et al. 2004). In yeast, eIF3a binds to 18S rRNA near the ribosomal A site and eIF3a and eIF3c bind ribosomal proteins rpS0A (on the back of the 40S subunit) and rpS10 (Valasek et al. 2003). Mammalian eIF3a, eIF3b, and eIF3d subunits bind mRNA when associated with the 40S subunit, which could stabilize eIF3's interaction with posttermination ribosomes (Unbehaun et al. 2004).

TERNARY COMPLEX FORMATION

The first step in initiation is assembly of eIF2, GTP, and Met-tRNA$_i^{Met}$ into a ternary complex that is responsible for delivering Met-tRNA$_i^{Met}$ to the 40S subunit and has a key role in responding to the identification of the initiation codon in the mRNA.

eIF2

eIF2 is a heterotrimer comprising a large γ subunit and smaller α and β subunits (Fig. 3A). eIF2γ has binding sites for GTP and Met-tRNA$_i^{Met}$ and is homologous to elongation factor EF-Tu/eEF1A, another translational GTPase that binds aminoacylated tRNAs and delivers them to the ribosome. By analogy with eEF1A, it is thought that the 3' end of Met-tRNA$_i^{Met}$ binds in a cleft in eIF2γ, formed in part by the switch 1 region of the GTPase active site (Schmitt et al. 2002; Roll-Mecak et al. 2004; Yatime et al. 2006). Structural changes in the switch 1 and 2 regions of eIF2γ caused by GTP hydrolysis and Pi release likely account for the resulting

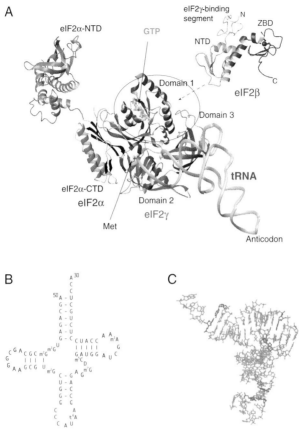

Figure 3. Structure of the eIF2·GTP·Met-tRNA$_i^{Met}$ ternary complex (TC). (*A*) Model for the organization of the TC, in which human eIF2α (PDB 1Q8K) is shown without its carboxy-terminal tail, with the amino-terminal domain (NTD) (which is mobile with respect to the carboxy-terminal domain [CTD]) arbitrarily shown in a single unique orientation and with the CTD oriented as in Ito et al. (2004); archaeal eIF2γ (PDB 1KK1) is shown in *red* and *orange* with bound GTP in *green* and labeled to show the three domains. eIF2β from *Methanobacterium thermoautotrophicum* (PDB 1NEE) is shown with the NTD (*violet*), zinc-binding domain (ZBD) (*purple*), and the amino terminus (*beige*). The NTD is involved in binding eIF2γ and likely contacts it near the nucleotide-binding site (*circled*); it and the ZBD are mobile with respect to each other. Met-tRNA$_i^{Met}$ is modeled as in Schmitt et al. (2002) and Roll-Mecak et al. (2004). Domain 2 of eIF2γ was aligned to domain 2 of EF-Tu in a complex with GTP and Cys-tRNA^{cys} (PDB 1B23). (*B*) Secondary structure of human initiator tRNA; (*C*) tertiary structure of yeast initiator tRNA (PDB 1YFG; Basavappa and Sigler 1991). (*Red*) Key initiator-specific bases required for function. For clarity, bases that function to block binding to eEF1A (e.g., A50:U64 and U51:A63 in mammals) are not highlighted. (*A*, Generously provided by A. Marintchev and G. Wagner and modified, with permission, from Marintchev and Wagner 2004 [©Cambridge University Press].)

decreased affinity for Met-tRNA$_i^{Met}$ (Yatime et al. 2006). Mutations of amino acids in eIF2γ proposed to be important for tRNA binding result in reduced affinity for Met-tRNA$_i^{Met}$, supporting the structural model (Erickson and Hannig 1996; Schmitt et al. 2002; Roll-Mecak et al. 2004). Interactions with the methionine moiety on the 3' end of Met-tRNA$_i^{Met}$ are broken when GTP is replaced by GDP on eIF2, supporting the proposal that changes in the 3'-end-binding site upon GTP hydrolysis are responsible for releasing Met-tRNA$_i^{Met}$ from eIF2 (Kapp and Lorsch 2004b).

The role of eIF2β is less clear. It consists of three regions, defined on the basis of sequence motifs and proposed functions. The amino-terminal region contains three lysine-repeat tracts that contribute to eIF2β's RNA-binding activity and interacts with eIF5 and eIF2B (Asano et al. 1999; Laurino et al. 1999; Das and Maitra 2000). The central region interacts with eIF2γ, and the carboxy-terminal region contains a zinc finger and is required for the subunit's RNA-binding activity (Thompson et al. 2000; Cho and Hoffman 2002; Gutierrez et al. 2004). Several mutations in and around this zinc finger allow initiation at UUG codons, implicating it in the mechanism of start codon selection (Donahue et al. 1988).

eIF2α is critical for regulation of eIF2's activity (Chapter 12), but it is dispensable in yeast if eIF2β, eIF2γ, and tRNA$_i^{Met}$ are overexpressed (Erickson et al. 2001). Purified eIF2βγ also binds Met-tRNA$_i^{Met}$ and GTP in vitro, although genetic analysis in yeast and structural studies of the aIF2αγ heterodimer suggest that eIF2α affects Met-tRNA$_i^{Met}$ binding to the a/eIF2 trimer (Mouat and Manchester 1998; Erickson et al. 2001; Nika et al. 2001; Roll-Mecak et al. 2004; Yatime et al. 2006). Structural analysis of a/eIF2α revealed an amino-terminal OB fold, an adjacent α-helical subdomain, and a carboxy-terminal α/β domain (Ito et al. 2004; Yatime et al. 2005, 2006). The carboxy-terminal domain is structurally similar to that of eEF1Bα, the GTP/GDP exchange factor for eEF1A, and this has suggested, given that eIF2γ is structurally similar to eEF1A, that eIF2α and eIF2γ interact in a structurally analogous manner to eEF1A and eEF1Bα (Ito et al. 2004). Mutagenesis experiments with ar-chaeal and yeast eIF2 have indicated that the eIF2α-binding site on eIF2γ is adjacent to the Met-tRNA$_i^{Met}$-binding site, consistent with a role for eIF2α in tRNA recognition. On the basis of structural model and mu-tagenesis data, the carboxy-terminal domain of eIF2α was proposed to interact with eIF2γ and, directly or indirectly, stabilize the interaction of eIF2γ with Met-tRNA$_i^{Met}$ (Ito et al. 2004; Yatime et al. 2004). These predictions are consistent with a recent crystal structure of the archaeal αγ complex (Yatime et al. 2006). eIF2α and eIF2β bind at independent sites on eIF2γ and do not seem to interact with each other (Schmitt et al. 2002).

Initiator tRNA

Two different methionyl tRNAs are used in protein synthesis in all three kingdoms of life: elongator methionyl tRNA (Met-tRNA$_m^{Met}$) inserts methionine at internal sites in the growing polypeptide during elongation, whereas Met-tRNA$_i^{Met}$ initiates protein synthesis. There are several possible reasons that distinct methionyl tRNAs are used. First, Met-tRNA$_i^{Met}$ must bind to the P site of the small ribosomal subunit rather than in the A site, which may require specific structural features that contribute to binding affinity or that are uniquely compatible with the mechanism of initiation. eEF1A is abundant, so that the use of a specific tRNA for initiation that does not bind to eEF1A ensures a regulatable supply of the ternary complex for the rate-limiting initiation process (Astrom et al. 1999). Finally, levels of the two tRNAs may be regulated independently, suggesting that the use of distinct tRNAs creates a mechanism for separate control of initiation and/or elongation rates (see, e.g., Kanduc 1997).

Initiator methionyl tRNA has unique sequence and structural features that enable it to be specifically recognized by eIF2 or excluded from binding to eEF1A and that are important for productive binding in the P site (Fig. 3B,C). The A:U base pair at the top of the acceptor stem is critical for binding of Met-tRNA$_i^{Met}$ to eIF2·GTP (Farruggio et al. 1996; von Pawel-Rammingen et al. 1992) and is required for formation of the key stabilizing contact between the methionine moiety and the factor (Kapp and Lorsch 2004b). Two initiator-specific G:C base pairs in the anticodon stem are important for the binding of the eIF2·GTP·Met-tRNA$_i^{Met}$ ternary complex to the 40S subunit (Astrom et al. 1993). Finally, several conserved bases in the D and T loops and T stem of the initiator tRNA form a unique substructure (Basavappa and Sigler 1991) and are important for the function of tRNA, at least in part by preventing binding to eEF1A (von Pawel-Rammingen et al. 1992; Drabkin et al. 1998).

Ternary Complex Formation

After each round of initiation, eIF2 is released as an inactive complex with GDP (Raychaudhuri et al. 1985). Release of GDP from eIF2 is very slow (≤ 0.2 min^{-1}), and to function in multiple rounds of initiation, eIF2 therefore requires a guanine nucleotide exchange factor (GEF). This exchange factor, called eIF2B, is unusually ornate for a GEF, consisting of five different subunits, and accelerates the replacement of GDP with GTP by at least tenfold (Nika et al. 2000; Williams et al. 2001). eIF2B and its key role in the control of translation initiation are described in greater detail in Chapter 9.

After GDP has been replaced with GTP on eIF2, the factor switches into a form with high affinity for Met-tRNA$_i^{Met}$ (Erickson and Hannig 1996; Kapp and Lorsch 2004b). This switch is mediated, at least in part, by turning on an interaction between eIF2 and the methionine moiety of the Met-tRNA$_i^{Met}$ upon GTP binding (Kapp and Lorsch 2004b). GTP-dependent methionine recognition may help to prevent unacylated tRNA from entering the initiation pathway.

ASSEMBLY OF THE 43S PREINITIATION COMPLEX

The interaction of the ternary complex with a 40S subunit to form a 43S preinitiation complex is facilitated by eIF1, eIF1A, and eIF3 (Trachsel et al. 1977; Benne and Hershey 1978; Peterson et al. 1979; Thomas et al. 1980a,b; Chaudhuri et al. 1999; Battiste et al. 2000; Majumdar et al. 2003). Some or all of these factors bind the 40S subunit before the ternary complex, likely as a result of their prior role in subunit dissociation. Genetic and biochemical data have confirmed the importance of corresponding yeast factors in 43S complex formation and also implicated eIF5 in this process (Chapter 9). eIF5 interacts with subunits of eIF2 and eIF3 and is likely recruited to the 40S subunit with these factors (Das and Maitra 2001). The mechanisms by which these factors cooperatively promote 43S complex formation include direct interactions with eIF2 and the 40S subunit, interactions between factors, and induced conformational changes in the 40S subunit (see, e.g., Fletcher et al. 1999; Valasek et al. 2002; Maag and Lorsch 2003; Kolupaeva et al. 2005). The roles of individual factors in this process have been studied in the greatest detail in *Saccharomyces cerevisiae* as a result of extensive genetic studies of its translation apparatus, and these are discussed in Chapter 9.

A clear understanding of how these factors promote recruitment of the ternary complex and function in subsequent steps in initiation requires accurate knowledge of their location within the 43S complex. eIF1 has been mapped to the interface surface of the ribosome, between the platform and Met-tRNA$_i^{Met}$ (Fig. 2B) (Lomakin et al. 2003), and eIF1A is thought to bind in the A site, by analogy with IF1 (Battiste et al.

2000; Carter et al. 2001). Modeling based on cryo-EM data indicates that eIF3 binds primarily to the solvent surface of the 40S subunit (Fig. 2C) (Siridechadilok et al. 2005); its location and structure suggest that conformational changes may be required for it to bind to eIF1 (Fletcher et al. 1999). Early immuno-EM reconstructions suggested that eIF2 binds between the head and body of the 40S subunit (Bommer et al. 1991). In this model, eIF2β is close to the beak of the 40S subunit and eIF2α is near the platform, with eIF2γ between them. More recently, modeling of the ternary complex, based on the structure of eIF2's subunits and data concerning its ligand interactions, also suggested that it binds on the platform, contacting the head (Marintchev and Wagner 2004). This model proposes that the orientation of Met-tRNA$_i^{Met}$ in the 43S complex differs from that in the final 80S complex so that the acceptor end of Met-tRNA$_i^{Met}$ may rotate slightly from the E site toward the canonical P-site position seen in prokaryotic 70S ribosomes (Yusupov et al. 2001) as a consequence of initiation codon recognition in a manner that might displace eIF1 and promote coordinated hydrolysis of eIF2-bound GTP (see below).

RECRUITMENT OF THE 43S PREINITIATION COMPLEX TO mRNA

The 43S complex selects the initiation codon by first binding to the capped 5' end of a mRNA and then scanning downstream on the 5' leader until it encounters the initiation codon. Although 43S complexes are intrinsically capable of 5'-end-dependent attachment to mRNA that has an unstructured 5'UTR (untranslated region) (Pestova and Kolupaeva 2002), the 5'UTRs of almost all mRNAs contain some secondary structure, and binding of 43S complexes to them involves the cooperative activities of eIF4F, eIF4B, and possibly PABP, which are thought to unwind the 5' cap-proximal region of a mRNA to prepare an unstructured binding site for the 43S complex. mRNA binds to a cleft on the 40S subunit, which is closed by a noncovalent "latch" formed by helices 16 and 34 of 18S rRNA (Spahn et al. 2004).

Binding of the 43S complex to a capped mRNA begins with recognition of the m^{7}G cap by the eIF4E subunit of eIF4F, which is a heterotrimer that also contains eIF4A (an ATPase/RNA helicase of the DEAD-box family; see Fig. 6A) and eIF4G (a large modular protein; see Fig. 7). Recent biophysical data favor a model in which eIF4F is assembled before it binds to the cap and modulates the structure of adjacent mRNA after attachment, permitting binding of the 43S complex to the cap-proximal region.

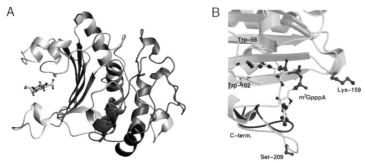

Figure 4. Structural features of the eIF4E/eIF4G cap-binding complex and its interaction with mRNA. (*A*) Lateral view of the cap-bound yeast eIF4E/eIF4G$_{393-490}$ complex, showing eIF4E as well as an amino-terminal helical tail (*yellow*) and helical elements of eIF4G (*blue*) whose folding is mutually induced by the interaction. The m^{7}GDP cap binds in a narrow cap-binding slot and is stacked between W58 and W104. (*B*) The nucleotide adjacent to the m^{7}Gppp moiety of the cap structure may regulate the flexibility of the carboxy-terminal loop region in eIF4E: comparison between this loop region in complexes of human eIF4E bound to m^{7}GpppA (*green*) and of mouse eIF4E bound to m^{7}GDP (*red line*). (*A*, Generously provided by T. von der Haar and reprinted, with permission, from von der Haar et al. 2004 [© Nature Publishing Group]; *B*, generously provided by K. Tomoo and reprinted, with permission, from Tomoo et al. 2002 [© The Biochemical Society].)

eIF4E consists of a curved β-sheet of eight antiparallel strands backed by three long helices (Fig. 4A) (Marcotrigiano et al. 1997; Matsuo et al. 1997). The cap stacks between two tryptophan residues on the concave surface; additional contacts between eIF4E and the nucleotide adjacent to the cap stabilize binding of eIF4E to capped mRNA, alter eIF4E's conformation slightly, and may have functional consequences for initiation (see below) (Fig. 4B) (Tomoo et al. 2002; Slepenkov et al. 2006; for review, see von der Haar et al. 2004). Binding of eIF4E to the cap serves to mediate binding of the eIF4A and eIF4G subunits of eIF4F to the 5′ end of mRNA. Although a conserved Y(X$_4$)Lϕ motif in eIF4G is required for its binding to eIF4E, and oligopeptides containing this motif bind eIF4E tightly (K_d in the nanomolar range), such peptides neither alter the conformation of eIF4E nor induce the strong enhancement of cap binding that is characteristic of the eIF4G–eIF4E interaction (Hagighat and Sonenberg 1997; Ptushkina et al. 1998; Marcotrigiano et al. 1999; von der Haar et al. 2000). The interaction of yeast eIF4E and a fragment of eIF4G (eIF4G$_{348-514}$) that induces this enhanced binding (see Fig. 7B) triggers a coupled folding transition, such that eIF4G$_{348-514}$ forms an annular structure that wraps around the "wrist"

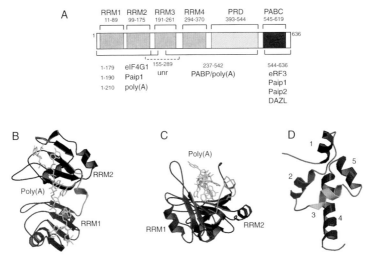

Figure 5. Structure and interactions of human poly(A)-binding protein. (*A*) Schematic diagram showing the RRM1–RRM4 domains, the PRD proline-rich domain, the carboxy-terminal PABC domain, and sites of interaction with eIF4GI, unr, poly(A) RNA, PABP, eRF3, Paip1, Paip2, and DAZL (Bushell et al. 2001; Chang et al. 2004; Collier et al. 2005; Mangus et al. 2003 and references therein). (*B,C*) Crystal structure of RRM1 and RRM2 (residues 1–190) bound to A$_{11}$ (Deo et al. 1999) showing the extended RNA-binding surface (*B*) and the RNA-binding trough formed by RRMs 1 and 2 (*C*). (*D*) Secondary structure of the PABC domain (Kozlov et al. 2001), showing the five carboxy-terminal helices (numbered from the amino terminus). (*B–D*, Generously provided by A. Jacobson and reprinted, with permission, from Mangus et al. 2003 [BioMed Central Ltd.].)

of a fist-like structure formed by induced folding of the amino terminus of eIF4E (Gross et al. 2003). The resulting complex is stable and long-lived ($t_{1/2}$ = 6 minutes), has high affinity for the cap, and dissociates from it slowly, in contrast to the rapid cycle of weak binding to and release from the cap of eIF4E alone (Ptushkina et al. 1998; von der Haar et al. 2000; Gross et al. 2003). The folded domain of yeast eIF4G and the corresponding region of mammalian eIF4G are homologous, so it is likely that the eIF4E–eIF4G interaction in mammalian eIF4F is stabilized in a similar way. PABP (Fig. 5), which binds to eIF4G (see Fig. 7), further enhances the interaction of eIF4F with the cap, possibly due to additional induced conformational changes (Borman et al. 2000; von der Haar et al. 2000).

Like other helicases, eIF4A alternates between open and closed conformations during its working cycle. Its two domains are connected by an extended linker in an open "dumbbell" conformation in the only

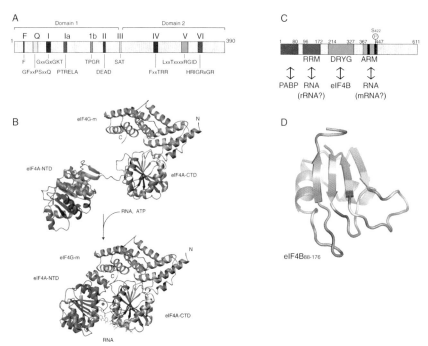

Figure 6. Structures, domain organization, and interactions of eIF4A and eIF4B. (*A*) Schematic diagram of human eIF4A showing its two-domain structure and the ten conserved motifs characteristic of DEAD-box proteins (Cordin et al. 2006). The Q, I, II, and VI motifs participate in binding and hydrolysis of ATP; motifs 1a, 1b, IV, and V are thought to be involved in RNA binding; and motif III links ATP binding/ hydrolysis to conformational changes in eIF4A required for its helicase activity. (*B*) Model of the transition from open (inactive) to closed (active) conformations of eIF4A, stabilized by binding of the middle domain of eIF4G (eIF4G-m) to the eIF4A carboxy-terminal domain based on the crystal structure of yeast eIF4A and mjDEAD (Caruthers et al. 2000; Oberer et al. 2005). (*Gold*) Residues in the eIF4G-m-binding surface of the eIF4A carboxy-terminal domain. Surface mutations in eIF4G-m that impair binding of eIF4A and RNA are colored *blue* and *green*, respectively. (*C*) Schematic diagram of human eIF4B showing RRM, DRYG, and arginine-rich motif (ARM) domains and the site of S6 kinase-mediated phosphorylation, and sites of interaction with PABP, RNA, and eIF4B (Méthot et al. 1996; Fleming et al. 2003; Raught et al. 2004 and references therein). (*D*) Ribbon diagram of the solution structure of the RNA-recognition motif (RRM) domain of human eIF4B, colored to show residues that undergo large (*red*), medium (*orange*), and small (*yellow*) ^{1}H/^{15}N chemical-shift perturbations on binding RNA (Fleming et al. 2003). (*B*, Generously provided by G. Wagner and reprinted, with permission, from Oberer et al. 2005; *D*, generously provided by S. Curry and modified, with permission, from Fleming et al. 2003 [© American Chemical Society].)

determined eIF4A structure (Caruthers et al. 2000), but it can be modeled onto the structures of other helicases in the closed, active conformation (Oberer et al. 2005; Sengoku et al. 2006). The functions of consensus sequence motifs in eIF4A (Fig. 6A) in binding and hydrolysis of ATP, in binding RNA, and in coupling ATP hydrolysis with helicase activity have been elucidated (for review, see Cordin et al. 2006). The structure of the core of the *Drosophila* DEAD-box protein Vasa with U_{10} RNA and an ATP analog reveals interactions of these sequence motifs at the ATPase and RNA-binding sites that are consistent with their known functions (Sengoku et al. 2006). Binding and hydrolysis of ATP lead to changes in the conformation and RNA affinity of eIF4A that could trans-duce the derived energy to disrupt base-pairing in RNA helices (see, e.g., Lorsch and Herschlag 1998). Some eIF4A mutants are *trans*-dominant in-hibitors of initiation and likely act by inhibiting the activity of eIF4F, which suggests that eIF4A functions primarily as a subunit of eIF4F dur-ing initiation (Pause et al. 1994). eIF4A is a weak, nonprocessive heli-case but is activated by eIF4B and by incorporation into eIF4F (Rozen et al. 1990; Korneeva et al. 2005; for review, see Kapp and Lorsch 2004a). eIF4B (Fig. 6C,D) (Fleming et al. 2003) is a cofactor for eIF4A that may potentiate its RNA helicase activity by increasing its affinity for RNA (Abramson et al. 1988). eIF4B's association with initiation complexes may be regulated by phosphorylation by S6 kinase (Holz et al. 2005). A related factor, eIF4H, is homologous to the amino-terminal region of eIF4B but is less effective than eIF4B in activating eIF4A and eIF4F and cannot substitute for it in all circumstances; it is not known whether eIF4B and eIF4H act in a mRNA-specific manner or under different physiological circumstances (Rogers et al. 2001; Dmitriev et al. 2003). The central HEAT-repeat domain of eIF4G (Fig. 7C,D) enhances RNA cross-linking to eIF4A and is thought to stabilize the closed "active" orientation of eIF4A's domains (see Fig. 6B) (Pestova et al. 1996; Oberer et al. 2005). During ribosomal recruitment to mRNA, eIF4G coordinates the RNA helicase activity of eIF4A and the cap-binding activity of eIF4E as well as binding directly to mRNA and to the eIF3 component of the 43S complex. Elucidation of these activities and various other observa-tions (described below) have together led to a model for the action of cap-bound eIF4F in initiation in which it both prepares the cap-proximal region of mRNA for binding of the 43S complex by unwinding local secondary structure and actively promotes binding of this complex to the mRNA (see Rhoads 1991; Gingras et al. 1999). Binding of eIF4E to the cap and of eIF4G with immediately adjacent mRNA is ATP-independent, whereas cross-linking of eIF4A and eIF4B to these nucleotides

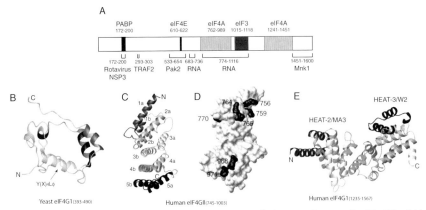

Figure 7. Structure, domain organization, and interactions of eIF4GI. (*A*) Schematic representation of human eIF4GI (AY082886) showing the relative positions and amino acid numbering of binding domains for RNA (Lomakin et al. 2000), initiation factors as indicated (Prévôt et al. 2003 and references therein), and the regulatory proteins Pak2 kinase, Mnk1 kinase, TRAF2, and rotavirus NSP3 (Morino et al. 2000; Groft and Burley 2002; Kim et al. 2005; Ling et al. 2005). (*B*) Solution structure of the eIF4E-binding domain of yeast eIF4G (393–490), derived from the complex of eIF4E with this domain and viewed from the dorsal surface of eIF4E (Gross et al. 2003). The helices before the conserved Y(X)$_4$Lφ motif helix are colored from *dark blue* to *light blue*; the conserved Y(X)$_4$Lφ motif helix is in *yellow*; and the helix after it is in *dark orange-red*. (*C,D*) Ribbon drawing (*C*) and surface representation (*D*) of the conserved central eIF4A-binding domain of eIF4GII (eIF4GII$_{745-1003}$) viewed perpendicular to the α-helix axes, with the concave surface in the foreground. (*C*) Pairs of helices are color-coded and numbered from the amino terminus. (*D*) The surface is color-coded to show the locations of mutations that affect binding to eIF4A (*blue*). (*E*) Ribbon drawing of the HEAT-2/MA3 and HEAT-3/W2 domains of human eIF4GI (1235–1567) (Bellsolell et al. 2006), showing color-coded pairs of helices. (*B*, Generously provided by A. Marintchev and G. Wagner; *D*, generously provided by A.V. Pisarev and modified from Marcotrigiano et al. 2001; *E*, generously prepared by A.V. Pisarev and used by generous permission of S.K. Burley.)

is ATP-dependent (Sonenberg 1988). Incorporation of eIF4A into eIF4F enhances its RNA helicase activity more than 20-fold, and eIF4F and eIF4B together form active helicase complexes on mRNA that unwind cap-proximal mRNA in an ATP-dependent manner (Ray et al. 1985; Rozen et al. 1990; Jaramillo et al. 1991). The interactions of eIF4A and eIF4B are inhibited by cap-proximal secondary structure in a manner that parallels its inhibition of translation (Pelletier and Sonenberg 1985; Lawson et al. 1988). These observations are consistent with findings that the requirement for the cap, eIF4E, eIF4A, and ATP for ribosomal binding increases in proportion to the degree of secondary structure in the 5'UTR (Jackson 1991; Svitkin et al. 2001; for reviews, see Rhoads 1991; Gingras et al. 1999).

As well as binding to and enhancing the activity of eIF4A, eIF4G is thought to have a direct role in mediating recruitment of the 43S complex to the unwound "landing site" on the mRNA via its interaction with eIF3 (Lamphear et al. 1995; Imataka and Sonenberg 1997; Korneeva et al. 2005). This model for eIF4G's activity is supported by the observation that tethering the central eIF3- and eIF4A-binding domains of eIF4G to an internal site in an mRNA is sufficient to promote ribosomal recruitment to a downstream location in vivo (De Gregorio et al. 1999). Ribosomal recruitment following structural alteration of cap-proximal mRNA likely involves multiple additional bridging interactions between eIF3, eIF4B, and the 40S subunit, as well as of eIF4A, eIF4B, and possibly eIF2 with mRNA (Lamphear et al. 1995; Methot et al. 1996; for discussion, see Nygard and Nilsson 1990; Rhoads 1991). There is undoubtedly some redundancy within these interactions, which may account for observations that yeast eIF3 does not bind to eIF4G directly but does interact with it indirectly via eIF5 and eIF1 components of a yeast eIF3c/eIF5/eIF1 complex (Yamamoto et al. 2005) and that in some circumstances, eIF4F's activity in 5'-end-dependent initiation can be partially complemented by eIF4A and a fragment of eIF4G that does not bind PABP, eIF4E, or eIF3 (Ali and Jackson 2001). Despite the identification of many interactions that might mediate end-dependent ribosomal recruitment to mRNA, the functional importance of many of them has not been proved, and important questions about this process remain unanswered.

Two sets of observations suggest that components of the eIF4F/eIF4B complex are bound to mRNA on the E-site side of the 40S subunit. Footprinting of 48S complexes revealed factor protection of mRNA on the 5' but not the 3' side of ribosome-bound mRNA (see, e.g., Lazarowitz and Robertson 1977), and cryo-EM-based models of eIF4G bound to eIF3 on the 40S subunit placed it very close to the E site (Siridechadilok et al. 2005). These results immediately raise two questions: Is the 43S complex recruited by these factors threaded onto the mRNA or does it bind to a segment of the 5' leader immediately downstream from the cap? From which nucleotide does the 43S complex begin inspection of the 5' leader for the initiation codon during the scanning process? Initiation as a result of the latter mechanism could provide an explanation for the rarity of eukaryotic mRNAs with very short 5' leaders, and the leakiness of initiation that occurs on them (Kozak 1991), in addition to the susceptibility to dissociation by eIF1 of 43S complexes bound to them (Pestova and Kolupaeva 2002).

RIBOSOMAL SCANNING ON mRNA

The scanning model provides the most coherent account for the mechanism of ribosomal start-site selection, but almost 30 years after it was proposed (Kozak 1978), knowledge of the mechanistic details of the scanning process remains fragmentary. According to this model, the 40S subunit carrying Met-tRNA$_i^{Met}$ and a set of initiation factors migrates in a 5' to 3' direction from its attachment site at the 5' end of an mRNA to the first AUG triplet, which is recognized as the start site for initiation. The processivity of scanning in a yeast system has been reported to be absolute, without appreciable ribosomal drop-off (Berthelot et al. 2004). Scanning consists of several linked processes: unwinding of structured RNA in the 5' leader, ribosomal movement, rejection of potential mismatches between the anticodon of Met-tRNA$_i^{Met}$ and noncognate triplets in the 5' leader, and, finally, selection of the initiation codon.

Attachment of 43S complexes to capped and uncapped mRNAs might occur by substantially different mechanisms, which might account for some differences in the subsequent transition to scanning. Thus, only aberrant ribosomal complexes formed in the absence of eIF1 on native capped β-globin mRNA, near the 5' end, whereas 43S complexes were able to scan downstream along the 5'UTR of in-vitro-transcribed uncapped β-globin mRNA to form aberrant 48S initiation complexes on near-cognate GUG triplets and even to some extent to reach the AUG initiation codon (Pestova et al. 1998; Pestova and Kolupaeva 2002). Thus, eIF1-dependent disruption of one or more of the interactions that promoted attachment or induction of the necessary conformational changes in ribosomal complexes is particularly important for permitting ribosomal movement from the initial binding site in the case of cap-dependent initiation.

The question of if and when the relay of interactions among the cap, eIF4F, and the 43S complex is disrupted, and whether potential dissociation involves the interactions of the cap and eIF4E, of eIF4E and eIF4G, or of eIF4F and eIF3 has not been resolved (for discussion, see Jackson 2000). The requirement of eIF4G and eIF4A for scanning on structured 5' leaders and association of eIF4G with scanning ribosomal complexes (Pestova and Kolupaeva 2002; Pöyry et al. 2004), as well as the stability of the cap–eIF4F interaction (Gross et al. 2003), suggest that association of eIF4E/eIF4G with the cap and eIF4G with eIF3 is most likely retained during scanning, which would imply looping out of the 5' leader as the cap-bound ribosomal complex scans away from the initial binding site. However, if eIF4F must be dissociated from the cap, there is some evidence that mammalian eIF4B may enhance its release (Ray et al. 1986), but this is contradicted by a report that wheat eIF4B enhances this interaction (Khan and Goss 2005).

Ribosomal 43S complexes containing eIF1, eIF1A, eIF3, and the ternary complex are capable of scanning on unstructured mRNA without ATP or eIFs 4A, 4B, and 4F (Pestova and Kolupaeva 2002). Scanning thus reflects an intrinsic ability of the 43S complex to move on mRNA, but the contribution of individual factors to this process is not known. Some factors that contribute to formation of the 43S complex may combine with the 40S subunit to form a closed channel through which the 5' leader moves that may limit dissociation and thus enhance processivity (RNA probing studies indicate that factors associated with the 40S subunit may cover as many as 18 additional nucleotides upstream of the 40S subunit [see, e.g., Lazarowitz and Robertson 1977]), whereas other factors may induce the necessary conformational changes in the 43S complex. Although 43S complexes are able to recognize the 5' end of unstructured 5'UTR, bind to it, and scan along it to the initiation codon without ATP and factors associated with ATP hydrolysis, ATP and the eIF4 group of factors are required for scanning if the 5' leader contains even weak internal secondary structure (Pestova and Kolupaeva 2002). This observation accounts for earlier proposals that scanning is ATP-dependent (Kozak 1980) and is consistent with the findings that the requirement for ATP and eIF4A is proportional to the degree of secondary structure in the 5' leader (Jackson 1991; Svitkin et al. 2001). Recent data suggest that eIF4G remains associated with the 40S subunit after it begins scanning (Pöyry et al. 2004), and this association could provide a mechanism to ensure processive eIF4A-dependent unwinding of RNA in the leader and coupling of this process to ribosomal movement.

The location of the eIF4G/eIF4A complex on the scanning ribosomal complex would give valuable insights into its mechanism of action during scanning; as noted above, it may bind to the E-site side of the 40S subunit, which would imply that it "pulls" mRNA through the mRNA-binding cleft, rather than unwinding it at the leading edge of the 40S subunit (Siridechadilok et al. 2005). This location may impose directionality on the scanning process by preventing ribosomal movement in the 3' to 5' direction, which is important bearing in mind that ribosomes are intrinsically able to scan in this direction, as is apparent from translation of mRNAs on which reinitiation occurs on an initiation codon that is upstream of the termination codon (Meyers 2003). Such a mechanism would in some way likely also rely on unwinding of mRNA at the leading edge of the ribosomal complex, most likely by the intrinsic RNA-unwinding activity of the ribosome itself (Takyar et al. 2005), but possibly by an additional RNA helicase. eIF1 and eIF1A contribute to the processivity of scanning in a mechanistically distinct way from eIF4F, likely by altering the structure of the mRNA-binding cleft of the 40S subunit or by altering the positions or interactions of associated ligands, including initiator tRNA (see below).

A second DEAD-box protein, Ded1p, which has RNA helicase and strand annealing activities (Iost et al. 1999; Yang and Jankowsky 2005), has been implicated in translation initiation in yeast, but it is not known whether it acts primarily in unwinding cap-proximal mRNA structure or in scanning (Chuang et al. 1997; de la Cruz et al. 1997; Berthelot et al. 2004). Mutations in Ded1p are synthetic lethals with mutations in *TIF1* (eIF4A) and *cdc33* (eIF4E) and with deletions of *TIF4631* (eIF4G) or the nonessential *STM1/TIF3* (eIF4B) (de la Cruz et al. 1997). Ded1p can be replaced by the mouse PL10 homolog, one of a family of eukaryotic DEAD-box RNA helicases that might also be involved in translation (Linder 2003).

RECOGNITION OF THE INITIATION CODON

Recognition of the initiation codon is principally determined by its complementarity with the anticodon of Met-tRNA$_i^{Met}$ (Cigan et al. 1988). To ensure the fidelity of initiation, scanning ribosomes must have a discriminatory mechanism that can distinguish AUG from non-AUG codons and that destabilizes premature, partial base-pairing of triplets in the 5′UTR with the Met-tRNA$_i^{Met}$ anticodon. Biochemical analyses indicate that eIF1 is critical in both these processes, as well as in discriminating against initiation at AUG codons that have unfavorable nucleotide context or are located within approximately 8 nucleotides of the 5′ terminus (Pestova and Kolupaeva 2002). Importantly, eIF1 can also induce dissociation of aberrantly preassembled ribosomal complexes (Pestova et al. 1998; Pestova and Kolupaeva 2002; Lomakin et al. 2006). These results are consistent with earlier genetic analyses in yeast, which indicated that mutations in a defined region of eIF1 enhance initiation on a UUG triplet (Yoon and Donahue 1992). These activities of eIF1 are similar to those of prokaryotic IF3 (Petrelli et al. 2001). These factors are not homologs, but both bind to the small ribosomal subunit between the platform and initiator tRNA in a manner that precludes direct inspection of base-paired codon–anticodon triplets, and both can perform their discriminatory functions in heterologous systems (Dallas and Noller 2001; Lomakin et al. 2003, 2006). These observations strongly favor eIF1 acting indirectly to monitor the fidelity of initiation codon selection by inducing conformational changes in the ribosomal complex. Recent data suggest that like IF3, eIF1 may also participate in selection of Met-tRNA$_i^{Met}$, discriminating against mutations in the three consecutive universally conserved G-C pairs in the anticodon stem (see Fig. 2A) (Lomakin et al. 2006), which could occur by a mechanism involving eukaryotic equivalents of *Escherichia coli* G_{1338} and A_{1339} in 18S rRNA (Lancaster and Noller 2005).

A model to integrate the role of eIF1 in promoting scanning and ensuring the fidelity of initiation codon selection postulates that in its presence, ribosomal complexes are in an "open" conformation in which the positions of tRNA and mRNA on the 40S subunit are favorable for scanning, but that to form stable 48S complexes at the AUG codon in an mRNA, ribosomal complexes must undergo conformational changes induced by codon–anticodon base-pairing that lead to their adopting a "closed" conformation (Pestova and Hellen 2000; Pestova and Kolupaeva 2002). eIF1 could antagonize this rearrangement and induce dissociation of preassembled aberrant ribosomal complexes that lack stabilizing interactions involving the base-paired codon and anticodon, upstream RNA, or context nucleotides (Pestova and Kolupaeva 2002).

Experiments in a reconstituted yeast system support the "open-closed" isomerization model and its importance in start codon recognition and indicate that eIF1A and eIF5 also have key roles in the AUG-dependent shift from the "open" to the "closed" complex (Maag et al. 2006). An implication of this model is that the key context nucleotides at −3 and +4 positions must interact in a specific manner with components of the 48S complex, stabilizing it against the dissociating activity of eIF1: Recent data indicate that a G at the +4 position interacts specifically with $AA_{1818-1819}$ of 18S rRNA, whereas a G at the −3 position interacts with eIF2α (Pisarev et al. 2006). The interactions of A at these positions were not investigated but are likely similar. In addition to eIF1, genetic experiments in yeast have identified all three subunits of eIF2 and eIF5 as the principal components of the yeast translation apparatus involved in identifying the AUG codon and mediating the response to its identification (Donahue 2000). These mutations are thought to act, at least in part, by allowing GTP hydrolysis (or P_i release) in response to non-AUG codons or by causing premature release of eIF2 from Met-tRNA$_i^{Met}$ during scanning (Huang et al. 1997). More recently, in vivo experiments in yeast have indicated that eIF1A, eIF4G, and eIF3 also influence start codon recognition, possibly via their interactions with eIFs 1, 2, or 5 (He et al. 2003; Nielsen et al. 2004; Valasek et al. 2004; Fekete et al. 2005). In the case of eIF1A, in vitro data suggest that a fidelity-compromising mutation disrupts an AUG-dependent interaction between yeast eIF1A and eIF5 that may influence the equilibrium between the "open" and "closed" states of the ribosomal complex (Maag et al. 2006). A mutation in the eIF1-binding domain of yeast eIF4G that reduces the fidelity of start-site selection is suppressed by overexpressing eIF1, suggesting that disruption of the eIF1-eIF4G interaction might lead to premature release of eIF1 at non-AUG codons (see below) and thus the observed phenotype (He et al. 2003). One striking behavior of the majority of mutations that reduce the fidelity of initiation codon selection in yeast is that they appear to specifically permit initiation at UUG codons in preference to other non-AUG codons (see, e.g., Huang et al. 1997). This is not a priori an expected result; given the possibility of forming a U:G wobble base pair between the anticodon and a GUG codon, GUG would be expected to be the preferred triplet. Although the reason for the UUG preference has not been elucidated, recent in vitro data are consistent with the notion that UUG is a "special" codon (Maag et al. 2006). It is possible that the U at position 1 makes a specific contact with some group within the 43S complex that determines its preferential use.

RESPONSE TO AUG RECOGNITION AND RIBOSOMAL SUBUNIT JOINING

The eIFs 1, 1A, 2, and 3 occlude surfaces of the 40S subunit that form key intersubunit bridges with the 60S subunit in the 80S ribosome (see above), and these factors (or in the case of eIF3, the domain of it that prevents the B4 intersubunit contact) must therefore be displaced from the interface of the 40S subunit to permit subunit joining. This process requires two additional factors, eIF5, a GTPase-activating protein (GAP) specific for eIF2, and eIF5B, a ribosome-dependent GTPase that is homologous to the prokaryotic initiation factor IF2 (Pestova et al. 2000a; Das and Maitra 2001). The current model for this process is that eIF5 induces hydrolysis of eIF2-bound GTP, which leads to a reduction in eIF2's affinity for Met-tRNA$_i^{Met}$. eIF2-GDP and other initiation factors are then displaced from the 40S subunit by the combined action of eIF5B and the 60S subunit during the actual subunit joining event.

eIF5 binds to the eIF2β and eIF2γ subunits, but these interactions alone are not sufficient to stimulate the GTPase activity of the eIF2γ subunit. Full activation by eIF5 only occurs when the eIF2-GTP/Met-tRNA$_i^{Met}$ complex is bound to the 40S subunit, so it is likely that the 40S subunit stabilizes the active conformation of eIF2γ (Das and Maitra 2001; Algire et al. 2005). Thus, the GTPase activity of eIF2 in the 48S complex is 10^6-fold greater in the presence of eIF5 than when eIF2 is part of an isolated ternary complex or of a ribosomal 43S complex (Algire et al. 2005). Therefore, unlike all other translational GTPases, eIF2 has a dedicated GAP and does not rely on the GTPase-activating center of the 60S subunit. Activation by eIF5 has generally been considered to be dependent on establishment of codon–anticodon base-pairing in the P site (Raychaudhuri et al. 1985; Unbehaun et al. 2004) so that recognition of the initiation codon by the scanning 43S complex and establishment of base-pairing with Met-tRNA$_i^{Met}$ acts as a switch that triggers irreversible changes in the 48S complex, committing it to initiation at the correct site. However, in a reconstituted yeast system containing eIFs 1, 1A, 2, and 5, Pi release rather than GTP hydrolysis was the step most strongly dependent on initiation codon recognition (Algire et al. 2005). It should be noted that because release of phosphate drives GTP hydrolysis to completion, and thus the two processes are coupled, the reported effects on GTP hydrolysis and Pi release are most likely not inconsistent with one another.

eIF5 contains domains that are homologous to the amino-terminal and zinc-finger domains of eIF2β, an atypical HEAT-repeat domain that binds eIF2β (and in yeast, eIF1 and eIF3c), and in higher eukaryotes, a negatively charged carboxy-terminal tail that binds eIF5B (see Fig. 8A,B) (Das and Maitra 2001; Cho and Hoffman 2002; Marintchev and Wagner 2004; Yamamoto et al. 2005; Conte et al. 2006). Surprisingly, the amino-terminal domain of eIF5 is structurally similar to that of eIF1 as well, defining a new class of fold—the eIF12S fold—suggesting that these proteins may have arisen via gene duplication (Conte et al. 2006). The similar structures of eIF1 and the eIF5 amino-terminal domain suggest a model in which eIF1 antagonizes the proper interaction of eIF5 with eIF2 (Conte et al. 2006), consistent with the proposal that one or more conformational changes within the preinitiation complex take place prior to AUG-dependent irreversible hydrolysis of GTP. At least some of the interactions between eIF5 and other factors may be regulated: Mammalian and yeast eIF5 are phosphorylated by casein kinase 2 (CK2) in the immediate vicinity of the binding site for the amino-terminal tail of eIF2β in a manner that is required for their efficient interaction; CK2 binds directly to eIF2β and also phosphorylates its amino-terminal domain (Llorens et al. 2003; Maiti et al. 2003; Homma et al. 2005).

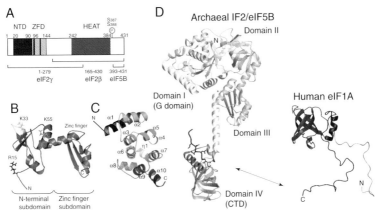

Figure 8. Structures, domain organization, and interactions of eIF5 and eIF5B. (*A*) Schematic representation of human eIF5, showing the relative positions and amino acid numbering of the amino-terminal domain (NTD), zinc-finger domain (ZFD), and HEAT-repeat domain (Cho and Hoffmann 2002; Conte et al. 2006), two CK2 phosphorylation sites, and domains of eIF5 that bind specifically to eIF2β and to eIF5B (Das and Maitra 2000; Marintchev and Wagner 2004). The interaction of eIF5 with eIF2γ was reported for yeast factors (Alone and Dever 2006). (*B*) Solution structure of the amino-terminal region of human eIF5, showing the amino-terminal subdomain (NTD) and the zinc-finger subdomain (ZFD). The structure shown is a representative, but the relative orientation of these subdomains could not be unambiguously determined. The NTD contains residues involved in the GAP function, including the "primary" arginine R15 and the putative "secondary" lysines, K33 and K55. (*Yellow*) The imperfect Walker A-box motif. (*C*) Crystal structure of the carboxy-terminal domain (residues 241–395) of yeast eIF5 (PDB 2FUL), which contains ten α-helices (α1–α10) and a 3_{10}-helix (η1). These helices are labeled and shown in different colors, and form four antiparallel helical repeats (repeat I consists of α1 and α2; repeat II consists of α3, α4, and α5; repeat III consists of α6, η1, and α7; and repeat IV consists of α8, α9, and α10). These repeats pack against each other in a counterclockwise manner, so that the eIF5 carboxy-terminal domain has a left-handed twist. (*D*) Model for the structures and interactions of eIF1A and eIF5B. Shown are the structures of human eIF1A (Battiste et al. 2000) and of the archaeal IF2/eIF5B homolog (Roll-Mecak et al. 2004), showing the segment of eIF1A that is absent from prokaryotic IF1. The arrow indicates the interactions between eIF1A and eIF5B (Marintchev et al. 2003), and a peptide corresponding to the binding segment of human eIF1A (shown, with side chains, in *blue*) has been modeled onto the structure of domain IV of archaeal IF2/eIF5B. (*B*, Generously provided by M.R. Conte and modified from Conte et al. 2006; *C*, generously provided by W. Gong and modified from Wei et al. 2006; *D*, generously provided by A. Marintchev and G. Wagner and modified, with permission, from Marintchev et al. 2003 [© National Academy of Sciences].)

It has been proposed that eIF5 acts as a classic GAP by providing an "arginine finger" (Arg-15) that may stabilize the transition state for GTP hydrolysis *in trans;* Lys-33 (which is part of an imperfect Walker A-box motif) and Lys-55 (or both) may have a secondary role in eIF5's GAP function (Das et al. 2001; Paulin et al. 2001). The amino-terminal domain (residues 1–279) of yeast eIF5 is necessary and sufficient for its binding to the G domain of eIF2γ and is as active as full-length eIF5 in activating eIF2's ribosome-dependent GTPase activity (Alone and Dever 2006). An implication of these observations is that the direct interaction of the eIF5 carboxy-terminal domain with eIF2β does not act as an allosteric regulator of eIF2γ G-domain function and instead may serve to recruit eIF5 to ribosome-bound eIF2. An alternate hypothesis suggests that eIF5 normally functions by interacting with eIF2β to derepress eIF2γ's GTPase activity (Marintchev and Wagner 2004). This hypothesis is based on observations concerning the effects of mutations in two domains of yeast eIF2β and probably in eIF2γ that weaken their mutual interactions and that yield eIF2 that has intrinsic (eIF5-independent) GTPase activity (Huang et al. 1997; Hashimoto et al. 2002). Interestingly, archaea lack a homolog of eIF5 and archaeal eIF2β lacks domains corresponding to those to which eIF5 binds in eukaryotes; it is not known whether archaeal eIF2 has intrinsic GTPase activity or if it requires an accessory GAP unrelated to eIF5 (Londei 2005). The fact that initiation in archaea does not involve ribosomal scanning and instead either occurs on leaderless mRNAs or is mediated by Shine-Dalgarno interactions raises the interesting possibility that the appearance of eIF5 in eukaryotes is directly connected to the use by eukaryotic ribosomes of scanning to locate the initiation codon, either to provide an additional structural component (eIF5) of the apparatus required for scanning or to ensure the integrity of the ternary complex during scanning by preventing premature GTP hydrolysis prior to AUG recognition. In the absence of eIF1, AUG recognition is uncoupled from activation of GTP hydrolysis (Unbehaun et al. 2004) or subsequent Pi release (Algire et al. 2005); premature GTPase activation (or Pi release) is therefore repressed by eIF1. Derepression occurs in response to AUG recognition by the 43S complex, by reducing the affinity of eIF1 for the complex, resulting in either its release or displacement from its original location on the 40S subunit, likely as a result of a conformational change in the ribosomal complex (Unbehaun et al. 2004; Valasek et al. 2004; Maag et al. 2005). eIF1 therefore has at least two significant roles in initiation codon recognition: (1) destabilizing incorrect codon–anticodon pairings and correct pairings in poor sequence contexts in the P site and (2) mediating the response to the identification of the correct AUG codon.

Hydrolysis of eIF2-bound GTP was until recently thought to lead directly to immediate loss of eIF2-GDP and other factors from the 40S subunit (Trachsel et al. 1977; Benne and Hershey 1978; Das and Maitra 2001). This model was based on data from assays done using AUG triplets as surrogates for mRNA, a reduced set of factors, and prolonged incubation, and involved purification of ribosomal complexes by sucrose density gradient centrifugation (which imposes significant shear forces on them). It has now been refined. The interaction of eIF2-GDP with the 40S subunit is weakened upon GTP hydrolysis, but it does not dissociate immediately. Importantly, retention of eIF2-GDP on the 40S subunit is dependent on interactions with bound mRNA, including a specific interaction with the key -3 context nucleotide (Pisarev et al. 2006). This observation is consistent with the facts that the affinity of Met-tRNA$_i^{Met}$ for eIF2-GDP is only an order of magnitude less than for eIF2-GTP (which is a thousand times less than the equivalent difference for EF-Tu) and that GTP hydrolysis does not induce the large conformational changes in the switch 1 and switch 2 elements of eIF2γ that are characteristic of GTPases such as EF-Tu (Schmitt et al. 2002; Kapp and Lorsch 2004b; Roll-Mecak et al. 2004). However, even if eIF2-GDP is displaced artificially (e.g., as a result of sucrose density gradient centrifugation), its loss does not lead to dissociation of other factors from the 40S subunit, particularly eIF3 (Unbehaun et al. 2004; Kolupaeva et al. 2005), which prevent subunit joining from occurring (Pestova et al. 2000b). Continued binding of eIF3 (and possibly of associated factors such as eIF4G) is stabilized by the presence of mRNA on the 40S subunit (Unbehaun et al. 2004; Kolupaeva et al. 2005), and studies on reinitiation suggest that eIF3 may in fact not even be completely released from the ribosome for several elongation cycles after initiation (Pöyry et al. 2004). As noted above, AUG recognition may result in either release or displacement of eIF1 from its original location on the 40S subunit (e.g., moving fully onto eIF3). eIF1A is displaced from 48S complexes during sucrose density gradient centrifugation (so that its displacement from the 40S subunit during subunit joining has not been analyzed), but the fact that it binds to eIF5B has prompted speculation that these factors may be released in a coordinate manner (see Fig. 8C) (Choi et al. 2000; Marintchev et al. 2003).

Displacement of factors from the 40S subunit after eIF5-mediated hydrolysis of eIF2-bound GTP is mediated by eIF5B (Pestova et al. 2000a,b), a universally conserved factor that is known as IF2 in prokaryotes (Choi et al. 1998). Like IF2, eIF5B is a ribosome-dependent GTPase; both factors bind GTP to achieve an active conformation and mediate subunit joining without hydrolysis of bound GTP but must complete this step to dissociate from the assembled ribosome (Pestova et al. 2000b; Antoun et al. 2003). Without GTP hydrolysis, eIF5B occludes the ribosomal A site and prevents elongation (Pestova et al. 2000b). Mutations that impair eIF5B's GTPase activity are suppressed by second-site mutations that lower its affinity for the ribosome, suggesting that GTP hydrolysis is not required to perform an essential mechanical function (Shin et al. 2002). The conformational changes in prokaryotic IF2 that occur upon hydrolysis of bound GTP and prepare it for release (as visualized by cryo-EM) have been described as a rotation that leads to partial retraction of domain IV from the A site and consequent loss of interaction with the P-site tRNA and a reduction in interactions with the decoding region and the GTPase-associated center of the large subunit (Myasnikov et al. 2005). These conformational changes and changes observed in the crystal structures of archaeal eIF5B in GTP- and GDP-bound states (see below) are consistent.

eIF5B has a long, charged, and poorly conserved amino-terminal domain and a short, helical carboxy-terminal subdomain flanking four domains (I–IV) that together form a chalice shape (Fig. 8C) (Roll-Mecak et al. 2000). Domain I contains the five sequence motifs (G1–G5) characteristic of GTP-binding translation factors, as well as the switch 1 and switch 2 effector regions; the β-barrel domain IV is connected to domain III by a long helix and flanked by two additional helices not present in prokaryotic IF2. The switch 2 element makes several contacts with GTP-Mg^{2+} (but not with GDP), and the interaction induces a minor conformational change in switch 2 that causes domain II to rotate; this is amplified and transmitted via domain III to domain IV, which undergoes a greater rigid-body movement about the long helix that connects it to domain III and effectively acts as a lever arm (Roll-Mecak et al. 2000). Whereas domain IV of prokaryotic IF2 binds fMet-tRNA$_f^{Met}$ specifically and with the same affinity as full-length IF2, consistent with a role in stimulating its binding to the 30S subunit, eukaryotic eIF5B binds Met-tRNA$_i^{Met}$ with low affinity and almost without specificity (Guillon et al. 2005). Domain IV of eukaryotic eIF5B also binds eIF1A in an interaction that involves the eukaryote-specific carboxy-terminal α-helices of eIF5B (Marintchev et al. 2003), and this interaction likely positions domain IV in the ribosomal A site near to the acceptor arm of Met-tRNA$_i^{Met}$. This location has been confirmed for prokaryotic IF2 by cryo-EM analysis (Myasnikov et al. 2005), and the high degree of sequence conservation between eIF5B and prokaryotic IF2 suggests that a similar location for domain IV of eIF5B is likely. The interaction between the carboxyl termini of eIF1A and eIF5B is important for efficient ribosomal subunit joining and subsequent GTP hydrolysis by eIF5B in the reconstituted yeast system (Acker et al. 2006), most likely because it either promotes the proper orientation of the two factors on the 40S subunit or induces a conformational rearrangement in eIF5B. It is possible that the interaction between the carboxyl terminus of eIF1A and eIF5 that is thought to have a role in the response to start codon recognition (Maag et al. 2006) is released upon GTP hydrolysis by eIF2, allowing the carboxyl terminus of eIF1A to interact with eIF5B. If so, it would provide a switch that promotes subunit joining and related events only after start codon recognition.

If eIF5B's putative interaction with Met-tRNA$_i^{Met}$ has any functional relevance, it is likely to be at the level of the ribosome and could involve either a mechanical role (to displace eIF2-GDP from the ribosome and/or to adjust the position of Met-tRNA$_i^{Met}$) analogously to a proposed activity of IF2 (La Teana et al. 1996) or a regulatory role as part of a checkpoint to ensure accurate 80S ribosome formation (Shin et al. 2002). These functions could occur sequentially and are thus not necessarily mutually exclusive. Adjustment of the position of Met-tRNA$_i^{Met}$ on the 40S subunit during subunit joining has been proposed to accommodate its likely change from having a high degree of mobility in the P site (to allow inspection of potential initiation codons during scanning) to adopting a unique and rigid position (which is necessary for peptide formation and is required by the structure of the large ribosomal subunit) (Marintchev and Wagner 2004). Consistent with the former role, partial displacement of eIF2-GDP by eIF5B was observed to occur in a context-dependent manner, such that eIF2-GDP was displaced more readily from 48S complexes assembled on initiation codons in suboptimal context (Pisarev et al. 2006). The displacement of eIF1 and eIF3 and the complete displacement of eIF2 requires the 60S subunit in addition to eIF5B (Unbehaun et al. 2004).

The mRNA poly(A) tail and PABP have also been implicated in the subunit joining process in yeast cells and in mammalian cell-free extracts (Sachs and Davis 1989; Kahvejian et al. 2005). In yeast, PABP has been proposed to derepress inhibition of eIF5B by Slh1p and Ski2p, putative RNA helicases (Searfoss et al. 2001); although these helicases have human homologs, it is not known whether PABP influences the subunit joining process in higher eukaryotes in the same way as in yeast.

PERSPECTIVES

The outlines of the mechanism of end-dependent translation initiation and most of the initiation factors required for this process have been known for almost 30 years (see, e.g., Trachsel et al. 1977; Benne and Hershey 1978; Peterson et al. 1979). In this chapter, we have emphasized recent genetic, biochemical, and biophysical developments that have enhanced understanding of some steps in this process, including formation of the 43S complex, initiation codon selection, and subunit joining. There is clearly still much to be learned about these steps. Moreover, the mechanisms of several other steps in initiation, such as attachment of the 43S complex to mRNA and subsequent scanning, are barely understood, and knowledge of how the individual biochemical activities of many factors are integrated into the initiation process is rudimentary.

Detailed understanding of the initiation process will be impossible without detailed architectural models of ribosomal initiation complexes at different stages during initiation. In addition to the many high-resolution structures of individual components of the translation apparatus that are already or will soon become available, such models will require detailed structures of the individual ribosomal subunits and sophisticated mapping, cryo-EM, or even crystallographic approaches to map the precise locations of ligands on them.

ACKNOWLEDGMENTS

We apologize to those whose work was not cited or discussed here because of the broad scope of this review and space limitations. Research in the authors' laboratories is supported by grants from the National Institutes of Health to T.V.P. (GM-63940 and GM-59660), to C.U.T.H. (AI-51340), and to J.R.L. (GM-62128) and from the American Cancer Society (RSG-03-156-01-GMC) and the American Heart Association (0555466U) to J.R.L. We thank M. Conte, S. Curry, W. Gong, A. Marintchev, A. Pisarev, and B. Siridechadilok for preparing parts of several of the figures.

REFERENCES

Abramson R.D., Dever T.E., and Merrick W.C. 1988. Biochemical evidence supporting a mechanism for cap-independent and internal initiation of eukaryotic mRNA. *J. Biol. Chem.* **263:** 6016–6019.

Acker M.G., Shin B.-S., Dever T.E., and Lorsch J.R. 2006. Interaction between eukaryotic initiation factors 1A and 5B is required for efficient ribosomal subunit joining. *J. Biol. Chem.* **281:** 8469–8477.

Algire M.A., Maag D., and Lorsch J.R. 2005. Pi release from eIF2, not GTP hydrolysis, is the step controlled by start-site selection during eukaryotic translation initiation. *Mol. Cell* **20:** 251–262.

Ali I.K. and Jackson R.J. 2001. The translation of capped mRNAs has an absolute requirement for the central domain of eIF4G but not for the cap-binding initiation factor eIF4E. *Cold Spring Harbor Symp. Quant. Biol.* **66:** 377–387.

Alone P.V. and Dever T.E. 2006. Direct binding of translation initiation factor eIF2γ-G domain to its GTPase-activating and GDP-GTP exchange factors eIF5 and eIF2Bε. *J. Biol. Chem.* **281:** 12636–12644.

Antoun A., Pavlov M.Y., Andersson K., Tenson T., and Ehrenberg M. 2003. The roles of initiation factor 2 and guanosine triphosphate in initiation of protein synthesis. *EMBO J.* **22:** 5593–5601.

Asano K., Krishnamoorthy T., Phan L., Pavitt G.D., and Hinnebusch A.G. 1999. Conserved bipartite motifs in yeast eIF5 and eIF2Bε, GTPase-activating and GDP-GTP exchange factors in translation initiation, mediate binding to their common substrate eIF2. *EMBO J.* **18:** 1673–1688.

Astrom S.U., von Pawel-Rammingen U., and Bystrom A.S. 1993. The yeast initiator tRNAMet can act as an elongator tRNA(Met) in vivo. *J. Mol. Biol.* **233:** 43–58.

Astrom S.U., Nordlund M.E., Erickson F.L., Hannig E.M., and Bystrom A.S. 1999. Genetic interactions between a null allele of the RIT1 gene encoding an initiator tRNA-specific modification enzyme and genes encoding translation factors in *Saccharomyces cerevisiae*. *Mol. Gen. Genet.* **261:** 967–976.

Basavappa R. and Sigler P.B. 1991. The 3 Å crystal structure of yeast initiator tRNA: Functional implications in initiator/elongator discrimination. *EMBO J.* **10:** 3105–3111.

Battiste J.L., Pestova T.V., Hellen C.U.T., and Wagner G. 2000. The eIF1A solution structure reveals a large RNA-binding surface important for scanning function. *Mol. Cell* **5:** 109–119.

Bellsolell L., Cho-Park P.F., Poulin F., Sonenberg S., and Burley S.K. 2006. Two structurally atypical HEAT domains in the C-terminal portion of human eIF4G support binding to eIF4A and Mnk1. *Structure* **14:** 913–923.

Benne R. and Hershey J.W. 1978. The mechanism of action of protein synthesis initiation factors from rabbit reticulocytes. *J. Biol. Chem.* **253:** 3078–3087.

Berthelot K., Muldoon M., Rajkowitsch L., Hughes J., and McCarthy J.E. 2004. Dynamics and processivity of 40S ribosome scanning on mRNA in yeast. *Mol. Microbiol.* **51:** 987–1001.

Bommer U.A., Lutsch G., Stahl J., and Bielka H. 1991. Eukaryotic initiation factors eIF-2 and eIF-3: Interactions, structure and localization in ribosomal initiation complexes. *Biochimie* **73:** 1007–1019.

Borman A.M., Michel Y.M., and Kean K.M. 2000. Biochemical characterisation of cap-poly(A) synergy in rabbit reticulocyte lysates: The eIF4G-PABP interaction increases the functional affinity of eIF4E for the capped mRNA 5′-end. *Nucleic Acids Res.* **28:** 4068–4075.

Bushell M., Wood W., Carpenter G., Pain V.M., Morley S.J., and Clemens M.J. 2001. Disruption of the interaction of mammalian protein synthesis eukaryotic initiation factor 4B with the poly(A)-binding protein by caspase- and viral protease-mediated cleavages. *J. Biol. Chem.* **276:** 23922–23928.

Carter A.P., Clemons W.M., Jr., Brodersen D.E., Morgan-Warren R.J., Hartsch T., Wimberly B.T., and Ramakrishnan V. 2001. Crystal structure of an initiation factor bound to the 30S ribosomal subunit. *Science* **291:** 498–501.

Caruthers J.M., Johnson E.R., and McKay D.B. 2000. Crystal structure of yeast initiation factor 4A, a DEAD-box RNA helicase. *Proc. Natl. Acad. Sci.* **97:** 13080–13085.

Chang T.C., Yamashita A., Chen C.Y., Yamashita Y., Zhu W., Durdan S., Kahvejian A., Sonenberg N., and Shyu A.B. 2004. UNR, a new partner of poly(A)-binding protein, plays a key role in translationally coupled mRNA turnover mediated by the c-fos major coding-region determinant. *Genes Dev.* **18:** 2010–2023.

Chaudhuri J., Chowdhury D., and Maitra U. 1999. Distinct functions of eukaryotic translation initiation factors eIF1A and eIF3 in the formation of the 40 S ribosomal preinitiation complex. *J. Biol. Chem.* **274:** 17975–17980.

Cho S. and Hoffman D.W. 2002. Structure of the beta subunit of translation initiation factor 2 from the archaeon *Methanococcus jannaschii:* A representative of the eIF2beta/eIF5 family of proteins. *Biochemistry* **41:** 5730–5742.

Choi S.K., Lee J.H., Zoll W.L., Merrick W.C., and Dever T.E. 1998. Promotion of met-tRNAiMet binding to ribosomes by yIF2, a bacterial IF2 homolog in yeast. *Science* **280:** 1757–1760.

Choi S.K., Olsen D.S., Roll-Mecak A., Martung A., Remo K.L., Burley S.K., Hinnebusch A.G., and Dever T.E. 2000. Physical and functional interaction between the eukaryotic orthologs of prokaryotic translation initiation factors IF1 and IF2. *Mol. Cell. Biol.* **20:** 7183–7191.

Chuang R.Y., Weaver P.L., Liu Z., and Chang T.H. 1997. Requirement of the DEAD-Box protein Ded1p for messenger RNA translation. *Science* **275:** 1468–1471.

Churbanov A., Rogozin I.B., Babenko V.N., Ali H., and Koonin E.V. 2005. Evolutionary conservation suggests a regulatory function of AUG triplets in 5′-UTRs of eukaryotic genes. *Nucleic Acids Res.* **33:** 5512–5520.

Cigan A.M., Feng L., and Donahue T.F. 1988. tRNA$_i^{(met)}$ functions in directing the scanning ribosome to the start site of translation. *Science* **242:** 93–97.

Collier B., Gorgoni B., Loveridge C., Cooke H.J., and Gray N.K. 2005. The DAZL family proteins are PABP-binding proteins that regulate translation in germ cells. *EMBO J.* **24:** 2656–2666.

Conte M.R., Kelly G., Babon J., Sanfelice D., Youell J., Smerdon S.J., and Proud C.G. 2006. Structure of the eukaryotic initiation factor (eIF) 5 reveals a fold common to several translation factors. *Biochemistry* **45:** 4550–4558.

Cordin O., Banroques J., Tanner N.K., and Linder P. 2006. The DEAD-box protein family of RNA helicases. *Gene* **367:** 17–37.

Dallas A. and Noller H.F. 2001. Interaction of translation initiation factor 3 with the 30S ribosomal subunit. *Mol. Cell* **8:** 855–864.

Das S. and Maitra U. 2000. Mutational analysis of mammalian translation initiation factor 5 (eIF5): Role of interaction between the beta subunit of eIF2 and eIF5 in eIF5 function in vitro and in vivo. *Mol. Cell. Biol.* **20:** 3942–3950.

———. 2001. Functional significance and mechanism of eIF5-promoted GTP hydrolysis in eukaryotic translation initiation. *Prog. Nucleic Acid Res. Mol. Biol.* **70:** 207–231.

Das S., Ghosh R., and Maitra U. 2001. Eukaryotic translation initiation factor 5 functions as a GTPase-activating protein. *J. Biol. Chem.* **276:** 6720–6726.

De Gregorio E., Preiss T., and Hentze M.W. 1999. Translation driven by an eIF4G core domain *in vivo. EMBO J.* **18:** 4865–4874.

de la Cruz J., Iost I., Kressler D., and Linder P. 1997. The p20 and Ded1 proteins have antagonistic roles in eIF4E-dependent translation in *Saccharomyces cerevisiae. Proc. Natl. Acad. Sci.* **94:** 5201–5206.

Deo R.C., Bonanno J.B., Sonenberg N., and Burley S.K. 1999. Recognition of polyadenylate RNA by the poly(A)-binding protein. *Cell* **98:** 835–845.

Donahue T.F. 2000. Genetic approaches to translation initiation in *Saccharomyces cerevisiae.* In *Translational control of gene expression* (ed. N. Sonenberg et al.), pp. 487–502. Cold Spring Harbor Laboratory Press, Cold Spring Harbor, New York.

Donahue T.F., Cigan A.M., Pabich E.K., and Valavicius B.C. 1988. Mutations at a Zn(II) finger motif in the yeast eIF-2 beta gene alter ribosomal start-site selection during the scanning process. *Cell* **54:** 621–632.

Dmitriev S.E., Terenin I.M., Dunaevsky Y.E., Merrick W.C., and Shatsky I.N. 2003. Assembly of 48S translation initiation complexes from purified components with mRNAs that have some base pairing within their 5′ untranslated regions. *Mol. Cell. Biol.* **23:** 8925–8933.

Drabkin H.J., Estrella M., and Rajbhandary U.L. 1998. Initiator-elongator discrimination in vertebrate tRNAs for protein synthesis. *Mol. Cell. Biol.* **18:** 1459–1466.

Erickson F.L. and Hannig E.M. 1996. Ligand interactions with eukaryotic translation initiation factor 2: Role of the gamma-subunit. *EMBO J.* **15:** 6311–6320.

Erickson F.L., Nika J., Rippel S., and Hannig E.M. 2001. Minimum requirements for the function of eukaryotic translation initiation factor 2. *Genetics* **158:** 123–132.

Farruggio D., Chaudhuri J., Maitra U., and RajBhandary U.L. 1996. The A1 × U72 base pair conserved in eukaryotic initiator tRNAs is important specifically for binding to the eukaryotic translation initiation factor eIF2. *Mol. Cell. Biol.* **16:** 4248–4256.

Fekete C.A., Applefield D.J., Blakely S.A., Shirokikh N., Pestova T., Lorsch J.R., and Hinnebusch A.G. 2005. The eIF1A C-terminal domain promotes initiation complex assembly, scanning and AUG selection in vivo. *EMBO J.* **24:** 3588–3601.

Fleming K., Ghuman J., Yuan X., Simpson P., Szendroi A., Matthews S., and Curry S. 2003. Solution structure and RNA interactions of the RNA recognition motif from eukaryotic translation initiation factor 4B. *Biochemistry* **42:** 8966–8975.

Fletcher C.M., Pestova T.V., Hellen C.U.T., and Wagner G. 1999. Structure and interactions of the translation initiation factor eIF1. *EMBO J.* **18:** 2631–2637.

Fraser C.S., Lee J.Y., Mayeur G.L., Bushell M., Doudna J.A., and Hershey J.W. 2004. The j-subunit of human translation initiation factor eIF3 is required for the stable binding of eIF3 and its subcomplexes to 40 S ribosomal subunits in vitro. *J. Biol. Chem.* **279:** 8946–8956.

Freienstein C. and Blobel G. 1975. Nonribosomal proteins associated with eukaryotic native small ribosomal subunits. *Proc. Natl. Acad. Sci.* **72:** 3392–3396.

Geballe A.P. and Sachs M.S. 2000. Translational control by upstream open reading frames. In *Translational control of gene expression* (ed. N. Sonenberg et al.), pp. 595–614. Cold Spring Harbor Laboratory Press, Cold Spring Harbor, New York.

Gingras A.C., Raught B., and Sonenberg N. 1999. eIF4 initiation factors: Effectors of mRNA recruitment to ribosomes and regulators of translation. *Annu. Rev. Biochem.* **68:** 913–963.

Groft C.M. and Burley S.K. 2002. Recognition of eIF4G by rotavirus NSP3 reveals a basis for mRNA circularization. *Mol. Cell* **9:** 1273–1283.

Gross J.D., Moerke N.J., von der Haar T., Lugovskoy A.A., Sachs A.B., McCarthy J.E., and Wagner G. 2003. Ribosome loading onto the mRNA cap is driven by conformational coupling between eIF4G and eIF4E. *Cell* **115:** 739–750.

Guillon L., Schmitt E., Blanquet S., and Mechulam Y. 2005. Initiator tRNA binding by e/aIF5B, the eukaryotic/archaeal homologue of bacterial initiation factor IF2. *Biochemistry* **44:** 15594–15601.

Gunnery S., Maivali U., and Mathews M.B. 1997. Translation of an uncapped mRNA involves scanning. *J. Biol. Chem.* **272:** 21642–21646.

Gutierrez P., Osborne M.J., Siddiqui N., Trempe J.F., Arrowsmith C., and Gehring K. 2004. Structure of the archaeal translation initiation factor aIF2 beta from *Methanobacterium thermoautotrophicum:* Implications for translation initiation. *Protein Sci.* **13:** 659–667.

Haghighat A. and Sonenberg N. 1997. eIF4G dramatically enhances the binding of eIF4E to the mRNA 5′-cap structure. *J. Biol. Chem.* **272:** 21677–21680.

Hashimoto N.N., Carnevalli L.S., and Castilho B.A. 2002. Translation initiation at non-AUG codons mediated by weakened association of eukaryotic initiation factor (eIF) 2 subunits. *Biochem J.* **367:** 359–368.

He H., von der Haar T., Singh C.R., Ii M., Li B., Hinnebusch A.G., McCarthy J.E., and Asano K. 2003. The yeast eukaryotic initiation factor 4G (eIF4G) HEAT domain interacts with eIF1 and eIF5 and is involved in stringent AUG selection. *Mol. Cell. Biol.* **23:** 5431–5445.

Holz M.K., Ballif B.A., Gygi S.P., and Blenis J. 2005. mTOR and S6K1 mediate assembly of the translation preinitiation complex through dynamic protein interchange and ordered phosphorylation events. *Cell* **123:** 569–580.

Homma M.K., Wada I., Suzuki T., Yamaki J., Krebs E.G., and Homma Y. 2005. CK2 phosphorylation of eukaryotic translation initiation factor 5 potentiates cell cycle progression. *Proc. Natl. Acad. Sci.* **102:** 15688–15693.

Huang H.K., Yoon H., Hannig E.M., and Donahue T.F. 1997. GTP hydrolysis controls stringent selection of the AUG start codon during translation initiation in *Saccharomyces cerevisiae*. *Genes Dev.* **11:** 2396–2413.

Imataka H. and Sonenberg N. 1997. Human eukaryotic translation initiation factor 4G (eIF4G) possesses two separate and independent binding sites for eIF4A. *Mol. Cell. Biol.* **17:** 6940–6947.

Iost I., Dreyfus M., and Linder P. 1999. Ded1p, a DEAD-box protein required for translation initiation in *Saccharomyces cerevisiae*, is an RNA helicase. *J. Biol. Chem.* **274:** 17677–17683.

Ito T., Marintchev A., and Wagner G. 2004. Solution structure of human initiation factor eIF2alpha reveals homology to the elongation factor eEF1B. *Structure* **12:** 1693–1704.

Jacobson A. 2000. Poly(A) metabolism and translation: The closed-loop model. In *Translational control of gene expression* (ed. N. Sonenberg et al.), pp. 451–480. Cold Spring Harbor Laboratory Press, Cold Spring Harbor, New York.

Jackson R.J. 1991. The ATP requirement for initiation of eukaryotic translation varies according to the mRNA species. *Eur. J. Biochem.* **200:** 285–294.

———. 2000. Comparative view of initiation site selection mechanisms. In *Translational control of gene expression* (ed. N. Sonenberg et al.), pp. 127–183. Cold Spring Harbor Laboratory Press, Cold Spring Harbor, New York.

Jaramillo M., Dever T.E., Merrick W.C., and Sonenberg N. 1991. RNA unwinding in translation: Assembly of helicase complex intermediates comprising eukaryotic initiation factors eIF-4F and eIF-4B. *Mol. Cell. Biol.* **11:** 5992–5997.

Kahvejian A., Svitkin Y.V., Sukarieh R., M'Boutchou M.N., and Sonenberg N. 2005. Mammalian poly(A)-binding protein is a eukaryotic translation initiation factor, which acts via multiple mechanisms. *Genes Dev.* **19:** 104–113.

Kanduc D. 1997. Changes of tRNA population during compensatory cell proliferation: Differential expression of methionine-tRNA species. *Arch. Biochem. Biophys.* **342:** 1–5.

Kapp L.D. and Lorsch J.R. 2004a. The molecular mechanics of eukaryotic translation. *Annu. Rev. Biochem.* **73:** 657–704.

———. 2004b. GTP-dependent recognition of the methionine moiety on initiator tRNA by translation factor eIF2. *J. Mol. Biol.* **335:** 923–936.

Khan M.A. and Goss D.J. 2005. Translation initiation factor (eIF) 4B affects the rates of binding of the mRNA m7G cap analogue to wheat germ eIFiso4F and eIFiso4F.PABP. *Biochemistry* **44:** 4510–4516.

Kim W.J., Back S.H., Kim V., Ryu I., and Jang S.K. 2005. Sequestration of TRAF2 into stress granules interrupts tumor necrosis factor signaling under stress conditions. *Mol. Cell. Biol.* **25:** 2450–2462.

Kisselev L.L. and Buckingham R.H. 2000. Translational termination comes of age. *Trends Biochem. Sci.* **25:** 561–566.

Kolupaeva V.G., Unbehaun A., Lomakin I.B., Hellen C.U., and Pestova T.V. 2005. Binding of eukaryotic initiation factor 3 to ribosomal 40S subunits and its role in ribosomal dissociation and anti-association. *RNA* **11:** 470–486.

Korneeva N.L., First E.A., Benoit C.A., and Rhoads R.E. 2005. Interaction between the NH2-terminal domain of eIF4A and the central domain of eIF4G modulates RNA-stimulated ATPase activity. *J. Biol. Chem.* **280:** 1872–1881.

Kozak M. 1978. How do eucaryotic ribosomes select initiation regions in messenger RNA? *Cell* **15:** 1109–1123.

———. 1980. Role of ATP in binding and migration of 40S ribosomal subunits. *Cell* **22:** 459–467.

———. 1983. Translation of insulin-related polypeptides from messenger RNAs with tandemly reiterated copies of the ribosome binding site. *Cell* **34:** 971–978.

———. 1991. Structural features in eukaryotic mRNAs that modulate the initiation of translation. *J. Biol. Chem.* **266:** 19867–19870.

Kozlov G., Trempe J.F., Khaleghpour K., Kahvejian A., Ekiel I., and Gehring K. 2001. Structure and function of the C-terminal PABC domain of human poly(A)-binding protein. *Proc. Natl. Acad. Sci.* **98:** 4409–4413.

Lamphear B.J., Kirchweger R., Skern T., and Rhoads R.E. 1995. Mapping of functional domains in eukaryotic protein synthesis initiation factor 4G (eIF4G) with picornaviral proteases. Implications for cap-dependent and cap-independent translational initiation. *J. Biol. Chem.* **270:** 21975–21983.

Lancaster L. and Noller H.F. 2005. Involvement of 16S rRNA nucleotides G1338 and A1339 in discrimination of initiator tRNA. *Mol. Cell* **20:** 623–632.

La Teana A., Pon C.L., and Gualerzi C.O. 1996. Late events in translation initiation. Adjustment of fMet-tRNA in the ribosomal P-site. *J. Mol. Biol.* **256:** 667–675.

Laurino J.P., Thompson G.M., Pacheco E., and Castilho B.A. 1999. The beta subunit of eukaryotic translation initiation factor 2 binds mRNA through the lysine repeats and a region comprising the C2-C2 motif. *Mol. Cell. Biol.* **19:** 173–181.

Lawson T.G., Cladaras M.H., Ray B.K., Lee K.A., Abramson R.D., Merrick W.C., and Thach R.E. 1988. Discriminatory interaction of purified eukaryotic initiation factors 4F plus 4A with the 5′ ends of reovirus messenger RNAs. *J. Biol. Chem.* **263:** 7266–7276.

Lazarowitz S.G. and Robertson H.D. 1977. Initiator regions from the small size class of reovirus messenger RNA protected by rabbit reticulocyte ribosomes. *J. Biol. Chem.* **252:** 7842–7849.

Linder P. 2003. Yeast RNA helicases of the DEAD-box family involved in translation initiation. *Biol. Cell* **95:** 157–167.

Ling J., Morley S.J., and Traugh J.A. 2005. Inhibition of cap-dependent translation via phosphorylation of eIF4G by protein kinase Pak2. *EMBO J.* **24:** 4094–4105.

Llorens F., Roher N., Miro F.A., Sarno S., Ruiz F.X., Meggio F., Plana M., Pinna L.A., and Itarte E. 2003. Eukaryotic translation-initiation factor eIF2b binds to protein kinase CK2: Effects on CK2a activity. *Biochem. J.* **375:** 623–631.

Lomakin I.B., Hellen C.U.T., and Pestova T.V. 2000. Physical association of eukaryotic initiation factor 4G (eIF4G) with eIF4A strongly enhances binding of eIF4G to the internal ribosomal entry site of encephalomyocarditis virus and is required for internal initiation of translation. *Mol. Cell. Biol.* **20:** 6019–6029.

Lomakin I.B., Kolupaeva V.G., Marintchev A., Wagner G., and Pestova T.V. 2003. Position of eukaryotic initiation factor eIF1 on the 40S ribosomal subunit determined by directed hydroxyl radical probing. *Genes Dev.* **17:** 2786–2797.

Lomakin I.B., Shirokikh N.E., Yusupov M.M., Hellen C.U.T., and Pestova T.V. 2006. The fidelity of translation initiation: Reciprocal activities of eIF1, IF3 and YciH. *EMBO J.* **25:** 196–210.

Londei P. 2005. Evolution of translational initiation: New insights from the archaea. *FEMS Microbiol. Rev.* **29:** 185–200.

Lorsch J.R. and Herschlag D. 1998. The DEAD box protein eIF4A. 1. A minimal kinetic and thermodynamic framework reveals coupled binding of RNA and nucleotide. *Biochemistry* **37:** 2180–2193.

Maag D. and Lorsch J.R. 2003. Communication between eukaryotic translation initiation factors 1 and 1A on the yeast small ribosomal subunit. *J. Mol. Biol.* **330:** 917–924.

Maag D., Algire M.A., and Lorsch J.R. 2006. Communication between eukaryotic translation initiation factors 5 and 1A within the ribosomal pre-initiation complex plays a role in start site selection. *J. Mol. Biol.* **356:** 724–737.

Maag D., Fekete C.A., Gryczynski Z., and Lorsch J.R. 2005. A conformational change in the eukaryotic translation preinitiation complex and release of eIF1 signal recognition of the start codon. *Mol. Cell* **17:** 265–275.

Maiti T., Bandyopadhyay A., and Maitra U. 2003. Casein kinase II phosphorylates translation initiation factor 5 (eIF5) in *Saccharomyces cerevisiae*. *Yeast* **20:** 97–108.

Majumdar R., Bandyopadhyay A., and Maitra U. 2003. Mammalian translation initiation factor eIF1 functions with eIF1A and eIF3 in the formation of a stable 40 S preinitiation complex. *J. Biol. Chem.* **278:** 6580–6587.

Mangus D.A., Evans M.C., and Jacobson A. 2003. Poly(A)-binding proteins: Multifunctional scaffolds for the post-transcriptional control of gene expression. *Genome Biol.* **4:** 223.

Marcotrigiano J., Gingras A.C., Sonenberg N., and Burley S.K. 1997. Cocrystal structure of the messenger RNA 5′ cap-binding protein (eIF4E) bound to 7-methyl-GDP. *Cell* **89:** 951–961.

———. 1999. Cap-dependent translation initiation in eukaryotes is regulated by a molecular mimic of eIF4G. *Mol. Cell* **3:** 707–716.

Marcotrigiano J., Lomakin I.B., Sonenberg N., Pestova T.V., Hellen C.U.T., and Burley S.K. 2001. A conserved HEAT domain within eIF4G directs assembly of the translation initiation machinery. *Mol. Cell* **7:** 193–203.

Marintchev A. and Wagner G. 2004. Translation initiation: Structures, mechanisms and evolution. *Q. Rev. Biophys.* **37:** 197–284.

Marintchev A., Kolupaeva V.G., Pestova T.V., and Wagner G. 2003. Mapping the binding interface between human eukaryotic initiation factors 1A and 5B: A new interaction between old partners. *Proc. Natl. Acad. Sci.* **100:** 1535–1540.

Matsuo H., Li H., McGuire A.M., Fletcher C.M., Gingras A.C., Sonenberg N., and Wagner G. 1997. Structure of translation factor eIF4E bound to m7GDP and interaction with 4E-binding protein. *Nat. Struct. Biol.* **4:** 717–724.

Mayeur G.L., Fraser C.S., Peiretti F., Block K.L., and Hershey J.W. 2003. Characterization of eIF3k: A newly discovered subunit of mammalian translation initiation factor eIF3. *Eur. J. Biochem.* **270:** 4133–4139.

Méthot N., Pickett G., Keene J.D., and Sonenberg N. 1996. In vitro RNA selection identifies RNA ligands that specifically bind to eukaryotic translation initiation factor 4B: The role of the RNA recognition motif. *RNA* **2:** 38–50.

Meyers G. 2003. Translation of the minor capsid protein of a calicivirus is initiated by a novel termination-dependent reinitiation mechanism. *J. Biol. Chem.* **278:** 34051–34060.

Michel Y.M., Poncet D., Piron M., Kean K.M., and Borman A.M. 2000. Cap-Poly(A) synergy in mammalian cell-free extracts. Investigation of the requirements for poly(A)-mediated stimulation of translation initiation. *J. Biol. Chem.* **275:** 32268–32276.

Moldave K. 1985. Eukaryotic protein synthesis. *Annu. Rev. Biochem.* **54:** 1109–1149.

Morino S., Imataka H., Svitkin Y.V., Pestova T.V., and Sonenberg N. 2000. Eukaryotic translation initiation factor 4E (eIF4E) binding site and the middle one-third of eIF4GI constitute the core domain for cap-dependent translation, and the C-terminal one-third functions as a modulatory region. *Mol. Cell. Biol.* **20:** 468–477.

Mouat M.F. and Manchester K. 1998. An alpha subunit-deficient form of eukaryotic protein synthesis initiation factor eIF-2 from rabbit reticulocyte lysate and its activity in ternary complex formation. *Mol. Cell. Biochem.* **183:** 69–78.

Myasnikov A.G., Marzi S., Simonetti A., Giuliodori A.M., Gualerzi C.O., Yusupova G., Yusupov M., and Klaholz B.P. 2005. Conformational transition of initiation factor 2 from the GTP- to GDP-bound state visualized on the ribosome. *Nat. Struct. Mol. Biol.* **12:** 1145–1149.

Nielsen K.H., Szamecz B., Valasek L., Jivotovskaya A., Shin B.S., and Hinnebusch A.G. 2004. Functions of eIF3 downstream of 48S assembly impact AUG recognition and GCN4 translational control. *EMBO J.* **23:** 1166–1177.

Nika J., Rippel S., and Hannig E.M. 2001. Biochemical analysis of the eIF2beta gamma complex reveals a structural function for eIF2alpha in catalyzed nucleotide exchange. *J. Biol. Chem.* **276:** 1051–1056.

Nika J., Yang W., Pavitt G.D., Hinnebusch A.G., and Hannig E.M. 2000. Purification and kinetic analysis of eIF2B from *Saccharomyces cerevisiae*. *J. Biol. Chem.* **275:** 26011–26017.

Nygard O. and Nilsson L. 1990. Translational dynamics. Interactions between the translational factors, tRNA and ribosomes during eukaryotic protein synthesis. *Eur. J. Biochem.* **191:** 1–17.

Oberer M., Marintchev A., and Wagner G. 2005. Structural basis for the enhancement of eIF4A helicase activity by eIF4G. *Genes Dev.* **19:** 2212–2223.

Palmiter R.D. 1972. Regulation of protein synthesis in chick oviduct. II. Modulation of polypeptide elongation and initiation rates by estrogen and progesterone. *J. Biol. Chem.* **247:** 6770–6780.

Paulin F.E., Campbell L.E., O'Brien K., Loughlin J., and Proud C.G. 2001. Eukaryotic translation initiation factor 5 (eIF5) acts as a classical GTPase-activator protein. *Curr. Biol.* **11:** 55–59.

Pause A., Methot N., Svitkin Y., Merrick W.C., and Sonenberg N. 1994. Dominant negative mutants of mammalian translation initiation factor eIF-4A define a critical role for eIF-4F in cap-dependent and cap-independent initiation of translation. *EMBO J.* **13:** 1205–1215.

Peabody D.S. 1989. Translation initiation at non-AUG triplets in mammalian cells. *J. Biol. Chem.* **264:** 5031–5035.

Pelletier J. and Sonenberg N. 1985. Insertion mutagenesis to increase secondary structure within the 5′ noncoding region of a eukaryotic mRNA reduces translational efficiency. *Cell* **40:** 515–526.

Pesole G., Gissi C., Grillo G., Licciulli F., Liuni S., and Saccone C. 2000. Analysis of oligonucleotide AUG start codon context in eukariotic mRNAs. *Gene* **261**: 85–91.

Pestova T.V. and Hellen C.U.T. 2000. The structure and function of initiation factors in eukaryotic protein synthesis. *Cell. Mol. Life Sci.* **57**: 651–674.

Pestova T.V. and Kolupaeva V.G. 2002. The roles of individual eukaryotic translation initiation factors in ribosomal scanning and initiation codon selection. *Genes Dev.* **16**: 2906–2922.

Pestova T.V., Borukhov S.I., and Hellen C.U.T. 1998. Eukaryotic ribosomes require initiation factors 1 and 1A to locate initiation codons. *Nature* **394**: 854–859.

Pestova T.V., Dever T.E., and Hellen C.U.T. 2000a. Ribosomal subunit joining. In *Translational control of gene expression* (ed. N. Sonenberg et al.), pp. 425–445. Cold Spring Harbor Laboratory Press, Cold Spring Harbor, New York.

Pestova T.V., Shatsky I.N., and Hellen C.U.T. 1996. Functional dissection of eukaryotic initiation factor 4F: The 4A subunit and the central domain of the 4G subunit are sufficient to mediate internal entry of 43S preinitiation complexes. *Mol. Cell. Biol.* **16**: 6870–6878.

Pestova T.V., Lomakin I.B., Lee J.H., Choi S.K., Dever T.E., and Hellen C.U.T. 2000b. The joining of ribosomal subunits in eukaryotes requires eIF5B. *Nature* **403**: 332–335.

Peterson D.T., Merrick W.C., and Safer B. 1979. Binding and release of radiolabeled eukaryotic initiation factors 2 and 3 during 80 S initiation complex formation. *J. Biol. Chem.* **254**: 2509–2516.

Petrelli D., LaTeana A., Garofalo C., Spurio R., Pon C.L., and Gualerzi C.O. 2001. Translation initiation factor IF3: Two domains, five functions, one mechanism? *EMBO J.* **20**: 4560–4569.

Pisarev A.V., Kolupaeva V.G., Pisareva V.P., Merrick W.C., Hellen C.U.T., and Pestova T.V. 2006. Specific functional interactions of nucleotides at key −3 and +4 positions flanking the initiation codon with components of the mammalian 48S translation initiation complex. *Genes Dev.* **20**: 4624–4636.

Pöyry T.A., Kaminski A., and Jackson R.J. 2004. What determines whether mammalian ribosomes resume scanning after translation of a short upstream open reading frame? *Genes Dev.* **18**: 62–75.

Prévôt D., Darlix J.L., and Ohlmann T. 2003. Conducting the initiation of protein synthesis: The role of eIF4G. *Biol. Cell* **95**: 141–156.

Ptushkina M., von der Haar T., Vasilescu S., Frank R., Birkenhager R., and McCarthy J.E. 1998. Cooperative modulation by eIF4G of eIF4E-binding to the mRNA 5′ cap in yeast involves a site partially shared by p20. *EMBO J.* **17**: 4798–4808.

Raught B., Peiretti F., Gingras A.C., Livingstone M., Shahbazian D., Mayeur G.L., Polakiewicz R.D., Sonenberg N., and Hershey J.W. 2004. Phosphorylation of eucaryotic translation initiation factor 4B Ser422 is modulated by S6 kinases. *EMBO J.* **23**: 1761–1769.

Ray B.K., Lawson T.G., Abramson R.D., Merrick W.C., and Thach R.E. 1986. Recycling of messenger RNA cap-binding proteins mediated by eukaryotic initiation factor 4B. *J. Biol. Chem.* **261**: 11466–11470.

Ray B.K., Lawson T.G., Kramer J.C., Cladaras M.H., Grifo J.A., Abramson R.D., Merrick W.C., and Thach R.E. 1985. ATP-dependent unwinding of messenger RNA structure by eukaryotic initiation factors. *J. Biol. Chem.* **260**: 7651–7658.

Raychaudhuri P., Chaudhuri A., and Maitra U. 1985. Eukaryotic initiation factor 5 from calf liver is a single polypeptide chain protein of Mr = 62,000. *J. Biol. Chem.* **260**: 2132–2139.

Rhoads R.E. 1991. Initiation: mRNA and 60S subunit binding. In *Translation in eukaryotes* (ed. H. Trachsel), pp. 109–148. CRC Press, Boca Raton, Florida.

Rogers G.W., Jr., Richter N.J., Lima W.F., and Merrick W.C. 2001. Modulation of the helicase activity of eIF4A by eIF4B, eIF4H, and eIF4F. *J. Biol. Chem.* **276:** 30914–30922.

Roll-Mecak A., Cao C., Dever T.E., and Burley S.K. 2000. X-ray structures of the universal translation initiation factor IF2/eIF5B: Conformational changes on GDP and GTP binding. *Cell* **103:** 781–792.

Roll-Mecak A., Alone P., Cao C., Dever T.E., and Burley S.K. 2004. X-ray structure of translation initiation factor eIF2gamma: Implications for tRNA and eIF2alpha binding. *J. Biol. Chem.* **279:** 10634–10642.

Rozen F., Edery I., Meerovitch K., Dever T.E., Merrick W.C., and Sonenberg N. 1990. Bidirectional RNA helicase activity of eucaryotic translation initiation factors 4A and 4F. *Mol. Cell. Biol.* **10:** 1134–1144.

Sachs A.B. and Davis R.W. 1989. The poly(A) binding protein is required for poly(A) shortening and 60S ribosomal subunit-dependent translation initiation. *Cell* **58:** 857–867.

Schmitt E., Blanquet S., and Mechulam Y. 2002. The large subunit of initiation factor aIF2 is a close structural homologue of elongation factors. *EMBO J.* **21:** 1821–1832.

Searfoss A., Dever T.E., and Wickner R. 2001. Linking the 3′ poly(A) tail to the subunit joining step of translation initiation: Relations of Pab1p, eukaryotic translation initiation factor 5b (Fun12p), and Ski2p-Slh1p. *Mol. Cell. Biol.* **21:** 4900–4908.

Sengoku T., Nureki O., Nakamura A., Kobayashi S., and Yokoyama S. 2006. Structural basis for RNA unwinding by the DEAD-box protein *Drosophila* Vasa. *Cell* **125:** 287–300.

Shabalina S.A., Ogurtsov A.Y., Rogozin I.B., Koonin E.V., and Lipman D.J. 2004. Comparative analysis of orthologous eukaryotic mRNAs: Potential hidden functional signals. *Nucleic Acids Res.* **32:** 1774–1782.

Shin B.S., Maag D., Roll-Mecak A., Arefin M.S., Burley S.K., Lorsch J.R., and Dever T.E. 2002. Uncoupling of initiation factor eIF5B/IF2 GTPase and translational activities by mutations that lower ribosome affinity. *Cell* **111:** 1015–1025.

Siridechadilok B., Fraser C.S., Hall R.J., Doudna J.A., and Nogales E. 2005. Structural roles for human translation factor eIF3 in initiation of protein synthesis. *Science* **310:** 1513–1515.

Slepenkov S.V., Darzynkiewicz, E., and Rhoads R.E. 2006. Stopped-flow kinetic analysis of eIF4E and phosphorylated eIF4E binding to cap analogs and capped oligoribonucleotides. Evidence of a one-step binding mechanism. *J. Biol. Chem.* **281:** 14927–14938.

Sonenberg N. 1988. Cap-binding proteins of eukaryotic messenger RNA: Functions in initiation and control of translation. *Prog. Nucleic Acid Res. Mol. Biol.* **35:** 173–207.

Spahn C.M., Jan E., Mulder A., Grassucci R.A., Sarnow P., and Frank J. 2004. Cryo-EM visualization of a viral internal ribosome entry site bound to human ribosomes: The IRES functions as an RNA-based translation factor. *Cell* **118:** 465–475.

Spahn C.M., Beckmann R., Eswar N., Penczek P.A., Sali A., Blobel G., and Frank J. 2001. Structure of the 80S ribosome from *Saccharomyces cerevisiae*—tRNA-ribosome and subunit-subunit interactions. *Cell* **107:** 373–386.

Svitkin Y.V., Pause A., Haghighat A., Pyronnet S., Witherell G., Belsham G.J., and Sonenberg N. 2001. The requirement for eukaryotic initiation factor 4A (elF4A) in translation is in direct proportion to the degree of mRNA 5′ secondary structure. *RNA* **7:** 382–394.

Takyar S., Hickerson R.P., and Noller H.F. 2005. mRNA helicase activity of the ribosome. *Cell* **120:** 49–58.

Thomas A., Goumans H., Voorma H.O., and Benne R. 1980a. The mechanism of action of eukaryotic initiation factor 4C in protein synthesis. *Eur. J. Biochem.* **107:** 39–45.

Thomas A., Spaan W., van Steeg H., Voorma H.O., and Benne R. 1980b. Mode of action of protein synthesis initiation factor eIF-1 from rabbit reticulocytes. *FEBS Lett.* **116:** 67–71.

Thompson G.M., Pacheco E., Melo E.O., and Castilho B.A. 2000. Conserved sequences in the β subunit of archaeal and eukaryal translation initiation factor 2 (eIF2), absent from eIF5, mediate interaction with eIF2γ. *Biochem. J.* **347:** 703–709.

Tomoo K., Shen X., Okabe K., Nozoe Y., Fukuhara S., Morino S., Ishida T., Taniguchi T., Hasegawa H., Terashima A., et al. 2002. Crystal structure of 7-methylguanosine 5′-triphosphate (m7GTP)- and P1-7-methylguanosine P3-adenosine-5′, 5′-triphosphate (m7 GpppA)-bound human full length eukaryotic initiation factor 4E: Biological importance of the C-terminal flexible region. *Biochem. J.* **362:** 539–544.

Trachsel H., Erni B., Schreier M.H., and Staehelin T. 1977. Initiation of mammalian protein synthesis. II. The assembly of the initiation complex with purified initiation factors. *J. Mol. Biol.* **116:** 755–767.

Uchida N., Hoshino S., Imataka H., Sonenberg N., and Katada T. 2002. A novel role of the mammalian GSPT/eRF3 associating with poly(A)-binding protein in Cap/Poly(A)-dependent translation. *J. Biol. Chem.* **277:** 50286–50292.

Unbehaun A., Borukhov S.I., Hellen C.U.T., and Pestova T.V. 2004. Release of initiation factors from 48S complexes during ribosomal subunit joining and the link between establishment of codon-anticodon base-pairing and hydrolysis of eIF2-bound GTP. *Genes Dev.* **18:** 3078–3093.

Valasek L., Nielsen K.H., and Hinnebusch A.G. 2002. Direct eIF2-eIF3 contact in the multifactor complex is important for translation initiation in vivo. *EMBO J.* **21:** 5886–5898.

Valasek L., Nielsen K.H., Zhang F., Fekete C.A., and Hinnebusch A.G. 2004. Interactions of eukaryotic translation initiation factor 3 (eIF3) subunit NIP1/c with eIF1 and eIF5 promote preinitiation complex assembly and regulate start codon selection. *Mol. Cell. Biol.* **24:** 9437–9455.

Valasek L., Mathew A.A., Shin B.S., Nielsen K.H., Szamecz B., and Hinnebusch A.G. 2003. The yeast eIF3 subunits TIF32/a, NIP1/c, and eIF5 make critical connections with the 40S ribosome in vivo. *Genes Dev.* **17:** 786–799.

von der Haar T., Ball P.D., and McCarthy J.E. 2000. Stabilization of eukaryotic initiation factor 4E binding to the mRNA 5′-Cap by domains of eIF4G. *J. Biol. Chem.* **275:** 30551–30555.

von der Haar T., Gross J.D., Wagner G., and McCarthy J.E. 2004. The mRNA cap-binding protein eIF4E in post-transcriptional gene expression. *Nat. Struct. Mol. Biol.* **11:** 503–511.

von Pawel-Rammingen U., Astrom S., and Bystrom A.S. 1992. Mutational analysis of conserved positions potentially important for initiator tRNA function in *Saccharomyces cerevisiae*. *Mol. Cell. Biol.* **12:** 1432–1442.

Wei Z., Xue Y., Xu H., and Gong W. 2006. Crystal structure of the C-terminal domain of *S. cerevisiae* eIF5. *J. Mol. Biol.* **359:** 1–9.

Williams D.D., Price N.T., Loughlin A.J., and Proud C.G. 2001. Characterization of the mammalian initiation factor eIF2B complex as a GDP dissociation stimulator protein. *J. Biol. Chem.* **276:** 24697–24703.

Yamamoto Y., Singh C.R., Marintchev A., Hall N.S., Hannig E.M., Wagner G., and Asano K. 2005. The eukaryotic initiation factor (eIF) 5 HEAT domain mediates multifactor assembly and scanning with distinct interfaces to eIF1, eIF2, eIF3, and eIF4G. *Proc. Natl. Acad. Sci.* **102:** 16164–16169.

Yang Q. and Jankowsky E. 2005. ATP- and ADP-dependent modulation of RNA unwinding and strand annealing activities by the DEAD-Box protein DED1. *Biochemistry* **44:** 13591–13601.

Yatime L., Mechulam Y., Blanquet S., and Schmitt E. 2006. Structural switch of the γ subunit in an archaeal aIF2αγ heterodimer. *Structure* **14:** 119–128.

Yatime L., Schmitt E., Blanquet S., and Mechulam Y. 2004. Functional molecular mapping of archaeal translation initiation factor 2. *J. Biol. Chem.* **279:** 15984–15993.

———. 2005. Structure-function relationships of the intact aIF2alpha subunit from the archaeon *Pyrococcus abyssi*. *Biochemistry* **44:** 8749–8756.

Yoon H.J. and Donahue T.F. 1992. The suil suppressor locus in *Saccharomyces cerevisiae* encodes a translation factor that functions during tRNA(iMet) recognition of the start codon. *Mol. Cell. Biol.* **12:** 248–260.

Yusupov M.M., Yusupova G.Z., Baucom A., Lieberman K., Earnest T.N., Cate J.H., and Noller H.F. 2001. Crystal structure of the ribosome at 5.5 Å resolution. *Science* **292:** 883–896.

Zhou C., Arslan F., Wee S., Krishnan S., Ivanov A.R., Oliva A., Leatherwood J., and Wolf D.A. 2005. PCI proteins eIF3e and eIF3m define distinct translation initiation factor 3 complexes. *BMC Biol.* **3:** 14.

5

Translation Initiation by Viral Internal Ribosome Entry Sites

Jennifer A. Doudna
Departments of Molecular and Cell Biology and Chemistry
Howard Hughes Medical Institute
University of California, Berkeley, California 94720

Peter Sarnow
Department of Microbiology and Immunology
Stanford University School of Medicine
Stanford, California 94305

THE INITIATION OF PROTEIN SYNTHESIS IS A COMPLEX and highly regulated process in all organisms. In eukaryotic cells, the 40S ribosomal subunit must be recruited to an mRNA and correctly positioned at the initiation codon prior to joining with the 60S subunit to form a translationally active complex. In most mRNAs, this step is initiated by an interaction of the cap-binding protein complex eukaryotic initiation factor 4F (eIF4F), composed of factors eIF4E, eIF4A, and eIF4G, with the m^7GpppN cap structure, which is located at the 5′ end of all polymerase-II-transcribed mRNAs. Subsequently, the 40S subunit that carries the initiator methionyl-tRNA–eIF2–GTP complex is thought to attach at or near the 5′ end of the mRNA, aided by an interaction of 40S-associated factor eIF3 with eIF4G. The 40S subunit then scans the mRNA in a 5′ to 3′ direction until an appropriate AUG start codon is encountered where the 60S subunit joins to form a translation-competent 80S ribosome with the AUG positioned in the ribosomal P site (for review, see Kozak 1989; Dever 2002; Chapter 4). The mRNA is thought to have a passive role in this process, and typically lacks significant secondary structure upstream of the AUG start codon that might interfere with scanning.

In intriguing contrast, certain viral mRNAs and cellular mRNAs that encode products involved in growth control, differentiation, apoptosis, and oncogenesis contain untranslated regions that are often highly conserved, may extend for several hundred nucleotides, and appear to contain extensive secondary and tertiary structures. As predicted, many such mRNAs are poorly translated (for review, see Kozak 1991). However, in some instances, mRNAs utilize an alternative initiation mechanism of internal ribosome binding in which the 5′-untranslated region (5′UTR) of the mRNA has an active role in 40S subunit recruitment. The 5′-untranslated RNA segments responsible for such internal initiation of translation are known as internal ribosome entry sites (IRESs) (see Fig. 1) (for review, see Hellen and Sarnow 2001; Jackson 2005; Komar and Hatzoglou 2005; Pisarev et al. 2005; Spriggs et al. 2005). A detailed understanding of internal translation initiation is important both for elucidating a key mechanism in the control of eukaryotic gene expression and for developing effective antiviral therapies.

In this chapter, we discuss viral IRES mechanisms, with a focus on mechanistic insights learned from both structural and biochemical studies of picornaviral, hepatitis C, and cricket paralysis viral IRES-containing RNAs. We also describe translational control mechanisms of cellular mRNAs containing IRES functionalities.

DISCOVERY OF INTERNAL RIBOSOME ENTRY SITES

Although viral genomes may encode some or even all of the macromolecular machinery that is needed to amplify their genomes, viral mRNAs are generally dependent on the host translation apparatus in infected cells (Chapter 20). Thus, dependent on the outcome of the infectious cycle, viruses have evolved a variety of mechanisms to usurp host ribosomes for viral mRNA translation. As a consequence, viruses that inhibit translation of host-cell mRNA have frequently been used to study translation mechanisms in mammalian cells (for review, see Gale et al. 2000). For example, it has been known for a long time that infection of cells with poliovirus, a cytoplasmic RNA virus belonging to the Picornaviridae, results in the selective translational inhibition of host but not of viral RNAs, concomitant with the cleavage and modification of several canonical translation initiation factors (for review, see Ehrenfeld 1996). The efficient translation of the viral genome was puzzling, because the 5′UTRs in the three poliovirus genotypes are approximately 750 nucleotides in length and are burdened with RNA structures and multiple unused AUG start codons. A model to explain the mechanism of selective viral mRNA translation in infected cells was based on the finding that poliovirus mRNA contains a 5′-terminal pU residue, instead of a cap structure (Nomoto et al. 1976), arguing that poliovirus mRNA translation must proceed by a 5′ cap-independent mechanism. Intriguingly, insertion of the 5′UTRs of the poliovirus or encephalomyocarditis virus (EMCV) genome between two adjacent coding regions in a bicistronic mRNA construct enabled translation of the downstream cistron independent of the upstream cistron. This finding suggested that cap-independent translation of picornavirus mRNAs proceeds by a mechanism whereby ribosomes are recruited to RNA sequence elements that function as IRESs (Jang et al. 1988; Pelletier and Sonenberg 1988). Indeed, circular RNAs containing the EMCV IRES were subsequently shown to be translated by ribosomes (Chen and Sarnow 1995), whereas RNA circles lacking IRES elements could not be translated (Kozak 1979; Konarska et al. 1981; Chen and Sarnow 1995). Subsequent studies have shown that picornavirus IRES elements possess structures that must be maintained, because small deletions, insertions, or point mutations within the IRES elements dramatically reduced their activities (for review, see Jang 2005).

By using the bicistronic mRNA assay described above, numerous RNA viruses and at least two DNA viruses (Bieleski and Talbot 2001; Grundhoff and Ganem 2001; Griffiths and Coen 2005) have been shown to utilize IRES mechanisms for translation initiation (Table 1) (for review, see Hellen and Sarnow 2001; Jang 2006). In most cases studied in detail to date, these viral IRESs require a specific RNA structure to be active, but they differ in their requirements for host translation initiation factors. Except within families of related viruses, however, there are few similarities in sequence, size, or secondary structure among different viral IRES elements. Thus, it is not yet clear whether different IRES structures reflect different modes of interaction of the IRES with ribosomes or IRES *trans*-acting factors (ITAFs) or whether IRES elements contain functions that are important in viral RNA replication and, perhaps, in viral RNA packaging. Indeed, biochemical and genetic evidence supports both scenarios. For example, cryo-electron microscopy (cryo-EM) studies of the IRESs from hepatitis C virus (HCV) and cricket paralysis virus (CrPV) associated with 40S ribosomal subunits have revealed that these IRES elements occupy distinct sites on the ribosomal subunit, yet induce similar conformational changes in the 40S subunits, implying that binding by these diverse IRES elements produces similar effects (Spahn et al. 2001, 2004). On the other hand, it has been documented that RNA sequences residing within the poliovirus IRES also modulate replication and packaging efficiency of the viral genome (Borman et al. 1994; Johansen and Morrow 2000).

Because the poliovirus IRES functions efficiently in extracts made from uninfected cells (Pelletier and Sonenberg 1989), the question arose whether certain cellular mRNAs harbored IRES elements as well. Indeed, using bicistronic mRNA-IRES assays (for review, see Hellen and Sarnow

Table 1. Representative examples of IRESs in animal viral genomes

Virus	Reference
Picornaviridae	
Poliovirus	Pelletier and Sonenberg (1988)
Rhinovirus	Borman and Jackson (1992)
Encephalomyocarditis virus	Jang et al. (1988)
Foot-and-mouth disease virus	Kuhn et al. (1990)
Porcine teschovirus 1 Talfan	Kaku et al. (2002)
Flaviviridae	
Hepatitis C virus	Tsukiyama-Kohara et al. (1992)
Classical swine fever virus	Rijnbrand et al. (1997)
Bovine virus diarrhea virus	Poole et al. (1995)
GB virus B	Grace et al. (1999)
GB virus A	Simons et al. (1996)
GB virus C	Simons et al. (1996)
Dicistroviridae	
Plautia stali intestine virus	Sasaki and Nakashima (1999)
Rhopalosiphum padi virus	Domier et al. (2000)
Cricket paralysis virus	Wilson et al. (2000b)
Taura syndrome virus	Cevallos and Sarnow (2005)
Herpesviridae	
Kaposi's sarcoma-associated herpesvirus	Bieleski and Talbot (2001); Grundhoff and Ganem (2001); Low et al. (2001)
Herpes simplex virus	Griffiths and Coen (2005)
Retroviridae	
Friend murine leukemia virus *gag* mRNA	Berlioz and Darlix (1995)
Moloney murine leukemia virus *gag* mRNA	Vagner et al. (1995)
Rous sarcoma virus	Deffaud and Darlix (2000)
Human immunodeficiency virus 1	Buck et al. (2001); Brasey et al. (2003)
Human immunodeficiency virus 2	Herbreteau et al. (2005)

Order of entry follows the description in the text.

2001; Komar and Hatzoglou 2005; Pisarev et al. 2005) or genome-wide screens to identify cellular mRNAs that can be actively translated during poliovirus infection when translation of the bulk of host-cell mRNAs is inhibited (Johannes et al. 1999), a variety of cellular mRNAs have been discovered that contain IRESs. Cellular conditions and *trans*-acting factors that regulate cellular IRES elements are being discovered at a rapid pace (Chapter 6), but little is known about the structure of cellular IRES elements or the mechanistic details by which these RNA elements mediate internal initiation. In contrast, a wealth of information has been obtained about IRES function from studying the picornavirus IRES elements. More recently, structural information about HCV–ribosome and CrPV–ribosome complexes has provided a window through which one can watch the subversion of ribosomes by IRES elements. These exciting findings are described in this chapter.

PICORNAVIRUSES SET A PARADIGM FOR IRES FUNCTION

On the basis of sequence information, it became clear that the picornavirus IRES elements (Table 1) can be divided into at least two types: type-1 IRES elements that are located in the genomes of enteroviruses (e.g., poliovirus) and rhinoviruses, and type-2 IRES elements that reside in the genomes of aphtoviruses (e.g., foot-and-mouth disease virus) and cardioviruses (e.g., EMCV) (Fig. 1). Hepatitis A and C viral IRESs belong to a subclass within the type-2 IRES elements, and the recently described porcine teschovirus (Kaku et al. 2002; Chard et al. 2006) encodes an IRES element that is structurally and functionally similar to that of the HCV IRES. It is likely that this IRES was acquired during a mixed infection via an RNA recombination event.

Although there is high conservation of predicted RNA structure within each type of IRES, few similarities exist between the two IRES types, with the exception of a conserved oligopyrimidine tract sequence element that resides approximately 25 nucleotides upstream of an AUG codon in each IRES type (Fig. 1) (for review, see Jang 2006). In type-1 IRESs, this pyrimidine-AUG motif is an important element for IRES function, but it does not include the translation start codon that resides 40–150 nucleotides further downstream (Fig. 1), dependent on the type-1 IRES. Therefore, after recruitment onto type-1 IRES elements, the 40S subunit must traverse, by an as yet unknown mechanism, the viral genome to identify the authentic AUG start codon. In contrast, type-2 IRES elements bind 40S ribosomal subunits at the pyrimidine-AUG motif, which includes the authentic start-site codon (Fig. 1). The sequence motifs in the IRES that guide the binding of 40S subunits have been delineated in translation assays using bicistronic mRNAs that contain IRES elements with defined mutations (for review, see Jang 2006). These studies revealed that the presence and integrity of domain II through domain VI in type-1 IRESs and domain H through domain L in type-2 IRESs are important for IRES function (Fig. 1).

The minimum set of canonical translation factors that are essential to recruit 40S subunits to these IRES domains has been identified in pioneering studies by Pestova, Hellen, and coworkers. Using the EMCV

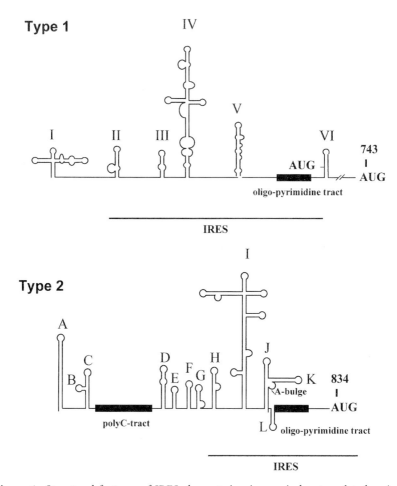

Figure 1. Structural features of IRES elements in picornaviral untranslated regions. (*Top panel*) Type-1 IRES, exemplified in the poliovirus RNA genome; (*bottom panel*) type-2 IRES, exemplified by the encephalomyocarditis virus genome. (*Black rectangles*) Locations of the oligopyrimidine tract and poly(C) tract sequences. The A-bulge loop in the J-K domain of the type-2 IRES is marked.

IRES as a model system, it was determined in reconstitution experiments that factors eIF2, eIF3, eIF4G, eIF4A, eIF4B, and initiator tRNA^{Met} were sufficient to recruit and to position 43S complexes at the authentic start codon in an ATP-dependent reaction and that the cap-binding protein eIF4E was dispensable for 43S recruitment (Pestova et al. 1996a,b). Further experiments revealed that it was the central domain of eIF4G, which retains the binding sites for eIF4A and eIF3, that interacted with the J-K domains of the IRES, including the A-bulge loop (Fig. 1) (Lomakin et al. 2000). It is thought that similar canonical factors aid in the recruitment of 43S subunits to A-bulge-loop-lacking type-1 IRES elements, but evidence supporting this presumption has been sparse (Ochs et al. 2002). Curiously, the HAV IRES requires intact eIF4F for its function. It has been speculated that the eIF4E component of eIF4F may interact with a "flipped out" G residue somewhere within the HAV IRES to modulate the binding of eIF4F to the IRES (Jackson 2005).

That noncanonical translation factors may have roles in IRES-mediated translation was already suspected in 1984 when Dorner et al. (1984) noted that translation of poliovirus RNA in a rabbit reticulocyte lysate (RRL) occurred at numerous nonauthentic internal start codons. Translation at the authentic start codon could be maximized after addition of extracts prepared from cultured cells, arguing that a canonical or a noncanonical translation factor was limiting in the RRL. Although numerous IRES-binding proteins have been identified over the years, strong experimental evidence for roles in modulating picornavirus IRESs has been obtained for three proteins: The Unr (upstream of N-*ras*) factor, an RNA-binding protein with five cold-shock domains, strongly stimulates the type-1 rhinovirus IRES when added to a reaction in the RRL (Hunt et al. 1999); the RNA-binding protein PCBP-2 (poly[C]-binding protein 2) greatly stimulates the type-1 poliovirus IRES (Blyn et al. 1997; Gamarnik and Andino 1997); and ITAF$_{45}$ (IRES *trans*-acting factor, 45 kD) is a cell-cycle-regulated protein that modulates the type-2 FMDV IRES (Pilipenko et al. 2000). Roles for picornavirus IRES-modulating factors have also been reported for the La autoantigen and the polypyrimidine-tract-binding protein PTB; however, these reports have been somewhat controversial. Addition of recombinant La stimulates the poliovirus IRES in the RRL (Meerovitch et al. 1993), but the relevance of this finding was questioned because nonphysiological concentrations of La protein were needed to stimulate viral RNA translation. However, recent studies have shown that short interfering RNA (siRNA)-mediated knockdown of La in cultured cells causes a significant inhibition of the poliovirus IRES, arguing that La is a relevant IRES regulator in vivo (Costa-Mattioli et al. 2004).

Similarly, depending on the translation assay used, PTB has been found to display different effects on picornavirus type-2 IRES and HCV IRES function (Ali and Siddiqui 1995; Kaminski et al. 1995). The importance of PTB in type-2 IRES activity has been demonstrated by the requirement for nPTB, a neural homolog of PTB, in the assembly of 43S complexes onto the IRES from the virulent GDVII strain of Theiler's murine encephalomyocarditis virus (TMEV) (Pilipenko et al. 2001). Mutations in the IRES that abolished binding of nPTB generated viral mutants that were less virulent in mice, suggesting a role for the TMEV IRES and nPTB in viral pathogenesis. The notion that PTB acts as an RNA chaperone that may influence IRES structure comes from studies with the EMCV IRES. It was found that IRES activity was modulated by PTB and the size of the A-rich bulge in the J-K loop (Fig. 1). For example, an IRES with a bulge of 7 As was more dependent on PTB than an IRES with a 6-A bulge; curiously, the EMCV genome contains a 6-A bulge and its IRES activity is independent of PTB (Kaminski and Jackson 1998).

So far, the roles of many ITAFs in IRES functions have been examined only in translation-competent extracts. With the advent of RNA interference (RNAi) technology (for review, see Sandy et al. 2005; Zamore and Haley 2005), ITAF-encoding genes are being knocked down in mammalian cells and their effects on IRES function are being examined (Costa-Mattioli et al. 2004). Furthermore, the finding that some IRES elements can function in *Saccharomyces cerevisiae* (Thompson et al. 2001; Rosenfeld and Racaniello 2005) suggests that genetic approaches toward studying IRES function have become available.

HCV IRES–RIBOSOME COMPLEXES REVEAL ROLES OF THE RIBOSOME IN THE INTERNAL INITIATION MECHANISM

Although the picornavirus IRES elements set paradigms for understanding RNA structures and factors that aid in ribosome recruitment, little was learned about ribosomal functions that are needed for formation of IRES–ribosome complexes. With the finding that the HCV IRES can form high-affinity binary complexes with 40S subunits (see below), structural biologists entered the field. Their work resulted in the purification and biochemical characterization of IRES–ribosome complexes, and their stunning visualization by cryo-EM, and made the HCV IRES the best-characterized of the viral IRESs studied to date.

Like the related classical swine fever virus (CSFV), bovine viral diarrhea virus (BVDV), and GB virus B (GBV-B), HCV utilizes an IRES element distinct in sequence and secondary structure from those of the picornaviruses (Tsukiyama-Kohara et al. 1992; Wang et al. 1993; Poole et al. 1995; Rijnbrand et al. 1997; Grace et al. 1999). In notable contrast to EMCV and poliovirus IRESs, the HCV IRES RNA was found to bind directly to 40S ribosomal subunits without requiring any of the host translation initiation factors (Pestova et al. 1998; Kieft et al. 2001a). This property has also been observed for the intergenic region IRES in dicistroviruses (for review, see Jan 2006). Toeprinting experiments to map precise positioning of the 40S subunit on the HCV IRES RNA showed that the AUG initiation codon in the domain IV loop is positioned at or near the P site in the ribosome (Pestova et al. 1998). Functional preinitiation complexes could be assembled from those IRES-40S particles by the addition of just two additional factors, eIF3 and eIF2, bound to GTP and initiator tRNA (Pestova et al. 1998). The HCV IRES thus bypasses requirements for the large cap-binding complex eIF4F.

The 341-nucleotide HCV 5′UTR is highly conserved among viral isolates and shares secondary structural features with the IRESs of certain flaviviruses. Mutagenesis and chemical modification studies led to a model for the secondary structural fold of the RNA consisting of four domains (I–IV) in which the AUG initiation codon is located near the 3′ end (Fig. 2)

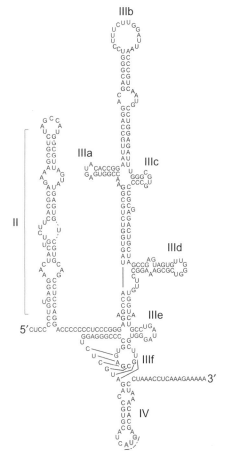

Figure 2. Secondary structure of the HCV IRES. Domains are labeled and the AUG start codon is shaded.

(Honda et al. 1996; Reynolds et al. 1996; Zhao and Wimmer 2001). Sequence differences among the IRESs of different HCV genotypes are primarily compensatory nucleotide changes that maintain the proposed base-pairing interactions. Within the HCV 5′UTR, domain I located at the extreme 5′ end is not part of the IRES element. However, it is part of a binding site for a liver-specific microRNA (miRNA), miR-122, that is essential for maintaining viral RNA abundance in cultured liver cells (Jopling et al. 2005). Domain II is important for IRES function, but its deletion does not completely abolish activity (Tsukiyama-Kohara et al. 1992; Reynolds et al. 1995, 1996; Kolupaeva et al. 2000a,b). Domain III includes a four-way junction (IIIabc) and two stem-loops (IIId and IIIe) that are essential for function. Structures of the isolated II, IIId, and IIIe hairpins have been determined by solution nuclear magnetic resonance (NMR) spectroscopy (Klinck et al. 2000; Lukavsky et al. 2000, 2003), whereas the structure of the IIIabc helical junction was solved by X-ray crystallography (Kieft et al. 2002). Domain II adopts a distorted L-shaped structure, an overall conformation in the free form that is markedly similar to its 40S subunit-bound form (Spahn et al. 2001; Lukavsky et al. 2003). The IIId structure revealed a loop-E motif similar to that found in the sarcin-ricin loop of 28S rRNA, suggesting that it may be a site of inter-molecular interaction. In the IIIe structure, loop bases are oriented toward the major groove, creating a unique tetraloop fold (Lukavsky et al. 2000). The structure of the IIIabc helical junction consists of two coaxial helical stacks, one with canonical A-form geometry that buttresses the other, which has a distorted helical shape and contains most of the nucleotides implicated in binding to the translation factor eIF3 (Kieft et al. 2002). It is not clear whether these domain III substructures represent the conformations found in the intact IRES or IRES–40S subunit complex (Melcher et al. 2003). Toeprinting and direct binding assays demonstrated that eIF3 binds in the IIIabc region, whereas the 40S subunit contacts domain II and the IIId, IIIe, and pseudoknot structures (Pestova et al. 1998; Kolupaeva et al. 2000a; Kieft et al. 2001a). The domain IV hairpin contains the AUG start codon and is apparently destabilized during translation initiation (Reynolds et al. 1995).

In 48S preinitiation complexes formed by interaction of the HCV IRES with the 40S subunit and factors, toeprinting experiments have shown that the initiator tRNA anticodon is base-paired to the mRNA initiator codon embedded within the IRES (Pestova et al. 1998). The 48S complex then recruits the 60S ribosomal subunit in a GTP hydrolysis-dependent step, releasing eIF2 and forming the translationally competent 80S ribosome. In many ways, this mechanism is similar to the canonical cap-dependent mechanism of translation initiation, differing only at the step of 40S particle recruitment.

STRUCTURAL AND MECHANISTIC STUDIES OF HCV IRES COMPLEXES

The HCV IRES binds the 40S ribosomal subunit and eIF3 in the absence of any additional protein factors, interactions conferred by the three-dimensional structure of the IRES RNA. Quantitative biochemical and biophysical analyses revealed that 40S subunit binding to the HCV IRES RNA involves intermolecular contacts to most of the IRES sequence (Pestova et al. 1998; Kieft et al. 2001b), whereas eIF3 binding is localized to the stem-loops IIIa–IIIc (Buratti et al. 1998; Pestova et al. 1998; Sizova et al. 1998; Odreman-Macchioli et al. 2000; Kieft et al. 2001b). The IRES–40S subunit interaction is a high-affinity complex with a dissociation constant of 2 nM, whereas the IRES-eIF3 complex forms with a dissociation constant of approximately 35 nM (Kieft et al. 2001b). These high-affinity binding constants raise the possibility that a conformational change or cellular factors are necessary to dissociate the IRES once translation begins.

A major focus of mechanistic investigation of the HCV IRES has been structural determination of HCV IRES–ribosome complexes by single-particle cryo-EM. The first structures of HCV IRES–40S ribosomal subunit complexes, solved by cryo-EM at a resolution of about 20 Å, revealed that the IRES binds to the "head" and "platform" regions of the 40S subunit in a single extended conformation (Fig. 3) (Spahn et al. 2001).

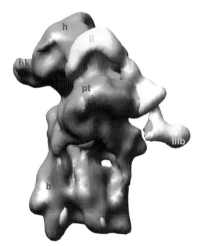

Figure 3. Structure of the HCV IRES–40S ribosomal subunit complex determined at a resolution of about 20 Å by cryo-EM.

Interestingly, the HCV IRES RNA induces a conformational change in the 40S subunit that converts the mRNA-binding cleft into a tunnel. On the basis of structural and biochemical studies with IRES mutants, domain II was found to be responsible for the observed conformational change, possibly through its interaction near the E site on the 40S subunit (Spahn et al. 2001). This implies the fascinating possibility that the HCV IRES recruits the mammalian ribosome by inducing or stabilizing a 40S conformation optimal for positioning the viral mRNA initiation codon in the ribosome-decoding center.

Subsequent biochemical studies led to a more detailed model for the HCV IRES-induced assembly of active 80S ribosomes through the ordered interaction of the IRES with the 40S subunit and the required initiation factors (Ji et al. 2004; Otto and Puglisi 2004). The HCV IRES can form a binary complex with an initiation-factor-free 40S ribosomal subunit, followed by preinitiation complex assembly at the AUG initiation codon upon association of eIF3 and other factors. 80S complexes form upon GTP-dependent association of the 60S subunit. Efficient assembly of the 48S-like and 80S complexes on the IRES mRNA is dependent on maintenance of the highly conserved HCV IRES structure, and different domains of the IRES appear to have specific roles in the ribosome assembly process.

The high affinity of the HCV IRES RNA for 40S subunits suggests that the IRES remains associated after 60S subunit association, enabling the in vitro assembly of 80S–IRES complexes in a HeLa cell extract translation system. Using cycloheximide, a drug that blocks ribosome elongation, to stall translation initiation complexes after subunit joining, human 80S–HCV IRES complexes were affinity-purified, and the structure was determined at about 20 Å resolution by single-particle cryo-EM (Boehringer et al. 2005). Interestingly, the structure of the 40S ribosomal subunit in this complex differs significantly from its structure both within elongating ribosomes and in the 40S–HCV IRES particle. The conformation of the mRNA entry channel resembles that observed in empty 40S subunits, whereas the mRNA exit (E) site is similar in all cases. The structure of the HCV IRES as determined at this resolution is virtually the same as that in the 40S and 80S complexes. Curiously, no initiator tRNA was present in the IRES–80S sample, and the occupancy of the 40S-associated protein RACK1, a kinase recruitment factor, was diminished relative to that in the 40S subunit alone. The significance of these properties of the sample remain to be elucidated.

STRUCTURAL ROLES OF INITIATION FACTOR eIF3 IN HCV IRES-MEDIATED TRANSLATION INITIATION

The two-subunit composition of the ribosome is an important aspect of its regulation. When not actively synthesizing proteins, the ribosomal subunits can dissociate from each other and remain separated in part through the action of the approximately 750-kD initiation factor eIF3 (Trachsel and Staehelin 1979; Bommer et al. 1991; Hershey et al. 1996; Asano et al. 1997; Chaudhuri et al. 1999). The mammalian eIF3 is composed of at least 12 proteins, and it exists in the cytoplasm in complex with the 40S ribosomal subunit (Westermann and Nygard 1983). eIF3 is apparently released sometime after association of the 40S and 60S subunits to form active 80S ribosomes (Benne and Hershey 1976). Interestingly, the specific binding interaction observed between the HCV IRES and eIF3 is essential to the translational activity of the IRES (Pestova et al. 1998; Kieft et al. 2001b).

Structural studies of human eIF3 and its complexes with the HCV IRES and translational machinery suggest how and why this interaction contributes to viral translation initiation, and how the IRES functionally replaces the initiation factor eIF4F complex in positioning the 5′ end of the mRNA on the 40S ribosomal subunit (Siridechadilok et al. 2005). The approximately 30-Å reconstruction of human eIF3 revealed a distinctive architecture consisting of five discrete domains. In the orientation presented in Figure 4, eIF3 shows anthropomorphic features that were used to name the different domains according to body parts that include a head, arms, and legs.

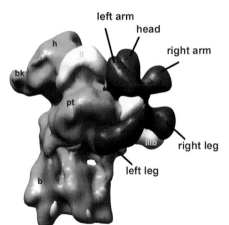

Figure 4. Structural model of the human eIF3–40S–HCV IRES complex derived from cryo-EM structures and difference mapping.

This modular structure must interact with RNA and protein factors to organize the assembly of the translation initiation complex on the 40S ribosomal subunit. As mentioned above, the HCV IRES RNA structure binds directly to eIF3 to form part of the ternary complex necessary for viral protein synthesis. The IRES somehow circumvents the need for eIF3 binding to initiation factor eIF4G, a component of the 5′-cap-binding complex eIF4F, which is essential to position the 5′ end of the message on the ribosome during translation initiation on most cellular mRNAs. The extended interaction surface modeled between the IRES and eIF3 agrees well with biochemical studies in which at least four subunits of eIF3 (eIF3a, -b/c, -d, and -f) could be cross-linked to the IRES.

Superposition of the eIF3–IRES structure onto a previous cryo-EM reconstruction of the 40S–IRES using the similar IRES density observed in each case produced a model of the ternary 40S–eIF3–IRES complex that illustrates the interaction of eIF3 with the 40S subunit. The bulk of eIF3 mass in the ternary complex localizes to the solvent-exposed side of 40S subunit, where the front surface of eIF3 partially contacts the 40S. The front of the eIF3 left leg sits below the platform near the 60S subunit interface, whereas the left arm points toward the tRNA E site. The extensive interactions between the 40S subunit, eIF3, and the HCV IRES in the model, in which the IRES appears to thread through a cleft on the back side of eIF3 and clamp eIF3 with 40S subunit, raise the possibility that IRES binding to eIF3 or to the 40S subunit precedes ternary complex assembly.

These insights into the structural role of eIF3 in HCV IRES-dependent translation initiation also provide clues to its activities during cap-dependent initiation. First, the position of eIF3 in this 40S–eIF3–IRES model suggests a plausible explanation for its role in preventing premature joining of 40S and 60S ribosomal subunits by blocking a critical intersubunit contact between the small-subunit protein S15/rpS13 and helix 34 of the large subunit (Siridechadilok et al. 2005). Second, the 40S–eIF3-IRES model and additional structural data for an eIF3–eIF4G complex provide a means to infer the placement of eIF4G relative to the 40S and other initiation factors, locating eIF4G very close to the tRNA exit (E) site of the 40S subunit (Siridechadilok et al. 2005). Because eIF4G is a component of eIF4F, this analysis predicts the position of the cap-binding complex and the approximate location of the 5′ m^7G cap of an mRNA. Comparison of the 40S–eIF3–IRES and 40S–eIF3–eIF4G models suggests an intriguing similarity in the function of the HCV IRES and eIF4G to anchor the attached mRNA strand, in either case, near the exit site on the head of the 40S ribosomal subunit. The structurally analogous positioning of the IRES RNA and the eIF4G-containing protein complex implies mechanistic overlap in their activities, possibly explaining why HCV does not need eIF4F for viral protein synthesis.

IRES ELEMENTS IN DICISTROVIRIDAE: A STREAMLINED MECHANISM FOR RIBOSOME RECRUITMENT

Study of several different viral IRESs has revealed distinct mechanisms for 40S ribosomal recruitment. In contrast to the HCV and poliovirus IRESs, a more recently discovered class of divergent viral IRES elements can assemble elongation-competent ribosomes in the absence of any canonical translation initiation factors and initiator tRNA. These IRES elements have been discovered in the Dicistroviridae family of insect viruses (for review, see Jan 2006). For example, CrPV contains a single positive-strand RNA genome that encodes two nonoverlapping reading frames, each initiated by an IRES (Wilson et al. 2000b). It was shown that the intergenic region (IGR) IRES elements in CrPV, related *Plautia stali* intestine virus (PSIV), and Taura syndrome virus (TSV) did not require initiator tRNA for translational initiation (Nakashima et al. 1999; Wilson et al. 2000a,b; Cevallos and Sarnow 2005). In the case of the CrPV IGR IRES, mutagenesis and toeprinting analyses of IRES–ribosome complexes revealed that the ribosomal P site was occupied by a CCU triplet (Wilson et al. 2000a), which is base-paired with 5′ upstream sequences to generate a pseudoknot-type structural element. The integrity of the pseudoknot-type structure was essential for IRES activity, because mutations that disrupted this element abolished IGR IRES activity, whereas compensatory mutations rescued IRES activity (Wilson et al. 2000a; Jan and Sarnow 2002). The ribosomal A site was occupied by a GCU triplet that encodes alanine, the amino-terminal amino acid of the second open reading frame, encoding the structural precursor protein (Wilson et al. 2000a).

Like the HCV IRES, the CrPV IRES can recruit 40S ribosomal subunits in the absence of initiator tRNA^{Met}/eIF2–GTP complexes with high affinity ($K_d = 24$ nM) (Jan and Sarnow 2002; Costantino and Kieft 2005). Unlike the HCV IRES, however, neither GTP hydrolysis nor initiator methionyl-tRNA/eIF2-GTP complexes were required for the assembly of 80S ribosomes after addition of purified 60S subunits (Jan et al. 2003). The reconstituted 80S–IRES complexes contained the CCU triplet of the IRES in the ribosomal P site and the GCU triplet in the ribosomal A site. Similarly, preformed 80S subunits were able to be directly recruited by the IRES without the aid of additional canonical translation factors

(Pestova et al. 2004). Remarkably, the in-vitro-assembled 80S–CrPV IRES RNA complexes could support translation elongation after addition of tRNAs and elongation factors eEF1A and eEF2 (Jan et al. 2003; Pestova and Hellen 2003; Cevallos and Sarnow 2005). In contrast, the EMCV IRES, which requires canonical initiation factors to start translation initiation (Pestova et al. 1996a), did not direct the synthesis of small peptides (Jan et al. 2003; Cevallos and Sarnow 2005). Inclusion of cycloheximide in the translation reactions abolished the accumulation of the peptides, arguing that the peptides were synthesized by elongating ribosomes (Jan et al. 2003; Cevallos and Sarnow 2005). Overall, these data show that the CrPV and TSV IGR IRES elements can start protein biosynthesis from the A site of the ribosome without initiator tRNA and canonical initiation factors known to mediate cap-dependent translation. Why such a diverse IRES element has been selected is unknown. The finding that the CrPV IGR IRES can function in cells when initiation factor eIF2 is phosphorylated and, as a consequence, cap-dependent translation is inhibited (Fernandez et al. 2002) suggests that the IGR IRES can escape certain antiviral innate immune responses.

To investigate the molecular mechanism by which the CrPV IGR IRES recruits the ribosome, three-dimensional reconstructions of the CrPV IRES bound to human 40S subunits and 80S ribosomes were generated by single-particle cryo-EM (Spahn et al. 2004). A defined CrPV IRES structure was observed at the ribosome intersubunit space, where it contacts both the 40S and 60S subunits (Fig. 5). The CrPV IRES occupies the region involved in tRNA binding to the ribosomal P and

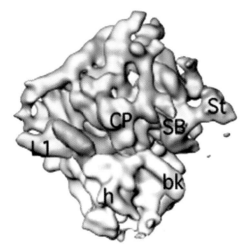

Figure 5. Structure of the CrPV IRES–human 80S ribosome complex determined at a resoluton of about 20 Å by cryo-EM.

E sites, in striking contrast to the HCV IRES that binds primarily to the more evolutionarily divergent solvent side of the 40S subunit. However, binding of the CrPV IRES induces a conformational change at the mRNA entry channel similar to that observed upon HCV IRES binding, suggesting that these changes are part of a general mechanism for translation initiation. Such conformational manipulations may contribute to threading of the 3′ segment of the mRNA into the mRNA entry channel, implying that structural rearrangements of the ribosome are integral to the general mechanism of translation initiation. Interestingly, defined movements of the head domain of prokaryotic 30S ribosomal subunits are also thought to be crucial during initiation (Lata et al. 1996; Carter et al. 2001).

ALTERNATIVE MECHANISMS OF RIBOSOME RECRUITMENT BY VIRAL IRES RNAs

As discussed above, most viral IRES elements reside upstream of the mRNA-coding sequence. However, increasing evidence indicates that there are other mechanistic paradigms for viral IRES activity. For example, the IRES in the monopartite, double-stranded RNA genome of giardiavirus extends into the mRNA-coding region (Garlapati and Wang 2004). Two recent discoveries in herpes simplex virus (HSV) and in human immunodeficiency virus (HIV) suggest that IRES elements can function entirely from within the coding region of an mRNA.

The only DNA virus that had been shown to contain an IRES element is the Kaposi's sarcoma-associated herpesvirus (Bieleski and Talbot 2001; Grundhoff and Ganem 2001; Low et al. 2001). However, recent studies with HSV revealed the presence of an unusual IRES element that directs low-level translation of the viral thymidine kinase (*tk*) gene under some conditions. The most common mutations observed in HSV strains resistant to the antiviral drug acyclovir are frameshift lesions in the viral *tk* gene that would be expected to abolish TK expression (Gaudreau et al. 1998). Analysis using a cell-free translation system led to the conclusion that such lesions are adjacent to a 12-nucleotide sequence in the coding region of the gene that functions as an IRES to initiate translation independent of an AUG codon (Griffiths and Coen 2005). Like the CrPV IRES, the *tk* IRES is active in the presence of low concentrations of edeine, an antibiotic that disrupts the activity of the eIF2–GTP/Met-tRNAi complex (Wilson et al. 2000a). Results from experiments in which stop codons were introduced into a reporter gene construct suggested that translation initiates from the *tk* IRES at or just downstream from a CUG codon (Griffiths and Coen 2005). It will be interesting to learn how a sequence as short as the *tk* IRES can recruit the translational machinery to the adjacent mRNA, apparently without requiring a large RNA structure.

The genomic RNAs of both HIV-1 and HIV-2 contain IRES elements that appear to control translation initiation of viral genes by distinct mechanisms. In HIV-1, an IRES element within the 5′UTR is active at the G_2/M phase of the host cell cycle (Brasey et al. 2003), whereas a second IRES located entirely within the *gag*-coding region drives translation of a 40-kD amino-terminally truncated Gag isoform by recruiting ribosomes to an internal AUG codon (Buck et al. 2001). In HIV-2, three isoforms of the Gag protein found within immature viral particles are thought to be produced by a combination of 5′-end-dependent and internal translation initiation that may help balance protein production during viral replication and packaging (Herbreteau et al. 2005). In this case, however, the IRES sequence located within the *gag*-coding region is large (>400 nucleotides) and apparently forms a tertiary structure important for recruiting ribosomes to two in-frame AUG codons internal to the message. Interplay between cap-dependent and cap-independent translation initiation is proposed to be a key regulator of the HIV-2 life cycle (Herbreteau et al. 2005).

CONCLUDING REMARKS

The discovery of viral IRES elements has led to new insights into the molecular basis for eukaryotic translation initiation. It is now clear that there are multiple distinct mechanisms by which RNA structures or sequences recruit ribosomes to a translation start site in the absence of canonical translation initiation factors. Structural studies of the HCV and CrPV IRES RNAs in complex with the ribosome support the idea that despite different sequences, structures, and host-factor requirements, these IRESs stabilize the same conformational state of the 40S ribosomal subunit. This finding implies that the observed rearrangement of the mRNA entry channel is a fundamental step in the initiation process, perhaps induced by one or more initiation factors during cap-dependent translation initiation. Future studies should reveal whether the molecular underpinnings of translation initiation are the same regardless of the means by which ribosomes are recruited to translation initiation sites and also whether the properties of the viral IRESs studied to date are shared with the IRES elements of other viral and cellular RNAs.

ACKNOWLEDGMENTS

J.D. thanks C. Fraser, W. Gilbert, H. Ji, A. Mehle, and B. Siridechadilok for many helpful discussions and comments on the manuscript, and B. Siridechadilok, A. Mehle, and C. Fraser for preparation of figures. Research in her laboratory was supported by the Howard Hughes Medical Institute and the National Institutes of Health. P.S. was supported by grants from the National Institues of Health (GM069007, AI47365).

REFERENCES

Ali N. and Siddiqui A. 1995. Interaction of polypyrimidine tract-binding protein with the 5′ noncoding region of the hepatitis C virus RNA genome and its functional requirement in internal initiation of translation. *J. Virol.* **69:** 6367–6375.

Asano K., Kinzy T.G., Merrick W.C., and Hershey J.W.B. 1997. Conservation and diversity of eukaryotic translation initiation factor eIF3. *J. Biol. Chem.* **272:** 1101–1109.

Benne R. and Hershey J.W. 1996. Purification and characterization of initiation factor IF-E3 from rabbit reticulocytes. *Proc. Natl. Acad. Sci.* **73:** 3005–3009.

Berlioz C. and Darlix J.L. 1995. An internal ribosomal entry mechanism promotes translation of murine leukemia virus gag polyprotein precursors. *J. Virol.* **69:** 2214–2222.

Bieleski L. and Talbot S.J. 2001. Kaposi's sarcoma-associated herpesvirus vCyclin open reading frame contains an internal ribosome entry site. *J. Virol.* **75:** 1864–1869.

Blyn L.B., Towner J.S., Semler B.L., and Ehrenfeld E. 1997. Requirement of poly(rC) binding protein 2 for translation of poliovirus RNA. *J. Virol.* **71:** 6243–6246.

Boehringer D., Thermann R., Ostareck-Lederer A., Lewis J.D., and Stark H. 2005. Structure of the hepatitis C virus IRES bound to the human 80S ribosome: Remodeling of the HCV IRES. *Structure* **13:** 1695–1706.

Bommer U.A., Lutsch G., Stahl J., and Bielka H. 1991. Eukaryotic initiation factors eIF-2 and eIF-3: Interactions, structure and localization in ribosomal initiation complexes. *Biochimie* **73:** 1007–1019.

Borman A. and Jackson R.J. 1992. Initiation of translation of human rhinovirus RNA: Mapping the internal ribosome entry site. *Virology* **188:** 685–696.

Borman A.M., Deliat F.G., and Kean K.M. 1994. Sequences within the poliovirus internal ribosome entry segment control viral RNA synthesis. *EMBO J.* **13:** 3149–3157.

Brasey A., Lopez-Lastra M., Ohlmann T., Beerens N., Berkhout B., Darlix J.L., and Sonenberg N. 2003. The leader of human immunodeficiency virus type 1 genomic RNA harbors an internal ribosome entry segment that is active during the G2/M phase of the cell cycle. *J. Virol.* **77:** 3939–3949.

Buck C.B., Shen X., Egan M.A., Pierson T.C., Walker C.M., and Siliciano R.F. 2001. The human immunodeficiency virus type 1 gag gene encodes an internal ribosome entry site. *J. Virol.* **75:** 181–191.

Buratti E., Tisminetzky S., Zotti M., and Baralle F.E. 1998. Functional analysis of the interaction between HCV 5′UTR and putative subunits of eukaryotic translation initiation factor eIF3. *Nucleic Acids Res.* **26:** 3179–3187.

Carter A.P., Clemons W.M., Jr., Brodersen D.E., Morgan-Warren R.J., Hartsch T., Wimberly B.T., and Ramakrishnan V. 2001. Crystal structure of an initiation factor bound to the 30S ribosomal subunit. *Science* **291:** 498–501.

Cevallos R.C. and Sarnow P. 2005. Factor-independent assembly of elongation-competent ribosomes by an internal ribosome entry site located in an RNA virus that infects penaeid shrimp. *J. Virol.* **79:** 677–683.

Chard L.S., Kaku Y., Jones B., Nayak A., and Belsham G.J. 2006. Functional analyses of RNA structures shared between the internal ribosome entry sites of hepatitis C virus and the picornavirus porcine teschovirus 1 Talfan. *J. Virol.* **80:** 1271–1279.

Chaudhuri J., Chowdhury D., and Maitra U. 1999. Distinct functions of eukaryotic translation initiation factors eIF1A and eIF3 in the formation of the 40 S ribosomal preinitiation complex. *J. Biol. Chem.* **274:** 17975–17980.

Chen C. and Sarnow P. 1995. Initiation of protein synthesis by the eukaryotic translational apparatus on circular RNAs. *Science* **268:** 415–417.

Costa-Mattioli M., Svitkin Y., and Sonenberg N. 2004. La autoantigen is necessary for optimal function of the poliovirus and hepatitis C virus internal ribosome entry site in vivo and in vitro. *Mol. Cell. Biol.* **24:** 6861–6870.

Costantino D. and Kieft J.S. 2005. A preformed compact ribosome-binding domain in the cricket paralysis-like virus IRES RNAs. *RNA* **11:** 332–343.

Deffaud C. and Darlix J.L. 2000. Rous sarcoma virus translation revisited: Characterization of an internal ribosome entry segment in the 5′ leader of the genomic RNA. *J. Virol.* **74:** 11581–11588.

Dever T.E. 2002. Gene-specific regulation by general translation factors. *Cell* **108:** 545–556.

Domier L.L., McCoppin N.K., and D'Arcy C.J. 2000. Sequence requirements for translation initiation of Rhopalosiphum padi virus ORF2. *Virology* **268:** 264–271.

Dorner A.J., Semler B.L., Jackson R.J., Hanecak R., Duprey E., and Wimmer E. 1984. In vitro translation of poliovirus RNA: Utilization of internal initiation sites in reticulocyte lysate. *J. Virol.* **50:** 507–514.

Ehrenfeld E. 1996. Initiation of translation by picornavirus RNAs. In *Translational control* (ed. J.W.B. Hershey et al.), pp. 549–573. Cold Spring Harbor Laboratory Press, Cold Spring Harbor, New York.

Fernandez J., Yaman I., Sarnow P., Snider M.D., and Hatzoglou M. 2002. Regulation of internal ribosomal entry site-mediated translation by phosphorylation of the translation initiation factor eIF2alpha. *J. Biol. Chem.* **277:** 19198–19205.

Gale M., Jr., Tan S.L., and Katze M.G. 2000. Translational control of viral gene expression in eukaryotes. *Microbiol. Mol. Biol. Rev.* **64:** 239–280.

Gamarnik A.V. and Andino R. 1997. Two functional complexes formed by KH domain containing proteins with the 5′ noncoding region of poliovirus RNA. *RNA* **3:** 882–892.

Garlapati S. and Wang C.C. 2004. Identification of a novel internal ribosome entry site in giardiavirus that extends to both sides of the initiation codon. *J. Biol. Chem.* **279:** 3389–3397.

Gaudreau A., Hill E., Balfour H.H., Jr., Erice A., and Boivin G. 1998. Phenotypic and genotypic characterization of acyclovir-resistant herpes simplex viruses from immunocompromised patients. *J. Infect. Dis.* **178:** 297–303.

Grace K., Gartland M., Karayiannis P., McGarvey M.J., and Clarke B. 1999. The 5′ untranslated region of GB virus B shows functional similarity to the internal ribosome entry site of hepatitis C virus. *J. Gen. Virol.* **80:** 2337–2341.

Griffiths A. and Coen D.M. 2005. An unusual internal ribosome entry site in the herpes simplex virus thymidine kinase gene. *Proc. Natl. Acad. Sci.* **102:** 9667–9672.

Grundhoff A. and Ganem D. 2001. Mechanisms governing expression of the v-FLIP gene of Kaposi's sarcoma-associated herpesvirus. *J. Virol.* **75:** 1857–1863.

Hellen C.U. and Sarnow P. 2001. Internal ribosome entry sites in eukaryotic mRNA molecules. *Genes Dev.* **15:** 1593–1612.

Herbreteau C.H., Weill L., Decimo D., Prevot D., Darlix J.L., Sargueil B., and Ohlmann T. 2005. HIV-2 genomic RNA contains a novel type of IRES located downstream of its initiation codon. *Nat. Struct. Mol. Biol.* **12:** 1001–1007.

Hershey J.W.B., Asano K., Naranda T., Vornlocher H.P., Hanachi P., and Merrick W.C. 1996. Conservation and diversity in the structure of translation initiation factor EIF3 from humans and yeast. *Biochimie* **78:** 903–907.

Honda M., Ping L.H., Rijnbrand R.C., Amphlett E., Clarke B., Rowlands D., and Lemon S.M. 1996. Structural requirements for initiation of translation by internal ribosome entry within genome-length hepatitis C virus RNA. *Virology* **222:** 31–42.

Hunt S.L., Hsuan J.J., Totty N., and Jackson R.J. 1999. unr, a cellular cytoplasmic RNA-binding protein with five cold-shock domains, is required for internal initiation of translation of human rhinovirus RNA. *Genes Dev.* **13:** 437–448.

Jackson R.J. 2005. Alternative mechanisms of initiating translation of mammalian mRNAs. *Biochem. Soc. Trans.* **33:** 1231–1241.

Jan E. 2006. Divergent IRES elements in invertebrates. *Virus Res.* **119:** 16–28.

Jan E. and Sarnow P. 2002. Factorless ribosome assembly on the internal ribosome entry site of cricket paralysis virus. *J. Mol. Biol.* **324:** 889–902.

Jan E., Kinzy T.G., and Sarnow P. 2003. Divergent tRNA-like element supports initiation, elongation, and termination of protein biosynthesis. *Proc. Natl. Acad. Sci.* **100:** 15410–15415.

Jang S.K. 2006. Internal initiation: IRES elements of picornaviruses and hepatitis c virus. *Virus Res.* **119:** 2–15.

Jang S.K., Krausslich H.G., Nicklin M.J., Duke G.M., Palmenberg A.C., and Wimmer E. 1988. A segment of the 5′ nontranslated region of encephalomyocarditis virus RNA directs internal entry of ribosomes during in vitro translation. *J. Virol.* **62:** 2636–2643.

Ji H., Fraser C.S., Yu Y., Leary J., and Doudna J.A. 2004. Coordinated assembly of human translation initiation complexes by the hepatitis C virus internal ribosome entry site RNA. *Proc. Natl. Acad. Sci.* **101:** 16990–16995.

Johannes G., Carter M.S., Eisen M.B., Brown P.O., and Sarnow P. 1999. Identification of eukaryotic mRNAs that are translated at reduced cap binding complex eIF4F concentrations using a cDNA microarray. *Proc. Natl. Acad. Sci.* **96:** 13118–13123.

Johansen L.K. and Morrow C.D. 2000. The RNA encompassing the internal ribosome entry site in the poliovirus 5′ nontranslated region enhances the encapsidation of genomic RNA. *Virology* **273:** 391–399.

Jopling C.L., Yi M., Lancaster A.M., Lemon S.M., and Sarnow P. 2005. Modulation of hepatitis C virus RNA abundance by a liver-specific MicroRNA. *Science* **309:** 1577–1581.

Kaku Y., Chard L.S., Inoue T., and Belsham G.J. 2002. Unique characteristics of a picornavirus internal ribosome entry site from the porcine teschovirus-1 talfan. *J. Virol.* **76:** 11721–11728.

Kaminski A. and Jackson R.J. 1998. The polypyrimidine tract binding protein (PTB) requirement for internal initiation of translation of cardiovirus RNAs is conditional rather than absolute. *RNA* **4:** 626–638.

Kaminski A., Hunt S.L., Patton J.G., and Jackson R.J. 1995. Direct evidence that polypyrimidine tract binding protein (PTB) is essential for internal initiation of translation of encephalomyocarditis virus RNA. *RNA* **1:** 924–938.

Kieft J.S., Grech A., Adams P., and Doudna J.A. 2001a. Mechanisms of internal ribosome entry in translation initiation. *Cold Spring Harbor Symp. Quant. Biol.* **66:** 277–283.

Kieft J.S., Zhou K., Jubin R., and Doudna J.A. 2001b. Mechanism of ribosome recruitment by hepatitis C IRES RNA. *RNA* **7:** 194–206.

Kieft J.S., Zhou K., Grech A., Jubin R., and Doudna J.A. 2002. Crystal structure of an RNA tertiary domain essential to HCV IRES-mediated translation initiation. *Nat. Struct. Biol.* **9:** 370–374.

Klinck R., Westhof E., Walker S., Afshar M., Collier A., and Aboul-Ela F. 2000. A potential RNA drug target in the hepatitis C virus internal ribosomal entry site. *RNA* **6:** 1423–1431.

Kolupaeva V.G., Pestova T.V., and Hellen C.U. 2000a. An enzymatic footprinting analysis of the interaction of 40S ribosomal subunits with the internal ribosomal entry site of hepatitis C virus. *J. Virol.* **74:** 6242–6250.

———. 2000b. Ribosomal binding to the internal ribosomal entry site of classical swine fever virus. *RNA* **6:** 1791–1807.

Komar A.A. and Hatzoglou M. 2005. Internal ribosome entry sites in cellular mRNAs: Mystery of their existence. *J. Biol. Chem.* **280:** 23425–23428.

Konarska M., Filipowicz W., Domdey H., and Gross H. 1981. Binding of ribosomes to linear and circular forms of the 5′-terminal leader fragment of tobacco-mosaic-virus RNA. *Eur. J. Biochem.* **114:** 221–227.

Kozak M. 1979. Inability of circular mRNA to attach to eukaryotic ribosomes. *Nature* **280:** 82–85.

———. 1989. The scanning model for translation: An update. *J. Cell Biol.* **108:** 229–241.

———. 1991. An analysis of vertebrate mRNA sequences: Intimations of translational control. *J. Cell Biol.* **115:** 887–903.

Kuhn R., Luz N., and Beck E. 1990. Functional analysis of the internal translation initiation site of foot-and-mouth disease virus. *J. Virol.* **64:** 4625–4631.

Lata K.R., Agrawal R.K., Penczek P., Grassucci R., Zhu J., and Frank J. 1996. Three-dimensional reconstruction of the *Escherichia coli* 30 S ribosomal subunit in ice. *J. Mol. Biol.* **262:** 43–52.

Lomakin I.B., Hellen C.U., and Pestova T.V. 2000. Physical association of eukaryotic initiation factor 4G (eIF4G) with eIF4A strongly enhances binding of eIF4G to the internal ribosomal entry site of encephalomyocarditis virus and is required for internal initiation of translation. *Mol. Cell. Biol.* **20:** 6019–6029.

Low W., Harries M., Ye H., Du M.Q., Boshoff C., and Collins M. 2001. Internal ribosome entry site regulates translation of Kaposi's sarcoma-associated herpesvirus FLICE inhibitory protein. *J. Virol.* **75:** 2938–2945.

Lukavsky P.J., Kim I., Otto G.A., and Puglisi J.D. 2003. Structure of HCV IRES domain II determined by NMR. *Nat. Struct. Biol.* **10:** 1033–1038.

Lukavsky P.J., Otto G.A., Lancaster A.M., Sarnow P., and Puglisi J.D. 2000. Structures of two RNA domains essential for hepatitis C virus internal ribosome entry site function. *Nat. Struct. Biol.* **7:** 1105–1110.

Meerovitch K., Svitkin Y.V., Lee H.S., Lejbkowicz F., Kenan D.J., Chan E.K., Agol V.I., Keene J.D., and Sonenberg N. 1993. La autoantigen enhances and corrects aberrant translation of poliovirus RNA in reticulocyte lysate. *J. Virol.* **67:** 3798–3807.

Melcher S.E., Wilson T.J., and Lilley D.M. 2003. The dynamic nature of the four-way junction of the hepatitis C virus IRES. *RNA* **9:** 809–820.

Nakashima N., Sasaki J., and Toriyama S. 1999. Determining the nucleotide sequence and capsid-coding region of himetobi P virus: A member of a novel group of RNA viruses that infect insects. *Arch. Virol.* **144:** 2051–2058.

Nomoto A., Lee Y.F., and Wimmer E. 1976. The 5' end of poliovirus mRNA is not capped with m7G(5')ppp(5')Np. *Proc. Natl. Acad. Sci.* **73:** 375–380.

Ochs K., Saleh L., Bassili G., Sonntag V.H., Zeller A., and Niepmann M. 2002. Interaction of translation initiation factor eIF4B with the poliovirus internal ribosome entry site. *J. Virol.* **76:** 2113–2122.

Odreman-Macchioli F.E., Tisminetzky S.G., Zotti M., Baralle F.E., and Buratti E. 2000. Influence of correct secondary and tertiary RNA folding on the binding of cellular factors to the HCV IRES. *Nucleic Acids Res.* **28:** 875–885.

Otto G.A. and Puglisi J.D. 2004. The pathway of HCV IRES-mediated translation initiation. *Cell* **119:** 369–380.

Pelletier J. and Sonenberg N. 1988. Internal initiation of translation of eukaryotic mRNA directed by a sequence derived from poliovirus RNA. *Nature* **334:** 320–325.

———. 1989. Internal binding of eucaryotic ribosomes on poliovirus RNA: Translation in HeLa cell extracts. *J. Virol.* **63:** 441–444.

Pestova T.V. and Hellen C.U. 2003. Translation elongation after assembly of ribosomes on the cricket paralysis virus internal ribosomal entry site without initiation factors or initiator tRNA. *Genes Dev.* **17:** 181–186.

Pestova T.V., Hellen C.U., and Shatsky I.N. 1996a. Canonical eukaryotic initiation factors determine initiation of translation by internal ribosomal entry. *Mol. Cell. Biol.* **16:** 6859–6869.

Pestova T.V., Lomakin I.B., and Hellen C.U. 2004. Position of the CrPV IRES on the 40S subunit and factor dependence of IRES/80S ribosome assembly. *EMBO Rep.* **5:** 906–913.

Pestova T.V., Shatsky I.N., and Hellen C.U. 1996b. Functional dissection of eukaryotic initiation factor 4F: The 4A subunit and the central domain of the 4G subunit are sufficient to mediate internal entry of 43S preinitiation complexes. *Mol. Cell. Biol.* **16:** 6870–6878.

Pestova T.V., Shatsky I.N., Fletcher S.P., Jackson R.J., and Hellen C.U. 1998. A prokaryotic-like mode of cytoplasmic eukaryotic ribosome binding to the initiation codon during internal translation initiation of hepatitis C and classical swine fever virus RNAs. *Genes Dev.* **12:** 67–83.

Pilipenko E.V., Viktorova E.G., Guest S.T., Agol V.I., and Roos R.P. 2001. Cell-specific proteins regulate viral RNA translation and virus-induced disease. *EMBO J.* **20:** 6899–6908.

Pilipenko E.V., Pestova T.V., Kolupaeva V.G., Khitrina E.V., Poperechnaya A.N., Agol V.I., and Hellen C.U. 2000. A cell cycle-dependent protein serves as a template-specific translation initiation factor. *Genes Dev.* **14:** 2028–2045.

Pisarev A.V., Shirokikh N.E., and Hellen C.U. 2005. Translation initiation by factor-independent binding of eukaryotic ribosomes to internal ribosomal entry sites. *C.R. Biol.* **328:** 589–605.

Poole T.L., Wang C., Popp R.A., Potgieter L.N., Siddiqui A., and Collett M.S. 1995. Pestivirus translation initiation occurs by internal ribosome entry. *Virology* **206:** 750–754.

Reynolds J.E., Kaminski A., Carroll A.R., Clarke B.E., Rowlands D.J., and Jackson R.J. 1996. Internal initiation of translation of hepatitis C virus RNA: The ribosome entry site is at the authentic initiation codon. *RNA* **2:** 867–878.

Reynolds J.E., Kaminski A., Kettinen H.J., Grace K., Clarke B.E., Carroll A.R., Rowlands D.J., and Jackson R.J. 1995. Unique features of internal initiation of hepatitis C virus RNA translation. *EMBO J.* **14:** 6010–6020.

Rijnbrand R., van der Straaten T., van Rijn P.A., Spaan W.J., and Bredenbeek P.J. 1997. Internal entry of ribosomes is directed by the 5′ noncoding region of classical swine fever virus and is dependent on the presence of an RNA pseudoknot upstream of the initiation codon. *J. Virol.* **71:** 451–457.

Rosenfeld A.B. and Racaniello V.R. 2005. Hepatitis C virus internal ribosome entry site-dependent translation in *Saccharomyces cerevisiae* is independent of polypyrimidine tract-binding protein, poly(rC)-binding protein 2, and La protein. *J. Virol.* **79:** 10126–10137.

Sandy P., Ventura A., and Jacks T. 2005. Mammalian RNAi: A practical guide. *Biotechniques* **39:** 215–224.

Sasaki J. and Nakashima N. 1999. Translation initiation at the CUU codon is mediated by the internal ribosome entry site of an insect picorna-like virus in vitro. *J. Virol.* **73:** 1219–1226.

Simons J.N., Desai S.M., Schultz D.E., Lemon S.M., and Mushahwar I.K. 1996. Translation initiation in GB viruses A and C: Evidence for internal ribosome entry and implications for genome organization. *J. Virol.* **70:** 6126–6135.

Siridechadilok B., Fraser C.S., Hall R.J., Doudna J.A., and Nogales E. 2005. Structural roles for human translation factor eIF3 in initiation of protein synthesis. *Science* **310:** 1513–1515.

Sizova D.V., Kolupaeva V.G., Pestova T.V., Shatsky I.N., and Hellen C.U. 1998. Specific interaction of eukaryotic translation initiation factor 3 with the 5′ nontranslated regions of hepatitis C virus and classical swine fever virus RNAs. *J. Virol.* **72:** 4775–4782.

Spahn C.M., Jan E., Mulder A., Grassucci R.A., Sarnow P., and Frank J. 2004. Cryo-EM visualization of a viral internal ribosome entry site bound to human ribosomes: The IRES functions as an RNA-based translation factor. *Cell* **118:** 465–475.

Spahn C.M., Kieft J.S., Grassucci R.A., Penczek P.A., Zhou K., Doudna J.A., and Frank J. 2001. Hepatitis C virus IRES RNA-induced changes in the conformation of the 40s ribosomal subunit. *Science* **291:** 1959–1962.

Spriggs K.A., Bushell M., Mitchell S.A., and Willis A.E. 2005. Internal ribosome entry segment-mediated translation during apoptosis: The role of IRES-trans-acting factors. *Cell Death Differ.* **12:** 585–591.

Thompson S.R., Gulyas K.D., and Sarnow P. 2001. Internal initiation in *Saccharomyces cerevisiae* mediated by an initiator tRNA/eIF2-independent internal ribosome entry site element. *Proc. Natl. Acad. Sci.* **98:** 12972–12977.

Trachsel H. and Staehelin T. 1979. Initiation of mammalian protein synthesis. The multiple functions of the initiation factor eIF-3. *Biochim. Biophys. Acta* **565:** 305–314.

Tsukiyama-Kohara K., Iizuka N., Kohara M., and Nomoto A. 1992. Internal ribosome entry site within hepatitis C virus RNA. *J. Virol.* **66:** 1476–1483.

Vagner S., Waysbort A., Marenda M., Gensac M.C., Amalric F., and Prats A.C. 1995. Alternative translation initiation of the Moloney murine leukemia virus mRNA controlled by internal ribosome entry involving the p57/PTB splicing factor. *J. Biol. Chem.* **270:** 20376–20383.

Wang C., Sarnow P., and Siddiqui A. 1993. Translation of human hepatitis C virus RNA in cultured cells is mediated by an internal ribosome-binding mechanism. *J. Virol.* **67:** 3338–3344.

Westermann P. and Nygard O. 1983. The spatial arrangement of the complex between eukaryotic initiation factor eIF-3 and 40 S ribosomal subunit. Cross-linking between factor and ribosomal proteins. *Biochim. Biophys. Acta* **741:** 103–108.

Wilson J.E., Pestova T.V., Hellen C.U., and Sarnow P. 2000a. Initiation of protein synthesis from the A site of the ribosome. *Cell* **102:** 511–520.

Wilson J.E., Powell M.J., Hoover S.E., and Sarnow P. 2000b. Naturally occurring dicistronic cricket paralysis virus RNA is regulated by two internal ribosome entry sites. *Mol. Cell. Biol.* **20:** 4990–4999.

Zamore P.D. and Haley B. 2005. Ribo-gnome: The big world of small RNAs. *Science* **309:** 1519–1524.

Zhao W.D. and Wimmer E. 2001. Genetic analysis of a poliovirus/hepatitis C virus chimera: New structure for domain II of the internal ribosomal entry site of hepatitis C virus. *J. Virol.* **75:** 3719–3730.

6

Translation Initiation Via Cellular Internal Ribosome Entry Sites

Orna Elroy-Stein

Department of Cell Research & Immunology
George S. Wise Faculty of Life Science
Tel Aviv University, 69978 Tel Aviv, Israel

William C. Merrick

Department of Biochemistry, School of Medicine
Case Western Reserve University
Cleveland, Ohio 44106-4935

INTERNAL RIBOSOME ENTRY SITES (IRESs) in eukaryotic mRNAs were first discovered in viral mRNAs in 1988 (Jang et al. 1988; Pelletier and Sonenberg 1988; for review, see Chapter 5). The first cellular IRES was documented a few years later in the mRNA encoding the human immunoglobulin heavy-chain binding protein, BiP (Macejak and Sarnow 1991). Since then, several dozen cellular IRESs have been reported, although the authenticity of some of them has been called into question. In this chapter, we undertake the definition and description of cellular IRESs, their modes of regulation, and their biological significance.

It has long been known that some cellular proteins continue to be expressed under conditions where cap-dependent translation is severely compromised, such as during poliovirus infection, stress, and mitosis (Sarnow 1989; Johannes and Sarnow 1998; Johannes et al. 1999; Clemens 2001; Qin and Sarnow 2004). Such observations led to the hypothesis that these proteins might be expressed from mRNAs under the control of an IRES. Following this reasoning, microarray analysis of polysomes from poliovirus-infected cells, where the cap-binding complex eIF4F is disrupted by cleavage of eIF4G, indicated that up to 3% of eukaryotic mRNAs might contain IRES elements (Johannes et al. 1999; Qin and Sarnow 2004). The mRNAs suggested to have IRES elements are generally not translated efficiently under normal conditions, and they appear to require down-regulation of cap-dependent translation for their expression (Merrick 2004; Qin and Sarnow 2004). Furthermore, many of these mRNAs encode proteins that are known or expected to facilitate recovery from stress and are generally not abundantly expressed.

HOW TO DETECT A CELLULAR IRES

The difficulty in characterizing cellular IRES elements that confer a relatively low efficiency of expression has contributed to skepticism about their credibility (Kozak 2005), but it should be noted that the controversy initially aroused by the discovery of viral IRESs has now evaporated (Schneider et al. 2001). Conclusive demonstration that a given cellular mRNA uses internal ribosome entry, rather than some other mechanism, requires careful experimentation. The standard test, used in the majority of studies of both viral and cellular IRESs, is the bicistronic assay (Jang et al. 1988; Pelletier and Sonenberg 1988). A more rigorous test employing a circular mRNA has been used to confirm the function of a viral IRES (Chen and Sarnow 1995). In the bicistronic assay, the bicistronic mRNA is engineered to contain two cistrons with the putative IRES element inserted between them. The first cistron is translated by the cap-dependent scanning mechanism, whereas translation of the second cistron does not happen unless internal initiation at the IRES element occurs. A panel of controls is required to exclude possible confounders that fall into two general classes: (1) The bi-cistronic-derived mRNA length and structure are not as expected or (2) the mRNA is translated by some means other than via an IRES. In the first class, several studies have revealed the existence of cryptic promoters and cryptic splice sites that can give rise to new capped 5′ ends within some of the published IRESs (for review, see Kozak 2005). To rule out cryptic promoter activity, promoter-less vectors are often used. However, cryptic promoters may not be detected unless the vector contains an enhancer element (Bert et al. 2006). The commonly used pRF bicistronic vector, which contains the *Renilla* luciferase and firefly luciferase (Luc) coding regions as the first and second cistrons, respectively, was shown to generate transcripts that are substrates for aberrant splicing due to the use of an alternate splice acceptor site within the IRES sequence, leading to misinterpretations regarding IRES function and strength (Van Eden et al. 2004a). It is increasingly recognized that direct transfection of RNA into cells, as well as the use of in vitro translation systems, provides important controls for alternative splicing and transcriptional start sites. In a number of published cases, however, some of these controls are lacking.

In the main, problems of the second class (alternative translational strategies) have been dealt with by employing controls to exclude leaky scanning, reinitiation, and shunting. These involve the insertion of hairpins near the 5′ terminus or AUGs upstream of the first cistron to repress the translation of the first cistron, with the expectation that translation of the second cistron through internal initiation will not be affected.

The discovery of the IRES in BiP mRNA was based on study of a construct in which its 220-nucleotide 5′UTR was inserted between the chloramphenicol acetyltransferase (CAT) and Luc coding regions of a bicistronic plasmid vector; i.e., 5′-CAT/BiP 5′UTR/Luc. The BiP 5′UTR sequence enabled translation of the downstream cistron independent of the upstream sequence. This strategy followed that employed to define the poliovirus IRES (Pelletier and Sonenberg 1988; Chapter 5). Northern blot analysis established the integrity of the bicistronic mRNA, transcribed from the transfected plasmid. Inhibition of cap-dependent translation by a strong hairpin placed in the 5′UTR or by proteolytic inactivation of eIF4F led to decreased translation of the first cistron but not of the second cistron. These observations were attributed to the function of an IRES in the BiP sequence that drives 5′ cap-independent translation of the second cistron (Macejak and Sarnow 1991; Yang and Sarnow 1997). Following the BiP IRES report, a handful of cellular IRESs were proposed and reported based on similar criteria (a list can be found at http://www.iresite.org [Mokrejs et al. 2006]). Among these are genes involved in cell proliferation, differentiation, and cell death. Some examples include growth factors such as fibroblast growth factor-2 (FGF-2; Vagner et al. 1995) and vascular endothelial growth factor (VEGF; Akiri et al. 1998; Stein et al. 1998), oncogenes such as c-myc (Nanbru et al. 1997; Stoneley et al. 1998), and inhibitors of apoptosis such as X-linked inhibitor of apoptosis protein (XIAP; Holcik et al. 1999) and c-IAP1 (Van Eden et al. 2004b). It is noteworthy that these proteins are biologically effective at rather small concentrations and thus do not need to be expressed at high levels. In the following sections, the current state of the field is discussed with emphasis on particularly illustrative examples.

TRANSLATION DURING MITOSIS AND DIFFERENTIATION

Global translation in mammals is subjected to transient inhibition under specific physiological conditions, in keeping with cellular requirements. For example, translation rates are robust in the G_1 phase of the cell cycle but are low during mitosis (Bonneau and Sonenberg 1987; Datta et al. 1999; Heesom et al. 2001; Pyronnet et al. 2001). Nonetheless, specific cellular mRNAs remain associated with polysomes, raising the possibility that cap-independent translation initiation is involved (Qin and Sarnow 2004). The mRNA encoding the PITSLRE kinase, a master mitotic CDK p34^{cdc2}-related kinase, serves as an example (Cornelis et al. 2000; Tinton et al. 2005; Petretti et al. 2006). The PITSLRE mRNA produces two isoforms, p110^{PITSLRE} and p58^{PITSLRE}. Translation of p58^{PITSLRE} is initiated at an in-frame AUG (AUG18) within the coding region of p110^{PITSLRE}, downstream from 17 AUG codons, some of which are in a more favorable context than AUG18. Mutational analysis of AUG18 confirmed that p58^{PITSLRE} is the product of alternative start-site selection rather than proteolytic cleavage of the p110^{PITSLRE} protein. Insertion of regions of PITSLRE sequence into a bicistronic plasmid demonstrated that a 380-nucleotide segment located upstream of AUG18 enabled translation of the second cistron in transfected cells. This observation led to the deduction that an IRES element drives the translation of the p58^{PITSLRE} isoform. No expression of β-galactosidase (βgal) from the second cistron was observed using segments upstream of the 380-nucleotide segment, ruling out ribosomal readthrough. Quantitative correlation of βgal activity with the amount of bicistronic βgal RNA detected by northern blots ruled out the possibility that the protein was derived from aberrant short mRNA species generated from the bicistronic plasmid. Moreover, synchronization of a stable cell line expressing a reporter construct showed that PITSLRE IRES activity was induced 4-fold in the G$_2$/M phase of the cell cycle, consistent with a role for p58^{PITSLRE} in mitotic spindle formation (Petretti et al. 2006).

The mRNA encoding ornithine decarboxylase (ODC), the rate-limiting enzyme in the biosynthesis of polyamines, serves as a second example of an IRES-containing mRNA (Pyronnet et al. 2000). Active translation of ODC mRNA during mitosis ensures elevated levels of polyamines for mitotic spindle formation and chromatin condensation. Translation of the ODC mRNA in the rabbit reticulocyte lysate in vitro translation system pretreated with rhinovirus 2A protease (2A^{pro}), which cleaves eIF4G, suggested the existence of an IRES element within a segment of the ODC 5'UTR, downstream of a stable hairpin structure that acts as a repressor for ribosomal scanning. The IRES element was functional in HeLa cells transfected with a bicistronic plasmid, 5'-CAT/ODC IRES/Luc. Long exposure of northern blots ruled out the presence of short RNA molecules containing only the Luc coding regions, allowing the conclusion that translation of the second cistron involves an IRES. Moreover, rapamycin treatment, which blocks cap-dependent, but not cap-independent, translation (Beretta et al. 1996a,b), prevented ODC induction during the G$_1$/S phase but not the G$_2$/M phase of the cell cycle.

Analysis of synchronized HeLa cells revealed that the ODC IRES, located in the 3′ proximal half of the 5′UTR, is preferentially active during G_2/M.

Similar to the transiently lower global rate of translation that occurs during mitosis, down-regulation of global protein synthesis is experienced by cells undergoing terminal differentiation, which involves extensive changes in the pattern of gene expression. The transient inhibition of 5′ cap-dependent translation initiation increases the availability of free ribosomes required for efficient translation of the new repertoire of mRNAs made in cells undergoing specialization. One possible scenario is that a transient shift to IRES-mediated translation occurs during the time of translational inhibition as a result of the decreased competition with cap-dependent translation. The ability to utilize IRES-mediated initiation would allow the ongoing synthesis of proteins needed for the differentiation process. Evidence supporting this idea comes from studies of the initial phases of megakaryocytic differentiation (Gerlitz et al. 2002).

PROPOSED IRES FUNCTION DURING APOPTOSIS

The induction of apoptosis is associated with substantial inhibition of cap-dependent protein synthesis due to specific fragmentation of eIF4GI and eIF4GII and alterations in the phosphorylation state of eIF2α, eIF4E, and 4E-BPs (Chapter 16). Yet, both death and survival require de novo protein synthesis from a subclass of mRNAs. Several observations suggest that this is achieved by a switch from cap-dependent to cap-independent translation. Whereas the overall translation rate of Fas-treated cells is 30–40% that of naive cells, specific proteins are preferentially synthesized. For example, the rate of β-tubulin synthesis was reduced by 88%, but that of death-associated protein 5 (DAP5, also known as p97 and NAT1) was only marginally affected. The mRNA levels of both proteins dropped to the same extent, demonstrating that DAP5 synthesis is due to regulation at the translational level (Henis-Korenblit et al. 2000). Similarly, despite inhibition of global protein synthesis, the level of c-myc protein is maintained during TRAIL-mediated apoptosis without any change in the half-life of the c-myc protein or the level of its mRNA (Stoneley et al. 2000). Depending on the apoptosis scenario, the life/death decision is dictated by the balance between the levels of proapoptotic proteins (such as c-myc, DAP5, Apaf-1) and antiapoptotic proteins (such as XIAP, HIAP2, c-IAP, Bag-1, Bcl-2), which remain actively synthesized. The 5′UTRs of mRNAs encoding the pro- and antiapoptotic proteins were termed "death IRESs" based on results from bicistronic vector transfection assays which evince increased translation of the second cistron relative to the first cistron under apoptotic conditions (for review, see Holcik and Sonenberg 2005; Lewis and Holcik 2005; Marash and Kimchi 2005). Although alternative interpretations of these bicistronic tests have not been rigorously ruled out for some of the proposed death IRESs, IRES activity in the mRNA encoding the antiapoptotic protein Bcl-2 was confirmed both in vitro and in vivo (Sherrill et al. 2004). The major Bcl-2 transcript harbors a 1400-nucleotide 5′UTR that includes 10 AUG codons upstream of the translation initiation start site. Insertion of the 5′UTR into the intercistronic region of a bicistronic construct enabled translation of the second cistron independently of the upstream sequence in vitro. Transfection of cells with bicistronic RNA harboring the Bcl-2 5′UTR in the intercistronic region resulted in an 8-fold increase of second cistron translation compared to a control RNA lacking the Bcl-2 5′UTR. This observation is consistent with the presence of an IRES in the Bcl-2 5′UTR. The IRES activity was induced 3- to 6-fold under apoptotic conditions triggered by either etoposide or sodium arsenite, which lead to inhibition of 5′-cap-dependent initiation (Sherrill et al. 2004).

Of particular interest in the context of internal translation initiation during apoptosis is the DAP5 protein, a member of the eIF4G family. DAP5 harbors binding sites for eIF3 and eIF4A but, unlike eIF4G, not for PABP and eIF4E (Imataka et al. 1997; Levy-Strumpf et al. 1997; Morley et al. 1997; Yamanaka et al. 1997; Henis-Korenblit et al. 2000). During apoptosis, DAP5 mRNA remains actively translated via a proposed IRES as mentioned above (Henis-Korenblit et al. 2000), and the protein is proposed to serve as a positive *trans*-acting factor to promote the IRES function of mRNAs encoding Apaf-1, c-myc, HIAP2, and XIAP as well as its own mRNA. Similar to other members of the eIF4G protein family, DAP5 (97 kD) undergoes caspase cleavage near its carboxyl terminus upon induction of apoptosis, yielding a p86 cleavage product that maintains its eIF4A and eIF3 binding sites (Henis-Korenblit et al. 2000). eIF4GI is cleaved simultaneously by caspase-3 at two sites to generate products designated N-FAG, M-FAG/p76, and C-FAG (the amino-terminal, middle, and carboxy-terminal Fragments of Apoptotic cleavage of eIF4G) (Bushell et al. 2000; Chapter 16). M-FAG/p76 contains the binding sites for eIF4E, eIF3, and eIF4A. This raised the question of which eIF4G or DAP5 fragment or full-length form is more potent in activating the proposed death IRESs. In an eIF4GI-depleted HeLa cell-free translation extract, both M-FAG/p76 and intact DAP5 stimulated the proposed IRES activities of c-myc, XIAP, and c-IAP (Hundsdoerfer et al. 2005). Cotransfection experiments revealed that, depending on the experimental system used, the p86 form of DAP5 is a more potent activator compared to intact DAP5 and to M-FAG/p76 (Henis-Korenblit et al. 2000, 2002; Nevins et al. 2003; Warnakulasuriyarachchi et al. 2004).

PRIMARY SEQUENCE AND *cis*-ELEMENTS OF PROPOSED CELLULAR IRESs

It is not yet understood how the translation machinery binds internally to cellular mRNAs. As with viral IRESs, there are no obvious primary sequence similarities that define a consensus sequence. In addition, synthetic sequences often confer up-regulated expression of a reporter relative to a control RNA sequence (Anthony and Merrick 1991; Owens et al. 2001), and yet not all of these are real IRES elements (i.e., RNA sequences/structures that can be found in authentic eukaryotic mRNAs). However, polypyrimidine tracts (PPT) are a common feature of many proposed cellular IRES elements. It is suggested that the polypyrimidine tract binding protein (PTB) mediates IRES activity by binding to the PPT (see next section), but this has not been confirmed as yet.

Many proposed cellular IRES elements are functionally modular in that their activity increases or decreases when various segments are deleted (Yang and Sarnow 1997; Akiri et al. 1998; Huez et al. 1998; Stein et al. 1998; Chappell et al. 2000; Coldwell et al. 2000; Le Quesne et al. 2001; Chappell and Mauro 2003; Cencig et al. 2004; Jopling et al. 2004; Dobson et al. 2005). It is believed that such modules correspond to distinct structures and/or factor-binding sites that function together to yield optimal or regulated activity.

Small segments complementary to 18S rRNA have been suggested to function as ribosome recruiters in some picornaviral IRESs (Scheper et al. 1994) and are also found in short synthetic IRESs selected from libraries of random oligonucleotides (Owens et al. 2001). Because the efficiency of rRNA–mRNA interactions depends on flanking sequences and secondary structures (Mauro and Edelman 2002), it is tempting to speculate that, in some cases, induced changes in the RNA structure by the binding of a *trans*-acting RNA-binding protein may govern the ability of the IRES to interact with rRNA. Such direct mRNA–rRNA interactions are used for selection of initiator regions in prokaryotic mRNAs (Shine and Dalgarno 1974; Steitz and Jakes 1975). Criteria similar to those used to establish the interaction of 16S rRNA with the Shine-Dalgarno sequence in bacteria were recently applied to provide evidence for an mRNA–rRNA base-pairing mechanism of translation initiation in eukaryotes. Manipulation of both mRNA and rRNA sequences demonstrated a requirement for sequence complementary between the 9-nucleotide element found in the mouse *Gtx* homeodomain mRNA and the 18S rRNA for reporter gene expression (Dresios et al. 2005).

Although structural models have been postulated based on enzymatic and chemical probing for the proposed IRESs of c-myc (Le Quesne et al. 2001), L-myc (Jopling et al. 2004), FGF-1 (Martineau et al. 2004), FGF-2 (Bonnal et al. 2003), Apaf-1 (Mitchell et al. 2003), and the p36 isoform of Bag-1 (Pickering et al. 2004), there is no convincing evidence that higher-ordered structures are significant or required for cellular IRES function. Furthermore, common structural features have not been found, even when the closely related c-myc and L-myc 5′UTRs are compared. One possibility is that cellular IRES elements are dynamic structures that can undergo conformational switches that affect their activity. The leader of the mRNA encoding the cat-1 Arg/Lys transporter, which confers stress-induced translational regulation (Fernandez et al. 2002a,b, 2001), serves as an example. Under conditions of amino acid starvation, it undergoes a structural switch that is induced by cap-dependent translation of an inhibitory upstream open reading frame (uORF), acting as a zipper to remodel and activate the downstream IRES. Analysis of systematic 5′ end truncations of the leader in a monocistronic mRNA construct used for in vitro translation demonstrated that the 5′ end of the uORF is inhibitory for initiation of Cat-1. Starvation of cells transfected with bicistronic plasmids containing a series of mutations in the uORF showed that induction of the IRES-mediated translation is independent of the uORF peptide sequence but dependent on the process of protein synthesis. Deletion mapping combined with enzymatic and chemical probing led to the conclusion that translation of the uORF mediates dynamic RNA rearrangements within the Cat-1 5′UTR, possibly by enabling the formation of an active IRES structure dependent on its binding to an as-yet-unknown factor (Yaman et al. 2003). Slowing translation elongation through the uORF, either by cycloheximide or by the introduction of rare codons, stimulated the translation of the IRES-mediated downstream ORF, further supporting the role of uORF translation in remodeling and enhancing IRES activity (Fernandez et al. 2005). This type of regulatory control has been well characterized as translational attenuation for the *trp* operon and *ermC* in bacterial systems (see Kolter and Yanofsky 1982 and Hahn et al. 1982, respectively).

IRES *TRANS*-ACTING FACTORS

As with viral IRES elements (Chapter 5), additional proteins (beyond the canonical initiation factors) are often required for optimal expression from cellular IRESs. These IRES *trans*-acting factors (ITAFs) are frequently normal cellular proteins, but they could in theory be viral gene products. It is hypothesized that ITAFs may remodel IRES structures by stabilizing a conformation that is active for ribosome recruitment. A second hypothesis considers ITAF function as a bridge between the RNA and the translating ribosomes. UV cross-linking experiments and electrophoretic mobility shift assays (EMSA) led to the discovery that several known RNA-binding proteins also bind to specific regions shown to be important for the activity of the proposed IRES elements. Examples include La autoantigen (Holcik and Korneluk 2000; Kim et al. 2001), heterogeneous nuclear ribonucleoprotein (hnRNP)-C1/C2 (Holcik et al. 2003; Kim et al. 2003), Upstream of N-ras (Unr) (Mitchell et al. 2001; Tinton et al. 2005), and proteins belonging to the PTB protein family (also known as hnRNP-I) (see below). ITAF function is commonly assayed in cell-free translation systems by varying the concentration of the candidate ITAF protein. Increasingly, assays are performed in vivo under conditions of overexpression and/or siRNA-mediated knockdown of proposed ITAFs. At present, the observed influence of the proposed ITAFs on IRES activity is usually a 1.5- to 3-fold effect, suggesting that their effect may not be a major one. However, there are technical limitations to these experiments, and the existence of functionally redundant factors has not been excluded, so it is premature to draw firm conclusions.

One of the best-studied ITAFs is PTB. This protein serves as a splicing repressor by binding to a single-stranded UCUU sequence within a PPT (Wagner and Garcia-Blanco 2001), but its role as an ITAF has been proposed to work by binding to double-stranded RNA structures (Mitchell et al. 2005). Decreased PTB binding through deletion or mutation of IRES PPTs correlated with reduced activity of the proposed IRES elements of Apaf-1 and p27Kip (Mitchell et al. 2003; Cho et al. 2005). The observations that PTB enhances the proposed IRES activity of Apaf-1 (Mitchell et al. 2003), Bag-1 (Pickering et al. 2003), p27Kip (Cho et al. 2005), and HIF1α (Schepens et al. 2005) are in contrast to the finding that PTB inhibits the function of the BiP and Unr IRES elements (Kim et al. 2000; Cornelis et al. 2005). These findings suggest that a given ITAF may act either as a positive or a negative factor, depending on the IRES element under consideration (e.g., Apaf-1 vs. Bip) or on other possible ITAFs that associate with the IRES (e.g., Unr, which binds together with PTB to the Apaf-1 IRES).

The ability of an ITAF to modulate cellular IRES function could depend on its intracellular level, subcellular localization, and/or posttranslational modifications in response to physiological signals. For example, posttranslational modifications of hnRNP-C correlated with its redistribution from the nucleus to the cytoplasm and with its ability to modulate the function of c-myc and XIAP IRES elements (Holcik et al. 2003; Kim et al. 2003). Currently available data are mostly of a correlative nature, so the relevance of ITAF modifications to actual ITAF function remains to be experimentally proven.

PHYSIOLOGICAL RELEVANCE OF WEAK CELLULAR IRESs

As mentioned above, one of the challenges in studying cellular IRES elements stems from their low activity when compared to viral IRESs or cap-dependent translation. From a biological point of view, however, there is no compelling reason to expect cellular IRESs to be as active as viral IRESs; many of the latter form part of a viral strategy to take over the cell's translational machinery (Chapter 20), whereas many cellular IRES-containing mRNAs express proteins that have biological impact at low concentrations. Induction of the transcription factors GCN4 and ATF4 in response to cellular stress is mediated at the translational level (Chapters 9 and 13) and affords powerful examples of the physiological relevance of a mild translational effect. Enhanced ATF4 translation, via a 5′ cap-dependent mechanism, correlates with a shift of its mRNA to fractions that are engaged with only two/three ribosomes compared to one ribosome under normal conditions (Lu et al. 2004). The key feature is that expression from the ATF4 mRNA can be regulated, not that the 5′UTR sequence or structure leads to maximal levels of ATF4 production.

From an evolutionary point of view, it is tempting to speculate that IRES elements have evolved by random genomic events followed by natural selection when a cellular advantage was provided. Therefore, weak IRESs may become stronger in the future depending on selection pressures. Similar to cryptic promoters and alternative splice sites which constantly evolve in the ever-changing genome (Ast 2004; Soreq et al. 2004), weak cellular IRES elements may represent an additional mechanism used to confer options to enhance physiological adaptability.

STIMULATION OF IRES ACTIVITY UNDER STRESS

Numerous stress signals increase the phosphorylation of eIF2α by activation of one of the four stress-activated eIF2α kinases, leading to a reduction in eIF2·GTP levels and reduced rates of global translation (Chapters 4, 12, and 13). Yet, a subset of mRNAs remains actively translated by a mechanism that involves uORF utilization for leaky scanning and reinitiation (Lu et al. 2004; Vattem and Wek 2004), or IRES function (for review, see Komar and Hatzoglou 2005). For example, the bulk of VEGF mRNA remains associated with translating polysomes under hypoxic conditions (Stein et al. 1998), suggesting that its IRES element is resistant to reduced levels of ternary complex ($eIF2 \cdot GTP \cdot Met-tRNA_i$) due to $eIF2\alpha$ phosphorylation. In vitro analysis has shown that the EMCV IRES is also insensitive to a reduction in ternary complex levels (Hui et al. 2003). The observation of IRES activity under reduced levels of $eIF2 \cdot GTP \cdot Met-tRNA_i$ presents a paradox, because standard model pathways for 80S initiation complex formation require binding of the ternary complex to the 40S subunit prior to mRNA binding. Thus, a reduction in 43S complexes (containing ternary complexes) would be expected to decrease expression from all mRNAs.

A direct connection between $eIF2\alpha$ phosphorylation and activation of the IRES elements of VEGF, c-myc, cat-1, p58[PITSLRE] kinase, and ODC has been demonstrated by overexpression of $eIF2\alpha$ kinases, kinase inhibitors, or the nonphosphorylatable $eIF2\alpha$-S51A variant (Fernandez et al. 2002a,b; Gerlitz et al. 2002; Tinton et al. 2005). A possible explanation is that the ternary complex is recruited more efficiently to IRES-containing mRNAs. The key feature of this model is that there must be some difference between the normal 80S initiation pathway and the pathway for IRES-containing mRNAs to allow the efficient use of ternary complexes at low concentration. The difference might be due to a change in the sequence of events (e.g., the binding of mRNA to the 40S subunit prior to the binding of the ternary complex) or a change in the kinetic rates that relate to the many intermediate steps in the 80S pathway. Another hypothetical mechanism is to pre-bind the multifactor complex (MFC) (Chapter 9) to the 40S subunit before the ternary complex and the mRNA bind. The explanation may lie in a combination of these possibilities, depending on the specific IRES in question.

In sum, a number of IRES-containing mRNAs appear to be resistant to reduced levels of ternary complexes. This phenomenon makes biological sense because it allows many stress response proteins to be synthesized in the face of a global reduction in protein synthesis, thereby facilitating the cell's recovery from stress. However, further work is necessary to determine the mechanism whereby expression from IRES-containing mRNAs continues despite reduced levels of ternary complexes.

SUMMARY AND PERSPECTIVES

The discovery of cellular IRES elements has led to important new insights into the regulation of translation initiation as an additional level of control of gene expression in response to physiological and environmental signals. Protein synthesis from IRES-containing mRNAs frequently occurs at times when there is a global reduction in cap-dependent translation. Furthermore, the use of IRES elements is not necessarily a stand-alone process, because their use can be coupled to other mechanisms of regulation. For example, expression from an IRES element could be regulated by a metabolically sensitive binding protein, by initial cap-dependent translation of a uORF followed by IRES-mediated initiation (as noted for cat-1 mRNA), or by ribosomal shunting.

It is likely that we are only beginning to understand the biochemical and biological mechanisms that are associated with the use and regulation of expression from mRNAs containing IRES elements. This is illustrated by two recent sets of findings. First, eIF2A appears to be an inhibitor of Ure2p IRES-mediated expression (Komar et al. 2005) and of other cellular mRNAs in yeast that are utilized during glucose deprivation (W. Gilbert and J.A. Doudna, pers. comm.). On the other hand, recent data indicate that reduction of eIF5B levels by siRNA methodology has a greater effect on IRES-mediated expression than on cap-dependent translation under stress conditions in mammalian cells (O. Elroy-Stein, unpubl.). These two observations may be coupled, because although either eIF5 or eIF5B can trigger hydrolysis of the GTP in the ternary complexes associated with 48S complexes (Pestova et al. 2000), only eIF5B has been shown to function with eIF2A and to lead to its release (Adams et al. 1975). Second, mice lacking the DKC1 gene display an impaired ability to utilize IRES-containing mRNAs (Yoon et al. 2006). The DKC1 gene, which is related to premature aging and cancer in humans, encodes an enzyme responsible for the conversion of uridine to pseudouridine in rRNA (Ruggero et al. 2003; Chapter 15). Although it is not anticipated that pseudouridylization is a reversible and regulatory phenomenon, this observation suggests that ribosomal modifications may regulate its activity or mRNA specificity. Most previous studies had failed to distinguish such differences between modified and unmodified ribosomes, but changes in the less common initiation pathways (reinitiation, shunting, or IRES-mediated initiation) might have been obscured by the predominant translation of cap-dependent mRNAs. Thus, it should be worthwhile to reinvestigate the role of ribosome modifications with sensitive reporters that report on the utilization of these minor pathways for translation initiation.

Future studies in this field will undoubtedly provide a deeper understanding of cellular IRES function and reveal the extent to which the cellular IRES elements share properties with viral IRES elements. Although the complexity of the cellular regulatory networks, coupled with the relatively weak expression of cellular IRESs, demands especially rigorous experimentation, the potential utility of cellular IRES-mediated translation as an alternative way to modulate gene expression clearly warrants further investigation.

ACKNOWLEDGMENTS

We thank Jennifer A. Doudna and Peter Sarnow for their instructive criticism, useful comments, and suggestions. We apologize to all the cellular IRES researchers whose work was not directly cited in this chapter. This work was supported in part by grants to O. E.-S. from the Israel Science Foundation and the US–Israel Binational Science Foundation, and to W.C.M. from the National Institutes of Health (GM26796 and GM68079).

REFERENCES

Adams S.L., Safer B., Anderson W.F., and Merrick W.C. 1975. Eukaryotic initiation complex formation. Evidence for two distinct pathways. *J. Biol. Chem.* **250:** 9083–9089.

Akiri G., Nahari D., Finkelstein Y., Le S.Y., Elroy-Stein O., and Levi B.Z. 1998. Regulation of vascular endothelial growth factor (VEGF) expression is mediated by internal initiation of translation and alternative initiation of transcription. *Oncogene* **17:** 227–236.

Anthony D.D. and Merrick W.C. 1991. Eukaryotic initiation factor (eIF)-4F. Implications for a role in internal initiation of translation. *J. Biol. Chem.* **266:** 10218–10226.

Ast G. 2004. How did alternative splicing evolve? *Nat. Rev. Genet.* **5:** 773–782.

Beretta L., Svitkin Y.V., and Sonenberg N. 1996a. Rapamycin stimulates viral protein synthesis and augments the shutoff of host protein synthesis upon picornavirus infection. *J. Virol.* **70:** 8993–8996.

Beretta L., Gingras A.C., Svitkin Y.V., Hall M.N., and Sonenberg N. 1996b. Rapamycin blocks the phosphorylation of 4E-BP1 and inhibits cap-dependent initiation of translation. *EMBO J.* **15:** 658–664.

Bert A.G., Grepin R., Vadas M.A., and Goodall G.J. 2006. Assessing IRES activity in the HIF1-α and other cellular 5′UTRs. *RNA* **12:** 1–10.

Bonnal S., Schaeffer C., Creancier L., Clamens S., Moine H., Prats A.C., and Vagner S. 2003. A single internal ribosome entry site containing a G quartet RNA structure drives fibroblast growth factor 2 gene expression at four alternative translation initiation codons. *J. Biol. Chem.* **278:** 39330–39336.

Bonneau A.M. and Sonenberg N. 1987. Involvement of the 24-kDa cap-binding protein in regulation of protein synthesis in mitosis. *J. Biol. Chem.* **262:** 11134–11139.

Bushell M., Poncet D., Marissen W.E., Flotow H., Lloyd R.E., Clemens M.J., and Morley S.J. 2000. Cleavage of polypeptide chain initiation factor eIF4GI during apoptosis in lymphoma cells: Characterization of an internal fragment generated by caspase-3-mediated cleavage. *Cell Death Differ.* **7:** 628–636.

Cencig S., Nanbru C., Le S.Y., Gueydan C., Huez G., and Kruys V. 2004. Mapping and characterization of the minimal internal ribosome entry segment in the human c-myc mRNA 5′ untranslated region. *Oncogene* **23:** 267–277.

Chappell S.A. and Mauro V.P. 2003. The internal ribosome entry site (IRES) contained within the RNA-binding motif protein 3 (Rbm3) mRNA is composed of functionally distinct elements. *J. Biol. Chem.* **278:** 33793–33800.

Chappell S.A., Edelman G.M., and Mauro V.P. 2000. A 9-nt segment of a cellular mRNA can function as an internal ribosome entry site (IRES) and when present in linked multiple copies greatly enhances IRES activity. *Proc. Natl. Acad. Sci.* **97:** 1536–1541.

Chen C.Y. and Sarnow P. 1995. Initiation of protein synthesis by the eukaryotic translational apparatus on circular RNAs. *Science* **268:** 415–417.

Cho S., Kim J.H., Back S.H., and Jang S.K. 2005. Polypyrimidine tract-binding protein enhances the internal ribosomal entry site-dependent translation of p27Kip1 mRNA and modulates transition from G1 to S phase. *Mol. Cell. Biol.* **25:** 1283–1297.

Clemens M.J. 2001. Translational regulation in cell stress and apoptosis. Roles of the eIF4E binding proteins. *J. Cell. Mol. Med.* **5:** 221–239.

Coldwell M.J., Mitchell S.A., Stoneley M., MacFarlane M., and Willis A.E. 2000. Initiation of Apaf-1 translation by internal ribosome entry. *Oncogene* **19:** 899–905.

Cornelis S., Tinton S.A., Schepens B., Bruynooghe Y., and Beyaert R. 2005. UNR translation can be driven by an IRES element that is negatively regulated by polypyrimidine tract binding protein. *Nucleic Acids Res.* **33:** 3095–3108.

Cornelis S., Bruynooghe Y., Denecker G., Van Huffel S., Tinton S., and Beyaert R. 2000. Identification and characterization of a novel cell cycle-regulated internal ribosome entry site. *Mol. Cell* **5:** 597–605.

Datta B., Datta R., Mukherjee S., and Zhang Z. 1999. Increased phosphorylation of eukaryotic initiation factor 2alpha at the G2/M boundary in human osteosarcoma cells correlates with deglycosylation of p67 and a decreased rate of protein synthesis. *Exp. Cell Res.* **250:** 223–230.

Dobson T., Minic A., Nielsen K., Amiott E., and Krushel L. 2005. Internal initiation of translation of the TrkB mRNA is mediated by multiple regions within the 5′ leader. *Nucleic Acids Res.* **33:** 2929–2941.

Dresios J., Chappell S.A., Zhou W., and Mauro V.P. 2005. An mRNA-rRNA base-pairing mechanism for translation initiation in eukaryotes. *Nat. Struct. Mol. Biol.* **13:** 30–34.

Fernandez J., Bode B., Koromilas A., Diehl J.A., Krukovets I., Snider M.D., and Hatzoglou M. 2002a. Translation mediated by the internal ribosome entry site of the cat-1 mRNA is regulated by glucose availability in a PERK kinase-dependent manner. *J. Biol. Chem.* **277:** 11780–11787.

Fernandez J., Yaman I., Mishra R., Merrick W.C., Snider M.D., Lamers W.H., and Hatzoglou M. 2001. Internal ribosome entry site-mediated translation of a mammalian mRNA is regulated by amino acid availability. *J. Biol. Chem.* **276:** 12285–12291.

Fernandez J., Yaman I., Merrick W.C., Koromilas A., Wek R.C., Sood R., Hensold J., and Hatzoglou M. 2002b. Regulation of internal ribosome entry site-mediated translation by eukaryotic initiation factor-2alpha phosphorylation and translation of a small upstream open reading frame. *J. Biol. Chem.* **277:** 2050–2058.

Fernandez J., Yaman I., Huang C., Liu H., Lopez A.B., Komar A.A., Caprara M.G., Merrick W.C., Snider M.D., Kaufman R.J., et al. 2005. Ribosome stalling regulates IRES-mediated translation in eukaryotes, a parallel to prokaryotic attenuation. *Mol. Cell* **17:** 405–416.

Gerlitz G., Jagus R., and Elroy-Stein O. 2002. Phosphorylation of initiation factor-2 alpha is required for activation of internal translation initiation during cell differentiation. *Eur. J. Biochem.* **269:** 2810–2819.

Hahn J., Grandi G., Gryczan T.J., and Dubnau D. 1982. Translational attenuation of *ermC:*

A deletion analysis. *Mol. Gen. Genet.* **186:** 204–216.

Heesom K.J., Gampel A., Mellor H., and Denton R.M. 2001. Cell cycle-dependent phosphorylation of the translational repressor eIF-4E binding protein-1 (4E-BP1). *Curr. Biol.* **11:** 1374–1379.

Henis-Korenblit S., Strumpf N.L., Goldstaub D., and Kimchi A. 2000. A novel form of DAP5 protein accumulates in apoptotic cells as a result of caspase cleavage and internal ribosome entry site-mediated translation. *Mol. Cell. Biol.* **20:** 496–506.

Henis-Korenblit S., Shani G., Sines T., Marash L., Shohat G., and Kimchi A. 2002. The caspase-cleaved DAP5 protein supports internal ribosome entry site-mediated translation of death proteins. *Proc. Natl. Acad. Sci.* **99:** 5400–5405.

Holcik M. and Korneluk R.G. 2000. Functional characterization of the X-linked inhibitor of apoptosis (XIAP) internal ribosome entry site element: Role of La autoantigen in XIAP translation. *Mol. Cell. Biol.* **20:** 4648–4657.

Holcik M. and Sonenberg N. 2005. Translational control in stress and apoptosis. *Nat. Rev. Mol. Cell Biol.* **6:** 318–327.

Holcik M., Gordon B.W., and Korneluk R.G. 2003. The internal ribosome entry site-mediated translation of anti-apoptotic protein XIAP is modulated by the heterogeneous nuclear ribonucleoproteins C1 and C2. *Mol. Cell. Biol.* **23:** 280–288.

Holcik M., Lefebvre C., Yeh C., Chow T., and Korneluk R.G. 1999. A new internal-ribosome-entry-site motif potentiates XIAP-mediated cytoprotection. *Nat. Cell Biol.* **1:** 190–192.

Huez I., Creancier L., Audigier S., Gensac M.C., Prats A.C., and Prats H. 1998. Two independent internal ribosome entry sites are involved in translation initiation of vascular endothelial growth factor mRNA. *Mol. Cell. Biol.* **18:** 6178–6190.

Hui D.J., Bhasker C.R., Merrick W.C., and Sen C.G. 2003. Viral stress inducible protein p56 inhibits translation by blocking the interaction of eIF3 with the ternary complex (eIF2·GTP·Met-tRNAi). *J. Biol. Chem.* **278:** 39477–39482.

Hundsdoerfer P., Thoma C., and Hentze M.W. 2005. Eukaryotic translation initiation factor 4GI and p97 promote cellular internal ribosome entry sequence-driven translation. *Proc. Natl. Acad. Sci.* **102:** 13421–13426.

Imataka H., Olsen H.S., and Sonenberg N. 1997. A new translational regulator with homology to eukaryotic translation initiation factor 4G. *EMBO J.* **16:** 817–825.

Jang S.K., Krausslich H.G., Nicklin M.J., Duke G.M., Palmenberg A.C., and Wimmer E. 1988. A segment of the 5′ nontranslated region of encephalomyocarditis virus RNA directs internal entry of ribosomes during in vitro translation. *J. Virol.* **62:** 2636–2643.

Johannes G. and Sarnow P. 1998. Cap-independent polysomal association of natural mRNAs encoding c-myc, BiP, and eIF4G conferred by internal ribosome entry sites. *RNA* **4:** 1500–1513.

Johannes G., Carter M.S., Eisen M.B., Brown P.O., and Sarnow P. 1999. Identification of eukaryotic mRNAs that are translated at reduced cap binding complex eIF4F concentrations using a cDNA microarray. *Proc. Natl. Acad. Sci.* **96:** 13118–13123.

Jopling C.L., Spriggs K.A., Mitchell S.A., Stoneley M., and Willis A.E. 2004. L-Myc protein synthesis is initiated by internal ribosome entry. *RNA* **10:** 287–298.

Kim J.H., Paek K.Y., Choi K., Kim T.D., Hahm B., Kim K.T., and Jang S.K. 2003. Heterogeneous nuclear ribonucleoprotein C modulates translation of c-myc mRNA in a cell cycle phase-dependent manner. *Mol. Cell. Biol.* **23:** 708–720.

Kim Y.K., Hahm B., and Jang S.K. 2000. Polypyrimidine tract-binding protein inhibits translation of BiP mRNA. *J. Mol. Biol.* **304:** 119–133.

Kim Y.K., Back S.H., Rho J., Lee S.H., and Jang S.K. 2001. La autoantigen enhances translation of BiP mRNA. *Nucleic Acids Res.* **29:** 5009–5016.

Komar A.A. and Hatzoglou M. 2005. Internal ribosome entry sites in cellular mRNAs: Mystery of their existence. *J. Biol. Chem.* **280:** 23425–23428.

Komar A.A., Gross S.R., Barth-Baus D., Strachan R., Hensold J.O., Goss Kinzy T., and Merrick W.C. 2005. Novel characteristics of the biological properties of the yeast *Saccharomyces cerevisiae* eukaryotic initiation factor 2A. *J. Biol. Chem.* **280:** 15601–15611.

Kolter R. and Yanofsky C. 1982. Attenuation in amino acid biosynthetic operons. *Annu. Rev. Genet.* **16:** 113–134.

Kozak M. 2005. A second look at cellular mRNA sequences said to function as internal ribosome entry sites. *Nucleic Acids Res.* **33:** 6593–6602.

Le Quesne J.P., Stoneley M., Fraser G.A., and Willis A.E. 2001. Derivation of a structural model for the c-myc IRES. *J. Mol. Biol.* **310:** 111–126.

Levy-Strumpf N., Deiss L.P., Berissi H., and Kimchi A. 1997. DAP-5, a novel homolog of eukaryotic translation initiation factor 4G isolated as a putative modulator of gamma interferon-induced programmed cell death. *Mol. Cell. Biol.* **17:** 1615–1625.

Lewis S.M. and Holcik M. 2005. IRES in distress: Translational regulation of the inhibitor of apoptosis proteins XIAP and HIAP2 during cell stress. *Cell Death Differ.* **12:** 547–553.

Lu P.D., Harding H.P., and Ron D. 2004. Translation reinitiation at alternative open reading frames regulates gene expression in an integrated stress response. *J. Cell Biol.* **167:** 27–33.

Macejak D.G. and Sarnow P. 1991. Internal initiation of translation mediated by the 5′ leader of a cellular mRNA. *Nature* **353:** 90–94.

Marash L. and Kimchi A. 2005. DAP5 and IRES-mediated translation during programmed cell death. *Cell Death Differ.* **12:** 554–562.

Martineau Y., Le Bec C., Monbrun L., Allo V., Chiu I.M., Danos O., Moine H., Prats H., and Prats A.C. 2004. Internal ribosome entry site structural motifs conserved among mammalian fibroblast growth factor 1 alternatively spliced mRNAs. *Mol. Cell. Biol.* **24:** 7622–7635.

Mauro V.P. and Edelman G.M. 2002. The ribosome filter hypothesis. *Proc. Natl. Acad. Sci.* **99:** 12031–12036.

Merrick W.C. 2004. Cap-dependent and cap-independent translation in eukaryotic systems. *Gene* **332:** 1–11.

Mitchell S.A., Brown E.C., Coldwell M.J., Jackson R.J., and Willis A.E. 2001. Protein factor requirements of the Apaf-1 internal ribosome entry segment: Roles of polypyrimidine tract binding protein and upstream of N-ras. *Mol. Cell. Biol.* **21:** 3364–3374.

Mitchell S.A., Spriggs K.A., Coldwell M.J., Jackson R.J., and Willis A.E. 2003. The Apaf-1 internal ribosome entry segment attains the correct structural conformation for function via interactions with PTB and unr. *Mol. Cell* **11:** 757–771.

Mitchell S.A., Spriggs K.A., Bushell M., Evans J.R., Stoneley M., Le Quesne J.P., Spriggs R.V., and Willis A.E. 2005. Identification of a motif that mediates polypyrimidine tract-binding protein-dependent internal ribosome entry. *Genes Dev.* **19:** 1556–1571.

Mokrejs M., Vopalensky V., Kolenaty O., Masek T., Feketova Z., Sekyrova P., Skaloudova B., Kriz V., and Pospisek M. 2006. IRESite: The database of experimentally verified IRES structures (www.iresite.org). *Nucleic Acids Res.* **34:** D125–130.

Morley S.J., Curtis P.S., and Pain V.M. 1997. eIF4G: Translation's mystery factor begins to yield its secrets. *RNA* **3:** 1085–1104.

Nanbru C., Lafon I., Audigier S., Gensac M.C., Vagner S., Huez G., and Prats A.C. 1997. Alternative translation of the proto-oncogene c-myc by an internal ribosome entry site. *J. Biol. Chem.* **272:** 32061–32066.

Nevins T.A., Harder Z.M., Korneluk R.G., and Holcik M. 2003. Distinct regulation of internal ribosome entry site-mediated translation following cellular stress is mediated by apoptotic fragments of eIF4G translation initiation factor family members eIF4GI and p97/DAP5/NAT1. *J. Biol. Chem.* **278:** 3572–3579.

Owens G.C., Chappell S.A., Mauro V.P., and Edelman G.M. 2001. Identification of two short internal ribosome entry sites selected from libraries of random oligonucleotides. *Proc. Natl. Acad. Sci.* **98:** 1471–1476.

Pelletier J. and Sonenberg N. 1988. Internal initiation of translation of eukaryotic mRNA directed by a sequence derived from poliovirus RNA. *Nature* **334:** 320–325.

Pestova T.V., Lomakin I.B., Lee J.H., Choi S.K., Dever T.E., and Hellen C.U.T. 2000. The joining of ribosomal subunits in eukaryotes requires eIF5B. *Nature* **403:** 332–335.

Petretti C., Savoian M., Montembault E., Glover D.M., Prigent C., and Giet R. 2006. The PITSLRE/CDK11p58 protein kinase promotes centrosome maturation and bipolar spindle formation. *EMBP Rep.* **7:** 418–424.

Pickering B.M., Mitchell S.A., Evans J.R., and Willis A.E. 2003. Polypyrimidine tract binding protein and poly r(C) binding protein 1 interact with the BAG-1 IRES and stimulate its activity in vitro and in vivo. *Nucleic Acids Res.* **31:** 639–646.

Pickering B.M., Mitchell S.A., Spriggs K.A., Stoneley M., and Willis A.E. 2004. Bag-1 internal ribosome entry segment activity is promoted by structural changes mediated by poly(rC) binding protein 1 and recruitment of polypyrimidine tract binding protein 1. *Mol. Cell. Biol.* **24:** 5595–5605.

Pyronnet S., Dostie J., and Sonenberg N. 2001. Suppression of cap-dependent translation in mitosis. *Genes Dev.* **15:** 2083–2093.

Pyronnet S., Pradayrol L., and Sonenberg N. 2000. A cell cycle-dependent internal ribosome entry site. *Mol. Cell* **5:** 607–616.

Ruggero D., Grisendi S., Piazza F., Rego E., Mari F., Rao P.H., Cordon-Cardo C., and Pandolfi P.P. 2003. Dyskeratosis congenita and cancer in mice deficient in ribosomal RNA modification. *Science* **299:** 259–262.

Qin X. and Sarnow P. 2004. Preferential translation of internal ribosome entry site-containing mRNAs during the mitotic cycle in mammalian cells. *J. Biol. Chem.* **279:** 13721–13728.

Sarnow P. 1989. Translation of glucose-regulated protein 78/immunoglobulin heavy-chain binding protein mRNA is increased in poliovirus-infected cells at a time when cap-dependent translation of cellular mRNAs is inhibited. *Proc. Natl. Acad. Sci.* **86:** 5795–5799.

Schepens B., Tinton S.A., Bruynooghe Y., Beyaert R., and Cornelis S. 2005. The polypyrimidine tract-binding protein stimulates HIF-1alpha IRES-mediated translation during hypoxia. *Nucleic Acids Res.* **33:** 6884–6894.

Scheper G.C., Voorma H.O., and Thomas A.A. 1994. Basepairing with 18S ribosomal RNA in internal initiation of translation. *FEBS Lett.* **352:** 271–275.

Schneider R., Agol V.I., Andino R., Bayard F., Cavener D.R., Chappell S.A., Chen J.J., Darlix J.L., Dasgupta A., Donze O., et al. 2001. New ways of initiating translation in eukaryotes. *Mol. Cell. Biol.* **21:** 8238–8246.

Sherrill K.W., Byrd M.P., Van Eden M.E., and Lloyd R.E. 2004. BCL-2 translation is mediated via internal ribosome entry during cell stress. *J. Biol. Chem.* **279:** 29066–29074.

Shine J. and Dalgarno L. 1974. The 3′-terminal sequence of *Escherichia coli* 16S ribosomal

RNA: Complementarity to nonsense triplets and ribosome binding sites. *Proc. Natl. Acad. Sci.* **71:** 1342–1346.

Sorek R., Shamir R., and Ast G. 2004. How prevalent is functional alternative splicing in the human genome? *Trends Genet.* **20:** 68–71.

Stein I., Itin A., Einat P., Skaliter R., Grossman Z., and Keshet E. 1998. Translation of vascular endothelial growth factor mRNA by internal ribosome entry: Implications for translation under hypoxia. *Mol. Cell. Biol.* **18:** 3112–3119.

Steitz J.A. and Jakes K. 1975. How ribosomes select initiator regions in mRNA: Base pair formation between the 3′ terminus of 16S rRNA and the mRNA during initiation of protein synthesis in *Escherichia coli. Proc. Natl. Acad. Sci.* **72:** 4734–4738.

Stoneley M., Paulin F.E., Le Quesne J.P., Chappell S.A., and Willis A.E. 1998. C-Myc 5′ untranslated region contains an internal ribosome entry segment. *Oncogene* **16:** 423–428.

Stoneley M., Chappell S.A., Jopling C.L., Dickens M., MacFarlane M., and Willis A.E. 2000. c-Myc protein synthesis is initiated from the internal ribosome entry segment during apoptosis. *Mol. Cell. Biol.* **20:** 1162–1169.

Tinton S.A., Schepens B., Bruynooghe Y., Beyaert R., and Cornelis S. 2005. Regulation of the cell-cycle-dependent internal ribosome entry site of the PITSLRE protein kinase: Roles of Unr (upstream of N-ras) protein and phosphorylated translation initiation factor eIF-2alpha. *Biochem. J.* **385:** 155–163.

Vagner S., Gensac M.C., Maret A., Bayard F., Amalric F., Prats H., and Prats A.C. 1995. Alternative translation of human fibroblast growth factor 2 mRNA occurs by internal entry of ribosomes. *Mol. Cell. Biol.* **15:** 35–44.

Van Eden M.E., Byrd M.P., Sherrill K.W., and Lloyd R.E. 2004a. Demonstrating internal ribosome entry sites in eukaryotic mRNAs using stringent RNA test procedures. *RNA* **10:** 720–730.

———. 2004b. Translation of cellular inhibitor of apoptosis protein 1 (c-IAP1) mRNA is IRES mediated and regulated during cell stress. *RNA* **10:** 469–481.

Vattem K.M. and Wek R.C. 2004. Reinitiation involving upstream ORFs regulates ATF4 mRNA translation in mammalian cells. *Proc. Natl. Acad. Sci.* **101:** 11269–11274.

Wagner E.J. and Garcia-Blanco M.A. 2001. Polypyrimidine tract binding protein antagonizes exon definition. *Mol. Cell. Biol.* **21:** 3281–3288.

Warnakulasuriyarachchi D., Cerquozzi S., Cheung H.H., and Holcik M. 2004. Translational induction of the inhibitor of apoptosis protein HIAP2 during endoplasmic reticulum stress attenuates cell death and is mediated via an inducible internal ribosome entry site element. *J. Biol. Chem.* **279:** 17148–17157.

Yaman I., Fernandez J., Liu H., Caprara M., Komar A.A., Koromilas A.E., Zhou L., Snider M.D., Scheuner D., Kaufman R.J., and Hatzoglou M. 2003. The zipper model of translational control: A small upstream ORF is the switch that controls structural remodeling of an mRNA leader. *Cell* **113:** 519–531.

Yamanaka S., Poksay K.S., Arnold K.S., and Innerarity T.L. 1997. A novel translational repressor mRNA is edited extensively in livers containing tumors caused by the transgene expression of the apoB mRNA-editing enzyme. *Genes Dev.* **11:** 321–333.

Yang Q. and Sarnow P. 1997. Location of the internal ribosome entry site in the 5′ noncoding region of the immunoglobulin heavy-chain binding protein (BiP) mRNA: Evidence for specific RNA-protein interactions. *Nucleic Acids Res.* **25:** 2800–2807.

Yoon A., Peng G., Brandenburg Y., Zollo O., Xu W., Rego E., and Ruggero D. 2006. Impaired control of IRES-mediated translation in X-linked dyskeratosis congenita. *Science* **312:** 902–906.

7

Translation Termination, the Prion [*PSI*⁺], and Ribosomal Recycling

Måns Ehrenberg and Vasili Hauryliuk

Department of Cell and Molecular Biology
Uppsala University, BMC, Box 576, SE-751 24
Uppsala, Sweden

Colin G. Crist,[1] and Yoshikazu Nakamura

Department of Basic Medical Sciences
Institute of Medical Science
Tokyo 108-8639, Japan

FOLLOWING COMPLETION OF THE ELONGATION PHASE of protein synthesis (Chapters 2 and 3), the ribosome is brought into its pretermination complex when a stop (nonsense) codon is translocated into its A site (Fig. 1A). Stop codons are normally not read by tRNAs, but by class 1 release factor (RF) proteins (Scolnick et al. 1968; Caskey et al. 1970; Capecchi and Klein 1970). There are two RFs (RF1 and RF2) in eubacteria, one (eRF1) in eukaryotes and one (aRF1) in archaea. RF1 recognizes the stop codons UAG and UAA, RF2 recognizes UAA and UGA (Scolnick et al. 1968), and eRF1 and aRF1 recognize all three stop codons (Konecki et al. 1977). Although there are sequence homologies between RF1 and RF2, as well as between eRF1 and aRF1, there is only a single universal homology among all class 1 RFs: the GGQ (Gly Gly Gln) motif (Frolova et al. 1999). Binding of a class 1 RF to a ribosome programmed with its cognate stop codon (Fig. 1B) induces hydrolysis of the ester bond that links the completed protein with the tRNA bound to the P site (Fig. 1C), leading to rapid dissociation of the protein from the ribosome. This general scheme for the first phase of termination of protein synthesis (Caskey et al. 1977) is still valid, but the molecular details

---

[1]Present address: Deptartment of Developmental Biology, Pasteur Institute, 75015 Paris, France.

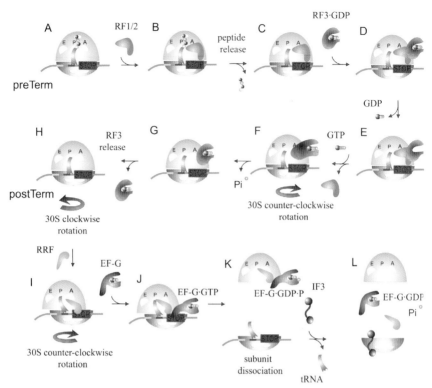

Figure 1. A model for translation termination and ribosomal recycling in prokaryotes. (*A*) Pretermination ribosomal complex; (*B*) binding of class 1 termination factor RF1/2; (*C*) peptide release; (*D*) binding of class 2 termination factor RF3 in complex with GDP; (*E,F*) Exchange of GDP to GTP on RF3 induces counterclockwise ratcheting-like rotation of the 30S subunit relative to the 50S subunit, driving class 1 termination factor dissociation from the ribosomal complex; (*G,H*) GTP hydrolysis by RF3 promotes factor dissociation, followed by 30S relaxation; (*I*) RRF binding to the posttermination ribosomal complex induces counterclockwise rotation of ribosomal subunits; (*J,K*) concerted action of RRF and EF-G leads to GTP-dependent subunit dissociation; (*L*) IF3 prevents reassociation of 70S and accelerates mRNA and tRNA dissociation from 30S.

of stop codon recognition and induction of ester bond hydrolysis remain to be clarified.

In eubacteria and eukaryotes, but not in archaea, there exists a class 2 RF: the GTPases RF3 in eubacteria and eRF3 in eukaryotes (Grentzmann et al. 1994; Mikuni et al. 1994; Zhouravleva et al. 1995). The function of RF3 is to remove class 1 RFs from the ribosome after release of the peptide chain from the P-site tRNA (Fig. 1D–H) (Freistroffer et al. 1997; Zavialov et al. 2001, 2002). The main role of eRF3 in termination of eukaryotic protein synthesis is still unknown (Kisselev et al. 2003; Nakamura and Ito 2003).

After termination of protein synthesis by a class 1 RF followed by its subsequent dissociation from the ribosome, the resulting ribosomal posttermination complex still contains the mRNA and a deacylated tRNA in the P site (Fig. 1H). This complex is quite stable in eubacteria (Hirashima and Kaji 1970, 1972a,b), requiring that ribosomes be recycled in order to participate in a new round of initiation (Fig. 1I–L). The ribosomal recycling factor RRF, elongation factor EF-G (Janosi et al. 1996; Inokuchi et al. 2000; Agrawal et al. 2004), and initiation factor IF3 (Karimi et al. 1999) serve this function in a GTP-dependent manner (Fig. 1) (Karimi et al. 1999). How ribosomal recycling occurs in eukaryotes has remained obscure.

CLASS 1 RELEASE FACTORS

Mimicry in Termination of Protein Synthesis

The modes of action of tRNAs during peptide elongation and of class 1 RFs during termination are similar. In both cases, codons in the ribosomal A site are recognized with high precision, which results in a distal chemical event: either the transfer of a peptide from the P-site to the A-site tRNA or the disruption of the bond between a finished peptide and the P-site tRNA. Accordingly, RFs can be viewed as functional mimics of tRNAs (Moffat and Tate 1994). This view is further supported by the observation that deacylated tRNAs can rapidly terminate protein synthesis in the presence of ethanol (Caskey et al. 1971; Zavialov et al. 2002). Given the similarity of function, it has been proposed that class 1 RFs are *structural* mimics of tRNAs (Ito et al. 1996; Nakamura et al. 2000; Nakamura and Ito 2003).

Three implications follow from this idea. First, the structures of class 1 RFs (Fig. 2) should closely resemble the universal structure of the tRNA molecule (Nakamura et al. 2000, 2001; Nakamura 2001). Second, the RFs should have anticodon motifs that recognize stop codons by direct contact in the ribosomal A site. Such anticodon motifs were identified in domain 2 (Vestergaard et al. 2001) of the eubacterial RFs through genetic complementation and biochemical approaches (Ito et al. 2000). For RF1, the tripeptide motif PAT recognizes the UAG and UAA stop codons. For RF2, the tripeptide motif SPF recognizes the stop codons UAA and UGA. Furthermore, NIKS and YXCXXXF were identified as anticodon motifs in eRF1 and aRF1, respectively (Song et al. 2000; Frolova et al. 2002; Seit-

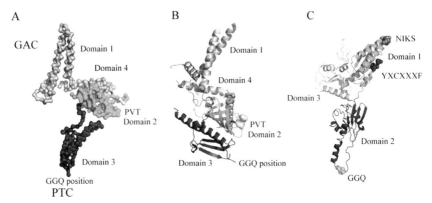

Figure 2. Structures of class 1 RFs on and off the ribosome. (*A*) Cryo-EM of *E. coli* RF1 bound to the posttermination ribosomal complex, accession number 2FVO (Rawat et al. 2006); (*B*) crystal structure of *Streptococcus mutans* RF1, accession number 1ZBT (Joint Center for Structural Genomics [JCSG]); (*C*) crystal structure of *Homo sapiens* eRF1, accession number 1DT9 (Song et al. 2000).

Nebi et al. 2002; Kisselev et al. 2003). Third, the RFs should have aminoacyl-like motifs that induce hydrolysis of the ester bond in the P-site-bound peptidyl-tRNA. On the basis of the experimental observation that substitutions of the Gly residues in the universally conserved GGQ motif of eRF1 resulted in loss of termination activity but not ribosome binding of the factor, it was suggested that the GGQ triplet in domain 3 of RF1 and RF2 (Vestergaard et al. 2001; Shin et al. 2004) and domain 2 of eRF1 (Song et al. 2000) induces hydrolysis of the ester bond in peptidyl-tRNA (Frolova et al. 1999). According to the mimicry hypothesis, the distance between the PAT/SPF and GGQ motifs in eubacterial class 1 RFs should correspond to the distance between the anticodon and CCA ends of a ribosome-bound tRNA molecule and thus be close to 75 Å (Yusupov et al. 2001; Klaholz et al. 2003; Rawat et al. 2003). Corroboration of the mimicry hypothesis and its molecular implications has been the driving force for intense research efforts during the last decade.

Structures of Class 1 Release Factors On and Off the Ribosome

The first crystal structure of a class 1 RF was that of human eRF1 (Fig. 2C) (Song et al. 2000). It has three domains, and the GGQ motif is at the tip of domain 2, whereas domain 3 interacts with eRF3 (Merkulova et al. 1999). According to the mimicry hypothesis, the NIKS motif at the tip of domain 1 corresponds to the tRNA anticodon and recognizes the three stop codons (Song et al. 2000). However, the distance between the NIKS and GGQ motifs is significantly greater than 75 Å, suggesting either that the mimicry hypothesis is wrong or that a conformational change of eRF1 occurs as it binds to a ribosome programmed with a stop codon and ready for termination of mRNA translation (Fig. 1A,B) (Kononenko et al. 2004). Surprisingly, the similar crystal structures of RF2 (Vestergaard et al. 2001) and RF1 (Fig. 2B) (Shin et al. 2004) (accession number 1ZBT) turned out to be very different from that of eRF1 (Song et al. 2000). RF2 and RF1 each have four domains with the GGQ motif at the tip of domain 3, and the SPF or PAT motif at the tip of domain 2. Domains 2, 3, and 4 are tightly bound together, so that the distance between the GGQ and SPF or PAT motifs is only about 25 Å, which is much shorter than the distance between the GGQ and NIKS motifs in eRF1 and the 75 Å distance required by the tRNA mimicry hypothesis. The structural data led Vestergaard et al. (2001) to propose that eRF1 and RF2 terminate protein synthesis by different molecular mechanisms.

The inconsistency between the mimicry hypothesis and the crystal structures of RF2 and RF1 was resolved by cryo-electron microscopy (cryo-EM) reconstructions of the structures of RF2 (Klaholz et al. 2003; Rawat et al. 2003) and RF1 (Fig. 2A) (Rawat et al. 2006) in complex with the ribosome (Fig. 1B,C). The cryo-EM reconstructions show that GGQ on the tip of domain 3 and SPF or PAT on the tip of domain 2 span a distance of 73 Å, thereby allowing simultaneous contact of the SPF or PAT motif with the mRNA in the decoding center and the GGQ motif with the peptidyl-transfer center. In the crystal, domains 2, 3, and 4 of RF2 and RF1 form a cluster, but in the ribosome complex, domain 3 has moved away from the cluster toward the peptidyl-transfer center. On the assumption that the crystal and solution structures of RF2 and RF1 are equivalent, it was suggested that each factor enters the ribosome in a compact form and unfolds when either the SPF or PAT motif contacts one of its cognate stop codons, thereby allowing GGQ-induced ester bond hydrolysis in the peptidyl-transfer center (Rawat et al. 2003).

Solution structures may in principle be different from crystal structures. Therefore, small-angle X-ray scattering (SAXS) data were recorded for native RF1 and for a truncated variant of RF1 that lacks domain 1 but is still active in termination of protein synthesis (Mora et al. 2003). The experimentally recorded SAXS data from these authentic solution structures were compared with simulated SAXS data derived from the ribosome-bound cryo-EM structures of RF1 and RF2 and the crystal structures of free RF1 (Shin et al. 2004) and RF2 (Vestergaard et al. 2001). The experimentally recorded SAXS data were also used for ab initio reconstructions of the native and truncated forms of RF1 (Vestergaard et al. 2005). All results in this SAXS study suggest that the authentic solution structure of RF1 is open, as in the cryo-EM (Klaholz et al. 2003; Rawat et al. 2003, 2006) and low-resolution crystal (Petry et al. 2005) reconstructions of ribosome-bound RF2 and RF1, and not closed, as in the crystal reconstructions of free RF2 (Vestergaard et al. 2001) and free RF1 from *Thermotoga maritime* (Shin et al. 2004) and *Streptococcus mutans* (Joint Center for Structural Genomics [JCSG], accession number 1ZBT).

These SAXS data suggest that the crystal structures of the free RFs are different from their solution structures. However, the compact crystal form of bacterial class 1 RFs appears all the same to be biologically relevant in an important modification reaction. The Q (Gln) in the universal GGQ motif of class 1 RFs is modified to N^5-methyl-glutamine in eubacteria (Dincbas-Renqvist et al. 2000), eukaryotes (Heurgue-Hamard et al. 2005; Polevoda et al. 2005), and probably also in archaea (R. Buckingham, pers. comm.). In *Escherichia coli*, the modification is catalyzed by the *prmC*-encoded methyltransferase PrmC and results in an increase in the termination efficiencies of RF1 and RF2 (Dincbas-Renqvist et al. 2000; Nakahigashi et al. 2002; Heurgue-Hamard et al. 2005). In the crystal structure of the PrmC·RF1 complex, RF1 is in the same compact structure (Graille et al. 2005) as in the crystal structures of free RF1 (Shin et al. 2004) and RF2 (Vestergaard et al. 2001). Furthermore, the interaction between the aminoterminal domain of PrmC and domain 2 of RF1 requires the compact form of RF1. This suggests that PrmC first binds to a class 1 RF in its open form (Vestergaard et al. 2005). Then the RF changes conformation to the compact form, as seen in the crystals of the free RFs (Vestergaard et al. 2001; Shin et al. 2004), and becomes methylated. Subsequently, the methylated RF returns to the free state and its open conformation.

Stop-codon Recognition by Class 1 Release Factors

According to the mimicry hypothesis, stop codons should be recognized by direct contact with anticodon-like peptide motifs in the class 1 RFs (Nakamura and Ito 2002). Another possibility, the "termination adaptor hypothesis," states that stop codons are recognized by complementary sequences in rRNA and that, subsequently, the resulting RNA duplexes are recognized by class 1 RFs (Arkov and Murgola 1999). This hypothesis was recently revived with support from rRNA sequence data (Ivanov et al. 2001). However, photocross-linking studies show that eRF1, RF1, and RF2 are in close contact with their cognate stop codons on the ribosome (Tate et al. 1990; Brown and Tate 1994; Chavatte et al. 2001, 2002; Volkov et al. 2002; Liang et al. 2005). In addition, patterns of cross-linking between rRNA and mRNA with amino-acid-encoding (sense) or stop codons in the ribosomal A site are very similar, making selective duplex formation between stop codons and rRNA triplets unlikely. Furthermore, cryo-EM (Klaholz et al. 2003; Rawat et al. 2003, 2006) and the crystal structures (Petry et al. 2005) show the ribosome-bound RF1 PAT and RF2 SPF motifs in very close contact with their stop codons, leaving little room for mRNA:rRNA duplex formation.

Finally, by using rabbit ribosomes, it was demonstrated that eRF1 from the ciliate *Euplotes aediculatus*, in which UAA and UAG are the only stop codons, with UGA being reassigned to Cys (Kervestin et al. 2001), can terminate at UAA and UAG but not UGA. A control experiment showed that mammalian eRF1 can terminate at all three stop codons, implying that the codon specificity resides in the RF and not in rRNA. Together, these experimental observations appear to rule out the termination adaptor hypothesis and give support to the notion that the PAT/SPF motifs have anticodon-like properties. Nevertheless, it is still unclear as to how RF1 and RF2 recognize U while discriminating against C, G, and A in the first position of all stop codons. Interestingly, the crystallographic reconstructions of ribosome-bound RF1 and RF2 suggest that an additional motif, at the tip of the α5 helix in domain 2 of both factors, may be involved in stop-codon, and possibly first position U, recognition (Petry et al. 2005). However, clarification of this issue requires much higher-resolution structures than those obtained so far.

The suggestion that the NIKS motif in eRF1 directly interacts with stop codons (Fig. 2C) (Song et al. 2000) receives support from cross-linking data showing that a photo-activatable s4U in the first position of UAG, UAA, UGA, or UGG (Trp) in the ribosomal A site cross-links to residues in the NIKS loop (Song et al. 2000; Chavatte et al. 2001, 2002). This could mean that elements in the NIKS loop recognize the omnipresent U in the first position of all stop codons.

In organisms like *Schizosaccharomyces pombe*, where eRF1 terminates at all three stop codons, the TASNIKS peptide sequence at the tip of domain 1 is highly conserved, but not in ciliates with reassigned stop codons, suggesting that the TASNIKS motif is directly involved in stop-codon recognition. In the ciliate *Tetrahymena thermophila* where UAA and UAG are reassigned to Gln, TASNIKS is replaced by the KATNIKD hepta peptide. To probe the role of the TASNIK motif in stop-codon recognition, hybrids were constructed between, on the one hand, domain 1 of *Tetrahymena* eRF1 containing the wild-type (KATNIKD) or mutated (TASNIKD or TASNIKS) sequence and, on the other hand, domains 2 and 3 of eRF1 from *S. pombe* (Ito et al. 2002). The wild-type eRF1 with the KATNIKD sequence could not complement a temperature-sensitive fission yeast phenotype (*sup45ts*) caused by an alteration in eRF1, but the TASNIKD and TASNIKS mutants restored viability at the elevated temperature. From these observations, it was suggested that the TAS tripeptide in the TASNIKS motif of eRF1 recognizes AA and AG in the last two positions of a stop codon. However, these in vivo results could not be reproduced in vitro, where none of the eRF1 mutants terminated at UAG or UAA, although they had full release activity at the UGA codon. It should be noted, however, that the signal-to-background ratio ranged between two and eight in these in vitro experiments, meaning that large variations in the efficiency of UAG or UAA reading among the eRF1 variants could have been hidden in the background, whereas these variations could, in principle, account for the positive complementation in vivo.

The data discussed so far are compatible with an anticodon-like peptide sequence in eRF1, but other observations indicate that codon recognition may depend on scattered motifs in the eRF1 structure. In vitro analysis of the termination efficiency of selected mutants of eRF1 showed that the motifs YXCXXXF (positions 125–131 in human eRF1) (Seit-Nebi et al. 2002; Kolosov et al. 2005) and NIKS (positions 61–64) (Chavatte et al. 2002; Frolova et al. 2002; K. Bulygin et al., pers. comm.), separated by approximately 15 Å (Kong et al. 2004), are important for stop-codon recognition. In addition, amino acid substitutions between positions 51 and 132 in domain 1 of eRF1 from yeast affect stop-codon recognition. Recently, two more invariant residues, Glu-55 and Tyr-125 (human eRF1 numbering), potentially involved in codon recognition, have been identified, suggesting that a three-dimensional network of amino acids may be responsible for stop-codon reading by eRF1 (Kolosov et al. 2005).

INDUCTION OF ESTER BOND HYDROLYSIS BY CLASS 1 RELEASE FACTORS

Mutations in the GGQ motif greatly reduce termination efficiency and cell viability (Song et al. 2000; Mora et al. 2003; Kong et al. 2004). It has been suggested that GGQ induces ester bond hydrolysis of the P-site-bound peptidyl-tRNA by activating a catalytic reaction carried out by the peptidyl-transferase center of the ribosome (Frolova et al. 1999). It has also been proposed that the glutamine in GGQ directly catalyzes the hydrolytic reaction by activating a water molecule and orienting it toward the peptidyl-tRNA (Kong et al. 2004). The first proposal is attractive because of the similarity of the *trans*-peptidation and peptide release reactions. In peptidyl transfer, a carbonyl carbon atom in the ester bond that links the peptidyl residue to the 3′ end of the tRNA in the P site is attacked by the α-amino group of the aminoacyl-tRNA in the A site, whereas in ester bond hydrolysis, the attack is made by a water molecule. The proposal is further supported by the observations that a deacylated tRNA in the A site can induce ester bond hydrolysis in a reaction stimulated by ethanol (Caskey et al. 1971; Zavialov et al. 2002) and that many antibiotics inhibit both peptidyl transfer and termination by class 1 RFs (Scolnick et al. 1968; Capecchi and Klein 1970; Caskey et al. 1971; Menninger 1971; V. Hauryliuk and M. Ehrenberg, unpubl.). The second proposal is contradicted by experimental observations that a GGQ to GGA mutant of RF1 (Zavialov et al. 2002) and several glutamine substitutions in GGQ of eRF1 (Seit-Nebi et al. 2000, 2001) retain significant termination activity in vitro. Genetic experiments suggest that the conserved base A2603 in helix 93 of 23S rRNA is important for peptide release, by excluding water during elongation and allowing water to enter the peptidyl-transfer center during termination of protein synthesis (Polacek et al. 2003).

ACCURACY OF TERMINATION OF PROTEIN SYNTHESIS IN EUBACTERIA

Termination of peptide elongation at sense codons by class 1 RFs is costly for the cell (Kurland and Ehrenberg 1985) and occurs at very low frequency in *E. coli* (Jorgensen et al. 1993). The accuracy (A) of termination catalyzed by RF1 and RF2 has been studied biochemically, where the Michaelis-Menten parameters k_{cat} and K_m were measured for termination at cognate stop codons, as well as at all sense codons differing from nonsense codons by a single base change (Freistroffer et al. 2000). A is defined as the ratio between the efficiency (k_{cat}/K_m) of termination at the cognate nonsense ($(k_{cat}/K_m)^c$) and at the near-cognate sense ($(k_{cat}/K_m)^{nc}$) codon. In the absence of the class 2 RF3, the accuracy is in the range between one thousand and one million for both RFs, with the UAU (Tyr) or UGG (Trp) codon as hot spots for erroneous termination by RF1 ($A = 1100$) or RF2 ($A = 2400$), respectively. Recognition of U in the first codon position is remarkably precise for both RFs. For instance, RF1 discriminates against a C replacing the U in the first codon position with an A value of almost one million. The accuracy of termination originates in K_m values that are about one thousand times larger, and k_{cat} values that are between two and one thousand times smaller, for termination at sense codons than at nonsense codons. These data suggest that there is a codon-independent positioning of a class 1 RF in the A site, where it is unable to induce hydrolysis of the peptidyl-tRNA. The low affinity of this binding mode leads to the similar and high K_m values for termination at sense codons. There is also a codon-dependent positioning of the factor, where it is able to induce ester bond hydrolysis. It is the standard free-energy difference between the codon-dependent and -independent states of the RF that leads to the largely varying k_{cat} values for termination at the different sense codons. These parameters were obtained in the absence of the class 2 RF3, and since addition of RF3 reduced, rather than enhanced, the accuracy of termination, it was concluded that there is no proofreading (Hopfield 1974; Ninio 1975) in termination of protein synthesis (Freistroffer et al. 2000).

The accuracy of codon reading by tRNAs during the elongation phase is, in contrast, greatly amplified by proofreading (Hopfield 1974; Ninio 1975) driven by GTP hydrolysis on elongation factor EF-Tu (Thompson and Stone 1977; Ruusala et al. 1982). In spite of this, recent biochemical data suggest that the accuracy of termination (Freistroffer et al. 2000) is much larger than the accuracy for tRNA selection (Gromadski and Rodnina 2004) by about one thousand. One reason for this appears to be that although class 1 RF-dependent termination at sense codons is characterized by relatively large K_m values, near-cognate codon reading by aminoacyl-tRNAs in ternary complexes with EF-Tu and GTP is characterized by K_m values smaller than those associated with cognate codon reading. A common feature of codon reading by class 1 RFs and tRNAs is that the forward rate constant, i.e., for ester bond hydrolysis in the former case and GTP hydrolysis in the latter case, is much smaller for incorrect codon reading than for correct codon reading.

The accuracy of codon reading by the eukaryotic class 1 eRF1 has not been quantified. The intriguing observation that eRF1 forms a ternary complex with eRF3 and GTP (V. Hauryliuk, pers. comm.), which resembles that formed by aminoacyl-tRNA with EF-Tu and GTP, could mean that the accuracy of termination in eukaryotes is amplified by proofreading.

CLASS 2 RELEASE FACTORS

Class 2 Release Factors Are Small GTPases

The class 2 RF3 in bacteria and eRF3 in eukaryotes are members of the GTPase family (Grentzmann et al. 1994; Mikuni et al. 1994; Zhouravleva et al. 1995). Such enzymes display a transition from their GDP-bound state to their GTP-bound state, in which they accomplish their task. This is followed by GTP hydrolysis and return to their GDP-bound form. The GDP-to-GTP exchange is often aided by a guanine nucleotide exchange factor (GEF), and GTP hydrolysis is sometimes triggered by a GTPase-activating protein (GAP) (Bourne 1993, 1995; Sprang 1997a,b, 2001). The primary sequences of RF3 and eRF3 have the consensus elements GXXXXGK(S/T) implicated in binding the α- and β-phosphates of the nucleotide, (T/G)(C/S)A, accommodating the guanine ring and the NKXD tetrapeptide, and coordinating the γ-phosphate of GTP and Mg^{2+}. The crystal structures of eRF3 in the GDP- or GDPNP-bound forms are very similar (Kong et al. 2004), as are the crystal structures of EF-G in the GDP-bound (Czworkowski et al. 1994; Czworkowski and Moore 1997; Hansson et al. 2005) and GDPNP-bound (Czworkowski et al. 1994; Czworkowski and Moore 1997; Hansson et al. 2005) forms. The crystal structure of RF3 has not been published, but there is a cryo-EM reconstruction of the ribosome-bound factor (Klaholz et al. 2004), albeit at low resolution and without support from crystal coordinates of the factor domains.

Although eRF3 is essential for viability (Kushnirov et al. 1988; Wilson and Culbertson 1988), RF3 is not (Ter-Avanesyan et al. 1993; Mikuni et al. 1994; Grentzmann et al. 1994). There is a ribosome-dependent GTPase activity of RF3, which is further stimulated by RF1 or RF2 (Zavialov et al. 2001), whereas the GTPase activity of eRF3 depends strictly on the presence of both the ribosome and eRF1 (Zhouravleva et al. 1995; Frolova et al. 1996). Free eRF3 forms a complex with eRF1 (Stansfield et al. 1995; Zhouravleva et al. 1995; Frolova et al. 1998; Ito et al. 1998; Merkulova et al. 1999; Kobayashi et al. 2004), which is stabilized by the presence of GTP, but not GDP (Kobayashi et al. 2004), whereas no such free complex exists for RF1 or RF2 and RF3 (Ito et al. 1996; Pel et al. 1998).

Functional Roles of RF3 and eRF3

In the eubacterial cell, RF3 rapidly removes RF1 or RF2 from the ribosome after release of a polypeptide from the P-site-bound tRNA (Freistroffer et al. 1997). Its recently clarified working cycle (see Fig. 1) (Zavialov et al. 2001, 2002; Zavialov and Ehrenberg 2003) starts with the free factor in the GDP form, since it binds GTP with very low affinity compared to GDP, and its dissociation kinetics are too slow to account for its cycling in vivo. Experiments with class 1 RF mutants that are active in ribosome binding but deficient in ester bond hydrolysis reveal that rapid exchange of radiolabeled GDP with unlabeled GDP occurs when RF3 has bound to a ribosome in complex with RF1 or RF2 and with either a peptidyl- or deacylated-tRNA in the P site. However, rapid GDP-to-GTP exchange and rapid release of the class 1 RF can occur only after termination, when there is a deacylated tRNA in the P site. This mechanism prevents premature class 1 RF removal by RF3 and relies on three features of the system: (1) the ability of the P-site-bound deacylated tRNA, but not the peptidyl-tRNA (Lill et al. 1989), to accommodate into the P/E site (Moazed and Noller 1989) of the ratcheted ribosome conformation (Frank and Agrawal 2000, 2001); (2) the high affinity of the GTP-bound form of RF3 for the ratcheted ribosome conformation (Zavialov and Ehrenberg 2003), implying that GDP-to-GTP exchange on RF3 requires a transition from the relaxed state to the ratcheted state of the ribosome (Fig. 1E,F); and (3) the low affinity of class 1 RFs to the ratcheted ribosome conformation, which leads to their rapid dissociation from this ribosome state. Then, GTP is hydrolyzed, followed by rapid dissociation of RF3 from the ribosome.

Although eRF3 is an essential factor, its mode of action has not been clarified (Kisselev and Buckingham 2000; Kisselev et al. 2003; Agaphonov et al. 2005). It may be that eRF3 removes eRF1 from the ribosome after termination, in analogy with the function of RF3. As discussed above, eRF3 could be involved in a putative proofreading mechanism in eukaryote termination.

Prion State of Yeast Class 2 Release Factors

In yeast, the class 2 eRF3 (Sup35) has been characterized as a member of a group of proteins that behave as "yeast prions," exhibiting remarkable transmission properties that fulfill the criteria for prions enunciated by Prusiner in 1982 (Prusiner 1982; Wickner 1994; Wickner et al. 2004).

eRF3 proteins from distantly related yeast (Santoso et al. 2000; Nakayashiki et al. 2001) have an amino-terminal prion domain (PrD) containing five or more tandem oligopeptide repeats that are similar to those of the mammalian prion protein (PrP), as well as a high concentration of polar glutamine and asparagine residues. The PrD drives a prion conformation that reduces the efficiency of translation termination, providing an easily detectable nonsense suppressor phenotype termed $[PSI^+]$. A highly charged middle (M) domain separates the carboxyl and amino termini of eRF3 in *Saccharomyces cerevisiae*. Although the function of the M domain is unknown, similar positively charged residues in *S. pombe* block the eRF1-binding site in the carboxyl terminus of eRF3, potentially regulating eRF1 binding to eRF3 in a competitive manner (Kong et al. 2004). It is intriguing to think of this M-domain/C-domain interaction as a mechanism by which the amino-terminal PrD is anchored, thereby inhibiting the conformational freedom it may require to adopt the $[PSI^+]$

conformation when eRF3 is not engaged in productive interactions with eRF1. The N domain is also important for several protein–protein interactions. The most thoroughly characterized are its interactions with the poly(A)-binding protein (PABP) and the highly charged M domain of eRF3. The former interaction is evolutionarily conserved throughout eukaryotes and is thought to link translation termination with translation initiation in protein biosynthesis (Inge-Vechtomov et al. 2003).

   S. cerevisiae cells lacking the N and M domains are viable, but they are unable to become [*PSI*⁺]. [*PSI*⁺] cells are marked by a conformational change in the PrD of eRF3 in which the otherwise unstructured PrD adopts a β-sheet-rich conformation consistent with amyloid. As with PrP, this prion conformation then nucleates remaining soluble eRF3 proteins to adopt the same conformation and join the growing amyloid by a mechanism termed nucleated conformational conversion (Serio et al. 2000). As with any heritable element, [*PSI*⁺] requires a replication mechanism without which cells containing the large cytoplasmic [*PSI*⁺] aggregate would simply be diluted out of a growing culture. The replication machinery comes in the form of heat shock protein Hsp104, which functions to break up large aggregates to be further resolubilized, an activity essential for acquired thermotolerance. Hsp104 is thought to disrupt large [*PSI*⁺] aggregates to produce small, diffusible seeds that are transmissible to daughter cells during cell division and is thus also required for [*PSI*⁺] propagation (Ness et al. 2002).

Physiological Relevance of the Yeast Release Factor Prion

It is of great interest to determine whether or not "nonsense suppression" as a phenotype of [*PSI*⁺] cells has a true physiological role. One of the more intriguing postulates is that [*PSI*⁺] cells can take advantage of pre-existing genomic diversity to establish complex traits. In a comprehensive screen of the *S. cerevisiae* genome, Gerstein and colleagues identified open reading frames (ORFs) of two gene classes that could theoretically be influenced by [*PSI*⁺] (Harrison et al. 2002). The first class contains ORFs that are disabled by the presence of a premature stop codon and the second class contains two ORFs that could potentially become a single ORF by the readthrough of a single stop codon. Using homology with annotated yeast ORFs and non-yeast proteins, many of the ORFs identified in this screen were determined to encode proteins involved in growth inhibition, flocculation, and stress induction (Harrison et al. 2002), suggesting that nonsense suppression of premature stop codons within pre-existing ORFs could be beneficial in diverse growth conditions.

How might $[PSI^+]$ cells utilize such preexisting genomic variation? One might argue that such genomic variation is irrelevant in light of the nonsense-mediated RNA decay (NMD) pathway. After all, any mRNA transcript of an ORF containing a premature stop codon would be degraded, thereby avoiding buildup of possibly deleterious truncations (Gonzalez et al. 2001). Nevertheless, eRF3 does interact with components of the NMD machinery via its carboxy-terminal domain (Wang et al. 2001), and it is known that readthrough of premature stop codons (afforded by $[PSI^+]$) antagonizes NMD in *S. cerevisiae* (Keeling et al. 2004). Therefore, $[PSI^+]$ provides an additional mechanism to antagonize the NMD pathway, thereby allowing the exploitation of preexisting genomic diversity in the form of ORFs with premature stop codons.

It has been shown that $[PSI^+]$ cells outperformed isogenic eRF3 prion-free ($[psi^-]$) cells in response to ethanol stress in three independent genotypes and in response to heat stress in two independent genotypes (Eaglestone et al. 1999). To determine whether $[PSI^+]$ could be responsible for diverse growth phenotypes responding to a variety of external parameters, isogenic $[psi^-]$ and $[PSI^+]$ cells were assayed for their response to different carbon and nitrogen sources, in the presence of various inhibitors of cellular processes, stress conditions, and temperature. In a strain-dependent manner, conspicuous differences in either growth rate or colony morphology were observed for isogenic $[PSI^+]$ and $[psi^-]$ cells in response to many of the growth conditions tested (True and Lindquist 2000). Similar $[PSI^+]$-like growth phenotypes also could be achieved by targeted mutations in the carboxyl terminus of eRF3 that enhanced nonsense suppression, demonstrating that the $[PSI^+]$ effects are independent of diverse effects on the cell caused by the $[PSI^+]$ aggregate (True et al. 2004).

There are several characteristics, unique to $[PSI^+]$, that may confer on *S. cerevisiae* a unique ability to adapt to a fluctuating environment. First, $[PSI^+]$ is epigenetic. This would allow *S. cerevisiae* to adapt to environmental conditions without the need for permanent genetic change. Since environments may fluctuate rapidly, a nonpermanent response may be beneficial. Second, $[PSI^+]$ is metastable. The de novo genesis and loss of the $[PSI^+]$ phenotype occur at low frequency; about 1 in 10^6–10^7 $[psi^-]$ cells are $[PSI^+]$. Therefore, in any substantial population of $[psi^-]$ cells, $[PSI^+]$ cells will also exist with the capacity to thrive under different growth conditions and vice versa. Third, $[PSI^+]$ is variable with differing levels of nonsense suppression and stability. This would allow certain subpopulations of *S. cerevisiae* to outperform other populations in response to variable environments. The stable nature of strong $[PSI^+]$

may even allow [*PSI*$^+$] surviving cells to adopt the necessary genetic mutations to thrive under prolonged diverse growth conditions, and this has prompted Lindquist and colleagues to describe [*PSI*$^+$] as a capacitor for evolutionary change (True and Lindquist 2000; True et al. 2004).

RECYCLING OF EUBACTERIAL RIBOSOMES TO A NEW ROUND OF INITIATION

After termination of eubacterial protein synthesis and removal of a class 1 RF by the class 2 RF, the ribosome is in a stable, posttermination state, with the mRNA retained and a deacylated tRNA in the P site (Fig. 1H). Kaji and collaborators (Hirashima and Kaji 1970, 1972a,b) showed by sucrose gradient experiments that *E. coli* ribosomes purified as mRNA-bound polysome structures can be dissociated from the mRNA by puromycin treatment. This involves removal of the nascent peptides from the P-site tRNAs and requires the addition of elongation factor EF-G and a previously unknown protein, the ribosome recycling factor (RRF). Furthermore, these authors found that the presence of initiation factor IF3 is necessary for formation of stably dissociated ribosomal subunits (Hirashima and Kaji 1973; Hirokawa et al. 2002).

The first demonstration that EF-G and RRF do not need IF3 to disassemble the posttermination ribosome into subunits (Fig. 1I–K) was provided by experiments (Karimi et al. 1999) in which small peptides were synthesized at a fixed concentration of 30S preinitiation complexes and a limiting concentration of 50S subunits. It was shown (in the absence of IF3) that EF-G, RRF, and GTP, but not the noncleavable analog GDPNP, are sufficient to promote rapid peptide synthesis by recycling 50S subunits through the excess 30S preinitiation complexes. However, IF3 is needed for ribosomal recycling at 1:1 subunit stochiometry, suggesting that IF3 is required only to prevent premature association of ribosomal subunits. In 2005, three laboratories independently confirmed the EF-G- and RRF-dependent dissociation of posttermination ribosomes into subunits (Hirokawa et al. 2005; Peske et al. 2005; Zavialov et al. 2005b). The notion that ribosome dissociation depends on translocation of RRF from the A to P site and of the deacylated P-site tRNA to the E site (Selmer et al. 1999) is rendered unlikely by the finding that EF-G variants, active in GTP hydrolysis but defective in tRNA translocation, fully activate ribosome dissociation by RRF in vivo and in vitro (Fujiwara et al. 2004).

It has been suggested that the posttermination ribosome dissociation reaction is driven by the positively cooperative binding of the GTP form of EF-G with RRF to the 50S ribosomal subunit, as compared to the negatively cooperative binding of these factors to the 70S subunit (Kiel et al. 2003; Zavialov et al. 2005b). A detailed kinetic description of the process (Peske et al. 2005) and a structural corollary (Gao et al. 2005) were provided recently. However, the proposed mechanism does not account for the requirements of GTP *as well as GTP hydrolysis* for posttermination ribosome splitting (Fig. 1I–K), similar to these two requirements for translocation of mRNA and tRNAs by EF-G (Zavialov et al. 2005a).

There remains some controversy regarding mRNA and tRNA release from the posttermination ribosome by the actions of RRF, EF-G, and IF3. It has been suggested that RRF ejects deacylated tRNA from the P site of the posttermination ribosome and that after ribosome splitting, mRNA dissociates from the 30S subunit aided by IF3 (Hirokawa et al. 2005, 2006). According to other models, the primary event in ribosomal recycling is subunit dissociation, followed by spontaneous (Zavialov et al. 2005b) or IF3-aided (Peske et al. 2005) release of tRNA and mRNA. The discrepancy regarding spontaneous or IF3-aided release of mRNA may be accounted for by the strong Shine-Dalgarno sequence in the mRNA used by Zavialov et al. (2005b) that may mask the IF3 effect on mRNA release, and by the absence of a Shine-Dalgarno sequence in the mRNA used by Peske et al. (2005).

CONCLUDING REMARKS

Despite extensive research over decades, several fundamental questions regarding termination and ribosome recycling remain unanswered. Although codon reading by tRNAs during protein elongation has been clarified in molecular detail by high-resolution crystal structures (Ogle and Ramakrishnan 2005), we still do not know even the elementary facts regarding codon recognition by class 1 RFs. As emphasized in this review, the accuracy of codon reading by RF1 and RF2 is impressively high despite the absence of proofreading and remains virtually unexplained. To solve this problem, crystal structures of ribosome-bound class 1 RFs with higher resolution than so far available (Petry et al. 2005) will be necessary, combined with extended mutational analyses and improved biochemical assays.

It is also not clear how the GGQ motif and region affect the peptidyl-transfer center for rapid hydrolysis of the ester bond in the P-site-bound peptidyl-tRNA. However, the mechanism of peptidyl transfer has been elucidated recently by standard free-energy calculations (Trobro and Aqvist 2005), based on an early crystal structure of the 50S subunit (Hansen et al. 2002), confirmed with a later 50S subunit structure (Schmeing et al. 2005). It is hoped that similar approaches may help clarify the accuracy of codon reading by class 1 RFs, as well as how they induce ester bond hydrolysis of peptidyl-tRNA.

Another riddle relates to the primary function of the class 2 eRF3. This essential factor may indeed remove class 1 RF from the ribosome, but it could also be involved in an accuracy-enhancing proofreading mechanism during stop-codon reading by eRF1. To move on from mere speculation regarding its function, it will be necessary to develop the biochemistry of eukaryotic protein synthesis much further. The essential role of EF-G-promoted GTP hydrolysis has not been clarified, either during prokaryotic ribosomal recycling or during translocation of mRNA and tRNAs in the elongation phase. Ribosomal recycling among eukaryotes is another mystery, since no RRF analog has been found. There, subunit dissociation might be catalyzed by the concerted action of initiation factors, eRF1 and eRF3 (L. Kisselev, pers. comm.), but experimental proof for this is lacking. Clearly, providing the answers to these fundamental questions requires a focused and sustained future research effort.

ACKNOWLEDGMENTS

We thank Lev Kisselev for sharing unpublished results and Bente Vestergaard for comments on Figure 2. This work was supported by grants to M.E. from the Swedish Research Council and the National Institutes of Health (GM70768) and to Y.N. from the Ministry of Education, Sports, Culture, Science, and Technology of Japan and CREST Japan Science and Technology Agency.

REFERENCES

Agaphonov M., Romanova N., Sokolov S., Iline A., Kalebina T., Gellissen G., and Ter-Avanesyan M. 2005. Defect of vacuolar protein sorting stimulates proteolytic processing of human urokinase-type plasminogen activator in the yeast *Hansenula polymorpha*. *FEMS Yeast Res.* **5:** 1029–1035.

Agrawal R.K., Sharma M.R., Kiel M.C., Hirokawa G., Booth T.M., Spahn C.M., Grassucci R.A., Kaji A., and Frank J. 2004. Visualization of ribosome-recycling factor on the *Escherichia coli* 70S ribosome: Functional implications. *Proc. Natl. Acad. Sci.* **101:** 8900–8905.

Arkov A.L. and Murgola E.J. 1999. Ribosomal RNAs in translation termination: Facts and hypotheses. *Biochemistry* **64:** 1354–1359.

Bourne H.R. 1993. GTPases. A turn-on and a surprise. *Nature* **366:** 628–629.

———. 1995. GTPases: A family of molecular switches and clocks. *Philos. Trans. R. Soc. Lond. B Biol. Sci.* **349:** 283–289.

Brown C.M. and Tate W.P. 1994. Direct recognition of mRNA stop signals by *Escherichia coli* polypeptide chain release factor two. *J. Biol. Chem.* **269:** 33164–33170.

Capecchi M.R. and Klein H.A. 1970. Characterization of three proteins involved in polypeptide chain termination. *Cold Spring Harbor Symp. Quant. Biol.* **34:** 469–477.

Caskey C.T., Bosch L., and Konecki D.S. 1977. Release factor binding to ribosome requires an intact 16 S rRNA 3′ terminus. *J. Biol. Chem.* **252:** 4435–4437.

Caskey C.T., Beaudet A.L., Scolnick E.M., and Rosman M. 1971. Hydrolysis of fMet-tRNA by peptidyl transferase. *Proc. Natl. Acad. Sci.* **68:** 3163–3167.

Caskey T., Scolnick E., Tompkins R., Goldstein J., and Milman G. 1970. Peptide chain termination, codon, protein factor, and ribosomal requirements. *Cold Spring Harbor Symp. Quant. Biol.* **34:** 479–488.

Chavatte L., Frolova L., Kisselev L., and Favre A. 2001. The polypeptide chain release factor eRF1 specifically contacts the s(4)UGA stop codon located in the A site of eukaryotic ribosomes. *Eur. J. Biochem.* **268:** 2896–2904.

Chavatte L., Seit-Nebi A., Dubovaya V., and Favre A. 2002. The invariant uridine of stop codons contacts the conserved NIKSR loop of human eRF1 in the ribosome. *EMBO J.* **21:** 5302–5311.

Czworkowski J. and Moore P.B. 1997. The conformational properties of elongation factor G and the mechanism of translocation. *Biochemistry* **36:** 10327–10334.

Czworkowski J., Wang J., Steitz T.A., and Moore P.B. 1994. The crystal structure of elongation factor G complexed with GDP, at 2.7 Å resolution. *EMBO J.* **13:** 3661–3668.

Dincbas-Renqvist V., Engstrom A., Mora L., Heurgue-Hamard V., Buckingham R., and Ehrenberg M. 2000. A post-translational modification in the GGQ motif of RF2 from *Escherichia coli* stimulates termination of translation. *EMBO J.* **19:** 6900–6907.

Eaglestone S.S., Cox B.S., and Tuite M.F. 1999. Translation termination efficiency can be regulated in *Saccharomyces cerevisiae* by environmental stress through a prion-mediated mechanism. *EMBO J.* **18:** 1974–1981.

Frank J. and Agrawal R.K. 2000. A ratchet-like inter-subunit reorganization of the ribosome during translocation. *Nature* **406:** 318–322.

———. 2001. Ratchet-like movements between the two ribosomal subunits: Their implications in elongation factor recognition and tRNA translocation. *Cold Spring Harbor Symp. Quant. Biol.* **66:** 67–75.

Freistroffer D.V., Kwiatkowski M., Buckingham R.H., and Ehrenberg M. 2000. The accuracy of codon recognition by polypeptide release factors. *Proc. Natl. Acad. Sci.* **97:** 2046–2051.

Freistroffer D.V., Pavlov M.Y., MacDougall J., Buckingham R.H., and Ehrenberg M. 1997. Release factor RF3 in *E. coli* accelerates the dissociation of release factors RF1 and RF2 from the ribosome in a GTP-dependent manner. *EMBO J.* **16:** 4126–4133.

Frolova L., Seit-Nebi A., and Kisselev L. 2002. Highly conserved NIKS tetrapeptide is functionally essential in eukaryotic translation termination factor eRF1. *RNA* **8:** 129–136.

Frolova L., Le Goff X., Zhouravleva G., Davydova E., Philippe M., and Kisselev L. 1996. Eukaryotic polypeptide chain release factor eRF3 is an eRF1- and ribosome-dependent guanosine triphosphatase. *RNA* **2:** 334–341.

Frolova L.Y., Tsivkovskii R.Y., Sivolobova G.F., Oparina N.Y., Serpinsky O.I., Blinov V.M., Tatkov S.I., and Kisselev L.L. 1999. Mutations in the highly conserved GGQ motif of class 1 polypeptide release factors abolish ability of human eRF1 to trigger peptidyl-tRNA hydrolysis. *RNA* **5:** 1014–1020.

Frolova L.Y., Simonsen J.L., Merkulova T.I., Litvinov D.Y., Martensen P.M., Rechinsky V.O., Camonis J.H., Kisselev L.L., and Justesen J. 1998. Functional expression of eukaryotic polypeptide chain release factors 1 and 3 by means of baculovirus/insect cells and complex formation between the factors. *Eur. J. Biochem.* **256:** 36–44.

Fujiwara T., Ito K., Yamami T., and Nakamura Y. 2004. Ribosome recycling factor disassembles the post-termination ribosomal complex independent of the ribosomal translocase activity of elongation factor G. *Mol. Microbiol.* **53:** 517–528.

Gao N., Zavialov A.V., Li W., Sengupta J., Valle M., Gursky R.P., Ehrenberg M., and Frank J. 2005. Mechanism for the disassembly of the posttermination complex inferred from cryo-EM studies. *Mol. Cell* **18:** 663–674.

Gonzalez C.I., Wang W., and Peltz S.W. 2001. Nonsense-mediated mRNA decay in *Saccharomyces cerevisiae:* A quality control mechanism that degrades transcripts harboring premature termination codons. *Cold Spring Harbor Symp. Quant. Biol.* **66:** 321–328.

Graille M., Heurgue-Hamard V., Champ S., Mora L., Scrima N., Ulryck N., van Tilbeurgh H., and Buckingham R.H. 2005. Molecular basis for bacterial class I release factor methylation by PrmC. *Mol. Cell* **20:** 917–927.

Grentzmann G., Brechemier-Baey D., Heurgue-Hamard V., and Buckingham R.H. 1995. Function of polypeptide chain release factor RF-3 in *Escherichia coli.* RF-3 action in termination is predominantly at UGA-containing stop signals. *J. Biol. Chem.* **270:** 10595–10600.

Grentzmann G., Brechemier-Baey D., Heurgue V., Mora L. and Buckingham R.H. 1994. Localization and characterization of the gene encoding release factor RF3 in *Escherichia coli. Proc. Natl. Acad. Sci.* **91:** 5848–5852.

Gromadski K.B. and Rodnina M.V. 2004. Kinetic determinants of high-fidelity tRNA discrimination on the ribosome. *Mol. Cell* **13:** 191–200.

Hansen J.L., Schmeing T.M., Moore P.B., and Steitz T.A. 2002. Structural insights into peptide bond formation. *Proc. Natl. Acad. Sci.* **99:** 11670–11675.

Hansson S., Singh R., Gudkov A.T., Liljas A., and Logan D.T. 2005. Crystal structure of a mutant elongation factor G trapped with a GTP analogue. *FEBS Lett.* **579:** 4492–4497.

Harrison P., Kumar A., Lan N., Echols N., Snyder M., and Gerstein M. 2002. A small reservoir of disabled ORFs in the yeast genome and its implications for the dynamics of proteome evolution. *J. Mol. Biol.* **316:** 409–419.

Heurgue-Hamard V., Champ S., Mora L., Merkulova-Rainon T., Kisselev L.L., and Buckingham R.H. 2005. The glutamine residue of the conserved GGQ motif in *Saccharomyces cerevisiae* release factor eRF1 is methylated by the product of the YDR140w gene. *J. Biol. Chem.* **280:** 2439–2445.

Hirashima A. and Kaji A. 1970. Factor dependent breakdown of polysomes. *Biochem. Biophys. Res. Commun.* **41:** 877–883.

———. 1972a. Factor-dependent release of ribosomes from messenger RNA. Requirement for two heat-stable factors. *J. Mol. Biol.* **65:** 43–58.

———. 1972b. Purification and properties of ribosome-releasing factor. *Biochemistry* **11:** 4037–4044.

———. 1973. Role of elongation factor G and a protein factor on the release of ribosomes from messenger ribonucleic acid. *J. Biol. Chem.* **248:** 7580–7587.

Hirokawa G., Demeshkina N., Iwakura N., Kaji H., and Kaji A. 2006. The ribosome-recycling step: Consensus or controversy? *Trends Biochem. Sci.* **31:** 143–149.

Hirokawa G., Nijman R.M., Raj V.S., Kaji H., Igarashi K., and Kaji A. 2005. The role of ribosome recycling factor in dissociation of 70S ribosomes into subunits. *RNA* **11:** 1317–1328.

Hirokawa G., Kiel M.C., Muto A., Selmer M., Raj V.S., Liljas A., Igarashi K., Kaji H., and Kaji A. 2002. Post-termination complex disassembly by ribosome recycling factor, a functional tRNA mimic. *EMBO J.* **21:** 2272–2281.

Hopfield J.J. 1974. Kinetic proofreading: A new mechanism for reducing errors in biosynthetic processes requiring high specificity. *Proc. Natl. Acad. Sci.* **71:** 4135–4139.

Inge-Vechtomov S., Zhouravleva G., and Philippe M. 2003. Eukaryotic release factors (eRFs) history. *Biol. Cell* **95:** 195–209.

Inokuchi Y., Hirashima A., Sekine Y., Janosi L., and Kaji A. 2000. Role of ribosome recycling factor (RRF) in translational coupling. *EMBO J.* **19:** 3788–3798.

Ito K., Ebihara K., and Nakamura Y. 1998. The stretch of C-terminal acidic amino acids of translational release factor eRF1 is a primary binding site for eRF3 of fission yeast. *RNA* **4:** 958–972.

Ito K., Uno M., and Nakamura Y. 2000. A tripeptide "anticodon" deciphers stop codons in messenger RNA. *Nature* **403:** 680–684.

Ito K., Ebihara K., Uno M., and Nakamura Y. 1996. Conserved motifs in prokaryotic and eukaryotic polypeptide release factors: tRNA-protein mimicry hypothesis. *Proc. Natl. Acad. Sci.* **93:** 5443–5448.

Ito K., Frolova L., Seit-Nebi A., Karamyshev A., Kisselev L., and Nakamura Y. 2002. Omnipotent decoding potential resides in eukaryotic translation termination factor eRF1 of variant-code organisms and is modulated by the interactions of amino acid sequences within domain 1. *Proc. Natl. Acad. Sci.* **99:** 8494–8499.

Ivanov V., Beniaminov A., Mikheyev A., and Minyat E. 2001. A mechanism for stop codon recognition by the ribosome: A bioinformatic approach. *RNA* **7:** 1683–1692.

Janosi L., Hara H., Zhang S., and Kaji A. 1996. Ribosome recycling by ribosome recycling factor (RRF)—An important but overlooked step of protein biosynthesis. *Adv. Biophys.* **32:** 121–201.

Jorgensen F., Adamski F.M., Tate W.P., and Kurland C.G. 1993. Release factor-dependent false stops are infrequent in *Escherichia coli*. *J. Mol. Biol.* **230:** 41–50.

Karimi R., Pavlov M.Y., Buckingham R.H., and Ehrenberg M. 1999. Novel roles for classical factors at the interface between translation termination and initiation. *Mol. Cell* **3:** 601–609.

Keeling K.M., Lanier J., Du M., Salas-Marco J., Gao L., Kaenjak-Angeletti A., and Bedwell D.M. 2004. Leaky termination at premature stop codons antagonizes nonsense-mediated mRNA decay in *S. cerevisiae*. *RNA* **10:** 691–703.

Kervestin S., Frolova L., Kisselev L., and Jean-Jean O. 2001. Stop codon recognition in ciliates: Euplotes release factor does not respond to reassigned UGA codon. *EMBO Rep.* **2:** 680–684.

Kiel M.C., Raj V.S., Kaji H., and Kaji A. 2003. Release of ribosome-bound ribosome recycling factor by elongation factor G. *J. Biol. Chem.* **278:** 48041–48050.

Kisselev L.L. and Buckingham R.H. 2000. Translational termination comes of age. *Trends Biochem. Sci.* **25:** 561–566.

Kisselev L., Ehrenberg M., and Frolova L. 2003. Termination of translation: Interplay of mRNA, rRNAs and release factors? *EMBO J.* **22:** 175–182.

Klaholz B.P., Myasnikov A.G., and Van Heel M. 2004. Visualization of release factor 3 on the ribosome during termination of protein synthesis. *Nature* **427:** 862–865.

Klaholz B.P., Pape T., Zavialov A.V., Myasnikov A.G., Orlova E.V., Vestergaard B., Ehrenberg M., and van Heel M. 2003. Structure of the *Escherichia coli* ribosomal termination complex with release factor 2. *Nature* **421:** 90–94.

Kobayashi T., Funakoshi Y., Hoshino S., and Katada T. 2004. The GTP-binding release factor eRF3 as a key mediator coupling translation termination to mRNA decay. *J. Biol. Chem.* **279:** 45693–45700.

Kolosov P., Frolova L., Seit-Nebi A., Dubovaya V., Kononenko A., Oparina N., Justesen J., Efimov A., and Kisselev L. 2005. Invariant amino acids essential for decoding function of polypeptide release factor eRF1. *Nucleic Acids Res.* **33:** 6418–6425.

Konecki D.S., Aune K.C., Tate W., and Caskey C.T. 1977. Characterization of reticulocyte release factor. *J. Biol. Chem.* **252:** 4514–4520.

Kong C., Ito K., Walsh M.A., Wada M., Liu Y., Kumar S., Barford D., Nakamura Y., and Song H. 2004. Crystal structure and functional analysis of the eukaryotic class II release factor eRF3 from *S. pombe. Mol. Cell* **14:** 233–245.

Kononenko A.V., Dembo K.A., Kiselev L.L., and Volkov V.V. 2004. Molecular morphology of eukaryotic class I translation termination factor eRF1 in solution (transl.). *Mol. Biol.* **38:** 303–311.

Kushnirov V.V., Ter-Avanesyan M.D., Telckov M.V., Surguchov A.P., Smirnov V.N., and Inge-Vechgtomov S.G. 1988. Nucleotide sequence of the SUP2 (SUP35) gene of *Saccharomyces cerevisiae. Gene* **66:** 45–54.

Kurland C.G. and Ehrenberg M. 1985. Constraints on the accuracy of messenger RNA movement. *Q. Rev. Biophys.* **18:** 423–450.

Liang H., Wong J.Y., Bao Q., Cavalcanti A.R., and Landweber L.F. 2005. Decoding the decoding region: Analysis of eukaryotic release factor (eRF1) stop codon-binding residues. *J. Mol. Evol.* **60:** 337–344.

Lill R., Robertson J.M., and Wintermeyer W. 1989. Binding of the 3′ terminus of tRNA to 23S rRNA in the ribosomal exit site actively promotes translocation. *EMBO J.* **8:** 3933–3938.

Menninger J.R. 1971. A simple assay for protein chain termination using natural peptidyl-tRNA. *Biochim. Biophys. Acta* **240:** 237–243.

Merkulova T.I., Frolova L.Y., Lazar M., Camonis J., and Kisselev L.L. 1999. C-terminal domains of human translation termination factors eRF1 and eRF3 mediate their in vivo interaction. *FEBS Lett.* **443:** 41–47.

Mikuni O., Ito K., Moffat J., Matsumura K., McCaughan K., Nobukuni T., Tate W., and Nakamura Y. 1994. Identification of the *prfC* gene, which encodes peptide-chain-release factor 3 of *Escherichia coli. Proc. Natl. Acad. Sci.* **91:** 5798–5802.

Moazed D. and Noller H.F. 1989. Intermediate states in the movement of transfer RNA in the ribosome. *Nature* **342:** 142–148.

Moffat J.G. and Tate W.P. 1994. A single proteolytic cleavage in release factor 2 stabilizes ribosome binding and abolishes peptidyl-tRNA hydrolysis activity. *J. Biol. Chem.* **269:** 18899–18903.

Mora L., Zavialov A., Ehrenberg M., and Buckingham R.H. 2003. Stop codon recognition and interactions with peptide release factor RF3 of truncated and chimeric RF1 and RF2 from *Escherichia coli. Mol. Microbiol.* **50:** 1467–1476.

Nakahigashi K., Kubo N., Narita S., Shimaoka T., Goto S., Oshima T., Mori H., Maeda M., Wada C., and Inokuchi H. 2002. HemK, a class of protein methyl transferase with similarity to DNA methyl transferases, methylates polypeptide chain release factors, and hemK knockout induces defects in translational termination. *Proc. Natl. Acad. Sci.* **99:** 1473–1478.

Nakamura Y. 2001. Molecular mimicry between protein and tRNA. *J. Mol. Evol.* **53:** 282–289.

Nakamura Y. and Ito K. 2002. A tripeptide discriminator for stop codon recognition. *FEBS Lett.* **514:** 30–33.

———. 2003. Making sense of mimic in translation termination. *Trends Biochem. Sci.* **28:** 99–105.

Nakamura Y., Ito K., and Ehrenberg M. 2000. Mimicry grasps reality in translation termination. *Cell* **101:** 349–352.

Nakamura Y., Uno M., Toyoda T., Fujiwara T., and Ito K. 2001. Protein tRNA mimicry in translation termination. *Cold Spring Harbor Symp. Quant. Biol.* **66:** 469–475.

Nakayashiki T., Ebihara K., Bannai H., and Nakamura Y. 2001. Yeast [PSI$^+$] "prions" that are crosstransmissible and susceptible beyond a species barrier through a quasi-prion state. *Mol. Cell* **7:** 1121–1130.

Ness F., Ferreira P., Cox B.S., and Tuite M.F. 2002. Guanidine hydrochloride inhibits the generation of prion "seeds" but not prion protein aggregation in yeast. *Mol. Cell. Biol.* **22:** 5593–5605.

Ninio J. 1975. Kinetic amplification of enzyme discrimination. *Biochimie* **57:** 587–595.

Ogle J.M. and Ramakrishnan V. 2005. Structural insights into translational fidelity. *Annu. Rev. Biochem.* **74:** 129–177.

Pel H.J., Moffat J.G., Ito K., Nakamura Y., and Tate W.P. 1998. *Escherichia coli* release factor 3: Resolving the paradox of a typical G protein structure and atypical function with guanine nucleotides. *RNA* **4:** 47–54.

Peske F., Rodnina M.V., and Wintermeyer W. 2005. Sequence of steps in ribosome recycling as defined by kinetic analysis. *Mol. Cell* **18:** 403–412.

Petry S., Brodersen D.E., Murphy F.V., IV, Dunham C.M., Selmer M., Tarry M.J., Kelley A.C., and Ramakrishnan V. 2005. Crystal structures of the ribosome in complex with release factors RF1 and RF2 bound to a cognate stop codon. *Cell* **123:** 1255–1266.

Polacek N., Gomez M.J., Ito K., Xiong L., Nakamura Y., and Mankin A. 2003. The critical role of the universally conserved A2602 of 23S ribosomal RNA in the release of the nascent peptide during translation termination. *Mol. Cell* **11:** 103–112.

Polevoda B., Span L., and Sherman F. 2005. The yeast translation release factors Mrf1p and Sup45p (eRF1) are methylated, respectively, by the methyltransferases Mtq1p and Mtq2p. *J. Biol. Chem.* **281:** 2562–2571.

Prusiner S.B. 1982. Novel proteinaceous infectious particles cause scrapie. *Science* **216:** 136–144.

Rawat U., Gao H., Zavialov A., Gursky R., Ehrenberg M., and Frank J. 2006. Interactions of the release factor RF1 with the ribosome as revealed by cryo-EM. *J. Mol. Biol.* **357:** 1144–1153.

Rawat U.B., Zavialov A.V., Sengupta J., Valle M., Grassucci R.A., Linde J., Vestergaard B., Ehrenberg M., and Frank J. 2003. A cryo-electron microscopic study of ribosome-bound termination factor RF2. *Nature* **421:** 87–90.

Ruusala T., Ehrenberg M., and Kurland C.G. 1982. Is there proofreading during polypeptide synthesis? *EMBO J.* **1:** 741–745.

Santoso A., Chien P., Osherovich L.Z., and Weissman J.S. 2000. Molecular basis of a yeast prion species barrier. *Cell* **100:** 277–288.

Schmeing T.M., Huang K.S., Kitchen D.E., Strobel S.A., and Steitz T.A. 2005. Structural insights into the roles of water and the 2′ hydroxyl of the P site tRNA in the peptidyl transferase reaction. *Mol. Cell* **20:** 437–448.

Scolnick E., Tompkins R., Caskey T. and Nirenberg M. 1968. Release factors differing in specificity for terminator codons. *Proc. Natl. Acad. Sci.* **61:** 768–774.

Seit-Nebi A., Frolova L. and Kisselev L. 2002. Conversion of omnipotent translation termination factor eRF1 into ciliate-like UGA-only unipotent eRF1. *EMBO Rep.* **3:** 881–886.

Seit-Nebi A., Frolova L., Justesen J., and Kisselev L. 2001. class 1 translation termination factors: Invariant GGQ minidomain is essential for release activity and ribosome binding but not for stop codon recognition. *Nucleic Acids Res.* **29:** 3982–3987.

Seit Nebi A., Frolova L., Ivanova N., Poltaraus A., and Kiselev L. 2000. Mutation of a glutamine residue in the universal tripeptide GGQ in human eRF1 termination factor does not cause complete loss of its activity (transl.). *Mol. Biol.* **34:** 899–900.

Selmer M., Al-Karadaghi S., Hirokawa G., Kaji A., and Liljas A. 1999. Crystal structure of *Thermotoga maritima* ribosome recycling factor: A tRNA mimic. *Science* **286:** 2349–2352.

Serio T.R., Cashikar A.G., Kowal A.S., Sawicki G.J., Moslehi J.J., Serpell L., Arnsdorf M.F., and Lindquist S.L. 2000. Nucleated conformational conversion and the replication of conformational information by a prion determinant. *Science* **289:** 1317–1321.

Shin D.H., Brandsen J., Jancarik J., Yokota H., Kim R., and Kim S.H. 2004. Structural analyses of peptide release factor 1 from *Thermotoga maritima* reveal domain flexibility required for its interaction with the ribosome. *J. Mol. Biol.* **341:** 227–239.

Song H., Mugnier P., Das A.K., Webb H.M., Evans D.R., Tuite M.F., Hemmings B.A., and Barford D. 2000. The crystal structure of human eukaryotic release factor eRF1—Mechanism of stop codon recognition and peptidyl-tRNA hydrolysis. *Cell* **100:** 311–321.

Sprang S.R. 1997a. G protein mechanisms: Insights from structural analysis. *Annu. Rev. Biochem.* **66:** 639–678.

———. 1997b. G proteins, effectors and GAPs: Structure and mechanism. *Curr. Opin. Struct. Biol.* **7:** 849–856.

———. 2001. GEFs: Master regulators of G-protein activation. *Trends Biochem. Sci.* **26:** 266–267.

Stansfield I., Jones K.M., Kushnirov V.V., Dagkesamanskaya A.R., Poznyakovski A.I., Paushkin S.V., Nierras C.R., Cox B.S., Ter-Avanesyan M.D., and Tuite M.F. 1995. The products of the SUP45 (eRF1) and SUP35 genes interact to mediate translation termination in *Saccharomyces cerevisiae*. *EMBO J.* **14:** 4365–4373.

Tate W., Greuer B., and Brimacombe R. 1990. Codon recognition in polypeptide chain termination: Site directed crosslinking of termination codon to *Escherichia coli* release factor 2. *Nucleic Acids Res.* **18:** 6537–6544.

Ter-Avanesyan M.D., Kushnirov V.V., Dagkesamanskaya A.R., Didichenko S.A., Chernoff Y.O., Inge-Vechtomov S.G., and Smirnov V.N. 1993. Deletion analysis of the SUP35 gene of the yeast *Saccharomyces cerevisiae* reveals two non-overlapping functional regions in the encoded protein. *Mol. Microbiol.* **7:** 683–692.

Thompson R.C. and Stone P.J. 1977. Proofreading of the codon-anticodon interaction on ribosomes. *Proc. Natl. Acad. Sci.* **74:** 198–202.

Trobro S. and Aqvist J. 2005. Mechanism of peptide bond synthesis on the ribosome. *Proc. Natl. Acad. Sci.* **102:** 12395–12400.

True H.L. and Lindquist S.L. 2000. A yeast prion provides a mechanism for genetic variation and phenotypic diversity. *Nature* **407:** 477–483.

True H.L., Berlin I., and Lindquist S.L. 2004. Epigenetic regulation of translation reveals hidden genetic variation to produce complex traits. *Nature* **431:** 184–187.

Vestergaard B., Van L.B., Andersen G.R., Nyborg J., Buckingham R.H., and Kjeldgaard M.

2001. Bacterial polypeptide release factor RF2 is structurally distinct from eukaryotic eRF1. *Mol. Cell* **8:** 1375–1382.

Vestergaard B., Sanyal S., Roessle M., Mora L., Buckingham R.H., Kastrup J.S., Gajhede M., Svergun D.I., and Ehrenberg M. 2005. The SAXS solution structure of RF1 differs from its crystal structure and is similar to its ribosome bound cryo-EM structure. *Mol. Cell* **20:** 929–938.

Volkov K.V., Aksenova A.Y., Soom M.J., Osipov K.V., Svitin A.V., Kurischko C., Shkundina I.S., Ter-Avanesyan M.D., Inge-Vechtomov S.G., and Mironova L.N. 2002. Novel non-Mendelian determinant involved in the control of translation accuracy in *Saccharomyces cerevisiae*. *Genetics* **160:** 25–36.

Wang W., Czaplinski K., Rao Y., and Peltz S.W. 2001. The role of Upf proteins in modulating the translation read-through of nonsense-containing transcripts. *EMBO J.* **20:** 880–890.

Wickner R.B. 1994. [URE3] as an altered URE2 protein: Evidence for a prion analog in *Saccharomyces cerevisiae*. *Science* **264:** 566–569.

Wickner R.B., Edskes H.K., Roberts B.T., Baxa U., Pierce M.M., Ross E.D., and Brachmann A. 2004. Prions: Proteins as genes and infectious entities. *Genes Dev.* **18:** 470–485.

Wilson P.G. and Culbertson M.R. 1988. SUF12 suppressor protein of yeast. A fusion protein related to the EF-1 family of elongation factors. *J. Mol. Biol.* **199:** 559–573.

Yusupov M.M., Yusupova G.Z., Baucom A., Lieberman K., Earnest T.N., Cate J.H., and Noller H.F. 2001. Crystal structure of the ribosome at 5.5 Å resolution. *Science* **292:** 883–896.

Zavialov A.V. and Ehrenberg M. 2003. Peptidyl-tRNA regulates the GTPase activity of translation factors. *Cell* **114:** 113–122.

Zavialov A.V., Buckingham R.H., and Ehrenberg M. 2001. A posttermination ribosomal complex is the guanine nucleotide exchange factor for peptide release factor RF3. *Cell* **107:** 115–124.

Zavialov A.V., Hauryliuk V.V., and Ehrenberg M. 2005a. Guanine-nucleotide exchange on ribosome-bound elongation factor G initiates the translocation of tRNAs. *J. Biol.* **4:** 9.

———. 2005b. Splitting of the posttermination ribosome into subunits by the concerted action of RRF and EF-G. *Mol. Cell* **18:** 675–686.

Zavialov A.V., Mora L., Buckingham R.H., and Ehrenberg M. 2002. Release of peptide promoted by the GGQ motif of class 1 release factors regulates the GTPase activity of RF3. *Mol. Cell* **10:** 789–798.

Zhouravleva G., Frolova L., Le Goff X., Le Guellec R., Inge-Vechtomov S., Kisselev L., and Philippe M. 1995. Termination of translation in eukaryotes is governed by two interacting polypeptide chain release factors, eRF1 and eRF3. *EMBO J.* **14:** 4065–4072.

8

Coupled Termination-Reinitiation Events in mRNA Translation

Richard J. Jackson, Ann Kaminski, and Tuija A.A. Pöyry
Department of Biochemistry
University of Cambridge
Cambridge CB2 1GA, United Kingdom

Bicistronic (and polycistronic) mRNA translation is commonplace in eubacteria, although translation of the downstream cistrons may not invariably involve coupled termination-reinitiation events. In contrast, efficient translation of the downstream cistron of a bicistronic mRNA is exceedingly rare in eukaryotes, unless there is an internal ribosome entry site (IRES) or unless the upstream cistron is very short. In this chapter, we first review the eubacterial mechanism of coupled termination-reinitiation and then discuss the reasons that similar events do not generally occur in eukaryotes. We then review the mechanism underlying one of the few exceptional cases of reinitiation on a eukaryotic bicistronic mRNA with a long upstream open reading frame (ORF), and we conclude by examining what it is that is so very different when the ORF is short that it results in reinitiation being the normal (default) outcome.

TRANSLATIONAL COUPLING AND TERMINATION-REINITIATION IN PROKARYOTES

Each cistron of a eubacterial polycistronic mRNA usually has its own Shine-Dalgarno (SD) motif, which should, in principle, allow it to be translated independently. However, in a great many polycistronic mRNAs, there is translational coupling: Initiation of translation of a downstream cistron (cistron $n + 1$) is quite strictly dependent on translation of an upstream cistron, generally the immediate upstream cistron (cistron n). The usual explanation is that the initiation site of cistron $n + 1$ is

occluded by secondary structure base-pairing with the coding region of
cistron *n* (often the far upstream coding region), and so it requires translation of cistron *n* to unwind this secondary structure in order to reveal
the SD-AUG motif of cistron *n* + 1. However, this does not imply an
obligatory termination-reinitiation mechanism, in which the ribosome
that initiates translation of cistron *n* + 1 must necessarily be the ribosome that has just completed translation of cistron *n*.

Bacteriophage RNAs of the MS2, R17, f2, and fr family provide a
useful example of general translational coupling and its distinction from
an obligatory termination-reinitiation event. The only initiation site accessible in the virion RNA (whether on addition to a cell-free system or in
intact *Escherichia coli* cells immediately following infection) is that of
the central (130-codon) coat protein cistron (Fig. 1A). The 5′-proximal
"maturation protein" cistron initiation site is occluded in local secondary
structure (Groenveld et al. 1995); the initiation site of the 3′-proximal
synthetase cistron is also occluded, but in this case via long-range base-
pairing with the codon 24–32 region of the coat protein cistron (Min Jou
et al. 1972; Berkhout and van Duin 1985). Consequently, it requires translation of the coat protein cistron past codon 30 to unwind this long-range
interaction to allow ribosome access to the synthetase initiation site.
Thus, translation of viral RNA with a nonsense mutation (premature
termination codon) at codon 6 of the coat protein cistron fails to show

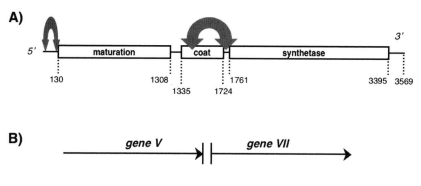

Figure 1. (*A*) Diagrammatic representation of bacteriophage MS2 virion RNA
cistrons, with the overlapping lysis gene (nucleotides 1678–1902) omitted for clarity.
The arrows show the RNA–RNA base-pairing interactions that occlude the initiation
sites of the maturation protein and RNA synthetase cistrons. (*B*) Sequence of the
junction between bacteriophage f1 mRNA cistrons coding for gene V and gene VII
proteins, with the gene V termination codon shown in italics, the gene VII initiation
codon in bold, and the vestigial Shine-Dalgarno motif underlined.

any expression of the downstream RNA synthetase cistron, but synthetase is expressed efficiently from a viral RNA with a nonsense mutation at codon 50 (Lodish and Zinder 1966; Tooze and Weber 1967; Min Jou et al. 1972). In the latter case, it is intuitively improbable that the ribosome (or its 30S subunit) which initiates translation at the synthetase initiation site is invariably the self-same 70S ribosome/30S subunit that had just previously terminated translation no less than approximately 240 nucleotides further upstream, and much more likely to be initiated by ribosomes recruited de novo from the free ribosome pool.

There are examples, however, of translational coupling which have a different explanation that is likely to require obligatory termination-reinitiation. In these cases, the downstream cistron has an extremely weak, almost vestigial, SD interaction and is unable to recruit ribosomes de novo at even low efficiency. The classic example is found in bacterio-phage f1 mRNA, where expression of the downstream gene VII cistron is absolutely dependent on translation of the upstream gene V cistron (Ivey-Hoyle and Steege 1989, 1992), whose termination codon is extremely close to the gene VII AUG initiation codon (Fig. 1B). In this case, it seems unlikely that secondary structure is the major reason for the obligatory coupling. Rather, the evidence suggests that the extremely weak SD motif of gene VII (..GG..) is only able to capture 30S subunits if they are delivered close by via termination of gene V translation. The efficiency of such capture is nevertheless low, <10% (Ivey-Hoyle and Steege 1992). The low expression of gene VII is reduced even further if the gene V termination codon is displaced further upstream (Ivey-Hoyle and Steege 1989). Conversely, mutations that progressively improve the gene VII SD motif toward the consensus increase gene VII expression (especially with additional mutations reducing the GC content of the rest of the initia-tion site) and make it more and more independent of gene V translation (Ivey-Hoyle and Steege 1992).

A similar type of translational coupling required for the synthesis of the lysis protein of MS2 and fr RNA bacteriophages has been studied in detail by van Duin's group (Adhin and van Duin 1990). These authors concluded that reinitiation (1) could occur anywhere within a window of about 40 nucleotides on either side of the stop codon, with a slight pref-erence for downstream sites rather than upstream; (2) showed a strong preference for the nearest suitable site, even to the extent of preferring a nearby UUG or GUG codon over a more distant AUG; and (3) needed at least a vestigial SD motif appropriately close to the reinitiation codon. This has echoes of the reinitiation on *lacI* mRNA with nonsense mutations in the 5′ part of the gene, which likewise occurred in some cases at UUG and GUG codons with minimal SD motifs (2–3 residues) that were silent as initiation sites in the absence of the nearby stop codon (Steege 1977).

POSTTERMINATION EVENTS IN PROKARYOTIC SYSTEMS

Adhin and van Duin (1990) suggested that following termination, the ribosome (or more probably its small subunit) could undergo bidirectional diffusion for a very brief period of time, but it would soon dissociate from the mRNA unless some stabilizing interaction were brought into play. An SD motif, even a weak one, could provide such a stabilizing effect and pause the diffusing ribosome for sufficient time for it to acquire fMet-tRNA$_f$ if there was a potential initiation codon at an appropriate distance (Spanjaard and van Duin 1989). Many years previously, Sarabhai and Brenner (1967), in a veritable intellectual tour de force based solely on genetic evidence, had proposed that following termination (at a nonsense mutant site in bacteriophage T4 rII mRNA), ribosomes underwent "phaseless wandering" on the mRNA for a limited distance and could reinitiate translation within the window covered by this wandering. They deduced that reinitiation was probably very inefficient (1–3%) and could occur in any frame either upstream or downstream from the stop codon and that the reinitiation site was silent in the absence of the nearby termination codon.

How far is the Adhin and van Duin model (based mainly on the preferences for the site of reinitiation) consistent with what is now known about the mechanism of disassembly of the posttermination 70S ribosome (for a more detailed discussion, see Chapter 7)? According to a kinetic analysis using fluorescent ligands and stopped-flow plus quench-flow methods, this disassembly takes place in the following sequence of events (Peske et al. 2005): (1) In Step 1, the combined action of ribosome release factor (RRF) and EFG-GTP causes relatively rapid dissociation of the 50S subunit, leaving the 30S subunit/deacylated tRNA complex still bound to the same site on the mRNA; (2) in Step 2, which is approximately tenfold slower, the binding of IF3 to the 30S subunit ejects the deacylated tRNA; and (3) in Step 3, the 30S subunit, presumably still with bound IF3, dissociates from the mRNA at a rate that was only marginally, and perhaps not significantly, slower than the Step 2 rate.

What do these kinetic analyses say about the Adhin and van Duin bidirectional diffusion model? If the deacylated tRNA remains in the 30S subunit P site throughout this process, it seems most unlikely that any bidirectional diffusion could occur before Step 2 had been completed, as the codon–anticodon interaction in the 30S subunit P site would likely act as a strong restraint against any movement. If this assumption is correct, then the near equality of the Step 2 and Step 3 rates does not necessarily refute the Adhin and van Duin model, but does strongly imply that any diffusion can only be extremely limited in time (and presumably therefore also in distance).

If posttermination dissociation of the ribosome into subunits were to be slow, for example, if RRF were limiting, could the ribosome remain mRNA-associated and reinitiate at a nearby site as a 70S ribosome? De novo initiation by 70S ribosomes certainly can occur in special circumstances, such as on leaderless mRNAs (O'Donnell and Janssen 2002; Moll et al. 2004), but for reinitiation by posttermination 70S ribosomes, there would need to be a special mechanism to eject the deacylated tRNA.

WHY DOES TERMINATION-REINITIATION AFTER A LONG ORF GENERALLY NOT OCCUR IN EUKARYOTES?

The conventional wisdom is that the eukaryotic cytoplasmic translation system cannot execute termination-reinitiation after translation of a long ORF. Despite this conventional wisdom, in bicistronic mRNA assays for IRESs, the downstream cistron expression in the control lacking an IRES is not absolutely zero, even though it is extremely low. Moreover, the fact that this low level can be reduced by inserting a large nonfunctional mutant picornavirus IRES between the two cistrons (Johannes et al. 1999) suggests that the low level of downstream cistron expression may be largely due to ribosomes that reinitiate translation after terminating upstream cistron translation. The efficiency of this putative termination-reinitiation event is unknown, since absolute (i.e., molar) yields of the two translation products are seldom reported, but the general impression is that it is more in the 0.1−1.0% range than in the 1−10% range. In the remainder of this chapter, we shall ignore the embarrassing fact of low expression of the downstream cistron while trying not to completely deny its existence.

In our attempts to explain why termination-reinitiation after translation of a long ORF in eukaryotes is generally so rare (and/or so extremely inefficient) compared to eubacteria, we have the huge disadvantage of knowing virtually nothing about the mechanism of disassembly of the eukaryotic posttermination ribosome. There is no obvious homolog of RRF in eukaryotic cytoplasm, and although it has been speculated that eRF3 (eukaryotic release factor 3) might do the same job (Buckingham et al. 1997), this remains unproven. This speculation is based on no more than the fact that, like RRF, eRF3 is essential for viability, whereas prokaryotic RF3 is not and that eRF3 is larger than either RRF or RF3 and so might conceivably have both functions. However, most of the essential domains of the approximately 630-amino acid eRF3 are already accounted for. The amino-terminal approximately 200-amino acid domain not only is very different between mammals and yeast, where it has prion-like properties (for review, see Chapter 7), but also even shows variation between the two mammalian eRF3 isoforms (Hoshino et al. 1998) and is absent from *Giardia lamblia* eRF3 (Inagaki and Ford Doolittle 2000). Despite the fact that this domain in both yeast and mammalian eRF3 interacts with the poly(A)-binding protein (Hoshino et al. 1999), there is good evidence that it is not essential for viability in yeast (Ter-Avanesyan et al. 1993; Zhouravleva et al. 1995).

The essential part of eRF3 is thus the carboxy-terminal approximately 430-amino acid domain that is very highly conserved between the two mammalian eRF3 isoforms and strongly conserved between mammalian and yeast factors (Hoshino et al. 1998), and it has strong sequence and crystal structure homology with eEF1A (Kong et al. 2004). This carboxy-terminal domain interacts with eRF1 and also with the nonsense-mediated decay factors Upf1-3. Despite the strong homology with eEF1A, there is no evidence that eRF3 binds an aminoacyl-tRNA, but it does exhibit GTP-dependent interaction with eRF1 (Kobayashi et al. 2004), which has structural mimicry with tRNA (Song et al. 2000). Given the well-known fact that EF-Tu and EF-G (or eEF1A and eEF2) cannot simultaneously bind to the same ribosome (because they share a common binding site), it seems very probable that eRF3 and eEF2 cannot bind simultaneously to the eukaryotic ribosome in a manner that RRF and EFG evidently can with bacterial ribosomes. Thus, the fine details of the mechanism of post-termination release of ribosomes from the mRNA may be different in eukaryotes and prokaryotes. We also do not know whether initiation factors are involved in disassembly of the posttermination 80S ribosome, similar to the role of IF3 in prokaryotes. Possible candidates would be eIF3, eIF1A, and perhaps eIF1.

Even if the fine details may, however, be different between prokaryotes and eukaryotes, it seems likely that the final outcome is very similar. In particular, there is good evidence that in reticulocyte lysate systems capable of multiple cycles of translation on endogenous globin mRNA, the ribosomes released at termination do not pass through the 80S monomeric ribosome pool before reinitiating translation (Adamson et al. 1970). Therefore, the disassembly of the posttermination ribosome is likely to involve sequential release of the two ribosomal subunits, leaving a (probably) short-lived association of the 40S subunit with the mRNA in a state where it could undergo bidirectional diffusion, albeit only for a very brief period of time. Since reinitiation in prokaryotes requires at least a vestigial SD motif, presumably to stabilize the 30S/mRNA interaction, thereby delaying 30S subunit dissociation from the mRNA and to pause the diffusing 30S subunit at an appropriate position for reinitiation once fMet-tRNA$_f$ has been acquired, the apparent lack of reinitiation in eukaryotes is likely to be due to the absence of anything akin to the SD interaction to retain and restrain the 40S subunit on the mRNA. Thus, dissociation of the subunit from the mRNA is likely to be the normal default outcome in eukaryotes. It follows that in the rare cases where reinitiation does occur at reasonable efficiency, there must almost certainly be some mechanism, most likely requiring a specific *cis*-acting RNA element, that fulfills a role similar to that of the prokaryotic SD interaction.

EXCEPTIONAL CASES OF TERMINATION-REINITIATION AFTER TRANSLATION OF A LONG EUKARYOTIC ORF

The subgenomic RNAs (sgRNA) of caliciviruses probably represent the best-characterized example of reinitiation by mammalian ribosomes after translating a long ORF. Caliciviruses fall into two classes according to whether the full-length genomic (virion) RNA is tricistronic or bicistronic (Clarke and Lambden 2000). In the former case, exemplified by feline calicivirus (FCV), it is thought that only the 5′-proximal cistron A is expressed from the virion RNA (Fig. 2A). This encodes a polyprotein with a protease module that processes the polyprotein to give all the individual nonstructural proteins, including an RNA replicase, and also processes the major capsid precursor. The RNA replicase generates not only additional copies of full-length viral RNA, but also a single species of subgenomic RNA, which is bicistronic, with the upstream sgORF-1 (identical to ORF-B of the virion RNA) encoding an approximately 70-kD major capsid precursor, whereas sgORF-2 (identical to ORF-C of the virion RNA) encodes a putative minor capsid protein produced in about 15% relative molar yield (Herbert et al. 1996; Wirblich et al. 1996). FCV sgORFs 1 and 2 overlap by 4 nucleotides (..**AUGA**..), whereas in other caliciviruses, the overlap ranges from 1 to 8 nucleotides (Fig. 2A).

In the bicistronic class, exemplified by rabbit hemorrhagic disease virus (RHDV), ORF-B, including its potential initiation codon, has become fused in-frame to ORF-A (Clarke and Lambden 2000), and translation of the full-length virion RNA results in the synthesis of both the AB fusion protein and the ORF-C product (Wirblich et al. 1996; Konig

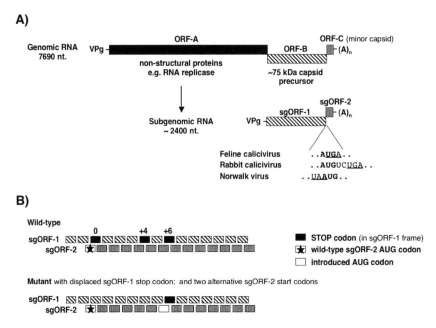

Figure 2. (*A*) Depiction of the ORFs in FCV genomic (virion) RNA and the single subgenomic (sg)RNA generated during FCV infection, with the overlap region between sgORFs 1 and 2 found in other species of calicivirus also shown. (*B*) Explanatory diagram showing the disposition of termination codons in the wild-type sgORF-1 frame; and the mutant in which the ORF-1 termination codon was displaced six codons downstream, and an additional potential (re)initiation codon introduced into the sgORF-2 frame overlapping the displaced sgORF-1 termination codon.

et al. 1998). Despite this difference, RHDV infection leads to the production of a subgenomic RNA that is an exact counterpart of FCV sgRNA, and it is likely that capsid protein synthesis in the infected cell is overwhelmingly from translation of this sgRNA rather than the full-length genomic RNA. An unusual feature of both the full-length and subgenomic RNAs of caliciviruses is that they do not have a 5′ cap, but instead a covalently linked approximately 10-kD VPg that is encoded within ORF-A (Herbert et al. 1997). Nevertheless, it is thought that initiation on these RNAs is by a ribosome scanning mechanism virtually identical to that operating with conventional capped mRNAs, although obviously there must be some difference in the details of how the 5′ end is recognized by initiation factors (Goodfellow et al. 2005).

The translation of RHDV sgRNA has been studied in transfection assays by Meyers (2003), and we have examined FCV sgRNA translation in vitro (T.A.A. Pöyry et al., unpubl.), where time-course assays provided direct evidence for a termination-reinitiation event for FCV sgORF-2 expression. The approximately 70-kD sgORF-1 product first appeared long before the much smaller sgORF-2 product, and addition of an initiation inhibitor (edeine) at various times showed that initiation of sgORF-2 translation started only after the first ribosomes had fully completed translation of the upstream sgORF-1.

In the RHDV system, sgORF-2 expression required the terminal 84 nucleotides of sgORF-1, but no other virus-coding sequences (Meyers 2003). Detailed mapping revealed a remarkably sharp boundary: sgORF-2 expression was completely abolished if only the terminal 75 nucleotides of sgORF-1 were retained, and it was reduced fivefold with just 81 or 78 nucleotides of sgORF-1. Translation through this essential 84-nucleotide element was needed, as displacing the sgORF-1 stop codon toward the 5′ end of the mRNA abolished sgORF-2 expression (Meyers 2003). However, engineering a frameshift through most of the 84-nucleotide segment had no effect on reinitiation efficiency, implying that it is required by virtue of its RNA sequence or structure, rather than for its encoded peptide (Meyers 2003). Also of interest is the fact that the efficiency of the putative reinitiation event leading to sgORF-2 expression from RHDV sgRNA was reduced only threefold when the AUG (re)initiation codon was mutated to ACG (Meyers 2003). FCV sgORF-2 expression shows very similar requirements for just the terminal 84 nucleotides of sgORF-1 (the terminal 72 nucleotides are insufficient) and a similar tolerance to mutation of the sgORF-2 initiation codon to non-AUG codons.

Although FCV sgORF-2 expression was abolished if the sgORF-1 termination codon was displaced toward the 5′ end of the sgRNA, a limited downstream displacement by up to about 10 codons was tolerated, and in this background, we could introduce an alternative (re)initiation codon in the sgORF-2 frame (Fig. 2B) overlapping the displaced sgORF-1 stop codon (..**AUG**A..). Strikingly, in these circumstances, all (re)initiation of sgORF-2 expression still occurred at the original sgORF-2 AUG (re)initiation codon, rather than at the introduced AUG. Thus, the reinitiation codon must be selected by virtue of its juxtaposition to the critical 84-nucleotide element at the end of sgORF-1, rather than by virtue of its juxtaposition to the sgORF-1 stop codon (Fig. 2B). By UV cross-linking assays coupled with immunoprecipitations, the 84-nucleotide segment was shown to bind eIF3, and when reticulocyte lysates were supplemented with purified HeLa cell eIF3, there was a quite significant increase in sgORF-2 expression but little effect on sgORF-1 translation, with the result that the sgORF-2/ORF-1 product yield ratio increased.

As we had predicted in the previous section, there *is* a requirement for a specific *cis*-acting RNA element for reinitiation on the FCV and RHDV sgRNAs. This element seems to bind eIF3, which could serve to capture the posttermination 40S subunit, via the well-known eIF3–40S subunit interaction (Korneeva et al. 2000), and restrain it in an appropriate position relative to the (re)initiation site. Capture by eIF3 (bound to the 84-nucleotide element) would also provide a plausible explanation for the observation that reinitiation in the FCV system did not require eIF4G and was refractory to dominant negative eIF4A mutants.

Another potential case of reinitiation dependent on a specific *cis*-acting RNA element is the bicistronic M2 RNA of respiratory syncytial virus (Ahmadian et al. 2000; Gould and Easton 2005), which is a member of the Pneumoviridae family and unrelated to caliciviruses. In this case, the overlap between the two ORFs is larger, 26 or 32 nucleotides (there are two potential AUGs for reinitiation of translation), and the required *cis*-acting element, identified as the 3′-terminal 150–200 nucleotides of the upstream ORF, is longer. As in the case of FCV subgenomic RNA, downstream displacement of the termination codon by 14 codons severely compromises reinitiation.

Other examples of bicistronic mRNAs where the downstream cistron is possibly translated by a termination-reinitiation event are the M RNA (segment 7) of the influenza B viruses, where the overlap is ..UA**AUG**.. (Horvath et al. 1990), and non-long terminal repeat retrotransposons, some of which have ..UA**A**UG.. overlaps, and others ..**AUG**A.. (Kojima et al. 2005). In both cases, the proximity of the termination and restart codons was critical for downstream cistron expression. In addition, a downstream secondary structure element in the retrotransposon RNA was suggested to be important for the putative termination-reinitiation event. Thus far, the only candidate bicistronic mRNA of nonviral origin where a similar termination-reinitiation might conceivably operate is an embryonic alternative splice variant of the murine glutamic acid decarboxylase GAD 67 mRNA, where there is a ..UG**AUG**.. overlap (Szabo et al. 1994).

Although all the above examples concern bicistronic mRNAs, there is just one situation where translation of polycistronic mRNAs has been claimed to occur independent of any *cis*-acting RNA element (including IRESs) but dependent on a *trans*-acting protein. These functionally polycistronic mRNAs are transcribed from the cauliflower mosaic virus genome, and their translation by what are likely to be termination-reinitiation events requires the immediate-early virus-encoded protein TAV (*trans*-activator/viroplasmin). It has been suggested that TAV interacts with eIF3, specifically with the eIF3g subunit, leading to stabilization of the eIF3–ribosome interaction, so that it persists throughout the elongation phase of translation and thereby facilitates reinitiation (Ryabova et al. 2004).

REINITIATION AFTER A SHORT ORF IN EUKARYOTES: SIMILARITIES AND DIFFERENCES BETWEEN YEAST AND HIGHER EUKARYOTES

In contrast to reinitiation after translation of a long ORF, which seems an extremely rare event, reinitiation after a short ORF (denoted here as uORF for upstream ORF) is quite common in higher eukaryotes, although it does not necessarily occur with every uORF. When it does occur, this reinitiation involves resumption of scanning by the ribosome (or more likely its 40S subunit) following translation of the uORF. No less than approximately 35% of mammalian mRNAs have at least one short upstream ORF (Peri and Pandey 2001; Iacono et al. 2005). On the other hand, notwithstanding the fact that yeast GCN4 mRNA, which has four uORFs, has been the focus of a massive amount of research, ORFs are very much rarer in yeast and occur in possibly less than 1% of all yeast mRNAs (Zhang and Dietrich 2005). Indeed, the Sherman group has argued that with uORFs in yeast mRNAs, rescanning and subsequent reinitiation (as occurs with GCN4 uORF-1) is the exception rather than the rule, whereas the converse seems true in higher eukaryotes (Yun et al. 1996). Therefore, some caution should be exercised in making uncritical extrapolations from yeast to mammalian reinitiation mechanisms. It is not clear whether such differences as occur are due to differences in the details of the termination and posttermination events in the two systems or to differences in initiation factors; for example, the fact that yeast eIF3 consists of only six polypeptides (Valásek et al. 2001) and only makes indirect interactions with eIF4G, bridged by eIF1 and eIF5 (He et al. 2003), whereas mammalian eIF3 has an additional seven polypeptides (Kolupaeva et al. 2005) and can bind directly to the central domain of eIF4G (Korneeva et al. 2000).

One feature of yeast GCN4 mRNA that has not been recorded with mammalian mRNAs is that resumption of scanning after translation of the first uORF is strongly influenced by 5′UTR sequences extending as much as 160 nucleotides upstream of uORF-1 (Grant et al. 1995), for reasons that remain obscure. Of the other three uORFs, uORF-4 is certainly nonpermissive for rescanning, as is also likely to be the case with uORF-3, whereas the status of uORF-2 is somewhat uncertain. Moreover, uORF-4 retains its restrictive property even when it and its immediate flanking sequences are transferred to the privileged position of uORF-1.

A detailed mutagenesis screen showed that it is the combination of the last sense codon and the ten residues downstream from the stop codon that determine the nonpermissive character of uORF-4. For the downstream element, a whole range of largely AU-rich sequences were found to allow resumption of scanning, but introduction of sufficient G and C residues could prevent it; for the last sense codon, it was likewise codons consisting mainly or exclusively of G and C residues that were most restrictive, with all four proline codons among the worst (Grant and Hinnebusch 1994). It was suggested that these restrictive sequences might cause a delay in the termination process (or perhaps in the disassembly of the posttermination ribosome), leading to a failure to resume scanning (Grant and Hinnebusch 1994). In general, changes in the sequence immediately downstream from the uORF have not been found to exert such a strong influence on reinitiation efficiency in mammalian systems, although there is one report where they did (Lincoln et al. 1998).

One very important feature of rescanning, common to both yeast and higher eukaryotes, is that after termination of translation of a permissive uORF, the 40S subunits that resume scanning actually start migrating without any associated eIF2/GTP/Met-tRNA$_i$ ternary complex, and they acquire a ternary complex during the course of scanning. Thus, an associated eIF2 ternary complex is not a prerequisite for the actual migration of the rescanning 40S subunit, but it is, of course, obligatory for recognition of a downstream initiation codon, an event that is largely dependent on codon–anticodon pairing. Thus, if the spacing between the uORF stop codon and the reinitiation codon is very short, as in the HER-2 mRNA where it is just 5 nucleotides (Child et al. 1999), reinitiation efficiency is extremely low, but it can be increased by expanding the spacing (Kozak 1987; Child et al. 1999). Surprisingly, the low efficiency of (re)initiation at the HER-2 ORF start site proper is significantly augmented, especially in tumor cells, if the construct includes the HER-2 3′UTR (Mehta et al. 2006). The HER-2 3′UTR was shown to bind HuR, heterogeneous nuclear ribonucleoprotein (hnRNP) A1, and hnRNP C1/C2, but it remains a complete mystery how this 3′UTR–protein complex could increase the apparent efficiency of reinitiation at a site such a short distance downstream from the uORF stop codon (Mehta et al. 2006; Sachs and Geballe 2006).

As acquisition of an eIF2/GTP/Met-tRNA$_i$ ternary complex occurs during the course of 40S subunit rescanning, a decrease in the availability of ternary complexes (as a result of either eIF2α phosphorylation, or debilitating point mutations in eIF2B, or a reduction in initiator tRNA gene copy number) delays ternary complex acquisition with the consequence that the rescanning 40S subunit will have traveled further before

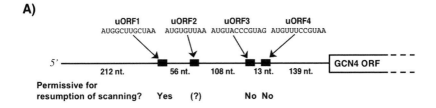

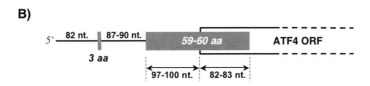

Figure 3. (*A*) Schematic diagram of the 5′UTR of *S. cerevisiae GCN4* mRNA, showing the sequences of the four uORFs and the spacing between the different ORFs and between the 5′ end and uORF-1. (*B*) Schematic diagram of mammalian *ATF4* mRNA, with upstream uORFs shown as *gray* boxes, and the *ATF4* ORF as an *open rectangle*. Based on the sequences of the human, mouse, rat, and bovine *ATF4* mRNAs, the coding potential (sense codons) of the two uORFs is shown, as are the distances between the 5′ end and uORF-1, between uORF-1 and uORF-2, and between the start of ORF-2 and the start of the *ATF4* ORF (from Vattem and Wek 2004). A very similar configuration, with only slight differences in the length of uORF-2 and the spacing between uORFs 1 and 2, occurs in chicken and zebra fish *ATF4* mRNA.

regaining competence for initiation codon recognition. The kinetics of reacquisition of a ternary complex are the key to the approximately tenfold up-regulation of yeast Gcn4p synthesis in response to amino acid starvation, which activates the *GCN2* eIF2 kinase (Chapter 9), and the fivefold increase in *ATF4* mRNA translation in response to thapsigargin-induced endoplasmic reticulum (ER) stress that results in activation of the PERK eIF2 kinase (Chapter 13).

Whereas yeast *GCN4* mRNA has four uORFs (Fig. 3A), vertebrate *ATF4* mRNA has just two upstream ORFs, highly conserved in size and position across different species (Fig. 3B). The opposing characteristics of these two upstream ORFs is shown by the fact that mutation of the initiation codon of the 5′-proximal uORF results in constitutively low *ATF4* expression, unresponsive to thapsigargin, whereas mutation of the second ORF start codon results in very high constitutive expression (Chapter 13). Because this second ORF is relatively long and overlaps the *ATF4* ORF quite extensively, ribosomes that translate it will be unable to access the *ATF4* ORF initiation site on the same round of translation. A reduction in the intracellular concentration of eIF2 ternary complex will decrease the probability of any rescanning 40S subunits acquiring a ternary complex before reaching the second ORF initiation codon, and consequently, there will be an increase in the probability of reinitiation at the *ATF4* ORF site (Vattem and Wek 2004). Thus, with both *GCN4* and *ATF4* mRNAs, the essential determinants of translational control are (1) the distances between the various potential initiation codons; (2) the fact that the 5′-proximal uORF is not only recognized efficiently by scanning 40S subunits, but is also permissive for resumption of scanning; and (3) the existence of one or more short (or fairly short) ORFs further downstream whose translation results in the ribosome dissociating from the mRNA without engaging the main ORF initiation codon.

WHAT IS THE MECHANISM OF REINITIATION AFTER TRANSLATION OF A uORF IN MAMMALIAN SYSTEMS?

Reinitiation efficiency in mammalian systems falls off quite abruptly if the length of the uORF is increased beyond approximately 20 codons (Luukkonen et al. 1995; Kozak 2001). Moreover, it is abolished if a uORF that would otherwise be short enough to allow a reasonable level of reinitiation includes a pseudoknot structure that would be expected to cause the ribosomes translating the uORF to pause (Somogyi et al. 1993; Kontos et al. 2001; Kozak 2001). The efficiency of reinitiation is also decreased by low concentrations of cycloheximide sufficient to slow the rate of elongation, but not to cause complete arrest (Pöyry et al. 2004).

These results suggest that the critical parameter is not the length of the uORF per se, but the time taken to complete its translation. This, in turn, leads to the idea that resumption of scanning might depend on some of the factor–ribosome interactions that promoted the primary initiation event at the uORF AUG persisting for the few seconds necessary to complete translation of the uORF (including the termination step). However, even if we knew which are the critical initiation factors that might have to maintain contact with the ribosome during translation of the uORF, there is currently no satisfactory way of proving that their interactions with the ribosome do in fact persist for this very brief period. It would probably require the development of fluorescence methods for detecting factor–ribosome interactions in a real-time kinetic analysis by stopped-flow methods.

It occurred to us that it might be possible to solve at least part of this conundrum by reformulating the question to ask whether reinitiation is dependent on precisely which factors were involved in the initiation of translation of the uORF, the primary initiation event. To this end, we took a construct with a short ORF and a reasonably efficient downstream reinitiation site and replaced the 5′UTR with various viral IRESs or other 5′UTR sequences with different initiation factor requirements (Pöyry et al. 2004). We found that efficient reinitiation (in the range 25–35%) occurred only if eIF4F participated in this primary initiation event, although the full-length eIF4G polypeptide could be replaced by the central one-third eIF4G fragment (p50) or the carboxy-terminal two-thirds fragment (p100), both of which include the critical central domain that has sites for interaction with eIF4A and eIF3 (Korneeva et al. 2000).

These observations suggest that for reinitiation to occur after translation of a uORF, the interactions between the ribosome and eIF4F must persist, perhaps somewhat rearranged, throughout the time taken to complete translation of the uORF, and as eIF4F appears to have only indirect interactions with the 40S subunit, bridged by eIF3 (Korneeva et al. 2000), it seems likely that some of the eIF3–40S subunit contacts would also need to be very briefly maintained. If these interactions are still in place by the time translation of the uORF (including the termination step) has been completed, they could promote resumption of scanning; but if the interactions had been completely disrupted by the time termination of uORF translation has been completed, then ribosome release from the mRNA would be the likely default outcome.

This model is something of a heresy, since conventional wisdom has it that all the initiation factors dissociate from the 40S subunit at, or momentarily before, the ribosomal subunit joining step. The most recent analysis actually suggested that only eIF2–GDP dissociates following eIF5-triggered GTP hydrolysis, but that all the other factors dissociate at the final eIF5B-dependent step (Unbehaun et al. 2004). However, this and all the earlier investigations into this issue have involved much longer timescales than the few seconds that we are postulating to be the length of time that the critical factor–ribosome interactions are retained following initiation at the uORF AUG.

Clearly, it is quite impossible for *all* the initiation factors to be retained on the ribosome during translation of the uORF. Those factors that are thought to bind to the 40S subunit on the subunit interface must surely be all released to allow ribosome subunit joining. There is direct structural evidence that eIF1 is in this category (Lomakin et al. 2003), and eIF1A can also be included by analogy with its prokaryotic paralog (Carter et al. 2001). In addition, biochemical evidence suggests that both eIF2 and eIF5 bind to the subunit interface side of the 40S subunit

(Valásek et al. 2003; Unbehaun et al. 2004). As for eIF5B, by analogy with its prokaryotic paralog, IF2 (Marzi et al. 2003), it may well remain ribosome-associated until after subunit joining has occurred, but it almost certainly has to dissociate before elongation can commence, so as to allow elongation factors access to the ribosomal A site.

The two factors with interaction sites on the 40S subunits that make them reasonable candidates for very brief retention after initiation of uORF translation are eIF3 and eIF4G (or the complete eIF4F complex), precisely those that we think would be most important in our model. Biochemical "docking" assays suggest that the bulk of the six-subunit yeast eIF3 binds to the back, or solvent face, of the 40S subunit, with only small incursions into the subunit interface side (Valásek et al. 2003). A similar picture emerges from recent cryo-electron microscopy (cryo-EM) studies, where mammalian eIF3 (which consists of 13 polypeptides) was seen as resembling the shape of a (five-pointed) starfish, most of which was interacting with the solvent face behind the platform, with just one of the limbs (the so-called "left leg") wrapping round and slightly invading the subunit interface (Siridechadilok et al. 2005). This suggests that with a small displacement of this limb, interactions of the other limbs of eIF3 with the solvent face of the 40S subunit might allow eIF3 to remain briefly associated with the 40S subunit after ribosomal subunit joining. This postinitiation eIF3–40S subunit interaction would likely be metastable, but that is all that our model requires.

The cryo-EM reconstructions also placed eIF4G in a location slightly behind and above the platform and 40S-bound eIF3, and not significantly intruding into the subunit interface (Siridechadilok et al. 2005), suggesting that it, too, could possibly be retained after subunit joining, albeit somewhat rearranged and in a metastable direct interaction with eIF3, and hence interacting indirectly with the 40S subunit. An interesting implication of this eIF4G location on the trailing (5′) side of the ribosome is that scanning is likely to involve the eIF4F complex "pushing" the 40S subunit along the mRNA from the rear, rather than "leading" from the front.

Put in the most generalized terms possible, our results show that the ribosome (or probably just its 40S subunit) that initiates translation of the uORF is somehow "licensed" for reinitiation, but only if the complete eIF4F complex (or at least p50/eIF4A or p100/eIF4A) was involved in this primary initiation event, and that the 40S subunit rapidly loses this "license" over time. Within these rather lax constraints, it is obvious that there are many possible alternative explanations, but we consider our model to be the simplest consistent with the facts. It also has the appealing feature that the brief retention of some eIF3–ribosome interactions during uORF translation could not only help retain the 40S subunit on the mRNA, but also lead to the re-recruitment (during the resumed 40S subunit-scanning stage) of eIF1, eIF5, and an eIF2/GTP/Met-tRNA$_i$ ternary complex, which have all been shown to interact directly or indirectly with eIF3 (Asano et al. 2001).

CIRCUMSTANCES IN WHICH A uORF MAY TRIGGER RIBOSOME SHUNTING

Although translation of a uORF is normally followed by strictly linear 40S subunit scanning, investigations of cauliflower mosaic virus 35S mRNA translation have revealed that in some circumstances, it can result in ribosome shunting, a form of discontinuous scanning in which the ribosome (or more likely just the 40S subunit) bypasses a large segment of the 5'UTR and reengages with the mRNA at a far downstream site, the shunt-landing (or shunt-acceptor) site. The first 78 nucleotides of this RNA is considered to be relatively unstructured and includes a 5'-proximal uORF at nucleotides 61–72. This unstructured segment is followed by an approximately 580-nucleotide irregular stem-loop with several uORFs, which in turn is followed by an approximately 65-nucleotide relatively unstructured region (the putative shunt-landing site) before the start of the main ORF, ORF-VII. In broad terms, this configuration is conserved in the homologous mRNAs of all other plant pararetroviruses (Pooggin et al. 1999). Forced evolution experiments and extensive mutagenesis screens have shown that the essential features for shunting are (1) the existence of the large stem-loop, (2) the 5'-proximal uORF preceding this stem, and (3) the short spacing between the uORF termination codon and the base of the stem-loop (Pooggin et al. 1998, 2000; Ryabova and Hohn 2000; Hohn et al. 2001). However, neither the size of the irregular stem-loop nor the presence of many uORFs within it seems to be critical, as efficient shunting still occurs if it is replaced in its entirety by an 18-bp perfect hairpin (-45 kcal/mole) with no uORFs (Hemmings-Mieszczak and Hohn 1999).

Shunting therefore appears to be triggered when a ribosome is positioned at the uORF termination codon with its leading edge engaging the base of the stem and possibly starting to unwind this base-paired stem. It remains completely unknown why this particular set of circumstances should induce shunting (presumably by the 40S subunit, rather than an 80S ribosome, although this remains unproven) instead of resumption of strictly linear scanning, nor do we know how or when a shunting ribosome acquires a new eIF2/GTP/Met-tRNA$_i$ ternary complex. Although there is clear evidence that the particular configuration in the plant pararetrovirus RNAs triggers ribosome shunting, this remains the only such example of uORF-dependent shunting. It also needs to be emphasized that there are other completely different circumstances which can lead to ribosome shunting, as illustrated by the other paradigm of the adenovirus late mRNAs transcribed from the major late promoter, which have no upstream uORFs and no extensive stem-loops of high stability (Zhang et al. 1989; Yueh and Schneider 2000; Xi et al. 2004).

THE SPECIAL CASE OF SEQUENCE-SPECIFIC uORFs

Apart from the sequence surrounding the termination codon of *GCN4* uORF-4 (and probably also uORFs 2 and 3) being nonpermissive for resumption of scanning (which may be a peculiarity of yeast), the implicit assumption in the discussion so far has been that it is the length rather than the sequence of the uORF that is the determinant of how much it down-regulates translation of the main ORF. However, there are three well-known cases of inhibitory uORFs in eukaryotic mRNAs where (part of) the amino acid sequence is critical (Morris and Geballe 2000): *Neurospora crassa* arg-2 mRNA and other fungal homologs such as *Saccharomyces cerevisiae* CPA1 mRNA which code for arginine-specific carbamoyl phosphate synthase, a key enzyme in arginine biosynthesis; mammalian *S*-adenosyl-methionine decarboxylase (AdoMetDC) which provides the precursor for the conversion of putrescine to spermidine and spermidine to spermine; and uORF-2 of human cytomegalovirus gp48 mRNA which encodes a protein of unknown function. The inhibitory effect of the arg-2 and AdoMetDC uORFs is enhanced by elevated concentrations of arginine and polyamines (spermidine or spermine), respectively, in a type of negative feedback.

Because the uORF initiation codon is close to the 5′ cap in AdoMetDC mRNA and has suboptimal context in the other two cases, these uORFs capture only a fraction of the 40S subunits scanning from the 5′-proximal loading site, leaving the rest able to access the main ORF initiation site by leaky scanning. However, those ribosomes that do translate the uORF undergo a prolonged pause or stall at its termination codon, and this stalling creates a strong roadblock that prevents any following 40S subunits from reaching the main ORF by leaky scanning. There is probably very little, if any, termination-reinitiation, and, indeed, the model presented in the previous section would predict that the prolonged stall would abrogate subsequent resumption of scanning by ribosomes that eventually overcome the stall. As termination-reinitiation is very unlikely in these cases, they are, strictly speaking, outside the remit of this chapter; for further information, see reviews by Morris and Geballe (2000) and Geballe and Sachs (2000).

Subsequent work has shown that the peptide encoded by the arg-2 uORF seems to block the ribosomal nascent protein tunnel (as can also occur in eubacterial systems; Tenson and Ehrenberg 2002), for although the ribosomes appear to stall at the termination codon, they still stall at the same site (the 25th codon) even if the stop codon is displaced further downstream, or if the uORF (minus its termination codon) is placed in the middle of a larger cistron (Fang et al. 2000, 2004). In contrast, in the other two cases, the position of the uORF termination codon is critical, and the nascent peptide seems to interfere specifically with the translation termination process (Cao and Geballe 1996, 1998; Law et al. 2001; Janzen et al. 2002; Raney et al. 2002).

The current view is that only a minority of uORFs are sequence-dependent, although direct tests of this issue have been applied to only a very few of the approximate 35% of mammalian mRNAs with upstream uORFs. This is an area where a bioinformatics approach of comparing the relative conservation of nucleotide and encoded amino acid sequences of uORFs, and non-uORF 5′UTR sequences, in homologous mRNAs across different species could perhaps allow uORFs to be sorted into those that are possibly sequence-dependent and others that are probably sequence-independent. Such analyses will need to take into account the findings obtained with the three prototypic sequence-specific uORFs, namely, that the amino acid sequence need not be conserved in every position and the location of the uORF within the 5′ leader need not necessarily be conserved.

A good illustrative example is provided by plant AdoMetDC mRNA, which, like its mammalian counterpart, is subject to translational repression by polyamines (Hanfrey et al. 2005). Although the details have yet to be worked out, it is clear that this control is dependent on the configuration of two highly conserved uORFs that overlap by one residue (..<u>UR</u>**AUG**..): a 5′-proximal minute uORF of 2–3 sense codons (depending on the plant species) and a 48- to 54-codon uORF-2 in which no less than 31 positions are absolutely conserved in amino acid sequence across different plant species that are thought to have diverged some 400 million years ago (Hanfrey et al. 2005). This is a rather compelling indication that uORF-2, at least, falls into the sequence-specific category. What still remains completely unknown, however, is whether each and every sequence-specific uORF exerts its influence by causing ribosome stalling.

DO POSTTERMINATION BACKWARD SCANNING AND RETROREINITIATION REALLY OCCUR IN EUKARYOTES?

Some 20 years ago, there were two claims of reinitiation occurring in mammalian cells at sites located upstream of the termination codon (Peabody and Berg 1986; Peabody et al. 1986; Thomas and Capecchi 1986), a scenario that has since been named retroreinitiation (Amrani et al. 2004). These original claims were criticized on the grounds that the experiments had not distinguished true retroscanning/reinitiation from primary initiation due to leaky scanning (Kozak 1989, 2001). Whether this explanation is correct or not, suffice it to say that neither claim was followed up and, until quite recently, there have been no further claims of retroscanning. In fact, in a direct test for such backward scanning in vitro, an extremely low frequency of reinitiation, too low for accurate quantification, was seen when there were four residues between the stop and restart codons, but no reinitiation whatsoever with a ten-residue spacer (Kozak 2001). On the other hand, some apparent backward scanning has been seen in transfected cell assays when the overlap of the two ORFs was 10 nucleotides, but not when it was much longer (Gunnery et al. 1997).

Recently, retroscanning/reinitiation has been claimed to occur following termination at a premature termination codon (PTC) in yeast cell-free extracts (Amrani et al. 2004). Given the Sherman group conclusion that yeast ribosomes do not generally reinitiate after translation of a short ORF (Yun et al. 1996), and the fact that there is no proven case of efficient backward scanning and reinitiation even in mammalian systems, it is pertinent to ask whether leaky scanning might not be the true explanation for the putative retroreinitiation in this case, too.

The mRNA construct consisted of the first approximately 50 codons of the *CAN1* mRNA fused out of frame to a luciferase ORF, with a PTC at codon 47 of the *CAN1* sequence, and potential reinitiation codons in the luciferase frame either at +5 (5 nucleotides downstream from the PTC), or −11, −21, or −32 nucleotides upstream of the PTC (Amrani et al. 2004). The conclusion was that termination at the PTC is abnormally slow in comparison with termination at a natural stop codon and that when the ribosomes are eventually released from PTC following termination, an unspecified fraction of them can reinitiate at a nearby AUG, with a preference for reinitiation at an upstream site (−11, or even −32), rather than at an AUG at +5. Much of the evidence consists of toeprint assays, and there is no disputing that these show an abnormally long ribosomal pause at the PTC. Once the PTC toeprint has disappeared, toeprints caused by ribosomes stalled at the reinitiation site can be seen if cycloheximide has been added in the meantime. However, trapping the ribosomes at the reinitiation sites using cycloheximide makes it hard to estimate the efficiency of any reinitiation and to deduce how the ribosome reached that site. It is also important not to overlook the fact that if there are several ribosomes stalled at specific positions on the same mRNA, a toeprint assay can only detect the ribosome nearest to the primer-binding site, in this case the ribosome at the PTC, and can give no information on whether there are any stalled at upstream sites (let alone define the actual binding sites of such ribosomes).

The limited data on luciferase expression suggest that reinitiation at either upstream or downstream sites is rather inefficient, and although some of these expression results are indicative of a leaky scanning explanation, others are not (see supplementary Fig. 2 of Amrani et al. 2004). The critical test would seem to be to examine what happens in a translation assay (rather than a toeprint assay) when there are two potential upstream (re)initiation sites, say at −11 and −32: If luciferase expression comes predominantly from the −32 site, this would be strong support for the leaky scanning explanation, whereas preferential use of the −11 AUG would favor retroscanning/reinitiation; if they were both used equally, we would be none the wiser! If there are strong indications of retroreinitiation, it would be important (1) to determine its efficiency with respect to the absolutely minimal levels seen by Kozak (2001) and (2) to test whether this retroreinitiation is a general phenomenon, or whether the *CAN1* segment fortuitously includes a motif that mimics the function of the 84-nucleotide *cis*-acting element essential for reinitiation on the RHDV and FCV subgenomic mRNAs, as discussed above. Until these issues have been thoroughly investigated, we consider it premature to discard the view that even though posttermination 40S subunits may undergo very limited bidirectional random diffusion for an extremely brief time period, systematic rescanning over relatively long distances occurs only in the 5′→3′ direction, not 3′→5′.

CONCLUDING REMARKS

The balance of the evidence suggests that the normal default posttermination event in both eubacterial and eukaryotic systems is complete dissociation of both ribosomal subunits from the mRNA. However, in the course of disassembly of the posttermination ribosome–mRNA complex, the small ribosomal subunit may remain mRNA-associated for a very brief period of time, during which it can undergo random bidirectional diffusion on the mRNA. For reinitiation to occur during this very short time window, special events are required to retain and restrain the small subunit on the mRNA in a position and status permissive for reinitiation. In prokaryotic systems, this requires an appropriate SD-AUG (NUG) tandem within the window of bidirectional diffusion. Except when the 5′-proximal ORF is short, reinitiation in eukaryotic (cytoplasmic) systems is rare and requires a fairly extensive *cis*-acting RNA element that fulfills the role of the much smaller prokaryotic SD element by retaining and restraining the small ribosomal subunit on the mRNA, either by direct binding or by indirect interactions bridged by initiation factors, at a site that is in appropriate juxtaposition to the reinitiation codon. In the numerous cases of reinitiation after a relatively very short 5′-proximal ORF in eukaryotic mRNAs, it is probably short-lived persistence of critical interactions between the ribosome and some of the initiation factors (notably eIF4F and probably eIF3) that was instrumental in promoting initiation at the uORF initiation codon which not only keeps the post-termination 40S ribosomal subunit on the mRNA, but also potentiates resumption of scanning to the next downstream initiation site and rerecruitment of the other necessary initiation factors.

ACKNOWLEDGMENTS

Work from our laboratory described in this chapter was supported by a Wellcome Trust Programme grant.

REFERENCES

Adamson S.D., Howard G.A., and Herbert E. 1970. The ribosome cycle in a reconstituted cell-free system from reticulocytes. *Cold Spring Harbor Symp. Quant. Biol.* **34:** 547–554.

Adhin M.R. and van Duin J. 1990. Scanning model for translational reinitiation in eubacteria. *J. Mol. Biol.* **213:** 811–818.

Ahmadian G., Randhawa J.S., and Easton A.J. 2000. Expression of the ORF-2 protein of the human respiratory syncytial virus M2 gene is initiated by a ribosomal termination-dependent reinitiation mechanism. *EMBO J.* **19:** 2681–2689.

Amrani N., Ganesan R., Kervestin S., Mangus D.A., Ghosh S., and Jacobson A. 2004. A *faux* 3′-UTR promotes aberrant termination and triggers nonsense-mediated mRNA decay. *Nature* **432:** 112–118.

Asano K., Phan L., Valásek L., Schoenfeld L.W., Shalev A., Clayton J., Nielsen K., Donahue T.F., and Hinnebusch A.G. 2001. A multifactor complex of eIF1, eIF2, eIF3, eIF5, and tRNA$_i^{Met}$ promotes initiation complex assembly and couples GTP hydrolysis to AUG recognition. *Cold Spring Harbor Symp. Quant. Biol.* **66:** 403–415.

Berkhout B. and van Duin J. 1985. Mechanism of translational coupling between coat protein and replicase genes of RNA bacteriophage MS2. *Nucleic Acids Res.* **13:** 6955–6967.

Buckingham R.H., Grentzmann G., and Kisselev L. 1997. Polypeptide chain release factors. *Mol. Microbiol.* **24:** 449–456.

Cao J. and Geballe A.P. 1996. Inhibition of nascent-peptide release at translation termination. *Mol. Cell. Biol.* **16:** 7109–7114.

———. 1998. Ribosomal release without peptidyl tRNA hydrolysis at translation termination in a eukaryotic system. *RNA* **4:** 181–188.

Carter A.P., Clemons W.M. Jr., Brodersen D.E., Morgan-Warren R.J., Hartsch T., Wimberly B.T., and Ramakrishnan V. 2001. Crystal structure of an initiation factor bound to the 30S ribosomal subunit. *Science* **291:** 498–501.

Child S.J., Miller M.K., and Geballe A.P. 1999. Translational control by an upstream open reading frame in the HER-2/neu transcript. *J. Biol. Chem.* **274:** 24335–24341.

Clarke I.N. and Lambden P.R. 2000. Organization and expression of calicivirus genes. *J. Infect. Dis.* **181:** S309-S316.

Fang P., Wang Z., and Sachs M.S. 2000. Evolutionarily conserved features of the arginine attenuator peptide provide the necessary requirements for its function in translational regulation. *J. Biol. Chem.* **275:** 26710–26719.

Fang P., Spevak C.C., Wu C., and Sachs M.S. 2004. A nascent polypeptide domain that can regulate translation elongation. *Proc. Natl. Acad. Sci.* **101:** 4059–4064.

Geballe A.P. and Sachs M.S. 2000. Translational control by upstream open reading frames. In *Translational control of gene expression* (ed. N. Sonenberg et al.), pp. 595–614. Cold Spring Harbor Laboratory Press, Cold Spring Harbor, New York.

Goodfellow I., Chaudhry Y., Gioldasi I., Gerondopoulos A., Natoni A., Labrie L., Laliberte J.F., and Roberts L. 2005. Calicivirus translation initiation requires an interaction between VPg and eIF 4E. *EMBO Rep.* **6:** 968–972.

Gould P.S. and Easton A.J. 2005. Coupled translation of the respiratory syncytial virus M2 open reading frames requires upstream sequences. *J. Biol. Chem.* **280:** 21972–21980.

Grant C.M. and Hinnebusch A.G. 1994. Effect of sequence context at stop codons on efficiency of reinitiation in GCN4 translational control. *Mol. Cell. Biol.* **14:** 606–618.

Grant C.M., Miller P.F., and Hinnebusch A.G. 1995. Sequences 5′ of the first upstream open reading frame in GCN4 mRNA are required for efficient translational reinitiation. *Nucleic Acids Res.* **23:** 3980–3988.

Groenveld H., Thimon K., and van Duin J. 1995. Translational control of maturation-protein synthesis in phage MS2: A role for the kinetics of RNA folding? *RNA* **1:** 79–88.

Gunnery S., Mäivali Ü., and Mathews M.B. 1997. Translation of an uncapped mRNA involves scanning. *J. Biol. Chem.* **272:** 21642–21646.

Hanfrey C., Elliott K.A., Franceschetti M., Mayer M.J., Illingworth C., and Michael A.J. 2005. A dual upstream open reading frame-based autoregulatory circuit controlling polyamine-responsive translation. *J. Biol. Chem.* **280:** 39229–39237.

He H., von der Haar T., Singh C.R., Ii M., Li B., Hinnebusch A.G., McCarthy J.E., and Asano K. 2003. The yeast eukaryotic initiation factor 4G (eIF4G) HEAT domain interacts with eIF1 and eIF5 and is involved in stringent AUG selection. *Mol. Cell. Biol.* **23:** 5431–5445.

Hemmings-Mieszczak M. and Hohn T. 1999. A stable hairpin preceded by a short open reading frame promotes nonlinear ribosome migration on a synthetic mRNA leader. *RNA* **5:** 1149–1157.

Herbert T.P., Brierley I., and Brown T.D.K. 1996. Detection of the ORF3 polypeptide of feline calicivirus in infected cells and evidence for its expression from a single, functionally bicistronic, subgenomic mRNA. *J. Gen. Virol.* **77:** 123–127.

―――. 1997. Identification of a protein linked to the genomic and subgenomic mRNAs of feline calicivirus and its role in translation. *J. Gen. Virol.* **78:** 1033–1040.

Hohn T., Park H.S., Guerra-Peraza O., Stavolone L., Pooggin M.M., Kobayashi K., and Ryabova L.A. 2001. Shunting and controlled reinitiation: The encounter of cauliflower mosaic virus with the translational machinery. *Cold Spring Harbor Symp. Quant. Biol.* **66:** 269–276.

Horvath C.M., Williams M.A., and Lamb R.A. 1990. Eukaryotic coupled translation of tandem cistrons: Identification of the influenza B virus BM2 polypeptide. *EMBO J.* **9:** 2639–2647.

Hoshino S., Imai M., Kobayashi T., Uchida N., and Katada T. 1999. The eukaryotic polypeptide chain releasing factor (eRF3/GSPT) carrying the translation termination signal to the 3′-poly(A) tail of mRNA. Direct association of eRF3/GSPT with polyadenylate binding protein. *J. Biol. Chem.* **274:** 16677–16680.

Hoshino S., Imai M., Mizutani M., Kikuchi Y., Hanaoka F., Ui M., and Katada T. 1998. Molecular cloning of a novel member of the eukaryotic polypeptide chain releasing factors (eRF). Its identification as eRF3 interacting with eRF1. *J. Biol. Chem.* **273:** 22254–22259.

Iacono M., Mignone F., and Pesole G. 2005. uAUG and uORFs in human and rodent 5′ untranslated mRNAs. *Gene* **349:** 97–105.

Inagaki Y. and Ford Doolittle W. 2000. Evolution of the eukaryotic translation termination system: Origins of release factors. *Mol. Biol. Evol.* **17:** 882–889.

Ivey-Hoyle M. and Steege D.A. 1989. Translation of phage f1 gene VII occurs from an inherently defective initiation site made functional by coupling. *J. Mol. Biol.* **208:** 233–244.

―――. 1992. Mutational analysis of an inherently defective translation initiation site. *J. Mol. Biol.* **224:** 1039–1054.

Janzen D.M., Frolova L., and Geballe A.P. 2002. Inhibition of translation termination mediated by an interaction of eukaryotic release factor 1 with a nascent peptidyl-tRNA. *Mol. Cell. Biol.* **22:** 8562–8570.

Johannes G., Carter M.S., Eisen M.B., Brown P.O., and Sarnow P. 1999. Identification of eukaryotic mRNAs that are translated at reduced cap binding complex eIF4F concentrations using a cDNA microarray. *Proc. Natl. Acad. Sci.* **96:** 13118–13123.

Kobayashi T., Funakoshi Y., Hoshino S., and Katada T. 2004. The GTP-binding release factor eRF3 as a key mediator coupling translation termination to mRNA decay. *J. Biol. Chem.* **279:** 45693–45700.

Kojima K.K., Matsumoto T., and Fujiwara H. 2005. Eukaryotic translational coupling in UAAUG stop-start codons for the bicistronic RNA translation of the non-long terminal repeat retrotransposon SART1. *Mol. Cell. Biol.* **25:** 7675–7686.

Kolupaeva V.G., Unbehaun A., Lomakin I.B., Hellen C.U.T., and Pestova T.V. 2005. Binding of eukaryotic initiation factor 3 to ribosomal 40S subunits and its role in ribosomal dissociation and anti-association. *RNA* **11:** 470–486.

Kong C., Ito K., Walsh M.A., Wada M., Liu Y., Kumar S., Barford D., Nakamura Y., and Song H. 2004. Crystal structure and functional analysis of the eukaryotic class II release factor eRF3 from *S. pombe*. *Mol. Cell* **14:** 233–245.

Konig M., Thiel H.J., and Meyers G. 1998. Detection of viral proteins after infection of cultured hepatocytes with rabbit hemorrhagic disease virus. *J. Virol.* **72:** 4492–4497.

Kontos H., Napthine S., and Brierley I. 2001. Ribosomal pausing at a frameshifter RNA pseudoknot is sensitive to reading phase but shows little correlation with frameshift efficiency. *Mol. Cell. Biol.* **21:** 8657–8670.

Korneeva N.L., Lamphear B.J., Hennigan F.L., and Rhoads R.E. 2000. Mutually cooperative binding of eukaryotic translation initiation factor (eIF) 3 and eIF4A to human eIF4G-1. *J. Biol. Chem.* **275:** 41369–41376.

Kozak M. 1987. Effects of intercistronic length on the efficiency of reinitiation by eucaryotic ribosomes. *Mol. Cell. Biol.* **7:** 3438–3445.

———. 1989. The scanning model for translation: An update. *J. Cell Biol.* **108:** 229–241.

———. 2001. Constraints on reinitiation of translation in mammals. *Nucleic Acids Res.* **29:** 5226–5232.

Law G.L., Raney A., Heusner C., and Morris D.R. 2001. Polyamine regulation of ribosome pausing at the upstream open reading frame of S-adenosylmethionine decarboxylase. *J. Biol. Chem.* **276:** 38036–38043.

Lincoln A.J., Monczak Y., Williams S.C., and Johnson P.F. 1998. Inhibition of CCAAT/enhancer-binding protein alpha and beta translation by upstream open reading frames. *J. Biol. Chem.* **273:** 9552–9560.

Lodish H.F. and Zinder N.D. 1966. Mutants of the bacteriophage f2. 8. Control mechanisms for phage-specific syntheses. *J. Mol. Biol.* **19:** 333–348.

Lomakin I.B., Kolupaeva V.G., Marintchev A., Wagner G., and Pestova T.V. 2003. Position of eukaryotic initiation factor eIF1 on the 40S ribosomal subunit determined by directed hydroxyl radical probing. *Genes Dev.* **17:** 2786–2797.

Luukkonen B.G.M., Tan W., and Schwartz S. 1995. Efficiency of reinitiation of translation on human immunodeficiency virus type 1 mRNAs is determined by the length of the upstream open reading frame and by intercistronic distance. *J. Virol.* **69:** 4086–4094.

Marzi S., Knight W., Brandi L., Caserta E., Soboleva N., Hill W.E., Gualerzi C.O., and Lodmell J.S. 2003. Ribosomal localization of translation initiation factor IF2. *RNA* **9:** 958–969.

Mehta A., Trotta C.R., and Peltz S.W. 2006. Derepression of the Her-2 uORF is mediated by a novel post-transcriptional control mechanism in cancer cells. *Genes Dev.* **20:** 939–953.

Meyers G. 2003. Translation of the minor capsid protein of a calicivirus is initiated by a novel termination-dependent reinitiation mechanism. *J. Biol. Chem.* **278:** 34051–34056.

Min Jou W., Haegeman G., Ysebaert M., and Fiers W. 1972. Nucleotide sequence of the gene coding for the bacteriophage MS2 coat protein. *Nature* **237:** 82–88.

Moll I., Hirokawa G., Kiel M.C., Kaji A., and Blasi U. 2004. Translation initiation with 70S ribosomes: An alternative pathway for leaderless mRNAs. *Nucleic Acids Res.* **32:** 3354–3363.

Morris D.R. and Geballe A.P. 2000. Upstream open reading frames as regulators of mRNA translation. *Mol. Cell. Biol.* **20:** 8635–8642.

O'Donnell S.M. and Janssen G.R. 2002. Leaderless mRNAs bind 70S ribosomes more strongly than 30S ribosomal subunits in *Escherichia coli*. *J. Bacteriol.* **184:** 6730–6733.

Peabody D.S. and Berg P. 1986. Termination-reinitiation occurs in the translation of mammalian cell mRNAs. *Mol. Cell. Biol.* **6:** 2695–2703.

Peabody D.S., Subramani S., and Berg P. 1986. Effect of upstream reading frames on translation efficiency in simian virus 40 recombinants. *Mol. Cell. Biol.* **6:** 2704–2711.

Peri S. and Pandey A. 2001. A reassessment of the translation initiation codon in vertebrates. *Trends Genet.* **17:** 685–687.

Peske F., Rodnina M.V., and Wintermeyer W. 2005. Sequence of steps in ribosome recycling as defined by kinetic analysis. *Mol. Cell* **18:** 403–412.

Pooggin M.M., Hohn T., and Futterer J. 1998. Forced evolution reveals the importance of short open reading frame A and secondary structure in the cauliflower mosaic virus 35S RNA leader. *J. Virol.* **72:** 4157–4169.

———. 2000. Role of a short open reading frame in ribosome shunt on the cauliflower mosaic virus RNA leader. *J. Biol. Chem.* **275:** 17288–17296.

Pooggin M.M., Futterer J., Skryabin K.G., and Hohn T. 1999. A short open reading frame terminating in front of a stable hairpin is the conserved feature in pregenomic RNA leaders of plant pararetroviruses. *J. Gen. Virol.* **80:** 2217–2228.

Pöyry T.A.A., Kaminski A., and Jackson R.J. 2004. What determines whether mammalian ribosomes resume scanning after translation of a short upstream open reading frame? *Genes Dev.* **18:** 62–75.

Raney A., Law G.L., Mize G.J., and Morris D.R. 2002. Regulated translation termination at the upstream open reading frame in S-adenosylmethionine decarboxylase mRNA. *J. Biol. Chem.* **277:** 5988–5994.

Ryabova L.A. and Hohn T. 2000. Ribosome shunting in the cauliflower mosaic virus 35S RNA leader is a special case of reinitiation of translation functioning in plant and animal systems. *Genes Dev.* **14:** 817–829.

Ryabova L., Park H.S., and Hohn T. 2004. Control of translation reinitiation on the cauliflower mosaic virus (CaMV) polycistronic RNA. *Biochem. Soc. Trans.* **32:** 592–596.

Sachs M.S. and Geballe A.P. 2006. Downstream control of upstream open reading frames. *Genes Dev.* **20:** 915–921.

Sarabhai A. and Brenner S. 1967. A mutant which reinitiates the polypeptide chain after chain termination. *J. Mol. Biol.* **27:** 145–162.

Siridechadilok B., Fraser C.S., Hall R.J., Doudna J.A., and Nogales E. 2005. Structural roles for human translation factor eIF3 in initiation of protein synthesis. *Science* **310:** 1513–1515.

Somogyi P., Jenner A.J., Brierley I., and Inglis S.C. 1993. Ribosomal pausing during translation of an RNA pseudoknot. *Mol. Cell. Biol.* **13:** 6931–6940.

Song H., Mugnier P., Das A.K., Webb H.M., Evans D.R., Tuite M.F., Hemmings B.A., and Barford D. 2000. The crystal structure of human eukaryotic release factor eRF1—Mechanism of stop codon recognition and peptidyl-tRNA hydrolysis. *Cell* **100:** 311–321.

Spanjaard R.A. and van Duin J. 1989. Translational reinitiation in the presence and absence of a Shine and Dalgarno sequence. *Nucleic Acids Res.* **17:** 5501–5507.

Steege D.A. 1977. 5′-Terminal nucleotide sequence of *Escherichia coli* lactose repressor mRNA: Features of translational initiation and reinitiation sites. *Proc. Natl. Acad. Sci.* **74:** 4163–4167.

Szabo G., Katarova Z., and Greenspan R. 1994. Distinct protein forms are produced from alternatively spliced bicistronic glutamic acid decarboxylase mRNAs during development. *Mol. Cell. Biol.* **14:** 7535–7545.

Tenson T. and Ehrenberg M. 2002. Regulatory nascent peptides in the ribosomal tunnel. *Cell* **108:** 591–594.

Ter-Avanesyan M.D., Kushnirov V.V., Dagkesamanskaya A.R., Didichenko S.A., Chernoff Y.O., Inge-Vechtomov S.G., and Smirnov V.N. 1993. Deletion analysis of the SUP35 gene of the yeast *Saccharomyces cerevisiae* reveals two non-overlapping functional regions in the encoded protein. *Mol. Microbiol.* **7:** 683–692.

Thomas K.R. and Capecchi M.R. 1986. Introduction of homologous DNA sequences into mammalian cell lines induces mutations in the cognate gene. *Nature* **324:** 34–38.

Tooze J. and Weber K. 1967. Isolation and characterization of amber mutants of bacteriophage R17. *J. Mol. Biol.* **28:** 311–330.

Unbehaun A., Borukhov S.I., Hellen C.U.T., and Pestova T.V. 2004. Release of initiation factors from 48S complexes during ribosomal subunit joining and the link between establishment of codon-anticodon base-pairing and hydrolysis of eIF2-bound GTP. *Genes Dev.* **18:** 3078–3093.

Valásek L., Phan L., Schoenfeld L.W., Valaskova V., and Hinnebusch A.G. 2001. Related eIF3 subunits TIF32 and HCR1 interact with an RNA recognition motif in PRT1 required for eIF3 integrity and ribosome binding. *EMBO J.* **20:** 891–904.

Valásek L., Mathew A.A., Shin B.S., Nielsen K.H., Szamecz B., and Hinnebusch A.G. 2003. The yeast eIF3 subunits TIF32/a, NIP1/c, and eIF5 make critical connections with the 40S ribosome in vivo. *Genes Dev.* **17:** 786–799.

Vattem K.M. and Wek R.C. 2004. Reinitiation involving upstream ORFs regulates ATF4 mRNA translation in mammalian cells. *Proc. Natl. Acad. Sci.* **101:** 11269–11274.

Wirblich C., Thiel H.J., and Meyers G. 1996. Genetic map of the calicivirus rabbit hemorrhagic disease virus as deduced from in vitro translation studies. *J. Virol.* **70:** 7974–7983.

Xi Q., Cuesta R., and Schneider R.J. 2004. Tethering of eIF4G to adenoviral mRNAs by viral 100k protein drives ribosome shunting. *Genes Dev.* **18:** 1997–2009.

Yueh A. and Schneider R.J. 2000. Translation by ribosome shunting on adenovirus and hsp70 mRNAs facilitated by complementarity to 18S rRNA. *Genes Dev.* **14:** 414–421.

Yun D.F., Laz T.M., Clements J.M., and Sherman F. 1996. mRNA sequences influencing translation and the selection of AUG initiator codons in the yeast *Saccharomyces cerevisiae*. *Mol. Microbiol.* **19:** 1225–1239.

Zhang Y., Dolph P.J., and Schneider R.J. 1989. Secondary structure analysis of adenovirus tripartite leader. *J. Biol. Chem.* **264:** 10679–10684.

Zhang Z. and Dietrich F.S. 2005. Identification and characterization of upstream open reading frames (uORF) in the 5′ untranslated regions (UTR) of genes in *Saccharomyces cerevisiae*. *Curr. Genet.* **48:** 77–87.

Zhouravleva G., Frolova L., Le Goff X., Le Guellec R., Inge-Vechtomov S., Kisselev L., and Philippe M. 1995. Termination of translation in eukaryotes is governed by two interacting polypeptide chain release factors, eRF1 and eRF3. *EMBO J.* **14:** 4065–4072.

9

Mechanism of Translation Initiation in the Yeast *Saccharomyces cerevisiae*

Alan G. Hinnebusch and Thomas E. Dever

Laboratory of Gene Regulation and Development
National Institute of Child Health and Human Development
National Institutes of Health
Bethesda, Maryland 20892

Katsura Asano

Molecular Cellular Developmental Biology Program
Division of Biology, Kansas State University
Manhattan, Kansas 66506

OUR UNDERSTANDING OF THE TRANSLATION INITIATION PATHWAY is based largely on biochemical analysis of purified mammalian factors, as described in Chapter 4. Genetic and biochemical analysis of the yeast *Saccharomyces cerevisiae* is helping to elucidate molecular mechanisms in the pathway, and genetic studies in yeast provide the best opportunity to establish the physiological relevance of conclusions reached using cell-free systems, thus providing the impetus for this review. According to the current model (Fig. 1), the pathway begins with binding to the free 40S subunit of a ternary complex (TC) comprising eIF2, GTP, and Met-tRNA$_i^{Met}$, in a reaction promoted by eIF1, eIF1A, and eIF3. The 43S preinitiation complex (PIC) thus formed binds to the mRNA, with the assistance of eIF4F, -4A, -4B, and poly(A)-binding protein (PABP), and scans the mRNA for the start codon. Base-pairing of Met-tRNA$_i^{Met}$ with AUG stimulates hydrolysis of the GTP bound to eIF2, dependent on the GTPase-activating protein (GAP) eIF5 and 40S subunit, the eIF2-GDP and other factors are released, and eIF5B-GTP stimulates 60S subunit joining. The eIF2-GDP must be recycled to eIF2-GTP by the guanine nucleotide

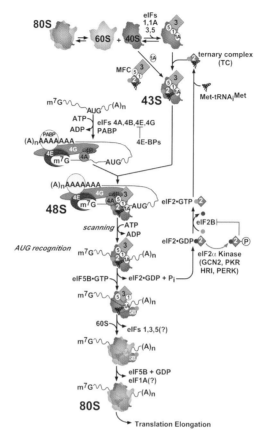

Figure 1. Eukaryotic translation initiation pathway. See text for details. (MFC) Multifactor complex. (Modified from Hinnebusch et al. 2004.)

exchange factor (GEF) eIF2B for a new round of initiation, and this reaction is inhibited under stress conditions by phosphorylation of the α subunit of eIF2 on Ser-51.

Many advances in our knowledge of the mechanism and regulation of Met-tRNA$_i^{Met}$ recruitment have come from genetic analysis of translational control of yeast *GCN4* mRNA by the eIF2α kinase GCN2 (*gen*-eral *c*ontrol *n*on-derepressible 2). Activation of GCN2 by amino acid starvation produces eIF2α phosphorylation (eIF2[αP]) at a level that does not fully inhibit the recycling of eIF2-GDP and protein synthesis, and specifically increases translation of *GCN4* mRNA. Because GCN4 transcriptionally activates amino acid biosynthetic enzymes subject to general amino acid control, cells can increase amino acid production while reducing general protein synthesis in nutrient-poor environments. Induction of *GCN4* translation is mediated by four short open reading frames (uORFs) in its mRNA leader. In the current model (Fig. 2A,B), ribosomes scanning from the cap translate uORF1, and about 50% resume scanning as 40S subunits. Under nonstarvation conditions, these reinitiating ribosomes rebind the TC and reinitiate at uORFs 2–4, then dissociate from the mRNA without translating the *GCN4* ORF. Inhibition of eIF2B by eIF2[αP] in starved cells lowers the TC concentration so that about 50% of the 40S subunits scanning from uORF1 reach uORF4 before rebinding TC and, lacking Met-tRNA$_i^{Met}$, bypass uORF4. Most of these ribosomes rebind the TC before reaching the *GCN4* AUG codon and reinitiate there instead. Thus, reducing TC levels by phosphorylating eIF2α induces *GCN4* translation by allowing a fraction of scanning 40S subunits to bypass the inhibitory uORFs 2–4 (Hinnebusch 2005).

Because *GCN4* translation is a sensitive indicator of TC formation, mutations in eIF2γ and the β, γ, δ, and ϵ subunits of eIF2B were first isolated on the basis of constitutive derepression of *GCN4* translation (*general control derepressed*, or Gcd$^-$ phenotype). These mutations also produce a slow-growth phenotype (Slg$^-$) and reduce general protein synthesis on rich medium, indicating nonlethal impairment of the essential functions of eIF2 or eIF2B. The derepression of *GCN4* conferred by these mutations is maintained in *gcn2*Δ cells, indicating that they reduce TC levels independently of eIF2α phosphorylation (Fig. 2C,E). Overexpressing eIF2, and regulatory mutations in certain subunits of eIF2B and eIF2α, prevent derepression of *GCN4* in starved wild-type cells (Gcn$^-$ phenotype) by reducing the inhibitory effect of eIF2[αP] on eIF2B function (Fig. 2D). The Gcn$^-$ phenotype is manifested by sensitivity to amino acid analogs that inhibit biosynthesis (Hinnebusch 2005).

Donahue and colleagues established another valuable genetic assay that reveals defects in AUG selection by scanning ribosomes (Donahue 2000). These workers isolated mutations that allow expression of the defective *his4–303* allele, lacking an AUG start codon, by enabling translation to begin at an in-frame UUG near the beginning of the gene. Characterizing mutations with this Sui$^-$ (*suppressor of initiation codon*) phenotype in the anticodon of tRNA$_i^{Met}$, the eIF2 subunits, eIF5, and eIF1 first implicated these factors in AUG selection during scanning.

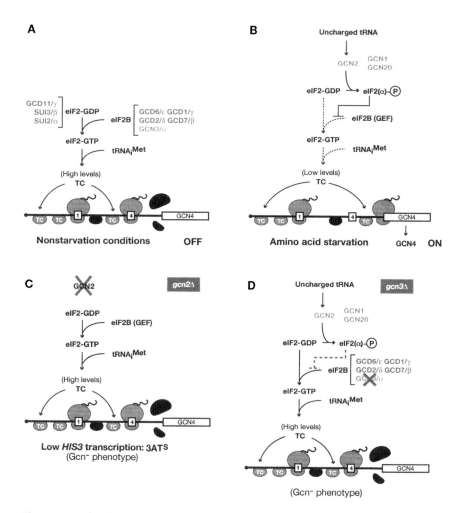

Figure 2. Molecular model for *GCN4* translational control. Following translation of uORF1, the 40S subunit remains attached to the mRNA and resumes scanning. Under nonstarvation conditions (*A*), the 40S subunit quickly rebinds TC and reinitiates at uORF4, and the 80S ribosome dissociates after terminating at uORF4. Under amino acid starvation conditions (*B*), many 40S ribosomes fail to rebind TC until scanning past uORF4 because the TC concentration is low and reinitiate at *GCN4* instead. TC levels are reduced in starved cells due to phosphorylation of eIF2 by GCN2, converting eIF2 to an inhibitor of eIF2B. (*C*) *GCN4* translation is constitutively repressed in *gcn2Δ* cells owing to the inability to phosphorylate eIF2 in response to starvation. The GCN4 target gene *HIS3*, encoding a histidine biosynthetic enzyme, is not induced by the histidine starvation imposed by 3-aminotriazole (3AT), conferring 3AT-sensitivity (3AT^s) indicative of Gcn⁻ mutants. (*D*) *gcn3Δ* mutants exhibit a Gcn⁻ phenotype because eIF2B lacking GCN3 is not inhibited by eIF2[αP].

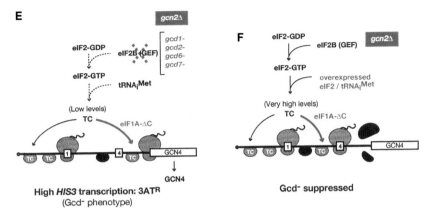

Figure 2. *(Continued)* *(E)* Nonlethal recessive mutations in essential subunits of eIF2B reduce GEF function and lower TC formation, constitutively derepressing *GCN4* translation in *gcn2Δ* cells (Gcd⁻ phenotype). The constitutive induction of *HIS3* transcription suppresses the (3ATˢ) phenotype in *gcn2Δ* cells. Truncating the CTD of eIF1A (ΔC) produces a Gcd⁻ phenotype differently, by decreasing the rate of TC binding to 40S subunits. *(F)* The defect in TC loading produced by eIF1A-ΔC is suppressed by overproducing TC. (Modified from Hinnebusch 2005.)

FORMATION OF TC

The Met-tRNA$_i^{Met}$ is delivered to the free 40S subunit by GTP-bound eIF2 in the TC. Yeast eIF2-GTP has about 10-fold higher affinity than eIF2-GDP for Met-tRNA$_i^{Met}$, and Met-tRNA$_i^{Met}$ similarly increases the affinity of eIF2 for GTP, reducing the GTP dissociation rate (Erickson and Hannig 1996; Kapp and Lorsch 2004). Thermodynamic coupling between GTP and Met-tRNA$_i^{Met}$ in eIF2 binding is dependent on the methionine moiety, and it was proposed that GTP elicits a conformational change that allows methionine to access its binding pocket in eIF2 (Kapp and Lorsch 2004).

Eukaryotic tRNA$_i^{Met}$ has sequence and structural characteristics that allow eIF2 to distinguish it from elongator tRNAs (Hinnebusch 2000), including the A1:U72 bp at the top of the acceptor stem and G:C bps in the anticodon stem, both implicated in eIF2 binding, and (for yeast tRNA$_i^{Met}$) A54 in loop IV (von Pawel-Rammingen et al. 1992; Farruggio et al. 1996). Mutating A1:U72 to either G:C, A:C, or G:U decreased the affinity of yeast Met-tRNA$_i^{Met}$ for eIF2-GTP and reduced thermodynamic coupling between GTP and methionine. Thus, A1:U72 likely promotes GTP-stimulated interaction between eIF2 and methionine (Kapp and Lorsch 2004).

eIF2γ Plays the Central Role in Binding Guanine Nucleotides and Initiator tRNA

eIF2γ is a GTP-binding protein closely related to the elongation factor EF-Tu, which binds aminoacylated tRNAs to the ribosome during elongation. The crystal structure of an archaeal eIF2γ homolog (aIF2γ) from *Pyrococcus abyssi* revealed three domains similar to domains I–III of EF-Tu, including the guanine nucleotide-binding pocket in the G domain (Schmitt et al. 2002). The γ subunit alone and the aIF2 holoprotein reconstituted from *Sulfolobus solfataricus* have similar affinities for GTP, indicating that γ is sufficient for binding nucleotide (Pedulla et al. 2005). *P. abyssi* aIF2γ also displays weak Met-tRNA$_i^{Met}$ binding and discriminates against tRNA$_i^{Met}$ acylated with valine (Yatime et al. 2004), showing that γ also contains binding determinants for tRNA$_i^{Met}$ and methionine.

Most substitutions in yeast eIF2γ of the conserved lysine in the NK̲XD motif of the GDP-binding pocket are lethal (Erickson and Hannig 1996), but the viable arginine substitution (*gcd11-K250R*) leads to Slg$^-$ and Gcd$^-$ phenotypes suppressible by overexpressing tRNA$_i^{Met}$ (Fig. 3A). In vitro, the *K250R* mutation increased off-rates for GDP and GTP, and adding Met-tRNA$_i^{Met}$ overcame the GTP-binding defect (Erickson and Hannig 1996). Thus, eliminating an important contact between eIF2γ and GTP decreases TC formation in vivo, with derepression of *GCN4* translation and diminished protein synthesis. These results demonstrate the physiological relevance of thermodynamic coupling between GTP and Met-tRNA$_i^{Met}$ in binding eIF2.

In the bacterial EF-Tu/GDPNP/Phe-tRNA^{Phe} complex, the switch-1 and switch-2 regions contact the γ-phosphate of GDPNP, while domain II is disengaged from the G domain in the inactive GDP-bound structure. In contrast, both GDP- and GDPNP-bound forms of *P. abyssi* aIF2γ show close packing of domains II and III against the G domain, with little difference in switch-1 and switch-2 conformations (Schmitt et al. 2002), making it unclear how GTP stimulates Met-tRNA$_i^{Met}$ binding. In the aIF2 α/g heterodimer of *S. solfataricus*, however, the switch regions are moved closer to positions they occupy in the EF-Tu/GDPNP/Phe-tRNA^{Phe} complex, and in a model docking Met-tRNA$_i^{Met}$ to the α/γ heterodimer, the altered location of switch 1 opens a channel between the G domain and domain II of the γ subunit to accommodate methionine (Yatime et al. 2006). Y51 in switch 1 stacks against the methionine and, consistently, the *gcd11-Y142H* mutation in the corresponding residue of yeast eIF2γ produces Gcd$^-$ and Slg$^-$ phenotypes (Fig. 3A) suppressible by overproducing tRNA$_i^{Met}$ (Harashima and Hinnebusch 1986; Dorris et al. 1995). Moreover, purified eIF2

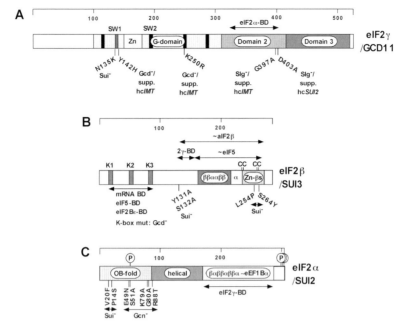

Figure 3. Important domains and residues in subunits of yeast eIF2. (*A*) The primary sequence of eIF2γ is depicted with motifs characteristic of G proteins as black bars, including switch 2 (SW2). Locations and phenotypes of selected mutations are below, indicating suppression (supp.) by overexpression of tRNA$_i^{Met}$ (hcIMT) or eIF2α (hcSUI2) where relevant. Binding domains (BDs) for factors or mRNA are delimited with double-headed arrows. (*B*) Primary sequence of eIF2β showing the K-boxes 1–3, ββαββ domain, and Zn-binding β-sheet (Zn-βs) predicted from the aIF2β structure. Regions of similarity to other factors (e.g., ~eIF5) are delimited with double-headed arrows. (*C*) Primary sequence of eIF2α is depicted showing the phosphorylation site for GCN2 at position 51. (Modified from Hinnebusch et al. 2004.)

containing the *gcd11-Y142H* subunit shows reduced Met-tRNA$_i^{Met}$ binding but normal off-rates for GDP and GTP (Erickson and Hannig 1996). The docking model for aIF2 α/γ also rationalizes the Gcd⁻ (Harashima and Hinnebusch 1986) and Slg⁻ phenotypes, suppressible by tRNA$_i^{Met}$ overexpression, for the G397A substitution in domain II (Dorris et al. 1995) and lethality of an isoleucine substitution at the same residue (Roll-Mecak et al. 2004) in yeast eIF2γ (A296 in *S. solfataricus* aIF2γ; see Fig. 4B).

The α and β subunits of yeast and archaeal eIF2 interact directly with the γ subunit, but not with each other (Thompson et al. 2000; Schmitt et al. 2002), suggesting that γ is the core subunit of eIF2. The *S. solfataricus* γ/α heterodimer contains two loops in domain III (CTD, carboxy-terminal domain) of the α subunit inserted in two pockets of domain II of γ

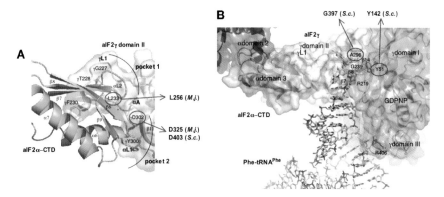

Figure 4. Binding of aIF2γ to tRNA and eIF2α-CTD. (*A*) Close-up of the interface between aIF2α-CTD (*blue*) and domain II of aIF2γ (*yellow*) from *S. solfataricus*. Residues corresponding to those in *M. jannaschii* (*M.j.*) aIF2γ or yeast (*S.c.*) eIF2γ implicated in binding to aIF2α/eIF2α are circled. (*B*) Model of Phe-tRNA^Phe (in stick figure) docked on the *S. solfataricus* aIF2αγ heterodimer. Residues corresponding to those in *M. jannaschii* (*M.j.*) aIF2γ or yeast (*S.c.*) eIF2γ implicated in tRNA binding are circled. (Modified, with permission, from Yatime et al. 2006.)

(Fig. 4A). Consistent with this, the D403A mutation in yeast eIF2γ, corresponding to D302 in the second α-binding pocket of *S. solfataricus aIF2γ*, produced a Slg⁻ phenotype suppressible by overexpressing eIF2α/SUI2 but not Met-tRNA$_i^{Met}$. The equivalent mutation in *Methanococcus jannaschii* aIF2γ (D325A) impaired binding to aIF2α in vitro, as did the L256D mutation (Roll-Mecak et al. 2004) corresponding to L233 in the first α-binding pocket of *S. solfataricus aIF2γ* (Fig. 4A). Thus, the crystal structure of the *S. solfataricus aIF2γ/α* heterodimer likely provides a good model for the physiological interface between γ and α subunits in both aIF2 and eIF2.

In vitro assays suggest that association with the aIF2α CTD increases the affinity of aIF2γ for Met-tRNA$_i^{Met}$ by two orders of magnitude (Yatime et al. 2004, 2006). Presumably, α acts indirectly by adjusting the tRNA-binding channel of γ (Fig. 4B) (Yatime et al. 2006). In vitro analysis of a yeast eIF2 β/γ complex showed that eliminating the α subunit from eIF2 produced only a 5-fold reduction in affinity for Met-tRNA$_i^{Met}$ (Nika et al. 2001). Thus, α seems to play a much smaller role in Met-tRNA$_i^{Met}$ binding to eIF2 versus aIF2.

eIF2β: Interactions with Met-tRNA$_i^{Met}$, mRNA, and eIF2Bε

Whereas the *P. abyssi* αγ heterodimer and aIF2 holoproteins bind Met-tRNA$_i^{Met}$ with indistinguishable affinities (Yatime et al. 2004; Pedulla et al. 2005), a mammalian eIF2$\alpha\gamma$ dimer binds nucleotide but cannot form a stable complex with Met-tRNA$_i^{Met}$ (Flynn et al. 1993), suggesting that β is more critical than α for Met-tRNA$_i^{Met}$ binding in eukaryotes. Consistent with this, the *SUI3-S264Y* mutation in yeast eIF2β impairs Met-tRNA$_i^{Met}$ binding by eIF2 (Huang et al. 1997).

M. jannaschii aIF2β contains a $\beta\beta\alpha\alpha\beta\beta$ domain connected by a linker to a carboxy-terminal β-sheet stabilized by Zn^{2+} (Cho and Hoffman 2002). eIF2β contains an additional amino-terminal domain (NTD) containing three polylysine stretches (K-boxes 1–3) not found in archaea (Fig. 3B). The cysteine residues in the Zn-binding pocket of yeast eIF2β are essential for viability (Castilho-Valavicius et al. 1992), and a *SUI3* allele lacking the Zn pocket has a dominant Gcd$^-$ phenotype, suggesting that eIF2 containing the mutant protein is defective in formation or 40S binding of TC. Numerous dominant Sui$^-$ alleles in *SUI3* alter conserved residues around the Zn pocket (Donahue et al. 1988; Castilho-Valavicius et al. 1992), including *S264Y* and *L254P* that increase GTP hydrolysis by purified TC. *S264Y* also increases Met-tRNA$_i^{Met}$ dissociation from eIF2-GTP. Both defects could increase the probability that TC dissociates and leaves Met-tRNA$_i^{Met}$ base-paired with UUG in the P site, accounting for the Sui$^-$ phenotypes (Huang et al. 1997). A segment of yeast eIF2β is necessary and sufficient for eIF2γ binding (amino acids 128–159) (Fig. 3B), and substitutions of conserved Tyr-131 and Ser-132 impair binding of eIF2β to eIF2γ in vitro and to the eIF2 $\gamma\alpha$ heterodimer in vivo. The *SUI3-YS* allele containing both substitutions confers Ts$^-$ and Sui$^-$ phenotypes and is synthetically lethal with the Sui$^-$ *SUI3-S264Y* allele. By weakening β–γ interaction, *SUI3-YS* may exacerbate the hyperactive GTPase function or the defect in Met-tRNA$_i^{Met}$ binding conferred by *S264Y* (Hashimoto et al. 2002).

eIF2 can bind mRNA via the β subunit, and this activity may be a determinant of translational efficiency (Hinnebusch 2000). The binding of mRNA by yeast eIF2β is dependent partly on its Zn-binding motif and more strongly on its K-boxes (Fig. 3B). Deletion of all three K-boxes is lethal, but *SUI3* alleles with any single K-box are viable (Laurino et al. 1999). Removing K-boxes 1 and 2 abolishes the Sui$^-$ phenotype of the *SUI3-S264Y* allele, possibly by weakening interaction of TC with the UUG start codon, or by impairing eIF2 interaction with eIF5, which also involves the K-boxes (see below). Although the K-boxes are not essential for TC formation or recruitment, a *SUI3* allele lacking all three K-boxes confers dominant Slg$^-$ and Gcd$^-$ phenotypes (Laurino et al. 1999), and all viable K-box mutations produce Gcd$^-$ phenotypes (Asano et al. 1999). Consistent with this, the K-boxes stabilize interaction of eIF2β or eIF2 holoprotein with the catalytic ϵ subunit of eIF2B (GCD6), and most likely promote recycling of eIF2-GDP and TC formation. Archaea lack orthologs of eIF2Bε and eIF5, suggesting that the eIF2β K-boxes evolved to facilitate the interactions of eIF2 with its GEF and GAP (Asano et al. 1999).

Eukaryotic eIF2α Promotes and Regulates GDP–GTP Exchange by eIF2B

The 3D structures of archaeal and human eIF2α reveal three distinct domains (Fig. 3C). As mentioned above, the carboxy-terminal domain III of aIF2α interacts with aIF2γ and promotes Met-tRNA$_i^{Met}$ binding (Yatime et al. 2004, 2006). The amino-terminal domain I contains a β-barrel "OB fold" (Nonato et al. 2002; Dhaliwal and Hoffman 2003; Ito et al. 2004) with nonspecific RNA-binding activity (Yatime et al. 2004) and interacts with the central, helical domain of eIF2α. Two Sui$^-$ mutations were isolated in domain I of yeast eIF2α, *P13S* and *V19F* (Cigan et al. 1989), but the mechanism of their Sui$^-$ phenotypes is unknown.

Domain I contains the regulatory phosphorylation site at Ser-51, and Gcn$^-$ mutations of single residues in this region of yeast eIF2α (Fig. 3D) reduce the inhibitory effect of eIF2(αP) on eIF2B activity without reducing eIF2α phosphorylation (Vazquez de Aldana et al. 1993; Krishnamoorthy et al. 2001; Dey et al. 2005). These residues form a contiguous surface on one face of eIF2α that is thought to be the binding site for the eIF2B regulatory subcomplex (see below). In fact, genetic data indicate that yeast eIF2α functions primarily to promote and regulate GDP–GTP exchange by eIF2B. *sui2Δ* cells lacking eIF2α can survive if eIF2(βγ) and tRNA$_i^{Met}$ are overexpressed, and a more complete bypass of eIF2α function is achieved by overexpressing eIF2γ-K250R, wild-type eIF2β, and tRNA$_i^{Met}$, as *K250R* enhances spontaneous GDP–GTP exchange by eIF2. Overexpressing eIF2 subunits (with the K250R mutation) and tRNA$_i^{Met}$ also suppresses the lethality of deleting all four essential eIF2B subunits. Consistent with this, there is a ~20-fold increase in the K_m for eIF2[βγ]-GDP versus eIF2[αβγ]-GDP in eIF2B-catalyzed nucleotide exchange (Nika et al. 2000), suggesting that eIF2α contributes to eIF2 binding by eIF2B. Perhaps the critical role of the α subunit in Met-tRNA$_i^{Met}$ binding by aIF2 was transferred to the β subunit during eukaryotic evolution as the α subunit acquired new functions in nucleotide exchange.

eIF2B REGULATES TC FORMATION

eIF2B is required to displace the GDP bound to eIF2 for its replacement with GTP. eIF2B contains five conserved subunits, ε/GCD6, δ/GCD2, γ/GCD1, β/GCD7, α/GCN3, and all but α/GCN3 are essential proteins.

Recessive mutations in the four essential subunits confer Ts$^-$ and Gcd$^-$ phenotypes (Hinnebusch 1996) consistent with impaired eIF2 recycling. However, eIF2Bε exhibits significant GEF activity on its own (Fabian et al. 1997; Pavitt et al. 1998; Gomez and Pavitt 2000), lodged in the last ~25% (195 amino acids) of the protein (Gomez et al. 2002). The extreme carboxyl terminus of this segment contains conserved *a*romatic and *a*cidic (AA) boxes 1 and 2 required for binding eIF2. A 7-Ala substitution in box 2 (*gcd6–7A*) impairs eIF2–eIF2B interaction in vivo and confers a Gcd$^-$ phenotype suppressible by overexpressing the four macromolecules in TC (eIF2α, -β, -γ, and tRNA$_i^{Met}$) from a high-copy (hc) plasmid (hc-TC). *gcd6–7A* also impairs binding of recombinant eIF2β-NTD to eIF2Bε in vitro (Asano et al. 1999). Deleting both AA boxes destroys the eIF2-binding and GEF functions of solitary ε/GCD6 and reduces the GEF activity of eIF2B holoprotein (Gomez and Pavitt 2000; Gomez et al. 2002). As noted above, the K-boxes in eIF2β-NTD are required for tight binding of eIF2 (and recombinant eIF2β) to recombinant eIF2Bε, suggesting that the K-box domain interacts with the eIF2Bε CTD (Asano et al. 1999). A region amino-terminal to the AA boxes in eIF2Bε/GCD6 (residues 518–581) is required for GEF activity but not eIF2 binding (Gomez et al. 2002), and Glu-569 appears to be the critical catalytic residue in this region (Boesen et al. 2004). The 3D structure of the yeast eIF2Bε CTD comprises four hairpins of antiparallel α helices (HEAT repeats) (see Fig. 6B) (Boesen et al. 2004) with Glu-569 in helix II, AA box 1 spanning helices VI–VII, and AA box 2 in helix VIII.

Because deleting the AA boxes in ε/GCD6 does not abolish GEF activity of eIF2B holoprotein, there must be additional contacts between eIF2B and eIF2 subunits. Indeed, the α, β, and δ eIF2B subunits form a subcomplex that can bind eIF2α (see below). Point mutations in a conserved Asn-Phe-Asp motif in ε/GCD6 (amino acids 249–251) have no effect on GEF activity of the isolated subunit, but reduce eIF2B function to nearly that of ε/GCD6 alone. These mutations do not impair eIF2B assembly or eIF2 binding and thus appear to disrupt a stimulatory effect of other eIF2B subunits on ε/GCD6 catalytic function (Gomez and Pavitt 2000).

Inhibition of eIF2B by Phosphorylated eIF2

Phosphorylated eIF2 can form the TC and function efficiently in PIC assembly but is a poor substrate for nucleotide exchange by eIF2B. Phosphorylation of eIF2 increases its affinity for eIF2B (Proud 1992; Pavitt et al. 1998), and it is frequently assumed that eIF2(αP)-GDP forms a nondissociable complex with eIF2B; however, other evidence indicates that eIF2(αP)-GDP acts as a competitive inhibitor (Rowlands et al. 1988; Dever et al. 1995).

eIF2α phosphorylation does not induce *GCN4* translation in *gcn3Δ* mutants lacking eIF2Bα (Gcn$^-$ phenotype; Fig. 2D), and *gcn3Δ* does not impair growth in rich medium (Hannig and Hinnebusch 1988; Dever et al. 1993), thus indicating that α/GCN3 is required only for inhibition of eIF2B by eIF2(αP). Consistent with this, yeast eIF2B lacking α/GCN3 shows full GEF activity that is resistant to eIF2(αP) in vitro (Pavitt et al. 1998). Point mutations conferring Gcn$^-$ phenotypes also were isolated in the essential β/GCD7 and δ/GCD2 subunits of eIF2B, and both proteins are similar in sequence to GCN3. The majority of eIF2 can be phosphorylated in such mutants without impairing translation. Hence, inhibition of eIF2B by eIF2(αP) depends on the α/β/δ subunits of eIF2B (Vazquez de Aldana and Hinnebusch 1994; Pavitt et al. 1997). It was demonstrated biochemically that *gcn3Δ* and the Gcn$^-$ alleles *GCD7-S119P* and *GCD7-I118T,-D178Y* rescue eIF2B from eIF2(αP) by allowing nucleotide exchange on both phosphorylated and unphosphorylated eIF2-GDP (Pavitt et al. 1998).

When overexpressed, α/GCN3, β/GCD7, and δ/GCD2 form a stable subcomplex that binds purified eIF2 in a manner stimulated by Ser-51 phosphorylation (Pavitt et al. 1998) and reduces the toxicity of hyperactivated GCN2 by sequestering eIF2(αP) (Yang and Hinnebusch 1996). Because the eIF2B α/β/δ subcomplex, but not individual subunits, binds recombinant eIF2α-P in vitro, the three subunits likely interact to form a binding pocket for eIF2α (Krishnamoorthy et al. 2001). All Gcn$^-$ mutations tested in eIF2Bβ or the eIF2α NTD decrease binding of the eIF2B α/β/δ subcomplex or holoprotein to eIF2α-P in vitro (Krishnamoorthy et al. 2001; Dey et al. 2005). This suggests that tight binding of the phosphorylated α subunit of eIF2 to the α/β/δ regulatory subcomplex impedes interaction of the catalytic subunit of eIF2B (ϵ/GCD6) with the GDP-binding pocket of eIF2γ in the manner required for GDP release. The *GCD7* mutations also decrease interaction between eIF2B and unphosphorylated eIF2, indicating that contacts between the α/β/δ subcomplex and eIF2α contribute to productive interaction with nonphosphorylated eIF2-GDP as well (Krishnamoorthy et al. 2001). Presumably, phospho-Ser51 provides an extra contact with the α/β/δ subcomplex that is critical for disrupting interaction of GCD6 with the GDP-binding pocket in eIF2γ. The eIF2B regulatory subunits show similarity to a methionine salvage enzyme whose 3D structure (Bumann et al. 2004) and that of a closely related archaeal protein (Kakuta et al. 2004) have been determined. The biological significance of this unexpected similarity is currently unknown.

FORMATION OF THE 43S PIC

In reconstituted mammalian systems, binding of TC to 40S subunits is promoted independently by eIF1, eIF1A, and eIF3 (Fig. 1), with the greatest stimulation occurring when all three are present (Majumdar et al. 2003; Kolupaeva et al. 2005). In a reconstituted yeast system, eIF1A and eIF1 function in the absence of eIF3 to stimulate TC binding to 43S–mRNA complexes, dependent on AUG (Algire et al. 2002; Maag et al. 2005b). However, yeast eIF3 and eIF5 enhance TC binding to the 43S PIC (J. Lorsch, pers. comm.), which has >50-fold lower affinity for TC than the 43S–mRNA complex (Maag et al. 2005b). Consistent with this, mutations in the b/PRT1 subunit of eIF3 (*prt1-1*) (Danaie et al. 1995; Phan et al. 1998) and the eIF5-CTD (*tif5-7A*) (Asano et al. 2001) reduce binding of Met-tRNA$_i^{Met}$ to 40S subunits in yeast extracts in a manner rescued by the purified factors.

The additive effects of mammalian eIF1, eIF1A, and eIF3 in TC recruitment can be understood at least partly by their interdependence in 40S binding (Majumdar et al. 2003). Similarly, there is approximately 10-fold thermodynamic coupling in 40S binding of yeast eIF1 and eIF1A in vitro (Maag and Lorsch 2003). TC also stabilizes 40S binding of mammalian eIF3 (lacking the j subunit) in the absence of mRNA (Kolupaeva et al. 2005), and TC increases the affinity of yeast eIF1 for 40S ribosomes in the absence of mRNA (Maag et al. 2005b). The discovery that yeast eIF3, eIF1, eIF5, and TC can be isolated in a multifactor complex (MFC) (Asano et al. 2000) raised the possibility that these factors are recruited to the 40S as a preformed unit, as discussed below.

eIF3 Resides in an MFC with eIFs 1, 2, and 5

eIF3 from budding yeast contains as stoichiometric subunits homologs of only 5 of the 13 subunits of mammalian eIF3 (eIF3a/TIF32, eIF3b/PRT1, eIF3c/NIP1, eIF3g/TIF35, and eIF3i/TIF34) (Phan et al. 1998), which are all essential for translation initiation in yeast (Hinnebusch 2000). A sixth yeast homolog, eIF3j/HCR1, is a nonessential subunit that enhances interactions between eIF3 and other eIFs (Valášek et al. 1999, 2001) and promotes 40S binding of eIF3 (Nielsen et al. 2006). Because no other eIF3 subunits are found, the 5 essential subunits present in budding yeast may execute the critical functions of the more complicated mammalian factor. Interestingly, 5 of the 7 additional eIF3 subunits missing in budding yeast are present in the fission yeast *Schizosaccharomyces pombe* and reside in a complex with epitope-tagged versions of the core eIF3 subunits (Akiyoshi et al. 2000; Zhou et al. 2005; and references therein). Further genetic analyses are required to define the functions of these other subunits. Pairwise interactions among the yeast eIF3 subunits led to a subunit interaction model for the complex (Fig. 5A) (Verlhac et al. 1997; Asano et al. 1998; Phan et al. 1998; Vornlocher et al. 1999; Valášek et al. 2001). Many aspects of this model have been confirmed in vivo by making deletions of pre-dicted binding domains and determining the compositions of the result-ing purified subcomplexes (Valášek et al. 2002). Analyses of subunit in-teractions in mammalian eIF3 are consistent with the yeast model (Methot et al. 1997; Shalev et al. 2001; Fraser et al. 2004).

PRT1/b forms distinct subcomplexes in vivo with TIF32/a and NIP1/c or with TIF34/i and TIF35/g, in accordance with the model (Fig. 5A). Whereas the larger a/b/c subcomplex can restore 40S binding of both Met-tRNA$_i^{Met}$ and mRNA and the translation of reporter mRNA in a *prt1-1* extract, the smaller b/i/g subcomplex is relatively inert (Phan et al. 2001). Consistent with this, expression of amino-terminally truncated PRT1/b lacking the predicted RRM (ΔRRM, Fig. 5D) sequesters TIF34/i and TIF35/g in an inactive subcomplex lacking TIF32/a and NIP1/c that cannot bind to ribosomes in vivo and has a dominant-negative Slg$^-$ phe-notype (Evans et al. 1995; Valášek et al. 2001). Moreover, substitution of the RNP1 motif of the PRT1 RRM (*prt1-rnp1*) impairs direct interactions of PRT1 with TIF32/a and HCR1/j and decreases eIF3 recruitment by 40S-bound HCR1/j in vivo. Overexpressing the HCR1-R215I variant decreases the Ts$^-$ phenotype of the *prt1-rnp1* mutation and elevates 40S binding of eIF3 in *prt1-rnp1* cells, and the *hcr1Δ* mutation reduces 40S binding by wild-type eIF3 (Nielsen et al. 2006). Thus, PRT1-RNP1/HCR1 and HCR1–40S interactions promote ribosome binding of yeast eIF3 in vivo.

Yeast eIF3 copurifies with both eIF1 (Naranda et al. 1996; Phan et al. 1998) and eIF5 (Phan et al. 1998). In vitro, eIF1 and the eIF5-CTD bind simultaneously to the NTD of eIF3 subunit NIP1/c (Phan et al. 1998; Asano et al. 1999, 2000), and consistent with this, eIF1 and eIF5 copu-rify with the eIF3 a/b/c subcomplex (Phan et al. 2001). In vitro, the yeast eIF5-CTD interacts simultaneously with the eIF2β-NTD (K-box do-main), the NIP1/c-NTD, and eIF1 (Asano et al. 1999, 2000; Singh et al. 2004b), suggesting that eIF5-CTD can link together eIF2, eIF3, and eIF1 in the MFC (Fig. 5A). The MFC containing eIFs 1, 2, 3, and 5 was puri-fied free of ribosomes and found to contain tRNA$_i^{Met}$ in 1:1 stoichiome-try with eIF2 (i.e., the TC) (Asano et al. 2000). Mutating the AA boxes in the eIF5-CTD (*tif5-7A* allele) impairs interaction of eIF5 with eIF2β-NTD and NIP1/c-NTD in vitro and reduces association of eIF2, eIF3, and eIF5 in extracts. That *tif5-7A* impairs translation and cell growth in a manner partially suppressed by hc-TC suggests that interac-

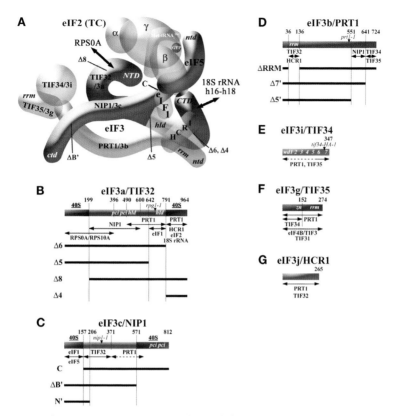

Figure 5. Schematic representations of the multifactor complex (MFC) and primary structures of yeast eIF3. (*A*) Model of the yeast MFC based on binary interactions between isolated recombinant subunits and affinity purification of MFC subcomplexes produced by His$_8$-tagged subunits harboring deletions of predicted binding domains for other components of the complex. eIF3 subunits (*orange, red,* and *purple shapes*) are labeled with yeast (e.g., TIF32) and universal (e.g., 3a) designations. The three subunits of eIF2 (*green*) with GTP and Met-tRNA$_i^{Met}$ bound (primarily) to eIF2γ comprise the TC. The protein subunits and Met-tRNA$_i^{Met}$ are shown roughly in proportion to their molecular weights. The eIF5, NIP1, and termini of TIF32 are depicted as solid rather than transparent shapes to emphasize their importance in binding to 40S ribosomes. Interactions of eIF3 subunits with RPS0A and helices 16–18 of 18S rRNA are depicted (Valášek et al. 2003), as are locations of relevant deletion endpoints. (*NTD*) Amino-terminal domain; (*CTD*) carboxy-terminal domain; (*hld*) HCR1-like domain; (*rrm*) RNA recognition motif. (*B–G*) The primary sequences of yeast eIF3 depicted using the color scheme in *A* with amino acid positions and locations of point mutations above, and locations of deletions and binding domains for other factors indicated below. The locations of PCI homology domains in TIF32 and NIP1 (*pci*), the HCR1-like domain in TIF32 (*hld*), WD repeats in TIF34 (*wd 1,2, ... 7*), predicted in PRT1 and TIF35, and a predicted Zn-binding domain in TIF35 (*zn*) are also indicated in the colored rectangles. See text for further details. (Modified from Hinnebusch et al. 2004.)

tion of eIF2 with eIF5-CTD in the MFC enhances translation initiation in vivo (Asano et al. 1999, 2000). eIF1 can bind to eIF2β-NTD in addition to eIF5-CTD and NIP1/c-NTD, expanding the network of interactions among these factors in the MFC (Singh et al. 2004b).

The prediction that NIP1/c-NTD interacts with eIF1 and mediates eIF2–eIF3 interaction via the eIF5-CTD was confirmed in vivo by showing that a His$_8$-tagged version of NIP1/c-lacking the NTD (NIP1-C, Fig. 5C) copurified with all eIF3 subunits but not with eIFs 1, 2, or 5, whereas the tagged NTD of NIP1/c copurified with eIFs 1, 2, and 5 but not with eIF3 subunits. A similar analysis of TIF32/a truncations confirmed binding domains for NIP1/c and PRT1/b and revealed additional binding domains for eIF2β and eIF1 in the CTD. The extreme CTD of TIF32/a (Δ4, Fig. 5B) can bind eIF2β in vitro and eIF2 in vivo in a subcomplex lacking other MFC components. Deletion of the TIF32 CTD is lethal, and overexpressing this truncated allele (hc *tif32-Δ6*, Fig. 5B), generating an otherwise intact MFC lacking eIF2, confers a dominant Slg$^-$ phenotype that is partially suppressed by hc-TC. Thus, direct eIF2–eIF3 contact through the TIF32-CTD also seems to be required for optimal initiation in vivo (Valášek et al. 2002).

Interaction of eIF2β-NTD with eIF5 enhances association of eIF2β-NTD with NIP1/c-NTD in vitro, suggesting that the eIF2β–eIF5 interaction nucleates the eIF2-eIF5-eIF3 subassembly of the MFC. Consistent with this, hc-TC in *tif5-7A* cells restores both eIF3–eIF2 and eIF2–eIF5 interactions. Furthermore, deleting the last 10 residues of eIF5 (*tif5-W391Δ*) selectively reduces binding of eIF5-CTD to eIF2β-NTD in vitro but additionally impairs eIF3–eIF5 and eIF2–eIF3 interactions in vivo (Singh et al. 2004a), and hc-TC partially suppresses the Ts$^-$ phenotype of *tif5-W391Δ* (Singh et al. 2005). From a homology model of the eIF5-CTD constructed from the HEAT-repeat structure of eIF2Bε-CTD (Boesen et al. 2004), it was proposed that the *tif5-W391Δ* mutation eliminates an eIF2β contact in the eIF5-CTD without disrupting connections between helices 6 and 8 required for NIP1/c-NTD interaction (Fig. 6B,C) (Singh et al. 2004a). Mutating a patch of acidic residues adjacent to the carboxyl terminus (the *AN1* allele) also selectively impairs binding of eIF2β-NTD to eIF5-CTD, extending the boundaries of the predicted eIF2β-binding surface (Fig. 6B,C). Mutating a cluster of basic residues on a different face of eIF5-CTD (substituted by the *BN1* mutation) identified a potential binding surface for NIP1/c-NTD and eIF1. The effects of the *AN1* and *BN1* mutations on eIF2–eIF3–eIF5 interactions in vivo are more complex, presumably because they disrupt allosteric coupling between eIF2β-NTD and NIP1/c-NTD in binding eIF5-CTD. As discussed below, the *AN1* and *BN1* mutations have distinct effects on *GCN4* translational

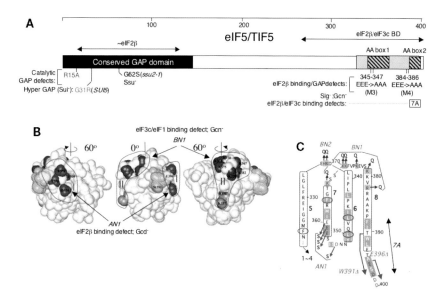

Figure 6. Primary and predicted tertiary structures of eIF5. (*A*) Primary structure of yeast eIF5 depicting conserved domains, region of similarity to eIF2β, and binding domains (BD) for eIF2β and eIF3c/NIP1. The locations and phenotypes/effects of point mutations, including *tif5-7A* (7A), are shown below. (*B*) Different views of the homology model of the eIF5-CTD HEAT domain constructed from the crystal structure of eIF2Bε-CTD (Boesen et al. 2004) showing residues deleted or substituted by the *W391D* (*orange* and *yellow*), *AN1* (*red*), and *BN1* (*dark blue*) mutations. (*C*) Primary sequence of eIF5-CTD in helices 5–8 (boxed) arranged according to their topology, showing the *W391Δ*, *AN1*, and *BN1* mutations, residues in the AA boxes (highlighted in *yellow*), and residues mutated in *tif5–7A* (*7A*). (*A*, Modified from Hinnebusch et al. 2004; *B*, *C*, modified, with permission, from Yamamoto et al. 2005 [©National Academy of Sciences].)

control consistent with TC recruitment (*AN1*) and scanning (*BN1*) defects in vivo (Yamamoto et al. 2005).

Formation of the MFC Stimulates 43S PIC Assembly in Yeast Cells

If assembly of eIFs 1, 3, 5 and TC in the MFC allows cooperative binding of these factors to the 40S subunit, then mutations that disrupt the connections between constituent factors should impair 40S binding of multiple MFC components. There is increasing evidence supporting this prediction. In cell extracts, the *prt1–1* mutation in eIF3b impairs 40S binding of all MFC components (Phan et al. 2001), and the *tif5-7A* mutation in eIF5 reduces TC binding to 40S subunits (Asano et al. 2001). In vivo, *tif5-7A* only affects 40S binding of eIF5 itself (Asano et al. 2001), but overexpressing *tif32-Δ6* (which eliminates the TIF32-CTD/eIF2β interaction) in *tif5-7A* cells exacerbates the Slg⁻ phenotype and reduces 40S binding of eIF2 and eIF3 in native PICs (Nielsen et al. 2004). Thus, the contacts between eIF2 and eIF3 via the eIF5-CTD and TIF32-CTD (Fig. 5A) make additive contributions to 40S binding of eIF2 and eIF3 in vivo. Consistent with this, the *tif5-AN1* mutation (Fig. 6B) disrupts eIF2–eIF3 association in the MFC and produces a Gcd⁻ phenotype in vivo (Yamamoto et al. 2005). Interactions mediated by the NIP1/c-NTD also promote efficient 40S binding of MFC components. The defective subcomplex formed by NIP1-NTD, eIFs 2, 1, and 5 cannot bind efficiently to 40S subunits (Valášek et al. 2003), confers a Gcd⁻ phenotype that is suppressed by hc-TC (Valášek et al. 2002), and lowers 40S binding of eIF2 in vivo (Nielsen et al. 2004). Moreover, mutations in two 10-amino acid stretches of the NIP1/c-NTD also produce Gcd⁻ phenotypes suppressed by hc-TC, reduce binding of NIP1-NTD to eIF1 and eIF5 in vitro, and decrease 40S binding of eIF2 and eIF5 in vivo (Valášek et al. 2004).

It was shown recently that three yeast "degron" mutants endowed with conditional expression of eIF2β, eIF3a plus eIF3b, or eIF5 exhibit reduced 40S binding of all MFC constituents following depletion of the relevant factor in each strain. None of these mutants showed complete loss of MFC recruitment, indicating that eIF2, eIF3, and eIF5 are not fully interdependent for 40S binding in vivo, although the kinetics of binding may have been reduced more dramatically than the steady-state levels of binding to 40S subunits by MFC constituents (Jivotovskaya et al. 2006). Finally, eIF1 also makes an important contribution to MFC recruitment in vivo, because attaching the FLAG epitope to its carboxyl terminus (eIF1-FL) is lethal, produces a dominant Gcd⁻ phenotype, and reduces 40S binding of MFC components when eIF1-FL is overexpressed in wild-type cells (Singh et al. 2004b).

As noted above, *tif5-7A* does not reduce steady-state levels of 40S-bound MFC in vivo. In fact, it confers a Gcn⁻ phenotype, indicating impaired induction of *GCN4* translation in response to eIF2α phosphorylation. Interestingly, the Gcn⁻ phenotype is suppressed by hc *tif32-Δ6* which, as noted above, reduces 40S binding of eIF2 and eIF3 in *tif5-7A* cells. Hence, it was proposed that *tif5-7A* impairs scanning or AUG recognition in a manner that prevents reinitiating ribosomes from bypassing the *GCN4* uORF4 when TC levels are reduced by eIF2α phosphorylation. When *tif5-7A* is combined with hc *tif32-Δ6*, the additional impairment of TC recruitment outweighs the scanning defect and restores the bypass of uORF4 and ensuing translation of *GCN4*, suppressing the Gcn⁻ phenotype of *tif5-7A* (Nielsen et al. 2004). A similar explanation applies to Ts⁻ point mutations in the eIF5-CTD that destabilize the MFC and produce Gcd⁻ phenotypes at 30°C but confer Gcn⁻ phenotypes and more severe impairment of translation at 36°C. Apparently, the more severe disruption of the MFC at 36°C elicits a scanning defect (leaky scanning of uORF1 in this case) that outweighs the impaired TC recruitment responsible for the Gcd⁻ phenotype at 30°C (Singh et al. 2005). The *tif5-BN1* allele, which disrupts interaction of eIF5-CTD with eIF3c/NIP1 and eIF1, belongs to this class of mutations, suggesting that these connections are important for AUG selection during scanning (Yamamoto et al. 2005).

40S Binding Sites for MFC Components

Mammalian eIF1 binds to the 40S interface surface near the P site, similar to that of bacterial IF3 on the 30S subunit (Lomakin et al. 2003). Domains in yeast eIF3 core subunits that bind to 40S ribosomes were identified by determining whether subcomplexes formed by mutant versions of a/TIF32 or c/NIP1 compete with native MFC for 40S binding in vivo. Deleting the amino and carboxyl termini of NIP1 and the TIF32-NTD impairs 40S binding by otherwise intact MFC complexes containing these truncated proteins (*TIF32-Δ8* mutation, Fig. 5B; *NIP1-ΔB'*, Fig. 5C), suggesting that these segments make direct contact with the 40S subunit. Indeed, full-length NIP1, the amino-terminal half of TIF32, and eIF5 comprise a minimal 40S binding unit (MBU) sufficient for 40S binding in vivo and in vitro, albeit at levels reduced from that with intact MFC. The eIF5 is necessary for 40S binding of eIF3 subunits only when the TIF32-CTD is deleted, as in the case of the MBU formed by TIF32-Δ5 (Fig. 5B) (Valášek et al. 2003). Interestingly, the TIF32-NTD and NIP1 interact with isolated 40S subunit proteins RPS0A and RPS10A, suggesting that eIF3 contacts the solvent-exposed back side of the 40S. TIF32-CTD also interacts with helices 16–18 of 18S rRNA, prompting the suggestion that TIF32 gains access to the interface side of the 40S near the A site (Valášek et al. 2003). A cryo-EM model of mammalian eIF3 bound to the 40S subunit (Siridechadilok et al. 2005) shows the bulk of eIF3 binding to the back side of the 40S, but an extension of eIF3 reaches around to the interface side near the E site, not A site, of the 40S. There is little information about the 40S-binding sites for eIF5 and TC, but Met-tRNA$_i^{Met}$ is expected to occupy the P site.

Functions of eIF1A in 43S PIC Assembly

eIF1A is about 20% identical in sequence to bacterial IF1, and both proteins contain the β-barrel OB-fold (Battiste et al. 2000), and eIF1A also contains an α-helical domain and unstructured amino- and carboxy-terminal extensions (Fig. 7). Because IF1 binds to the A site of the 30S subunit, the OB-fold in eIF1A is thought to occupy the same location in 40S subunits. Nuclear magnetic resonance (NMR) analysis identified residues in the OB-fold and helical domains of mammalian eIF1A that may contact RNA, and mutations of several such residues reduced RNA binding and impaired scanning in a reconstituted system. The *K67D* mutation in the OB-fold also impaired TC binding to 40S subunits in vitro (Fig. 7B) (Battiste et al. 2000).

Yeast eIF1A is essential for translation initiation in vivo (Kainuma and Hershey 2001). Deleting the unstructured CTD of yeast eIF1A (ΔC) impairs TC recruitment in vivo, conferring a Gcd$^-$ phenotype that is suppressed by hc-TC (Fig. 7A) (Olsen et al. 2003). Consistent with this, ΔC cells show reduced levels of 40S-bound eIF2, eIF3, and eIF5, despite higher 40S binding by the mutant protein itself, and the defect in PIC assembly is partially rescued by hc-TC. Assays with purified components showed that ΔC decreases the rate of TC loading while increasing the affinity of eIF1A for 40S subunits. Substitution of amino acids 66–70 in helix $\alpha 1$ of the OB-fold (mutant *66-70*) reduces 40S binding of eIF1A in vivo and increases the K_d for the 40S–eIF1A complex by about 100-fold in vitro. This mutation also impairs TC recruitment in vivo, conferring a Gcd$^-$ phenotype (Fig. 7A), and decreases 40S binding of eIF2, eIF3, and eIF5, and these defects are diminished by increasing the concentration of the mutant protein to restore its 40S association. Thus, $\alpha 1$ in eIF1A likely contacts the 40S A site (Fekete et al. 2005).

Substitution of amino acids 98–101 in the helical domain of eIF1A also impairs MFC binding to the 40S in vivo in a manner suppressed by hc-TC, but it produces a Gcn$^-$ rather than Gcd$^-$ phenotype, apparently by decreasing the rate of ribosomal scanning between uORFs 1 and 4 on *GCN4* mRNA. In accord with this explanation, *98-101* reduces the ability of yeast eIF1A to promote scanning to the AUG codon in reconstituted mammalian 48S PICs, and it confers a Sui$^-$ phenotype in yeast (Fig. 7A). Similar to the Gcn$^-$ mutations in eIF5 discussed above, the impairment of scanning by *98-101* probably masks its defect in TC recruitment during translation of *GCN4* mRNA. The ΔC mutation in eIF1A also impairs scanning in vitro and confers a Sui$^-$ phenotype in vivo, but its defect in TC recruitment must outweigh the scanning deficiency to yield a net Gcd$^-$ phenotype (Fekete et al. 2005).

FORMATION OF THE 48S PIC

Secondary structure in the 5'UTR of mRNA impedes translation by interfering with recruitment of the 43S PIC to the mRNA. mRNA recruitment is promoted by eIF4E and PABP, which bind to the cap and poly(A)

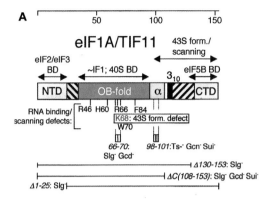

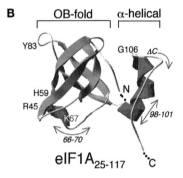

Figure 7. Important domains and 3D structures of eIF1A. (*A*) Primary structure of yeast eIF1A is depicted showing the α-helical (α), 3_{10} helix, and other relevant domains, based on the 3D structure of human eIF1A in *B*. The region of similarity to IF1 (~IF1) predicted to be the 40S-binding domain (BD) is indicated, as are binding domains for other factors and the carboxy-terminal region implicated by mutations in 43S PIC assembly (43S form.) and scanning in vivo. Locations and phenotypes of selected mutations are shown below, as are yeast residues corresponding to those in human eIF1A whose mutation produced defects in RNA binding and scanning (R46, H60, R66, K68, W70, F84) or 43S PIC formation (K68). (*B*) Solution structure of residues 25–117 of human eIF1A (Battiste et al. 2000) (PDB ID: 1D7Q) showing the residues implicated in RNA binding, scanning, or 43S PIC assembly for the human protein, and the predicted locations of selected mutations in yeast eIF1A, as described in *A*. (Modified from Hinnebusch et al. 2004.)

tail, respectively, and interact with separate domains of eIF4G in the eIF4F complex. The eIF4G can potentially form a protein bridge in the 43S PIC, and eIF4G also recruits the RNA helicase eIF4A to remove secondary structure from the 5' end of the mRNA (Fig. 1). Yeast eIF4E is encoded by the single essential gene *CDC33* (Altmann et al. 1985), whereas two isoforms of eIF4G (-1 and -2), about 50% identical, are encoded by *TIF4631* and *TIF4632*, respectively, either of which is sufficient for viability (Goyer et al. 1993; de la Cruz et al. 1997). eIF4A is also encoded by two genes, *TIF1* and *TIF2*, that specify identical proteins, and either is sufficient for wild-type growth (Tanner and Linder 2001). Unlike the case in mammals, eIF4A is not stably associated with eIF4E and eIF4G in yeast extracts (Goyer et al. 1989). The single-copy essential gene *PAB1* encodes yeast PABP (Sachs 2000).

Role of the eIF4E–eIF4G Complex in Binding the m⁷GpppN Cap

Yeast eIF4E is a globular cap-binding protein with an unstructured amino-terminal extension (Matsuo et al. 1997). *cdc33* mutations have been described that impair cap binding by eIF4E or its interaction with eIF4G, including *cdc33-4-2*, which confers Ts⁻ translation both in vivo and in cell extracts in a manner rescued by recombinant CDC33 (Altmann et al. 1988). eIF4E binds to a segment in the amino-terminal domain of eIF4G containing the conserved motif $Y(X_4)L\Phi$ (where Φ is a hydrophobic amino acid) (Fig. 8) (Mader et al. 1995). In strains containing only one eIF4G isoform, alanine substitution of the tyrosine residue of the motif is lethal, whereas substituting the last two leucine residues (*tif4631–459* and *tif4632–430*) confers a Ts⁻ phenotype suppressible by eIF4E overexpression and reduces eIF4G–eIF4E association (Tarun and Sachs 1997; Tarun et al. 1997). eIF4E induces the segment in eIF4G1 surrounding the $Y(X)_4L\Phi$ motif (amino acids 393–490) to fold into a "molecular bracelet" of α helices encircling the unstructured amino terminus of eIF4E (Hershey et al. 1999; Gross et al. 2003). Deletion of the yeast eIF4E NTD (Δ*30* mutation) decreases its affinity for eIF4G$_{393–490}$, abolishes induced folding of the eIF4G1 segment, and reduces affinity for m⁷GDP. In vivo, Δ*30* confers a Ts⁻ phenotype and lowers polysome content, suggesting that the molecular bracelet is required for optimal initiation.

These last results explain earlier findings that interaction with eIF4G increases affinity of eIF4E for capped mRNA (Haghighat and Sonenberg 1997; Ptushkina et al. 1998), and that a ~150-amino acid eIF4G1 fragment, but not a 17-amino acid peptide containing the $Y(X)_4L\Phi$ motif, harbors this stimulatory activity (von Der Haar et al. 2000). Mutations on the dorsal surface of yeast eIF4E (e.g., W75R) that simultaneously impair eIF4E–eIF4G and eIF4E–cap interactions (Ptushkina et al. 1998) may disrupt binding of the eIF4G bracelet to eIF4E. A larger amino-terminal segment of eIF4G confers greater stimulation of m⁷G–eIF4E interaction than does the eIF4E-binding domain alone, in a manner enhanced by yeast PABP, so PABP may also stimulate eIF4E–cap interaction (von Der Haar et al. 2000).

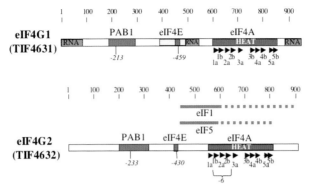

Figure 8. Primary structures of yeast eIF4G1 and eIF4G2. Binding domains for PAB1, eIF4E, eIF4A, and RNA depicted with colored segments, the five pairs of helices in the HEAT domain as arrowheads, binding domains in eIF4G2 for eIF1 and eIF5 above, and locations of selected mutations (e.g., *-213*) below the schematics.

Simultaneous Binding of eIF4E and PABP to eIF4G Can Circularize mRNA

The cap and poly(A) tail stimulate translation synergistically (Gallie 1991; Iizuka et al. 1994; Tarun and Sachs 1995), and yeast PAB1 is essential for translation initiation in vivo (Sachs 2000). Tarun and Sachs showed that immunoneutralization of PAB1 abolished the ability of the poly(A) tail to stimulate initiation in extracts and obtained in vitro evidence that eIF4E–cap and PAB1–poly(A) tail interactions provide independent means of recruiting the 40S ribosome to mRNA (Tarun and Sachs 1995, 1997). Consistent with this, translation of reporter mRNA introduced into yeast by electroporation is stimulated by a poly(A) tail (Preiss and Hentze 1998) in wild-type but not in *pab1Δ* cells (carrying a suppressor mutation *spb2-1* that allows cells to survive without PAB1) (Searfoss et al. 2001).

Tarun and Sachs (1996) also showed that the association of PAB1 with eIF4E is dependent on both eIF4G and the poly(A) tail and thus capable of circularizing mRNA (Wells et al. 1998). PAB1 binds in vitro to ~110-residue segments in eIF4G located amino-terminal to the eIF4E-binding sites (Fig. 8) in a manner dependent on poly(A) (Tarun and Sachs 1996; Tarun et al. 1997). The *tif4631-213* and *tif4632-233* mutations altering (K/R)LRK residues in the PAB1-binding domains (Fig. 8) reduce PAB1 binding and impair translation of uncapped, polyadenylated mRNA but not capped mRNA lacking poly(A) (Tarun et al. 1997), establishing the functional importance of PAB1–eIF4G interaction in vitro. The second of the four RRM domains of PAB1 is essential for poly(A)-stimulated translation and binding to eIF4G in vitro (Deardorff and Sachs 1997; Kessler and Sachs 1998), and RRM2 point mutations

K180E,E181R and *D184E,A185K,L186M* impair both activities. Interestingly, PAB1 can stimulate translation of capped mRNAs lacking poly(A), and this "*trans*-activation" requires residues in RRM2 and the amino terminus of eIF4G, but is independent of the eIF4G–PAB1 interaction. Thus, the PAB1 RRM2 may interact with an unidentified PIC component in a manner promoted by the eIF4G NTD (Otero et al. 1999).

Despite the in vitro evidence implicating PAB1–eIF4G interaction in translation, the *tif4631-213* and *tif4632-233* mutations in PAB1-binding sites, and even deletion of the amino-terminal 400 amino acids of the eIF4G proteins, all have little effect on cell growth (Tarun and Sachs 1996). Similarly, deletion of RRM2 of PAB1 reduces the growth rate by only 50% (Kessler and Sachs 1998). However, the *tif4631-Δ300* allele (removing the PAB1-binding domain) is synthetically lethal in combination with mutations in the eIF4E-binding site in eIF4G1 (*tif4631-459*). Similarly, a mutation affecting RRM domains 1 and 2 in PAB1 (*pab1-16*) is synthetically lethal with the *cdc33*-1 mutation in eIF4E. Thus, PAB1–eIF4G interaction is likely critical in vivo only when the eIF4E–eIF4G1 interaction is impaired (Tarun et al. 1997).

Functions of eIF4A, Other DEAD-box Helicases, and eIF4B in mRNA Recruitment

eIF4A belongs to the DEAD-box RNA helicase family (Tanner and Linder 2001), and both mammalian and yeast eIF4A exhibit RNA-dependent ATPase activity and bidirectional RNA helicase activity. The crystal structure of yeast eIF4A reveals a dumbbell shape with two globular domains connected by a linker (Caruthers et al. 2000), which is likely an inactive conformation, as the ATP-binding cleft should be formed by interaction of the two domains (Oberer et al. 2005).

The lethal *A66V* substitution in yeast eIF4A increases the K_m for ATP 5-fold, impairs RNA helicase activity in vitro, and abolishes the rescue of translation in an eIF4A-depleted extract by recombinant eIF4A, proving that ATP hydrolysis by eIF4A is crucial for translation initiation (Blum et al. 1992). A yeast extract harboring eIF4A-A66V is defective in translating an mRNA containing AUG only 8 nucleotides from the cap, for which scanning should not be required, suggesting that eIF4A is required for mRNA recruitment even without secondary structure in the 5'UTR (Blum et al. 1992). However, it is unknown whether eIF4A is essential for mRNA binding to the 43S PIC in vivo. In fact, yeast contains a second DEAD-box helicase, DED1, also required for translation initiation in vivo and in vitro (Chuang et al. 1997), and *DED1* interacts genetically with the eIF4 factors (de la Cruz et al. 1997). Furthermore, overexpression of DBP1, a third yeast DEAD-box helicase 72% identical to DED1, suppresses the lethality of *ded1Δ* (de la Cruz et al. 1997). Hence, multiple RNA helicases may function in mRNA recruitment in vivo. Because DED1 also participates in ribosome biogenesis (Yarunin et al. 2005), at least some of its effects on translation could be secondary to defects in ribosome structure or abundance.

The helicase activity of eIF4A is stimulated by mammalian eIF4B. The yeast eIF4B homolog, encoded by *TIF3* (Altmann et al. 1993), is nonessential but required for normal cell growth. TIF3 overexpression suppresses an eIF4A mutation, and *tif3Δ* exacerbates growth defects of eIF4E and eIF4G mutations (Coppolecchia et al. 1993), consistent with TIF3 stimulating eIF4A in vivo. Moreover, *tif3Δ* confers cold-sensitive translation in extracts that is exacerbated by insertion of stem-loop structures 22 nucleotides from the cap, suggesting an inability to melt secondary structures in the 5'UTR (Altmann et al. 1993). Ribosome binding of a reporter mRNA containing a short 5'UTR also is impaired in a *tif3Δ* extract, indicating that TIF3 is required for mRNA recruitment even without secondary structure near the cap. Curiously, unlike mammalian eIF4B, TIF3 does not stimulate the helicase activity of eIF4A in vitro (Altmann et al. 1993), making it unclear how the helicase activity of eIF4A is stimulated in yeast.

Both TIF3 and mammalian eIF4B exhibit ATP- and eIF4A-independent RNA annealing activities which might promote mRNA–rRNA base-pairing during scanning (Altmann et al. 1995). A predicted RRM in the TIF3 NTD has RNA-binding activity and promotes both annealing in vitro and TIF3 function in vivo. The CTD also binds RNA but is dispensable in vivo when TIF3 is overexpressed (Niederberger et al. 1998). TIF3 interacts with the CTD of TIF35/eIF3g (Vornlocher et al. 1999), and mammalian eIF4B binds to eIF3a (Methot et al. 1996). Thus, TIF3/eIF4B may stimulate mRNA recruitment as an adapter between mRNA and eIF3.

Complex Formation between eIF4A and the HEAT Domain of eIF4G

The helicase activity of mammalian eIF4A is also enhanced by eIF4G, and eIF4A binds eIF4G at one (yeast) or two (mammalian) locations in the carboxy-terminal portion of the molecule. The amino-proximal eIF4A-binding domain in human eIF4GII (the middle or eIF4G-m domain) forms a stack of five helical hairpins (HEAT repeats) (Fig. 8). NMR chemical shift mapping and mutational analysis suggested that the amino-terminal hairpins of eIF4G-m contact the eIF4A-CTD adjacent to DEAD-box motifs implicated in binding ATP and RNA, so that mRNA and eIF4G-m could bind eIF4A simultaneously (Oberer et al. 2005). Because mutations in the carboxy-terminal hairpins of eIF4G-m also affected eIF4A–eIF4Gm interaction (Marcotrigiano et al. 2001), it was proposed that eIF4G-m acts as a "soft clamp" spanning the two eIF4A domains to stabilize the closed, active conformation of eIF4A (Oberer et al. 2005).

The predicted interfaces of mammalian eIF4G-m and eIF4A-CTD are conserved in the yeast proteins, and an internal fragment of yeast eIF4G1 binds eIF4A in vitro and impairs translation in a manner overcome by increasing eIF4A concentration (Dominguez et al. 1999). Ts⁻ mutations in one or more residues in the amino-terminal HEAT repeats of yeast eIF4G2 (*tif4632-1*, *-6*, and *-8*) (Fig. 8, *-6*) reduce the affinity for eIF4A in vitro and are suppressed in vivo and in a *tif4631-8* extract by eIF4A overexpression (Neff and Sachs 1999). Ts⁻ mutations in the HEAT domain of eIF4G1 also reduce affinity for eIF4A in vitro (Dominguez et al. 2001), some of which map in the carboxy-terminal HEAT repeats. Thus, eIF4G and eIF4A likely interact in vivo in the manner proposed in the soft clamp model of Oberer et al. eIF4G1 also contains three domains with RNA-binding activity that make overlapping contributions to eIF4G function in vivo and could mediate interactions with rRNA or mRNA (Berset et al. 2003).

Functions of eIF3, eIF1, eIF1A, and eIF5 in mRNA Binding

Mammalian eIF3, eIF1, and eIF1A stimulate mRNA binding to the 40S subunit in vitro. Yeast eIF3 was implicated in this reaction by showing that the *prt1-1* mutation impairs 40S binding of mRNA in extracts in a manner rescued by eIF3 or the TIF32/a-PRT1/b-NIP1/c eIF3 subcomplex (Phan et al. 2001). Furthermore, co-depletion of eIF3a and eIF3b in yeast degron mutants impairs mRNA recruitment in vivo (Jivotovskaya et al. 2006). Because TC stimulates mRNA binding to 40S subunits (Maag et al. 2005b; Jivotovskaya et al. 2006), eIF3 may promote mRNA recruitment at least partly via TC loading.

A more direct mechanism is envisioned for mammalian eIF3 based on its physical interaction with a segment in eIF4G carboxy-terminal to the middle HEAT domain, enabling eIF3 to bridge the 40S subunit and the eIF4F–mRNA complex. However, there is no direct evidence that the eIF3–eIF4G interaction is crucial for mRNA recruitment. In fact, the eIF3-binding domain is not conserved in yeast eIF4G (Marintchev and Wagner 2005), and eIF3–eIF4G interaction is difficult to observe in yeast extracts. Moreover, depletion of eIF4G from yeast cells leads to accumulation of at least two different mRNAs in native 48S PICs. Thus, eIF4G is not essential for recruitment of certain mRNAs in yeast, although it could be critical for a wild-type rate of mRNA binding (Jivotovskaya et al. 2006). eIF3 may also bind mRNA directly, considering that multiple subunits of mammalian eIF3 are cross-linked to mRNA in 48S PICs and mRNA–eIF3 complexes in vitro (Unbehaun et al. 2004; Kolupaeva et al. 2005).

There is evidence that the eIF5 CTD bridges eIF4G and eIF3 to promote mRNA recruitment in yeast. eIF5-CTD can bind the carboxy-terminal half of eIF4G2 in vitro, dependent on multiple surfaces of the predicted eIF5 HEAT domain (Fig. 6A) (Asano et al. 2001; Yamamoto et al. 2005). An eIF3–eIF4G interaction observed in vivo was eliminated by the *tif5-7A* mutation, which also reduces 40S binding of mRNA in cell extracts (Asano et al. 2001). On the other hand, depletion of eIF5 in a yeast degron mutant leads to accumulation of mRNA in 48S PICs, suggesting that the GAP function of eIF5 in subunit-joining is more rate-limiting than its role in mRNA recruitment (Jivotovskaya et al. 2006).

RIBOSOMAL SCANNING AND AUG RECOGNITION

The predominant role of scanning in mammalian translation initiation was established primarily by the work of Kozak (1989). Genetic studies by Sherman et al. played a key role in discovering the scanning process in yeast, demonstrating that the first AUG from the 5' end is selected as the start codon independent of sequence context (Sherman and Stewart 1982). Genetic experiments by Donahue et al. proved that base-pairing between the initiator anticodon and the start codon, regardless of their exact sequences, is a fundamental requirement for initiation in yeast, implying that the anticodon of tRNA$_i^{Met}$ inspects the mRNA during scanning (Cigan et al. 1988). As noted above, these workers also implicated eIF1, eIF2 subunits, and eIF5 in AUG selection by isolating Sui$^-$ mutations in each protein (Donahue 2000). Biochemical analyses of Sui$^-$ mutations in eIF2β, eIF2γ, and eIF5 suggest that elevated rates of GTPase hydrolysis by eIF2, or more rapid dissociation of the TC, can increase utilization of UUG triplets as start codons (Huang et al. 1997). These findings imply that GTP hydrolysis by the TC is normally suppressed at non-AUG triplets during scanning.

Critical Role of eIF1 in Scanning and AUG Selection

Using a reconstituted mammalian system, Pestova and colleagues showed that the 43S PIC can bind to mRNA bearing an unstructured 5'UTR independent of eIF4F, eIF4A, and ATP, but requires eIF1 or eIF4G to scan to the AUG codon. AUG recognition on mRNA with a structure-laden 5'UTR additionally requires eIF4F, eIF4B, ATP, and eIF1A (Pestova et al. 1998; Pestova and Kolupaeva 2002). These results provide biochemical evidence that eIF4F and eIF4A helicase activity facilitates scanning by unwinding secondary structure in the 5'UTR, and that eIF1 and eIF1A promote a scanning-permissive conformation of the PIC. However, the role of eIF4F/eIF4A in scanning has not been demonstrated in vivo. In fact, there is evidence that the helicases DED1 and DBP1 are more critical than eIF4A for efficient scanning through a long 5'UTR in yeast cells (Berthelot et al. 2004). On the other hand, a subset of yeast mRNAs accumulate in 48S complexes in yeast depleted of eIF4G, consistent with a rate-limiting defect in scanning after 43S binding to the mRNA (Jivotovskaya et al. 2006).

Pestova et al. also found that elimination of eIF1 leads to aberrant 48S PIC assembly in vitro at near-cognate start codons (e.g., AUU), at AUG triplets in a suboptimal sequence context, and at AUGs located within 4 nuceotides of the 5' end (Pestova and Kolupaeva 2002). These results, and their finding that mammalian eIF1 binds near the P site (Lomakin et al. 2003), led them to propose that eIF1 induces an open conformation of the 40S subunit that rejects codon–anticodon mismatches in the P site and promotes continued scanning. Defects in these activities could explain the increased selection of UUG start codons produced by the *sui1-1* mutation in yeast eIF1 (Yoon and Donahue 1992), as it greatly reduces the level of yeast eIF1 found in native 43S PICs (Valášek et al. 2004) and may simply eliminate eIF1 from the scanning complex.

Interestingly, a mutation that reduces the GAP function of eIF5 (*ssu2-1*) is a suppressor of *sui1-1* (Donahue 2000; Asano et al. 2001), suggesting that *sui1–1* elevates eIF5 GAP function. This would imply that eIF1 normally inhibits eIF5 function at non-AUG codons. Consistent with this, overexpressing wild-type eIF1 suppresses the Sui⁻ phenotype of the *TIF5-G31R* mutation in eIF5 (*SUI5*), which increases GAP function in vitro, and *sui1-1* is synthetically lethal with *TIF5-G31R* (Valášek et al. 2004). Pestova et al. provided biochemical support for this model by showing that eIF1 decreases the rate of eIF5-dependent GTP hydrolysis in 48S PICs lacking an AUG. Furthermore, Lorsch and coworkers showed that AUG in the P site triggers dissociation of eIF1 from yeast PICs in vitro, leading them to propose that AUG recognition ejects eIF1 from its location near the P site, promoting the closed conformation that impedes scanning and derepressing eIF5-stimulated GTP hydrolysis (Maag et al. 2005b).

Surprisingly, the rate and extent of eIF5-dependent GTP hydrolysis in the reconstituted yeast PICs are stimulated only about 3-fold by AUG, whereas the rate of P_i release from eIF2-GDP is 100-fold lower when the mRNA contained CUC versus AUG. P_i release may be triggered by dissociation of eIF1, as the rates of P_i release at AUG and eIF1 dissociation are both reduced by the *mof2-1* mutation in eIF1 (Algire et al. 2005). Thus, eIF1-regulated P_i release from eIF2-GDP, rather than GTP hydrolysis, may be the rate-limiting reaction in AUG selection.

Interestingly, alanine substitutions in the NIP1/c NTD exhibit a Sui⁻ phenotype, suppress Sui⁻ mutations in eIF1, eIF2β, or eIF5, and impair NIP1/c-NTD binding to eIF1 and eIF5 in vitro. Hence, the NIP1-NTD may coordinate interactions between eIF1 and eIF5 involved in AUG selection (Valášek et al. 2004). eIF1 function in AUG selection also seems to depend on its interaction with eIF4G, as the *tif4632-1* mutation reduces its binding to eIF1 in vitro and confers a Sui⁻ phenotype suppressed by eIF1 overexpression (He et al. 2003).

Point mutations on the RNA-binding surface of mammalian eIF1A lead to accumulation of incorrect 48S complexes upstream of the AUG codon in the reconstituted mammalian system (Battiste et al. 2000). The same defect is observed for mutations in the helical (*98-101*) or unstructured CTD (Δ*C*) of yeast eIF1A (Fig. 7A) when it replaces the mammalian factor. As noted above, both *98-101* and Δ*C* confer Sui⁻ phenotypes in yeast cells, and *98-101* also impairs derepression of *GCN4* translation (Gcn⁻ phenotype) by a mechanism that likely involves a decreased rate of scanning between uORF1 and uORF4 (Fekete et al. 2005). Perhaps a defect in scanning increases the probability of selecting non-AUG triplets by increasing the "dwell time" of the triplet in the P site. Inefficient scanning was also evoked to explain the Gcn⁻ phenotype of the eIF3b mutant *prt1-1*, which accumulates 48S complexes in vivo (Nielsen et al. 2004).

eIF5 Functions in Start Codon Selection as a GTPase Activating Protein for eIF2

Depletion of eIF5 from yeast cells impairs translation initiation and leads to accumulation of 48S PICs (Jivotovskaya et al. 2006), supporting the idea that eIF5 GAP function is a prerequisite for 60S subunit joining in vivo. The effects of the *TIF5-G31R* and *ssu2-1/tif5-G62S* mutations on eIF5 GAP function (Huang et al. 1997; Donahue 2000) implicated the eIF5 NTD in this activity (Fig. 6A). Consistent with this, the yeast eIF5 NTD binds to the G domain of eIF2γ and stimulates GTP hydrolysis by eIF2 in vitro (Alone and Dever 2006). There is genetic and biochemical evidence that invariant Arg-15 functions as an "arginine finger" to stabilize the transition state of the GTP hydrolysis reaction (Das et al. 2001; Paulin et al. 2001; Algire et al. 2005). eIF5 GAP function is strongly dependent on the 40S subunit (Raychaudhuri et al. 1985), with the yeast factor showing >2000-fold higher activity for 48S PICs versus free TC (Algire et al. 2005). Hence, a segment of the 40S ribosome may interact with the switch-1 or switch-2 segments in eIF2γ to trigger GTP hydrolysis. Mutations in the AA boxes of the CTD of mammalian eIF5 that impair interaction with eIF2β (*M3* and *M4*, Fig. 6A) reduce the GAP activity of eIF5 in vitro (Das and Maitra 2000), but this is not observed for comparable mutations in yeast eIF5 (*tif5-7A* and *tif5-12A*) (Asano et al. 2001). Thus, interaction of the eIF5-CTD with eIF2β may serve primarily to recruit eIF5 to eIF2.

It appears that eIF5 plays an additional role in AUG selection that involves a switch in eIF1A binding to the 40S subunit (Maag et al. 2005a). Lorsch et al. showed that AUG and eIF5 collaborate to produce tighter association of eIF1A with the 48S PIC by overcoming a destabilizing effect of the eIF1A CTD on 40S binding. Truncation of the eIF1A CTD by the ΔC mutation and the *SUI5* mutation in eIF5 both stabilize eIF1A binding to PICs containing UUG in the P site, consistent with their Sui[-] phenotypes (Fekete et al. 2005). Thus, pairing of Met-tRNA$_i^{Met}$ with AUG in the P site evokes multiple conformational changes in the PIC involving eIF1, eIF1A, and eIF5 that arrest scanning and stimulate GTP hydrolysis and P$_i$ release by the TC.

eIF5B CATALYZES SUBUNIT JOINING IN THE SECOND GTP-DEPENDENT STEP OF TRANSLATION INITIATION

Reconstitution of the 60S joining reaction with mammalian factors showed that the reaction requires eIF5B (Pestova et al. 2000), an ortholog of prokaryotic IF2 containing a GTP-binding domain (Fig. 9A) (Lee et al. 1999). Both yeast and mammalian eIF5B have GTPase activities dependent on 40S and 60S subunits (Merrick et al. 1975; Pestova et al. 2000; Shin et al. 2002), and analysis of the D759N substitution in the NKxD motif of human eIF5B demonstrated that GTP binding is required for the subunit-joining activity of eIF5B (Fig. 9A) (Lee et al. 2002). eIF5B is nonessential in yeast, but *fun12Δ* strains lacking the factor have a severe Slg[-] phenotype (Choi et al. 1998).

The crystal structure of archaeal aIF5B reveals a 4-domain protein resembling a chalice (Roll-Mecak et al. 2000). The G domain and domain II can be superimposed on domains I and II of other translation GTPases and, together with domain III, form the cup of the chalice, which is con-

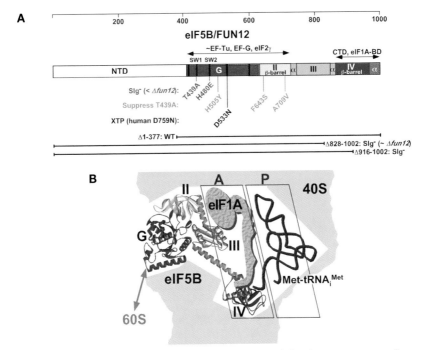

Figure 9. Structural and functional properties of eIF5B. (*A*) Primary structure of yeast eIF5B/FUN12 is depicted showing the locations of structural domains (NTD, G, II, III, and IV), eIF1A-binding domain (BD), and locations and properties of key point and deletion mutations. Dominant-negative *T439A* mutation in switch 1 (SW1) confers a Slg⁻ phenotype in *fun12Δ* cells and is suppressed by the mutations shown in green. (*B*) Hypothetical model of eIF5B binding and interactions on the 40S subunit. See text for details. The aIF5B images were generated using the program WebLab Viewer Lite (v. 3.2, Accelrys, Inc.). (Modified from Hinnebusch et al. 2004.)

nected to domain IV by a 40-Å α-helical stem (Fig. 9). Comparing aIF5B in GTP- and GDP-bound states revealed a ~5-Å lever-type motion of the helical stem and domain IV (Roll-Mecak et al. 2000), which is thought to trigger release of eIF5B from the 80S initiation complex.

In vitro analysis of mammalian eIF5B showed that GTP hydrolysis is not required for 60S subunit joining, but for conversion of the 80S complex to a translation-competent state by the release of eIF5B (Pestova et al. 2000). Consistent with this, mutation of Thr-439 in switch 1 of yeast eIF5B, or His-480 in switch 2 of yeast eIF5B and the corresponding His-706 in human eIF5B, destroys the ribosome-dependent GTPase activity and its function in protein synthesis, but not subunit-joining activity. Moreover, the eIF5B-H706E mutant remains bound to the 80S product of the subunit-joining reaction (Lee et al. 2002; Shin et al. 2002), and overexpressing the mutant factors confers dominant Slg$^-$ phenotypes in *fun12Δ* cells. Remarkably, the H505Y mutation suppresses the Slg$^-$ phenotype and translation defect conferred by the eIF5B-T439A mutation (Fig. 9, A,B, left panel) without restoring eIF5B GTPase activity, by reducing the affinity of eIF5B for the 80S ribosome (Shin et al. 2002). Thus, eIF5B GTPase activity is required primarily for release of eIF5B from the 80S initiation complex. Yeast eIF5B mutants lacking GTPase activity exhibit a Gcn$^-$ phenotype that can be attributed to leaky scanning of *GCN4* uORF1, a defect consistent with the subunit-joining function of eIF5B (Shin et al. 2002).

Bacterial IF2 binds fMet-tRNA$_i^{Met}$ to the 30S subunit, most likely through binding of domain IV to the fMet and acceptor stem of fMet-tRNA$_i^{Met}$ (Boelens and Gualerzi 2002). Although yeast eIF5B is deficient in this activity (Guillon et al. 2005), there is biochemical evidence that it stabilizes Met-tRNA$_i^{Met}$ binding to the 80S complex in vitro (Shin et al. 2002). eIF5B may also stabilize 48S PICs following GTP hydrolysis by the TC, as polysomes containing 40S halfmers are eliminated in *fun12* mutants lacking eIF5B (Lee et al. 2002). Moreover, the Slg$^-$ phenotype of *fun12Δ* strains is partially suppressed by overexpressing tRNA$_i^{Met}$ (Choi et al. 1998).

Yeast eIF5B and eIF1A interact in vitro and are found associated in cell extracts in a manner requiring the last 24 residues in eIF1A and the carboxy-terminal 153 residues of eIF5B (Choi et al. 2000; Olsen et al. 2003). NMR analysis reveals that the carboxy-terminal 14 residues of human eIF1A lie in a shallow hydrophobic groove between helices 13 and 14 in domain IV of eIF5B (Marintchev et al. 2003). Mutational analysis indicates that eIF1A–eIF5B association through their carboxyl termini is important in vivo (Choi et al. 2000; Kainuma and Hershey 2001; Olsen et al. 2003) and promotes subunit joining and GTP hydrolysis by eIF5B in vitro (Acker et al. 2006). Overexpression of eIF1A exacerbates the growth defect of *fun12* mutants lacking the eIF1A-binding domain in eIF5B, and it was proposed that eIF1A is dependent on eIF5B for efficient release from the A site, so that eIF1A overexpression in *fun12* mutants impedes entry of the first eEF1A-GTP-aminoacyl-tRNA complex (Choi et al. 2000).

Wagner and colleagues proposed an intriguing model for eIF5B and eIF1A bound to the 40S ribosome (Fig. 9B), in which the OB-fold of eIF1A in the A site contacts domain II of eIF5B in the manner proposed for bacterial IF1–IF2 interaction. The extreme carboxyl terminus of eIF1A binds domain IV in eIF5B, as described above, with the remainder of the unstructured eIF1A CTD spanning the 50 Å between eIF5B domains II and IV. Domain IV in eIF5B additionally contacts the methionine and acceptor stem of Met-tRNA$_i^{Met}$ in the P site (Marintchev et al. 2003). This model is largely consistent with the recent cryo-EM analysis of IF2 bound to 70S ribosomes (Allen et al. 2005; Myasnikov et al. 2005), which also predicts interaction of the G domain of eIF5B with the GTPase-activating center of the large subunit and domain II with the small subunit (Roll-Mecak et al. 2001; Ramakrishnan 2002; Allen et al. 2005). It is reasonable to propose that binding of eIF5B to the 40S subunit alters 40S structure to facilitate 60S subunit joining with attendant release of eIF3 and eIF1 (Unbehaun et al. 2004). Joining of the 60S subunit would trigger GTP hydrolysis by eIF5B, producing a conformational change that leads to dissociation of eIF5B and eIF1A and leaves the 80S ribosome ready to enter the elongation phase of protein synthesis.

BEYOND PIC ASSEMBLY: NEW HORIZONS IN YEAST TRANSLATIONAL CONTROL

Whereas translation of cellular mRNA is enhanced by the poly(A) tail, the L-A and M1 viral mRNAs are not polyadenylated but are still translated efficiently during virus propagation. Wickner et al. showed that deletion of the nonessential DEAD-box helicases *SKI2* and *SLH1* allows mRNAs lacking poly(A) to be translated as efficiently as poly(A)-containing mRNAs in yeast cells (Searfoss and Wickner 2000). They propose that poly(A)/PAB1 is required to overcome translational inhibition by SKI2 and SLH1 of nonpolyadenylated mRNAs. Other results suggest that poly(A)/PAB1 stimulates translation by overcoming the inhibitory effect of SKI2/SLH1 on eIF5B function in subunit joining (Searfoss et al. 2001). Thus, their genetic analysis not only uncovered a new function for PAB1 independent of its role in mRNA recruitment, but also suggested that poly(A)-dependent translation in vivo is the result of the active inhibition of nonpolyadenylated mRNA translation by the protein complex containing SKI2 and SLH2.

Fascinating work by Parker and colleagues recently identified a new level of translational control involving sequestration of translationally repressed mRNAs in P-bodies, cellular foci of mRNA decapping and degradation. It appears that effective translational repression in nutrient-starved yeast requires the exclusion of ribosomes from mRNAs by the formation of mRNP complexes with DEAD-box helicase DHH1 and the PAT1 protein, which then aggregate in P-bodies. It seems that mRNAs sent to P-bodies are not necessarily degraded and can return to the translated pool when nutrients are resupplied. Interestingly, DHH1 homologs are required for translational repression in fruit flies, nematodes, and frogs, suggesting that this general mechanism of repression is widespread in eukaryotes (Coller and Parker 2005).

PERSPECTIVES

Combining genetic, biochemical, and structural analysis of the yeast initiation factors will continue to provide valuable information about the physiological importance of their interactions and activities in living cells, which should be highly relevant to their closely conserved mammalian counterparts. In-depth structure–function studies of the factors will continue, and suppressor analysis will likely uncover new interactions/functions of the known factors or identify novel accessory proteins. Genetic analysis can now be extended easily to the rRNA, using yeast strains harboring a single rDNA allele (Wai et al. 2000), and should complement efforts to obtain high-resolution structures of initiation complexes. Besides revealing the mechanisms of PIC assembly and function, genetic, biochemical, and cytological approaches will likely uncover an unexpected network of translational regulators that integrate mRNA sequestration and turnover with translation.

ACKNOWLEDGMENTS

We thank Regina Renk for help in preparing the manuscript and Jon Lorsch for helpful discussions.

REFERENCES

Acker M.G., Shin B.S., Dever T.E., and Lorsch J.R. 2006. Interaction between eukaryotic initiation factors 1A and 5B is required for efficient ribosomal subunit joining. *J. Biol. Chem.* **281:** 8469–8475.

Akiyoshi Y., Clayton J., Phan L., Yamamoto M., Hinnebusch A.G., Watanabe Y., and Asano K. 2000. Fission yeast homolog of murine Int-6 protein, encoded by mouse mammary tumor virus integration site, is associated with the conserved core subunits of eukaryotic translation initiation factor 3. *J. Biol. Chem.* **276:** 10056–10062.

Algire M.A., Maag D., and Lorsch J.R. 2005. Pi release from eIF2, not GTP hydrolysis, is the step controlled by start-site selection during eukaryotic translation initiation. *Mol. Cell* **20:** 251–262.

Algire M.A., Maag D., Savio P., Acker M.G., Tarun S.Z., Jr., Sachs A.B., Asano K., Nielsen K.H., Olsen D.S., Phan L., et al. 2002. Development and characterization of a reconstituted yeast translation initiation system. *RNA* **8:** 382–397.

Allen G.S., Zavialov A., Gursky R., Ehrenberg M., and Frank J. 2005. The cryo-EM structure of a translation initiation complex from *Escherichia coli*. *Cell* **121:** 703–712.

Alone P.V. and Dever T.E. 2006. Direct binding of translation initiation factor eIF2gamma-G domain to its GTPase-activating and GDP-GTP exchange factors eIF5 and eIF2B epsilon. *J. Biol. Chem.* **281:** 12636–12644.

Altmann M., Edery I., Sonenberg N., and Trachsel H. 1985. Purification and characterization of protein synthesis initiation factor eIF-4E from the yeast *Saccharomyces cerevisiae*. *Biochemistry* **24:** 6085–6089.

Altmann M., Edery I., Trachsel H., and Sonenberg N. 1988. Site-directed mutagenesis of the tryptophan residues in yeast eukaryotic initiation factor 4E. *J. Biol. Chem.* **263:** 17229– 17232.

Altmann M., Wittmer B., Methot N., Sonenberg N., and Trachsel H. 1995. The *Saccharomyces cerevisiae* translation initiation factor Tif3 and its mammalian homologue, eIF-4B, have RNA annealing activity. *EMBO J.* **14:** 3820–3827.

Altmann M., Muller P.P., Wittmer B., Ruchti F., Lanker S., and Trachsel H. 1993. A *Saccharomyces cerevisiae* homologue of mammalian translation initiation factor 4B contributes to RNA helicase activity. *EMBO J.* **12:** 3997–4003.

Asano K., Clayton J., Shalev A., and Hinnebusch A.G. 2000. A multifactor complex of eukaryotic initiation factors eIF1, eIF2, eIF3, eIF5, and initiator tRNA[Met] is an important translation initiation intermediate in vivo. *Genes Dev.* **14:** 2534–2546.

Asano K., Phan L., Anderson J., and Hinnebusch A.G. 1998. Complex formation by all five homologues of mammalian translation initiation factor 3 subunits from yeast *Saccharomyces cerevisiae. J. Biol. Chem.* **273:** 18573–18585.

Asano K., Krishnamoorthy T., Phan L., Pavitt G.D., and Hinnebusch A.G. 1999. Conserved bipartite motifs in yeast eIF5 and eIF2Bε, GTPase-activating and GDP-GTP exchange factors in translation initiation, mediate binding to their common substrate eIF2. *EMBO J.* **18:** 1673–1688.

Asano K., Shalev A., Phan L., Nielsen K., Clayton J., Valášek L., Donahue T.F., and Hinnebusch A.G. 2001. Multiple roles for the carboxyl terminal domain of eIF5 in translation initiation complex assembly and GTPase activation. *EMBO J.* **20:** 2326–2337.

Battiste J.B., Pestova T.V., Hellen C.U.T., and Wagner G. 2000. The eIF1A solution structure reveals a large RNA-binding surface important for scanning function. *Mol. Cell* **5:** 109–119.

Berset C., Zurbriggen A., Djafarzadeh S., Altmann M., and Trachsel H. 2003. RNA-binding activity of translation initiation factor eIF4G1 from *Saccharomyces cerevisiae. RNA* **9:** 871–880.

Berthelot K., Muldoon M., Rajkowitsch L., Hughes J., and McCarthy J.E. 2004. Dynamics and processivity of 40S ribosome scanning on mRNA in yeast. *Mol. Microbiol.* **51:** 987–1001.

Blum S., Schmid S.R., Pause A., Buser P., Linder P., Sonenberg N., and Trachsel H. 1992. ATP hydrolysis by initiation factor 4A is required for translation initiation in *Saccharomyces cerevisiae. Proc. Natl. Acad. Sci.* **89:** 7664–7668.

Boelens R. and Gualerzi C.O. 2002. Structure and function of bacterial initiation factors. *Curr. Protein Pept. Sci.* **3:** 107–119.

Boesen T., Mohammad S.S., Pavitt G.D., and Andersen G.R. 2004. Structure of the catalytic fragment of translation initiation factor 2B and identification of a critically important catalytic residue. *J. Biol. Chem.* **279:** 10584–10592.

Bumann M., Djafarzadeh S., Oberholzer A.E., Bigler P., Altmann M., Trachsel H., and Baumann U. 2004. Crystal structure of yeast Ypr118w, a methylthioribose-1-phosphate isomerase related to regulatory eIF2B subunits. *J. Biol. Chem.* **279:** 37087–37094.

Caruthers J.M., Johnson E.R., and McKay D.B. 2000. Crystal structure of yeast initiation factor 4A, a DEAD-box RNA helicase. *Proc. Natl. Acad. Sci.* **97:** 13080–13085.

Castilho-Valavicius B., Thompson G.M., and Donahue T.F. 1992. Mutation analysis of the Cys-X_2-Cys-X_{19}-Cys-X_2-Cys motif in the α subunit of eukaryotic translation factor 2. *Gene Expr.* **2:** 297–309.

Cho S. and Hoffman D.W. 2002. Structure of the beta subunit of translation initiation factor 2 from the archaeon *Methanococcus jannaschii:* A representative of the eIF2beta/eIF5 family of proteins. *Biochemistry* **41:** 5730–5742.

Choi S.K., Lee J.H., Zoll W.L., Merrick W.C., and Dever T.E. 1998. Promotion of Met-tRNA$_i^{Met}$ binding to ribosomes by yIF2, a bacterial IF2 homolog in yeast. *Science* **280:** 1757–1760.

Choi S.K., Olsen D.S., Roll-Mecak A., Martung A., Remo K.L., Burley S.K., Hinnebusch A.G., and Dever T.E. 2000. Physical and functional interaction between the eukaryotic orthologs of prokaryotic translation initiation factors IF1 and IF2. *Mol. Cell. Biol.* **20:** 7183–7191.

Chuang R.Y., Weaver P.L., Liu Z., and Chang T.H. 1997. Requirement of the DEAD-Box protein ded1p for messenger RNA translation. *Science* **275:** 1468–1471.

Cigan A.M., Feng L., and Donahue T.F. 1988. tRNA$_i^{Met}$ functions in directing the scanning ribosome to the start site of translation. *Science* **242:** 93–97.

Cigan A.M., Pabich E.K., Feng L., and Donahue T.F. 1989. Yeast translation initiation suppressor *sui2* encodes the α subunit of eukaryotic initiation factor 2 and shares identity with the human α subunit. *Proc. Natl. Acad. Sci.* **86:** 2784–2788.

Coller J. and Parker R. 2005. General translational repression by activators of mRNA decapping. *Cell* **122:** 875–886.

Coppolecchia R., Buser P., Stotz A., and Linder P. 1993. A new yeast translation initiation factor suppresses a mutation in the eIF-4A RNA helicase. *EMBO J.* **12:** 4005–4011.

Danaie P., Wittmer B., Altmann M., and Trachsel H. 1995. Isolation of a protein complex containing translation initiation factor Prt1 from *Saccharomyces cerevisiae*. *J. Biol. Chem.* **270:** 4288–4292.

Das S. and Maitra U. 2000. Mutational analysis of mammalian translation initiation factor 5 (eIF5): Role of interaction between the beta subunit of eIF2 and eIF5 in eIF5 function in vitro and in vivo. *Mol. Cell. Biol.* **20:** 3942–3950.

Das S., Ghosh R., and Maitra U. 2001. Eukaryotic translation initiation factor 5 functions as a GTPase-activating protein. *J. Biol. Chem.* **276:** 6720–6726.

Deardorff J.A. and Sachs A.B. 1997. Differential effects of aromatic and charged residue substitutions in the RNA binding domains of the yeast poly(A)-binding protein. *J. Mol. Biol.* **269:** 67–81.

de la Cruz J., Iost I., Kressler D., and Linder P. 1997. The p20 and Ded1 proteins have antagonistic roles in eIF4E-dependent translation in *Saccharomyces cerevisiae*. *Proc. Natl. Acad. Sci.* **94:** 5201–5206.

Dever T.E., Yang W., Åström S., Byström A.S., and Hinnebusch A.G. 1995. Modulation of tRNA$_i^{Met}$, eIF-2 and eIF-2B expression shows that *GCN4* translation is inversely coupled to the level of eIF-2·GT·Met-tRNA$_i^{Met}$ ternary complexes. *Mol. Cell. Biol.* **15:** 6351–6363.

Dever T.E., Chen J.J., Barber G.N., Cigan A.M., Feng L., Donahue T.F., London I.M., Katze M.G., and Hinnebusch A.G. 1993. Mammalian eukaryotic initiation factor 2α kinases functionally substitute for GCN2 in the *GCN4* translational control mechanism of yeast. *Proc. Natl. Acad. Sci.* **90:** 4616–4620.

Dey M., Trieselmann B.A., Locke E.G., Lu J., Cao C., Dar A.C., Krishnamoorthy T., Dong J., Sicheri F., and Dever T.E. 2005. PKR and GCN2 kinases and guanine nucleotide exchange factor eIF2B recognize overlapping surfaces on translation factor eIF2α. *Mol. Cell. Biol.* **25:** 3063–3075.

Dhaliwal S. and Hoffman D.W. 2003. The crystal structure of the N-terminal region of the alpha subunit of translation initiation factor 2 (eIF2α) from *Saccharomyces cerevisiae* provides a view of the loop containing serine 51, the target of the eIF2α-specific kinases. *J. Mol. Biol.* **334:** 187–195.

Dominguez D., Kislig E., Altmann M., and Trachsel H. 2001. Structural and functional similarities between the central eukaryotic initiation factor (eIF)4A-binding domain of mammalian eIF4G and the eIF4A-binding domain of yeast eIF4G. *Biochem. J.* **355:** 223–230.

Dominguez D., Altmann M., Benz J., Baumann U., and Trachsel H.. 1999. Interaction of translation initiation factor eIF4G with eIF4A in the yeast *Saccharomyces cerevisiae*. *J. Biol. Chem.* **274:** 26720–26726.

Donahue T. 2000. Genetic approaches to translation initiation in *Saccharomyces cerevisiae*. In *Translational control of gene expression* (ed. N. Sonenberg et al.), pp. 487–502. Cold Spring Harbor Laboratory Press, Cold Spring Harbor, New York.

Donahue T.F., Cigan A.M., Pabich E.K., and Castilho-Valavicius B. 1988. Mutations at a Zn(II) finger motif in the yeast eIF-2β gene alter ribosomal start-site selection during the scanning process. *Cell* **54:** 621–632.

Dorris D.R., Erickson F.L., and Hannig E.M. 1995. Mutations in *GCD11*, the structural gene for eIF-2γ in yeast, alter translational regulation of *GCN4* and the selection of the start site for protein synthesis. *EMBO J.* **14:** 2239–2249.

Erickson F.L. and Hannig E.M. 1996. Ligand interactions with eukaryotic translation initiation factor 2: Role of the γ-subunit. *EMBO J.* **15:** 6311–6320.

Evans D.R.H., Rasmussen C., Hanic-Joyce P.J., Johnston G.C., Singer R.A., and Barnes C.A. 1995. Mutational analysis of the Prt1 protein subunit of yeast translation initiation factor 3. *Mol. Cell. Biol.* **15:** 4525–4535.

Fabian J.R., Kimball S.R., Heinzinger N.K., and Jefferson L.S. 1997. Subunit assembly and guanine nucleotide exchange activity of eukaryotic initiation factor-2B expressed in Sf9 cells. *J. Biol. Chem.* **272:** 12359–12365.

Farruggio D., Chaudhuri J., Maitra U., and RajBhandary U.L. 1996. The A1 U72 base pair conserved in eukaryotic initiator tRNAs is important specifically for binding to the eukaryotic translation initiation factor eIF2. *Mol. Cell. Biol.* **16:** 4248–4256.

Fekete C.A., Applefield D.J., Blakely S.A., Shirokikh N., Pestova T., Lorsch J.R., and Hinnebusch A.G. 2005. The eIF1A C-terminal domain promotes initiation complex assembly, scanning and AUG selection in vivo. *EMBO J.* **24:** 3588–3601.

Flynn A., Oldfield S., and Proud C.G. 1993. The role of the β-subunit of initiation factor eIF2 in initiation complex formation. *Biochim. Biophys. Acta* **1174:** 117–121.

Fraser C.S., Lee J.Y., Mayeur G.L., Bushell M., Doudna J.A., and Hershey J.W. 2004. The j-subunit of human translation initiation factor eIF3 is required for the stable binding of eIF3 and its subcomplexes to 40 S ribosomal subunits in vitro. *J. Biol. Chem.* **279:** 8946–8956.

Gallie D.R. 1991. The cap and poly(A) tail function synergistically to regulate mRNA translational efficiency. *Genes Dev.* **5:** 2108–2116.

Gomez E. and Pavitt G.D. 2000. Identification of domains and residues within the epsilon subunit of eukaryotic translation initiation factor 2B (eIF2Bε) required for guanine nucleotide exchange reveals a novel activation function promoted by eIF2B complex formation. *Mol. Cell. Biol.* **20:** 3965–3976.

Gomez E., Mohammad S.S., and Pavitt G.D. 2002. Characterization of the minimal catalytic domain within eIF2B: The guanine-nucleotide exchange factor for translation initiation. *EMBO J.* **21:** 5292–5301.

Goyer C., Altmann M., Trachsel H., and Sonenberg N. 1989. Identification and characterization of cap binding proteins from yeast. *J. Biol. Chem.* **264:** 7603–7610.

Goyer C., Altmann M., Lee H.S., Blanc A., Deshmukh M., Woolford J.L., Trachsel H., and Sonenberg N. 1993. *TIF4631* and *TIF4632*: Two yeast genes encoding the high-molecular-weight subunits of the cap-binding protein complex (eukaryotic initiation factor 4F) contain an RNA recognition motif-like sequence and carry out an essential function. *Mol. Cell. Biol.* **13:** 4860–4874.

Gross J.D., Moerke N.J., von der Haar T., Lugovskoy A.A., Sachs A.B., McCarthy J.E., and Wagner G. 2003. Ribosome loading onto the mRNA cap is driven by conformational coupling between eIF4G and eIF4E. *Cell* **115:** 739–750.

Guillon L., Schmitt E., Blanquet S., and Mechulam Y. 2005. Initiator tRNA binding by e/aIF5B, the eukaryotic/archaeal homologue of bacterial initiation factor IF2. *Biochemistry* **44:** 15594–15601.

Haghighat A. and Sonenberg N. 1997. eIF4G dramatically enhances the binding of eIF4E to the mRNA 5'-cap structure. *J. Biol. Chem.* **272:** 21677–21680.

Hannig E.M. and Hinnebusch A.G. 1988. Molecular analysis of *GCN3*, a translational activator of *GCN4*: Evidence for posttranslational control of *GCN3* regulatory function. *Mol. Cell. Biol.* **8:** 4808–4820.

Harashima S. and Hinnebusch A.G. 1986. Multiple *GCD* genes required for repression of *GCN4*, a transcriptional activator of amino acid biosynthetic genes in *Saccharomyces cerevisiae*. *Mol. Cell. Biol.* **6:** 3990–3998.

Hashimoto N.N., Carnevalli L.S., and Castilho B.A. 2002. Translation initiation at non-AUG codons mediated by weakened association of eIF2 subunits. *Biochem. J.* **367:** 359–368.

He H., von der Haar T., Singh C.R., Ii M., Li B., Hinnebusch A.G., McCarthy J.E., and Asano K. 2003. The yeast eukaryotic initiation factor 4G (eIF4G) HEAT domain interacts with eIF1 and eIF5 and is involved in stringent AUG selection. *Mol. Cell. Biol.* **23:** 5431–5445.

Hershey P.E.C., McWhirter S.M., Gross J., Wagner G., Alber T., and Sachs A.B. 1999. The cap binding protein eIF4E promotes folding of a functional domain of yeast translation initiation factor eIF4G1. *J. Biol. Chem.* **274:** 21297–21304.

Hinnebusch A.G. 1996. Translational control of *GCN4*: Gene-specific regulation by phosphorylation of eIF2. In *Translational control* (ed. J.W.B. Hershey et al.), pp. 199–244. Cold Spring Harbor Laboratory Press, Cold Spring Harbor, New York.

———. 2000. Mechanism and regulation of initiator methionyl-tRNA binding to ribosomes. In *Translational control of gene expression* (ed. N. Sonenberg et al.), pp. 185–243. Cold Spring Harbor Laboratory Press, Cold Spring Harbor, New York.

———. 2005. Translational regulation of gcn4 and the general amino acid control of yeast. *Annu. Rev. Microbiol.* **59:** 407–450.

Hinnebusch A.G., Dever T.D., and Sonenberg N. 2004. Mechanism and regulation of protein synthesis initiation in eukaryotes. In *Protein synthesis and ribosome structure* (ed. N. Nierhaus and D.N. Wilson), pp. 241–322. Wiley-VCH, Weinheim, Germany.

Huang H., Yoon H., Hannig E.M., and Donahue T.F. 1997. GTP hydrolysis controls stringent selection of the AUG start codon during translation initiation in *Saccharomyces cerevisiae*. *Genes Dev.* **11:** 2396–2413.

Iizuka N., Najita L., Franzusoff A., and Sarnow P. 1994. Cap-dependent and cap-independent translation by internal initiation of mRNAs in cell extracts prepared from *Saccharomyces cerevisiae*. *Mol. Cell. Biol.* **14:** 7322–7330.

Ito T., Marintchev A., and Wagner G. 2004. Solution structure of human initiation factor eIF2alpha reveals homology to the elongation factor eEF1B. *Structure* **12:** 1693–1704.

Jivotovskaya A.V., Valasek L., Hinnebusch A.G., and Nielsen K.H. 2006. Eukaryotic translation initiation factor 3 (eIF3) and eIF2 can promote mRNA binding to 40S subunits independently of eIF4G in yeast. *Mol. Cell. Biol.* **26:** 1355–1372.

Kainuma M. and Hershey J.W.B. 2001. Depletion and deletion analyses of eucaryotic translation initiation factor 1A *Saccharomyces cerevisiae*. *Biochimie* **83**: 505–514.

Kakuta Y., Tahara M., Maetani S., Yao M., Tanaka I., and Kimura M. 2004. Crystal structure of the regulatory subunit of archaeal initiation factor 2B (aIF2B) from hyperthermophilic archaeon *Pyrococcus horikoshii* OT3: A proposed structure of the regulatory subcomplex of eukaryotic IF2B. *Biochem. Biophys. Res. Commun.* **319**: 725–732.

Kapp L.D. and Lorsch J.R. 2004. GTP-dependent recognition of the methionine moiety on initiator tRNA by translation factor eIF2. *J. Mol. Biol.* **335**: 923–936.

Kessler S.H. and Sachs A.B. 1998. RNA recognition motif 2 of yeast Pab1p is required for its functional interaction with eukaryotic translation initiation factor 4G. *Mol. Cell. Biol.* **18**: 51–57.

Kolupaeva V.G., Unbehaun A., Lomakin I.B., Hellen C.U., and Pestova T.V. 2005. Binding of eukaryotic initiation factor 3 to ribosomal 40S subunits and its role in ribosomal dissociation and anti-association. *RNA* **11**: 470–486.

Kozak M. 1989. The scanning model for translation: An update. *J. Cell Biol.* **108**: 229–241.

Krishnamoorthy T., Pavitt G.D., Zhang F., Dever T.E., and Hinnebusch A.G. 2001. Tight binding of the phosphorylated α subunit of initiation factor 2 (eIF2α) to the regulatory subunits of guanine nucleotide exchange factor eIF2B is required for inhibition of translation initiation. *Mol. Cell. Biol.* **21**: 5018–5030.

Laurino J.P., Thompson G.M., Pacheco E., and Castilho B.A. 1999. The β subunit of eukaryotic translation initiation factor 2 binds mRNA through the lysine repeats and a region comprising the C_2-C_2 motif. *Mol. Cell. Biol.* **19**: 173–181.

Lee J.H., Choi S.K., Roll-Mecak A., Burley S.K., and Dever T.E. 1999. Universal conservation in translation initiation revealed by human and archaeal homologs of bacterial translation initiation factor IF2. *Proc. Natl. Acad. Sci.* **96**: 4342–4347.

Lee J.H., Pestova T.V., Shin B.S., Cao C., Choi S.K., and Dever T.E. 2002. Initiation factor eIF5B catalyzes second GTP-dependent step in eukaryotic translation initiation. *Proc. Natl. Acad. Sci.* **99**: 16689–16694.

Lomakin I.B., Kolupaeva V.G., Marintchev A., Wagner G., and Pestova T.V. 2003. Position of eukaryotic initiation factor eIF1 on the 40S ribosomal subunit determined by directed hydroxyl radical probing. *Genes Dev.* **17**: 2786–2797.

Maag D. and Lorsch J.R. 2003. Communication between eukaryotic translation initiation factors 1 and 1A on the yeast small ribosomal subunit. *J. Mol. Biol.* **330**: 917–924.

Maag D., Algire M.A., and Lorsch J.R. 2005a. Communication between eukaryotic translation initiation factors 5 and 1A within the ribosomal pre-initiation complex plays a role in start site selection. *J. Mol. Biol.* **356**: 724–737.

Maag D., Fekete C.A., Gryczynski Z., and Lorsch J.R. 2005b. A conformational change in the eukaryotic translation preinitiation complex and release of eIF1 signal recognition of the start codon. *Mol. Cell* **17**: 265–275.

Mader S., Lee H., Pause A., and Sonenberg N. 1995. The translation initiation factor eIF-4E binds to a common motif shared by the translation factor eIF-4γ and the translational repressors 4E-binding proteins. *Mol. Cell. Biol.* **15**: 4990–4997.

Majumdar R., Bandyopadhyay A., and Maitr U. 2003. Mammalian translation initiation factor eIF1 functions with eIF1A and eIF3 in the formation of a stable 40 S preinitiation complex. *J. Biol. Chem.* **278**: 6580–6587.

Marcotrigiano J., Lomakin I.B., Sonenberg N., Pestova T.V., Hellen C.U.T., and Burley S.K. 2001. A conserved HEAT domain within eIF4G directs assembly of the translation initiation machinery. *Mol. Cell* **7**: 193–203.

Marintchev A. and Wagner G. 2005. eIF4G and CBP80 share a common origin and similar domain organization: Implications for the structure and function of eIF4G. *Biochemistry* **44:** 12265–12272.

Marintchev A., Kolupaeva V.G., Pestova T.V., and Wagner G. 2003. Mapping the binding interface between human eukaryotic initiation factors 1A and 5B: A new interaction between old partners. *Proc. Natl. Acad. Sci.* **100:** 1535–1540.

Matsuo H., Li H., McGuire A.M., Fletcher C.M., Gingras A.-C., Sonenberg N., and Wagner G. 1997. Structure of translation factor eIF4E bound to m7GDP and interaction with 4E-binding protein. *Nat. Struct. Biol.* **4:** 717–724.

Merrick W.C., Kemper W.M., and Anderson W.F. 1975. Purification and characterization of homogeneous initiation factor M2A from rabbit reticulocytes. *J. Biol. Chem.* **250:** 5556–5562.

Methot N., Song M.S., and Sonenberg N. 1996. A region rich in aspartic acid, arginine, tyrosine, and glycine (DRYFG) mediates eukaryotic initiation factor 4B (eIF4B) self-association and interaction with eIF3. *Mol. Cell. Biol.* **16:** 5328–5334.

Methot N., Rom E., Olsen H., and Sonenberg N. 1997. The human homologue of the yeast Prt1 protein is an integral part of the eukaryotic initiation factor 3 complex and interacts with p170. *J. Biol. Chem.* **272:** 1110–1116.

Myasnikov A.G., Marzi S., Simonetti A., Giuliodori A.M., Gualerzi C.O., Yusupova G., Yusupov M., and Klaholz B.P. 2005. Conformational transition of initiation factor 2 from the GTP- to GDP-bound state visualized on the ribosome. *Nat. Struct. Mol. Biol.* **12:** 1145–1149.

Naranda T., MacMillan S.E., Donahue T.F., and Hershey J.W. 1996. SUI1/p16 is required for the activity of eukaryotic translation initiation factor 3 in *Saccharomyces cerevisiae*. *Mol. Cell. Biol.* **16:** 2307–2313.

Neff C.L. and Sachs A.B. 1999. Eukaryotic translation initiation factors eIF4G and eIF4A from *Saccharomyces cerevisiae* physically and functionally interact. *Mol. Cell. Biol.* **19:** 5557–5564.

Niederberger N., Trachsel H., and Altmann M. 1998. The RNA recognition motif of yeast translation initiation factor Tif3/eIF4B is required but not sufficient for RNA strand-exchange and translational activity. *RNA* **4:** 1259–1267.

Nielsen K.H., Valášek L., Sykes C., Jivotovskaya A., and Hinnebusch A.G. 2006. Interaction of the RNP1 motif in PRT1 with HCR1 promotes 40S binding of eukaryotic initiation factor 3 in yeast. *Mol. Cell. Biol.* **26:** 2984–2998.

Nielsen K.H., Szamecz B., Valášek L., Jivotovskaya A., Shin B.S., and Hinnebusch A.G. 2004. Functions of eIF3 downstream of 48S assembly impact AUG recognition and GCN4 translational control. *EMBO J.* **23:** 1166–1177.

Nika J., Rippel S., and Hannig E.M. 2000. Biochemical analysis of the eIF2βγ complex reveals a structural function for eIF2α in catalyzed nucleotide exchange. *J. Biol. Chem.* **276:** 1051–1056.

———. 2001. Biochemical analysis of the eIF2beta gamma complex reveals a structural function for eIF2alpha in catalyzed nucleotide exchange. *J. Biol. Chem.* **276:** 1051–1056.

Nonato M.C., Widom J., and Clardy J. 2002. Crystal structure of the N-terminal segment of human eukaryotic translation initiation factor 2alpha. *J. Biol. Chem.* **277:** 17057–17061.

Oberer M., Marintchev A., and Wagner G. 2005. Structural basis for the enhancement of eIF4A helicase activity by eIF4G. *Genes Dev.* **19:** 2212–2223.

Olsen D.S., Savner E.M., Mathew A., Zhang F., Krishnamoorthy T., Phan L., and Hinnebusch A.G. 2003. Domains of eIF1A that mediate binding to eIF2, eIF3 and eIF5B and promote ternary complex recruitment *in vivo*. *EMBO J.* **22:** 193–204.

Otero L.J., Ashe M.P., and Sachs A.B. 1999. The yeast poly(A)-binding protein Pab1p stimulates *in vitro* poly(A)-dependent and cap-dependent translation by distinct mechanisms. *EMBO J.* **18:** 3153–3163.

Paulin F.E., Campbell L.E., O'Brien K., Loughlin J., and Proud C.G. 2001. Eukaryotic translation initiation factor 5 (eIF5) acts as a classical GTPase-activator protein. *Curr. Biol.* **11:** 55–59.

Pavitt G.D., Yang W., and Hinnebusch A.G. 1997. Homologous segments in three subunits of the guanine nucleotide exchange factor eIF2B mediate translational regulation by phosphorylation of eIF2. *Mol. Cell. Biol.* **17:** 1298–1313.

Pavitt G.D., Ramaiah K.V.A., Kimball S.R., and Hinnebusch A.G. 1998. eIF2 independently binds two distinct eIF2B subcomplexes that catalyze and regulate guanine-nucleotide exchange. *Genes Dev.* **12:** 514–526.

Pedulla N., Palermo R., Hasenohrl D., Blasi U., Cammarano P., and Londei P. 2005. The archaeal eIF2 homologue: Functional properties of an ancient translation initiation factor. *Nucleic Acids Res.* **33:** 1804–1812.

Pestova T.V. and Kolupaeva V.G. 2002. The roles of individual eukaryotic translation initiation factors in ribosomal scanning and initiation codon selection. *Genes Dev.* **16:** 2906–2922.

Pestova T.V., Borukhov S.I., and Hellen C.U.T. 1998. Eukaryotic ribosomes require initiation factors 1 and 1A to locate initiation codons. *Nature* **394:** 854–859.

Pestova T.V., Lomakin I.B., Lee J.H., Choi S.K., Dever T.E., and Hellen C.U.T. 2000. The joining of ribosomal subunits in eukaryotes requires eIF5B. *Nature* **403:** 332–335.

Phan L., Schoenfeld L.W., Valášek L., Nielsen K.H., and Hinnebusch A.G. 2001. A subcomplex of three eIF3 subunits binds eIF1 and eIF5 and stimulates ribosome binding of mRNA and tRNA$_i^{Met}$. *EMBO J.* **20:** 2954–2965.

Phan L., Zhang X., Asano K., Anderson J., Vornlocher H.P., Greenberg J.R., Qin J., and Hinnebusch A.G. 1998. Identification of a translation initiation factor 3 (eIF3) core complex, conserved in yeast and mammals, that interacts with eIF5. *Mol. Cell. Biol.* **18:** 4935–4946.

Preiss T. and Hentze M.W. 1998. Dual function of the messenger RNA cap structure in poly(A)-tail-promoted translation in yeast. *Nature* **392:** 516–520.

Proud C.G. 1992. Protein phosphorylation in translational control. *Curr. Top. Cell. Regul.* **32:** 243–369.

Ptushkina M., von der Haar T., Vasilescu S., Frank R., Birkenhager R., and McCarthy J.E.G. 1998. Cooperative modulation by eIF4G of eIF4E binding to the mRNA 5'cap in yeast involves a site partially shared by p20. *EMBO J.* **17:** 4798–4808.

Ramakrishnan V. 2002. Ribosome structure and the mechanism of translation. *Cell* **108:** 557–572.

Raychaudhuri P., Chaudhuri A., and Maitra U. 1985. Eukaryotic initiation factor 5 from calf liver is a single polypeptide chain protein of Mr = 62,000. *J. Biol. Chem.* **260:** 2132–2139.

Roll-Mecak A., Cao C., Dever T.E., and Burley S.K. 2000. X-ray structures of the universal translation initiation factor IF2/eIF5B. Conformational changes on GDP and GTP binding. *Cell* **103:** 781–792.

Roll-Mecak A., Shin B.S., Dever T.E., and Burley S.K. 2001. Engaging the ribosome: Universal IFs of translation. *Trends Biochem. Sci.* **26:** 705–709.

Roll-Mecak A., Alone P., Cao C., Dever T.E., and Burley S.K. 2004. X-ray structure of translation initiation factor eIF2gamma: Implications for tRNA and eIF2alpha binding. *J. Biol. Chem.* **279:** 10634–10642.

Rowlands A.G., Panniers R., and Henshaw E.C. 1988. The catalytic mechanism of guanine nucleotide exchange factor action and competitive inhibition by phosphorylated eukaryotic initiation factor 2. *J. Biol. Chem.* **263:** 5526–5533.

Sachs A. 2000. Physical and functional interactions between the mRNA cap structure and the poly(A) tail. In *Translational control of gene expression* (ed. N. Sonenberg et al.), pp. 447–465. Cold Spring Harbor Laboratory Press, Cold Spring Harbor, New York.

Schmitt E., Blanquet S., and Mechulam Y. 2002. The large subunit of initiation factor aIF2 is a close structural homologue of elongation factors. *EMBO J.* **21:** 1821–1832.

Searfoss A.M. and Wickner R.B. 2000. 3' Poly(A) is dispensable for translation. *Proc. Natl. Acad. Sci.* **97:** 9133–9137.

Searfoss A., Dever T.E., and Wickner R. 2001. Linking the 3' poly(A) tail to the subunit joining step of translation initiation: Relations of Pab1p, eukaryotic translation initiation factor 5b (Fun12p), and Ski2p-Slh1p. *Mol. Cell. Biol.* **21:** 4900–4908.

Shalev A., Valášek L., Pise-Masison C.A., Radonovich M., Phan L., Clayton J., He H., Brady J.N., Hinnebusch A.G., and Asano K. 2001. *Saccharomyces cerevisiae* protein Pci8p and human protein eIF3e/Int-6 interact with the eIF3 core complex by binding to cognate eIF3b subunits. *J. Biol. Chem.* **276:** 34948–34957.

Sherman F. and Stewart J.W. 1982. Mutations altering initiation of translation of yeast iso-1-cytochrome *c*; contrasts between the eukaryotic and prokaryotic initiation process. In *The molecular biology of the yeast* Saccharomyces: *Metabolism and gene expression* (ed. J.N. Strathern et al.), pp. 301–334. Cold Spring Harbor Laboratory Press, Cold Spring Harbor, New York.

Shin B.S., Maag D., Roll-Mecak A., Arefin M.S., Burley S.K., Lorsch J.R., and Dever T.E. 2002. Uncoupling of initiation factor eIF5B/IF2 GTPase and translational activities by mutations that lower ribosome affinity. *Cell* **111:** 1015–1025.

Singh C.R., Yamamoto Y., and Asano K. 2004a. Physical association of eukaryotic initiation factor (eIF) 5 carboxyl-terminal domain with the lysine-rich eIF2beta segment strongly enhances its binding to eIF3. *J. Biol. Chem.* **279:** 49644–49655.

Singh C.R., He H., Ii M., Yamamoto Y., and Asano K. 2004b. Efficient incorporation of eukaryotic initiation factor 1 into the multifactor complex is critical for formation of functional ribosomal preinitiation complexes in vivo. *J. Biol. Chem.* **279:** 31910–31920.

Singh C.R., Curtis C., Yamamoto Y., Hall N.S., Kruse D.S., He H., Hannig E.M., and Asano K. 2005. Eukaryotic translation initiation factor 5 is critical for integrity of the scanning preinitiation complex and accurate control of GCN4 translation. *Mol. Cell. Biol.* **25:** 5480–5491.

Siridechadilok B., Fraser C.S., Hall R.J., Doudna J.A., and Nogales E. 2005. Structural roles for human translation factor eIF3 in initiation of protein synthesis. *Science* **310:** 1513–1515.

Tanner N.K. and Linder P. 2001. DExD/H box RNA helicases: From generic motors to specific dissociation functions. *Mol. Cell* **8:** 251–262.

Tarun S.Z. and. Sachs A.B. 1995. A common function for mRNA 5' and 3' ends in translation initiation in yeast. *Genes Dev.* **9**: 2997–3007.

———. 1996. Association of the yeast poly(A) tail binding protein with translation initiation factor eIF-4G. *EMBO J.* **15:** 7168–7177.

———. 1997. Binding of eukaryotic translation initiation factor 4E (eIF4E) to eIF4G represses translation of uncapped mRNA. *Mol. Cell. Biol.* **17:** 6876–6886.

Tarun S.Z., Wells S.E., Deardorff J.A., and Sachs A.B. 1997. Translation initiation factor eIF4G mediates in vitro poly (A) tail-dependent translation. *Proc. Natl. Acad. Sci.* **94:** 9046–9051.

Thompson G.M., Pacheco E., Melo E.O., and Castilho B.A. 2000. Conserved sequences in the beta subunit of archaeal and eukaryal translation initiation factor 2 (eIF2), absent from eIF5, mediate interaction with eIF2gamma. *Biochem J.* **347:** 703–709.

Unbehaun A., Borukhov S.I., Hellen C.U., and Pestova T.V. 2004. Release of initiation factors from 48S complexes during ribosomal subunit joining and the link between establishment of codon-anticodon base-pairing and hydrolysis of eIF2-bound GTP. *Genes Dev.* **18:** 3078–3093.

Valášek L., Nielsen K.H., and Hinnebusch A.G. 2002. Direct eIF2-eIF3 contact in the multifactor complex is important for translation initiation in vivo. *EMBO J.* **21:** 5886–5898.

Valášek L., Hašek J., Trachsel H., Imre E.M., and Ruis H. 1999. The *Saccharomyces cerevisiae HCRI* gene encoding a homologue of the p35 subunit of human translation eukaryotic initiation factor 3 (eIF3) is a high copy suppressor of a temperature-sensitive mutation in the Rpg1p subunit of yeast eIF3. *J. Biol. Chem.* **274:** 27567–27572.

Valášek L., Nielsen K.H., Zhang F., Fekete C.A., and Hinnebusch A.G. 2004. Interactions of eukaryotic translation initiation factor 3 (eIF3) subunit NIP1/c with eIF1 and eIF5 promote preinitiation complex assembly and regulate start codon selection. *Mol. Cell. Biol.* **24:** 9437–9455.

Valášek L., Phan L., Schoenfeld L.W., Valásková V., and Hinnebusch A.G. 2001. Related eIF3 subunits TIF32 and HCR1 interact with an RNA recognition motif in PRT1 required for eIF3 integrity and ribosome binding. *EMBO J.* **20:** 891–904.

Valášek L., Mathew A., Shin B.S., Nielsen K.H., Szamecz B., and Hinnebusch A.G. 2003. The yeast eIF3 subunits TIF32/a and NIP1/c and eIF5 make critical connections with the 40S ribosome in vivo. *Genes Dev.* **17:** 786–799.

Vazquez de Aldana C.R. and Hinnebusch A.G. 1994. Mutations in the GCD7 subunit of yeast guanine nucleotide exchange factor eIF-2B overcome the inhibitory effects of phosphorylated eIF-2 on translation initiation. *Mol. Cell. Biol.* **14:** 3208–3222.

Vazquez de Aldana C.R., Dever T.E., and Hinnebusch A.G. 1993. Mutations in the α subunit of eukaryotic translation initiation factor 2 (eIF-2α) that overcome the inhibitory effects of eIF-2α phosphorylation on translation initiation. *Proc. Natl. Acad. Sci.* **90:** 7215–7219.

Verlhac M.-H., Chen R.-H., Hanachi P., Hershey J.W.B., and Derynck R. 1997. Identification of partners of TIF34, a component of the yeast eIF3 complex, required for cell proliferation and translation initiation. *EMBO J.* **16:** 6812–6822.

von Der Haar T., Ball P.D., and McCarthy J.E. 2000. Stabilization of eukaryotic initiation factor 4E binding to the mRNA 5'-cap by domains of eIF4G. *J. Biol. Chem.* **275:** 30551–30555.

von Pawel-Rammingen U., Åström S., and Byström A.S. 1992. Mutational analysis of conserved positions potentially important for initiator tRNA function in *Saccharomyces cerevisiae*. *Mol. Cell. Biol.* **12:** 1432–1442.

Vornlocher H.P., Hanachi P., Ribeiro S., and Hershey J.W.B. 1999. A 110-kilodalton subunit of translation initiation factor eIF3 and an associated 135-kilodalton protein are encoded by the *Saccharomyces cerevisiae TIF32* and *TIF31* genes. *J. Biol. Chem.* **274:** 16802–16812.

Wai H.H., Vu L., Oakes M., and Nomura M. 2000. Complete deletion of yeast chromosomal rDNA repeats and integration of a new rDNA repeat: Use of rDNA deletion strains for functional analysis of rDNA promoter elements *in vivo*. *Nucleic Acids Res.* **28:** 3524–3534.

Wells S.E., Hillner P.E., Vale R.D., and Sachs A.B. 1998. Circularization of mRNA by eukaryotic translation initiation factors. *Mol. Cell* **2:** 135–140.

Yamamoto Y., Singh C.R., Marintchev A., Hall N.S., Hannig E.M., Wagner G., and Asano K. 2005. The eukaryotic initiation factor (eIF) 5 HEAT domain mediates multifactor assembly and scanning with distinct interfaces to eIF1, eIF2, eIF3, and eIF4G. *Proc. Natl. Acad. Sci.* **102:** 16164–16149.

Yang W. and Hinnebusch A.G. 1996. Identification of a regulatory subcomplex in the guanine nucleotide exchange factor eIF2B that mediates inhibition by phosphorylated eIF2. *Mol. Cell. Biol.* **16:** 6603–6616.

Yarunin A., Panse V.G., Petfalski E., Dez C., Tollervey D., and Hurt E. 2005. Functional link between ribosome formation and biogenesis of iron-sulfur proteins. *EMBO J.* **24:** 580–588.

Yatime L., Mechulam Y., Blanquet S., and Schmitt E. 2006. Structural switch of the gamma subunit in an archaeal aIF2alphagamma heterodimer. *Structure* **14:** 119–128.

Yatime L., Schmitt E., Blanquet S., and Mechulam Y. 2004. Functional molecular mapping of archaeal translation initiation factor 2. *J. Biol. Chem.* **279:** 15984–15993.

Yoon H.J. and Donahue T.F. 1992. The *sui1* suppressor locus in *Saccharomyces cerevisiae* encodes a translation factor that functions during tRNA$_i^{Met}$ recognition of the start codon. *Mol. Cell. Biol.* **12:** 248–260.

Zhou C., Arslan F., Wee S., Krishnan S., Ivanov A.R., Oliva A., Leatherwood J., and Wolf D.A. 2005. PCI proteins eIF3e and eIF3m define distinct translation initiation factor 3 complexes. *BMC Biol.* **3:** 14.

10

cis-Regulatory Sequences and *trans*-Acting Factors in Translational Control

Matthias W. Hentze

Gene Expression Unit
European Molecular Biology Laboratory
69117 Heidelberg, Germany

Fátima Gebauer

Gene Regulation Program
Centre de Regulació Genómica (CRG-UPF)
08003 Barcelona, Spain

Thomas Preiss

Molecular Genetics Program
Victor Chang Cardiac Research Institute
Darlinghurst (Sydney), NSW 2010, Australia

ALL NUCLEUS-ENCODED EUKARYOTIC MRNAS possess a 5′ cap structure (m^7GpppN) and, with the exception of the metazoan histone mRNAs, a 3′ poly(A) tail as well. These modifications are added cotranscriptionally in the nucleus (Maniatis and Reed 2002; Proudfoot et al. 2002), and they influence many aspects of mRNA metabolism, including splicing, transport, and stability. The cap structure and the poly(A) tail also promote translation initiation, and, as detailed below, their functions in initiation are frequently the direct or indirect targets of regulatory intervention.

Canonical cap-dependent translation initiation can be divided into four substeps: (1) formation of a 43S preinitiation complex, (2) recruitment of the 43S complex to the 5′ end of the mRNA, (3) scanning of the 5′-untranslated region (UTR) and start codon recognition, and (4) assembly of the 80S ribosome (for a detailed description, see Preiss and

Hentze 2003; Sonenberg and Dever 2003; Chapter 4). The interplay between *cis*-acting elements and *trans*-acting factors can control mRNA translation at all of these steps.

POINTS OF GLOBAL CONTROL DURING TRANSLATION INITIATION

Although all stages of translation can be the target of regulation, most known control mechanisms affect the initiation stage. A broad distinction can be made between global and mRNA-specific translational control processes (Gebauer and Hentze 2004). mRNA-specific control is usually exerted through regulatory elements in the 5′UTR or 3′UTR of the mRNA (Fig. 1; for a more detailed description, see below). Global control frequently results from changes in the phosphorylation state of initiation factors or the regulators that interact with them and affects the translation of most cellular mRNAs in a similar way (Dever 2002). Common targets are the cap-binding complex, eukaryotic initiation factor 4F (eIF4F), or the ternary complex component eIF2.

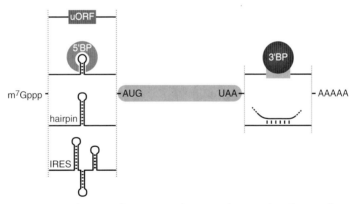

Figure 1. *cis*-Acting elements that can regulate translation. The 5′UTR of an mRNA may contain upstream open reading frames (uORF) and, more generally, upstream start codons (uAUG) that can affect downstream initiation. 5′UTR RNA motifs bound by cognate binding proteins or RNA secondary structures (i.e., hairpins) may also hinder translation initiation, for instance, by sterically interfering with access of initiation complexes to the mRNA. Certain viral and some cellular mRNAs exhibit internal ribosome entry sites (IRES) in their 5′UTR that can direct translation initiation complexes to the mRNA by a cap-independent mechanism. The 3′UTRs of mRNAs frequently contain elements that regulate their translation and/or other aspects of their posttranscriptional fate. Such elements serve to assemble specific binding proteins or interact with regulatory RNAs, such as microRNAs.

Four cellular kinases are known to phosphorylate the α subunit of eIF2, leading to the sequestration of the guanine nucleotide exchange factor eIF2B and accumulation of inactive GDP-bound eIF2 (Chapters 12 and 13). The main effect of eIF2α phosphorylation is a global decrease of cellular translation, but paradoxically, it can also activate the translation of specific mRNAs by means of regulatory upstream open reading frames (uORFs; see Fig. 1) in their 5′UTR. The best-characterized example of this type of control is the yeast *GCN4* mRNA (Hinnebusch 1997; Hinnebusch and Natarajan 2002). This transcript has four short uORFs, which collectively mediate the induction of translation at the main *GCN4* ORF in response to amino acid starvation by a reinitiation mechanism. Ribosomes nearly all translate the first uORF. About half of the ribosomes (or 40S subunits) then resume scanning and are thought to regain initiation competence by rebinding active ternary complex (eIF2, GTP, initiator tRNA) from the cellular pool. If levels of active eIF2 are high, this happens quickly and most ribosomes reinitiate translation at the third or fourth uORF, resulting in low levels of Gcn4p translation. In amino-acid-starved cells, active eIF2 levels are low, and this allows more ribosomes to skip over the other uORFs before regaining competence for initiation, thus leading to activation of initiation at the main *GCN4* ORF and enhanced Gcn4p translation. Translational control of the mammalian ATF4 mRNA has been shown to operate in a similar way (Lu et al. 2004; Vattem and Wek 2004).

The alternate use of several in-frame start codons can be considered as a special form of uAUG-mediated regulation leading to the expression of several proteins from one mRNA, often by "leaky scanning." Human fibroblast growth factor 2 (FGF-2) occurs in five isoforms with different amino-terminal extensions that are translated from a common mRNA by alternative use of in-frame start codons (four CUGs and one AUG) (Bonnal et al. 2003; Touriol et al. 2003). Translation from the 5′ proximal CUG codon is initiated by the cap-dependent mechanism, whereas the other four initiation codons are served by an internal ribosome entry site (IRES, see below). The FGF-2 protein isoforms have different subcellular localizations and functions. A sizeable proportion of eukaryotic mRNAs possess 5′UTRs that contain uAUGs or uORFs, and a recent survey found conserved uAUGs in at least 20–30% of mammalian genes (Churbanov et al. 2005), suggesting that they represent a common *cis*-acting element to generate protein diversity and to regulate protein expression.

The eIF4F complex is a major target of translational control by extracellular stimuli. The MAP kinase pathway promotes phosphorylation of eIF4E on Ser-209 by the kinase Mnk1, correlating with increased translation (Chapter 14). A family of eIF4E-binding proteins (4E-BPs) act in their hypophosphorylated form as competitive inhibitors of eIF4G binding to eIF4E (Gingras et al. 2004; Proud 2004). Growth stimulatory signals through the PI3K/AKT/TOR pathway lead to hyperphosphorylation of the 4E-BPs, their dissociation from eIF4E, and enhanced global translation (Chapter 14). Despite the universal presence of a cap structure at the 5′ end of cellular mRNAs, it is nevertheless frequently observed that translation of specific mRNAs reacts to changes in the level of active eIF4F in nonstandard ways. Some mRNAs appear to require higher eIF4F activity for their translation owing to their highly structured 5′UTRs (see Fig. 1) (Zimmer et al. 2000). Several such mRNAs encode proteins that promote cell growth and proliferation, and a selective up-regulation of these mRNAs could explain the well-documented involvement of eIF4E in cancer (Huang and Houghton 2003; De Benedetti and Graff 2004; Mamane et al. 2004; Rosenwald 2004; Chapter 15). On the other end of the spectrum are mRNAs that can be translated in a cap-independent fashion. Again, the proteins encoded by such mRNAs are frequently involved in the regulation of cell growth and proliferation, differentiation, or survival. The implication is that cells need to maintain translation of these mRNAs under stress conditions, when canonical cap-dependent translation is compromised (Chapter 16). The 5′UTRs of many of these mRNAs were reported to contain IRES elements (Fig. 1) (Holcik et al. 2000; Vagner et al. 2001; Stoneley and Willis 2004; Komar and Hatzoglou 2005). IRES elements are structural RNA elements, first discovered in picornaviral RNAs, that recruit initiation complexes directly to internal sites on the mRNA. They do this without a need for the mRNA cap structure and—at least in the case of viral IRES elements—often require only a subset of the canonical initiation factors (Hellen and Sarnow 2001; Bushell and Sarnow 2002; Chapter 5).

In summary, a major response of cells to stress or growth stimuli is to alter the availability of pivotal initiation factors and thereby modulate their global translation activity. Subsets of cellular mRNAs harbor *cis*-acting elements (uORFs, secondary structures, or IRES elements) in their 5′UTRs that can specify an mRNA-specific response (hypersensitivity, resistance) to such general changes (Dever 2002).

POLY(A) TAIL FUNCTION IN TRANSLATION

Despite its position at the 3′ end of the mRNA molecule, the poly(A) tail represents a critical *cis*-acting element for translation initiation (Jacobson 1996; Sachs 2000; Preiss 2002). The known *trans*-acting factors for poly(A) tail function are the poly(A)-binding proteins (PABPs)

(Mangus et al. 2003; Gorgoni and Gray 2004). They exist in cytoplasmic and nuclear forms, which bear little resemblance to each other. Unicellular eukaryotes typically possess a single gene encoding a cytoplasmic PABP (Pab1p in the yeast nomenclature), and several PABP genes are found in metazoans and plants, affording additional means for cell-type/stage-specific control of poly(A) tail function. Humans have four PABP genes that have recognizable homology with each other and with yeast Pab1p: the ubiquitously expressed PABPC1 (also called PABP1, PABP, PAB1, PAB), PABPC3 (or testis [t]PABP), PABPC4 (variously referred to as inducible [i]PABP, or APP-1), and PABPC5 (also known as X-linked PABP or PABP5). The amino termini of all these PABPs contain four conserved RNA recognition motifs (RRMs). PABPC1, -3, and -4 further exhibit a conserved carboxy-terminal domain (PABC), linked to the remainder of the protein via a region rich in proline and methionine residues.

To date, most functional analyses have been done with proteins of the PABP1 type, i.e., human PABPC1 or yeast Pab1p. In the following discussion, we refer to these simply as PABP. Several molecules of PABP can bind to poly(A) tails with a periodicity of about 25 adenosine residues, although 12 adenosines are in principle sufficient for binding. RRMs 1 and 2 bind to poly(A) with high affinity and specificity, whereas RRMs 3 and 4 exhibit more generic RNA-binding activity.

Role of the Poly(A) Tail in Cap-dependent Initiation

Much progress in understanding the function of the poly(A) tail has come from studies in cell-free translation extracts derived from *Saccharomyces cerevisiae*, which are capable of recapitulating the synergistic interactions between the cap structure and the poly(A) tail in translation. The observed synergy results at least in part from a competition between mRNAs for limiting components of the translation machinery (Tarun and Sachs 1997; Preiss and Hentze 1998; for comprehensive reviews, see Sachs 2000; Preiss 2002). The poly(A) tail, like the cap structure, promotes the recruitment of 43S preinitiation complexes to the mRNA (Tarun and Sachs 1995). Poly(A)-mediated initiation and the cap/poly(A) synergy require PABP (Tarun and Sachs 1995) and its interaction with eIF4G (Tarun and Sachs 1996; Tarun et al. 1997). The interacting regions were mapped to RRMs 1 and 2 of PABP and the amino terminus of eIF4G (Kessler and Sachs 1998). This suggested that concurrent interactions of eIF4E and PABP with eIF4G could bridge the opposite ends of an mRNA to form a "closed-loop configuration" during translation (Fig. 2a), consistent with functional interactions between the cap structure and the poly(A) tail (Preiss and Hentze 1998) and the occurrence of circular polyribosomes in electron micrographs (Christensen et al. 1987). Indeed, atomic force microscopy revealed that recombinant eIF4E, eIF4G, PABP, and a capped and polyadenylated template can form pseudo-circular complexes in vitro (Wells et al. 1998). Potential advantages of a pseudo-circular structure of mRNAs are easily recognized (for summary, see Sachs 2000). At a minimum, multiple mRNA–protein and protein–protein contacts could help stabilize translation initiation complexes on the mRNA. Cooperative binding interactions within the cap-eIF4E-eIF4G-PABP-poly(A) assembly (Haghighat and Sonenberg 1997; Le et al. 1997; Ptushkina et al. 1998, 1999; Wei et al. 1998; Borman et al. 2000) suggest that these effects are not simply additive but may represent at least part of the molecular basis for the cap/poly(A) synergy. Such a mechanism also helps ensure that only intact mRNAs act as efficient templates. Furthermore, the association of both mRNA ends with the translation machinery may antagonize mRNA decay via the deadenylation/decapping pathway (but see also below).

The orthologs of PABP1 and eIF4G also interact with each other in plant (Le et al. 1997) and mammalian cells (Imataka et al. 1998; Piron et al. 1998). The interaction is important for poly(A)-stimulated translation in vitro and in vivo (Wakiyama et al. 2000). Interestingly, the nonstructural rotavirus protein NSP3 binds to the amino-terminal region of human eIF4G and competitively displaces PABP (Piron et al. 1998). NSP3 further interacts with the (nonadenylated) 3′ end of rotavirus mRNA and provides the virus with a translational advantage in two ways: blocking PABP-dependent cellular translation and bridging the two ends of rotavirus mRNAs (Imataka et al. 1998; Vende et al. 2000).

The eIF4G–PABP interaction is thus important for the participation of the 3′ poly(A) tail in translation initiation at the 5′ end, but additional 5′ to 3′ contacts are likely to be involved. Further evidence implicates the poly(A) tail in 60S subunit recruitment (Fig. 2b). Early work in reticulocyte lysates attributed the modest stimulatory effect of a poly(A) tail on translation to enhanced 60S subunit joining (Munroe and Jacobson 1990). A recent study supports this concept (Kahvejian et al. 2005). Biochemical depletion of PABP reduced translation in a poly(A)-dependent mammalian cell-free translation system and showed a stronger inhibition of 80S versus 48S initiation complex formation. In yeast, suppressor mutations of a deletion of the essential *PAB1* gene show alterations in the level of 60S ribosomal subunits (Sachs and Davis 1989, 1990). Further genetic studies suggested that the poly(A)/PABP complex derepresses 60S subunit joining by inhibiting the function of the RNA helicases Ski2p and

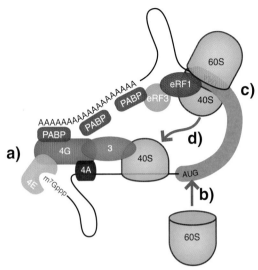

Figure 2. mRNA looping model. This schematic is designed to highlight models for an involvement of the poly(A) tail in the different stages of translation (initiation). (*a*) eIF4E bound to the cap structure and PABP bound to the poly(A) tail jointly recruit eIF4G and the 43S preinitiation complex to the mRNA. (*b*) The poly(A) tail also stimulates the 60S joining step at the start codon. (*c*) The eRF3–PABP interaction provides a means to couple translation termination and poly(A) tail function. This may serve to link translation to mRNA deadenylation/decay and/or to facilitate (*d*) a recycling of terminating ribosomal subunits back to the 5′UTR for a new round of initiation. For clarity, only the most relevant initiation factors are shown, and they are labeled in the diagram without the usual "eIF" prefix.

Slh1p, which in turn inhibit eIF5B and eIF5 (Masison et al. 1995; Benard et al. 1998, 1999; Searfoss and Wickner 2000; Searfoss et al. 2001).

There are also indications for contacts of the PABPs with the initiation machinery other than to eIF4G. Different, nonoverlapping portions of PABP can stimulate translation in tethered-function assays. Stimulation by the RRM 1 and 2 portion requires an interaction with eIF4G; however, RRMs 3 and 4 can also stimulate translation in isolation (Gray et al. 2000), although none of the known PABP interactors are known to bind this region. Several mutations in yeast eIF4G or PABP that decrease or abolish poly(A)-dependent translation do not have the same deleterious effect on translational synergy between cap structure and poly(A) tail in vitro and do not cause cell inviability (Tarun et al. 1997). A mutation in PABP was isolated that inhibits poly(A)-dependent translation but does not abolish eIF4G binding (Kessler and Sachs 1998).

An interaction between PABP and eIF4B, a factor that stimulates eIF4A helicase activity, was shown in plants (Le et al. 1997) and in mammalian cells (Bushell et al. 2001). Mammalian PABP contacts eIF4B through its PABC region (Deo et al. 2001; Kozlov et al. 2001). This domain binds specifically to a 12-amino acid peptide motif (termed PAM-2, for PABP-interacting motif 2) present in several interaction partners, including the PABP-interacting proteins, Paip1 and Paip2, and the eukaryotic release factor (eRF) 3 (Uchida et al. 2002; see below). In addition to PAM-2, both Paip1 and Paip2 harbor a second PABP-binding site termed PAM-1, through which they can interact with the RRM 1 – 3 regions of PABP. Paip1 exhibits homology with the central third of eIF4G (Craig et al. 1998). It forms complexes of 1:1 stoichiometry with PABP (Roy et al. 2002), interacts with eIF4A, and coactivates cap-dependent translation, although it lacks a canonical eIF4E-binding motif (Craig et al. 1998). In contrast, Paip2 is a small acidic protein that acts as a translational repressor. Two molecules of Paip2 can simultaneously bind PABP and reduce its affinity for oligo(A). Paip1 and Paip2 compete with each other for binding to PABP (Khaleghpour et al. 2001a, b).

The PABC domain also interacts with the translation termination factor eRF3 of yeast, *Xenopus,* and mammalian cells (Hoshino et al. 1999; Cosson et al. 2002). This interaction is thought to have two possible consequences: It could serve to loop out the 3′UTR of the mRNA and bring terminating ribosomes into the vicinity of the poly(A) tail and, by virtue of the PABP/eIF4G/eIF4E bridging interaction, also close to the cap structure and the 5′UTR (Fig. 2c). It is tempting to speculate that these interactions may limit the dissociation of ribosomal subunits from the mRNA after termination and instead aid their recycling or channeling back to the 5′ end of the same mRNA molecule to initiate a new round of translation (Fig. 2d) (Uchida et al. 2002). So far, there is no direct evidence for this attractive concept of ribosome recycling. Alternatively, the PABP–eRF3 interaction could relay a signal from translation to mRNA decay (Uchida et al. 2002; Hosoda et al. 2003; Kobayashi et al. 2004). eRF3 reportedly inhibits PABP multimerization, thus potentially contributing to its dissociation from the poly(A) tail and linking translation termination to poly(A) shortening and, ultimately, mRNA decay.

Role of the Poly(A) Tail in IRES-mediated Initiation

The effects of the poly(A) tail on IRES-driven translation introduce further complexity. Picornaviral RNAs are polyadenylated, and the translation of all three classes of picornaviral IRES elements is enhanced by a poly(A) tail (Bergamini et al. 2000; Svitkin et al. 2001). Using either cleavage of eIF4GI by viral proteases (to separate the amino-terminal PABP- and eIF4E-interacting regions from the remainder of the protein) or Paip2-mediated disruption of PABP/poly(A) binding to eIF4G as experimental tools in vitro, it was shown that this translational enhancement requires an interaction between eIF4G and PABP (Svitkin et al. 2001). This result suggests that IRES/poly(A) cooperativity could be important not only in cases where eIF4G is not proteolytically cleaved (the cardioviruses), but also during the early stages of infection with viruses that induce eIF4G cleavage. Viral RNA translation after eIF4G cleavage is expected to be PABP/poly(A) tail-independent. These observations further support a central role of the eIF4G/PABP interaction during translation initiation.

Surprisingly, similar analyses with IRES elements derived from cellular mRNAs paint a different picture. c-*myc* and several other cellular IRES elements are all sensitive to eIF4GI depletion in vitro, and their translation can be reconstituted by addition of the middle region of eIF4G or, alternatively, by the eIF4G homolog DAP5/p97/NAT-1 (Hundsdoerfer et al. 2005). c-*myc* and BiP-IRES-driven translation are enhanced by a poly(A) tail, but poly(A) stimulation of the c-*myc* and BiP IRES elements persists after cleavage of eIF4G and even following efficient depletion of PABP from the translation extracts (Thoma et al. 2004). This indicates that the poly(A) tail can stimulate translation in these cases by a mechanism that is independent of intact eIF4G and that may not even require PABP at all.

MRNA-SPECIFIC REGULATORY SEQUENCES AND THEIR *TRANS*-ACTING FACTORS

Translational regulation is often used to control the expression of specific mRNAs at particular times and/or locations within the cell. "mRNA-specific" translational control is achieved by *trans*-acting factors (proteins or miRNAs; Chapter 11) binding to sequence elements usually located in the UTRs of the mRNA (Fig. 1), often the 3'UTR. Very commonly, factor binding to its target site on the mRNA leads to translational repression. mRNA-specific translational regulation is widely used in biology to modulate processes such as metabolism, cell differentiation, embryonic patterning, and synaptic transmission (for review, see Kuersten and Goodwin 2003; Chapters 18, 19, and 24). Although many examples have been described, the molecular mechanisms of translational control have been elucidated only in a few cases. We

Table 1. *cis*-Acting regulatory sequences and corresponding *trans*-acting factors

mRNA	*cis*-Element	Sequence	Binding protein	Mechanism
Ferritin	IRE	combined sequence/structure motif	IRP	steric block to 43S binding
Cyclin B1[a]	CPE	UUUUAU	CPEB	inhibition of eIF4G binding to eIF4E
Caudal	BBR	positions 1350–1470 of caudal mRNA[b]	Bicoid	inhibition of eIF4E binding to cap
LOX	DICE	$(C_4PuC_3UCUUC_4 AAG)_{10}$	hnRNPs K and E	inhibition of 60S subunit joining
Oskar	BRE	UUg/aUa/gUg/aUU	Bruno	inhibition of 43S binding via RNA oligomerization and block of eIF4G binding to eIF4E
msl-2	poly(U)	U_{7-16}	SXL	inhibition of 43S binding and scanning
Nanos	TCE/SRE	CUGGC	Smaug	inhibition of eIF4G binding to eIF4E

[a]One of the several CPE-containing mRNAs.
[b]Position 1 is the first nucleotide of the start codon.

focus on those examples of translational control for which mechanistic data are available (see Table 1). For further information, see several recent reviews (Mendez and Richter 2001; Gebauer and Hentze 2004; Kindler et al. 2005; Richter and Sonenberg 2005; Wilhelm and Smibert 2005). We structure this section by focusing on the steps of the translation initiation pathway that are controlled by the respective regulatory factors.

Targeting 40S Ribosomal Subunit Binding to the mRNA

Most mRNAs in the cell appear to be translated via a cap-dependent mechanism. As explained above, the cap structure helps recruit the small ribosomal subunit to the mRNA during translation initiation. This step is inhibited by a variety of mechanisms (Fig. 3).

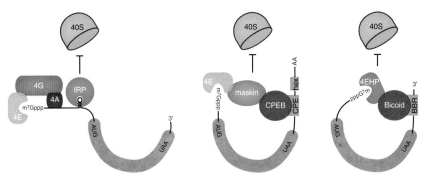

Figure 3. Mechanisms blocking 43S ribosomal subunit recruitment. Three distinct examples are shown: (1) IRP binds to a 5′ stem loop and sterically blocks the binding of the 43S preinitiation complex to eIF4F, (2) the CPEB/Maskin complex binds to the CPE in the 3′UTR of the mRNA and competitively inhibits the association of eIF4G with eIF4E, and (3) Bicoid binds to the Bicoid-binding region (BBR) in the 3′UTR and recruits the translation-incompetent cap-binding factor 4EHP, thereby inhibiting the association of eIF4F to the 5′ cap structure.

IRE/IRP

Iron regulatory proteins (IRP1 and IRP2) modulate cellular iron storage via the translational regulation of ferritin heavy- and light-chain mRNAs (Hentze et al. 2004). Under conditions of low iron, IRP binds to a cap-proximal stem loop in the 5′UTR of ferritin mRNA termed the iron-responsive element (IRE) and inhibits 43S ribosomal recruitment (Gray and Hentze 1994). The same mechanism controls additional mRNAs that encode proteins important in iron metabolism (Hentze et al. 2004). Translational repression by IRP binding does not affect the association of the cap-binding complex eIF4F to the mRNA, but it does affect the joining of the 43S preinitiation complex to this factor (Muckenthaler et al. 1998). It is relieved when the IRE is moved to a cap-distal position, suggesting that the IRE/IRP complex impedes 43S recruitment sterically (Goossen et al. 1990). This conclusion is consistent with the finding that RNA-binding proteins with no role in translational control can act as translational repressors when bound to structures located within approximately 40 nucleotides of the cap (Stripecke et al. 1994).

CPE/CPEB-Maskin

During *Xenopus* oocyte maturation (meiosis) and early embryonic divisions, many maternal mRNAs become cytoplasmically polyadenylated and translationally activated. These mRNAs share the presence of the cytoplasmic polyadenylation element (CPE) in their 3′UTRs. The CPE is a U-rich sequence that serves as the binding site for CPEB, a bifunctional regulator that stimulates poly(A) tail elongation during maturation and mediates repression of the translation of CPE-containing mRNAs prior to maturation (Chapters 18 and 19). To repress translation, CPEB recruits Maskin, a protein that contains an eIF4E interaction motif similar to that found in eIF4G (Stebbins-Boaz et al. 1999). A Maskin peptide that includes the eIF4E-binding sequence can inhibit translation in vivo, suggesting that Maskin competes with eIF4G for eIF4E binding and therefore inhibits the formation of the cap-binding complex (Stebbins-Boaz et al. 1999), although this has not yet been shown directly. In this respect, the CPEB–Maskin complex can be considered as a "message-specific 4E-BP" (Chapter 14).

Caudal mRNA/Bicoid

Anterior body patterning in *Drosophila* requires the localized activity of the homeodomain transcription factor Bicoid (Bcd). In addition to its function in transcription, Bcd binds to a region in the 3′UTR of Caudal (cad) mRNA termed the Bicoid-binding region (BBR) and inhibits its translation (Dubnau and Struhl 1996; Rivera-Pomar et al. 1996). Similar to Maskin, Bcd contains an eIF4E-binding motif (Niessing et al. 2002), leading to the initial hypothesis that Bcd inhibits translation by a Maskin-type mechanism. However, the eIF4E-binding motif of Bcd overlaps with that for binding d4EHP (*Drosophila* 4E-homologous protein), a 4E-like protein that cannot interact with eIF4G and therefore blocks the cap structure in an unproductive way (Cho et al. 2005). Mutations in the d4EHP-binding domain of Bcd cause anterior expression of cad and developmental defects. d4EHP interacts with both Bcd and the cap structure of cad mRNA, and this set of interactions maintains cad mRNA translationally repressed (Cho et al. 2005). Thus, Bcd inhibits small ribosomal subunit binding by recruiting a translationally incompetent 4E-like protein to the cap structure.

Inhibition of 60S Ribosomal Subunit Joining

Translation can also be inhibited after 43S complex recruitment. For example, *GCN4* mRNA translation is regulated at the level of reinitiation, once the small ribosomal subunit has been recruited to the mRNA (Chapters 12 and 13; see above). In principle, post-recruitment regulation could be achieved by modulating any of the steps subsequent to 40S ribosomal subunit binding; namely, scanning, AUG recognition, 60S ribosomal subunit joining, elongation, termination, or even the stability of the nascent polypeptide. However, relatively few examples of post-recruitment regulation have been described so far.

LOX mRNA/hnRNPs K and E

15-Lipoxygenase (LOX) is an enzyme involved in the degradation of mitochondria occurring during terminal erythroid differentiation. LOX mRNA remains translationally silenced in erythroid precursor cells by the binding of heterogeneous nuclear ribonucleoproteins (hnRNPs) K and E1/2 to an element in its 3′UTR, termed the differentiation control element (DICE) (Ostareck-Lederer and Ostareck 2004). Translational repression of LOX mRNA is independent of the poly(A) tail and occurs efficiently when translation is driven by the EMCV (encephalomyocarditis virus) or CSVF (classical swine fever virus) IRES elements, indicating that repression is also independent of the cap structure and implicating a step in translation initiation that is shared by the initiation pathways of cap-dependent and IRES-mediated translation (Ostareck et al. 1997, 2001). Indeed, sucrose gradient and toeprint analyses showed that in the presence of hnRNPs K and E, the 43S preinitiation complex can reach the initiation codon, but the 80S complex formation is inhibited (Ostareck et al. 2001). These data show that 60S ribosomal subunit joining is blocked (Fig. 4). Hypothetically, the regulators could target either the

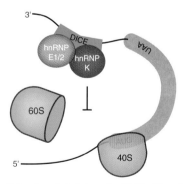

Figure 4. Inhibition of 60S subunit joining. hnRNPs K and E bind to DICE in the 3′UTR of LOX mRNA and block the joining of the 60S ribosomal subunit to the 40S subunit placed at the initiation codon. Whether the hnRNP K/E complex targets an initiation factor or a ribosomal component is unknown.

ribosome or an initiation factor required for 60S subunit joining. The observation that translation of a DICE-containing mRNA driven by the CrPV (cricket paralysis virus) IRES—which does not require any known initiation factor—is not inhibited by hnRNPs K and E suggests indirectly that hnRNPs K and E may target an initiation factor (Ostareck et al. 2001), but further experiments are required to clarify this point.

Multistep Mechanisms

A theme that has begun to emerge most recently is that mRNA-binding proteins can regulate translation by using more than one mechanism. Multistep mechanisms allow the tight redundant (fail-safe) translational control of mRNAs that encode critical proteins whose misexpression would be deleterious. However, the different steps of multistep regulation are not necessarily simultaneous with one another: The same regulatory protein may act in different ways and with different partners at distinct locations and/or stages of development.

oskar mRNA/Bruno

Oskar protein acts as a morphogen essential for the formation of germ cells and abdominal structures in the fruit fly. oskar mRNA is synthesized by nurse cells and then deployed at the anterior of the oocyte and transported to the posterior, where it is translated. Prior to its localization at the posterior pole, translation of oskar mRNA is repressed via the ovarian RNA-binding protein Bruno. Failure to repress oskar translation results in the disruption of the embryonic body plan and lethality (Kim-Ha et al. 1995). Research on the mechanism of oskar translational regulation has been complicated by the fact that localization of the oskar transcript is tightly coupled to translational control. Factors involved in oskar mRNA localization can also affect translation indirectly.

Bruno binds to defined, repeated sequence elements in the 3′UTR of oskar mRNA, referred to as the Bruno response elements (BREs) (Table 1). Mutating the BREs results in premature translation of oskar mRNA but does not interfere with mRNA localization, suggesting that Bruno regulates the translation of Oskar directly (Kim-Ha et al. 1995). A direct role for Bruno in translational repression was demonstrated in vitro by the use of cell-free translation extracts from *Drosophila* ovaries and embryos (Lie and Macdonald 1999; Castagnetti et al. 2000). These studies also showed that BRE-mediated repression is independent of the cap structure and the poly(A) tail (Lie and Macdonald 1999; Castagnetti and Ephrussi 2003).

Sucrose density gradient analysis shows that the repression mechanism targets the recruitment of the 43S preinitiation complex (Chekulaeva et al. 2006). This result is at odds with another report suggesting that the repressed oskar mRNA is associated with polysomes, arguing that Bruno inhibits the elongation/termination of translation (Braat et al. 2004). However, the experiments of Braat and colleagues involve changes of several experimental variables (puromycin addition, Mg^{2+} concentration, duration and temperature of incubation) at the same time, which may explain this discrepancy. Consistent with Bruno inhibiting translation initiation, Bruno binds the protein Cup, which contains a canonical eIF4E-binding site and competes with eIF4G for eIF4E binding (Wilhelm et al. 2003; Nakamura et al. 2004; Zappavigna et al. 2004). This result is in apparent contradiction to the demonstration that translational repression can occur independent of the cap structure (Lie and Macdonald 1999), but a second, cap- and Cup-independent mechanism to control oskar mRNA translation was uncovered recently. This second mechanism consists of Bruno-driven mRNA oligomerization and the formation of repressed higher-order "silencing particles" (up to 80S) that are inaccessible for ribosomes (Chekulaeva et al. 2006). Such oligomer formation seems to be particularly suitable for coupling translational repression with the localization of mRNAs, which are usually transported to their final destinations in the form of large particles. Thus, the translation of oskar mRNA is regulated by at least two Bruno-dependent mechanisms: the inhibition of 43S complex recruitment in a cap-dependent way via Cup, and Bruno-driven oskar mRNA oligomerization into "silencing particles" that are inaccessible to ribosome binding. Possibly, the two mechanisms are linked by their requirement for silencing particle formation (Chekulaeva et al. 2006).

msl-2 mRNA/SXL

Sex-lethal (SXL) is a female-specific RNA-binding protein that regulates sex determination in *Drosophila* via the modulation of the alternative splicing of downstream genes (Forch and Valcárcel 2003). SXL also regulates dosage compensation by inhibiting the expression of male-specific-lethal 2 (*msl-2*) mRNA that encodes a critical component of the dosage compensation complex. Expression of MSL-2 in females causes the inappropriate assembly of dosage compensation complexes on the two female X chromosomes and lethality (Kelley et al. 1995). In *Drosophila melanogaster*, SXL inhibits *msl-2* expression by first promoting the retention of a facultative intron in the 5′UTR of *msl-2* mRNA and then repressing its translation (Bashaw and Baker 1997; Kelley et al. 1997; Gebauer et al. 1998). Because corresponding splice sites are absent in *Drosophila virilis*,

SXL-mediated splicing regulation of *msl-2* mRNA is dispensable in this species. To inhibit translation, SXL must bind to specific sites located in both the (intronic) 5′UTR and the 3′UTR of *msl-2* mRNA (Table 1) (Bashaw and Baker 1997; Kelley et al. 1997). Mutational studies further support the notion that the sole purpose of retaining the intron is to provide 5′ SXL-binding sites for subsequent translational repression. Sucrose density gradient and toeprint assays uncovered that SXL functions via a dual mechanism: SXL bound to the 3′UTR inhibits the initial recruitment of the 43S ribosomal complex to *msl-2* mRNA, whereas SXL bound to the 5′UTR can block the scanning of 43S complexes that escaped the first inhibitory mechanism (Fig. 5) (Beckmann et al. 2005). As for oskar, repressed *msl-2* mRNPs have higher density than the common mRNPs, but in contrast to Bruno, no mRNA oligomerization activity by SXL has been detected (K. Beckmann and M. W. Hentze, unpubl.).

Scanning inhibition by SXL bound to the 5′UTR does not seem to operate by a simple steric hindrance mechanism, because the conserved SXL homolog from *Musca domestica*, which binds to the *msl-2* 5′UTR sites with an affinity similar to that of the *Drosophila* protein, neither represses translation nor inhibits 43S scanning (Grskovic et al. 2003; Beckmann et al. 2005). This result indirectly suggests that 5′UTR-bound

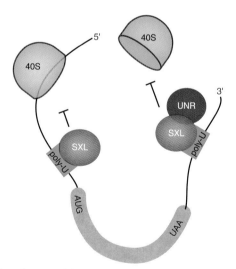

Figure 5. Translational control by a multistep mechanism. SXL binds to uridine stretches in the 5′UTR and 3′UTR of *msl-2* mRNA. 3′-Bound SXL recruits UNR, and the complex inhibits the initial association of the 43S preinitiation complex with the mRNA. SXL bound to the 5′UTR inhibits the scanning of those 43S complexes that may have escaped the 3′-mediated control.

SXL interferes with the function of an initiation factor required for scanning. It is also possible that a higher-order complex of SXL molecules, perhaps in conjunction with other proteins, blocks 43S complex transit. Acting from the 3'UTR, SXL requires the RNA-binding protein UNR (upstream of N-*ras*) as a corepressor to efficiently block translation (Grskovic et al. 2003; Abaza et al. 2006; Duncan et al. 2006). It will be intriguing to decipher how the 3'UTR SXL/corepressor assembly inhibits 43S complex recruitment.

Nanos mRNA/Smaug

Nanos is a posterior determinant of *D. melanogaster* whose primary function is to repress the translation of maternal hunchback mRNA. As for oskar, nanos mRNA is produced by nurse cells and deposited into the oocyte, where a small fraction of it localizes to the posterior pole and is translated during late oogenesis and embryogenesis (Chapter 24). Some 96% of the nanos mRNA remains unlocalized and translationally repressed (Gavis and Lehmann 1994; Bergsten and Gavis 1999). A translational control element (TCE) in the 3'UTR of nanos mRNA is necessary and sufficient for repression (Dahanukar and Wharton 1996; Gavis et al. 1996; Smibert et al. 1996). TCE is a bipartite RNA element (Gavis et al. 1996; Crucs et al. 2000) that harbors a CUGGC motif which is bound by the protein Smaug and thus called the Smaug recognition element (SRE) (Smibert et al. 1996). Smaug is required for the translational repression of nanos mRNA in the embryo (Smibert et al. 1996, 1999; Dahanukar et al. 1999; Crucs et al. 2000). In addition to the SRE, the TCE also bears a distal stem that is not bound by Smaug, yet it is required for translational repression during late oogenesis (Crucs et al. 2000; Forrest et al. 2004). Smaug interacts with the TCE via a SAM (sterile α motif) domain, a motif normally involved in protein–protein interactions (Aviv et al. 2003; Green et al. 2003). A search for proteins interacting with the SAM domain of Smaug identified Cup, the eIF4E-binding protein that also contributes to oskar mRNA translational regulation (Nelson et al. 2004; see above). Indeed, Cup appears to repress nanos mRNA translation, because embryos that overexpress a Cup mutant protein unable to interact with eIF4E show reduced levels of Smaug-mediated repression (Nelson et al. 2004). These results suggest that Smaug blocks translation initiation via eIF4E. Surprisingly, however, about 50% of the unlocalized, repressed nanos mRNA in the embryo has been reported to be associated with polysomes (Clark et al. 2000). This apparent contradiction may be resolved by temporal considerations. Nanos mRNA is produced and actively translated in nurse cells, and it is subsequently deployed in the oocyte where it is translationally repressed in a Smaug-independent fashion (Forrest et al. 2004). Because the mRNA was translated in the nurse cells, it seems plausible that this first Smaug-independent repression could occur on polysome-associated mRNA. During the first hours of embryogenesis, translational repression becomes Smaug-dependent (Smibert et al. 1996; Dahanukar et al. 1999), so nanos mRNA must switch from a Smaug-independent (elongation?) to a Smaug-dependent (initiation) block. It is conceivable that ribosomes remain bound to the mRNA after the switch to a translation initiation block.

Interesting Examples with Currently Unknown Inhibitory Mechanisms

In finishing this chapter, we include two interesting examples of translational control where the regulatory mechanisms are not yet clear.

Ceruloplasmin mRNA/GAIT Complex

Ceruloplasmin (Cp) is a glycoprotein with important functions in inflammation and iron metabolism. Cp synthesis is transiently induced by interferon-γ (IFN-γ) and ceases after 24 hours of IFN-γ treatment by a mechanism that involves the translational repression of Cp mRNA (Mazumder and Fox 1999). A 29-nucleotide *cis*-acting element in the 3′UTR of Cp mRNA, termed the IFN-γ-activated inhibitor of translation (GAIT) element, is necessary for translational repression (Sampath et al. 2003). This element binds a complex of four proteins: glyceraldehyde 3-phosphate dehydrogenase (GAPDH), NS1-associated protein 1 (NSAP1), the large ribosomal subunit protein L13a, and glutamyl-prolyl-tRNA synthetase (GluProRS), all of which are involved in translation inhibition of Cp mRNA (Mazumder et al. 2003; Sampath et al. 2004). Both GluProRS and L13a are phosphorylated upon IFN-γ treatment and released from other cellular complexes (the aminoacyl-tRNA multisynthetase complex [MSC] and the 60S ribosomal subunit, respectively) to become functional components of the GAIT complex. Phosphorylation and release of these proteins from their respective compartments does not affect global protein synthesis (Mazumder et al. 2003; Sampath et al. 2004). Although the mechanism of translational repression by the GAIT complex is unknown, repressed Cp mRNA is found in nonpolysomal fractions, suggesting that translation initiation is inhibited (Mazumder and Fox 1999). In addition, translation inhibition requires PABP, eIF4G, and the poly(A) tail, which led to the proposal that mRNA circularization via the PABP/eIF4G interaction is necessary to bring the GAIT complex in close proximity to the 5′ end of Cp mRNA and, thus, for translational repression (Mazumder et al. 2001).

HAC1 mRNA

The endoplasmic reticulum (ER) is the site of maturation and folding of secreted and membrane-bound proteins. Upon ER stress, unfolded proteins accumulate, and cells activate a signaling pathway termed the unfolded protein response (UPR) to counteract the toxic effects of this accumulation. Central to the UPR pathway in *Saccharomyces cerevisiae* is the translational activation of the mRNA encoding the transcription factor Hac1p. Two forms of *HAC1* mRNA can be detected, which differ by the presence or absence of an unusual intron in the ORF region. In unstressed cells, unspliced *HAC1* mRNA associates with polysomes, but no Hac1p is detected, suggesting that translation elongation and/or termination is blocked (Chapman and Walter 1997). Repressed *HAC1* mRNA can be immunoprecipitated with an antibody that recognizes the nascent amino terminus of the protein, showing that ribosomes have indeed initiated translation (Chapman and Walter 1997). Base-pairing between the 3′ intron and the 5′UTR of *HAC1* mRNA is crucial for translation inhibition (Ruegsegger et al. 2001). Upon UPR induction, removal of the intron by an unconventional cytoplasmic splicing and ligation mechanism leads to *HAC1* mRNA translational activation (Ruegsegger et al. 2001). It remains to be resolved how long-range base-pairing can inhibit ribosome transit, particularly because translation elongation is known to be generally quite processive.

PERSPECTIVES

As we outlined in this chapter, much exciting progress has been made during the last few years in understanding the interplay among *cis*-acting elements within mRNAs, *trans*-acting regulatory factors, and the translation (initiation) machinery. Our picture of the latter is becoming more and more refined thanks to outstanding biochemical and structural work on translation initiation factors and the ribosome, but much challenge lies ahead in comprehending the messenger RNP (the mRNA with its associated binding proteins) as a template for translation. The understanding of transcription has greatly benefited from extension of the consideration of promoter/enhancer sequences and DNA-binding proteins to the complex chromatin template. Similar progress in understanding mRNAs and mRNPs beyond their immediate *cis*-acting elements and RNA-binding proteins should be very beneficial to the field. From several experimental angles, we will need to study the composition of native mRNPs and their dynamic rearrangements during translation, translational repression, mRNA transport, and degradation.

It will be very interesting to determine whether a transient inhibition of translation induces changes in the mRNA/mRNP template (e.g., P-body formation), which further solidify a translational block ("chromatinization of the mRNA"). Recent results in yeast can be considered from such a viewpoint (Coller and Parker 2005). In several cases of mRNA-specific control via *trans*-acting factors, the inhibited (target) step of the translation (initiation) pathway has now been uncovered, and we await a better understanding of the connections between the regulatory factors and the translation machinery. The Maskin-type regulators and d4EHP (Stebbins-Boaz et al. 1999; Cho et al. 2005) can serve as examples, although curiously and with the exception of Cup-mediated regulation of oskar mRNA translation (Chekulaeva et al. 2006), the predicted target steps have not yet been directly confirmed experimentally.

ACKNOWLEDGMENTS

M.W.H. is supported by the European Molecular Biology Laboratory and grants from the Deutsche Forschungsgemeinschaft. F.G. acknowledges support from the Spanish Ministry of Science and Technology, the La Caixa Foundation, and the Catalonian Research Department (DURSI). T.P. is supported by The Sylvia & Charles Viertel Charitable Foundation, the National Health and Medical Research Council, and the Australian Research Council.

REFERENCES

Abaza I., Coll O., Patalano S., and Gebauer F. 2006. *Drosophila* UNR is required for translational repression of male-specific-lethal 2 mRNA during regulation of X chromosome dosage compensation. *Genes Dev.* **20:** 380–389.

Aviv T., Lin Z., Lau S., Rendl L.M., Sicheri F., and Smibert C.A. 2003. The RNA-binding SAM domain of Smaug defines a new family of post-transcriptional regulators. *Nat. Struct. Biol.* **10:** 614–621.

Bashaw G.J. and Baker B.S. 1997. The regulation of the *Drosophila* msl-2 gene reveals a function for Sex-lethal in translational control. *Cell* **89:** 789–798.

Beckmann K., Grskovic M., Gebauer F., and Hentze M.W. 2005. A dual inhibitory mechanism restricts msl-2 mRNA translation for dosage compensation in *Drosophila*. *Cell* **122:** 529–540.

Benard L., Carroll K., Valle R.C., and Wickner R.B. 1998. Ski6p is a homolog of RNA-processing enzymes that affects translation of non-poly(A) mRNAs and 60S ribosomal subunit biogenesis. *Mol. Cell. Biol.* **18:** 2688–2696.

Benard L., Carroll K., Valle R.C., Masison D.C., and Wickner R.B. 1999. The ski7 antiviral protein is an EF1-alpha homolog that blocks expression of non-poly(A) mRNA in *Saccharomyces cerevisiae. J. Virol.* **73:** 2893–2900.

Bergamini G., Preiss T., and Hentze M.W. 2000. Picornavirus IRESes and the poly(A) tail jointly promote cap-independent translation in a mammalian cell-free system. *RNA* **6:** 1781–1790.

Bergsten S.E. and Gavis E.R. 1999. Role for mRNA localization in translational activation but not spatial restriction of nanos RNA. *Development* **126:** 659–669.

Bonnal S., Schaeffer C., Creancier L., Clamens S., Moine H., Prats A.C., and Vagner S. 2003. A single internal ribosome entry site containing a G quartet RNA structure drives fibroblast growth factor 2 gene expression at four alternative translation initiation codons. *J. Biol. Chem.* **278:** 39330–39336.

Borman A.M., Michel Y.M., and Kean K.M. 2000. Biochemical characterisation of cap-poly(A) synergy in rabbit reticulocyte lysates: The eIF4G-PABP interaction increases the functional affinity of eIF4E for the capped mRNA 5′-end. *Nucleic Acids Res.* **28:** 4068–4075.

Braat A.K., Yan N., Arn E., Harrison D., and Macdonald P.M. 2004. Localization-dependent oskar protein accumulation; control after the initiation of translation. *Dev. Cell.* **7:** 125–131.

Bushell M. and Sarnow P. 2002. Hijacking the translation apparatus by RNA viruses. *J. Cell Biol.* **158:** 395–399.

Bushell M., Wood W., Carpenter G., Pain V.M., Morley S.J., and Clemens M.J. 2001. Disruption of the interaction of mammalian protein synthesis eukaryotic initiation factor 4B with the poly(A)-binding protein by caspase- and viral protease-mediated cleavages. *J. Biol. Chem.* **276:** 23922–23928.

Castagnetti S. and Ephrussi A. 2003. Orb and a long poly(A) tail are required for efficient oskar translation at the posterior pole of the *Drosophila* oocyte. *Development* **130:** 835–843.

Castagnetti S., Hentze M.W., Ephrussi A., and Gebauer F. 2000. Control of oskar mRNA translation by Bruno in a novel cell-free system from *Drosophila* ovaries. *Development* **127:** 1063–1068.

Chapman R.E. and Walter P. 1997. Translational attenuation mediated by an mRNA intron. *Curr. Biol.* **7:** 850–859.

Chekulaeva M., Hentze M.W., and Ephrussi A. 2006. Bruno acts as a dual repressor of oskar translation, promoting mRNA oligomerization and formation of silencing particles. *Cell* **124:** 521–533.

Cho P.F., Poulin F., Cho-Park Y.A., Cho-Park I.B., Chicoine J.D., Lasko P., and Sonenberg N. 2005. A new paradigm for translational control: Inhibition via 5′-3′ mRNA tethering by Bicoid and the eIF4E cognate 4EHP. *Cell* **121:** 411–423.

Christensen A.K., Kahn L.E., and Bourne C.M. 1987. Circular polysomes predominate on the rough endoplasmic reticulum of somatotropes and mammotropes in the rat anterior pituitary. *Am. J. Anat.* **178:** 1–10.

Churbanov A., Rogozin I.B., Babenko V.N., Ali H., and Koonin E.V. 2005. Evolutionary conservation suggests a regulatory function of AUG triplets in 5′-UTRs of eukaryotic genes. *Nucleic Acids Res.* **33:** 5512–5520.

Clark I.E., Wyckoff D., and Gavis E.R. 2000. Synthesis of the posterior determinant Nanos is spatially restricted by a novel cotranslational regulatory mechanism. *Curr. Biol.* **10:** 1311–1314.

Coller J. and Parker R. 2005. General translational repression by activators of mRNA decapping. *Cell* **122:** 875–886.

Cosson B., Berkova N., Couturier A., Chabelskaya S., Philippe M., and Zhouravleva G. 2002. Poly(A)-binding protein and eRF3 are associated in vivo in human and *Xenopus* cells. *Biol. Cell* **94:** 205–216.

Craig A.W., Haghighat A., Yu A.T., and Sonenberg N. 1998. Interaction of polyadenylate-binding protein with the eIF4G homologue PAIP enhances translation. *Nature* **392:** 520–523.

Crucs S., Chatterjee S., and Gavis E.R. 2000. Overlapping but distinct RNA elements control repression and activation of nanos translation. *Mol. Cell* **5:** 457–467.

Dahanukar A. and Wharton R.P. 1996. The Nanos gradient in *Drosophila* embryos is generated by translational regulation. *Genes Dev.* **10:** 2610–2620.

Dahanukar A., Walker J.A., and Wharton R.P. 1999. Smaug, a novel RNA-binding protein that operates a translational switch in *Drosophila*. *Mol. Cell* **4:** 209–218.

De Benedetti A. and Graff J.R. 2004. eIF-4E expression and its role in malignancies and metastases. *Oncogene* **23:** 3189–3199.

Deo R.C., Sonenberg N., and Burley S.K. 2001. X-ray structure of the human hyperplastic discs protein: An ortholog of the C-terminal domain of poly(A)-binding protein. *Proc. Natl. Acad. Sci.* **98:** 4414–4419.

Dever T.E. 2002. Gene-specific regulation by general translation factors. *Cell* **108:** 545–556.

Dubnau J. and Struhl G. 1996. RNA recognition and translational regulation by a homeodomain protein. *Nature* **379:** 694–699.

Duncan K., Grskovic M., Strein C., Beckmann K., Niggeweg R., Abaza I., Gebauer F., Wilm M., and Hentze M.W. 2006. Sex-lethal imparts a sex-specific function to UNR by recruiting it to the msl-2 mRNA 3′UTR: Translational repression for dosage compensation. *Genes Dev.* **20:** 368–379.

Forch P. and Valcárcel J. 2003. Splicing regulation in *Drosophila* sex determination. *Prog. Mol. Subcell. Biol.* **31:** 127–151.

Forrest K.M., Clark I.E., Jain R.A., and Gavis E.R. 2004. Temporal complexity within a translational control element in the nanos mRNA. *Development* **131:** 5849–5857.

Gavis E.R. and Lehmann R. 1994. Translational regulation of nanos by RNA localization. *Nature* **369:** 315–318.

Gavis E.R., Lunsford L., Bergsten S.E., and Lehmann R. 1996. A conserved 90 nucleotide element mediates translational repression of nanos RNA. *Development* **122:** 2791–2800.

Gebauer F. and Hentze M.W. 2004. Molecular mechanisms of translational control. *Nat. Rev. Mol. Cell Biol.* **5:** 827–835.

Gebauer F., Merendino L., Hentze M.W., and Valcarcel J. 1998. The *Drosophila* splicing regulator sex-lethal directly inhibits translation of male-specific-lethal 2 mRNA. *RNA* **4:** 142–150.

Gingras A.C., Raught B., and Sonenberg N. 2004. mTOR signaling to translation. *Curr. Top. Microbiol. Immunol.* **279:** 169–197.

Goossen B., Caughman S.W., Harford J.B., Klausner R.D., and Hentze M.W. 1990. Translational repression by a complex between the iron-responsive element of ferritin mRNA and its specific cytoplasmic binding protein is position-dependent in vivo. *EMBO J.* **9:** 4127–4133.

Gorgoni B. and Gray N.K. 2004. The roles of cytoplasmic poly(A)-binding proteins in regulating gene expression: A developmental perspective. *Brief. Funct. Genomics Proteomics* **3:** 125–141.

Gray N.K. and Hentze M.W. 1994. Iron regulatory protein prevents binding of the 43S translation pre-initiation complex to ferritin and eALAS mRNAs. *EMBO J.* **13:** 3882–3891.

Gray N.K., Coller J.M., Dickson K.S., and Wickens M. 2000. Multiple portions of poly(A)-binding protein stimulate translation in vivo. *EMBO J.* **19:** 4723–4733.

Green J.B., Gardner C.D., Wharton R.P., and Aggarwal A.K. 2003. RNA recognition via the SAM domain of Smaug. *Mol. Cell* **11:** 1537–1548.

Grskovic M., Hentze M.W., and Gebauer F. 2003. A co-repressor assembly nucleated by Sexlethal in the 3′UTR mediates translational control of *Drosophila* msl-2 mRNA. *EMBO J.* **22:** 5571–5781.

Haghighat A. and Sonenberg N. 1997. eIF4G dramatically enhances the binding of eIF4E to the mRNA 5′-cap structure (erratum in *J. Biol. Chem.* [1997] **272:** 29398). *J. Biol. Chem.* **272:** 21677–21680.

Hellen C.U. and Sarnow P. 2001. Internal ribosome entry sites in eukaryotic mRNA molecules. *Genes Dev.* **15:** 1593–1612.

Hentze M.W., Muckenthaler M.U., and Andrews N.C. 2004. Balancing acts: Molecular control of mammalian iron metabolism. *Cell* **117:** 285–297.

Hinnebusch A.G. 1997. Translational regulation of yeast GCN4. A window on factors that control initiator-tRNA binding to the ribosome. *J. Biol. Chem.* **272:** 21661–21664.

Hinnebusch A.G. and Natarajan K. 2002. Gcn4p, a master regulator of gene expression, is controlled at multiple levels by diverse signals of starvation and stress. *Eukaryot. Cell* **1:** 22–32.

Holcik M., Sonenberg N., and Korneluk R.G. 2000. Internal ribosome initiation of translation and the control of cell death. *Trends Genet.* **16:** 469–473.

Hoshino S., Imai M., Kobayashi T., Uchida N., and Katada T. 1999. The eukaryotic polypeptide chain releasing factor (eRF3/GSPT) carrying the translation termination signal to the 3′-poly(A) tail of mRNA. Direct association of eRF3/GSPT with polyadenylate-binding protein. *J. Biol. Chem.* **274:** 16677–16680.

Hosoda N., Kobayashi T., Uchida N., Funakoshi Y., Kikuchi Y., Hoshino S., and Katada T. 2003. Translation termination factor eRF3 mediates mRNA decay through the regulation of deadenylation. *J. Biol. Chem.* **278:** 38287–38291.

Huang S. and Houghton P.J. 2003. Targeting mTOR signaling for cancer therapy. *Curr. Opin. Pharmacol.* **3:** 371–377.

Hundsdoerfer P., Thoma C., and Hentze M.W. 2005. Eukaryotic translation initiation factor 4GI and p97 promote cellular internal ribosome entry sequence-driven translation. *Proc. Natl. Acad. Sci.* **102:** 13421–13426.

Imataka H., Gradi A., and Sonenberg N. 1998. A newly identified N-terminal amino acid sequence of human eIF4G binds poly(A)-binding protein and functions in poly(A)-dependent translation. *EMBO J.* **17:** 7480–7489.

Jacobson A. 1996. Poly(A) metabolism and translation: The closed-loop model. In *Translational control* (ed. J.W.B. Hershey et al.), pp. 451–480. Cold Spring Harbor Laboratory Press, Cold Spring Harbor, New York.

Kahvejian A., Svitkin Y.V., Sukarieh R., M'Boutchou M.N., and Sonenberg N. 2005. Mammalian poly(A)-binding protein is a eukaryotic translation initiation factor, which acts via multiple mechanisms. *Genes Dev.* **19:** 104–113.

Kelley R.L., Wang J., Bell L., and Kuroda M.I. 1997. Sex lethal controls dosage compensation in *Drosophila* by a non-splicing mechanism. *Nature* **387:** 195–199.

Kelley R.L., Solovyeva I., Lyman L.M., Richman R., Solovyev V., and Kuroda M.I. 1995. Expression of msl-2 causes assembly of dosage compensation regulators on the X chromosomes and female lethality in *Drosophila*. *Cell* **81**: 867–877.

Kessler S.H. and Sachs A.B. 1998. RNA recognition motif 2 of yeast Pab1p is required for its functional interaction with eukaryotic translation initiation factor 4G. *Mol. Cell. Biol.* **18**: 51–57.

Khaleghpour K., Svitkin Y.V., Craig A.W., DeMaria C.T., Deo R.C., Burley S.K., and Sonenberg N. 2001a. Translational repression by a novel partner of human poly(A) binding protein, Paip2. *Mol. Cell* **7**: 205–216.

Khaleghpour K., Kahvejian A., De Crescenzo G., Roy G., Svitkin Y.V., Imataka H., O'Connor-McCourt M., and Sonenberg N. 2001b. Dual interactions of the translational repressor Paip2 with poly(A) binding protein. *Mol. Cell. Biol.* **21**: 5200–5213.

Kim-Ha J., Kerr K., and Macdonald P.M. 1995. Translational regulation of oskar mRNA by bruno, an ovarian RNA-binding protein, is essential. *Cell* **81**: 403–412.

Kindler S., Wang H., Richter D., and Tiedge H. 2005. RNA transport and local control of translation. *Annu. Rev. Cell Dev. Biol.* **21**: 223–245.

Kobayashi T., Funakoshi Y., Hoshino S., and Katada T. 2004. The GTP-binding release factor eRF3 as a key mediator coupling translation termination to mRNA decay. *J. Biol. Chem.* **279**: 45693–45700.

Komar A.A. and Hatzoglou M. 2005. Internal ribosome entry sites in cellular mRNAs: Mystery of their existence. *J. Biol. Chem.* **280**: 23425–23428.

Kozlov G., Trempe J.F., Khaleghpour K., Kahvejian A., Ekiel I., and Gehring K. 2001. Structure and function of the C-terminal PABC domain of human poly(A)-binding protein. *Proc. Natl. Acad. Sci.* **98**: 4409–4413.

Kuersten S. and Goodwin E.B. 2003. The power of the 3′UTR: Translational control and development. *Nat. Rev. Genet.* **4**: 626–637.

Le H., Tanguay R.L., Balasta M.L., Wei C.C., Browning K.S., Metz A.M., Goss D.J., and Gallie D.R. 1997. Translation initiation factors eIF-iso4G and eIF-4B interact with the poly(A)-binding protein and increase its RNA binding activity. *J. Biol. Chem.* **272**: 16247–16255.

Lie Y.S. and Macdonald P.M. 1999. Translational regulation of oskar mRNA occurs independent of the cap and poly(A) tail in *Drosophila* ovarian extracts. *Development* **126**: 4989–4996.

Lu P.D., Harding H.P., and Ron D. 2004. Translation reinitiation at alternative open reading frames regulates gene expression in an integrated stress response. *J. Cell Biol.* **167**: 27–33.

Mamane Y., Petroulakis E., Rong L., Yoshida K., Ler L.W., and Sonenberg N. 2004. eIF4E—From translation to transformation. *Oncogene* **23**: 3172–3179.

Mangus D.A., Evans M.C., and Jacobson A. 2003. Poly(A)-binding proteins: Multifunctional scaffolds for the post-transcriptional control of gene expression. *Genome Biol.* **4**: 223.

Maniatis T. and Reed R. 2002. An extensive network of coupling among gene expression machines. *Nature* **416**: 499–506.

Masison D.C., Blanc A., Ribas J.C., Carroll K., Sonenberg N., and Wickner R.B. 1995. Decoying the cap-mRNA degradation system by a double-stranded RNA virus and poly(A)-mRNA surveillance by a yeast antiviral system. *Mol. Cell. Biol.* **15**: 2763–2771.

Mazumder B. and Fox P.L. 1999. Delayed translational silencing of ceruloplasmin transcript in gamma interferon-activated U937 monocytic cells: Role of the 3′ untranslated region. *Mol. Cell. Biol.* **19**: 6898–6905.

Mazumder B., Seshadri V., Imataka H., Sonenberg N., and Fox P.L. 2001. Translational silencing of ceruloplasmin requires the essential elements of mRNA circularization: Poly(A) tail, poly(A)-binding protein, and eukaryotic translation initiation factor 4G. *Mol. Cell. Biol.* **21:** 6440–6449.

Mazumder B., Sampath P., Seshadri V., Maitra R.K., DiCorleto P.E., and Fox P.L. 2003. Regulated release of L13a from the 60S ribosomal subunit as a mechanism of transcript-specific translational control. *Cell* **115:** 187–198.

Mendez R. and Richter J.D. 2001. Translational control by CPEB: A means to the end. *Nat. Rev. Mol. Cell Biol.* **2:** 521–529.

Muckenthaler M., Gray N.K., and Hentze M.W. 1998. IRP-1 binding to ferritin mRNA prevents the recruitment of the small ribosomal subunit by the cap-binding complex eIF4F. *Mol. Cell* **2:** 383–388.

Munroe D. and Jacobson A. 1990. mRNA poly(A) tail, a 3′ enhancer of translational initiation. *Mol. Cell. Biol.* **10:** 3441–3455.

Nakamura A., Sato K., and Hanyu-Nakamura K. 2004. *Drosophila* cup is an eIF4E binding protein that associates with Bruno and regulates oskar mRNA translation in oogenesis. *Dev. Cell* **6:** 69–78.

Nelson M.R., Leidal A.M., and Smibert C.A. 2004. *Drosophila* Cup is an eIF4E-binding protein that functions in Smaug-mediated translational repression. *EMBO J.* **23:** 150–159.

Niessing D., Blanke S., and Jackle H. 2002. Bicoid associates with the 5′-cap-bound complex of caudal mRNA and represses translation. *Genes Dev.* **16:** 2576–2582.

Ostareck D.H., Ostareck-Lederer A., Shatsky I.N., and Hentze M.W. 2001. Lipoxygenase mRNA silencing in erythroid differentiation: The 3′UTR regulatory complex controls 60S ribosomal subunit joining. *Cell* **104:** 281–289.

Ostareck D.H., Ostareck-Lederer A., Wilm M., Thiele B.J., Mann M., and Hentze M.W. 1997. mRNA silencing in erythroid differentiation: hnRNP K and hnRNP E1 regulate 15-lipoxygenase translation from the 3′ end. *Cell* **89:** 597–606.

Ostareck-Lederer A. and Ostareck D.H. 2004. Control of mRNA translation and stability in haematopoietic cells: The function of hnRNPs K and E1/E2. *Biol. Cell* **96:** 407–411.

Piron M., Vende P., Cohen J., and Poncet D. 1998. Rotavirus RNA-binding protein NSP3 interacts with eIF4GI and evicts the poly(A) binding protein from eIF4F. *EMBO J.* **17:** 5811–5821.

Preiss T. 2002. The end in sight: Poly(A), translation and mRNA stability in eukaryotes. In *Translation mechanisms* (ed. L. Brakier-Gingras and J. Lapointe), pp. 197–212. Landes Bioscience, Georgetown, Texas. Web edition: http://www.eurekah.com/chapter.php?chapid= 840&bookid=59&catid=54.

Preiss T. and Hentze M.W. 1998. Dual function of the messenger RNA cap structure in poly(A)-tail-promoted translation in yeast. *Nature* **392:** 516–520.

———. 2003. Starting the protein synthesis machine: Eukaryotic translation initiation. *Bioessays* **25:** 1201–1211.

Proud C.G. 2004. Role of mTOR signalling in the control of translation initiation and elongation by nutrients. *Curr. Top. Microbiol. Immunol.* **279:** 215–244.

Proudfoot N.J., Furger A., and Dye M.J. 2002. Integrating mRNA processing with transcription. *Cell* **108:** 501–512.

Ptushkina M., von der Haar T., Karim M.M., Hughes J.M.X., and McCarthy J.E. 1999. Repressor binding to a dorsal regulatory site traps human eIF4E in a high cap-affinity state. *EMBO J.* **18:** 4068–4075.

Ptushkina M., von der Haar T., Vasilescu S., Frank R., Birkenhager R., and McCarthy J.E. 1998. Cooperative modulation by eIF4G of eIF4E-binding to the mRNA 5′ cap in yeast involves a site partially shared by p20. *EMBO J.* **17:** 4798–4808.

Richter J.D. and Sonenberg N. 2005. Regulation of cap-dependent translation by eIF4E inhibitory proteins. *Nature* **433:** 477–480.

Rivera-Pomar R., Niessing D., Schmidt-Ott U., Gehring W.J., and Jackle H. 1996. RNA binding and translational suppression by bicoid. *Nature* **379:** 746–749.

Rosenwald I.B. 2004. The role of translation in neoplastic transformation from a pathologist's point of view. *Oncogene* **23:** 3230–3247.

Roy G., De Crescenzo G., Khaleghpour K., Kahvejian A., O'Connor-McCourt M., and Sonenberg N. 2002. Paip1 interacts with poly(A) binding protein through two independent binding motifs. *Mol. Cell. Biol.* **22:** 3769–3782.

Ruegsegger U., Leber J.H., and Walter P. 2001. Block of HAC1 mRNA translation by long-range base pairing is released by cytoplasmic splicing upon induction of the unfolded protein response. *Cell* **107:** 103–114.

Sachs A.B. 2000. Physical and functional interactions between the mRNA cap structure and the poly(A) tail. In *Translational control of gene expression* (ed. N. Sonenberg et al.), pp. 447–465. Cold Spring Harbor Laboratory Press, Cold Spring Harbor, New York.

Sachs A.B. and Davis R.W. 1989. The poly(A) binding protein is required for poly(A) shortening and 60S ribosomal subunit-dependent translation initiation. *Cell* **58:** 857–867.

———. 1990. Translation initiation and ribosomal biogenesis: Involvement of a putative rRNA helicase and RPL46. *Science* **247:** 1077–1079.

Sampath P., Mazumder B., Seshadri V., and Fox P.L. 2003. Transcript-selective translational silencing by gamma interferon is directed by a novel structural element in the ceruloplasmin mRNA 3′ untranslated region. *Mol. Cell. Biol.* **23:** 1509–1519.

Sampath P., Mazumder B., Seshadri V., Gerber C.A., Chavatte L., Kinter M., Ting S.M., Dignam J.D., Kim S., Driscoll D.M., and Fox P.L. 2004. Noncanonical function of glutamyl-prolyl-tRNA synthetase: Gene-specific silencing of translation. *Cell* **119:** 195–208.

Searfoss A.M. and Wickner R.B. 2000. 3′ poly(A) is dispensable for translation. *Proc. Natl. Acad. Sci.* **97:** 9133–9137.

Searfoss A., Dever T.E., and Wickner R. 2001. Linking the 3′ poly(A) tail to the subunit joining step of translation initiation: Relations of Pab1p, eukaryotic translation initiation factor 5b (Fun12p), and Ski2p-Slh1p. *Mol. Cell. Biol.* **21:** 4900–4908.

Smibert C.A., Wilson J.E., Kerr K., and Macdonald P.M. 1996. Smaug protein represses translation of unlocalized nanos mRNA in the *Drosophila* embryo. *Genes Dev.* **10:** 2600–2609.

Smibert C.A., Lie Y.S., Shillinglaw W., Henzel W.J., and Macdonald P.M. 1999. Smaug, a novel and conserved protein, contributes to repression of nanos mRNA translation *in vitro*. *RNA* **5:** 1535–1547.

Sonenberg N. and Dever T.E. 2003. Eukaryotic translation initiation factors and regulators. *Curr. Opin. Struct. Biol.* **13:** 56–63.

Stebbins-Boaz B., Cao Q., de Moor C.H., Mendez R., and Richter J.D. 1999. Maskin is a CPEB-associated factor that transiently interacts with elF-4E. *Mol. Cell* **4:** 1017–1027.

Stoneley M. and Willis A.E. 2004. Cellular internal ribosome entry segments: Structures, trans-acting factors and regulation of gene expression. *Oncogene* **23:** 3200–3207.

Stripecke R., Oliveira C.C., McCarthy J.E., and Hentze M.W. 1994. Proteins binding to 5′ untranslated region sites: A general mechanism for translational regulation of mRNAs in human and yeast cells. *Mol. Cell. Biol.* **14:** 5898–5909.

Svitkin Y.V., Imataka H., Khaleghpour K., Kahvejian A., Liebig H.D., and Sonenberg N. 2001. Poly(A)-binding protein interaction with eIF4G stimulates picornavirus IRES-dependent translation. *RNA* **7:** 1743–1752.

Tarun S.Z., Jr. and Sachs A.B. 1995. A common function for mRNA 5′ and 3′ ends in translation initiation in yeast. *Genes Dev.* **9:** 2997–3007.

———. 1996. Association of the yeast poly(A) tail binding protein with translation initiation factor eIF-4G. *EMBO J.* **15:** 7168–7177.

———. 1997. Binding of eukaryotic translation initiation factor 4E (eIF4E) to eIF4G represses translation of uncapped mRNA. *Mol. Cell. Biol.* **17:** 6876–6886.

Tarun S.Z., Jr., Wells S.E., Deardorff J.A., and Sachs A.B. 1997. Translation initiation factor eIF4G mediates *in vitro* poly(A) tail-dependent translation. *Proc. Natl. Acad. Sci.* **94:** 9046–9051.

Thoma C., Bergamini G., Galy B., Hundsdoerfer P., and Hentze M.W. 2004. Enhancement of IRES-mediated translation of the c-myc and BiP mRNAs by the poly(A) tail is independent of intact eIF4G and PABP. *Mol. Cell* **15:** 925–935.

Touriol C., Bornes S., Bonnal S., Audigier S., Prats H., Prats A.C., and Vagner S. 2003. Generation of protein isoform diversity by alternative initiation of translation at non-AUG codons. *Biol. Cell* **95:** 169–178.

Uchida N., Hoshino S., Imataka H., Sonenberg N., and Katada T. 2002. A novel role of the mammalian GSPT/eRF3 associating with poly(A)-binding protein in Cap/poly(A)-dependent translation. *J. Biol. Chem.* **277:** 50286–50292.

Vagner S., Galy B., and Pyronnet S. 2001. Irresistible IRES. Attracting the translation machinery to internal ribosome entry sites. *EMBO Rep.* **2:** 893–898.

Vattem K.M. and Wek R.C. 2004. Reinitiation involving upstream ORFs regulates ATF4 mRNA translation in mammalian cells. *Proc. Natl. Acad. Sci.* **101:** 11269–11274.

Vende P., Piron M., Castagne N., and Poncet D. 2000. Efficient translation of rotavirus mRNA requires simultaneous interaction of NSP3 with the eukaryotic translation initiation factor eIF4G and the mRNA 3′ end. *J. Virol.* **74:** 7064–7071.

Wakiyama M., Imataka H., and Sonenberg N. 2000. Interaction of eIF4G with poly(A)-binding protein stimulates translation and is critical for *Xenopus* oocyte maturation. *Curr. Biol.* **10:** 1147–1150.

Wei C.C., Balasta M.L., Ren J., and Goss D.J. 1998. Wheat germ poly(A) binding protein enhances the binding affinity of eukaryotic initiation factor 4F and (iso)4F for cap analogues. *Biochemistry* **37:** 1910–1916.

Wells S.E., Hillner P.E., Vale R.D., and Sachs A.B. 1998. Circularization of mRNA by eukaryotic translation initiation factors. *Mol. Cell* **2:** 135–140.

Wilhelm J.E. and Smibert C.A. 2005. Mechanisms of translational regulation in *Drosophila*. *Biol. Cell* **97:** 235–252.

Wilhelm J.E., Hilton M., Amos Q., and Henzel W.J. 2003. Cup is an eIF4E binding protein required for both the translational repression of oskar and the recruitment of Barentsz. *J. Cell Biol.* **163:** 1197–1204.

Zappavigna V., Piccioni F., Villaescusa J.C., and Verrotti A.C. 2004. Cup is a nucleocytoplasmic shuttling protein that interacts with the eukaryotic translation initiation factor 4E to modulate *Drosophila* ovary development. *Proc. Natl. Acad. Sci.* **101:** 14800–14805.

Zimmer S.G., DeBenedetti A., and Graff J.R. 2000. Translational control of malignancy: The mRNA cap-binding protein, eIF-4E, as a central regulator of tumor formation, growth, invasion and metastasis. *Anticancer Res.* **20:** 1343–1351.

11

Regulation of mRNA Molecules by MicroRNAs

Karen A. Wehner and Peter Sarnow
Department of Microbiology and Immunology
Stanford University School of Medicine
Stanford, California 94305

Twenty-two years before microRNAs (miRNAs) would become *Science's* "the 2001 molecule of the year," the first miRNA, *lin-4*, was discovered in *Caenorhabditis elegans*. Since then, these small noncoding RNA molecules, approximately 22 nucleotides in length, have been detected in many eukaryotic organisms, and it is estimated that they control approximately one-fourth of all cellular mRNAs at the posttranscriptional level. This chapter focuses primarily on animal miRNAs and their roles in posttranscriptional control of gene expression.

HISTORICAL BACKGROUND ON ANIMAL miRNAs

lin-4 was identified as a loss-of-function mutation that gave rise to reiterations in cell lineages, retarded development, and abnormal morphology in *C. elegans* (Horvitz and Sulston 1980; Chalfie et al. 1981). Additional genetic screens identified semidominant gain-of-function mutations in the *lin-14* gene that, like the *lin-4* mutation, led to cell lineage reiteration and retarded development (Ambros and Horvitz 1984, 1987). On the basis of the similar phenotypes of the *lin-4* loss-of-function and *lin-14* gain-of-function mutants, it was postulated that the two genes interact genetically (Ambros 1989). Indeed, epistasis experiments revealed that the *lin-4* mutant phenotype was dependent on *lin-14*, suggesting that the role of the *lin-4* gene product is to act as a negative regulator of *lin-14* (Ambros 1989).

The mechanism by which *lin-4* negatively regulates *lin-14* was unknown until 1991, when it was demonstrated that the *lin-4* gene product down-regulated *lin-14* protein levels by interacting with the *lin-14* 3′ untranslated region (3′UTR) (Arasu et al. 1991; Wightman et al. 1991). Subsequently, the *lin-4* gene product was discovered to be a small RNA molecule, rather than a protein, with complementarity to multiple regions of the *lin-14* 3′UTR (Lee et al. 1993; Wightman et al. 1993). These regions of complementarity, located in highly conserved segments of the 3′UTR, are discontinuous and are predicted to form *lin-4/lin-14* RNA duplexes that contain a central bulge (Wightman et al. 1993).

Early efforts to determine the level at which *lin-4* controls *lin-14* expression employed RNase protection assays (RPA) and western blots to examine RNA and protein levels, respectively (Wightman et al. 1993; Olsen and Ambros 1999). Results from these assays indicated that *lin-14* mRNA levels were not noticeably altered in the presence of the *lin-4* RNA. In contrast, *lin-14*-encoded protein levels were dramatically reduced in the presence of *lin-4* (Wightman et al. 1993; Olsen and Ambros 1999). The finding that regulation of the *lin-14* mRNA by *lin-4* did not involve transcript stability suggested that translation of *lin-14* mRNA was inhibited by the small *lin-4* RNA. The *lin-14* mRNA was found to cosediment with polysomes in both the absence and presence of *lin-4* (Olsen and Ambros 1999). Because *lin-14* protein levels dramatically decrease when the *lin-4* RNA is associated with *lin-14* mRNAs, it was hypothesized that the *lin-4* RNA blocked *lin-14* mRNA translation at a step following initiation. These experiments provided the first detailed look at the mechanism by which the *lin-4* RNA controlled expression of *lin-14* mRNA and established the basis for future experimentation in the field.

The second small regulatory RNA to be discovered in *C. elegans* was *let-7*. *let-7* was identified through a genetic screen designed to find genes that are involved in developmental timing (Reinhart et al. 2000). Like the *lin-4* RNA, the *let-7* RNA is 21 nucleotides in length, and is expressed at a specific time during development. In addition, *let-7* inhibits the expression of mRNAs that contain sites with complementarity to *let-7* in their 3′UTR, such as the *lin-41* mRNA (Reinhart et al. 2000; Slack et al. 2000). However, unlike *lin-4*, the *let-7* RNA is highly conserved at the sequence level throughout the eukaryotic kingdom (Pasquinelli et al. 2000). With the identification and characterization of *let-7*, it became obvious that this type of small RNA molecule was not restricted to *C. elegans*, but was likely to be involved in the temporal regulation of development throughout the eukaryotic kingdom.

Recent research on *let-7/lin-41* and *lin-4/lin-14* interactions, using northern analyses to monitor full-length RNA species, revealed that contrary to the initial findings (Olsen and Ambros 1999), target mRNA levels are in fact reduced in the presence of the small RNAs (Bagga et al. 2005), suggesting that perhaps not all miRNAs inhibit target gene expression solely by a translational inhibition mechanism.

miRNA BIOGENESIS

Mammalian miRNA genes are genomically encoded as monocistronic and polycistronic gene clusters and are found within intronic regions of protein-coding genes, within intronic regions of noncoding genes, and as independent transcriptional units (Fig. 1) (Lagos-Quintana et al. 2001; Lau et al. 2001; Lee et al. 2002). Transcription of miRNA genes by RNA polymerase II results in the production of large primary miRNA precursors (pri-miRNAs) that contain hairpin structures harboring the mature miRNA (Cai et al. 2004; Lee et al. 2004). To yield mature, functional miRNAs, the miRNA sequence must be excised from the pri-miRNA by a maturation process that involves both nuclear and cytoplasmic cleavage events (Fig. 1).

miRNA processing occurs in a stepwise manner that requires two RNase III enzymes: Drosha for initiation of miRNA processing and Dicer for completion of miRNA processing (Fig. 1) (Kim 2005). In the nucleus, cleavage of pri-miRNAs by Drosha occurs rapidly to yield an approximately 70-nucleotide-long RNA stem-loop structure, termed the pre-miRNA. Following nuclear export by Exportin 5, Dicer rapidly cleaves pre-miRNAs to produce miRNA duplexes of approximately 22 nucleotides in length. Because both cleavage events are mediated by RNase III enzymes, both ends of the duplex possess characteristic 5′ monophosphates, 3′ hydroxyl moieties, and 2-nucleotide 3′ overhangs.

miRNP COMPOSITION AND ASSEMBLY

miRNA effector complexes, also known as the miRNA-containing ribonucleoprotein particle (miRNP) or miRNA-containing RNA-induced silencing complex (miRISC), contain a number of protein components and the miRNA guide strand. Following Dicer cleavage, one strand of the 22-nucleotide miRNA duplex is selected for miRNP incorporation, resulting in a single-stranded miRNA molecule that can hybridize to target mRNAs with complementary sequences. It is thought that strand

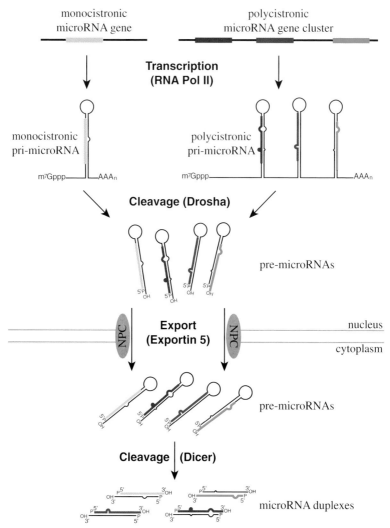

Figure 1. miRNA biogenesis pathway. Transcription of the miRNA genes by RNA polymerase II results in the production of monocistronic or polycistronic miRNA precursors (pri-miRNAs) that possess both 5′ 7-methylguanosine caps and 3′ poly(A) tails. Cleavage of the pri-miRNAs by the RNase III enzyme Drosha results in the production of the pre-miRNA hairpins that possess 2-nucleotide 3′ overhangs, 5′ monophosphates, and 3′ hydroxyl groups. The pre-miRNA hairpins are subsequently exported from the nucleus to the cytoplasm via the nuclear pore complex (NPC) in an Exportin-5-dependent manner. Once in the cytoplasm, the pre-miRNAs are further processed by the RNase III enzyme Dicer. Cleavage by Dicer yields mature miRNA duplexes of approximately 22 nucleotides in length with both sets of termini possessing 2-nucleotide 3′ overhangs, 5′ monophosphates, and 3′ hydroxyl groups.

selection is based on the thermodynamic properties of the miRNA duplex ends; in particular, it is the RNA strand displaying the weakest thermodynamic stability at its 5′ end that is incorporated into the miRNP (Fig. 2A) (Khvorova et al. 2003).

Although strand selection and RNP incorporation are not well characterized for the miRNP, they have been extensively studied in the assembly of the *Drosophila* short interfering RISC (siRISC) (Liu et al. 2003; Tomari et al. 2004; Matranga et al. 2005). Results from these studies have revealed that Dicer requires protein cofactors for its role in strand selection. In mammalian cells, Dicer has been shown to interact with two proteins, TRBP (*trans*-activating region [TAR] RNA-binding protein) (Chendrimada et al. 2005; Forstemann et al. 2005; Haase et al. 2005) and PACT (PKR-activating protein) (Patel and Sen 1998; Lee et al. 2006), that have the potential to serve as cofactors for strand selection. Although TRBP and PACT are both required for accumulation of the mature miRNA guide strand, their precise role in the strand selection process still remains to be elucidated.

The single-stranded miRNA guide strand is aided by proteins both to find its target mRNA and to regulate the function of the target mRNA. A myriad of protein factors have been predicted to be components of the miRNP. These proteins include, but are not limited to: FMRP, FXR1, FXR2, Dicer, Ago1-4, Gemin3, Gemin4, MOV10, TNRC6B, and PRMT5 (Mourelatos et al. 2002; Sasaki et al. 2003; Jin et al. 2004; Liu et al. 2004; Meister et al. 2004a,b; Nelson et al. 2004). The four Argonaute proteins have all been found to coimmunoprecipitate with Dicer (Sasaki et al. 2003; Gregory et al. 2004; Chendrimada et al. 2005). In the case of Ago2, association with Dicer has been demonstrated during the final stages of strand selection and initial stages of miRNP formation (Gregory et al. 2004; Chendrimada et al. 2005). Ago2 also appears to be a component of the mature miRNP, as it has been shown to possess the endonucleolytic activity responsible for miRNA-mediated cleavage in perfectly complementary miRNA/mRNA targets (Liu et al. 2004; Meister et al. 2004b). At this point, it is not known what roles the additional three Argonaute proteins have in miRNP biogenesis or miRNA function, but it is interesting to speculate that they may form alternate miRNPs by replacing Ago2 in the assembly process. Because these proteins do not have the endonucleolytic activity of Ago2, it is likely that their presence would result in an miRNP with a different regulatory function. Indeed, Ago2$^{-/-}$ cells are not capable of supporting miRNA-mediated cleavage of perfectly complementary targets, but they are capable of supporting miRNA-mediated regulation of imperfectly matched targets (Liu et al. 2004). Finally, the four

Argonaute proteins display distinct tissue-type-specific expression patterns (Sasaki et al. 2003), arguing for a role of these proteins as cell-specific modulators of miRNA function.

Components of the miRNA/RNP biogenesis pathway have also been found to interact with the fragile X mental retardation protein (FMRP), an RNA-binding protein that negatively regulates initiation of translation (Laggerbauer et al. 2001; Jin et al. 2004) of mRNAs that contain G quartet structures (Darnell et al. 2001; Schaeffer et al. 2001). FMRP has two autosomal paralogs, FXR1 and FXR2, that also bind mRNA substrates, but unlike FMRP, they do not negatively regulate translation (Laggerbauer et al. 2001). In addition to binding mRNA substrates and to each other, FMRP, FXR1, and FXR2 also have the ability to interact with components

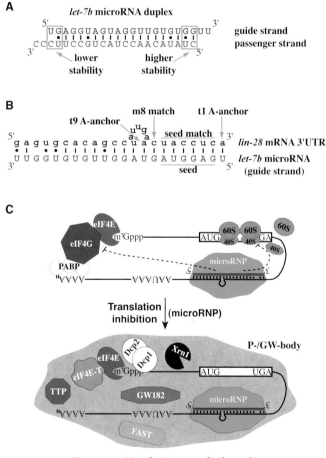

Figure 2. (*See facing page for legend.*)

of the miRNA/RNP biogenesis machinery (Jin et al. 2004). Recently, it has been demonstrated that FMRP, FXR1, and FXR2 associate with Ago2, miRNAs, and pre-miRNAs (Jin et al. 2004), suggesting that these proteins may be present in the miRNP both during guide strand selection and during the final stages of miRNP maturation. Loss of FMRP function causes the fragile X mental retardation syndrome (Kremer et al. 1991; Oberle et al. 1991; Pieretti et al. 1991). Although the exact roles of the fragile X proteins in miRNP assembly and function have yet to be determined, one wonders whether loss of FMRP-mediated regulation of translation via the miRNA pathway may contribute to disease outcome. Overall, the precise step at which each of these proteins joins the miRNP and the mechanistic role each protein has in the miRNP have yet to be determined, but will likely reveal important insights into miRNP assembly and function. The possible role of FMRP in translational control is also discussed in Chapter 18.

RULES FOR ENGAGEMENT: A HAPPY UNION

One of the biggest tasks for miRNP complexes is the recognition of target mRNAs. The rules governing miRNA/mRNA target interactions were initially inferred from the validated *C. elegans* miRNA/mRNA targets *lin-*

Figure 2. miRNA strand selection, targeting, and mechanism. (*A*) miRNA guide strand selection is based on the thermodynamic stability at the ends of the miRNA duplex. The strand destined to become the guide strand has lower thermodynamic stability at its 5′ end (unpaired U) than the passenger strand (Watson-Crick paired G). The *let-7b* miRNA guide strand is shown in uppercase red letters and the *let-7b* miRNA passenger strand is shown in uppercase blue letters. Vertical lines (|) indicate Watson-Crick base-pairing. Dots (·) indicate non-Watson-Crick base-pairing. (*B*) The *lin-28* mRNA/*let-7b* miRNA target pair illustrates several well-conserved features of an mRNA/miRNA target pair. These features include imperfect complementarity between the two RNA strands, the potential for large internal bulges, a 6-nucleotide seed:seed match interaction, a t1 A-anchor, an m8 match, and a t9 A-anchor. The *lin-28* mRNA is shown in lowercase black letters. (*C*) There are currently two mechanistic models for miRNA-mediated control of gene expression. One model suggests that the miRNA/miRNP interacts with the cap-binding protein, eIF4E, preventing eIF4E and PABP from cooperatively recruiting eIF4G. Thus, binding of the miRNA/miRNP would lead to a block in the initiation of translation. A second model suggests that the mRNA-associated miRNP can lead to premature translation termination by causing a drop-off of translating ribosomes. Transcripts regulated by miRNAs are targeted to P/GW bodies either concurrent with or subsequent to a block in translation. P/GW-body proteins Dcp1, Dcp2, Xrn1, GW182, TTP, and FAST are then likely to be involved in the degradation of the miRNA targets.

4/lin-14 and *let-7/lin-41*. Comprehensive analyses of 3′UTRs which contain miRNA-binding sites that are conserved in several species have subsequently revealed the defining characteristics of an miRNA/mRNA interaction (Fig. 2B) (Lewis et al. 2005). Watson-Crick base-pair complementarity between six consecutive nucleotides in the mRNA (the so-called seed match) with corresponding nucleotides 2–7 of the miRNA (the so-called miRNA seed) was found to be essential. The nucleotide following the seed match at position 8 (t8) is highly conserved, as is its tendency to form a Watson-Crick base pair with nucleotide 8 of the miRNA (m8) (Fig. 2B) (Lewis et al. 2005). The interaction between t8 and m8 has been designated an "m8 match" (Lewis et al. 2005). The residue one nucleotide downstream from the seed match (t1) is also highly conserved and is often an A residue (t1 A-anchor). Although not as highly conserved as the other positions mentioned above, position 9 of the miRNA target site (t9) also tends to be conserved and is often also an A (t9 A-anchor). Besides these characteristics, conservation of the remaining portions of the target site and regions immediately flanking the seed match is fairly low (Lewis et al. 2005). Exceptions to these generalizations exist and some miRNA/mRNA complexes do not display all of these characteristics (see below) (Lewis et al. 2005), resulting in the need for each putative miRNA/mRNA interaction to be evaluated on its own and verified by functional assays (see below).

THE MULTIPLE FACETS OF miRNA/mRNA INTERACTIONS

As described at the beginning of this chapter, *lin-4* and *let-7* provided paradigms on target recognition and gene regulation by miRNAs in *C. elegans*. More recently, numerous animal miRNA/mRNA complexes have been identified and their functional roles have been elucidated. Although some miRNA/mRNA interactions lead to the regulation of gene expression via RNA degradation mechanisms, others lead to the control of gene expression at the level of translation.

Pancreatic Islet-specific miRNA *miR-375*

miR-375 has been identified as a pancreatic-islet-specific miRNA and has been shown to negatively regulate glucose-stimulated insulin secretion (Poy et al. 2004). A group of 64 computationally predicted targets for *miR-375* included 5 with potential roles in insulin secretion and islet cell function. Of the five targets, myotrophin (*Mtpn*) was the only target whose protein levels decreased in response to excess levels of *miR-375* and increased in the absence of *miR-375* (Poy et al. 2004). Furthermore, the putative *Mtpn miR-375*-binding site was found to be necessary and sufficient for *miR-375* to negatively regulate the expression of a luciferase reporter (Poy et al. 2004). These results indicate that *Mtpn* may negatively regulate insulin secretion by controlling *Mtpn* gene expression. Furthermore, the *miR-375:Mtpn* target pair is similar to the classic *lin-4* and *let-7* examples from *C. elegans* where binding of the miRNAs to target 3'UTRs down-regulates expression of the mRNA target.

HOXB8 mRNA Is Regulated by miRNA *miR-196*

miR-196 miRNAs are encoded at three positions in the Hox A, B, and C clusters in vertebrates (Yekta et al. 2004). Intriguingly, *miR-196* exhibits near-perfect sequence complementarity to conserved regions of three genes involved in vertebrate development: HOXB8, HOXC8, and HOXD8. It has been demonstrated that *miR-196* negatively regulates the expression of these genes by directing cleavage of their transcripts (Yekta et al. 2004). These results suggest that *miR-196* restricts HOX gene expression during vertebrate development through mRNA cleavage. This finding was unusual, because until then, miRNA-mediated cleavage had only been observed in plants (Llave et al. 2002).

miRNA-mediated Degradation of mRNAs Containing AU-rich Elements

Similar to *miR-196*, miRNA *miR-16* has also been shown to lead to the decay of its target mRNAs (Jing et al. 2005). Unlike *miR-196*, *miR-16* has unusually little complementarity to its target mRNAs. The seed-like region of *miR-16* is 8 nucleotides long and begins at nucleotide 5 of the miRNA, rather than nucleotide 1, as is the case with most other miRNAs. The *miR-16* seed-like region has complementarity to an AU-rich element (ARE), which resides in the 3'UTRs of unstable mRNAs. It appears that *miR-16* is required for ARE-dependent RNA decay, because siRNA-mediated knockdown of *miR-16* or Dicer led to stabilization of ARE-containing mRNAs (Jing et al. 2005). The interaction between *miR-16* and the ARE appears to be sequence-specific and requires the ARE-binding protein tristetraprolin (TTP), which also interacts with Ago2 and Ago4 (Jing et al. 2005). These results argue that TTP is required to assist in the targeting of the miRNP to the ARE-containing messages because the weak seed-like sequence in *miR-16* may not be sufficient to mediate the assembly of *miR-16*/mRNA complexes (Jing et al. 2005).

Functional Consequences of Sequestration
of Liver-specific miRNA *miR-122*

Antagomirs represent a new class of modified oligonucleotides that are stable in cultured cells and sequester intracellular miRNAs with high efficiency (Hutvagner et al. 2004; Meister et al. 2004a; Krutzfeldt et al. 2005). Krutzfeldt and colleagues injected mice intravenously with antagomirs complementary to liver-specific miRNA *miR-122*, which resulted in reduced levels of endogenous *miR-122* and plasma cholesterol (Krutzfeldt et al. 2005). Furthermore, microarray analysis revealed that many cholesterol biosynthesis pathway genes that are normally repressed in the liver were derepressed in the absence of *miR-122*. The 3′UTRs of many of the transcripts from these genes had at least one site of complementarity to the *miR-122* seed sequence, suggesting that they are *miR-122* targets (Krutzfeldt et al. 2005). This study showed that miRNAs can be targeted by antagomirs in an organism without causing obvious side effects, which opens the possibility to perform gene therapy by controlling the expression of miRNAs.

In a curious and unanticipated twist, *miR-122* was noted to have sequence complementarity to two regions of the hepatitis C virus (HCV) RNA genome, one at its 5′ end and one at its 3′ end (Jopling et al. 2005). Sequestration of *miR-122*, using 2′O-methyl antisense oligonucleotides, resulted in decreased viral RNA abundance, suggesting that *miR-122* has a vital role in HCV gene amplification. Because all validated animal miRNA targets have been found in the 3′UTRs of mRNAs, it was surprising that only the 5′ target site was responsive to *miR-122*-mediated regulation. Mutations in the 5′ *miR-122*-binding site abolished viral genome replication, but ectopic expression of miRNAs with compensatory mutations restored replication, demonstrating a genetic interaction between *miR-122* and the HCV RNA genome (Jopling et al. 2005). This finding provides an example of a cellular miRNA that positively regulates its target by binding to the 5′UTR.

Misexpression of a miRNA Reprograms Gene Expression

Many miRNAs are expressed in a tissue-specific manner. What would happen if a brain-specific miRNA were expressed in nonneuronal cells? Lim et al. (2005) noted that expression of brain-specific miRNA *miR-124* in cervix-carcinoma-derived HeLa cells led to changes in transcript expression that resembled more a brain-specific transcript expression profile. Similarly, expression of muscle-specific miRNA, *miR-1*, led to changes in transcript expression toward the profile observed in muscle. In both instances, more than 100 transcripts were down-regulated, and fusion of the 3'UTRs of candidate target genes to luciferase reporter mRNAs revealed that many of the 3'UTRs were indeed responsive to miRNA-mediated regulation (Lim et al. 2005). These results led to the suggestion that specific miRNAs regulate tissue-specific gene expression and that many more targets than once thought define tissue-specific gene expression.

Numerous additional miRNA targets have been predicted computationally through a variety of means, including the use of computer programs such as miRanda and PicTar (John et al. 2004; Krek et al. 2005). Although many of the bioinformatically predicted targets remain to be verified, the most significant outcome from these studies has been the realization that the expression of hundreds, and perhaps thousands, of genes may be regulated by miRNAs.

Identification of mRNAs That Are Translationally Controlled by miRNAs

In an effort to identify mRNAs that are translationally controlled by miRNA, Nakamoto et al. (2005) knocked down miRNA *miR-30a-3p* by siRNA and fractionated cell extracts on a sucrose gradient. mRNAs cosedimenting with polyribosomes in the presence or absence of *miR-30a-3p* were analyzed by cDNA microarray. Eight mRNAs that sedimented with larger polyribosome fractions in the absence of *miR-30a-3p* were further examined for the presence of *miR-30a-3p*-binding sites. Each of the eight potential targets had at least one seed match to *miR-30a-3p*: Five mRNA targets had perfect seed matches and three had seed matches containing a G-U wobble pairing (Nakamoto et al. 2005). Protein levels from two potential targets, cystine-rich angiogenic inducer 61 (*CYR61*) and cyclin-dependent kinase 6 (*CDK6*), were found to increase when *miR-30a-3p* was knocked down and to decrease when excess *miR-30a-3p* was transfected into cells (Nakamoto et al. 2005). Both *CYR61* and *CDK6* have multiple putative miRNA-binding sites, and it is not yet known which site(s) is essential for miRNA-mediated regulation. It is important to note that none of the targets identified here had been identified by computational algorithms. Thus, identification of miRNA/mRNA interactions by functional means may be an important tool in the identification of new miRNA targets.

MECHANISMS OF TARGET mRNA REGULATION BY miRNAs

Interaction of miRNAs with target mRNAs can lead to up-regulation or down-regulation of target mRNAs. So far, miRNA-mediated up-regulation has only been observed for HCV RNA by *miR-122* (Jopling et al. 2005), but its mechanism is not yet known. Many inroads have been made, however, toward understanding the mechanism of miRNA-mediated down-regulation of target mRNAs. It appears that this process involves two mechanisms that may not always be exclusive: degradation and/or translational inhibition of target mRNAs.

Reporter Systems to Study miRNA-mediated Regulation of Target mRNAs

Several reporter assays have been developed that aid in the study of the mechanism of miRNA-mediated regulation of target mRNAs. Most reporter systems employ either a luciferase reporter or a green fluorescent protein (GFP) reporter, each of which has its own unique appeal. The luciferase reporter system allows the fast and quantitative analysis of reporter mRNA expression. The GFP reporter system allows easy visual analysis of transfection efficiencies and is amenable to western blot analysis. Both reporter systems have been used with a variety of promoters directing reporter mRNA expression. Either the relatively strong cytomegalovirus (CMV) promoter and/or the relatively weak herpes simplex virus (HSV)-TK promoter has been used. Promoter selection may be one of the most important considerations, because it has been reported that relative overexpression of either the miRNA or the mRNA target can lead to the erroneous situations in which a miRNA exhibits off-target effects or does not appear to regulate an authentic target (Doench and Sharp 2004).

miRNAs Can Lead to Degradation of Target mRNAs

As discussed above with *miR-196* (Yekta et al. 2004) and *miR-16* (Jing et al. 2005), miRNAs can induce the degradation of target mRNAs. To explore the mechanism of miRNA-induced degradation of target mRNAs, Wu and colleagues (2006) examined the kinetics of degradation of reporter mRNAs containing *miR-125-* or *let-7*-binding sites in their 3'UTRs. It was discovered that both miRNAs accelerated the decay of the target mRNAs by removal of the 3'-terminal polyadenosine nucleotides in the mRNAs (Wu et al. 2006). Clearly, this mechanism is distinct from the endonucleolytic RNA cleavage employed by siRNAs. Importantly, poly(A) shortening was not the result of translational repression of the miRNA-targeted mRNAs (Wu et al. 2006). Thus, miRNAs can modulate mRNA gene expression by RNA degradation or translational inhibition.

miRNAs Can Inhibit Translation of Target mRNAs

Recently, three reports have demonstrated that miRNAs can control mRNA expression by inhibiting either translation initiation (Humphreys et al. 2005; Pillai et al. 2005) or translation elongation (Petersen et al. 2006).

Pillai et al. (2005) used luciferase reporters, expressed from CMV promoters, containing either one perfectly complementary or three imperfectly complementary *let-7* targets to examine the mechanism of miRNA-mediated regulation. Expression of both reporter transcripts was found to be tenfold down-regulated in the presence of *let-7*, but only fivefold when *let-7* was sequestered by 2′O-Me oligonucleotides. Furthermore, in the presence of *let-7*, the transcripts containing bulged *let-7* target sites were found to sediment in lighter fractions of a sucrose gradient, indicating that they were not being actively translated. However, upon addition of 2′O-Me oligonucleotides against *let-7*, the transcripts shifted into heavier polysome-containing fractions, suggesting that *let-7* inhibits translation of target mRNAs. Notably, the addition of the initiation inhibitor, NaF, did not alter the sedimentation profile of the transcripts containing bulged *let-7* target sites, suggesting that the *let-7* may inhibit translation at the step of initiation (Pillai et al. 2005). Next, these investigators used in-vitro-synthesized RNA to study the role of cap structure and poly(A) sequences in miRNA-mediated translational control. It was found that the cap structure, but not the poly(A) tail, had a significant role in miRNA-dependent regulation (Pillai et al. 2005). This finding predicted that internal ribosome entry site (IRES)-containing mRNAs, which do not require the cap-binding protein eIF4E for their activities (Hellen and Sarnow 2001), should be refractory to miRNA-mediated translational inhibition. Indeed, this predicted result was obtained with reporter mRNAs containing IRESs from encephalomyocarditis virus (EMCV) or HCV. Because the EMCV IRES requires eIF4G but not eIF4E, and the HCV IRES does not require either initiation factor to recruit ribosomes, these results support a role for eIF4E in miRNA-mediated translational control.

In a second study, Humphreys et al. employed a system in which effects of short, partially complementary RNAs on target mRNAs were examined after cotransfection of a synthetic siRNA, termed CXCR4

(Doench et al. 2003; Doench and Sharp 2004), and a plasmid expressing a luciferase reporter mRNA, containing four partially complementary sites to the CXCR4 siRNA (Humphreys et al. 2005). Similarly to Pillai et al., the authors noted an approximately 12-fold reduction of protein expression from mRNAs containing CXCR4-binding sites and noted that IRES-containing mRNAs were refractory to CXCR4-mediated inhibition. In contrast to studies by Pillai et al., these authors noted that both the 5′cap structure and 3′ poly(A) sequences were required for CXCR4-mediated regulation (Humphreys et al. 2005). Taken together, both studies suggest that miRNAs block translation initiation by interfering with the binding of eIF4E to the 5′-terminal cap structure (Fig. 2C). Although these findings are in contrast to the initial findings that in *C. elegans*, *lin-4* may lead to inhibition of *lin-14* mRNA translation at a postinitiation step (Olsen and Ambros 1999), a miRNA-induced block of translation initiation is further supported by the finding that miRNP components and miRNA-targeted mRNAs are found in structures known as processing bodies (P-bodies)/GW-bodies that contain components of translation complexes stalled at the initiation step but not the elongation step (Eystathioy et al. 2003; Andrei et al. 2005; Jakymiw et al. 2005; Kedersha et al. 2005; Liu et al. 2005a; Pillai et al. 2005).

Using the same CXCR4 system, Peterson et al. (2006) noted that translation of CXCR4-containing mRNAs was repressed by approximately 30-fold in the presence of siCXCR4 duplex RNAs, yet repressed mRNAs associated with polysomes that were translationally active. It was concluded that CXCR4-mediated repression occurred at a postinitiation step, because both cap-dependent and IRES-dependent translation of mRNAs, containing CXCR4-binding sites, was sensitive to inhibitors of translation initiation (Petersen et al. 2006). Experimental approaches that monitored the sizes of newly synthesized peptides and the efficiency of translational readthrough at stop codons in CXCR4-targeted mRNAs suggested that the short CXCR4 RNAs repressed mRNA translation by inducing a drop-off of translating ribosomes (Fig. 2C) (Petersen et al. 2006). The mechanism underlying this ribosomal drop-off is not known, but it could involve miRNP-induced alterations of ribosomal elongation rates or changes in elongation or termination factor concentrations on siRNA-targeted mRNAs. One of the main differences between the Petersen et al. study and the studies by Pillai et al. and Humphreys et al. is that the reporter mRNAs in the Petersen et al. study were transcribed in the nucleus by RNA polymerase II. In contrast, the other two studies obtained mechanistic details by monitoring translation of transfected, capped, and uncapped reporter mRNAs. Thus, future studies should focus on whether the origin, location, and number of target sites in mRNAs or the specificity for miRNAs and associated proteins determines the mechanism of translational inhibition and the rate of turnover of targeted mRNAs.

Fate of mRNAs Translationally Inhibited by miRNA

The intracellular fate of miRNA-targeted reporter RNAs was followed by Pillai et al. (2005), who noted that *let-7*-targeted and translationally inhibited mRNAs localize to P/GW-bodies and adjacent cytoplasmic structures. P-bodies, also known as GW-bodies, are cytoplasmic foci where decapping and 5′–3′ exonucleolytic decay of mRNAs occur (Eystathioy et al. 2003; Sheth and Parker 2003; Kedersha et al. 2005). Accordingly, the decapping factors Dcp1 and Dcp2, and the 5′–3′ exoribonuclease Xrn1, are found in P/GW-bodies. In mammalian cells, P-bodies are closely related in location to stress granules (SGs), which form in response to cellular stresses (Kedersha et al. 2005). Both types of cytoplasmic foci are induced by stress and dispersed by translation elongation inhibitors, but not by translation initiation inhibitors (Kedersha et al. 2005). P-bodies and SGs share common proteins, such as Fas-activated serine/threonine phosphoprotein (FAST), XRN1, eIF4E, and TTP (Kedersha et al. 2005); the latter is the ARE-binding protein tristetraprolin that also interacts with Ago2 and Ago4 (Jing et al. 2005). However, both types of foci also contain unique protein components: For example, SGs contain T-cell internal antigen-1 (TIA-1), a phosphorylated form of eIF2α, eIF3, G3BP, eIF4G, and PABP-1 (Kedersha et al. 2005). P-bodies, on the other hand, contain GW182, Dcp1a, Dcp2, and the eIF4E-binding protein 4E-T (Andrei et al. 2005; Ferraiuolo et al. 2005; Kedersha et al. 2005). GW182 is an RNA-binding protein that contains a putative nuclear localization signal, glycine-tryptophan (GW) repeats, and one RNA recognition motif (Eystathioy et al. 2002). Taken together, it seems that both SGs and P-bodies are cytoplasmic sites at which translationally inactive mRNAs can be sorted for storage, prepared for reinitiation, or degraded.

The notion that miRNA-targeted mRNAs are destined to migrate to these foci is substantiated by the findings that components of the miRNP associate with components of P/GW-bodies. For example, Ago2 coimmunoprecipitates with GW182, Dcp1, Dcp2, and Xrn1, whereas Ago1 coimmunoprecipitates with Dcp1 and Dcp2 (Jakymiw et al. 2005; Liu et al. 2005a,b; Pillai et al. 2005). Furthermore, Ago2 localizes to P/GW-bodies (Liu et al. 2005a; Sen and Blau 2005), and *Drosophila* P-body components GW182, Dcp1, and Dcp2 are required for miRNA-mediated repression of targeted reporter mRNAs (Liu et al. 2005b; Rehwinkel et al. 2005). Finally, the remarkable result that tethering of Ago2 to 3′UTRs of reporter mRNAs mimicked translational inhibition by miRNAs (Pillai et al. 2004) might be explained if tethered mRNAs localized to P-bodies. Overall, these results suggest a link between miRNA function and P/GW-bodies and argue that inhibition of translation initiation by miRNPs localizes mRNAs to P/GW-bodies, most likely as a consequence of translation inhibition (Fig. 2C). Of course, it will be very interesting to explore whether mRNAs that are inhibited at a postinitiation step by miRNPs (Petersen et al. 2006) will localize as well to P/GW-bodies.

miRNAs AND CANCER

Considering the prevalence of miRNAs and their roles in regulating genes that are involved in cell growth and differentiation, it is not surprising that many miRNAs have been implicated to function as tumor suppressors or oncogenes. We briefly summarize some of these findings and refer to an excellent review on the roles for miRNAs in cancer (Esquela-Kerscher and Slack 2006).

One of the first examples that pointed to tumor suppressor function for miRNAs came from two clustered miRNA genes, *miR-15a* and *miR-16-1*, which were often found to be down-regulated or deleted in patients suffering from B-cell chronic lymphocytic leukemia (Calin et al. 2002). Subsequently, it was shown that both miRNAs can down-regulate the anti-apoptotic gene *BCL2* that is expressed in many cancer cells (Cimmino et al. 2005). Similarly, some members of the human *let-7* family were found to be down-regulated in human lung cancer cells, correlating with a shortened postoperative survival for cancer patients (Takamizawa et al. 2004). Recently, *RAS* genes that encode membrane-associated GTPase signaling proteins were found to contain multiple binding sites for *let-7* in their 3′UTRs, resulting in translational down-regulation of *RAS* mRNAs. RAS protein levels were highly elevated in human lung cancer cells compared to normal cells, concomitant with low levels of *let-7*, suggesting that miRNAs could be useful as therapeutic agents (Johnson et al. 2005).

On the other hand, two recent findings point to roles for miRNAs in tumorigenesis (He et al. 2005; O'Donnell et al. 2005). It was noted that the 13q31 locus is preferentially amplified in B-cell lymphomas. Curiously, this locus encodes a cluster of seven miRNAs, known as the *miR-17-92* cluster. Transcription of this cluster is up-regulated by the *myc* oncogene, and both the miRNA cluster and *myc* function cooperatively as oncogenes to cause tumorigenesis. One of the identified targets for the *miR-17-92* cluster is the transcription factor *E2F1* gene, whose product is hypothesized to modulate the expression of apoptotic genes (He et al. 2005; O'Donnell et al. 2005). Thus, a complex network of miRNA expression and target gene regulation can control the onset of cancer.

As a final remark in this short summary, an astonishing report by Lu et al. (2005) noted that an expression profile, obtained from as few as 217 miRNA genes, could accurately pinpoint the lineages and differentiation states of human cancer cells. Such a prediction could not be accomplished when expression profiles of 16,000 protein-coding genes were used in the analysis (Lu et al. 2005). This study suggests that miRNA profiling might provide an invaluable tool to aid in the classification, prognosis, and treatment of cancer.

CONCLUDING REMARKS

miRNA-targeted mRNAs are being discovered by computational analyses at a staggering rate (see, e.g., Lewis et al. 2003). Although it seems clear that some miRNAs trigger degradation of target mRNAs, the scope of translational inhibition of target mRNAs is just beginning to emerge. Most likely, the turnover and translational efficiency of targeted mRNAs can both be affected, and the predominant mechanism may reflect the time that targeted mRNAs reside in SGs and P-bodies. It is also important to remember that endogenous miRNAs can also up-regulate targeted mRNAs, exemplified by *miR-122* and HCV RNA. Thus, miRNAs may function in diverse and unexpected ways.

ACKNOWLEDGMENTS

We thank Christopher Potter for critical reading of this chapter. Work in the authors' laboratory was supported by grants from the National Institutes of Health (GM069007, AI47365) and DRG1775 from the Damon Runyon Cancer Research Foundation (K.A.W.).

REFERENCES

Ambros V. 1989. A hierarchy of regulatory genes controls a larva-to-adult developmental switch in *C. elegans*. *Cell* **57:** 49–57.

Ambros V. and Horvitz H.R. 1984. Heterochronic mutants of the nematode *Caenorhabditis elegans*. *Science* **226:** 409–416.

———. 1987. The *lin-14* locus of *Caenorhabditis elegans* controls the time of expression of specific postembryonic developmental events. *Genes Dev.* **1:** 398–414.

Andrei M.A., Ingelfinger D., Heintzmann R., Achsel T., Rivera-Pomar R., and Luhrmann R. 2005. A role for eIF4E and eIF4E-transporter in targeting mRNPs to mammalian processing bodies. *RNA* **11:** 717–727.

Arasu P., Wightman B., and Ruvkun G. 1991. Temporal regulation of *lin-14* by the antagonistic action of two other heterochronic genes, *lin-4* and *lin-28*. *Genes Dev.* **5:** 1825–1833.

Bagga S., Bracht J., Hunter S., Massirer K., Holtz J., Eachus R., and Pasquinelli A.E. 2005. Regulation by *let-7* and *lin-4* miRNAs results in target mRNA degradation. *Cell* **122:** 553–563.

Cai X., Hagedorn C.H., and Cullen B.R. 2004. Human microRNAs are processed from capped, polyadenylated transcripts that can also function as mRNAs. *RNA* **10:** 1957–1966.

Calin G.A., Dumitru C.D., Shimizu M., Bichi R., Zupo S., Noch E., Aldler H., Rattan S., Keating M., Rai K., et al. 2002. Frequent deletions and down-regulation of micro-RNA genes miR15 and miR16 at 13q14 in chronic lymphocytic leukemia. *Proc. Natl. Acad. Sci.* **99:** 15524–15529.

Chalfie M., Horvitz H.R., and Sulston J.E. 1981. Mutations that lead to reiterations in the cell lineages of *C. elegans*. *Cell* **24:** 59–69.

Chendrimada T.P., Gregory R.I., Kumaraswamy E., Norman J., Cooch N., Nishikura K., and Shiekhattar R. 2005. TRBP recruits the Dicer complex to Ago2 for microRNA processing and gene silencing. *Nature* **436:** 740–744.

Cimmino A., Calin G.A., Fabbri M., Iorio M.V., Ferracin M., Shimizu M., Wojcik S.E., Aqeilan R.I., Zupo S., Dono M., et al. 2005. miR-15 and miR-16 induce apoptosis by targeting BCL2. *Proc. Natl. Acad. Sci.* **102:** 13944–13949.

Darnell J.C., Jensen K.B., Jin P., Brown V., Warren S.T., and Darnell R.B. 2001. Fragile X mental retardation protein targets G quartet mRNAs important for neuronal function. *Cell* **107:** 489–499.

Doench J.G. and Sharp P.A. 2004. Specificity of microRNA target selection in translational repression. *Genes Dev.* **18:** 504–511.

Doench J.G., Petersen C.P., and Sharp P.A. 2003. siRNAs can function as miRNAs. *Genes Dev.* **17:** 438–442.

Esquela-Kerscher A. and Slack F.J. 2006. Oncomirs—microRNAs with a role in cancer. *Nat. Rev. Cancer* **6:** 259–269.

Eystathioy T., Chan E.K., Tenenbaum S.A., Keene J.D., Griffith K., and Fritzler M.J. 2002. A phosphorylated cytoplasmic autoantigen, GW182, associates with a unique population of human mRNAs within novel cytoplasmic speckles. *Mol. Biol. Cell* **13:** 1338–1351.

Eystathioy T., Jakymiw A., Chan E.K., Seraphin B., Cougot N., and Fritzler M.J. 2003. The GW182 protein colocalizes with mRNA degradation associated proteins hDcp1 and hLSm4 in cytoplasmic GW bodies. *RNA* **9:** 1171–1173.

Ferraiuolo M.A., Basak S., Dostie J., Murray E.L., Schoenberg D.R., and Sonenberg N. 2005. A role for the eIF4E-binding protein 4E-T in P-body formation and mRNA decay. *J. Cell Biol.* **170:** 913–924.

Forstemann K., Tomari Y., Du T., Vagin V.V., Denli A.M., Bratu D.P., Klattenhoff C., Theurkauf W.E., and Zamore P.D. 2005. Normal microRNA maturation and germ-line stem cell maintenance requires Loquacious, a double-stranded RNA-binding domain protein. *PLoS Biol.* **3:** e236.

Gregory R.I., Yan K.P., Amuthan G., Chendrimada T., Doratotaj B., Cooch N., and

Shiekhattar R. 2004. The Microprocessor complex mediates the genesis of microRNAs. *Nature* **432:** 235–240.

Haase A.D., Jaskiewicz L., Zhang H., Laine S., Sack R., Gatignol A., and Filipowicz W. 2005. TRBP, a regulator of cellular PKR and HIV-1 virus expression, interacts with Dicer and functions in RNA silencing. *EMBO Rep.* **6:** 961–967.

He L., Thomson J.M., Hemann M.T., Hernando-Monge E., Mu D., Goodson S., Powers S., Cordon-Cardo C., Lowe S.W., Hannon G.J., and Hammond S.M. 2005. A microRNA polycistron as a potential human oncogene. *Nature* **435:** 828–833.

Hellen C.U. and Sarnow P. 2001. Internal ribosome entry sites in eukaryotic mRNA molecules. *Genes Dev.* **15:** 1593–1612.

Horvitz H.R. and Sulston J.E. 1980. Isolation and genetic characterization of cell-lineage mutants of the nematode *Caenorhabditis elegans. Genetics* **96:** 435–454.

Humphreys D.T., Westman B.J., Martin D.I., and Preiss T. 2005. MicroRNAs control translation initiation by inhibiting eukaryotic initiation factor 4E/cap and poly(A) tail function. *Proc. Natl. Acad. Sci.* **102:** 16961–16966.

Hutvagner G., Simard M.J., Mello C.C., and Zamore P.D. 2004. Sequence-specific inhibition of small RNA function. *PLoS Biol.* **2:** E98.

Jakymiw A., Lian S., Eystathioy T., Li S., Satoh M., Hamel J.C., Fritzler M.J., and Chan E.K. 2005. Disruption of GW bodies impairs mammalian RNA interference. *Nat. Cell Biol.* **7:** 1167–1174.

Jin P., Zarnescu D.C., Ceman S., Nakamoto M., Mowrey J., Jongens T.A., Nelson D.L., Moses K., and Warren S.T. 2004. Biochemical and genetic interaction between the fragile X mental retardation protein and the microRNA pathway. *Nat. Neurosci.* **7:** 113–117.

Jing Q., Huang S., Guth S., Zarubin T., Motoyama A., Chen J., Di Padova F., Lin S.C., Gram H., and Han J. 2005. Involvement of microRNA in AU-rich element-mediated mRNA instability. *Cell* **120:** 623–634.

John B., Enright A.J., Aravin A., Tuschl T., Sander C., and Marks D.S. 2004. Human MicroRNA targets. *PLoS Biol.* **2:** e363.

Johnson S.M., Grosshans H., Shingara J., Byrom M., Jarvis R., Cheng A., Labourier E., Reinert K.L., Brown D., and Slack F.J. 2005. RAS is regulated by the *let-7* microRNA family. *Cell* **120:** 635–647.

Jopling C.L., Yi M., Lancaster A.M., Lemon S.M., and Sarnow P. 2005. Modulation of hepatitis C virus RNA abundance by a liver-specific MicroRNA. *Science* **309:** 1577–1581.

Kedersha N., Stoecklin G., Ayodele M., Yacono P., Lykke-Andersen J., Fitzler M.J., Scheuner D., Kaufman R.J., Golan D.E., and Anderson P. 2005. Stress granules and processing bodies are dynamically linked sites of mRNP remodeling. *J. Cell Biol.* **169:** 871–884.

Khvorova A., Reynolds A., and Jayasena S.D. 2003. Functional siRNAs and miRNAs exhibit strand bias. *Cell* **115:** 209–216.

Kim V.N. 2005. MicroRNA biogenesis: Coordinated cropping and dicing. *Nat. Rev. Mol. Cell Biol.* **6:** 376–385.

Krek A., Grun D., Poy M.N., Wolf R., Rosenberg L., Epstein E.J., MacMenamin P., da Piedade I., Gunsalus K.C., Stoffel M., and Rajewsky N. 2005. Combinatorial microRNA target predictions. *Nat. Genet.* **37:** 495–500.

Kremer E.J., Pritchard M., Lynch M., Yu S., Holman K., Baker E., Warren S.T., Schlessinger D., Sutherland G.R., and Richards R.I. 1991. Mapping of DNA instability at the fragile X to a trinucleotide repeat sequence p(CCG)n. *Science* **252:** 1711–1714.

Krutzfeldt J., Rajewsky N., Braich R., Rajeev K.G., Tuschl T., Manoharan M., and Stoffel

M. 2005. Silencing of microRNAs in vivo with "antagomirs". *Nature* **438:** 685–689.

Laggerbauer B., Ostareck D., Keidel E.M., Ostareck-Lederer A., and Fischer U. 2001. Evidence that fragile X mental retardation protein is a negative regulator of translation. *Hum. Mol. Genet.* **10:** 329–338.

Lagos-Quintana M., Rauhut R., Lendeckel W., and Tuschl T. 2001. Identification of novel genes coding for small expressed RNAs. *Science* **294:** 853–858.

Lau N.C., Lim L.P., Weinstein E.G., and Bartel D.P. 2001. An abundant class of tiny RNAs with probable regulatory roles in *Caenorhabditis elegans. Science* **294:** 858–862.

Lee R.C., Feinbaum R.L., and Ambros V. 1993. The *C. elegans* heterochronic gene *lin-4* encodes small RNAs with antisense complementarity to *lin-14. Cell* **75:** 843–854.

Lee Y., Jeon K., Lee J.T., Kim S., and Kim V.N. 2002. MicroRNA maturation: Stepwise processing and subcellular localization. *EMBO J.* **21:** 4663–4670.

Lee Y., Hur I., Park S.Y., Kim Y.K., Suh M.R., and Kim V.N. 2006. The role of PACT in the RNA silencing pathway. *EMBO J.* **25:** 522–532.

Lee Y., Kim M., Han J., Yeom K.H., Lee S., Baek S.H., and Kim V.N. 2004. MicroRNA genes are transcribed by RNA polymerase II. *EMBO J.* **23:** 4051–4060.

Lewis B.P., Burge C.B., and Bartel D.P. 2005. Conserved seed pairing, often flanked by adenosines, indicates that thousands of human genes are microRNA targets. *Cell* **120:** 15–20.

Lewis B.P., Shih I.H., Jones-Rhoades M.W., Bartel D.P., and Burge C.B. 2003. Prediction of mammalian microRNA targets. *Cell* **115:** 787–798.

Lim L.P., Lau N.C., Garrett-Engele P., Grimson A., Schelter J.M., Castle J., Bartel D.P., Linsley P.S., and Johnson J.M. 2005. Microarray analysis shows that some microRNAs downregulate large numbers of target mRNAs. *Nature* **433:** 769–773.

Liu J., Valencia-Sanchez M.A., Hannon G.J., and Parker R. 2005a. MicroRNA-dependent localization of targeted mRNAs to mammalian P-bodies. *Nat. Cell Biol.* **7:** 719–723.

Liu J., Rivas F.V., Wohlschlegel J., Yates J.R., Parker R., and Hannon G.J. 2005b. A role for the P-body component GW182 in microRNA function. *Nat. Cell Biol.* **7:** 1161–1166.

Liu J., Carmell M.A., Rivas F.V., Marsden C.G., Thomson J.M., Song J.J., Hammond S.M., Joshua-Tor L., and Hannon G.J. 2004. Argonaute2 is the catalytic engine of mammalian RNAi. *Science* **305:** 1437–1441.

Liu Q., Rand T.A., Kalidas S., Du F., Kim H.E., Smith D.P., and Wang X. 2003. R2D2, a bridge between the initiation and effector steps of the *Drosophila* RNAi pathway. *Science* **301:** 1921–1925.

Llave C., Xie Z., Kasschau K.D., and Carrington J.C. 2002. Cleavage of Scarecrow-like mRNA targets directed by a class of *Arabidopsis* miRNA. *Science* **297:** 2053–2056.

Lu J., Getz G., Miska E.A., Alvarez-Saavedra E., Lamb J., Peck D., Sweet-Cordero A., Ebert B.L., Mak R.H., Ferrando A.A., et al. 2005. MicroRNA expression profiles classify human cancers. *Nature* **435:** 834–838.

Matranga C., Tomari Y., Shin C., Bartel D.P., and Zamore P.D. 2005. Passenger-strand cleavage facilitates assembly of siRNA into Ago2-containing RNAi enzyme complexes. *Cell* **123:** 607–620.

Meister G., Landthaler M., Dorsett Y., and Tuschl T. 2004a. Sequence-specific inhibition of microRNA- and siRNA-induced RNA silencing. *RNA* **10:** 544–550.

Meister G., Landthaler M., Patkaniowska A., Dorsett Y., Teng G., and Tuschl T. 2004b. Human Argonaute2 mediates RNA cleavage targeted by miRNAs and siRNAs. *Mol. Cell* **15:** 185–197.

Mourelatos Z., Dostie J., Paushkin S., Sharma A., Charroux B., Abel L., Rappsilber J.,

Mann M., and Dreyfuss G. 2002. miRNPs: A novel class of ribonucleoproteins containing numerous microRNAs. *Genes Dev.* **16:** 720–728.

Nakamoto M., Jin P., O'Donnell W.T., and Warren S.T. 2005. Physiological identification of human transcripts translationally regulated by a specific microRNA. *Hum. Mol. Genet.* **14:** 3813–3821.

Nelson P.T., Hatzigeorgiou A.G., and Mourelatos Z. 2004. miRNP:mRNA association in polyribosomes in a human neuronal cell line. *RNA* **10:** 387–394.

Oberle I., Rousseau F., Heitz D., Kretz C., Devys D., Hanauer A., Boue J., Bertheas M.F., and Mandel J.L. 1991. Instability of a 550-base pair DNA segment and abnormal methylation in fragile X syndrome. *Science* **252:** 1097–1102.

O'Donnell K.A., Wentzel E.A., Zeller K.I., Dang C.V., and Mendell J.T. 2005. c-Myc-regulated microRNAs modulate E2F1 expression. *Nature* **435:** 839–843.

Olsen P.H. and Ambros V. 1999. The *lin-4* regulatory RNA controls developmental timing in *Caenorhabditis elegans* by blocking LIN-14 protein synthesis after the initiation of translation. *Dev. Biol.* **216:** 671–680.

Pasquinelli A.E., Reinhart B.J., Slack F., Martindale M.Q., Kuroda M.I., Maller B., Hayward D.C., Ball E.E., Degnan B., Muller P., et al. 2000. Conservation of the sequence and temporal expression of *let-7* heterochronic regulatory RNA. *Nature* **408:** 86–89.

Patel R.C. and Sen G.C. 1998. PACT, a protein activator of the interferon-induced protein kinase, PKR. *EMBO J.* **17:** 4379–4390.

Petersen C.P., Bordeleau M.E., Pelletier J., and Sharp P.A. 2006. Short RNAs repress translation after initiation in mammalian cells. *Mol. Cell* **21:** 533–542.

Pieretti M., Zhang F.P., Fu Y.H., Warren S.T., Oostra B.A., Caskey C.T., and Nelson D.L. 1991. Absence of expression of the FMR-1 gene in fragile X syndrome. *Cell* **66:** 817–822.

Pillai R.S., Artus C.G., and Filipowicz W. 2004. Tethering of human Ago proteins to mRNA mimics the miRNA-mediated repression of protein synthesis. *RNA* **10:** 1518–1525.

Pillai R.S., Bhattacharyya S.N., Artus C.G., Zoller T., Cougot N., Basyuk E., Bertrand E., and Filipowicz W. 2005. Inhibition of translational initiation by Let-7 MicroRNA in human cells. *Science* **309:** 1573–1576.

Poy M.N., Eliasson L., Krutzfeldt J., Kuwajima S., Ma X., Macdonald P.E., Pfeffer S., Tuschl T., Rajewsky N., Rorsman P., and Stoffel M. 2004. A pancreatic islet-specific microRNA regulates insulin secretion. *Nature* **432:** 226–230.

Rehwinkel J., Behm-Ansmant I., Gatfield D., and Izaurralde E. 2005. A crucial role for GW182 and the DCP1:DCP2 decapping complex in miRNA-mediated gene silencing. *RNA* **11:** 1640–1647.

Reinhart B.J., Slack F.J., Basson M., Pasquinelli A.E., Bettinger J.C., Rougvie A.E., Horvitz H.R., and Ruvkun G. 2000. The 21-nucleotide *let-7* RNA regulates developmental timing in *Caenorhabditis elegans*. *Nature* **403:** 901–906.

Sasaki T., Shiohama A., Minoshima S., and Shimizu N. 2003. Identification of eight members of the Argonaute family in the human genome. *Genomics* **82:** 323–330.

Schaeffer C., Bardoni B., Mandel J.L., Ehresmann B., Ehresmann C., and Moine H. 2001. The fragile X mental retardation protein binds specifically to its mRNA via a purine quartet motif. *EMBO J.* **20:** 4803–4813.

Sen G.L. and Blau H.M. 2005. Argonaute 2/RISC resides in sites of mammalian mRNA decay known as cytoplasmic bodies. *Nat. Cell Biol.* **7:** 633–636.

Sheth U. and Parker R. 2003. Decapping and decay of messenger RNA occur in cytoplasmic processing bodies. *Science* **300:** 805–808.

Slack F.J., Basson M., Liu Z., Ambros V., Horvitz H.R., and Ruvkun G. 2000. The *lin-41* RBCC gene acts in the *C. elegans* heterochronic pathway between the *let-7* regulatory RNA and the LIN-29 transcription factor. *Mol. Cell* **5:** 659–669.

Takamizawa J., Konishi H., Yanagisawa K., Tomida S., Osada H., Endoh H., Harano T., Yatabe Y., Nagino M., Nimura Y., et al. 2004. Reduced expression of the *let-7* microRNAs in human lung cancers in association with shortened postoperative survival. *Cancer Res.* **64:** 3753–3756.

Tomari Y., Matranga C., Haley B., Martinez N., and Zamore P.D. 2004. A protein sensor for siRNA asymmetry. *Science* **306:** 1377–1380.

Wightman B., Ha I., and Ruvkun G. 1993. Posttranscriptional regulation of the heterochronic gene *lin-14* by *lin-4* mediates temporal pattern formation in *C. elegans*. *Cell* **75:** 855–862.

Wightman B., Burglin T.R., Gatto J., Arasu P., and Ruvkun G. 1991. Negative regulatory sequences in the *lin-14* 3′-untranslated region are necessary to generate a temporal switch during *Caenorhabditis elegans* development. *Genes Dev.* **5:** 1813–1824.

Wu L., Fan J., and Belasco J.G. 2006. MicroRNAs direct rapid deadenylation of mRNA. *Proc. Natl. Acad. Sci.* **103:** 4034–4039.

Yekta S., Shih I.H., and Bartel D.P. 2004. MicroRNA-directed cleavage of HOXB8 mRNA. *Science* **304:** 594–596.

12

The eIF2α Kinases

Thomas E. Dever
Laboratory of Gene Regulation and Development
National Institute of Child Health and Human Development
Bethesda, Maryland 20892

Arvin C. Dar and Frank Sicheri
Program in Molecular Biology and Cancer
Samuel Lunenfeld Research Institute
Mount Sinai Hospital, Toronto, Ontario
M5G 1X5 Canada; Department of
Molecular and Medical Genetics
University of Toronto, Ontario
M5S 1A8 Canada

PERHAPS THE BEST-CHARACTERIZED MECHANISM of translational control in eukaryotic cells involves phosphorylation of eukaryotic translation initiation factor eIF2. As described in Chapters 4 and 9, eIF2, consisting of three subunits, α, β, and γ, specifically binds the initiator methionyl-tRNA (Met-tRNA$_i^{Met}$) in a GTP-dependent manner and delivers this essential component of translation initiation to the small ribosomal subunit. The γ-subunit of eIF2 is responsible for GTP binding, and like other GTP-binding proteins, eIF2 cycles between its GTP-bound state and its GDP-bound state. The recycling of inactive eIF2·GDP to active eIF2·GTP is catalyzed by the guanine nucleotide exchange factor eIF2B. It is this recycling reaction that is regulated by phosphorylation of eIF2. Phosphorylation of Ser-51 in mature eIF2α converts eIF2 from a substrate to a competitive inhibitor of eIF2B. (It is noteworthy that according to the DNA sequence, the phosphorylation site in eIF2α is Ser-52. However, because the initiating Met of eIF2α is posttranslationally cleaved, the phosphorylated residue is Ser-51 in the mature protein.) This phosphorylation of eIF2α enhances its interaction with a trimeric regulatory eIF2Bαβδ subcomplex that can be biochemically separated from the pentameric eIF2B complex (Chapter 9). In all cells examined, the amount of eIF2B is limiting compared to the amount of eIF2. As a consequence, phosphorylation of a small percentage of eIF2α results in the apparent sequestration of eIF2B in inactive phosphorylated eIF2·eIF2B complexes and in the inhibition of protein synthesis.

Initially, eIF2 phosphorylation was linked to the shut-off of protein synthesis in heme-deprived or double-stranded RNA (dsRNA)-treated rabbit reticulocyte lysates (Farrell et al. 1977). Further study revealed that these translational control mechanisms were functional in a variety of cells and that distinct, but related, protein kinases phosphorylate eIF2α in response to heme deprivation and dsRNA treatment. Later, it was found that various other cellular stress conditions including amino acid starvation in yeast and endoplasmic reticulum (ER) stress in mammalian cells activate additional eIF2α kinases (for information on the physiological functions of eIF2α phosphorylation, see Chapter 13). Although eIF2α phosphorylation was first described about 30 years ago, recent molecular, biochemical, and structural studies have provided fresh insights into the activation mechanisms and substrate recognition properties of the kinases that phosphorylate eIF2α. In this chapter, we provide an overview of the four well-characterized eIF2α kinases, summarize their activation mechanisms including kinase dimerization and autophosphorylation, and detail the molecular mechanism of eIF2α kinase function as revealed from structural studies on the kinases.

eIF2α KINASES: CONSERVED KINASE DOMAIN WITH DISTINCT REGULATORY DOMAINS

Four eIF2α kinases have been characterized: HRI (HUGO gene symbol: EIF2AK1), PKR (EIF2AK2), PERK (EIF2AK3), and GCN2 (EIF2AK4). The conserved eIF2α kinase domains in these four proteins share approximately 25–37% amino acid sequence identity, but only about 17% identity with the kinase domain in the cAMP-dependent protein kinase PKA. Sequence alignments of a large grouping of diverse protein kinases reveal that 26 of approximately 300 residues are preferentially conserved among the eIF2α kinases (and not conserved in the diverse kinases) (see Dar et al. 2005; Dey et al. 2005b). In addition to these conserved residues, two hallmarks of the eIF2α kinase domains (see below) are (1) the presence of an insert ranging in size between 30 and >200 residues between sheets β4 and β5 in the amino-terminal lobe of the kinases and (2) a shortened linker connecting helices αF and αG as well as an extended helix αG in the carboxy-terminal lobe of the kinase domain. As described later below, the unique features of helix αG directly contribute to the specific recognition of eIF2α by these kinases.

The ability of the four eIF2α kinases to respond to distinct stimuli is determined by the presence of unique regulatory domains in the proteins. As detailed below and depicted in Figure 1, the regulatory elements are appended to the amino terminus, carboxyl terminus, or within the conserved eIF2α kinase domain.

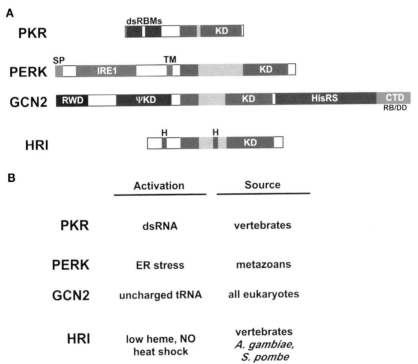

Figure 1. (A) Architecture of the eIF2α kinases. (Blue) Conserved kinase domains (KD); (gray) kinase insert. Also shown are the dsRBMs in PKR; the signal peptide (SP), IRE1 homology region, and transmembrane domain (TM) in PERK; the amino-terminal RWD domain, pseudo-kinase domain (ΨKD), histidyl-tRNA synthetase-related domain (HisRS), and the ribosome-binding (RB) and dimerization domain (DD) in the carboxy-terminal domain (CTD) of GCN2; and the two heme-binding sites (H) in HRI. (B) The activating condition or ligand and the known sources for each kinase are also indicated.

PKR: Amino-terminal dsRNA-binding Domains Govern Kinase Activation in Response to Viral Infection

PKR is thought to be present in all vertebrates. A protein that has homology with the PKR kinase domain has been identified in frogs and fish; however, as described below, the activation of the fish kinase may differ slightly from that of mammalian PKR. Expression of PKR is induced by interferon, and the latent enzyme has a molecular mass of about 68 kD. In human PKR, two dsRNA-binding motifs (dsRBMs) located between residues 6–79 and 96–169 precede the protein kinase domain (residues 258–551) (Fig. 1). The dsRBMs in PKR are connected by a flexible linker, and both domains resemble similar domains in Staufen, ADAR1, and *Escherichia coli* RNase III (for a recent review, see Tian et al. 2004).

It has been proposed that the dsRBMs of PKR prevent kinase activity in the absence of dsRNA and promote dimerization and kinase activation in the presence of dsRNA activator. Whereas dsRBM1 bound dsRNA with micromolar affinity, dsRBM2 bound dsRNA weakly, but cooperatively, with dsRBM1 (Green and Mathews 1992; Schmedt et al. 1995). The amino-terminal region of PKR containing the dsRBMs interacted with the kinase domain in a yeast two-hybrid assay (Sharp et al. 1998), and nuclear magnetic resonance (NMR) data revealed that dsRBM2, but not dsRBM1, interacted with the kinase domain (Nanduri et al. 2000). Because dsRNA binding induces a conformational change in PKR (Manche et al. 1992; Carpick et al. 1997), it is proposed that dsRBM2 inhibits the kinase domain in latent PKR and that binding of dsRNA to dsRBM2 relieves this inhibition.

In addition to binding dsRNA, the dsRBMs in PKR mediate dimerization of the protein. Numerous studies have revealed the critical importance of dsRNA binding for PKR activation. Mutations in dsRBM1 that impair binding of dsRNA to the isolated dsRBMs impaired PKR function (Cosentino et al. 1995; Carpick et al. 1997; Patel and Sen 1998). Surprisingly, these mutations did not significantly impair dimerization of the isolated dsRBMs (Ortega et al. 1996; Wu and Kaufman 1996), suggesting that the dsRBMs can dimerize independently of dsRNA binding. In contrast to the apparently limited importance of dsRNA for dimerization of the isolated dsRBMs, dimerization of full-length PKR is strongly dependent on dsRNA binding (Wu and Kaufman 1997; Zhang et al. 2001). Mutations in the dsRBMs that impair binding of dsRNA likewise impair dimerization and activation of full-length PKR by dsRNA.

The binding of dsRNA to the PKR dsRBMs is independent of the RNA sequence, but the dsRNAs must be at least 16 bp in length to bind (Schmedt et al. 1995). Although an individual dsRBM protects approximately 11 bp (Schmedt et al. 1995), activation of PKR requires longer dsRNA molecules at least 40 bp in length (Manche et al. 1992; Zheng and Bevilacqua 2004). Whereas low concentrations of dsRNA activate PKR, higher dsRNA concentrations inhibit the kinase (Hunter et al. 1975). This bell-shaped curve for activation of PKR by dsRNA is consistent with the idea that two molecules of PKR must bind to the same dsRNA molecule to activate the kinase. Higher dsRNA concentrations would favor binding of inactive PKR monomers to different dsRNA molecules, and thus prevent PKR dimerization. Furthermore, PKR activation displays second-order kinetics consistent with active PKR functioning as a dimer (Kostura and Mathews 1989; Lemaire et al. 2005). Interestingly, the dsRBMs can bind imperfect dsRNA structures, including molecules containing bulges, loops, and even pseudoknots. In addition, highly structured RNAs, including viral RNAs, viral dsRNA genomes, and mRNAs with extensive secondary structures, can bind and activate PKR (Davis and Watson 1996; Bevilacqua et al. 1998; Ben-Asouli et al. 2002). The small highly structured adenoviral VA (virus-associated) RNA$_I$ binds to the PKR dsRBMs and blocks PKR activation (Chapter 20), perhaps because only one molecule of PKR can bind to a single molecule of VA RNA$_I$. The dsRBMs in PKR have a preference for binding A-form dsRNA (Bevilacqua et al. 1998); however, it is intriguing that a PKR ortholog, known as PKZ, in zebra fish and goldfish contains two Z-DNA-binding domains in place of the dsRBMs in mammalian PKR (Rothenburg et al. 2005). It remains to be determined whether PKZ is activated by Z-form dsRNA and whether PKZ represents the PKR ortholog in fish.

A consensus model for PKR activation (Fig. 2) proposes that dsRBM2 masks the kinase domain in latent monomeric PKR. Binding of dsRNA to the dsRBMs of PKR causes a conformational change exposing the kinase domain. In addition, binding of two PKR molecules to the same molecule of dsRNA enables PKR dimerization and kinase activation.

PERK: Regulatory Domain Localized to the ER Lumen Senses Unfolded Proteins and ER Stress

PERK (PKR-like ER kinase) homologs have been identified in both vertebrates (mammals, chicken, frog) and invertebrates (*Drosophila melanogaster, Caenorhabditis elegans*). The four elements of the approximately 1100-residue protein from amino to carboxyl termini are a signal peptide, amino-terminal region, transmembrane domain, and kinase domain (Fig. 1). It is tempting to speculate that the novel eIF2α kinases present in *Toxoplasma gondii* (TgIF2K-A; Sullivan et al. 2004) and *Plasmodium falciparum* (PfPK4; Mohrle et al. 1997) are orthologs of PERK given the presence of a transmembrane domain in these proteins. PERK was independently discovered as a protein sharing sequence homology with the

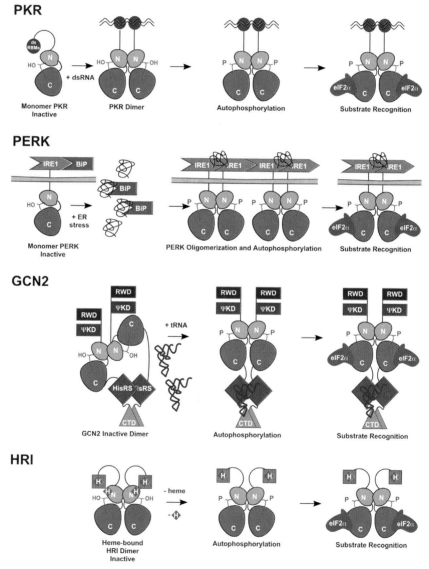

Figure 2. (*See facing page for legend.*)

ER stress-responsive IRE1 kinase (Harding et al. 1999) and as a pancreatic threonine kinase (where it was referred to as PEK) (Shi et al. 1998). The homology between PERK and IRE1 is restricted to the amino-terminal regulatory regions of the proteins, which are inserted into the lumen of the ER. Under conditions of ER stress, IRE1 oligomerizes, which activates a carboxy-terminal endonuclease and triggers the unfolded protein response (UPR), an ER stress-responsive pathway (Credle et al. 2005; Chapter 13).

Like IRE1, PERK is activated under ER stress conditions. Interestingly, PERK responsiveness to ER stress was maintained when the amino-terminal and transmembrane domains were replaced by the corresponding regions of IRE1 (Bertolotti et al. 2000). These results suggest a common activation mechanism for IRE1 and PERK. In unstressed cells, PERK interacts with the ER chaperones BiP (GRP78) and GRP94 in a manner dependent on its amino-terminal regulatory region (Bertolotti et al. 2000; Ma et al. 2002). Consistent with this idea, PERK is present in a complex of approximately 210–230 kD in unstressed cells. Under conditions of ER stress, the chaperones dissociate from PERK and the kinase shifts to a much larger complex ($\sim$320 to >600 kD) (Bertolotti et al. 2000). Two models have been proposed for PERK and IRE1 activation under conditions of ER stress (Fig. 2). The first model proposes

Figure 2. Mechanism of activation of the eIF2α kinases. (*Blue*) Amino- and carboxy-terminal lobes of the conserved eIF2α kinase domains. The conserved phospho-regulatory site in the activation segment (*purple*) is depicted in its nonphosphorylated (OH) and phosphorylated (P) states. The regulatory domains of the four kinases are depicted in the same manner as in Fig. 1, and the eIF2α is shown in *green*. Briefly, binding of dsRNA (*wavy lines*) to monomeric, inactive PKR promotes dimerization triggering autophosphorylation and substrate recognition. Binding of ER chaperones including BiP to the luminal IRE1 homology domain (R) is thought to maintain PERK in a monomeric, inactive state. Under ER stress conditions, unfolded proteins accumulate (*squiggly line*) and either titrate the ER chaperones or bind to the IRE1 homology domains, leading to PERK oligomerization, autophosphorylation, and eIF2α recognition. Interactions among the CTD, HisRS, and kinase domains maintain inactive GCN2 as a dimer. Binding of uncharged tRNA to the HisRS plus CTD is thought to induce a conformational change in the protein promoting autophosphorylation and eIF2α recognition. In the presence of heme, the activity of dimeric HRI is repressed perhaps through interactions between the amino-terminal heme-binding site and the kinase domain. Dissociation of heme from the site in the insertion sequence in the amino-terminal lobe of the kinase domain activates the kinase to autophosphorylate leading to eIF2α recognition. See text for further details.

that binding of BiP and other ER chaperones to the amino-terminal region of PERK prevents kinase dimerization and maintains the kinase in an inactive state. Upon ER stress, unfolded proteins accumulate in the ER and titrate the chaperones away from PERK, enabling kinase dimerization and activation (Bertolotti et al. 2000). Consistent with this model, deletion of the GRP78-binding segment from the amino-terminal region of PERK resulted in activation of the kinase in the absence of an ER stress signal (Ma et al. 2002). The second model, based in part on the presence of a putative peptide-binding domain in the IRE1 homolog region, proposes that binding of unfolded proteins to the IRE1 domain promotes kinase oligomerization (Credle et al. 2005). Consistently, deletion of the IRE1 homology region from the PERK amino terminus blocked oligomerization and prevented PERK activation in response to ER stress (Ma et al. 2002). Interestingly, both models reveal a similarity to PKR in that regulated dimerization of the amino-terminal region of PERK promotes kinase activation.

GCN2: HisRS Domain Binds Uncharged tRNA to Sense Amino Acid Deficiency

GCN2 is present in all eukaryotes including fungi, plants, invertebrates, and vertebrates including mammals. First identified in yeast, GCN2 is activated and phosphorylates eIF2α upon binding the uncharged cognate tRNAs that accumulate in cells starved for any single amino acid. The architecture of GCN2 is complex, consisting of (1) an amino-terminal charged region (RWD, Fig. 1) that binds the GCN1/GCN20 complex; (2) a pseudokinase domain that resembles authentic protein kinases except that key catalytic residues are lacking; (3) an eIF2α kinase domain; (4) a domain resembling histidyl-tRNA synthetase (HisRS); and (5) a carboxy-terminal domain (CTD) that dimerizes, enhances tRNA binding, and mediates ribosome binding in cell extracts. Consistent with activation of GCN2 in response to starvation for any single amino acid, deacylated tRNA^{Phe}, tRNA^{Tyr}, tRNA^{Lys}, and tRNA^{Val} bound to GCN2 with similar affinities, whereas aminoacylated Phe-tRNA^{Phe} bound with lower affinity (Dong et al. 2000). Efficient tRNA binding to GCN2 requires both the HisRS and carboxy-terminal domains (Dong et al. 2000) and is dependent on the HisRS m2 motif—a conserved element required for tRNA binding to authentic class II tRNA synthetases. Consistently, mutation of the conserved Tyr-1119 and Arg-1120 in the m2 motif blocked yeast GCN2 function in vivo and prevented eIF2α phosphorylation in vitro (Wek et al. 1995; Dong et al. 2000).

GCN2 is thought to be a constitutive dimer mediated by interactions between the CTD, HisRS, and protein kinase domains of two GCN2 protomers (Qiu et al. 1998). In addition to these dimeric interactions, the CTD interacts with the HisRS and protein kinase domains and the HisRS and protein kinase domains interact as well (Qiu et al. 2001). These interdomain interactions may regulate GCN2 kinase activity as binding of a fragment containing the HisRS + CTD to the protein kinase domain is competed by tRNA (Dong et al. 2000). In addition, a hyperactivating E803V mutation in the GCN2 kinase domain that impaired binding of the isolated HisRS + CTD fragment to the protein kinase domain enhanced tRNA binding to full-length GCN2 (Dong et al. 2000). Thus, it has been proposed that in the absence of a starvation signal, GCN2 is maintained in an OFF state by interdomain interactions between the protein kinase domain and the flanking HisRS and CTD region(s). Under conditions of amino acid starvation, uncharged tRNAs accumulate and bind to the HisRS (+CTD) regions, which releases these domains from their inhibitory interactions with the protein kinase domain (Fig. 2). It is proposed that tRNA binding to the HisRS domain in addition enables realignment of active-site residues in the protein kinase domain to enable conversion of GCN2 to its ON state.

In addition to amino acid starvation conditions, GCN2 is activated in yeast cells subjected to purine nucleotide limitation, high salinity, glucose limitation, rapamycin, and methylmethane sulfonate (MMS) and in mammalian cells by ultraviolet (UV) light (see Deng et al. 2002; Hinnebusch 2005). Interestingly, in all cases examined, the activation of GCN2 by these various stress conditions is dependent on the tRNA-binding properties (m2 motif) of the HisRS domain and on GCN2 dimerization mediated by the CTD. This has led to the idea that uncharged tRNA is the common activator of GCN2. Thus, the various stress conditions either indirectly activate GCN2 by altering the abundance of uncharged tRNA in the cell or modulate GCN2 sensitivity to uncharged tRNA such that the basal levels of uncharged tRNA normally present in the cell now activate the kinase. Consistent with this latter possibility, the hyperactivating E803V mutation in the kinase domain enhanced tRNA binding to full-length GCN2 (Dong et al. 2000). Moreover, rapamycin reduced phosphorylation of Ser-577 in yeast GCN2, and substitution of Ser-577 by Ala-577 enhanced tRNA binding to GCN2 (Garcia-Barrio et al. 2002; Cherkasova and Hinnebusch 2003). Thus, rapamycin activation of yeast GCN2 is likely to result from dephosphorylation of Ser-577 and the resultant increased binding affinity for uncharged tRNA. The notion that diverse stresses all activate GCN2 via uncharged tRNA binding to the HisRS domain is consistent with the common requirement for GCN1 and GCN20 for activation. These proteins form a complex that binds to both the ribosome and the amino terminus (RWD) of GCN2, and they are proposed to transfer the activating ligand, uncharged tRNA, from the ribosomal A site to GCN2 (for review, see Hinnebusch 2005).

HRI: Heme Regulation via Binding Sites Both Within and Amino-terminal to the Kinase Domain

HRI homologs have been identified in vertebrates and in *Schizosaccharomyces pombe* (Zhan et al. 2002), but not other yeasts. In addition, HRI-like proteins were identified in the genome sequences of *Anopheles gambiae* and *Bombyx mori*, but not other insects (Zhan et al. 2004). HRI is the principal kinase in erythroid cells where it is activated under conditions of heme deprivation (Lu et al. 2001). It is thought that the primary function of HRI in erythroid cells is to coordinate globin synthesis with available iron by blocking protein synthesis when heme is scarce. Interestingly, HRI is present in other cell types, and it has been reported to respond to other stress conditions including arsenite and cadmium exposure, heat shock, osmotic stress, and nitric oxide (Lu et al. 2001; McEwen et al. 2005).

Mammalian HRI undergoes a complex maturation process requiring the activity of Hsp90 and other chaperones (see Chen 2000). Prior to maturation, HRI is unresponsive to changes in heme levels, but following maturation, the kinase is repressed by heme. Two heme-binding sites have been mapped on HRI (Rafie-Kolpin et al. 2000). The first heme-binding site is located in the amino-terminal portion of HRI preceding the kinase domain (Fig. 1). This binding site appears to bind heme constitutively. In contrast, the second heme-binding site mapped to the kinase insertion sequence between β strands 4 and 5 in the amino-terminal kinase lobe. Reversible binding of heme to this latter site is associated with the regulation of the kinase activity. Mature HRI is a homodimer, and inhibition of HRI kinase activity by heme has been linked to intermolecular disulfide-bond formation between the two HRI protomers in a dimer (Yang et al. 1992). A recent study found that in the presence of heme, the amino-terminal domain interacts with the kinase domain, and this interaction is disrupted under conditions of heme deficiency that activate the kinase (Yun et al. 2005). Thus, it is proposed that in the presence of heme, HRI is maintained in an OFF state because of the interdomain interactions as well as intermolecular disulfide bonds that repress the kinase (Fig. 2). Under conditions of heme deprivation, the heme dissociates from the kinase insert region, the interdomain interactions as well as the intermolecular disulfide bonds are broken, and the kinase converts to its ON state.

eIF2α KINASE DOMAIN DIMERIZATION AND AUTOPHOSPHORYLATION

A common requirement for activation of the four eIF2α kinases is dimerization (Fig. 2). Both PKR and PERK are monomers in their latent state, and kinase activation is linked to dimerization or oligomerization, respectively, of their regulatory domains. GCN2 is a constitutive dimer, and following maturation, the dimeric form of HRI is responsive to regulation by heme. The importance of dimerization for PKR and PERK function is best illustrated by the substitution of heterologous dimerization domains in place of the regulatory domains of these proteins. Whereas the isolated kinase domain of PKR is nonfunctional for eIF2α phosphorylation both in vitro and in cells, a chimeric protein consisting of the dimerization domain from glutathione-S-transferase (GST) and the PKR kinase domain phosphorylated eIF2α both in vitro and in yeast cells (Ung et al. 2001; Dar and Sicheri 2002). Similarly, substitution of the amino-terminal region of PERK by a constitutive dimerization domain yielded a functional kinase (Liu et al. 2000). Moreover, fusion of the PKR kinase domain to the regulated dimerization domain from *E. coli* GyrB resulted in activation of eIF2α kinase activity in mammalian cells by the bifunctional ligand coumermycin that generates a dimeric linkage of two GyrB domains (Ung et al. 2001). Finally, fusion of the PERK kinase domain to the extracellular domain of CD4 or to an FK506-binding domain conferred PERK kinase activation by anti-CD4 antibodies and a dimeric form of FK506 (AP20187), respectively (Bertolotti et al. 2000; Lu et al. 2004). Thus, activation and catalytic activities of the eIF2α kinases are apparently dependent on dimerization mediated by domains extrinsic to the kinase domain. As recent studies, discussed below, revealed the critical importance of PKR kinase domain dimerization, it seems likely that the dimeric contacts intrinsic to the kinase domain are of insufficient strength to activate the kinase and that additional dimerization contacts are critical for eIF2α kinase activity. Thus, the regulatory domains of the eIF2α kinases may serve both to maintain the kinase in an inactive state under nonstress conditions, perhaps through autoinhibitory contacts in some cases, and to activate the kinase by providing critical dimeric contacts under activating conditions.

Protein kinase activation is commonly coupled with kinase domain autophosphorylation. When overexpressed in yeast or bacterial cells, PKR, PERK, and HRI were phosphorylated on a large number of sites (Zhang et al. 1998; Ma et al. 2001; Rafie-Kolpin et al. 2003). Mass spectrometry and phospho-specific antibody analyses mapped many of the phosphorylation sites on PKR and PERK; however, mutation of most of the PKR autophosphorylation sites did not affect kinase function (Taylor et al. 1996; Romano et al. 1998; Zhang et al. 2001; Su et al. 2006). It seems likely that many of these sites are phosphorylated at substoichiometric levels and thus may not represent bona fide phosphoregulatory sites on the kinases. Although mutation of some of the phosphorylation sites in PKR and HRI impaired kinase activity (e.g., Thr-451 or Tyr-293 in PKR and Thr-490 in HRI) (Zhang et al. 1998; Ma et al. 2001; Rafie-Kolpin et al. 2003; Su et al. 2006), it is possible that the mutated residues are critical for catalytic function rather than serving as a phosphorylation site. In the PKR crystal structure, the only stoichiometrically phosphorylated residue is Thr-446 (Dar et al. 2005). This residue, located in the kinase domain activation segment, is also a site of autophosphorylation in GCN2, PERK, and HRI (Romano et al. 1998; Ma et al. 2001; Rafie-Kolpin et al. 2003). The corresponding Thr-882 appears to be the predominant autophosphorylation site on GCN2 in vitro (Romano et al. 1998). Autophosphorylation of PKR on Thr-446 is linked to kinase activation, as it is dependent on the dsRBMs and correlates with kinase domain dimerization (Dey et al. 2005a). Mutation of Thr-446 in PKR to alanine severely impairs kinase activity (Romano et al. 1998; Dey et al. 2005a), and likewise, mutation of the corresponding Thr-882 residue in GCN2 (Romano et al. 1998) and that of the Thr-485 residue in HRI (Rafie-Kolpin et al. 2003) impair translational regulation and kinase function. Thus, autophosphorylation of the eIF2α kinases on the conserved threonine in the activation segment is a critical step in the activation pathway for the kinases.

SUBSTRATE RECOGNITION BY PKR

Peptides Encompassing the Ser-51 Acceptor Site of eIF2α Are Poor PKR Substrates

The peptide sequence encompassing the Ser-51 site in eIF2α does not appear to contribute greatly to the overall ability of PKR to phosphorylate eIF2α. Most compelling is the finding that PKR phosphorylates full-length eIF2α approximately 1000 times more efficiently than short model peptide substrates derived from eIF2α (Mellor and Proud 1991). In further support, PKR readily phosphorylates eIF2α mutants in which the Ser-51 position is replaced by threonine and more unexpectedly tyrosine (Lu et al. 1999). Notwithstanding, a phospho-acceptor site consensus for basic residues, especially arginines, three and four residues carboxy-terminal to serine was delineated for PKR (Proud et al. 1991). In addition, discrimination against basic residues amino-terminal to Ser-51 was observed (Proud et al. 1991). This preference closely resembles the native sequence of eIF2α around the Ser-51 site (SELS$_{51}$RRR), which is highly conserved among eIF2α orthologs. Demonstrating that the fold of eIF2α and not just a linear peptide epitope is an important determinant for recognition is the observation that only native and not heat-denatured eIF2α is efficiently phosphorylated by the PKR-related kinase HRI (Kramer and Hardesty 1981; Dey et al. 2005b).

Structural Analysis of eIF2α and Virus-encoded Pseudosubstrate K3L

The structure of an amino-terminal fragment (residues 3–182) of human eIF2α revealed a single autonomously folding domain composed of a five-stranded β-barrel buttressed to a helical bundle (Nonato et al. 2002). As a result of a protease contaminant, a labile site in the vicinity of Ser-51 was cleaved, and this region could not be visualized. In a second structure of eIF2α (residues 1–175 from *Saccharomyces cerevisiae*), the Ser-51 acceptor site was resolved and appeared to be hidden from protein kinase attack through its integration into a well-defined 3–10 helix, turn, 3–10 helix fold (see eIF2α in pink in Fig. 3) (Dhaliwal and Hoffman 2003). A recent NMR structure of full-length human eIF2α has revealed the same well-defined elements within the amino-terminal domain and, additionally, the presence of an autonomously folding carboxy-terminal domain (Ito et al. 2004). Comparison of the NMR spectra of wild type and a phospho-mimetic Ser-51Asp mutant of eIF2α did not reveal significant conformational differences. This suggested that a conformational change in eIF2α, resulting solely from Ser-51 phosphorylation, is not likely to have a role in the downstream signaling response (Ito et al. 2004).

The vaccinia virus protein, K3L, shares striking similarity to a portion of the amino-terminal domain of eIF2α. This observation is consistent with K3L functioning as a competitive inhibitor of eIF2α binding to PKR. The structure of K3L consists solely of a five-stranded β-barrel with a helix insert between β strands 3 and 4 (Dar and Sicheri 2002). Comparison of isolated K3L and eIF2α structures revealed major differences in the conformations of the helix inserts and the β1–β2 linker, which together lie close in three-dimensional space. These differences likely

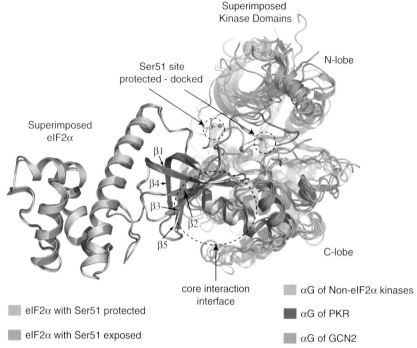

Figure 3. Bipartite substrate recognition of eIF2α by PKR. Helix αG in PKR binds the five-stranded β-barrel core of eIF2α (labeled *1–5*), and the active site of PKR interacts with the Ser-51 site within eIF2α. Displayed are the structures of isolated eIF2α and several other kinase domains superimposed onto the PKR–eIF2α complex. The structures of eIF2α with Ser-51 in its protected state (1Q46) and eIF2α with Ser-51 exposed for phosphorylation are shown in *pink* and *blue*, respectively. The positions of Ser-51 in its protected state and its anticipated position for phosphorylation within the active site of PKR are depicted in a *yellow/red* ball and stick representation. The structures of the kinase domains of PKR (2A1A and 2A19) and GCN2 (1ZY5) are superimposed with the non-eIF2α kinases, PKA (1JBP), Aurora A (1MQ4), and the Death-associated protein kinase (1IG1). As can be deduced, helix αG in PKR (*red*) and GCN2 (*green*) adopts a noncanonical position from that which has been observed in other protein kinases (*orange*).

reflect the fact that K3L has evolved to function as an inhibitor, rather than as a substrate, of PKR.

Residues conserved between the K3L and eIF2α protein families map to a surface of the β-barrel core that was hypothesized to function as a PKR docking motif (Dar and Sicheri 2002). Whereas most kinase targeting motifs are highly flexible linear sequences, the proposed PKR docking motif in eIF2α and K3L is assembled from dispersed residues in the K3L/eIF2α primary structure, which converge in space through the globular fold of the protein and are far removed ($\gg 20$ Å) from the Ser-51 acceptor site in eIF2α (Dar and Sicheri 2002).

PKR Recognition Determinants Map to a Common Surface on K3L and eIF2α

Mutational analyses of eIF2α and K3L revealed two distinct regions as important determinants for a functional interaction with PKR. The region of strongest similarity between K3L and eIF2α resides within strands β4, β5, and the intervening linker (Fig. 3). Mutations within this conserved K3L/eIF2α surface blocked the ability of K3L to suppress the slow growth/toxic phenotype resulting from PKR expression in yeast (Kawagishi-Kobayashi et al. 1997). Furthermore, substitution of Asp-83 with alanine in the corresponding surface of eIF2α eliminated Ser-51 phosphorylation by PKR and GCN2 both in vitro and in vivo (Dey et al. 2005b). Intriguingly, mutation of His-47 in K3L to arginine (in the projected vicinity of the Ser-51 acceptor site of eIF2α) had the opposite effect and created a super potent inhibitor (Kawagishi-Kobayashi et al. 1997).

Detailed kinetic analyses revealed that K3L inhibits PKR phosphorylation of model peptide substrates through a noncompetitive mechanism, but the phosphorylation of eIF2α by a competitive mechanism (Dar and Sicheri 2002). Analysis of K3L mutants revealed two categories of behavior. One class of mutants greatly reduced the ability of K3L to bind and inhibit PKR (as reflected by higher K_i and lower I_{max} values). These mutants map to the K3L-eIF2α conserved surface. The second class of mutants had little to no effect on the ability of K3L to bind PKR but reduced the ability of K3L to maximally inhibit PKR phosphorylation of peptide substrate. This class of mutants mapped to the helix insert region of K3L. Taken together, the kinetic data can be reconciled by a model in which the helix insert, housing the Ser-51 site in eIF2α and the specialized inhibitory structure in K3L, engages the active site of PKR, whereas the eIF2α/K3L conserved surface engages a remote site on the kinase domain. Interestingly, these same structural elements in eIF2α were found to be important for the binding and inhibition of eIF2B by phosphorylated eIF2, suggesting that eIF2B and the eIF2α kinases recognize the same or overlapping surfaces on eIF2α (Dey et al. 2005b).

Structural Analysis of the PKR–eIF2α Complex

Crystal structures of active PKR kinase domains in complex with eIF2α have resolved the molecular basis for bipartite recognition (see Fig. 3) (Dar et al. 2005). The complexes also revealed a mode of kinase domain dimerization that is of likely relevance to the functioning of the entire eIF2α kinase family. The catalytic domain of PKR adopts a symmetrical N-lobe to N-lobe dimer configuration, with the eIF2α-binding site located on the opposite facing surface of the carboxy-terminal lobe (see Fig. 2 for a cartoon representation of the PKR–eIF2α complex). The PKR kinase domain possesses a bilobal structure characteristic of all eukaryotic protein kinases (for review, see Johnson et al. 1996; Huse and Kuriyan 2002). The active site is located at a cleft between a β-rich amino-terminal lobe and an α-helical carboxy-terminal lobe (see Fig. 4) (Huse and Kuriyan 2002). In an ATP analog (AMP-PNP)-bound form of PKR, the nucleotide is coordinated in a productive manner, consistent with the finding that the PKR construct is competent for autophosphorylation and eIF2α phosphorylation in vitro. In the PKR–eIF2α complexes, Thr-446 appears to be stoichiometrically

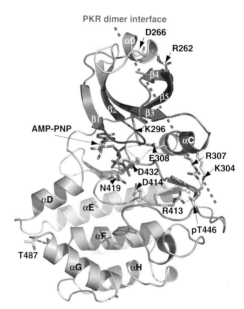

Figure 4. Frontal view of the PKR kinase domain. Ribbon representation of the PKR kinase domain highlighting functionally important residues within the catalytic cleft and the eIF2α-binding site. (*Purple*) N-lobe of PKR; (*green*) C-lobe of PKR. (*Yellow*) Helix α0 (involved in dimerization); (*red*) G-loop (involved in ATP coordination); (*orange*) activation segment (the regulatory element containing the T446 autophosphorylation site). Residues involved in Mg-ATP binding, the phosphotransfer reaction (K296, E308, D414, N419, D432), T446–P coordination (K304, R307, R413), and dimerization (R262, D266) are highlighted in ball and stick representations. (*Pink balls*) Mg ions. PKR dimerization is mediated by the surface highlighted by the *red dashed line*.

phosphorylated, and the activation segments adopt a fully ordered productive conformation.

Structures of active and repressed states of protein kinases have revealed that the modulation of the position of helix αC within the N-lobe and the activation segment conformation within the C-lobe can serve to regulate catalytic function (for review, see Huse and Kuriyan 2002). In the PKR crystal structures, helix αC composes an integral part of the dimerization interface; in addition, the αC residues Lys-304 and Arg-307, together with Arg-413 of the RD sequence motif, coordinate the phosphomoiety of Thr-446 in the kinase activation segment (Fig. 4). These structural features raise the intriguing possibility that dimerization-mediated modulation of helix αC and activation segment conformation serve to regulate catalytic function. A full understanding of how the absence of dimerization and activation segment phosphorylation directly impinges on PKR catalytic function awaits the structure of the down-regulated form of PKR.

Comparison of PKR in its active conformation with structures of other active kinases in the protein data bank (there are currently 85 kinase structures with less than 70% sequence identity deposited in the Protein Data Bank or PDB) reveals two gross differences: a unique conformation of the P + 1 loop (a subelement of the activation segment) and a noncanonical position and length of helix αG. The P + 1 loop of protein kinases generally serves as a platform for the phospho-acceptor sequence of the substrate and imparts a preference for serine/threonine versus tyrosine (Nolen et al. 2004). In PKR, the P + 1 portion of the activation segment deviates from the orientations characteristic of either serine/threonine or tyrosine kinases (Dar et al. 2005). As the Ser-51 acceptor site of eIF2α was disordered in the PKR–eIF2α complexes, it is unclear how its mode of interaction with the unique P + 1 loop conformation relates to the ability of PKR to phosphorylate both serine/threonine and tyrosine containing peptides.

Helix αG in PKR adopts a unique conformation not observed previously in other protein kinase structures (see Fig. 3). Specifically, helix αG in PKR is one turn longer and tilted 40° from its expected orientation (Dar et al. 2005). These features enhance the accessibility of the amino-terminal end of helix αG, providing a binding site for the highly conserved surface of eIF2α and K3L described above. In contrast, the reciprocal surface on PKR is not highly conserved with only 2 of 13 contact residues invariant among the eIF2α kinases. The absence of conservation on the contact surface of PKR may be explained in part by the fact that many of the interactions involve main-chain atoms of PKR.

In addition to the noncanonical position of helix αG, which is the result of a short αF-αG linker, the eIF2α kinases share the αG sequence motif $T_{487}xxE_{490}xxxx\phi_{495}xx\phi_{498}R_{499}$ (where ϕ represents hydrophobic residues phenylalanine, valine, or leucine). An in-depth mutational analysis revealed that whereas positions 490 and 495 can tolerate several nonconservative substitutions, mutation of Thr-487 (except to serine) blocks eIF2α phosphorylation (Dey et al. 2005a). Thus, the key eIF2α kinase determinants appear to be the short αF-αG linker in combination with a threonine (or perhaps serine) at the 487 position (see Fig. 4 for the location of Thr-487 in PKR). The other conserved determinants within helix αG are not essential but most likely reveal some insight into the common ancestry of the eIF2α kinases.

It is noteworthy that in the isolated eIF2α structures, the Ser-51 site is not accessible to an attacking protein kinase, and if docked onto PKR, Ser-51 would lie 17 Å away from the active site of the kinase. Although the noncanonical position of helix αG brings the Ser-51 residue into the vicinity of the PKR active site (Fig. 3), the fact that the Ser-51-encompassing region was disordered in the two PKR–eIF2α complexes, but ordered in the two apo eIF2α structures, suggested that PKR binding allosterically induced a conformational change in eIF2α (Dar et al. 2005). This would enable Ser-51 to engage the kinase active site in an extended conformation. If true, this could serve to prevent inappropriate phosphorylation of Ser-51 by protein kinases with matching peptide phosphorylation preferences.

COUPLING AMONG PKR DIMERIZATION, AUTOPHOSPHORYLATION, AND SUBSTRATE RECOGNITION

Mapping of residues selectively conserved among eIF2α kinases onto the PKR structure revealed a striking degree of conservation at the dimer interface and an unexpectedly low level of conservation within the eIF2α substrate-binding site. Mutational analyses of PKR revealed that dimerization, activation segment phosphorylation, and substrate recognition were mechanistically interconnected (Dey et al. 2005a). Because the PKR dimerization and substrate-binding sites are physically remote, a plausible "coupling" role for the phosphorylated activation segment was proposed. Analytical centrifugation experiments demonstrated that phosphorylation of Thr-446 promoted a transition of the kinase domain from a monomeric state to a dimeric state, and surface plasmon resonance (SPR) experiments demonstrated that Thr-446 phosphorylation was the primary regulatory determinant for eIF2α/K3L binding. In the absence of Thr-446 phosphorylation, the PKR kinase domain does not bind eIF2α even when dimerized by fusion to GST.

Mutations identified as having the ability to activate an isolated PKR kinase domain mapped in close proximity to the dimer interface observed in the crystal structures of PKR. These mutations enhanced kinase domain dimerization (Dey et al. 2005a). The results supported the notion that the promotion of a specific dimer conformation as visualized by the PKR crystal structure is mechanistically important for the PKR activation mechanism. In further support of this idea, other dimer interface mutations in full-length PKR were identified that disrupted autophosphorylation on Thr-446 and blocked PKR toxicity in yeast and eIF2α phosphorylation in vitro (Dey et al. 2005a). These results demonstrated the essential requirement for catalytic domain dimerization in the PKR activation mechanism. Further data in support of the back-to-back kinase domain arrangement observed in the crystal structures were provided by the PKR R262D/D266R mutant in which the polarity of a conserved intermolecular salt bridge was reversed (see helix α0 in Fig. 4 for residue positions). In these experiments, both wild type and charge-reversed double mutants, but not single-site mutants that abolish salt bridge formation, displayed wild-type function (Dey et al. 2005a).

COMPARISON OF PKR AND GCN2 CATALYTIC DOMAIN STRUCTURES

Comparison of an isolated GCN2 kinase domain structure from yeast (Padyana et al. 2005) with the structure of the human PKR kinase domain in complex with eIF2α (Dar et al. 2005) reveals several insights of potential relevance to the regulation of the eIF2α kinase family (see Fig. 5). Like PKR, the GCN2 kinase domain adopts a dimer configuration in several different space groups and crystallographic environments. Both the GCN2 and PKR dimers are formed through interactions between residues on the same planar surface of the N-lobes. However, the orientation of the kinase domain dimer in GCN2 is dissimilar from that observed for PKR. In particular, the protomers in GCN2 are arranged antiparallel to each other relative to the parallel arrangement of PKR (the two dimer configurations differ by a 180° rotation normal to the planar contact surface; Fig. 5A,B).

In the PKR structures, the absence of an eIF2α-binding partner was correlated with major disorder in the C-lobe, including the substrate-binding site within helix αG (Dar et al. 2005). In contrast, the isolated GCN2 structures were fully modeled with the exception of sections within the activation segment including the phosphoregulatory site, Thr-882, and a region within the β4–β5 linker that corresponds to the site of

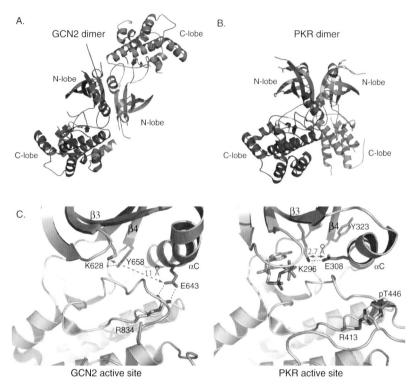

Figure 5. Repressed and activated states of GCN2 and PKR, respectively. (*A*) Ribbon diagram of the GCN2 dimer (1ZY5); protomers are colored red and blue. (*B*) Ribbon diagram of the PKR dimer (2A19); protomers are colored red and blue. (*C*) Active sites of GCN2 (*left*) and PKR (*right*). Equivalent residues from GCN2 and PKR, based on structural alignment, have been colored identically. The GCN2 active site is characterized by interactions between Lys-628 and Tyr-658 and also Glu-643 and Arg-834. Tyr-323 in PKR (equivalent to Tyr-658 in GCN2) is flipped away from the active site and forms a key dimerization contact. Arg-413 in PKR (equivalent to Arg-834 in GCN2) serves to coordinate Thr-446–P. The conserved salt bridge, which is indicative of active kinase, is formed in PKR (K296-E308 distance is 2.7 Å) but is not evident in the GCN2 structure (K628-E643 distance is 11.0 Å). Switching from the repressed to the active states in GCN2 and PKR may involve conversion between the left and right images, respectively, such that the highlighted interactions are transitory.

an engineered deletion (Padyana et al. 2005). Despite relatively high atomic B factors, helix αG of GCN2 was fully modeled in the identical noncanonical position as that observed for helix αG in the eIF2α-bound PKR structures (Fig. 3). This suggested that in GCN2, helix αG displacement occurs independent of eIF2α binding. Hence, unlike PKR, the ability of GCN2 to bind eIF2α may not be regulated.

Each of the GCN2 structures was deficient for activation segment phosphorylation (Thr-882 and portions of the activation segment were disordered and hence were not modeled in the GCN2 structures), and the corresponding active sites displayed features nonconducive to optimal catalytic efficiency. The most pronounced feature includes the absence of a salt bridge formed between subdomains 2 and 3 corresponding to the kinase invariant residues Lys-628 and Glu-643 in GCN2 (the two residues are 11 Å apart; Fig. 5C). These residues project from strand β3 and helix αC of the kinase N-lobe, respectively. In the active state of protein kinases, the glutamic acid–lysine salt bridge functions critically to laterally coordinate the α and β phosphates of ATP (Johnson et al. 1996) (in PKR, the two residues Lys-296 and Glu-308 are 2.7 Å apart; Fig. 5C). In the GCN2 structures, Lys-628 instead interacts with Tyr-658 through a water-mediated H-bond (Padyana et al. 2005). Interestingly, in the PKR structure, the Tyr-658 equivalent (Tyr-323) is involved in a dimer interface contact (Dar et al. 2005). As the GCN2 conformation appears to be nonproductive, a rearrangement must be achieved to attain a catalytically competent state. On the basis of its close relation to PKR, this rearrangement in GCN2 may occur in response to autophosphorylation of the activation segment residue Thr-882 in conjunction with a transition from an antiparallel to a parallel dimer configuration. Specifically, a transition to a parallel dimer configuration would displace Tyr-658 from an inward orientation to an outward orientation (Fig. 5C). This would break its interaction with Lys-628, leaving the lysine side chain available for alternate interactions. Coincidentally, activation segment phosphorylation on Thr-882 would break the interaction between Arg-834 and Glu-643, by favoring an alternate interaction of Thr-882–P with Arg-834. This would leave Glu-643 available for interaction with Lys-628, thereby achieving a catalytically competent state (as in PKR, Fig. 5C). A structure of GCN2 in its fully active state would go a long way to proving whether the aforementioned hypothetical model is correct. However, precedents for this sort of behavior have been observed for other kinase systems including the Src and cyclin-dependent families of protein kinases (De Bondt et al. 1993; Jeffrey et al. 1995; Sicheri et al. 1997).

SUMMARY

With these recent structural studies on GCN2 and PKR, we have gained new insights into the nature of both the repressed (GCN2) and the activated (PKR) states of the eIF2α kinase domain, and we can propose a consensus model for eIF2α kinase activation. For activation of PKR, binding of dsRNA presumably relieves an autoinhibitory interaction and promotes kinase dimerization. In contrast, activation of GCN2 (and the heme-sensitive form of HRI) involves binding of the activating ligand to a constitutive kinase dimer. Following dimerization (or binding of the activating ligand), specific intermolecular contacts as revealed in the PKR crystal structure mediate kinase domain dimerization and trigger autophosphorylation on the kinase domain activation segment. Finally, binding of eIF2α to the activated kinase domain apparently induces a conformational change that exposes Ser-51 to the active site of the kinase. Goals for future investigations include determining how well the active and inactive conformations observed in the PKR and GCN2 structures are conserved among the four eIF2α kinases; testing whether the activation and substrate recognition mechanisms outlined for PKR and GCN2 are shared by the other eIF2α kinases and identifying any changes in the mechanisms; determining how ligand interactions with the various eIF2α kinase regulatory domains evoke the necessary structural rearrangements to activate the kinases; and finally, determining how phosphorylation of eIF2α alters its interaction with eIF2B.

ACKNOWLEDGMENTS

This work was supported in part by the Intramural Research Program of the NICHD, National Institutes of Health (T.E.D.) and by grants from the National Cancer Institute of Canada and the Canadian Institutes of Health Research (F.S.).

REFERENCES

Ben-Asouli Y., Banai Y., Pel-Or Y., Shir A., and Kaempfer R. 2002. Human interferon-gamma mRNA autoregulates its translation through a pseudoknot that activates the interferon-inducible protein kinase PKR. *Cell* **108**: 221–232.

Bertolotti A., Zhang Y., Hendershot L.M., Harding H.P., and Ron D. 2000. Dynamic interaction of BiP and ER stress transducers in the unfolded-protein response. *Nat. Cell Biol.* **2**: 326–332.

Bevilacqua P.C., George C.X., Samuel C.E., and Cech T.R. 1998. Binding of the protein kinase PKR to RNAs with secondary structure defects: Role of the tandem A-G mismatch and noncontiguous helixes. *Biochemistry* **37**: 6303–6316.

Carpick B.W., Graziano V., Schneider D., Maitra R.K., Lee X., and Williams B.R. 1997. Characterization of the solution complex between the interferon-induced, double-stranded RNA-activated protein kinase and HIV-I trans-activating region RNA. *J. Biol. Chem.* **272**: 9510–9516.

Chen J.-J. 2000. Heme-regulated eIF2a kinase. In *Translational control of gene expression* (ed.

N. Sonenberg et al.), pp. 529–546. Cold Spring Harbor Laboratory Press, Cold Spring Harbor, New York.

Cherkasova V.A. and Hinnebusch A.G. 2003. Translational control by TOR and TAP42 through dephosphorylation of eIF2alpha kinase GCN2. *Genes Dev.* **17:** 859–872.

Cosentino G.P., Venkatesan S., Serluca F.C., Green S.R., Mathews M.B., and Sonenberg N. 1995. Double-stranded-RNA-dependent protein kinase and TAR RNA-binding protein form homo- and heterodimers in vivo. *Proc. Natl. Acad. Sci.* **92:** 9445–9449.

Credle J.J., Finer-Moore J.S., Papa F.R., Stroud R.M., and Walter P. 2005. On the mechanism of sensing unfolded protein in the endoplasmic reticulum. *Proc. Natl. Acad. Sci.* **102:** 18773–18784.

Dar A.C. and Sicheri F. 2002. X-ray crystal structure and functional analysis of vaccinia virus K3L reveals molecular determinants for PKR subversion and substrate recognition. *Mol. Cell* **10:** 295–305.

Dar A.C., Dever T.E., and Sicheri F. 2005. Higher-order substrate recognition of eIF2alpha by the RNA-dependent protein kinase PKR. *Cell* **122:** 887–900.

Davis S. and Watson J.C. 1996. In vitro activation of the interferon-induced, double-stranded RNA-dependent protein kinase PKR by RNA from the 3′ untranslated regions of human alpha-tropomyosin. *Proc. Natl. Acad. Sci.* **93:** 508–513.

De Bondt H.L., Rosenblatt J., Jancarik J., Jones H.D., Morgan D.O., and Kim S.H. 1993. Crystal structure of cyclin-dependent kinase 2. *Nature* **363:** 595–602.

Deng J., Harding H.P., Raught B., Gingras A.C., Berlanga J.J., Scheuner D., Kaufman R.J., Ron D., and Sonenberg N. 2002. Activation of GCN2 in UV-irradiated cells inhibits translation. *Curr. Biol.* **12:** 1279–1286.

Dey M., Cao C., Dar A.C., Tamura T., Ozato K., Sicheri F., and Dever T.E. 2005a. Mechanistic link between PKR dimerization, autophosphorylation, and eIF2alpha substrate recognition. *Cell* **122:** 901–913.

Dey M., Trieselmann B., Locke E.G., Lu J., Cao C., Dar A.C., Krishnamoorthy T., Dong J., Sicheri F., and Dever T.E. 2005b. PKR and GCN2 kinases and guanine nucleotide exchange factor eukaryotic translation initiation factor 2B (eIF2B) recognize overlapping surfaces on eIF2alpha. *Mol. Cell. Biol.* **25:** 3063–3075.

Dhaliwal S. and Hoffman D.W. 2003. The crystal structure of the N-terminal region of the alpha subunit of translation initiation factor 2 (eIF2alpha) from *Saccharomyces cerevisiae* provides a view of the loop containing serine 51, the target of the eIF2alpha-specific kinases. *J. Mol. Biol.* **334:** 187–195.

Dong J., Qiu H., Garcia-Barrio M., Anderson J., and Hinnebusch A.G. 2000. Uncharged tRNA activates GCN2 by displacing the protein kinase moiety from a bipartite tRNA-binding domain. *Mol. Cell* **6:** 269–279.

Farrell P.J., Balkow K., Hunt T., Jackson R.J., and Trachsel H. 1977. Phosphorylation of initiation factor eIF-2 and the control of reticulocyte protein synthesis. *Cell* **11:** 187–200.

Garcia-Barrio M., Dong J., Cherkasova V.A., Zhang X., Zhang F., Ufano S., Lai R., Qin J., and Hinnebusch A.G. 2002. Serine 577 is phosphorylated and negatively affects the tRNA binding and eIF2alpha kinase activities of GCN2. *J. Biol. Chem.* **277:** 30675–30683.

Green S.R. and Mathews M.B. 1992. Two RNA-binding motifs in double-stranded RNA-activated protein kinase, DAI. *Genes Dev.* **6:** 2478–2490.

Harding H.P., Zhang Y., and Ron D. 1999. Protein translation and folding are coupled by an endoplasmic-reticulum-resident kinase. *Nature* **397:** 271–274.

Hinnebusch A.G. 2005. Translational regulation of GCN4 and the general amino acid control of yeast. *Annu. Rev. Microbiol.* **59:** 407–450.

Hunter T., Hunt T., Jackson R.J., and Robertson H.D. 1975. The characteristics of inhibition of protein synthesis by double-stranded ribonucleic acid in reticulocyte lysates. *J. Biol. Chem.* **250:** 409–417.

Huse M. and Kuriyan J. 2002. The conformational plasticity of protein kinases. *Cell* **109:** 275–282.

Ito T., Marintchev A., and Wagner G. 2004. Solution structure of human initiation factor eIF2alpha reveals homology to the elongation factor eEF1B. *Structure* **12:** 1693–1704.

Jeffrey P.D., Russo A.A., Polyak K., Gibbs E., Hurwitz J., Massague J., and Pavletich N.P. 1995. Mechanism of CDK activation revealed by the structure of a cyclinA-CDK2 complex. *Nature* **376:** 313–320.

Johnson L.N., Noble M.E., and Owen D.J. 1996. Active and inactive protein kinases: Structural basis for regulation. *Cell* **85:** 149–158.

Kawagishi-Kobayashi M., Silverman J.B., Ung T.L., and Dever T.E. 1997. Regulation of the protein kinase PKR by the vaccinia virus pseudosubstrate inhibitor K3L is dependent on residues conserved between the K3L protein and the PKR substrate eIF2alpha. *Mol. Cell. Biol.* **17:** 4146–4158.

Kostura M. and Mathews M.B. 1989. Purification and activation of the double-stranded RNA-dependent eIF-2 kinase DAI. *Mol. Cell. Biol.* **9:** 1576–1586.

Kramer G. and Hardesty B. 1981. Phosphorylation reactions that influence the activity of eIF-2. *Curr. Top. Cell. Regul.* **20:** 185–203.

Lemaire P.A., Lary J., and Cole J.L. 2005. Mechanism of PKR activation: Dimerization and kinase activation in the absence of double-stranded RNA. *J. Mol. Biol.* **345:** 81–90.

Liu C.Y., Schroder M., and Kaufman R.J. 2000. Ligand-independent dimerization activates the stress response kinases IRE1 and PERK in the lumen of the endoplasmic reticulum. *J. Biol. Chem.* **275:** 24881–24885.

Lu J., O'Hara E.B., Trieselmann B.A., Romano P.R., and Dever T.E. 1999. The interferon-induced double-stranded RNA-activated protein kinase PKR will phosphorylate serine, threonine, or tyrosine at residue 51 in eukaryotic initiation factor 2alpha. *J. Biol. Chem.* **274:** 32198–32203.

Lu L., Han A.P., and Chen J.J. 2001. Translation initiation control by heme-regulated eukaryotic initiation factor 2alpha kinase in erythroid cells under cytoplasmic stresses. *Mol. Cell. Biol.* **21:** 7971–7980.

Lu P.D., Jousse C., Marciniak S.J., Zhang Y., Novoa I., Scheuner D., Kaufman R.J., Ron D., and Harding H.P. 2004. Cytoprotection by pre-emptive conditional phosphorylation of translation initiation factor 2. *EMBO J.* **23:** 169–179.

Ma K., Vattem K.M., and Wek R.C. 2002. Dimerization and release of molecular chaperone inhibition facilitate activation of eukaryotic initiation factor-2 kinase in response to endoplasmic reticulum stress. *J. Biol. Chem.* **277:** 18728–18735.

Ma Y., Lu Y., Zeng H., Ron D., Mo W., and Neubert T.A. 2001. Characterization of phosphopeptides from protein digests using matrix-assisted laser desorption/ionization time-of-flight mass spectrometry and nanoelectrospray quadrupole time-of-flight mass spectrometry. *Rapid Commun. Mass Spectrom.* **15:** 1693–1700.

Manche L., Green S.R., Schmedt C., and Mathews M.B. 1992. Interactions between double-stranded RNA regulators and the protein kinase DAI. *Mol. Cell. Biol.* **12:** 5238–5248.

McEwen E., Kedersha N., Song B., Scheuner D., Gilks N., Han A., Chen J.J., Anderson P., and Kaufman R.J. 2005. Heme-regulated inhibitor kinase-mediated phosphorylation of eukaryotic translation initiation factor 2 inhibits translation, induces stress granule formation, and mediates survival upon arsenite exposure. *J. Biol. Chem.* **280:** 16925–16933.

Mellor H. and Proud C.G. 1991. A synthetic peptide substrate for initiation factor-2 kinases. *Biochem. Biophys. Res. Commun.* **178:** 430–437.

Mohrle J.J., Zhao Y., Wernli B., Franklin R.M., and Kappes B. 1997. Molecular cloning, characterization and localization of PfPK4, an eIF-2alpha kinase-related enzyme from the malarial parasite *Plasmodium falciparum*. *Biochem. J.* **328:** 677–687.

Nanduri S., Rahman F., Williams B.R., and Qin J. 2000. A dynamically tuned double-stranded RNA binding mechanism for the activation of antiviral kinase PKR. *EMBO J.* **19:** 5567–5574.

Nolen B., Taylor S., and Ghosh G. 2004. Regulation of protein kinases; controlling activity through activation segment conformation. *Mol. Cell* **15:** 661–675.

Nonato M.C., Widom J., and Clardy J. 2002. Crystal structure of the N-terminal segment of human eukaryotic translation initiation factor 2alpha. *J. Biol. Chem.* **21:** 21.

Ortega L.G., McCotter M.D., Henry G.L., McCormack S.J., Thomis D.C., and Samuel C.E. 1996. Mechanism of interferon action. Biochemical and genetic evidence for the intermolecular association of the RNA-dependent protein kinase PKR from human cells. *Virology* **215:** 31–39.

Padyana A.K., Qiu H., Roll-Mecak A., Hinnebusch A.G., and Burley S.K. 2005. Structural basis for autoinhibition and mutational activation of eukaryotic initiation factor 2alpha protein kinase GCN2. *J. Biol. Chem.* **280:** 29289–29299.

Patel R.C. and Sen G.C. 1998. Requirement of PKR dimerization mediated by specific hydrophobic residues for its activation by double-stranded RNA and its antigrowth effects in yeast. *Mol. Cell. Biol.* **18:** 7009–7019.

Proud C.G., Colthurst D.R., Ferrari S., and Pinna L.A. 1991. The substrate specificity of protein kinases which phosphorylate the alpha subunit of eukaryotic initiation factor 2. *Eur. J. Biochem.* **195:** 771–779.

Qiu H., Garcia-Barrio M.T., and Hinnebusch A.G. 1998. Dimerization by translation initiation factor 2 kinase GCN2 is mediated by interactions in the C-terminal ribosome-binding region and the protein kinase domain. *Mol. Cell. Biol.* **18:** 2697–2711.

Qiu H., Dong J., Hu C., Francklyn C.S., and Hinnebusch A.G. 2001. The tRNA-binding moiety in GCN2 contains a dimerization domain that interacts with the kinase domain and is required for tRNA binding and kinase activation. *EMBO J.* **20:** 1425–1438.

Rafie-Kolpin M., Han A.P., and Chen J.J. 2003. Autophosphorylation of threonine 485 in the activation loop is essential for attaining eIF2alpha kinase activity of HRI. *Biochemistry* **42:** 6536–6544.

Rafie-Kolpin M., Chefalo P.J., Hussain Z., Hahn J., Uma S., Matts R.L., and Chen J.J. 2000. Two heme-binding domains of heme-regulated eukaryotic initiation factor-2alpha kinase. N terminus and kinase insertion. *J. Biol. Chem.* **275:** 5171–5178.

Romano P.R., Garcia-Barrio M.T., Zhang X., Wang Q., Taylor D.R., Zhang F., Herring C., Mathews M.B., Qin J., and Hinnebusch A.G. 1998. Autophosphorylation in the activation loop is required for full kinase activity in vivo of human and yeast eukaryotic initiation factor 2alpha kinases PKR and GCN2. *Mol. Cell. Biol.* **18:** 2282–2297.

Rothenburg S., Deigendesch N., Dittmar K., Koch-Nolte F., Haag F., Lowenhaupt K., and Rich A. 2005. A PKR-like eukaryotic initiation factor 2alpha kinase from zebrafish contains Z-DNA binding domains instead of dsRNA binding domains. *Proc. Natl. Acad. Sci.* **102:** 1602–1607.

Schmedt C., Green S.R., Manche L., Taylor D.R., Ma Y., and Mathews M.B. 1995. Functional characterization of the RNA-binding domain and motif of the double-stranded RNA-dependent protein kinase DAI (PKR). *J. Mol. Biol.* **249:** 29–44.

Sharp T.V., Moonan F., Romashko A., Joshi B., Barber G.N., and Jagus R. 1998. The vaccinia virus E3L gene product interacts with both the regulatory and the substrate binding regions of PKR: Implications for PKR autoregulation. *Virology* **250:** 302–315.

Shi Y., Vattem K.M., Sood R., An J., Liang J., Stramm L., and Wek R.C. 1998. Identification and characterization of pancreatic eukaryotic initiation factor 2 alpha-subunit kinase, PEK, involved in translational control. *Mol. Cell. Biol.* **18:** 7499–7509.

Sicheri F., Moarefi I., and Kuriyan J. 1997. Crystal structure of the Src family tyrosine kinase Hck. *Nature* **385:** 602–609.

Su Q., Wang S., Baltzis D., Qu L.-K., Wong A.H.-T., and Koromilas A.E. 2006. Tyrosine phosphorylation acts as a molecular switch to full-scale activation of the eIF2α RNA-dependent protein kinase. *Proc. Natl. Acad. Sci.* **103:** 63–68.

Sullivan W.J., Jr., Narasimhan J., Bhatti M.M., and Wek R.C. 2004. Parasite-specific eIF2 (eukaryotic initiation factor-2) kinase required for stress-induced translation control. *Biochem J.* **380:** 523–531.

Taylor D.R., Lee S.B., Romano P.R., Marshak D.R., Hinnebusch A.G., Esteban M., and Mathews M.B. 1996. Autophosphorylation sites participate in the activation of the double-stranded-RNA-activated protein kinase PKR. *Mol. Cell. Biol.* **16:** 6295–6302.

Tian B., Bevilacqua P.C., Diegelman-Parente A., and Mathews M.B. 2004. The double-stranded-RNA-binding motif: Interference and much more. *Nat. Rev. Mol. Cell Biol.* **5:** 1013–1023.

Ung T.L., Cao C., Lu J., Ozato K., and Dever T.E. 2001. Heterologous dimerization domains functionally substitute for the double-stranded RNA binding domains of the kinase PKR. *EMBO J.* **20:** 3728–3737.

Wek S.A., Zhu S., and Wek R.C. 1995. The histidyl-tRNA synthetase-related sequence in the eIF-2 alpha protein kinase GCN2 interacts with tRNA and is required for activation in response to starvation for different amino acids. *Mol. Cell. Biol.* **15:** 4497–4506.

Wu S. and Kaufman R.J. 1996. Double-stranded (ds) RNA binding and not dimerization correlates with the activation of the dsRNA-dependent protein kinase (PKR). *J. Biol. Chem.* **271:** 1756–1763.

———. 1997. A model for the double-stranded RNA (dsRNA)-dependent dimerization and activation of the dsRNA-activated protein kinase PKR. *J. Biol. Chem.* **272:** 1291–1296.

Yang J.M., London I.M., and Chen J.J. 1992. Effects of hemin and porphyrin compounds on intersubunit disulfide formation of heme-regulated eIF-2 alpha kinase and the regulation of protein synthesis in reticulocyte lysates. *J. Biol. Chem.* **267:** 20519–20524.

Yun B.G., Matts J.A., and Matts R.L. 2005. Interdomain interactions regulate the activation of the heme-regulated eIF 2 alpha kinase. *Biochim. Biophys. Acta* **1725:** 174–181.

Zhan K., Narasimhan J., and Wek R.C. 2004. Differential activation of eIF2 kinases in response to cellular stresses in *Schizosaccharomyces pombe*. *Genetics* **168:** 1867–1875.

Zhan K., Vattem K.M., Bauer B.N., Dever T.E., Chen J.J., and Wek R.C. 2002. Phosphorylation of eukaryotic initiation factor 2 by heme-regulated inhibitor kinase-related protein kinases in *Schizosaccharomyces pombe* is important for resistance to environmental stresses. *Mol. Cell. Biol.* **22:** 7134–7146.

Zhang F., Romano P.R., Nagamura-Inoue T., Tian B., Dever T.E., Mathews M.B., Ozato K., and Hinnebusch A.G. 2001. Binding of double-stranded RNA to protein kinase PKR is required for dimerization and promotes critical autophosphorylation events in the activation loop. *J. Biol. Chem.* **276:** 24946–24958.

Zhang X., Herring C.J., Romano P.R., Szczepanowska J., Brzeska H., Hinnebusch A.G., and Qin J. 1998. Identification of phosphorylation sites in proteins separated by polyacrylamide gel electrophoresis. *Anal. Chem.* **70:** 2050–2059.

Zheng X. and Bevilacqua P.C. 2004. Activation of the protein kinase PKR by short double-stranded RNAs with single-stranded tails. *RNA* **10:** 1934–1945.

13

---

eIF2α Phosphorylation in Cellular Stress Responses and Disease

David Ron and Heather P. Harding
Skirball Institute
New York University School of Medicine
New York, New York 10016

PHOSPHORYLATION OF THE α SUBUNIT OF EUKARYOTIC translation initiation factor 2 (eIF2α) is a highly conserved regulatory event activated in response to diverse stresses (Chapter 12). It elicits translational reprogramming as its primary consequence and secondarily affects the transcriptional profile of cells (Chapter 9). Together, these two strands of the eIF2α phosphorylation-dependent integrated stress response (ISR) broadly affect gene expression, amino acid and energy metabolism, and the protein-folding environment in the cell. Rare human mutations and transgenic mice, in which components of the ISR have been severely altered, reveal the pathway's importance to mammalian pathophysiology. Here, we review the components of the mammalian ISR and consider their function in the context of the cellular adaptation to protein misfolding, nutrient deprivation, and other stresses. We address the potential importance of the ISR to such common human diseases as diabetes mellitus, the metabolic syndrome, osteoporosis, neurodegeneration, and demyelination. Special emphasis is placed on instances suggesting that failure of homeostasis in the ISR contributes to disease, and these are considered in the context of the hypothetical therapeutic opportunities they present.

BACKGROUND

Molecular and Physiological Principles That Determine the Consequences of eIF2α Phosphorylation

Phosphorylation on serine 51 of its α subunit converts eIF2 from a substrate to an inhibitor of its guanine nucleotide exchange factor, eIF2B. Thus, the level of phosphorylated eIF2α regulates the rate at which eIF2 can be recycled to the GTP-bound form to join in a ternary complex with charged initiator methionyl-tRNA and promote the initiation of mRNA translation (Chapter 4). Consequently, the reversible phosphorylation of eIF2α has evolved as a potent means for regulating translation initiation rates. This phosphorylation event is purely regulatory; it has no direct effect on the function of eIF2 as an initiation factor. Indeed, the side chain of residue 51 is relatively unimportant, as translation initiation proceeds very well with an eIF2α^{S51A} mutant protein, a point we return to shortly.

Experiments in yeast suggest that the measurable effects of eIF2α phosphorylation can be suppressed by mutations in components of eIF2B (Chapter 9). Therefore, the consequences of eIF2α phosphorylation are likely mediated through their effects on translation. Despite this narrow convergence on a single molecular target, the moderating effect of eIF2(αP) on protein synthesis has surprisingly broad biological consequences.

Protein synthesis consumes much energy and competes with other metabolic processes for the utilization of free amino acids. For example, it is estimated that more than half the energy consumed by a prominent secretory cell such as the hepatocyte is devoted to protein synthesis (Pannevis and Houlihan 1992). Thus, the ability to reversibly attenuate protein synthesis is an important adaptation of cells to diverse stressful conditions that threaten energy homeostasis or require the diversion of amino acids to other metabolic pathways (Chapters 14 and 17).

Newly synthesized proteins pose a significant burden on the chaperone machinery of the cell, which assists nascent chains to attain their proper three-dimensional structure. The reserve capacity of the protein-folding machinery in the various compartments is relatively limited, as attested to by the dire consequences of (cis) mutations affecting the ability of abundantly expressed proteins to fold (Carrell and Lomas 1997). Therefore, eIF2(αP)-mediated attenuation of protein synthesis protects chaperone networks from client protein overload. It may be important in this regard that eIF2(αP) targets the initiation phase of protein synthesis and does not attenuate elongation. It is now appreciated that the ribosome nascent chain complex is associated with cellular chaperones that function as components of a coupled translation and folding machine (Bukau 2005). Attenuated processivity of this machine would tend to prolong the engagement of chaperones with incomplete nascent chains, whereas attenuated initiation would allow chaperones to recycle efficiently. Similar considerations apply to the pool of ribosomes and translation factors; by attenuating the initiation phase of translation, eIF2(αP) signaling facilitates the reprogramming of translation around the changing complement of mRNAs in stressed cells.

Although ternary complexes of eIF2 + GTP + charged $tRNA_i^{met}$ are required for translation initiation at all AUG codons (Chapter 4), the effects of limiting their availability (by eIF2α phosphorylation) has surprisingly gene-specific consequences. The mRNA encoding the yeast transcription factor Gcn4p and its metazoan counterpart, ATF4, are both subject to paradoxical translational up-regulation when levels of ternary complexes decline (Dever et al. 1992; Harding et al. 2000b). Translation initiation at the protein coding AUG of these unusual mRNAs is normally repressed by the presence of several conserved inhibitory upstream open reading frames. However, stress conditions that lead to eIF2α phosphorylation and declining ternary complex formation derepress ATF4 and GCN4 translation by a mechanism of regulated translation reinitiation (Lu et al. 2004a; Vattem and Wek 2004; Chapter 9).

Translational up-regulation of these transcription factor(s) activates a gene expression program whose targets adapt the cell to the specific stresses that promote eIF2α phosphorylation. In yeast, mutations that abolish eIF2(αP) signaling by inhibiting kinase activity or preventing eIF2α phosphorylation *in cis* have consequences similar to those affecting Gcn4p, suggesting that this linear signaling pathway evolved predominantly to control transcription rather than protein synthesis (Dever 2002). The situation in mammals is less clear. The phenotypic overlap of ATF4 deletion and various mutations that inactivate signaling upstream in the pathway are incomplete (Harding et al. 2003); however, ATF4 is not the only effector of regulated gene expression by the eIF2(αP) signaling pathway (Jiang et al. 2003; Deng et al. 2004). Therefore, the relative contributions of translational attenuation and transcriptional activation to the downstream consequences of signaling by eIF2(αP) remain to be defined. It is important to note, in this regard, that despite considerable effort, no other mRNA has been clearly demonstrated to be subjected to eIF2(αP)-regulated translation reinitiation, and the mechanisms linking eIF2(αP) signaling to the ATF4-independent gene activation program of mammals remain obscure.

Regulating Cellular Levels of Phosphorylated eIF2α

The eIF2(αP)-dependent signaling nexus is regulated by kinases and phosphatases. There are four known eIF2α kinases in vertebrates: GCN2 responds predominantly to amino acid deprivation, PERK responds to the imbalance between the load of endoplasmic reticulum (ER) client proteins and chaperones (so-called ER stress), and HRI and PKR are activated, respectively, by unbalanced synthesis of heme and globin in the erythroid lineage or the presence of double-stranded RNA in virally infected cells (Chapter 20).

Despite a diversity in upstream activating events, these four kinases have similar catalytic domains and share a single known effector, eIF2(αP) (Chapter 12). Formal evidence for this last point is provided by the shared phenotype of mutations in yeast that eliminate GCN2, or block the ability of eIF2α to be phosphorylated by the kinase (replacement of eIF2α's serine 51 by an alanine, $eIF2\alpha^{S51A}$) (Dever et al. 1992). The situation in mammals is complicated by the presence of more than one kinase. Nonetheless, the overlap in phenotype between homozygosity for $eIF2\alpha^{S51A}$ and loss of PERK (the dominant eIF2α kinase under basal conditions, in many cells) suggests that in mammals, too, these kinases have a single cellular target (Harding et al. 2001; Scheuner et al. 2001; Zhang et al. 2002a). Furthermore, the $eIF2\alpha^{S51A}$ mutation abolishes all measurable consequences of PERK activation (Lu et al. 2004a). Thus, claims to the contrary notwithstanding (Cullinan et al. 2003), we believe that most if not all signaling by this family of kinases is mediated by a single substrate, eIF2α.

Phosphorylation of eIF2α is highly dynamic and regulated. In addition to the aforementioned kinases, there are two known phosphatases that dephosphorylate eIF2(αP). A constitutive complex, consisting of an eIF2α-specific regulatory subunit (CReP) and the catalytic subunit of protein phosphatase I (PP1c), contributes to the high basal eIF2(αP)-directed phosphatase activity of mammalian cells (Jousse et al. 2003). In stressed cells, the eIF2α phosphorylation-dependent gene expression program activates a second related phosphatase regulatory subunit encoded by the *GADD34* gene (Novoa et al. 2001; Brush et al. 2003). *GADD34*, too, forms a complex with PP1c and significantly increases the eIF2(αP)-directed phosphatase activity of cells, ensuring recovery of protein synthesis during the later phases of stress responses that activate the eIF2α kinases (Kojima et al. 2003; Novoa et al. 2003). We do not know whether, between them, *GADD34* and CReP account for all eIF2(αP) dephosphorylation, nor do we know whether the four known kinases account for all the eIF2α-directed kinase activity of cells.

Given the diversity in upstream activators, all of which are channeled to a single downstream event, we propose that the consequences of eIF2α phosphorylation, in metazoans, be referred to as the integrated stress response (ISR, Fig. 1). As we shall see, the ISR, with its translational and transcriptional components, forms an important strand of several stress-responsive signal transduction pathways. Much of what we know about the role of eIF2α phosphorylation in health and disease has been

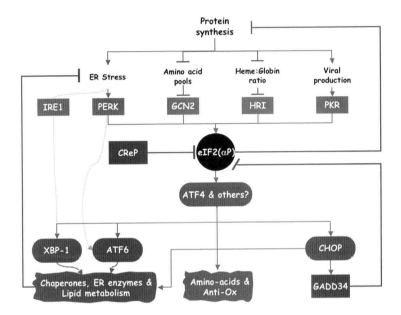

Figure 1. Schematic overview of the eIF2α-phosphorylation-dependent integrated stress response (ISR). The four eIF2α kinases—PERK, GCN2, HRI, and PKR—are modulated by distinct stress signals, which in turn are influenced by rates of protein synthesis. eIF2α phosphorylation and attenuated protein synthesis are especially important in limiting the levels of ER stress. Expression of ATF4 and possibly other transcription factors is increased by eIF2α phosphorylation, and these activate downstream genes. The latter include the genes encoding the transcription factors ATF6 and XBP-1, which are also controlled by parallel ER stress signals (*dashed arrows*) and genes downstream from that which encode chaperones, ER enzymes, enzymes involved in lipid metabolism, amino acid transporters, and enzymes involved in amino acid metabolism and anti-oxidative stress response genes. Also induced is the gene encoding *CHOP*, a transcription factor that activates *GADD34*, which encodes an eIF2(αP)-specific regulatory subunit of a phosphatase that inhibits signaling in the ISR. CReP is a constitutive regulatory subunit of a phosphatase that dephosphorylates eIF2(αP). (*Arrows*) Activating signals; (*blunted lines*) inhibitory signals.

derived by analysis of mutations that affect this pathway. Next, we consider these mutations from the perspective of their effects on defined physiological processes.

PATHOPHYSIOLOGICAL MECHANISMS: eIF2α PHOSPHORYLATION AND PROTEIN FOLDING

PERK and HRI Couple Protein Synthesis to the Rate of Posttranslational Protein Metabolism

Evidence for the role of eIF2α phosphorylation in preventing protein misfolding is provided by the analysis of mutations in PERK and HRI. PERK is an ER-localized type-I *trans*-membrane protein whose lumenal domain senses the balance between unfolded/misfolded lumenal client proteins and the host of chaperones in the organelle. An imbalance, also referred to as ER stress, leads to PERK activation that occurs by oligomerization in the plane of the membrane, *trans*-autophosphorylation, and a marked enhancement in kinase activity and in the kinase's affinity for its substrate, which take place on the cytosolic face of the ER membrane (Bertolotti et al. 2000; Marciniak et al. 2006). The resultant increase in eIF2(αP) levels attenuates protein synthesis and diminishes the load of newly synthesized unfolded ER client proteins. PERK thus matches the rate of client protein synthesis on the cytoplasmic side of the ER to the folding environment on the lumenal side (Harding et al. 1999; Sood et al. 2000).

In the absence of PERK, cells are unable to modulate protein synthesis in response to protein misfolding in the ER lumen, and *PERK*$^{-/-}$ cells experience higher levels of ER stress when subjected to manipulations that adversely affect the folding environment in the ER (Harding et al. 2000a). PERK deletion in mice and homozygosity for loss-of-function mutations in humans (the Wolcott Rallison syndrome) are associated with profound dysfunction in a number of tissues constituted of mainly secretory cells. Thus, the mutant mice and humans develop a syndromatic form of diabetes mellitus with onset in infancy, exocrine pancreatic dysfunction, and a severe bone defect (Delepine et al. 2000; Harding et al. 2001; Zhang et al. 2002a). As expected, hypersensitivity to conditions that promote protein misfolding in the ER is also observed in the eIF2α^{S51A} mutation that abolishes responsiveness to PERK (Scheuner et al. 2001).

A conceptually similar scenario exists with regard to HRI. This eIF2α kinase is expressed prominently in erythroid precursors, where it is repressed by free heme. When heme is limiting, HRI phosphorylates eIF2α, thereby attenuating protein synthesis, which in erythroid precursors amounts mostly to inhibition of globin-chain production. This translational repression maintains a balance between heme and globin production to ensure that the newly synthesized globin chains are incorporated into hemoglobin molecules. As long as heme biosynthesis is maintained, HRI is dispensable. However, when heme biosynthesis becomes limiting (commonly a consequence of nutritional iron deficiency), HRI is rendered essential. In its absence, unchecked protein synthesis leads to the accumulation of unliganded globin chains that misfold and are converted to dangerous proteotoxins. What would otherwise be a mild red cell production defect in wild-type animals is converted to a life-threatening anemia due to destruction of red cell precursors in the bone marrow of $HRI^{-/-}$ mice (Chen 2000; Han et al. 2001).

Recently, HRI has been implicated in a different context in which translational control protects cells from proteotoxicity. Arsenite, an abundant environmental toxin that leads to protein misfolding in the cytoplasm, was noted to promote eIF2α phosphorylation through the activation of HRI. It is likely, although unproven, that the hypersensitivity of $HRI^{-/-}$ cells to arsenite is due, at least in part, to their inability to attenuate the load of newly synthesized proteins in response to the threat of toxin-induced misfolding (McEwen et al. 2005).

These dramatic phenotypes of loss-of-function mutations in upstream activators of the ISR point to the critical role of eIF2α phosphorylation in controlling protein synthesis under specific stressful circumstances. It is tempting to speculate that modulation of protein synthesis may have a broad role in protecting cells from physiological and environmental use-dependent attrition; i.e., that eIF2α phosphorylation counteracts processes that contribute to aging. The insulin-producing β cells of the pancreas is one scenario where this process is believed to have an important role in pathophysiology.

Obesity in humans and mice leads to insulin resistance, which is counteracted by increased production of pro-insulin in the ER of the pancreatic β cells. This adaptation is promoted by a dramatic glucose-dependent translation up-regulation of insulin and other secreted proteins in the β cell (Skelly et al. 1996). PERK-dependent eIF2α phosphorylation protects the β cell from the consequences of the elevated client protein load and, at the expense of imposing transient limitation of insulin biosynthesis, ensures the long-term survival of the tissue (Fig. 2). The ultimate failure of this protective mechanism is believed to contribute to the onset of diabetes mellitus in the obese, a theory that is supported by the observation that minor defects in the ISR accelerate the onset of glucose intolerance (Scheuner et al. 2005).

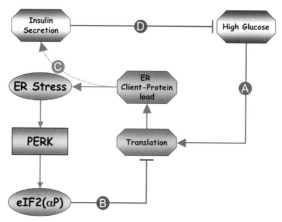

Figure 2. eIF2α phosphorylation in the pancreatic β cell. High-serum glucose (*A*) stimulates translation in the β cell, increasing ER client protein load and promoting a physiological ER stress. The latter is modulated by PERK-mediated eIF2α phosphorylation (*B*), which protects the β cell from ER stress at the expense of insulin synthesis (*C*). However, this compromise is highly advantageous to β cells, which are rapidly destroyed in the absence of PERK. Insulin resistance, which decreases signaling in *D*, places extra stress on the β cell by increased signaling in *A*. The survival of the cell depends, in part, on the adequacy of the protective PERK-dependent mechanism described here.

Unfolded Protein Load and eIF2(αP) Dephosphorylation

The important role of eIF2(αP)-mediated translational control in checking the risk of chaperone overload and misfolding is also supported by the consequences of pharmacological and genetic manipulation of eIF2(αP)-directed phosphatase activity. A screen for chemical compounds that protect cultured cells from the lethal effects of tunicamycin (an agent that perturbs the folding environment in the ER by blocking glycosylation) led to the identification of salubrinal, an in vivo inhibitor of CReP and GADD34 (Boyce et al. 2005). Furthermore, CReP knockdown by RNA interference (RNAi) (Jousse et al. 2003) and deletion of GADD34 (Marciniak et al. 2004) are both protective against experimental models of ER stress. These findings are consistent with a simple model whereby diminished phosphatase activity sustains eIF2(αP) levels and attenuates newly synthesized unfolded protein load and with it levels of ER stress.

The transcription factor CHOP has an especially interesting role in linking eIF2α phosphorylation and ER stress. The *CHOP* gene is strongly induced by ER stress, as part of the PERK→eIF2(αP)→ATF4-dependent gene expression program mentioned above (Fig. 1) (Harding et al. 2000b). *CHOP* deletion has been noted to protect against the lethal affects of ER stress (Zinszner et al. 1998), whereas CHOP overexpression sensitizes cells (McCullough et al. 2001). GADD34 is a direct transcriptional target of CHOP, such that eIF2(αP)-directed phosphatase activity is diminished in stressed $CHOP^{-/-}$ cells. As a consequence, mutant cells synthesize fewer proteins and, compared with the wild type, confront their chaperones with a diminished load of unfolded client proteins. Mutant cells are thus partially protected from the lethal consequences of perturbations to the protein-folding environment in the ER (Marciniak et al. 2004).

This protection extends beyond experimental pharmacological inducers of protein misfolding to other paradigms with disease relevance. The dominant C96Y mutation in mouse insulin 2 (the so-called Akita mutation) causes the protein to misfold, and the ensuing compromised protein-folding environment in the ER greatly shortens β-cell life span, leading to the development of diabetes mellitus in the affected mice. A $Ddit3^{-/-}$ ($CHOP^{-/-}$) genotype is partially protective against β-cell death and delays the onset of diabetes mellitus in $Ins2^{C96Y/+}$ (Akita) mice (Oyadomari et al. 2002).

CHOP deletion has also been noted to protect dopaminergic neurons of mice against 6-hydroxydopamine injection (Silva et al. 2005). This compound is believed to mimic endogenous and environmental agents that contribute to dopaminergic cell death, the underlying cellular substrate for the development of Parkinson's disease. Interestingly, exposure to 6-hydroxydopamine causes ER stress in cultured dopaminergic neurons (Ryu et al. 2002; Holtz and O'Malley 2003), suggesting that the effect of CHOP deletion might be mediated through the ability of sustained elevation of eIF2(αP) levels to protect against ER stress.

In both models, CHOP is induced by the ISR, which is a PERK→eIF2(αP)→ATF4-dependent strand of an evolutionarily conserved response to protein misfolding in the ER known as the unfolded protein response (UPR). The UPR has two other well-characterized strands which also contribute to the activation of genes that enhance the ability of the ER to cope with unfolded and misfolded client proteins: One mediated by the ER-localized signal transducer IRE1 and its downstream effector, the transcription factor XBP-1, activation of which involves a noncanonical mRNA splicing event (Patil and Walter 2001; also see Chapter 10). The other is mediated by the release of the transcription factor ATF6 from its ER tether in stressed cells (for review, see Mori

2000). This three-stranded UPR protects cells from ER stress by attenuating client protein load and by augmenting ER capacity (Harding et al. 2002; Kaufman 2002).

CHOP constitutes a relatively downstream component of the ISR/UPR that contributes to *GADD34* activation, which, in turn, counteracts the effects of PERK in the UPR and serves as a negative feedback loop that maintains protein synthesis. This eIF2(αP)$\rightarrow$ATF4$\rightarrow$CHOP$\rightarrow$GADD34$\dashv$eIF2(αP) negative feedback loop presumably evolved in response to the inherent reciprocal relationship between protein synthesis and chaperone reserve. High levels of eIF2(αP) favor chaperone reserve over protein synthesis and, under some circumstances, survival of the individual cell. However, this adaptation tends to compromise the ability of the stressed cell to fulfill its physiological mission (e.g., synthesize insulin, in the example described above) and, over time, might even compromise the cell's ability to synthesize proteins required for its own survival.

Metazoans need to balance these contradictory influences and set the thresholds for activating the various components of the ISR/UPR at their appropriate level. The data fit a model whereby thresholds for activation that proved adaptive under the selective pressures that existed when the system evolved might not be appropriate for all circumstances. According to this theory, the protection against agents that perturb folding in the ER afforded by CHOP deletion or GADD34 inhibition (by salubrinal) likely reflects a failure of homeostasis operating at the level of client protein synthesis (Fig. 3). However, it is not possible to exclude an alternative hypothesis whereby CHOP and GADD34 evolved primarily to kill cells experiencing insurmountable levels of ER stress. The killing of severely damaged ER stressed cells could also have a homeostatic role, as removal of the damaged cells would provide space for regeneration; it is interesting in this regard that CHOP and GADD34 are relatively late additions, being found in vertebrates, but not in simpler metazoans that do not engage in tissue regeneration. This second hypothesis, too, fits a model of failure of homeostasis, but one that is based primarily on the control of cell death rather than protein synthesis.

Besides being intuitively appealing, the model, whereby the eIF2(αP)$\rightarrow$ATF4$\rightarrow$CHOP$\rightarrow$GADD34$\dashv$eIF2(αP) loop provides advantages to the organism that are independent of cell death, is supported by at least two observations: (1) the impaired survival of GADD34 mutant cells exposed to high doses of thapsigargin (an agent that strongly activates PERK and promotes a sustained shutdown of protein synthesis in the mu-

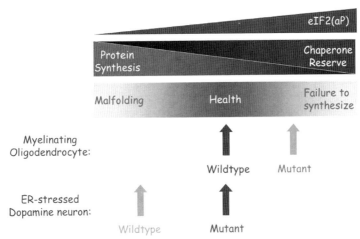

Figure 3. Failure of homeostasis in the ISR. CHOP-mediated GADD34 expression affects the balance between protein synthesis and chaperone reserve (see Fig. 1). The set point for this component of the ISR maintains homeostasis of the wild type under conditions that prevailed during selection of the phenotype. In the hypothetical example considered here, the set point allows the oligodendrocyte to cope with the physiologically stressful process of myelination. The wild-type cell strikes an appropriate balance between too much eIF2α phosphorylation, which risks failure to synthesize enough myelin constituents, and too little eIF2α phosphorylation, which risks protein misfolding in the ER. However, in certain unusual situations, such as the 6-hydroxydopamine-stressed dopamineric neuron, the set point of the wild type is maladaptive, favoring survival of the mutant. This reflects a failure of homeostasis.

tant cells [Novoa et al. 2003]) and (2) the impaired survival of CHOP mutant mice with a confounding second mutation in proteolipid protein (PLP); specifically, a mutation that causes the protein to misfold in the ER of oligodendrocytes and induces a model of Pelizaeus-Merzbacher leukodystrophy. The reduced survival of the PLP^{mut}; CHOP knockout mice is associated with more, not less, oligodendrocyte cell death (Southwood et al. 2002), arguing that the ability of CHOP to promote protein synthesis rather than cell death accounts for its survival advantage in that context.

The observations made on mice and cells with mutations in ISR components suggest that the pathway could be manipulated pharmacologically to beneficial ends. Specifically, inhibitors of eIF2α dephosphorylation might protect cells against ER stress. However, the safety of such manipulation will need to be considered carefully. In the next section, we discuss further complexity introduced into the ISR by its downstream gene expression program.

PATHOPHYSIOLOGICAL MECHANISMS: eIF2α PHOSPHORYLATION AND GENE EXPRESSION

The well-recognized consequences (i.e., scoreable phenotypes) of eIF2α phosphorylation in yeast are attributed to activation of a gene expression program mediated by the translational up-regulation of the transcription factor Gcn4p. The upstream eIF2α kinase, Gcn2p, is activated by uncharged tRNAs and thus responds to decreased availability of amino acids. The gene expression program activated by Gcn4p promotes the import and synthesis of multiple amino acids and is thus referred to as the general (amino acid) control (GC) response (Chapter 9). Amino acid transporters and biosynthetic enzymes are also among the genes prominently up-regulated by the mammalian ISR, but this pathway has diversified to include target genes of other classes, as described below.

Integrated Stress Response and Oxidative Stress

Impaired signaling in the ISR by mutations at any level in the pathway is associated with enhanced accumulation of reactive oxygen species (Scheuner et al. 2001; Harding et al. 2003). These observations reveal that the mammalian ISR has evolved a significant role in combating oxidative stress. This is played out partially at the level of enhanced import of cysteine, a precursor of the cell's major redox buffers, but also includes up-regulation of genes directly involved in redox homeostasis. This aspect of the ISR is especially important during ER stress and likely reflects the role of the ER as a producer of H_2O_2 and a consumer of reduced glutathione (for review, see Tu and Weissman 2004), the consequences of which are counteracted by the gene expression program induced by the ISR (Harding et al. 2003).

The challenge to redox homeostasis posed by ER activity is intrinsic to the biochemistry of the major ER oxidase, ERO1 (Tu and Weissman 2004). Interestingly, *ERO1a*, encoding one of the two mammalian isoforms of this enzyme, is activated by CHOP. This connection can be rationalized as an attempt by stressed cells to up-regulate their capacity to process ER client proteins, whose folding depends on disulfide bond formation. However, in the context of severe ER stress, the up-regulation of *ERO1a* poses a significant challenge to the ability of the cell to maintain a reduced cytoplasm, accounting for the tendency of CHOP to promote oxidative stress (McCullough et al. 2001; Ikeyama et al. 2003; Marciniak et al. 2004).

ISR and Conventional UPR Target Genes

The induction of ERO1α is but one example of the role of ISR in enhancing the capacity of the ER to process client proteins. To achieve this goal, it also collaborates with other strands of the UPR to activate genes that enhance the cell's capacity to cope with ER client protein load. This is mediated in part by transcriptional activation of key UPR regulators such as XBP-1 and ATF6, transcription factors that transduce ER stress signals to the nucleus (Calfon et al. 2002; Harding et al. 2003). It is unclear how PERK-dependent eIF2(αP)-mediated ER stress signaling transcriptionally up-regulates *XBP-1* and *ATF6α*, as these genes are induced normally in *ATF4*$^{-/-}$ cells (Harding et al. 2003).

It is apparent from the above that the prosurvival effects of the ISR are attributed to regulation of both protein synthesis and gene expression. The latter conclusion has received direct support from an experimental system that allows conditional activation of PERK by an artificial ligand, independently of the induction of ER stress. Transient, preemptive activation of PERK (and induction of the ISR) markedly protected HT22 cells from subsequent challenges that involve oxidative stress, although the latter were applied *after* the translational effects of PERK activation had worn off (Lu et al. 2004b). Further support for the protective role of attenuated eIF2 activity (the predicted consequence of eIF2α phosphorylation) was provided by the results of a somatic cell genetic screen in which interference with eIF2α expression protected HT22 cells from oxidative stress (Tan et al. 2001).

ATF4 activation is unlikely to account for all of these protective effects, as the contribution of ATF4 to the survival of ER-stressed cells can be substituted by supplementing the medium with nonessential amino acids and reducing substances, whereas that of the upstream kinase PERK cannot be substituted by this simple expedient (Harding et al. 2003). It is therefore likely that eIF2α phosphorylation activates other, yet to be identified, transcription factors, as exemplified recently by the case of NF-κB (Jiang et al. 2003; Deng et al. 2004). Establishing the relative contribution of the ISR's translational and transcriptional arms to its cytoprotective actions will therefore require a more-detailed understanding of the mediators of the transcriptional program and their genetic manipulation.

The prosurvival effects of the ISR may have an important role in cancer biology. The survival of cancer cells in vivo requires that they negotiate the ischemic environment of tumors. The attendant oxygen and nutrient deprivation challenge the protein-folding environment in the ER and induce a protective UPR of which PERK activation and the ISR are prominent features (Koumenis et al. 2002). Remarkably, transformed cells lacking either PERK or its downstream effector, ATF4, are markedly impaired in their ability to survive as tumors in vivo (Bi et al. 2005). These observations suggest that inhibition of the ISR may have a role in cancer therapy.

The discussion, so far, has centered on the role of ISR in cell survival, specifically in the context of misfolded protein stress. However, there are reasons to believe that ISR signaling, induced by physiological levels of ER stress, also contributes to cell function. This aspect is especially interesting with regard to the intersection of the ISR with calcium signaling in excitatory cells.

The ER is a major repository of intracellular calcium, and its stores are tapped in a variety of excitation-coupled phenomena (e.g., secretion and muscle contraction). The most important outcome of this is to raise cytosolic calcium levels and activate cytosolic effector mechanisms; inevitably, however, such signaling is associated with a measure of ER calcium store depletion. Severe calcium store depletion is a powerful activator of PERK (Harding et al. 1999), presumably a consequence of the role of ER calcium in chaperone function. There are reasons to believe that ER calcium excursions in the physiological range might also activate PERK, as physiological agonists that deplete ER calcium stores have been noted to activate the ISR (Kimball and Jefferson 1990; Alcazar et al. 1997). More recently, analysis of $PERK^{-/-}$ mice has revealed a broad defect in excitation-coupled secretion in the pancreas and contraction in smooth muscle cells (Huang et al. 2006). This suggests that physiological levels of excitatory activity lead to PERK activation and ISR signaling, which are, in turn, required for the proper function of the excitatory apparatus.

PATHOPHYSIOLOGICAL MECHANISMS: eIF2α PHOSPHORYLATION AND NUTRITION

Availability of amino acids is the primary regulator of eIF2α phosphorylation in yeast. In mammals, too, nutrient availability regulates eIF2α phosphorylation, but the connection has acquired a surprising measure of complexity, and eIF2α phosphorylation has emerged as an important signaling node in integrating nutritional and metabolic cues. Below, we consider several examples with potential pathophysiological implications.

GCN2 and the Adaptation to a Diet with Imbalanced Amino Acid Content

Remarkably, $GCN2^{-/-}$ mice and worms exhibit no obvious impairment when maintained on a normal diet. Even when placed on a diet deficient in an essential amino acid, adult $GCN2^{-/-}$ mice fare no worse than their wild-type littermates (D. Ron and H.P. Harding, unpubl.), although recent evidence suggests that the mutant animals are unable to conserve muscle mass under these circumstances (Anthony et al. 2004). Furthermore, when pregnant dams are placed on an amino-acid-deficient diet, their developing $GCN2^{-/-}$ pups are at a disadvantage compared to the wild type, but this too is not an especially strong phenotype (Zhang et al. 2002b), all suggesting that mammals have redundant physiological pathways for dealing with amino acid deprivation.

It has long been known that omnivorous animals such as rats will avoid otherwise palatable and nutritious foods that are deficient in merely one essential amino acid. The aversion to imbalanced foods is presumably highly adaptive as it directs the foraging animal to avoid dangerous micro-nutrient deficiency. It is also relatively easy to measure this response by comparing the consumption of two otherwise identical meals, one of which is deficient in a single amino acid. Bilateral lesions in the anterior piriform cortex abolish the bias against the imbalanced foods (Gietzen 1993). Interestingly, the consumption of an imbalanced meal leads to eIF2α phosphorylation in neurons in that nucleus (Gietzen et al. 2004), and such phosphorylation and the associated aversive response are missing in mice lacking GCN2 (Hao et al. 2005; Maurin et al. 2005).

GCN2-dependent eIF2α phosphorylation is both a necessary and sufficient signal, as localized instillation of an inhibitor of a single tRNA synthetase into the anterior piriform cortex elicited an aversive response in animals fed a balanced diet (Hao et al. 2005). It is likely that a transient decline in serum levels of a specific amino acid is sufficient to trigger GCN2 activation in neurons of the piriform cortex and that this upstream event initiates the aversive response. Given the prominent role of eIF2α phosphorylation in GCN2 activity in yeast, it is likely that in mammals, too, it is the mediator of kinase action, although this remains to be proven. Nor it is known how GCN2 activation and eIF2α phosphorylation trigger the necessary neuronal activity; the short latency of the response suggests that the effects are directly translational, but it is impossible at present to exclude a contribution at other levels of gene expression. Despite these uncertainties, the role of GCN2 in the mammalian aversive response to imbalanced diet resembles its ancestral role in adapting unicellular organisms to an amino-acid-deficient environment.

GCN2 also regulates other neuronal activities, as brain explants of $GCN2^{-/-}$ mice exhibit a complex defect in long-term potentiation, and the knockout mice have a prominent defect in memory (Costa-Mattioli et al. 2005). Unlike the response to an imbalanced diet, it is unclear how neuronal stimulation in long-term potentiation experiments or during normal memory formation activates GCN2. It is possible that ion fluxes during neuronal activity affect amino acid import and thereby activate GCN2 by the canonical mechanism of uncharged tRNAs, but this is merely a speculation that the authors of this chapter find attractive.

GCN2, Tryptophan Metabolism, and Immunomodulation

An especially interesting example of the evolution of the role of GCN2 as sensor of uncharged tRNAs is provided by its role in the immune response. The tryptophan degrading enzyme indolamine 2,3-dioxygenase (IDO) is activated in certain antigen-presenting cells and its activity has an important role in modulating adjacent T cells, to attenuate TH1 cytotoxic responses. IDO is believed to have a role in such important phenomena as tumor anergy or preventing fetal allograph rejection. The role of tryptophan metabolism in IDO signaling is well documented in vitro, where supplementation with the amino acid prevents IDO expression in dendritic cells from affecting T-cell action. This led Mellor and Munn (2004) to hypothesize that a localized environment of tryptophan depletion contributes to IDO's immunomodulatory role.

Given the ability of GCN2 to respond to depletion of even a single amino acid, it was natural to inquire whether IDO expression led to kinase activation and an ISR in the responding T cell. This proved to be the case. Furthermore, T cells explanted from $GCN2^{-/-}$ mice were profoundly deficient in their response to IDO-expressing cells in coculture experiments, suggesting that GCN2 signaling has an important role in mediating the effects of IDO in the responding T cell (Munn et al. 2005).

These intriguing observations are clearly consistent with Munn and Mellor's hypothesis, but they are also open to an alternative interpretation whereby IDO leads to the conversion of tryptophan or another precursor to metabolites that diffuse to the T cell and activate GCN2 (e.g., by inhibiting tryptophan uptake into the cells or the tryptophanyl tRNA synthetase). The reversal of IDO's effects by tryptophan supplementation would then be explained by competition at the level of IDO (if the precursor were other than tryptophan) or at the level of the metabolites' target. Either way, this study suggests that canonical GCN2 signaling and the downstream ISR have been co-opted for immunomodulation in T cells.

ISR and Intermediary Metabolism

The development of diabetes mellitus, exocrine pancreatic insufficiency, and bone disease in mice and humans lacking PERK is readily explained by impaired function and survival of secretory cells (β cells, acinar cells, and osteoblasts) deprived of PERK's modulatory role on client protein synthesis and deprived of the pro-survival benefits of the ISR. The profound metabolic consequences likely reflect this basic problem of ER homeostasis. However, analysis of more subtle mutations in *PERK* and *eIF2α* suggests a more complex role for the ISR in regulating intermediary metabolism.

The first clue came from mice homozygous for the *eIF2α*^{S51A} mutation, which precludes regulated phosphorylation. The embryos died of hypoglycemia shortly after birth, due to their inability to maintain blood glucose levels following disruption of the feto-maternal circulation. This defect appeared to be multifactorial, as it correlated with lower levels of gluconeogenic enzymes and lower glycogen stores in the liver of the mutant mice (Scheuner et al. 2001). It is presently unclear whether this defect represents a failure of ISR signaling in the postnatal liver and, if so, what are the kinases involved in such signaling normally? The *eIF2α*$^{S51A/S51A}$ mice also have a profound defect in pancreatic islet development, which may have contributed to the hypoglycemia by impaired glucagon production.

More recently, Kaufman and colleagues have evaluated the heterozygous *eIF2α*$^{S51A/+}$ mice, and these too have metabolic defects, albeit more subtle: When placed on a high-fat diet, the *eIF2α*$^{S51A/+}$ mice are more glucose-intolerant than wild-type mice. This phenotype is due in part to compromised insulin secretion, which is consistent with the role of the ISR in preserving β-cell function (Scheuner et al. 2005). However, the mutant mice have other metabolic abnormalities, such as increased body weight and hyperlipidemia, which are not easily attributed to defective β-cell function. Instead, these characteristics are consistent with a role for the ISR in regulating aspects of lipid metabolism.

At present, we are unable to offer a unified explanation for all these effects of altered signaling in the eIF2(αP) pathways, but interesting clues have surfaced recently. Hotamisligil and colleagues have documented higher levels of ER stress signaling in the adipose tissue and liver of mice placed on a high-fat diet (Ozcan et al. 2004). The mechanism for such activation remains unclear, but the phenomenon appears to be physiologically significant, as mutations affecting the IRE1→XBP-1 arm of the UPR profoundly reduce insulin sensitivity in mice fed a high-fat diet. PERK and IRE1 are activated by similar molecular mechanisms. Therefore, it is likely that mice on a high-fat diet (and obese people too) have enhanced PERK signaling in their liver and fat, and it follows that the $eIF2\alpha^{S51A/+}$ mice are impaired in conveying that signal.

The IRE1→XBP-1 arm of the UPR up-regulates phospholipid biosynthesis in yeast and mammals (for review, see Ron and Hampton 2004). The sterol-regulated transcription factors, SREBPs, are responsive to the lipid composition of the ER membrane. Furthermore, the machinery for their activation by regulated intramembrane proteolysis is shared by ATF6, a component of the UPR. We have recently discovered that the ISR inhibits SREBP target gene expression by attenuating SREBP activation (Harding et al. 2005). This inhibition could be mediated indirectly by an effect of the ISR on ER lipid composition or through cross-talk between the three arms of the UPR or, perhaps more directly, through the translational or transcriptional components of the ISR. Either way, the accumulating evidence for regulation of lipid metabolism by ER stress-mediated eIF2α phosphorylation suggests the need for further studies to clarify the role it has in the metabolic syndrome of obesity and diabetes mellitus.

PATHOPHYSIOLOGICAL MECHANISMS: TRANSLATIONAL REPRESSION CELL DEATH AND SURVIVAL

The discussion so far has emphasized the benefits of eIF2α phosphorylation and the downstream response, both to individual cells and to the organism as a whole. There are, however, a number of circumstances in which ISR activation correlated with more, not less, cell death. These will be considered next.

PKR, an eIF2α kinase activated by viral infection, is clearly important to vertebrate innate immunity. Evidence for this is provided by the observation that most animal cell viruses have evolved specific mechanisms for blocking eIF2α phosphorylation in their host cell (Chapter 20). It appears, however, that PKR provides its benefit to the organism by promoting the death of virally infected cells (Srivastava et al. 1998). Thus, in the context of viral infection, translational repression synergizes with other signals to promote apoptosis (Chapter 16).

Animal models of stroke are associated with conspicuous levels of eIF2α phosphorylation in the affected neurons (DeGracia et al. 1997), which correlate with PERK activation (Kumar et al. 2001). However, in these experimental systems, eIF2α phosphorylation correlates not with cell survival, but rather with cell death (DeGracia et al. 2002). This has led to the speculation that impaired recovery of protein synthesis during the reperfusion phase of ischemic injury compromises neuronal survival (Paschen 2003). At the same time, however, hibernation, one of the most stress-resistant states of the central nervous system, is associated with very high levels of eIF2α phosphorylation (Frerichs et al. 1998). It therefore would appear that, like many regulated phenomena, the effects of eIF2α phosphorylation on nervous tissue survival are biphasic. A rare human genetic syndrome, childhood ataxia with cerebral hypomyelination (CACH, also known as vanishing white matter), is further instructive in this regard.

CACH is a severe disorder of the white matter associated with abnormalities in the myelin-producing oligodendrocytes. It is caused by loss-of-function mutations in any one of several subunits of eIF2B, the nucleotide exchange factor for eIF2 (Leegwater et al. 2001; Richardson et al. 2004). The CACH-associated mutations mimic the consequences of eIF2α phosphorylation and, to a first approximation, can be considered ISR-activating (Kantor et al. 2005). Furthermore, it is possible that eIF2α phosphorylation contributes directly to the pathophysiology of the CACH syndrome, as patients are reported to experience catastrophic deterioration in their condition following acute stressful conditions such as febrile illness or head trauma, which might compromise eIF2B function further through the induction of eIF2α phosphorylation.

The severe consequences of the CACH mutations point to the dangers of ISR hyperactivation, especially as it pertains to the myelin-producing oligodendrocyte. These cells seem to be perched in an especially precarious position as they are sensitive to both mild defects in mounting an ISR (Lin et al. 2005) and to excesses in eIF2α phosphorylation (Southwood et al. 2002) or to mutations that mimic its consequences (Fogli et al. 2004).

SUMMARY

The phosphorylation of eIF2α and signaling downstream represent an ancient and conserved adaptation to cell stress, which we refer to as the integrated stress response. The ISR influences the balance of precursors and macromolecular end products in protein, lipid, and carbohydrate metabolism, and in higher eukaryotes, this pathway has acquired a prominent role in regulating the protein-folding environment in the ER. Signaling in the ISR has biphasic and tissue-specific effects on cell survival under various stressful conditions. Because the pathway is regulated by specific kinases and phosphatases that are, in principle, amenable to pharmacological manipulation, the ISR is a potential drug target for the treatment of a variety of common disorders.

ACKNOWLEDGMENTS

Work in the authors' lab has been supported by grants from the National Institutes of Health, the Juvenile Diabetes Research Foundation, and the Ellison Medical Research Foundation.

REFERENCES

Alcazar A., Martin de la Vega C., Bazan E., Fando J.L., and Salinas M. 1997. Calcium mobilization by ryanodine promotes the phosphorylation of initiation factor 2 alpha subunit and inhibits protein synthesis in cultured neurons. *J. Neurochem.* **69:** 1703–1708.

Anthony T.G., McDaniel B.J., Byerley R.L., McGrath B.C., Cavener D.R., McNurlan M.A., and Wek R.C. 2004. Preservation of liver protein synthesis during dietary leucine deprivation occurs at the expense of skeletal muscle mass in mice deleted for eIF2 kinase GCN2. *J. Biol. Chem.* **279:** 36553–36561.

Bertolotti A., Zhang Y., Hendershot L., Harding H., and Ron D. 2000. Dynamic interaction of BiP and the ER stress transducers in the unfolded protein response. *Nat. Cell Biol.* **2:** 326–332.

Bi M., Naczki C., Koritzinsky M., Fels D., Hu N., Harding H., Novoa I., Varia M., Raleigh J., Scheuner D., et al. 2005. ER stress-regulated translation increases tolerance to extreme hypoxia and promotes tumor growth. *EMBO J.* **24:** 3470–3481.

Boyce M., Bryant K.F., Jousse C., Long K., Harding H.P., Scheuner D., Kaufman R.J., Ma D., Coen D., Ron D., and Yuan J. 2005. A selective inhibitor of eIF2α dephosphorylation protects cells from ER stress. *Science* **307:** 935–939.

Brush M.H., Weiser D.C., and Shenolikar S. 2003. Growth arrest and DNA damage-inducible protein GADD34 targets protein phosphatase 1alpha to the endoplasmic reticulum and promotes dephosphorylation of the alpha subunit of eukaryotic translation initiation factor 2. *Mol. Cell. Biol.* **23:** 1292–1303.

Bukau B. 2005. Ribosomes catch Hsp70s. *Nat. Struct. Mol. Biol.* **12:** 472–473.

Calfon M., Zeng H., Urano F., Till J.H., Hubbard S.R., Harding H.P., Clark S.G., and Ron D. 2002. IRE1 couples endoplasmic reticulum load to secretory capacity by processing the *XBP-1* mRNA. *Nature* **415:** 92–96.

Carrell R.W. and Lomas D.A. 1997. Conformational disease. *Lancet* **350:** 134–138.

Chen J.-J. 2000. Heme-regulated eIF2α kinase. In *Translational control of gene expression* (ed. N. Sonenberg et al.), pp. 529–546. Cold Spring Harbor Laboratory Press, Cold Spring Harbor, New York.

Costa-Mattioli M., Gobert D., Harding H.P., Herdy B., Azzi M., Bruno M., Ben Mamou C., Marcinkiewicz E., Yoshida M., Imataka H., et al. 2005. Translational control of hippocampal synaptic plasticity and memory by an eIF2 kinase, GCN2. *Nature* **436:** 1166–1173.

Cullinan S.B., Zhang D., Hannink M., Arvisais E., Kaufman R.J., and Diehl J.A. 2003. Nrf2 is a direct PERK substrate and effector of PERK-dependent cell survival. *Mol. Cell. Biol.* **23:** 7198–7209.

DeGracia D.J., Kumar R., Owen C.R., Krause G.S., and White B.C. 2002. Molecular pathways of protein synthesis inhibition during brain reperfusion: Implications for neuronal survival or death. *J. Cereb. Blood Flow Metab.* **22:** 127–141.

DeGracia D.J., Sullivan J.M., Neumar R.W., Alousi S.S., Hikade K.R., Pittman J.E., White B.C., Rafols J.A., and Krause G.S. 1997. Effect of brain ischemia and reperfusion on the localization of phosphorylated eukaryotic initiation factor 2 alpha. *J. Cereb. Blood Flow Metab.* **17:** 1291–1302.

Delepine M., Nicolino M., Barrett T., Golamaully M., Lathrop G.M., and Julier C. 2000. EIF2AK3, encoding translation initiation factor 2-alpha kinase 3, is mutated in patients with Wolcott-Rallison syndrome. *Nat. Genet.* **25:** 406–409.

Deng J., Lu P.D., Zhang Y., Scheuner D., Kaufman R.J., Sonenberg N., Harding H.P., and Ron D. 2004. Translational repression mediates activation of nuclear factor kappa B by phosphorylated translation initiation factor 2. *Mol. Cell. Biol.* **24:** 10161–10168.

Dever T.E. 2002. Gene-specific regulation by general translation factors. *Cell* **108:** 545–556.

Dever T.E., Feng L., Wek R.C., Cigan A.M., Donahue T.F., and Hinnebusch A.G. 1992. Phosphorylation of initiation factor 2 alpha by protein kinase GCN2 mediates gene-specific translational control of GCN4 in yeast. *Cell* **68:** 585–596.

Fogli A., Schiffmann R., Bertini E., Ughetto S., Combes P., Eymard-Pierre E., Kaneski C.R., Pineda M., Troncoso M., Uziel G., et al. 2004. The effect of genotype on the natural history of eIF2B-related leukodystrophies. *Neurology* **62:** 1509–1517.

Frerichs K.U., Smith C.B., Brenner M., DeGracia D.J., Krause G.S., Marrone L., Dever T.E., and Hallenbeck J.M. 1998. Suppression of protein synthesis in brain during hibernation involves inhibition of protein initiation and elongation. *Proc. Natl. Acad. Sci.* **95:** 14511–14516.

Gietzen D.W. 1993. Neural mechanisms in the responses to amino acid deficiency. *J. Nutr.* **123:** 610–625.

Gietzen D.W., Ross C.M., Hao S., and Sharp J.W. 2004. Phosphorylation of eIF2alpha is involved in the signaling of indispensable amino acid deficiency in the anterior piriform cortex of the brain in rats. *J. Nutr.* **134:** 717–723.

Han A.P., Yu C., Lu L., Fujiwara Y., Browne C., Chin G., Fleming M., Leboulch P., Orkin S.H., and Chen J.J. 2001. Heme-regulated eIF2alpha kinase (HRI) is required for translational regulation and survival of erythroid precursors in iron deficiency. *EMBO J.* **20:** 6909–6918.

Hao S., Sharp J.W., Ross-Inta C.M., McDaniel B.J., Anthony T.G., Wek R.C., Cavener D.R., McGrath B.C., Rudell J.B., Koehnle T.J., and Gietzen D.W. 2005. Uncharged tRNA and sensing of amino acid deficiency in mammalian piriform cortex. *Science* **307:** 1776–1778.

Harding H., Zhang Y., and Ron D. 1999. Translation and protein folding are coupled by an endoplasmic reticulum resident kinase. *Nature* **397:** 271–274.

Harding H.P., Calfon M., Urano F., Novoa I., and Ron D. 2002. Transcriptional and translational control in the mammalian unfolded protein response. *Annu. Rev. Cell Dev. Biol.* **18:** 575–599.

Harding H., Zhang Y., Bertolotti A., Zeng H., and Ron D. 2000a. Perk is essential for translational regulation and cell survival during the unfolded protein response. *Mol. Cell* **5:** 897–904.

Harding H., Novoa I., Zhang Y., Zeng H., Wek R.C., Schapira M., and Ron D. 2000b. Regulated translation initiation controls stress-induced gene expression in mammalian cells. *Mol. Cell* **6:** 1099–1108.

Harding H., Zeng H., Zhang Y., Jungreis R., Chung P., Plesken H., Sabatini D., and Ron D. 2001. Diabetes mellitus and exocrine pancreatic dysfunction in Perk−/− mice reveals a role for translational control in survival of secretory cells. *Mol. Cell* **7:** 1153–1163.

Harding H.P., Zhang Y., Khersonsky S., Marciniak S., Scheuner D., Kaufman R.J., Javitt N., Chang Y.T., and Ron D. 2005. Bioactive small molecules reveal antagonism between the integrated stress response and sterol regulated gene expression. *Cell Metab.* **2:** 361–371.

Harding H., Zhang Y., Zeng H., Novoa I., Lu P., Calfon M., Sadri N., Yun C., Popko B.,

Paules R., et al. 2003. An integrated stress response regulates amino acid metabolism and resistance to oxidative stress. *Mol. Cell* **11:** 619–633.

Holtz W.A. and O'Malley K.L. 2003. Parkinsonian mimetics induce aspects of unfolded protein response in death of dopaminergic neurons. *J. Biol. Chem.* **278:** 19367–19377.

Huang G., Yao J., Zeng W., Mizuno Y., Kamm K.E., Stull J.T., Harding H.P., Ron D., and Muallem S. 2006. ER stress disrupts Ca2+-signaling complexes and Ca2+ regulation in secretory and muscle cells from PERK-knockout mice. *J. Cell Sci.* **119:** 153–161.

Ikeyama S., Wang X.T., Li J., Podlutsky A., Martindale J.L., Kokkonen G., Van Huizen R., Gorospe M., and Holbrook N.J. 2003. Expression of the pro-apoptotic gene gadd153/chop is elevated in liver with aging and sensitizes cells to oxidant injury. *J. Biol. Chem.* **278:** 16726–16731.

Jiang H.Y., Wek S.A., McGrath B.C., Scheuner D., Kaufman R.J., Cavener D.R., and Wek R.C. 2003. Phosphorylation of the alpha subunit of eukaryotic initiation factor 2 is required for activation of NF-kappaB in response to diverse cellular stresses. *Mol. Cell. Biol.* **23:** 5651–5663.

Jousse C., Oyadomari S., Novoa I., Lu P.D., Zhang Y., Harding H.P., and Ron D. 2003. Inhibition of a constitutive translation initiation factor 2α phosphatase, CReP, promotes survival of stressed cells. *J. Cell Biol.* **163:** 767–775.

Kantor L., Harding H.P., Ron D., Schiffmann R., Kaneski C.R., Kimball S.R., and Elroy-Stein O. 2005. Heightened stress response in primary fibroblasts expressing mutant eIF2B genes from CACH/VWM leukodystrophy patients. *Hum. Genet.* **118:** 99–106.

Kaufman R.J. 2002. Orchestrating the unfolded protein response in health and disease. *J. Clin. Invest.* **110:** 1389–1398.

Kimball S.R. and Jefferson L.S. 1990. Mechanism of the inhibition of protein synthesis by vasopressin in rat liver. *J. Biol. Chem.* **265:** 16794–16798.

Kojima E., Takeuchi A., Haneda M., Yagi F., Hasegawa T., Yamaki K.I., Takeda K., Akira S., Shimokata K., and Isobe K.I. 2003. The function of GADD34 is a recovery from a shut-off of protein synthesis induced by ER stress-elucidation by GADD34-deficient mice. *FASEB J.* **17:** 1573–1575.

Koumenis C., Naczki C., Koritzinsky M., Rastani S., Diehl A., Sonenberg N., Koromilas A., and Wouters B.G. 2002. Regulation of protein synthesis by hypoxia via activation of the endoplasmic reticulum kinase PERK and phosphorylation of the translation initiation factor eIF2alpha. *Mol. Cell. Biol.* **22:** 7405–7416.

Kumar R., Azam S., Sullivan J., Owen C., Cavener D., Zhang P., Ron D., Harding H., Chen J., Han A., et al. 2001. Brain ischemia and reperfusion activates the eukaryotic initiation factor 2α kinase, PERK. *J. Neurochem.* **77:** 1418–1421.

Leegwater P.A., Vermeulen G., Konst A.A., Naidu S., Mulders J., Visser A., Kersbergen P., Mobach D., Fonds D., van Berkel C.G., et al. 2001. Subunits of the translation initiation factor eIF2B are mutant in leukoencephalopathy with vanishing white matter. *Nat. Genet.* **29:** 383–388.

Lin W., Harding H., Ron D., and Popko B. 2005. Endoplasmic reticulum stress modulates the response of myelinating oligodendrocytes to the immune cytokine interferon-γ. *J. Cell Biol.* **169:** 603–612.

Lu P.D., Harding H.P., and Ron D. 2004a. Translation re-initiation at alternative open reading frames regulates gene expression in an integrated stress response. *J. Cell Biol.* **167:** 27–33.

Lu P.D., Jousse C., Marciniak S.J., Zhang Y., Novoa I., Scheuner D., Kaufman R.J., Ron D., and Harding H.P. 2004b. Cytoprotection by pre-emptive conditional phosphorylation of translation initiation factor 2. *EMBO J.* **23:** 169–179.

Marciniak S.J., Garcia-Bonilla L., Hu J., Harding H.P., and Ron D. 2006. Activation-dependent substrate recruitment by the eukaryotic translation initiation factor 2 kinase PERK. *J. Cell Biol.* **172:** 201–209.

Marciniak S.J., Yun C.Y., Oyadomari S., Novoa I., Zhang Y., Jungreis R., Nagata K., Harding H.P., and Ron D. 2004. CHOP induces death by promoting protein synthesis and oxidation in the stressed endoplasmic reticulum. *Genes Dev.* **18:** 3066–3077.

Maurin A., Jousse C., Averous J., Parry L., Bruhat A., Cherasse Y., Zeng H., Zhang Y., Harding H.P., Ron D., and Fafournoux P. 2005. The GCN2 kinase biases feeding behavior to maintain amino acid homeostasis in omnivores. *Cell Metab.* **1:** 273–277.

McCullough K.D., Martindale J.L., Klotz L.O., Aw T.Y., and Holbrook N.J. 2001. Gadd153 sensitizes cells to endoplasmic reticulum stress by down-regulating Bcl2 and perturbing the cellular redox state. *Mol. Cell. Biol.* **21:** 1249–1259.

McEwen E., Kedersha N., Song B., Scheuner D., Gilks N., Han A., Chen J.J., Anderson P., and Kaufman R.J. 2005. Heme-regulated inhibitor (HRI) kinase-mediated phosphorylation of eukaryotic translation initiation factor 2 (eIF2) inhibits translation, induces stress granule formation, and mediates survival upon arsenite exposure. *J. Biol. Chem.* **280:** 16925–16933.

Mellor A.L. and Munn D.H. 2004. IDO expression by dendritic cells: Tolerance and tryptophan catabolism. *Nat. Rev. Immunol.* **4:** 762–774.

Mori K. 2000. Tripartite management of unfolded proteins in the endoplasmic reticulum. *Cell* **101:** 451–454.

Munn D.H., Sharma M.D., Baban B., Harding H.P., Zhang Y., Ron D., and Mellor A.L. 2005. GCN2 kinase in T cells mediates proliferative arrest and anergy induction in response to indoleamine 2,3-dioxygenase. *Immunity* **22:** 633–642.

Novoa I., Zeng H., Harding H., and Ron D. 2001. Feedback inhibition of the unfolded protein response by GADD34-mediated dephosphorylation of eIF2α. *J. Cell Biol.* **153:** 1011–1022.

Novoa I., Zhang Y., Zeng H., Jungreis R., Harding H.P., and Ron D. 2003. Stress-induced gene expression requires programmed recovery from translational repression. *EMBO J.* **22:** 1180–1187.

Oyadomari S., Koizumi A., Takeda K., Gotoh T., Akira S., Araki E., and Mori M. 2002. Targeted disruption of the Chop gene delays endoplasmic reticulum stress-mediated diabetes. *J. Clin. Invest.* **109:** 525–532.

Ozcan U., Cao Q., Yilmaz E., Lee A.H., Iwakoshi N.N., Ozdelen E., Tuncman G., Gorgun C., Glimcher L.H., and Hotamisligil G.S. 2004. Endoplasmic reticulum stress links obesity, insulin action, and type 2 diabetes. *Science* **306:** 457–461.

Pannevis M.C. and Houlihan D.F. 1992. The energetic cost of protein synthesis in isolated hepatocytes of rainbow trout (*Oncorhynchus mykiss*). *J. Comp. Physiol. B* **162:** 393–400.

Paschen W. 2003. Shutdown of translation: Lethal or protective? Unfolded protein response versus apoptosis. *J. Cereb. Blood Flow Metab.* **23:** 773–779.

Patil C. and Walter P. 2001. Intracellular signaling from the endoplasmic reticulum to the nucleus: The unfolded protein response in yeast and mammals. *Curr. Opin. Cell Biol.* **13:** 349–355.

Richardson J.P., Mohammad S.S., and Pavitt G.D. 2004. Mutations causing childhood ataxia with central nervous system hypomyelination reduce eukaryotic initiation factor 2B complex formation and activity. *Mol. Cell. Biol.* **24:** 2352–2363.

Ron D. and Hampton R. 2004. Membrane biogenesis and the unfolded protein response. *J. Cell Biol.* **167:** 23–25.

Ryu E.J., Harding H.P., Angelastro J.M., Vitolo O.V., Ron D., and Greene L.A. 2002. Endoplasmic reticulum stress and the unfolded protein response in cellular models of Parkinson's disease. *J. Neurosci.* **22:** 10690–10698.

Scheuner D., Song B., McEwen E., Gillespie P., Saunders T., Bonner-Weir S., and Kaufman R.J. 2001. Translational control is required for the unfolded protein response and in-vivo glucose homeostasis. *Mol. Cell* **7:** 1165–1176.

Scheuner D., Mierde D.V., Song B., Flamez D., Creemers J.W., Tsukamoto K., Ribick M., Schuit F.C., and Kaufman R.J. 2005. Control of mRNA translation preserves endoplasmic reticulum function in beta cells and maintains glucose homeostasis. *Nat. Med.* **11:** 757–764.

Silva R., Ries V., Oo T., Yarygina O., Jackson-Lewis V., Ryu E., Lu P., Marciniak S., Ron D., Przedborski S., et al. 2005. CHOP/GADD153 is a mediator of apoptotic death in substantia nigra dopamine neurons in an in vivo neurotoxin model of parkinsonism. *J. Neurochem.* **95:** 974–986.

Skelly R.H., Schuppin G.T., Ishihara H., Oka Y., and Rhodes C.J. 1996. Glucose-regulated translational control of proinsulin biosynthesis with that of the proinsulin endopeptidases PC2 and PC3 in the insulin-producing MIN6 cell line. *Diabetes* **45:** 37–43.

Sood R., Porter A.C., Ma K., Quilliam L.A., and Wek R.C. 2000. Pancreatic eukaryotic initiation factor-2alpha kinase (PEK) homologues in humans, *Drosophila melanogaster* and *Caenorhabditis elegans* that mediate translational control in response to endoplasmic reticulum stress. *Biochem. J.* **346:** 281–293.

Southwood C.M., Garbern J., Jiang W., and Gow A. 2002. The unfolded protein response modulates disease severity in Pelizaeus-Mezbacher disease. *Neuron* **36:** 585–596.

Srivastava S.P., Kumar K.U., and Kaufman R.J. 1998. Phosphorylation of eukaryotic translation initiation factor 2 mediates apoptosis in response to activation of the double-stranded RNA-dependent protein kinase. *J. Biol. Chem.* **273:** 2416–2423.

Tan S., Somia N., Maher P., and Schubert D. 2001. Regulation of antioxidant metabolism by translation initiation factor 2alpha. *J. Cell Biol.* **152:** 997–1006.

Tu B.P. and Weissman J.S. 2004. Oxidative protein folding in eukaryotes: Mechanisms and consequences. *J. Cell Biol.* **164:** 341–346.

Vattem K.M. and Wek R.C. 2004. Reinitiation involving upstream ORFs regulates ATF4 mRNA translation in mammalian cells. *Proc. Natl. Acad. Sci.* **101:** 11269–11274.

Zhang P., McGrath B., Li S., Frank A., Zambito F., Reinert J., Gannon M., Ma K., McNaughton K., and Cavener D.R. 2002a. The PERK eukaryotic initiation factor 2 alpha kinase is required for the development of the skeletal system, postnatal growth, and the function and viability of the pancreas. *Mol. Cell. Biol.* **22:** 3864–3874.

Zhang P., McGrath B.C., Reinert J., Olsen D.S., Lei L., Gill S., Wek S.A., Vattem K.M., Wek R.C., Kimball S.R., et al. 2002b. The GCN2 eIF2alpha kinase is required for adaptation to amino acid deprivation in mice. *Mol. Cell. Biol.* **22:** 6681–6688.

Zinszner H., Kuroda M., Wang X., Batchvarova N., Lightfoot R.T., Remotti H., Stevens J.L., and Ron D. 1998. CHOP is implicated in programmed cell death in response to impaired function of the endoplasmic reticulum. *Genes Dev.* **12:** 982–995.

14

Signaling to Translation Initiation

Brian Raught
University Health Network
Ontario Cancer Institute and McLaughlin
Centre for Molecular Medicine
Toronto, Ontario M5G 1L7, Canada

Anne-Claude Gingras
Samuel Lunenfeld Research Institute
Mount Sinai Hospital
Toronto, Ontario M5G 1X5, Canada

FOLLOWING TRANSCRIPTION AND SEVERAL NUCLEAR processing events (including splicing, a quality control step, the "pioneer" round of translation, and nucleocytoplasmic transport), eukaryotic mRNAs are competent for translation. However, ribosomes alone lack the ability to identify and position themselves at an initiation codon. Instead, the protein synthetic machinery must be recruited to the mRNA 5′ end via the concerted action of the eukaryotic translation initiation factors (eIFs). This complex recruitment process, also referred to as the initiation phase of translation, culminates in the positioning of a charged ribosome (i.e., an 80S ribosome loaded with an initiator methionyl-tRNA in its P site) at an initiation codon (for a more complete description of this process, see Chapter 4). The initiation process is rate-limiting for translation in many cases and is subject to exquisite regulation.

Early studies indicated that translation rates are primarily regulated at the initiation phase and are tightly controlled in response to extracellular stimuli and stresses or changes in local environmental conditions: for example, hormone/growth factor signals, amino acid or nutrient availability, and environmental stresses such as heat or osmotic shock (see Chapter 13). Moreover, precise (often localized) regulation of the translation of specific mRNAs or mRNA classes is critical for the proper progression of a variety of physiological and developmental processes (e.g., to establish protein gradients in developing embryos or learning and memory formation). Dysregulation of translational control may also be an important component of cellular transformation (Chapter 15).

Many of the translation initiation factors were demonstrated to be phosphoproteins more than 25 years ago (see, e.g., Traugh et al. 1976; Duncan and Hershey 1985, 1987). The same stimuli and stresses observed to modulate translation rates were also found to effect changes in the levels of phosphate incorporated into these translation factors and/or to alter the number of isoforms observed via two-dimensional isoelectric focusing (IEF)/SDS-polyacrylamide gel electrophoresis (PAGE) analyses. More recent studies have confirmed and extended these critical early observations.

In this chapter, we discuss the methods employed to determine the location of phosphorylation sites, describe the major signaling pathways involved in regulating translation factor phosphorylation, and summarize current knowledge regarding functional consequences of phosphorylation on individual translation factors and inhibitors.

IDENTIFICATION OF PHOSPHORYLATION SITES ON TRANSLATION FACTORS

Small-scale Targeted Phosphorylation Studies

Until recently, phosphorylation site identification on a protein of interest was generally conducted using small-scale targeted approaches. These types of studies have yielded a detailed understanding of the regulation of the phosphorylation events on several different translation factors (and other proteins involved in the regulation of translation; Fig. 1, Table 1).

Targeted studies have traditionally utilized in vitro or in vivo ^{32}P labeling followed by two-dimensional phosphopeptide mapping (see, e.g., Gingras et al. 1999b), phosphoamino acid analysis, and/or IEF/SDS-PAGE to monitor the phosphorylation state of a protein(s) of interest. Mass spectrometry (and/or Edman degradation) has often been used to identify ^{32}P-labeled phosphoproteins or phosphopeptides isolated by SDS-PAGE or thin-layer chromatography. Phosphospecific antibodies, combined with mutational analysis, have been utilized to confirm phosphorylation site identifications and allow additional follow-up functional studies. These targeted approaches are extremely well suited for the characterization of the in vivo regulation of modification sites in a protein of interest. The same methods have also been used to characterize the intracellular signaling pathways that modulate the phosphorylation state of

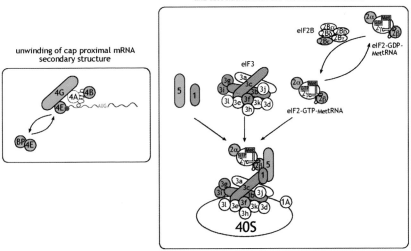

Figure 1. Multiple phosphoproteins prepare the mRNA and the 40S ribosomal subunit for translation initiation. Factors are color-coded. (*Green*) High-confidence phosphorylation sites identified by small-scale targeted studies (possibly in addition to large-scale identification); (*orange*) phosphorylation sites detected only in one or more large-scale phosphoproteomic studies; (*unshaded*) no phosphorylation sites known or detected. (*Left*) Factors involved in unwinding of cap-proximal mRNA secondary structure. eIF4G is a high-molecular-weight scaffold that interacts with eIF4E and eIF4A to form the eIF4F complex. The ubiquitous cofactor eIF4B is required for eIF4A RNA unwinding activity. eIF4F formation is regulated by the 4E-BPs. (*Right*) Priming of the 40S ribosome. eIF2 forms a ternary complex with Met-tRNA$_i^{Met}$ and GTP; eIF2B is a multisubunit guanine exchange factor necessary for GDP to GTP exchange on eIF2. These factors, as well as the ribosome-associated multisubunit complex eIF3 and the eIF1 protein, are nucleated by the eIF5 scaffolding GAP protein. (For a review of translation initiation, see Chapter 4.)

specific residues and for ordering phosphorylation events on a given protein (see, e.g., Martin-Perez and Thomas 1983; Krieg et al. 1988; Gingras et al. 2001). Targeted analyses have resulted in the identification of more than 25 serine/threonine phosphorylation sites on mammalian translation initiation factors (Table 1). However, the bulk of these sites are confined to just a few proteins (e.g., eIF2Bε, 6 sites; 4E-BP1, 7 sites; and eIF4GI, 3 sites).

It is clear that many translation factor phosphorylation sites remain unidentified, even on proteins that have been the subject of intense scrutiny in small-scale experiments. For example, two-dimensional phosphopeptide maps of eIF4GI yield at least nine highly reproducible phos-

Table 1. In vivo phosphorylation sites identified on mammalian translation initiation factors and inhibitors

Factor	In vivo phosphorylation sites[a]
eIF1	*Y30[1]*
eIF2α	**S48[2]; S51[3]**
eIF2β	**S2[4]; S67[5]**
eIF2Bε	**S466[6]; S469[6]; S540[6,7,8]**; *S544[6,8,9]*; **S717[6]; S718[6]**
eIF3b (p116)	*S83[9]; S85[9]; S125[9]*
eIF3c (p110)	*S39[10]; S166[9]; T524[9]; S909[11]*
eIF3f (p47)	**S46[12]**
eIF3g (p44)	*T41[13]; S42[9,13]*
eIF3i (p36)	*Y445[1]*
eIF4E	**S209[14,15]; S210[15]**
4E-BP1	**T37[16]; T46[16]; S65[16]; T70[16]; S84[16]; S101[17]; S112[18]**
4E-BP2	**T37[19]; T46[19]; S65[20]; T70[20]**
eIF4GI	*Y594[1]; S1146[11]*; **S1148[21]**; *S1188[21,22]*; S1210[9]; *S1232[9,21]*
eIF4GII	**S1156[23]**
p97	*T508[9,13]*
eIF4B	*S93[9,13]; Y211[22]; Y266[22]; Y316[22]; S406[22]*; **S422[24,22,11]**; *S425[11]; S498[9]; S504[9,11]*
eIF4H	*Y12[1,25]; Y45[1]; Y101[1]*
eIF5	*S389[13,26]; S390[13,26]*
eIF5B	*S107[9]; S113[9]; S135[9]; S137[11]; S164[9]; S182[9]; S183[11]; S186[9]; S190[9]; S214[13]; S1168[9]*

[a]Numbering was done according to human nomenclature. Sites identified in small-scale targeted experiments are highlighted in bold; sites identified in high-throughput shotgun sequencing are in *italics*. For small-scale experiments, only the original publications are referenced; all identifications from large-scale experiments are cited. References: (1) Rush et al. (2005); (2) Wettenhall et al. (1986); (3) Kudlicki et al. (1987) and Pathak et al. (1988); (4) Llorens et al. (2003); (5) Welsh et al. (1994); (6) Wang et al. (2001); (7) Welsh et al. (1998); (8) Woods et al. (2001); (9) Beausoleil et al. (2004); (10) Gevaert et al. (2005); (11) Kim et al. (2005); (12) Shi et al. (2003); (13) Ballif et al. (2005); (14) Joshi et al. (1995); (15) Whalen et al. (1996); (16) Fadden et al. (1997); (17) Wang et al. (2003); (18) Heesom et al. (1998); (19) Wang et al. (2005); (20) B. Raught et al. (unpubl.); (21) Raught et al. (2000b); (22) Cutillas et al. (2005); (23) Qin et al. (2003); (24) Raught et al. (2004); (25) Tao et al. (2005); (26) Majumdar et al. (2002).

phopeptides, only four of which were accounted for by the sequenced phosphorylation sites in small-scale studies (Raught et al. 2000b). In addition, eIF4B generally migrates as more than five isoforms on IEF/SDS-PAGE (Duncan and Hershey 1985), yet only a single regulated site was identified in small-scale targeted experiments (Raught et al. 2004). Thus, additional complementary approaches will be extremely useful in identifying and characterizing the full range of phosphorylation sites in translation factors.

Large-scale Phosphoproteomics

The advent of large-scale phosphoproteomics has resulted in the recent identification of a large number of putative phosphorylation sites, including many in translation initiation factors (Table 1). Although the methodologies used in large-scale approaches are outside the scope of this chapter (for reviews, see Mann et al. 2002; Kalume et al. 2003; Steen and Mann 2004; Gingras et al. 2005a), it is important to note that not all mass-spectrometric-based phosphopeptide assignments are equally reliable. Peptide assignments are made by search engines that simply compare an observed fragmentation pattern with theoretical fragmentation patterns generated in silico from a user-defined protein database. Matches are associated with a series of scores indicating how well the observed peptide fragmentation pattern matches the theoretical fragmentation pattern for a given peptide. Different laboratories utilize different software and scoring standards, and peptide assignments (as well as phosphorylation site assignments) can therefore differ dramatically in their accuracy. No community-wide standards for peptide assignment have been adopted by proteomics researchers. Thus, without directly examining the raw data for each experiment, it remains extremely difficult at this time to determine a confidence value for a given peptide/phosphopeptide assignment. Even though many of the data are undoubtedly of very high quality, here we refer to all mass-spectrometric-derived phosphopeptide assignments as "putative," to recognize this fact.

Phosphoproteomic studies have thus far been generally undirected, i.e., global in scope, with little or no enrichment for any particular functional class of protein. Most phosphoproteomic studies have employed some type of broad-spectrum enrichment strategy to isolate all phosphoproteins (or phosphopeptides) from highly complex mixtures (such as cell or tissue lysates); the phosphoproteins/peptides are then generally further separated (via liquid chromatography or other biochemical method) prior to analysis by mass spectrometry. Such techniques will likely prove to be extremely valuable in the identification of translation initiation factor modification sites in the future: Many putative phosphorylation sites have already been identified on eIF4B (see Table 1), consistent with the multiple isoforms detected by IEF/SDS-PAGE (Duncan and Hershey 1985). Several putative phosphorylation sites on eIF3b (p116), eIF3c (p110), and eIF5B have also been identified (Table 1).

Importantly, mass spectrometric technologies offer the possibility of almost unlimited semidirected approaches, in which phosphorylation sites on partially purified proteins or protein fractions (e.g., translation factor multiprotein complexes or ribosomal salt washes) may be identified. As more laboratories initiate phosphoproteome studies of this type, it is likely that the quantity—and quality—of the data will increase dramatically.

SIGNALING PATHWAYS REGULATING TRANSLATION INITIATION

It is clear that phosphorylation status can regulate the activity of translation factors. It was less clear until recently, however, which intracellular signaling pathways were responsible for effecting the phosphorylation of these proteins. Recent advances have provided abundant evidence that two signaling cascades in particular have critical roles in this process: the mTOR/PI3K/Akt and MAPK signaling modules.

TOR

Translation is a demanding process in terms of energy consumption; in fact, cells expend more energy on protein synthesis than any other activity (Schmidt 1999). Also important is the fact that mammalian cells do not have the ability to synthesize all amino acids and must therefore acquire many of them from the environment. Thus, one critical decision that cells must make in response to a "request" for increased translation by a stimulatory hormone/growth factor (or other type of) signal is whether sufficient amino acids are available in the environment to accommodate the demand. If so, the cell may proceed to synthesize protein, as required. If not, however, the cell must degrade endogenous protein to recycle into new polypeptides. Similarly, if a cell senses that energy supplies are scarce, or the cell is exposed to environmental stress, survival may require that limited energy supplies are directed away from use for protein synthesis and toward use in repair processes or protective measures. This complex "decision" process appears to be mediated (in part) by the target of rapamycin (TOR) proteins.

The TOR proteins are evolutionarily conserved high-molecular-weight protein kinases that have key roles in cell growth, proliferation, and survival (for review, see Lorberg and Hall 2004). The yeast TOR proteins were discovered in genetic screens for resistance to the antibiotic, immunosuppressant, and anticancer drug rapamycin (Heitman et al. 1991). Following cell entry, rapamycin binds to FK506-binding proteins (FKBPs); this drug-receptor gain-of-function complex can efficiently inhibit some aspects of TOR signaling. The mammalian TOR protein (mTOR) is thought to act in a checkpoint control capacity for translation, accepting signaling input from both hormone/growth factor receptors and nutrient sensors, and in this way allows translation to occur only in the presence of sufficient nutrients to fuel protein synthesis (Fig. 2).

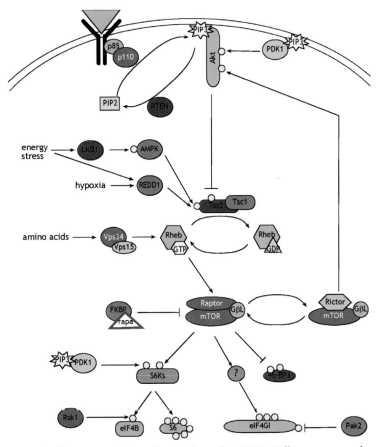

Figure 2. Signaling to translation factors through mTOR. Cells can accurately sense the levels of nutrients and energy available in their local environment, and they can quickly respond to many types of extracellular stresses. mTOR complexes act at translation checkpoints. mTOR associates with GβL; this dimer can interact with Raptor to form a rapamycin-sensitive complex that signals to translation factors. Alternatively, mTOR-GβL associates with Rictor to create a rapamycin-resistant complex signaling to Akt and the cytoskeleton. Many growth factor or hormone receptors activate PI3K (p85/p110 subunits shown)/Akt signaling to repress TSC1/2 (a Rheb GAP), and thereby allow the Rheb GTPase to signal to mTOR. Available energy levels (derived from the intracellular AMP/ATP ratio) are relayed to mTOR via an LKB1/AMPK signaling module; high levels of AMP activate TSC1/2 to repress Rheb signaling. Similarly, hypoxic stress signals via REDD to inhibit Rheb signaling activity. Amino acids signal via an unknown mechanism through hVps34/hVps15 to activate Rheb. (*Green shading*) Protein transmitting a positive signal to mTOR and/or a positive affect on translation initiation; (*red shading*) negative signaling to mTOR and/or a negative affect on translation initiation; (*yellow circles*) phosphorylation on a given substrate. (FKBP) Any of the many FK506-binding proteins in mammalian cells; (Rapa) rapamycin. Please note that this pathway has been simplified: Feedback loops and cross-talk have deliberately been excluded for clarity. See text for details.

In the absence of sufficient environmental amino acids, mTOR signaling can instruct the cell to break down endogenous proteins (primarily ribosomes) via a process called autophagy (for review, see Meijer and Codogno 2004) to generate free amino acids. Growth factors, nutrients (branched chain amino acids and glucose, in particular), and high ATP concentrations each transmit a positive signal to mTOR. Conversely, low nutrient or energy levels, and many forms of environmental stress, negatively modulate TOR signaling.

Much has been learned recently regarding how these multiple, complex inputs modulate mTOR signaling to translation (Fig. 2). mTOR can assemble into two alternative multiprotein complexes (for review, see Martin and Hall 2005). An mTOR–GβL subcomplex (GβL is the mammalian ortholog of the yeast LST8 protein) forms mutually exclusive interactions with Raptor (the yeast KOG protein ortholog) or Rictor (orthologous to yeast Avo3). These two complexes display distinct signaling activities. mTOR–GβL–Raptor (also known as mTORC1) signaling is sensitive to rapamycin and can directly phosphorylate the translation regulatory proteins S6K (the ribosomal protein S6 kinase; Chapter 17) and 4E-BP1 (discussed further below) and indirectly effects the phosphorylation of other translation factors (Kim et al. 2002; Loewith et al. 2002). Conversely, the mTOR–GβL–Rictor (mTORC2) complex is insensitive to rapamycin, but it phosphorylates Ser-473 on the serine/threonine kinase Akt (a proto-oncogene, itself an upstream component in the mTOR activation cascade; see below) and perhaps other AGC kinases (Sarbassov et al. 2004, 2005; Ali and Sabatini 2005). The mTOR–GβL–Rictor complex also signals to the actin cytoskeleton via protein kinase Cα (PKCα) and Rho (Jacinto et al. 2004; Sarbassov et al. 2004).

Intracellular signals emanating from various types of growth factor receptors are relayed through phosphoinositide 3-kinase (PI3K; for review, see Hay 2005). PI3K phosphorylates 4,5 phosphatidyl inositol bisphosphate (PIP2) at the 3′ position of the inositol ring to produce PIP3 (a process opposed by the PTEN tumor suppressor). PIP3 acts as a second messenger to recruit and activate PDK1 (phospholipid-dependent kinase 1) and Akt. PDK1 phosphorylates sites in the activation loop of AGC kinases such as Akt, GSK-3, and the S6Ks. Activated Akt signals to mTOR by phosphorylating multiple sites on the TSC2 protein (for review, see Kwiatkowski and Manning 2005). TSC2 is a component of the tuberous sclerosis complex (TSC), a tumor suppressor and GAP for the small GTPase Rheb (Stocker et al. 2003). Akt-mediated phosphorylation of TSC2 inhibits its ability to act as a Rheb GAP. Rheb was reported to interact with the mTOR–GβL–Raptor complex: Increased Rheb-GTP levels apparently activate mTOR signaling to translation factors via an ill-defined GTP-dependent mechanism (Long et al. 2005). Importantly, extensive cross-talk appears to occur between the PI3K/mTOR and Ras/MAP kinase pathways (also see below). Ras itself can activate PI3K, and mutations in neurofibromin (encoded by the tumor suppressor gene NF1, a RasGAP) lead to hyperactivation of the PI3K/mTOR pathway (Johannessen et al. 2005). Another entry point for cross-talk is at the level of TSC2, which is phosphorylated and inactivated by ERK and the 90-kD ribosomal S6 protein kinase Rsk (in addition to Akt; for review, see Kwiatkowski and Manning 2005). Additional cross-talk at the level of the translation initiation factor eIF4B is described below. Such cross-talk is likely to be cell-type- and stimulus-dependent and may allow increased flexibility in signaling to translation.

In contrast to signaling by growth factors or hormones, how nutrients signal to mTOR is less clear. The upstream components of the growth factor pathway, PI3K, PDK1, and Akt, do not appear to be involved in this process. However, Rheb appears to be required for the amino acid signaling input to mTOR, in that forced Rheb expression protects mTOR signaling from down-regulation by amino acid withdrawal (for review, see Kwiatkowski and Manning 2005). Intriguingly, increasing the levels of endogenous Rheb-GTP by decreasing the levels of the TSC GAP (GTPase-activating proteins) proteins is not sufficient to protect mTOR signaling from amino acid withdrawal, and amino acids do not regulate GTP loading on Rheb. Recently, a novel upstream component of amino acid signaling to mTOR was identified: The class III PI3K member hVps34 (which is localized to endosomes) is activated by amino acid addition, leading to an increase in endosomal PIP3. hVps34 and its binding partner hVps15 are required for amino acid signaling to mTOR (Nobukuni et al. 2005). It remains to be determined how amino acids activate hVps34 and how activated hVps34 or endosomal PIP3 signals through mTOR.

Negative environmental signals also modulate mTOR–GβL–Raptor signaling via direct regulation of TSC1/TSC2. Energy deprivation increases the AMP/ATP ratio, leading to activation of the AMP-activated protein kinase (AMPK) which phosphorylates and activates TSC2 and thereby elicits a down-regulation of signaling to mTOR (Hahn-Windgassen et al. 2005). Similarly, hypoxia leads to HIF (hypoxia-induced factor)-dependent activation of REDD1 (regulated in development and DNA damage) and REDD2, proteins that act as TSC activators (Brugarolas et al. 2004; Reiling and Hafen 2004). The AMPK and Akt inputs to the TSC proteins were previously thought to be independent, but a recent report indicates that Akt down-regulates AMPK activity by decreasing the AMPATP ratio, allowing a more complete inhibition of the TSC complex and better activation of the mTOR–GβL–Raptor complex (Hahn-Windgassen et al. 2005).

A plethora of recent studies has indicated that phosphorylation of the S6Ks, several translation factors (eIF4B, eIF4GI, and eEF2), and the 4E-BPs is mediated by mTOR signaling. However, as opposed to the upstream components of the mTOR cascade, signaling downstream from mTOR remains relatively poorly understood: As discussed further below, recent evidence suggests that control of translation initiation may be mediated (in part) by a mitogen and nutrient-regulated interaction between mTOR and eIF3 (Holz et al. 2005). Importantly, although mTOR immunoprecipitates (likely containing mTOR–GβL–Raptor, and perhaps other proteins) can phosphorylate both S6K1 and 4E-BP1 in vitro, the sequence context of the phosphorylation sites in these proteins differs dramatically, suggesting that mTOR itself may possess little substrate specificity and that associated proteins are responsible for directing the kinase to the proper substrate or site. This hypothesis is consistent with the presence of a common binding moiety (the TOS motif) in both S6K1 and 4E-BP1 that mediates an interaction with Raptor (Schalm et al. 2003). Alternatively, it remains possible that S6K and/or 4E-BP1 is not directly phosphorylated by mTOR at all, but by other kinases present in mTOR immunoprecipitates.

The Ras-MAP Kinase Pathway

Several distinct mitogen-activated protein kinase (MAPK) cascades have evolved in metazoans (Fig. 3) (for review, see Roux and Blenis 2004). Within each of these cascades, a MAPK kinase kinase (MAPKKK) is activated, which phosphorylates and activates a MAPK kinase (MAPKK), which in turn phosphorylates and activates a MAPK. MAPK can itself phosphorylate and activate additional kinases, known collectively as the MKs (MAPK-activated protein kinases).

The classic cascade consists of Raf (MAPKKK), MEK1 or 2 (Map or Erk kinase; MAPKK), and ERK1 or 2 (extracellular signal-regulated kinase; MAPK). Raf couples the ERK cascade to the small GTPase Ras: Activation of Ras by growth factors or phorbol esters activates the entire cascade, yielding activated ERK. In addition to directly phosphorylating several nuclear substrates (e.g., Elk-1, c-Fos, and STAT3), activated ERKs can also phosphorylate and activate three distinct MK families: the Rsk, Msk, and Mnk families (see below).

Other MAPK cascades are specialized for transmitting stress signals (for review, see Johnson and Lapadat 2002). The stress-activated protein kinases include the JNK and p38MAPK families, which are activated by common MAPKKK enzymes (including MEKK1-4), but divergent MAPKKs (MEK3 and MEK6 for p38MAPK, and MEK4 and MEK7 for

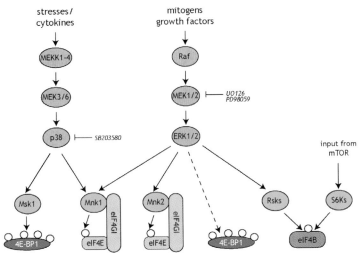

Figure 3. Signaling to translation initiation via MAP kinase cascades. MAPK modules are composed of a MAPKKK (e.g., Raf), a MAPKK (e.g., MEK1), and a MAPK (e.g., ERK1). The ERK signaling cascade is activated principally downstream from Ras by growth factors and mitogens, whereas the p38MAPK cascade responds to cellular stresses. MAPKs phosphorylate and activate downstream kinases, known as MKs, a group including the Rsks, Mnks, and Msks. Some MKs (e.g., Mnk1) can be activated by both stresses (via p38MAPK) and growth factors (through ERK); other MKs exhibit specificity for one pathway. Mnk1 and Mnk2 interact with eIF4GI, which brings them to the vicinity of their substrate, eIF4E. The Rsks participate in the phosphorylation of eIF4B at Ser 422. Msk1 mediates UVB-induced phosphorylation of 4E-BP1; mitogens can also induce 4E-BP1 hyperphosphorylation through the ERK pathway, although the kinase directly responsible remains to be identified. See text for details.

JNK). Although activation of MKs has not been demonstrated thus far for the JNKs, p38MAPK does activate MKs, some of which are also activated by ERK (Mnk1 and the Msks), in addition to p38MAPK-specific MKs, Mk2, and Mk3 (for review, see Roux and Blenis 2004).

Specificity of signaling by the MAPK cascades is partially achieved by docking interactions among MAPKKs, MAPKs, and MAPK substrates, as well as regulatory proteins and scaffolding molecules (for review, see Enslen and Davis 2001). For example, the D domain found in many MAPK substrates (including almost all MKs) is characterized by a cluster of positively charged residues flanked by hydrophobic residues and interacts with a domain on the carboxyl terminus of ERK and p38MAPK that contains negatively charged and hydrophobic amino acids (Tanoue et al. 2000). Small changes in the number of contiguous basic residues, as well as in the amount and location of the hydrophobic residues within D domains of the MKs, explain their activation by ERK and/or p38MAPK.

Several MK families now appear to be important for translational control: (1) MAPK-interacting kinases (Mnks) are phosphorylated and activated by both ERKs (Mnk1 and Mnk2) and p38MAPKs (Mnk1) and thus transduce signals emanating from both growth factor stimulation and stress (Fukunaga and Hunter 1997; Wang et al. 1998). Mnks connect the MAPKs to translation initiation via phosphorylation of eIF4E (see below). (2) The 90-kD ribosomal S6 kinase (Rsk1-4) family members are activated downstream from the ERK cascade (Roux and Blenis 2004). They were initially thought to phosphorylate the ribosomal S6 protein in vivo, an event that was later reassigned to S6K1 and S6K2, but were recently implicated in the phosphorylation of eIF4B (see below). (3) Mitogen- and stress-activated protein kinase 1 (Msk1) mediates UVB-induced 4E-BP1 phosphorylation (Liu et al. 2002). (4) Mk2 and Mk3, activated by p38MAPK, are involved in controlling the translation and/or stability of tumor necrosis factor α (TNF-α) mRNA, bind and/or phosphorylate several RNA-binding proteins, and participate in the phosphorylation of the translation elongation factor 2 kinase (eEF2 kinase) (Knebel et al. 2002).

Other Pathways

Although not as well characterized, other signaling modules have also been implicated in the control of translation initiation. p21-activated protein kinase 2 (Pak2) is activated in response to various types of cellular stresses and (as discussed below) may be responsible for an inhibitory signal to translation via phosphorylation of eIF4GI (Ling et al. 2005). Ca^{2+}/CaM signaling also appears to have a role in translational control, as phosphorylation of the eIF4GII protein is mediated by calmodulin-dependent protein kinase I (CaMKI) (Qin et al. 2003). Gsk-3, itself a substrate of Akt, phosphorylates and inhibits eIF2Bε (Welsh et al. 1998). Casein kinase 2 (CK2), which has long been used to phosphorylate translation initiation factors in vitro, has recently been implicated in the phosphorylation of eIF2B in HeLa cells (Llorens et al. 2003) and eIF5 (Majumdar et al. 2002; Homma et al. 2005). Finally, the cell-cycle-regulated kinase Cdk11 was also recently demonstrated to modulate phosphorylation of eIF3f (p47; Shi et al. 2003). It is clear that even more intracellular signaling modules will be implicated in translational control as this process is studied in more cell and tissue types and under more varied types of physiological conditions.

Among other factors, these signaling pathways likely provide fine control of specific mRNAs or mRNA classes and/or increased flexibility and control under conditions that are specific to certain cell/tissue types or developmental processes.

REGULATION (AND CONSEQUENCES) OF PHOSPHORYLATION

Many directed studies have characterized how the phosphorylation state of individual residues in translation initiation factors is regulated and, in some cases, have highlighted the biological function of phosphorylation events on a given site. One particularly well-studied example, eIF2α phosphorylation, is described in detail in Chapters 12 and 13. Here, we highlight our current state of knowledge regarding the phosphorylation of the other mammalian translation initiation factors and the 4E-BPs.

4E-BPs

One of the best understood translation regulatory mechanisms is the control of eIF4F formation by the 4E-BP family of translation repressors (for review, see Gingras et al. 1999a; Raught et al. 2000a). The 4E-BPs compete with the eIF4Gs for an overlapping binding site on eIF4E: 4E-BP and eIF4G binding to eIF4E are mutually exclusive. 4E-BP binding to eIF4E abrogates eIF4F formation and thereby inhibits cap-dependent translation (whereas cap-independent translation may actually be enhanced by the 4E-BPs) (Svitkin et al. 2005). 4E-BP binding to eIF4E is regulated by phosphorylation (Lin et al. 1994; Pause et al. 1994): Whereas hypophosphorylated 4E-BPs bind avidly to eIF4E, hyperphosphorylated 4E-BPs do not. Due to their central role in the regulation of translation initiation, the 4E-BPs have been the subject of intensive study.

Although the function of 4E-BPs is well understood at the molecular level, the physiological role of the 4E-BPs is only now beginning to emerge. Several earlier studies indicated that overexpression of mammalian 4E-BPs in transformed cell lines leads to reversion of the transformed phenotype (Rousseau et al. 1996; Polunovsky et al. 2000). Similarly, forced expression of 4E-BP mutants that bind constitutively to eIF4E (human 4E-BP1 in cell lines or *Drosophila* 4E-BP in the wing) (Miron et al. 2001; Fingar et al. 2002) yields a decrease in cell size. Taken together, these studies pointed to a role for the 4E-BPs as negative growth regulators.

Flies lacking d4E-BP (the sole 4E-BP isoform in this organism), however, develop normally, with no reported differences in cell size. Instead, d4E-BP null flies are hypersensitive to bacterial infection, wounding, oxidative stress (Bernal and Kimbrell 2000; Tettweiler et al. 2005), and starvation (Teleman et al. 2005; Tettweiler et al. 2005). In particular, d4E-BP$^{-/-}$ animals demonstrate a marked inability to down-regulate fat metabolism in response to starvation; starved d4E-BP$^{-/-}$ flies thus lose fat stores and die at dramatically increased rates compared to wild-type animals (Teleman et al. 2005). Conversely, forced d4E-BP overexpression or long-term rapamycin treatment (which constitutively activates endogenous d4E-BP) elicits increased fat levels and increased survival time under starvation conditions. These data are all consistent with a model in which the 4E-BPs act as a metabolic "brake" in response to various types of cellular stress (Teleman et al. 2005; Tettweiler et al. 2005). For example, in cells lacking sufficient environmental amino acids and/or energy to fuel protein synthesis (nutritional stress), or in cells exposed to environmental stresses (e.g., infection or high oxidant levels), the 4E-BPs can rapidly decrease cap-dependent translation to concentrate limited resources on an appropriate stress response.

Mice (and other mammals) possess three 4E-BPs: 4E-BP1 is the predominant isoform in adipocytes, and 4E-BP2 is more abundant in brain (Tsukiyama-Kohara et al. 2001). The respective phenotypes of the knockout animals are consistent with this tissue distribution. Similar to that observed in flies, although viable and apparently healthy, the 4E-BP1 knockout mouse is leaner than its wild-type counterparts, displaying smaller white fat pads and a somewhat increased metabolic rate (at least in males) (Tsukiyama-Kohara et al. 2001). The 4E-BP2 knockout mouse is impaired in some aspects of learning and memory (Banko et al. 2005). Clearly, much work remains on the characterization of double- and triple-knockout animals. In the more complex mammalian system, it will also be critical to address the effects of various types of stress.

4E-BP1 phosphorylation was observed long before its gene was cloned: In rodent adipocytes, a small heat- and acid-stable protein was reported to be one of the most highly phosphorylated proteins in response to insulin stimulation (see, e.g., Blackshear et al. 1982). This protein was later cloned independently as (1) PHAS-I, a phosphorylated heat- and acid-stable insulin-stimulated protein (Lin et al. 1994), and (2) 4E-BP1, an interaction partner of eIF4E (Pause et al. 1994). Numerous reports have established that 4E-BP phosphorylation is up-regulated following treatment of cells, tissues, or organisms with a vast array of stimuli, including hormones, growth factors, cytokines, and nutrients (for summary, see Gingras et al. 1999a, 2004; Raught et al. 2000a). Conversely, 4E-BPs are dephosphorylated following infection with some viruses, in response to cellular stress or following nutrient deprivation. The primary signaling pathway controlling 4E-BP1 phosphorylation is the PI3K/mTOR module, although some reports have implicated RAS/ERK and p38MAPK in the phosphorylation of some 4E-BP1 residues under specific conditions (see, e.g., Herbert et al. 2002; Liu et al. 2002; Naegele and Morley 2004).

The functional significance of global 4E-BP1 hyperphosphorylation is well understood—a decrease in affinity for eIF4E, leading to a relief of translational repression. 4E-BP1 phosphorylation occurs on multiple sites: Thr-37, Thr-46, Ser-65, Thr-70, Ser-83, Ser-101 (all lying in a Ser/ThrPro consensus), and Ser-112 (a SerGln site). Phosphorylation on these sites proceeds in a hierarchical manner: Phosphorylation on Thr-37 and Thr-46 is required as a priming event for subsequent phospho-rylation of Thr-70, which precedes phosphorylation of Ser-65 and release from eIF4E (Gingras et al. 1999b, 2001; Mothe-Satney et al. 2000). In a recent report, phosphorylation of Ser-101 was found to be constitutive but required for efficient phosphorylation of Ser-65 in vivo (Wang et al. 2003). The function of phosphorylation at Ser-84 and Ser-112 is less clear: Ser-112 phosphorylation was postulated to act as a priming event (Heesom et al. 1998), although subsequent studies have found no effect of mutating this site to an alanine on the phosphorylation of other sites (Ferguson et al. 2003).

Due to the hierarchical nature of 4E-BP1 phosphorylation, it has been somewhat difficult to dissect the function of each particular phos-phorylation event and to determine the specific signaling inputs that modulate the modification of each residue (e.g., affecting Thr-37 or Thr-46 phosphorylation can indirectly affect the phosphorylation of other sites). It has been convincingly demonstrated, however, that the sensitiv-ity of the different sites to rapamycin varies: In the presence of serum, phosphorylation of Thr-37 and Thr-46 is only modestly affected by ra-pamycin treatment, whereas phosphorylation on Ser-65 and Thr-70 is nearly abrogated (Gingras et al. 2001; Wang et al. 2005). Ser-101 and Ser-112 are apparently not rapamycin-sensitive (Heesom et al. 1998; Wang et al. 2003). Similarly, Thr-37 and Thr-46 are only mildly affected by the presence of serum or activation of the PI3K/Akt pathway, whereas Ser-65 and Thr-70 phosphorylation are exquisitely sensitive to these treat-ments (see, e.g., Gingras et al. 1998). The identity of the kinases respon-sible for phosphorylating each of the sites in vivo remains unclear: mTOR immunoprecipitates have been reported to phosphorylate five of these sites in some studies (Brunn et al. 1997a,b), whereas other reports have indicated a clear preference for phosphorylation of the priming sites, Thr-37 and Thr-46 (Burnett et al. 1998; Gingras et al. 1999b). The identification of 4E-BP kinases, as well as the quantitative impact of the phosphorylation of individual sites on binding to eIF4E, will undoubtedly continue to be intensively pursued.

Two critical sequence elements contribute to 4E-BP1 phosphorylation. The TOS (TOR signaling) motif, identified in both the carboxyl terminus of all 4E-BPs and the amino terminus of the S6Ks, mediates an interaction with Raptor (either directly or through another molecule) (Schalm and Blenis 2002; Schalm et al. 2003). Mutations in this binding site dramatically reduce phosphorylation of 4E-BP1 on Thr-37/Thr-46, Ser-65, and Thr-70, but inhibition of phosphorylation at Ser-65 and Thr-70 is more profound than that at Thr-37/Thr-46. Competition between the S6K and 4E-BP TOS motifs for an interaction with Raptor can explain earlier observations indicating that 4E-BP1 phosphorylation is inhibited by S6K1 overexpression (von Manteuffel et al. 1997). Overexpression of an S6K1 TOS mutant does not inhibit 4E-BP1 phosphorylation (Schalm and Blenis 2002). A second sequence motif located in the amino-terminal portion of 4E-BP1, RAIP (named according to its amino acid sequence), is also required for 4E-BP1 phosphorylation (Tee and Proud 2002; Beugnet et al. 2003; Choi et al. 2003). Mutation of this motif has profound effects on the phosphorylation of the Thr-37/Thr-46 priming sites. Although a report indicated that mutations in RAIP also influenced 4E-BP1 binding to Raptor (Choi et al. 2003), this has not been confirmed in other studies (see, e.g., Beugnet et al. 2003), and the mechanism by which RAIP sequences influence 4E-BP1 phosphorylation remains unclear.

Importantly, mTOR also appears to signal to phosphatases to regulate 4E-BP1 phosphorylation. A calyculin-A-inhibitable phosphatase that exhibited activity toward 4E-BP1 coprecipitates with mTOR (Peterson et al. 1999). In addition, pulse-chase studies clearly indicate that rapamycin treatment rapidly activates a phosphatase which is specifically active against Ser-65 and Thr-70, the two sites that are most sensitive to rapamycin treatment (B. Raught et al., unpubl.). The *Saccharomyces cerevisiae* TOR proteins signal through PP2A-type phosphatases, via the phosphatase-binding proteins Tap42 and Tip41 (Di Como and Arndt 1996; Jacinto et al. 2001). Although these proteins are conserved in mammalian cells, and both are involved in binding to PP2A-type phosphatases (Chen et al. 1998; Gingras et al. 2005b), it remains controversial whether they are involved in mTOR signaling to translation.

eIF4E

Phosphorylation of mammalian eIF4E occurs principally on a single residue, Ser-209 (Joshi et al. 1995; Whalen et al. 1996), and is mediated by the kinases Mnk1 and Mnk2 (Fukunaga and Hunter 1997; Wang et al. 1998; Pyronnet et al. 1999; Waskiewicz et al. 1999). The Mnks physically interact with eIF4G, effectively localizing the kinases to their substrate (Pyronnet et al. 1999). There has been much debate recently as to the physiological function of phosphorylation at eIF4E Ser-209 (Scheper and Proud 2002 and references therein). The effects of eIF4E phosphorylation on cap-dependent translation rates in vitro are insignificant. However, forced hyperphosphorylation of eIF4E through overexpression of the Mnks in mammalian cells or in *Drosophila* leads to a reduction in translation and/or growth rates (Knauf et al. 2001). Phosphorylation also appears to elicit a modest decrease in the affinity of eIF4E for capped mRNA or capped nucleotide analogs (for review, see Scheper and Proud 2002), a caveat being that most experiments have evaluated the association of capped mRNA with prephosphorylated eIF4E, an approach that could miss dynamically regulated interactions. Mutational analyses have indicated that eIF4E phosphorylation is required for efficient nucleocytoplasmic transport of the cyclin D1 mRNA and for cellular transformation (Topisirovic et al. 2004).

A fly expressing a *Drosophila* eIF4E phosphorylation mutant exhibits developmental delays and reduced cell and organism size in adults (Lachance et al. 2002). Whether this is due to a lack of eIF4E phosphorylation per se is unknown, as the effects of the amino acid substitution on eIF4E function are difficult to assess. In this regard, mice lacking both Mnk1 and Mnk2 are deficient in eIF4E phosphorylation, yet global translation rates are unaltered, and the mice develop normally (Ueda et al. 2004). Similarly, *Drosophila* mutants in the Mnk ortholog Lk6 develop normally under standard culture conditions (Arquier et al. 2005; Reiling et al. 2005). Importantly, however, when nutrients are limiting, a reduction in organism size is observed in the Lk6 mutant fly (Reiling et al. 2005). Diet-dependent effects of the Lk6 mutation on growth could reconcile many of the contradictory findings regarding the role of eIF4E phosphorylation: Mnk activity may become limiting in response to a poor nutrient supply (or other types of environmental stresses). Taken together, these data suggest that phosphorylation of eIF4E is not critical for growth under normal conditions. In the presence of adequate nutrients, 4E-BP1 is phosphorylated downstream from mTOR, resulting in a sufficient pool of eIF4E for incorporation into eIF4F. Under starvation conditions, however, 4E-BP1 is dephosphorylated and sequesters a proportion of eIF4E to decrease the abundance of active eIF4F. In such cases, loss of eIF4E phosphorylation may lead to decreased translation efficiency and growth reduction.

eIF4Gs

In mammals, the eIF4G family is composed of three large scaffolding proteins (eIF4GI, eIF4GII, and p97) that have critical roles in both cap-dependent and cap-independent translation initiation. eIF4GI and eIF4GII appear to be functionally equivalent; both proteins bind to the same complement of translation factors, and both proteins display translation stimulatory activity in vitro and in vivo (Gradi et al. 1998). Interestingly, however, the phosphorylation state of the eIF4G proteins does not appear to be regulated by the same intracellular signaling pathways (discussed below). Adding even more complexity, multiple eIF4GI mRNAs may be expressed in mammalian cells, some of which can contain five or more translation start sites. More study will be required to characterize functional differences between the various eIF4GI isoforms (Byrd et al. 2002, 2005). p97 shares homology only with the carboxy-terminal two-thirds of the two "full-length" eIF4Gs. p97 can also interact with eIF3, eIF4A, and Mnk1, but it lacks the amino-terminal eIF4E- and PABP-binding sites present in the eIF4Gs. In vitro translation assays suggested that p97 may have a negative role in cap-dependent translation and/or a positive role in cap-independent translation (Imataka et al. 1997). Gene knockout studies indicate that although global translation rates are relatively unaffected by the absence of p97, this protein is required for differentiation and embryogenesis (Yamanaka et al. 2000).

eIF4GI and eIF4GII share 46% overall identity, but this similarity is not evenly distributed throughout the proteins; the middle and carboxy-terminal regions of eIF4GI and eIF4GII, responsible for eIF3/eIF4A and eIF4A/Mnk1 binding, respectively, are highly homologous (>60% identity). In contrast, the region that lies between these protein-binding sites (the so-called "hinge") displays only 17% identity. Two-dimensional phosphopeptide mapping and mass spectrometry analysis showed that three phosphorylation sites, Ser-1148, Ser-1188, and Ser-1232 (numbered according to the 1600-amino acid human protein and originally reported as Ser-1108, Ser-1148, and Ser-1192 in a 1560-amino acid amino-terminally truncated protein), were identified in the eIF4GI hinge region (Raught et al. 2000b). These residues are not conserved in eIF4GII or p97. PI3K/mTOR signaling regulates the phosphorylation of all three of these residues in response to hormonal or mitogenic stimulation. In whole animals, increased eIF4F complex assembly following feeding was associated with a tenfold increase in eIF4GI Ser-1148 phosphorylation (Vary and Lynch 2005). The Pak2 kinase was recently reported to interact with eIF4GI. In response to cellular stress, Pak2 phosphorylates eIF4GI Ser-896, which lies in the middle eIF3/eIF4A-binding region. Consistent with decreased translation following cellular stress, this phosphorylation event was observed to inhibit translation, possibly via inhibition of cap binding or dissociation of the eIF4E-eIF4G complex (Ling et al. 2005). This disruption may, however, also be due to binding of Pak2 to the eIF4E-binding region, rather than phosphorylation per se. Large-scale proteomic studies recently identified additional phosphorylation sites on eIF4GI: Tyr-594 (Rush et al. 2005), Ser-1146 (Kim et al. 2005), and Ser-1210 (Beausoleil et al. 2004).

eIF4GII phosphorylation does not appear to be regulated by the same intracellular signaling pathways as eIF4GI. In HEK293 cells, eIF4GII phosphorylation is not responsive to serum or hormonal stimulation, nor is it sensitive to chemical inhibitors of the PI3K/mTOR signaling modules (Raught et al. 2000b). However, phosphorylation screening with CaMKI, a Ca^{2+}/calmodulin-dependent protein kinase for which very few substrates are known, identified eIF4GII (Qin et al. 2003). Consistent with these data, eIF4GII phosphorylation was found to be sensitive to the calcium ionophore ionomycin. The CaMKI phosphorylation site was localized to Ser-1156, located in the "hinge" region of eIF4GII. Thus, although the known protein-binding regions (as well as all of the known scaffolding/recruiting functions) have been conserved in the two eIF4G proteins, the "hinge" regions appear to have undergone divergent evolution, apparently resulting in the two proteins acquiring responsiveness to two different intracellular signaling pathways. Further study of signaling to the eIF4G family may shed light on this intriguing idea.

eIF4B

The DEAD box RNA helicase eIF4A (a component of the eIF4F complex) displays only very low levels of helicase activity in isolation (Rogers et al. 2001). This activity is significantly stimulated, however, by the ubiquitous eIF4B protein (Trachsel et al. 1977; Benne and Hershey 1978; Pestova et al. 1996; Morino et al. 2000). eIF4B possesses no autonomous catalytic activity (Grifo et al. 1984), but instead, it appears to enhance the affinity of eIF4A for ATP and mRNA (Rogers et al. 1999), thereby increasing its processivity (Rogers et al. 2001). For more in-depth discussions of eIF4A/eIF4B functions, see Chapters 4 and 9.

eIF4B is a phosphoprotein that migrates as multiple isoelectric variants (consistent with multiple phosphorylation site) in the IEF/SDS-PAGE system (Duncan and Hershey 1984). eIF4B hyperphosphorylation, as elicited by mitogens and phorbol esters, correlates with cellular growth status (Duncan and Hershey 1985) and an increase in the translation rates of mRNAs possessing long structured 5'-untranslated regions (5'UTRs). In fact, eIF4B is required for 48S translation initiation complex formation in vitro on 5'UTRs with even moderate amounts of secondary structure (Dmitriev et al. 2003). Interestingly, recombinant eIF4B protein cannot substitute for native eIF4B in this system (Dmitriev et al. 2003), suggesting that one or more posttranslational modifications occurring in eukaryotic cells are required for eIF4B activity.

We identified a serum- and mitogen-inducible phosphorylation site on eIF4B, Ser-422 (Raught et al. 2004). Further study suggested that this residue is an S6K phosphorylation site; to date, eIF4B is the only translation initiation factor demonstrated to be downstream from the S6Ks. Interestingly, however, some degree of eIF4B Ser-422 phosphorylation was also observed to be rapamycin-resistant in some cell and tissue types (Shahbazian et al. 2006). Residual, rapamycin-resistant eIF4B Ser-422 phosphorylation was also observed in mice lacking both the S6K1 and S6K2 genes. The MEK inhibitor U0126 and the MKK1/ERK1/ERK2 inhibitor PD184352 both blocked the residual serum-stimulated, rapamycin-resistant eIF4B Ser-422 phosphorylation, suggesting that modification of this residue is also effected by MAPKs. Finally, RNA interference (RNAi) directed against the MAPK-regulated Rsk proteins (the 90-kD ribosomal S6 kinases, a family of four isoforms in mammals) indicated that one or more members of this protein family are responsible for the rapamycin-resistant eIF4B Ser-422 phosphorylation. eIF4B Ser-422 phosphorylation thus appears to be modulated via both the PI3K/mTOR pathway (via the S6Ks) and the MAPK pathway (via the Rsks).

Ser-422 phosphorylation is also one of the few translation initiation factor modification events for which we have a putative function: regulation of eIF3 binding (it is not known, however, whether this interaction is direct or indirect). Two separate studies (Holz et al. 2005; Shahbazian et al. 2006) have demonstrated that (1) an eIF4B–eIF3 interaction is modulated by nutrients and mitogens, (2) mutation of eIF4B Ser-422 to alanine leads to a significant decrease in the interaction with eIF3, and (3) mutation of Ser-422 to aspartate or glutamate (to create a phosphomimetic protein) increases the eIF4B–eIF3 interaction.

Large-scale phosphoproteome studies identified Ser-93 (Ballif et al. 2004; Beausoleil et al. 2004), Tyr-211, Tyr-266, Tyr-316, Ser-406 (Cutillas et al. 2005), Ser-422 (Cutillas et al. 2005; Kim et al. 2005), Ser-425 (Kim et al. 2005), Ser-498 (Beausoleil et al. 2004), Ser-504 (Beausoleil et al. 2004; Kim et al. 2005), and Ser-597 (Ballif et al. 2004) as phosphorylation sites in eIF4B. Further targeted studies will be required to validate the identifications and to determine the functional significance of the large-scale observations.

eIF3

The eIF3 complex (consisting of at least 12 proteins in mammals) (Mayeur et al. 2003) appears to be critical for the organization of higher-order structures on and around the 40S ribosomal subunit, and recent evidence suggests that it may act as a signaling scaffold for S6K/mTOR signaling (Holz et al. 2005). eIF3 was one of the first translation factors demonstrated to be a phosphoprotein, as well as one of the earliest shown to undergo regulated phosphorylation in response to insulin and phorbol esters. Earlier studies indicated that the eIF3a (p170; Pincheira et al. 2001) and eIF3b (p116; Morley and Traugh 1990) subunits are phosphorylated in vivo under various conditions, whereas large-scale phosphoproteomic studies have putatively identified phosphorylation sites in subunits b, c, f, g, and i (see Table 1).

Blenis and coworkers (Holz et al. 2005) have suggested that eIF3 acts as a dynamic scaffold for mTOR and S6K1. These researchers demonstrated that inactive S6K1 associates with eIF3. Stimulation of serum-starved cells with hormones, mitogens, or phorbol esters, or addition of amino acids to nutrient-deprived cells, stimulates the docking of the mTOR–GβL–Raptor complex to eIF3 in a rapamycin-sensitive manner. mTOR–GβL–Raptor can phosphorylate and activate the eIF3-bound S6K1. At the same time, hormone/nutrient stimulation leads to an association of eIF3 with S6K1 substrates such as eIF4B and the ribosomal S6 protein. Activation of S6K1 by mTOR leads to its release from the eIF3/mTOR complex, allowing it to phosphorylate its translational targets. Further mechanistic details will be sure to follow. It will be extremely interesting to determine how the phosphorylation status of eIF3 may modulate its interaction with these (and other) signaling molecules.

eIF2

eIF2 is a heterotrimer required for the loading of methionyl-initiator tRNA onto the 40S ribosomal subunit (Chapter 9). eIF2α phosphorylation, which leads to sequestration of its guanine exchange factor (GEF) eIF2B, represents a major checkpoint in translational control and is discussed in depth in Chapters 12 and 13.

eIF2B

Reloading of GTP on eIF2 is regulated at the level of the activity of its GEF, eIF2B (a five-subunit protein complex), a positive effector of translation initiation. Phosphorylation of the catalytic eIF2Bε subunit is regulated by insulin and amino acids and has an important regulatory role. eIF2Bε is multiply phosphorylated in vivo on Ser-466, Ser-469, Ser-540, Ser-544, Ser-717, and Ser-718 (Wang et al. 2001; Beausoleil et al. 2004; Gevaert et al. 2005). eIF2Bε can also be modified in vitro on these sites by several different kinases (including GSK-3, CKI, and CKII; see Wang et al. 2001 and references therein). Phosphorylation (in vitro) by CKI or CKII stimulates eIF2B activity, but phosphorylation by GSK-3 (in vitro and in vivo) has the opposite effect (for review, see Kimball and Jefferson 2000). Phosphorylation of eIF2B by GSK-3 in vivo on Ser-540 is an important mechanism in mediating the action of insulin on eIF2B GEF activity: Insulin activates Akt, which inhibits GSK-3, resulting in the dephosphorylation of Ser-540 and GEF activation. The kinases responsible for phosphorylating the other eIF2Bε sites in vivo, as well as the mechanism(s) by which amino acids control eIF2B activity, remain to be identified.

eIF5

eIF5 is a GTPase-activating protein (GAP) specific for eIF2 and additionally serves as a scaffold molecule for the assembly of the multifactor complex (MFC, composed of the initiation factors eIF1, eIF2, eIF3, and eIF5, and Met-tRNA$_i^{Met}$). The MFC is an important intermediate in translation and is reviewed in detail in Chapters 4 and 9. Due to its important enzymatic and scaffolding roles, phosphorylation of eIF5 is likely to affect translation. Two groups have reported that casein kinase 2 (CK2) interacts with and phosphorylates eIF5 on two carboxy-terminal serine residues (corresponding to Ser-389/390 in the human molecule) (Majumdar et al. 2002; Homma et al. 2005). (These phospho-sites were also identified in a large-scale phosphoproteomic study of developing brain; Ballif et al. 2004). CK2-mediated phosphorylation is posited to be required for the association of eIF5 with eIF2 and is therefore critical for eIF5 GAP function. Expression of an eIF5 protein in which these phosphorylation sites were mutated to alanine residues resulted in a decrease in mature MFC formation and a resultant decrease in the transition from

40S to 80S translation initiation complexes on mRNA. In addition, the mutant eIF5 protein elicited perturbation of cell cycle progression and a significant reduction in growth rate. The possible regulatory roles of these phosphorylation events have not yet been addressed.

eIF5B

eIF5B is the mammalian homolog of the prokaryotic IF2 protein, and it (along with eIF1A) mediates the docking of the 60S ribosomal subunit onto the mRNA-associated 40S subunit to form functional 80S ribosomes. Interestingly, large-scale phosphoproteomic studies have identified 11 putative eIF5B phosphorylation sites (see Table 1), with all but one located in the amino terminus.

CONCLUSIONS

Significant progress has been made recently in (1) the identification of phosphorylation sites in translation initiation factors, (2) the characterization of how these phosphorylation sites are regulated in response to extracellular stimuli and stress, and (3) our understanding of the intracellular signaling cascades that mediate the phosphorylation of translation initiation factors. In particular, the importance of the PI3K/mTOR and MAPK pathways has been highlighted recently, and our understanding of mTOR signaling has dramatically improved in the last couple of years. The very recent findings concerning signaling downstream from mTOR (as it is docked on eIF3) provide a foundation for further study in this extremely interesting area.

Advances in mass spectrometric techniques have dramatically improved our ability to identify phosphorylation sites (and other types of PTMs). Indeed, undirected, large-scale shotgun proteomics studies have already identified more putative phosphorylation sites in the translation initiation factors than were identified in the previous 25 years using standard directed approaches. Further improvement in mass spectrometer technology, coupled with advancements in our ability to utilize mass spectrometers to quantitate phosphopeptides, will lead to a new era in our understanding of the regulation of translation via intracellular signaling pathways. Coupled with standard smaller-scale targeted studies, we expect much progress to be made in this regard in the next 5 years.

The remaining major hurdle is to determine at a mechanistic level how these modifications affect translation factor activity and then to clarify how these changes in activity relate to the control of global translation rates or the translation of specific mRNAs or mRNA classes. This is not a simple task: For example, studies on the single phosphorylation site of eIF4E have reported both positive and negative effects on translation, depending on the system used and the parameters measured. Ideally, a mutant protein defective in phosphorylation should be expressed under its own promoter in the absence of the endogenous protein (using a knockin approach), which is not always straightforward in higher eukaryotes. Changes in structure imparted by the serine (or threonine) to alanine substitution may also cloud results. Similarly, mutation of one phosphoresidue can adversely affect the phosphorylation state of other residues, e.g., in the case of hierarchical phosphorylation events.

Finally, although we have focused solely on phosphorylation in this review, we posit that other posttranslational modifications will also be found to have critical roles in translational control; e.g., the L28 ribosomal protein is ubiquitinated in a cell-cycle-regulated manner (Spence et al. 2000), but how this might affect translation is unknown. Fortunately, modern mass spectrometric techniques, combined with traditional cell biological and biochemical approaches, should allow us to study the effects of these types of modifications on translational control.

REFERENCES

Ali S.M. and Sabatini D.M. 2005. Structure of S6 kinase 1 determines whether raptor-mTOR or rictor-mTOR phosphorylates its hydrophobic motif site. *J. Biol. Chem.* **280:** 19445–19448.

Arquier N., Bourouis M., Colombani J., and Leopold P. 2005. *Drosophila* Lk6 kinase controls phosphorylation of eukaryotic translation initiation factor 4E and promotes normal growth and development. *Curr. Biol.* **15:** 19–23.

Ballif B.A., Villen J., Beausoleil S.A., Schwartz D., and Gygi S.P. 2004. Phosphoproteomic analysis of the developing mouse brain. *Mol. Cell. Proteomics* **3:** 1093–1101.

Ballif B.A., Roux P.P., Gerber S.A., MacKeigan J.P., Blenis J., and Gygi S.P. 2005. Quantitative phosphorylation profiling of the ERK/p90 ribosomal S6 kinase-signaling cassette and its targets, the tuberous sclerosis tumor suppressors. *Proc. Natl. Acad. Sci.* **102:** 667–672.

Banko J.L., Poulin F., Hou L., DeMaria C.T., Sonenberg N., and Klann E. 2005. The translation repressor 4E-BP2 is critical for eIF4F complex formation, synaptic plasticity, and memory in the hippocampus. *J. Neurosci.* **25:** 9581–9590.

Beausoleil S.A., Jedrychowski M., Schwartz D., Elias J.E., Villen J., Li J., Cohn M.A., Cantley L.C., and Gygi S.P. 2004. Large-scale characterization of HeLa cell nuclear phosphoproteins. *Proc. Natl. Acad. Sci.* **101:** 12130–12135.

Benne R. and Hershey J.W. 1978. The mechanism of action of protein synthesis initiation factors from rabbit reticulocytes. *J. Biol. Chem.* **253:** 3078–3087.

Bernal A. and Kimbrell D.A. 2000. *Drosophila* Thor participates in host immune defense and connects a translational regulator with innate immunity. *Proc. Natl. Acad. Sci.* **97:** 6019–6024.

Beugnet A., Wang X., and Proud C.G. 2003. Target of rapamycin (TOR)-signaling and RAIP motifs play distinct roles in the mammalian TOR-dependent phosphorylation of initiation factor 4E-binding protein 1. *J. Biol. Chem.* **278:** 40717–40722.

Blackshear P.J., Nemenoff R.A., and Avruch J. 1982. Preliminary characterization of a heat-stable protein from rat adipose tissue whose phosphorylation is stimulated by insulin. *Biochem. J.* **204:** 817–824.

Brugarolas J., Lei K., Hurley R.L., Manning B.D., Reiling J.H., Hafen E., Witters L.A., Ellisen L.W., and Kaelin W.G., Jr. 2004. Regulation of mTOR function in response to hypoxia by REDD1 and the TSC1/TSC2 tumor suppressor complex. *Genes Dev.* **18:** 2893–2904.

Brunn G.J., Fadden P., Haystead T.A., and Lawrence J.C., Jr. 1997a. The mammalian target of rapamycin phosphorylates sites having a (Ser/Thr)-Pro motif and is activated by antibodies to a region near its COOH terminus. *J. Biol. Chem.* **272:** 32547–32550.

Brunn G.J., Hudson C.C., Sekulic A., Williams J.M., Hosoi H., Houghton P.J., Lawrence J.C., Jr., and Abraham R.T. 1997b. Phosphorylation of the translational repressor PHAS-I by the mammalian target of rapamycin. *Science* **277:** 99–101.

Burnett P.E., Barrow R.K., Cohen N.A., Snyder S.H., and Sabatini D.M. 1998. RAFT1 phosphorylation of the translational regulators p70 S6 kinase and 4E-BP1. *Proc. Natl. Acad. Sci.* **95:** 1432–1437.

Byrd M.P., Zamora M., and Lloyd R.E. 2002. Generation of multiple isoforms of eukaryotic translation initiation factor 4GI by use of alternate translation initiation codons. *Mol. Cell. Biol.* **22:** 4499–4511.

———. 2005. Translation of eukaryotic translation initiation factor 4GI (eIF4GI) proceeds from multiple mRNAs containing a novel cap-dependent internal ribosome entry site (IRES) that is active during poliovirus infection. *J. Biol. Chem.* **280:** 18610–18622.

Chen J., Peterson R.T., and Schreiber S.L. 1998. Alpha 4 associates with protein phosphatases 2A, 4, and 6. *Biochem. Biophys. Res. Commun.* **247:** 827–832.

Choi K.M., McMahon L.P., and Lawrence J.C., Jr. 2003. Two motifs in the translational repressor PHAS-I required for efficient phosphorylation by mammalian target of rapamycin and for recognition by raptor. *J. Biol. Chem.* **278:** 19667–19673.

Cutillas P.R., Geering B., Waterfield M.D., and Vanhaesebroeck B. 2005. Quantification of gel-separated proteins and their phosphorylation sites by LC-MS using unlabeled internal standards: Analysis of phosphoprotein dynamics in a B cell lymphoma cell line. *Mol. Cell. Proteomics* **4:** 1038–1051.

Di Como C.J. and Arndt K.T. 1996. Nutrients, via the Tor proteins, stimulate the association of Tap42 with type 2A phosphatases. *Genes Dev.* **10:** 1904–1916.

Dmitriev S.E., Terenin I.M., Dunaevsky Y.E., Merrick W.C., and Shatsky I.N. 2003. Assembly of 48S translation initiation complexes from purified components with mRNAs that have some base pairing within their 5′ untranslated regions. *Mol. Cell. Biol.* **23:** 8925–8933.

Duncan R. and Hershey J.W. 1984. Heat shock-induced translational alterations in HeLa cells. Initiation factor modifications and the inhibition of translation. *J. Biol. Chem.* **259:** 11882–11889.

———. 1985. Regulation of initiation factors during translational repression caused by serum depletion. Covalent modification. *J. Biol. Chem.* **260:** 5493–5497.

———. 1987. Initiation factor protein modifications and inhibition of protein synthesis.

Mol. Cell. Biol. **7:** 1293–1295.

Enslen H. and Davis R.J. 2001. Regulation of MAP kinases by docking domains. *Biol. Cell* **93:** 5–14.

Fadden P., Haystead T.A., and Lawrence J.C., Jr. 1997. Identification of phosphorylation sites in the translational regulator, PHAS-I, that are controlled by insulin and rapamycin in rat adipocytes. *J. Biol. Chem.* **272:** 10240–10247.

Ferguson G., Mothe-Satney I., and Lawrence J.C., Jr. 2003. Ser-64 and Ser-111 in PHAS-I are dispensable for insulin-stimulated dissociation from eIF4E. *J. Biol. Chem.* **278:** 47459–47465.

Fingar D.C., Salama S., Tsou C., Harlow E., and Blenis J. 2002. Mammalian cell size is controlled by mTOR and its downstream targets S6K1 and 4EBP1/eIF4E. *Genes Dev.* **16:** 1472–1487.

Fukunaga R. and Hunter T. 1997. MNK1, a new MAP kinase-activated protein kinase, isolated by a novel expression screening method for identifying protein kinase substrates. *EMBO J.* **16:** 1921–1933.

Gevaert K., Staes A., Van Damme J., De Groot S., Hugelier K., Demol H., Martens L., Goethals M., and Vandekerckhove J. 2005. Global phosphoproteome analysis on human HepG2 hepatocytes using reversed-phase diagonal LC. *Proteomics* **5:** 3589–3599.

Gingras A.C., Aebersold R., and Raught B. 2005a. Advances in protein complex analysis using mass spectrometry. *J. Physiol.* **563:** 11–21.

Gingras A.C., Raught B., and Sonenberg N. 1999a. eIF4 initiation factors: Effectors of mRNA recruitment to ribosomes and regulators of translation. *Annu. Rev. Biochem.* **68:** 913–963.

Gingras A.C., Raught B., and Sonenberg N. 2004. mTOR signaling to translation. *Curr. Top. Microbiol. Immunol.* **279:** 169–197.

Gingras A.C., Kennedy S.G., O'Leary M.A., Sonenberg N., and Hay N. 1998. 4E-BP1, a repressor of mRNA translation, is phosphorylated and inactivated by the Akt(PKB) signaling pathway. *Genes Dev.* **12:** 502–513.

Gingras A.C., Gygi S.P., Raught B., Polakiewicz R.D., Abraham R.T., Hoekstra M.F., Aebersold R., and Sonenberg N. 1999b. Regulation of 4E-BP1 phosphorylation: A novel two-step mechanism. *Genes Dev.* **13:** 1422–1437.

Gingras A.C., Caballero M., Zarske M., Sanchez A., Hazbun T.R., Fields S., Sonenberg N., Hafen E., Raught B., and Aebersold R. 2005b. A novel, evolutionarily conserved protein phosphatase complex involved in Cisplatin sensitivity. *Mol. Cell. Proteomics* **4:** 1725–1740.

Gingras A.C., Raught B., Gygi S.P., Niedzwiecka A., Miron M., Burley S.K., Polakiewicz R.D., Wyslouch-Cieszynska A., Aebersold R., and Sonenberg N. 2001. Hierarchical phosphorylation of the translation inhibitor 4E-BP1. *Genes Dev.* **15:** 2852–2864.

Gradi A., Imataka H., Svitkin Y.V., Rom E., Raught B., Morino S., and Sonenberg N. 1998. A novel functional human eukaryotic translation initiation factor 4G. *Mol. Cell. Biol.* **18:** 334–342.

Grifo J.A., Abramson R.D., Satler C.A., and Merrick W.C. 1984. RNA-stimulated ATPase activity of eukaryotic initiation factors. *J. Biol. Chem.* **259:** 8648–8654.

Hahn-Windgassen A., Nogueira V., Chen C.C., Skeen J.E., Sonenberg N., and Hay N. 2005. Akt activates the mammalian target of rapamycin by regulating cellular ATP level and AMPK activity. *J. Biol. Chem.* **280:** 32081–32089.

Hay N. 2005. The Akt-mTOR tango and its relevance to cancer. *Cancer Cell* **8:** 179–183.

Heesom K.J., Avison M.B., Diggle T.A., and Denton R.M. 1998. Insulin-stimulated kinase from rat fat cells that phosphorylates initiation factor 4E-binding protein 1 on the rapamycin-insensitive site (serine-111). *Biochem. J.* **336:** 39–48.

Heitman J., Movva N.R., and Hall M.N. 1991. Targets for cell cycle arrest by the immunosuppressant rapamycin in yeast. *Science* **353:** 905–909.

Herbert H.P., Tee A.R., and Proud C.G. 2002. The extracellular signal-regulated kinase pathway regulates the phosphorylation of 4E-BP1 at multiple sites. *J. Biol. Chem.* **277:** 11591–11596.

Holz M.K., Ballif B.A., Gygi S.P., and Blenis J. 2005. mTOR and S6K1 mediate assembly of the translation preinitiation complex through dynamic protein interchange and ordered phosphorylation events. *Cell* **123:** 569–580.

Homma M.K., Wada I., Suzuki T., Yamaki J., Krebs E.G., and Homma Y. 2005. CK2 phosphorylation of eukaryotic translation initiation factor 5 potentiates cell cycle progression. *Proc. Natl. Acad. Sci.* **102:** 15688–15693.

Imataka H., Olsen H.S., and Sonenberg N. 1997. A new translational regulator with homology to eukaryotic translation initiation factor 4G. *EMBO J.* **16:** 817–825.

Jacinto E., Guo B., Arndt K.T., Schmelzle T., and Hall M.N. 2001. TIP41 interacts with TAP42 and negatively regulates the TOR signaling pathway. *Mol. Cell* **8:** 1017–1026.

Jacinto E., Loewith R., Schmidt A., Lin S., Ruegg M.A., Hall A., and Hall M.N. 2004. Mammalian TOR complex 2 controls the actin cytoskeleton and is rapamycin insensitive. *Nat. Cell Biol.* **6:** 1122–1128.

Johannessen C.M., Reczek E.E., James M.F., Brems H., Legius E., and Cichowski K. 2005. The NF1 tumor suppressor critically regulates TSC2 and mTOR. *Proc. Natl. Acad. Sci.* **102:** 8573–8578.

Johnson G.L. and Lapadat R. 2002. Mitogen-activated protein kinase pathways mediated by ERK, JNK, and p38 protein kinases. *Science* **298:** 1911–1912.

Joshi B., Cai A.L., Keiper B.D., Minich W.B., Mendez R., Beach C.M., Stepinski J., Stolarski R., Darzynkiewicz E., and Rhoads R.E. 1995. Phosphorylation of eukaryotic protein synthesis initiation factor 4E at Ser-209. *J. Biol. Chem.* **270:** 14597–14603.

Kalume D.E., Molina H., and Pandey A. 2003. Tackling the phosphoproteome: Tools and strategies. *Curr. Opin. Chem. Biol.* **7:** 64–69.

Kim D.H., Sarbassov D.D., Ali S.M., King J.E., Latek R.R., Erdjument-Bromage H., Tempst P., and Sabatini D.M. 2002. mTOR interacts with raptor to form a nutrient-sensitive complex that signals to the cell growth machinery. *Cell* **110:** 163–175.

Kim J.E., Tannenbaum S.R., and White F.M. 2005. Global phosphoproteome of HT-29 human colon adenocarcinoma cells. *J. Proteome Res.* **4:** 1339–1346.

Kimball S.R. and Jefferson L.S. 2000. Regulation of translation initiation in mammalian cells by amino acids. In *Translational control of gene expression* (ed. N. Sonenberg et al.), pp. 561–579. Cold Spring Harbor Laboratory Press, Cold Spring Harbor, New York.

Knauf U., Tschopps C., and Gram H. 2001. Negative regulation of protein translation by mitogen-activated protein kinase-interacting kinases 1 and 2. *Mol. Cell. Biol.* **21:** 5500–5511.

Knebel A., Haydon C.E., Morrice N., and Cohen P. 2002. Stress-induced regulation of eukaryotic elongation factor 2 kinase by SB203580-sensitive and insensitive pathways. *Biochem. J.* **367:** 525–532.

Krieg J., Hofsteenge J., and Thomas G. 1988. Identification of the 40 S ribosomal protein S6 phosphorylation sites induced by cycloheximide. *J. Biol. Chem.* **263:** 11473–11477.

Kudlicki W., Wettenhall R.E., Kemp B.E., Szyszka R., Kramer G., and Hardesty B. 1987. Evidence for a second phosphorylation site on eIF-2 alpha from rabbit reticulocytes. *FEBS Lett.* **215:** 16–20.

Kwiatkowski D.J. and Manning B.D. 2005. Tuberous sclerosis: A GAP at the crossroads of multiple signaling pathways. *Hum. Mol. Genet.* **14:** R251–R258.

Lachance P.E., Miron M., Raught B., Sonenberg N., and Lasko P. 2002. Phosphorylation of eukaryotic translation initiation factor 4E is critical for growth. *Mol. Cell. Biol.* **22:** 1656–1663.

Lin T.A., Kong X., Haystead T.A., Pause A., Belsham G., Sonenberg N., and Lawrence J.C., Jr. 1994. PHAS-I as a link between mitogen-activated protein kinase and translation initiation. *Science* **266:** 653–656.

Ling J., Morley S.J., and Traugh J.A. 2005. Inhibition of cap-dependent translation via phosphorylation of eIF4G by protein kinase Pak2. *EMBO J.* **24:** 4094–4105.

Liu G., Zhang Y., Bode A.M., Ma W.Y., and Dong Z. 2002. Phosphorylation of 4E-BP1 is mediated by the p38/MSK1 pathway in response to UVB irradiation. *J. Biol. Chem.* **277:** 8810–8816.

Llorens F., Roher N., Miro F.A., Sarno S., Ruiz F.X., Meggio F., Plana M., Pinna L.A., and Itarte E. 2003. Eukaryotic translation-initiation factor eIF2beta binds to protein kinase CK2: Effects on CK2alpha activity. *Biochem. J.* **375:** 623–631.

Loewith R., Jacinto E., Wullschleger S., Lorberg A., Crespo J.L., Bonenfant D., Oppliger W., Jenoe P., and Hall M.N. 2002. Two TOR complexes, only one of which is rapamycin sensitive, have distinct roles in cell growth control. *Mol. Cell* **10:** 457–468.

Long X., Lin Y., Ortiz-Vega S., Yonezawa K., and Avruch J. 2005. Rheb binds and regulates the mTOR kinase. *Curr. Biol.* **15:** 702–713.

Lorberg A. and Hall M.N. 2004. TOR: The first 10 years. *Curr. Top. Microbiol. Immunol.* **279:** 1–18.

Majumdar R., Bandyopadhyay A., Deng H., and Maitra U. 2002. Phosphorylation of mammalian translation initiation factor 5 (eIF5) in vitro and in vivo. *Nucleic Acids Res.* **30:** 1154–1162.

Mann M., Ong S.E., Gronborg M., Steen H., Jensen O.N., and Pandey A. 2002. Analysis of protein phosphorylation using mass spectrometry: Deciphering the phosphoproteome. *Trends Biotechnol.* **20:** 261–268.

Martin D.E. and Hall M.N. 2005. The expanding TOR signaling network. *Curr. Opin. Cell Biol.* **17:** 158–166.

Martin-Perez J. and Thomas G. 1983. Ordered phosphorylation of 40S ribosomal protein S6 after serum stimulation of quiescent 3T3 cells. *Proc. Natl. Acad. Sci.* **80:** 926–930.

Mayeur G.L., Fraser C.S., Peiretti F., Block K.L., and Hershey J.W. 2003. Characterization of eIF3k: A newly discovered subunit of mammalian translation initiation factor eIF3. *Eur. J. Biochem.* **270:** 4133–4139.

Meijer A.J. and Codogno P. 2004. Regulation and role of autophagy in mammalian cells. *Int. J. Biochem. Cell Biol.* **36:** 2445–2462.

Miron M., Verdu J., Lachance P.E., Birnbaum M.J., Lasko P.F., and Sonenberg N. 2001. The translational inhibitor 4E-BP is an effector of PI(3)K/Akt signalling and cell growth in *Drosophila*. *Nat. Cell Biol.* **3:** 596–601.

Morino S., Imataka H., Svitkin Y.V., Pestova T.V., and Sonenberg N. 2000. Eukaryotic translation initiation factor 4E (eIF4E) binding site and the middle one-third of eIF4GI constitute the core domain for cap-dependent translation, and the C-terminal one-third functions as a modulatory region. *Mol. Cell. Biol.* **20:** 468–477.

Morley S.J. and Traugh J.A. 1990. Differential stimulation of phosphorylation of initiation factors eIF-4F, eIF-4B, eIF-3, and ribosomal protein S6 by insulin and phorbol esters. *J. Biol. Chem.* **265:** 10611–10616.

Mothe-Satney I., Yang D., Fadden P., Haystead T.A., and Lawrence J.C., Jr. 2000. Multiple mechanisms control phosphorylation of PHAS-I in five (S/T)P sites that govern translational repression. *Mol. Cell. Biol.* **20:** 3558–3567.

Naegele S. and Morley S.J. 2004. Molecular cross-talk between MEK1/2 and mTOR signaling during recovery of 293 cells from hypertonic stress. *J. Biol. Chem.* **279:** 46023–46034.

Nobukuni T., Joaquin M., Roccio M., Dann S.G., Kim S.Y., Gulati P., Byfield M.P., Backer J.M., Natt F., Bos J.L., et al. 2005. Amino acids mediate mTOR/raptor signaling through activation of class 3 phosphatidylinositol 3OH-kinase. *Proc. Natl. Acad. Sci.* **102:** 14238–14243.

Pathak V.K., Schindler D., and Hershey J.W. 1988. Generation of a mutant form of protein synthesis initiation factor eIF-2 lacking the site of phosphorylation by eIF-2 kinases. *Mol. Cell. Biol.* **8:** 993–995.

Pause A., Belsham G.J., Gingras A.C., Donze O., Lin T.A., Lawrence J.C., Jr., and Sonenberg N. 1994. Insulin-dependent stimulation of protein synthesis by phosphorylation of a regulator of 5′-cap function. *Nature* **371:** 762–767.

Pestova T.V., Hellen C.U., and Shatsky I.N. 1996. Canonical eukaryotic initiation factors determine initiation of translation by internal ribosomal entry. *Mol. Cell. Biol.* **16:** 6859–6869.

Peterson R.T., Desai B.N., Hardwick J.S., and Schreiber S.L. 1999. Protein phosphatase 2A interacts with the 70-kDa S6 kinase and is activated by inhibition of FKBP12-rapamycinassociated protein. *Proc. Natl. Acad. Sci.* **96:** 4438–4442.

Pincheira R., Chen Q., Huang Z., and Zhang J.T. 2001. Two subcellular localizations of eIF3 p170 and its interaction with membrane-bound microfilaments: Implications for alternative functions of p170. *Eur. J. Cell Biol.* **80:** 410–418.

Polunovsky V.A., Gingras A.C., Sonenberg N., Peterson M., Tan A., Rubins J.B., Manivel J.C., and Bitterman P.B. 2000. Translational control of the antiapoptotic function of Ras. *J. Biol. Chem.* **275:** 24776–24780.

Pyronnet S., Imataka H., Gingras A.C., Fukunaga R., Hunter T., and Sonenberg N. 1999. Human eukaryotic translation initiation factor 4G (eIF4G) recruits mnk1 to phosphorylate eIF4E. *EMBO J.* **18:** 270–279.

Qin H., Raught B., Sonenberg N., Goldstein E.G., and Edelman A.M. 2003. Phosphorylation screening identifies translational initiation factor 4GII as an intracellular target of Ca(2+)/calmodulin-dependent protein kinase I. *J. Biol. Chem.* **278:** 48570–48579.

Raught B., Gingras A.C., and Sonenberg N. 2000a. Regulation of ribosomal recruitment in eukaryotes. In *Translational control of gene expression* (ed. N. Sonenberg et al.), pp. 245–293. Cold Spring Harbor Laboratory Press, Cold Spring Harbor, New York.

Raught B., Gingras A.C., Gygi S.P., Imataka H., Morino S., Gradi A., Aebersold R., and Sonenberg N. 2000b. Serum-stimulated, rapamycin-sensitive phosphorylation sites in the eukaryotic translation initiation factor 4GI. *EMBO J.* **19:** 434–444.

Raught B., Peiretti F., Gingras A.C., Livingstone M., Shahbazian D., Mayeur G.L., Polakiewicz R.D., Sonenberg N., and Hershey J.W. 2004. Phosphorylation of eucaryotic translation initiation factor 4B Ser422 is modulated by S6 kinases. *EMBO J.* **23:** 1761–1769.

Reiling J.H. and Hafen E. 2004. The hypoxia-induced paralogs Scylla and Charybdis inhibit growth by down-regulating S6K activity upstream of TSC in *Drosophila. Genes Dev.* **18:** 2879–2892.

Reiling J.H., Doepfner K.T., Hafen E., and Stocker H. 2005. Diet-dependent effects of the *Drosophila* Mnk1/Mnk2 homolog Lk6 on growth via eIF4E. *Curr. Biol.* **15:** 24–30.

Rogers G.W., Jr., Richter N.J., and Merrick W.C. 1999. Biochemical and kinetic characterization of the RNA helicase activity of eukaryotic initiation factor 4A. *J. Biol. Chem.* **274:** 12236–12244.

Rogers G.W., Jr., Richter N.J., Lima W.F., and Merrick W.C. 2001. Modulation of the helicase activity of eIF4A by eIF4B, eIF4H, and eIF4F. *J. Biol. Chem.* **276:** 30914–30922.

Rousseau D., Gingras A.C., Pause A., and Sonenberg N. 1996. The eIF4E-binding proteins 1 and 2 are negative regulators of cell growth. *Oncogene* **13:** 2415–2420.

Roux P.P. and Blenis J. 2004. ERK and p38 MAPK-activated protein kinases: A family of protein kinases with diverse biological functions. *Microbiol. Mol. Biol. Rev.* **68:** 320–344.

Rush J., Moritz A., Lee K.A., Guo A., Goss V.L., Spek E.J., Zhang H., Zha X.M., Polakiewicz R.D., and Comb M.J. 2005. Immunoaffinity profiling of tyrosine phosphorylation in cancer cells. *Nat. Biotechnol.* **23:** 94–101.

Sarbassov D.D., Guertin D.A., Ali S.M., and Sabatini D.M. 2005. Phosphorylation and regulation of Akt/PKB by the rictor-mTOR complex. *Science* **307:** 1098–1101.

Sarbassov D.D., Ali S.M., Kim D.H., Guertin D.A., Latek R.R., Erdjument-Bromage H., Tempst P., and Sabatini D.M. 2004. Rictor, a novel binding partner of mTOR, defines a rapamycin-insensitive and raptor-independent pathway that regulates the cytoskeleton. *Curr. Biol.* **14:** 1296–1302.

Schalm S.S. and Blenis J. 2002. Identification of a conserved motif required for mTOR signaling. *Curr. Biol.* **12:** 632–639.

Schalm S.S., Fingar D.C., Sabatini D.M., and Blenis J. 2003. TOS motif-mediated raptor binding regulates 4E-BP1 multisite phosphorylation and function. *Curr. Biol.* **13:** 797–806.

Scheper G.C. and Proud C.G. 2002. Does phosphorylation of the cap-binding protein eIF4E play a role in translation initiation? *Eur. J. Biochem.* **269:** 5350–5359.

Schmidt E.V. 1999. The role of c-myc in cellular growth control. *Oncogene* **18:** 2988–2996.

Shahbazian D., Roux P.P., Mieulet V., Cohen M.S., Raught B., Taunton J., Hershey J.W.B., Blenis J., Pende M., and Sonenberg N. 2006. The mTOR/PI3K and MAPK pathways converge on eIF4B to control its phosphorylation and activity. *EMBO J.* **25:** 2781–2791.

Shi J., Feng Y., Goulet A.C., Vaillancourt R.R., Sachs N.A., Hershey J.W., and Nelson M.A. 2003. The p34cdc2-related cyclin-dependent kinase 11 interacts with the p47 subunit of eukaryotic initiation factor 3 during apoptosis. *J. Biol. Chem.* **278:** 5062–5071.

Spence J., Gali R.R., Dittmar G., Sherman F., Karin M., and Finley D. 2000. Cell cycle-regulated modification of the ribosome by a variant multiubiquitin chain. *Cell* **102:** 67–76.

Steen H. and Mann M. 2004. The ABC's (and XYZ's) of peptide sequencing. *Nat. Rev. Mol. Cell Biol.* **5:** 699–711.

Stocker H., Radimerski T., Schindelholz B., Wittwer F., Belawat P., Daram P., Breuer S., Thomas G., and Hafen E. 2003. Rheb is an essential regulator of S6K in controlling cell growth in *Drosophila*. *Nat. Cell Biol.* **5:** 559–565.

Svitkin Y.V., Herdy B., Costa-Mattioli M., Gingras A.C., Raught B., and Sonenberg N. 2005. Eukaryotic translation initiation factor 4E availability controls the switch between cap-dependent and internal ribosomal entry site-mediated translation. *Mol. Cell. Biol.* **25:** 10556–10565.

Tanoue T., Adachi M., Moriguchi T., and Nishida E. 2000. A conserved docking motif in MAP kinases common to substrates, activators and regulators. *Nat. Cell Biol.* **2:** 110–116.

Tao W.A., Wollscheid B., O'Brien R., Eng J.K., Li X.J., Bodenmiller B., Watts J.D., Hood L., and Aebersold R. 2005. Quantitative phosphoproteome analysis using a dendrimer conjugation chemistry and tandem mass spectrometry. *Nat. Methods* **2:** 591–598.

Tee A.R. and Proud C.G. 2002. Caspase cleavage of initiation factor 4E-binding protein 1 yields a dominant inhibitor of cap-dependent translation and reveals a novel regulatory motif. *Mol. Cell. Biol.* **22:** 1674–1683.

Teleman A.A., Chen Y.W., and Cohen S.M. 2005. 4E-BP functions as a metabolic brake used under stress conditions but not during normal growth. *Genes Dev.* **19:** 1844–1848.

Tettweiler G., Miron M., Jenkins M., Sonenberg N., and Lasko P.F. 2005. Starvation and oxidative stress resistance in *Drosophila* are mediated through the eIF4E-binding protein, d4E-BP. *Genes Dev.* **19:** 1840–1843.

Topisirovic I., Ruiz-Gutierrez M., and Borden K.L. 2004. Phosphorylation of the eukaryotic translation initiation factor eIF4E contributes to its transformation and mRNA transport activities. *Cancer Res.* **64:** 8639–8642.

Trachsel H., Erni B., Schreier M.H., and Staehelin T. 1977. Initiation of mammalian protein synthesis. II. The assembly of the initiation complex with purified initiation factors. *J. Mol. Biol.* **116:** 755–767.

Traugh J.A., Tahara S.M., Sharp S.B., Safer B., and Merrick W.C. 1976. Factors involved in initiation of haemoglobin synthesis can be phosphorylated in vitro. *Nature* **263:** 163–165.

Tsukiyama-Kohara K., Poulin F., Kohara M., DeMaria C.T., Cheng A., Wu Z., Gingras A.C., Katsume A., Elchebly M., Spiegelman B.M., et al. 2001. Adipose tissue reduction in mice lacking the translational inhibitor 4E-BP1. *Nat. Med.* **7:** 1128–1132.

Ueda T., Watanabe-Fukunaga R., Fukuyama H., Nagata S., and Fukunaga R. 2004. Mnk2 and Mnk1 are essential for constitutive and inducible phosphorylation of eukaryotic initiation factor 4E but not for cell growth or development. *Mol. Cell. Biol.* **24:** 6539–6549.

Vary T.C. and Lynch C.J. 2005. Meal feeding enhances formation of eIF4F in skeletal muscle: Role of increased eIF4E availability and eIF4G phosphorylation. *Am. J. Physiol. Endocrinol. Metab.* **290:** E631-E642.

von Manteuffel S.R., Dennis P.B., Pullen N., Gingras A.C., Sonenberg N., and Thomas G. 1997. The insulin-induced signalling pathway leading to S6 and initiation factor 4E binding protein 1 phosphorylation bifurcates at a rapamycin-sensitive point immediately upstream of p70s6k. *Mol. Cell. Biol.* **17:** 5426–5436.

Wang X., Beugnet A., Murakami M., Yamanaka S., and Proud C.G. 2005. Distinct signaling events downstream of mTOR cooperate to mediate the effects of amino acids and insulin on initiation factor 4E-binding proteins. *Mol. Cell. Biol.* **25:** 2558–2572.

Wang X., Li W., Parra J.L., Beugnet A., and Proud C.G. 2003. The C terminus of initiation factor 4E-binding protein 1 contains multiple regulatory features that influence its function and phosphorylation. *Mol. Cell. Biol.* **23:** 1546–1557.

Wang X., Paulin F.E., Campbell L.E., Gomez E., O'Brien K., Morrice N., and Proud C.G. 2001. Eukaryotic initiation factor 2B: Identification of multiple phosphorylation sites in the epsilon-subunit and their functions in vivo. *EMBO J.* **20:** 4349–4359.

Wang X., Flynn A., Waskiewicz A.J., Webb B.L., Vries R.G., Baines I.A., Cooper J.A., and Proud C.G. 1998. The phosphorylation of eukaryotic initiation factor eIF4E in response to phorbol esters, cell stresses, and cytokines is mediated by distinct MAP kinase pathways. *J. Biol. Chem.* **273:** 9373–9377.

Waskiewicz A.J., Johnson J.C., Penn B., Mahalingam M., Kimball S.R., and Cooper J.A. 1999. Phosphorylation of the cap-binding protein eukaryotic translation initiation factor 4E by protein kinase Mnk1 in vivo. *Mol. Cell. Biol.* **19:** 1871–1880.

Welsh G.I., Miller C.M., Loughlin A.J., Price N.T., and Proud C.G. 1998. Regulation of eukaryotic initiation factor eIF2B: Glycogen synthase kinase-3 phosphorylates a conserved serine which undergoes dephosphorylation in response to insulin. *FEBS Lett.* **421:** 125–130.

Welsh G.I., Price N.T., Bladergroen B.A., Bloomberg G., and Proud C.G. 1994. Identification of novel phosphorylation sites in the beta-subunit of translation initiation factor eIF-2. *Biochem. Biophys. Res. Commun.* **201:** 1279–1288.

Wettenhall R.E., Kudlicki W., Kramer G., and Hardesty B. 1986. The NH2-terminal sequence of the alpha and gamma subunits of eukaryotic initiation factor 2 and the phosphorylation site for the heme-regulated eIF-2 alpha kinase. *J. Biol. Chem.* **261:** 12444–12447.

Whalen S.G., Gingras A.C., Amankwa L., Mader S., Branton P.E., Aebersold R., and Sonenberg N. 1996. Phosphorylation of eIF-4E on serine 209 by protein kinase C is inhibited by the translational repressors, 4E-binding proteins. *J. Biol. Chem.* **271:** 11831–11837.

Woods Y.L., Cohen P., Becker W., Jakes R., Goedert M., Wang X., and Proud C.G. 2001. The kinase DYRK phosphorylates protein-synthesis initiation factor eIF2Bepsilon at Ser539 and the microtubule-associated protein tau at Thr212: Potential role for DYRK as a glycogen synthase kinase 3-priming kinase. *Biochem. J.* **355:** 609–615.

Yamanaka S., Zhang X.Y., Maeda M., Miura K., Wang S., Farese R.V., Jr., Iwao H., and Innerarity T.L. 2000. Essential role of NAT1/p97/DAP5 in embryonic differentiation and the retinoic acid pathway. *EMBO J.* **19:** 5533–5541.

15

Translational Control in Cancer Development and Progression

Robert J. Schneider

Department of Microbiology and NYU Cancer Institute
New York University School of Medicine
New York, New York 10016

Nahum Sonenberg

Department of Biochemistry and McGill Cancer Centre
University of McGill, Montréal
Québec H3G 1Y6, Canada

TRANSLATIONAL CONTROL HAS AN IMPORTANT ROLE in key physiological pathways that have a direct impact on cancer development and progression. These include pathways for cell proliferation and growth, cellular responses to stresses such as hypoxia and nutritional deprivation, and stimulation by mitogenic signals (for previous reviews, see Dua et al. 2001; Meric and Hunt 2002; Rosenwald 2004; Holcik and Sonenberg 2005). Consequently, regulation of protein synthesis has emerged as an important component of cancer etiology, both at the level of global control of the proteome and for selective translation of specific mRNAs and classes of mRNAs. What is surprising is how long it has taken to appreciate the central importance and elucidate the key mechanisms of translational control in cancer development and progression. Despite the infancy of this field of research, it is already apparent that translational control of cancer is multifaceted, presenting modifications unique to different types of cancers, as well as different stages and grades of disease. Changes in translation associated with cancer development and progression observed to date involve altered expression of translation components, including translation factors, ribosomes, translation factor regulatory proteins, and tRNAs; altered expression and translation of specific mRNAs; and altered activity of signal transduction pathways that control the activity of protein synthesis, both overall and of individual mRNAs. These changes are manifested in a variety of ways, including up-regulation of global protein synthesis, increased translation of individual mRNAs, and selective translation of antiapoptotic, proangiogenic, proproliferative, and hypoxia-mediated mRNAs. Other transformation-associated changes in translation are directed to uncoupling of signal transduction and translational control pathways that suppress translation during physiological stresses and impair cell growth (cell mass) and cellular proliferation (cell division).

The complexity of translational control in cancer results in part from the fact that even within a single tumor, cancer is often a heterogeneous disease, with tumors containing cells at different stages of transformation and sometimes derived from different origins. In addition, tumor cells orchestrate alterations to the extracellular matrix which becomes an integral part of the tumor. The matrix consists of a variety of stromal cells (fibroblasts and epithelial cells), the vasculature composed of endothelial and smooth-muscle cells (pericytes), and immune cells such as dendritic cells and lymphocytes. These cell types have essential roles in solid tumor growth, invasiveness (penetration of tissue basement membrane), and malignancy, including metastatic spread and growth of cancer cells to distant sites (Bissell and Radisky 2001; Bissell and Labarge 2005). All of these cells can be in different states of growth and proliferation, ranging from highly hypoxic or dead tumor cells at the necrotic core of the tumor to an actively dividing polyglot of cells in the tumor and its microenvironment. In describing alterations in the translational machinery at the molecular level, studies therefore need to distinguish between the different cell types in a tumor. Moreover, given the crucial importance of the extracellular matrix in tumor suppression and promotion properties (Bissell and Labarge 2005), the focus of translational control solely on tumor cells to the exclusion of cells of the stroma and vasculature can only partially describe the role of translation in cancer development and progression.

HISTORICAL EVIDENCE FOR TRANSLATIONAL CONTROL IN CANCER ETIOLOGY

Because protein synthesis can account for up to 40% of the total energy and oxygen requirements of the cell (Buttgereit and Brand 1995), it stands to reason that protein synthesis might be uncoupled from its control by the physiological events that limit cancer development, such as hypoxia, cell cycling, nutrient deprivation, and constitutive stimulation of mitogenic and oncogenic events that promote cellular transformation. It has been known for more than 100 years that nucleoli, the sites of rRNA synthesis and ribosome assembly, are considerably enlarged and more numerous in highly transformed cells (Pianese 1896). The link between expanded size and function of nucleoli (and therefore increased ribosome content) and malignant transformation was identified more than a century ago and ultimately utilized as a component of the pathological index for cancer malignancy (Gani 1976), and remains in use today (for review, see Koss 1982). In addition, there has long been an appreciation of the importance of protein synthesis in cell growth and proliferation, as the cell needs to double its protein mass prior to mitosis in order to divide (Pardee 1989). Thus, the rate of protein synthesis and its control should influence the growth of cancer cells.

The connection between increased rates of protein synthesis and cell proliferation was recognized in the 1970s (Johnson et al. 1976; Baxter and Stanners 1978). For example, engineered reduction of protein synthesis by half is sufficient to drive cells into the quiescent G_0 resting phase of the cell cycle, causing growth arrest and reduction of ribosome content until full protein synthesis resumes (for review, see Zetterberg et al. 1995; Rosenwald 2004). No such arrest or reduction in ribosome content occurs in transformed cells (Stanners et al. 1979). Moreover, cell cycle progression and rRNA synthesis by RNA polymerase I are coupled so as to achieve the lowest level of rRNA synthesis in M phase and the highest level in S phase (Grummt 1999). Control of rRNA synthesis is also operative during nutrient starvation (Mayer et al. 2004). Such reductions do not typically occur in highly transformed cells (for review, see Ruggero and Pandolfi 2003). Although not widely investigated, in some studies, in situ rates of protein synthesis suggest increased rates in breast and colon tumors compared to normal tissue (Heys et al. 1991, 1992). More recently, studies have begun to identify the molecular connections between the translation machinery and control of cell growth and proliferation, permitting determination of the role of components of the translation machinery and its regulators in cancer biogenesis, which is the subject of this review.

ALTERATION OF RIBOSOMAL PROTEINS AND RNAs ASSOCIATED WITH CANCER DEVELOPMENT

Dysregulation of ribosome production (rRNA, ribosomal proteins) is associated with cellular transformation and cancer progression, although the connection and its effects are not straightforward, likely involving tumor suppressor genes and proto-oncogenes (for review, see Ruggero and Pandolfi 2003). Mutation of the *DKC1* gene causes dyskeratosis congenita, a disease of premature aging and increased tumor susceptibility. The DKC1 protein is a pseudouridine synthase that modifies rRNA and telomerase RNA and is involved in ribosome biosynthesis (Ruggero and Pandolfi 2003). Impaired rRNA modification and ribosome function are associated with increased cancer risk (Ruggero and Pandolfi 2003), which was recently linked to impaired translation of internal ribosome entry site (IRES) mRNAs, including tumor suppressor p27 (Kip1), although the mechanism is not yet understood (Yoon et al. 2006). More directly, the transcription factor UBF (upstream binding factor), which acts on RNA polymerase I that synthesizes rRNAs, is a primary regulator of rRNA synthesis and is regulated by mitogenic signals and a number of proto-oncogenes and tumor suppressors (Ruggero and Pandolfi 2003). Whether increased rRNA synthesis is a secondary effect of cellular transformation or drives it has not yet been resolved. However, practically all rRNA becomes incorporated into ribosomes, suggesting that production of rRNA drives ribosome abundance and overall levels of protein synthesis (Zetterberg and Killander 1965; Liebhaber et al. 1978). Moreover, UBF activity is up-regulated by the extracellular signal-regulated kinases (ERKs) (Stefanovsky et al. 2001), PI3K (phosphatidylinositol 3-kinase), Akt, and mTOR (mammalian target of rapamycin), which are often activated in transformation and may also be increased in some cancers (Martin et al. 2004; for review, see Hay and Sonenberg 2004). These events provide molecular connections between carcinogenesis and translation activity. Additionally, the tumor suppressors retinoblastoma (Rb) protein and p53, which down-regulate rRNA synthesis by acting on UBF, are often mutated in many advanced cancers (Cavanaugh et al. 1995; Zhai and Comai 2000).

As noted above, a striking feature of many types of cancer is the increased expression of ribosomal proteins for both subunits, particularly in hematological malignancies such as leukemias, but in some solid tumors as well (Ferrari et al. 1990; Zhang et al. 1997; Bassoe et al. 1998; Loging and Reisman 1999; Uechi et al. 2001). An increase in ribosomal proteins and rRNA synthesis promotes the assembly of ribosomes. Furthermore, there is compelling evidence that ribosome production is stimulated by the proto-oncogene *myc* when overexpressed during cellular transformation (for review, see Schmidt 2004). Among the genes activated by *myc* are ribosomal protein genes and factors involved in ribosome biogenesis (Coller et al. 2000; Boon et al. 2001; Menssen and Hermeking 2002). It is also well established that *myc* promotes cell growth, which is associated with increased transcription of eIF4E mRNA (Schmidt 2004). An increase in ribosome biogenesis is therefore an important component of the ability to increase cell mass and has specific links to the control of protein synthesis (Iritani and Eisenman 1999; Iritani et al. 2002).

Evidence for an association between ribosome content and cancer development derives more directly from two studies. Mutation in *Drosophila* of a tumor suppressor gene known as *brat* (brain tumor), which controls rRNA and tRNA synthesis, results in increased rRNA levels, an enlarged brain, and development of malignant neuroblastomas (Frank et al. 2002). In contrast, a genome-wide screen in zebra fish for tumor suppressor genes by establishment of heterozygous mutations demonstrated an astonishing incidence of tumors with loss-of-function mutations in ribosomal protein genes (Amsterdam et al. 2004). Of the 12 lines of zebra fish identified with elevated cancer incidence, 11 were heterozygous for ribosomal protein gene mutations that resulted in decreased levels of 18S rRNA and ribosomes. The major tumor type consisted of malignant peripheral nerve sheet tumors (80%), although a variety of other tumor types developed as well. Collectively, these two studies may indicate that normal levels of ribosomal proteins are the important feature, possibly because they maintain translation of tumor suppressors.

A relatively underappreciated connection between the control of tRNA synthesis or abundance and cancer etiology also exists. In addition to its role in activating rRNA synthesis by RNA polymerase I, ERK and mitogen-inducible UBF also activate transcription of tRNAs and 5S rRNA by RNA polymerase III (Stefanovsky et al. 2001). Because ERK activity is up-regulated in approximately one-third of advanced human cancers due to upstream activating *ras* mutations (Downward 2003), this results in a coordinate increase in both rRNA and tRNA levels in a large number of human tumors. In addition, loss of p53 and elevated Rb and Myc proteins, which are common to a variety of cancers, stimulate RNA polymerase III activity and therefore increase tRNA synthesis, providing a link to cancer cell development (for review, see White 2005). Moreover, certain tRNA modifications are statistically associated with a higher incidence of several types of human cancers, including lung and ovarian adenocarcinomas (Dirheimer et al. 1995; Spinola et al. 2005), although cause and effect have not been established.

TRANSLATIONAL REGULATION BY uORFS AND ALTERNATE 5′UTRs

mRNAs with long and significantly structured 5′-untranslated regions (5′UTRs) are typically inefficiently translated. Strong 5′UTR secondary structure (in excess of about −40 kcal/mole) acts as an impediment to initiation by impairing eIF4F function or ribosome scanning (Pelletier and Sonenberg 1985; Chapter 4). As a general rule, the average well-translated cellular mRNA has a 5′UTR of 20–50 nucleotides in length, and approximately 90% of vertebrate mRNAs have 5'UTRs 10–200 nucleotides in length (for review, see Willis 1999; Chapter 6). A small group of mRNAs contain long 5'UTRs (several hundred to a thousand or more nucleotides), some with numerous upstream open reading frames (uORFs), that are poorly translated (Willis 1999). This group includes many mRNAs that encode proto-oncogenes, antiapoptotic proteins (Chapter 16), and growth factors involved in regulation of the cell cycle, cellular proliferation, DNA metabolism, and DNA repair. The products of these mRNAs can promote transformation if overexpressed. In the absence of alternate means of initiation, such as internal ribosome entry or ribosome shunting, mRNAs with highly structured 5'UTRs are generally impaired in translation and benefit disproportionately from increased expression of eIF4E and increased abundance of the cap-initiation complex (described later in this review). However, there are a number of examples in which alteration of mRNA sequence itself likely participates in cellular transformation, as reviewed next.

The *mdm2* gene (murine double minute gene 2) encodes an E3 ubiquitin ligase that targets the p53 and Rb proteins for proteasome-mediated decay, thereby controlling the level of p53 and affecting transformation (Haupt et al. 1997; Kubbutat et al. 1997). Mdm2 protein acts as an oncoprotein and is overexpressed in a significant number of human tumors, although it is most frequently found in soft tissue sarcomas, gliomas, and osteosarcomas (for review, see Bond et al. 2005). In some tumors, Mdm2 protein is overexpressed at the translational level without alteration of *mdm2* mRNA abundance (Landers et al. 1997; Capoulade et al. 1998). The *mdm2* gene expresses two mRNAs from two promoters, giving rise to mRNAs that differ only in their 5'UTRs. The long mRNA form that contains two uORFs is constitutively expressed in normal cells and is poorly translated (Landers et al. 1997; Piette et al. 1997; Brown et al. 1999; Jin et al. 2003). In contrast, the smaller *mdm2* mRNA contains a short 5'UTR with no uORFs and is expressed in a p53-inducible manner in advanced tumors and some tumor-derived cell lines where it is well translated (Landers et al. 1997; Piette et al. 1997; Brown et al. 1999; Jin et al. 2003).

Utilizing another *cis*-acting mechanism, the cyclin-dependent kinase inhibitor 2A gene (*CDKN2A*, also known as p16[INK4a], p14[ARF], and MST1) encodes an inhibitor of *cdk4* and *cdk6* that participates in G_1 checkpoint control and acts as a tumor suppressor through regulation of p53 and Rb proteins (for review, see Zhang and Xiong 2001; Sharpless 2005). Mutation of the *CDKN2A* gene has been observed in a wide variety of human tumors, including approximately 30% of familial cases of malignant melanomas. Some of these cases have been shown to involve mutation of

CDKN2A, creating a novel uORF that effectively captures scanning ribosomes, thereby reducing translation of CDKN2A protein and G_1 checkpoint control (Liu et al. 1999).

A number of mechanisms have been described in which mutation of specific mRNAs alters translational efficiency, associated with either cancer development or progression (for review, see Meric and Hunt 2002). One of the best-studied examples is alteration of *BRCA1* mRNA in breast cancer. Inherited mutations of the *BRCA1* gene have been identified in 3% of all breast cancers (and some ovarian cancers) and 50% of familial breast cancers and confer a high level of susceptibility (Miki et al. 1994). In addition, the majority of high-grade breast ductal carcinomas, the major presentation of breast cancer, express very low levels of BRCA1 protein (Wilson et al. 1999). BRCA1 interacts with a number of proteins, including oncogenes, tumor suppressors, and DNA-damage-repair proteins, and has effects on practically all phases of the cell cycle, broadly increasing genetic instability when mutated (Deng 2006). BRCA1 expression levels are decreased in many cancers by mutations that introduce a premature stop codon and promote nonsense-mediated mRNA decay (NMD) (Perrin-Vidoz et al. 2002). Of the 3% of sporadic breast cancer cases that involve BRCA1 mutations, some involve alteration of the BRCA1 5′UTR (Sobczak and Krzyzosiak 2002). In the mammary gland, BRCA1 is encoded by an mRNA containing a short 5′UTR and is therefore well translated, which promotes cell cycle control. One study suggests that in BRCA1-associated sporadic cases of breast cancer, a second promoter may be utilized that produces an mRNA with a much longer 5′UTR containing three uORFs, reducing BRCA1 translation (Sobczak and Krzyzosiak 2002).

A very similar mechanism has emerged to explain the overexpression of the retinoic acid receptor β2 (RARβ2) isoform. RARβ2 is a tumor suppressor that is lost in some tumors (Chen et al. 2002). RARβ2 in conjunction with retinoic acids provides protection against breast and other cancers (Chen et al. 2002). The RARβ2 mRNA contains a long (~470 nucleotides) and structured 5′UTR with multiple uORFs that strongly impair its translation (Peng et al. 2005). Variant RARβ2 transcripts with short 5′UTRs lacking the uORFs have been identified in cells that display greater sensitivity to retinoic acid (Peng et al. 2005). The greater sensitivity of tumor cells to retinoic acid inhibition is associated with increased translation of the variant RARβ2 mRNA.

Finally, many investigators have begun to explore the role of inherited polymorphisms in 5′UTR sequences, as well as polymorphisms in translation factors and translation regulatory factors in the origin of different cancers. In one particularly compelling example, which has direct implications in clinical treatment, it was shown that the thymidylate syn-thase (TS) 5′UTR contains a polymorphic tandem repeat sequence that elevates TS expression at the level of translation by three- to fourfold (Kawakami et al. 2001). TS is the target enzyme of the anticancer agent 5-fluorouracil (5FU). As demonstrated by Danenberg and coworkers, quantitation of the TS mRNA in colorectal cancers showed that those mRNAs containing a three-repeat sequence rather than a two-repeat sequence in the 5′UTR were strongly increased in TS protein but not mRNA (Kawakami et al. 2001). TS protein levels are predictive of patient response to 5FU. It was suggested that the two-repeat sequence likely sta-bilizes an inhibitory hairpin immediately downstream from the initiating AUG, whereas the hairpin is destabilized by the three-repeat sequence.

TRANSLATION INITIATION FACTORS IN CANCER DEVELOPMENT AND PROGRESSION

As described above, studies have not always distinguished translation factor alterations in the tumor itself from those of tumor-associated stro-mal cells and the endothelial and smooth-muscle cells in the tumor vasculature, nor have they accounted for the variable abundance of can-cer cells within tumors. It is noteworthy that although the development and progression of different types of solid tumors may face similar bar-riers, such as adaptation to hypoxia, nutritional deprivation, and the need to promote tumor-directed angiogenesis, the mechanisms by which translation is altered in tumors of different tissue origins and at differ-ent stages of transformation can differ substantially.

Overexpression of eIF4E

Early studies demonstrated that overexpression of eIF4E promotes frank transformation of immortalized NIH-3T3 and CHO (Chinese hamster ovary) cells (Lazaris-Karatzas et al. 1990; De Benedetti et al. 1994). Enor-mous attention has subsequently been focused on the important role that translation initiation factor abundance can have in cancer etiology. This was followed by the demonstration that Myc oncogene-dependent trans-formation of primary rodent fibroblasts requires the cooperation of eIF4E (Lazaris-Karatzas and Sonenberg 1992; De Benedetti et al. 1994) and that eIF4E overexpression stimulates *ras* oncogene activity (Lazaris-Karatzas et al. 1992) Overexpression of eIF4E has also been shown to con-fer solid-tumor growth properties to immortalized cells in subcutaneous xenotransplant nude mouse models (Lazaris-Karatzas et al. 1990; De Benedetti et al. 1994). Transformation of immortalized cells was attributed to direct overexpression of eIF4E, as antisense inhibition of eIF4E partially reversed in vitro transformation parameters such as growth in soft agar, increased cellular proliferation rates, and higher levels of mRNA translation (De Benedetti et al. 1991; Rinker-Schaeffer et al. 1992; Graff et al. 1995). More recently, it was shown that small interfering RNA (siRNA) knockdown of eIF4E in a head and neck squamous carcinoma cell line impairs in vitro parameters of tumor cell growth and transformation (Oridate et al. 2005). Conversely, overexpression of eIF4E in conjunction with Myc in murine hematopoietic stem cells enhances development of B-cell lymphomas (Wendel et al. 2004). Similarly, strong eIF4E overexpression in transgenic mice leads to the development of several types of carcinomas, predominantly B-cell lymphomas, followed by angiosarcomas, hepatocellular carcinomas, and lung adenocarcinomas. All carcinomas emerge fairly late in the life span of the mouse and require *myc* activation (Ruggero et al. 2004). Promotion of tumorigenesis by overexpressed eIF4E in multiple settings and in murine models is highly compelling, raising two questions that have been surprisingly difficult to answer clearly. Is eIF4E overexpressed in human cancers, and if so, what is its role in cancer development and progression?

eIF4E protein is overexpressed severalfold in a wide variety of cell lines derived from different human tumors (Miyagi et al. 1995). Overexpression of eIF4E in human tumors has been demonstrated, but it is still unresolved whether eIF4E is overexpressed in as wide a variety of tumor types as reported and at the levels often cited. At issue is whether immunoblot analysis of eIF4E in whole-tumor protein extracts can be considered representative of eIF4E expression levels in cancer cells in tumors. As noted earlier, a significant fraction of the tumor composition can consist of other cell types, including rapidly dividing endothelial cells and pericytes that comprise the vasculature, infiltrating immune cells, and stromal fibroblasts, all of which would be expected to express increased levels of eIF4E in accord with their significant levels of proliferation. To ensure that immunoblot studies are representative of the factor levels in tumors, it is imperative that studies utilize immunohistochemical (IHC) staining of the factor and in a statistically relevant number of tumor cell fields in pathologically defined tumor specimens. For hematological malignancies such as leukemias and lymphomas, cell sorting of peripheral blood lymphocytes using tumor cell markers followed by immunoblotting studies can be highly informative. Taking this as a gold standard, eIF4E has been found to be significantly overexpressed in the majority of non-Hodgkin's lymphomas compared to normal lymphocytes, and eIF4E overexpression is associated with disease progression from an indolent to a moderately aggressive phenotype (Wang et al. 1999). In normal cells, up to 25% of eIF4E can be found in the nucleus (Lejbkowicz et al. 1992). It was postulated that in hematological malignancies, the nuclear fraction of eIF4E contributes to disease progression (Lai and Borden 2000), although this concept has not been tested and remains unresolved.

In solid tumors, eIF4E is overexpressed in a majority of colon cancers, displaying a striking up-regulation early in disease development (Rosenwald et al. 1999; Berkel et al. 2001). eIF4E levels in lung cancer show an interesting pattern of up-regulation. eIF4E is strongly overexpressed in most bronchioloalveolar lung carcinomas, but surprisingly, not in squamous cell lung carcinomas, despite the fact that both are aggressive cancers in which tumor cells proliferate rapidly (Rosenwald et al. 2001). Some head and neck squamous carcinomas demonstrate increased expression of eIF4E (Nathan et al. 1997; Sorrells et al. 1999), where it may correlate with eIF4E gene amplification (Haydon et al. 2000). The association of eIF4E overexpression and progression to malignant disease is also observed in thyroid cancer. eIF4E is overexpressed in the majority of aggressive thyroid carcinomas but not in nonmalignant papillary carcinomas of the thyroid (Wang et al. 2001). Similarly, eIF4E is overexpressed in invasive (aggressive) bladder cancers but not in superficial (indolent) bladder cancers, although in this particular study, only eIF4E mRNA was analyzed (Sorrells et al. 1999). On the other hand, despite the fact that malignant melanomas are highly aggressive and fast-growing tumors, they do not overexpress eIF4E (Rosenwald et al. 2003). Thus, although overexpression of eIF4E is strongly associated with aggressive or malignant disease in certain cancers, this is not a universal finding.

As for the levels of eIF4E protein expression in human breast cancers, this has proved to be a complex issue and it is not yet resolved. In part, this reflects the complexity and heterogeneity of breast cancer itself, which is composed of a group of different tumor types arising from several different types of cells in the breast. In part, it also reflects problems inherent in the use of immunoblot analysis to determine protein factor levels in tumors. The analysis of tumor homogenates by immunoblot can be influenced by the density of cancer cells in the tumors, as well as the presence of other types of cells that can be quite abundant. For example, IHC analysis demonstrated a moderate twofold increased level of eIF4E in some very early-stage precancerous lesions known as ductal carcinomas in situ (DCIS) and some early-stage invasive ductal carcinomas (Nathan et al. 1997). In tissue culture, isolated human primary mammary epithelial cells express much lower levels of eIF4E than are found in breast cancer cell lines, and these primary cells demonstrate in vitro signs of transformation if eIF4E is engineered to be overexpressed (Avdulov et al. 2004). In contrast, most studies that have reported greatly elevated levels of eIF4E (five- to tenfold) in human breast cancers and an associated increased risk of disease progression or recurrence have utilized immunoblot analysis of tumor homogenates (Kerekatte et al. 1995; Li et al. 1998, 2002; McClusky et al. 2005). Other groups did not observe these remarkable increases in eIF4E protein levels in a variety of breast cancers when investigated by IHC (Rosenwald 2004). Similarly, in unpublished immunohistochemistry studies (R.J. Schneider, unpubl.), only a two- to threefold increase in eIF4E levels was observed in a variety of breast cancer types of different stages and histopathological grades (locally advanced, inflammatory, invasive ductal, and lobular carcinomas). As noted above, an increased concentration of tumor cells or the vasculature in the specimen could account for the difference in results among the studies. Moreover, both increased tumor angiogenesis and cell density, which might be the source of greatly elevated eIF4E levels, are associated with poor survival. Clearly, additional studies using IHC and other approaches need to be conducted to analyze the state of eIF4E in a variety of breast cancers at different stages and grades to resolve this issue.

What Is the Role of eIF4F Overexpression in Cancer Development and Progression?

Evidence supports the view that cap-dependent mRNAs compete for eIF4E binding, which facilitates eIF4F-dependent translation initiation (for review, see Gingras et al. 1999; Mamane et al. 2004). High levels of eIF4E only modestly increase global levels of mRNA translation (Koromilas et al. 1992a). In contrast, high levels of eIF4E selectively and strongly increase translation of a small subset of mRNAs that contain structured 5′UTRs (Koromilas et al. 1992a), reflecting the keen competition for this factor. mRNAs that are strongly increased in translation by elevated levels of eIF4E include ornithine decarboxylase (ODC), vascular endothelial growth factor (VEGF), cyclin D1, c-Myc, and fibroblast growth factor 2 (FGF-2) (Koromilas et al. 1992a; Rosenwald et al. 1993, 1995; Kevil et al. 1995, 1996; Rousseau et al. 1996; Shantz et al. 1996; Scott et al. 1998; Zimmer et al. 2000). These and other mRNAs are severely depleted from polysomes following eIF4E inhibition by rapamycin treatment and activation of 4E-BP1 (Grolleau et al. 2002). Thus, overexpression of eIF4E can selectively and strongly increase translation of key mRNAs whose products can promote cellular proliferation and partici-pate in tumorigenesis.

Overexpression of eIF4E is also associated with inhibition of apo-ptosis by nutrient and serum deprivation (Polunovsky et al. 1996) and by overexpression of c-Myc (Tan et al. 2000; Ruggero et al. 2004) or Ras activation (Polunovsky et al. 2000). Although eIF4E overexpression se-lectively enhances translation of certain mRNAs, it stimulates the trans-port of cyclin D1 mRNA from the nucleus to the cytoplasm (Rousseau et al. 1996). In fact, overexpression of an eIF4E mutant protein that re-tains binding to m^{7}G-cap structures but not eIF4G still promotes high levels of cyclin D1 mRNA transport (Cohen et al. 2001). These data sug-gest that overexpression of eIF4E likely participates in transformation through increased translation of some structured growth-promoting mRNAs, as well as the increased transport to the cytoplasm of cyclin D1 mRNA, and possibly other growth and proliferation stimulatory mRNAs. It should also be noted that the potential function of eIF4E phosphory-lation has not been resolved in protein synthesis, and studies have not rigorously examined eIF4E phosphorylation in different cancers. In ad-dition, the importance of 4E-BP1 abundance and activity in cancers has not been significantly investigated. Several studies have reported *increased* expression of 4E-BP1 with increased cancer progression in a large frac-tion of prostate, head and neck, and gastrointestinal cancers (Martin et al. 2000; Nathan et al. 2004; Kremer et al. 2006). The phosphorylation status of 4E-BP1 has been reported as elevated in some human breast (Zhou et al. 2004) and gastrointestinal cancers (Kremer et al. 2006). In the latter case, total 4E-BP1 protein levels were not examined, although we have also found 4E-BP1 protein levels to be strongly elevated in lo-cally advanced breast cancers (R.J. Schneider, unpubl.).

Overexpression and Phosphorylation of the eIF2α Subunit in Human Cancers

As described elsewhere in this volume (Chapters 4, 9, and 12), eIF2 con-sists of three subunits (α, β, γ), which carry the initiator methionyl-tRNA to the 40S ribosome subunit with a molecule of GTP. The eIF2α subunit is a target of negative regulation and translation inhibition, as its phos-phorylation by a group of related protein kinases inhibits the recycling of GTP on eIF2 by the GTP exchange factor eIF2B, thereby blocking ini-tiation (Chapter 12). Overexpression in NIH-3T3 cells of a dominant-in-terfering form of eIF2α kinase, a mutant form of eIF2α that cannot be phosphorylated, is transforming and promotes tumorigenesis in a xenotransplant nude mouse model (Koromilas et al. 1992b; Donze et al. 1995; Raught et al. 1996). These data are consistent with the negative effects of eIF2α phosphorylation on cell proliferation (Wek 1994). The eIF2α subunit is overexpressed (demonstrated by IHC) in adenomas and carcinomas of the colon, in benign melanocytic nevi (moles), malignant melanomas (Rosenwald et al. 2003), and bronchioloalveolar carcinomas, but not other lung cancers (Rosenwald et al. 2001). In hematological malignancies, eIF2α is overexpressed in the majority of aggressive non-Hodgkin's lymphomas compared to nonactivated lymphocytes, but not in diffuse large B-cell lymphomas (Rosenwald 2004), whereas normal proliferating B and T cells (compared to dormant cells) also express very high levels of both eIF4E and eIF2α (Wang et al. 2001).

Reciprocally, up-regulation of PKR, which opposes eIF2 function, was suggested to have tumor suppressor activity in experimental systems (Meurs et al. 1993; Barber et al. 1995). Increased PKR levels have been observed with increased cellular differentiation in certain cancers, particularly some head and neck squamous carcinomas (Haines et al. 1993a,b, 1998) and in some indolent papillary thyroid carcinomas, but not in aggressive nonpapillary thyroid cancers (Terada et al. 2000). Importantly, increased expression of PKR was not associated with diminished parameters of cancer progression or increased survival benefit, only lower cancer cell proliferation. This is also consistent with the decreased levels of PKR reported in some preleukemic and leukemic human myeloid cells that display decreased levels of proliferation (Beretta et al. 1996). The paradoxically higher levels of PKR in high-grade viral hepatocellular carcinomas (Shimada et al. 1998) might result from production of type I interferons in response to viral infection, which can induce PKR expression. However, elevated levels of active (phosphorylated) PKR and phosphorylated eIF2α were found in melanoma lymph node metastases and colon carcinomas (Kim et al. 2002). In breast cancer, the expression level of PKR is complex and in some studies also contrary to that expected. Low levels of PKR were reported in preneoplastic breast lesions such as ductal and lobular hyperplasias and lobular carcinoma in situ (LCIS), whereas higher levels were found in invasive ductal carcinomas that are more aggressive (Haines et al. 1996; Kim et al. 2000, 2002). In breast and certain other cancers, it is possible that PKR activity is atypically impaired by interaction with an inhibitor (Savinova et al. 1999). Alternatively, there is some evidence that cancer cells may have elevated levels of eIF2B, the eIF2α GTP exchange factor, which would counter translation inhibition by eIF2α phosphorylation and PKR activation (Kim et al. 2000). It is clear that considerably more research needs to be devoted to understanding the role of the eIF2α phosphorylation pathway in human malignancies.

Overexpression of Other Initiation Factors in Human Cancer

Other initiation factors have also been implicated in human cancers, although they are not nearly as well investigated as eIF4E and eIF2. Engineered overexpression of eIF4G, the large scaffolding component of eIF4F, in immortalized NIH-3T3 cells can be transforming, as measured by in vitro assays and tumorigenesis in nude mice (Fukuchi-Shimogori et al. 1997). Transformation by eIF4G in this system occurs in the absence of eIF4E overexpression, suggesting that it is mediated by eIF4G itself. With the recognition that studies have not yet surveyed a large number of human tumors, eIF4G has been found to be overexpressed at the protein level in approximately 15–30% of squamous cell lung cancers (Brass et al. 1997; Keiper et al. 1999; Bauer et al. 2002; Prevot et al. 2003) and a subset of late-stage prostate cancers (Wang et al. 2005). In both lung and prostate cancers, the overexpression of eIF4G is immunogenic and may be associated with gene amplification. eIF4A, the ATP-dependent RNA helicase component of eIF4F, is reportedly overexpressed in some human malignant melanoma cells in culture (Eberle et al. 1997) and in some primary hepatocellular carcinomas (Shuda et al. 2000). The function and extent of eIF4G or eIF4A overexpression in human cancer is unknown and will require more extensive investigation. There is some indirect evidence, however, that up-regulation of eIF4A could be important to the progression of some cancers. A tumor suppressor protein known as Pdcd4 has been shown to bind eIF4A and inhibit its helicase activity, impairing its involvement in translation and providing a possible link between tumor suppression and translation inhibition (Yang et al. 2003). Recent studies show that Pdcd4 is degraded in a S6K1-dependent manner by a specific ubiquitin ligase, which is linked to cell growth (Dorello et al. 2006). It is possible that degradation of Pdcd4 and increased eIF4A activity, rather than abundance, play a role in development or progression of certain cancers.

Several subunits of the multiprotein initiation factor eIF3 have been shown to be elevated in some human cancers and derived cell lines at the protein level. The 170-kD eIF3 protein (eIF3a) is overexpressed in some breast, esophageal, stomach, lung, and cervical cancers of different grades (Bachmann et al. 1997; Dellas et al. 1998; Chen and Burger 1999). The correlation of eIF3a protein overexpression and development of these cancers is quite significant. eIF3a is also overexpressed during S phase in cultured cells and may promote the translation of a subset of mRNAs involved in cell proliferation including the p27 mRNA, providing a possible molecular link to malignancy and translational control (Dong and Zhang 2003; Dong et al. 2004). The p40 subunit of eIF3 (eIF3h) is reportedly up-regulated in some breast and prostate cancers (Nupponen et al. 1999), whereas the p48 subunit of eIF3 (eIF3e) is decreased in a fraction of breast and lung carcinomas (Marchetti et al. 2001), and the p110 subunit (eIF3c) is elevated in one form of human testicular cancer (seminomas) (Rothe et al. 2000). As with all observations of altered initiation factor expression, studies need to determine the functional implications (if any) in cancer development and progression using experimental models. It is worth noting that mutation of the yeast homolog of the p110 subunit of eIF3c results in an inability to transit the G_1 cell cycle checkpoint (Kovarik et al. 1998), and overexpression of a truncated p48 subunit (eIF3e) can transform NIH-3T3 cells (Mayeur and Hershey 2002). It is not obvious how increased expression of a truncated form of eIF3e promotes cell proliferation and transformation. There is evidence that some eIF3 genes are amplified commensurate with their overexpression, particularly the eIF3h (S3) gene (Saramaki et al. 2001). With the sequencing of the human genome, analysis of regions that are amplified in certain human cancers have revealed that translation factors reside within loci that are frequently amplified and rearranged. For instance, region 3q26 is frequently amplified in human lung, ovarian, esophageal, breast, prostate, and nasopharyngeal cancers. One study has found that initiation factor eIF5A2, which resides within this domain, is indeed amplified (Guan et al. 2004). Decreased expression of eIF5A2 using antisense RNA in an ovarian cancer cell line decreased specific parameters of transformation, such as growth in soft agar and tumorigenesis in nude mice, implicating eIF5A2 overexpression in promotion of advanced ovarian cancer (Guan et al. 2004).

Overexpression of Elongation Factors in Human Cancer

Although translational regulation is typically equated with the regulation of initiation, there is a considerable amount of regulation that occurs at the level of elongation as well (Chapter 21). It is therefore not surprising that elongation factors and their regulators are also subject to altered expression in human cancers and can be shown to promote transformation when overexpressed in model systems. As with overexpression of specific translation initiation factors, studies need to determine whether the factor participates in transformation at the level of translation or through other functions of these factors. For instance, elongation factor eEF1A is overexpressed in some metastatic breast adenocarcinomas (Edmonds et al. 1996), some pancreatic, colon, lung, and stomach cancers (Grant et al. 1992), and 25% of human ovarian cancers (Anand et al. 2002). Its overexpression causes anchorage-independent growth in NIH-3T3 cells and increased tumor growth of ovarian carcinoma cell lines in nude mice (Anand et al. 2002). Nevertheless, it is not yet resolved whether eEF1A acts on translation to increase tumorigenesis since it can be utilized for its G-protein activity by other proteins, including cytoskeletal components such as actins and tubulin (for review, see Lamberti et al. 2004). This is an important issue that will require additional studies. mRNA levels for the eEF1γ protein are also reportedly overexpressed in some pancreatic, breast, colon, and stomach cancers (Lew et al. 1992; Ender et al. 1993; Mimori et al. 1995).

The elongation factor eEF2 kinase is overexpressed and more active in several human cancers, including some malignant breast cancers and glioblastomas (Parmer et al. 1997, 1999). eEF2 kinase activity is reportedly elevated in S phase and in proliferating cells in cancers (Calberg et al. 1991; Bagaglio et al. 1993; Bagaglio and Hait 1994; Parmer et al. 1999). As noted above for eEF1A, eEF2 kinase likely has targets other than eEF2, which might explain its elevated activity, since it should decrease the rate of protein synthesis at the step of elongation. However, the enhanced expression of eEF2 observed in highly transformed breast cancer cell lines indicates that higher eEF2 kinase activity is likely negated (Connolly et al. 2006). Studies need to determine whether the elevated level and activity of eEF2 kinase (which is controlled by mTOR; Chapter 21) are important in cancer progression, and if so, whether it promotes transformation through altered translational control or other targets.

SIGNAL TRANSDUCTION CONTROL OF TRANSLATION IN CANCER

It should not be surprising that aberrant activation of specific arms of the translational machinery and its regulators by altered signal transduction pathways has a key role in the etiology of human cancers. After all, this is precisely the well-established paradigm by which oncogenic transformation functions to transcriptionally activate genes that drive cancer development and progression, and many of these same signaling pathways also regulate translation. Translational control by signal transduction pathways and its impact on transformation is itself a large area of research and is reviewed in detail in Chapter 14. Here, we provide a summary of the key features of this important area of translational control and refer readers to several recent reviews for an in-depth discussion (Gingras et al. 2004; Holland et al. 2004; Bader et al. 2005; Hay 2005).

Activation of the GTPase Ras, and hyperactivation of kinase PI3K and the serine-threonine kinase Akt, promote development of human cancers. Ras, PI3K, and Akt are also key regulators of protein synthesis. PI3K is a multisubunit protein kinase consisting of a 110-kD catalytic subunit (p110) and an 85-kD regulatory subunit that blocks p110 kinase activity (for review, see Vivanco and Sawyers 2002). p110 activity is stimulated by Ras activation and by receptor tyrosine kinases, including the ErbB2/ErbB3 (Her2/neu) receptor that is overexpressed in a subset of breast cancers (Vivanco and Sawyers 2002; Hay 2005). The activity of PI3K is opposed by the phosphatase PTEN (phosphatases and tensin homolog deleted on chromosome 10), which is a tumor suppressor. PTEN and p85 are mutationally inactivated to different extents in a wide variety of malignant human cancers, but predominantly in prostate and endometrial cancers, glioblastomas, and melanomas, thereby constitutively activating PI3K (Vivanco and Sawyers 2002; Bader et al. 2005). Increased activation of Akt isoforms 1, 2, or 3 by PIP_3 and any of several protein kinases, including PDK1 and PDK2, promotes cell growth and proliferation by several different pathways including inactivation of FOXO transcription factors, which block cell proliferation and impair protein synthesis (for review, see Bader et al. 2005). Hyperactivated Akt isoforms have been reported in many human cancers (for review, see Bader et al. 2005). Activation of PI3K and Akt are linked to activation of cap-dependent mRNA translation through the serine-threonine PI3K-related protein kinase known as mTOR.

mTOR is indirectly activated by Ras, Akt, and a variety of receptor tyrosine kinases, and mitogenic stimuli acting through the PI3K-Akt pathway (see Fig. 1). mTOR is also activated by amino acids and the increased energy state of the cell and is a major regulator of protein synthesis (for review, see Gingras et al. 2004). mTOR inhibition blocks cap-dependent mRNA translation, causes cells to arrest in the G_1 phase of the cell cycle, and impairs RNA polymerases I and III (Gingras et al. 2004). The increased levels of rRNA and ribosomal proteins associated with transformation are to some extent a result of increased mTOR activity in cancer cells (Martin et al. 2004). Activated Akt phosphorylates and inactivates negative regulators of mTOR known as the tuberous sclerosis complex proteins 1 and 2 (TSC1/2) (Inoki et al. 2002), which act as tumor suppressors (van Slegtenhorst et al. 1998). TSC2 is a GTPase-activating protein (GAP), which inhibits the G protein Rheb. Rheb is thought to be the effector of TSC1/2, typically preventing inhibition of (i.e., activating) mTOR under mitogenic (pro-proliferative) conditions (for review, see Holland et al. 2004; Parsa and Holland 2004). Conse-

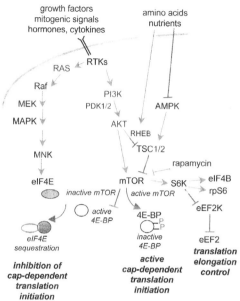

Figure 1. Regulatory pathways for translational control. Key signaling pathways and targets for the translational control of mRNA initiation and elongation are shown. Nutrients and amino acids inhibit AMP kinase (AMPK), an activator of the TSC1/2 complex, and stimulate the small G-protein RHEB, an inhibitor of TSC1/2 complex, which inhibits mTOR. Mitogenic signals (growth factors, hormones, cytokines) act through receptor tyrosine kinases (RTKs) to stimulate RAS, leading to activation of MNK1 and eIF4E phosphorylation and activation of PI3 kinase (PI3K), which stimulates AKT, an inhibitor of TSC1/2 and an activator of mTOR. PTEN, a PI3K phosphatase, opposes activation of the pathway. mTOR activity is therefore stimulated by mitogenic and growth-promoting signals and suppressed by growth-inhibiting signals. Activated mTOR stimulates ribosomal protein S6 kinase, which phosphorylates ribosomal protein S6 (rpS6) that is involved in regulating cell size, and initiation factor eIF4B that promotes initiation complex function or assembly (Raught et al. 2004; Holz et al 2005; Shahbazian et al. 2006). S6 kinase also phosphorylates elongation factor eEF2 kinase. Down-regulation of mTOR activity removes the negative inhibitory signal and results in suppression of elongation. Activated mTOR phosphorylates and maintains 4E-BPs in the inactive state. Down-regulation of mTOR activity results in derepression of 4E-BPs through decreased phosphorylation and sequestration of eIF4E, thereby reducing cap-dependent mRNA translation initiation.

quently, inactivating mutations in TSC1 and TSC2 result in unopposed activation of mTOR by upstream signals acting on PI3K and Akt. Surprisingly, mutations in TSC1/2 are quite rare in human cancers (Kwiatkowski 2003). The only example involves tuberous sclerosis complex patients in which TSC1 or TSC2 is mutated, resulting in development of hamartomas (benign tumors) in multiple organs (Kwiatkowski 2003). The fact that Rheb can also bind and act on mTOR (Long et al. 2005a,b) might account for the lack of TSC1/2 mutations in human cancer. Mutation of mTOR has also never been described in human cancers. It is likely that neither inactivation of TSC1/2 nor activation of mTOR is sufficient for transformation. How then does activation of Ras and the PI3K-Akt pathway promote oncogenic transformation at the level of translational control?

The major downstream targets that are phosphorylated and controlled by mTOR are the S6 ribosomal protein kinases S6K1 and S6K2 (for review, see Avruch et al. 2001; Chapter 17) and the eIF4E sequestering proteins 4E-BP1, 2, and 3 (for review, see Gingras et al. 2004). The 4E-BP proteins are typically hyperphosphorylated by mTOR in mitogen-stimulated and/or proliferating cells, which inhibits their ability to compete for eIF4E binding to eIF4G and increases cap-dependent mRNA translation. With down-regulation of the PI3K/Akt pathway, mTOR is no longer activated and 4E-BP1 becomes dephosphorylated, unmasking its eIF4E-binding site and resulting in sequestration of eIF4E. Thus, cap-dependent mRNA translation can be promoted by pro-transforming growth factors and mutations such as those in PTEN that hyperactivate the PI3K-Akt signaling pathway. The drug rapamycin and its analogs (CCI-779, RAD001) inhibit mTOR activation. They are particularly potent in inhibiting growth of tumors that harbor PTEN mutations or tumors in which PI3K and Akt are constitutively activated (Geoerger et al. 2001; Neshat et al. 2001; Podsypanina et al. 2001). The increase in mTOR activity resulting from constitutive activation of the PI3K-Akt pathway does not strongly increase overall protein synthesis. Rather, as described earlier in this chapter, there is a profound and selective increase in translation of a group of rapamycin-sensitive mRNAs, accounting for approximately 5–7% of those of the cell, which correspond to regulators of cell proliferation, hypoxia responsiveness, and angiogenesis (Grolleau et al. 2002). In this regard, rapamycin was shown to inhibit tumor cell growth in an experimental xenotransplant mouse model by impairing angiogenesis, possibly by acting directly on endothelial cell translation or by reducing the production of tumor-derived VEGF (Guba et al. 2002). Endothelial cells may in fact be acutely sensitive to translational control. In a mouse model system, a natural collagen fragment known as tumstatin, which binds $\alpha_v\beta_3$ integrin and blocks its activation of the PI3K-Akt pathway, impairs stimulation of cap-dependent mRNA translation and profoundly limits tumor-directed vascular formation (Maeshima et al. 2002).

Selectively increased translation of specific mRNAs might also account in part for the ability of Ras and Akt to act in concert to increase oncogenic transformation and tumorigenesis. In a murine model of glioblastoma, which develops as a result of combined introduction of activated Ras and Akt into glial progenitor cells, only small changes in the overall transcription profile were observed (Rajasekhar et al. 2003). However, there was a significant increase in recruitment into polyribosomes of a small number of mRNAs that function in cell growth, proliferation, and tumorigenesis such as VEGF, which was shown to require, in part, the increased availability of eIF4E.

CONCLUDING REMARKS

It is clear that translational control has an important role in cancer etiology. Many of the mutations in key signal transduction pathways that have been linked to altered transcriptional control and DNA replication in cancer development and progression are also involved in altered translational control. The same driving forces that promote altered gene expression in selection for transformed cells, such as hypoxia, angiogenesis, inhibition of apoptosis, increased cell growth rates, and proliferation, are also the driving forces in altered translational control and translation of specific mRNAs. Although the overexpression of certain factors has been widely associated with transformed phenotypes, cancer is a complex and heterogeneous disease, which can vary enormously at different stages and grades even within the same disease. The molecular and cellular diversity inherent in cancer development is likely reflected in the variety of mechanisms by which the translational machinery is altered; particularly the overexpression of different initiation factors in different types of cancers. Future studies need to carefully study the changes in the translational machinery with special attention to the many different types of cancers and their histopathological presentation, rather than broadly grouping together all cancers of a specific tissue. Moreover, the translational changes in the tumor microenvironment are probably as important as those that occur in the tumor itself and should become an important area of investigation in the future.

ACKNOWLEDGMENTS

R.J.S. received funding from the National Institutes of Health, National Cancer Institute, Department of Defense Breast Cancer Research Program, Estee Lauder–Breast Cancer Research Foundation, Avon Breast Cancer Research Foundation, and the Komen Breast Cancer Research Foundation.

N.S. received funding from the National Institutes of Health, National Cancer Institute of Canada, Canadian Institute of Health Research, and a Howard Hughes Medical Institute International Scholar grant.

REFERENCES

Amsterdam A., Sadler K.C., Lai K., Farrington S., Bronson R.T., Lees J.A., and Hopkins N. 2004. Many ribosomal protein genes are cancer genes in zebrafish. *PLoS Biol.* **2:** E139.

Anand N., Murthy S., Amann G., Wernick M., Porter L.A., Cukier I.H., Collins C., Gray J.W., Diebold J., Demetrick D.J., and Lee J.M. 2002. Protein elongation factor EEF1A2 is a putative oncogene in ovarian cancer. *Nat. Genet.* **31:** 301–305.

Avdulov S., Li S., Michalek V., Burrichter D., Peterson M., Perlman D.M., Manivel J.C., Sonenberg N., Yee D., Bitterman P.B., and Polunovsky V.A. 2004. Activation of translation complex eIF4F is essential for the genesis and maintenance of the malignant phenotype in human mammary epithelial cells. *Cancer Cell* **5:** 553–563.

Avruch J., Belham C., Weng Q., Hara K., and Yonezawa K. 2001. The p70 S6 kinase integrates nutrient and growth signals to control translational capacity. *Prog. Mol. Subcell. Biol.* **26:** 115–154.

Bachmann F., Banziger R., and Burger M.M. 1997. Cloning of a novel protein overexpressed in human mammary carcinoma. *Cancer Res.* **57:** 988–994.

Bader A.G., Kang S., Zhao L., and Vogt P.K. 2005. Oncogenic PI3K deregulates transcription and translation. *Nat. Rev. Cancer* **5:** 921–929.

Bagaglio D.M. and Hait W.N. 1994. Role of calmodulin-dependent phosphorylation of elongation factor 2 in the proliferation of rat glial cells. *Cell Growth Differ.* **5:** 1403–1408.

Bagaglio D.M., Cheng E.H., Gorelick F.S., Mitsui K., Nairn A.C., and Hait W.N. 1993. Phosphorylation of elongation factor 2 in normal and malignant rat glial cells. *Cancer Res.* (suppl. 10) **53:** 2260–2264.

Barber G.N., Jagus R., Meurs E.F., Hovanessian A.G., and Katze M.G. 1995. Molecular mechanisms responsible for malignant transformation by regulatory and catalytic domain variants of the interferon-induced enzyme RNA-dependent protein kinase. *J. Biol. Chem.* **270:** 17423–17428.

Bassoe C.F., Bruserud O., Pryme I.F., and Vedeler A. 1998. Ribosomal proteins sustain morphology, function and phenotype in acute myeloid leukemia blasts. *Leuk. Res.* **22:** 329–339.

Bauer C., Brass N., Diesinger I., Kayser K., Grasser F.A., and Meese E. 2002. Overexpression of the eukaryotic translation initiation factor 4G (eIF4G-1) in squamous cell lung carcinoma. *Int. J. Cancer* **98:** 181–185.

Baxter G.C. and Stanners C.P. 1978. The effect of protein degradation on cellular growth characteristics. *J. Cell. Physiol.* **96:** 139–145.

Beretta L., Gabbay M., Berger R., Hanash S.M., and Sonenberg N. 1996. Expression of the protein kinase PKR is modulated by IRF-1 and is reduced in 5q- associated leukemias. *Oncogene* **12:** 1593–1596.

Berkel H.J., Turbat-Herrera E.A., Shi R., and de Benedetti A. 2001. Expression of the translation initiation factor eIF4E in the polyp-cancer sequence in the colon. *Cancer Epidemiol. Biomark. Prev.* **10:** 663–666.

Bissell M.J. and Labarge M.A. 2005. Context, tissue plasticity, and cancer: Are tumor stem cells also regulated by the microenvironment? *Cancer Cell* **7:** 17–23.

Bissell M.J. and Radisky D. 2001. Putting tumours in context. *Nat. Rev. Cancer* **1**: 46–54.

Bond G.L., Hu W., and Levine A.J. 2005. MDM2 is a central node in the p53 pathway: 12 years and counting. *Curr. Cancer Drug Targets* **5**: 3–8.

Boon K., Caron H.N., van Asperen R., Valentijn L., Hermus M.C., van Sluis P., Roobeek I., Weis I., Voute P.A., Schwab M., and Versteeg R. 2001. N-myc enhances the expression of a large set of genes functioning in ribosome biogenesis and protein synthesis. *EMBO J.* **20**: 1383–1393.

Brass N., Heckel D., Sahin U., Pfreundschuh M., Sybrecht G.W., and Meese E. 1997. Translation initiation factor eIF-4gamma is encoded by an amplified gene and induces an immune response in squamous cell lung carcinoma. *Hum. Mol. Genet.* **6**: 33–39.

Brown C.Y., Mize G.J., Pineda M., George D.L., and Morris D.R. 1999. Role of two upstream open reading frames in the translational control of oncogene mdm2. *Oncogene* **18**: 5631–5637.

Buttgereit F. and Brand M.D. 1995. A hierarchy of ATP-consuming processes in mammalian cells. *Biochem. J.* **312**: 163–167.

Calberg U., Nilsson A., Skog S., Palmquist K., and Nygard O. 1991. Increased activity of the eEF-2 specific, Ca^{2+} and calmodulin dependent protein kinase III during the S-phase in Ehrlich ascites cells. *Biochem. Biophys. Res. Commun.* **180**: 1372–1376.

Capoulade C., Bressac-de Paillerets B., Lefrere I., Ronsin M., Feunteun J., Tursz T., and Wiels J. 1998. Overexpression of MDM2, due to enhanced translation, results in inactivation of wild-type p53 in Burkitt's lymphoma cells. *Oncogene* **16**: 1603–1610.

Cavanaugh A.H., Hempel W.M., Taylor L.J., Rogalsky V., Todorov G., and Rothblum L.I. 1995. Activity of RNA polymerase I transcription factor UBF blocked by Rb gene product. *Nature* **374**: 177–180.

Chen G. and Burger M.M. 1999. p150 expression and its prognostic value in squamous-cell carcinoma of the esophagus. *Int. J. Cancer* **84**: 95–100.

Chen L.I., Sommer K.M., and Swisshelm K. 2002. Downstream codons in the retinoic acid receptor beta-2 and beta-4 mRNAs initiate translation of a protein isoform that disrupts retinoid-activated transcription. *J. Biol. Chem.* **277**: 35411–35421.

Cohen N., Sharma M., Kentsis A., Perez J.M., Strudwick S., and Borden K.L. 2001. PML RING suppresses oncogenic transformation by reducing the affinity of eIF4E for mRNA. *EMBO J.* **20**: 4547–4559.

Coller H.A., Grandori C., Tamayo P., Colbert T., Lander E.S., Eisenman R.N., and Golub T.R. 2000. Expression analysis with oligonucleotide microarrays reveals that MYC regulates genes involved in growth, cell cycle, signaling, and adhesion. *Proc. Natl. Acad. Sci.* **97**: 3260–3265.

Connolly E., Braunstein S., Formenti S.C., and Schneider R.J. 2006. Hypoxia inhibits translation through a 4E-BP1 and eEF2 kinase pathway controlled by mTOR and uncoupled in breast cancer cells. *Mol. Cell. Biol.* **26**: 3955–3965.

De Benedetti A., Joshi B., Graff J.R., and Zimmer S.G. 1994. CHO cells transformed by the translation factor eIF4E display increased c-myc expression but require overexpression of Max for tumorigenicity. *Mol. Cell. Differ.* **2**: 347–371.

De Benedetti A., Joshi-Barve S., Rinker-Schaeffer C., and Rhoads R.E. 1991. Expression of antisense RNA against initiation factor eIF-4E mRNA in HeLa cells results in lengthened cell division times, diminished translation rates, and reduced levels of both eIF-4E and the p220 component of eIF-4F. *Mol. Cell. Biol.* **11**: 5435–5445.

Dellas A., Torhorst J., Bachmann F., Banziger R., Schultheiss E., and Burger M.M. 1998. Expression of p150 in cervical neoplasia and its potential value in predicting survival. *Cancer* **83**: 1376–1383.

Deng C.X. 2006. BRCA1: Cell cycle checkpoint, genetic instability, DNA damage response and cancer evolution. *Nucleic Acids Res.* **34:** 1416–1426.

Dirheimer G., Baranowski W., and Keith G. 1995. Variations in tRNA modifications, particularly of their queuine content in higher eukaryotes. Its relation to malignancy grading. *Biochimie* **77:** 99–103.

Dong Z. and Zhang J.T. 2003. EIF3 p170, a mediator of mimosine effect on protein synthesis and cell cycle progression. *Mol. Biol. Cell.* **14:** 3942–3951.

Dong Z., Liu L.H., Han B., Pincheira R., and Zhang J.T. 2004. Role of eIF3 p170 in controlling synthesis of ribonucleotide reductase M2 and cell growth. *Oncogene* **23:** 3790–3801.

Donze O., Jagus R., Koromilas A.E., Hershey J.W., and Sonenberg N. 1995. Abrogation of translation initiation factor eIF-2 phosphorylation causes malignant transformation of NIH 3T3 cells. *EMBO J.* **14:** 3828–3834.

Dorello N.V., Peschiarol A., Guardavaccaro D., Colburn N.H., Sherman N.E., and Pagano M. 2006. SCFBTRCP- and S6K1-mediated degradation of PDCD4 promotes protein translation and cell growth in response to mitogens. *Science* (in press).

Downward J. 2003. Targeting RAS signalling pathways in cancer therapy. *Nat. Rev. Cancer* **3:** 11–22.

Dua K., Williams T.M., and Beretta L. 2001. Translational control of the proteome: Relevance to cancer. *Proteomics* **1:** 1191–1199.

Duncan R., Milburn S.C., and Hershey J.W. 1987. Regulated phosphorylation and low abundance of HeLa cell initiation factor eIF-4F suggest a role in translational control. Heat shock effects on eIF-4F. *J. Biol. Chem.* **262:** 380–388.

Eberle J., Krasagakis K., and Orfanos C.E. 1997. Translation initiation factor eIF-4A1 mRNA is consistently overexpressed in human melanoma cells in vitro. *Int. J. Cancer* **71:** 396–401.

Edmonds B.T., Wyckoff J., Yeung Y.G., Wang Y., Stanley E.R., Jones J., Segall J., and Condeelis J. 1996. Elongation factor-1 alpha is an overexpressed actin binding protein in metastatic rat mammary adenocarcinoma. *J. Cell Sci.* **109:** 2705–2714.

Ender B., Lynch P., Kim Y.H., Inamdar N.V., Cleary K.R., and Frazier M.L. 1993. Overexpression of an elongation factor-1 gamma-hybridizing RNA in colorectal adenomas. *Mol. Carcinog.* **7:** 18–20.

Ferrari S., Manfredini R., Tagliafico E., Rossi E., Donelli A., Torelli G., and Torelli U. 1990. Noncoordinated expression of S6, S11, and S14 ribosomal protein genes in leukemic blast cells. *Cancer Res.* **50:** 5825–5828.

Frank D.J. and Roth M.B. 1998. ncl-1 is required for the regulation of cell size and ribosomal RNA synthesis in *Caenorhabditis elegans. J. Cell Biol.* **140:** 1321–1329.

Frank D.J., Edgar B.A., and Roth M.B. 2002. The *Drosophila melanogaster* gene brain tumor negatively regulates cell growth and ribosomal RNA synthesis. *Development* **129:** 399–407.

Fukuchi-Shimogori T., Ishii I., Kashiwagi K., Mashiba H., Ekimoto H., and Igarashi K. 1997. Malignant transformation by overproduction of translation initiation factor eIF4G. *Cancer Res.* **57:** 5041–5044.

Gani R. 1976. The nucleoli of cultured human lymphocytes. I. Nucleolar morphology in relation to transformation and the DNA cycle. *Exp. Cell Res.* **97:** 249–258.

Geoerger B., Kerr K., Tang C.B., Fung K.M., Powell B., Sutton L.N., Phillips P.C., and Janss A.J. 2001. Antitumor activity of the rapamycin analog CCI-779 in human primitive neuroectodermal tumor/medulloblastoma models as single agent and in combination chemotherapy. *Cancer Res.* **61:** 1527–1532.

Gingras A.-C., Raught B., and Sonenberg N. 1999. eIF4 initiation factors: Effectors of mRNA recruitment to ribosomes and regulators of translation. *Annu. Rev. Biochem.* **68:** 913–963.

———. 2004. mTOR signaling to translation. *Curr. Top. Microbiol. Immunol.* **279:** 169–197.

Graff J.R., Boghaert E.R., De Benedetti A., Tudor D.L., Zimmer C.C., Chan S.K., and Zimmer S.G. 1995. Reduction of translation initiation factor 4E decreases the malignancy of ras-transformed cloned rat embryo fibroblasts. *Int. J. Cancer* **60:** 255–263.

Grant A.G., Flomen R.M., Tizard M.L., and Grant D.A. 1992. Differential screening of a human pancreatic adenocarcinoma lambda gt11 expression library has identified increased transcription of elongation factor EF-1 alpha in tumour cells. *Int. J. Cancer* **50:** 740–745.

Grolleau A., Bowman J., Pradet-Balade B., Puravs E., Hanash S., Garcia-Sanz J.A., and Beretta L. 2002. Global and specific translational control by rapamycin in T cells uncovered by microarrays and proteomics. *J. Biol. Chem.* **277:** 22175–22184.

Grummt I. 1999. Regulation of mammalian ribosomal gene transcription by RNA polymerase I. *Prog. Nucleic Acid Res. Mol. Biol.* **62:** 109–154.

Guan X.Y., Fung J.M., Ma N.F., Lau S.H., Tai L.S., Xie D., Zhang Y., Hu L., Wu Q.L., Fang Y., and Sham J.S. 2004. Oncogenic role of eIF-5A2 in the development of ovarian cancer. *Cancer Res.* **64:** 4197–4200.

Guba M., von Breitenbuch P., Steinbauer M., Koehl G., Flegel S., Hornung M., Bruns C.J., Zuelke C., Farkas S., Anthuber M., et al. 2002. Rapamycin inhibits primary and metastatic tumor growth by antiangiogenesis: Involvement of vascular endothelial growth factor. *Nat. Med.* **8:** 128–135.

Haines G.K., III, Becker S., Ghadge G., Kies M., Pelzer H., and Radosevich J.A. 1993a. Expression of the double-stranded RNA-dependent protein kinase (p68) in squamous cell carcinoma of the head and neck region. *Arch. Otolaryngol. Head Neck Surg.* **119:** 1142–1147.

Haines G.K., Cajulis R., Hayden R., Duda R., Talamonti M., and Radosevich J.A. 1996. Expression of the double-stranded RNA-dependent protein kinase (p68) in human breast tissues. *Tumour Biol.* **17:** 5–12.

Haines G.K., Ghadge G.D., Becker S., Kies M., Pelzer H., Thimmappaya B., and Radosevich J.A. 1993b. Correlation of the expression of double-stranded RNA-dependent protein kinase (p68) with differentiation in head and neck squamous cell carcinoma. *Virchows Arch. B Cell Pathol. Incl. Mol. Pathol.* **63:** 289–295.

Haines G.K., III, Panos R.J., Bak P.M., Brown T., Zielinski M., Leyland J., and Radosevich J.A. 1998. Interferon-responsive protein kinase (p68) and proliferating cell nuclear antigen are inversely distributed in head and neck squamous cell carcinoma. *Tumour Biol.* **19:** 52–59.

Haupt Y., Maya R., Kazaz A., and Oren M. 1997. Mdm2 promotes the rapid degradation of p53. *Nature* **387:** 296–299.

Hay N. 2005. The Akt-mTOR tango and its relevance to cancer. *Cancer Cell* **8:** 179–183.

Hay N. and Sonenberg N. 2004. Upstream and downstream of mTOR. *Genes Dev.* **18:** 1926–1945.

Haydon M.S., Googe J.D., Sorrells D.S., Ghali G.E., and Li B.D. 2000. Progression of eIF4E gene amplification and overexpression in benign and malignant tumors of the head and neck. *Cancer* **88:** 2803–2810.

Heys S.D., Park K.G., McNurlan M.A., Keenan R.A., Miller J.D., Eremin O., and Garlick

P.J. 1992. Protein synthesis rates in colon and liver: Stimulation by gastrointestinal pathologies. *Gut* **33:** 976–981.

Heys S.D., Park K.G., McNurlan M.A., Calder A.G., Buchan V., Blessing K., Eremin O., and Garlick P.J. 1991. Measurement of tumour protein synthesis in vivo in human colorectal and breast cancer and its variability in separate biopsies from the same tumour. *Clin. Sci.* **80:** 587–593.

Holcik M. and Sonenberg N. 2005. Translational control in stress and apoptosis. *Nat. Rev. Mol. Cell Biol.* **6:** 318–327.

Holland E.C., Sonenberg N., Pandolfi P.P., and Thomas G. 2004. Signaling control of mRNA translation in cancer pathogenesis. *Oncogene* **23:** 3138–3144.

Holz M.K., Ballif B.A., Gygi S.P., and Blenis J. 2005. mTOR and S6K1 mediate assembly of the translation preinitiation complex through dynamic protein interchange and ordered phosphorylation events. *Cell* **123:** 569–580.

Inoki K., Li Y., Zhu T., Wu J., and Guan K.L. 2002. TSC2 is phosphorylated and inhibited by Akt and suppresses mTOR signalling. *Nat. Cell Biol.* **4:** 648–657.

Iritani B.M. and Eisenman R.N. 1999. c-Myc enhances protein synthesis and cell size during B lymphocyte development. *Proc. Natl. Acad. Sci.* **96:** 13180–13185.

Iritani B.M., Delrow J., Grandori C., Gomez I., Klacking M., Carlos L.S., and Eisenman R.N. 2002. Modulation of T-lymphocyte development, growth and cell size by the Myc antagonist and transcriptional repressor Mad1. *EMBO J.* **21:** 4820–4830.

Jin X., Turcott E., Englehardt S., Mize G.J., and Morris D.R. 2003. The two upstream open reading frames of oncogene mdm2 have different translational regulatory properties. *J. Biol. Chem.* **278:** 25716–25721.

Johnson L.F., Levis R., Abelson H.T., Green H., and Penman S. 1976. Changes in RNA in relation to growth of the fibroblast. IV. Alterations in the production and processing of mRNA and rRNA in resting and growing cells. *J. Cell Biol.* **71:** 933–938.

Kawakami K., Salonga D., Park J.M., Danenberg K.D., Uetake H., Brabender J., Omura K., Watanabe G., and Danenberg P.V. 2001. Different lengths of a polymorphic repeat sequence in the thymidylate synthase gene affect translational efficiency but not its gene expression. *Clin. Cancer Res.* **7:** 4096–4101.

Keiper B.D., Gan W., and Rhoads R.E. 1999. Protein synthesis initiation factor 4G. *Int. J. Biochem. Cell Biol.* **31:** 37–41.

Kerekatte V., Smiley K., Hu B., Smith A., Gelder F., and De Benedetti A. 1995. The proto-oncogene/translation factor eIF4E: A survey of its expression in breast carcinomas. *Int. J. Cancer* **64:** 27–31.

Kevil C., Carter P., Hu B., and De Benedetti A. 1995. Translational enhancement of FGF-2 by eIF-4 factors, and alternate utilization of CUG and AUG codons for translation initiation. *Oncogene* **11:** 2339–2348.

Kevil C.G., De Benedetti A., Payne D.K., Coe L.L., Laroux F.S., and Alexander J.S. 1996. Translational regulation of vascular permeability factor by eukaryotic initiation factor 4E: Implications for tumor angiogenesis. *Int. J. Cancer* **65:** 785–790.

Kim S.H., Forman A.P., Mathews M.B., and Gunnery S. 2000. Human breast cancer cells contain elevated levels and activity of the protein kinase, PKR. *Oncogene* **19:** 3086–3094.

Kim S.H., Gunnery S., Choe J.K., and Mathews M.B. 2002. Neoplastic progression in melanoma and colon cancer is associated with increased expression and activity of the interferon-inducible protein kinase, PKR. *Oncogene* **21:** 8741–8748.

Koromilas A.E., Lazaris-Karatzas A., and Sonenberg N. 1992a. mRNAs containing extensive secondary structure in their 5′ non-coding region translate efficiently in cells overexpressing initiation factor eIF-4E. *EMBO J.* **11:** 4153–4158.

Koromilas A.E., Roy S., Barber G.N., Katze M.G., and Sonenberg N. 1992b. Malignant transformation by a mutant of the IFN-inducible dsRNA-dependent protein kinase. *Science* **257:** 1685–1689.

Koss L.G. 1982. Analytical and quantitative cytology. A historical perspective. *Anal. Quant. Cytol.* **4:** 251–256.

Kovarik P., Hasek J., Valasek L., and Ruis H. 1998. RPG1: An essential gene of *Saccharomyces cerevisiae* encoding a 110-kDa protein required for passage through the G1 phase. *Curr. Genet.* **33:** 100–109.

Kremer C.L., Klein R.R., Mendelson J., Browne W., Samadzedeh L.K., Vanpatten K., Highstrom L., Pestano G.A., and Nagle R.B. 2006. Expression of mTOR signaling pathway markers in prostate cancer progression. *Prostate* **66:** 1203–1212.

Kubbutat M.H., Jones S.N., and Vousden K.H. 1997. Regulation of p53 stability by Mdm2. *Nature* **387:** 299–303.

Kwiatkowski D.J. 2003. Tuberous sclerosis: From tubers to mTOR. *Ann. Hum. Genet.* **67:** 87–96.

Lai H.K. and Borden K.L. 2000. The promyelocytic leukemia (PML) protein suppresses cyclin D1 protein production by altering the nuclear cytoplasmic distribution of cyclin D1 mRNA. *Oncogene* **19:** 1623–1634.

Lamberti A., Caraglia M., Longo O., Marra M., Abbruzzese A., and Arcari P. 2004. The translation elongation factor 1A in tumorigenesis, signal transduction and apoptosis (review article). *Amino Acids* **26:** 443–448.

Landers J.E., Cassel S.L., and George D.L. 1997. Translational enhancement of mdm2 oncogene expression in human tumor cells containing a stabilized wild-type p53 protein. *Cancer Res.* **57:** 3562–3568.

Lazaris-Karatzas A. and Sonenberg N. 1992. The mRNA 5′ cap-binding protein, eIF-4E, cooperates with v-myc or E1A in the transformation of primary rodent fibroblasts. *Mol. Cell. Biol.* **12:** 1234–1238.

Lazaris-Karatzas A., Montine K.S., and Sonenberg N. 1990. Malignant transformation by a eukaryotic initiation factor subunit that binds to mRNA 5′ cap. *Nature* **345:** 544–547.

Lazaris-Karatzas A., Smith M.R., Frederickson R.M., Jaramillo M.L., Liu Y.L., Kung H.F., and Sonenberg N. 1992. Ras mediates translation initiation factor 4E-induced malignant transformation. *Genes Dev.* **6:** 1631–1642.

Lejbkowicz F., Goyer C., Darveau A., Neron S., Lemieux R., and Sonenberg N. 1992. A fraction of the mRNA 5′ cap-binding protein, eukaryotic initiation factor 4E, localizes to the nucleus. *Proc. Natl. Acad. Sci.* **89:** 9612–9616.

Lew Y., Jones D.V., Mars W.M., Evans D., Byrd D., and Frazier M.L. 1992. Expression of elongation factor-1 gamma-related sequence in human pancreatic cancer. *Pancreas* **7:** 144–152.

Li B.D., McDonald J.C., Nassar R., and De Benedetti A. 1998. Clinical outcome in stage I to III breast carcinoma and eIF4E overexpression. *Ann. Surg.* **227:** 756–763.

Li B.D., Gruner J.S., Abreo F., Johnson L.W., Yu H., Nawas S., McDonald J.C., and De Benedetti A. 2002. Prospective study of eukaryotic initiation factor 4E protein elevation and breast cancer outcome. *Ann. Surg.* **235:** 732–738.

Liebhaber S.A., Wolf S., and Schlessinger D. 1978. Differences in rRNA metabolism of primary and SV40-transformed human fibroblasts. *Cell* **13:** 121–127.

Liu L., Dilworth D., Gao L., Monzon J., Summers A., Lassam N., and Hogg D. 1999. Mutation of the CDKN2A 5′UTR creates an aberrant initiation codon and predisposes to melanoma. *Nat. Genet.* **21:** 128–132.

Loging W.T. and Reisman D. 1999. Elevated expression of ribosomal protein genes L37,

RPP-1, and S2 in the presence of mutant p53. *Cancer Epidemiol. Biomark. Prev.* **8:** 1011–1016.

Long X., Ortiz-Vega S., Lin Y., and Avruch J. 2005a. Rheb binding to mammalian target of rapamycin (mTOR) is regulated by amino acid sufficiency. *J. Biol. Chem.* **280:** 23433–23436.

Long X., Lin Y., Ortiz-Vega S., Yonezawa K., and Avruch J. 2005b. Rheb binds and regulates the mTOR kinase. *Curr. Biol.* **15:** 702–713.

Maeshima Y., Sudhakar A., Lively J.C., Ueki K., Kharbanda S., Kahn C.R., Sonenberg N., Hynes R.O., and Kalluri R. 2002. Tumstatin, an endothelial cell-specific inhibitor of protein synthesis. *Science* **295:** 140–143.

Mamane Y., Petroulakis E., Rong L., Yoshida K., Ler L.W., and Sonenberg N. 2004. eIF4E — From translation to transformation. *Oncogene* **23:** 3172–3179.

Marchetti A., Buttitta F., Pellegrini S., Bertacca G., and Callahan R. 2001. Reduced expression of INT-6/eIF3-p48 in human tumors. *Int. J. Oncol.* **18:** 175–179.

Martin D.E., Soulard A., and Hall M.N. 2004. TOR regulates ribosomal protein gene expression via PKA and the Forkhead transcription factor FHL1. *Cell* **119:** 969–979.

Martin M.E., Perez M.I., Redondo C., Alvarez M.I., Salinas M., and Fando J.L. 2000. 4E binding protein 1 expression is inversely correlated to the progression of gastrointestinal cancers. *Int. J. Biochem. Cell Biol.* **32:** 633–642.

Mayer C., Zhao J., Yuan X., and Grummt I. 2004. mTOR-dependent activation of the transcription factor TIF-IA links rRNA synthesis to nutrient availability. *Genes Dev.* **18:** 423–434.

Mayeur G.L. and Hershey J.W. 2002. Malignant transformation by the eukaryotic translation initiation factor 3 subunit p48 (eIF3e). *FEBS Lett.* **514:** 49–54.

McClusky D.R., Chu Q., Yu H., De Benedetti A., Johnson L.W., Meschonat C., Turnage R., McDonald J.C., Abreo F., and Li B.D. 2005. A prospective trial on initiation factor 4E (eIF4E) overexpression and cancer recurrence in node-positive breast cancer. *Ann. Surg.* **242:** 584–590.

Menssen A. and Hermeking H. 2002. Characterization of the c-MYC-regulated transcriptome by SAGE: Identification and analysis of c-MYC target genes. *Proc. Natl. Acad. Sci.* **99:** 6274–6279.

Meric F. and Hunt K.K. 2002. Translation initiation in cancer: A novel target for therapy. *Mol. Cancer Ther.* **1:** 971–979.

Meurs E.F., Galabru J., Barber G.N., Katze M.G., and Hovanessian A.G. 1993. Tumor suppressor function of the interferon-induced double-stranded RNA-activated protein kinase. *Proc. Natl. Acad. Sci.* **90:** 232–236.

Miki Y., Swensen J., Shattuck-Eidens D., Futreal P.A., Harshman K., Tavtigian S., Liu Q., Cochran C., Bennett L.M., Ding W., et al. 1994. A strong candidate for the breast and ovarian cancer susceptibility gene BRCA1. *Science* **266:** 66–71.

Mimori K., Mori M., Tanaka S., Akiyoshi T., and Sugimachi K. 1995. The overexpression of elongation factor 1 gamma mRNA in gastric carcinoma. *Cancer* (suppl. 6) **75:** 1446–1449.

Miyagi Y., Sugiyama A., Asai A., Okazaki T., Kuchino Y., and Kerr S.J. 1995. Elevated levels of eukaryotic translation initiation factor eIF-4E, mRNA in a broad spectrum of transformed cell lines. *Cancer Lett.* **91:** 247–252.

Nathan C.A., Carter P., Liu L., Li B.D., Abreo F., Tudor A., Zimmer S.G., and De Benedetti A. 1997. Elevated expression of eIF4E and FGF-2 isoforms during vascularization of breast carcinomas. *Oncogene* **15:** 1087–1094.

Nathan C.A., Amirghahari N., Abreo F., Rong X., Caldito G., Jones M.L., Zhou H., Smith

M., Kimberly D., and Glass J. 2004. Overexpressed eIF4E is functionally active in surgical margins of head and neck cancer patients via activation of the Akt/mammalian target of rapamycin pathway. *Clin. Cancer Res.* **10:** 5820–5827.

Neshat M.S., Mellinghoff I.K., Tran C., Stiles B., Thomas G., Petersen R., Frost P., Gibbons J.J., Wu H., and Sawyers C.L. 2001. Enhanced sensitivity of PTEN-deficient tumors to inhibition of FRAP/mTOR. *Proc. Natl. Acad. Sci.* **98:** 10314–10319.

Nupponen N.N., Porkka K., Kakkola L., Tanner M., Persson K., Borg A., Isola J., and Visakorpi T. 1999. Amplification and overexpression of p40 subunit of eukaryotic translation initiation factor 3 in breast and prostate cancer. *Am. J. Pathol.* **154:** 1777–1783.

Oridate N., Kim H.J., Xu X., and Lotan R. 2005. Growth inhibition of head and neck squamous carcinoma cells by small interfering RNAs targeting eIF4E or cyclin D1 alone or combined with cisplatin. *Cancer Biol. Ther.* **4:** 318–323.

Pardee A.B. 1989. G1 events and regulation of cell proliferation. *Science* **246:** 603–608.

Parmer T.G., Ward M.D., and Hait W.N. 1997. Effects of rottlerin, an inhibitor of calmodulin-dependent protein kinase III, on cellular proliferation, viability, and cell cycle distribution in malignant glioma cells. *Cell Growth Differ.* **8:** 327–334.

Parmer T.G., Ward M.D., Yurkow E.J., Vyas V.H., Kearney T.J., and Hait W.N. 1999. Activity and regulation by growth factors of calmodulin-dependent protein kinase III (elongation factor 2-kinase) in human breast cancer. *Br. J. Cancer* **79:** 59–64.

Parsa A.T. and Holland E.C. 2004. Cooperative translational control of gene expression by Ras and Akt in cancer. *Trends Mol. Med.* **10:** 607–613.

Pelletier J. and Sonenberg N. 1985. Insertional mutagenesis to increase secondary structure within the 5′ noncoding region of a eukaryotic mRNA reduces translational efficiency. *Cell* **40:** 515–526.

Peng X., Mehta R.G., Tonetti D.A., and Christov K. 2005. Identification of novel RAR-beta2 transcript variants with short 5′-UTRs in normal and cancerous breast epithelial cells. *Oncogene* **24:** 1296–1301.

Perrin-Vidoz L., Sinilnikova O.M., Stoppa-Lyonnet D., Lenoir G.M., and Mazoyer S. 2002. The nonsense-mediated mRNA decay pathway triggers degradation of most BRCA1 mRNAs bearing premature termination codons. *Hum. Mol. Genet.* **11:** 2805–2814.

Pianese G. 1896. Beitrag zur histologie und Aetiologie der carcinoma histologische und experimentelle. *Beitr. Pathol. Anat. Allg. Pathol.* **142:** 1–193.

Piette J., Neel H., and Marechal V. 1997. Mdm2: Keeping p53 under control. *Oncogene* **15:** 1001–1010.

Podsypanina K., Lee R.T., Politis C., Hennessy I., Crane A., Puc J., Neshat M., Wang H., Yang L., Gibbons J., et al. 2001. An inhibitor of mTOR reduces neoplasia and normalizes p70/S6 kinase activity in Pten+/− mice. *Proc. Natl. Acad. Sci.* **98:** 10320–10325.

Polunovsky V.A., Rosenwald I.B., Tan A.T., White J., Chiang L., Sonenberg N., and Bitterman P.B. 1996. Translational control of programmed cell death: Eukaryotic translation initiation factor 4E blocks apoptosis in growth-factor-restricted fibroblasts with physiologically expressed or deregulated Myc. *Mol. Cell. Biol.* **16:** 6573–6581.

Polunovsky V.A., Gingras A.C., Sonenberg N., Peterson M., Tan A., Rubins J.B., Manivel J.C., and Bitterman P.B. 2000. Translational control of the antiapoptotic function of Ras. *J. Biol. Chem.* **275:** 24776–24780.

Prevot D., Darlix J.L., and Ohlmann T. 2003. Conducting the initiation of protein synthesis: The role of eIF4G. *Biol. Cell.* **95:** 141–156.

Rajasekhar V.K., Viale A., Socci N.D., Wiedmann M., Hu X., and Holland E.C. 2003. Oncogenic Ras and Akt signaling contribute to glioblastoma formation by differential recruitment of existing mRNAs to polysomes. *Mol. Cell* **12:** 889–901.

Raught B., Gingras A.C., James A., Medina D., Sonenberg N., and Rosen J.M. 1996. Expression of a translationally regulated, dominant-negative CCAAT/enhancer-binding protein beta isoform and up-regulation of the eukaryotic translation initiation factor 2alpha are correlated with neoplastic transformation of mammary epithelial cells. *Cancer Res.* **56:** 4382–4386.

Raught B., Peiretti F., Gingras A.C., Livingstone M., Shahbazian D., Mayeur G.L., Polakiewicz R.D., Sonenberg N., and Hershey J.W. 2004. Phosphorylation of eucaryotic translation initiation factor 4B Ser422 is modulated by S6 kinases. *EMBO J.* **23:** 1761–1769.

Rinker-Schaeffer C.W., Austin V., Zimmer S., and Rhoads R.E. 1992. *ras*-transformation of cloned rat embryo fibroblasts results in increased rates of protein synthesis and phosphorylation of eukaryotic initiation factor 4E. *J. Biol. Chem.* **267:** 10659–10664.

Rosenwald I.B. 2004. The role of translation in neoplastic transformation from a pathologist's point of view. *Oncogene* **23:** 3230–3247.

Rosenwald I.B., Lazaris-Karatzas A., Sonenberg N., and Schmidt E.V. 1993. Elevated levels of cyclin D1 protein in response to increased expression of eukaryotic initiation factor 4E. *Mol. Cell. Biol.* **13:** 7358–7363.

Rosenwald I.B., Hutzler M.J., Wang S., Savas L., and Fraire A.E. 2001. Expression of eukaryotic translation initiation factors 4E and 2alpha is increased frequently in bronchioloalveolar but not in squamous cell carcinomas of the lung. *Cancer* **92:** 2164–2171.

Rosenwald I.B., Wang S., Savas L., Woda B., and Pullman J. 2003. Expression of translation initiation factor eIF-2alpha is increased in benign and malignant melanocytic and colonic epithelial neoplasms. *Cancer* **98:** 1080–1088.

Rosenwald I.B., Chen J.J., Wang S., Savas L., London I.M., and Pullman J. 1999. Upregulation of protein synthesis initiation factor eIF-4E is an early event during colon carcinogenesis. *Oncogene* **18:** 2507–2517.

Rosenwald I.B., Kaspar R., Rousseau D., Gehrke L., Leboulch P., Chen J.J., Schmidt E.V., Sonenberg N., and London I.M. 1995. Eukaryotic translation initiation factor 4E regulates expression of cyclin D1 at transcriptional and post-transcriptional levels. *J. Biol. Chem.* **270:** 21176–21180.

Rothe M., Ko Y., Albers P., and Wernert N. 2000. Eukaryotic initiation factor 3 p110 mRNA is overexpressed in testicular seminomas. *Am. J. Pathol.* **157:** 1597–1604.

Rousseau D., Kaspar R., Rosenwald I., Gehrke L., and Sonenberg N. 1996. Translation initiation of ornithine decarboxylase and nucleocytoplasmic transport of cyclin D1 mRNA are increased in cells overexpressing eukaryotic initiation factor 4E. *Proc. Natl. Acad. Sci.* **93:** 1065–1070.

Ruggero D. and Pandolfi P.P. 2003. Does the ribosome translate cancer? *Nat. Rev. Cancer* **3:** 179–192.

Ruggero D., Montanaro L., Ma L., Xu W., Londei P., Cordon-Cardo C., and Pandolfi P.P. 2004. The translation factor eIF-4E promotes tumor formation and cooperates with c-Myc in lymphomagenesis. *Nat. Med.* **10:** 484–486.

Saramaki O., Willi N., Bratt O., Gasser T.C., Koivisto P., Nupponen N.N., Bubendorf L., and Visakorpi T. 2001. Amplification of EIF3S3 gene is associated with advanced stage in prostate cancer. *Am. J. Pathol.* **159:** 2089–2094.

Savinova O., Joshi B., and Jagus R. 1999. Abnormal levels and minimal activity of the dsRNA-activated protein kinase, PKR, in breast carcinoma cells. *Int. J. Biochem. Cell Biol.* **31:** 175–189.

Schmidt E.V. 2004. The role of c-myc in regulation of translation initiation. *Oncogene* **23:** 3217–3221.

Scott P.A., Smith K., Poulsom R., De Benedetti A., Bicknell R., and Harris A.L. 1998. Differential expression of vascular endothelial growth factor mRNA vs protein isoform expression in human breast cancer and relationship to eIF-4E. *Br. J. Cancer* **77:** 2120–2128.

Shahbazian D., Roux P.P., Mieulet V., Cohen M.S., Raught B., Taunton J., Hershey J.W., Blenis J., Pende M., and Sonenberg N. 2006. The mTOR/PI3K and MAPK pathways converge on eIF4B to control its phosphorylation and activity. *EMBO J.* **25:** 2781–2791.

Shantz L.M., Hu R.H., and Pegg A.E. 1996. Regulation of ornithine decarboxylase in a transformed cell line that overexpresses translation initiation factor eIF-4E. *Cancer Res.* **56:** 3265–3269.

Sharpless N.E. 2005. INK4a/ARF: A multifunctional tumor suppressor locus. *Mutat. Res.* **576:** 22–38.

Shimada A., Shiota G., Miyata H., Kamahora T., Kawasaki H., Shiraki K., Hino S., and Terada T. 1998. Aberrant expression of double-stranded RNA-dependent protein kinase in hepatocytes of chronic hepatitis and differentiated hepatocellular carcinoma. *Cancer Res.* **58:** 4434–4438.

Shuda M., Kondoh N., Tanaka K., Ryo A., Wakatsuki T., Hada A., Goseki N., Igari T., Hatsuse K., Aihara T., et al. 2000. Enhanced expression of translation factor mRNAs in hepatocellular carcinoma. *Anticancer Res.* **20:** 2489–2494.

Sobczak K. and Krzyzosiak W.J. 2002. Structural determinants of BRCA1 translational regulation. *J. Biol. Chem.* **277:** 17349–17358.

Sorrells D.L., Meschonat C., Black D., and Li B.D. 1999. Pattern of amplification and overexpression of the eukaryotic initiation factor 4E gene in solid tumor. *J. Surg. Res.* **85:** 37–42.

Spinola M., Galvan A., Pignatiello C., Conti B., Pastorino U., Nicander B., Paroni R., and Dragani T.A. 2005. Identification and functional characterization of the candidate tumor suppressor gene TRIT1 in human lung cancer. *Oncogene* **24:** 5502–5509.

Stanners C.P., Adams M.E., Harkins J.L., and Pollard J.W. 1979. Transformed cells have lost control of ribosome number through their growth cycle. *J. Cell. Physiol.* **100:** 127–138.

Stefanovsky V.Y., Pelletier G., Hannan R., Gagnon-Kugler T., Rothblum L.I., and Moss T. 2001. An immediate response of ribosomal transcription to growth factor stimulation in mammals is mediated by ERK phosphorylation of UBF. *Mol. Cell* **8:** 1063–1073.

Tan A., Bitterman P., Sonenberg N., Peterson M., and Polunovsky V. 2000. Inhibition of Myc-dependent apoptosis by eukaryotic translation initiation factor 4E requires cyclin D1. *Oncogene* **19:** 1437–1447.

Terada T., Maeta H., Endo K., and Ohta T. 2000. Protein expression of double-stranded RNA-activated protein kinase in thyroid carcinomas: Correlations with histologic types, pathologic parameters, and Ki-67 labeling. *Hum. Pathol.* **31:** 817–821.

Uechi T., Tanaka T., and Kenmochi N. 2001. A complete map of the human ribosomal protein genes: Assignment of 80 genes to the cytogenetic map and implications for human disorders. *Genomics* **72:** 223–230.

van Slegtenhorst M., Nellist M., Nagelkerken B., Cheadle J., Snell R., van den Ouweland A., Reuser A., Sampson J., Halley D., and van der Sluijs P. 1998. Interaction between hamartin and tuberin, the TSC1 and TSC2 gene products. *Hum. Mol. Genet.* **7:** 1053–1057.

Vivanco I. and Sawyers C.L. 2002. The phosphatidylinositol 3-kinase AKT pathway in human cancer. *Nat. Rev. Cancer* **2:** 489–501.

Wang S., Lloyd R.V., Hutzler M.J., Rosenwald I.B., Safran M.S., Patwardhan N.A., and Khan A. 2001. Expression of eukaryotic translation initiation factors 4E and 2alpha correlates with the progression of thyroid carcinoma. *Thyroid* **11:** 1101–1107.

Wang S., Rosenwald I.B., Hutzler M.J., Pihan G.A., Savas L., Chen J.J., and Woda B.A. 1999. Expression of the eukaryotic translation initiation factors 4E and 2alpha in non-Hodgkin's lymphomas. *Am. J. Pathol.* **155:** 247–255.

Wang X., Yu J., Sreekumar A., Varambally S., Shen R., Giacherio D., Mehra R., Montie J.E., Pienta K.J., Sanda M.G., et al. 2005. Autoantibody signatures in prostate cancer. *N. Engl. J. Med.* **353:** 1224–1235.

Wek R.C. 1994. eIF-2 kinases: Regulators of general and gene-specific translation initiation. *Trends Biochem. Sci.* **19:** 491–496.

Wendel H.G., De Stanchina E., Fridman J.S., Malina A., Ray S., Kogan S., Cordon-Cardo C., Pelletier J., and Lowe S.W. 2004. Survival signalling by Akt and eIF4E in oncogenesis and cancer therapy. *Nature* **428:** 332–337.

White R.J. 2005. RNA polymerases I and III, growth control and cancer. *Nat. Rev. Mol. Cell Biol.* **6:** 69–78.

Willis A.E. 1999. Translational control of growth factor and proto-oncogene expression. *Int. J. Biochem. Cell Biol.* **31:** 73–86.

Wilson C.A., Ramos L., Villasenor M.R., Anders K.H., Press M.F., Clarke K., Karlan B., Chen J.J., Scully R., Livingston D., et al. 1999. Localization of human BRCA1 and its loss in high-grade, non-inherited breast carcinomas. *Nat. Genet.* **21:** 236–240.

Yang H.S., Jansen A.P., Komar A.A., Zheng X., Merrick W.C., Costes S., Lockett S.J., Sonenberg N., and Colburn N.H. 2003. The transformation suppressor Pdcd4 is a novel eukaryotic translation initiation factor 4A binding protein that inhibits translation. *Mol. Cell. Biol.* **23:** 26–37.

Yoon A., Peng G., Brandenburg Y., Zollo O., Xu W., Rego E., and Ruggero D. 2006. Impaired control of IRES-mediated translation in X-linked dyskeratosis congenita. *Science* **312:** 902–906.

Zetterberg A. and Killander D. 1965. Quantitative cytochemical studies on interphase growth. II. Derivation of synthesis curves from the distribution of DNA, RNA and mass values of individual mouse fibroblasts in vitro. *Exp. Cell Res.* **39:** 22–32.

Zetterberg A., Larsson O., and Wiman K.G. 1995. What is the restriction point? *Curr. Opin. Cell Biol.* **7:** 835–842.

Zhai W. and Comai L. 2000. Repression of RNA polymerase I transcription by the tumor suppressor p53. *Mol. Cell. Biol.* **20:** 5930–5938.

Zhang L., Zhou W., Velculescu V.E., Kern S.E., Hruban R.H., Hamilton S.R., Vogelstein B., and Kinzler K.W. 1997. Gene expression profiles in normal and cancer cells. *Science* **276:** 1268–1272.

Zhang Y. and Xiong Y. 2001. Control of p53 ubiquitination and nuclear export by MDM2 and ARF. *Cell Growth Differ.* **12:** 175–186.

Zhou X., Tan M., Stone Hawthorne V., Klos K.S., Lan K.H., Yang Y., Yang W., Smith T.L., Shi D., and Yu D. 2004. Activation of the Akt/mammalian target of rapamycin/4E-BP1 pathway by ErbB2 overexpression predicts tumor progression in breast cancers. *Clin. Cancer Res.* **10:** 6779–6788.

Zimmer S.G., DeBenedetti A., and Graff J.R. 2000. Translational control of malignancy: The mRNA cap-binding protein, eIF-4E, as a central regulator of tumor formation, growth, invasion and metastasis. *Anticancer Res.* **20:** 1343–1351.

16

Matters of Life and Death: Translation Initiation during Apoptosis

Simon J. Morley and Mark J. Coldwell

Department of Biochemistry, School of
Life Sciences, University of Sussex
Falmer, Brighton, BN1 9QG, United Kingdom

RECENT STUDIES HAVE IDENTIFIED SEVERAL MECHANISTIC LINKS between the regulation of translation and the process of apoptosis induced via either receptor-dependent or receptor-independent mechanisms. Rates of protein synthesis are controlled by a wide range of agents that induce cell death, with many changes that occur to the translational machinery preceding overt apoptosis and loss of cell viability. In this chapter, we summarize the temporal regulation of translation initiation in response to the activation of apoptosis focusing on (1) early changes in protein phosphorylation, (2) specific proteolytic cleavage of initiation factors, (3) selective maintenance of populations of mRNA associated with the translational machinery, and (4) potential role for the reported increases in the cleavage of ribosomal RNA and increased turnover rates of mRNA. Any one event or combination of such events influences the translational capacity of the cell, allowing it to make a critical decision between survival and a commitment to die. Posttranscriptional control has a central role in this choice as the level of expression and activity of many effector proteins required for this decision are regulated at the translational level.

APOPTOSIS

Apoptosis as a phenomenon of programmed cell death by a suicide mechanism was first described by Kerr et al. (1972), with the morphological characteristics of apoptosis, which are distinct from those of a necrotic cell, being defined a year later (Schweichel and Merker 1973). The first noticeable physical change in a cell undergoing apoptosis is the condensation of the chromatin within the nucleus. The cytoplasm of the cell also condenses, and then begins to bud away from the outer surface of the cell in a process known as blebbing (Wyllie 1993). Intracellular organelles remain morphologically unaffected and the dead cell is consumed by phagocytosis. Apoptosis is a process that is important in numerous biological systems including tissue development (Cecconi 1999; Hengartner 2000) and the maintenance of correct cellular turnover in adults, such as the deletion of self-reacting cell types in the immune system. The breakdown of this mechanism is thought to be important in the development of cancers, AIDS, and various neurodegenerative disorders including Alzheimer's disease and multiple sclerosis (Flirski and Slobow 2005; Huang et al. 2005).

Components of the Apoptotic Pathways

The study of apoptosis began to advance rapidly with the discovery that morphogenetic apoptosis during the development of the model nematode, *Caenorhabditis elegans*, required the products of three essential genes, *ced-9*, *ced-4*, and *ced-3* (Horvitz et al. 1994). The genes of the Bcl-2 family (Reed 1998) are the mammalian homologs of *ced-9*, and the many members of this family can be either antiapoptotic (e.g., Bcl-2, Bcl-X_L, Bcl-w, and Diva/Boo) or proapoptotic (e.g., Bax, Bad, and Bid). The mammalian *ced-3* homologs comprise a group of related cysteine proteases, termed caspases, since they cleave proteins after a specific aspartate residue (Adams 2003). The human homolog of *ced-4* acts upstream of caspase-3 in the apoptotic protease cascade and was thus named the Apoptotic protease activating factor or Apaf-1 (Zou et al. 1997; see below). All caspases exist as inactive zymogens, are cleaved and activated in response to apoptotic signals (for review, see Cryns and Yuan 1998), and can be subdivided into the apical (or initiator) and the effector (or executioner) caspases. Activated caspase-3 is considered to be the central effector caspase and has been found to cleave many substrates (for review, see Fischer et al. 2003), including several components of the translation initiation pathway (see below).

Receptor-mediated Extrinsic Pathway of Apoptosis

The molecular mechanisms of apoptosis have been the subject of intense research (for review, see Adams 2003) and hence are only covered superficially here. Cell death can be directly induced following the stimulation of specific cell surface death receptors (Fig. 1). These receptors contain a globular intracellular protein interaction domain known as the death domain

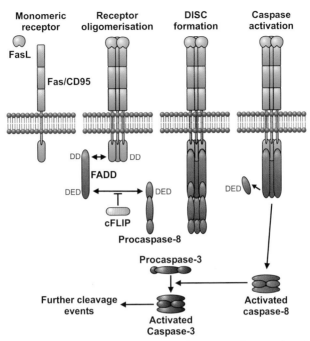

Figure 1. Activation of the extrinsic pathway of apoptosis by Fas signaling. Binding of the Fas ligand to the Fas receptor leads to the formation of the death-inducing signaling complex (DISC), which contains FADD and procaspase-8. Procaspase-8 becomes activated by autoproteolysis, forming a homodimer, which is then able to activate the effector caspase, caspase-3 (see text for details). (DD) Death domain; (DED) death effector domain.

(DD). They include those of the tumor necrosis factor (TNF)/nerve growth factor (NGF) superfamily such as TNFR1, death receptors DR-3, DR-4/TNF-related apoptosis-inducing ligand (TRAIL-R1), DR-5/TRAIL-R2, and the CD95 (Apo-1/Fas) antigen (for review, see Budihardjo et al. 1999). Upon binding of its cognate ligand, activated Fas receptors trimerize and recruit the Fas-associated death domain (FADD) adapter protein containing both DD and the death-effector domain (DED). The current models suggest that FADD binds directly to the receptor through DD interactions and to procaspase-8/procaspase-10 through DED interaction (Adams 2003; Riedl and Shi 2004), resulting in the assembly of the death-inducing complex (DISC) at the plasma membrane (Fig. 1). Similarly, the TRAIL molecule binds to the death receptors DR4/DR5, which in their bound form interact with FADD and procaspase-8 (Riedl and Shi 2004), although other adapter proteins interact with the receptor (Wajant et al. 2003; Ho and Hawkins 2005). An endogenous inhibitor, c-FLIP, which exists in a long and short form from alternative mRNA splicing (Krueger et al. 2001), competes with procaspase-8 for binding to the DISC. Although low levels of the c-FLIP$_{long}$ may be required for DISC formation (Chang et al. 2002), c-FLIP$_{short}$ abrogates caspase-8 activation. Indeed, the de novo synthesis of c-FLIP$_{short}$ is required for the resistance of short-term-activated T cells to apoptosis (Schmitz et al. 2004). Recruitment of procaspases 8 and 10 to activated receptor complexes results in a conformational change and their autoactivation by proteolysis, allowing the processing and activation of the effector caspase-3 (Fig. 1). As well as activating the caspase cascade, caspase-8 cleaves Bid, generating tBid, which affects the integrity of the mitochondrial membrane, controlling multiple death-promoting factors residing in the intermembrane space (Fig. 2).

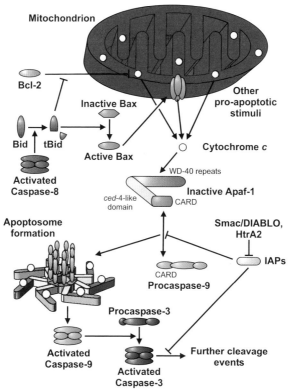

Figure 2. Activation of the intrinsic pathway of apoptosis. Cleavage of Bid by caspase-8 leads to activation and oligomerization of Bax. This and other stimuli cause the release of cytochrome *c* from the mitochondria, where it binds to Apaf-1 monomers. Assembly of Apaf-1 and procaspase-9 molecules into the apoptosome causes the activation of caspase-9, which is then able to activate caspase-3 (see text for details). (CARD) Caspase recruitment domain; (WD-40 repeats) repeats rich in tryptophan and aspartic acid.

Mitochondrially Induced Intrinsic Pathway of Apoptosis

Most proapoptotic signaling pathways activated in response to DNA damage, hypoxia, growth factor depletion, or calcium overload appear to converge at the level of the mitochondria (Fig. 2), regulating molecules that control their integrity (for review, see Adams 2003; Yousefi et al. 2003). A number of apoptotic stimuli tip the fine balance between the endoplasmic reticulum (ER) and mitochondrially associated pro- and antiapoptotic factors of the Bcl-2 family (Danial and Korsmeyer 2004). For example, tBid (Fig. 2) counteracts Bcl-2, directly activates the proapoptotic protein, Bax, and also promotes the oligomerization of Bak, resulting in the permeabilization of the outer mitochondrial membrane and release of proteins from the intermembrane space. One of these, cytochrome *c*, promotes the assembly of the apoptosome, an ATP-dependent complex between Apaf-1 and procaspase-9 (Wang 2001). This results in oligomerization of Apaf-1 monomers into a wheel-shaped structure (Acehan et al. 2002) promoting autoactivation of caspase-9 (Fig. 2). This initiator caspase can then process procaspase-3 into its active form. The inhibitor of apoptosis proteins (IAPs; XIAP, cIAP1, cIAP2) can prevent inappropriate caspase-9 activation by interfering with procaspase-9 dimerization (Fig. 2). However, in a multilayered and finely tuned response, families of IAP proteins synthesized to modulate procaspase-9 cleavage (Van Eden et al. 2004; Holcik and Sonenberg 2005; Lewis and Holcik 2005) are themselves regulated by such proteins as Smac/DIABLO and HtrA2 (Fig. 2) (Bergmann et al. 2003; Holcik and Sonenberg 2005).

Signaling for Cell Survival Versus Cell Death

A number of cytokines and growth factors including insulin-like growth factor (IGF), NGF, and platelet-derived growth factor (PDGF) can sustain cell survival by stimulating signaling through the phosphatidylinositol 3-kinase (PI3K), Akt (protein kinase B), mammalian target of rapamycin (mTOR) pathway (Chapter 14). Akt phosphorylates several targets involved in the regulation of apoptosis including the proapoptotic protein, Bad (Zhao et al. 2004), and also regulates the activity of transcription factors that direct the expression of a number of cell death genes. Phosphorylation of Bad causes its dissociation from Bcl-X_L and its sequestration by 14-3-3 proteins, thereby preventing it from translocating to the mitochondria, an event required for its proapoptotic function (Kennedy et al. 1999; Yang et al. 2001). Phosphorylation of forkhead (FKHRL1) by Akt prevents its translocation to the nucleus (again via 14-3-3 proteins) and represses the expression of mRNAs encoding the Fas ligand and Bim (Linseman et al. 2002; Gilley et al. 2003). Akt also phosphorylates IκB, allowing the relocalization of NF-κB to the nucleus and transcription of prosurvival genes (Beere 2004) such as cIAP1 (You et al. 1997) and cIAP2 (Chu et al. 1997; Holcik and Sonenberg 2005). Up-regulated expression of these antiapoptotic proteins also has a role in modulating the fine balance between cell death and cancer (Chapter 15). For instance, unregulated Akt signaling promotes c-FLIP$_{short}$ levels in TRAIL-resistant glioblastomas (Panner et al. 2005), thereby preventing TRAIL-induced procaspase-8 activation but promoting NF-κB-mediated transcription of prosurvival genes. As part of the apoptotic program, activated death receptors have also been reported to attenuate survival signaling events, thereby potentiating cell death (Yousefi et al. 2003). This may, in part, be mediated by the recruitment of phosphatases (such as SHP-1) to the DISC, which crosstalk with survival signaling events, abrogating their output by dephosphorylating and inactivating key upstream kinases such as Akt (Beere 2004). In other examples, signaling molecules such as Raf-1 and MEK kinase 1 (MEKK1) are direct targets for the activated caspases (Widmann et al. 1998; Cornelis et al. 2005). With Raf-1, caspase-9-mediated cleavage causes the catalytic fragment to relocalize to the mitochondria where it may be partially activated as part of a final attempt to promote cell survival. To a similar end, MEKK1, which functions as part of the antiapoptotic survival pathway with NF-κB, is first activated during genotoxic stress. However, MEKK1 is then cleaved by caspase-3, generating a potent proapoptotic fragment that can further propagate caspase activation in an amplification loop (Widmann et al. 1998).

APOPTOSIS-MEDIATED MODIFICATIONS OF FACTORS INVOLVED IN mRNA TRANSLATION

One of the early responses to the induction of apoptosis is a rapid, but incomplete, inhibition of the rate of protein synthesis (Deckwerth and Johnson 1993; Morley et al. 1998; Zhou et al. 1998; Scott and Adebodun 1999). This is mediated by changes in the phosphorylation of translation initiation factors, the regulation of their assembly into functional complexes (for review, see Clemens et al. 2000; Morley 2001), and the targeted cleavage of factors by activated caspases (Figs. 3 and 4). However, some mRNAs (described below) remain translated under apoptotic conditions, and it is still unclear as to what role, if any, the cleavage of initiation factors has in the inhibition of translation (Morley et al. 2005). For example, treatment of Jurkat T cells with anti-Fas antiserum results in a 60–70% decrease in the rate of protein synthesis within 2–4 hours (Morley et al. 1998, 2000; Saelens et al. 2001; Jeffrey et al. 2002). The

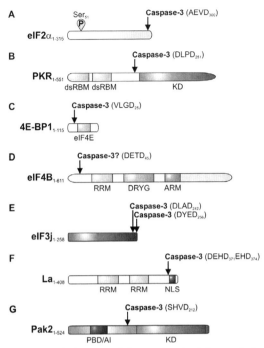

Figure 3. Caspase-mediated cleavage of factors involved in regulation of translation. For details, see text and references therein. (RRM) RNA recognition motif; (KD) kinase domain; (DRYG) region rich in aspartate, tyrosine, and glycine; (ARM) arginine-rich motif; (NLS) nuclear localization signal; (PDB/AI) p21-binding domain/autoinhibitory domain; (dsRBM) double-stranded RNA-binding motif.

z.VAD.fmk-sensitive (therefore caspase-dependent) inhibition of protein synthesis was associated with a substantial decrease in heavy polysomes (Morley et al. 1998; Zhou et al. 1998), indicative of a block at the level of initiation. However, under some conditions (e.g., TRAIL treatment of HeLa cells), mRNA degradation may also be responsible, in part, for the observed disaggregation of polysomes (Bushell et al. 2004; discussed below). It is likely that this pleiotropic response reflects a general switch from cap-dependent to cap-independent translation during the apoptotic program, inevitably influencing the translational capacity of the cell, allowing it to make critical decisions between death and survival.

PKR, eIF2α Phosphorylation, and DISC Assembly

The interferon-induced, double-stranded RNA-dependent protein kinase, PKR, can play an important part in regulating apoptosis (for review, see Barber 2005; Chapter 12). Studies with recombinant vaccinia viruses ex-

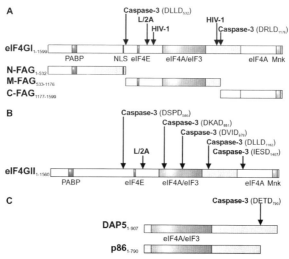

Figure 4. Caspase-mediated cleavage of proteins of the eIF4G family. (*A*) Proteolysis of eIF4GI by caspase-3 generates three discrete "fragments of apoptotic cleavage of eIF4G" (FAGs). These are different from the fragments produced after cleavage by picornaviral (L or 2A) proteases and HIV-1 protease. (*B*) Additional caspase cleavage sites are found in the eIF4GII molecule. (*C*) Cleavage of the DAP5 molecule by caspase-3 generates a fragment designated as p86.

pressing PKR suggested that the active kinase was able to promote apoptosis in HeLa cells (Lee and Esteban 1994) and that murine fibroblasts lacking PKR were resistant to TNF-α-induced apoptosis (Der et al. 1997). The most studied substrate of activated PKR is the α subunit of eIF2 with expression of a nonphosphorylatable form of eukaryotic initiation factor 2α (eIF2α) (Ser-51Ala) rendering cells resistant to apoptosis (Jagus et al. 1999; Clemens et al. 2000; Proud 2005; Chapter 13). However, it is not clear whether the effect PKR can have on translation is sufficient to induce apoptosis. It is possible that the down-regulation of general protein synthesis can lead to a reduction in the translation of mRNAs that encode proteins with an antiapoptotic role (e.g., c-FLIP$_{short}$; Fig. 1) (Fulda et al. 2000) or, conversely, the preferential translation of those mRNAs encoding proapoptotic proteins (e.g., Bax; Tan and Katze 1999). The findings that overexpression of a kinase-dead PKR ablated expression of mRNAs encoding Fas, TNFR-1, FADD, FLICE, Bad, and Bax and that PKR can indirectly activate the transcription of several proapoptotic genes (Balachandran et al. 1998) suggest a role for PKR at the level of transcription. PKR also induces the activation of caspase-8 in a manner that is dependent on FADD but independent of Apaf-1 and the Fas/CD95 and TNF-α receptors (Gil and Esteban 2000). These results suggest either that PKR stimulates other death receptors to form a complex similar to the DISC or that PKR is able to cause the FADD–procaspase-8 interaction to occur independently in the cytoplasm.

eIF2 appears to be regulated during apoptosis in two ways. The extent of phosphorylation of the α subunit (Fig. 3A) is often increased in cells following exposure to a variety of proapoptotic stimuli (Morley et al. 1998, 2000; Saelens et al. 2001; Jeffrey et al. 2002), with PKR being the strongest candidate kinase due to its important role in promoting cell death (Jeffrey et al. 2002; Donze et al. 2004; Hsu et al. 2004). In fact, the effects of the TNF-α family on translation appear to require the participation of both PKR-mediated and caspase-dependent events. Consistent with these observations, eIF2α becomes highly phosphorylated in cells exposed to TNF-α or TRAIL at times preceding any loss of cell viability or the onset of cell death (I.W. Jeffrey and M.J. Clemens, in prep.). TRAIL did not induce the activation of PKR but did result in the appearance of a PKR cleavage product of approximately 43 kD, which may correspond to a catalytically active fragment of PKR (Fig. 3B). Such a fragment, generated by caspase-3 cleavage at $DLPD_{251}$, is highly active as an eIF2α kinase arising as a result of the removal of the inhibitory double-stranded RNA-binding domain from PKR (Saelens et al. 2001). This is consistent with the finding that in TRAIL-treated MCF-7 cells, the phosphorylation of eIF2α is itself a caspase-dependent process that can be prevented by treatment of the cells with z.VAD-fmk or the caspase-8-specific inhibitor, z.IETD-fmk (I.W. Jeffrey and M.J. Clemens, unpubl.). In contrast, activation of the Fas (CD95) receptor in Jurkat cells induced a biphasic activation of eIF2α phosphorylation, yet there was no requirement for caspase-8 (Morley et al. 2000; Morley and Pain 2001; Saelens et al. 2001). In addition, a small fraction of eIF2α was also often cleaved at $AEVD_{300}$ (Fig. 3A) to give rise to a carboxy-terminally truncated product (Satoh et al. 1999; Marissen et al. 2000b; Saelens et al. 2001). The physiological significance of this cleavage is unclear; however, the truncated form of eIF2α showed very rapid exchange of GTP for GDP on eIF2, in a process that no longer required eIF2B (Marissen et al. 2000b), with expression of this cleaved protein overcoming PKR-mediated translational suppression (Satoh et al. 1999).

Overall, the caspase-mediated cleavage of eIF2α may render a population of eIF2 constitutively active in a manner independent of its state of phosphorylation, thereby protecting protein synthesis from complete down-regulation. This activity could be especially relevant for the continued translation of IRES-containing mRNAs (see below), since although the utilization of these mRNAs is independent of changes in eIF4G or 4E-BP function, it does require eIF2 activity. However, such a possibility has not yet been demonstrated experimentally. It is also not known whether there is any functional relationship between eIF2α phosphorylation and the susceptibility of the protein to cleavage by caspases.

eIF4GI, eIF4GII, and DAP5

The mRNA-binding stage of translation is a major site of regulation, requiring the activity of eIF4F (for review, see Morley et al. 2005), a heterotrimeric complex comprising eIF4E, eIF4GI, or eIF4GII and eIF4A (Chapter 4). The short-term availability of eIF4E for interaction with eIF4G is controlled by a family of small eIF4E-binding proteins (the 4E-BPs; see below), which act as competitive inhibitors of the interaction between eIF4E and eIF4G (Chapter 14). Many studies (for review, see Morley et al. 2005) have demonstrated that both eIF4GI and eIF4GII are targets for specific degradation during apoptosis (Fig. 4A,B). However, to date, it is still unclear what the exact role for eIF4GI and eIF4GII cleavage is during early apoptosis. Activation of the extrinsic pathway (by deprivation of serum growth factors, activation of the Fas receptor, TNF-α, or TRAIL) or the intrinsic pathway of apoptosis (by treatment with cycloheximide, the proteasome inhibitor MG132, the chemotherapeutic agent cisplatin, or the DNA-damaging agent etoposide) led to the inhibition of translation and the progressive degradation of eIF4GI and eIF4GII (for review, see Morley 2001; Holcik and Sonenberg 2005; Morley et al. 2005). In contrast to the differential cleavage rates of eIF4GI and eIF4GII during picornavirus infection (Gradi et al. 1998), the regulated destruction of eIF4GI and eIF4GII during apoptosis occurred with similar kinetics (Clemens et al. 1998; Bushell et al. 2000a; Marissen et al. 2000a; Morley et al. 2000). However, the inhibition of protein synthesis and the cleavage of eIF4GI are not necessarily directly correlated. Indeed, in a number of cell lines, cisplatin caused a decrease in translation rate well after the total cleavage of eIF4GI, and in MCF-7 cells, decreased translation occurs without the cleavage of eIF4GI at all (Bushell et al. 2000a; Jeffrey et al. 2002; Marissen et al. 2004). Under the same conditions, there were no major decreases in the levels of several other initiation factors, including eIF4E and eIF4A (Bushell et al. 2000b; Morley et al. 2000), suggesting that it will be important to investigate the level of eIF2α phosphorylation in these cells. PABP can be cleaved by a partially caspase-dependent mechanism. However, it is not a substrate for caspase-3 itself; the cleavage site remains undefined and the role for such cleavage in the inhibition of translation is not known (Marissen et al. 2004).

In most (but not all) cases, both the inhibition of translation and the cleavage of eIF4G can be prevented by cell-permeable caspase inhibitors, with the activity of caspase-3 being both necessary and sufficient for the proteolysis of eIF4GI and eIF4GII in vitro and in vivo (for review, see Morley et al. 2005). In contrast to the cleavage of eIF4GI and eIF4GII by picornaviral protease which bifurcates the protein, caspases produce considerably different fragments. Caspase-3 cleaves eIF4GI at two sites: one at $DLLD_{532}$ and the other at $DRLD_{1176}$ (Fig. 4A). The three distinct breakdown products of eIF4GI resulting from the two cleavages have been termed fragments of apoptotic cleavage of eIF4G (FAGs) (Clemens et al. 1998; Bushell et al. 2000b) and designated N-FAG, M-FAG, and C-FAG. The possible role for these fragments in controlling translation initiation is addressed in more detail below. In contrast, eIF4GII is cleaved at five sites ($DSPD_{568}$, $DKAD_{851}$, $DVID_{978}$, $DLLD_{1162}$, $IESD_{1407}$) (Fig. 4B), essentially destroying all of the known functional domains of the protein and rendering them unlikely to participate in translation (Marissen et al. 2000a; Morley 2001; Morley and Pain 2001).

In addition, a gene variously called *p97, DAP5,* or *NAT1* (Imataka et al. 1997; Levy-Strumpf et al. 1997; Yamanaka et al. 1997) encodes a protein that is homologous to the central and carboxy-terminal parts of eIF4G (Fig. 4C). *DAP5* is also cleaved during apoptosis, at a single site downstream from the sequence $DETD_{790}$. The resulting amino-terminal fragment, termed p86, is still capable of binding eIF3 and eIF4A, but it is distinct from M-FAG as it lacks the eIF4E-binding site (Henis-Korenblit et al. 2000). This truncated version of the protein may have a critical role in translational control during apoptosis (see below).

4E-BP1

The extent of sequestration of eIF4E by 4E-BP1 is determined by its phosphorylation state, which is regulated by a number of signaling pathways. As discussed in more detail in Chapter 14, phosphorylation at multiple sites strongly reduces the affinity of the 4E-BPs for eIF4E, releasing eIF4E to participate in the initiation process. On the contrary, under conditions that block cell proliferation or induce apoptosis, 4E-BP phosphorylation decreases and the 4E-BPs associate avidly with eIF4E. Prominent among such are staurosporine, TRAIL, Fas ligand, DNA-damage inducers, and activated p53 (Morley et al. 2000; Tee and Proud 2000, 2001, 2002; Morley 2001; Jeffrey et al. 2002; Clemens 2004; Constantinou and Clemens 2005). Both dephosphorylation and caspase-mediated cleavage of 4E-BP1 at $VLGD_{25}$ (see Fig. 3C) occur in cells exposed to staurosporine, etoposide

(Tee and Proud 2000, 2001, 2002; Clemens 2004), and TRAIL (Jeffrey et al. 2002). Inhibition of protein synthesis caused by the activation of the tumor suppressor protein p53 is also associated with dephosphorylation and proteolytic cleavage of 4E-BP1, with the latter bound to eIF4E in preference to the remaining full-length 4E-BP1 (Constantinou and Clemens 2005). Down-regulation of cap-dependent translation is inevitably a consequence of these changes, but again there is the possibility of differential effects on different mRNAs, depending on the requirement for the level of eIF4F in the cell. Moreover, IRES-driven translation escapes the inhibition altogether because it is an eIF4E-independent process. Consistent with this, reports suggest that DNA-damaging agents, such as staurosporine, etoposide, cisplatin, and mitomycin C, all result in the dephosphorylation of 4E-BP1 before apoptosis is evident, conditions that promote the IRES-driven translation of Bcl-2 (Sherrill et al. 2004) and cIAP1 (Van Eden et al. 2004), possibly delaying the onset of cell death.

eIF4B

eIF4B is required for mRNA binding to ribosomes because it stimulates the RNA helicase activity of eIF4A in vitro (Chapter 4). Cleavage of eIF4B has been observed during apoptosis (see Fig. 3D), although this caspase-3-mediated event occurs with delayed kinetics relative to that seen for eIF4GI (Bushell et al. 2000a). eIF4B is also cleaved during apoptosis in MCF-7 cells lacking caspase-3, suggesting that other caspases may be involved (Jeffrey et al. 2002). Cleavage of eIF4B occurs in its amino-terminal domain in a region that is required for its interaction with PABP; the consequence of such cleavage remains unclear, as truncated eIF4B still co-isolates with eIF4F (Bushell et al. 2000a) and still contains the DRYG motif, which is essential for self-association and for its interaction with the largest subunit of eIF3 (eIF3a; Methot et al. 1996).

eIF3

Mammalian eIF3, which binds directly to 40S ribosomal subunits (Chapter 4), contains at least 12 nonidentical protein subunits, designated a–l in order of decreasing molecular weight. Recently, fission yeast has been shown to possess two distinct eIF3 complexes and one of them contains a further subunit, designated eIF3m (Zhou et al. 2005). Specific functions for mammalian eIF3 subunits have been identified by a variety of in vitro experiments (Hershey and Merrick 2000; Fraser et al. 2004; Kolupaeva et al. 2005) and are discussed in more detail in Chapters 4 and 9. Cleavage of the p35 (j) subunit of eIF3 with caspase-3 was observed during apoptosis (see Fig. 3E), with none of the other subunits of eIF3 being affected (Bushell et al. 2000a). The appearance of the ΔeIF3j fragment is evident within 4 hours of cycloheximide-induced apoptosis of BJAB cells, a slower rate than the cleavage of eIF4GI, with complete conversion to the truncated form evident at later times (Bushell et al. 2000a). In vitro experiments indicated that caspase-3 had the ability to cleave eIF3j at $DLAD_{242}$ and $DYED_{256}$ (Bushell et al. 2000a). The exact function of the eIF3j subunit in the eIF3 multisubunit complex remained unresolved for a number of years before it was determined that it has a central role in mediating the stable association of eIF3 with the 40S ribosome (Fraser et al. 2004). eIF3j that has been processed to the shorter form by caspase-3 shows a dramatically reduced affinity for the 40S ribosome and consequently does not efficiently recruit eIF3 to the 40S ribosome (Fraser et al. 2004), suggesting a role for this cleavage in the general inhibition of translation at later times during apoptosis.

Other Factors Affecting Translation Initiation

Polypyrimidine-tract-binding Protein

Cellular IRES sequences are relatively inefficiently utilized in vitro, but their activity can be stimulated by IRES *trans*-activating factors (ITAFs) such as polypyrimidine-tract-binding protein (PTB) (Spriggs et al. 2005; Chapters 5 and 6). Recent work has now shown that transient production of the antiapoptotic proteins Apaf-1 and BAG-1 through IRES-mediated translation requires PTB binding (Mitchell et al. 2005; Spriggs et al. 2005) and that levels of PTB are up-regulated during apoptosis (A.E. Willis and M. Bushell, in prep.).

La Autoantigen

Another ITAF, the La (SS-B) autoantigen, is a target for caspase-mediated cleavage (Fig. 3F) and dephosphorylation during apoptosis (Rutjes et al. 1999; Ayukawa et al. 2000). Cleavage at the carboxyl terminus of the protein, downstream from $DEHD_{371}$ or $DEHD_{374}$, results in the loss of a nuclear localization signal, redeploying the truncated protein to the cytoplasm (Ayukawa et al. 2000). This relocalization of La may promote its ability to influence translation initiation, including effects on internal initiation (Meerovitch et al. 1993) and the inhibition of PKR activity (Xiao et al. 1994; James et al. 1999).

Pak2

Pak2 (also termed hPAK65 and γ-Pak) is one of several mammalian p21-activated kinases (Bokoch 2003) and is cleaved by caspase-3 during apoptosis after SHVD$_{212}$ (Fig. 3G). This separates the amino-terminal regulatory domain, which binds the p21-GTPases and acts to repress Pak2 activity in their absence, from the carboxy-terminal catalytic domain (Lee et al. 1997). Once cleaved in this way, the autoactivated carboxy-terminal domain of Pak2 phosphorylates the eIF4E kinase, Mnk1, at Thr-22 and Ser-27 and, without influencing its kinase activity, reduces the binding of Mnk1 to eIF4GI (Orton et al. 2004). This prevents eIF4E phosphorylation prior to the cleavage of eIF4GI. In addition, Pak2 can phosphorylate eIF4GI directly on Ser-896 (close to the central eIF4A-binding site), thereby reducing its activity (Ling et al. 2005). Because Pak2 and eIF4E compete for binding to eIF4G, this suggests that Pak2 activated during apoptosis may contribute to translational inhibition by decreasing levels of eIF4F.

CONSEQUENCES OF INITIATION FACTOR MODIFICATIONS FOR mRNA TRANSLATION DURING APOPTOSIS

The progress of apoptosis is characterized by a general, but incomplete, inhibition of cap-dependent protein synthesis and the maintenance of some cap-independent translation. Apoptosis-associated modifications in the translation machinery include the poorly understood up-regulation of 4E-BP1 (I.W. Jeffrey and M.J. Clemens, in prep.) and PTB levels (A.E. Willis and M. Bushell, in prep.). Although the former would contribute to an inhibition of translation (Chapter 14), the latter would be predicted to maintain IRES-mediated translation of Apaf-1 and BAG-1 (Mitchell et al. 2005; Spriggs et al. 2005). Such changes are superimposed upon the specific fragmentation of proteins (eIF4GI/II, eIF4B, eIF3j), alterations in the state of phosphorylation of initiation factors (eIF2α, eIF4E, 4E-BP1), and interference with protein–protein interactions. An example of the latter is the caspase-mediated activation of Pak2, which impinges on eIF4F complex assembly (Ling et al. 2005). Any or all of the above events could potentially contribute to the observed modulation of protein synthesis, and it is likely that the relative importance of the various changes may be different with alternative stimuli and at distinct stages of the ongoing apoptotic response. For example, in anti-Fas-treated Jurkat cells, the caspase-8-independent increase in eIF2α phosphorylation is associated with a reversible, general inhibition of protein synthesis and polysome disaggregation at early times (Morley and Pain 2001). At slightly later times, this is followed by the irreversible cleavage of eIF4GI, the loss of p70^{S6K} activity and an increase in the binding of 4E-BP1 to eIF4E, predicted to facilitate cap-independent translation (Morley 2001).

Because several factors with different functions in translation are modified before and during the early stages of apoptosis, it is difficult to dissect the individual contributions of the changes observed to the overall regulation of protein synthesis. As with many other aspects of apoptosis, the mechanisms of translational down-regulation are almost certainly multifactorial (Fig. 5). In addition, changes in mRNA utilization and/or stability can also have a major impact on global protein synthesis and/or the translation of specific mRNAs during apoptosis (Del Prete et al. 2002; Bushell et al. 2004). For example, TRAIL treatment of HeLa cells, which reduces translation rates by 75% within 2 hours, induces the partial cleavage of eIF4GI and the dephosphorylation of 4E-BP1, with a concomitant loss of mRNA encoding ribosomal protein S6 and actin (Bushell et al. 2004). However, microarray analyses indicate that although the majority of mRNAs are lost from polysomes during apoptosis, about 3% remain associated with actively translating ribosomes under these conditions. Many of those mRNAs lost from polysomes encode translation initiation and elon-

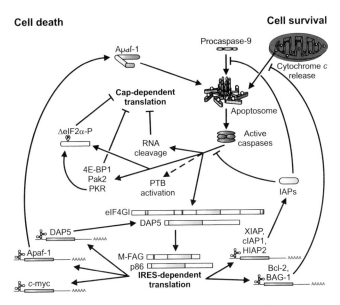

Figure 5. Translation initiation during apoptosis. An overview of the complex interplay between events leading to cell death and those required for cell survival. As described in detail in the text, emphasis is placed upon how these events are believed to impinge upon cap-dependent and IRES-dependent translation.

gation factors, ribosomal proteins, glycolytic enzymes, and regulators of cell cycle progression. In contrast, one-fourth of the mRNAs that remain associated with polysomes encode proteins involved in transcriptional control (including c-Myc), chromosome remodeling, some members of the Notch signaling pathway, and p70 S6 kinase (M. Bushell, in prep.). An added consideration is the finding that there is also cleavage of 18S and 28S rRNAs in ribosomes in apoptotic cells, although this may be a somewhat later event (King et al. 2000). These findings raise the question as to whether the modified ribosomes have a specialized function in apoptosing cells or whether rRNA cleavage is incidental. Despite these uncertainties, with our knowledge of the roles of individual initiation factors, it is possible to identify several specific effects that can be predicted from the changes induced by the phosphorylation and/or cleavage of these proteins.

Effects of Modifications of the eIF4F Complex

We have shown the existence of a modified form of eIF4F during apoptosis containing dephosphorylated eIF4E, eIF4A, but with the central M-FAG fragment in place of full-length eIF4GI (Bushell et al. 2000a, b; Morley 2001; Morley et al. 2005). It is distinct from the steady-state eIF4F complex, remaining stable for several hours in apoptosing cells before M-FAG is further degraded, with the loss of the eIF4E-binding site (Morley 2001; Morley and Pain 2001). As such, this complex may still be able to support either cap-dependent (Morino et al. 2000) or -independent initiation (Fig. 6) (De Gregorio et al. 1999). Indeed, when HeLa cells were exposed to TRAIL, apoptosis was rapidly induced and cap-dependent translation rates were severely but not completely inhibited (Stoneley et al. 2000). The presence of either the c-Myc or HRV2 IRES between the cistrons of a dicistronic luciferase reporter mRNA (Stoneley et al. 1998) was able to maintain the translation of the downstream firefly luciferase sequence under these conditions. c-Myc mRNA itself can initiate translation in either a cap-dependent manner or an IRES-dependent manner (Stoneley et al. 2000), and the continued expression of the c-Myc protein from the IRES may be important for the later stages of apoptosis (Bushell et al. 2004). c-Myc has a very short half-life (~20 min; Hann and Eisenman 1984; West et al. 1998) and thus, it would have to be continuously synthesized to maintain suitable levels of the protein. In addition, the IRES present in the mRNA encoding XIAP was utilized during apoptosis (Holcik et al. 1999, 2000), whereas the *DAP5* IRES was used to maintain translation initiation during Fas-mediated apoptosis (Henis-Korenblit et al. 2000). Reaper mRNA, which encodes a proapoptotic pro-

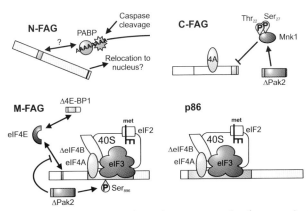

Figure 6. Modified translation complexes form as a result of apoptosis. A summary of changes in interactions and in the phosphorylation status of components of the translation initiation pathway observed during apoptosis resulting from caspase-mediated cleavage and kinase activation (see text for details).

tein from *Drosophila melanogaster*, is also translated in a cap-independent manner. The 5′UTR (untranslated region) of reaper displays IRES activity in vitro and in vivo, with the mRNA associated with polysomes in response to apoptosis or thermal stress (Hernandez et al. 2004). Studies using RNA interference (RNAi) and cross-linking suggest that maintained reaper mRNA translation may, in part, be mediated by its interaction with La (Vazquez-Pianzola et al. 2005). Interestingly, IRES activity has been identified within the 5′UTR of the La mRNA (Carter and Sarnow 2000), and it is tempting to hypothesize that the La protein may be able to regulate expression of both itself and reaper at the level of translation initiation. These results suggest that this may be a common mechanism in ensuring maintenance of expression of ITAFs to allow translation of other IRES-driven mRNAs under conditions when cap-dependent translation is compromised (Spriggs et al. 2005). Not all mRNAs that can be translated by an IRES-dependent mechanism continue to be translated during apoptosis, however. Indeed, in HeLa cells, the Apaf-1 and BAG-1 IRES elements were unable to maintain translation of a reporter gene beyond the first 2 hours of TRAIL treatment (Coldwell 2001).

The question still remains as to what actually drives the selective translation of these IRES-containing mRNAs during the apoptotic process. Is it cap-independent translation mediated by truncated initiation factors (Fig. 5) or is it a modified form of cap-dependent translation? Recent unpublished work has suggested that cIAP2, which is an E3 ubiquitin ligase that targets Smac/DIABLO and RIP for degradation (Hu and Yang 2003; Park et al. 2004), may be translated by stress-mediated shunting (Ryabova et al. 2002; R.E. Lloyd, unpubl.), indicating that such a mechanism is also a possibility. As discussed recently (Morley et al. 2005), there is still some confusion over any possible role for the cleavage fragments of eIF4GI/II in modulating translation rates during early apoptosis (Fig. 6). N-FAG, which can be stable in apoptosing cells for at least 24 hours (Marissen and Lloyd 1998), had no effect on the translation of a number of dicistronic reporter genes expressed in 293 cells (Nevins et al. 2003). Upon eIF4GI cleavage, N-FAG may accumulate in the nucleus where it is unlikely to influence translation rates directly (Fig. 6) (Coldwell et al. 2004). However, it cannot be discounted that N-FAG may impinge upon translation following nuclear envelope breakdown at later stages of the apoptotic program. C-FAG (Fig. 6), which is predominantly cytoplasmic, does not influence translation rates when expressed in vivo or when added to in vitro translation systems (S.J. Morley, unpubl.). Effects on IRES-driven translation are also unlikely to be a result of the cleavage of eIF4GII during apoptosis. The N-FAG equivalent of eIF4GII, which lacks the KRRRK nuclear localization signal found in eIF4GI (Coldwell et al. 2004), remains intact and is predicted to be cytoplasmic; whether this can influence translation rates is unknown at this time. Further cleavage of eIF4GII (Fig. 4B) destroys the eIF4A- and eIF3-binding sites of the M-FAG equivalent and cleaves the eIF4GII counterpart of C-FAG, making it unlikely that they directly function during apoptosis.

There has been a report that M-FAG influenced the translation of reporter mRNAs containing the Apaf-1 IRES. In cells that were not undergoing apoptosis, translation was responsive to the expression of M-FAG (Nevins et al. 2003). However, both our own unpublished work and an independent study have concluded that overexpression of M-FAG alone had no effect on c-Myc, XIAP, or Apaf-1 IRES-driven translation (Henis-Korenblit et al. 2002). The reasons for these differences are unclear. Therefore, the most plausible candidate for modulating selective IRES-driven translation during apoptosis is *DAP5/NAT1/p97* (Marash and Kimchi 2005). *DAP5* is a caspase-activated translation factor (Figs. 4C and 6), able to maintain the translation of its own mRNA during apoptosis (Henis-Korenblit et al. 2000). Removal of the carboxyl terminus of the protein allows it to stimulate IRES-driven translation and abrogates its ability to inhibit cap-dependent translation. As such, the cleaved form of *DAP5* (p86) has been demonstrated to stimulate translation from a reporter mRNA containing Apaf-1, c-Myc, or XIAP IRES sequences both in vivo (Henis-Korenblit et al. 2002; Nevins et al. 2003) and in vitro (Hundsdoerfer et al. 2005). In addition, translation of a further member of the IAP family, HIAP2, is regulated by IRES-driven translation, which requires cleaved *DAP5* produced in response to ER stress (Warnakula-suriyarachchi et al. 2004). Consequently, it seems likely that *DAP5* has a central role in allowing the cell to rapidly and simultaneously up-regulate expression of both pro- and antiapoptotic proteins for a short period of time, generating a fine balance between apoptosis and survival in the presence of the apoptotic trigger (Fig. 5).

SUMMARY

What is clear from published work is that following the induction of apoptosis or a severe stress that may lead down this pathway, a combination of events influences the translational capacity of the cell, allowing it to make a critical decision between survival or death (Fig. 5). In this model, the early inhibition of cap-dependent translation by inactivation of PI3K/AKT/mTOR signaling and a transient increase in eIF2α phosphorylation decrease the levels of rapidly turned over proteins, such as c-FLIP$_{short}$, mcl-1, and FADD. In contrast, increased eIF2α phosphorylation promotes the synthesis of Bad/Bax, influencing apical caspase activation and sensitizing the cell to incoming cues (Figs. 1 and 2). Selective mRNA degradation and/or the maintained translation of mRNAs encoding transcription factors and chromosome remodeling factors potentiate this effect. The cleavage and activation of *DAP5* (Figs. 5 and 6), elevated expression of PTB, and relocalization of cleaved La to the cytoplasm maintain the synthesis of c-Myc, Apaf-1, Bcl-2, cIAP1, BAG-1, and XIAP, proteins required to regulate the complex interplay between the pro- and antiapoptotic pathways while the cell decides its fate (Fig. 5). Once the death of the cell is inevitable, only those IRES elements required for maintenance of expression of short-lived proteins during the latter stages of apoptosis remain functional. De novo cap-dependent translation is severely inhibited as a result of these events, and the cleavage of eIF4GI, eIF4GII, the late cleavage of PABP, and the decreased ability of eIF4E to participate in the initiation process finally drive the cell down the death pathway. However, although this scenario fits the available data, the relative importance of initiation factor cleavages has yet to be proven experimentally.

ACKNOWLEDGMENTS

We thank all members of our laboratory, past and present, for helpful discussions, and we thank colleagues for providing us with their unpublished data. Research in our laboratory has been funded by grants (40800, 067517, 56778, 58915) from The Wellcome Trust, during which time S.J.M. was a senior research fellow of The Wellcome Trust.

REFERENCES

Acehan D., Jiang X., Morgan D.G., Heuser J.E., Wang X., and Akey C.W. 2002. Three-dimensional structure of the apoptosome: Implications for assembly, procaspase-9 binding, and activation. *Mol. Cell* **9:** 423–432.

Adams J.M. 2003. Ways of dying: Multiple pathways to apoptosis. *Genes Dev.* **17:** 2481–2495.

Ayukawa K., Taniguchi S., Masumoto J., Hashimoto S., Sarvotham H., Hara A., Aoyama T., and Sagara J. 2000. La autoantigen is cleaved in the COOH terminus and loses the nuclear localization signal during apoptosis. *J. Biol. Chem.* **275:** 34465–34470.

Balachandran S., Kim C.N., Yeh W.C., Mak T.W., Bhalla K., and Barber G.N. 1998. Activation of the dsRNA-dependent protein kinase, PKR, induces apoptosis through FADD-mediated death signaling. *EMBO J.* **17:** 6888–6902.

Barber G.N. 2005. The dsRNA-dependent protein kinase, PKR and cell death. *Cell Death Differ.* **12:** 563–570.

Beere H.M. 2004. "The stress of dying": The role of heat shock proteins in the regulation of apoptosis. *J. Cell Sci.* **117:** 2641–2651.

Bergmann A., Yang A.Y., and Srivastava M. 2003. Regulators of IAP function: Coming to grips with the grim reaper. *Curr. Opin. Cell Biol.* **15:** 717–724.

Bokoch G.M. 2003. Biology of the p21-activated kinases. *Annu. Rev. Biochem.* **72:** 743–781.

Budihardjo I., Oliver H., Lutter M., Luo X., and Wang X. 1999. Biochemical pathways of caspase activation during apoptosis. *Annu. Rev. Cell Dev. Biol.* **15:** 269–290.

Bushell M., Stoneley M., Sarnow P., and Willis A.E. 2004. Translation inhibition during the induction of apoptosis: RNA or protein degradation? *Biochem. Soc. Trans.* **32:** 606–610.

Bushell M., Wood W., Clemens M.J., and Morley S.J. 2000a. Changes in integrity and association of eukaryotic protein synthesis initiation factors during apoptosis. *Eur. J. Biochem.* **267:** 1083–1091.

Bushell M., Poncet D., Marissen W.E., Flotow H., Lloyd R.E., Clemens M.J., and Morley S.J. 2000b. Cleavage of polypeptide chain initiation factor eIF4GI during apoptosis in lymphoma cells: Characterisation of an internal fragment generated by caspase-3-mediated cleavage. *Cell Death Differ.* **7:** 628–636.

Carter M.S. and Sarnow P. 2000. Distinct mRNAs that encode La autoantigen are differentially expressed and contain internal ribosome entry sites. *J. Biol. Chem.* **275:** 28301–28307.

Cecconi F. 1999. Apaf1 and the apoptotic machinery. *Cell Death Differ.* **6:** 1087–1098.

Chang D.W., Xing Z., Pan Y., Algeciras-Schimnich A., Barnhart B.C., Yaish-Ohad S., Peter M.E., and Yang X. 2002. c-FLIP(L) is a dual function regulator for caspase-8 activation and CD95-mediated apoptosis. *EMBO J.* **21:** 3704–3714.

Chu Z.L., McKinsey T.A., Liu L., Gentry J.J., Malim M.H., and Ballard D.W. 1997. Suppression of tumor necrosis factor-induced cell death by inhibitor of apoptosis c-IAP2 is under NF-kappaB control. *Proc. Natl. Acad. Sci.* **94:** 10057–10062.

Clemens M.J. 2004. Targets and mechanisms for the regulation of translation in malignant transformation. *Oncogene* **23:** 3180–3188.

Clemens M.J., Bushell M., and Morley S.J. 1998. Degradation of eukaryotic polypeptide chain initiation factor (eIF) 4G in response to induction of apoptosis in human lymphoma cell lines. *Oncogene* **17:** 2921–2931.

Clemens M.J., Bushell M., Jeffrey I.W., Pain V.M., and Morley S.J. 2000. Translation initiation factor modifications and the regulation of protein synthesis in apoptotic cells. *Cell Death Differ.* **7:** 603–615.

Coldwell M.J. 2001. "Identification and functional analysis of internal ribosome entry segments in Apaf-1 and BAG-1." Ph.D. thesis, University of Leicester, Leicester, United Kingdom.

Coldwell M.J., Hashemzadeh-Bonehi L., Hinton T.M., Morley S.J., and Pain V.M. 2004. Expression of fragments of translation initiation factor eIF4GI reveals a nuclear localisation signal within the N-terminal apoptotic cleavage fragment N-FAG. *J. Cell Sci.* **117:** 2545–2555.

Constantinou C. and Clemens M.J. 2005. Regulation of the phosphorylation and integrity of protein synthesis initiation factor eIF4GI and the translational repressor 4E-BP1 by p53. *Oncogene* **24:** 4839–4850.

Cornelis S., Bruynooghe Y., Van Loo G., Saelens X., Vandenabeele P., and Beyaert R. 2005. Apoptosis of hematopoietic cells induced by growth factor withdrawal is associated with caspase-9 mediated cleavage of Raf-1. *Oncogene* **24:** 1552–1562.

Cryns V. and Yuan J. 1998. Proteases to die for. *Genes Dev.* **12:** 1551–1570.

Danial N.N. and Korsmeyer S.J. 2004. Cell death: Critical control points. *Cell* **116:** 205–219.

Deckwerth T.L. and Johnson E.M. 1993. Temporal analysis of events associated with programmed cell death (apoptosis) of sympathetic neurons deprived of nerve growth factor. *J. Cell Biol.* **123:** 1207–1222.

De Gregorio E., Preiss T., and Hentze M.W. 1999. Translation driven by an eIF4G core domain *in vivo. EMBO J.* **18:** 4865–4874.

Del Prete M.J., Robles M.S., Guao A., Martinez-A C., Izquierdo M., and Garcia-Sanz J.A. 2002. Degradation of cellular mRNA is a general early apoptosis-induced event. *FASEB J.* **16:** 2003–2005.

Der S.D., Yang Y.L., Weissmann C., and Williams B.R.G. 1997. A double-stranded RNA-activated protein kinase-dependent pathway mediating stress-induced apoptosis. *Proc. Natl. Acad. Sci.* **94:** 3279–3283.

Donze O., Deng J., Curran J., Sladek R., Picard D., and Sonenberg N. 2004. The protein kinase PKR: A molecular clock that sequentially activates survival and death programs. *EMBO J.* **23:** 564–571.

Fischer U., Janicke R.U., and Schulze-Osthoff K. 2003. Many cuts to ruin: A comprehensive update of caspase substrates. *Cell Death Differ.* **10:** 76–100.

Flirski M. and Slobow T. 2005. Biochemical markers and risk factors of Alzheimer's disease. *Curr. Alzheimer Res.* **2:** 47–64.

Fraser C.S., Lee J.Y., Mayeur G.L., Bushell M., Doudna J.A., and Hershey J.W. 2004. The j-subunit of human translation initiation factor eIF3 is required for the stable binding of eIF3 and its subcomplexes to 40 S ribosomal subunits in vitro. *J. Biol. Chem.* **279:** 8946–8956.

Fulda S., Meyer E., and Debatin K.M. 2000. Metabolic inhibitors sensitize for CD95 (APO-1/Fas)-induced apoptosis by down-regulating Fas-associated death domain-like interleukin 1-converting enzyme inhibitory protein expression. *Cancer Res.* **60:** 3947–3956.

Gil J. and Esteban M. 2000. The interferon-induced protein kinase (PKR), triggers apoptosis through FADD-mediated activation of caspase 8 in a manner independent of Fas and TNF-alpha receptors. *Oncogene* **19:** 3665–3674.

Gilley J., Coffer P.J., and Ham J. 2003. FOXO transcription factors directly activate bim gene expression and promote apoptosis in sympathetic neurons. *J. Cell Biol.* **162:** 613–622.

Gradi A., Imataka H., Svitkin Y.V., Rom E., Raught B., Morino S., and Sonenberg N. 1998. A novel functional human eukaryotic translation initiation factor 4G. *Mol. Cell. Biol.* **18:** 334–342.

Hann S.R. and Eisenman R.N. 1984. Proteins encoded by the human c-myc oncogene: Differential expression in neoplastic cells. *Mol. Cell. Biol.* **4:** 2486–2497.

Hengartner M.O. 2000. The biochemistry of apoptosis. *Nature* **407:** 770–776.

Henis-Korenblit S., Levy-Strumpf N., Goldstaub D., and Kimchi A. 2000. A novel form of DAP5 protein accumulates in apoptotic cells as a result of caspase cleavage and internal ribosome entry site-mediated translation. *Mol. Cell. Biol.* **20:** 496–506.

Henis-Korenblit S., Shani G., Sines T., Marash L., Shohat G., and Kimchi A. 2002. The caspase-cleaved DAP5 protein supports internal ribosome entry site-mediated translation of death proteins. *Proc. Natl. Acad. Sci.* **99:** 5400–5405.

Hernandez G., Vazquez-Pianzola P., Sierra J.M., and Rivera-Pomar R. 2004. Internal ribosome entry site drives cap-independent translation of reaper and heat shock protein 70 mRNAs in *Drosophila* embryos. *RNA* **10:** 1783–1797.

Hershey J.W.B. and Merrick W.C. 2000. The pathway and mechanism of initiation of protein synthesis. In *Translational control of gene expression* (ed. N. Sonenberg et al.), pp. 33–88. Cold Spring Harbor Laboratory Press, Cold Spring Harbor, New York.

Ho P.K. and Hawkins C.J. 2005. Mammalian initiator apoptotic caspases. *FEBS J.* **272:** 5436–5453.

Holcik M. and Sonenberg N. 2005. Translational control in stress and apoptosis. *Nat. Rev. Mol. Cell Biol.* **6:** 318–327.

Holcik M., Yeh C., Korneluk R.G., and Chow T. 2000. Translational upregulation of X-linked inhibitor of apoptosis (XIAP) increases resistance to radiation induced cell death. *Oncogene* **19:** 4174–4177.

Holcik M., Lefebvre C., Yeh C., Chow T., and Korneluk R.G. 1999. A new internal-ribosome-entry-site motif potentiates XIAP-mediated cytoprotection. *Nat. Cell Biol.* **1:** 190–192.

Horvitz H.R., Shaham S., and Hengartner M.O. 1994. The genetics of programmed cell death in the nematode *Caenorhabditis elegans*. *Cold Spring Harbor Symp. Quant. Biol.* **59:** 377–385.

Hsu L.C., Park J.M., Zhang K., Luo J.L., Maeda S., Kaufman R.J., Eckmann L., Guiney D.G., and Karin M. 2004. The protein kinase PKR is required for macrophage apoptosis after activation of Toll-like receptor 4. *Nature* **428:** 341–345.

Hu S. and Yang X. 2003. Cellular inhibitor of apoptosis 1 and 2 are ubiquitin ligases for the apoptosis inducer Smac/DIABLO. *J. Biol. Chem.* **278:** 10055–10060.

Huang Y., Erdmann N., Zhao J., and Zheng J. 2005. The signaling and apoptotic effects of TNF-related apoptosis-inducing ligand in HIV-1 associated dementia. *Neurotox. Res.* **8:** 135–148.

Hundsdoerfer P., Thoma C., and Hentze M.W. 2005. Eukaryotic translation initiation factor 4GI and p97 promote cellular internal ribosome entry sequence-driven translation. *Proc. Natl. Acad. Sci.* **20:** 13421–13426.

Imataka H., Olsen H.S., and Sonenberg N. 1997. A new translational regulator with homology to eukaryotic translation initiation factor 4G. *EMBO J.* **16:** 817–825.

Jagus R., Joshi B., and Barber G.N. 1999. PKR, apoptosis and cancer. *Int. J. Biochem. Cell Biol.* **31:** 123–138.

James M.C., Jeffrey I.W., Pruijn G.J., Thijssen J.P., and Clemens M.J. 1999. Translational control by the La antigen. Structure requirements for rescue of the double-stranded RNA-mediated inhibition of protein synthesis. *Eur J. Biochem.* **266:** 151–162.

Jeffrey I.W., Bushell M., Tilleray V.J., Morley S.J., and Clemens M.J. 2002. Inhibition of protein synthesis in apoptosis: Differential requirements by the tumour necrosis factor α family and a DNA damaging agent for caspases and the double-stranded RNA-dependent protein kinase. *Cancer Res.* **62:** 2272–2280.

Kennedy S.G., Kandel E.S., Cross T.K., and Hay N. 1999. Akt/Protein kinase B inhibits cell death by preventing the release of cytochrome c from mitochondria. *Mol. Cell. Biol.* **19:** 5800–5810.

Kerr J.F., Wyllie A.H., and Currie A.R. 1972. Apoptosis: A basic biological phenomenon with wide-ranging implications in tissue kinetics. *Br. J. Cancer.* **26:** 239–257.

King K.L., Jewell C.M., Bortner C.D., and Cidlowski J.A. 2000. 28S ribosome degradation in lymphoid cell apoptosis: Evidence for caspase and Bcl-2-dependent and -independent pathways. *Cell Death Differ.* **7:** 994–1001.

Kolupaeva V.G., Unbehaun A., Lomakin I.B., Hellen C.U., and Pestova T.V. 2005. Binding of eukaryotic initiation factor 3 to ribosomal 40S subunits and its role in ribosomal dissociation and anti-association. *RNA* **11:** 470–486.

Krueger A., Schmitz I., Baumann S., Krammer P.H., and Kirchhoff S. 2001. Cellular FLICE-inhibitory protein splice variants inhibit different steps of caspase-8 activation at the CD95 death-inducing signaling complex. *J. Biol. Chem.* **276:** 20633–20640.

Lee N., MacDonald H., Reinhard C., Halenbeck R., Roulston A., Shi T., and Williams L.T. 1997. Activation of hPAK65 by caspase cleavage induces some of the morphological and biochemical changes of apoptosis. *Proc. Natl. Acad. Sci.* **94:** 13642–13647.

Lee S.B. and Esteban M. 1994. The interferon-induced double-stranded RNA-activated protein kinase induces apoptosis. *Virology* **199:** 491–496.

Levy-Strumpf N., Deiss L.P., Berissi H., and Kimchi A. 1997. DAP-5, a novel homolog of eukaryotic translation initiation factor 4G isolated as a putative modulator of gamma interferon-induced programmed cell death. *Mol. Cell. Biol.* **17:** 1615–1625.

Lewis S.M. and Holcik M. 2005. IRES in distress: Translational regulation of the inhibitor of apoptosis proteins XIAP and HIAP2 during cell stress. *Cell Death Differ.* **12:** 547–553.

Ling J., Morley S.J., and Traugh J.A. 2005. Inhibition of cap-dependent translation via phosphorylation of eIF4G by protein kinase Pak2. *EMBO J.* **24:** 4094–4105.

Linseman D.A., Phelps R.A., Bouchard R.J., Le S.S., Laessig T.A., McClure M.L., and Heidenreich K.A. 2002. Insulin-like growth factor-I blocks Bcl-2 interacting mediator of cell death (Bim) induction and intrinsic death signaling in cerebellar granule neurons. *J. Neurosci.* **22:** 9287–9297.

Marash L. and Kimchi A. 2005. DAP5 and IRES-mediated translation during programmed cell death. *Cell Death Differ.* **12:** 554–562.

Marissen W.E. and Lloyd R.E. 1998. Eukaryotic translation initiation factor 4G is targeted for proteolytic cleavage by caspase 3 during inhibition of translation in apoptotic cells. *Mol. Cell. Biol.* **18:** 7565–7574.

Marissen W.E., Gradi A., Sonenberg N., and Lloyd R.E. 2000a. Cleavage of eukaryotic translation initiation factor 4GII correlates with translation inhibition during apoptosis. *Cell Death Differ.* **7:** 1234–1243.

Marissen W.E., Triyoso D., Younan P., and Lloyd R.E. 2004. Degradation of poly(A)-binding protein in apoptotic cells and linkage to translation regulation. *Apoptosis* **9:** 67–75.

Marissen W.E., Guo Y., Thomas A.A., Matts R.L., and Lloyd R.E. 2000b. Identification of caspase 3-mediated cleavage and functional alteration of eukaryotic initiation factor 2α in apoptosis. *J. Biol. Chem.* **275:** 9314–9323.

Meerovitch K., Svitkin Y.V., Lee H.S., Lejbkowicz F., Kenan D.J., Chan E.K., Agol V.I.,

Keene J.D., and Sonenberg N. 1993. La autoantigen enhances and corrects aberrant translation of poliovirus RNA in reticulocyte lysate. *J. Virol.* **67:** 3798–3807.

Methot N., Song M.S., and Sonenberg N. 1996. A region rich in aspartic acid, arginine, tyrosine, and glycine (DRYG) mediates eukaryotic initiation factor 4B (eIF4B) self-association and interaction with eIF3. *Mol. Cell. Biol.* **16:** 5328–5334.

Mitchell S.A., Spriggs K.A., Bushell M., Evans J.R., Stoneley M., Le Quesne J.P.C., Spriggs R.V., and Willis A.E. 2005. Identification of a motif that mediates polypyrimidine tract-binding protein-dependent internal ribosome entry. *Genes Dev.* **19:** 1556–1571.

Morino S., Imataka H., Svitkin Y.V., Pestova T.V., and Sonenberg N. 2000. Eukaryotic translation initiation factor 4E (eIF4E) binding site and the middle one-third of eIF4GI constitute the core domain for cap-dependent translation, and the C-terminal one-third functions as a modulatory region. *Mol. Cell. Biol.* **20:** 468–477.

Morley S.J. 2001. The regulation of eIF4F during cell growth and cell death. *Prog. Mol. Subcell. Biol.* **27:** 1–37.

Morley S.J. and Pain V.M. 2001. Proteasome inhibitors and immunosuppressive drugs promote the cleavage of eIF4GI and eIF4GII by caspase-8-independent mechanisms in Jurkat T cell lines. *FEBS Lett.* **503:** 206–212.

Morley S.J., Coldwell M.J., and Clemens M.J. 2005. Initiation factor modifications in the preapoptotic phase. *Cell Death Differ.* **12:** 571–584.

Morley S.J., McKendrick L., and Bushell M. 1998. Cleavage of translation initiation factor 4G (eIF4G) during anti-Fas IgM-induced apoptosis does not require signalling through p38 mitogen-activated protein (MAP) kinase. *FEBS Lett.* **438:** 41–48.

Morley S.J., Jeffrey I., Bushell M., Pain V.M., and Clemens M.J. 2000. Differential requirements for caspase-8 activity in the mechanism of phosphorylation of eIF2alpha, cleavage of eIF4GI and signaling events associated with the inhibition of protein synthesis in apoptotic Jurkat T cells. *FEBS Lett.* **477:** 229–236.

Nevins T.A., Harder Z.M., Korneluk R.G., and Holcik M. 2003. Distinct regulation of internal ribosome entry site-mediated translation following cellular stress is mediated by apoptotic fragments of eIF4G translation initiation factor family members eIF4GI and p97/DAP5/NAT1. *J. Biol. Chem.* **278:** 3572–3579.

Orton K.C., Ling J., Waskiewicz A.J., Cooper J.A., Merrick W.C., Korneeva N.L., Rhoads R.E., Sonenberg N., and Traugh J.A. 2004. Phosphorylation of Mnk1 by caspase-activated Pak2/gamma-PAK inhibits phosphorylation and interaction of eIF4G with Mnk. *J. Biol. Chem.* **279:** 38649–38657.

Panner A., James C.D., Berger M.S., and Pieper R.O. 2005. mTOR controls FLIP$_S$ translation and TRAIL sensitivity in glioblastoma multiforme cells. *Mol. Cell. Biol.* **25:** 8809–8823.

Park S.M., Yoon J.B., and Lee T.H. 2004. Receptor interacting protein is ubiquitinated by cellular inhibitor of apoptosis proteins (c-IAP1 and c-IAP2) in vitro. *FEBS Lett.* **566:** 151–156.

Proud C.G. 2005. eIF2 and the control of cell physiology. *Semin. Cell Dev. Biol.* **16:** 3–12.

Reed J.C. 1998. Bcl-2 family proteins. *Oncogene* **17:** 3225–3236.

Riedl S.J. and Shi Y. 2004. Molecular mechanisms of caspase regulation during apoptosis. *Nat. Rev. Mol. Cell Biol.* **5:** 897–907.

Rutjes S.A., Utz P.J., van der Heijden A., Broekhuis C., van Venrooij W.J., and Pruijn G.J. 1999. The La (SS-B) autoantigen, a key protein in RNA biogenesis, is dephosphorylated and cleaved early during apoptosis. *Cell Death Differ.* **6:** 976–986.

Ryabova L.A., Pooggin M.M., and Hohn T. 2002. Viral strategies of translation initiation: Ribosomal shunt and reinitiation. *Prog. Nucleic Acid Res. Mol. Biol.* **72:** 1–39.

Saelens X., Kalai M., and Vandenabeele P. 2001. Translation inhibition in apoptosis: Caspase-dependent PKR activation and eIF2-alpha phosphorylation. *J. Biol. Chem.* **276:** 41620–41628.

Satoh S., Hijikata M., Handa H., and Shimotohno K. 1999. Caspase-mediated cleavage of eukaryotic translation initiation factor subunit 2alpha. *Biochem. J.* **342:** 65–70.

Schmitz I., Weyd H., Krueger A., Baumann S., Fas S.C., Krammer P.H., and Kirchhoff S. 2004. Resistance of short term activated T cells to CD95-mediated apoptosis correlates with de novo protein synthesis of c-FLIPshort. *J. Immunol.* **172:** 2194–2200.

Schweichel J.U. and Merker H.J. 1973. The morphology of various types of cell death in prenatal tissues. *Teratology* **7:** 253–266.

Scott C.E. and Adebodun F. 1999. 13C-NMR investigation of protein synthesis during apoptosis in human leukemic cell lines. *J. Cell. Physiol.* **181:** 147–152.

Sherrill K.W., Byrd M.P., Van Eden M.E., and Lloyd R.E. 2004. BCL-2 translation is mediated via internal ribosome entry during cell stress. *J. Biol. Chem.* **279:** 29066–29074.

Spriggs K.A., Bushell M., Mitchell S.A., and Willis A.E. 2005. Internal ribosome entry segment-mediated translation during apoptosis: The role of IRES-trans-acting factors. *Cell Death Differ.* **12:** 585–591.

Stoneley M., Paulin F.E.M., Le Quesne J.P.C., Chappell S.A., and Willis A.E. 1998. C-Myc 5′ untranslated region contains an internal ribosome entry segment. *Oncogene* **16:** 423–428.

Stoneley M., Chappell S.A., Jopling C.L., Dickens M., MacFarlane M., and Willis A.E. 2000. c-Myc protein synthesis is initiated from the internal ribosome entry segment during apoptosis. *Mol. Cell. Biol.* **20:** 1162–1169.

Tan S.L. and Katze M.G. 1999. The emerging role of the interferon-induced PKR protein kinase as an apoptotic effector: A new face of death? *J. Interferon Cytokine Res.* **19:** 543–554.

Tee A.R. and Proud C.G. 2000. DNA-damaging agents cause inactivation of translational regulators linked to mTOR signalling. *Oncogene* **19:** 3021–3031.

———. 2001. Staurosporine inhibits phosphorylation of translational regulators linked to mTOR. *Cell Death Differ.* **8:** 841–849.

———. 2002. Caspase cleavage of initiation factor 4E-binding protein 1 yields a dominant inhibitor of cap-dependent translation and reveals a novel regulatory motif. *Mol. Cell. Biol.* **22:** 1674–1683.

Van Eden M.E., Byrd M.P., Sherrill K.W., and Lloyd R.E. 2004. Translation of cellular inhibitor of apoptosis protein 1 (c-IAP1) mRNA is IRES mediated and regulated during cell stress. *RNA* **10:** 469–481.

Vazquez-Pianzola P., Urlaub H., and Rivera-Pomar R. 2005. Proteomic analysis of reaper 5′ untranslated region-interacting factors isolated by tobramycin affinity-selection reveals a role for La antigen in reaper mRNA translation. *Proteomics* **5:** 1645–1655.

Wajant H., Pfizenmaier K., and Scheurich P. 2003. Tumor necrosis factor signaling. *Cell Death Differ.* **10:** 45–65.

Wang X. 2001. The expanding role of mitochondria in apoptosis. *Genes Dev.* **15:** 2922–2933.

Warnakulasuriyarachchi D., Cerquozzi S., Cheung H.H., and Holcik M. 2004. Translational induction of the inhibitor of apoptosis protein HIAP2 during endoplasmic reticulum stress attenuates cell death and is mediated via an inducible internal ribosome entry site element. *J. Biol. Chem.* **279:** 17148–17157.

West M.J., Stoneley M., and Willis A.E. 1998. Translational induction of the c-myc oncogene via activation of the FRAP/TOR signalling pathway. *Oncogene* **17:** 769–780.

Widmann C., Gerwins P., Johnson N.L., Jarpe M.B., and Johnson G.L. 1998. MEK kinase 1, a substrate for DEVD-directed caspases, is involved in genotoxin-induced apoptosis. *Mol. Cell. Biol.* **18:** 2416–2429.

Wyllie A.H. 1993. Apoptosis (the 1992 Frank Rose Memorial Lecture). *Br. J. Cancer.* **67:** 205–208.

Xiao Q., Sharp T.V., Jeffrey I.W., James M.C., Pruijn G.J., van Venrooij W.J., and Clemens M.J. 1994. The La antigen inhibits the activation of the interferon-inducible protein kinase PKR by sequestering and unwinding double-stranded RNA. *Nucleic Acids Res.* **22:** 2512–2518.

Yamanaka S., Poksay K.S., Arnold K.S., and Innerarity T.L. 1997. A novel translational repressor mRNA is edited extensively in livers containing tumours caused by the transgene expression of the apoB mRNA-editing enzyme. *Genes Dev.* **11:**321–333.

Yang H., Masters S.C., Wang H., and Fu H. 2001. The proapoptotic protein Bad binds the amphipathic groove of 14-3-3zeta. *Biochim. Biophys. Acta* **1547:** 313–319.

You M., Ku P.T., Hrdlickova R., and Bose H.R., Jr. 1997. ch-IAP1, a member of the inhibitor-of-apoptosis protein family, is a mediator of the antiapoptotic activity of the v-Rel oncoprotein. *Mol. Cell. Biol.* **17:** 7328–7341.

Yousefi S., Conus S., and Simon H.U. 2003. Cross-talk between death and survival pathways. *Cell Death Differ.* **10:** 861–863.

Zhao S., Konopleva M., Cabreira-Hansen M., Xie Z., Hu W., Milella M., Estrov Z., Mills G.B., and Andreeff M. 2004. Inhibition of phosphatidylinositol 3-kinase dephosphorylates BAD and promotes apoptosis in myeloid leukemias. *Leukemia* **18:** 267–275.

Zhou B.B., Li H.L., Yuan J.Y., and Kirschner M.W. 1998. Caspase-dependent activation of cyclin-dependent kinases during Fas-induced apoptosis in Jurkat cell lines. *Proc. Natl. Acad. Sci.* **95:** 6785–6790.

Zhou C., Arslan F., Wee S., Krishnan S., Ivanov A.R., Oliva A., Leatherwood J., and Wolf D.A. 2005. PCI proteins eIF3e and eIF3m define distinct translation initiation factor 3 complexes. *BMC Biol.* **3:** 14–30.

Zou H., Henzel W.J., Liu X., Lutschg A., and Wang X. 1997. Apaf-1, a human protein homologous to *C. elegans* CED-4, participates in cytochrome c-dependent activation of caspase-3. *Cell.* **90:** 405–413.

17

Translational Control in Metabolic Diseases: The Role of mTOR Signaling in Obesity and Diabetes

Sara C. Kozma, Sung Hee Um, and George Thomas
Genome Research Institute
University of Cincinnati
Cincinnati, Ohio 45237

FOR SOME TIME IT HAS BEEN RECOGNIZED not only that protein synthesis is regulated by growth factors and hormonal signaling (Shi et al. 2003), but that the translation machinery is also specifically affected by nutrient levels (Clemens et al. 1980; Pain et al. 1980). These stimuli modulate both the global synthesis of proteins and the selective translation of specific mRNAs. Thus, given the impact of nutrient supply and endocrine signaling on protein synthesis, it is logical to presume that pathological conditions affecting nutrient homeostasis would result in major deregulation of protein synthesis. Currently, the most prevalent homeostatic disorder is the metabolic syndrome, defined as a cluster of pathologies that always includes obesity, plus at least two of the following factors: raised serum triglyceride levels, reduced high-density-lipoprotein cholesterol levels, raised blood pressure, and raised fasting plasma glucose.

The recent dramatic increase in the incidence of obesity has strongly contributed to an escalation of the metabolic syndrome manifestations in Western societies. It is believed that the increase in obesity derives from the fact that during evolution, food scarcity led to the development of dominant genetic traits to secure and manage caloric intake (Neel 1999). In Western societies, food availability, which increased dramatically in the 1950s, began to reveal these calorie-securing traits, and obesity emerged as a prevalent disorder that has since reached epidemic proportions. The nutrient overload resulting from increased food intake is being further accentuated by a decrease in physical activity and a demographic shift to an aging population (Pi-Sunyer 2002). As a result, more than 60% of adults in the United States are diagnosed as either clinically overweight or obese (Must et al. 1999; Marx 2003).

Many deleterious health effects are associated with obesity. For example, at the present rate, obesity will supersede smoking as the main cause of cardiovascular disease in the very near future (Morgan et al. 2004; Reynolds and He 2005). In addition, recent epidemiological studies show that the effects of metabolic syndrome also include a high risk for development of cancer (Calle et al. 2003; El-Serag 2004), with 15–20% of cancer deaths attributed each year to obesity (Calle and Kaaks 2004). However, the major outcome of the metabolic syndrome progression is insulin resistance in peripheral tissues, including skeletal muscle, liver, and adipose tissue, followed by development of type-2 diabetes (Kahn et al. 2005).

Diabetes mellitus is a group of metabolic diseases characterized by high blood glucose levels, which result from defects in insulin secretion, insulin action, or both. Type-2 diabetes is defined as a deficient response to insulin, which translates into impaired use of carbohydrate and resulting hyperglycemia. In the early stages of the disease, the pancreatic insulin-secreting β cells are challenged to increase insulin secretion levels. Type-2 diabetes comes into play when insulin secretion fails to compensate for insulin resistance, and almost always involves the failure of β cells to secrete insulin. A rarer form of diabetes, which shares hyperglycemia as a common feature with type-2 diabetes, is type-1 diabetes which arises from diseases that cause extensive destruction of pancreatic islets, including pancreatitis, cancer, or surgical excision. Whereas these two types of diabetes represent distinct pathogenic processes, they both are chronic disorders that affect carbohydrate, fat, and protein metabolism. In addition, the long-term complications of type-1 and type-2 diabetes are similar and consist of morphologic changes in arteries (atherosclerosis), the basement membrane of small vessels (microangiopathy), kidneys (diabetic nephropathy), retina (retinopathy), and nerves (neuropathy). Alterations in translational control are entrenched in the development of these pathological states. Indeed, increased levels of food intake leading to nutrient overload place a higher demand on the translational machinery of the cells of many organs. As mentioned above, insulin controls glucose homeostasis by regulating glucose utilization in peripheral tissues, but it also modulates its own production and secretion in pancreatic β cells (Kahn 1998; Leibiger et al. 1998; Kulkarni et al. 1999). In the peripheral tissues, insulin induces positive anabolic responses, most notably ribosome biogenesis and protein synthesis, that are dependent on nutritional state (Hay and Sonenberg 2004). An emerging concept is that the signaling components that control the translational apparatus are also intimately linked to those that control energy balance. This idea is based on the large amounts of energy consumed through the biogenesis and output of the protein synthetic machinery. Here, we have concentrated on the role of the mammalian target of rapamycin (mTOR) signaling pathway in this response.

mTOR: AT THE CROSSROADS OF INSULIN AND NUTRIENT SIGNALING

With the advent of metazoans and the emergence of humoral systems, signaling networks consisting of sensors, signal transducers, and effector molecules have developed to maintain cellular homeostasis. Recent studies on insulin signaling and cellular energy homeostasis have revealed that the mTOR signal transduction pathway integrates with the humorally controlled class-1 phosphatidylinositide-3OH kinase (PI3K) signaling pathway to maintain normal growth during development and organ homeostasis in the adult (Manning and Cantley 2003; Jaeschke et al. 2004; Um et al. 2006; Chapter 14). mTOR is a large 300,000-molecular-weight protein kinase that belongs to the PI3 kinase-related protein kinase family (PIKK) (Keith and Schreiber 1995), which also includes ataxia telangiectasia-mutated (ATM), ataxia telangiectasia and Rad3-related (ATR), and DNA-dependent protein kinase (DNA-PK) (Dennis et al. 1999; Abraham 2002). mTOR resides on an evolutionarily conserved nutrient effector pathway, which unicellular eukaryotes utilize in a cell-autonomous manner (Matsuo et al. 2003). As with β-cell function (Swenne 1992; Hugl et al. 1998), mTOR activity is extremely sensitive to nutrients, including glucose and amino acids, as well as mitogens, such as insulin (Hara et al. 1998; Iiboshi et al. 1999; Kim et al. 2002; McDaniel et al. 2002). Genetic studies in *Saccharomyces cerevisiae, Caenorhabditis elegans* (Long et al. 2002), *Drosophila melanogaster* (Oldham et al. 2000; Zhang et al. 2000), and *Mus musculus* (Gangloff et al. 2004; Murakami et al. 2004) showed that mTOR and its orthologs are highly conserved genes that have an essential role in cell growth and development. Whereas the other PIKK family members function as checkpoints in the control of DNA damage repair (Khanna and Jackson 2001), mTOR acts as a positive effector in nutrient/energy signaling. Genetic studies in *C. elegans* (Long et al. 2002) and *D. melanogaster* (Oldham et al. 2000; Zhang et al. 2000) have confirmed that the mTOR orthologs, *cTOR* and *dTOR*, respectively, also have essential roles in cell growth and development, which are, likewise, tightly linked to nutritional status.

The yeast ortholog of mTOR, TOR, was initially identified in *S. cerevisiae* in a genetic screen for mutants resistant to the cytostatic effects of rapamycin (Heitman et al. 1991), an antibiotic produced by the soil bacterium *Streptomyces hygroscopicus* (Dennis and Thomas 2002; see Chapter 14). In mammals, rapamycin blocks mTOR function by forming an inhibitory complex with the immunophilin FKBP12, which binds to and attenuates mTOR signaling (Abraham 2002; Dennis and Thomas 2002). mTOR mediates its downstream anabolic effects through direct phosphorylation of key substrates, including eukaryotic initiation factor 4E binding protein 1 (4E-BP1) (von Manteuffel et al. 1996, 1997; Brunn et al. 1997) and ribosomal protein S6 kinase 1 (S6K1) (Fig. 1) (Burnett et al. 1998; Isotani et al. 1999; Dennis et al. 2001; Saitoh et al. 2002). Interestingly, these proteins have mTOR recognition motifs that are quite distinct. In S6K1, a critical residue is T389, which resides in a hydrophobic motif, whereas in 4E-BP1, the critical sites are S65 and T70, which display serine or threonine followed by proline (S/TP) phosphorylation motifs. Such critical S/TP motifs have also been identified in S6K1, including S371 (Saitoh et al. 2002; Chapter 14).

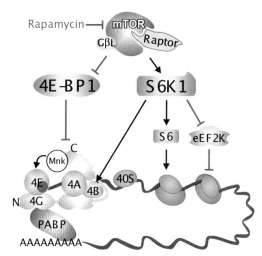

Figure 1. Downstream effectors of mTOR signaling: Nutrients and insulin induce mTOR phosphorylation of 4E-BP1 and S6K1 in a rapamycin-sensitive manner. Phosphorylation of 4E-BP1 decreases its affinity for eIF4E (4E). eIF4E can then interact with eIF4G (4G) to form the eIF4F complex with eIF4A (4A), the poly(A)-binding protein (PABP), and MAPK-activated protein kinase 1 (Mnk). In this complex, eIF4E is phosphorylated by Mnk and interacts directly with the m^{7}G cap of mRNA to facilitate cap-dependent translation. mTOR phosphorylation and activation of S6K1 lead to the subsequent phosphorylation of eIF4B, S6, and eEF2K.

Targets of mTOR: 4E-BPs

A key event in the initiation of translation is the recruitment of mRNA to the preinitiation 40S ribosomal complex and the scanning of the 5'UTR (untranslated region) until the first AUG initiation codon is encountered (Chapter 4). This event is mediated by the multiprotein eukaryotic initiation factor 4F (eIF4F) complex and regulated in part by phosphorylation of individual components of the complex. The binding of the m^7G cap of mRNA to the eIF4F complex is mediated by the cap-binding protein, eIF4E (Fig. 1). eIF4E interacts with the eIF4F complex through eIF4G. The latter acts as a protein scaffold for the binding of the other proteins to the complex, including the mRNA helicase, eIF4A; the poly(A)-binding protein (PABP); and the mitogen-activated protein kinase-interacting kinase 1 (Mnk1), which mediates eIF4E phosphorylation (Fig. 1) (Pyronnet et al. 1999; Chapter 14). The motif through which eIF4G interacts with eIF4E is shared with 4E-BP1-3 (4E-BPs), such that when the 4E-BPs are bound to eIF4E, they prevent its interaction with eIF4G and the formation of a competent preinitiation complex. To generate a competent signaling complex, the 4E-BPs must be released from eIF4E, a step mediated by multiple phosphorylations of the 4E-BPs (Gingras et al. 1999b). Moreover, only when bound within the eIF4F complex can eIF4E become phosphorylated by Mnk1 (Pyronnet et al. 1999), and this event is indicative of eIF4F complex formation (Gingras et al. 1999a).

Targets of mTOR: S6Ks

S6K1 is a member of the A, G, and C family of serine/threonine protein kinases (Hanks and Hunter 1995; G. Manning et al. 2002). Two isoforms of S6K1 have been identified (Banerjee et al. 1990; Kozma et al. 1990; Grove et al. 1991; Reinhard et al. 1992); the larger, termed S6K1L (Um et al. 2006) has an amino-terminal 23-amino acid extension (Grove et al. 1991; Reinhard et al. 1992), generated from an alternative weak translational initiation start site (Kozak 1991). The 23-amino acid extension contains a nuclear localization signal, which targets S6K1L to the nucleus (Reinhard et al. 1994). In contrast, the shorter S6K1S appears to be localized to the cytoplasm (Reinhard et al. 1994). The finding that S6 phosphorylation was largely intact and still sensitive to rapamycin in $S6K1^{-/-}$ mice (Shima et al. 1998) led to the discovery of S6K2 (Gout et al. 1998; Saitoh et al. 1998; Shima et al. 1998). There are also two S6K2 isoforms: S6K2L, the larger, and S6K2S, the smaller, apparently derived by alternate splicing (Gout et al. 1998; Lee-Fruman et al. 1999; Um et al.

2006). S6K1 and S6K2 show high identity in sequence and structure, which extends to invertebrates where single-copy orthologs for S6K1 and S6K2 have been found in a number of species (Montagne and Thomas 2004). The current model of S6K1 activation, which may also apply to S6K2, proceeds with the phosphorylation of the S/TP sites in the autoinhibitory domain, which disrupts an interaction between the amino-terminal TOR signaling (TOS) motif (see below, mTOR Signaling Complexes) and the carboxyl autoinhibitory domain to allow phosphorylation of T389 (Schalm et al. 2005). These two events then allow phosphatidylinositol-dependent kinase (PDK1) to dock on phosphorylated T389 and phosphorylate T229 in the activation loop, resulting in S6K1 activation (Biondi et al. 2001; Frodin et al. 2002). Consistent with this model, S6K1 T229 and T389 phosphorylation are severely reduced in embryonic stem (ES) cells from mice deficient for PDK1 (Williams et al. 2000) or mTOR (Gangloff et al. 2004), respectively. That S6K1 and S6K2 are kinases allows mTOR to further amplify signaling downstream to additional substrates (see below).

mTOR Signaling Complexes

The signaling of mTOR to substrates such as S6K1 and 4E-BP1 requires two associated proteins: regulatory associated protein of mTOR (raptor) (Hara et al. 2002; Kim et al. 2002; Loewith et al. 2002), and G protein β-subunit-like protein (GβL) (Kim et al. 2003). This complex of three subunits is referred to as Complex 1 (Fig. 1) (Chapter 14). Complex 1 interacts with downstream substrates through raptor, which recognizes mTOR substrates through their TOS motifs (Schalm and Blenis 2002), whereas GβL is required to make a competent signaling complex that can respond to nutrients such as the branched-chain amino acid leucine and glucose (Kim et al. 2003). Raptor interacts with both the amino- and carboxy-terminal domains of mTOR, although the binding is strongest to the amino terminus (Kim et al. 2002). In contrast, GβL binds specifically to the catalytic domain at the carboxyl end of mTOR. Moreover, whereas the binding between mTOR and raptor is either weakened or disrupted by nutrient addition or rapamycin, respectively, such treatment has no effect on GβL's ability to bind to mTOR (Kim et al. 2003). The ability of Complex 1 to bind to substrates such as S6K1 was first suggested by the finding that overexpression of either a wild-type, kinase-dead, or activated S6K1 inhibited mTOR signaling toward 4E-BP1 (von Manteuffel et al. 1997). This effect was traced to the conserved TOS motif (Schalm and Blenis 2002; Schalm et al. 2003) in S6K1, S6K2, and the 4E-BPs (Fig. 1).

Consistent with the role of TOR in regulating nutrient-mediated cell growth in yeast, S6K1 and 4E-BP1 phosphorylation is modulated by amino acid availability through Complex 1, as scored by increased phosphorylation at specific residues (Kim et al. 2002). In addition, it recently has been shown that intracellular ATP levels (Dennis et al. 2001) and AMP-dependent protein kinase (AMPK) (Inoki et al. 2003; Shaw et al. 2004), as well as phosphatidic acid (Fang et al. 2001), regulate mTOR activity in a manner similar to that of amino acids.

A second mTOR complex, Complex 2, has been identified, which also contains GβL, but where raptor is replaced by a large adaptor protein, termed rapamycin-independent companion of mTOR (rictor) or AVO3 (Loewith et al. 2002; Sarbassov et al. 2004). This second complex does not signal to either S6K1 or 4E-BP1, is resistant to rapamycin, and was initially shown to control actin cytoskeleton dynamics (Jacinto et al. 2004; Sarbassov et al. 2004). However, recent studies also show that Complex 2 phosphorylates and activates PKB (Sarbassov et al. 2005), in a manner analogous to phosphorylation and activation of S6K1 by Complex 1 (Fig. 2). Consistent with this finding, the activities of Complex 1 and Complex 2 are sensitive to wortmannin, an inhibitor of class 1 PI3K (see below). Presumably, rictor binds to mTOR at a location similar to that of raptor, as both compete for binding to mTOR (Sarbassov et al. 2004).

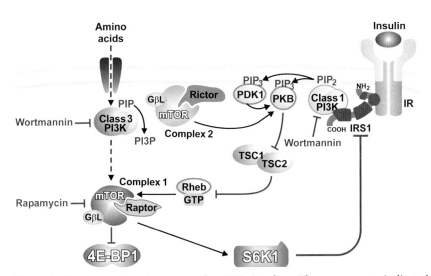

Figure 2. Insulin and nutrients control mTOR signaling: The components indicated in the insulin- and nutrient-mediated regulation of mTOR signaling are described in the text.

Insulin Induction of mTOR Signaling

Growth factors binding to their respective receptors induce intermolecular receptor autophosphorylation at specific tyrosine residues. These residues then act as docking sites for proteins containing either phosphotyrosine-binding (PTB) domains or phosphotyrosine-binding Src homology 2 (SH2) domains. One of these tyrosine phosphorylation-docking sites is known to recruit the p85 adaptor molecule that together with the p110 catalytic subunit forms PI3K (Fruman et al. 1998). The recruitment of PI3K to the receptor stimulates the production of the lipid second messengers, phosphatidylinositide-3, 4, 5-P_3 (PIP$_3$), and phosphatidylinositide-3, 4-P_2 (PIP$_2$) (Fig. 2). Increased PIP$_3$ binds to the amino-terminal plextrin homology (PH) domain of PKB, recruiting PKB to the membrane where it is phosphorylated by PDK1 at T308 in the PKB activation loop site. This step is argued to be preceded by the Complex-2-mediated phosphorylation of PKB S473, which resides in a conserved hydrophobic motif found in all members of the AGC family of serine/threonine protein kinases (Pearson et al. 1995; Sarbassov et al. 2005). Given that PKB S473 phosphorylation is blocked by wortmannin, a PI3K inhibitor, which also blocks growth-factor-induced activation of Complex 2 (Sarbassov et al. 2005), the assumption is that activation of Complex 2 is also mediated by class 1 PI3K. In contrast, activation of PKB by PIP$_3$ and PIP$_2$ is counteracted by the lipid phosphatase and tensin homolog deleted from chromosome 10 (PTEN), a tumor suppressor gene that is either absent or mutated in a large number of cancers (Maehama and Dixon 1999).

Activated PKB is then predicted to phosphorylate tuberous sclerosis complex (TSC) protein 2 (Fig. 2), which, together with TSC1, makes up a tumor suppressor complex associated with the autosomal-dominant genetic disorder, TSC (Montagne et al. 2001). Phosphorylation of TSC2 at a set of specific residues either targets TSC2 for destruction (Nellist et al. 2002), for being sequestered by 14-3-3 (Li et al. 2002), or disrupts its interaction with TSC1 (Fig. 2) (B.D. Manning et al. 2002). The sequence of TSC1 contains a predicted transmembrane domain and TSC2 comprises a small-GTPase-activating protein (GAP) domain (Hay and Sonenberg 2004). In *Drosophila*, it was shown that *dS6K* is epistatic to *dTsc1* and *dTsc2* (Gao and Pan 2001; Potter et al. 2001; Tapon et al. 2001), consistent with overexpression of *dTsc1/2* in larvae repressing dS6K activity, and with dS6K activity being highly elevated in *dTsc1* or *dTsc2* mutant larvae (Gao et al. 2002; Inoki et al. 2002; Radimerski et al. 2002a). Similarly, S6K1 and 4E-BP1 phosphorylation are elevated in mouse embryonic fibroblasts (MEFs) derived from TSC2-deficient mice, or in cell lines derived from humans having point mutations or lesions in the sequence of either TSC1 or TSC2 (Gao et al. 2002; Goncharova et al. 2002; Inoki et al. 2002; Jaeschke et al. 2002; B.D. Manning et al. 2002; Tee et al. 2002). Conversely, overexpression of TSC1 and TSC2 suppressed insulin-induced S6K1 T389 phosphorylation and S6K1 activation. These studies led our laboratory and others to the finding that the inhibitory effects of TSC1/TSC2 in mammals or *dTsc1/dTsc2* in *Drosophila* are elicited through the GTPase-stimulating activity of TSC2, which acts to drive a Ras family member, termed Ras homolog enriched in brain (Rheb) (Yamagata et al. 1994), into the inactive GDP state (Fig. 2) (Garami et al. 2003; Saucedo et al. 2003; Zhang et al. 2003). There are two genes for *Rheb*, *Rheb1* and *Rheb2*, but most studies to date concern *Rheb1*. Consistent with its role in insulin action, it was shown that insulin induces an increase in *Rheb1* GTP levels (Garami et al. 2003), and recent studies strongly support the idea that overexpressed Rheb directly interacts and stimulates the ability of Complex 1 to signal downstream to S6K1 and 4E-BP1 (Fig. 2) (Long et al. 2005a,b).

Nutrient Effector Pathway

Recent evidence suggests that nutrient input into the Complex 1 signaling pathway is not mediated by either TSC1/TSC2 or *Rheb* (Nobukuni et al. 2005). In the absence of TSC1/TSC2, S6K1 T389 phosphorylation is elevated and refractory to mitogen stimulation, such as insulin, but can still be regulated by amino acids (Nobukuni et al. 2005). However, this is not the case for *Rheb*, as short interfering RNA (siRNA) knockdown of Rheb protein levels blocks both the insulin and amino acid input to S6K1 T389 phosphorylation (Nobukuni et al. 2005). Nonetheless, withdrawal of amino acids, which immediately triggers S6K1 T389 dephosphorylation, has no effect on elevated levels of Rheb-GTP, leading to the hypothesis that Rheb-GTP is necessary but not sufficient to drive S6K1 activation in the absence of amino acids (Nobukuni et al. 2005). These findings suggested that the nutrient input to S6K1 resides on a pathway parallel to that of the TSC1/2-Rheb axis. Earlier studies demonstrated that wortmannin, a potent class 1 PI3K inhibitor, blocks amino acid-induced T389 phosphorylation (Hara et al. 1998; Iiboshi et al. 1999). Given that amino acids do not induce PKB activation (Nobukuni et al. 2005), these findings suggested that the wortmannin-sensitive target responsible for mediating the nutrient input to S6K1 was distinct from class 1 PI3K.

The observations above led to the identification of class 3 PI3K, hVps34, standing for human equivalent of vacuolar protein-sorting 34, in *S. cerevisiae* (Schu et al. 1993) as the target by which these responses were mediated (Fig. 2). This was demonstrated by studies showing that ectopic expression of hVps34 drives S6K1 T389 phosphorylation, but only in the presence of amino acids, and that this effect is blocked by specific siRNAs directed against hVps34 (Byfield et al. 2005; Nobukuni et al. 2005). Moreover, stimulation of cells with amino acids increases hVps34 activity as measured by the production of phosphatidylinositol 3-phosphate (PI3P), the product of hVps34 (Byfield et al. 2005; Nobukuni et al. 2005). Consistent with the finding that amino acids do not induce PKB activation, siRNA depletion of hVps34 has no effect on PKB activation (Byfield et al. 2005; Nobukuni et al. 2005). PI3P mediates the recruitment of proteins containing Fab1/TOTB/Vac1/EEA1 (FYVE) or PI3K-targeting phox homology (Px) domains (Lemmon 2003) to endosomal membranes, with PI3P-rich microdomains acting as signaling platforms (Burda et al. 2002; Gonzalez-Gaitan and Stenmark 2003; Miaczynska et al. 2004). Consistent with hVps34 mediating the amino acid input to S6K1, this response is attenuated by ectopic expression of a cDNA containing two FYVE domains, which bind to PI3P and prevent binding of proteins containing either FYVE or PX domains, and preventing S6K1 T389 phosphorylation (Nobukuni et al. 2005).

Integration of Insulin and Nutrient Signaling

The studies described above do not answer, as yet, the mechanism by which the insulin and nutrient inputs are integrated to coordinate mTOR signaling. Earlier studies have shown that nutrients such as amino acids and glucose can mediate an interaction between mTOR and raptor, which reflects their ability to mediate downstream signaling to S6K1 and 4E-BP1 (Kim et al. 2002). These authors found that raptor binds less tightly to mTOR in the presence of either amino acids or glucose, whereas the binding of GβL is unaffected (Kim et al. 2003). The mechanism responsible for altering the interaction between mTOR and raptor is unknown; however, this "active conformation" correlates with the inability of the complex to signal downstream to S6K1 (Kim et al. 2002). Moreover, the presence or absence of insulin apparently has no effect on the mTOR–raptor interaction (Kim and Sabatini 2004). In contrast, it has been shown that insulin drives Rheb into the GTP-bound state (Garami et al. 2003), whereas amino acids apparently have little effect on Rheb GTP levels (Nobukuni et al. 2005). However, overexpressed Rheb binds tightly to the mTOR catalytic domain in the presence, but not the absence, of amino acids, regardless of whether Rheb is GTP- or GDP-loaded (Long et al. 2005a,b). Thus, it may be that hVps34 mediates the ability of nutrients to induce mTOR Complex 1 into a conformation that permits

Rheb to bind to mTOR and induce downstream signaling. In summary, although this model is speculative, it allows a framework to understand how distinct inputs from nutrients and insulin may converge to regulate Complex 1 signaling (Fig. 2).

Role of 4E-BP1 in Adipose Tissue

Among the three known mammalian 4E-BPs, 4E-BP1 is the predominant isoform in adipose tissue; 4E-BP2 is more abundant in brain, whereas 4E-BP3 is present in colon, white adipose tissue, and liver (Tsukiyama-Kohara et al. 2001; Banko 2005). The most studied of the 4E-BPs to date is 4E-BP1, which undergoes multiple phosphorylation at a number of sites in a hierarchal manner (Gingras et al. 1999b). Rapamycin blocks the phosphorylation of these sites, indicating that mTOR activity is required to release 4E-BP1 from eIF4E (Chapter 14). Mitogen withdrawal (von Manteuffel et al. 1996) and amino acid deprivation (Hara et al. 1998; Wang et al. 1998) lead to a decrease in 4E-BP1 phosphorylation and consequent inhibition of cap-dependent translation. Because 4E-BP1 was initially identified in rat adipose tissue as a protein that became highly phosphorylated in response to insulin treatment (Belsham and Denton 1980; Belsham et al. 1982), it was reasoned that its deletion would have a pronounced effect on adipose development and, potentially, glucose homeostasis. The subsequent generation of mice deficient for 4E-BP1 led to the finding that they were hypoglycemic, although insulin levels appeared to be normal (Tsukiyama-Kohara et al. 2001). This raises the possibility that these mice are hypersensitive to insulin in their peripheral tissues (see below).

Given the positive role of eIF4E in translation, and the fact that the 4E-BP1 gene is a negative effector of translation (Gingras et al. 2001), it might be expected that deletion of 4E-BP1 would lead to an increase in the size of mice. Unexpectedly, male mice were moderately smaller in size, despite the fact that the absence of 4E-BP1 led to an increase in eIF4E phosphorylation, from 6% in wild-type mouse embryo fibroblasts (MEFs) to 45% in 4E-BP1-deficient MEFs, an increase that is negated by the reintroduction of 4E-BP1 (Tsukiyama-Kohara et al. 2001). This difference in weight of male mice could not be accounted for by hypophagia; instead, it was found that this difference was due to a 60% reduction in white adipose tissue (WAT) mass, which was consistent with a severe reduction in circulating leptin levels (Tsukiyama-Kohara et al. 2001). In analyzing inguinal and retroperitoneal WAT, these authors found a sharp increase in multilocular adipocytes, a hallmark of brown adipose tissue (BAT). This indicated an increase in energy expenditure, which was consistent with a 15% increase in metabolic rate. This led to the finding that uncoupling protein 1 (UCP1), which uncouples oxidative phosphorylation from ATP synthesis in the inner mitochondrial membrane of BAT, is dramatically increased in WAT. Overexpression of UCP1 in WAT by the tissue-specific aP2 promoter has been shown to prevent genetic obesity in mice (Kopecky et al. 1995). Because UCP1 transcription, but not transcription of either UCP2 or UCP3, is induced by the peroxisome proliferator-activated receptor-γ (PPAR-γ) coactivator 1 (PGC1), the expression of PGC1 was examined. Interestingly, PGC1 levels were increased threefold in WAT of 4E-BP1-depleted mice. The fact that this increase was only observed at the protein level suggested that its expression was mediated at the translational level in WAT. These results demonstrate a role for 4E-BP1 in adipose tissue development and energy homeostasis and suggest that small-molecule inhibitors that disrupt the 4E-BP1/eIF4E interaction would be effective in the treatment of obesity and insulin resistance.

S6K1 and Anabolic Processes

A role for S6K1 in the insulin-induced anabolic responses associated with cell growth was revealed originally through its first known substrate, ribosomal protein S6 (Volarevic et al. 2000), whose phosphorylation correlates with increased rates of protein synthesis (Fumagalli and Thomas 2000). Recently, four additional substrates have been identified that imply a role for S6K1 in the insulin-induced anabolic response, including eIF4B, which regulates the unwinding of the 5′ ends of mRNA (Raught et al. 2004); eukaryotic elongation factor-2 kinase (eEF2K), which regulates the phosphorylation of eukaryotic elongation factor 2 (eEF2), involved in controlling ribosomal transit rates (Wang et al. 2001; Chapter 21); the proapoptotic protein Bcl-2-associated death protein (BAD) (Harada et al. 2001); and most recently, S6K1 Aly/REF-like target (SKAR) a nuclear protein proposed to couple transcription with pre-mRNA splicing and mRNA export (Richardson et al. 2004). Consistent with a role in cell growth, S6K1-deficient mice are viable, although they are 15–20% smaller than wild-type mice at birth and display a lean phenotype (Shima et al. 1998). However, in all tissues examined in $S6K1^{-/-}$ mice, S6K2 is up-regulated, suggesting a compensatory response for the loss of S6K1 (Shima et al. 1998). Consistent with this, 30% perinatal lethality is observed in mice deficient for both S6K1 and S6K2 (Pende et al. 2004). Surprisingly, body size does not correlate with the extent of S6 phosphorylation, because in $S6K1^{-/-}$ mice, S6 phosphorylation is affected very little, whereas in $S6K2^{-/-}$ mice, which exhibit normal body size, S6 phosphorylation is significantly reduced (Pende et al. 2004). These studies led to the conclusion that the reduced body size of S6K1-deficient mice is due to phosphorylation of a distinct substrate (Pende et al. 2004). Such a role was recently proposed for SKAR, whose depletion has a pronounced effect on cell size (Richardson et al. 2004). However, it should be noted that in mice in which the wild-type S6 gene was replaced with a gene harboring alanines in the five phosphorylation sites of S6, cell size is affected in the pancreas (Ruvinsky et al. 2005). This appears to result from an increase in the rate of cell division, because the size of the mice, as well as total tissue mass, is not affected in the S6 mutant mice. Despite this observation, MEFs derived from these mice are small in size and exhibit accelerated cell cycle progression (Ruvinsky et al. 2005). Interestingly, this condition is not observed in MEFs from $S6K1^{-/-}:S6K2^{-/-}$ mice, although S6 phosphorylation is virtually abolished (Pende et al. 2004).

Earlier studies also implicated S6K1 in the translational up-regulation of a family of mRNAs, which contain a polypyrimidine tract at their 5′ transcriptional start site, or 5′ TOP (Jefferies et al. 1997; Kawasome et al. 1998; Schwab et al. 1999). The 5′ TOP mRNAs largely encode components of the translational apparatus, most notably ribosomal proteins (Meyuhas et al. 1996; Fumagalli and Thomas 2000), and thus have a critical role in cell growth (Thomas 2000; Rudra and Warner 2004). Previous studies showed that the translation of 5′ TOP mRNAs is under selective translational control (Jefferies et al. 1994a) and is inhibited by rapamycin (Jefferies et al. 1994b; Terada et al. 1994), with their selective translation requiring an intact 5′ TOP tract (Jefferies et al. 1997; Schwab et al. 1999). Moreover, a dominant interfering cDNA of S6K1 inhibited the translational up-regulation of 5′ TOP mRNAs to the same extent as rapamycin, whereas an S6K1 mutant, which is rapamycin resistant, largely protected 5′ TOP mRNA translation from the inhibitory effects of rapamycin (Jefferies et al. 1997; Schwab et al. 1999). However, rapamycin still abolished the translation of 5′ TOP mRNAs in MEFs and ES cells deficient for both S6K1 and S6K2, where S6 phosphorylation is largely ablated (Pende et al. 2004). It should be noted that Stolovich et al. (2002) reported that 5′ TOP mRNA translation was still regulated in MEFs deficient in S6K1, which they argued were also deficient for S6 phosphorylation. However, other investigators showed that S6 phosphorylation was intact in these same cells and mediated by S6K2 (Lee-Fruman et al. 1999), in agreement with the findings above (Pende et al. 2004). The reason for this discrepancy remains unclear. Also consistent with the results in MEFs deficient for S6K1 and S6K2, MEFs harboring the five-phosphorylation-site mutant of S6 are unaffected in their ability to up-regulate the translation of 5′ TOP mRNAs in response to serum (Ruvinsky et al. 2005). How then does one reconcile these results with the earlier findings that S6K1 regulates the translation of 5′ TOP mRNAs (Jefferies et al. 1997)? The effect of the dominant-negative S6K1 allele can be rationalized by its ability to titrate Complex 1, inhibiting its ability to phosphorylate a key downstream target of Complex 1 involved in regulating 5′ TOP translation (von Manteuffel et al. 1997). However, that the rapamycin-resistant cDNA of S6K1 protects the translation of these transcripts from the inhibitory effects of rapamycin (Jefferies et al. 1997; Schwab et al. 1999) is less easily explained. It is possible (1) that S6K1 is not involved in this response, but that when overexpressed, it induces the phosphorylation of a protein involved in 5′ TOP mRNA translation, whose phosphorylation is not mediated by endogenous S6K1, or (2) that in the absence of S6K1 and S6K2, a rapamycin-sensitive, compensatory pathway is activated. To resolve these issues will require a molecular understanding of the mechanisms that control 5′ TOP mRNA translation.

S6K1 in Hypoinsulinemia

The positive effect of S6K1 on anabolic processes that drive cell growth is consistent with diminished β-cell size in $S6K1^{-/-}$ mice, which leads to reduced pancreatic insulin content, an impairment in insulin secretion, and glucose intolerance (Pende et al. 2000). The observed phenotype closely parallels those of preclinical type-2 diabetes in which malnutrition-induced hypoinsulinemia predisposes individuals to glucose intolerance (Swenne et al. 1992; Phillips 1996; DeFronzo 1997). As stated above, this may be due to the fact that β-cell growth, like S6K1 activity, is extremely sensitive to mitogens, such as insulin and insulin-like growth factors, as well as to nutrients, including glucose and amino acids (Swenne 1992; Hugl et al. 1998). Given that mice harboring the five-phosphorylation-site mutant of S6 also exhibit reduced β-cell size are hypoinsulinemic, and display impaired glucose tolerance, it has been suggested that the S6K1-deficient phenotype is attributed to decreased S6 phosphorylation in β cells (Ruvinsky et al. 2005). However, in all tissues examined from $S6K1^{-/-}$ mice, including β cells, we have not noted a reduction in S6 phosphorylation (M. Pende et al., unpubl.), arguing that the effects observed in the S6 alanine mutant mice are distinct from those elicited by loss of S6K1. Given this difference, it will be of interest to determine whether the effects on β-cell size are more severe in $S6K1^{-/-}:S6K2^{-/-}$ mice than in $S6K1^{-/-}$ mice.

Several attractive S6K1 targets have arisen that could explain the effects of S6K1 deficiency on β-cell size. For example, recent studies have shown that eIF4B, which stimulates the helicase activity of the DEAD box protein eIF4A to unwind secondary structures in the 5′UTR of mRNAs, is a substrate of S6K1 (Raught et al. 2004). In vitro S6K1 phosphorylates eIF4B S422, which displays the R(R)RXXSX S6K1 substrate-recognition motif (Flotow and Thomas 1992) and is located in an RNA-binding region required for eIF4A helicase-promoting activity (Raught et al. 2004). These authors showed that this site is sensitive to rapamycin and that the rapamycin-resistant cDNA of S6K1 (Jefferies et al. 1997) confers rapamycin resistance upon eIF4B S422 phosphorylation (Raught et al. 2004). Critically, substitution of an alanine at S422 results in a loss of activity in an in vivo translation assay, indicating that phosphorylation of this site has an important role in eIF4B function. Moreover, recent studies have shown that S6K1 in the inactive state is associated with eukaryotic translation initiation factor 3 (eIF3) and, upon phosphorylation by Complex 1, is released from eIF3 such that it can phosphorylate eIF4B S422, leading to enhanced rates of translation (Holz et al. 2005). It also has been demonstrated that rapamycin inhibits protein synthesis transit rates (Redpath et al. 1996) through blocking the phosphorylation of eEF2K (Wang et al. 2001). The only identified substrate for eEF2K is eEF2, which mediates the translocation step of elongation, with phosphorylation of eEF2 T56 by eEF2K preventing eEF2 from binding to the ribosome (Proud 2004; Chapter 21). In response to insulin, S6K1 and S6K2 mediate the phosphorylation of eEF2K S366, inhibiting eEF2K activity and eEF2 T56 phosphorylation, which then act in concert to increase elongation rates of translation (Wang et al. 2001). It will be of interest to monitor the phosphorylation status of either eIF4B or eEF2K in S6K1-deficient mice to determine whether they contribute to the reduced β-cell size and hypoinsulinemia induced by loss of S6K1.

S6K1 Is a Negative Effector of Insulin Signaling

Despite the positive anabolic role of S6K1 in insulin signaling described above, fasting and feeding glucose homeostasis were normal in $S6K1^{-/-}$ mice, leading to the speculation that such mice may be hypersensitive to insulin signaling in their peripheral tissues, which could dampen hyperglycemia caused by decreased circulating insulin (Pende et al. 2000). This finding suggested that activation of S6K1 was also involved in a negative anabolic response that suppressed insulin signaling. The observation was consistent with parallel studies showing that an insulin-induced negative feedback loop, mediated by serine/threonine phosphorylation of insulin receptor substrate 1 (IRS1), was blocked by rapamycin (Zick 2004). A role for S6K1 in negatively regulating insulin signaling was first inferred from studies demonstrating that the activity of the PKB *Drosophila* ortholog, dPKB, is elevated in *Drosophila* larvae lacking dS6K or in *Drosophila Kc167* where dS6K protein levels were reduced with double-stranded interfering RNA (Radimerski et al. 2002a,b). Consistent with this finding, removal of either *dTsc1* or *dTsc2*, negative effectors of dTOR/dS6K signaling, leads to constitutive dS6K activation and the coordinate suppression of dPKB activity. This negative effect of removing either *dTsc1* or *dTsc2* on dPKB activity is rescued in a *dS6K*-deficient background (Radimerski et al. 2002b), arguing that the inhibition of dPKB activity in a *dTsc1* or *dTsc2* mutant is mediated by *dS6K*. Consistent with these findings in MEFs lacking TSC2 or in mammalian cells overexpressing Rheb, S6K1 activity is constitutively activated and PKB activity is suppressed (Jaeschke et al. 2002; Garami et al. 2003). Recent studies have shown that in the case of insulin signaling, the effect of S6K1 on PKB is mediated by a negative feedback loop from S6K1 to IRS1, which suppresses PI3K activation and, consequently, PKB activation (Harrington et al. 2004; Shah et al. 2004; Um et al. 2004). As described above, S6K1 activation is also mediated by nutrients, such as amino acids, independent of mitogens (Hara et al. 1998; Patti et al. 1998; Dennis et al. 2001; Kim et al. 2002). As with insulin signaling, this response also leads to suppression of PI3K-mediated PKB activation through IRS1 phosphorylation (Um et al. 2004). That these effects are mediated by Complex 1 signaling and S6K1 activation is supported by the observation that amino acids inhibit insulin-induced class 1 PI3K signaling, an inhibitory response that is reversed by the inhibition of Complex 1/S6K1 signaling by rapamycin (Tremblay and Marette 2001). Here, it is important to note that in insulin-resistant states of obesity, circulating concentrations of amino acids are elevated, particularly branched-chain amino acids (Felig et al. 1969, 1970, 1974), and that the infusion of amino acids in humans, similar to what is observed for lipids, induces insulin resistance in experimental settings (Krebs et al. 2002, 2003). The understanding that S6K1 is involved in this response comes from three recent observations. First, overexpression of a kinase-dead *S6K1*, but not wild-type *S6K1*, cDNA blocks Rheb-induced PKB activation, showing that these effects are mediated by S6K1 (Shah et al. 2004). Second, insulin resistance, induced in rats by chronic insulin treatment (hyperinsulinemic-euglycemic clamp), is reversed by administration of rapamycin, an effect that is paralleled by a loss of IRS1/IRS2 phosphorylation and S6K1 inactivation (Ueno et al. 2005). Finally, a novel S6K1 phosphorylation site in IRS1 was recently identified. The phosphorylation of this site is blocked by siRNAs directed against S6K1, and mutation of the site to alanine potentiates IRS1 tyrosine phosphorylation and PKB activation (A. Marette, pers. comm.). Thus, a signaling component intimately involved in controlling translation is also involved in regulating insulin signaling in response to nutrient levels, providing one example of how such metabolic processes are coupled.

SUMMARY

The initial discovery that increased S6 phosphorylation was associated with increased growth following hepatectomy (Gressner and Wool 1974) has led during the last 30 years to the elucidation of a complex signaling pathway that is intimately associated with cell metabolism and energy balance (Um et al. 2006). Consistent with this, these pathways are closely involved in regulating protein synthesis and ribosome biogenesis, two of the most energy-consuming anabolic processes in a growing cell (Rudra and Warner 2004). In the future, it will be critical to further explore the molecular mechanisms that link the translational machinery with the metabolic state and to potentially exploit this knowledge in the treatment of pathologies such as those associated with metabolic syndrome, in which these mechanisms have gone awry.

REFERENCES

Abraham R.T. 2002. Identification of TOR signaling complexes: More TORC for the cell growth engine. *Cell* **111**: 9–12.

Banerjee P., Ahamad M.F., Grove J.R., Kozlosky C., Price D.J., and Avruch J. 1990. Molecular structure of a major insulin/mitogen-activated 70kDa S6 protein kinase. *Proc. Natl. Acad. Sci.* **87**: 8550–8554.

Banko J.L., Poulin F., Hou L., DeMaria C.T., Sonenberg N., and Klann E. 2005. The translation repressor 4E-BP2 is critical for eIF4F complex formation, synaptic plasticity, and memory in the hippocampus. *J. Neurosci.* **25**: 9581–9590.

Belsham G.J. and Denton R.M. 1980. The effect of insulin and adrenaline on the phosphorylation of a 22 000-molecular weight protein within isolated fat cells; possible identification as the inhibitor-1 of the "general phosphatase" (proceedings). *Biochem. Soc. Trans.* **8**: 382–383.

Belsham G.J., Brownsey R.W., and Denton R.M. 1982. Reversibility of the insulin-stimulated phosphorylation of ATP citrate lyase and a cytoplasmic protein of subunit Mr 22000 in adipose tissue. *Biochem. J.* **204**: 345–352.

Biondi R.M., Kieloch A., Currie R.A., Deak M., and Alessi D.R. 2001. The PIF-binding pocket in PDK1 is essential for activation of S6K and SGK, but not PKB. *EMBO J.* **20**: 4380–4390.

Brunn G.J., Hudson C.C., Sekulic A., Williams J.M., Hosoi H., Houghton P.J., Lawrence J.C., Jr., and Abraham R.T. 1997. Phosphorylation of the translational repressor PHAS-I by the mammalian target of rapamycin. *Science* **277**: 99–101.

Burda P., Padilla S.M., Sarkar S., and Emr S.D. 2002. Retromer function in endosome-to-Golgi retrograde transport is regulated by the yeast Vps34 PtdIns 3-kinase. *J. Cell Sci.* **115**: 3889–3900.

Burnett P.E., Barrow R.K., Cohen N., Snyder S.H., and Sabatini D.M. 1998. RAFT1 phosphorylation of the translational regulators p70 S6 kinase and 4E-BP1. *Proc. Natl. Acad. Sci.* **95**: 1432–1437.

Byfield M.P., Murray J.T., and Backer J.M. 2005. hVps34 is a nutrient-regulated lipid kinase required for activation of p70 S6 kinase. *J. Biol. Chem.* **280**: 33076–33082.

Calle E.E. and Kaaks R. 2004. Overweight, obesity and cancer: Epidemiological evidence and proposed mechanisms. *Nat. Rev. Cancer* **4:** 579–591.

Calle E.E., Rodriguez C., Walker-Thurmond K., and Thun M.J. 2003. Overweight, obesity, and mortality from cancer in a prospectively studied cohort of U.S. adults. *N. Engl. J. Med.* **348:** 1625–1638.

Clemens M.J., Henshaw E.C., and Pain V.M. 1980. Regulation of polypeptide-chain initiation in Ehrlich ascites-tumour cells by amino acid starvation (proceedings). *Biochem. Soc. Trans.* **8:** 350–351.

DeFronzo R.A. 1997. Pathogenesis of type 2 diabetes: Metabolic and molecular implications for identifying diabetes genes. *Diabetes Rev.* **5:** 177–269.

Dennis P.B. and Thomas G. 2002. Quick guide: Target of rapamycin. *Curr. Biol.* **12:** R269.

Dennis P.B., Fumagalli S., and Thomas G. 1999. Target of rapamycin (TOR): Balancing the opposing forces of protein synthesis and degradation. *Curr. Opin. Genet. Dev.* **9:** 49–54.

Dennis P.B., Jaeschke A., Saitoh M., Fowler B., Kozma S.C., and Thomas G. 2001. Mammalian TOR: A homeostatic ATP sensor. *Science* **294:** 1102–1105.

El-Serag H.B. 2004. Hepatocellular carcinoma: Recent trends in the United States. *Gastroenterology* **127:** S27–S34.

Fang Y., Vilella-Bach M., Bachmann R., Flanigan A., and Chen J. 2001. Phosphatidic acid-mediated mitogenic activation of mTOR signaling. *Science* **294:** 1942–1945.

Felig P., Marliss E., and Cahill G.F., Jr. 1969. Plasma amino acid levels and insulin secretion in obesity. *N. Engl. J. Med.* **281:** 811–816.

———. 1970. Are plasma amino acid levels elevated in obesity? *N. Engl. J. Med.* **282:** 166.

Felig P., Wahren J., Hendler R., and Brundin T. 1974. Splanchnic glucose and amino acid metabolism in obesity. *J. Clin. Invest.* **53:** 582–590.

Flotow H. and Thomas G. 1992. Substrate recognition determinants of the mitogen-activated 70K S6 kinase from rat liver. *J. Biol. Chem.* **267:** 3074–3078.

Frodin M., Antal T.L., Dummler B.A., Jensen C.J., Deak M., Gammeltoft S., and Biondi R.M. 2002. A phosphoserine/threonine-binding pocket in AGC kinases and PDK1 mediates activation by hydrophobic motif phosphorylation. *EMBO J.* **21:** 5396–5407.

Fruman D.A., Meyers R.E., and Cantley L.C. 1998. Phosphoinositide kinases. *Annu. Rev. Biochem.* **67:** 481–507.

Fumagalli S. and Thomas G. 2000. S6 phosphorylation and signal transduction. In *Translational control of gene expression* (ed. N. Sonenberg et al.), pp. 695–717. Cold Spring Harbor Laboratory Press, Cold Spring Harbor, New York.

Gangloff Y.G., Mueller M., Dann S.G., Svoboda P., Sticker M., Spetz J.F., Um S.H., Brown E.J., Cereghini S., Thomas G., and Kozma S.C. 2004. Disruption of the mouse mTOR gene leads to early postimplantation lethality and prohibits embryonic stem cell development. *Mol. Cell. Biol.* **24:** 9508–9516.

Gao X. and Pan D. 2001. TSC1 and TSC2 tumor suppressors antagonize insulin signaling in cell growth. *Genes Dev.* **15:** 1383–1392.

Gao X., Zhang Y., Arrazola P., Hino O., Kobayashi T., Yeung R.S., Ru B., and Pan D. 2002. Tsc tumour suppressor proteins antagonize amino-acid-TOR signalling. *Nat. Cell Biol.* **4:** 699–704.

Garami A., Zwartkruis F.J., Nobukuni T., Joaquin M., Roccio M., Stocker H., Kozma S.C., Hafen E., Bos J.L., and Thomas G. 2003. Insulin activation of Rheb, a mediator of mTOR/S6K/4E-BP signaling, is inhibited by TSC1 and 2. *Mol. Cell* **11:** 1457–1466.

Gingras A.C., Raught B., and Sonenberg N. 1999a. eIF4 initiation factors: Effectors of mRNA recruitment to ribosomes and regulators of translation. *Annu. Rev. Biochem.* **68:** 913–963.

————. 2001. Regulation of translation initiation by FRAP/mTOR. *Genes Dev.* **15:** 807–826.

Gingras A.C., Gygi S.P., Raught B., Polakiewicz R.D., Abraham R.T., Hoekstra M.F., Aebersold R., and Sonenberg N. 1999b. Regulation of 4E-BP1 phosphorylation: A novel two-step mechanism. *Genes Dev.* **13:** 1422–1437.

Goncharova E.A., Goncharov D.A., Eszterhas A., Hunter D.S., Glassberg M.K., Yeung R.S., Walker C.L., Noonan D., Kwiatkowski D.J., Chou M.M., et al. 2002. Tuberin regulates p70 S6 kinase activation and ribosomal protein S6 phosphorylation. A role for the TSC2 tumor suppressor gene in pulmonary lymphangioleiomyomatosis (LAM). *J. Biol. Chem.* **277:** 30958–30967.

Gonzalez-Gaitan M. and Stenmark H. 2003. Endocytosis and signaling: A relationship under development. *Cell* **115:** 513–521.

Gout I., Minami T., Hara K., Tsujishita Y., Filonenko V., Waterfield M.D., and Yonezawa K. 1998. Molecular cloning and characterization of a novel p70 S6 kinase, p70 S6 kinase b containing a proline-rich region. *J. Biol. Chem.* **273:** 30061–30064.

Gressner A.M. and Wool I.G. 1974. The phosphorylation of liver ribosomal proteins *in vivo*. Evidence that only a single small subunit (S6) is phosphorylated. *J. Biol. Chem.* **249:** 6917–6925.

Grove J.R., Banerjee P., Balasubramanyam A., Coffer P.J., Price D.J., Avruch J., and Woodgett J.R. 1991. Cloning and expression of two human p70 S6 kinase polypeptides differing only at their amino termini. *Mol. Cell. Biol.* **11:** 5541–5550.

Hanks S.K. and Hunter T. 1995. The eukaryotic protein kinase superfamily: Kinase (catalytic) domain structure and classification. *FASEB J.* **9:** 576–596.

Hara K., Yonezawa K., Weng Q.P., Kozlowski M.T., Belham C., and Avruch J. 1998. Amino acid sufficiency and mTOR regulate p70 S6 kinase and eIF-4E BP1 through a common effector mechanism. *J. Biol. Chem.* **273:** 14484–14494.

Hara K., Maruki Y., Long X., Yoshino K., Oshiro N., Hidayat S., Tokunaga C., Avruch J., and Yonezawa K. 2002. Raptor, a binding partner of target of rapamycin (TOR), mediates TOR action. *Cell* **110:** 177–189.

Harada H., Andersen J.S., Mann M., Terada N., and Korsmeyer S.J. 2001. p70S6 kinase signals cell survival as well as growth, inactivating the pro-apoptotic molecule BAD. *Proc. Natl. Acad. Sci.* **98:** 9666–9670.

Harrington L.S., Findlay G.M., Gray A., Tolkacheva T., Wigfield S., Rebholz H., Barnett J., Leslie N.R., Cheng S., Shepherd P.R., et al. 2004. The TSC1-2 tumor suppressor controls insulin-PI3K signaling via regulation of IRS proteins. *J. Cell Biol.* **166:** 213–223.

Hay N. and Sonenberg N. 2004. Upstream and downstream of mTOR. *Genes Dev.* **18:** 1926–1945.

Heitman J., Movva N.R., and Hall M.N. 1991. Targets for cell cycle arrest by the immunosuppressant rapamycin in yeast. *Science* **253:** 905–909.

Holz M.K., Ballif B.A., Gygi S.P., and Blenis J. 2005. mTOR and S6K1 mediate assembly of the translation preinitiation complex through dynamic protein interchange and ordered phosphorylation events. *Cell* **123:** 569–580.

Hugl S.R., White M.F., and Rhodes C.J. 1998. Insulin-like growth factor I (IGF-I)-stimulated pancreatic beta-cell growth is glucose-dependent. Synergistic activation of insulin receptor substrate-mediated signal transduction pathways by glucose and IGF-I in INS-1 cells. *J. Biol. Chem.* **273:** 17771–17779.

Iiboshi Y., Papst P.J., Kawasome H., Hosoi H., Abraham R.T., Houghton P.J., and Terada N. 1999. Amino acid-dependent control of p70(s6k). Involvement of tRNA aminoacylation in the regulation. *J. Biol. Chem.* **274:** 1092–1099.

Inoki K., Zhu T., and Guan K.L. 2003. TSC2 mediates cellular energy response to control cell growth and survival. *Cell* **115**: 577–590.

Inoki K., Li Y., Zhu T., Wu J., and Guan K.L. 2002. TSC2 is phosphorylated and inhibited by Akt and suppresses mTOR signalling. *Nat. Cell Biol.* **4**: 648–657.

Isotani S., Hara K., Tokunaga C., Inoue H., Avruch J., and Yonezawa K. 1999. Immunopurified mammalian target of rapamycin phosphorylates and activates p70 S6 kinase alpha in vitro. *J. Biol. Chem.* **274**: 34493–34498.

Jacinto E., Loewith R., Schmidt A., Lin S., Ruegg M.A., Hall A., and Hall M.N. 2004. Mammalian TOR complex 2 controls the actin cytoskeleton and is rapamycin insensitive. *Nat. Cell Biol.* **6**: 1122–1128.

Jaeschke A., Dennis P.B., and Thomas G. 2004. mTOR: A mediator of intracellular homeostasis. *Curr. Top. Microbiol. Immunol.* **279**: 283–298.

Jaeschke A., Hartkamp J., Saitoh M., Roworth W., Nobukuni T., Hodges A., Sampson J., Thomas G., and Lamb R. 2002. Tuberous sclerosis complex tumor suppressor-mediated S6 kinase inhibition by phosphatidylinositide-3-OH kinase is mTOR independent. *J. Cell Biol.* **159**: 217–224.

Jefferies H.B.J., Thomas G., and Thomas G. 1994a. Elongation factor-1a mRNA is selectively translated following mitogenic stimulation. *J. Biol. Chem.* **269**: 4367–4372.

Jefferies H.B.J., Reinhard C., Kozma S.C., and Thomas G. 1994b. Rapamycin selectively represses translation of the "polypyrimidine tract" mRNA family. *Proc. Natl. Acad. Sci.* **91**: 4441–4445.

Jefferies H.B.J., Fumagalli S., Dennis P.B., Reinhard C., Pearson R.B., and Thomas G. 1997. Rapamycin suppresses 5′TOP mRNA translation through inhibition of p70[s6k]. *EMBO J.* **12**: 3693–3704.

Kahn B.B. 1998. Type 2 diabetes: When insulin secretion fails to compensate for insulin resistance. *Cell* **92**: 593–596.

Kahn R., Buse J., Ferrannini E., and Stern M. 2005. The metabolic syndrome: Time for a critical appraisal. Joint statement from the American Diabetes Association and the European Association for the Study of Diabetes. *Diabetologia* **48**: 1684–1699.

Kawasome K., Papst P., Webb S., Keller G.M., Johnson G.L., Gelfand E.W., and Terada N. 1998. Targeted disruption of p70[s6k] defines its role in protein synthesis and rapamycin sensitivity. *Proc. Natl. Acad. Sci.* **95**: 5033–5038.

Keith C.T. and Schreiber S.L. 1995. PIK-related kinases: DNA repair, recombination, and cell cycle checkpoints. *Science* **270**: 50–51.

Khanna K.K. and Jackson S.P. 2001. DNA double-strand breaks: Signaling, repair and the cancer connection. *Nat. Genet.* **27**: 247–254.

Kim D.H. and Sabatini D.M. 2004. Raptor and mTOR: Subunits of a nutrient-sensitive complex. *Curr. Top. Microbiol. Immunol.* **279**: 259–270.

Kim D.H., Sarbassov D.D., Ali S.M., King J.E., Latek R.R., Erdjument-Bromage H., Tempst P., and Sabatini D.M. 2002. mTOR interacts with raptor to form a nutrient-sensitive complex that signals to the cell growth machinery. *Cell* **110**: 163–175.

Kim D.H., Sarbassov D.D., Ali S.M., Latek R.R., Guntur K.V., Erdjument-Bromage H., Tempst P., and Sabatini D.M. 2003. GbetaL, a positive regulator of the rapamycin-sensitive pathway required for the nutrient-sensitive interaction between raptor and mTOR. *Mol. Cell* **11**: 895–904.

Kopecky J., Clarke G., Enerback S., Spiegelman B., and Kozak L.P. 1995. Expression of the mitochondrial uncoupling protein gene from the aP2 gene promoter prevents genetic obesity. *J. Clin. Invest.* **96**: 2914–2923.

Kozak M. 1991. An analysis of vertebrate mRNA sequences: Intimations of translational control. *J. Cell Biol.* **115:** 887–903.

Kozma S.C., Ferrari S., Bassand P., Siegmann M., Totty N., and Thomas G. 1990. Cloning of the mitogen-activated S6 kinase from rat liver reveals an enzyme of the second messenger subfamily. *Proc. Natl. Acad. Sci.* **87:** 7365–7369.

Krebs M., Krssak M., Bernroider E., Anderwald C., Brehm A., Meyerspeer M., Nowotny P., Roth E., Waldhausl W., and Roden M. 2002. Mechanism of amino acid-induced skeletal muscle insulin resistance in humans. *Diabetes* **51:** 599–605.

Krebs M., Brehm A., Krssak M., Anderwald C., Bernroider E., Nowotny P., Roth E., Chandramouli V., Landau B.R., Waldhausl W., and Roden M. 2003. Direct and indirect effects of amino acids on hepatic glucose metabolism in humans. *Diabetologia* **46:** 917–925.

Kulkarni R.N., Bruning J.C., Winnay J.N., Postic C., Magnuson M.A., and Kahn C.R. 1999. Tissue-specific knockout of the insulin receptor in pancreatic beta cells creates an insulin secretory defect similar to that in type 2 diabetes. *Cell* **96:** 329–339.

Lee-Fruman K.K., Kuo C.J., Lippincott J., Terada N., and Blenis J. 1999. Characterization of S6K2, a novel kinase homologous to S6K1. *Oncogene* **18:** 5108–5114.

Leibiger I.B., Leibiger B., Moede T., and Berggren P.O. 1998. Exocytosis of insulin promotes insulin gene transcription via the insulin receptor/PI-3 kinase/p70 s6 kinase and CaM kinase pathways. *Mol. Cell* **1:** 933–938.

Lemmon M.A. 2003. Phosphoinositide recognition domains. *Traffic* **4:** 201–213.

Li Y., Inoki K., Yeung R., and Guan K.L. 2002. Regulation of TSC2 by 14-3-3 binding. *J. Biol. Chem.* **277:** 44593–44596.

Loewith R., Jacinto E., Wullschleger S., Lorberg A., Crespo J.L., Bonenfant D., Oppliger W., Jenoe P., and Hall M.N. 2002. Two TOR complexes, only one of which is rapamycin sensitive, have distinct roles in cell growth control. *Mol. Cell* **10:** 457–468.

Long X., Ortiz-Vega S., Lin Y., and Avruch J. 2005a. Rheb binding to mammalian target of rapamycin (mTOR) is regulated by amino acid sufficiency. *J. Biol. Chem.* **280:** 23433–23436.

Long X., Lin Y., Ortiz-Vega S., Yonezawa K., and Avruch J. 2005b. Rheb binds and regulates the mTOR kinase. *Curr. Biol.* **15:** 702–713.

Long X., Spycher C., Han Z.S., Rose A.M., Muller F., and Avruch J. 2002. TOR deficiency in *C. elegans* causes developmental arrest and intestinal atrophy by inhibition of mRNA translation. *Curr. Biol.* **12:** 1448–1461.

Maehama T. and Dixon J.E. 1999. PTEN: A tumour suppressor that functions as a phospholipid phosphatase. *Trends Cell Biol.* **9:** 125–128.

Manning B.D. and Cantley L.C. 2003. Rheb fills a GAP between TSC and TOR. *Trends Biochem. Sci.* **28:** 573–576.

Manning B.D., Tee A.R., Logsdon M.N., Blenis J., and Cantley L.C. 2002. Identification of the tuberous sclerosis complex-2 tumor suppressor gene product tuberin as a target of the phosphoinositide 3-kinase/akt pathway. *Mol. Cell* **10:** 151–162.

Manning G., Whyte D.B., Martinez R., Hunter T., and Sudarsanam S. 2002. The protein kinase complement of the human genome. *Science* **298:** 1912–1934.

Marx J. 2003. Cellular warriors at the battle of the bulge. *Science* **299:** 846–849.

Matsuo T., Kubo Y., Watanabe Y., and Yamamoto M. 2003. *Schizosaccharomyces pombe* AGC family kinase Gad8p forms a conserved signaling module with TOR and PDK1-like kinases. *EMBO J.* **22:** 3073–3083.

McDaniel M.L., Marshall C.A., Pappan K.L., and Kwon G. 2002. Metabolic and autocrine regulation of the mammalian target of rapamycin by pancreatic beta-cells. *Diabetes* **51:** 2877–2885.

Meyuhas O., Avni D., and Shama S. 1996. Translational control of ribosomal protein mRNAs in eukaryotes. In *Translational control* (ed. J.W.B. Hershey et al.), pp. 363–388. Cold Spring Harbor Laboratory Press, Cold Spring Harbor, New York.

Miaczynska M., Pelkmans L., and Zerial M. 2004. Not just a sink: Endosomes in control of signal transduction. *Curr. Opin. Cell Biol.* **16:** 400–406.

Montagne J. and Thomas G. 2004. S6K integrates nutrient and mitogen signals to control cell growth. In *Cell growth: Control of cell size* (ed. M.N. Hall et al.), pp. 265–298. Cold Spring Harbor Laboratory Press, Cold Spring Harbor, New York.

Montagne J., Radimerski T., and Thomas G. 2001. Insulin signaling: Lessons from the *Drosophila* tuberous sclerosis complex, a tumor suppressor. *Sci. STKE* **2001:** E36.

Morgan K.P., Kapur A., and Beatt K.J. 2004. Anatomy of coronary disease in diabetic patients: An explanation for poorer outcomes after percutaneous coronary intervention and potential target for intervention. *Heart* **90:** 732–738.

Murakami M., Ichisaka T., Maeda M., Oshiro N., Hara K., Edenhofer F., Kiyama H., Yonezawa K., and Yamanaka S. 2004. mTOR is essential for growth and proliferation in early mouse embryos and embryonic stem cells. *Mol. Cell. Biol.* **24:** 6710–6718.

Must A., Spadano J., Coakley E.H., Field A.E., Colditz G., and Dietz W.H. 1999. The disease burden associated with overweight and obesity. *J. Am. Med. Assoc.* **282:** 1523–1529.

Neel J.V. 1999. Diabetes mellitus: A "thrifty" genotype rendered detrimental by "progress"? 1962. *Bull. World Health Org.* **77:** 694–703.

Nellist M., Goedbloed M.A., de Winter C., Verhaaf B., Jankie A., Reuser A.J., van den Ouweland A.M., van der Sluijs P., and Halley D.J. 2002. Identification and characterization of the interaction between tuberin and 14-3-3zeta. *J. Biol. Chem.* **277:** 39417–39424.

Nobukuni T., Joaquin M., Roccio M., Dann S.G., Kim S.Y., Gulati P., Byfield M.P., Backer J.M., Natt F., Bos J.L., Zwartkruis F.J., and Thomas G. 2005. Amino acids mediate mTOR/raptor signaling through activation of class 3 phosphatidylinositol 3OH-kinase. *Proc. Natl. Acad. Sci.* **102:** 14238–14243.

Oldham S., Montagne J., Radimerski T., Thomas G., and Hafen E. 2000. Genetic and biochemical characterization of dTOR, the *Drosophila* homolog of the target of rapamycin. *Genes Dev.* **14:** 2689–2694.

Pain V.M., Lewis J.A., Huvos P., Henshaw E.C., and Clemens M.J. 1980. The effects of amino acid starvation on regulation of polypeptide chain initiation in Ehrlich ascites tumor cells. *J. Biol. Chem.* **255:** 1486–1491.

Patti M.E., Brambilla E., Luzi L., Landaker E.J., and Kahn C.R. 1998. Bidirectional modulation of insulin action by amino acids. *J. Clin. Invest.* **101:** 1519–1529.

Pearson R.B., Dennis P.B., Han J.W., Williamson N.A., Kozma S.C., Wettenhall R.E., and Thomas G. 1995. The principal target of rapamycin-induced p70[s6k] inactivation is a novel phosphorylation site within a conserved hydrophobic domain. *EMBO J.* **21:** 5279–5287.

Pende M., Kozma S.C., Jaquet M., Oorschot V., Burcelin R., Le Marchand-Brustel Y., Klumperman J., Thorens B., and Thomas G. 2000. Hypoinsulinaemia, glucose intolerance and diminished beta-cell size in S6K1-deficient mice. *Nature* **408:** 994–997.

Pende M., Um S.H., Mieulet V., Sticker M., Goss V.L., Mestan J., Mueller M., Fumagalli S., Kozma S.C., and Thomas G. 2004. S6K1($-/-$)/S6K2($-/-$) mice exhibit perinatal lethality and rapamycin-sensitive 5′-terminal oligopyrimidine mRNA translation and reveal a mitogen-activated protein kinase-dependent S6 kinase pathway. *Mol. Cell. Biol.* **24:** 3112–3124.

Phillips D.I. 1996. Insulin resistance as a programmed response to fetal undernutrition (comments). *Diabetologia* **39:** 1119–1122.

Pi-Sunyer F.X. 2002. The obesity epidemic: Pathophysiology and consequences of obesity. *Obesity Res.* (suppl. 2) **10:** 97S–104S.

Potter C.J., Huang H., and Xu T. 2001. *Drosophila* tsc1 functions with tsc2 to antagonize insulin signaling in regulating cell growth, cell proliferation, and organ size. *Cell* **105:** 357–368.

Proud C.G. 2004. mTOR-mediated regulation of translation factors by amino acids. *Biochem. Biophys. Res. Commun.* **313:** 429–436.

Pyronnet S., Imataka H., Gingras A.C., Fukunaga R., Hunter T., and Sonenberg N. 1999. Human eukaryotic translation initiation factor 4G (eIF4G) recruits mnk1 to phosphorylate eIF4E. *EMBO J.* **18:** 270–279.

Radimerski T., Montagne J., Hemmings-Mieszczak M., and Thomas G. 2002a. Lethality of *Drosophila* lacking TSC tumor suppressor function rescued by reducing dS6K signaling. *Genes Dev.* **16:** 2627–2632.

Radimerski T., Montagne J., Rintelen F., Stocker H., van Der Kaay J., Downes C.P., Hafen E., and Thomas G. 2002b. dS6K-regulated cell growth is dPKB/dPI(3)K-independent, but requires dPDK1. *Nat. Cell Biol.* **4:** 251–255.

Raught B., Peiretti F., Gingras A.C., Livingstone M., Shahbazian D., Mayeur G.L., Polakiewicz R.D., Sonenberg N., and Hershey J.W. 2004. Phosphorylation of eucaryotic translation initiation factor 4B Ser422 is modulated by S6 kinases. *EMBO J.* **23:** 1761–1769.

Redpath N.T., Foulstone E.J., and Proud C.G. 1996. Regulation of translation elongation factor-2 by insulin via a rapamycin-sensitive signalling pathway. *EMBO J.* **15:** 2291–2297.

Reinhard C., Thomas G., and Kozma S.C. 1992. A single gene encodes two isoforms of the p70 S6 kinase: Activation upon mitogenic stimulation. *Proc. Natl. Acad. Sci.* **89:** 4052–4056.

Reinhard C., Fernandez A., Lamb N.J.C., and Thomas G. 1994. Nuclear localization of p85[s6k]: Functional requirement for entry into S phase. *EMBO J.* **13:** 1557–1565.

Reynolds K. and He J. 2005. Epidemiology of the metabolic syndrome. *Am. J. Med. Sci.* **330:** 273–279.

Richardson C.J., Broenstrup M., Fingar D.C., Julich K., Ballif B.A., Gygi S., and Blenis J. 2004. SKAR is a specific target of S6 kinase 1 in cell growth control. *Curr. Biol.* **14:** 1540–1549.

Rudra D. and Warner J.R. 2004. What better measure than ribosome synthesis? *Genes Dev.* **18:** 2431–2436.

Ruvinsky I., Sharon N., Lerer T., Cohen H., Stolovich-Rain M., Nir T., Dor Y., Zisman P., and Meyuhas O. 2005. Ribosomal protein S6 phosphorylation is a determinant of cell size and glucose homeostasis. *Genes Dev.* **19:** 2199–2211.

Saitoh M., ten Dijke P., Miyazono K., and Ichijo H. 1998. Cloning and characterization of p70[s6k]b defines a novel family of p70 S6 kinases. *Biochem. Biophys. Res. Commun.* **253:** 470–476.

Saitoh M., Pullen N., Brennan P., Cantrell D., Dennis P.B., and Thomas G. 2002. Regulation of an activated S6 kinase 1 variant reveals a novel mammalian target of rapamycin phosphorylation site. *J. Biol. Chem.* **277:** 20104–20112.

Sarbassov D.D., Guertin D.A., Ali S.M., and Sabatini D.M. 2005. Phosphorylation and regulation of Akt/PKB by the rictor-mTOR complex. *Science* **307:** 1098–1101.

Sarbassov D.D., Ali S.M., Kim D.H., Guertin D.A., Latek R.R., Erdjument-Bromage H., Tempst P., and Sabatini D.M. 2004. Rictor, a novel binding partner of mTOR, defines a rapamycin-insensitive and raptor-independent pathway that regulates the cytoskeleton. *Curr. Biol.* **14:** 1296–1302.

Saucedo L.J., Gao X., Chiarelli D.A., Li L., Pan D., and Edgar B.A. 2003. Rheb promotes cell growth as a component of the insulin/TOR signalling network. *Nat. Cell Biol.* **5:** 566–571.

Schalm S.S. and Blenis J. 2002. Identification of a conserved motif required for mTOR signaling. *Curr. Biol.* **12:** 632–639.

Schalm S.S., Tee A.R., and Blenis J. 2005. Characterization of a conserved C-terminal motif (RSPRR) in ribosomal protein S6 kinase 1 required for its mammalian target of rapamycin-dependent regulation. *J. Biol. Chem.* **280:** 11101–11106.

Schalm S.S., Fingar D.C., Sabatini D.M., and Blenis J. 2003. TOS motif-mediated raptor binding regulates 4E-BP1 multisite phosphorylation and function. *Curr. Biol.* **13:** 797–806.

Schu P.V., Takegawa K., Fry M.J., Stack J.H., Waterfield M.D., and Emr S.D. 1993. Phosphatidylinositol 3-kinase encoded by yeast VPS34 gene essential for protein sorting. *Science* **260:** 88–91.

Schwab M.S., Kim S.H., Terada N., Edfjall C., Kozma S.C., Thomas G., and Maller J.L. 1999. p70(S6K) controls selective mRNA translation during oocyte maturation and early embryogenesis in *Xenopus laevis. Mol. Cell. Biol.* **19:** 2485–2494.

Shah O.J., Wang Z., and Hunter T. 2004. Inappropriate activation of the TSC/Rheb/mTOR/S6K cassette induces IRS1/2 depletion, insulin resistance, and cell survival deficiencies. *Curr. Biol.* **14:** 1650–1656.

Shaw R.J., Bardeesy N., Manning B.D., Lopez L., Kosmatka M., DePinho R.A., and Cantley L.C. 2004. The LKB1 tumor suppressor negatively regulates mTOR signaling. *Cancer Cell* **6:** 91–99.

Shi Y., Taylor S.I., Tan S.L., and Sonenberg N. 2003. When translation meets metabolism: Multiple links to diabetes. *Endocr. Rev.* **24:** 91–101.

Shima H., Pende M., Chen Y., Fumagalli S., Thomas G., and Kozma S.C. 1998. Disruption of the p70^{s6k}/p85^{s6k} gene reveals a small mouse phenotype and a new functional S6 kinase. *EMBO J.* **17:** 6649–6659.

Stolovich M., Tang H., Hornstein E., Levy G., Cohen R., Bae S.S., Birnbaum M.J., and Meyuhas O. 2002. Transduction of growth or mitogenic signals into translational activation of TOP mRNAs is fully reliant on the phosphatidylinositol 3-kinase-mediated pathway but requires neither S6K1 nor rpS6 phosphorylation. *Mol. Cell. Biol.* **22:** 8101–8113.

Swenne I. 1992. Pancreatic beta-cell growth and diabetes mellitus. *Diabetologia* **35:** 193–201.

Swenne I., Borg L.A., Crace C.J., and Schnell Landstrom A. 1992. Persistent reduction of pancreatic beta-cell mass after a limited period of protein-energy malnutrition in the young rat. *Diabetologia* **35:** 939–945.

Tapon N., Ito N., Dickson B.J., Treisman J.E., and Hariharan I.K. 2001. The *Drosophila* tuberous sclerosis complex gene homologs restrict cell growth and cell proliferation. *Cell* **105:** 345–355.

Tee A.R., Fingar D.C., Manning B.D., Kwiatkowski D.J., Cantley L.C., and Blenis J. 2002. Tuberous sclerosis complex-1 and -2 gene products function together to inhibit mammalian target of rapamycin (mTOR)-mediated downstream signaling. *Proc. Natl. Acad. Sci.* **99:** 13571–13576.

Terada N., Patel H.R., Takase K., Kohno K., Narin A.C., and Gelfand E.W. 1994. Rapamycin selectively inhibits translation of mRNAs encoding elongation factors and ribosomal proteins. *Proc. Natl. Acad. Sci.* **91:** 11477–11481.

Thomas G. 2000. An "encore" for ribosome biogenesis in cell proliferation. *Nat. Cell Biol.* **2:** E71–E72.

Tremblay F. and Marette A. 2001. Amino acid and insulin signaling via the mTOR/p70 S6 kinase pathway. A negative feedback mechanism leading to insulin resistance in skeletal muscle cells. *J. Biol. Chem.* **276:** 38052–38060.

Tsukiyama-Kohara K., Poulin F., Kohara M., DeMaria C.T., Cheng A., Wu Z., Gingras A.C., Katsume A., Elchebly M., Spiegelman B.M., et al. 2001. Adipose tissue reduction in mice lacking the translational inhibitor 4E-BP1. *Nat. Med.* **7:** 1128–1132.

Ueno M., Carvalheira J.B., Tambascia R.C., Bezerra R.M., Amaral M.E., Carneiro E.M., Folli F., Franchini K.G., and Saad M.J. 2005. Regulation of insulin signalling by hyperinsulinaemia: Role of IRS-1/2 serine phosphorylation and the mTOR/p70 S6K pathway. *Diabetologia* **48:** 506–518.

Um S.H., D'Alessio D., and Thomas G. 2006. Nutrient overload, insulin resistance and ribosomal protein S6 kinase 1, S6K1. *Cell Metab.* **3:** 393–402.

Um S.H., Frigerio F., Watanabe M., Picard F., Joaquin M., Sticker M., Fumagalli S., Allegrini P.R., Kozma S.C., Auwerx J., and Thomas G. 2004. Absence of S6K1 protects against age- and diet-induced obesity while enhancing insulin sensitivity. *Nature* **431:** 200–205.

Volarevic S., Stewart M.J., Ledermann B., Zilberman F., Terracciano L., Montini E., Grompe M., Kozma S.C., and Thomas G. 2000. Proliferation, but not growth, blocked by conditional deletion of 40S ribosomal protein S6. *Science* **288:** 2045–2047.

von Manteuffel S.R., Gingras A.C., Ming X.F., Sonenberg N., and Thomas G. 1996. 4E-BP1 phosphorylation is mediated by the FRAP-p70[s6k] pathway and is independent of mitogen-activated protein kinase. *Proc. Natl. Acad. Sci.* **93:** 4076–4080.

von Manteuffel S.R., Dennis P.B., Pullen N., Gingras A.C., Sonenberg N., and Thomas G. 1997. The insulin-induced signalling pathway leading to S6 and initiation factor 4E binding protein 1 phosphorylation bifurcates at a rapamycin-sensitive point immediately upstream of p70s6k. *Mol. Cell. Biol.* **17:** 5426–5436.

Wang X., Campbell L.E., Miller C.M., and Proud C.G. 1998. Amino acid availability regulates p70 S6 kinase and multiple translation factors. *Biochem. J.* **334:** 261–267.

Wang X., Li W., Williams M., Terada N., Alessi D.R., and Proud C.G. 2001. Regulation of elongation factor 2 kinase by p90(RSK1) and p70 S6 kinase. *EMBO J.* **20:** 4370–4379.

Williams M.R., Arthur J.S., Balendran A., van der Kaay J., Poli V., Cohen P., and Alessi D.R. 2000. The role of 3-phosphoinositide-dependent protein kinase 1 in activating AGC kinases defined in embryonic stem cells. *Curr. Biol.* **10:** 439–448.

Yamagata K., Sanders L.K., Kaufmann W.E., Yee W., Barnes C.A., Nathans D., and Worley P.F. 1994. rheb, a growth factor- and synaptic activity-regulated gene, encodes a novel Ras-related protein. *J. Biol. Chem.* **269:** 16333–16339.

Zhang H., Stallock J.P., Ng J.C., Reinhard C., and Neufeld T.P. 2000. Regulation of cellular growth by the *Drosophila* target of rapamycin dTOR. *Genes Dev.* **14:** 2712–2724.

Zhang Y., Gao X., Saucedo L.J., Ru B., Edgar B.A., and Pan D. 2003. Rheb is a direct target of the tuberous sclerosis tumour suppressor proteins. *Nat. Cell Biol.* **5:** 578–581.

Zick Y. 2004. Uncoupling insulin signalling by serine/threonine phosphorylation: A molecular basis for insulin resistance. *Biochem. Soc. Trans.* **32:** 812–816.

18

Translational Control of Synaptic Plasticity and Learning and Memory

Eric Klann

Departments of Molecular Physiology and Biophysics
and Neuroscience Baylor College of Medicine
Houston, Texas 77030

Joel D. Richter

Program in Molecular Medicine
University of Massachusetts Medical School
Worcester, Massachusetts 01605

ONE HALLMARK OF LONG-TERM MEMORY CONSOLIDATION is the requirement for new gene expression. Although memory formation has largely focused on transcriptional control (Kandel 2001), it has been known for more than four decades that it also requires protein synthesis (Flexner et al. 1963). This and other early studies offered little in the way of molecular mechanisms because they relied mostly on the injection of general translation inhibitors into animals. The last 10 years, however, have witnessed major advances in our understanding of translational control of memory and its cellular foundation, synaptic plasticity. In this chapter, we discuss the most salient aspects of translational control of these essential brain activities and present our thoughts on some of the key issues remaining to be elucidated.

TEMPORAL PHASES OF SYNAPTIC PLASTICITY AND MEMORY

How are memories stored at the cellular level? Most neuroscientists hypothesize that memory involves changes in the strength of synaptic connections between neurons (i.e., synaptic transmission). These changes in synaptic efficacy are referred to as *synaptic plasticity* and are manifested as either an increase (potentiation) or decrease (depression) in strength.

Long-term potentiation (LTP) and long-term depression (LTD) have been intensively studied in the rodent hippocampus, a brain structure that is critical for processing information about space, time, and the relationship between objects. Both LTP and LTD can be induced routinely in vitro with distinct patterns of electrical stimulation delivered to synapses in preparations of hippocampal slices.

More than 20 years ago, hippocampal LTP was shown to require new protein synthesis in vivo (Krug et al. 1984). A number of subsequent studies using transcription and translation inhibitors demonstrated that LTP has distinct temporal phases (Frey et al. 1988; Huang and Kandel 1994; Nguyen et al. 1994): The phase that is induced by a single train of high-frequency stimulation (HFS) lasting for 1–2 hours is referred to as the "early phase" LTP (E-LTP), whereas the phase that is induced by multiple, spaced trains of HFS persisting for 3 or more hours is referred to as the "late phase" LTP (L-LTP). Mechanistically, these two forms of LTP differ with respect to their requirement for new mRNA and protein synthesis. L-LTP is sensitive to both transcription and translation inhibitors, whereas E-LTP is not sensitive to either. Temporal phases of hippocampal LTP also have been observed when LTP is induced with θ-burst stimulation, a type of stimulation protocol that more closely resembles hippocampal neuron activity in vivo (Nguyen and Kandel 1997). Finally, distinct temporal phases have been described for synaptic plasticity in invertebrate systems such as long-term facilitation (LTF) at the sensorimotor synapse in *Aplysia* (Kandel 2001).

The induction of both E-LTP and L-LTP requires activation of the N-methyl-D-aspartate (NMDA) subtype of glutamate receptor. Additional protein-synthesis-dependent forms of LTP can be induced in hippocampal slices with a number of receptor agonists, including the neurotrophins brain-derived neurotrophic factor (BDNF) and neurotrophin-3 (Kang and Schuman 1996), activators of the cAMP signaling pathway (Frey et al. 1993), and those that activate dopamine receptors (Huang and Kandel 1995). Moreover, long-lasting protein-synthesis-dependent forms of hippocampal LTD have also been described. Activation of group I metabotropic glutamate receptors (mGluRs) via direct pharmacological activation with the agonist 3,5-dihydroxyphenylglycine (DHPG) or with paired-pulse low-frequency stimulation (PP-LFS) results in protein-synthesis-dependent L-LTD (Huber et al. 2000). Protein-synthesis-dependent LTD also can be triggered by insulin (Huang et al. 2004). Finally, LFS induces NMDA-receptor-dependent LTD (Dudek and Bear 1992; Mulkey and Malenka 1992) that is blocked by protein synthesis inhibitors (Kauderer and Kandel 2000; Sajikumar and Frey 2004). Thus, protein-synthesis-dependent LTP and LTD can be induced with via activation of multiple receptors in the hippocampus.

Similar to synaptic plasticity, memory also exhibits distinct temporal phases that can be differentiated by transcription and translation inhibitors. A good illustration of this type of differentiation is the gill-withdrawal reflex in *Aplysia*, which is associated with synaptic facilitation. A single sensitizing stimulus results in short-term memory that depends on covalent modifications to existing proteins and is insensitive to transcription and translation inhibitors; conversely, multiple, spaced sensitizing stimulation results in long-term memory that is blocked by those inhibitors (Castellucci et al. 1989). Similar observations have been made in rodents where, in general, new mRNA and protein synthesis is required for long-term memory but not short-term memory (McGaugh 2000; Kandel 2001). Later in this chapter, we discuss specific mechanisms of translational control that mediate long-term memory in rodents.

CONTRIBUTION OF LOCAL PROTEIN SYNTHESIS TO SYNAPTIC PLASTICITY

Neuronal dendrites and dendritic spines contain polyribosomes (Steward and Levy 1982), translation factors (Tang et al. 2002), and mRNA that can be translated into protein at synapses (Crino and Eberwine 1996). Imaging studies with fluorescent reporters have shown that both BDNF and specific manipulations of neuronal activity trigger protein synthesis in dendrites (Aakulu et al. 2001; Ju et al. 2004; Sutton et al. 2004). Local (i.e., synaptic) protein synthesis also has been demonstrated to be necessary for BDNF-induced potentiation and mGluR-dependent LTD in the hippocampus (Kang and Schuman 1996; Huber et al. 2000). Protein synthesis inhibitors block both of these types of synaptic plasticities even when the cell bodies of the hippocampal neurons are severed from their dendrites. Local protein synthesis also is required for LTF in *Aplysia* sensory neurons and plasticity at nerve–muscle synapses in *Xenopus* (Martin et al. 1997; Zhang and Poo 2002). These findings demonstrate that local protein synthesis is required for numerous types of synaptic plasticities.

Emerging evidence indicates that local protein synthesis is required for hippocampal LTP. As mentioned earlier, L-LTP induced with multiple trains of HFS requires both transcription and translation. However, the translation inhibitors affect L-LTP at earlier times than the transcription inhibitors (Kelleher et al. 2004b; Banko et al. 2005), implying that L-LTP itself consists of an early translation-dependent phase that is independent of transcription followed by a transcription- and translation-dependent phase. The translation-dependent early phase of L-LTP is presumably due to local protein synthesis. Consistent with this idea, a translation-dependent, transcription-independent LTP has been shown in isolated dendrites in hippocampal slices (Cracco et al. 2005; Vickers et al. 2005). Furthermore, local application of a protein synthesis inhibitor to dendrites blocks L-LTP (Bradshaw et al. 2003). Thus, local protein synthesis is required for the earliest phase of L-LTP.

GENERAL TRANSLATIONAL CONTROL MECHANISMS DURING SYNAPTIC PLASTICITY

As described in Chapter 4, the integrity of the eukaryotic initiation factor 4F (eIF4F) cap-binding complex is modulated by eIF4E-binding proteins (4E-BPs). Hypophosphorylated 4E-BPs bind to eIF4E and inhibit translation whereas their hyperphosphorylated forms do not, thereby permitting eIF4F complex formation and translation initiation. Biochemical studies have shown that both LTP and mGluR-LTD are associated with enhanced 4E-BP phosphorylation (Kelleher et al. 2004b; Banko et al. 2005, 2006; Schmitt et al. 2005). 4E-BP phosphorylation is regulated by extracellular-signal-regulated kinase (ERK), phosphoinositide 3-kinase (PI3K), and mammalian target of rapamycin (mTOR) pathways, and it is known that mTOR phosphorylates 4E-BP (Gingras et al. 1998; Chapter 14). The ERK (Sweatt 2004; Thomas and Huganir 2004) and PI3K (Kelly and Lynch 2000; Beaumont et al. 2001; Sanna et al. 2002; Opazo et al. 2003) signaling cascades are required for a plethora of types of synaptic plasticities, and recent studies have demonstrated that mTOR activity is required for several protein-synthesis-dependent forms of synaptic plasticity. For example, the mTOR inhibitor rapamycin blocks LTF in *Aplysia* sensory neurons (Casadio et al. 1999) and at the crayfish neuromuscular junction (Beaumont et al. 2001). In the rodent hippocampus, rapamycin blocks mGluR-LTD (Hou and Klann 2004), insulin-induced LTD (Huang et al. 2004), BDNF-induced potentiation (Tang et al. 2002), and L-LTP (Tang et al. 2002; Cracco et al. 2005; Tsokas et al. 2005; Vickers et al. 2005). Thus, mTOR-dependent regulation of 4E-BP phosphorylation is a major component of protein-synthesis-dependent synaptic plasticity (Fig. 1).

In the mouse hippocampus, 4E-BP2 is the predominant 4E-BP isoform (Banko et al. 2005). LTP and mGluR-LTD have been examined in 4E-BP2 knockout mice; in hippocampal slices from these animals, protein synthesis-independent E-LTP is converted to protein-synthesis-dependent L-LTP and is correlated with increased eIF4F complex forma-

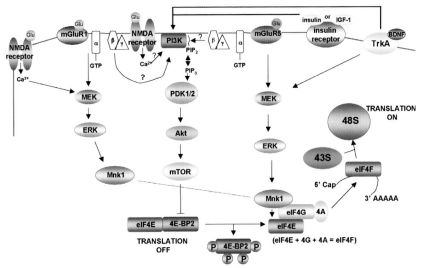

Figure 1. Signaling pathways that control translation initiation during various forms of LTP and LTD. Pharmacological, biochemical, electrophysiological, and genetic evidence indicates that activation of NMDA receptors and TrkA receptors results in LTP, whereas activation of group I mGluRs and insulin receptors results in LTD. Activation of MEK/ERK and PI3K/Akt is necessary for both types of synaptic plasticities. PI3K/Akt-dependent activation of mTOR results in the phosphorylation of 4E-BP2, which causes the dissociation of 4E-BP2 from eIF4E. This permits eIF4E to bind to eIF4G, resulting in the formation of the active eIF4F complex, which is a necessary component of the 48S initiation complex. Mnk1, which also binds eIF4G, is phosphorylated and activated by ERK. Mnk1 phosphorylates eIF4E, an event that is associated with increased translation rates. The eIF4F complex and the poly(A) tail act synergistically to stimulate mRNA translation. There is evidence that similar cascades are activated during LTF via serotonin-dependent signaling in *Aplysia* sensory neurons.

tion (Banko et al. 2005). In addition, enhanced mGluR-LTD is observed in the 4E-BP2 knockout mice (Banko et al. 2006). Taken together, these findings might suggest that the lack of 4E-BP2, which results in increased eIF4F complex formation, induces a robust but general synaptic plasticity in the hippocampus. On the other hand, L-LTP is abrogated in the 4E-BP2 knockout mice even though additional eIF4F complex formation is detected. Consequently, enhanced eIF4F complex formation and possibly excessive translation initiation may be detrimental to L-LTP (Banko et al. 2005). Interestingly, LTP-dependent increases in eIF4F complex formation also were observed in 4E-BP2 knockout mice, which strongly suggests that additional eIF4E-binding proteins regulate translation initiation during hippocampal LTP. In contrast, additional eIF4F complex formation was not observed in association with mGluR-LTD in 4E-BP2 knockout mice (Banko et al. 2006), suggesting that 4E-BP2 is the only eIF4E-binding protein regulated during mGluR-LTD.

eIF4E is a target for direct phosphorylation (Chapter 14), which could be important for synaptic plasticity. The neurotransmitter serotonin, which is required for LTF in *Aplysia* sensory neurons, induces a p38 MAP kinase-dependent increase in eIF4E phosphorylation (Dyer and Sossin 2000). ERK-dependent increases in eIF4E phosphorylation are induced in cultured neurons treated with BDNF (Kelleher et al. 2004b) and in hippocampal slices treated with NMDA (Banko et al. 2004). Moreover, increased eIF4E phosphorylation has been observed during several forms of synaptic plasticities; electrical stimulation that induces either E-LTP (Banko et al. 2005; Schmitt et al. 2005) or L-LTP (Kelleher et al. 2004b; Banko et al. 2005) triggers ERK-dependent increases in eIF4E phosphorylation. Finally, it recently was shown that mGluR-LTD is associated with an ERK-dependent increase in eIF4E phosphorylation (Banko et al. 2006). Although none of these studies directly demonstrates a requirement of eIF4E phosphorylation for protein-synthesis-dependent synaptic plasticity, they suggest that multiple forms of synaptic plasticities are associated with regulation of a critical translation factor required for cap-dependent translation.

Of course, translation initiation is regulated via other translation factors and protein kinases, such as eIF2α and its kinases, which are likely to be important for protein-synthesis-dependent synaptic plasticity. Regulation of eIF2α by the protein kinase GCN2 during hippocampal LTP and memory is discussed later in this chapter.

The data described above indicate that the signaling mechanisms used to regulate cap-dependent initiation are very similar for most forms of synaptic plasticity. How, then, can such signaling lead to translational control mechanisms that result in LTP versus LTD? One possibility is that there is mRNA-specific translational regulation, a topic discussed later in this chapter.

TRANSLATIONAL CONTROL OF LEARNING AND MEMORY

As noted earlier, long-term memory, but not short-term memory, requires mRNA and protein synthesis (McGaugh 2000; Kandel 2001). Translational control mechanisms that are involved in the formation of long-term memory are largely unknown, but several recent studies indicate that mechanisms that are required for hippocampal synaptic plasticity also are required for hippocampus-dependent long-term memory.

Two standard tests of hippocampal-dependent memory in rodents are platform location in the Morris water maze (spatial memory) and "freezing" in response to a mild foot shock (contextual fear memory). The Morris water maze measures the ability of an animal to learn and remember the relationship between multiple visual cues and the location of a hidden platform in a pool of opaque water (Morris 1984). A typical training protocol entails two blocks of four training trials per day with an interblock interval of 1 hour; this procedure would be repeated each day for 7 days. The time an animal takes to reach the platform is measured; it is referred to as escape latency and is one index of learning ability. Not surprisingly, the escape latency of an animal in the Morris water maze test usually decreases over the course of the training period. To assess spatial memory, the platform is removed from the pool and the animals are allowed to search for 60 seconds (usually 1 hour after the completion of the last training trial). In these probe trials, the time spent searching in each quadrant measures the spatial search strategy of the animal (Schenk and Morris 1985). In addition, platform crossings, the number of times an animal crosses the exact place where the platform had been located, are also noted.

The hippocampus is also involved in contextual fear memory. In this case, a conditioned fear paradigm is employed in which animals learn to fear a new environment because of its association with an aversive unconditioned stimulus such as a foot shock. When animals are exposed to the same context at a later time, those that are "conditioned" demonstrate several stereotypical fear responses such as freezing (Fanselow 1984; Phillips and LeDoux 1992).

Spatial memory has been examined in both the 4E-BP2- and GCN2-deficient mice. The 4E-BP2 knockout mice exhibit impaired spatial learning as measured by higher escape latencies during training, and impaired spatial memory as measured by both time spent in the target quadrant and the number of platform crossings (Banko et al. 2005). GCN2 knockout mice exhibit similar impairments in spatial learning and memory (Costa-Mattioli et al. 2005). Interestingly, when the GCN2 knockout mice were given a weak training regimen consisting of only one training trial per day, they exhibited enhanced learning and memory (Costa-Mattioli et al. 2005). These findings correlate nicely with the LTP studies in these animals where E-LTP was converted to L-LTP, but L-LTP was inhibited. Taken together, these findings suggest that proper translational control is required for hippocampus-dependent spatial learning and memory.

The 4E-BP2 knockout mice exhibit normal short-term contextual fear memory but impaired long-term contextual fear memory (Banko et al. 2005). Similarly, the GCN2 knockout mice also exhibit impaired long-term contextual fear memory (Costa-Mattioli et al. 2005). Thus, in addition to its role in spatial learning and memory, proper translational control also is required for contextual fear memory.

What are the upstream biochemical signaling mechanisms that regulate translation during memory formation? One possibility is the ERK signaling cascade, which is required for most forms of synaptic plasticities and long-term memory (Sweatt 2004; Thomas and Huganir 2004). Consistent with a role for ERK hippocampus-dependent memory, mice expressing a dominant-negative MEK (the upstream kinase that phosphorylates and activates ERK) have impaired long-term contextual fear memory (Kelleher et al. 2004b). Moreover, contextual fear conditioning results in increased phosphorylation of eIF4E that is absent in the dominant-negative MEK mice (Kelleher et al. 2004b). Many more studies will be required to delineate the signaling cascades that regulate translation factor function during both contextual fear memory and spatial memory in the hippocampus.

GENE-SPECIFIC TRANSLATIONAL CONTROL MECHANISMS IN SYNAPTIC PLASTICITY AND MEMORY

eIF2α Kinases and uORFs

As mentioned in Chapter 12, there are four eIF2α kinases, all of which have been found in the brain. The eIF2α kinases, which inhibit general protein synthesis during cellular stress by phosphorylating eIF2α, can also stimulate translation of mRNAs containing upstream open reading frames (uORFs), such as that encoding the transcription factor ATF4 (Harding et al. 2000; Vattem and Wek 2004). Interestingly, ATF4 is an inhibitor of synaptic plasticity and memory via its antagonism of the cyclic-AMP-response element (Yin et al. 1994; Bartch et al. 1995; Abel et al. 1998; Chen et al. 2003). Recent studies indicate that the eIF2α kinase GCN2 regulates ATF4 during synaptic plasticity in the hippocampus. GCN2-deficient mice exhibit LTP phenotypes similar to those of 4E-BP2-deficient mice; i.e., E-LTP is converted to protein-synthesis-dependent L-LTP and L-LTP is inhibited (Costa-Mattioli et al. 2005). The basal levels of eIF2α phosphorylation and the amount of ATF4 were decreased in the GCN2-deficient mice, and these decreases corresponded to enhanced CREB (cAMP-responsive element binding) function as measured by the expression of immediate-early genes regulated by CREB (Costa-Mattioli et al. 2005). These findings suggest that under resting conditions, GCN2 normally represses CREB-dependent transcription and that this suppression is removed when GCN2 is absent. In contrast to LTP, mGluR-LTD, which is enhanced in 4E-BP2-deficient mice (Banko et al. 2006), was unaffected in GCN2-deficient mice (Costa-Mattioli et al. 2005). Thus, the regulation of CREB-dependent transcription via GCN2-dependent ATF4 translation is specific to LTP.

CPEB and Cytoplasmic Polyadenylation

Cytoplasmic polyadenylation is one mechanism that governs specific gene expression in response to synaptic stimulation. Here, dormant mRNAs have relatively short poly(A) tails, usually about 20–40 nucleotides in length. In response to synaptic stimulation, the poly(A) tails on a number of mRNAs lengthen and translation ensues (Wu et al. 1998; Wells et al. 2001; Huang et al. 2002; Shin et al. 2004; Du and Richter 2005). Although most aspects of the molecular mechanism of polyadenylation-induced translation have been elaborated using *Xenopus* oocytes as they progress through the final stages of meiosis (maturation) in preparation for fertilization, most features of this process appear to be conserved in neurons. Two *cis* elements in the 3′-untranslated regions (UTRs) of responding mRNAs are necessary for polyadenylation: the cytoplasmic polyadenylation element (CPE, general structure of UUUUUAU) and the hexanucleotide AAUAAA, which is also important for nuclear pre-mRNA cleavage and polyadenylation (Fox et al. 1989; McGrew et al. 1989). CPEB is an RNA-recognition motif (RRM) and zinc-finger-containing protein that is the central player in cytoplasmic polyadenylation: It not only has a strong affinity for the CPE (Hake and Richter 1994; Hake et al. 1998), but also associates with a number of other key regulatory factors. These factors include symplekin, a scaffold-like protein upon which the polyadenylation machinery is assembled; cleavage and polyadenylation specificity factor (CPSF), a group of four proteins that binds the AAUAAA; and Gld2 (Barnard et al. 2004), an unusual poly(A) polymerase first discovered in yeast and *Caenorhabditis elegans* (Read et al. 2002; Saitoh et al. 2002; Wang et al. 2002). Polyadenylation is triggered by the phosphorylation of S174 or T171 (species-dependent) by the kinase Aurora A (Mendez et al. 2000b; Huang et al. 2002), although some evidence suggests that αCaMKII (calmodulin-dependent protein kinase II) can also phosphorylate this residue (Atkins et al. 2004). In mammalian hippocampal neurons, NMDA receptor activation triggers CPEB phosphorylation (Fig. 2) (Huang et al. 2002). Although phosphorylated CPEB has an enhanced affinity for CPSF (Mendez et al. 2000a), it is unclear

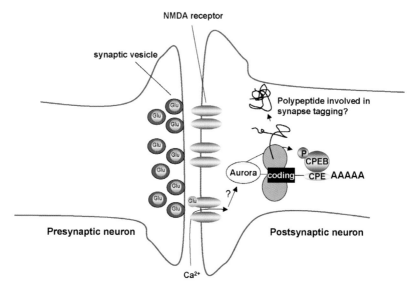

Figure 2. CPEB regulation of synaptic translation. Following synaptic stimulation, calcium enters the postsynaptic neuron via NMDA receptors. This process leads to the activation of the kinase Aurora A, which in turn phosphorylates CPEB and activates the cytoplasmic polyadenylation machinery. Polyadenylation induces the translation of specific (CPE-containing) mRNAs, whose products may be involved in synaptic tagging.

how this event stimulates polyadenylation. In any event, polyadenylation induces translation through other molecules, particularly Maskin, a CPEB-associated factor that also binds the cap-binding factor eIF4E (Stebbins-Boaz et al. 1999). The binding of Maskin to eIF4E precludes the assembly of the eIF4F initiation complex, thus preventing translation (Richter and Sonenberg 2005). The elongated poly(A) tail is bound by poly(A)-binding protein (PABP), which in turn binds eIF4G (Wakiyama et al. 2000); this interaction appears to enhance an eIF4G–eIF4E association at the expense of the Maskin–eIF4E association (Cao and Richter 2002). Consequently, eIF4F assembly on the cap takes place and translation proceeds. Another factor involved in CPEB-mediated translation is p54 (also referred to as DDX6), an RNA helicase (Minshall et al. 2001). Finally, at least a portion of CPEB is anchored to the plasma membrane through members of the amyloid precursor protein family (Cao et al. 2005) and possibly a guanine nucleotide exchange factor as well (Martinez et al. 2005). These proteins stimulate CPEB phosphorylation and resulting polyadenylation.

Aside from these molecular details, does this process influence local translation and synaptic plasticity? A number of experiments indicate that this is the case. First, CPEB and several of its associated factors (see above) are localized postsynaptically, where local translation most likely occurs (Huang et al. 2002). Second, a CPEB knockout mouse shows a deficit in hippocampal LTP induced by a θ-burst protocol (Alarcon et al. 2004). Third, CPEB knockout mice have a deficit in a form of memory known as extinction, which is a behavioral response that will diminish and gradually become extinct in the absence of reinforcement. Several studies make it clear that extinction is an active process that results in the formation of new memories and is not equivalent to forgetting. It also is clear that mechanisms underlying extinction can be distinct from those underlying acquisition and storage of memories (Abel and Lattal 2001). Extinction involves several brain areas including the hippocampus, amygdala, and prefrontal cortex, all of which contain CPEB (Berger-Sweeney et al. 2006). Furthermore, microarray analysis identified several mRNAs whose levels in the hippocampus differed in the CPEB knockout mouse compared to wild type (Berger-Sweeney et al. 2006); such mRNAs could be downstream from CPEB regulation. Finally, CPEB is present in invertebrate neurons as well, and these cells also support cytoplasmic polyadenylation (Liu and Schwartz 2003). In *Aplysia* sensory neurons in which CPEB RNA has been ablated by an antisense oligonucleotide, LTF is not properly maintained (Si et al. 2003b). Collectively, these results suggest that CPEB-mediated translation could result in the establishment of a "tag" at synapses and that the recognition of this tag by neurons is important for synaptic plasticity and, as a consequence, memory consolidation.

The form of CPEB in *Aplysia* and other invertebrate neurons differs from that in vertebrate neurons in that it contains a long stretch of polyglutamine. Polyglutamine is often found in proteins that have characteristics of a prion, i.e., an infectious agent consisting entirely of protein that is self-reproducing. This observation, plus the fact that CPEB RNA is detected in *Aplysia* neurons prompted Si et al. (2003a,b) to suggest that CPEB might assume a prion-like structure following synaptic stimulation, thereby forming an indelible mark at synapses. In such a scenario, CPEB itself might comprise the synaptic tag, rather than mediating the synthesis of a protein at synapses that constitutes the tag. These investigators showed that *Aplysia* CPEB had some features of a prion in vitro, such as resistance to protease and fast sedimentation rate in sucrose gradients. However, the most compelling evidence comes from work with yeast, where Si et al. (2003b) demonstrated that the *Aplysia* CPEB could assume two forms: One that is aggregated (i.e., prion-like) and one that is not. Surprisingly, these investigators showed that the aggregated form of CPEB is the one that binds RNA in vitro and that the aggregated form of the protein can convert the nonaggregated form into an aggregated form. Such epigenetic inheritance is a fundamental hallmark of prion formation. These authors speculate that synaptic stimulation causes CPEB to assume a prion-like form and that such a form either stimulates the translation of some RNAs, causes it to alter its substrate specificity, or releases some mRNAs from an inhibited state. Most importantly, the authors suggest that once in a prion-like form, CPEB needs no further stimulation (i.e., by kinases) to maintain its activity. Beyond doubt, a demonstration that CPEB does indeed form a prion-like structure in neurons would constitute a major advance in neuroscience.

If polyglutamine-containing CPEB forms a prion-like form in invertebrate neurons, then what about the polyglutamine-lacking CPEB in vertebrate neurons? Vertebrates contain three additional CPEB-like genes (Mendez and Richter 2001), all of which are expressed in the brain (Theis et al. 2003). Two of these other CPEB-like proteins do contain polyglutamine, although they are not nearly as long as that in the *Aplysia* CPEB. Whether these other CPEB-like proteins form prion-like structures, or indeed even regulate polyadenylation or translation, is unknown at this time.

FRAGILE X MENTAL RETARDATION SYNDROME AND FMRP

It has been estimated that one in approximately 4000 males and one in about 8000 females has the fragile X syndrome, a heritable form of mental retardation most obviously manifested by mild to severe mental retardation and connective tissue abnormalities. The fragile X gene is one that is subject to expansion by CGG triplets; fewer than 60 produce little evidence of the disease, whereas more than 200 lead to the most severe phenotypes. The repeat expansion occurs in an exon that gives rise to the mRNA 5'UTR (Gatchel and Zoghbi 2005) and the greater the number of repeats, the less FMRP (the protein derived from the fragile X gene) is produced. FMRP is widely expressed in animal tissues and shuttles between nucleus and cytoplasm; in neurons of the central nervous system (CNS), it is found postsynaptically. FMRP is a complex RNA-binding protein containing three KH domains (i.e., has homology with the RNA-binding region of heterogeneous nuclear RNP K) and an RGG (arginine-glycine-glycine) box. These observations suggest that FMRP might regulate local, synaptic mRNA translation and that the loss of this regulation could result in fragile X mental retardation. Perhaps the most compelling evidence that this is the case is the observation that an individual with a point mutation (I304N) in the second KH domain, which disrupts RNA binding in vitro (Siomi et al. 1994), has a severe fragile X phenotype (De Boulle et al. 1993; O'Donnell and Warren 2002; Bagni and Greenough 2005; Darnell et al. 2005).

How does FMRP regulate translation, and is the regulation positive or negative? The answers are not clear-cut, in part because of substantial discordance in the experimental data. First, the seemingly straightforward issue as to whether FMRP associates with polyribosomes is controversial. Some investigators find that the protein sediments with polysomes (Feng et al. 1997; Khandjian et al. 2004; Stefani et al. 2004), whereas others find that it sediments with nontranslating (i.e., slowly sedimenting, $\leq$80s) RNPs (Siomi et al. 1996; Zalfa et al. 2003), and yet others find that similar amounts are present in both fractions (Brown et al. 2001). Variables such as the type of detergent and/or ionic strength of the solutions used in cell fractionation, source of the biological material, and even age of the animals all contribute to this uncertainty; therefore, a general consensus as to the molecular function of this protein has not yet emerged. Second, several screens to identify mRNA targets of FMRP have yielded minimally overlapping sequence sets, again owing to differences in methodologies and sources of material used for the identification (Brown et al. 1998; Mayashiro et al. 2003).

Two observations suggest that FMRP regulation of translation might involve microRNAs (miRNAs) (for a discussion of miRNAs, see Chapter 11). First, immunoprecipitation of tandem affinity-tagged FMRP from *Drosophila* cells (notated as dFMR1) coimmunoprecipitated argonaute-2, a component of the RISC (RNA-induced silencing complex) (Ishizuka et al. 2002). Second, an analysis of RISC components in *Drosophila* cells revealed the presence of FMRP (Caudy et al. 2002). These observations do not necessarily imply a function for FMRP in the RISC complex, but it is compelling that several laboratories have reported that both FMRP (see above) and miRNAs are associated with polysomes (Olsen and Ambros 1999; Kim et al. 2004; Nelson et al. 2004), although this contention has also been disputed (Pillai et al. 2005). In *Drosophila*, Jin et al. (2004a,b) found that the neuromuscular junction is normal when dFMR-2 or argonaute-1 is heterozygous, but that there is a synaptic overgrowth phenotype when both genes are heterozygous in the same fly. Coupled with the observations that dFMR-1 is a component of RISC, this genetic interaction implies that these two proteins interact functionally and suggests that dFMR-1 might work with miRNAs to control translation.

Although FMRP-associated miRNAs have not yet been identified, another small RNA, unrelated to miRNA or other components of the RNA interference (RNAi) pathway, appears to be involved with FMRP-controlled translation. Zalfa et al. (2003) found that FMRP-mediated translational repression of αCaMKII mRNA at synapses of the mouse brain involves BC1 RNA, a small noncoding RNA long known to be dendritic (Chicurel et al. 1993). BC1 not only interacts with FMRP, but base-pairs with αCaMKII RNA, leading to the postulate that BC1-containing duplexes are promoted by FMRP (Bagni and Greenough 2005). How such an activity might cause translational repression is unknown. In contrast to these results, however, Wang et al. (2005) could detect no significant BC1 RNA–FMRP interaction. Rather, these authors found that BC1 RNA inhibits 48S initiation complex formation through associations with both eIF4A and poly(A)-binding protein. Under these circumstances, BC1 would probably act as a general inhibitor of translation, rather than an mRNA-specific inhibitor as suggested by Zalfa et al. (2003).

In summary, although most investigators in the fragile X field agree that one function of FMRP is to repress translation, the mechanism by which it does so has not been elucidated. On the other hand, that FMRP has a critical role in synaptic plasticity is seemingly beyond doubt. Huber et al. (2002) demonstrated that FMRP knockout mice have an enhanced mGluR-LTD, which, as noted previously, is protein-synthesis-dependent. How important are mGluRs in fragile X? Very important, at least as far as fruit flies are concerned. Mutant flies lacking the *dfmr1* gene have abnormal courtship behavior, poor memory, and structural anomalies in a part of the brain known as the mushroom bodies (Zhang et al. 2001; Dockendorff et al. 2002; Morales et al. 2002; Lee et al. 2003). The ingestion of drugs that antagonize mGluR signaling in mammals rescues all of these phenotypes (McBride et al. 2005)! Clearly, such results offer possible therapies for treating human fragile X mental retardation (Bear 2005; Dolen and Bear 2005).

PERSPECTIVES

One intriguing insight from the studies of translational control in synaptic plasticity conducted thus far is that the basic signal transduction cascades that are required to initiate translation are similar, regardless of whether the plasticity induced results in either LTP or LTD. In some ways, this makes sense. To synthesize the new proteins required for the expression of either LTP or LTD, translation must be initiated. Kelleher et al. (2004a) have suggested that long-lasting forms of LTP and LTD elicit a global regulation of protein synthesis via stimulation of these core signal transduction cascades that initiate translation. How then do LTP-specific and LTD-specific proteins get synthesized? As discussed earlier, one possibility is gene-specific regulation via mRNA-binding proteins such as CPEB and FMRP. Interestingly, CPEB (Huang et al. 2003) and FMRP (Rackham and Brown 2004; Antar et al. 2005), as well as the double-stranded RNA-binding protein staufen (Kiebler et al. 1999; Tang et al.

2001; Kanai et al. 2004), are all involved in RNA transport in dendrites. How these proteins transport mRNAs (presumably untranslated) is of considerable interest and may involve a large number of factors including molecular motors (Krichevsky and Kosik 2001; Kanai et al. 2004). A thorough review of RNA transport in neurons has been published recently (Kindler et al. 2005).

Whether either LTP or LTD is ultimately expressed must be due to either a difference in the synthesis of a small number of specific proteins or the creation of specific synaptic tags, one for LTP and another for LTD, that would permit the "capture" of distinct proteins from a large, homogeneous pool of newly synthesized plasticity proteins (Kelleher et al. 2004a). If the latter is true, one would predict that LTP and LTD would induce the synthesis of a similar pool of proteins that would be sufficient to enable the opposite type of plasticity (i.e., LTP would enable LTD, and LTD would enable LTP). In fact, this type of heterosynaptic associativity between LTP and LTD has been demonstrated (Sajikumar and Frey 2004). The idea that LTP and LTD result in the synthesis of similar sets of proteins remains to be determined, perhaps with either microarray studies of polysome fractions or proteomic analysis.

Reviews recount the past, and thus by their nature are a recitation of facts, or what pass for facts until superseded by new results that can destroy even central dogmas; imagining the future is the challenge. One now established "fact" is that local protein synthesis is required for synaptic plasticity and that modifiers of mRNA translation in neurons such as 4E-BP, CPEB, and FMRP are regulators of this process. What remains a black box, as mentioned above, are the identities of the proteins that are synthesized locally *and* that are modifiers of synaptic efficacy. Although several avenues of investigation will no doubt be necessary to reveal the full panoply of molecules involved, one intriguing route has recently been taken by Schratt et al. (2006), who demonstrate that a specific miRNA, mir-134, is not only localized to the synaptodendritic compartment, but that its regulation of Lim1 kinase (Limk1) mRNA expression controls synaptic spine development. Limk1 regulates the actin cytoskeleton by phosphorylating and inactivating cofilin, an actin-depolymerizing molecule; Limk1 knockout mice display enhanced hippocampal LTP (Meng et al. 2002). Given that there are nearly 100 miRNAs in neurons (Kim et al. 2004) that are regulated throughout development (Krichevsky et al. 2003), it seems reasonable to speculate that they will be found to be important modulators of local translation and synaptic plasticity. Moreover, the identities of the mRNAs to which the miRNAs anneal will help establish the molecular framework of tagging and plasticity, and learning and memory as well.

REFERENCES

Aakalu G., Smith W.B., Nguyen N., Jiang C., and Schuman E.M. 2001. Dynamic visualization of local protein synthesis in hippocampal neurons. *Neuron* **30:** 489–502.

Abel T. and Lattal K.M. 2001. Molecular mechanisms of memory acquisition, consolidation and retrieval. *Curr. Opin. Neurobiol.* **11:** 180–187.

Abel T., Martin K.C., Bartsch D., and Kandel E.R. 1998. Memory suppressor genes: Inhibitory constraints on the storage of long-term memory. *Science* **279:** 338–341.

Alarcon J.M., Hodgman R., Theis M., Huang Y.S., Kandel E.R., and Richter J.D. 2004. Selective modulation of some forms of schaffer collateral-CA1 synaptic plasticity in mice with a disruption of the CPEB-1 gene. *Learn. Mem.* **11:** 318–327.

Antar L.N., Dictenberg J.B., Plociniak M., Afroz R., and Bassell G.J. 2005. Localization of FMRP-associated mRNA granules and requirement of microtubules for activity-dependent trafficking in hippocampal neurons. *Genes Brain Behav.* **4:** 350–359.

Atkins C.M., Nozaki N., Shigeri Y., and Soderling T.R. 2004. Cytoplasmic polyadenylation element binding protein-dependent protein synthesis is regulated by calcium/calmodulin-dependent protein kinase II. *J. Neurosci.* **24:** 5193–5201.

Bagni C. and Greenough W.T. 2005. From mRNP trafficking to spine dysmorphogenesis: The roots of fragile X syndrome. *Nat. Rev. Neurosci.* **6:** 376–387.

Banko J.L., Hou L., and Klann E. 2004. NMDA receptor activation results in PKA- and ERK-dependent Mnk1 activation and increased eIF4E phosphorylation in hippocampal area CA1. *J. Neurochem.* **91:** 462–470.

Banko J.L., Hou L., Poulin F., Sonenberg N., and Klann E. 2006. Regulation of eukaryotic initiation factor 4E by converging signaling pathways during metabotropic glutamate receptor-dependent long-term depression. *J. Neurosci.* **26:** 2167–2173.

Banko J.L., Poulin F., Hou L., DeMaria C.T., Sonenberg N., and Klann E. 2005. The translation repressor 4E-BP2 is critical for eIF4F complex formation, synaptic plasticity, and memory in the hippocampus. *J. Neurosci.* **25:** 9581–9590.

Barnard D.C., Ryan K., Manley J.L., and Richter J.D. 2004. Symplekin and xGLD-2 are required for CPEB-mediated cytoplasmic polyadenylation. *Cell* **119:** 641–651.

Bartsch D., Ghirardi M., Skehel P.A., Karl K.A., Herder S.P., Chen M., Bailey C.H., and Kandel E.R. 1995. *Aplysia* CREB2 represses long-term facilitation: Relief of repression converts transient facilitation into long-term functional and structural change. *Cell* **83:** 979–992.

Bear M.F. 2005. Therapeutic implications of the mGluR theory of fragile X mental retardation. *Genes Brain Behav.* **4:** 393–398.

Beaumont V., Zhong N., Fletcher R., Froemke R.C., and Zucker R.S. 2001. Phosphorylation and local presynaptic protein synthesis in calcium- and calcineurin-dependent induction of crayfish long-term facilitation. *Neuron* **32:** 489–501.

Berger-Sweeney J., Zearfoss N.R., and Richter J.D. 2006. Reduced extinction of hippocampal-dependent memories in CPEB knockout mice. *Learn. Mem.* **13:** 4–7.

Bradshaw K.D., Emptage N.J., and Bliss T.V. 2003. A role for dendritic protein synthesis in hippocampal late LTP. *Eur. J. Neurosci.* **18:** 3150–3152.

Brown V., Small K., Lakkis L., Feng Y., Gunter C., Wilkinson K.D., and Warren S.T. 1998. Purified recombinant Fmrp exhibits selective RNA binding as an intrinsic property of the fragile X mental retardation protein. *J. Biol. Chem.* **273:** 15521–15527.

Brown V., Jin P., Ceman S., Darnell J.C., O'Donnell W.T., Tenenbaum S.A., Jin X., Feng Y., Wilkinson K.D., Keene J.D., et al. 2001. Microarray identification of FMRP-associated brain mRNAs and altered mRNA translational profiles in fragile X syndrome. *Cell* **107:** 477–487.

Cao Q. and Richter J.D. 2002. Dissolution of the maskin-eIF4E complex by cytoplasmic polyadenylation and poly(A)-binding protein controls cyclin B1 mRNA translation and oocyte maturation. *EMBO J.* **21:** 3852–3862.

Cao Q., Huang Y.S., Kan M.C., and Richter J.D. 2005. Amyloid precursor proteins anchor CPEB to membranes and promote polyadenylation-induced translation. *Mol. Cell. Biol.* **25:** 10930–10939.

Casadio A., Martin K.C., Giustetto M., Zhu H., Chen M., Bartsch D., Bailey C.H., and Kandel E.R. 1999. A transient neuron-wide form of CREB-mediated long-term facilitation can be stabilized at specific synapses by local protein synthesis. *Cell* **99:** 221–237.

Castellucci V.F., Blumenfeld H., Goelet P., and Kandel E.R. 1989. Inhibitor of protein synthesis blocks long-term behavioral sensitization in the isolated gill-withdrawal reflex of *Aplysia. J. Neurobiol.* **20:** 1–9.

Caudy A.A., Myers M., Hannon G.J., and Hammond S.M. 2002. Fragile X-related protein and VIG associate with the RNA interference machinery. *Genes Dev.* **16:** 2491–2496.

Chen A., Muzzio I.A., Malleret G., Bartsch D., Verbitsky M., Pavlidis P., Yonan A.L., Vronskaya S., Grody M.B., Cepeda I., et al. 2003. Inducible enhancement of memory storage and synaptic plasticity in transgenic mice expressing an inhibitor of ATF4 (CREB-2) and C/EBP proteins. *Neuron* **39:** 655–669.

Chicurel M.E., Terrian D.M., and Potter H. 1993. mRNA at the synapse: Analysis of a preparation enriched in hippocampal dendritic spines. *J. Neurosci.* **13:** 4054–4063.

Costa-Mattioli M., Gobert D., Harding H., Herdy B., Azzi M., Bruno M., Bidinosti M., Ben Mamou C., Marcinkiewicz E., Yoshida M., et al. 2005. Translational control of hippocampal synaptic plasticity and memory by the eIF2α kinase GCN2. *Nature* **436:** 1166–1170.

Cracco J.B., Serrano P., Moskowitz S.I., Bergold P.J., and Sacktor T.C. 2005. Protein synthesis-dependent LTP in isolated dendrites of CA1 pyramidal cells. *Hippocampus* **15:** 551–556.

Crino P.B. and Eberwine J. 1996. Molecular characterization of the dendritic growth cone: Regulated mRNA transport and local protein synthesis. *Neuron* **17:** 1173–1187.

Darnell J.C., Mostovetsky O., and Darnell R.B. 2005. FMRP RNA targets: Identification and validation. *Genes Brain Behav.* **4:** 341–349.

De Boulle K., Verkerk A.J., Reyniers E., Vits L., Hendrickx J., Van Roy B., Van den Bos F., de Graaff E., Oostra B.A., and Willems P.J. 1993. A point mutation in the FMR-1 gene associated with fragile X mental retardation. *Nat. Genet.* **3:** 31–35.

Dockendorff T.C., Su H.S., McBride S.M., Yang Z., Choi C.H., Siwicki K.K., Sehgal A., and Jongens T.A. 2002. *Drosophila* lacking dfmr1 activity show defects in circadian output and fail to maintain courtship interest. *Neuron* **34:** 973–984.

Dolen G. and Bear M.F. 2005. Courting a cure for fragile X. *Neuron* **45:** 642–644.

Du L. and Richter J.D. 2005. Activity-dependent polyadenylation in neurons. *RNA* **11:** 1340–1347.

Dudek S.M. and Bear M.F. 1992. Homosynaptic long-term depression in area CA1 of hippocampus and effects of N-methyl-D-aspartate blockade. *Proc. Natl. Acad. Sci.* **89:** 4363–4367.

Dyer J.R. and Sossin W.S. 2000. Regulation of eukaryotic initiation factor 4E phosphorylation in the nervous system of *Aplysia californica. J. Neurochem.* **75:** 872–881.

Faneslow M.S. 1984. Shock-induced analgesia on the formalin test: Effects of shock severity, naloxone, hypophysectomy, and associative variables. *Behav. Neurosci.* **98:** 79–95.

Feng Y., Absher D., Eberhart D.E., Brown V., Malter H.E., and Warren S.T. 1997. FMRP associates with polyribosomes as an mRNP, and the I304N mutation of severe fragile X syndrome abolishes this association. *Mol. Cell* **1:** 109–118.

Flexner J.B., Flexner L.B., and Stellar E. 1963. Memory in mice is affected by intracerebral puromycin. *Science* **141:** 57–59.

Fox C.A, Sheets M.D., and Wickens M.P. 1989. Poly(A) addition during maturation of frog oocytes: Distinct nuclear and cytoplasmic activities and regulation by the sequence UU-UUUAU. *Genes Dev.* **3:** 2151–2162.

Frey U., Huang Y.Y., and Kandel E.R. 1993. Effects of cAMP simulate a late stage of LTP in hippocampal CA1 neurons. *Science* **260:** 1661–1664.

Frey U., Krug M., Reymann K.G., and Matthies H. 1988. Anisomycin, an inhibitor of protein synthesis, blocks late phases of LTP phenomena in the hippocampal CA1 region *in vitro. Brain Res.* **452:** 57–65.

Gatchel J.R. and Zoghbi H.Y. 2005. Diseases of unstable repeat expansion: Mechanisms and common principles. *Nat. Rev. Genet.* **6:** 743–755.

Gingras A.C., Kennedy S.G., O'Leary M.A., Sonenberg N., and Hay N. 1998. 4E-BP1, a repressor of mRNA translation, is phosphorylated and inactivated by the Akt(PKB) signaling pathway. *Genes Dev.* **12:** 502–513.

Hake L.E. and Richter J.D. 1994. CPEB is a specificity factor that mediates cytoplasmic polyadenylation during *Xenopus* oocyte maturation. *Cell* **79:** 617–627.

Hake L.E., Mendez R., and Richter J.D. 1998. Specificity of RNA binding by CPEB: Requirement for RNA recognition motifs and a novel zinc finger. *Mol. Cell. Biol.* **18:** 685–693.

Harding H.P., Novoa I., Zhang Y., Zeng H., Wek R., Schapira M., and Ron D. 2000. Regulated translation initiation controls stress-induced gene expression in mammalian cells. *Mol. Cell* **6:** 1099–1108.

Hou L. and Klann E. 2004. Activation of the phosphoinositide 3-kinase-Akt-mammalian target of rapamycin signaling pathway is required for metabotropic glutamate receptor-dependent long-term depression. *J. Neurosci.* **24:** 6352–6361.

Huang C.C., Lee C.C., and Hsu K.S. 2004. An investigation into signal transduction mechanisms involved in insulin-induced long-term depression in the CA1 region of the hippocampus. *J. Neurochem.* **89:** 217–231.

Huang Y.S., Carson J.H., Barbarese E., and Richter J.D. 2003. Facilitation of dendritic mRNA transport by CPEB. *Genes Dev.* **17:** 638–653.

Huang Y.S., Jung M.Y., Sarkissian M., and Richter J.D. 2002. N-methyl-D-aspartate receptor signaling results in Aurora kinase-catalyzed CPEB phosphorylation and alpha CaMKII mRNA polyadenylation at synapses. *EMBO J.* **21:** 2139–2134.

Huang Y.Y. and Kandel E.R. 1994. Recruitment of long-lasting and protein kinase A-dependent long-term potentiation in the CA1 region of the hippocampus. *Learn. Mem.* **1:** 74–82.

———. 1995. D1/D5 receptor agonists induce a protein synthesis-dependent late potentiation in the CA1 region of the hippocampus. *Proc. Natl. Acad. Sci.* **92:** 2446–2450.

Huber K.M., Kayser M.S., and Bear M.F. 2000. Role for rapid dendritic protein synthesis in hippocampal mGluR-dependent long-term depression. *Science* **288:** 1254–1257.

Huber K.M., Gallagher S.M., Warren S.T., and Bear M.F. 2002. Altered synaptic plasticity in a mouse model of fragile X mental retardation. *Proc. Natl. Acad. Sci.* **99:** 7746–7750.

Ishizuka A., Siomi M.C., and Siomi H. 2002. A *Drosophila* fragile X protein interacts with components of RNAi and ribosomal proteins. *Genes Dev.* **16:** 2497–2508.

Jin P., Alisch R.S., and Warren S.T. 2004a. RNA and microRNAs in fragile X mental retardation. *Nat. Cell Biol.* **6:** 1048–1053.

Jin P., Zarnescu D.C., Ceman S., Nakamoto M., Mowrey J., Jongens T.A., Nelson D.L., Moses K., and Warren S.T. 2004b. Biochemical and genetic interaction between the fragile X mental retardation protein and the microRNA pathway. *Nat. Neurosci.* **7:** 113–117.

Ju W., Morishita W., Tsui J., Gaietta G., Deernick T.J., Adams S.R., Garner C.C., Tsien R.Y., Ellisman M.H., and Malenka R.C. 2004. Activity-dependent regulation of dendritic synthesis and trafficking of AMPA receptors. *Nat. Neurosci.* **7:** 244–253.

Kanai Y., Dohmae N., and Hirokawa N. 2004. Kinesin transports RNA: Isolation and characterization of an RNA-transporting granule. *Neuron* **43:** 513–525.

Kandel E.R. 2001. The molecular biology of memory storage: A dialogue between genes and synapses. *Science* **294:** 1030–1038.

Kang H. and Schuman E.M. 1996. A requirement for local protein synthesis in neurotrophin-induced hippocampal synaptic plasticity. *Science* **273:** 1402–1406.

Kauderer B.S. and Kandel E.R. 2000. Capture of a protein synthesis-dependent component of long-term depression. *Proc. Natl. Acad. Sci.* **97:** 13342–13347.

Kelleher R.J., III, Govindarajan A., and Tonegawa S. 2004a. Translational regulatory mechanisms in persistent forms of synaptic plasticity. *Neuron* **44:** 59–73.

Kelleher R.J., III, Govindarajan A., Jung H.Y., Kang H., and Tonegawa S. 2004b. Translational control by MAPK signaling in long-term synaptic plasticity and memory. *Cell* **116:** 1–20.

Kelly A. and Lynch M.A. 2000. Long-term potentiation in dentate gyrus of the rat is inhibited by the phosphoinositide 3-kinase inhibitor wortmannin. *Neuropharmacology* **39:** 643–651.

Khandjian E.W., Huot M.E., Tremblay S., Davidovic L., Mazroui R., and Bardoni B. 2004. Biochemical evidence for the association of fragile X mental retardation protein with brain polyribosomal ribonucleoparticles. *Proc. Natl. Acad. Sci.* **101:** 13357–13362.

Kiebler M.A, Hemraj I., Verkade P., Kohrmann M., Fortes P., Marion R.M., Ortin J., and Dotti C.G. 1999. The mammalian staufen protein localizes to the somatodendritic domain of cultured hippocampal neurons: Implications for its involvement in mRNA transport. *J. Neurosci.* **19:** 288–297.

Kim J., Krichevsky A., Grad Y., Hayes G.D., Kosik K.S., Church G.M., and Ruvkun G. 2004. Identification of many microRNAs that copurify with polyribosomes in mammalian neurons. *Proc. Natl. Acad. Sci.* **101:** 360–365.

Kindler S., Wang H., Richter D., and Tiedge H. 2005. RNA transport and local control of translation. *Annu. Rev. Cell Dev. Biol.* **21:** 223–245.

Krichevsky A.M. and Kosik K.S. 2001. Neuronal RNA granules: A link between RNA localization and stimulation-dependent translation. *Neuron* **32:** 683–696.

Krichevsky A.M., King K.S., Donahue C.P., Khrapko K., and Kosik K.S. 2003. A microRNA array reveals extensive regulation of microRNAs during brain development. *RNA* **9:** 1274–1281.

Krug M., Loessner B., and Ott T. 1984. Anisomycin blocks the late phase of long-term potentiation in the dentate gyrus of freely moving rats. *Brain Res. Bull.* **13:** 39–42.

Lee A., Li W., Xu K., Bogert B.A., Su K., and Gao F.B. 2003. Control of dendritic development by the *Drosophila* fragile X-related gene involves the small GTPase Rac1. *Development* **130:** 5543–5552.

Liu J. and Schwartz J.H. 2003. The cytoplasmic polyadenylation element binding protein and polyadenylation of messenger RNA in *Aplysia* neurons. *Brain Res.* **959:** 68–76.

Martin K.C., Casadio A., Zhu H., Yaping E., Rose J.C., Chen M., Bailey C.H., and Kandel E.R. 1997. Synapse-specific, long-term facilitation of *Aplysia* sensory to motor synapses: A function for local protein synthesis in memory storage. *Cell* **91:** 927–938.

Martinez S.E., Yuan L., Lacza C., Ransom H., Mahon G.M., Whitehead I.P., and Hake L.E. 2005. XGef mediates early CPEB phosphorylation during *Xenopus* oocyte meiotic maturation. *Mol. Biol. Cell* **16:** 1152–1164.

McBride S.M., Choi C.H., Wang Y., Liebelt D., Braunstein E., Ferreiro D., Sehgal A., Siwicki K.K., Dockendorff T.C., Nguyen H.T., et al. 2005. Pharmacological rescue of synaptic plasticity, courtship behavior, and mushroom body defects in a *Drosophila* model of fragile X syndrome. *Neuron* **45:** 753–764.

McGaugh J.L. 2000. Memory—A century of consolidation. *Science* **287:** 248–251.

McGrew L.L., Dworkin-Rastl E., Dworkin M.B., and Richter J.D. 1989. Poly(A) elongation during *Xenopus* oocyte maturation is required for translational recruitment and is mediated by a short sequence element. *Genes Dev.* **3:** 803–815.

Mendez R. and Richter J.D. 2001. Translational control by CPEB: A means to the end. *Nat. Rev. Mol. Cell Biol.* **2:** 521–529.

Mendez R., Murthy K.G., Ryan K., Manley J.L., and Richter J.D. 2000a. Phosphorylation of CPEB by Eg2 mediates the recruitment of CPSF into an active cytoplasmic polyadenylation complex. *Mol. Cell* **6:** 1253–1259.

Mendez R., Hake L.E., Andresson T., Littlepage L.E., Ruderman J.V., and Richter J.D. 2000b. Phosphorylation of CPE binding factor by Eg2 regulates translation of c-mos mRNA. *Nature* **404:** 302–307.

Meng Y., Zhang Y., Tregoubov V., Janus C., Cruz L., Jackson M., Lu W., MacDonald J.F., Wang J.Y., Falls D.L., and Jia Z. 2002. Abnormal spine morphology and enhanced LTP in LimK1 knockout mice. *Neuron* **35:** 121–133.

Minshall N., Thom G., and Standart N. 2001. A conserved role of a DEAD box helicase in mRNA masking. *RNA* **7:** 1728–1742.

Miyashiro K.Y., Beckel-Mitchener A., Purk T.P., Becker K.G., Barret T., Liu L., Carbonetto S., Weiler I.J., Greenough W.T., and Eberwine J. 2003. RNA cargoes associating with FMRP reveal deficits in cellular functioning in Fmr1 null mice. *Neuron* **37:** 417–431.

Morales J., Hiesinger P.R., Schroeder A.J., Kume K., Verstreken P., Jackson F.R., Nelson D.A., and Hassan B.A. 2002. *Drosophila* fragile X protein, DFXR, regulates neuronal morphology and function in the brain. *Neuron* **34:** 961–972.

Morris R. 1984. Developments of a water-maze procedure for studying spatial learning in the rat. *J. Neurosci. Methods* **11:** 47–60.

Mulkey R.M. and Malenka R.C. 1992. Mechanisms underlying induction of homosynaptic long-term depression in area CA1 of the hippocampus. *Neuron* **9:** 967–975.

Nelson P.T., Hatzigeorgiou A.G., and Mourelatos Z. 2004. miRNP:mRNA association in polyribosomes in a human neuronal cell line. *RNA* **10:** 387–394.

Nguyen P.V. and Kandel E.R. 1997. Brief theta burst stimulation induced a transcription-dependent late phase of LTP requiring cAMP in area CA1 of the mouse hippocampus. *Learn. Mem.* **4:** 230–243.

Nguyen P.V., Abel T., and Kandel E.R. 1994. Requirement of a critical period of transcription for induction of a late phase of LTP. *Science* **265:** 1104–1107.

O'Donnell W.T. and Warren S.T. 2002. A decade of molecular studies of fragile X syndrome. *Annu. Rev. Neurosci.* **25:** 315–338.

Olsen P.H. and Ambros V. 1999. The lin-4 regulatory RNA controls developmental timing in *Caenorhabditis elegans* by blocking LIN-14 protein synthesis after the initiation of translation. *Dev. Biol.* **216:** 671–680.

Opazo P., Watabe A.M., Grant S.G., and O'Dell T.J. 2003. Phosphatidylinositol 3-kinase regulates the induction of long-term potentiation through extracellular signal-regulated kinase-independent mechanisms. *J. Neurosci.* **23:** 3679–3688.

Phillips R.G. and LeDoux J.E. 1992. Differential contribution of amygdala and hippocampus to cued and contextual fear conditioning. *Behav. Neurosci.* **106:** 274–285.

Pillai R.S., Bhattacharyya S.N., Artus C.G., Zoller T., Cougot N., Basyuk E., Bertrand E., and Filipowicz W. 2005. Inhibition of translational initiation by let-7 microRNA in human cells. *Science* **309:** 1573–1576.

Rackham O. and Brown C.M. 2004. Visualization of RNA-protein interactions in living cells: FMRP and IMP1 interact on mRNAs. *EMBO J.* **23:** 3346–3355.

Read R.L., Martinho R.G., Wang S.W., Carr A.M., and Norbury C.J. 2002. Cytoplasmic poly(A) polymerases mediate cellular responses to S phase arrest. *Proc. Natl. Acad. Sci.* **99:** 12079–12084.

Richter J.D. and Sonenberg N. 2005. Regulation of cap-dependent translation by eIF4E inhibitory proteins. *Nature* **433:** 477–480.

Saitoh S., Chabes A., McDonald W.H., Thelander L., Yates J.R., and Russell P. 2002. Cid13 is a cytoplasmic poly(A) polymerase that regulates ribonucleotide reductase mRNA. *Cell* **109:** 563–573.

Sajikumar S. and Frey J.U. 2004. Late-associativity, synaptic tagging, and the role of dopamine during LTP and LTD. *Neurobiol. Learn. Mem.* **82:** 12–25.

Sanna P.P., Cammalleri M., Berton F., Simpson C., Lutjens R., Bloom F.E., and Francesconi W. 2002. Phosphatidylinositol 3-kinase is required for the expression but not for the induction or the maintenance of long-term potentiation in the hippocampal CA1 region. *J. Neurosci.* **22:** 3359–3365.

Schenk F. and Morris R.G. 1985. Dissociation between components of spatial memory in rats after recovery from the effects of retrohippocampal lesions. *Exp. Brain Res.* **58:** 11–28.

Schmitt J.M., Guire E.S., Saneyoshi T., and Soderling T.R. 2005. Calmodulin-dependent kinase kinase/calmodulin kinase I activity gates extracellular-regulated kinase-dependent long-term potentiation. *J. Neurosci.* **25:** 1281–1290.

Schratt G.M., Tuebing F., Nigh E.A., Kane C.G., Sabatini M.E., Kiebler M., and Greenberg M.E. 2006. A brain-specific microRNA regulates dendritic spine development. *Nature* **439:** 283–289.

Shin C.Y., Kundel M., and Wells D.G. 2004. Rapid, activity-induced increase in tissue plasminogen activator is mediated by metabotropic glutamate receptor-dependent mRNA translation. *J. Neurosci.* **24:** 9425–9433.

Si K., Lindquist S., and Kandel E.R. 2003a. A neuronal isoform of the *Aplysia* CPEB has prion-like properties. *Cell* **115:** 879–891.

Si K., Giustetto M., Etkin A., Hsu R., Janisiewicz A.M., Miniaci M.C., Kim J.H., Zhu H., and Kandel E.R. 2003b. A neuronal isoform of CPEB regulates local protein synthesis and stabilizes synapse-specific long-term facilitation in *Aplysia*. *Cell* **115:** 893–904.

Siomi H., Choi M., Siomi M.C., Nussbaum R.L., and Dreyfuss G. 1994. Essential role for KH domains in RNA binding: Impaired RNA binding by a mutation in the KH domain of FMR1 that causes fragile X syndrome. *Cell* **77:** 33–39.

Siomi M.C., Zhang Y., Siomi H., and Dreyfuss G. 1996. Specific sequences in the fragile X syndrome protein FMR1 and the FXR proteins mediate their binding to 60S ribosomal subunits and the interactions among them. *Mol. Cell. Biol.* **16:** 3825–3832.

Stebbins-Boaz B., Cao Q., de Moor C.H., Mendez R., and Richter J.D. 1999. Maskin is a CPEB-associated factor that transiently interacts with eIF-4E. *Mol. Cell* **4:** 1017–1027.

Stefani G., Fraser C.E., Darnell J.C., and Darnell R.B. 2004. Fragile X mental retardation protein is associated with translating polyribosomes in neuronal cells. *J. Neurosci.* **24:** 7272–7276.

Steward O. and Levy W.B. 1982. Preferential localization of polyribosomes under the base of dendrite spines in granule cells of the dentate gyrus. *J. Neurosci.* **2:** 284–291.

Sutton M.A., Wall N.R., Aakulu G.N., and Schuman E.M. 2004. Regulation of dendritic protein synthesis by miniature synaptic events. *Science* **304:** 1979–1983.

Sweatt J.D. 2004. Mitogen-activated protein kinases in synaptic plasticity and memory. *Curr. Opin. Neurobiol.* **14:** 311–317.

Tang S.J., Meulemans D., Vazquez L., Colaco N., and Schuman E. 2001. A role for a rat homolog of staufen in the transport of RNA to neuronal dendrites. *Neuron* **32:** 463–475.

Tang S.J., Reis G., Kang H., Gingras A.C., Sonenberg N., and Schuman E.M. 2002. A rapamycin-sensitive signaling pathway contributes to long-term synaptic plasticity in the hippocampus. *Proc. Natl. Acad. Sci.* **99:** 467–472.

Theis M., Si K., and Kandel E.R. 2003. Two previously undescribed members of the mouse CPEB family of genes and their inducible expression in the principal cell layers of the hippocampus. *Proc. Natl. Acad. Sci.* **100:** 9602–9607.

Thomas G.M. and Huganir R.L. 2004. MAPK cascade signalling and synaptic plasticity. *Nat. Rev. Neurosci.* **5:** 173–183.

Tsokas P., Grace E.A., Chan P., Ma T., Sealfon S.C., Iyengar R., Landau E.M., and Blitzer R.D. 2005. Local protein synthesis mediates a rapid increase in dendritic elongation factor 1A after induction of late long-term potentiation. *J. Neurosci.* **25:** 5833–5843.

Vattem K.M. and Wek R.C. 2004. Reinitiation involving upstream ORFs regulates ATF4 mRNA translation in mammalian cells. *Proc. Natl. Acad. Sci.* **101:** 11269–11274.

Vickers C.A., Dickson K.S., and Wyllie D.J. 2005. Induction and maintenance of late-phase long-term potentiation in isolated dendrites of rat hippocampal CA1 pyramidal neurons. *J. Physiol.* **568:** 803–813.

Wakiyama M., Imataka H., and Sonenberg N. 2000. Interaction of eIF4G with poly(A)-binding protein stimulates translation and is critical for *Xenopus* oocyte maturation. *Curr. Biol.* **10:** 1147–1150.

Wang H., Iacoangeli A., Lin D., Williams K., Denman R.B., Hellen C.U.T., and Tiedge H. 2005. Dendritic BC1 RNA in translational control mechanisms. *J. Cell Biol.* **171:** 811–821.

Wang L., Eckmann C.R., Kadyk L.C., Wickens M., and Kimble J. 2002. A regulatory cytoplasmic poly(A) polymerase in *Caenorhabditis elegans*. *Nature* **419:** 312–316.

Wells D.G., Dong X., Quinlan E.M., Huang Y.S., Bear M.F., Richter J.D., and Fallon J.R. 2001. A role for the cytoplasmic polyadenylation element in NMDA receptor-regulated mRNA translation in neurons. *J. Neurosci.* **21:** 9541–9548.

Wu L., Wells D., Tay J., Mendis D., Abbott M.A., Barnitt A., Quinlan E., Heynen A., Fallon J.R., and Richter J.D. 1998. CPEB-mediated cytoplasmic polyadenylation and the regulation of experience-dependent translation of alpha-CaMKII mRNA at synapses. *Neuron* **21:** 1129–1139.

Yin J.C., Wallach J.S., Del Vecchio M., Wilder E.L., Zhou H., Quinn W.G., and Tully T. 1994. Induction of a dominant negative CREB transgene specifically blocks long-term memory in *Drosophila*. *Cell* **79:** 49–58.

Zalfa F., Giorgi M., Primerano B., Moro A., Di Penta A., Reis S., Oostra B., and Bagni C. 2003. The fragile X syndrome protein FMRP associates with BC1 RNA and regulates the translation of specific mRNAs at synapses. *Cell* **112:** 317–327.

Zhang X. and Poo M.M. 2002. Localized synaptic potentiation by BDNF requires local protein synthesis in the developing axon. *Neuron* **36:** 675–688.

Zhang Y.Q., Bailey A.M., Matthies H.J., Renden R.B., Smith M.A., Speese S.D., Rubin G.M., and Broadie K. 2001. *Drosophila* fragile X-related gene regulates the MAP1B homolog Futsch to control synaptic structure and function. *Cell* **107:** 591–603.

19

Translational Control in Development

Beth Thompson,[1] Marvin Wickens,[1,2] and Judith Kimble[1,2,3]
[1]Cellular and Molecular Biology Program
[2]Department of Biochemistry
[3]Howard Hughes Medical Institute
University of Wisconsin
Madison, Wisconsin 53706

DEVELOPMENT REQUIRES THE COORDINATED EXPRESSION of selected genes at specific times and in specific cells. Such regulated expression controls the establishment of embryonic axes, the existence of stem cells, and the specification of individual cell fates. Translational control has a key role in this regulation. It is particularly important during early embryogenesis and in the germ line, where transcription is typically quiescent; there, control of translation and mRNA stability are the primary ways to regulate patterns of protein synthesis. Yet, translational regulation continues throughout development and in somatic tissues.

In this chapter, we view translational control from a developmental perspective. We discuss four major interfaces at which developmental biology meets molecular regulatory mechanisms: molecular switches, gradients, combinatorial control, and networks. These areas were chosen because they bear on fundamental processes of development. We emphasize instances in which sequence-specific regulatory factors control particular mRNAs, and we do not cover the role of general translation factors (e.g., eIF4E and eIF2α) on growth and differentiation (Chapter 4).

MECHANISMS OF TRANSLATIONAL CONTROL: A PRIMER

Translation is a multistep process and can be divided into three phases: initiation, elongation, and termination. In principle, translational control can be exerted in any of these phases. We focus in this chapter on initi-

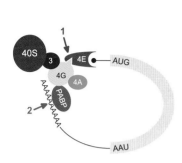

Figure 1. Common control points during translational initiation. Translation initiation occurs by assembly of a translational initiation complex, which in this simplified view includes the initiation factors eIF4E (*red*), eIF4G (*yellow*), eIF4A (*green*), and eIF3 (*dark blue*). PABP (*blue*) binds both the poly(A) tail and eIF4G, resulting in mRNA circularization and stabilization of the complex; eIF3 recruits the 40S ribosomal subunit (*purple*). Major points of regulation include (*1*) the eIF4E/eIF4G interaction (*red arrow*) and (*2*) length of the poly(A) tail (*blue arrow*).

ation, which appears to be the most common point of control during development.

Translational initiation involves more than 20 proteins, multiple biochemical complexes, and a series of separable steps (Chapter 4). A complex containing eukaryotic initiation factor 4E (eIF4E) (cap-binding protein) and eIF4G is crucial. At the 5′ end of the mRNA, eIF4E binds to both the m^7GpppN cap structure and the initiation factor, eIF4G; eIF4G in turn interacts with eIF3, which recruits the 40S subunit of the ribosome. For most mRNAs, the eIF4E–eIF4G–eIF3 complex is required for efficient translation initiation. At the 3′ end of the mRNA, poly(A)-binding protein (PABP) binds to both eIF4G and the 3′-terminal tract of polyadenosine (poly[A]), effectively circularizing the mRNA (Fig. 1) (Wells et al. 1998; Chapters 10 and 18).

The eIF4E–eIF4G interaction is a common target of control during development (Fig. 1, red arrow). Specifically, a regulatory protein binds eIF4E and prevents its interaction with eIF4G. That regulatory protein is tethered to a particular mRNA by means of a sequence-specific 3′-untranslated region (3′UTR)-binding protein (for review, see Gebauer and Hentze 2004). This mechanism echoes that of 4E-BP's global interference with eIF4E function (Richter and Sonenberg 2005).

The poly(A) tail is also a common site of control (Fig. 1, blue arrow). Indeed, changes in poly(A) length are a hallmark of translational regulation. Typically, long poly(A) tails are associated with increased translation, and short poly(A) tails are associated with decreased translation (Gray and Wickens 1998). These changes in poly(A) length are triggered on specific mRNAs by sequence-specific 3′UTR-binding proteins that recruit enzymes to that mRNA; those enzymes either add poly(A) or remove it.

TRANSLATIONAL REGULATORS: AN OVERVIEW

The translational regulators that mediate mRNA-specific controls in development include a multitude of proteins and microRNAs (miRNAs) (Chapters 10 and 11). An extensive review of all regulators would require its own chapter and perhaps its own volume. Instead, we focus on a few representative regulators, which have emerged as particularly important in developmental control.

PUF RNA-binding Proteins

The PUF (for *Pu*milio and *F*BF [*fem*-3-binding factor]) RNA-binding proteins control a wide range of developmental processes and are conserved in all eukarya (for review, see Wickens et al. 2002). *Drosophila* Pumilio and *Caenorhabditis elegans* FBF were its founding members. In general, PUF proteins bind to specific sequences in the 3'UTR of target mRNAs and either repress their translation or trigger their decay. In both instances, PUF activity is correlated with shorter poly(A) tails (Wreden et al. 1997; Goldstrohm et al. 2006).

PUF proteins bind RNA through eight consecutive PUF repeats, and flanking Csp regions. Crystal structures of both *Drosophila* and human Pumilio1 indicate that each repeat forms three α helices and that together, the repeats form a crescent-shaped three-dimensional structure (Edwards et al. 2001; Wang et al. 2001). The inner surface of the crescent binds RNA; consecutive α helices contact predominantly one nucleotide (Wang et al. 2001). Thus, the canonical PUF-binding elements are short segments of "single-stranded" RNA. Binding specificities among PUF proteins overlap but differ; for example, a single "extra" nucleotide in the binding site can switch specificity from one PUF protein to another (Opperman et al. 2005). The outer surface of the crescent-shaped PUF protein is important for protein–protein interactions (Wharton and Struhl 1991; Sonoda and Wharton 1999; Olivas and Parker 2000; Edwards et al. 2001, 2003).

PUF proteins control a variety of developmental processes, ranging from anterior–posterior patterning in *Drosophila* to sex determination in *C. elegans*. A common role appears to be promotion of proliferation at the expense of differentiation (Wickens et al. 2002). Indeed, both *Drosophila* Pumilio and *C. elegans* FBF are required for maintenance of germ-line stem cells (for review, see Wickens et al. 2002), and a PUF protein in planaria is essential for stem-cell maintenance during regeneration (Salvetti et al. 2005). Mammalian stem cells also possess PUF proteins, although their functions have not been determined to date (Moore et al. 2003).

CPEB RNA-binding Proteins

The CPEB (cytoplasmic polyadenylation element-binding protein) RNA-binding proteins control a wide range of developmental processes and are conserved in all metazoans. CPEB proteins contain two RNA recognition motifs (RRMs) and two zinc finger motifs, which are required for RNA binding (for review, see Mendez and Richter 2001). Within the 3′UTR of target mRNAs, CPEB binds to U-rich elements, called cytoplasmic polyadenylation elements (CPEs). Furthermore, CPEB can act as an activator or a repressor of target mRNAs (Mendez and Richter 2001; and see below). The crystal structure of CPEB has not yet been determined.

CPEB proteins have key roles in development of the germ line and early embryo. In *Xenopus*, CPEB is best known for its regulation of oocyte maturation (for review, see Mendez and Richter 2001; see below). Murine CPEB is critical for meiosis, which it controls at least in part through the polyadenylation of mRNAs that encode components of the synaptonemal complex (Tay and Richter 2001). In *Drosophila*, the CPEB homolog, *orb*, is required for multiple aspects of oogenesis and for the polyadenylation of specific target mRNAs (Huynh and St Johnston 2000; Chang et al. 2001; Castagnetti and Ephrussi 2003). *C. elegans* contains multiple CPEB homologs: CPB-1 and FOG-1 regulate spermatogenesis (Luitjens et al. 2000; Jin et al. 2001a), and CPB-3 controls germ-line apoptosis (Lettre et al. 2004; Boag et al. 2005). CPEB also acts in the nervous system (Chapter 18).

GLD-2 Poly(A) Polymerases

GLD-2 proteins are cytoplasmic poly(A) polymerases (PAPs) (Wang et al. 2002). Enzymatically active homologs have been identified in fission yeast (Read et al. 2002; Saitoh et al. 2002), *C. elegans* (Wang et al. 2002), *Xenopus* (Barnard et al. 2004; Rouhana et al. 2005), mice and humans (Kwak et al. 2004), and *Drosophila* (J.-E. Kwak and M. Wickens, unpubl.). The proteins are members of the β-nucleotidyl transferase superfamily, which includes enzymes that transfer nucleotides to a wide variety of substrates (Aravind and Koonin 1999). The GLD-2 enzymes have diverged extensively from canonical nuclear PAPs (Wang et al. 2002). Canonical nuclear PAPs possess an RRM-like motif that may bind RNA (Bard et al. 2000; Martin et al. 2000). Sequence inspection showed that GLD-2 proteins lack this motif. Instead, *C. elegans* GLD-2 physically interacts with specific

RNA-binding proteins that are thought to tether the polymerase to specific target mRNAs (Wang et al. 2002).

C. elegans GLD-2 controls multiple aspects of germ-line development and early embryogenesis (Kadyk and Kimble 1998; Wang et al. 2002). Vertebrate GLD-2 family members are present in both the germ line and soma, particularly in the nervous system (Barnard et al. 2004; Rouhana et al. 2005). In Xenopus oocytes, GLD-2 is the long-sought enzyme that adds poly(A) to stored maternal mRNAs and is associated with their activation (Barnard et al. 2004; Rouhana et al. 2005). GLD-2 has been implicated in meiotic maturation in C. elegans, Xenopus, and mice, suggesting a conserved role in that process. GLD-2 may also act in the nervous system (Rouhana et al. 2005; Chapter 18).

MicroRNAs

miRNAs are short, noncoding RNAs that have emerged as key regulators of many developmental processes (Bartel 2004; Alvarez-Garcia and Miska 2005; Chapters 10 and 11). miRNAs bind to complementary regions of target mRNAs, often in the 3'UTR, and down-regulate expression. Perfect pairing between the miRNA and its target generally results in endonucleolytic cleavage and subsequent degradation of the target mRNA. Imperfect pairing between the miRNA and its target generally results in translational repression of the target mRNA (Bartel 2004). A recent report indicates that imperfectly paired miRNAs might affect stability of their targets (Bagga et al. 2005), not just translational activity.

The first miRNA was discovered in C. elegans as a regulator of life cycle progression (Lee et al. 1993). Since that initial finding, the crucial role of miRNAs in development has expanded tremendously (Alvarez-Garcia and Miska 2005). In C. elegans, discovery of the let-7 miRNA provided another key breakthrough because of its conservation among metazoans (Pasquinelli et al. 2000). Also in C. elegans, lsy-6 and mir-273 control left–right neuronal asymmetry (Johnston and Hobert 2003; Chang et al. 2004); in Drosophila, miRNAs control cell death (Brennecke et al. 2003) and Notch signaling (Brennecke et al. 2005; Lai et al. 2005); in vertebrates, miRNAs control angiogenesis (Zhao et al. 2005) and limb development (Hornstein et al. 2005); and in Arabidopsis, miRNAs control leaf development (Mallory et al. 2004), among other events (Jones-Rhoades et al. 2006). These are but a few examples. For more targeted recent reviews of miRNAs in development, see Alvarez-Garcia and Miska (2005).

RNA REGULATORY SWITCHES AND DEVELOPMENTAL CHOICE

A molecular switch controls the transition from one state to another, such as "on" versus "off." In a developmental context, molecular switches govern myriad choices that direct the creation of a multicellular organism. One kind of molecular switch relies on a DNA-binding protein that recruits transcriptional repressors in one mode and transcriptional activators in another. A paradigmatic example is TCF/LEF, a DNA-binding protein that recruits the β-catenin transcriptional activator in response to Wnt signaling, but recruits the Groucho/HDAC transcriptional repressors in the absence of signaling (for review, see Bienz 1998). In this section, we focus on molecular switches that rely on RNA-binding proteins and operate on a translational level.

CPEB Phosphorylation Controls a Molecular Switch

CPEB is one of the first RNA-binding proteins recognized as a potential molecular switch. We focus here on the mechanism of CPEB-mediated translational control during frog oocyte maturation. *Xenopus* oocytes are arrested in prophase of meiosis I, until they are stimulated by progesterone to progress into metaphase of meiosis II, a process called "oocyte maturation." Prior to maturation, many maternal mRNAs are stored in the cytoplasm in a dormant repressed state. Upon progesterone stimulation, those stored mRNAs become polyadenylated and translationally active. Both repression and activation require cytoplasmic polyadenylation elements (CPEs) that are located in 3′UTRs near the cleavage and polyadenylation signal AAUAAA (Stutz et al. 1998; de Moor and Richter 1999; Minshall et al. 1999; Barkoff et al. 2000).

CPEB was initially identified as a CPE-binding protein (Hake and Richter 1994). Since then, it has been found to physically associate with both translational repressors and activators. The repressors include Maskin, Pumilio, and Nanos (Xcat-2) (Stebbins-Boaz et al. 1999; Nakahata et al. 2001, 2003). Maskin represses translation by binding the cap-binding protein, eIF4E, an interaction that sequesters eIF4E and prevents it from binding eIF4G (see Fig. 2A) (Stebbins-Boaz et al. 1999; Mendez and Richter 2001; Chapter 10). The mechanisms by which Pumilio and Nanos repress mRNAs are more poorly understood, but a yeast PUF protein recruits a deadenylase, Ccr4p, and an RNA helicase, Dhh1p, that can cause repression, and that mechanism may be conserved (Goldstrohm et al. 2006).

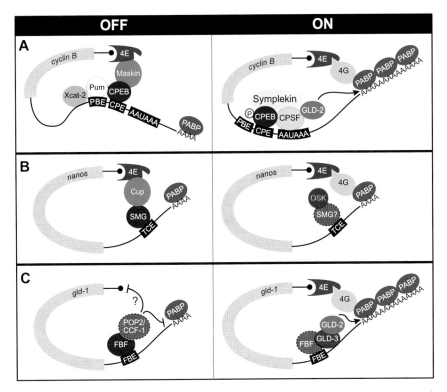

Figure 2. mRNA regulatory switches. mRNAs are depicted in either the repressed ("off") or active ("on") states. (*A*) CPEB can function as a molecular switch. As a repressor (*left*), CPEB (*dark purple*) associates with Maskin (*green*), which interferes with the eIF4E/eIF4G interaction. In addition, CPEB binds Pumilio (*light yellow*) and Xcat-2 (*orange*), which may also be involved in repression. As an activator (*right*), CPEB associates with CPSF (*light green*), Symplekin (*light blue*), and the poly(A) polymerase, GLD-2 (*brown*). The RNA is circularized by interactions between eIF4E (*red*), eIF4G (*gold*), and PABP (*blue*). CPEB phosphorylation switches it from a repressor to an activator. (*B*) Smaug may be able to function as a molecular switch. In this model, Smaug (*dark purple*) associates with the Maskin-like protein, Cup (*green*), when acting as a repressor (*left*), but with Oskar (*dark green*) to activate translation of *nanos* mRNA at the posterior pole. Control of protein presence in suspected as critical for the switch (*gray dashed circle*). See text for details. (*C*) FBF may be able to function as a molecular switch. In this model, FBF (*dark purple*) recruits the deadenylase complex containing POP2/CCF-1 (*gray*) to repress *gld-1* translation (*left*); FBF recruits GLD-3 (*light purple*) and GLD-2 (*brown*), which together function as a poly(A) polymerase, to activate *gld-1* translation (*right*). The switch mechanism is not understood. See text for details.

Translational activation of CPEB target mRNAs requires both alleviation of repression and activation. Oocyte maturation triggers CPEB phosphorylation, which is thought to weaken the interaction between eIF4E and Maskin, and thereby relieve repression (Mendez et al. 2000). CPEB phosphorylation also enhances binding of CPEB to the cleavage and polyadenylation specificity factor (CPSF) complex, which can promote CPE-dependent polyadenylation (Bilger et al. 1994; Mendez et al. 2000; Dickson et al. 2001). CPSF binds to the AAUAAA sequence, and that binding is enhanced by the presence of adjacent CPEs (Bilger et al. 1994). Importantly, CPSF also recruits the XlGLD-2 cytoplasmic PAP (Dickson et al. 2001; Barnard et al. 2004; Rouhana et al. 2005). In oocyte extracts, CPEB is present in a complex that contains CPSF, GLD-2, and the scaffolding protein, Symplekin (Fig. 2A, bottom) (Barnard et al. 2004; Rouhana et al. 2005). The role of GLD-2 in oocyte maturation may be conserved, since GLD-2 also functions during oocyte meiosis in *C. elegans* (Kadyk and Kimble 1998) and possibly in mice (Nakanishi et al. 2006).

The mechanism by which CPEB transitions from a translational repressor to a translational activator involves its phosphorylation, but aspects of the mechanism are not fully understood. CPEB interacts with translational activators, GLD-2 and CPSF, even in immature oocytes, when target mRNAs have short poly(A) tails (Barnard et al. 2004; Rouhana et al. 2005). Furthermore, although CPEB interacts with the translational repressors Maskin and Pumilio, GLD-2 does not interact with these proteins, suggesting the existence of distinct CPEB-containing complexes (Rouhana et al. 2005). Nonetheless, we emphasize that CPEB is capable of recruiting both translational activators and translational repressors, and this RNA-binding protein is therefore poised to mediate the switch between translational repression and activation. Such a CPEB-based molecular switch is clearly involved in developmental decisions and may also be a key element for controlling synaptic states of activity (Chapter 18).

Smaug, a Spatially Controlled Molecular Switch

The *Drosophila* RNA-binding protein, Smaug, is central to a molecular switch that controls translation of *nanos* mRNA. Smaug binds RNA in a sequence-specific manner through a sterile-α-motif (SAM) domain (Aviv et al. 2003); it is ubiquitously expressed in the embryo prior to cellularization (Smibert et al. 1999). Furthermore, Smaug family proteins are broadly conserved (Baez and Boccaccio 2005). In *Drosophila* embryos, *nanos* mRNA is translationally repressed during oogenesis and in the early embryo, except at the posterior pole (see Gradients below; for review, see Johnstone and Lasko 2001). Smaug protein binds to translational control elements (TCEs) in the *nanos* 3′UTR and represses translation of unlocalized *nanos* (Smibert et al. 1996). To this end, Smaug recruits the translational repressor, Cup. Cup uses a Maskin-like mechanism to inhibit translation: It binds eIF4E and precludes eIF4G binding (Nelson et al. 2004). On a different mRNA (*hsp83*), Smaug mediates translational repression by recruiting the Ccr4–Pop2–Not1 deadenylase complex (Semotok et al. 2005).

Smaug physically interacts with a translational activator, the Oskar protein (Dahanukar et al. 1999). Oskar is restricted to the posterior pole (Rongo et al. 1995), and Oskar, together with Tudor and Vasa, stimulates translation of *nanos* mRNA, at least in part by overcoming Smaug-mediated repression (Ephrussi and Lehmann 1992; Smith et al. 1992). Therefore, Smaug interacts with Cup to repress unlocalized *nanos* mRNA, but with Oskar to activate the localized mRNA (Fig. 2B). Importantly, Cup and Oskar associate with the same region of Smaug. Thus, the switch at the posterior pole may be triggered by simple displacement of Cup from *nanos* mRNA by localized Oskar protein (Nelson et al. 2004).

PUF Proteins Can Act as Translational Repressors or Activators

FBF was initially identified as a translational repressor of mRNAs controlling stem cells and sex determination in the *C. elegans* germ line (Zhang et al. 1997; Crittenden et al. 2002). In that germ line, the distal-most cells undergo mitosis, and more proximal germ cells enter meiosis (for review, see Kimble and Crittenden 2005). FBF promotes mitosis in the distal germ line (Crittenden et al. 2002), and three GLD proteins promote entry into meiosis (Kadyk and Kimble 1998; Eckmann et al. 2004; Hansen et al. 2004a,b). A key target of FBF repression is *gld-1* mRNA (Crittenden et al. 2002). Indirect evidence suggested that FBF might mediate repression by deadenylation of its target mRNAs (Ahringer and Kimble 1991).

Recent studies of the yeast PUF protein, Mpt5p, have implications for the mechanism of PUF-mediated repression during development. Mpt5p represses mRNAs, at least in part, by binding Pop2p, a component of the Ccr4–Not1 complex (Goldstrohm et al. 2006). This binding simultaneously recruits the deadenylase, Ccr4; the helicase Dhh1, which can repress mRNAs (Minshall et al. 2001; Coller and Parker 2005); and Dcp1, a decapping factor. The PUF–POP2 interaction is conserved, as it has been observed in *C. elegans* and human counterparts. Two homologs of

Dhh1, *Xenopus* p54 and *Drosophila* Me31B, are associated with and can control the translation of maternal mRNAs (Ladomery et al. 1997; Minshall et al. 2001; Nakamura et al. 2001). FBF, like other PUF proteins, thus may recruit a complex that both deadenylates and represses mRNAs. This idea can explain the otherwise enigmatic observation that *Drosophila* Pumilio acts through both deadenylation-dependent and -independent mechanisms (Wreden et al. 1997; Chagnovich and Lehmann 2001). Recruitment of a Ccr4 complex may be a common means of repression, as that complex is associated with *Drosophila* Smaug (Semotok et al. 2005) and mammalian TTP (Lykke-Andersen and Wagner 2005).

PUF proteins can also recruit translational activators. One example involves *gld-1* mRNA in *C. elegans*. In addition to being repressed by FBF, *gld-1* mRNA is activated by the GLD-2/GLD-3 PAP (N. Suh and J. Kimble, unpubl.). In *gld-2* mutants, *gld-1* mRNA is repressed and its poly(A) tail is shorter than in wild-type animals. Importantly, FBF has also been implicated in *gld-1* activation by certain genetic tests (Crittenden et al. 2002). Furthermore, in a yeast assay, FBF, GLD-2, and GLD-3 form a ternary complex on the *gld-1* 3′UTR, and FBF, GLD-2, and GLD-3 coimmunoprecipitate from worm extracts. Therefore, FBF appears to function as a molecular switch for control of *gld-1* RNA: FBF first represses *gld-1*, perhaps by binding the *C. elegans* Pop2 homolog, called CCF-1, and thereby recruiting a deadenylation–repression complex; later, FBF then activates *gld-1* translation by recruiting the GLD-2/GLD-3 PAP (Fig. 2C) (N. Suh and J. Kimble, unpubl.). The key event that triggers the switch has not been identified. A second mode of PUF-mediated activation may rely on DAZ: Vertebrate Pumilio binds to DAZ (Moore et al. 2003), which appears to translationally activate target mRNAs (Collier et al. 2005).

Sex-lethal

Drosophila sex determination and dosage compensation occur through an elegant series of posttranscriptional regulatory events (Cline and Meyer 1996). Regulation of splicing and translation combine to form a robust irreversible switch between male and female modes of development. The initial trigger for the switch is the ratio of X chromosomes to sets of autosomes, the X/A ratio.

The Sex-lethal (SXL) RNA-binding protein is required for dosage compensation and sex determination in flies. SXL expression is limited to females by a combination of transcriptional control and autoregulated splicing (for review, see Penalva and Sánchez 2003). Dosage compensation in flies is achieved by hyperactivating the X chromosome in males, a process that requires the MSL complex of proteins (Baker et al. 1994; Cline and Meyer 1996; Meller and Kuroda 2002; Nusinow and Panning 2005). The MSL complex is composed of five proteins, Mle and Msl-1, 2, 3, and 4, and two noncoding roX RNAs. The complex binds to multiple sites along the male X chromosome and alters chromatin structure. All proteins of the MSL complex are expressed in both males and females, except Msl-2, which is specifically repressed by SXL protein in females (Bashaw and Baker 1995). As a result, the MSL complex is functional only in males.

SXL regulates splicing of *msl-2* pre-mRNA and causes an intron to be retained within the 5′UTR of *msl-2*. The retained intron includes an SXL-binding site; four additional SXL-binding sites reside in the *msl-2* 3′UTR. Efficient translational repression of *msl-2* requires SXL binding to both the 5′UTR and 3′UTR (Gebauer et al. 1998). SXL bound to the 3′UTR inhibits the recruitment of the 43S ribosomal preinitiation complex to the mRNA, and SXL bound to the 5′UTR is proposed to interfere with ribosomal scanning for the initiation codon (Gebauer et al. 1998; Beckmann et al. 2005; Chapter 10).

The molecular event that switches translational activity is thus the presence of SXL protein exclusively in females, an event regulated by the X/A ratio. Once present, SXL represses *msl-2* mRNA, and the organism does not hyperactivate its X chromosome. When Sxl is missing, the MSL complex forms, and dosage compensation ensues.

Concluding Remarks

Common mechanisms have emerged among switches operating at a translational level. For example, both CPEB and Smaug direct translational repression by recruitment of a protein that interferes with binding of eIF4E to eIF4G: Maskin in the case of CPEB and Cup in the case of Smaug. In addition, Smaug and PUF proteins associate with the Ccr4–Pop2 deadenylation complex to direct translational repression. Thus, Smaug can use both of these common mechanisms. Since CPEB and PUF proteins associate, it also is possible that both mechanisms operate on common mRNA targets.

Activation mechanisms may also be common among different regulators. Both CPEB and PUF proteins recruit a GLD-2 cytoplasmic PAP for translational activation. In other instances, translational activation occurs simply by alleviating translational repression, as might be the case for Smaug and Oskar.

GRADIENTS OF TRANSLATIONAL REGULATORS

Gradients have emerged as a key molecular mechanism for the establishment of spatial and temporal coordinates that underlie many aspects of development. A gradient is defined as the graded distribution of a regulatory molecule that acts in a concentration-dependent manner to elicit multiple biological responses (Ashe and Briscoe 2006). One important feature of a gradient is the existence of thresholds: A concentration below a given threshold results in one response, and a concentration above the threshold results in another response (Fig. 3A). Classic gradients are formed by signaling morphogens, such as BMP, and by dose-dependent transcription factors, such as Bicoid (for reviews, see Gurdon and Bourillot 2001; Ephrussi and St Johnston 2004). Translational regulation can contribute to gradients in either of two ways: (1) Translational regulators can help form a gradient or (2) translational regulators can be distributed in a gradient and function in a concentration-dependent manner. Here, we review three examples of how translational regulation contributes to developmental gradients. Translational controls are a primary mechanism for establishing developmental gradients, but in addition, translational regulators can themselves provide a dose-dependent gradient of developmental regulation.

Anterior–Posterior Patterning in the *Drosophila* Embryo

The anterior–posterior axis is established in the *Drosophila* embryo by opposing gradients of two key transcription factors, called Caudal and Hunchback. The graded distribution of Caudal and Hunchback is accomplished by a combination of translational control and RNA localization. In this chapter, we focus on the translational regulators; RNA localization is reviewed elsewhere (Chapter 24). Figure 3, B and C, summarizes the critical gradients that determine the anterior–posterior axis of the *Drosophila* embryo.

Anterior Fates

Anterior determination begins during oogenesis with the anterior localization of *bicoid* mRNA (Berleth et al. 1988; St Johnston et al. 1989). After fertilization, localized *bicoid* mRNA is translated, and Bicoid protein diffuses posteriorly to generate a gradient of Bicoid protein (Fig. 3B). Bicoid has a dual role in anterior specification. In the nucleus, the Bicoid homeodomain transcription factor acts in a concentration-dependent

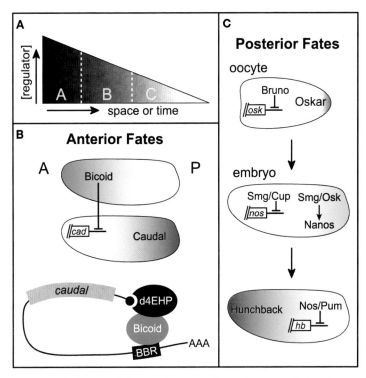

Figure 3. Control of *Drosophila* anterior–posterior axis by translational control and gradients. (*A*) A classic developmental gradient. A molecular regulator is present in a graded distribution, with respect to either space or time (*x*-axis). This regulator acts in a concentration-dependent manner to elicit specific biological fates. Here, a high concentration results in fate A, a medium concentration results in fate B, and a low concentration results in fate C. Broken lines indicate threshold concentrations. (*B*) Anterior fates are controlled by opposing gradients of Bicoid and Caudal proteins in the *Drosophila* embryo. Localized *bicoid* mRNA establishes a concentration gradient of Bicoid protein (*green*). The Bicoid translational regulator binds the *caudal* 3′UTR and represses translation, resulting in an opposing gradient of Caudal protein (*blue*). Bicoid interacts with d4EHP, which competes with eIF4E binding to the cap. d4EHP cannot interact with eIF4G and thus blocks translation initiation. (*C*) Posterior fates are controlled by a translational cascade. (*Top*) During oogenesis, *oskar* mRNA is localized to the posterior pole; unlocalized *oskar* mRNA is repressed by Bruno. *nanos* mRNA is also localized posteriorly during oogenesis (not shown). (*Middle*) Smaug and Cup repress unlocalized *nanos* mRNA in regions away from the posterior pole, but at the pole, Smaug associates with Oskar and *nanos* translation is activated. Nanos is distributed in a gradient emanating from the posterior pole (*yellow*). (*Bottom*) Nanos, together with Pumilio (Pum) and Brat, represses *hunchback* mRNA to establish an opposing gradient of Hunchback protein (*orange*).

manner to activate expression of anterior-specific genes (Berleth et al. 1988; Driever and Nüsslein-Volhard 1988a,b; Driever et al. 1989; Struhl et al. 1989). In the cytoplasm, the Bicoid translational regulator binds a regulatory element in the *caudal* 3′UTR called the BBR (for *bicoid-b*inding region) and represses its translation, resulting in an opposing gradient of Caudal protein (Fig. 3B, middle) (Mlodzik et al. 1985; Macdonald and Struhl 1986; Dubnau and Struhl 1996; Rivera-Pomar et al. 1996). Recently, the molecular mechanism by which Bicoid represses *caudal* translation has been elucidated (Cho et al. 2005). Bicoid binds d4EHP, a protein related to eIF4E cap-binding protein. d4EHP competes with eIF4E for cap recognition, but it cannot bind eIF4G; instead, it links the 5′ and 3′ ends of *caudal* mRNA and renders it translationally inactive (Fig. 3B).

Posterior Fates

Posterior determination involves a cascade of translational regulation that generates a gradient of Nanos protein, which in turn establishes a gradient of the Hunchback transcription factor (Fig. 3C). The cascade begins during oogenesis and ends in the early embryo (Fig. 3C).

Before *oskar* mRNA becomes localized at the posterior pole during oogenesis, its translation is repressed by the Bruno RNA-binding protein (Fig. 3C, top). Bruno contains three RRM motifs and represents a conserved family with broad roles in animal development (Good et al. 2000). The *oskar* 3′UTR contains three Bruno response elements, or BREs, which Bruno binds (Kim-Ha et al. 1995). Bruno functions as a dual repressor of *oskar* mRNA. First, Bruno recruits the Cup protein. As discussed in the previous section, Cup is an eIF4E-binding protein that is thought to interfere with translation initiation by interrupting the interaction between eIF4E and eIF4G (Wilhelm et al. 2003; Nakamura et al. 2004; Nelson et al. 2004). Second, Bruno acts by a novel Cup-independent mechanism, which involves multimerization of *oskar* mRNA into silencing particles (Chekulaeva et al. 2006). The mechanism by which *oskar* mRNA is translationally activated at the posterior pole remains unknown, but it appears to rely on binding of insulin growth factor II mRNA-binding protein (IMP) to IMP-binding elements in the *oskar* 3′UTR (Munro et al. 2006).

A gradient of Nanos protein is established by a combination of mRNA localization, translational repression, and translational activation (Fig. 3C, middle). Much (>90%) *nanos* mRNA remains unlocalized in the embryonic cytoplasm (Bergsten and Gavis 1999), and that mRNA must be repressed for proper development (Gavis and Lehmann 1992). Within the *nanos* 3′UTR are TCEs (Dahanukar and Wharton 1996) that bind the Smaug RNA-binding protein (Smibert et al. 1996). Smaug repression of *nanos* translation requires Cup protein (Nelson et al. 2004); its mechanism was described above (see RNA Regulatory Switches). Smaug also interacts with Oskar protein, which promotes Nanos translation at the posterior pole (Dahanukar et al. 1999). Therefore, Smaug, Cup, and Oskar collaborate to establish the graded distribution of the Nanos protein that emanates from the posterior (Fig. 3C, middle).

The Nanos gradient in turn establishes an opposing gradient of the Hunchback transcription factor (Fig. 3C, bottom). Nanos is a zinc finger RNA-binding protein that functions as part of a regulatory complex (Fig. 3C) (for review, see Johnstone and Lasko 2001). Specifically, Nanos represses *hunchback* mRNA at the translational level. Two elements within the 3'UTR of the *hunchback* mRNA, called the Nanos response elements, or NREs, mediate that repression (Wharton and Struhl 1991) together with at least three proteins. Pumilio, a PUF family RNA-binding protein, binds directly to the *hunchback* NREs (Murata and Wharton 1995). Although Pumilio is distributed uniformly in the embryo (Macdonald 1992), it only represses *hunchback* at the posterior where Nanos is localized. Nanos protein is recruited to *hunchback* mRNA by Pumilio (Sonoda and Wharton 1999), and then Brat, an NHL (for *NCL-1*, *HT2A*, and *LIN-41*) (Slack and Ruvkun 1998) homolog, is recruited by both Pumilio and Nanos (Sonoda and Wharton 2001). The mechanism by which the Pumilio–Nanos–Brat complex represses *hunchback* translation involves deadenylation (Wreden et al. 1997) as well as a deadenylation-independent mechanism (Fig. 3C, bottom) (Chagnovich and Lehmann 2001).

A Temporal Gradient of LIN-14 Controls Developmental Timing in *C. elegans*

Gradients can instruct patterning in time as well as in space. For example, a temporal gradient of the *C. elegans* protein, LIN-14, which is a novel transcription factor (Hristova et al. 2005), directs life cycle progression in somatic tissues (Ruvkun and Giusto 1989). After the embryo hatches, *C. elegans* develops through four larval stages, L1, L2, L3, and L4, and then transitions into adulthood. Genes in the heterochronic pathway govern progression through the larval stages (for review, see Slack and Ruvkun 1997). Multiple-component mRNAs of the heterochronic pathway are themselves translationally regulated, including *lin-14, lin-28,* and *lin-41.* LIN-14 and LIN-28 specify early larval fates and must be down-regulated to transition to later fates. LIN-41 is required for late larval fates and must be down-regulated for the L4-to-adult transition.

miRNA *lin-4* represses both *lin-14* and *lin-28* mRNAs (Lee et al. 1993; Wightman et al. 1993; Moss et al. 1997), and miRNA *let-7* represses *lin-41* mRNA (Reinhart et al. 2000; Slack et al. 2000). Forward genetics led to the discovery of *lin-4* and *let-7* as the first known miRNAs. However, the broader impact of miRNAs on gene regulation has now become clear (for review, see Pasquinelli et al. 2005; Chapter 11; this chapter, Translational Regulators).

Translational regulation of *lin-14* mRNA establishes a temporal gradient of LIN-14. LIN-14 abundance is high in L1 larvae but decreases in subsequent larval stages (Ruvkun and Giusto 1989). Genetic evidence suggests that LIN-14 may function in a concentration-dependent manner to achieve stage-specific fates: High LIN-14 promotes the L1 fate, medium LIN-14 promotes the L1 to L2 transition, and low LIN-14 promotes the L2 to L3 transition (Ha et al. 1996). The existence of multiple LIN-14 isoforms complicates the interpretation of these genetic experiments.

lin-4 miRNA governs the LIN-14 gradient (Lee et al. 1993; Wightman et al. 1993). In the absence of *lin-4*, LIN-14 remains abundant and early fates are reiterated (Chalfie et al. 1981; Arasu et al. 1991). *lin-4* interacts both in vivo and in vitro with complementary sites in the *lin-14* 3′UTR (Ha et al. 1996). Disruption of those sites leads to a high LIN-14 level and reiteration of early fates. The mechanism by which the heterochronic miRNAs control translation is a subject of debate. Early work indicated translational control, but more recent work suggests that *lin-4* and *let-7* affect mRNA stability (Bagga et al. 2005).

A FOG-1/CPEB Gradient Controls the *C. elegans* Germ Line

In the *C. elegans* germ line, the CPEB-related protein, FOG-1, functions in a dose-dependent manner that is graded in both space and time. FOG-1 is a key regulator of the sperm fate: Germ cells that normally differentiate as sperm become oocytes instead in *fog-1* null mutants (Barton and Kimble 1990). Further genetic tests showed that although high levels of FOG-1 promote spermatogenesis, low (but nonzero) levels of FOG-1 promote mitosis (Fig. 4A). The role of FOG-1 in mitosis was initially masked, because FOG-1 and FBF execute this function redundantly. Only when both are removed do early larval germ cells enter meiosis and differentiate as oocytes (Thompson et al. 2005). Importantly, a physical gradient of FOG-1 protein is established by FBF repression. In vivo, FOG-1 levels increase upon FBF removal, and FBF binds with high affinity and specificity to elements in the *fog-1* 3′UTR in vitro (Thompson et al. 2005). As *fog-1* mRNA escapes FBF repression, the abundance of FOG-1 increases dramatically. That dramatic

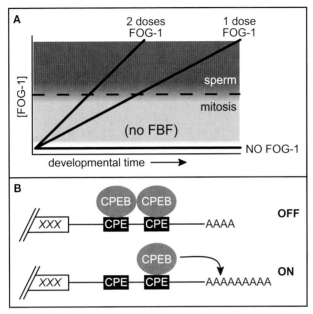

Figure 4. Dose-dependent regulation by CPEB-related proteins. (*A*) In *C. elegans*, low FOG-1 promotes mitosis (*light gray*), and high FOG-1 specifies the sperm fate (*dark gray*). Three situations are shown with respect to time; in all three cases, FBF has been removed (no FBF), which eliminates that redundancy of cell fate control. When no FOG-1 is present, no mitosis occurs and no sperm are made; instead, the few germ cells present become oocytes. When 1 dose of FOG-1 is present, the germ line develops at low levels of FOG-1 for longer time, and therefore more germ cells are made; however, at later stages of development, high FOG-1 levels are reached, and sperm are specified. When 2 doses of FOG-1 are present, the germ line accumulates high levels of FOG-1 quickly, fewer germ cells are made, and sperm are specified earlier. (*B*) In *Xenopus*, low CPEB promotes mitosis and high CPEB controls meiotic maturation. A possible mechanism involves the dose-dependent regulation of a hypothetical mRNA (Mendez et al. 2002). (*A*, Modified from Thompson et al. 2005.)

increase may rely on a proposed positive feedback loop, in which FOG-1 directly regulates its own mRNA (Jin et al. 2001b). Thus, the FOG-1 gradient controls multiple states of differentiation at different threshold levels, as is required of a developmentally meaningful gradient. With no FOG-1, oocytes are made; at low levels of FOG-1, germ cells divide mitotically; and at high FOG-1 levels, sperm are produced.

The graded distribution of FOG-1 and its dose-dependent properties are striking in light of a report that *Xenopus* CPEB also acts in a dose-dependent manner (Mendez et al. 2002). Remarkably, when present at a low level in the early embryo, CPEB promotes mitosis, whereas at a high level in the oocyte, it is responsible for progression through meiosis. The number of CPEB molecules bound to a particular 3′UTR may determine whether CPEB acts as a translational activator or repressor (Fig. 4B) (Mendez et al. 2002). In one model, CPEB multimerization represses translation, whereas a single CPEB molecule activates. Of course, other mechanisms are possible.

Concluding Remarks

Translational regulators of development exist in a graded distribution, either in space or in time. However, for most, we do not know whether the graded regulator functions by a classic dose-dependent mechanism. The FOG-1/CPEB gradient in *C. elegans* appears to be one exception, but virtually nothing is known about its regulation at a biochemical level. To understand how that gradient works, target mRNAs must be identified and analyzed for number and strength of FOG-1-binding sites and recruitment of additional factors.

COMBINATORIAL CONTROL

Translational regulators often function in unique combinations to elicit distinct developmental fates. This theme has long been appreciated in the world of transcriptional regulators. For example, the binding partners of a transcription factor can affect sequence specificity, binding affinity, and whether a protein activates or represses transcription. Moreover, the DNA-binding sites themselves can act as allosteric effectors. Combinatorial action enables relatively few regulators to possess diverse activities and respond uniquely to diverse stimuli.

Combinatorial control is also prominent in translational regulation during development. Multiple proteins can be associated in regulatory complexes on a single 3′UTR; conversely, individual proteins bind to and regulate multiple mRNAs. In this section, we emphasize the roles of protein partners in combinatorial control and draw on two well-characterized examples.

PUF Proteins and Combinatorial Control

PUF proteins act in concert with a variety of other proteins to control multiple mRNAs. Their protein partners have both distinct and overlapping roles in development. Different combinations of regulators elicit specific biological outcomes, and even the same regulators can act in concert or opposition depending on context.

PUF Proteins and Nanos

Nanos proteins are conserved among metazoa and characterized by two distinctive CCHC zinc fingers that bind RNA nonspecifically on their own (Curtis et al. 1995, 1997). Nanos proteins interact with PUF proteins to control sex determination in *C. elegans* and axis formation in *Drosophila*. In *C. elegans*, three Nanos paralogs are required for FBF-mediated repression of the *fem-3* RNA (Kraemer et al. 1999); in *Drosophila*, Nanos is required for Pumilio-mediated repression of *hunchback* mRNA (Murata and Wharton 1995; Sonoda and Wharton 1999). The two proteins physically interact in both systems, but although the worm proteins interact in the absence of RNA, the fly proteins interact only in the presence of an RNA-binding site (Kraemer et al. 1999; Sonoda and Wharton 1999).

The regulation of *hunchback* and *cyclin B* by *Drosophila* Pumilio reveals that two RNAs can assemble different, but related, complexes. Repression of *hunchback* mRNA by Pumilio and Nanos requires a third regulator, the NHL protein Brat (Sonoda and Wharton 2001). Repression of *cyclin B* mRNA is independent of Brat, yet requires Pumilio and Nanos (Sonoda and Wharton 2001). *hunchback* mRNA is repressed in the posterior of the embryo and is critical in pattern formation; *cyclin B* mRNA is repressed in primordial germ cells, where it is thought to slow the cell cycle (Asaoka-Taguchi et al. 1999). Thus, the same proteins, Pumilio and Nanos, regulate mRNAs vital in two different processes, and they recruit one key regulator on one mRNA, but not the other. In addition, Pumilio also regulates *bicoid* mRNA in the anterior embryo (Gamberi et al. 2002), where Nanos protein is not detected; an additional partner may be involved.

PUF–Nanos partnerships are pervasive. In *C. elegans*, multiple PUF proteins interact with their own Nanos partners and collaborate to regulate germ-line survival (Kraemer et al. 1999; Subramaniam and Seydoux 1999; D. Bernstein and M. Wickens, unpubl.). Moreover, even in the same tissue, PUF and Nanos proteins can function cooperatively or antagonistically. For example, in the *C. elegans* germ line, FBF and NOS-3 function together to promote the sperm–oocyte switch (Zhang et al. 1997; Kraemer et al. 1999), but antagonistically in the mitosis–meiosis decision (Crittenden et al. 2002; Eckmann et al. 2004; Hansen et al. 2004b). In *Drosophila*, both Pumilio and Nanos are required for germ-line stem-cell maintenance, but the phenotypes of the individual mutants are subtly different (Forbes and Lehmann 1998), suggesting distinct functions as well.

PUF Proteins and CPEB

CPEB proteins also interact with multiple partners to control mRNAs in a variety of ways (also see "Switches"). In *C. elegans*, a CPEB protein, CPB-1, interacts directly with the PUF protein, FBF (Luitjens et al. 2000). CPB-1 and FBF control spermatogenesis, although they appear to function at distinct steps (Luitjens et al. 2000). *Xenopus* CPEB also interacts with a PUF protein in frog oocytes. There, CPEB and Pumilio are both required for repression of specific maternal mRNAs before oocyte maturation, although the functional significance of the CPEB–Pumilio interaction remains unclear. The roles of both CPEB and Pumilio in the nervous system (Schweers et al. 2002; Dubnau et al. 2003; Mee et al. 2004; Menon et al. 2004; Ye et al. 2004; Chapter 18) suggest that their collaboration may be widespread and may extend to somatic tissues.

PUF Proteins and DAZ

DAZ (*d*eleted in *a*zoospermia) proteins contain an RRM motif and a variable number of DAZ motifs (Reijo et al. 1995). In humans, multiple DAZ-encoding genes reside on the Y chromosome, and deletions of this region are associated with spermatogenic failure (Reijo et al. 1995). Two autosomal DAZ homologs, Daz-Like (DAZL) and Boule (BOL), also have been characterized (Xu et al. 2001). Daz family members are present in germ cells and are required for fertility in evolutionarily divergent organisms, including flies (Eberhart et al. 1996), worms (Karashima et al. 2000; Maruyama et al. 2005), frogs (Houston and King 2000), and mammals (Reijo et al. 1995; Ruggiu et al. 1997).

   DAZ and PUF proteins physically interact. In humans, the PUF protein, PUM2, can interact with both DAZL and BOL (Moore et al. 2003; Urano et al. 2005). Interestingly, PUM2–DAZL and PUM2–BOL complexes appear to bind different RNAs (Urano et al. 2005). The PUM2–DAZL complex interacts with the NRE (a well-defined binding site for *Drosophila* Pumilio), whereas the PUM2–BOL complex binds the 3'UTR of another PUM2 target, *SDAD1*; neither complex binds the noncognate site.

   DAZL also interacts with PUM2 in *Xenopus* (Padmanabhan and Richter 2006). Here, PUM2 and DAZL regulate the *Ringo/Spy* RNA, an atypical activator of CDK1, which is required for CPEB-dependent polyadenylation during oocyte maturation. The 3′UTR of *Ringo/Spy* possesses two Pumilio-binding sites that are required for proper translational regulation. In addition to PUM2, DAZL also interacts with embryonic PABP (ePABP), a positive regulator of translation (Collier et al. 2005). Indeed, DAZL stimulates the translation of reporter RNAs in a tethered function assay (Collier et al. 2005). Prior to oocyte maturation when *Ringo/Spy* is translationally repressed, PUM2, DAZL, and ePABP all associate with the *Ringo/Spy* RNA. Upon stimulation by progesterone, when *Ringo/Spy* is translationally activated, ePABP and DAZL remain associated with the *Ringo/Spy* mRNA, but PUM2 does not.

hnRNPs in Erythropoiesis

Erythropoiesis provides another elegant example of combinatorial control (for review, see Ostareck-Lederer and Ostareck 2004). Translation of the reticulocyte-15-lipoxgenase (*r15-LOX*) mRNA is regulated during erythroid development. r15-LOX protein initiates mitochondrial breakdown, a final step in erythroid development. *r15-LOX* mRNA is present throughout erythropoiesis, but it is translationally repressed until reticulocytes mature into erythrocytes (Thiele et al. 1982). Translational repression of *r15-LOX* is mediated by a CU-rich element within the 3′UTR of the *r15-LOX* mRNA called the differential control element, or DICE. Heterogeneous nuclear ribonucleoprotein (hnRNP) K and hnRNP E1 bind DICE and repress *r15-LOX* translation (Ostareck et al. 1997), and the hnRNP K/E1 complex represses translation by blocking 80S ribosome assembly (Ostareck et al. 2001).

α-Globin mRNA is also controlled by hnRNP E1 during erythropoiesis (Kiledjian et al. 1995). In contrast to *r15-LOX*, α-globin mRNA is continuously translated during erythropoiesis and is stabilized by hnRNP E1. Here, hnRNP E1 interacts with hnRNP E2, AUF1, and PAB-C, which together comprise a "stabilization complex," or α-complex. The α-complex is thought to stabilize α-globin mRNA by protecting it from endonucleolytic cleavage. Thus, hnRNP E1 works with hnRNP K to repress one mRNA, but with the α-complex to stabilize another mRNA.

Concluding Remarks

The complexity of translational regulators and their interactions continues to increase, but principles are beginning to emerge: conserved partnerships, multiple partners of a single regulator, and context-dependent effects. The next key step will be identification of the full complement of complexes associated with particular target mRNAs under defined circumstances. The choice of target mRNAs is of critical importance, both for experimental accessibility and for developmental significance.

NETWORKS OF TRANSLATIONAL REGULATORS

A network describes connections between nodes. In a biological context, those nodes can act at a number of levels, including cells, synapses, or molecules. A central concept in the network field is that the behavior of the network is scale-free (for review, see Barabási and Oltvai 2004). In virtually all networks, a few nodes function as hubs and are connected by many links to other nodes.

The number of well-documented developmental regulators has vastly increased during the past two decades. Emphasis can now shift from simple pathways to complex networks of control. Transcriptional and protein–protein interaction networks are being compiled (for review, see Stathopoulos and Levine 2005) and used to assemble networks of control during development (Levine and Davidson 2005). To date, a similar effort has not been made for translational regulators.

In this section, we describe an emerging network of developmental control that relies largely on translational regulation. We have chosen regulation of the *C. elegans* germ line as our example for several reasons. First, virtually all regulators are broadly conserved, and their biological functions are also conserved, at least for some regulators. Thus, an understanding of this network will likely provide insight into fundamental stem-cell controls. Second, the details now available, although far from complete, make it possible to begin thinking on a network level. Third, the key regulators of this network have been introduced earlier in this chapter, which makes it easier to focus on network properties, rather than individual regulators and their relationships.

The *C. elegans* Germ Line: A Network Controlling Stem Cells and Differentiation

Germ-line stem cells (GSC) must decide whether to divide mitotically or to enter the meiotic cell cycle, and they must differentiate either as sperm or oocyte. Remarkably, the same regulators control these two decisions in the *C. elegans* germ line. Therefore, this system is useful for comparing two subnetworks and for learning how they are integrated into a larger network.

Mitosis–Meiosis Decision

A simplified regulatory pathway for the mitosis–meiosis decision is depicted in Figure 5A (for review, see Kimble and Crittenden 2005). Germ cells are signaled to remain in mitosis by GLP-1/Notch signaling from a

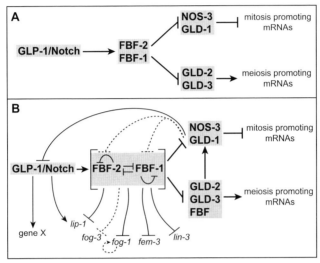

Figure 5. Regulation of the mitosis–meiosis decision in the *C. elegans* germ line. (*A*) A simple view of the pathway controlling the mitosis meiosis decision. The backbone of this pathway includes GLP-1/Notch signaling and FBF, which promote mitosis, and two redundant GLD branches that promote entry into meiosis. GLD-1 is a translational repressor and is likely to repress mRNAs, whereas GLD-2/GLD-3 is a poly(A) polymerase and is likely to activate mRNAs. (*B*) A more complete view of the emerging network that controls the mitosis–meiosis decision. (*Gray*) Backbone of the pathway. GLP-1/Notch signaling directly activates *fbf-2* transcription (Lamont et al. 2004) and *lip-1*, a putative MAPK phosphatase (Lee et al. 2006); GLP-1/Notch is likely to regulate additional target genes as well (gene *x*). FBF represses multiple RNAs, including mRNAs in both branches of the meiosis-promoting pathways (*gld-1*; Crittenden et al. 2002; *gld-3*; Eckmann et al. 2004). In addition, FBF directly represses *fog-1* (Thompson et al. 2005), *fem-3* (Zhang et al. 1997), *lip-1* (Lee et al. 2006), *lin-3* (Thompson et al. 2006), and possibly *fog-3* (Thompson et al. 2005). GLD-2 and GLD-3 function together as a novel poly(A) polymerase (Wang et al. 2002), and the *gld-1* mRNA itself appears to be a target of this poly(A) polymerase (N. Suh and J. Kimble, unpubl.). This network also includes multiple feedback loops: FBF-1 and FBF-2 inhibit both each other and themselves, a regulation that tightly controls FBF abundance (Lamont et al. 2004); GLD-1 represses *glp-1* mRNA (Marin and Evans 2003), which in effect turns down *fbf* expression. Also, the *fbf-1* and *fbf-2* 5′UTR regions contain GLD-1-binding sites, suggesting that GLD-1 might directly repress *fbf* expression (Ryder et al. 2004). Broken lines indicate that the interaction is hypothesized, but has not yet been experimentally validated.

somatic-stem-cell niche cell that resides at one end of the germ line. Downstream from GLP-1/Notch signaling, germ cells employ a series of RNA regulators to promote continued mitotic divisions or to enter the meiotic cell cycle. The PUF protein FBF is the primary regulator of stem cells and mitosis (Crittenden et al. 2002). Briefly, FBF controls each of two regulatory branches that promote entry into meiosis. In one branch is GLD-1, a STAR RNA-binding protein and translational repressor (Jones et al. 1996; Jan et al. 1999; Lee and Schedl 2001) and in the other branch are GLD-2 and GLD-3, which function together as a novel PAP (Wang et al. 2002). FBF promotes mitosis by repressing the activities of mRNAs in both branches of the meiotic entry pathway (Crittenden et al. 2002; Eckmann et al. 2002). GLD-1 is proposed to repress mitosis-promoting mRNAs, and the GLD-2/GLD-3 PAP is proposed to activate meiosis-promoting mRNAs.

Figure 5B presents a more detailed pathway, which verges on a network. This more complete picture includes more regulators, more functional interactions between regulators, and three levels of redundancy (Fig. 5B). The details of its complexity are not critical here and have been relegated to the figure legend for interested readers. Instead, we emphasize broad features of the network, which have parallels to other systems of regulation.

First, FBF emerges as a likely hub in the network. Multiple FBF target mRNAs are crucial regulators of the mitosis–meiosis decision (e.g., *gld-1, gld-3, lip-1; fog-1; fbf* itself) (Crittenden et al. 2002; Eckmann et al. 2002; Lamont et al. 2004; Thompson et al. 2005; Lee et al. 2006). A key feature of FBF is that it works not only as a repressor (Zhang et al. 1997; Crittenden et al. 2002), but also as an activator together with GLD-2 and GLD-3 (N. Suh and J. Kimble, unpubl.). The binding of FBF to a particular RNA can therefore mark it for either inactivity or activity, depending on prevailing conditions. The FBF hub therefore provides a seat for plasticity in the network. GLD-1 is likely to be another hub, but less is known about its mRNA targets that are relevant for this developmental decision. However, many potential GLD-1 target mRNAs have been identified (Lee and Schedl 2001; Ryder et al. 2004), consistent with hub status.

Second, the network includes several examples of redundancy: FBF is encoded by two nearly identical genes that are essentially redundant (Crittenden et al. 2002; Lamont et al. 2004), FBF and FOG-1 act redundantly to promote mitotic divisions in early larvae (Thompson et al. 2005), and the two branches of GLD activity are redundant for entry into meiosis.

Third, the network includes multiple feedback loops and cross-regulation. Negative feedback loops are crucial to ensure the transition from the FBF mitosis-promoting mode to the GLD meiosis-promoting mode. Mutual inhibition by the two FBFs is critical for maintenance of both FBF abundance and extent; a positive loop (GLD-2/GLD-3 directly activates *gld-1* expression) reinforces the transition into the GLD-dependent mode and makes that transition robust.

Finally, the intricacies of the network make it both robust and regulatable. For example, entry into meiosis can rely on either the GLD-1 or GLD-2/GLD-3 branches, which are linked by positive regulation. Therefore, the decision to enter meiosis, once made, is reinforced by a two-pronged coupled mechanism. In addition, the maintenance of germ cells in the mitotic cell cycle relies on redundant FBF proteins that are regulated to fine-tune the number of cells with the mitotic fate. Alteration of any one of the key components of this network does not eliminate the mitosis–meiosis decision, but instead changes the number of cells in mitosis and the timing of the switch into meiosis. This network therefore achieves a plasticity that may be one of its most crucial properties.

Sperm–Oocyte Decision

C. elegans hermaphrodites first produce sperm and then switch to making oocytes. The circuitry of this sperm–oocyte decision has been examined in detail (for review, see Ellis and Schedl 2006). Here, we discuss features of this circuit that are instructive for thinking about regulatory networks. An important feature is that the regulators of the sperm-oocyte decision are essentially the same as those of the mitosis–meiosis decision, including FBF, GLD-1, NOS-3, and GLD-3. Therefore, the regulatability inherent to this network, and which is used for determining the number of cells in mitosis, can also be used to regulate sperm number, a developmental attribute that is under strong genetic selection. One might think a priori that the circuits for the mitosis–meiosis and sperm–oocyte decisions would be the same. However, several regulators have distinct functional relationships in the two decisions. For example, both FBF and FOG-1 promote mitosis, but FBF is required for the switch to oogenesis, whereas FOG-1 specifies the sperm fate. Another example is GLD-3, which works upstream of FBF to promote continued spermatogenesis (Eckmann et al. 2002), but downstream from FBF to promote entry into meiosis (Eckmann et al. 2004). The take-home lessons are that the specifics of network function cannot be inferred from binding interactions and cannot be transferred from one subnetwork to another.

Parallels with Other Networks

The *C. elegans* germ-line network described briefly above has many parallels with other networks. Here, we briefly review parallels between the mitosis–meiosis network and other networks that are emerging at a translational level.

The existence of "hubs" is a major theme in network biology. In the mitosis–meiosis network, FBF stands out as a likely hub. Indeed, PUF proteins more generally appear to function as hubs. PUF proteins are associated with 100–200 RNAs in yeast extracts. In some cases, these mRNAs are clearly related in function. Yeast PUF3, for example, binds more than 100 nucleus-encoded mitochondrial mRNAs (Gerber et al. 2004). Coimmunoprecipitation experiments to identify mRNA targets of PUM2 and DAZL in human cells identified more than 60 potential target mRNAs, supporting the broad role of both PUF and DAZ family members in mRNA regulation (Fox et al. 2005). In other systems, RNA-binding proteins are also likely to function as hubs. Examples include Nova in the nervous system and ARE-binding proteins, where a single protein clearly recognizes and regulates multiple related mRNAs.

Redundancy is a prominent feature of the mitosis–meiosis network (see above). Other translational networks in the *C. elegans* germ line also exhibit redundancies. For example, two RNA regulators, GLD-1 and MEX-3, function redundantly to maintain totipotency in germ-line stem cells, a developmental phenomenon that is surely under rigorous control (Ciosk et al. 2006). Similarly, three Nanos homologs, NOS-1, NOS-2, and NOS-3, function redundantly to promote germ-line survival (Kraemer et al. 1999; Subramaniam and Seydoux 1999).

Other translational networks also employ extensive feedback regulation. For example, the translational regulation by miRNAs establishes a double-negative feedback loop that maintains a bistable state to control left/right asymmetry in the *C. elegans* nervous system (Johnston et al. 2005). In plants, two miRNAs target mRNAs that encode proteins implicated in miRNA biogenesis, resulting in a negative feedback loop (Dugas and Bartel 2004). The *Drosophila* CPEB homolog, Orb, autoregulates its own expression through a positive feedback loop (Tan et al. 2001), and some evidence suggests that the CPEB protein FOG-1 is also subject to positive autoregulation (Jin et al. 2001b).

Concluding Remarks

Translational networks have not yet been defined at a systems level. That enterprise will require a concerted effort of researchers from several fields, much as is currently happening with transcriptional networks. We suggest that elucidation of translational control networks is an important goal. One logical starting point is the network controlling development of the germ line, in part because translational control is so crucial in this tissue, in part because the first steps toward building that network have been taken, and in part because an understanding of the germ line has such clear intellectual and clinical significance.

PERSPECTIVES

Translational controls affect virtually all tissues during metazoan development, but they prevail in the germ line and early embryo. Remarkably, many conserved regulators control similar events in creatures as diverse as *C. elegans, Drosophila, Xenopus*, and mammals. For example, the maintenance of germ-line stem cells stands out as a conserved function of PUF RNA-binding proteins in animals as divergent as *Drosophila* and *C. elegans*, with hints that mammalian germ-line stem cells may also be controlled in the same way. Another striking example is the control of mitosis by low levels of CPEB in *Xenopus* and also by a CPEB homolog in *C. elegans*. These similarities across vast phylogenetic distances argue that the mechanisms represent ancient solutions for control of the metazoan cell cycle. As such, their detailed analysis will have broad significance, in terms of both fundamental principles and clinical applications.

Translational controls are not limited to the germ line and early embryo. Conserved translational regulators are also present in somatic tissues, and their developmental functions are beginning to emerge. Arguably, the best example is the use of miRNAs to control life cycle progression of somatic tissues in *C. elegans*. In addition, miRNAs have been implicated in various aspects of vertebrate somatic development, including the vasculature and limb. Protein regulators are also present in the soma, although their roles, for the most part, remain to be elucidated. It is clear that the impact of translational control on developing organisms is set to increase tremendously in the near future.

Translational regulators control development by the same basic principles as transcriptional regulators. Sequence-specific RNA-binding proteins (or miRNAs) recognize particular mRNAs to control their expression. They can activate or repress, and they do so by contact or interference with the basic machinery of translation. Regulatory proteins act as molecular switches, work in a combinatorial fashion to reinforce or antagonize complexes, and generate and function in gradients. All of these attributes are critical for their roles in development. In fact, several classic phenomena, such as establishment of the anterior–posterior axis in *Drosophila*, control of the sperm/oocyte decision in *C. elegans*, and control of germ-line stem cells in both organisms, are now understood to rely, in large part, on translational regulation.

Our understanding of how translational regulators are able to control, for example, stem-cell maintenance, body axis formation, and life cycle progression remains in its infancy, particularly with respect to biochemistry and systems biology. Nonetheless, the progress made to date provides a critical foundation for taking those next giant steps. More intensive molecular and bioinformatic analyses will surely reveal new biochemical complexes, new biochemical mechanisms, and new networks.

Translational controls of development have a remarkable parallel with translational controls of memory formation (Chapter 18 and this chapter). For example, CPEB and PUF proteins have been implicated in both, and GLD-2 colocalizes in the mammalian brain to regions dedicated to long-term memory. The critical link between development and neural functions underlying memory may be their inherent plasticity. Indeed, reliance on translational control circuits to accomplish plasticity takes advantage of the rapid and reversible changes that are typical of this level of gene control.

ACKNOWLEDGMENTS

We thank members of the Kimble and Wickens labs for ongoing discussions, Anne Helsley-Marchbanks for help preparing the manuscript, and Laura Vanderploeg for help with figures. M.W. and J.K. are supported by the National Institutes of Health; J.K. is an investigator of the Howard Hughes Medical Institute.

REFERENCES

Ahringer J. and Kimble J. 1991. Control of the sperm-oocyte switch in *Caenorhabditis elegans* hermaphrodites by the *fem-3* 3′ untranslated region. *Nature* **349:** 346–348.

Alvarez-Garcia I. and Miska E.A. 2005. MicroRNA functions in animal development and human disease. *Development* **132:** 4653–4662.

Arasu P., Wightman B., and Ruvkun G. 1991. Temporal regulation of *lin-14* by the antagonistic action of two other heterochronic genes, *lin-4* and *lin-28*. *Genes Dev.* **5:** 1825–1833.

Aravind L. and Koonin E.V. 1999. DNA polymerase β-like nucleotidyltransferase superfamily: Identification of three new families, classification and evolutionary history. *Nucleic Acids Res.* **27:** 1609–1618.

Asaoka-Taguchi M., Yamada M., Nakamura A., Hanyu K., and Kobayashi S. 1999. Maternal Pumilio acts together with Nanos in germline development in *Drosophila* embryos. *Nat. Cell Biol.* **1:** 431–437.

Ashe H.L. and Briscoe J. 2006. The interpretation of morphogen gradients. *Development* **133:** 385–394.

Aviv T., Lin Z., Lau S., Rendl L.M., Sicheri F., and Smibert C.A. 2003. The RNA-binding SAM domain of Smaug defines a new family of post-transcriptional regulators. *Nat. Struct. Biol.* **10:** 614–621.

Baez M.V. and Boccaccio G.L. 2005. Mammalian Smaug is a translational repressor that forms cytoplasmic foci similar to stress granules. *J. Biol. Chem.* **280:** 43131–43140.

Bagga S., Bracht J., Hunter S., Massirer K., Holtz J., Eachus R., and Pasquinelli A.E. 2005. Regulation by *let-7* and *lin-4* miRNAs results in target mRNA degradation. *Cell* **122:** 553–563.

Baker B.S., Gorman M., and Marín I. 1994. Dosage compensation in *Drosophila*. *Annu. Rev. Genet.* **28:** 491–521.

Barabási A.-L. and Oltvai Z.N. 2004. Network biology: Understanding the cell's functional organization. *Nat. Rev. Genet.* **5:** 101–113.

Bard J., Zhelkovsky A.M., Helmling S., Earnest T.N., Moore C.L., and Bohm A. 2000. Structure of yeast poly(A) polymerase alone and in complex with 3′-dATP. *Science* **289:** 1346–1349.

Barkoff A.F., Dickson K.S., Gray N.K., and Wickens M. 2000. Translational control of cyclin B1 mRNA during meiotic maturation: Coordinated repression and cytoplasmic polyadenylation. *Dev. Biol.* **220:** 97–109.

Barnard D.C., Ryan K., Manley J.L., and Richter J.D. 2004. Symplekin and xGLD-2 are required for CPEB-mediated cytoplasmic polyadenylation. *Cell* **119:** 641–651.

Bartel D.P. 2004. MicroRNAs: Genomics, biogenesis, mechanism, and function. *Cell* **116:** 281–297.

Barton M.K. and Kimble J. 1990. *fog-1*, a regulatory gene required for specification of spermatogenesis in the germ line of *Caenorhabditis elegans*. *Genetics* **125:** 29–39.

Bashaw G.J. and Baker B.S. 1995. The *msl-2* dosage compensation gene of *Drosophila* encodes a putative DNA binding protein whose expression is sex specifically regulated by *Sex-lethal*. *Development* **121:** 3245–3258.

Beckmann K., Grskovic M., Gebauer F., and Hentze M.W. 2005. A dual inhibitory mechanism restricts *msl-2* mRNA translation for dosage compensation in *Drosophila*. *Cell* **122:** 529–540.

Bergsten S.E. and Gavis E.R. 1999. Role for mRNA localization in translational activation but not spatial restriction of *nanos* RNA. *Development* **126:** 659–669.

Berleth T., Burri M., Thoma G., Bopp D., Richstein S., Frigerio G., Noll M., and Nüsslein-Volhard C. 1988. The role of localization of *bicoid* RNA in organizing the anterior pattern of the *Drosophila* embryo. *EMBO J.* **7:** 1749–1756.

Bienz M. 1998. TCF: Transcriptional activator or repressor? *Curr. Opin. Cell Biol.* **10:** 366–372.

Bilger A., Fox C.A., Wahle E., and Wickens M. 1994. Nuclear polyadenylation factors recognize cytoplasmic polyadenylation elements. *Genes Dev.* **8:** 1106–1116.

Boag P.R., Nakamura A., and Blackwell T.K. 2005. A conserved RNA-protein complex component involved in physiological germline apoptosis regulation in *C. elegans*. *Development* **132:** 4975–4986.

Brennecke J., Stark A., Russell R.B., and Cohen S.M. 2005. Principles of microRNA-target recognition. *PLoS Biol.* **3:** e85.

Brennecke J., Hipfner D.R., Stark A., Russell R.B., and Cohen S.M. 2003. *bantam* encodes a developmentally regulated microRNA that controls cell proliferation and regulates the proapoptotic gene *hid* in *Drosophila*. *Cell* **113:** 25–36.

Castagnetti S. and Ephrussi A. 2003. Orb and a long poly(A) tail are required for efficient *oskar* translation at the posterior pole of the *Drosophila* oocyte. *Development* **130:** 835–843.

Chagnovich D. and Lehmann R. 2001. Poly(A)-independent regulation of maternal *hunchback* translation in the *Drosophila* embryo. *Proc. Natl. Acad. Sci.* **98:** 11359–11364.

Chalfie M., Horvitz H.R., and Sulston J. 1981. Mutations that lead to reiterations in the cell lineages of *C. elegans*. *Cell* **24:** 59–69.

Chang J.S., Tan L., Wolf M.R., and Schedl P. 2001. Functioning of the *Drosophila orb* gene in *gurken* mRNA localization and translation. *Development* **128:** 3169–3177.

Chang S., Johnston R.J., Jr., Frøkjær-Jensen C., Lockery S., and Hobert O. 2004. MicroRNAs act sequentially and asymmetrically to control chemosensory laterality in the nematode. *Nature* **430:** 785–789.

Chekulaeva M., Hentze M.W., and Ephrussi A. 2006. Bruno acts as a dual repressor of *oskar* translation, promoting mRNA oligomerization and formation of silencing particles. *Cell* **124:** 521–533.

Cho P.F., Poulin F., Cho-Park Y.A., Cho-Park I.B., Chicoine J.D., Lasko P., and Sonenberg N. 2005. A new paradigm for translational control: Inhibition via 5'-3' mRNA tethering by Bicoid and the eIF4E cognate 4EHP. *Cell* **121:** 411–423.

Ciosk R., DePalma M., and Priess J.R. 2006. Translational regulators maintain totipotency in the *Caenorhabditis elegans* germline. *Science* **311:** 851–853.

Cline T.W. and Meyer B.J. 1996. Vive la différence: Males vs females in flies vs worms. *Annu. Rev. Genet.* **30:** 637–702.

Coller J. and Parker R. 2005. General translational repression by activators of mRNA decapping. *Cell* **122:** 875–886.

Collier B., Gorgoni B., Loveridge C., Cooke H.J., and Gray N.K. 2005. The DAZL family proteins are PABP-binding proteins that regulate translation in germ cells. *EMBO J.* **24:** 2656–2666.

Crittenden S.L., Bernstein D.S., Bachorik J.L., Thompson B.E., Gallegos M., Petcherski A.G., Moulder G., Barstead R., Wickens M., and Kimble J. 2002. A conserved RNA-binding protein controls germline stem cells in *Caenorhabditis elegans*. *Nature* **417:** 660–663.

Curtis D., Apfeld J., and Lehmann R. 1995. *nanos* is an evolutionarily conserved organizer of anterior-posterior polarity. *Development* **121:** 1899–1910.

Curtis D., Treiber D.K., Tao F., Zamore P.D., Williamson J.R., and Lehmann R. 1997. A CCHC metal-binding domain in Nanos is essential for translational regulation. *EMBO J.* **16:** 834–843.

Dahanukar A. and Wharton R.P. 1996. The Nanos gradient in *Drosophila* embryos is generated by translational regulation. *Genes Dev.* **10:** 2610–2620.

Dahanukar A., Walker J.A., and Wharton R.P. 1999. Smaug, a novel RNA-binding protein that operates a translational switch in *Drosophila*. *Mol. Cell* **4:** 209–218.

de Moor C.H. and Richter J.D. 1999. Cytoplasmic polyadenylation elements mediate masking and unmasking of cyclin B1 mRNA. *EMBO J.* **18:** 2294–2303.

Dickson K.S., Thompson S.R., Gray N.K., and Wickens M. 2001. Poly(A) polymerase and the regulation of cytoplasmic polyadenylation. *J. Biol. Chem.* **276:** 41810–41816.

Driever W. and Nüsslein-Volhard C. 1988a. The *bicoid* protein determines position in the *Drosophila* embryo in a concentration-dependent manner. *Cell* **54:** 95–104.

———. 1988b. A gradient of *bicoid* protein in *Drosophila* embryos. *Cell* **54:** 83–93.

Driever W., Thoma G., and Nüsslein-Volhard C. 1989. Determination of spatial domains of zygotic gene expression in the *Drosophila* embryo by the affinity of binding sites for the bicoid morphogen. *Nature* **340**: 363–367.

Dubnau J. and Struhl G. 1996. RNA recognition and translational regulation by a homeodomain protein. *Nature* **379**: 694–699.

Dubnau J., Chiang A.-S., Grady L., Barditch J., Gossweiler S., McNeil J., Smith P., Buldoc F., Scott R., Certa U., et al. 2003. The *staufen/pumilio* pathway is involved in *Drosophila* long-term memory. *Curr. Biol.* **13**: 286–296.

Dugas D.V. and Bartel B. 2004. MicroRNA regulation of gene expression in plants. *Curr. Opin. Plant Biol.* **7**: 512–520.

Eberhart C.G., Maines J.Z., and Wasserman S.A. 1996. Meiotic cell cycle requirement for a fly homologue of human *Deleted in Azoospermia*. *Nature* **381**: 783–785.

Eckmann C.R., Crittenden S.L., Suh N., and Kimble J. 2004. GLD-3 and control of the mitosis/meiosis decision in the germline of *Caenorhabditis elegans*. *Genetics* **168**: 147–160.

Eckmann C.R., Kraemer B., Wickens M., and Kimble J. 2002. GLD-3, a bicaudal-C homolog that inhibits FBF to control germline sex determination in *C. elegans*. *Dev. Cell* **3**: 697–710.

Edwards T.A., Pyle S.E., Wharton R.P., and Aggarwal A.K. 2001. Structure of Pumilio reveals similarity between RNA and peptide binding motifs. *Cell* **105**: 281–289.

Edwards T.A., Wilkinson B.D., Wharton R.P., and Aggarwal A.K. 2003. Model of the brain tumor–Pumilio translation repressor complex. *Genes Dev.* **17**: 2508–2513.

Ellis R. and Schedl T. 2006. Sex determination in the germ line (April 4, 2006). In *WormBook* (ed. The *C. elegans* Research Community). WormBook, doi/10.1895/wormbook.1.82.1, http://www.wormbook.org.

Ephrussi A. and Lehmann R. 1992. Induction of germ cell formation by *oskar*. *Nature* **358**: 387–392.

Ephrussi A. and St Johnston D. 2004. Seeing is believing: The bicoid morphogen gradient matures. *Cell* **116**: 143–152.

Forbes A. and Lehmann R. 1998. Nanos and Pumilio have critical roles in the development and function of *Drosophila* germline stem cells. *Development* **125**: 679–690.

Fox M., Urano J., and Reijo Pera R.A. 2005. Identification and characterization of RNA sequences to which human PUMILIO-2 (PUM2) and deleted in Azoospermia-like (DAZL) bind. *Genomics* **85**: 92–105.

Gamberi C., Peterson D.S., He L., and Gottlieb E. 2002. An anterior function for the *Drosophila* posterior determinant Pumilio. *Development* **129**: 2699–2710.

Gavis E. and Lehmann R. 1992. Localization of *nanos* RNA controls embryonic polarity. *Cell* **71**: 301–313.

Gebauer F. and Hentze M.W. 2004. Molecular mechanisms of translational control. *Nat. Rev. Mol. Cell Biol.* **5**: 827–835.

Gebauer F., Merendino L., Hentze M.W., and Valcarcel J. 1998. The *Drosophila* splicing regulator sex-lethal directly inhibits translation of *male-specific-lethal 2* mRNA. *RNA* **4**: 142–150.

Gerber A.P., Herschlag D., and Brown P.O. 2004. Extensive association of functionally and cytotopically related mRNAs with Puf family RNA-binding proteins in yeast. *PLoS Biol.* **2**: E79.

Goldstrohm A., Hook B., Seay D., and Wickens M. 2006. PUF proteins bind POP2p to regulate messenger RNAs. *Nat. Struct. Mol. Biol.* **13**: 533–539.

Good P.J., Chen Q., Warner S.J., and Herring D.C. 2000. A family of human RNA-binding proteins related to the *Drosophila* Bruno translational regulator. *J. Biol. Chem.* **275:** 28583–28592.

Gray N.K. and Wickens M. 1998. Control of translation initiation in animals. *Annu. Rev. Cell Dev. Biol.* **14:** 399–458.

Gurdon J.B. and Bourillot P.-Y. 2001. Morphogen gradient interpretation. *Nature* **413:** 797–803.

Ha I., Wightman B., and Ruvkun G. 1996. A bulged *lin-4/lin-14* RNA duplex is sufficient for *Caenorhabditis elegans lin-14* temporal gradient formation. *Genes Dev.* **10:** 3041–3050.

Hake L.E. and Richter J.D. 1994. CPEB is a specificity factor that mediates cytoplasmic polyadenylation during *Xenopus* oocyte maturation. *Cell* **79:** 617–627.

Hansen D., Hubbard E.J.A., and Schedl T. 2004a. Multi-pathway control of the proliferation versus meiotic development decision in the *Caenorhabditis elegans* germline. *Dev. Biol.* **268:** 342–357.

Hansen D., Wilson-Berry L., Dang T., and Schedl T. 2004b. Control of the proliferation versus meiotic development decision in the *C. elegans* germline through regulation of GLD-1 protein accumulation. *Development* **131:** 93–104.

Hornstein E., Mansfield J.H., Yekta S., Hu J.K.-H., Harfe B.D., McManus M.T., Baskerville S., Bartel D.P., and Tabin C.J. 2005. The microRNA *miR-196* acts upstream of Hoxb8 and Shh in limb development. *Nature* **438:** 671–674.

Houston D.W. and King M.L. 2000. A critical role for *Xdazl*, a germ plasm-localized RNA, in the differentiation of primordial germ cells in *Xenopus. Development* **127:** 447–456.

Hristova M., Birse D., Hong Y., and Ambros V. 2005. The *Caenorhabditis elegans* heterochronic regulator LIN-14 is a novel transcription factor that controls the developmental timing of transcription from the insulin/insulin-like growth factor gene *ins-33* by direct DNA binding. *Mol. Cell. Biol.* **25:** 11059–11072.

Huynh J.-R. and St Johnston D. 2000. The role of BicD, Egl, Orb and the microtubules in the restriction of meiosis to the *Drosophila* oocyte. *Development* **127:** 2785–2794.

Jan E., Motzny C.K., Graves L.E., and Goodwin E.B. 1999. The STAR protein, GLD-1, is a translational regulator of sexual identity in *Caenorhabditis elegans. EMBO J.* **18:** 258–269.

Jin S.W., Kimble J., and Ellis R.E. 2001a. Regulation of cell fate in *Caenorhabditis elegans* by a novel cytoplasmic polyadenylation element binding protein. *Dev. Biol.* **229:** 537–553.

Jin S.W., Arno N., Cohen A., Shah A., Xu Q., Chen N., and Ellis R.E. 2001b. In *Caenorhabditis elegans*, the RNA-binding domains of the cytoplasmic polyadenylation element binding protein FOG-1 are needed to regulate germ cell fates. *Genetics* **159:** 1617–1630.

Johnston R.J., Jr. and Hobert O. 2003. A microRNA controlling left/right neuronal asymmetry in *Caenorhabditis elegans. Nature* **426:** 845–849.

Johnston R.J., Jr., Chang S., Etchberger J.F., Ortiz C.O., and Hobert O. 2005. MicroRNAs acting in a double-negative feedback loop to control a neuronal cell fate decision. *Proc. Natl. Acad. Sci.* **102:** 12449–12454.

Johnstone O. and Lasko P. 2001. Translational regulation and RNA localization in *Drosophila* oocytes and embryos. *Annu. Rev. Genet.* **35:** 365–406.

Jones A.R., Francis R., and Schedl T. 1996. GLD-1, a cytoplasmic protein essential for oocyte differentiation, shows stage- and sex-specific expression during *Caenorhabditis elegans* germline development. *Dev. Biol.* **180:** 165–183.

Jones-Rhoades M.W., Bartel D.P., and Bartel B. 2006. MicroRNAs and their regulatory roles in plants. *Annu. Rev. Plant Biol.* **57:** 19–53.

Kadyk L.C. and Kimble J. 1998. Genetic regulation of entry into meiosis in *Caenorhabditis elegans*. *Development* **125:** 1803–1813.

Karashima T., Sugimoto A., and Yamamoto M. 2000. *Caenorhabditis elegans* homologue of the human azoospermia factor *DAZ* is required for oogenesis but not for spermatogenesis. *Development* **127:** 1069–1079.

Kiledjian M., Wang X., and Liebhaber S.A. 1995. Identification of two KH domain proteins in the α-globin mRNP stability complex. *EMBO J.* **14:** 4357–4364.

Kim-Ha J., Kerr K., and Macdonald P.M. 1995. Translational regulation of *oskar* mRNA by bruno, an ovarian RNA-binding protein, is essential. *Cell* **81:** 403–412.

Kimble J. and Crittenden S. 2005. Germline proliferation and its control (August 15, 2005). In *WormBook* (ed. The *C. elegans* Research Community). WormBook, doi/10.1895/wormbook.1.13.1, http://www.wormbook.org.

Kraemer B., Crittenden S., Gallegos M., Moulder G., Barstead R., Kimble J., and Wickens M. 1999. NANOS-3 and FBF proteins physically interact to control the sperm-oocyte switch in *Caenorhabditis elegans*. *Curr. Biol.* **9:** 1009–1018.

Kwak J.E., Wang L., Ballantyne S., Kimble J., and Wickens M. 2004. Mammalian GLD-2 homologs are poly(A) polymerases. *Proc. Natl. Acad. Sci.* **101:** 4407–4412.

Ladomery M., Wade E., and Sommerville J. 1997. Xp54, the *Xenopus* homologue of human RNA helicase p54, is an integral component of stored mRNP particles in oocytes. *Nucleic Acids Res.* **25:** 965–973.

Lai E.C., Tam B., and Rubin G.M. 2005. Pervasive regulation of *Drosophila* Notch target genes by GY-box-, Brd-box-, and K-box-class microRNAs. *Genes Dev.* **19:** 1067–1080.

Lamont L.B., Crittenden S.L., Bernstein D., Wickens M., and Kimble J. 2004. FBF-1 and FBF-2 regulate the size of the mitotic region in the *C. elegans* germline. *Dev. Cell* **7:** 697–707.

Lee M.-H. and Schedl T. 2001. Identification of in vivo mRNA targets of GLD-1, a maxi-KH motif containing protein required for *C. elegans* germ cell development. *Genes Dev.* **15:** 2408–2420.

Lee M.-H., Hook B., Lamont L.B., Wickens M., and Kimble J. 2006. LIP-1 phosphatase controls the extent of germline proliferation in *Caenorhabditis elegans*. *EMBO J.* **25:** 88–96.

Lee R.C., Feinbaum R.L., and Ambros V. 1993. The *C. elegans* heterochronic gene *lin-4* encodes small RNAs with antisense complementarity to *lin-14*. *Cell* **75:** 843–854.

Lettre G., Kritikou E.A., Jaeggi M., Calixto A., Fraser A.G., Kamath R.S., Ahringer J., and Hengartner M.O. 2004. Genome-wide RNAi identifies p53-dependent and -independent regulators of germ cell apoptosis in *C. elegans*. *Cell Death Differ.* **11:** 1198–1203.

Levine M. and Davidson E.H. 2005. Gene regulatory networks for development. *Proc. Natl. Acad. Sci.* **102:** 4936–4942.

Luitjens C., Gallegos M., Kraemer B., Kimble J., and Wickens M. 2000. CPEB proteins control two key steps in spermatogenesis in *C. elegans*. *Genes Dev.* **14:** 2596–2609.

Lykke-Andersen J. and Wagner E. 2005. Recruitment and activation of mRNA decay enzymes by two ARE-mediated decay activation domains in the proteins TTP and BRF-1. *Genes Dev.* **19:** 351–361.

Macdonald P. 1992. The *Drosophila pumilio* gene: An unusually long transcription unit and an unusual protein. *Development* **114:** 221–232.

Macdonald P.M. and Struhl G. 1986. A molecular gradient in early *Drosophila* embryos and its role in specifying the body pattern. *Nature* **324:** 537–545.

Mallory A.C., Reinhart B.J., Jones-Rhoades M.W., Tang G., Zamore P.D., Barton M.K., and Bartel D.P. 2004. MicroRNA control of PHABULOSA in leaf development: Importance of pairing to the microRNA 5′ region. *EMBO J.* **23:** 3356–3364.

Marin V.A. and Evans T.C. 2003. Translational repression of a *C. elegans* Notch mRNA by the STAR/KH domain protein GLD-1. *Development* **130:** 2623–2632.

Martin G., Keller W., and Doublie W. 2000. Crystal structure of mammalian poly(A) polymerase in complex with an analog of ATP. *EMBO J.* **19:** 4193–4203.

Maruyama R., Endo S., Sugimoto A., and Yamamoto M. 2005. *Caenorhabditis elegans* DAZ-1 is expressed in proliferating germ cells and directs proper nuclear organization and cytoplasmic core formation during oogenesis. *Dev. Biol.* **277:** 142–154.

Mee C.J., Pym E.C.G., Moffat K.G., and Baines R.A. 2004. Regulation of neuronal excitability through pumilio-dependent control of a sodium channel gene. *J. Neurosci.* **24:** 8695–8703.

Meller V.H. and Kuroda M.I. 2002. Sex and the single chromosome. *Adv. Genet.* **46:** 1–24.

Mendez R. and Richter J.D. 2001. Translational control by CPEB: A means to the end. *Nat. Rev. Mol. Cell Biol.* **2:** 521–529.

Mendez R., Barnard D., and Richter J.D. 2002. Differential mRNA translation and meiotic progression require Cdc2-mediated CPEB destruction. *EMBO J.* **21:** 1833–1844.

Mendez R., Murthy K.G.K., Ryan K., Manley J.L., and Richter J.D. 2000. Phosphorylation of CPEB by Eg2 mediates the recruitment of CPSF into an active cytoplasmic polyadenylation complex. *Mol. Cell* **6:** 1253–1259.

Menon K.P., Sanyal S., Habara Y., Sanchez R., Wharton R.P., Ramaswami M., and Zinn K. 2004. The translational repressor Pumilio regulates presynaptic morphology and controls postsynaptic accumulation of translation factor eIF-4E. *Neuron* **44:** 663–676.

Minshall N., Thom G., and Standart N. 2001. A conserved role of a DEAD box helicase in mRNA masking. *RNA* **7:** 1728–1742.

Minshall N., Walker J., Dale M., and Standart N. 1999. Dual roles of p82, the clam CPEB homolog, in cytoplasmic polyadenylation and translational masking. *RNA* **5:** 27–38.

Mlodzik M., Fjose A., and Gehring W.J. 1985. Isolation of *caudal*, a *Drosophila* homeo box-containing gene with maternal expression, whose transcripts form a concentration gradient at the pre-blastoderm stage. *EMBO J.* **4:** 2961–2969.

Moore F.L., Jaruzelska J., Fox M.S., Urano J., Firpo M.T., Turek P.J., Dorfman D.M., and Reijo Pera R.A. 2003. Human Pumilio-2 is expressed in embryonic stem cells and germ cells and interacts with DAZ (Deleted in AZoospermia) and DAZ-like proteins. *Proc. Natl. Acad. Sci.* **100:** 538–543.

Moss E.G., Lee R.C., and Ambros V. 1997. The cold shock domain protein LIN-28 controls developmental timing in *C. elegans* and is regulated by the *lin-4* RNA. *Cell* **88:** 637–646.

Munro T.P., Kwon S., Schnapp B.J., and St. Johnston D. 2006. A repeated IMP-binding motif controls *oskar* mRNA translation and anchoring independently of *Drosophila melanogaster* IMP. *J. Cell Biol.* **172:** 577–588.

Murata Y. and Wharton R. 1995. Binding of Pumilio to maternal *hunchback* mRNA is required for posterior patterning in *Drosophila* embryos. *Cell* **80:** 747–756.

Nakahata S., Katsu Y., Mita K., Inoue K., Nagahama Y., and Yamashita M. 2001. Biochemical identification of *Xenopus* Pumilio as a sequence-specific cyclin B1 mRNA-binding protein that physically interacts with a Nanos homolog, Xcat-2, and a cytoplasmic polyadenylation element-binding protein. *J. Biol. Chem.* **276:** 20945–20953.

Nakahata S., Kotani T., Mita K., Kawasaki T., Katsu Y., Nagahama Y., and Yamashita M. 2003. Involvement of *Xenopus* Pumilio in the translational regulation that is specific to cyclin B1 mRNA during oocyte maturation. *Mech. Dev.* **120:** 865–880.

Nakamura A., Sato K., and Hanyu-Nakamura K. 2004. *Drosophila* Cup is an eIF4E binding protein that associates with Bruno and regulates *oskar* mRNA translation in oogenesis. *Dev. Cell* **6:** 69–78.

Nakamura A., Amikura R., Hanyu K., and Kobayashi S. 2001. Me31B silences translation of oocyte-localizing RNAs through the formation of cytoplasmic RNP complex during *Drosophila* oogenesis. *Development* **128:** 3233–3242.

Nakanishi T., Kubota H., Ishibashi N., Kumagai S., Watanabe H., Yamashita M., Kashiwabara S., Miyado K., and Baba T. 2006. Possible role of mouse poly(A) polymerase mGLD-2 during oocyte maturation. *Dev. Biol.* **289:** 115–126.

Nelson M.R., Leidal A.M., and Smibert C.A. 2004. *Drosophila* Cup is an eIF4E-binding protein that functions in Smaug-mediated translational repression. *EMBO J.* **23:** 150–159.

Nusinow D.A. and Panning B. 2005. Recognition and modification of sex chromosomes. *Curr. Opin. Genet. Dev.* **15:** 206–213.

Olivas W. and Parker R. 2000. The Puf3 protein is a transcript-specific regulator of mRNA degradation in yeast. *EMBO J.* **19:** 6602–6611.

Opperman L., Hook B., DeFino M., Bernstein D.S., and Wickens M. 2005. A single spacer nucleotide determines the specificities of two mRNA regulatory proteins. *Nat. Struct. Mol. Biol.* **12:** 945–951.

Ostareck D.H., Ostareck-Lederer A., Shatsky I.N., and Hentze M.W. 2001. Lipoxygenase mRNA silencing in erythroid differentiation: The 3′UTR regulatory complex controls 60S ribosomal subunit joining. *Cell* **104:** 281–290.

Ostareck D.H., Ostareck-Lederer A., Wilm M., Thiele B.J., Mann M., and Hentze M.W. 1997. mRNA silencing in erythroid differentiation: hnRNP K and hnRNP E1 regulate 15-lipoxygenase translation from the 3′ end. *Cell* **89:** 597–606.

Ostareck-Lederer A. and Ostareck D.H. 2004. Control of mRNA translation and stability in haematopoietic cells: The function of hnRNPs K and E1/E2. *Biol. Cell* **96:** 407–411.

Padmanabhan K. and Richter J.D. 2006. Regulated Pumilio-2 binding controls RINGO/Spy mRNA translation and CPEB activation. *Genes Dev.* **20:** 199–209.

Pasquinelli A.E., Hunter S., and Bracht J. 2005. MicroRNAs: A developing story. *Curr. Opin. Genet. Dev.* **15:** 200–205.

Pasquinelli A.E., Reinhart B.J., Slack F., Martindale M.Q., Kuroda M.I., Maller B., Hayward D.C., Ball E.E., Degnan B., Müller P., et al. 2000. Conservation of the sequence and temporal expression of *let-7* heterochronic regulatory RNA. *Nature* **408:** 86–89.

Penalva L.O. and Sánchez L. 2003. RNA binding protein sex-lethal (Sxl) and control of *Drosophila* sex determination and dosage compensation. *Microbiol. Mol. Biol. Rev.* **67:** 343–359.

Read R.L., Martinho R.G., Wang S.-W., Carr A.M., and Norbury C.J. 2002. Cytoplasmic poly(A) polymerases mediate cellular responses to S phase arrest. *Proc. Natl. Acad. Sci.* **99:** 12079–12084.

Reijo R., Lee T.Y., Salo P., Alagappan R., Brown L.G., Rosenberg M., Rozen S., Jaffe T., Straus D., and Hovatta O., et al. 1995. Diverse spermatogenic defects in humans caused by Y chromosome deletions encompassing a novel RNA-binding protein gene. *Nat. Genet.* **10:** 383–393.

Reinhart B.J., Slack F.J., Basson M., Pasquinelli A.E., Bettinger J.C., Rougvie A.E., Horvitz H.R., and Ruvkun G. 2000. The 21-nucleotide *let-7* RNA regulates developmental timing in *Caenorhabditis elegans*. *Nature* **403:** 901–906.

Richter J.D. and Sonenberg N. 2005. Regulation of cap-dependent translation by eIF4E inhibitory proteins. *Nature* **433:** 477–480.

Rivera-Pomar R., Niessing D., Schmidt-Ott U., Gehring W.J., and Jäckle H. 1996. RNA binding and translational suppression by bicoid. *Nature* **379:** 746–749.

Rongo C., Gavis E.R., and Lehmann R. 1995. Localization of *oskar* RNA regulates *oskar* translation and requires Oskar protein. *Development* **121:** 2737–2746.

Rouhana L., Wang L., Buter N., Kwak J.E., Schiltz C.A., Gonzalez T., Kelley A.E., Landry C.F., and Wickens M. 2005. Vertebrate GLD2 poly(A) polymerases in the germline and the brain. *RNA* **11:** 1117–1130.

Ruggiu M., Speed R., Taggart M., McKay S.J., Kilanowski F., Saunders P., Dorin J., and Cooke H.J. 1997. The mouse *Dazla* gene encodes a cytoplasmic protein essential for gametogenesis. *Nature* **389:** 73–77.

Ruvkun G. and Giusto J. 1989. The *Caenorhabditis elegans* heterochronic gene *lin-14* encodes a nuclear protein that forms a temporal developmental switch. *Nature* **338:** 313–319.

Ryder S.P., Frater L.A., Abramovitz D.L., Goodwin E.B., and Williamson J.R. 2004. RNA target specificity of the STAR/GSG domain post-transcriptional regulatory protein GLD-1. *Nat. Struct. Mol. Biol.* **11:** 20–28.

Saitoh S., Chabes A., McDonald W.H., Thelander L., Yates J.R., III, and Russell P. 2002. Cid13 is a cytoplasmic poly(A) polymerase that regulates ribonucleotide reductase mRNA. *Cell* **109:** 563–573.

Salvetti A., Rossi L., Lena A., Batistoni R., Deri P., Rainaldi G., Locci M.T., Evangelista M., and Gremigni V. 2005. *DjPum*, a homologue of *Drosophila Pumilio*, is essential to planarian stem cell maintenance. *Development* **132:** 1863–1874.

Schweers B.A., Walters K.J., and Stern M. 2002. The *Drosophila melanogaster* translational repressor pumilio regulates neuronal excitability. *Genetics* **161:** 1177–1185.

Semotok J.L., Cooperstock R.L., Pinder B.D., Vari H.K., Lipshitz H.D., and Smibert C.A. 2005. Smaug recruits the CCR4/POP2/NOT deadenylase complex to trigger maternal transcript localization in the early *Drosophila* embryo. *Curr. Biol.* **15:** 284–294.

Slack F. and Ruvkun G. 1997. Temporal pattern formation by heterochronic genes. *Annu. Rev. Genet.* **31:** 611–634.

———. 1998. A novel repeat domain that is often associated with RING finger and B-box motifs. *Trends Biochem. Sci.* **23:** 474–475.

Slack F.J., Basson M., Liu Z., Ambros V., Horvitz H.R., and Ruvkun G. 2000. The *lin-41* RBCC gene acts in the *C. elegans* heterochronic pathway between the *let-7* regulatory RNA and the LIN-29 transcription factor. *Mol. Cell* **5:** 659–669.

Smibert C.A., Wilson J.E., Kerr K., and Macdonald P.M. 1996. smaug protein represses translation of unlocalized *nanos* mRNA in the *Drosophila* embryo. *Genes Dev.* **10:** 2600–2609.

Smibert C.A., Lie Y.S., Shillinglaw W., Henzel W.J., and Macdonald P.M. 1999. Smaug, a novel and conserved protein, contributes to repression of *nanos* mRNA translation in vitro. *RNA* **5:** 1535–1547.

Smith J.L., Wilson J.E., and Macdonald P.M. 1992. Overexpression of *oskar* directs ectopic activation of *nanos* and presumptive pole cell formation in *Drosophila* embryos. *Cell* **70:** 849–859.

Sonoda J. and Wharton R.P. 1999. Recruitment of Nanos to *hunchback* mRNA by Pumilio. *Genes Dev.* **13:** 2704–2712.

———. 2001. *Drosophila* Brain Tumor is a translational repressor. *Genes Dev.* **15:** 762–773.

Stathopoulos A. and Levine M. 2005. Genomic regulatory networks and animal development. *Dev. Cell* **9:** 449–462.

Stebbins-Boaz B., Cao Q., de Moor C.H., Mendez R., and Richter J.D. 1999. Maskin is a CPEB-associated factor that transiently interacts with elF-4E. *Mol. Cell* **4:** 1017–1027.

St Johnston D., Driever W., Berleth T., Richstein S., and Nüsslein-Volhard C. 1989. Multiple steps in the localization of *bicoid* RNA to the anterior pole of the *Drosophila* oocyte. *Development* (suppl.) **107:** 13–19.

Struhl G., Struhl K., and Macdonald P.M. 1989. The gradient morphogen *bicoid* is a concentration-dependent transcriptional activator. *Cell* **57:** 1259–1273.

Stutz A., Conne B., Huarte J., Gubler P., Volkel V., Flandin P., and Vassalli J.D. 1998. Masking, unmasking, and regulated polyadenylation cooperate in the translational control of a dormant mRNA in mouse oocytes. *Genes Dev.* **12:** 2535–2548.

Subramaniam K. and Seydoux G. 1999. *nos-1* and *nos-2*, two genes related to *Drosophila nanos*, regulate primordial germ cell development and survival in *Caenorhabditis elegans*. *Development* **126:** 4861–4871.

Tan L., Chang J.S., Costa A., and Schedl P. 2001. An autoregulatory feedback loop directs the localized expression of the *Drosophila* CPEB protein Orb in the developing oocyte. *Development* **128:** 1159–1169.

Tay J. and Richter J.D. 2001. Germ cell differentiation and synaptonemal complex formation are disrupted in CPEB knockout mice. *Dev. Cell* **1:** 201–213.

Thiele B.J., Andree H., Hohne M., and Rapoport S.M. 1982. Lipoxygenase mRNA in rabbit reticulocytes. Its isolation, characterization and translational repression. *Eur. J. Biochem.* **129:** 133–141.

Thompson B.E., Lamont L.B., and Kimble J. 2006. Germ-line induction of the *Caenorhabditis elegans* vulva. *Proc. Natl. Acad. Sci.* **103:** 620–625.

Thompson B.E., Bernstein D.S., Bachorik J.L., Petcherski A.G., Wickens M., and Kimble J. 2005. Dose-dependent control of proliferation and sperm specification by FOG-1/CPEB. *Development* **132:** 3471–3481.

Urano J., Fox M.S., and Reijo Pera R.A. 2005. Interaction of the conserved meiotic regulators, BOULE (BOL) and PUMILIO-2 (PUM2). *Mol. Reprod. Dev.* **71:** 290–298.

Wang L., Eckmann C.R., Kadyk L.C., Wickens M., and Kimble J. 2002. A regulatory cytoplasmic poly(A) polymerase in *Caenorhabditis elegans*. *Nature* **419:** 312–316.

Wang X., Zamore P.D., and Hall T.M.T. 2001. Crystal structure of a Pumilio homology domain. *Mol. Cell* **7:** 855–865.

Wells S.E., Hillner P.E., Vale R.D., and Sachs A.B. 1998. Circularization of mRNA by eukaryotic translation initiation factors. *Mol. Cell* **2:** 135–140.

Wharton R.P. and Struhl G. 1991. RNA regulatory elements mediate control of *Drosophila* body pattern by the posterior morphogen *nanos*. *Cell* **67:** 955–967.

Wickens M., Bernstein D.S., Kimble J., and Parker R. 2002. A PUF family portrait: 3′UTR regulation as a way of life. *Trends Genet.* **18:** 150–157.

Wightman B., Ha I., and Ruvkun G. 1993. Posttranscriptional regulation of the heterochronic gene *lin-14* by *lin-4* mediates temporal pattern formation in *C. elegans*. *Cell* **75:** 855–862.

Wilhelm J.E., Hilton M., Amos Q., and Henzel W.J. 2003. Cup is an eIF4E binding protein required for both the translational repression of *oskar* and the recruitment of Barentsz. *J. Cell Biol.* **163:** 1197–1204.

Wreden C., Verrotti A.C., Schisa J.A., Lieberfarb M.E., and Strickland S. 1997. *Nanos* and *pumilio* establish embryonic polarity in *Drosophila* by promoting posterior deadenylation of *hunchback* mRNA. *Development* **124:** 3015–3023.

Xu E.Y., Moore F.L., and Reijo Pera R.A. 2001. A gene family required for human germ cell development evolved from an ancient meiotic gene conserved in metazoans. *Proc. Natl. Acad. Sci.* **98:** 7414–7419.

Ye B., Petritsch C., Clark I.E., Gavis E.R., Jan L.Y., and Jan Y.N. 2004. *nanos* and *pumilio* are essential for dendrite morphogenesis in *Drosophila* peripheral neurons. *Curr. Biol.* **14:** 314–321.

Zhang B., Gallegos M., Puoti A., Durkin E., Fields S., Kimble J., and Wickens M.P. 1997. A conserved RNA-binding protein that regulates sexual fates in the *C. elegans* hermaphrodite germ line. *Nature* **390:** 477–484.

Zhao Y., Samal E., and Srivastava D. 2005. Serum response factor regulates a muscle-specific microRNA that targets *Hand2* during cardiogenesis. *Nature* **436:** 214–220.

20

Protein Synthesis and Translational Control during Viral Infection

Ian J. Mohr
Department of Microbiology and NYU Cancer Institute
New York University School of Medicine
New York, New York 10016

Tsafi Pe'ery[1,2] and Michael B. Mathews[1]
Departments of [1]Biochemistry and Molecular Biology and [2]Medicine
New Jersey Medical School, University of Medicine and Dentistry of New Jersey
Newark, New Jersey 07103

V IRUSES ARE OBLIGATE INTRACELLULAR PARASITES or symbionts and depend on cells for their replication. Nowhere is this dependency seen more clearly than in the translation system, as viruses—unlike cells and their endosymbiotic organelles, chloroplasts and mitochondria—lack a translational apparatus. Consequently, viruses must use the cellular apparatus for the synthesis of one of their principal components. Because they can be manipulated with relative ease, the study of viruses has been a preeminent source of information on the mechanism and regulation of the protein synthetic machinery (Table 1). Viruses do more than simply coopt the cellular machinery to produce viral proteins, however. Under extreme selection pressure, many viruses have evolved ways to gain a translational advantage for their mRNAs and to contend with potent host defense systems that affect protein synthesis.

Here we consider the interactions between viruses and the translation system of the cell under three headings:

1. *Translational mechanisms.* Viruses exploit a range of unorthodox mechanisms, most of which were discovered in viral systems. Many of them have proven to be used in the uninfected cell, albeit seem-

ingly less frequently or in special circumstances such as during apoptosis or in response to environmental stress.

2. *Modifications of the translation system.* Many viruses impose sweeping changes upon the cellular translation machinery and the signaling network that regulates it, modifying these systems to favor the synthesis of viral proteins at the cells' expense.

3. *Host defenses and viral countermeasures.* Host defenses impinge on translation at many levels, from direct effects on components of the translation system (e.g., initiation factors) to the level of the cell (as in apoptosis) and the whole organism (via interferon production and mobilization of the immune system). These defense mechanisms, aimed at obstructing viral polypeptide production and arresting the viral replicative program, are in turn blunted by viral countermeasures that aim to sustain viral propagation.

In this chapter, we focus chiefly on mammalian cells and their viruses (bacterial systems are discussed in Chapter 28), beginning with an introduction to virus–cell interactions from the standpoint of protein synthesis.

TRANSLATIONAL ASPECTS OF VIRUS INFECTION: AN OVERVIEW

Viruses are exceptionally heterogeneous, and their strategies for infecting cells and replicating within them are diverse. Correspondingly, the interactions of viruses with their hosts and the cellular protein synthesis machinery are rich and varied.

Viral Genomes

All viruses consist of a nucleic acid genome surrounded by a protective shell (capsid) of viral protein(s). Enveloped viruses, such as the retro-, herpes-, and influenza viruses, have an additional membranous covering containing viral proteins together with cellular components that is acquired as the virus "buds" through the cell membrane. During this process, some viruses incorporate cytosolic constituents such as tRNA (Kinzy and Goldman 2000). Virus particles vary in size over a large range. The smallest, such as picornaviruses and hepatitis delta virus (HDV), are comparable in size to a ribosome, whereas the largest mammalian viruses (poxviruses, such as vaccinia) approach the dimensions of a mitochondrion. Mimivirus, which infects *Acanthamoeba*, is the current record-holder, approximately the size of a mycoplasma cell.

Viral genomic complexity varies accordingly. The simplest viral genomes contain 3–4 kilobases (kb) of nucleic acid (e.g., RNA phages, par-

Table 1. Viruses and milestones in protein synthesis

Concept or discovery	Virus	References
Biochemical evidence for the existence of mRNA	phages T2 and T4	Volkin and Astrachan (1956); Gros et al. (1961); Brenner et al. (1961)
Faithfully initiating cell-free translation systems	phage f2; EMCV	Nathans et al. (1962); Kerr et al. (1966); Mathews and Korner (1970); Smith et al. (1970)
Breaking the genetic code	tobacco mosaic virus; phage T4	Wittmann and Wittmann-Liebold (1967); Barnett et al. (1967)
Identification of initiator tRNAs	phages R17 and f2; EMCV	Adams and Capecchi (1966); Webster et al. (1966); Smith and Marcker (1970)
Characterization of ribosome-binding sites	phages R17 and Qβ; brome mosaic virus; reovirus	Steitz (1969); Hindley and Staples (1969); Dasgupta et al. (1975); Lazarowitz and Robertson (1977)
Poly(A) tail	vaccinia virus	Kates (1971)
7-Methyl guanosine cap	reovirus; vaccinia virus	Furuichi et al. (1975); Wei and Moss (1975)
Scanning model for initiation site selection	several plant and animal viruses	Kozak (1978)
Frameshifting	Rous sarcoma virus	Jacks and Varmus (1985)
Internal ribosome entry site	poliovirus; EMCV	Pelletier and Sonenberg (1988); Jang et al. (1988)
Ribosome hopping	phage T4	Weiss et al. (1990)
Ribosome shunting	CaMV	Fütterer et al. (1993)
Ternary complex, AUG and eIF-independent initiation	CrPV and other dicistroviruses	Sasaki and Nakashima (1999); Pestova and Hellen (2003)

Abbreviations: (EMCV) encephalomyocarditis virus; (CaMV) cauliflower mosaic virus; (CrPV) cricket paralysis virus.

voviruses)—even less (1.7 kb) in the helper-dependent HDV, a satellite of hepatitis B virus (HBV)—whereas the most complex mammalian viral genomes are over 200 kb (herpes- and poxviruses). Mimiviruses have the largest known viral genome, ~1200 kb with >1200 predicted genes, more than twice as big as the smallest bacterial genomes (~500 kb and ~500 genes). Although virus genomes must contain all the *information* needed for their own multiplication, they do not encode all the *enzymes* that this entails. Instead, they rely on the cells that they infect to supply energy, chemicals, and most of the necessary biosynthetic machinery. Viruses typically encode one or more structural proteins together with a variable number of enzymes and regulatory products which function through interactions with both viral and host components, including components of the translation apparatus. They may also generate noncoding RNAs such as the adenovirus VA RNAs and Epstein-Barr virus EBERs (see below, Viral Countermeasures) and microRNAs (Cullen 2006).

Viral mRNA

The genetic material of viruses may be either DNA or RNA, single-stranded (ss) or double-stranded (ds), or partially both; if single-stranded, it may be of positive or negative polarity (i.e., equivalent to mRNA or to its complement). Correspondingly, viral strategies for replicating their genetic material and for generating mRNA are varied, involving enzymes of cellular origin, viral origin, or both (Pe'ery and Mathews 2006). Positive-stranded viral RNA genomes can serve directly as translation templates, but all other kinds of viral genomes use viral or cellular RNA polymerases to generate mRNA. Most DNA viruses and some RNA viruses (the retroviruses, HBV and HDV) exploit cellular RNA polymerase II (pol II) for this purpose. Accordingly, they produce mRNA in the nucleus, where they have access to the cellular RNA processing machinery. A few DNA viruses, notably the poxviruses, encode their own RNA polymerases, as well as enzymes for RNA capping and other modifications, and are based in the cytoplasm.

RNA viruses, apart from the retroviruses, HBV, and HDV, have to supply their own RNA-dependent RNA polymerases to produce mRNA, and most of them replicate in the cytoplasm. Numerous RNA viruses produce mRNAs that are not capped (e.g., picornaviruses and many plant viruses), necessitating the adoption of unconventional initiation strategies (see below, Viral mRNA Translation Strategies). Where the mRNA *is* capped (e.g., reo- and vesicular stomatitis virus [VSV]), the RNA viruses also furnish the enzymes for modifying its termini. Exceptionally, influenza viruses and their relatives replicate in the nucleus even though they encode their own RNA polymerase: They depend on host pol II transcripts as donors of capped 5′ oligonucleotides which serve as primers for viral mRNA synthesis.

Virus-encoded Translational Components

No known virus encodes a translation system, and mammalian viruses generally lack genes for any translational component, but the list of virus-encoded translation components is growing. It has been known since the early 1970s that some DNA bacteriophages encode tRNAs, eight in the case of phage T4, for example (McClain et al. 1972; Miller et al. 2003). Almost half of the phage genomes surveyed encoded at least one tRNA, and one was found to have as many as 26 tRNA genes with anticodons corresponding to 15 amino acids and a putative suppressor (Pedulla et al. 2003). The *Chlorella* viruses PBCV-1 and CVK2 encode 10–15 tRNAs, many of which are aminoacylated in infected cells (Nishida et al. 1999; Van Etten 2003). The murine herpesvirus MHV-68 contains eight tRNA-like genes, but many of them are defective and they do not give rise to charged tRNA (Bowden et al. 1997; Virgin et al. 1997). A few viruses encode some protein components of the translation apparatus. The *Chlorella* viruses specify a homolog of the elongation factor eEF3 (Yamada et al. 1993; Van Etten 2003), and mimivirus encodes nine translation factors as well as six tRNAs. The translation factors are homologs of three initiation factors (eIF1, eIF4A, and eIF4E), an elongation factor (eEF1A), a release factor (eRF1), and four aminoacyl-tRNA synthetases (Raoult et al. 2004).

What, then, are the biological roles of these viral tRNAs and translation factors? One hypothesis relates to host range and codon usage bias. The T4-encoded tRNAs are dispensable for infection of most bacterial strains under laboratory conditions, but they are required in certain hosts. It is possible, therefore, that the tRNAs contribute to the translation of a subset of phage mRNAs that are not well served by standard bacterial tRNA populations (Miller et al. 2003). Similarly, the anticodons of the *Chlorella* virus tRNAs correlate with viral gene codon usage, suggesting that the viral eEF3 and tRNAs might cooperate to facilitate translation of viral mRNAs. The variable spectrum of tRNAs found in these viral isolates may equip the viruses for replication in a range of host cells. However, another explanation must be found for the defective tRNA-like genes of MHV-68. Possibly they serve as decoys for an antiviral defense system analogous to the bacterial anticodon nuclease which cleaves cellular tRNAs (Pe'ery and Mathews 2000). In the same vein, the eIF2α homolog encoded by iridoviruses is thought to counteract a cellular defense mechanism (see Viral Countermeasures).

Host Cell Shutoff

Viral mRNAs must compete with cellular mRNAs for limiting quantities of translation factors in order to successfully reprogram the host machinery. Reprogramming also plays a pivotal role in development, differentiation, and stress responses (Schneider 2000; Chapters 13, 16, and 19). In some infections, viral mRNAs coexist with and are efficiently translated together with host mRNAs; in others, viral mRNAs dominate the scene, effectively excluding the production of host proteins. This phenomenon, known as host cell shutoff, may operate at several levels in the gene expression pathway (Pe'ery and Mathews 2006). Shutoff allows viruses to usurp the cellular machinery, thereby accelerating replication and perhaps enhancing virus yield, and it may preempt cellular antiviral defense mechanisms by compromising the synthesis of certain host proteins (see Regulation of mRNA Levels). Nonetheless, shutoff must be delicately balanced, because viruses need to keep their hosts alive and functioning so that the viral replicative cycle can be completed. The degree of host shutoff is variable, and related viruses may employ different mechanisms to accomplish the same end, or even do without it altogether, as illustrated by members of the picorna- and herpesvirus families.

Shutoff at the translational level generally entails two events: The virus compromises some aspect of the translation system and concomitantly engineers a bypass such that its protein synthesis escapes the restriction (Table 2). Thus, T-even phages cleave host tRNAs and replace them with phage-encoded isoaccepting species that are better attuned to phage mRNA codon usage (Mosig and Eiserling 1988). In eukaryotic cells, viruses often impede the cap-dependent translation of host mRNAs and circumvent this obstacle by initiation through alternative means such as internal ribosome entry or ribosome shunting (see Viral mRNA Translation Strategies). Viral mRNAs are at an advantage either because they do not require limiting initiation factors (e.g., poliovirus), or because they are "strong" mRNAs with *cis*-acting elements that compete effectively for the limiting factors (e.g., adenovirus). Alternatively, as with rotavirus, virus-coded proteins may set the viral mRNAs at an advantage (see Assembly of eIF4F Complexes). Another putative mechanism relies on a viral product, such as adenovirus VA RNA$_I$ and reovirus σ3, to selectively spare viral mRNAs from translational inhibition brought about by the RNA-regulated eIF2α kinase PKR (Sharpe and Fields 1982; O'Malley et al. 1989; Huang and Schneider 1990; Schmechel et al. 1997; Chapter 12). Preferential synthesis of viral proteins can also be achieved simply by mRNA competition: Viral mRNAs may dominate by their overwhelming preponderance at late times of infection (e.g., VSV; Lodish and Porter 1980, 1981) or translational efficiency (e.g., influenza virus; Kash et al. 2002).

Productive and Nonproductive Outcomes of Viral Infection

Many viral genomes, especially those of DNA viruses, are programmed to generate their products in an orderly fashion and temporal sequence, largely as a result of transcriptional controls. Examples of temporal switches operating at the translational level come from the ssRNA phages whose genomes serve directly as mRNAs (Pe'ery and Mathews 2000;

Chapter 28). In productive infections with most DNA viruses, the replicative cycle is divided into two phases, early and late, demarcated by the onset of viral DNA replication. During the early phase, a subset of the viral genes is expressed, including regulatory products, some of which suppress host defenses by interceding in processes such as apoptosis (see below, Viral Countermeasures). In more complex DNA viruses, such as herpes- and poxviruses, the early phase is subdivided into immediate early and intermediate stages. During the late phase, template number and transcriptional activity both increase, resulting in the abundant production of viral mRNAs encoding the coat protein(s) and other virion components, as well as proteins required for viral morphogenesis and related functions.

Productive infections of this kind underlie most of the discussion to this point, and translational control is more easily studied in such infections because the virus often comes to dominate all aspects of cell macromolecular synthesis. Other outcomes are possible, however, including abortive, persistent, and latent infections; cell transformation and neoplasia; and apoptosis (see below). For example, latently infected cells such as neurons infected with herpes simplex virus (HSV) contain the viral genome in a quiescent state, but virus production is undetectable until triggered by external stimuli. This might reflect a missing or limiting permissivity factor, or repression of viral gene expression.

Permissivity and virulence can be determined, at least in part, at the translational level. For example, in VSV infections of B lymphocytes, viral mRNA is associated with polysomes but is not translated without cellular activation by mitogens or phorbol esters (Schmidt et al. 1995). The $\gamma_1 34.5$ protein of HSV-1, which is required for neurovirulence, blocks the shutoff of host protein synthesis by PKR (He et al. 1997). The neuropathogenicity of poliovirus requires a sequence in its internal ribosome entry site (IRES) which is dispensable for virus growth in non-neuronal cells (Gromeier et al. 1999). Presumably, this sequence binds to a cellular IRES *trans*-acting factor (ITAF) that regulates its function. ITAFs with cell-type-specific distributions have been implicated in the tissue-specific IRES activity and virulence of several picornaviruses, including poliovirus (the polypyrimidine tract-binding protein, PTB), hepatitis A virus (PTB and glyceraldehyde 3-phosphate dehydrogenase), and human rhinovirus (DRBP76/NF90) (Yi et al. 2000; Guest et al. 2004; Merrill et al. 2006). Further examples of cell-specific permissivity factors will surely emerge in the future.

Antiviral Defenses

Not only must viruses commandeer the cellular translational factors necessary to conscript ribosomes, but they must also thwart cellular defense

Table 2. Virus–host interactions that regulate translation in infected cells

Virus (genome type)	Y/N	Suppression of host protein synthesis: mechanism	Viral strategies	
			to foster viral mRNA translation	to counteract eIF2α phosphorylation
Polio (+RNA)	Y	cleavage of eIF4GI and II and PABP by viral protease dephosphorylation of 4E-BP1 inhibition of host transcription by TBP cleavage	IRES	PKR degradation
EMCV (+RNA)	Y	dephosphorylation of 4E-BP1	IRES	none
HCV (+RNA)	?	unknown	IRES	NS5A inhibits PKR dimerization E2 protein is an eIF2α pseudosubstrate IRES inhibits PKR
CrPV (+RNA)	Y	unknown	eIFs not required for IRES	does not require eIF2
Sindbis (+RNA)	Y	inhibition of host mRNA transcription stimulation of eIF2α phosphorylation	ribosomes stall at viral mRNA element; eIF2A loads Met-tRNA$_i$	bypasses eIF2 by utilizing eIF2A
VSV (–RNA)	Y	inhibition of host mRNA transcription and nucleocytoplasmic transport (via M protein) dephosphorylation of eIF4E and 4E-BP1	unknown	none: very sensitive to interferon
HDV (–RNA)	N	none	unknown	viral genomic RNA can inhibit PKR activation
Influenza (–RNA)	Y	cap snatching by viral nuclease dephosphorylation of eIF4E	GRSF binds 5'UTR of viral mRNAs	activates cellular PKR inhibitor, P58[IPK] viral NS1 protein prevents PKR activation by dsRNA or PACT

Virus (Genome type)					
HIV (+RNA)	N	none		unknown	Tat and cellular TRBP bind TAR RNA TRBP binds to and inhibits PKR
Rotavirus (dsRNA)	Y	PABP displacement by rotavirus NSP3 protein binding to eIF4G		viral mRNA 3′ UTRs bind NSP3	8-kD NSP3 protein fragment binds dsRNA
Reovirus (dsRNA)	Y*	strain-dependent suppression involves PKR and RNase L		unknown	viral σ3 proteins bind dsRNA host-integrated stress response induced
SV40 (dsDNA)	Y	dephosphorylation of 4E-BP1 due to small t interacting with PP2A		IRES	large T allows translation despite eIF2α phosphorylation
HPV (dsDNA)	?	unknown		4E-BP1 phosphorylation in differentiating epithelial cells	E6 binds to GADD34/PP1α holoenzyme and causes eIF2α dephosphorylation
Adenovirus (dsDNA)	Y	dephosphorylation of eIF4E by viral 100K protein displacing Mnk from eIF4G		tripartite leader on late mRNAs causes ribosome shunting phosphorylation of 4E-BP1 by E4 ORF1 and ORF4-activated mTOR	VA RNA$_I$ binds PKR and prevents activation
HSV (dsDNA)	Y	reduction of host transcription inhibition of mRNA splicing enhancement of mRNA turnover by vhs (virus-encoded eIF4F-associated endonuclease)		promotes eIF4F assembly and eIF4E phosphorylation (no intrinsic preference for viral mRNAs demonstrated)	Us11 binds PKR and prevents activation by dsRNA or PACT $\gamma_1$34.5 protein binds cellular PP1α and targets its activity to eIF2α unidentified function inhibits PERK activation
HCMV (dsDNA)	N	none	as HSV	increases eIF4G and PABP abundance	TRS1 and IRS1 bind dsRNA and prevent PKR activation
EBV (dsDNA)	Y	unknown		lytic infection: unknown latency: LMP2A promotes 4E-BP1 phosphorylation	EBER RNAs bind PKR and prevent activation SM protein binds dsRNA and PKR
KSHV (dsDNA)	Y	acceleration of global mRNA turnover by ORF 37		unknown	virus-specified IRF-2 (vIRF-2) binds PKR and prevents its activation
Vaccinia (dsDNA)	Y	unknown		unknown	E3L binds dsRNA and PKR K3L is an eIF2α pseudosubstrate

See text for details. (Genome type) Nature and strandedness of the viral genomic nucleic acid. (KSHV) Kaposi's sarcoma-associated herpesvirus; (Y) yes; (N) no; (Y*) shutoff by reovirus is strain dependent; (?) unknown.

systems. Even single-celled organisms such as bacteria and yeast mount antiviral defenses—acting at the level of translation and of mRNA degradation, respectively—and their viruses have evolved corresponding specific antidotes (Benard et al. 1999; Kaufmann 2000; Pe'ery and Mathews 2000). Multicellular animals have sophisticated protective mechanisms which, in mammals, respond to viral invasion in three phases. First, cells are equipped with cell-autonomous innate mechanisms, such as RNA interference (RNAi), cytidine deamination, and apoptosis (gene-directed cell suicide or programmed cell death), that are constitutive and do not require induction by external agents. Second, there are innate mechanisms that are triggered by exposure to viruses and other stimuli, and are induced via soluble factors of the immune system such as interferons. Third is adaptive immunity, which is triggered by exposure to antigens and results in a long-lasting specific response mediated by antibodies and lymphocytes.

Apoptosis and interferon induction affect translation most directly and in related ways (see Host Defense Strategies and Mechanisms). Cellular translation initiation factors are prime targets for cleavage (apoptosis) and modification (interferon); interferon can trigger apoptosis by inducing mediators of the apoptotic pathway; and PKR plays a central role in both processes. By evoking cell death or blocking protein synthesis in infected cells, host defenses may sacrifice a small number of infected cells in order to contain virus replication and spread, thereby sparing the larger community of surrounding cells. Unchecked, these processes would extinguish viral propagation, but viruses have acquired a diverse array of functions that effectively counter the host defenses. In many cases, these countermeasures are critical determinants of viral pathogenesis, and several of them are deployed at the level of translation or have consequences for the protein synthesis system (see Viral Countermeasures).

VIRAL mRNA TRANSLATION STRATEGIES

Viruses depend on the cellular translational apparatus and have evolved to exploit it in unusual ways. These unorthodox mechanisms, which are depicted in Figure 1 and described briefly below, operate predominantly at the initiation step but also during elongation or termination. They represent departures from 5′ cap-dependent scanning (Kozak 1989; Chapter 4) and orderly decoding, directed by overriding or additional signals that specify deviations from the standard mechanism. Such mechanisms are often integral to viral translational and replicative strategies (e.g., internal ribosome entry) or regulatory mechanisms (e.g., ribosome shunting), or provide a means to expand the coding capacity within a confined genome (e.g., readthrough, frameshifting).

Polyprotein Synthesis

Many viral mRNAs encode a linked series of proteins as a single polypeptide that is translated from a unique start codon to a unique stop codon and then is proteolytically cleaved to liberate the mature proteins (Fig. 1A). Even though it employs standard translational mechanisms, this strategy deserves mention here because it is widely exploited by viruses to generate proteins in a fixed ratio with sparing use of transcription and translation signals. The primary translation product can be very long: For example, a precursor polypeptide of ~2200 amino acids is generated from the poliovirus genome and is processed by viral proteinases to yield about a dozen mature viral proteins (Villa-Komaroff et al. 1975). On the other hand, the primary translation product of the human immunodeficiency virus (HIV-1) *env* mRNA, the gp160 envelope precursor protein, gives rise to only two proteins, gp120 and gp41, after cleavage by cellular proteases (Robey et al. 1985). This strategy is also used by some cellular mRNAs, such as those for proopiomelanocortin (Nakanishi et al. 1979), ubiquitin (Wiborg et al. 1985), and fish antifreeze protein (Hsiao et al. 1990).

Internal Ribosome Entry

The entry of ribosomes at internal mRNA sites under the direction of an IRES, first observed in polio- and encephalomyocarditis virus (EMCV) (Jang et al. 1988; Pelletier and Sonenberg 1988), is a strategy employed by many viral RNAs (Hellen and Sarnow 2001; Chapter 5) and by some cellular mRNAs (Chapter 6). These picornaviral IRESs are located in the 5′-untranslated region (UTR) of the mRNA (Fig. 1B). No viral proteins have been shown to be essential for IRES function, although the poliovirus protease 2A plays a stimulatory role (Hambidge and Sarnow 1992), but a growing number of cellular ITAFs enhance IRES function.

Since IRESs function by recruiting ribosomes or key initiation factors, the requirements for canonical initiation factors are usually less stringent than those for cap-dependent translation (Chapter 5). These relaxed requirements free the viruses from the cap-dependent initiation mechanism and form the basis for some host shutoff mechanisms (discussed above and under Modification of Host Translation Factors in Virus-infected Cells). For poliovirus and EMCV, IRES-mediated initiation does not require the cap-binding protein eIF4E and is insensitive to cleavage of eIF4G by viral proteases. Hepatitis C virus (HCV) and the pestiviruses dispense with all of the eIF4 proteins. The insect dicistroviruses, such as cricket paralysis virus (CrPV) and *Plautia stali* intestine virus (PSIV), go to the extreme, dispensing with eIF2 and the initiator tRNA, Met-tRNA$_i$, as well

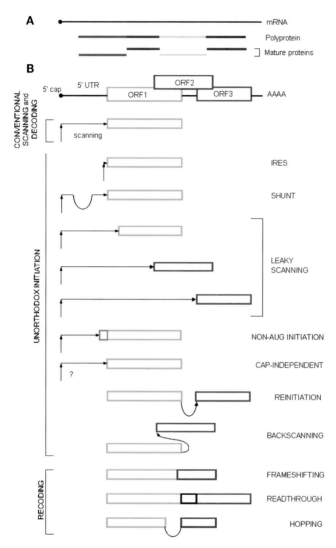

Figure 1. Viral translational strategies. (*A*) Polyprotein synthesis. The mRNA (*top line*) is translated to yield a precursor polypeptide (polyprotein; *second line*) that is proteolytically cleaved to give rise to mature viral protein products (represented as four individual segments). (*B*) Unconventional strategies. The top line represents a capped and polyadenylated mRNA with a 5′-untranslated region (5′UTR) and three open reading frames (ORFs 1–3). Various strategies for initiating protein synthesis on this mRNA and for decoding its information are depicted on the lines below. Although the 5′UTR of an uncapped mRNA is apparently scanned, the site of ribosome entry is not well defined, as indicated by the question mark (?).

as all other canonical initiation factors (Sasaki and Nakashima 1999; Pestova and Hellen 2003; Cevallos and Sarnow 2005). Hepatitis A virus (HAV), on the other hand, requires a functional eIF4G and is inhibited by 4E-BP1, which can sequester eIF4E (see Regulation of eIF4E Availability).

IRESs confer several additional advantages. First, they allow viral mRNAs to carry upstream sequences that are irrelevant to translation but play vital roles in the viral life cycle, such as in packaging or replication (Borman et al. 1994). Thus, AUG triplets and secondary structure that would inhibit initiation via the scanning mechanism can be tolerated in the 5′UTRs of these IRES-containing mRNAs. Second, IRES function obviates the need for capping, i.e., for the virus to replicate in the nucleus where the cellular capping enzymes reside, or to encode its own capping enzymes as many cytoplasmically replicating viruses do. A third benefit is that IRESs give access to internal initiation sites. HIV-1 and its simian relative, SIV, have two types of IRES elements even though their RNAs are capped and polyadenylated. One type is present in the 5′ leader, whereas the other is atypically placed internally within the *gag* gene where it can drive initiation at an internal AUG codon located some distance upstream (Buck et al. 2001; Brasey et al. 2003; Herbreteau et al. 2005; Nicholson et al. 2006). The role of these IRES elements in the viral replication is not yet clear. One possibility is that, like the internal IRES in the cellular PITSLRE protein kinase mRNA (Cornelis et al. 2000), they serve to produce a protein or isoform under specific conditions, such as during a particular phase of the cell cycle.

Ribosome Shunting

Shunting, also called discontinuous scanning, combines some features of linear scanning with certain characteristics of IRESs (Fig. 1B). First characterized in cauliflower mosaic virus (CaMV) 35S RNA (Fütterer et al. 1993), and related plant pararetroviruses, shunting has also been observed in adeno-, Sendai, papilloma- and duck hepatitis B viruses (Yueh and Schneider 1996; Latorre et al. 1998; Remm et al. 1999; Sen et al. 2004). Shunting allows 40S ribosomal subunits to bind to the cap, scan for a short distance from the 5′ terminus of the mRNA, then skip to a site up to hundreds of nucleotides downstream without scanning through the intervening sequences which may, therefore, be burdened with secondary structure and initiation codons that do not interfere. In CaMV, ribosomes translate a short upstream open reading frame (uORF) in a cap-dependent fashion, traverse an intervening highly structured region, then "land" near the initiation site for the major viral ORF, resume scanning, and initiate at the next AUG. In this case, and in Rice tungro bacilliform virus, shunting requires reinitiation after translation of the uORF (Ryabova and Hohn 2000; Pooggin et al. 2006). No plant or viral protein is essential, but the product of the CaMV ORF VI gene, transactivator/viroplasmin (TAV), is stimulatory.

In adenovirus late mRNAs, shunting occurs in the tripartite leader spliced onto the 5′ end of all late mRNAs. Ribosomal subunits appear to scan through the 5′-proximal region of the leader, skip a region of secondary structure, then land and continue scanning for a short distance. Leader sequences complementary to the 3′ end of 18S ribosomal RNA (rRNA) are required, although it is not known whether pairing occurs (Yueh and Schneider 2000). Initiation on such mRNAs takes place via conventional 5′ scanning and shunting at roughly equal levels in uninfected cells, but shunting predominates late in infection when the capbinding initiation complex is altered (see below, Regulated Phosphorylation of eIF4E). In the Sendai virus P/C mRNA, shunting is triggered by unidentified upstream signals, and landing is stimulated by a sequence located a short distance downstream of the start site (de Breyne et al. 2003). Ribosomes can initiate at the landing site even if it is mutated away from AUG, suggesting that it has an intrinsic affinity for ribosomes or host factors, as no viral proteins are required (Latorre et al. 1998).

Leaky Scanning

Leaky scanning permits a downstream AUG to be used for initiation in preference to, or in addition to, the first (i.e., 5′ proximal) AUG (Fig. 1B). The protein products may be in the same or different reading frames, and dicistronic mRNAs of this type are found in a large number of viruses. A principal feature defining the initiation site is the setting of the initiation codon. The most favorable context is GCCA/GCCAUGG, where the bases at positions −3 and +4 (relative to the A of the initiator AUG triplet, assigned +1) are the chief determinants of a strong initiation site (Kozak 1989). In these dicistronic mRNAs, the first AUG is often in a suboptimal context, suggesting that it functions as a weak initiator which is bypassed at a substantial frequency, allowing initiation at a subsequent site downstream.

This mechanism allows viruses to economize on coding space and signals for transcription and RNA processing. In HIV-1, for example, the Env protein is translated from mRNAs that contain an uORF encoding the accessory protein Vpu in a different reading frame. To permit Env synthesis, the *vpu* initiation site is required to be weak (Schwartz et al. 1992). When the two ORFs are in-frame with each other, the result is a nested pair of proteins with overlapping carboxy-terminal sequences and related functions as in simian virus 40 (SV40) coat proteins VP2 and VP3. On the other hand, when the two ORFs are in different reading frames, the resultant proteins need not be related either in structure or in function (as with HIV-1 Vpu and Env), although they sometimes are (e.g., adenovirus-5 E1B 19-kD and 55-kD proteins). Regulatory interactions have been reported between the two overlapping out-of-frame ORFs in the reovirus S1 transcript, where ribosomes translating the σ1 ORF may impede those translating the downstream σ1s ORF (Fajardo and Shatkin 1990; Belli and Samuel 1993).

Initiation at Non-AUG Codons

Initiation can take place at codons other than AUG, either instead of (Fütterer et al. 1996) or in addition to (Corcelette et al. 2000) initiation at the conventional signal. Ribosomes may be predisposed to start at such unorthodox sites when they use mechanisms such as IRESs or shunting that result in internal placement rather than scanning to locate the initiation codon (Latorre et al. 1998; Ryabova and Hohn 2000). In the dicistrovirus PSIV, for example, an IRES confers the ability to initiate with glutamine at a CAA codon (Sasaki and Nakashima 2000). The advantages accruing from this departure from conventional initiation potentially include increased protein-coding capacity and novel regulatory opportunities.

Translation of Uncapped RNA

The RNAs of some plant viruses, such as satellite tobacco necrosis virus (STNV) and barley yellow dwarf virus (BYDV; Wimmer et al. 1968), and the yeast virus L-A (Nemeroff and Bruenn 1987; Masison et al. 1995), lack 5′ cap and 3′ poly(A) structures. Unlike picornavirus RNA, which is also not capped, they function without an IRES element. In STNV and BYDV, a short 5′UTR sequence functions in concert with a longer, structured 3′UTR sequence termed a translational enhancer to recruit the cap-binding complex eIF4F (Timmer et al. 1993; Gazo et al. 2004). Plant potyvirus and feline calicivirus recruit eIF4E through the viral protein VPg that is covalently attached to the 5′ end of the RNA (Goodfellow et al. 2005; Grzela et al. 2006). The mechanisms used by plant viruses are described in detail in Chapter 26. In the case of the L-A virus, initiation depends on interactions with the yeast SKI (superkiller) gene products, which form part of a cellular defense system (Masison et al. 1995; Pe'ery and Mathews 2000).

Reinitiation

Multiple initiation sites and polycistronic mRNAs are the rule in bacteria and their viruses where internal ribosome entry directed by the Shine-Dalgarno sequence is the standard mechanism for initiation (Chapter 28). In phage ϕX174, two proteins are translated from the same sequence using separate ribosome-binding sites that give access to different reading frames (Ravetch et al. 1977). The individual sites are not necessarily utilized with equal efficiencies, however, and translation of a downstream cistron may depend on the translation of the upstream cistron, a phenomenon termed translational coupling (Pe'ery and Mathews 2000; Chapter 8).

Although it happens rarely, some eukaryotic mRNAs have more than one nonoverlapping ORF and reinitiation provides a means whereby two proteins can be made from a single mRNA as discussed in Chapter 8. For example, the rabbit calicivirus subgenomic mRNA is bicistronic, encoding major and minor capsid proteins from upstream and downstream ORFs, respectively. Access to the downstream ORF depends on reinitiation in response to an RNA signal (Meyers 2003). uORFs typically exert a profound negative influence over the translation of a downstream ORF, but in some mRNAs they confer inducibility upon the downstream ORF (Sachs and Geballe 2006). In most cases, such as yeast GCN4 and mammalian ATF4, short peptides encoded by the uORF do not play a part in the phenomenon (Chapters 9 and 13); in other cases, however, the peptide product itself mediates the inhibition. For example, the 22-residue uORF peptide represses production of the human cytomegalovirus (HCMV) gpUL4 (gp48) protein by blocking termination and stalling scanning ribosomes (Alderete et al. 1999). As discussed above, the short uORF of CaMV 35S RNA exerts a positive effect on the translation of downstream ORFs in the presence of TAV and is required for shunting.

Frameshifting

During the decoding of some mRNAs, the advancing ribosome slips forward or back by one nucleotide, resulting in a programmed +1 or –1 change of reading frame (Fig. 1B) (Chapter 22). Such recoding events were discovered in Rous sarcoma virus (Jacks and Varmus 1985) and are common in retroviruses. In HIV-1, for example, a –1 shift is an essential event in generating the Gag-Pol polyprotein (Fig. 1A) that is the precursor for reverse transcriptase. Frameshifting also occurs in many other viruses, e.g, the severe acute respiratory syndrome (SARS) virus (Baranov et al. 2005). The proportion of ribosomes that change reading frames is characteristic of each site, and is usually rather low so the translation products consist of a majority of the conventionally decoded polypeptide and a minority of the "recoded" form. In retroviruses, the Gag-Pol shifted product is about 5% of the unshifted Gag product. Most retroviruses translate *pol* this way, and similar events occur in coronaviruses and yeast L-A virus. There is little evidence that the proportion of frameshifting at a particular site is controlled; rather, this mechanism seems to be a device for increasing the coding capacity of the viral genome and for producing the two products in a fixed ratio. This ratio may be of critical importance, since a small change in the ratio is detrimental to virus assembly and proliferation (Dinman et al. 1998), raising the possibility of developing antiviral drugs acting at this level (Hung et al. 1998; Chapter 30).

Readthrough

The suppression or readthrough of stop codons is another recoding strategy used by many viruses to generate a carboxy-terminally extended protein at a fixed ratio to the conventionally translated product. Examples are found in viruses infecting plants, bacteria, and mammals. Thus, Moloney murine leukemia virus (MoMuLV) generates reverse transcriptase and integrase by suppressing termination at an amber codon, UAG, generating a Gag-Pol fusion protein which is subsequently cleaved in a strategy reminiscent of the polyprotein coding mechanism discussed above. Readthrough efficiency is regulated by the viral reverse transcriptase product as well as by the host release factor eRF1 (Orlova et al. 2003). The alphavirus RNA dependent RNA polymerase (P4) is produced by readthrough of an opal codon, UGA (Myles et al. 2006). As in frameshifting, the signal for termination suppression is often a downstream sequence or structure.

Hopping

Reminiscent of shunting during initiation, 50 nucleotides are bypassed during translation elongation to produce the topoisomerase subunit encoded by phage T4 gene *60* mRNA. In this solitary example, hopping is mediated by the nascent peptide and requires duplications flanking the bypassed sequence, as well as a stop codon and a hairpin structure containing the 5′ end of the gap sequence (Weiss et al. 1990). The hairpin requirement is reduced by a specific mutation in a 50S ribosomal subunit protein (Herbst et al. 1994). The regulatory significance of hopping is unknown, but not all ribosomes complete the hop by landing successfully. Their landing efficiency is influenced by the levels of initiation and release factors (Herr et al. 2001), hinting that this extraordinary event may serve a regulatory role in the production of gene *60* protein.

MODIFICATION OF HOST TRANSLATION FACTORS IN VIRUS-INFECTED CELLS

Prominent among the many changes to the translational machinery that occur in virus-infected cells are modifications of critical initiation factors, eIF2 and members of the eIF4 family, which mediate the entry of the initiator tRNA and the mRNA, respectively, into the preinitiation complex (Chapter 4). These alterations can influence the efficiency with which viral mRNAs are translated, the persistence of host protein synthesis, and the host antiviral response. In some instances, the modifications result from the direct action of viral proteins on host initiation factors, whereas in others they result from virus-induced activation of cellular signal transduction pathways that subsequently modify the initiation factor. The changes include covalent modifications, especially phosphorylation; association with viral or cellular regulators; and alterations in factor integrity and abundance.

Regulated Phosphorylation of eIF4E

The cap-binding protein eIF4E is phosphorylated on serine 209 by eIF4G-associated MAPK signal-integrating kinases, Mnk1 and Mnk2 (Fukunaga and Hunter 1997; Waskiewicz et al. 1997, 1999; Pyronnet et al. 1999; Scheper et al. 2001). eIF4E phosphorylation appears to play a regulatory role, although the details of how this is achieved and which mRNAs are most susceptible are poorly understood (Scheper and Proud 2002; Chapter 14). Nevertheless, eIF4E phosphorylation is strikingly increased or decreased upon infection with a variety of viruses (Fig. 2A).

Unphosphorylated eIF4E accumulates in conjunction with the inhibition of host protein synthesis following infection with adenovirus, VSV, or influenza virus (Huang and Schneider 1991; Feigenblum and Schneider 1993; Connor and Lyles 2002). In adenovirus-infected cells, decreased eIF4E phosphorylation is associated with host cell shutoff. The virus-encoded 100K protein binds to eIF4G, evicting Mnk from the eIF4F complex (Cuesta et al. 2000). Conditional alleles of the 100K protein that fail to shut off host protein synthesis at the restrictive temperature do not evict Mnk from the complex or induce the accumulation of unphosphorylated eIF4E. Translation of host mRNAs is impaired in the presence of the wild-type 100K protein, whereas adenoviral late mRNAs are translated effectively via ribosome shunting. Shunting is promoted by tyrosine phosphorylation of the 100K protein, which enhances its binding to mRNAs containing the adenoviral tripartite leader (Xi et al. 2004, 2005).

Conversely, eIF4E phosphorylation is *increased* in herpesvirus-infected cells as a result of signaling through the p38 kinase pathway (HSV-1) or both the p38 and ERK pathways (HCMV). Mnk inhibitors prevent eIF4E phosphorylation and reduce HSV-1 and HCMV replication in quiescent, primary human cells, suggesting that eIF4E phosphorylation enhances herpesvirus gene expression (Walsh and Mohr 2004; Walsh et al. 2005). Host cell shutoff, which is seen with HSV-1 but not HCMV, occurs via unrelated mechanisms (Mohr 2006b).

Regulation of eIF4E Availability

Members of the eIF4E-binding protein (4E-BP) family can act as translational repressors in a phosphorylation-dependent manner (Gingras et al. 2001; Chapter 14). In their hypophosphorylated form, 4E-BPs sequester eIF4E and inhibit cap-dependent translation. Hyperphosphorylation of the 4E-BPs, catalyzed by the mTOR kinase in response to environmental cues, results in the release of eIF4E, allowing it to associate with eIF4G and assemble, together with the RNA helicase eIF4A, into an active eIF4F complex (Fig. 2A) (Haghighat et al. 1995; Mader et al. 1995; Marcotrigiano et al. 1999).

The accumulation of hypophosphorylated 4E-BP1 and sequestration of eIF4E contribute to host shutoff by silencing cap-dependent translation (Gingras et al. 1996). This mechanism is operative in cells infected with picornaviruses such as EMCV and poliovirus, whose IRES elements are able to recruit ribosomes in the absence of a cap and eIF4E. Dephosphorylation of 4E-BP1 is implicated in the inhibition of host mRNA translation that occurs in VSV-infected cells (Connor and Lyles 2002). SV40 small-t antigen associates with the cellular protein phosphatase PP2A and promotes the accumulation of hypophosphorylated 4E-BP1, which correlates with eIF4E sequestration and the reduction in overall translation rates observed late in productively infected permissive cells (Yu et al. 2005). Because the decrease in available eIF4E does not alter the translation of VSV and SV40 mRNAs, both of which are capped, it is likely that they contain unidentified *cis*-acting elements that are critical for their translation.

Other viruses *promote* the phosphorylation of 4E-BP1, facilitating the release of eIF4E and its association with eIF4G (Fig. 2A) (Feigenblum and Schneider 1996; Gingras and Sonenberg 1997; Kudchodkar et al. 2004; Walsh and Mohr 2004; Moody et al. 2005; Walsh et al. 2005). In adenovirus-infected cells, this is achieved through the action of two proteins encoded by the E4 region. The E4 ORF1 protein activates PI3-kinase, mimicking the action of growth factor signaling, whereas the E4

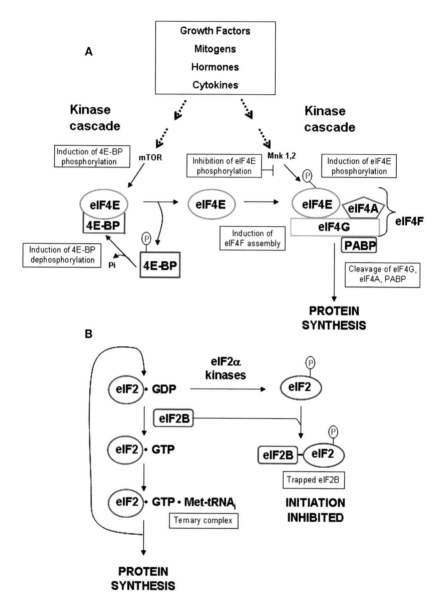

Figure 2. Regulation by phosphorylation of translation factors. (*A*) eIF4E and 4E-BPs. Mechanisms of viral intervention are indicated in boxes; for details, see text. (*B*) eIF2. Phosphorylation of the α subunit of eIF2 leads to sequestration of eIF2B and diminished ternary complex formation.

ORF4 product promotes phosphorylation of S6K1/p70^{S6K} through an unknown mechanism involving association with and relocalization of PP2A (O'Shea et al. 2005). Both pathways lead to activation of mTOR, which phosphorylates 4E-BP1. The increased 4E-BP1 phosphorylation observed in epithelial cells infected with human papilloma virus (HPV) appears to require mTOR and to be related to the differentiation process that occurs in infected, stratified epithelium (Oh et al. 2006). The 4E-BP1 translational repressor is also inactivated in quiescent primary cells infected with HSV-1, presumably facilitating cap-dependent initiation and infection (Walsh and Mohr 2004). This phosphorylation of 4E-BP1 is sensitive to rapamycin, an inhibitor of mTOR. In addition, steady-state levels of 4E-BP1 decline in HSV-1-infected cells via proteasome action, a response that is also prevented by rapamycin. Thus, 4E-BP1 phosphorylation results in its destruction as well as its release from eIF4E in HSV-1-infected cells. In HCMV-infected cells, 4E-BP1 phosphorylation is only partly sensitive to rapamycin (Kudchodkar et al. 2004; Walsh et al. 2005), reflecting the altered substrate specificity and rapamycin sensitivity of mTOR complexed with rictor and raptor induced by HCMV (Kudchodkar et al. 2006). Finally, in cells latently infected with EBV, the viral LMP2A protein stimulates mTOR through PI3-kinase and Akt (Moody et al. 2005). The resulting phosphorylation of 4E-BP1, together with increased translation of mRNAs such as c-myc, likely contributes to the proliferation of B cells latently infected with EBV.

eIF4F-associated Proteins and eIF4F Assembly

HSV-1 and HCMV induce the assembly of eIF4F complexes in virus-infected cells (Kudchodkar et al. 2004; Walsh and Mohr 2004; Walsh et al. 2005). Release of eIF4E from the 4E-BP1 repressor is insufficient to promote eIF4F assembly in HSV-1-infected cells. Instead, the ICP6 gene product binds to eIF4G and directly promotes the association of eIF4E with the amino terminus of eIF4G (Walsh and Mohr 2006). This is the first example of an initiation factor-associated protein that actively promotes complex assembly, and it defines a new controllable step in translation. Protein synthesis is slowed in cells infected with an ICP6 mutant where eIF4F assembly is deficient. Furthermore, phosphorylated eIF4E does not accumulate, presumably because eIF4E fails to interact with eIF4G and is less accessible to Mnk kinase. The ability to stimulate eIF4F complex assembly is likely to be important for productive viral growth in quiescent cells that have diminished ability to support high levels of protein synthesis (Stanners and Becker 1971).

The rotavirus NSP3 protein participates in an elegant mechanism that inhibits the translation of host mRNAs (Piron et al. 1998). NSP3 binds to eIF4GI and evicts cellular PABP, thereby disrupting the "closed-loop" mRNA topology that is believed to enhance translation of capped and polyadenylated mRNAs through the association of PABP with eIF4G (Wells et al. 1998; Kahvejian et al. 2001). Thus, translation of most cellular mRNAs is inhibited. Rotavirus mRNAs are not polyadenylated but, instead, have NSP3-binding sites at their 3′ ends (Poncet et al. 1993). NSP3 dimers bind to these sites and associate with eIF4G in a manner that could bridge the 3′ and 5′ ends of the viral mRNAs (Groft and Burley 2002), attaining functional circularization by imitating the PABP-bridged linkage.

Other viral proteins that associate with the eIF4F complex include adenovirus 100K (discussed above), influenza virus NS1 protein (Aragon et al. 2000; Burgui et al. 2003), and the HSV-1 virion host shutoff protein, vhs, which associates with eIF4F and enhances mRNA turnover (Feng et al. 2001). Selectivity of vhs for mRNA is presumably achieved through its association with eIF4A, eIF4B, or eIF4H (Doepker et al. 2004; Feng et al. 2005). Although vhs does not discriminate between host and viral mRNAs, viral mRNA production is sustained throughout infection whereas host transcription and splicing are impaired by other viral factors, resulting in the inhibition of host protein synthesis (Mohr 2006b). In addition, by enhancing viral mRNA turnover, vhs promotes the expression of successive classes of temporally transcribed genes (Read and Frenkel 1983).

Control of Initiation Factor Integrity and Abundance

Picornaviral proteinases that process their polyprotein translation products (Fig. 1A) also cleave certain initiation factors, thereby selectively impairing host cap-dependent translation (Fig. 2A). eIF4GI and eIF4GII are cleaved by the 2A proteinases of enteroviruses such as polio-, rhino-, and coxsackie B virus (Lloyd 2006), and by the leader (L) proteinase, together with the 3C proteinase, of foot and mouth disease virus (FMDV). The 3C proteinase also partially cleaves eIF4A (Kirchweger et al. 1994; Belsham et al. 2000; Gradi et al. 2004). Cleavage of eIF4G severs its amino-terminal eIF4E-binding domain from the rest of the molecule, but the carboxy-terminal proximal segment of eIF4G remains intact and is recruited to the viral IRES, where it engages eIF4A and eIF3 (Lloyd 2006). Although both eIF4GI and eIF4GII are substrates for the 2A proteinase, in human cells eIF4GII is cleaved with slower kinetics which correlate with the inhibition of host mRNA translation (Gradi et al. 1998; Svitkin et al. 1999). Retroviral proteinases also cleave eIF4GI and stimulate the translation of viral mRNAs even though their mRNAs are both capped and polyadenylated. The cleavage is only partial, however, and eIF4GII is unaffected, providing a means for maintaining eIF4F-dependent translation in HIV-1-infected cells (Ventoso et al. 2001; Ohlmann et al. 2002; Alvarez et al. 2003).

The enterovirus 3C proteinase, together with 2A, cleaves PABP such that its eIF4G interaction domain is severed from its RNA recognition motifs, thereby abrogating its ability to circularize mRNA (Kuyumcu-Martinez et al. 2004a). Only ribosome-associated PABP is targeted for cleavage, leaving the remaining pool of free PABP available for viral RNAs. Since enterovirus RNAs contain poly(A) tails that enhance IRES-mediated translation, there must be some mechanism to ensure that polyadenylated host transcripts are repressed prior to any effects on viral mRNA. Alternatively, cleavage of PABP bound to viral mRNAs might assist in clearing the RNA template of ribosomes, a prerequisite for RNA replication. A similar 3C-like proteinase cleaves PABP in calicivirus-infected cells (Kuyumcu-Martinez et al. 2004b). As caliciviral mRNAs are neither capped nor polyadenylated, viral mRNA translation is unimpeded.

Manifestly, such strategies are not open to viruses whose mRNAs require the cellular cap-dependent translation machinery. One such virus, HCMV, in fact induces an increase in the abundance of eIF4G, eIF4E, and PABP in infected cells (Walsh et al. 2005).

eIF4G Phosphorylation

Rapamycin-insensitive phosphorylation of eIF4GI occurs in cells infected with HCMV or SV40, both of which produce mRNAs that are capped and polyadenylated (Kudchodkar et al. 2004; Yu et al. 2005). Although such eIF4G phosphorylation might contribute to the reduction in translation rates observed at late times in SV40-infected cells, eIF4F complex assembly appears to be stimulated, and host shutoff does not occur in HCMV-infected cells. Enhanced eIF4G phosphorylation is also observed in influenza virus-infected cells, and the hyperphosphorylated form remains in the cap-binding complex (Feigenblum and Schneider 1993). Identification of the sites of phosphorylation may help illuminate the significance of this eIF4G modification in infected cells.

Phosphorylation of eIF2α Controls Translation

eIF2 is a heterotrimeric G protein which forms a ternary complex, eIF2·GTP·Met-tRNA$_i^{met}$, that supplies the initiator tRNA to the 40S ribosomal subunit, forming the 43S ribosomal complex (Schneider and Mohr 2003; Chapters 4 and 9). Upon engaging the 40S subunit, the ternary complex induces a structural reorganization that allows rapid GTP hydrolysis, stimulated by the GTPase-activating protein eIF5. Recognition of the AUG codon by the scanning ribosome complex triggers the release of P_i and eIF2·GDP, followed by 60S subunit joining (Algire et al. 2005). Participation in subsequent rounds of polypeptide chain initiation requires the activity of the guanine nucleotide exchange factor eIF2B to regenerate eIF2·GTP. Phosphorylation of eIF2 on serine 51 of its α subunit leads to a strong association between eIF2B and eIF2 that blocks the nucleotide exchange reaction (Fig. 2B). Cells have relatively low levels of eIF2B, therefore phosphorylation of small amounts of eIF2α can profoundly inhibit translation.

Of the four mammalian eIF2α kinases that are activated by different forms of cellular or environmental stress (Chapters 12 and 13), PKR and PERK (the PKR-like endoplasmic reticulum [ER] kinase, also called PEK) are pivotal in viral infection. Phosphorylation of eIF2α plays a key role in host defenses, apoptosis, and host cell shutoff. As a corollary, this process is a prime target for viral countermeasures that are described in the next section.

ANTIVIRAL DEFENSES AND VIRAL COUNTERMEASURES

Apoptosis and interferon are major elements of the cellular antiviral defense strategy that affect the translation system in overlapping ways. These include:

- activation of PKR
- phosphorylation of eIF2α
- induction of the 2′,5′-oligoadenylate synthetase/RNase L pathway
- proteolytic cleavage of initiation factors
- formation of stress granules
- inhibition of eIF3 function

Viral responses that tend to neutralize the first three of these defensive actions have been characterized, and some viruses even capitalize on eIF2α phosphorylation and stress granule formation. In this section, we describe the host defense mechanisms and the consequences of their deployment, then discuss viral countermeasures that act on the translation system.

Host Defense Strategies and Mechanisms

Apoptosis

Apoptosis is induced in a number of developmental and physiological situations (Chapter 16). When activated during viral infection, it is a radical defensive maneuver that eliminates virus-infected cells. Apoptotic signals are transmitted through signaling pathways and executed via the activation of caspases, which cleave many cellular proteins including translation factors eIF4GI, eIF4GII, eIF2α, eIF4B, and 4E-BP1 (Satoh et al. 1999; Bushell et al. 2000; Chapter 16). Along these pathways are many checkpoints where an antiapoptotic signal can reverse the process.

Most viral infections trigger the apoptotic program in mammalian cells either as a result of interactions between the virus and its receptors on the host cell surface, or as a result of viral manipulations of internal cellular systems such as transcription and translation. Synthesis of viral RNA and proteins during infection can activate PKR and PERK, respectively, and direct cells to apoptotic death. In addition, many viruses that infect quiescent cells prompt them to reenter the cell cycle. Cell cycle control proteins sense these manipulations and initiate apoptosis. Although few viruses (such as those causing latent infections) can prevent the induction of apoptosis, many have developed measures to retard the process by impeding it at various checkpoints, including those in the translation system.

The rate of protein synthesis is reduced in the early stage of apoptosis when an apoptotic cell is making both pro- and antiapoptotic proteins that affect its death decision. This partial inhibition of translation may benefit the virus, because the level of short-lived cellular antiviral proteins will decline. During this period, the virus can still use the translation machinery to generate its own proteins including antiapoptotic products. Viruses that manage to slow apoptosis—or, in some cases, take advantage of the process—succeed in producing progeny before their host cells die.

Interferon

The interferons are cytokines which are released by infected cells and induce effectors that act at the cellular level as defenses against viral infection. These antiviral activities include well-established direct effects on the cellular protein synthetic system. Virus infection is sensed through recognition of virus-specific molecules, including dsRNA and structured ssRNA, which are recognized by cellular transmembrane proteins in the toll-like receptor (TLR) family and cytoplasmic proteins including RIG and MDA5 as well as PKR (Yoneyama et al. 2004, 2005).

Recognition of viral infection leads to a cascade of signaling events that results in the secretion of type I interferons, interferons-α and -β, which alter the ability of the host to support viral replication (Levy et al. 2003). The interferons engage receptors on neighboring cells and "prime" them by inducing expression of more than 100 distinct interferon-stimulated genes (ISGs) that render the cells refractory to viral replication and spread, creating what has been termed an "antiviral state." Although the function of the interferon-induced proteome is far from completely understood, key elements that figure prominently in translational control are PKR, which is induced 3- to 5-fold by interferon treatment; 2′,5′-oligoadenylate synthetase (2-5 OAS), which makes an activator of RNase L; and the eIF3 inhibitor, P56. PKR and 2-5 OAS are activated as well as induced by dsRNA. Virus-encoded products are marshaled to obstruct the signaling pathways (Haller et al. 2006; Hiscott et al. 2006) and, as described below, to attenuate or neutralize the downstream effector enzymes, thereby alleviating translation shutoff as well as the onset of apoptosis.

Activation of PKR

PKR is the only known interferon-induced protein that exerts a direct effect on apoptosis via inhibition of the protein synthesis machinery. PKR is catalytically inert until it encounters an activating stimulus such as dsRNA. Profuse quantities of dsRNA and highly structured ssRNA accumulate in virus-infected cells. Some viral genomes are composed of dsRNA, whereas viruses with ssRNA genomes are likely to contain regions of secondary structure and proceed through a dsRNA intermediate in the course of their replication. DNA viruses, which accommodate large blocks of genetic information in a limited space, often have overlapping ORFs on opposing DNA strands and produce mRNA transcripts capable of associating to form dsRNA (Jacquemont and Roizman 1975; Maran and Mathews 1988). PKR contains two nonidentical dsRNA-binding motifs (dsRBMs; Tian et al. 2004; Chapter 12) and is activated when a PKR dimer assembles on a segment of dsRNA, resulting in a conformational change that leads to intermolecular autophosphorylation (see Fig. 3).

Virus-induced stress may also activate PKR and trigger apoptosis through PACT (protein activator of PKR), a cellular stress-induced protein that becomes activated through phosphorylation. PACT binds dsRNA, but this binding is not necessary for its interaction with PKR

(Patel and Sen 1998). Many ligands encoded by various viruses interfere with PKR activation and eIF2α phosphorylation and slow apoptosis, allowing the production of viral proteins and essential cellular proteins (see Viral Countermeasures).

PKR is necessary for the onset of translation inhibition and apoptosis. Thus, PKR-null mouse embryo fibroblasts are resistant to dsRNA-induced apoptosis (Der et al. 1997). Activated PKR phosphorylates eIF2α, its major substrate, leading to a global inhibition of translation. Interference with viral gene expression blocks the viral replicative program whereas effects on the host can lead to apoptosis. eIF2α phosphorylation correlates strongly with the onset of apoptosis. Thus, overexpression of a variant of eIF2α that resembles its phosphorylated form (S51D) can trigger apoptosis, whereas a non-phosphorylatable variant (S51A) partially protects against apoptosis (Srivastava et al. 1998). Increased eIF2α phosphorylation also correlates with the induction of apoptosis by tumor necrosis factor (TNF)-α and with viral infections that cause host protein synthesis shutoff (Clemens 2005). In addition, PKR can trigger apoptosis via phosphorylation of IKK (inhibitor of NF-κB kinase) and activation of NF-κB (Gil and Esteban 2000). Many pro- and antiapoptotic genes are responsive to NF-κB, and, therefore, the outcome of their induction is not clear-cut.

The cell cycle regulator p53, which is also induced by interferon, is translationally down-regulated via the PKR and 2-5 OAS pathways after infection with EMCV and human parainfluenza virus HPIV3. The reduction in p53 levels leads to apoptosis. Hence, p53 is a downstream effector of PKR- and 2-5 OAS-induced apoptosis (Marques et al. 2005). On the other hand, increased expression of many cellular genes in the apoptotic and stress-related pathways is observed when cells are infected with vaccinia virus recombinants expressing PKR (Guerra et al. 2006). Evidently, further work will be required to fully appreciate the downstream pathways of PKR-induced apoptosis.

Induction of the 2-5 OAS/RNase L Pathway

Production of oligoadenylate chains with a unique 2′-5′ linkage (2-5 OA) by three related dsRNA-activated synthetases (2-5 OASs) leads to activation of the latent cellular nuclease RNase L and degradation of both viral and cellular RNAs as well as rRNA (Iordanov et al. 2000). The 2-5 OASs lack dsRBMs but recognize dsRNA through a positively charged groove (Hartmann et al. 2003). Although the precise function of each isoform is poorly understood, they exhibit biochemical differences, including the size of the 2-5 OAs synthesized (Marie et al. 1990; Sarkar and Sen 1998).

The 2-5 OAs bind to monomeric RNase L subunits, promoting the formation of active homodimers that cleave RNA at the 3' side of UpXp sequences (Floyd-Smith et al. 1981, 1982; Wreschner et al. 1981; Silverman et al. 1988). Once set in motion, the response to 2-5 OAS is attenuated through the action of a phosphodiesterase that catalyzes the destruction of the activating signal (Kubota et al. 2004).

RNase L-mediated RNA cleavage slows translation in the infected cell and leads to apoptosis. Thus, RNase L-null mice are defective in apoptosis (Zhou et al. 1997), and apoptosis is induced by overexpressing RNase L; correspondingly, suppression of RNase L activity inhibits apoptosis in poliovirus-infected cells (Castelli et al. 1997). In PKR-null cells, the HCV ORF delivered by a vaccinia virus-based vector induced apoptosis via RNase L, demonstrating that the 2-5 OAS/RNase L pathway is independent of the PKR pathway (Diaz-Guerra et al. 1997; Gomez et al. 2005). Activation of the JNK (Jun N-terminal kinase) pathway, as a response to various stress stimuli, including viral infection, leads to apoptosis via the mitochondrial pathway. Both apoptosis and JNK activation by EMCV were attenuated in RNase L-null cells, suggesting that the JNK pathway is a direct link between RNase L and apoptosis (G. Li et al. 2004).

Proteolytic Cleavage of Initiation Factors

The induction of apoptosis by cellular factors such as TNF-α that are produced after viral infection causes the cleavage of various initiation factors. The best studied is the cleavage of initiation factors eIF4GI and eIF4GII by caspase-3, which takes place at sites that are distinct from those cleaved by viral proteases such as poliovirus 2A (Zamora et al. 2002; Chapter 16; see Control of Initiation Factor Integrity and Abundance). Indeed, overexpression of poliovirus 2A protease causes apoptotic cell death by inhibition of protein synthesis (Goldstaub et al. 2000). As a result of eIF4GI and eIF4GII cleavage, cap-dependent translation is greatly inhibited and only ~3% of mRNAs that are cap-independent can be translated (Johannes et al. 1999). The eIF4G homolog p97/DAP5/NAT1, which lacks the eIF4E-binding site, is cleaved by caspases in apoptotic cells, giving rise to a shorter protein that functions as a translation initiation factor for IRES-containing mRNAs. The p97 mRNA itself contains an IRES and is translated in apoptotic cells, as are mRNAs for several other proteins involved in apoptosis (Chapters 6 and 16). Poliovirus, and other viruses that use IRESs for initiation, exploit the inhibition of cap-dependent initiation and "hijack" the protein synthesis machinery for translation of their own mRNAs. Evidently, this task must be completed and viruses assembled before the final steps of apoptosis.

Formation of Stress Granules

Phosphorylation of eIF2α by PKR or another of its kinases can mediate stress granule (SG) formation. SGs contain modified 48S preinitiation complexes together with a selection of mRNAs whose translation is in abeyance. Depending on the cellular environment, the stalled preinitiation complexes are rerouted to produce proteins, or the mRNA is degraded, possibly after delivery to SG-linked processing (P-) bodies (Kedersha and Anderson 2002; Kedersha et al. 2005; Chapter 25). A link between apoptosis and SGs is suggested by the finding that mitochondrial apoptosis-inducing factor (AIF) inhibits SG formation (Cande et al. 2004). Whereas SG formation helps cells to adapt transiently to environmental stress, sustained stress such as viral infection results in cellular mRNA degradation and subsequent cell death. To counteract the stress response, Semliki Forest virus (SFV) and Sindbis virus (SV) mRNAs contain a translational enhancer element that confers resistance to high levels of phosphorylated eIF2α (see below). Additionally, SFV induces SG assembly early in infection, and this contributes to the inhibition of cellular protein synthesis. Selective SG disassembly in the vicinity of viral RNA replication coincides with SFV mRNA translation (McInerney et al. 2005). Because West Nile and Sendai viruses do not inhibit host protein synthesis, they might reduce SG formation by sequestering proteins TIA-1 and TIAR, which are essential for this process (Li et al. 2002; Wiegand et al. 2005).

Inhibition of eIF3 Function by P56

One of the most highly induced products of the interferon-stimulated gene family, P56, contains multiple tetratricopeptide repeats and interacts with eIF3. Remarkably, the human and murine orthologs are reported to have different actions. Whereas human P56 interacts with the eIF3e subunit and impairs the association of eIF3 with the eIF2 ternary complex (Guo et al. 2000), murine P56 interacts with eIF3c and inhibits the subsequent step, namely the association of eIF4F with 40S ribosomal subunits to form the 43S preinitiation complex (Hui et al. 2005). P56 inhibits cap-dependent translation and has a variable effect on IRESs (Hui et al. 2003; Wang et al. 2003). The EMCV IRES is relatively resistant to inhibition by P56, whereas HCV IRES is much more sensitive, perhaps reflecting the direct recruitment of eIF3 together with 40S subunits to this *cis*-acting element (Chapter 5). Although P56 induction can have potent effects on translation through eIF3, its activity does not appear to be responsive to dsRNA. The P56-mediated reduction of eIF3 activity may contribute to the antiproliferative effects of interferon. Presently, there is little information regarding the regulation of P56 activity or mechanisms whereby viruses might circumvent this impediment to mRNA translation.

Viral Countermeasures

Struggle for Control of the Translation Machinery

Surrendering initially infected cells for the benefit of the larger population is characteristic of the innate host response. Unopposed activation of PKR and RNase L would deplete levels of ternary complex, enhance mRNA decay, cleave rRNA, and obstruct viral and cellular protein synthesis in virus-infected cells. To the extent that host responses succeed in denying viruses access to the translation apparatus, the viral replication cycle is hindered and infection is effectively contained. Indeed, replication of the ssRNA genomes of picornaviruses (e.g., mengovirus and EMCV) and a flavivirus (West Nile virus) are particularly susceptible to the antiviral action of the 2-5 OAS/RNase L system (Chebath et al. 1987; Kajaste-Rudnitski et al. 2006), and picornaviral RNA has been isolated in association with 2-5 OAS (Gribaudo et al. 1991). The contest between viral and cellular functions to determine the abundance of unphosphorylated, active eIF2α is frequently a principal feature of viral pathogenesis (Leib et al. 1999, 2000; Brandt and Jacobs 2001; Mohr 2006a). Thus, viral mutants that are unable to prevent the accumulation of phosphorylated eIF2α are hypersensitive to interferon (Kitajewski et al. 1986; Beattie et al. 1996; Cheng et al. 2001; Cerveny et al. 2003) and attenuated in animals (Chou et al. 1990; Leib et al. 2000; Brandt and Jacobs 2001). Strikingly, although such mutant viruses are crippled in some cell types, they are capable of robust replication in tumor cells and are being evaluated as tumor-destroying oncolytic viruses, reflecting alterations to the translational control circuitry that are often found in neoplastic cells (Mohr 2005; Shmulevitz et al. 2005). In the therapeutic context, however, it is important to recognize that the cellular response to viral infection may include down-regulation of cellular stress pathways (e.g., those involving p53) that are critical for other cancer therapies (radiotherapy, chemotherapy).

Viral Gene Products Directed against the Activity of PKR

The profusion of viral products that counter the vigorous host response mediated by PKR (Fig. 3A and Table 2) includes adenovirus VA RNA$_I$ (Mathews and Shenk 1991) and the EBV EBERs (Sharp et al. 1993), abundant noncoding RNAs that bind to PKR but do not contain a dsRNA seg-ment capable of activating the enzyme. Incongruously, both VA RNA and EBER1 reportedly activate 2-5 OAS, although the significance of this find-ing remains unclear (Desai et al. 1995; Sharp et al. 1999). EBER expres-sion has been associated with resistance to the proapoptotic effects of in-terferon and Fas ligand (Nanbo et al. 2002, 2005), although the involvement of PKR activation in this interferon-induced apoptotic re-sponse has been questioned (Ruf et al. 2005). In addition, discrete ge-nomic elements, such as the HCV IRES, can also prevent PKR activation (Vyas et al. 2003). Some RNAs, such as the highly structured HDV genome and the HIV-1 transactivation response (TAR) RNA, are capable of acti-vating or inhibiting the enzyme under different conditions (Gunnery et al. 1992; Maitra et al. 1994; Robertson et al. 1996). TAR also activates 2-5 OAS, although HIV-1 Tat prevents activation of both 2-5 OAS and PKR by binding to TAR (Maitra et al. 1994) despite the fact that Tat is a sub-strate for PKR in vitro (Brand et al. 1997).

A prominent class of viral PKR inhibitors consists of dsRNA-bind-ing proteins that prevent PKR activation function, at least in part, by binding and sequestering dsRNA (Langland et al. 2006; Mohr 2006a). The two vaccinia virus E3L gene products, and the 8-kD protein en-coded by porcine group C rotavirus (Langland et al. 1994), bind RNA through their dsRBMs, but other members of this class of proteins (Table 2) recognize dsRNA through motifs that have yet to be described in cellular proteins (Wang et al. 1999; Olland et al. 2001; Khoo et al. 2002). In a variation on this theme, HIV-1 has engaged a cellular dsRNA-binding protein, TRBP, to bind the viral TAR element. TRBP is a homolog of PACT (Gupta et al. 2003) and a component of the RNA silencing complex RISC (Gregory et al. 2005). TRBP inhibits PKR ac-tivation through binding to both dsRNA and PKR (Park et al. 1994; Cosentino et al. 1995). On the other hand, PACT *activates* PKR, lead-ing to apoptosis, and this interaction is targeted by some viruses (Fig. 3B) (Patel and Sen 1998; Patel et al. 2000; Peters et al. 2002). For ex-ample, the influenza A virus NS1 protein binds to PKR and blocks its activation by PACT as well as by dsRNA (Li et al. 2006). NS1 also in-hibits the 2-5 OAS/RNase L pathway in infected cells (Min and Krug 2006) and associates with Staufen, another cellular dsRBM-containing protein, although the significance of this association is unclear (Falcon et al. 1999). Like TRBP, many of these virus-encoded dsRNA-binding proteins physically associate with PKR as well (Romano et al. 1998; Tan and Katze 1998; Poppers et al. 2003).

Although both vaccinia- and herpes simplex viruses are reportedly able to resist the antiviral activity of constitutively active 2-5 OAS, only the E3L

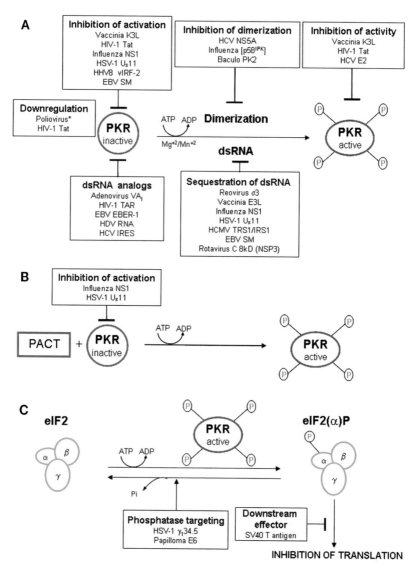

Figure 3. The activation and inhibition of PKR. PKR is activated by autophospho-rylation mediated by dsRNA (*A*) or PACT (*B*), and the activated enzyme phospho-rylates the α subunit of eIF2 (*C*). Steps in the pathway at which viral products in-terfere with PKR activation and the accumulation of phosphorylated eIF2 are shown schematically. Brackets indicate that the proximal effector is cellular; (*) indicates that the putative factor is cellular rather than viral. See text for details.

dsRNA-binding protein has a documented role in preventing rRNA decay due to activated 2-5 OAS and RNase L (Cayley et al. 1984; Rivas et al. 1998). In another strategy, certain HCV NS5A variants have been reported to bind PKR and inhibit dimerization (Gale et al. 1998). Finally, PKR activated in response to poliovirus infection is reportedly degraded by a cellular proteinase that requires an RNA cofactor (Black et al. 1993).

Combinatorial Tactics to Regulate eIF2α Phosphorylation

Vaccinia virus and HSV-1 each encode two proteins with targets downstream (Fig. 3C) as well as upstream of PKR activation, further highlighting the critical importance of preventing the accumulation of phosphorylated eIF2α. Acting early in the viral replicative cycle, the HSV-1 $\gamma_1$34.5 protein recruits the cellular protein phosphatase-1α (PP1α) to dephosphorylate eIF2 (He et al. 1997). The carboxy-terminal segment of the $\gamma_1$34.5 protein is homologous to the cellular GADD34 polypeptide, a PP1-binding protein that promotes eIF2α dephosphorylation during recovery from ER stress (Novoa et al. 2001). Late in the replicative cycle, when transcripts with the potential to form dsRNA accumulate, the Us11 eIF2α kinase-specific antagonist is produced. The Us11 protein binds to PKR and prevents its activation in response to both dsRNA and PACT (Peters et al. 2002; Mulvey et al. 2003, 2004). The vaccinia virus E3L protein binds dsRNA, whereas the K3L gene product is an eIF2α pseudosubstrate that binds PKR and prevents the kinase from engaging eIF2. Although both are present early in infection, E3L expression is more prolonged than K3L expression (Beattie et al. 1995). As with HSV-1, the dsRNA-binding antagonist acts later in the replicative program than the effector that targets a function downstream of kinase activation.

Impact of Viral Infection on Other eIF2α Kinases

As mentioned above, in addition to PKR, three other mammalian eIF2α kinases transduce stress signals to the translational machinery (Chapter 12). Downstream antagonists, such as HSV-1 $\gamma_1$34.5 phosphatase inhibitor and the vaccinia virus K3L pseudosubstrate, can counter the effects of a broad range of eIF2α kinases, including PERK (Sood et al. 2000; Cheng et al. 2005). The longer eIF2α homologs encoded by iridoviruses are also thought to act as pseudosubstrates (Yu et al. 1999; Essbauer et al. 2001), although the possibility that they are virus-specific eIF2 subunits has not been excluded. SV40 large T and HPV E6 have been reported to control eIF2α phosphorylation at a point subsequent to kinase activation (Rajan et al. 1995; Swaminathan et al. 1996; Kazemi et al. 2004), and an association of E6 with the GADD34/PP1α holoenzyme has been implicated. Other PKR effectors that have the potential to neutralize a broad range of eIF2α kinases include the E2 protein of the flavivirus HCV, which inhibits both PKR and PERK through a segment that contains homology with phosphorylation sites present in both PKR and eIF2α (Taylor et al. 1999; Pavio et al. 2002; Pavio et al. 2003). Not all flaviviruses can effectively counter PERK, however: Japanese encephalitis virus and cytopathic strains of bovine viral diarrhea virus (BVDV) both induce ER stress-mediated apoptosis (Jordan et al. 2002; Su et al. 2002). Similarly, infection with Mo-MuLV-ts1, a MoMuLV mutant that causes neuroimmunodegenerative syndrome in mice, induces apoptosis in astrocyte cultures by activating the ER stress response (Liu et al. 2004).

Controlling the response to ER stress is likely to be important in cells infected with enveloped viruses that either replicate on internal membranes (flaviviridae, picornavirus) or dramatically increase the burden of client proteins or particles that must transit through the ER as part of their exit strategy. Influenza virus exploits the cellular protein P58^{IPK}, as well as the viral NS1 protein, to prevent PKR activation. P58^{IPK} normally resides in an inactive complex with hsp40 (Melville et al. 1997; Tan et al. 1998). Upon its release from hsp40, P58^{IPK} can bind to both PKR and PERK and prevent their activation (Yan et al. 2002). Even though the mechanism by which influenza virus promotes the release of P58^{IPK} and its importance in viral pathogenesis remain to be determined, it has been reported that P58^{IPK} is required to prevent the accumulation of phosphorylated eIF2α during virus infection in plants (Bilgin et al. 2003). HCMV manipulates the unfolded protein response (UPR; Chapter 13), harnessing the desirable features, such as those that elevate the folding capacity of the ER, while suppressing those responses, such as translational arrest, that are clearly deleterious to viral replication (Isler et al. 2005a,b). Finally, HSV-1 specifies a function distinct from the γ$_1$34.5 phosphatase subunit and the Us11 PKR antagonist that confers resistance to ER stress, raising the possibility that viruses might target eIF2α kinases other than PKR to prevent eIF2α phosphorylation in response to many potential stresses (Mulvey et al. 2006).

Circumventing and Exploiting eIF2α Phosphorylation

Instead of combating the host response, some viruses have developed ways to capitalize on eIF2α phosphorylation to favor translation of viral mRNAs. In reovirus-infected cells, strains that impair host mRNA translation decrease the abundance of P58^{IPK} and induce eIF2α phosphorylation (Smith et al. 2006). This leads to the expression of a subset of genes, including ATF4, that are involved in the integrated stress response (including the UPR) which counteracts stress damage and suppresses the onset of apoptosis (Chapter 13). Phosphorylated eIF2α is beneficial for reovirus replication as the virus replicates better in cells that contain wild-type eIF2α than in cells containing its non-phosphorylatable variant which are defective in ATF4 induction. Because the IRES of the dicistrovirus CrPV functions in the absence of eIF2, as discussed above, viral mRNA translation can proceed under conditions that activate any of the eIF2α kinases (Wilson et al. 2000; Fernandez et al. 2002). A related, albeit less extreme, strategy has been evolved by Sindbis virus to ensure that viral structural proteins are produced despite the almost complete phosphorylation of eIF2α observed in infected cells. The viral 26S mRNA contains a stable hairpin element that promotes transient ribosome stalling immediately downstream of the AUG initiator codon. In the absence of functional eIF2α, Met-tRNA$_i$ is delivered to the stalled 40S subunits by eIF2A (Ventoso et al. 2006). Unlike the multisubunit factor eIF2, the activity of the 68-kD eIF2A polypeptide does not involve GTP binding, hydrolysis, or recycling, but it does require that the 40S subunit engage an AUG codon (Adams et al. 1975; Merrick et al. 1975).

Regulation of mRNA Levels

The synthesis of host defense mediators is obstructed by viral functions that target mRNA biogenesis, processing, localization, and metabolism (Table 2). Enterovirus proteinase 3C cleaves the TBP component of TFIID, inhibiting transcription from host defense genes (Clark et al. 1993). The VSV M protein blocks mRNA export by binding the nuclear export factor Rae1/mrnp41 (Faria et al. 2005). This prevents the cytoplasmic accumulation of host mRNAs, one of which encodes interferon-β (Stojdl et al. 2003), thereby preventing the expression of ISGs including PKR which confers resistance to VSV (Stojdl et al. 2000). The HSV-1-encoded ribonuclease vhs, which is important for pathogenesis in mice (Read and Frenkel 1983; Strelow and Leib 1995; Smith et al. 2002), destabilizes host mRNAs important for innate responses, such as those encoding type I interferons (Leib et al. 1999; Murphy et al. 2003; Duerst and Morrison 2004). The HSV-1 immediate-early protein ICP27 inhibits splicing, which is linked to nuclear export. In contrast, ICP27 promotes the export of viral mRNAs, the majority of which are unspliced (Sandri-Goldin 2004). Cellular pol II transcript abundance also decays rapidly in influenza virus-infected cells as viral mRNA 5' termini are procured from host mRNAs by a viral endonuclease in a process called "cap snatching" (Katze and Krug 1984). In addition, influenza virus NS1 protein associates with U6 snRNA to inhibit host pre-mRNA splicing (Lu et al. 1994) and interacts with the 30-kD subunit of CPSF (cleavage and polyadenylation specificity factor) to inhibit 3'-end formation of polyadenylated host mRNAs (Nemeroff et al. 1998). Improperly processed host mRNAs are retained in nuclei, whereas viral mRNAs, which do not require CPSF, are exported. The preferential translation of influenza mRNAs is thought to involve a cellular RNA-binding protein, GRSF-1, that recognizes a short *cis*-acting sequence in the 5'UTR of viral mRNAs (Kash et al. 2002).

The stability of viral mRNA is threatened by RNA interference (or posttranscriptional gene silencing) as a result of the accumulation of viral dsRNA. The cellular RNAi machinery processes dsRNA via the enzyme Dicer and uses the sequence specificity of the resulting products to selectively degrade target transcripts, viral mRNAs in this case. Not surprisingly, viruses are equipped with antidotes to this defense mechanism. The influenza- and vaccinia virus-encoded dsRNA-binding proteins NS1 and E3L, which both antagonize the interferon response as described above, suppress RNAi when expressed in insect cells, although their mechanism of action remains to be determined (W.X. Li et al. 2004). The B2 protein of nodamura virus, which infects both insect and mammalian hosts, inhibits Dicer-directed RNA processing by binding to its substrates and products (Li et al. 2002; Sullivan et al. 2005). The related flock house virus protein is required for viral RNA replication in nematodes, whereas B2-deficient mutants trigger a potent RNA-silencing activity (Lu et al. 2005). Adenovirus VA RNAs bind directly to Dicer and inhibit its function in infected cells by acting as competitive substrates (Lu and Cullen 2004; Gunnar-Andersson et al. 2005).

A new dimension of posttranscriptional regulation that can affect both cellular and viral target RNAs has been revealed by the discovery of cellular and virus-encoded microRNAs (miRNAs). Recently identified miRNAs encoded by DNA viruses have the potential to control the level and translation of select host mRNAs (Sullivan and Ganem 2005; Sullivan et al. 2005; Schutz and Sarnow 2006; Chapter 11). Strikingly, the HSV-1 latency-associated transcript (LAT) encodes no protein product but generates a miRNA that confers resistance to apoptosis by down-regulating cellular mRNAs involved in the initial stages of the apoptosis pathway (Gupta et al. 2006). Viral mRNAs can also be targets for cellular miRNAs, which may either positively or negatively affect virus replication. For example, the cellular miRNA miR-32 inhibits translation from target mRNAs produced by primate foamy virus-1, a retrovirus. This translation inhibition is partially relieved by the virus-encoded Tas polypeptide and is not limited to miR-32 (Lecellier et al. 2005). Conversely, a liver-specific miRNA up-regulates replication of the HCV genome in cultured liver cells (Jopling et al. 2005). Further study in cultured cells and animal models of infection will no doubt shed light on the biological functions of this control modality, which may have an especially critical role in plants and other organisms whose antiviral repertoire lacks cytokines and adaptive immunity.

SUMMARY AND PERSPECTIVES

The intimate relationships between viruses and the cellular translation system continue to provide insights into the mechanisms of protein synthesis and translational control. Reciprocally, increased knowledge of the mechanism and control of the translation process has deepened our understanding of viral infection strategies and of the interactions between viruses and their hosts. Intracellular stress engendered by viral infection affects the translation machinery through initiation factor cleavage, eIF2 kinase activation, and stress granule formation, each of which has the potential to trigger an apoptotic host response. Viral countermeasures include direct actions of viral products (e.g., PKR inhibitors) as well as indirect mechanisms that impinge on the stress pathway. For example, viral infection can harness cellular components (e.g., P58^{IPK}) to counteract stress, revealing the workings of host stress response pathways. The burgeoning RNAi field has profound and widespread implications for viral gene expression, including effects at the translational level. Viral miRNAs may function both to enhance viral protein synthesis and to inhibit cellular protein synthesis. Conversely, cells may use miRNAs to combat viral infection. First examples of these mechanisms have come to light, and we can confidently predict that many more will follow.

Recent findings have increased our understanding of the means by which viral products regulate host translation initiation factors and their associated proteins (e.g., eIF4F, 4E-BP1, and Mnk1) with effects on processes such as host cell shutoff and initiation complex assembly. Viral functions have been described which influence these processes in ways that have not been described in cells, but it is likely that cellular homologs or analogs of these viral products exist and await discovery. Correspondingly, numerous precedents (e.g., frameshifting, IRESs) suggest that what presently appear to be uniquely viral translation mechanisms will prove to have cellular counterparts. The *cis*-acting element newly discovered in Sindbis virus can direct ribosome pausing and the delivery of initiator tRNA by eIF2A, thereby evading the inhibition that would otherwise result from high levels of eIF2α phosphorylation. These findings imply a function for eIF2A that could have parallels in uninfected cells. Especially far-reaching consequences that could revise our notion of genomic coding capacity would flow from the finding of a cellular element, similar to that in dicistroviruses such as CrPV, capable of directing initiation-factor-independent initiation at non-AUG codons.

Although much has been learned, viruses have by no means yielded up all of their secrets. For example, the EBNA1 protein produced in cells latently infected with EBV has a glycine-alanine repetitive region (GAr) which inhibits protein synthesis *in cis*. Upon emerging from the ribosome, the GAr segment attenuates EBNA1 translation by an unknown mechanism, contributing to EBV's ability to evade host immune surveillance (Yin et al. 2003). Interference with this mechanism might allow a sustained immune response, which is one approach to antiviral therapy. Such potential interventions, and the emerging application of viruses in cancer and gene therapy, mandate an understanding of the interplay between viruses and cellular mechanisms in great depth. Given the many unsolved mysteries encrypted within their genomes and their reliance on the translational apparatus of their host cells, it is likely that viral systems will continue to illuminate protein synthesis and translational control paradigms for some time to come.

ACKNOWLEDGMENTS

The authors are supported by grants from the National Institute of General Medical Sciences (I.J.M.) and the National Institute of Allergy and Infectious Diseases (T.P. and M.B.M.) of the National Institutes of Health.

REFERENCES

Adams J.M. and Capecchi M. 1966. N-formylmethionyl-sRNA as the initiator of protein synthesis. *Proc. Natl. Acad. Sci.* **55:** 147–155.

Adams S.L., Safer B., Anderson W.F., and Merrick W.C. 1975. Eukaryotic initiation complex formation. Evidence for two distinct pathways. *J. Biol. Chem.* **250:** 9083–9089.

Alderete J.P., Jarrahian S., and Geballe A.P. 1999. Translational effects of mutations and polymorphisms in a repressive upstream open reading frame of the human cytomegalovirus UL4 gene. *J. Virol.* **73:** 8330–8337.

Algire M.A., Maag D., and Lorsch J.R. 2005. Pi release from eIF2, not GTP hydrolysis, is the step controlled by start-site selection during eukaryotic translation initiation. *Mol. Cell* **20:** 251–262.

Alvarez E., Menendez-Arias L., and Carrasco L. 2003. The eukaryotic translation initiation factor 4GI is cleaved by different retroviral proteases. *J. Virol.* **77:** 12392–12400.

Aragon T., de la Luna S., Novoa I., Carrasco L., Ortin J., and Nieto A. 2000. Eukaryotic translation initiation factor 4GI is a cellular target for NS1 protein, a translational activator of influenza virus. *Mol. Cell. Biol.* **20:** 6259–6268.

Baranov P.V., Henderson C.M., Anderson C.B., Gesteland R.F., Atkins J.F., and Howard M.T. 2005. Programmed ribosomal frameshifting in decoding the SARS-CoV genome. *Virology* **332:** 498–510.

Barnett L., Brenner S., Crick F.H.C., Schulman R.G., and Watts-Tobin R.J. 1967. Phase shift and other mutants in the first part of the *rII* cistron of phage T4. *Philos. Trans. R. Soc. Lond. B Biol. Sci.* **252:** 487–560.

Beattie E., Paoletti E., and Tartaglia J. 1995. Distinct patterns of IFN sensitivity observed in cells infected with vaccinia K3L- and E3L-mutant viruses. *Virology* **210:** 254–263.

Beattie E., Kauffman E.B., Martinez H., Perkus M.E., Jacobs B.L., Paoletti E., and Tartaglia J. 1996. Host-range restriction of vaccinia virus E3L-specific deletion mutants. *Virus Genes* **12:** 89–94.

Belli B.A. and Samuel C.E. 1993. Biosynthesis of reovirus-specified polypeptides: Identification of regions of the bicistronic reovirus S1 mRNA that affect the efficiency of translation in animal cells. *Virology* **193:** 16–27.

Belsham G.J., McInerney G.M., and Ross-Smith N. 2000. Foot-and-mouth disease virus 3C protease induces cleavage of translation initiation factors eIF4A and eIF4G within infected cells. *J. Virol.* **74:** 272–280.

Benard L., Carroll K., Valle R.C.P., Masison D.C., and Wickner R.B. 1999. The ski7 antiviral protein is an EF1-alpha homolog that blocks expression of non-Poly(A) mRNA in *Saccharomyces cerevisiae*. *J. Virol.* **73:** 2893–2900.

Bilgin D.D., Liu Y., Schiff M., and Dinesh-Kumar S.P. 2003. P58(IPK), a plant ortholog of double-stranded RNA-dependent protein kinase PKR inhibitor, functions in viral pathogenesis. *Dev. Cell* **4:** 651–661.

Black B.L., Rhodes R.B., McKenzie M., and Lyles D.S. 1993. The role of vesicular stomatitis virus matrix protein in inhibition of host-directed gene expression is genetically separable from its function in virus assembly. *J. Virol.* **67:** 4814–4821.

Borman A.M., Deliat F.G., and Kean K.M. 1994. Sequences within the poliovirus internal ribosome entry segment control viral RNA synthesis. *EMBO J.* **13:** 3149–3157.

Bowden R.J., Simas J.P., Davis A.J., and Efstathiou S. 1997. Murine gammaherpesvirus 68 encodes tRNA-like sequences which are expressed during latency. *J. Gen. Virol.* **78:** 1675–1687.

Brand S.R., Kobayashi R., and Mathews M.B. 1997. The Tat protein of human immunodeficiency virus type 1 is a substrate and inhibitor of the interferon-induced, virally activated protein kinase, PKR. *J. Biol. Chem.* **272:** 8388–8395.

Brandt T.A. and Jacobs B.L. 2001. Both carboxy- and amino-terminal domains of the vaccinia virus interferon resistance gene, E3L, are required for pathogenesis in a mouse model. *J. Virol.* **75:** 850–856.

Brasey A., Lopez-Lastra M., Ohlmann T., Beerens N., Berkhout B., Darlix J.L., and Sonenberg N. 2003. The leader of human immunodeficiency virus type 1 genomic RNA harbors an internal ribosome entry segment that is active during the G2/M phase of the cell cycle. *J. Virol.* **77:** 3939–3949.

Brenner S., Jacob F., and Meselson M. 1961. An unstable intermediate carrying information from genes to ribosomes for protein synthesis. *Nature* **190:** 576–581.

Buck C.B., Shen X., Egan M.A., Pierson T.C., Walker C.M., and Siliciano R.F. 2001. The human immunodeficiency virus type 1 gag gene encodes an internal ribosome entry site. *J. Virol.* **75:** 181–191.

Burgui I., Aragon T., Ortin J., and Nieto A. 2003. PABP1 and eIF4GI associate with influenza virus NS1 protein in viral mRNA translation initiation complexes. *J. Gen. Virol.* **84:** 3263–3274.

Bushell M., Wood W., Clemens M.J., and Morley S.J. 2000. Changes in integrity and association of eukaryotic protein synthesis initiation factors during apoptosis. *Eur. J. Biochem.* **267:** 1083–1091.

Cande C., Vahsen N., Metivier D., Tourriere H., Chebli K., Garrido C., Tazi J., and Kroemer G. 2004. Regulation of cytoplasmic stress granules by apoptosis-inducing factor. *J. Cell Sci.* **117:** 4461–4468.

Castelli J.C., Hassel B.A., Wood K.A., Li X.L., Amemiya K., Dalakas M.C., Torrence P.F., and Youle R.J. 1997. A study of the interferon antiviral mechanism: Apoptosis activation by the 2–5A system. *J. Exp. Med.* **186:** 967–972.

Cayley P.J., Davies J.A., McCullagh K.G., and Kerr I.M. 1984. Activation of the ppp(A2′p)nA system in interferon-treated, herpes simplex virus-infected cells and evidence for novel inhibitors of the ppp(A2′p)nA-dependent RNase. *Eur. J. Biochem.* **143:** 165–174.

Cerveny M., Hessefort S., Yang K., Cheng G., Gross M., and He B. 2003. Amino acid substitutions in the effector domain of the gamma(1)34.5 protein of herpes simplex virus 1 have differential effects on viral response to interferon-alpha. *Virology* **307:** 290–300.

Cevallos R.C. and Sarnow P. 2005. Factor-independent assembly of elongation-competent ribosomes by an internal ribosome entry site located in an RNA virus that infects penaeid shrimp. *J. Virol.* **79:** 677–683.

Chebath J., Benech P., Revel M., and Vigneron M. 1987. Constitutive expression of (2′-5′) oligo A synthetase confers resistance to picornavirus infection. *Nature* **330:** 587–588.

Cheng G., Feng Z., and He B. 2005. Herpes simplex virus 1 infection activates the endoplasmic reticulum resident kinase PERK and mediates eIF-2alpha dephosphorylation by the gamma(1)34.5 protein. *J. Virol.* **79:** 1379–1388.

Cheng G., Gross M., Brett M.E., and He B. 2001. AlaArg motif in the carboxyl terminus of the gamma(1)34.5 protein of herpes simplex virus type 1 is required for the formation of a high-molecular-weight complex that dephosphorylates eIF-2alpha. *J. Virol.* **75:** 3666–3674.

Chou J., Kern E.R., Whitley R.J., and Roizman B. 1990. Mapping of herpes simplex virus-1 neurovirulence to gamma 134.5, a gene nonessential for growth in culture. *Science* **250:** 1262–1266.

Clark M.E., Lieberman P.M., Berk A.J., and Dasgupta A. 1993. Direct cleavage of human TATA-binding protein by poliovirus protease 3C in vivo and in vitro. *Mol. Cell. Biol.* **13:** 1232–1237.

Clemens M.J. 2005. Translational control in virus-infected cells: Models for cellular stress responses. *Semin. Cell Dev. Biol.* **16:** 13–20.

Connor J.H. and Lyles D.S. 2002. Vesicular stomatitis virus infection alters the eIF4F translation initiation complex and causes dephosphorylation of the eIF4E binding protein 4E-BP1. *J. Virol.* **76:** 10177–10187.

Corcelette S., Masse T., and Madjar J.J. 2000. Initiation of translation by non-AUG codons in human T-cell lymphotropic virus type I mRNA encoding both Rex and Tax regulatory proteins. *Nucleic Acids Res.* **28:** 1625–1634.

Cornelis S., Bruynooghe Y., Denecker G., Van Huffel S., Tinton S., and Beyaert R. 2000. Identification and characterization of a novel cell cycle-regulated internal ribosome entry site. *Mol. Cell* **5:** 597–605.

Cosentino G.P., Venkatesan S., Serluca F.C., Green S.R., Mathews M.B., and Sonenberg N. 1995. Double-stranded-RNA-dependent protein kinase and TAR RNA-binding protein form homo- and heterodimers in vivo. *Proc. Natl. Acad. Sci.* **92:** 9445–9449.

Cuesta R., Xi Q., and Schneider R.J. 2000. Adenovirus-specific translation by displacement of kinase Mnk1 from cap-initiation complex eIF4F. *EMBO J.* **19:** 3465–3474.

Cullen B.R. 2006. Viruses and microRNAs. *Nat. Genet.* (suppl.) **38:** S25–S30.

Dasgupta R., Shih D.S., Saris C., and Kaesberg P. 1975. Nucleotide sequence of a viral RNA fragment that binds to eukaryotic ribosomes. *Nature* **256:** 624–628.

de Breyne S., Simonet V., Pelet T., and Curran J. 2003. Identification of a cis-acting element required for shunt-mediated translational initiation of the Sendai virus Y proteins. *Nucleic Acids Res.* **31:** 608–618.

Der S.D., Yang Y.L., Weissmann C., and Williams B.R.G. 1997. A double-stranded RNA-activated protein kinase-dependent pathway mediating stress-induced apoptosis. *Proc. Natl. Acad. Sci.* **94:** 3279–3283.

Desai S.Y., Patel R.C., Sen G.C., Malhotra P., Ghadge G.D., and Thimmapaya B. 1995. Activation of interferon-inducible 2′-5′ oligoadenylate synthetase by adenoviral VAI RNA. *J. Biol. Chem.* **270:** 3454–3461.

Diaz-Guerra M., Rivas C., and Esteban M. 1997. Activation of the IFN-inducible enzyme RNase L causes apoptosis of animal cells. *Virology* **236:** 354–363.

Dinman J.D., Ruiz-Echevarria M.J., and Peltz S.W. 1998. Translating old drugs into new treatments: Ribosomal frameshifting as a target for antiviral agents. *Trends Biotechnol.* **16:** 190–196.

Doepker R.C., Hsu W.L., Saffran H.A., and Smiley J.R. 2004. Herpes simplex virus virion host shutoff protein is stimulated by translation initiation factors eIF4B and eIF4H. *J. Virol.* **78:** 4684–4699.

Duerst R.J. and Morrison L.A. 2004. Herpes simplex virus 2 virion host shutoff protein interferes with type I interferon production and responsiveness. *Virology* **322:** 158–167.

Essbauer S., Bremont M., and Ahne W. 2001. Comparison of the eIF-2alpha homologous proteins of seven ranaviruses (Iridoviridae). *Virus Genes* **23:** 347–359.

Fajardo J.E. and Shatkin A.J. 1990. Translation of bicistronic viral mRNA in transfected cells: Regulation at the level of elongation. *Proc. Natl. Acad. Sci.* **87:** 328–332.

Falcon A.M., Fortes P., Marion R.M., Beloso A., and Ortin J. 1999. Interaction of influenza virus NS1 protein and the human homologue of Staufen in vivo and in vitro. *Nucleic Acids Res.* **27:** 2241–2247.

Faria P.A., Chakraborty P., Levay A., Barber G.N., Ezelle H.J., Enninga J., Arana C., van Deursen J., and Fontoura B.M. 2005. VSV disrupts the Rae1/mrnp41 mRNA nuclear export pathway. *Mol. Cell* **17:** 93–102.

Feigenblum D. and Schneider R.J. 1993. Modification of eukaryotic initiation factor 4F during infection by influenza virus. *J. Virol.* **67:** 3027–3035.

———. 1996. Cap-binding protein (eukaryotic initiation factor 4E) and 4E-inactivating protein BP-1 independently regulate cap-dependent translation. *Mol. Cell. Biol.* **16:** 5450–5457.

Feng P., Everly D.N., Jr., and Read G.S. 2001. mRNA decay during herpesvirus infections: Interaction between a putative viral nuclease and a cellular translation factor. *J. Virol.* **75:** 10272–10280.

———. 2005. mRNA decay during herpes simplex virus (HSV) infections: Protein-protein interactions involving the HSV virion host shutoff protein and translation factors eIF4H and eIF4A. *J. Virol.* **79:** 9651–9664.

Fernandez J., Yaman I., Sarnow P., Snider M.D., and Hatzoglou M. 2002. Regulation of internal ribosomal entry site-mediated translation by phosphorylation of the translation initiation factor eIF2alpha. *J. Biol. Chem.* **277:** 19198–19205.

Floyd-Smith G., Slattery E., and Lengyel P. 1981. Interferon action: RNA cleavage pattern of a (2′-5′)oligoadenylate-dependent endonuclease. *Science* **212:** 1030–1032.

Floyd-Smith G., Yoshie O., and Lengyel P. 1982. Interferon action. Covalent linkage of (2'-5')pppApApA(32P)pCp to (2'-5')(A)n-dependent ribonucleases in cell extracts by ultraviolet irradiation. *J. Biol. Chem.* **257:** 8584–8587.

Fukunaga R. and Hunter T. 1997. MNK1, a new MAP kinase-activated protein kinase, isolated by a novel expression screening method for identifying protein kinase substrates. *EMBO J.* **16:** 1921–1933.

Furuichi Y., Morgan M., Muthukrishnan S., and Shatkin A.J. 1975. Reovirus messenger RNA contains a methylated, blocked 5'-terminal structure: m^{7}G(5')ppp(5')G^{m}pCp-. *Proc. Natl. Acad. Sci.* **72:** 362–366.

Fütterer J., Kiss-László Z., and Hohn T. 1993. Nonlinear ribosome migration on cauliflower mosaic virus 35S RNA. *Cell* **73:** 789–802.

Fütterer J., Potrykus I., Bao Y., Li L., Burns T.M., Hull R., and Hohn T. 1996. Position-dependent ATT initiation during plant pararetrovirus rice tungro bacilliform virus translation. *J. Virol.* **70:** 2999–3010.

Gale M., Jr., Blakely C.M., Kwieciszewski B., Tan S.L., Dossett M., Tang N.M., Korth M.J., Polyak S.J., Gretch D.R., and Katze M.G. 1998. Control of PKR protein kinase by hepatitis C virus nonstructural 5A protein: Molecular mechanisms of kinase regulation. *Mol. Cell. Biol.* **18:** 5208–5218.

Gazo B.M., Murphy P., Gatchel J.R., and Browning K.S. 2004. A novel interaction of Cap-binding protein complexes eukaryotic initiation factor (eIF) 4F and eIF(iso)4F with a region in the 3'-untranslated region of satellite tobacco necrosis virus. *J. Biol. Chem.* **279:** 13584–13592.

Gil J. and Esteban M. 2000. Induction of apoptosis by the dsRNA-dependent protein kinase (PKR): Mechanism of action. *Apoptosis* **5:** 107–114.

Gingras A.C. and Sonenberg N. 1997. Adenovirus infection inactivates the translational inhibitors 4E-BP1 and 4E-BP2. *Virology* **237:** 182–186.

Gingras A., Svitkin Y., Belsham G.J., Pause A., and Sonenberg N. 1996. Activation of the translational suppressor 4E-BP1 following infection with encephalomyocarditis virus and poliovirus. *Proc. Natl. Acad. Sci.* **93:** 5578–5583.

Gingras A.C., Raught B., Gygi S.P., Niedzwiecka A., Miron M., Burley S.K., Polakiewicz R.D., Wyslouch-Cieszynska A., Aebersold R., and Sonenberg N. 2001. Hierarchical phosphorylation of the translation inhibitor 4E-BP1. *Genes Dev.* **15:** 2852–2864.

Goldstaub D., Gradi A., Bercovitch Z., Grosmann Z., Nophar Y., Luria S., Sonenberg N., and Kahana C. 2000. Poliovirus 2A protease induces apoptotic cell death. *Mol. Cell. Biol.* **20:** 1271–1277.

Goodfellow I., Chaudhry Y., Giodasi I., Gerondopoulos A., Natoni A., Labrie L., Laliberte J.F., and Roberts L. 2005. Calicivirus translation initiation requires an interaction between VPg and eIF4E. *EMBO Rep.* **6:** 968–972.

Gomez C.E., Vandermeeren A.M., Garcia M.A., Domingo-Gil E., and Esteban M. 2005. Involvement of PKR and RNase L in translational control and induction of apoptosis after Hepatitis C polyprotein expression from a vaccinia virus recombinant. *Virol. J.* **2:** 81.

Gradi A., Svitkin Y.V., Imataka H., and Sonenberg N. 1998. Proteolysis of human eukaryotic translation initiation factor eIF4GII, but not eIF4GI, coincides with the shut-off of host protein synthesis after poliovirus infection. *Proc. Natl. Acad. Sci.* **95:** 11089–11094.

Gradi A., Foeger N., Strong R., Svitkin Y.V., Sonenberg N., Skern T., and Belsham G.J. 2004. Cleavage of eukaryotic translation initiation factor 4GII within foot-and-mouth disease virus-infected cells: Identification of the L-protease cleavage site in vitro. *J. Virol.* **78:** 3271–3278.

Gregory R.I., Chendrimada T.P., Cooch N., and Shiekhattar R. 2005. Human RISC couples microRNA biogenesis and posttranscriptional gene silencing. *Cell* **123:** 631–640.

Gribaudo G., Lembo D., Cavallo G., Landolfo S., and Lengyel P. 1991. Interferon action: Binding of viral RNA to the 40-kilodalton 2′-5′-oligoadenylate synthetase in interferon-treated HeLa cells infected with encephalomyocarditis virus. *J. Virol.* **65:** 1748–1757.

Groft C.M. and Burley S.K. 2002. Recognition of eIF4G by rotavirus NSP3 reveals a basis for mRNA circularization. *Mol. Cell* **9:** 1273–1283.

Gromeier M., Bossert B., Arita M., Nomoto A., and Wimmer E. 1999. Dual stem loops within the poliovirus internal ribosomal entry site control neurovirulence. *J. Virol.* **73:** 958–964.

Gros F., Hiatt H., Gilbert W., Kurland G.G., Risebrough R.W., and Watson J.D. 1961. Unstable ribonucleic acid revealed by pulse labelling of *Escherichia coli*. *Nature* **190:** 581–585.

Grzela R., Strokovska L., Andrieu J.P., Dublet B., Zagorski W., and Chroboczek J. 2006. Potyvirus terminal protein VPg, effector of host eukaryotic initiation factor eIF4E. *Biochimie* **88:** 887–896.

Guerra S., Lopez-Fernandez L.A., Angel Garcia M., Zaballos A., and Esteban M. 2006. Human gene profiling in response to active protein kinase PKR in infected cells: Involvement of the transcription factor ATF-3 in PKR-induced apoptosis. *J. Biol. Chem.* **281:** 18734–18745.

Guest S., Pilipenko E., Sharma K., Chumakov K., and Roos R.P. 2004. Molecular mechanisms of attenuation of the Sabin strain of poliovirus type 3. *J. Virol.* **78:** 11097–11107.

Gunnar-Andersson M.G., Haasnoot P.C., Xu N., Berenjian S., Berkhout B., and Akusjarvi G. 2005. Suppression of RNA interference by adenovirus virus-associated RNA. *J. Virol.* **79:** 9556–9565.

Gunnery S., Green S.R., and Mathews M.B. 1992. Tat-responsive region RNA of human immunodeficiency virus type 1 stimulates protein synthesis in vivo and in vitro: Relationship between structure and function. *Proc. Natl. Acad. Sci.* **89:** 11557–11561.

Guo J., Hui D.J., Merrick W.C., and Sen G.C. 2000. A new pathway of translational regulation mediated by eukaryotic initiation factor 3. *EMBO J.* **19:** 6891–6899.

Gupta A., Gartner J.J., Sethupathy P., Hatzigeorgiou A.G., and Fraser N.W. 2006. Antiapoptotic function of a microRNA encoded by the HSV-1 latency-associated transcript. *Nature* **442:** 82–85.

Gupta V., Huang X., and Patel R.C. 2003. The carboxy-terminal, M3 motifs of PACT and TRBP have opposite effects on PKR activity. *Virology* **315:** 283–291.

Haghighat A., Mader S., Pause A., and Sonenberg N. 1995. Repression of cap-dependent translation by 4E-binding protein 1: Competition with p220 for binding to eukaryotic initiation factor-4E. *EMBO J.* **14:** 5701–5709.

Haller O., Kochs G., and Weber F. 2006. The interferon response circuit: Induction and suppression by pathogenic viruses. *Virology* **344:** 119–130.

Hambidge S. and Sarnow P. 1992. Translational enhancement of the poliovirus 5′ noncoding region mediated by virus-encoded polypeptide 2A. *Proc. Natl. Acad. Sci.* **89:** 10272–10276.

Hartmann R., Justesen J., Sarkar S.N., Sen G.C., and Yee V.C. 2003. Crystal structure of the 2′-specific and double-stranded RNA-activated interferon-induced antiviral protein 2′-5′-oligoadenylate synthetase. *Mol. Cell* **12:** 1173–1185.

He B., Gross M., and Roizman B. 1997. The gamma(1)34.5 protein of herpes simplex virus 1 complexes with protein phosphatase 1alpha to dephosphorylate the alpha subunit of the eukaryotic translation initiation factor 2 and preclude the shutoff of protein synthesis by double-stranded RNA-activated protein kinase. *Proc. Natl. Acad. Sci.* **94:** 843–848.

Hellen C.U. and Sarnow P. 2001. Internal ribosome entry sites in eukaryotic mRNA molecules. *Genes Dev.* **15:** 1593–1612.

Herbreteau C.H., Weill L., Decimo D., Prevot D., Darlix J.L., Sargueil B., and Ohlmann T. 2005. HIV-2 genomic RNA contains a novel type of IRES located downstream of its initiation codon. *Nat. Struct. Mol. Biol.* **12:** 1001–1007.

Herbst K.L., Nichols L.M., Gesteland R.F., and Weiss R.B. 1994. A mutation in ribosomal protein L9 affects ribosomal hopping during translation of gene 60 from bacteriophage T4. *Proc. Natl. Acad. Sci.* **91:** 12525–12529.

Herr A.J., Wills N.M., Nelson C.C., Gesteland R.F., and Atkins J.F. 2001. Drop-off during ribosome hopping. *J. Mol. Biol.* **311:** 445–452.

Hindley J. and Staples D.H. 1969. Sequence of a ribosome binding site in bacteriophage Qβ-RNA. *Nature* **224:** 964–967.

Hiscott J., Lin R., Nakhaei P., and Paz S. 2006. MasterCARD: A priceless link to innate immunity. *Trends Mol. Med.* **12:** 53–56.

Hsiao K.C., Cheng C.H., Fernandes I.E., Detrich H.W., and DeVries A.L. 1990. An antifreeze glycopeptide gene from the antarctic cod *Notothenia coriiceps neglecta* encodes a polyprotein of high peptide copy number. *Proc. Natl. Acad. Sci.* **87:** 9265–9269.

Huang J.T. and Schneider R.J. 1990. Adenovirus inhibition of cellular protein synthesis is prevented by the drug 2-aminopurine. *Proc. Natl. Acad. Sci.* **87:** 7115–7119.

———. 1991. Adenovirus inhibition of cellular protein synthesis involves inactivation of cap binding protein. *Cell* **65:** 271–280.

Hui D.J., Bhasker C.R., Merrick W.C., and Sen G.C. 2003. Viral stress-inducible protein p56 inhibits translation by blocking the interaction of eIF3 with the ternary complex eIF2.GTP.Met-tRNAi. *J. Biol. Chem.* **278:** 39477–39482.

Hui D.J., Terenzi F., Merrick W.C., and Sen G.C. 2005. Mouse p56 blocks a distinct function of eukaryotic initiation factor 3 in translation initiation. *J. Biol. Chem.* **280:** 3433–3440.

Hung M., Patel P., Davis S., and Green S.R. 1998. Importance of ribosomal frameshifting for human immunodeficiency virus type 1 particle assembly and replication. *J. Virol.* **72:** 4819–4824.

Iordanov M.S., Paranjape J.M., Zhou A., Wong J., Williams B.R., Meurs E.F., Silverman R.H., and Magun B.E. 2000. Activation of p38 mitogen-activated protein kinase and c-Jun NH(2)-terminal kinase by double-stranded RNA and encephalomyocarditis virus: Involvement of RNase L, protein kinase R, and alternative pathways. *Mol. Cell. Biol.* **20:** 617–627.

Isler J.A., Maguire T.G., and Alwine J.C. 2005a. Production of infectious human cytomegalovirus virions is inhibited by drugs that disrupt calcium homeostasis in the endoplasmic reticulum. *J. Virol.* **79:** 15388–15397.

Isler J.A., Skalet A.H., and Alwine J.C. 2005b. Human cytomegalovirus infection activates and regulates the unfolded protein response. *J. Virol.* **79:** 6890–6899.

Jacks T. and Varmus H.E. 1985. Expression of the Rous sarcoma virus *pol* gene by ribosomal frameshifting. *Science* **230:** 1237–1242.

Jacquemont B. and Roizman B. 1975. RNA synthesis in cells infected with herpes simplex virus. X. Properties of viral symmetric transcripts and of double-stranded RNA prepared from them. *J. Virol.* **15:** 707–713.

Jang S.K., Krausslich H.G., Nicklin M.J., Duke G.M., Palmenberg A.C., and Wimmer E. 1988. A segment of the 5′ nontranslated region of encephalomyocarditis virus RNA directs internal entry of ribosomes during in vitro translation. *J. Virol.* **62:** 2636–2643.

Johannes G., Carter M.S., Eisen M.B., Brown P.O., and Sarnow P. 1999. Identification of eukaryotic mRNAs that are translated at reduced cap binding complex eIF4F concentrations using a cDNA microarray. *Proc. Natl. Acad. Sci.* **96:** 13118–13123.

Jopling C.L., Yi M., Lancaster A.M., Lemon S.M., and Sarnow P. 2005. Modulation of hepatitis C virus RNA abundance by a liver-specific MicroRNA. *Science* **309:** 1577–1581.

Jordan R., Wang L., Graczyk T.M., Block T.M., and Romano P.R. 2002. Replication of a cytopathic strain of bovine viral diarrhea virus activates PERK and induces endoplasmic reticulum stress-mediated apoptosis of MDBK cells. *J. Virol.* **76:** 9588–9599.

Kahvejian A., Roy G., and Sonenberg N. 2001. The mRNA closed-loop model: The function of PABP and PABP-interacting proteins in mRNA translation. *Cold Spring Harbor Symp. Quant. Biol.* **66:** 293–300.

Kajaste-Rudnitski A., Mashimo T., Frenkiel M.P., Guenet J.L., Lucas M., and Despres P. 2006. The 2′,5′-oligoadenylate synthetase 1b is a potent inhibitor of West Nile virus replication inside infected cells. *J. Biol. Chem.* **281:** 4624–4637.

Kash J.C., Cunningham D.M., Smit M.W., Park Y., Fritz D., Wilusz J., and Katze M.G. 2002. Selective translation of eukaryotic mRNAs: Functional molecular analysis of GRSF-1, a positive regulator of influenza virus protein synthesis. *J. Virol.* **76:** 10417–10426.

Kates J. 1971. Transcription of the vaccinia virus genome and the occurrence of polyriboadenylic acid sequences in messenger RNA. *Cold Spring Harbor Symp. Quant. Biol.* **35:** 743–752.

Katze M.G. and Krug R.M. 1984. Metabolism and expression of RNA polymerase II transcripts in influenza virus-infected cells. *Mol. Cell. Biol.* **4:** 2198–2206.

Kaufmann G. 2000. Anticodon nucleases. *Trends Biochem. Sci.* **25:** 70–74.

Kazemi S., Papadopoulou S., Li S., Su Q., Wang S., Yoshimura A., Matlashewski G., Dever T.E., and Koromilas A.E. 2004. Control of alpha subunit of eukaryotic translation initiation factor 2 (eIF2 alpha) phosphorylation by the human papillomavirus type 18 E6 oncoprotein: Implications for eIF2 alpha-dependent gene expression and cell death. *Mol. Cell. Biol.* **24:** 3415–3429.

Kedersha N. and Anderson P. 2002. Stress granules: Sites of mRNA triage that regulate mRNA stability and translatability. *Biochem. Soc. Trans.* **30:** 963-969.

Kedersha N., Stoecklin G., Ayodele M., Yacono P., Lykke-Andersen J., Fitzler M.J., Scheuner D., Kaufman R.J., Golan D.E., and Anderson P. 2005. Stress granules and processing bodies are dynamically linked sites of mRNP remodeling. *J. Cell Biol.* **169:** 871–884.

Kerr I.M., Cohen N., and Work T.S. 1966. Factors controlling amino acid incorporation by ribosomes from Krebs II mouse ascites-tumour cells. *Biochem. J.* **98:** 826–835.

Khoo D., Perez C., and Mohr I. 2002. Characterization of RNA determinants recognized by the arginine- and proline-rich region of Us11, a herpes simplex virus type 1-encoded double-stranded RNA binding protein that prevents PKR activation. *J. Virol.* **76:** 11971–11981.

Kinzy T.G. and Goldman E. 2000. Nontranslational functions of components of the translational apparatus. In *Translational control of gene expression* (ed. N. Sonenberg et al.), pp. 973–997. Cold Spring Harbor Laboratory Press, Cold Spring Harbor, New York.

Kirchweger R., Ziegler E., Lamphear B.J., Waters D., Liebig H.D., Sommergruber W., Sobrino F., Hohenadl C., Blaas D., and Rhoads R.E., et al. 1994. Foot-and-mouth disease virus leader proteinase: Purification of the Lb form and determination of its cleavage site on eIF-4 gamma. *J. Virol.* **68:** 5677–5684.

Kitajewski J., Schneider R.J., Safer B., Munemitsu S.M., Samuel C.E., Thimmappaya B., and Shenk T. 1986. Adenovirus VAI RNA antagonizes the antiviral action of interferon by preventing activation of the interferon-induced eIF-2α kinase. *Cell* **45:** 195–200.

Kozak M. 1978. How do eucaryotic ribosomes select initiation regions in messenger RNA. *Cell* **15:** 1109–1123.

———. 1989. The scanning model for translation: An update. *J. Cell Biol.* **108:** 229–241.

Kubota K., Nakahara K., Ohtsuka T., Yoshida S., Kawaguchi J., Fujita Y., Ozeki Y., Hara A., Yoshimura C., Furukawa H., et al. 2004. Identification of 2′-phosphodiesterase, which plays a role in the 2-5A system regulated by interferon. *J. Biol. Chem.* **279:** 37832–37841.

Kudchodkar S.B., Yu Y., Maguire T.G., and Alwine J.C. 2004. Human cytomegalovirus infection induces rapamycin-insensitive phosphorylation of downstream effectors of mTOR kinase. *J. Virol.* **78:** 11030–11039.

———. 2006. Human cytomegalovirus infection alters the substrate specificities and rapamycin sensitivities of raptor- and rictor-containing complexes. *Proc. Natl. Acad. Sci.* **103:** 14182–14187.

Kuyumcu-Martinez N.M., Van Eden M.E., Younan P., and Lloyd R.E. 2004a. Cleavage of poly(A)-binding protein by poliovirus 3C protease inhibits host cell translation: A novel mechanism for host translation shutoff. *Mol. Cell. Biol.* **24:** 1779–1790.

Kuyumcu-Martinez M., Belliot G., Sosnovtsev S.V., Chang K.O., Green K.Y., and Lloyd R.E. 2004b. Calicivirus 3C-like proteinase inhibits cellular translation by cleavage of poly(A)-binding protein. *J. Virol.* **78:** 8172–8182.

Langland J.O., Pettiford S., Jiang B., and Jacobs B.L. 1994. Products of the porcine group C rotavirus NSP3 gene bind specifically to double-stranded RNA and inhibit activation of the interferon-induced protein kinase PKR. *J. Virol.* **68:** 3821–3829.

Langland J.O., Cameron J.M., Heck M.C., Jancovich J.K., and Jacobs B.L. 2006. Inhibition of PKR by RNA and DNA viruses. *Virus Res.* **119:** 100–110.

Latorre P., Kolakofsky D., and Curran J. 1998. Sendai virus Y proteins are initiated by a ribosomal shunt. *Mol. Cell. Biol.* **18:** 5021–5031.

Lazarowitz S.G. and Robertson H.D. 1977. Initiator regions from the small size class of reo-virus mRNA protected by rabbit reticulocyte ribosomes. *J. Biol. Chem.* **252:** 7842–7849.

Lecellier C.H., Dunoyer P., Arar K., Lehmann-Che J., Eyquem S., Himber C., Saib A., and Voinnet O. 2005. A cellular microRNA mediates antiviral defense in human cells. *Science* **308:** 557–560.

Leib D.A., Machalek M.A., Williams B.R., Silverman R.H., and Virgin H.W. 2000. Specific phenotypic restoration of an attenuated virus by knockout of a host resistance gene. *Proc. Natl. Acad. Sci.* **97:** 6097–6101.

Leib D.A., Harrison T.E., Laslo K.M., Machalek M.A., Moorman N.J., and Virgin H.W. 1999. Interferons regulate the phenotype of wild-type and mutant herpes simplex viruses in vivo. *J. Exp. Med.* **189:** 663–672.

Levy D.E., Marie I., and Prakash A. 2003. Ringing the interferon alarm: Differential regulation of gene expression at the interface between innate and adaptive immunity. *Curr. Opin. Immunol.* **15:** 52–58.

Li G., Xiang Y., Sabapathy K., and Silverman R.H. 2004. An apoptotic signaling pathway in the interferon antiviral response mediated by RNase L and c-Jun NH2-terminal kinase. *J. Biol. Chem.* **279:** 1123–1131.

Li S., Min J.Y., Krug R.M., and Sen G.C. 2006. Binding of the influenza A virus NS1 protein to PKR mediates the inhibition of its activation by either PACT or double-stranded RNA. *Virology* **349:** 13–21.

Li W., Li Y., Kedersha N., Anderson P., Emara M., Swiderek K.M., Moreno G.T., and Brinton M.A. 2002. Cell proteins TIA-1 and TIAR interact with the 3′ stem-loop of the West Nile virus complementary minus-strand RNA and facilitate virus replication. *J. Virol.* **76:** 11989–12000.

Li W.X., Li H., Lu R., Li F., Dus M., Atkinson P., Brydon E.W., Johnson K.L., Garcia-Sastre A., Ball L.A., et al. 2004. Interferon antagonist proteins of influenza and vaccinia viruses are suppressors of RNA silencing. *Proc. Natl. Acad. Sci.* **101:** 1350–1355.

Liu N., Kuang X., Kim H.T., Stoica G., Qiang W., Scofield V.L., and Wong P.K. 2004. Possible involvement of both endoplasmic reticulum- and mitochondria-dependent pathways in MoMuLV-ts1-induced apoptosis in astrocytes. *J. Neurovirol.* **10:** 189–198.

Lloyd R.E. 2006. Translational control by viral proteinases. *Virus Res.* **119:** 76–88.

Lodish H.F. and Porter M. 1980. Translational control of protein synthesis after infection by VSV. *J. Virol.* **36:** 719–733.

———. 1981. Vesicular stomatitis virus mRNA and inhibition of translation of cellular mRNA—Is there a P function in vesicular stomatitis virus? *J. Virol.* **38:** 504–517.

Lu R., Maduro M., Li F., Li H.W., Broitman-Maduro G., Li W.X., and Ding S.W. 2005. Animal virus replication and RNAi-mediated antiviral silencing in *Caenorhabditis elegans*. *Nature* **436:** 1040–1043.

Lu S. and Cullen B.R. 2004. Adenovirus VA1 noncoding RNA can inhibit small interfering RNA and MicroRNA biogenesis. *J. Virol.* **78:** 12868–12876.

Lu Y., Qian X.Y., and Krug R.M. 1994. The influenza virus NS1 protein: A novel inhibitor of pre-mRNA splicing. *Genes Dev.* **8:** 1817–1828.

Mader S., Lee H., Pause A., and Sonenberg N. 1995. The translation initiation factor eIF-4E binds to a common motif shared by the translation factor eIF-4γ and the translational repressors 4E-binding proteins. *Mol. Cell. Biol.* **15:** 4990–4997.

Maitra R.K., McMillan N.A., Desai S., McSwiggen J., Hovanessian A.G., Sen G., Williams B.R.G., and Silverman R.H. 1994. HIV-1 TAR RNA has an intrinsic ability to activate interferon-inducible enzymes. *Virology* **204:** 823–827.

Maran A. and Mathews M.B. 1988. Characterization of the double-stranded RNA implicated in the inhibition of protein synthesis in cells infected with a mutant adenovirus defective for VA RNA$_I$. *Virology* **164:** 106–113.

Marcotrigiano J., Gingras A.C., Sonenberg N., and Burley S.K. 1999. Cap-dependent translation initiation in eukaryotes is regulated by a molecular mimic of eIF4G. *Mol. Cell* **3:** 707–716.

Marie I., Svab J., Robert N., Galabru J., and Hovanessian A.G. 1990. Differential expression and distinct structure of 69- and 100-kDa forms of 2-5A synthetase in human cells treated with interferon. *J. Biol. Chem.* **265:** 18601–18607.

Marques J.T., Rebouillat D., Ramana C.V., Murakami J., Hill J.E., Gudkov A., Silverman R.H., Stark G.R., and Williams B.R. 2005. Down-regulation of p53 by double-stranded RNA modulates the antiviral response. *J. Virol.* **79:** 11105–11114.

Masison D.C., Blanc A., Ribas J.C., Carroll K., Sonenberg N., and Wickner R.B. 1995. Decoying the Cap⁻ mRNA degradation system by a double-stranded RNA virus and Poly(A)⁻ mRNA surveillance by a yeast antiviral system. *Mol. Cell. Biol.* **15:** 2763–2771.

Mathews M.B. and Korner A. 1970. The inhibitory action of a mammalian viral RNA on the initiation of protein synthesis in a reticulocyte cell-free system. *Nature* **228:** 661–663.

Mathews M.B. and Shenk T. 1991. Adenovirus virus-associated RNA and translation control. *J. Virol.* **65:** 5657–5662.

McClain W.H., Guthrie C., and Barrell B.G. 1972. Eight transfer RNAs induced by infection of *Escherichia coli* with bacteriophage T4. *Proc. Natl. Acad. Sci.* **69:** 3703–3707.

McInerney G.M., Kedersha N.L., Kaufman R.J., Anderson P., and Liljestrom P. 2005. Importance of eIF2alpha phosphorylation and stress granule assembly in alphavirus translation regulation. *Mol. Biol. Cell* **16:** 3753–3763.

Melville M.W., Hansen W.J., Freeman B.C., Welch W.J., and Katze M.G. 1997. The molecular chaperone hsp40 regulates the activity of P58IPK, the cellular inhibitor of PKR. *Proc. Natl. Acad. Sci.* **94:** 97–102.

Merrick W.C., Kemper W.M., Kantor J.A., and Anderson W.F. 1975. Purification and properties of rabbit reticulocyte protein synthesis elongation factor 2. *J. Biol. Chem.* **250:** 2620–2625.

Merrill M.K., Dobrikova E.Y., and Gromeier M. 2006. Cell-type-specific repression of internal ribosome entry site activity by double-stranded RNA-binding protein 76. *J. Virol.* **80:** 3147–3156.

Meyers G. 2003. Translation of the minor capsid protein of a calicivirus is initiated by a novel termination-dependent reinitiation mechanism. *J. Biol. Chem.* **278:** 34051–34060.

Miller E.S., Kutter E., Mosig G., Arisaka F., Kunisawa T., and Ruger W. 2003. Bacteriophage T4 genome. *Microbiol. Mol. Biol. Rev.* **67:** 86–156.

Min J.Y. and Krug R.M. 2006. The primary function of RNA binding by the influenza A virus NS1 protein in infected cells: Inhibiting the 2′-5′ oligo (A) synthetase/RNase L pathway. *Proc. Natl. Acad. Sci.* **103:** 7100–7105.

Mohr I. 2005. To replicate or not to replicate: Achieving selective oncolytic virus replication in cancer cells through translational control. *Oncogene* **24:** 7697–7709.

———. 2006a. Phosphorylation and dephosphorylation events that regulate viral mRNA translation. *Virus Res.* **119:** 89–99.

———. 2006b. Hailing the arrival of the messenger: Strategies for translational control in herpes simplex virus-infected cells. In *Alphaherpesvirus: Molecular and cellular biology* (ed. R. Sandri-Goldin), pp.105–120. Horizon Scientific Press, Norwich, United Kingdom.

Moody C.A., Scott R.S., Amirghahari N., Nathan C.A., Young L.S., Dawson C.W., and Sixbey J.W. 2005. Modulation of the cell growth regulator mTOR by Epstein-Barr virus-encoded LMP2A. *J. Virol.* **79:** 5499–5506.

Mosig G. and Eiserling F. 1988. Phage T4 structure and metabolism. In *The bacteriophages* (ed. R.L. Calendar), vol. 2, pp. 521–606. Plenum Press, New York.

Mulvey M., Arias C., and Mohr I. 2006. Resistance of mRNA translation to acute endoplasmic reticulum stress-inducing agents in herpes simplex virus type 1-infected cells requires multiple virus-encoded functions. *J. Virol.* **80:** 7354–7363.

Mulvey M., Camarena V., and Mohr I. 2004. Full resistance of herpes simplex virus type 1-infected primary human cells to alpha interferon requires both the Us11 and gamma(1)34.5 gene products. *J. Virol.* **78:** 10193–10196.

Mulvey M., Poppers J., Sternberg D., and Mohr I. 2003. Regulation of eIF2alpha phosphorylation by different functions that act during discrete phases in the herpes simplex virus type 1 life cycle. *J. Virol.* **77:** 10917–10928.

Murphy J.A., Duerst R.J., Smith T.J., and Morrison L.A. 2003. Herpes simplex virus type 2 virion host shutoff protein regulates alpha/beta interferon but not adaptive immune responses during primary infection in vivo. *J. Virol.* **77:** 9337–9345.

Myles K.M., Kelly C.L., Ledermann J.P., and Powers A.M. 2006. Effects of an opal termination codon preceding the nsP4 gene sequence in the O'Nyong-Nyong virus genome on *Anopheles gambiae* infectivity. *J. Virol.* **80:** 4992–4997.

Nakanishi S., Inoue A., Kita T., Nakamura M., Chang A.C., Cohen S.N., and Numa S. 1979. Nucleotide sequence of cloned cDNA for bovine corticotropin-beta-lipotropin precursor. *Nature* **278:** 423–427.

Nanbo A., Yoshiyama H., and Takada K. 2005. Epstein-Barr virus-encoded poly(A)⁻ RNA confers resistance to apoptosis mediated through Fas by blocking the PKR pathway in human epithelial intestine 407 cells. *J. Virol.* **79:** 12280–12285.

Nanbo A., Inoue K., Adachi-Takasawa K., and Takada K. 2002. Epstein-Barr virus RNA confers resistance to interferon-alpha-induced apoptosis in Burkitt's lymphoma. *EMBO J.* **21:** 954–965.

Nathans D., Notani G., Schwartz J.H., and Zinder N.D. 1962. Biosynthesis of the coat protein of coliphage f2 by *E. coli* extracts. *Proc. Natl. Acad. Sci.* **48:** 1424–1431.

Nemeroff M.E. and Bruenn J.A. 1987. Initiation by the yeast viral transcriptase *in vitro. J. Biol. Chem.* **262:** 6785–6787.

Nemeroff M.E., Barabino S.M., Li Y., Keller W., and Krug R.M. 1998. Influenza virus NS1 protein interacts with the cellular 30 kDa subunit of CPSF and inhibits 3′ end formation of cellular pre-mRNAs. *Mol. Cell* **1:** 991–1000.

Nicholson M.G., Rue S.M., Clements J.E., and Barber S.A. 2006. An internal ribosome entry site promotes translation of a novel SIV Pr55(Gag) isoform. *Virology* **349:** 325–334.

Nishida K., Kawasaki T., Fujie M., Usami S., and Yamada T. 1999. Aminoacylation of tRNAs encoded by *Chlorella* virus CVK2. *Virology* **263:** 220–229.

Novoa I., Zeng H., Harding H.P., and Ron D. 2001. Feedback inhibition of the unfolded protein response by GADD34-mediated dephosphorylation of eIF2alpha. *J. Cell Biol.* **153:** 1011–1022.

Oh K.J., Kalinina A., Park N.H., and Bagchi S. 2006. Deregulation of eIF4E: 4E-BP1 in differentiated human papillomavirus-containing cells leads to high levels of expression of the E7 oncoprotein. *J. Virol.* **80:** 7079–7088.

Ohlmann T., Prevot D., Decimo D., Roux F., Garin J., Morley S.J., and Darlix J.L. 2002. In vitro cleavage of eIF4GI but not eIF4GII by HIV-1 protease and its effects on translation in the rabbit reticulocyte lysate system. *J. Mol. Biol.* **318:** 9–20.

Olland A.M., Jane-Valbuena J., Schiff L.A., Nibert M.L., and Harrison S.C. 2001. Structure of the reovirus outer capsid and dsRNA-binding protein sigma3 at 1.8 Å resolution. *EMBO J.* **20:** 979–989.

O'Malley R.P., Duncan R.F., Hershey J.W.B., and Mathews M.B. 1989. Modification of protein synthesis initiation factors and shut-off of host protein synthesis in adenovirus-infected cells. *Virology* **168:** 112–118.

Orlova M., Yueh A., Leung J., and Goff S.P. 2003. Reverse transcriptase of Moloney murine leukemia virus binds to eukaryotic release factor 1 to modulate suppression of translational termination. *Cell* **115:** 319–331.

O'Shea C., Klupsch K., Choi S., Bagus B., Soria C., Shen J., McCormick F., and Stokoe D. 2005. Adenoviral proteins mimic nutrient/growth signals to activate the mTOR pathway for viral replication. *EMBO J.* **24:** 1211–1221.

Park H., Davies M.V., Langland J.O., Chang H.W., Nam Y.S., Tartaglia J., Paoletti E., Jacobs B.L., Kaufman R.J., and Venkatesan S. 1994. TAR RNA-binding protein is an inhibitor of the interferon-induced protein kinase PKR. *Proc. Natl. Acad. Sci.* **91:** 4713–4717.

Patel C.V., Handy I., Goldsmith T., and Patel R.C. 2000. PACT, a stress-modulated cellular activator of interferon-induced double-stranded RNA-activated protein kinase, PKR. *J. Biol. Chem.* **275:** 37993–37998.

Patel R.C. and Sen G.C. 1998. PACT, a protein activator of the interferon-induced protein kinase, PKR. *EMBO J.* **17:** 4379–4390.

Pavio N., Taylor D.R., and Lai M.M. 2002. Detection of a novel unglycosylated form of hepatitis C virus E2 envelope protein that is located in the cytosol and interacts with PKR. *J. Virol.* **76:** 1265–1272.

Pavio N., Romano P.R., Graczyk T.M., Feinstone S.M., and Taylor D.R. 2003. Protein synthesis and endoplasmic reticulum stress can be modulated by the hepatitis C virus envelope protein E2 through the eukaryotic initiation factor 2alpha kinase PERK. *J. Virol.* **77:** 3578–3585.

Pedulla M.L., Ford M.E., Houtz J.M., Karthikeyan T., Wadsworth C., Lewis J.A., Jacobs-Sera D., Falbo J., Gross J., Pannunzio N.R., et al. 2003. Origins of highly mosaic mycobacteriophage genomes. *Cell* **113:** 171–182.

Pe'ery T. and Mathews M.B. 2000. Viral translational strategies and host defense mechanisms. In *Translational control of gene expression* (ed. N. Sonenberg et al.), pp. 371–424. Cold Spring Harbor Laboratory Press, Cold Spring Harbor, New York.

———. 2006. Viral conquest of the host cell. In *Fields virology,* 5th edition (ed. D.M. Knipe et al.). Lippincott Williams & Wilkens, Philadelphia, Pennsylvania. (In press.)

Pelletier J. and Sonenberg N. 1988. Internal initiation of translation of eukaryotic mRNA directed by a sequence derived from poliovirus RNA. *Nature* **334:** 320–325.

Pestova T.V. and Hellen C.U. 2003. Translation elongation after assembly of ribosomes on the cricket paralysis virus internal ribosomal entry site without initiation factors or initiator tRNA. *Genes Dev.* **17:** 181–186.

Peters G.A., Khoo D., Mohr I., and Sen G.C. 2002. Inhibition of PACT-mediated activation of PKR by the herpes simplex virus type 1 Us11 protein. *J. Virol.* **76:** 11054–11064.

Piron M., Vende P., Cohen J., and Poncet D. 1998. Rotavirus RNA-binding protein NSP3 interacts with eIF4GI and evicts the poly(A) binding protein from eIF4F. *EMBO J.* **17:** 5811–5821.

Poncet D., Aponte C., and Cohen J. 1993. Rotavirus protein NSP3 (NS34) is bound to the 3′ end consensus sequence of viral mRNAs in infected cells. *J. Virol.* **67:** 3159–3165.

Pooggin M.M., Ryabova L.A., He X, Fütterer J., and Hohn T. 2006. Mechanism of ribosome shunting in Rice tungro bacilliform pararetrovirus. *RNA* **12:** 841–850.

Poppers J., Mulvey M., Perez C., Khoo D., and I. Mohr I. 2003. Identification of a lytic-cycle Epstein-Barr virus gene product that can regulate PKR activation. *J. Virol.* **77:** 228–236.

Pyronnet S., Imataka H., Gingras A.C., Fukunaga R., Hunter T., and Sonenberg N. 1999. Human eukaryotic translation initiation factor 4G (eIF4G) recruits mnk1 to phosphorylate eIF4E. *EMBO J.* **18:** 270–279.

Rajan P., Swaminathan S., Zhu J., Cole C.N., Barber G., Tevethia M.J., and Thimmapaya B. 1995. A novel translation regulation function for the simian virus 40 large-T antigen gene. *J. Virol.* **69:** 785–795.

Raoult D., Audic S., Robert C., Abergel C., Renesto P., Ogata H., La Scola B., Suzan M., and Claverie J.M. 2004. The 1.2-megabase genome sequence of Mimivirus. *Science* **306:** 1344–1350.

Ravetch J.V., Model P., and Robertson H.D. 1977. Isolation and characterization of phiX174 ribosome binding sites. *Nature* **265:** 698–702.

Read G.S. and Frenkel N. 1983. Herpes simplex virus mutants defective in the virion-associated shutoff of host polypeptide synthesis and exhibiting abnormal synthesis of alpha (immediate early) polypeptides. *J. Virol.* **46:** 498–512.

Remm M., Remm A., and Ustav M. 1999. Human papillomavirus type 18 E1 protein is translated from polycistronic mRNA by a discontinuous scanning mechanism. *J. Virol.* **73:** 3062–3070.

Rivas C., Gil J., Melkova Z., Esteban M., and Diaz-Guerra M. 1998. Vaccinia virus E3L protein is an inhibitor of the interferon (i.f.n.)-induced 2-5A synthetase enzyme. *Virology* **243:** 406–414.

Robertson H.D., Manche L., and Mathews M.B. 1996. Paradoxical interactions between human delta hepatitis agent RNA and the cellular protein kinase PKR. *J. Virol.* **70:** 5611–5617.

Robey W.G., Safai B., Oroszlan S., Arthur L.O., Gonda M.A., Gallo R.C., and Fischinger P.J. 1985. Characterization of envelope and core structural gene products of HTLV-III with sera from AIDS patients. *Science* **228:** 593–595.

Romano P.R., Zhang F., Tan S.L., Garcia-Barrio M.T., Katze M.G., Dever T.E., and Hinnebusch A.G. 1998. Inhibition of double-stranded RNA-dependent protein kinase PKR by vaccinia virus E3: Role of complex formation and the E3 N-terminal domain. *Mol. Cell. Biol.* **18:** 7304–7316.

Ruf I.K., Lackey K.A., Warudkar S., and Sample J.T. 2005. Protection from interferon-induced apoptosis by Epstein-Barr virus small RNAs is not mediated by inhibition of PKR. *J. Virol.* **79:** 14562–14569.

Ryabova L.A. and Hohn T. 2000. Ribosome shunting in the cauliflower mosaic virus 35S RNA leader is a special case of reinitiation of translation functioning in plant and animal systems. *Genes Dev.* **14:** 817–829.

Sachs M.S. and Geballe A.P. 2006. Downstream control of upstream open reading frames. *Genes Dev.* **20:** 915–921.

Sandri-Goldin R.M. 2004. Viral regulation of mRNA export. *J. Virol.* **78:** 4389–4396.

Sarkar S.N. and Sen G.C. 1998. Production, purification, and characterization of recombinant 2′, 5′-oligoadenylate synthetases. *Methods* **15:** 233–242.

Sasaki J. and Nakashima N. 1999. Translation initiation at the CUU codon is mediated by the internal ribosome entry site of an insect picorna-like virus in vitro. *J. Virol.* **73:** 1219–1226.

———. 2000. Methionine-independent initiation of translation in the capsid protein of an insect RNA virus. *Proc. Natl. Acad. Sci.* **97:** 1512–1515.

Satoh S., Hijikata M., Handa H., and Shimotohno K. 1999. Caspase-mediated cleavage of eukaryotic translation initiation factor subunit 2alpha. *Biochem. J.* **342:** 65–70.

Scheper G.C. and Proud C.G. 2002. Does phosphorylation of the cap-binding protein eIF4E play a role in translation initiation? *Eur. J. Biochem.* **269:** 5350–5359.

Scheper G.C., Morrice N.A., Kleijn M., and Proud C.G. 2001. The mitogen-activated protein kinase signal-integrating kinase Mnk2 is a eukaryotic initiation factor 4E kinase with high levels of basal activity in mammalian cells. *Mol. Cell. Biol.* **21:** 743–754.

Schmechel S., Chute M., Skinner P., Anderson R., and Schiff L. 1997. Preferential translation of reovirus mRNA by a sigma3-dependent mechanism. *Virology* **232:** 62–73.

Schmidt M.R., Gravel K.A., and Woodland R.T. 1995. Progression of a vesicular stomatitis virus infection in primary lymphocytes is restricted at multiple levels during B cell activation. *J. Immunol.* **155:** 2533–2544.

Schneider R.J. 2000. Translational control during heat shock. In *Translational control of gene expression* (ed. N. Sonenberg et al.), pp. 581–593. Cold Spring Harbor Laboratory Press, Cold Spring Harbor, New York.

Schneider R.J. and Mohr I. 2003. Translation initiation and viral tricks. *Trends Biochem. Sci.* **28:** 130–136.

Schutz S. and Sarnow P. 2006. Interaction of viruses with the mammalian RNA interference pathway. *Virology* **344:** 151–157.

Schwartz S., Felber B.K., and Pavlakis G.N. 1992. Mechanism of translation of monocistronic and multicistronic human immunodeficiency virus type 1 mRNAs. *Mol. Cell. Biol.* **12:** 207–219.

Sen N., Cao F., and Tavis J.E. 2004. Translation of duck hepatitis B virus reverse transcriptase by ribosomal shunting. *J. Virol.* **78:** 11751–11757.

Sharp T.V., Raine D.A., Gewert D.R., Joshi B., Jagus R., and Clemens M.J. 1999. Activation of the interferon-inducible (2′-5′) oligoadenylate synthetase by the Epstein-Barr virus RNA, EBER-1. *Virology* **257:** 303–313.

Sharp T.V., Schwemmle M., Jeffrey I., Laing K., Mellor H., Proud C.G., Hilse K., and Clemens M.J. 1993. Comparative analysis of the regulation of the interferon-inducible protein kinase PKR by Epstein-Barr virus RNAs EBER-1 and EBER-2 and adenovirus VAI RNA. *Nucleic Acids Res.* **21:** 4483–4490.

Sharpe A.H. and Fields B.N. 1982. Reovirus inhibition of cellular RNA and protein synthesis: Role of the S4 gene. *Virology* **122:** 381–391.

Shmulevitz M., Marcato P., and Lee P.W. 2005. Unshackling the links between reovirus oncolysis, Ras signaling, translational control and cancer. *Oncogene* **24:** 7720–7728.

Silverman R.H., Jung D.D., Nolan-Sorden N.L., Dieffenbach C.W., Kedar V.P., and SenGupta D.N. 1988. Purification and analysis of murine 2-5A-dependent RNase. *J. Biol. Chem.* **263:** 7336–7341.

Smith A.E. and Marcker K.A. 1970. Cytoplasmic methionine transfer RNAs from eukaryotes. *Nature* **226:** 607–610.

Smith A.E., Marcker K.A., and Mathews M.B. 1970. Translation of RNA from encephalomyocarditis virus in a mammalian cell-free system. *Nature* **225:** 184–187.

Smith J.A., Schmechel S.C., Raghavan A., Abelson M., Reilly C., Katze M.G., Kaufman R.J., Bohjanen P.R., and Schiff L.A. 2006. Reovirus induces and benefits from an integrated cellular stress response. *J. Virol.* **80:** 2019–2033.

Smith T.J., Morrison L.A., and Leib D.A. 2002. Pathogenesis of herpes simplex virus type 2 virion host shutoff (vhs) mutants. *J. Virol.* **76:** 2054–2061.

Sood R., Porter A.C., Ma K., Quilliam L.A., and Wek R.C. 2000. Pancreatic eukaryotic initiation factor-2alpha kinase (PEK) homologues in humans, *Drosophila melanogaster* and *Caenorhabditis elegans* that mediate translational control in response to endoplasmic reticulum stress. *Biochem. J.* **346:** 281–293.

Srivastava S.P., Kumar K.U., and Kaufman R.J. 1998. Phosphorylation of eukaryotic translation initiation factor 2 mediates apoptosis in response to activation of the double-stranded RNA-dependent protein kinase. *J. Biol. Chem.* **273:** 2416–2423.

Stanners C.P. and Becker H. 1971. Control of macromolecular synthesis in proliferating and resting Syrian hamster cells in monolayer culture. I. Ribosome function. *J. Cell. Physiol.* **77:** 31–42.

Steitz J.A. 1969. Polypeptide chain initiation: Nucleotide sequences of the three ribosomal binding sites in bacteriophage R17 RNA. *Nature* **224:** 957–964.

Stojdl D.F., Lichty B., Knowles S., Marius R., Atkins H., Sonenberg N., and Bell J.C. 2000. Exploiting tumor-specific defects in the interferon pathway with a previously unknown oncolytic virus. *Nat. Med.* **6:** 821–825.

Stojdl D.F., Lichty B.D., tenOever B.R., Paterson J.M., Power A.T., Knowles S., Marius R., Reynard J., Poliquin L., Atkins H., et al. 2003. VSV strains with defects in their ability to shutdown innate immunity are potent systemic anti-cancer agents. *Cancer Cell* **4:** 263–275.

Strelow L.I. and Leib D.A. 1995. Role of the virion host shutoff (vhs) of herpes simplex virus type 1 in latency and pathogenesis. *J. Virol.* **69:** 6779–6786.

Su H.L., Liao C.L., and Lin Y.L. 2002. Japanese encephalitis virus infection initiates endoplasmic reticulum stress and an unfolded protein response. *J. Virol.* **76:** 4162–4171.

Sullivan C.S. and Ganem D. 2005. MicroRNAs and viral infection. *Mol. Cell* **20:** 3–7.

Sullivan C.S., Grundhoff A.T., Tevethia S., Pipas J.M., and Ganem D. 2005. SV40-encoded microRNAs regulate viral gene expression and reduce susceptibility to cytotoxic T cells. *Nature* **435:** 682–686.

Svitkin Y.V., Gradi A., Imataka H., Morino S., and Sonenberg N. 1999. Eukaryotic initiation factor 4GII (eIF4GII), but not eIF4GI, cleavage correlates with inhibition of host cell protein synthesis after human rhinovirus infection. *J. Virol.* **73:** 3467–3472.

Swaminathan S., Rajan P., Savinova O., Jagus R., and Thimmapaya B. 1996. Simian virus 40 large-T bypasses the translational block imposed by the phosphorylation of eIF-2alpha. *Virology* **219:** 321–323.

Tan S.L. and Katze M.G. 1998. Biochemical and genetic evidence for complex formation between the influenza A virus NS1 protein and the interferon-induced PKR protein kinase. *J. Interferon Cytokine Res.* **18:** 757–766.

Tan S.L., Gale M.J., Jr., and Katze M.G. 1998. Double-stranded RNA-independent dimerization of interferon-induced protein kinase PKR and inhibition of dimerization by the cellular P58IPK inhibitor. *Mol. Cell. Biol.* **18:** 2431–2443.

Taylor D.R., Shi S.T., Romano P.R., Barber G.N., and Lai M.M. 1999. Inhibition of the interferon-inducible protein kinase PKR by HCV E2 protein. *Science* **285:** 107–110.

Tian B., Bevilacqua P.C., Diegelman-Parente A., and Mathews M.B. 2004. The double-stranded-RNA-binding motif: Interference and much more. *Nat. Rev. Mol. Cell Biol.* **5:** 1013–1023.

Timmer R.T., Benkowski L.A., Schodin D., Lax S.R., Metz A.M., Ravel J.M., and Browning K.S. 1993. The 5′ and 3′ untranslated regions of satellite tobacco necrosis virus RNA affect translational efficiency and dependence on a 5′ cap structure. *J. Biol. Chem.* **268:** 9504–9510.

Van Etten J.L. 2003. Unusual life style of giant *Chlorella* viruses. *Annu. Rev. Genet.* **37:** 153–195.

Ventoso I., Blanco R., Perales C., and Carrasco L. 2001. HIV-1 protease cleaves eukaryotic initiation factor 4G and inhibits cap-dependent translation. *Proc. Natl. Acad. Sci.* **98:** 12966–12971.

Ventoso I., Sanz M.A., Molina S., Berlanga J.J., Carrasco L., and Esteban M. 2006. Translational resistance of late alphavirus mRNA to eIF2alpha phosphorylation: A strategy to overcome the antiviral effect of protein kinase PKR. *Genes Dev.* **20:** 87–100.

Villa-Komaroff L., Guttman N., Baltimore D., and H.F. Lodishi. 1975. Complete translation of poliovirus RNA in a eukaryotic cell-tree system. *Proc. Natl. Acad. Sci.* **72:** 4157–4161.

Virgin H.W., IV, Latreille P., Wamsley P., Hallsworth K., Weck K.E., Dal Canto A.J., and Speck S.H. 1997. Complete sequence and genomic analysis of murine gammaherpesvirus 68. *J. Virol.* **71:** 5894–5904.

Volkin E. and Astrachan L. 1956. Phosphorus incorporation in *E. coli* ribonucleic acid after infection with bacteriophage T2. *Virology* **2:** 146–161.

Vyas J., Elia A., and Clemens M.J. 2003. Inhibition of the protein kinase PKR by the internal ribosome entry site of hepatitis C virus genomic RNA. *RNA* **9:** 858–870.

Walsh D. and Mohr I. 2004. Phosphorylation of eIF4E by Mnk-1 enhances HSV-1 translation and replication in quiescent cells. *Genes Dev.* **18:** 660–672.

———. 2006. Assembly of an active translation initiation factor complex by a viral protein. *Genes Dev.* **20:** 461–472.

Walsh D., Perez C., Notary J., and Mohr I. 2005. Regulation of the translation initiation factor eIF4F by multiple mechanisms in human cytomegalovirus-infected cells. *J. Virol.* **79:** 8057–8064.

Wang C., Pflugheber J., Sumpter R., Jr., Sodora D.L., Hui D., Sen G.C., and Gale M., Jr. 2003. Alpha interferon induces distinct translational control programs to suppress hepatitis C virus RNA replication. *J. Virol.* **77:** 3898–3912.

Wang W., Riedel K., Lynch P., Chien C.Y., Montelione G.T., and Krug R.M. 1999. RNA binding by the novel helical domain of the influenza virus NS1 protein requires its dimer structure and a small number of specific basic amino acids. *RNA* **5:** 195–205.

Waskiewicz A.J., Flynn A., Proud C.G., and Cooper J.A. 1997. Mitogen-activated protein kinases activate the serine/threonine kinases Mnk1 and Mnk2. *EMBO J.* **16:** 1909–1920.

Waskiewicz A.J., Johnson J.C., Penn B., Mahalingam M., Kimball S.R., and Cooper J.A. 1999. Phosphorylation of the cap-binding protein eukaryotic translation initiation factor 4E by protein kinase Mnk1 in vivo. *Mol. Cell. Biol.* **19:** 1871–1880.

Webster R.E., Engelhardt D., and Zinder N. 1966. In vitro protein synthesis: Chain initiation. *Proc. Natl. Acad. Sci.* **55:** 155–161.

Wei C.M. and Moss B. 1975. Methylated nucleotides block 5′-terminus of vaccinia virus messenger RNA. *Proc. Natl. Acad. Sci.* **72:** 318–322.

Weiss R.B., Huang W.M., and Dunn D.M. 1990. A nascent peptide is required for ribosomal bypass of the coding gap in bacteriophage T4 gene 60. *Cell* **62:** 117–126.

Wells S.E., Hillner P.E., Vale R.D., and Sachs A.B. 1998. Circularization of mRNA by eukaryotic translation initiation factors. *Mol. Cell* **2:** 135–140.

Wiborg O., Pedersen M.S., Wind A., Berglund L.E., Marcker K.A., and Vuust J. 1985. The human ubiquitin multigene family: Some genes contain multiple directly repeated ubiquitin coding sequences. *EMBO J.* **4:** 755–759.

Wiegand M., Bossow S., and Neubert W.J. 2005. Sendai virus trailer RNA simultaneously blocks two apoptosis-inducing mechanisms in a cell type-dependent manner. *J. Gen. Virol.* **86:** 2305–2314.

Wilson J.E., Pestova T.V., Hellen C.U., and Sarnow P. 2000. Initiation of protein synthesis from the A site of the ribosome. *Cell* **102:** 511–520.

Wimmer E., Chang A.Y., Clark J.M., Jr., and Reichmann M.E. 1968. Sequence studies of satellite tobacco necrosis virus RNA. Isolation and characterization of a 5′-terminal trinucleotide. *J. Mol. Biol.* **38:** 59–73.

Wittmann H.G. and Wittmann-Liebold B. 1967. Protein chemical studies of two RNA viruses and their mutants. *Cold Spring Harbor Symp. Quant. Biol.* **31:** 163–172.

Wreschner D.H., James T.C., Silverman R.H., and Kerr I.M. 1981. Ribosomal RNA cleavage, nuclease activation and 2-5A(ppp(A2′p)nA) in interferon-treated cells. *Nucleic Acids Res.* **9:** 1571–1581.

Xi Q., Cuesta R., and Schneider R.J. 2004. Tethering of eIF4G to adenoviral mRNAs by viral 100k protein drives ribosome shunting. *Genes Dev.* **18:** 1997–2009.

———. 2005. Regulation of translation by ribosome shunting through phosphotyrosine-dependent coupling of adenovirus protein 100k to viral mRNAs. *J. Virol.* **79:** 5676–5683.

Yamada T., Fukuda T., Tamura K., Furukawa S., and Songsri P. 1993. Expression of the gene encoding a translational elongation factor 3 homolog of *Chlorella* virus CVK2. *Virology* **197:** 742–750.

Yan W., Frank C.L., Korth M.J., Sopher B.L., Novoa I., Ron D., and Katze M.G. 2002. Control of PERK eIF2alpha kinase activity by the endoplasmic reticulum stress-induced molecular chaperone P58IPK. *Proc. Natl. Acad. Sci.* **99:** 15920–15925.

Yi M., Schultz D.E., and Lemon S.M. 2000. Functional significance of the interaction of hepatitis A virus RNA with glyceraldehyde 3-phosphate dehydrogenase (GAPDH): Opposing effects of GAPDH and polypyrimidine tract binding protein on internal ribosome entry site function. *J. Virol.* **74:** 6459–6468.

Yin Y., Manoury B., and Fahraeus R. 2003. Self-inhibition of synthesis and antigen presentation by Epstein-Barr virus-encoded EBNA1. *Science* **301:** 1371–1374.

Yoneyama M., Kikuchi M., Natsukawa T., Shinobu N., Imaizumi T., Miyagishi M., Taira K., Akira S., and Fujita T. 2004. The RNA helicase RIG-I has an essential function in double-stranded RNA-induced innate antiviral responses. *Nat. Immunol.* **5:** 730–737.

Yoneyama M., Kikuchi M., Matsumoto K., Imaizumi T., Miyagishi M., Taira K., Foy E., Loo Y.M., Gale M., Jr., Akira S., et al. 2005. Shared and unique functions of the DExD/H-box helicases RIG-I, MDA5, and LGP2 in antiviral innate immunity. *J. Immunol.* **175:** 2851–2858.

Yu Y., Kudchodkar S.B., and Alwine J.C. 2005. Effects of simian virus 40 large and small tumor antigens on mammalian target of rapamycin signaling: Small tumor antigen mediates hypophosphorylation of eIF4E-binding protein 1 late in infection. *J. Virol.* **79:** 6882–6889.

Yu Y.X., Bearzotti M., Vende P., Ahne W., and Bremont M. 1999. Partial mapping and sequencing of a fish iridovirus genome reveals genes homologous to the frog virus 3 p31, p40 and human eIF2alpha. *Virus Res.* **63:** 53–63.

Yueh A. and Schneider R.J. 1996. Selective translation initiation by ribosome jumping in adenovirus-infected and heat-shocked cells. *Genes Dev.* **10:** 1557–1567.

———. 2000. Translation by ribosome shunting on adenovirus and hsp70 mRNAs facilitated by complementarity to 18S rRNA. *Genes Dev.* **14:** 414–421.

Zamora M., Marissen W.E., and Lloyd R.E. 2002. Multiple eIF4GI-specific protease activities present in uninfected and poliovirus-infected cells. *J. Virol.* **76:** 165–177.

Zhou A., Paranjape J., Brown T.L., Nie H., Naik S., Dong B., Chang A., Trapp B., Fairchild R., Colmenares C., and Silverman R.H. 1997. Interferon action and apoptosis are defective in mice devoid of 2′,5′-oligoadenylate-dependent RNase L. *EMBO J.* **16:** 6355–6363.

21

Regulation of Translation Elongation and the Cotranslational Protein Targeting Pathway

Terence P. Herbert
Department of Cell Physiology and Pharmacology
Faculty of Medicine and Biological Sciences
University of Leicester
Leicester LE1 9HN, United Kingdom

Christopher G. Proud
Department of Biochemistry and Molecular Biology
University of British Columbia
Vancouver BC V6T 1Z3, Canada

THE FACTORS INVOLVED IN TRANSLATION ELONGATION are subject to sophisticated control mechanisms that come into play under a wide variety of conditions. Even though translation is most frequently controlled during the initiation phase (Chapter 1) and the regulatory mechanisms impinging on the initiation steps have received considerable attention (reviewed in several chapters of this book), accumulating information points to the elongation phase as a target for controls under defined circumstances. In this chapter, we focus on recent developments in understanding the control of elongation in mammalian cells. As a special case, we also discuss cotranslational protein targeting, a cellular process involving the control of elongation on an important class of mRNAs.

REGULATION OF TRANSLATION ELONGATION

The mechanism of peptide-chain elongation and the functions of translation elongation factors are described in Chapters 2 and 3. In addition, detailed aspects of the structure and function of eukaryotic elongation factor 2 (eEF2) are the subject of a recent informative review (Jorgensen et al. 2006). eEF2 is a phosphoprotein in mammalian cells, and most of the recent advances relate to the regulation of eEF2 and its cognate kinase, eEF2 kinase. eEF1A and eEF1B also are phosphoproteins and have been discussed in earlier reviews on this subject (Proud 2000; Traugh 2001; Browne and Proud 2002; Le Sourd et al. 2006).

Significance of eEF2 Phosphorylation for the Control of Protein Synthesis

Under a diverse range of conditions, the phosphorylation state of eEF2 changes in directions consistent with its having a role in regulating protein synthesis; i.e., its phosphorylation falls in response to a wide range of agents that activate protein synthesis and increases, for example, under conditions of adverse cellular energy status. Several studies have reported correlations between eEF2 phosphorylation and rates of protein synthesis and/or elongation (see, e.g., Redpath et al. 1996a; Diggle et al. 1998; Hovland et al. 1999; Scheetz et al. 2000; McLeod et al. 2001; Yan et al. 2003), but evidence that the changes in eEF2 phosphorylation are responsible for—or at least contribute to—the alterations in protein synthesis rates still remains limited.

The multiplicity of regulatory inputs into the enzyme that phosphorylates eEF2, eEF2 kinase (described below), suggests that it has a key role in cellular regulation, in particular the overall control of protein synthesis. Moreover, it has been proposed that controlling elongation may be important for modulating the translation of specific mRNAs. However, further work is required to improve our understanding of the physiological importance of the control of eEF2. One approach so far used in only one study (Terai et al. 2005) is to employ cells in which eEF2 kinase function has been ablated using small interfering RNA (siRNA) or by targeted transgenesis to ablate eEF2 kinase function.

eEF2

eEF2 mediates the translocation step of elongation where, following the addition of an amino acid to the nascent chain, the tRNA bearing that polypeptide moves from the A site into the P site on the ribosome and the ribosome moves by one codon along the mRNA. This is accompanied by hydrolysis of a GTP molecule (bound to eEF2). In fact, the addition of each amino acid requires the equivalent of at least four ATP or GTP molecules. In addition to the one required for the translocation step, at least one other GTP is hydrolyzed during the recruitment of the in-

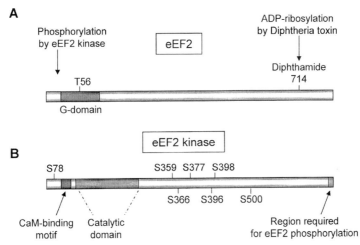

Figure 1. Principal features of eEF2 and eEF2 kinase. (*A*) The main features of eEF2 are indicated, including the major phosphorylation site (Thr-56) and the site of ADP ribosylation. (*B*) Important features of eEF2 kinase include its catalytic domain, CaM-binding motif, and known sites of phosphorylation. The region needed for phosphorylation of eEF2 is also shown.

coming aminoacyl-tRNA (mediated by eEF1A), and the equivalent of two ATPs are used by the aminoacyl-tRNA synthetase to charge the tRNA. Almost all the energy and amino acids consumed by protein synthesis are used during the elongation phase. As described below, the signal recognition particle (SRP) interferes with the function of eEF2 to arrest the elongating ribosome.

eEF2 undergoes three posttranslational modifications that contribute to its function and regulation. First, a completely conserved histidine residue is uniquely modified to diphthamide (H714 in human eEF2; Fig. 1A). The role of the diphthamide in eEF2 function remains unclear, although extensive mutagenesis studies indicate that it is not essential (for review, see Jorgensen et al. 2006). Second, the diphthamide is a substrate for ADP ribosylation by diphtheria toxin. This inhibits the activity of eEF2, thereby also inhibiting protein synthesis, and underlies the pathological effects of infection by *Corynebacterium diphtheriae*, as reviewed in Jorgensen et al. (2006). Third, mammalian eEF2 undergoes phosphorylation in vivo by eEF2 kinase at Thr-56 within its GTP-binding domain, which prevents its binding to ribosomes and thus inactivates eEF2 (Carlberg et al. 1990). Although eEF2 kinase can phosphorylate three adjacent residues in this region of eEF2 in vitro (Price et al. 1991), Thr-56 is the primary—and probably only—site that is phosphorylated in vivo.

eEF2 Kinase

eEF2 kinase is an unusual Ca^{2+}/calmodulin (CaM)-dependent enzyme (first termed Ca^{2+}/CaM-kinase III; Nairn et al. 1985; Ryazanov et al. 1988). Cloning of cDNAs for eEF2 kinase revealed a sequence that showed very little, if any, similarity to other protein kinases either of the main serine/threonine/tyrosine kinase superfamily or to the mitochondrial protein kinases (Redpath et al. 1996b; Ryazanov et al. 1997, 1999). Budding yeast do not contain any members of the α-kinase family. Yeast eEF2 is a substrate for the enzyme Rck2, which is linked to the HOG1 osmosensing MAP kinase pathway (Teige et al. 2001).

The human genome contains five other genes with sequences related to that of the catalytic domain of eEF2 kinase (Ryazanov et al. 1997; Drennan and Ryazanov 2004). Sequence comparisons with the predicted products of the other four genes indicate that the catalytic domain of eEF2 kinase lies within the region roughly bounded by residues 135–320 (Fig. 1B). One of the kinase genes encodes the transient receptor potential (TRP) channel ChaK1 (channel kinase 1). The cytoplasmic domain of this transmembrane protein shows sequence similarity to eEF2 kinase. The crystal structure of the corresponding region of ChaK1 has been solved (Yamaguchi et al. 2001) and is discussed briefly below. Biochemical studies from the laboratories of Ryazanov and Redpath (Diggle et al. 1999; Pavur et al. 2000) revealed several important features of other regions of the protein. For example, the CaM-binding domain is located close to the amino-terminal end of the catalytic region (Fig. 1B). This is rather unusual, as in other Ca^{2+}/CaM-dependent enzymes this feature lies carboxy-terminal to the catalytic domain. In such enzymes, Ca^{2+}/CaM binding is thought to result in a conformational change that causes a pseudosubstrate sequence to be removed from the protein-substrate-binding part of the active site, allowing the kinase to phosphorylate its true substrates. No pseudosubstrate region has yet been identified in the sequence of the eEF2 kinase, and its activation by Ca^{2+}/CaM presumably occurs by a different mechanism. CaM-binding motifs generally contain hydrophobic and basic (positively charged) residues that are important for CaM binding. This is also the case in eEF2 kinase, where the probable CaM-binding region contains several lysine or histidine residues and a tryptophan which is essential for CaM binding (Diggle et al. 1999).

eEF2 kinase activity is much higher at pH 6.6–6.8 than at pH 7.2–7.4 (Dorovkov et al. 2002), reflecting an increase in V_{max}. It was suggested that this may account for, or at least contribute to, the inhibition of protein synthesis that accompanies cellular acidosis that occurs, for example, as a consequence of hypoxia. Consistent with this idea, overexpression of eEF2 kinase led to inhibition of protein synthesis at acidic pH (Dorovkov et al.

2002). It did not do so at normal pH values, which could reflect insufficient eEF2 kinase activity under such conditions. Alternatively, elongation itself may not be limiting for the rate of protein synthesis at these pH values.

A second unusual feature of eEF2 kinase is that its extreme carboxyl terminus is required for efficient phosphorylation of eEF2 (Fig. 1B) (Diggle et al. 1999). The carboxyl terminus is not needed for autophosphorylation, showing that it is not required for catalytic activity, but may instead be involved in substrate binding. This idea is supported by the observation that the carboxy-terminal section of eEF2 kinase (lacking the catalytic domain) can be coimmunoprecipitated with eEF2 (Pavur et al. 2000).

Structure of the Catalytic Domain of the Related Protein, ChaK1

Some insight into the function of the eIF2 kinase can be obtained by an analysis of the cytosolic domain of the TRP channel, ChaK1, which shows homology with the eEF2 kinase. The three-dimensional structure of the catalytic fragment of ChaK1 has been solved crystallographically (Yamaguchi et al. 2001). As for the conventional protein kinases, it comprises two domains. The amino-terminal lobe is similar in overall topology to the amino-terminal lobe of conventional protein kinases and is likely to be involved in binding ATP. Despite the lack of obvious amino acid sequence resemblance, there are similarities between the three-dimensional structures of the ChaK1 amino-terminal lobe and other protein kinases (Ryazanov et al. 1997, 1999), especially in terms of the residues likely to be involved in catalysis, although there are also significant differences (Yamaguchi et al. 2001; Drennan and Ryazanov 2004). These residues are conserved in other members of this group of kinases including eEF2 kinase. As in conventional protein kinases, the carboxy-terminal lobe of ChaK1 appears to be involved in recognizing the protein substrate (Yamaguchi et al. 2001). However, in the eEF2 kinase, the extreme carboxyl terminus has a key role in the phosphorylation of eEF2, as noted above. Interestingly, the carboxy-terminal lobe of ChaK1 shows similarity to the ATP-grasp fold seen in metabolic enzymes, rather than that of classic protein kinases. Space constraints preclude further description of the structure of the catalytic domain of ChaK1 and its implications for eEF2 kinase. For an informative and detailed review, see Drennan and Ryazanov (2004).

Regulation of eEF2 Phosphorylation

The phosphorylation state of eEF2 changes under a range of different conditions. eEF2 phosphorylation at Thr-56 is decreased in response to a number of agents or conditions associated with the activation of protein synthesis. For example, stimuli that exert prohypertrophic effects in cardiac myocytes, such as angiotensin II in neonatal myocytes (Everett et al. 2001) or phenylephrine in adult myocytes (Wang and Proud 2002), bring about the dephosphorylation of eEF2 by mechanisms that involve signaling through the classic MAP kinase pathway. In pancreatic acini, several secretagogues (cholecystokinin, bombesin, and carbachol) induce rapid dephosphorylation of eEF2 and an acceleration in elongation rates (Sans et al. 2004). mTOR (mammalian target of rapamycin) signaling has been shown to control eEF2 phosphorylation in mammalian cells (Redpath et al. 1996a; see below). It is therefore interesting to note that in molluscan cells (*Aplysia* neurites), mTOR also regulates eEF2 phosphorylation: In particular, the ability of serotonin (5-hydroxytryptamine) to elicit dephosphorylation of eEF2 was blocked by rapamycin (Carroll et al. 2004). Because eEF2 is required for all cytoplasmic translation events, the serotonin-induced dephosphorylation of eEF2 may serve to enhance overall protein synthesis.

Control of eEF2 Kinase by Phosphorylation

In addition to acute control by Ca^{2+}/CaM and longer-term regulation via eEF2 kinase stability (see below), eEF2 kinase is also subjected to regulation through its own phosphorylation. The first such inputs to be identified were its modulation by autophosphorylation and by cAMP-dependent protein kinase (PKA), both of which lead to acquisition of Ca^{2+}/CaM-independent activity (for review, see Proud 2000). At low basal Ca^{2+}, this equates to activation of eEF2 kinase. Agents that raise cAMP do indeed increase eEF2 phosphorylation (Hovland et al. 1999; Diggle et al. 2001). Ser-500 in human eEF2 kinase is directly phosphorylated by PKA (Fig. 1B) (for review, see Proud 2000). (For consistency of presentation, residue numbers are given for human eEF2 kinase. For the rodent enzymes used in some studies, the numbering of all the residues mentioned here is lower by one.)

Regulation of eEF2 Kinase by Cellular Energy Status: A Role for AMPK?

Because elongation consumes a high proportion of metabolic energy in most cell types, one might anticipate the operation of a mechanism coupling rates of elongation to cellular energy supply. Conditions that deplete cellular ATP do indeed result in increased phosphorylation of eEF2. Incubation of cells in 2-deoxyglucose (which impairs glycolysis) or treatment with agents that interfere with mitochondrial function (uncouplers or oligomycin) causes increased phosphorylation of eEF2 (Horman et al. 2002, 2003; McLeod and Proud 2002; Browne et al. 2004). In the perfused heart, either ischemia or perfusion with medium lacking fatty acids (a major energy source in that tissue) elicited increased phosphorylation of eEF2 (Crozier et al. 2005). Reperfusion with glucose and palmitate decreased eEF2 phosphorylation to the level of the relevant control, but this was not so for organs perfused with medium containing only glucose. The observation that lack of fatty acids alone led to elevated eEF2 phosphorylation suggests that the phosphorylation of this factor is especially sensitive to cellular energy status. In the same study, alterations in protein factors involved in translation initiation were also observed but were less marked.

A candidate for linking cellular energy status with eEF2 phosphorylation is the AMP-activated protein kinase (AMPK), which is stimulated by the AMP that is generated from ADP (by adenylate kinase) when ATP levels drop. AMPK is a key regulator of the cellular energy economy (Hardie et al. 2003), a view consistent with the observation that treatment of cells with AICAR, which activates AMPK without altering adenine nucleotide levels, leads to increased phosphorylation of eEF2 (Horman et al. 2002). Two groups have now shown that AMPK directly phosphorylates eEF2 kinase (Horman et al. 2003; Browne et al. 2004). In vitro, phosphorylation occurs at three sites. Of these, only Ser-398 appears likely to be a physiological site for AMPK (Fig. 2A). For example, phosphorylation at Ser-398 is enhanced under conditions where AMPK is turned on, consistent with this being a physiological target of AMPK (Browne et al. 2004). Such conditions also increase the activity of eEF2 kinase (in parallel with increased phosphorylation of eEF2). However, it has not so far been shown directly that phosphorylation of eEF2 kinase by AMPK increases its activity.

Raising glucose levels activates protein synthesis in pancreatic β cells, which is likely to be important in increasing the synthesis of preproinsulin and other proteins. In pancreatic β cells or the related INS1-derived 832/13 cell line, elevated glucose led to dephosphorylation of eEF2 (Yan et al. 2003). This effect requires glucose metabolism and was independent of several signaling pathways including mTOR, extracellular-signal-regulated kinase (ERK), and phosphoinositide 3 (PI3) kinase. The data suggested that the effects of glucose concentrations on eEF2 might be mediated by AMPK. Indeed, treatment of 832/13 cells with AICAR (a cell permeant activator of AMPK) led to increased eEF2 phosphorylation. AMPK can also negatively control the mTOR pathway, which also feeds into the regulation of eEF2 kinase. TSC2, an upstream regulator of mTOR, is a

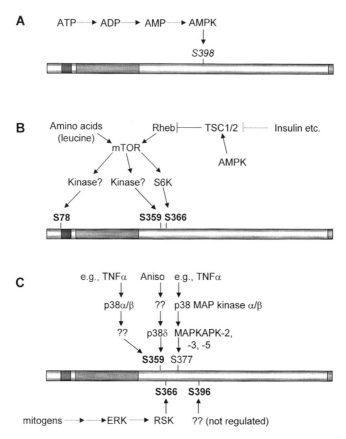

Figure 2. Regulatory inputs into eEF2 kinase. (*A*) Under conditions of ATP depletion, some of the resulting ADP is converted to AMP, activating AMPK that directly
phosphorylates eEF2 kinase at Ser-398, apparently activating it. (*B*) Inputs from
mTOR signaling include phosphorylation at three sites. The kinases acting at Ser-
78 and Ser-359 are so far unknown. Upstream control of mTOR signaling includes
its activation by amino acids (especially leucine) and by Rheb. The TSC1/2 complex
acts as a GTPase-activator protein for Rheb, and the function of this complex is
believed to be inactivated following phosphorylation of TSC2, e.g., in response to
insulin. Conversely, phosphorylation of TSC2 by AMPK leads to inhibition of
mTOR signaling. (*C*) Inputs from MAP kinase signaling include control by the ERK
pathway (via RSK, phosphorylation at Ser-366), by p38 MAP kinase α/β (indirectly,
at two sites, Ser-359 and Ser-377), and by p38 MAP kinase δ (also termed SAPK4;
Ser-359 and Ser-396). The available data indicate that Ser-396 is also phosphorylated by an additional unknown kinase. Phosphorylation at the sites shown in bold
inhibits eEF2 kinase in vitro, whereas phosphorylation at italicized sites likely
activates eEF2 kinase. (Aniso) Anisomycin; (MAPKAPK) MAP kinase-activated
protein kinase; (TNF) tumor necrosis factor.

direct substrate for AMPK (Fig. 2B) (Inoki et al. 2003). However, conditions that activate AMPK still increase eEF2 phosphorylation in cells lacking TSC2 (Smith et al. 2005), presumably due to the direct effects of AMPK on eEF2 kinase. The existence of two mechanisms for regulating eEF2 phosphorylation in response to ATP levels (Fig. 2) may reflect the importance of being able to shut down the energy-consuming process of peptide-chain elongation under adverse cellular conditions, in order to conserve nucleoside triphosphates for what are, at least temporarily, more essential cellular functions. A further example where phosphorylation of eEF2 may serve to conserve metabolic energy is during hibernation (Frerichs et al. 1998; Chen et al. 2001).

Regulation of eEF2 and eEF2 Kinase in the Heart

A recent study in neonatal cardiomyocytes suggests that eEF2 kinase protects these cells against hypoxic injury (Terai et al. 2005), via a mechanism linked to AMPK. AICAR normally protects myocytes from the proapoptotic effects of hypoxia, but small interfering RNA (siRNA)-mediated knockdown of eEF2 kinase attenuated this protective effect. This finding suggests that AMPK, via eEF2 kinase (and thus presumably, through the inhibition of translation), exerts a protective effect under conditions of oxygen deficiency.

As discussed by Proud (2000), Ser-500 is a substrate for PKA and phosphorylation here increases eEF2 kinase activity. Generally, this makes eEF2 kinase partly independent of Ca^{2+}/CaM. However, Ca^{2+} levels fluctuate constantly in the cardiac sarcoplasm, and the above mechanism would not be an appropriate one for controlling eEF2 kinase activity in cardiac myocytes (Horman et al. 2003). It is of interest to note that, for eEF2 kinase from adult cardiomyocytes, cAMP and PKA activation do not elicit calcium independence, but increase the maximal activity of the kinase instead (McLeod et al. 2001). This effect may serve to slow elongation and conserve energy in response to agents that stimulate its use by other processes (such as adrenalin, which stimulates contraction in cardiac myocytes). The data suggest that the form of eEF2 kinase found in adult cardiomyocytes may be distinct from that found in many other tissues (McLeod et al. 2001), but no cardiac-specific isoform of eEF2 kinase has yet been identified.

An interesting observation for the perfused whole heart is that increased workload actually causes the dephosphorylation of eEF2. This effect is independent of signaling through mTOR and does not appear to involve changes in eEF2 kinase activity (at least when measured in vitro under conditions of saturating Ca^{2+}/CaM) (Horman et al. 2003).

Increased workload is associated with activation of myocardial protein synthesis (as part of a hypertrophic program leading to cardiac growth): The dephosphorylation of eEF2 may contribute to this. It was noted above that hypertrophic agonists also elicit activation of eEF2 (Wang and Proud 2002). Horman et al. (2003) suggested that alterations in eEF2 phosphatase activity (PP-2A) may be responsible for the changes in eEF2 phosphorylation in the working heart, as also argued for the control of eEF2 by glucose or angiotensin, respectively (Everett et al. 2001; Yan et al. 2003).

This raises a related point; i.e., that it may be important to measure eEF2 kinase activity at subsaturating Ca^{2+}/CaM concentrations because mechanisms that control its activity may not modify its maximal activity but rather its sensitivity to activation by Ca^{2+}/CaM. A precedent for this is that the phosphorylation of eEF2 kinase by S6 kinase does not affect eEF2 kinase activity when measured at saturating Ca^{2+} concentrations, but markedly impairs its activity at submaximal concentrations (Wang et al. 2001). Thus, it is possible that some studies may not have detected alterations in eEF2 kinase activity that are due to shifts in its sensitivity to activation by Ca^{2+} ions, rather than to alterations in its maximal activity.

Regulation of eEF2 Kinase by Phosphorylation at Additional Sites

Significant advances have recently been made in understanding the control of eEF2 kinase by phosphorylation at other sites (as summarized in Fig. 2B,C). These mechanisms operate to regulate the activity of eEF2 kinase in response to amino acids and anabolic, mitogenic, or hypertrophic stimuli. For convenience, they may be divided into two categories.

Sites Modulated by mTOR Signaling (Fig. 2B)

eEF2 kinase is regulated by mTOR signaling, which decreases its overall activity (Redpath et al. 1996a) and impairs its ability to bind CaM (Browne and Proud 2004). However, eEF2 kinase polypeptide is not a direct substrate for phosphorylation by the protein kinase activity of mTOR (E.M. Smith and C.G. Proud, unpubl.). This may reflect the fact that eEF2 kinase lacks the TOR-signaling (TOS) motif that is found in several proteins that are phosphorylated by mTOR such as 4E-BP1 and the S6 kinases (Schalm and Blenis 2002). (The TOS motif binds the mTOR partner, raptor [Nojima et al. 2003; Schalm et al. 2003].) The first mTOR-regulated phosphorylation site in eEF2 kinase to be identified was Ser-366, which is phosphorylated in vitro by S6 kinase (Wang et al. 2001). Phosphorylation at Ser-366 decreases the activity of eEF2 kinase at submaximal concentrations of Ca^{2+}.

Phosphorylation at a second site in this region, Ser-359, also inhibits eEF2 kinase activity (Knebel et al. 2001). These workers first identified this site as a target for p38 MAP kinase δ (also termed stress-activated protein kinase 4 or SAPK4). Phosphorylation at Ser-359 can also be regulated via mTOR signaling, as illustrated by the observation that insulin-like growth factor 1 (IGF1) stimulates this and that the effect is inhibited by rapamycin (Knebel et al. 2001). As the activity of p38 MAP kinase δ is very low unless cells have been stimulated appropriately and is not regulated by IGF1 or mTOR, there must exist another protein kinase that phosphorylates this site. It is likely to be regulated by mTOR.

Ser-78 is the third known mTOR-controlled phosphorylation site in eEF2 kinase (Browne and Proud 2004). Phosphorylation at this site is markedly stimulated by insulin and this effect is blocked by rapamycin. It is also dependent on sufficient availability of amino acids, a second hallmark of mTOR-regulated events (Proud 2002). Ser-78 lies immediately next to the CaM-binding site. Phosphorylation at Ser-78 decreases the ability of eEF2 kinase to bind CaM in the presence of saturating concentrations of Ca^{2+} (Browne and Proud 2004). It is not known which kinase phosphorylates Ser-78.

Sites Phosphorylated by MAP Kinases or MAP Kinase-activated Kinases (Fig. 2C)

Ser-366 is also phosphorylated by members of the p90[RSK] group of protein kinases, which have a substrate specificity similar to that of the S6 kinases. These enzymes are activated by the classic MAP kinases (ERKs), and this site (which, as noted above, negatively regulates eEF2 kinase activity) is so far the only one known to be controlled by ERK signaling.

The stress-activated p38 MAP kinases (SAPK2a/b) and δ (SAPK4) phosphorylate eEF2 kinase at Ser-396 (Fig. 2C) (Knebel et al. 2002). However, the significance of phosphorylation here for the physiological control of eEF2 kinase activity is unclear, because phosphorylation was not altered in response to a range of conditions. Ser-377 is phosphorylated in vitro by several protein kinases that are activated by MAP kinases, including MAP-kinase-activated protein kinase 2 (MAPKAP-K2) (Knebel et al. 2002). This site is phosphorylated in vivo, and phosphorylation is enhanced by starving cells of amino acids and serum or by treating cells with anisomycin or tumor necrosis factor-α (TNF-α) (which both activate MAPKAP-K2). However, phosphorylation at Ser-377 does not affect the activity of eEF2 kinase in vitro (at least under the assay conditions used).

Stressful conditions often lead to increased phosphorylation of eEF2 (Knebel et al. 2002; McLeod and Proud 2002; Patel et al. 2002), so it is perhaps surprising that phosphorylation of eEF2 kinase by "stress-acti-vated" protein kinases leads to its inhibition. However, these kinases are also activated by cytokines (Karin 2005). Their links to eEF2 kinase may be part of a translation regulatory program to accelerate production of new proteins. Such control is currently not well understood, and further work is needed to study this.

Other Regulatory Mechanisms for eEF2 Kinase

An additional way by which eEF2 kinase activity may be regulated over the longer term is through its stability. The eEF2 kinase protein has been shown to have a short half-life and to be subjected to ubiquitination (Arora et al. 2003). Treatment of cells with the proteasome inhibitor MG132 blocked the degradation of eEF2 kinase and led to a build-up of ubiquitinated eEF2 kinase, consistent with a role for the proteasome in the stability of eEF2 kinase. eEF2 kinase binds Hsp90 (Nygård et al. 1991): Disruption of Hsp90 function with geldanamycin (previously shown to down-regulate eEF2 kinase; Yang et al. 2001) led to the increased appearance of ubiquitinated forms of the enzyme and its destabilization, resulting in a markedly decreased half-life.

A Role in Synaptic Plasticity?

Stimulation of a variety of receptors in neuronal cells also modulates eEF2 phosphorylation (Marin et al. 1997; Scheetz et al. 1997, 2000). Particularly interesting is the observation that activation of NMDA (N-methyl-D-as-partate) receptors increases the phosphorylation of eEF2 (Scheetz et al. 1997, 2000) and enhances the synthesis of Ca^{2+}/CaM kinase II, an enzyme known to be important in synaptic plasticity (Yamauchi 2005). Treatment with cycloheximide also enhanced the synthesis of Ca^{2+}/CaM kinase II, leading Yamauchi to suggest that the NMDA-induced phosphorylation of eEF2 may facilitate increased synthesis of Ca^{2+}/CaM kinase II because in-hibition of elongation actually enhances translation of the Ca^{2+}/CaM ki-nase II mRNA. Scheetz et al. (2000) suggested that partial inhibition of elongation may facilitate the translation of mRNAs that initiate translation inefficiently, by making elongation rather than initiation the rate-limiting process, as has previously been proposed (Walden and Thach 1986).

Other Points

It has been reported that eEF2 kinase activity is elevated in cancer cell lines and in human cancer tissue. These findings are rather unexpected given that eEF2 kinase inhibits protein synthesis. Arora et al. (2003) have reported that the imidazolium derivative NH125 inhibits eEF2 kinase activity and decreases the viability of a number of cancer cell lines. NH125 appears to block the G_1–S transition. Caution must always be exerted in interpreting such data: For example, rottlerin is also an inhibitor of eEF2 kinase and blocks glioma cells at the G_1–S point in the cell cycle (Parmer et al. 1997), but it is now known to inhibit a number of other protein kinases (Davies et al. 2000). However, in the case of NH125, overexpression of eEF2 kinase did diminish cellular sensitivity to this compound, suggesting that the effect on cell viability may indeed be mediated through this enzyme. Other inhibitors of eEF2 kinase have been reported (Cho et al. 2000), but the specificities of these compounds and of NH125 have not yet been established by reference to a broad panel of protein kinases.

COTRANSLATIONAL PROTEIN TARGETING

In eukaryotes, mRNAs encoding proteins destined for secretion or membrane insertion harbor a hydrophobic amino-terminal signal sequence that is required for the efficient targeting of these proteins to the endoplasmic reticulum (ER). This targeting reaction is mediated by two distinct pathways: the posttranslational and the cotranslational protein targeting pathways (for reviews, see Johnson and van Waes 1999; Keenan et al. 2001; Pool 2003; Egea et al. 2005). In the posttranslational targeting pathway, fully synthesized proteins are targeted to and translocated across the ER. This pathway is commonly used by yeast, but it is thought to be rarely in operation in mammalian cells, where the targeting of proteins to the ER is primarily mediated by the cotranslational targeting pathway (Fig. 3a). In cotranslational protein targeting to the ER, the SRP associates with the ribosome and is thought "to sample" the emerging nascent chain for the presence of a signal peptide (Ogg and Walter 1995). Upon signal peptide recognition, SRP then binds to the signal peptide to form a ribosome nascent chain–SRP (RNC–SRP) complex, which brings about the arrest of the elongation phase of protein synthesis, most likely by preventing the binding of elongation factors to the ribosome (Halic et al. 2004). The RNC–SRP complex then docks with the SRP receptor (SR) at the ER. Concomitantly with SRP dissociation from the complex and the resumption of elongation, the nascent chain is inserted into the translocon as a loop. Subsequent recognition of the signal peptide by the translocon results in the opening of the translocon. This allows the passage of the emerging nascent chain through the translocon, whereas the signal peptide remains associated with the translocon. Finally, the

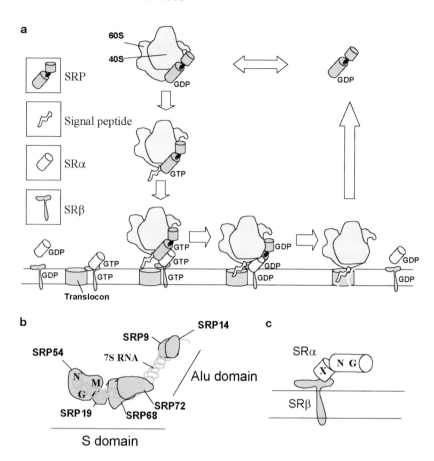

Figure 3. Cotranslational protein targeting to the ER. (*a*) Model of cotranslational protein targeting to the ER. SRP binds the ribosome. Upon emergence of the signal peptide, SRP undergoes a conformational rearrangement, resulting in an increased affinity for GTP. The resultant ribosome–nascent chain SRP (RNC–SRP) complex is then delivered to the ER via a GTP-dependent SR/SRP interaction. This leads to GAP activation and GTP hydrolysis, concomitant with the transfer of the signal peptide into the translocon (see text for further details). (*b*) Schematic representation of the signal recognition particle (SRP). (*c*) Schematic representation of the signal recognition particle receptor (SR).

completed polypeptide is released into the lumen of the ER through the action of the signal peptidase complex, which cleaves the signal peptide away from the completed polypeptide. The tight coupling of polypeptide synthesis with translocation of the nascent polypeptide across the ER ensures the prevention of protein folding prior to translocation.

SRP and Signal Peptide Recognition

Eukaryotic SRP consists of a 300-nucleotide 7S RNA and six protein subunits: SRP54, SRP19, SRP68, SRP72, SRP14, and SRP9 (Fig. 3b) (Walter and Blobel 1983a). Micrococcal nuclease treatment of SRP cleaves it into two domains, the Alu and S domains (Gundelfinger et al. 1983), each of which has a distinct function (Halic and Beckmann 2005). The Alu domain, made up of the 5' and 3' ends of the 7S RNA and a SRP14/SRP9 heterodimer, mediates peptide elongation arrest activity, whereas the S domain, comprising the remaining four proteins (SRP19, SRP54, and the SRP68/72 heterodimer) and the central portion of the 7S RNA ($\sim$100–250 nucleotides), is essential for signal sequence binding and SR docking. In fact, SRP54 and RNA helix 8 within the S domain, both of which are universally conserved, are sufficient to make a minimal SRP as found in *Escherichia coli*. Moreover, it is SRP54 that mediates signal sequence binding and GTP-dependent recognition of the SRP receptor.

SRP54 is a G-protein consisting of three domains, an N (amino-terminal), a G (GTPase), and an M (methionine-rich) domain (Fig. 3b) (Bernstein et al. 1989). This GTPase belongs to a subclass of a superfamily of small G-proteins defined by their unique insertion box (the I box, IBD), which results in low nucleotide affinity. However, the N domain is tightly associated with the G domain and has been suggested to have a role in regulating nucleotide affinity (Freymann et al. 1999; Montoya et al. 2000). SRP, via both RNA and protein interactions (Pool et al. 2002), can bind ribosomes actively involved in translation or those that are in an empty "inactive" state (Walter and Blobel 1983b; Flanagan et al. 2003). However, the affinity of SRP for the ribosome increases when the ribosome is in its active state (Flanagan et al. 2003). Cryo-electron microscopic (cryo-EM) studies have recently confirmed and extended previous findings demonstrating that the S domain interacts with ribosomal proteins L25 and L35 situated near the tunnel exit site of the large ribosomal subunit (Pool et al. 2002; Halic et al. 2004). This positioning is independent of the presence of a signal peptide and allows SRP to be orientated in such a way that SRP54 can scan the emerging nascent chain for the presence of a signal peptide (Pool et al. 2002). The M domain of SRP54 is thought to accommodate the hydrophobic signal sequence in a hydrophobic groove. Upon interaction of SRP54 with the emerging signal peptide, the affinity of SRP for the RNC complex increases dramatically (Flanagan et al. 2003). Thus, SRP, despite a 10- to 100-fold excess of ribosomes over SRP, is able to sample all emerging chains for the presence of a signal peptide (Ogg and Walter 1995).

SRP and Elongation Arrest

SRP is 230 Å in length and is long enough to make simultaneous contacts with the ribosomal exit site through its S domain and to bind the ribosome elongation-factor-binding site via its Alu domain (Andrews et al. 1985; Halic and Beckmann 2005). The binding of the Alu domain to the ribosome elongation-factor-binding site likely prevents access of elongation factors, thus mediating the "arrest" (or the slowing down) of elongation (Halic and Beckmann 2005). Elongation arrest has been proposed to allow sufficient time for targeting of the RNC complex to the SR (Walter and Blobel 1981). The structure of the elongation-arrested 80S ribosome with an emerging nascent signal sequence has been resolved at 12 Å resolution by cryo-EM (Halic et al. 2004). This showed that the first 48 nucleotides and the SRP9/14 heterodimer of the Alu domain make contacts with the ribosome. Both rRNA (helix 43 of 25S rRNA) and protein (rpL12) at the stalk base, probably through the universally conserved α-sarcin-ricin loop, contact the 5′ end of the 7S RNA. Contacts were also observed between the small ribosomal subunit through ribosomal 18S RNA and the SRP14/19 dimer. The sites of contact on the ribosome for the Alu domain are the same as those used by eEF2. Therefore, the Alu domain competes with eEF2 for binding to the elongation-factor-binding site on the ribosome by eEF2 mimicry, and it is this competition that provides a plausible explanation for SRP elongation arrest activity.

Targeting of the SRP–RNC Complex to the ER and Its Transfer to the Translocon

The binding of SRP54 to a nascent signal peptide emerging from the ribosome induces a dramatic conformational rearrangement in SRP54, which parallels an increase in its affinity for GTP (Bacher et al. 1999; Halic and Beckmann 2005). This increase in affinity of SRP54 for GTP is mediated by the ribosome through a mechanism that involves ribosomal protein L35 (Pool et al. 2002). GTP binding to SRP54 is a prerequisite for targeting of the RNC complex to the ER via the SR (Bacher et al. 1999). The mammalian SR is a heterodimer made up of SRα and SRβ (Fig. 3c). SRα contains three domains, an N (amino-terminal), G (GTPase), and X domain (for review, see Halic and Beckmann 2005). The N and G domains are structurally and functionally homologous to the NG domains of SRP54. The X domain interacts with SRβ, an Arf-like GTPase anchored to the ER membrane by a single transmembrane domain.

Upon docking of the RNC–SRP complex to the SR, interaction between SR and SRP54 results in an increase in both SRα and SRP54 GTP-binding affinities (Rapiejko and Gilmore 1992). The NG domains of SRP54 and SRα interact in their GTP-bound states. In the presence of the translocon, the nascent chain is released from the SRP/SR complex. This results in both SRα and SRP54 GAP activation and reciprocal GTP hydrolysis, leading to the dissociation of the complex (Connolly and Gilmore 1989; Rapiejko and Gilmore 1997). It is crucial that GTP hydrolysis is blocked until the signal sequence is released from SRP to ensure delivery of the nascent chain to the translocon. This targeting reaction is complicated by the fact that SRβ is also a GTPase and that interaction between SRα and SRβ via the X domain of SRα is nucleotide-dependent and requires SRβ to be in its GTP-bound state. Interestingly, the β subunit of the translocon (i.e., Sec61β) has been reported to act as the GEF for SRβ and therefore may have a role in sensing the availability of the translocon for nascent chain insertion (Fulga et al. 2001; Helmers et al. 2003). It has also been reported that the ribosome may act as the GAP for SRβ (Bacher et al. 1999). However, the details of the role and regulation of SRβ GTP hydrolysis are unclear.

Protein Translocation across the ER

The translocon, also referred to as the Sec61 complex, is a heterotrimeric complex composed of α, β, and γ subunits. The α subunit spans the outer ER membrane ten times, whereas the β and γ subunits have single spanning transmembrane domains. The complex is associated with a number of other protein complexes including the oligosaccharyl transferase complex; the TRAM complex, which is thought to act as a chaperone in membrane protein integration; the signal peptidase complex, which cleaves the signal peptide; and the TRAP complex (of unknown function).

The Sec61 complex forms a flexible channel with a hydrophilic interior. This channel is unusual because it not only opens across the membrane, but also opens laterally within the membrane. This allows transmembrane segments to move from the aqueous interior of the pore through a lateral gate into the hydrophobic environment of the lipid bilayer. A high-resolution 3.2 Å X-ray structure of the SecY complex, the bacterial homolog of the Sec61 complex, has been solved (Van den Berg et al. 2004). This revealed that the pore, which can be formed from a single copy of the SecY complex, resembles two aqueous funnels separated by a narrow constriction formed by a ring of six hydrophobic residues with branched side chains. The pore diameter is only 5–8 Å, but this widens, triggered by signal peptide insertion, to accommodate the translocating chain, through the lateral displacement of helices to which the pore residues are attached, analogous to the opening and closing of the diaphragm of a camera. The opening diaphragm accommodates passage of the translocating peptide through the pore and importantly, "like a gasket," molds itself around the translocating polypeptide while the polypeptide is actively translated. This mechanism inhibits the passage of small ions during the translocation process. Although a single SecY can form a pore, it is believed that oligomers of SecY are the active species in translocation.

Regulation of Targeting and Translation of mRNAs Encoding Secretory Proteins

Ribosomes engaged in the synthesis of secretory proteins are translocated to the ER to form a translocon-competent ribosome membrane junction. Upon termination of protein synthesis, only the small ribosomal subunit is released (Potter and Nicchitta 2000). The large ribosomal subunit remains stably associated with the ER (Potter et al. 2001). These membrane-bound 60S ribosomal subunits can participate in the de novo synthesis of both cytosolic and secretory proteins. However, RNC complexes synthesizing secretory proteins remain stably associated with the ER, whereas RNC complexes synthesizing cytosolic proteins are released from the ER to continue synthesis in the cytoplasm. Therefore, it has been proposed that the affinity of the elongating ribosome for the ER decreases unless it is actively engaged with the translocon. Additionally, reinitiation of translation of mRNAs encoding secretory proteins, which are ER-bound, can occur in the absence of SRP. This mechanism would clearly be more efficient than retargeting of the RNC–SRP complex to the ER upon each round of protein synthesis (Potter et al. 2001).

It is unclear whether mRNA targeting to the ER and protein translocation across the ER membrane are regulated by external stimuli. However, targeting of mRNA encoding secretory proteins to the ER is stimulated by an increase in glucose concentration in pancreatic β cells (Greenman et al. 2005). This potentially provides an important mechanism by which glucose can regulate the synthesis of secretory proteins, such as insulin. The actual mechanism by which glucose stimulates secretory protein synthesis is unknown. However, a number of components of the translocon/targeting apparatus are phosphorylated, including Sec61β, TRAM, SRα, and SRP72 (Utz et al. 1998; Gruss et al. 1999). Therefore, it is possible that phosphorylation may regulate mRNA targeting to the ER and subsequent protein translocation. Recently, a fungus-derived cyclopeptide was found to inhibit the translocation of specific proteins such as vascular cell adhesion molecule 1 (Besemer et al. 2005; Garrison et al.

2005). The specificity of this inhibition is conferred by the protein's signal peptide, indicating that signal sequence variation may modulate the functional expression of secreted or membrane proteins.

CONCLUSIONS AND PERSPECTIVES

Approximately one-third of all cellular protein synthesis occurs on the ER membrane, and for professional secretory cells, such as pancreatic β cells, more than 90% of the translated polypeptides are translocated into the ER. Much effort has been dedicated in determining the molecular mechanism of protein targeting and translocation with great success. Yet, rather surprisingly, very little is known regarding whether mRNA targeting and translocation are influenced by external stimuli or how, once the mRNA is targeted to the ER, subsequent rounds of protein synthesis are regulated at the ER.

It is now evident that both eEF2 and its kinase, eEF2 kinase, are subjected to sophisticated control mechanisms that allow the rapid modulation of eEF2 activity under diverse conditions. A clear priority now is to establish what contribution these regulatory mechanisms make to the control of protein synthesis, cell growth and survival, and the regulation of the translation of specific mRNAs. The exploitation of animals and cells devoid of eEF2 kinase activity would be a major step toward this goal. Furthermore, there remain major gaps in our understanding of the links between mTOR and the control of eEF2 kinase.

ACKNOWLEDGMENTS

Work in C.G.P.'s laboratory on eEF2 and eEF2 kinase has been supported by the Biotechnology and Biological Sciences Research Council and the British Heart Foundation, United Kingdom. Work in T.P.H.'s laboratory has been supported by Diabetes UK, Biotechnology and Biological Sciences Research Council, and the Wellcome Trust.

REFERENCES

Andrews D.W., Walter P., and Ottensmeyer F.P. 1985. Structure of the signal recognition particle by electron microscopy. *Proc. Natl. Acad. Sci.* **82:** 785–789.

Arora S., Yang J.M., Kinzy T.G., Utsumi R., Okamoto T., Kitayama T., Ortiz P.A., and Hait W.N. 2003. Identification and characterization of an inhibitor of eukaryotic elongation factor 2 kinase against human cancer cell lines. *Cancer Res.* **63:** 6894–6899.

Bacher G., Pool M., and Dobberstein B. 1999. The ribosome regulates the GTPase of the beta-subunit of the signal recognition particle receptor. *J. Cell Biol.* **146:** 723–730.

Bernstein H.D., Poritz M.A., Strub K., Hoben P.J., Brenner S., and Walter P. 1989. Model for signal sequence recognition from amino-acid sequence of 54K subunit of signal recognition particle. *Nature* **340:** 482–486.

Besemer J., Harant H., Wang S., Oberhauser B., Marquardt K., Foster C.A., Schreiner E.P., de Vries J.E., Dascher-Nadel C., and Lindley I.J. 2005. Selective inhibition of cotranslational translocation of vascular cell adhesion molecule 1. *Nature* **436:** 290–293.

Browne G.J. and Proud C.G. 2002. Regulation of peptide-chain elongation in mammalian cells. *Eur. J. Biochem.* **269:** 5360–5368.

———. 2004. A novel mTOR-regulated phosphorylation site in elongation factor 2 kinase modulates the activity of the kinase and its binding to calmodulin. *Mol. Cell. Biol.* **24:** 2986–2997.

Browne G.J., Finn S.G., and Proud C.G. 2004. Stimulation of the AMP-activated protein kinase leads to activation of eukaryotic elongation factor 2 kinase and to its phosphorylation at a novel site, serine 398. *J. Biol. Chem.* **279:** 12220–12231.

Carlberg U., Nilsson A., and Nygård O. 1990. Functional properties of phosphorylated elongation factor 2. *Eur. J. Biochem.* **191:** 639–645.

Carroll M., Warren O., Fan X., and Sossin W.S. 2004. 5-HT stimulates eEF2 dephosphorylation in a rapamycin-sensitive manner in *Aplysia* neurites. *J. Neurochem.* **90:** 1464–1476.

Chen Y., Matsushita M., Nairn A.C., Damuni Z., Cai D., Frerichs K.U., and Hallenbeck J.M. 2001. Mechanisms for increased levels of phosphorylation of elongation factor-2 during hibernation in ground squirrels. *Biochemistry* **40:** 11565–11570.

Cho S.I., Koketsu M., Ishihara H., Matsushita M., Nairn A.C., Fukazawa H., and Uehara Y. 2000. Novel compounds, "1,3-selenazine derivatives" as specific inhibitors of eukaryotic elongation factor-2 kinase. *Biochim. Biophys. Acta* **1475:** 207–215.

Connolly T. and Gilmore R. 1989. The signal recognition particle receptor mediates the GTP-dependent displacement of SRP from the signal sequence of the nascent polypeptide. *Cell* **57:** 599–610.

Crozier S.J., Vary T.C., Kimball S.R., and Jefferson L.S. 2005. Cellular energy status modulates translational control mechanisms in ischemic-reperfused rat hearts. *Am. J. Physiol. Heart Circ. Physiol.* **289:** H1242–H1250.

Davies S.P., Reddy H., Caivano M., and Cohen P. 2000. Specificity and mechanism of action of some commonly used protein kinase inhibitors. *Biochem. J.* **351:** 95–105.

Diggle T.A., Redpath N.T., Heesom K.J., and Denton R.M. 1998. Regulation of protein synthesis elongation factor-2 kinase by cAMP in adipocytes. *Biochem. J.* **336:** 525–529.

Diggle T.A., Seehra C.K., Hase S., and Redpath N.T. 1999. Analysis of the domain structure of elongation factor-2 kinase by mutagenesis. *FEBS Lett.* **457:** 189–192.

Diggle T.A., Subkhankulova T., Lilley K.S., Shikotra N., Willis A.E., and Redpath N.T. 2001. Phosphorylation of elongation factor-2 kinase on serine 499 by cAMP-dependent protein kinase induces Ca^{2+}/calmodulin-independent activity. *Biochem. J.* **353:** 621–626.

Dorovkov M.V., Pavur K.S., Petrov A.N., and Ryazanov A.G. 2002. Regulation of elongation factor-2 kinase by pH. *Biochemistry* **41:** 13444–13450.

Drennan D. and Ryazanov A.G. 2004. Alpha-kinases: Analysis of the family and comparison with conventional protein kinases. *Prog. Biophys. Mol. Biol.* **85:** 1–32.

Egea P.F., Stroud R.M., and Walter P. 2005. Targeting proteins to membranes: Structure of the signal recognition particle. *Curr. Opin. Struct. Biol.* **15:** 213–220.

Everett A.D., Stoops T.D., Nairn A.C., and Brautigan D. 2001. Angiotensin II regulates phosphorylation of translation elongation factor-2 in cardiac myocytes. *Am. J. Physiol.* **281:** H161–H167.

Flanagan J.J., Chen J.C., Miao Y., Shao Y., Lin J., Bock P.E., and Johnson A.E. 2003. Signal recognition particle binds to ribosome-bound signal sequences with fluorescence-detected subnanomolar affinity that does not diminish as the nascent chain lengthens. *J. Biol. Chem.* **278:** 18628–18637.

Frerichs K.U., Smith C.B., Brenner M., DeGracia D.J., Krause G.S., Marrone L., Dever T.E., and Hallenbeck J.M. 1998. Suppression of protein synthesis in brain during hibernation involves inhibition of protein initiation and elongation. *Proc. Natl. Acad. Sci.* **95:** 14511–14516.

Freymann D.M., Keenan R.J., Stroud R.M., and Walter P. 1999. Functional changes in the structure of the SRP GTPase on binding GDP and Mg2+GDP. *Nat. Struct. Biol.* **6:** 793–801.

Fulga T.A., Sinning I., Dobberstein B., and Pool M.R. 2001. SRbeta coordinates signal sequence release from SRP with ribosome binding to the translocon. *EMBO J.* **20:** 2338–2347.

Garrison J.L., Kunkel E.J., Hegde R.S., and Taunton J. 2005. A substrate-specific inhibitor of protein translocation into the endoplasmic reticulum. *Nature* **436:** 285–289.

Greenman I.C., Gomez E., Moore C.E., and Herbert T.P. 2005. The selective recruitment of mRNA to the ER and an increase in initiation are important for glucose-stimulated proinsulin synthesis in pancreatic beta-cells. *Biochem. J.* **391:** 291–300.

Gruss O.J., Feick P., Frank R., and Dobberstein B. 1999. Phosphorylation of components of the ER translocation site. *Eur. J. Biochem.* **260:** 785–793.

Gundelfinger E.D., Krause E., Melli M., and Dobberstein B. 1983. The organization of the 7SL RNA in the signal recognition particle. *Nucleic Acids Res.* **11:** 7363–7374.

Halic M. and Beckmann R. 2005. The signal recognition particle and its interactions during protein targeting. *Curr. Opin. Struct. Biol.* **15:** 116–125.

Halic M., Becker T., Pool M.R., Spahn C.M., Grassucci R.A., Frank J., and Beckmann R. 2004. Structure of the signal recognition particle interacting with the elongation-arrested ribosome. *Nature* **427:** 808–814.

Hardie D.G., Scott J.W., Pan D.A., and Hudson E.R. 2003. Management of cellular energy by the AMP-activated protein kinase system. *FEBS Lett.* **546:** 113–120.

Helmers J., Schmidt D., Glavy J.S., Blobel G., and Schwartz T. 2003. The beta-subunit of the protein-conducting channel of the endoplasmic reticulum functions as the guanine nucleotide exchange factor for the beta-subunit of the signal recognition particle receptor. *J. Biol. Chem.* **278:** 23686–23690.

Horman S., Beauloye C., Vertommen D., Vanoverschelde J.L., Hue L., and Rider M.H. 2003. Myocardial ischemia and increased heart work modulate the phosphorylation state of eukaryotic elongation factor-2. *J. Biol. Chem.* **278:** 41970–41976.

Horman S., Browne G.J., Krause U., Patel J.V., Vertommen D., Bertrand L., Lavoinne A., Hue L., Proud C.G., and Rider M.H. 2002. Activation of AMP-activated protein kinase leads to the phosphorylation of elongation factor 2 and an inhibition of protein synthesis. *Curr. Biol.* **12:** 1419–1423.

Hovland R., Eikhom T.S., Proud C.G., Cressey L.I., Lanotte M., Doskeland S.O., and Houge G. 1999. cAMP inhibits translation by inducing Ca2+/calmodulin-independent elongation factor 2 kinase activity in IPC-81 cells. *FEBS Lett.* **444:** 97–101.

Inoki K., Zhu T., and Guan K.L. 2003. TSC2 mediates cellular energy response to control cell growth and survival. *Cell* **115:** 577–590.

Johnson A.E. and van Waes M.A. 1999. The translocon: A dynamic gateway at the ER membrane. *Annu. Rev. Cell Dev. Biol.* **15:** 799–842.

Jorgensen R., Merrill A.R., and Andersen G.R. 2006. The life and death of translation elongation factor 2. *Biochem. Soc. Trans.* **34:** 1–6.

Karin M. 2005. Inflammation-activated protein kinases as targets for drug development. *Proc. Am. Thorac. Soc.* **2:** 386–390.

Keenan R.J., Freymann D.M., Stroud R.M., and Walter P. 2001. The signal recognition particle. *Annu. Rev. Biochem.* **70:** 755–775.

Knebel A., Morrice N., and Cohen P. 2001. A novel method to identify protein kinase substrates: eEF2 kinase is phosphorylated and inhibited by SAPK4/p38delta. *EMBO J.* **20:** 4360–4369.

Knebel A., Haydon C.E., Morrice N., and Cohen P. 2002. Stress-induced regulation of eEF2 kinase by SB203580-sensitive and -insensitive pathways. *Biochem. J.* **367:** 525–532.

Le Sourd F., Boulben S., Le Bouffant R., Cormier P., Morales J., Bellé R., and Mulner-Lorillon O. 2006. eFF1B: At the dawn of the 21st century. *Biochim. Biophys. Acta* **1759:** 13–31.

Marin P., Nastiuk K.L., Daniel N., Girault J., Czernik A.J., Glowinski J., Nairn A.C., and Premont J. 1997. Glutamate dependent phosphorylation of elongation factor 2 and inhibition of protein synthesis in neurons. *J. Neurosci.* **17:** 3445–3454.

McLeod L.E. and Proud C.G. 2002. ATP depletion increases phosphorylation of elongation factor eEF2 in adult cardiomyocytes independently of inhibition of mTOR signalling. *FEBS Lett.* **531:** 448–452.

McLeod L.E., Wang L., and Proud C.G. 2001. β-Adrenergic agonists increase phosphorylation of elongation factor 2 in cardiomyocytes without eliciting calcium-independent eEF2 kinase activity. *FEBS Lett.* **489:** 225–228.

Merrick W.C. and Nyborg J. 2000. The protein biosynthesis elongation cycle. In *Translational control of gene expression* (ed. N. Sonenberg et al.), pp. 89–125. Cold Spring Harbor Laboratory Press, Cold Spring Harbor, New York.

Montoya G., Kaat K., Moll R., Schafer G., and Sinning I. 2000. The crystal structure of the conserved GTPase of SRP54 from the archaeon *Acidianus ambivalens* and its comparison with related structures suggests a model for the SRP-SRP receptor complex. *Struct. Fold. Des.* **8:** 515–525.

Nairn A.C., Bhagat B., and Palfrey H.C. 1985. Identification of calmodulin-dependent protein kinase III and its major Mr 100,000 substrate in mammalian tissues. *Proc. Natl. Acad. Sci.* **82:** 7939–7943.

Nojima H., Tokunaga C., Eguchi S., Oshiro N., Hidayat S., Yoshino K., Hara K., Tanaka J., Avruch J., and Yonezawa K. 2003. The mTOR partner, raptor, binds the mTOR substrates, p70 S6 kinase and 4E-BP1, through their TOS (TOR signaling) motifs. *J. Biol. Chem.* **278:** 15461–15464.

Nygård O., Nilsson A., Carlberg U., Nilsson L., and Amons R. 1991. Phosphorylation regulates the activity of the eEF-2-specific Ca(2+)- and calmodulin-dependent protein kinase III. *J. Biol. Chem.* **266:** 16425–16430.

Ogg S.C. and Walter P. 1995. SRP samples nascent chains for the presence of signal sequences by interacting with ribosomes at a discrete step during translation elongation. *Cell* **81:** 1075–1084.

Parmer T.G., Ward M.D., and Hait W.N. 1997. Effects of rottlerin, an inhibitor of calmodulin-dependent protein kinase III, on cellular proliferation, viability, and cell cycle distribution in malignant glioma cells. *Cell Growth Differ.* **8:** 327–334.

Patel J., McLeod L.E., Vries R.G., Flynn A., Wang X., and Proud C.G. 2002. Cellular stresses profoundly inhibit protein synthesis and modulate the states of phosphorylation of multiple translation factors. *Eur. J. Biochem.* **269:** 3076–3085.

Pavur K.S., Petrov A.N., and Ryazanov A.G. 2000. Mapping the functional domains of elongation factor-2 kinase. *Biochemistry* **39:** 12216–12224.

Pool M.R. 2003. Getting to the membrane: How is co-translational protein targeting to the endoplasmic reticulum regulated? *Biochem. Soc. Trans.* **31:** 1232–1237.

Pool M.R., Stumm J., Fulga T.A., Sinning I., and Dobberstein B. 2002. Distinct modes of signal recognition particle interaction with the ribosome. *Science* **297:** 1345–1348.

Potter M.D. and Nicchitta C.V. 2000. Ribosome-independent regulation of translocon composition and Sec61alpha conformation. *J. Biol. Chem.* **275:** 2037–2045.

Potter M.D., Seiser R.M., and Nicchitta C.V. 2001. Ribosome exchange revisited: A mechanism for translation-coupled ribosome detachment from the ER membrane. *Trends Cell Biol.* **11:** 112–115.

Price N.T., Redpath N.T., Severinov K.V., Campbell D.G., Russell J.M., and Proud C.G. 1991. Identification of the phosphorylation sites in elongation factor-2 from rabbit reticulocytes. *FEBS Lett.* **282:** 253–258.

Proud C.G. 2000. Control of the elongation phase of protein synthesis. In *Translational control of gene expression* (ed. N. Sonenberg et al.), pp. 719–739. Cold Spring Harbor Laboratory Press, Cold Spring Harbor, New York.

———. 2002. Regulation of mammalian translation factors by nutrients. *Eur. J. Biochem.* **269:** 5338–5349.

Rapiejko P.J. and Gilmore R. 1992. Protein translocation across the ER requires a functional GTP binding site in the alpha subunit of the signal recognition particle receptor. *J. Cell Biol.* **117:** 493–503.

———. 1997. Empty site forms of the SRP54 and SR alpha GTPases mediate targeting of ribosome-nascent chain complexes to the endoplasmic reticulum. *Cell* **89:** 703–713.

Redpath N.T., Foulstone E.J., and Proud C.G. 1996a. Regulation of translation elongation factor-2 by insulin via a rapamycin-sensitive signalling pathway. *EMBO J.* **15:** 2291–2297.

Redpath N.T., Price N.T., and Proud C.G. 1996b. Cloning and expression of cDNA encoding protein synthesis elongation factor-2 kinase. *J. Biol. Chem.* **271:** 17547–17554.

Ryazanov A.G., Pavur K.S., and Dorovkov M.V. 1999. Alpha kinases: A new class of protein kinases with a novel catalytic domain. *Curr. Biol.* **9:** R43–R45.

Ryazanov A.G., Natapov P.G., Shestakova E.A., Severin F.F., and Spirin A.S. 1988. Phosphorylation of the elongation factor 2: The fifth Ca2+/calmodulin-dependent system of protein phosphorylation. *Biochimie* **70** 619–626.

Ryazanov A.G., Ward M.D., Mendola C.E., Pavur K.S., Dorovkov M.V., Wiedmann M., Erdjument-Bromage H., Tempst P., Parmer T.G., Prostko C.R., et al. 1997. Identification of a new class of protein kinases represented by eukaryotic elongation factor-2 kinase. *Proc. Natl. Acad. Sci.* **94:** 4884–4889.

Sans M.D., Xie Q., and Williams J.A. 2004. Regulation of translation elongation and phosphorylation of eEF2 in rat pancreatic acini. *Biochem. Biophys. Res. Commun.* **319:** 144–151.

Schalm S.S. and Blenis J. 2002. Identification of a conserved motif required for mTOR signaling. *Curr. Biol.* **12:** 632–639.

Schalm S.S., Fingar D.C., Sabatini D.M., and Blenis J. 2003. TOS motif-mediated raptor binding regulates 4E-BP1 multisite phosphorylation and function. *Curr. Biol.* **13:** 797–806.

Scheetz A.J., Nairn A.C., and Constantine-Paton M. 1997. N-methyl-D-aspartate receptor activation and visual activity induce elongation factor-2 phosphorylation in amphibian tecta: A role for N-methyl-D-aspartate receptors in controlling protein synthesis. *Proc. Natl. Acad. Sci.* **4:** 14770–14775.

———. 2000. NMDA-mediated control of protein synthesis at developing synapses. *Nat. Neurosci.* **3:** 1–7.

Smith E.M., Finn S.G., Tee A.R., Browne G.J., and Proud C.G. 2005. The tuberous sclerosis protein TSC2 is not required for the regulation of the mammalian target of rapamycin by amino acids and certain cellular stresses. *J. Biol. Chem.* **280:** 18717–18727.

Teige M., Scheikl E., Reiser V., Ruis H., and Ammerer G. 2001. Rck2, a member of the calmodulin-protein kinase family, links protein synthesis to high osmolarity MAP kinase signaling in budding yeast. *Proc. Natl. Acad. Sci.* **98:** 5625–5630.

Terai K., Hiramoto Y., Masaki M., Sugiyama S., Kuroda T., Hori M., Kawase I., and Hirota H. 2005. AMP-activated protein kinase protects cardiomyocytes against hypoxic injury through attenuation of endoplasmic reticulum stress. *Mol. Cell. Biol.* **25:** 9554–9575.

Traugh J.A. 2001. Insulin, phorbol ester and serum regulate the elongation phase of protein synthesis. *Prog. Mol. Subcell. Biol.* **26:** 33–48.

Utz P.J., Hottelet M., Le T.M., Kim S.J., Geiger M.E., Van Venrooij W.J., and Anderson P. 1998. The 72-kDa component of signal recognition particle is cleaved during apoptosis. *J. Biol. Chem.* **273:** 35362–35370.

Van den Berg B., Clemons W.M., Jr., Collinson I., Modis Y., Hartmann E., Harrison S.C., and Rapoport T.A. 2004. X-ray structure of a protein-conducting channel. *Nature* **427:** 36–44.

Walden W.E. and Thach R.E. 1986. Translational control of gene expression in a normal fibroblast. Characterization of a subclass of mRNAs with unusual kinetic properties. *Biochemistry* **25:** 2033–2041.

Walter P. and Blobel G. 1981. Translocation of proteins across the endoplasmic reticulum. III. Signal recognition protein (SRP) causes signal sequence-dependent and site-specific arrest of chain elongation that is released by microsomal membranes. *J. Cell Biol.* **91:** 557–561.

———. 1983a. Disassembly and reconstitution of signal recognition particle. *Cell* **34:** 525–533.

———. 1983b. Subcellular distribution of signal recognition particle and 7SL-RNA determined with polypeptide-specific antibodies and complementary DNA probe. *J. Cell Biol.* **97:** 1693–1699.

Wang L. and Proud C.G. 2002. Regulation of the phosphorylation of elongation factor 2 by MEK-dependent signalling in adult rat cardiomyocytes. *FEBS Lett.* **531:** 285–289.

Wang X., Li W., Williams M., Terada N., Alessi D.R., and Proud C.G. 2001. Regulation of elongation factor 2 kinase by p90[RSK1] and p70 S6 kinase. *EMBO J.* **20:** 4370–4379.

Yamaguchi H., Matsushita M., Nairn A.C., and Kuriyan J. 2001. Crystal structure of the atypical protein kinase domain of a TRP channel with phosphotransferase activity. *Mol. Cell* **7:** 1047–1057.

Yamauchi T. 2005. Neuronal Ca2+/calmodulin-dependent protein kinase II—Discovery, progress in a quarter of a century, and perspective: Implication for learning and memory. *Biol. Pharmacol. Bull.* **28:** 1342–1354.

Yan L., Nairn A.C., Palfrey H.C., and Brady M.J. 2003. Glucose regulates EF-2 phosphorylation and protein translation by a protein phosphatase-2A-dependent mechanism in INS-1-derived 832/13 cells. *J. Biol. Chem.* **278:** 18177–18183.

Yang J., Yang J.M., Iannone M., Shih W.J., Lin Y., and Hait W.N. 2001. Disruption of the EF-2 kinase/Hsp90 protein complex: A possible mechanism to inhibit glioblastoma by geldanamycin. *Cancer Res.* **61:** 4010–4016.

22

Regulation of Termination and Recoding

Jonathan D. Dinman
Department of Cell Biology and Molecular Genetics
University of Maryland, College Park, Maryland 20742

Marla J. Berry
Department of Cell and Molecular Biology
John A. Burns School of Medicine
University of Hawaii at Manoa, Honolulu
Hawaii 96813

PRECISE TRANSLATION OF MRNA SEQUENCE INTO PROTEIN is a universal requirement for all living organisms. Although this would appear to require strict adherence to the linear readout of mRNAs, the term "recoding" defines a group of molecular mechanisms that alter the interpretation of hard-wired sequences. Ribosomal frameshifting, termination readthrough, and incorporation of selenocysteine and pyrrolysine are examples of recoding, studies of which provide insights into the fundamental rules governing translational fidelity. Although monocistronic mRNAs are preferentially used by eukaryotes, the mRNAs of many RNA viruses encode multiple gene products. In the late 1970s, tryptic analyses of a number of retroviral proteins indicated that their Gag and Gag-Pol proteins had the same amino termini. Multiple mechanisms, including RNA splicing, termination suppression, and recoding were proposed to account for these observations. Subsequently, Gag-Pol production for a number of retroviruses and retrotransposons was shown to be due to recoding or frameshifting events (for review, see Jacks 1990). Similarly, selenocysteine was identified in glutathione peroxidase 1 in the early 1970s, but its incorporation via recoding of UGA was not revealed until 1986 (Chambers et al. 1986). Since then, translational recoding has been found to regulate gene expression in cellular mRNAs and a wide variety of RNA viruses, which may also present targets for the development of new antibiotic and antiviral therapies. A developing body of knowledge also suggests that these mechanisms may be used to posttranscriptionally regulate mRNAs of cellular origin, which has implications for expanding our understanding of developmental biology and cancer. Here, we review examples and mechanisms of the different modes of translational recoding as they occur in eukaryotic cells.

TRANSLATIONAL RECODING: PROGRAMMED FRAMESHIFTING

In response to specific *cis*-acting mRNA sequences, elongating ribosomes can be "programmed" to shift reading frame. A frameshift represents a final state that can stem from a myriad of processes. In the end, however, there are only two possible outcomes: The translational machinery can end up in either the -1 or $+1$ frame relative to the incoming reading frame. Thus, frameshifting is discussed by working backward from final effects to their underlying causes.

Programmed -1 Ribosomal Frameshifting

Programmed -1 ribosomal frameshifting (-1 PRF) is the result of a net shift of the translational reading frame by one base in the 5' direction. Although most known examples of frameshifting are of this class, almost all have been found in viruses. The locations of -1 PRF signals within viral genomes suggest two general strategies. The first is where a -1 PRF signal separates sequences encoding viral structural proteins from those encoding enzymes. The best-studied examples of this class are the L-A virus of yeast and human immunodeficiency virus type 1 (HIV-1). In these cases, frameshifting frequencies have evolved to produce the optimal ratios of structural to enzymatic proteins for virion assembly, and changes in -1 PRF efficiencies have strong inhibitory effects on virus particle morphogenesis (for review, see Dinman et al. 1998). The second is characterized by -1 PRF signals that separate sequences encoding immediate-early proteins from those encoding intermediate proteins. These have been proposed to act as switches between synthesis of proteins from the infectious ($+$) strand to the replicative phase of the viral life cycle (Barry and Miller 2002).

General Mechanism

The first mRNA motif identified in -1 PRF signals was a heptameric "slippery site," and requirements for 3' stimulatory elements and for "spacer" sequences separating the slippery site and downstream elements soon followed. These findings laid the foundation for the fundamental paradigm of −1 PRF: A strong mRNA secondary structure stimulates the simultaneous slippage of both the aminoacyl- (aa-) and peptidyl-tRNAs along the mRNA by one nucleotide in the 5′ direction on a special sequence that allows their non-wobble bases to pair with the −1 frame codons (for review, see Jacks 1990).

"Slippery Sites"

Directed mutagenesis approaches first suggested that frameshifting occurs at special heptameric "slippery sites," a hypothesis confirmed by protein sequence analyses. Parallel studies elucidated the rules defining the heptameric sequences (for review, see Brierley 1995). In general, the slippery site can be defined as N NNW WWH, where N is any three identical bases, W is A or U, and H is A, C, or U (the frame of the initiator AUG is indicated by the spacing). Exceptions to the rules governing acceptable slippery sites have helped to catalyze the development of alternative models. Slippage on the potato virus M (PVM) mRNA and in the measles virus P gene occurs while the A site is unoccupied (Gramstat et al. 1994; Liston and Briedis 1995). In contrast, A-site codon:anticodon interactions appear to drive −1 PRF at the unusual G UUA AAC slippery site of equine arteritis virus (Napthine et al. 2003). The actual and relative frameshift efficiencies promoted by different heptameric slippery sites can also vary significantly depending on the source of ribosomes and assay systems employed. A rigorous examination of this phenomenon using different slippery sites in the context of the SARS-CoV (severe acute respiratory syndrome–coronavirus) and HIV-1 frameshift signals reveals significant differences using translational systems derived from different taxa (animals, fungi, and plants), suggesting that −1 PRF can be used to tease out functional differences between translational machineries from among the different kingdoms (Plant and Dinman 2006). It also serves as a cautionary note against making general conclusions from data obtained by coupling specific −1 PRF signals with evolutionarily distant translational assay systems.

mRNA Structural Motifs

An early analysis of the bovine leukemia virus genome revealed the presence of overlapping reading frames followed by a predicted stem-loop, suggesting that such structures might act as physical barriers causing ribosomes to pause, thus allowing increased time for the ribosome-bound tRNAs to reach alternative configurations on mRNAs (Rice et al.

1985). After −1 PRF was confirmed, it became apparent that optimal frameshifting required a *cis*-acting element 3′ of the heptameric slippery site, initially thought to be a stem-loop (Jacks et al. 1988). The next breakthrough came with the demonstration that the frameshift stimulatory element in the avian infectious bronchitis virus (IBV) −1 PRF signal was an mRNA pseudoknot (Brierley et al. 1989). Many subsequent studies demonstrated that mRNA pseudoknots are widely used to stimulate −1 PRF and that these come in many different varieties (for review, see Giedroc et al. 2000). The most commonly employed structure is an H-type pseudoknot. There is considerable variation within this motif, for example, short and compact versus large and extended, helices being bent relative to one another versus coaxially stacked. More elaborate themes on the pseudoknot are becoming known, e.g., three stemmed structures (Baril et al. 2003; Plant et al. 2005) and long-range kissing loop interactions (Herold and Siddell 1993; Barry and Miller 2002).

Mutational analyses show a general relationship between thermodynamic stability and ability to stimulate −1 PRF (for review, see Brierley 1995). The stability of mRNA pseudoknot structures can be influenced by ions (Theimer and Giedroc 2000; Egli et al. 2002), base triple interactions (Kolk et al. 1998; Su et al. 1999; Michiels et al. 2001), pH (Nixon and Giedroc 2000), π-electron-stacking interactions, and even water (Sarkhel et al. 2003). Biophysical studies examining these types of considerations concluded that the global stability, ion-binding properties, and rates of pseudoknot unfolding were not the sole factors modulating −1 PRF efficiency (Theimer and Giedroc 2000), and a precise network of weak interactions nearest the helical junction of the two stems can significantly contribute to −1 PRF efficiency (Cornish et al. 2005).

Although mRNA pseudoknots tend to be preferred in −1 PRF signals, examples of −1 PRF stimulatory stem-loop structures do exist, e.g., in measles virus (Liston and Briedis 1995). The question of whether or not the stimulatory element in most strains of HIV-1 is a stem-loop or an mRNA pseudoknot remains controversial (for review, see Brierley and Dos Ramos 2006). It was first suggested that this signal required a stem-loop (Jacks et al. 1988), but this was disputed by experiments demonstrating that the HIV sequence lacking the proposed stem-loop sequence was capable of promoting efficient −1 PRF (Wilson et al. 1988; Moosmayer et al. 1991). However, later studies demonstrated that the presence of an intact stem-loop structure was required to promote efficient −1 PRF, a view further supported by phylogenetic studies (Brierley 1995; Brierley and Dos Ramos 2006). Three potential pseudoknots have been proposed (see Dinman et al. 2002), and other recent studies propose an extended stem-loop structure involving an upper helix, a bend-inducing bulge, and a lower helix formed by base-pairing of the 3′ sequence with the spacer region proximal to the slippery site (Gaudin et al. 2005; Staple and Butcher 2005). The very small distance (one base) between the 5′ end of the lower helix and the 3′ end of the slippery site makes it unlikely that the double-stranded RNA lower-helix structure exists inside of the downstream tunnel of the 80S ribosome when the slippery site is positioned at the ribosomal A and P sites. It is possible that the extended stem-loop structure may serve to initially pause the ribosome, so that subsequent denaturation of the lower helix would free sequence within it to participate in a more complex mRNA pseudoknot structure. Alternatively, there may simply be a variety of structures that stimulate −1PRF in HIV.

Mechanisms of −1 PRF

An early hypothesis suggested that the mRNA structural elements in the −1 PRF signals acted as binding sites for a ribosomal protein, RNA, or soluble elongation factor and that such interactions might then affect the fidelity of the ribosome–tRNA interaction at the decoding sites (Jacks et al. 1988). Although this hypothesis is difficult to disprove, in vitro competition experiments and gel-shift assays have not supported it (ten Dam et al. 1994; J.D. Dinman, unpubl.). The ability of frameshift signals from viruses specific for one host to function in evolutionarily distant species also suggests that enhancement is likely to be an intrinsic feature of the downstream element itself. Furthermore, the wide range of −1 PRF-stimulating mRNA structures makes it unlikely that there is a single protein (or RNA molecule) capable of directing frameshifting. However, the stimulatory effect of proteins on frameshifting cannot be completely disregarded. For example, an iron response element (IRE) could replace an mRNA pseudoknot (Kollmus et al. 1996), and −1 PRF was stimulated by annealing of complementary RNA sequences to mRNA templates downstream from slippery sites (Howard et al. 2004; Olsthoorn et al. 2004). The artificial nature of these systems supports the hypothesis that frameshifting can be stimulated by nonspecific steric interactions between the ribosome and downstream stimulatory elements.

PRF can be viewed within the context of the translation elongation cycle. Since −1 PRF occurs while the ribosomal A and P sites are occupied by tRNAs, it must happen after delivery of aa-tRNA to the A site by eukaryotic elongation factor 1A (eEF1A), but before translocation has been completed. Although attractive, a cotranslocational slippage model (Wilson et al. 1988; Yusupova et al. 2001) is not supported by genetic and biochemical studies (for review, see Harger et al. 2002). Rather, genetic, pharmacological, and biochemical studies suggest that the −1 PRF event happens during or after accommodation, and prior to peptidyltransfer (Dinman et al. 1997; Kinzy et al. 2002; Meskauskas et al. 2003a,b; Jacobs and Dinman 2004; Petrov et al. 2004). An alternative model has been proposed suggesting that −1 PRF occurs prior to accommodation while the aa-tRNA occupies the A/T hybrid state (Leger et al. 2004), although this is contradicted by a second study positing a posttranslocation simultaneous slippage mechanism while the slippery site occupies the E/P hybrid state (Horsfield et al. 1995). However, both of these studies are complicated by their employment of the controversial HIV-1 −1 PRF signal in a phylogenetically distant *Escherichia coli* translational system.

Although the actual −1 PRF event may not happen during translocation, the large movement by the ribosome during this process may still be important to the process. A "torsional restraint" model (Dinman 1995) begins with translocation of the ribosome into the 3′ stimulatory element, directing a ribosome to pause at the slippery site. As the ribosome unwinds the base of a stem-loop structure, unwinding of the top of the structure is restricted by the presence of the second stem, forcing ribosomes to pause over at the slippery site. A recent test of this model showed that oligonucleotides capable of forming "pseudo-pseudoknots" were able to greatly stimulate −1 PRF and to direct ribosomes to pause with their A and P sites positioned over the slippery site to significantly greater degrees than a simple stem-loop (Plant and Dinman 2005). A second model, the 9 Å solution, suggests how the −1 PRF mechanism is activated. During the process of accommodation, the aa-tRNA moves in the 5′ direction by the distance occupied by one nucleotide (Fredrick and Noller 2002; Noller et al. 2002), suggesting that sequence 3′ of the A-site codon of the mRNA is "pulled into" the ribosome by one base. Positioning of a bulky mRNA pseudoknot at the leading edge of the ribosome could impede this movement, stretching the mRNA between the slippery site and the pseudoknot. Decoupling the tRNAs from the mRNA allows the mRNA to slide back the distance of one nucleotide in the 3′ direction, providing a mechanism to relieve the tension, accounting for the −1 PRF event. This model can also account for the wide variety of mRNA pseudoknot structures and other elements that can stimulate −1 PRF because the global geometry of the elements themselves is not critical. Rather, proper positioning of ribosomes at the slippery site, and hindrance of the 9 Å displacement of the mRNA during accommodation, may determine the ability to enhance frameshifting. Combination of the torsional restraint and 9 Å models provides a possible explanation for how mRNA pseudoknots might stimulate −1 PRF (Fig. 1).

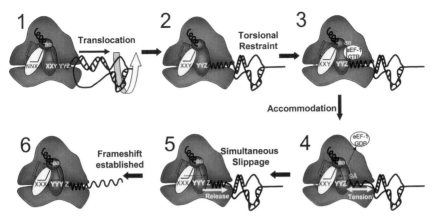

Figure 1. Torsional Restraint + the 9 Å Solution = Simultaneous Slippage. A possible mechanism of mRNA pseudoknot stimulated −1 PRF is diagrammed in a series of six steps. (*1*) During translocation, the ribosome begins to unwind the base of stem 1 of the mRNA pseudoknot. (*2*) Base-pairing in stem 2 limits the rotational freedom of stem 1, producing torsional restraint, positioning the ribosome's A and P sites at the heptameric slippery site (X XXY YYZ) of the mRNA. (*3*) The ternary complex delivers the aminoacyl-tRNA (aa-tRNA) to the 0-frame codon in the A site. (*4*) Accommodation of the aa-tRNA moves the anticodon loop, coupled with the A-site codon, by approximately 9 Å (one base) in the 5′ direction. The bulky pseudoknot resists this movement, creating a region of tension along the mRNA in the mRNA tunnel between the 3′ end of the A-site codon and the 5′ end of the pseudoknot. (*5*) Tension is released by unpairing of the A- and P-site tRNAs from their 0-frame codons, followed by slippage of a small segment of mRNA, and re-pairing of the non-wobble bases of the tRNAs with the −1 frame codons. (*6*) Eventually, the pseudoknot is denatured, the frameshift is established, and translation resumes in the new frame.

Regulation of Cellular Gene Expression by −1 PRF

Although −1 PRF was first discovered in viruses, it is logical to hypothesize that −1 PRF may be used by cellular genes. Indeed, there are a number of well-documented bacterial and archaeal examples (for review, see Baranov et al. 2002; Cobucci-Ponzano et al. 2005). Few such examples have been documented in eukaryotes, and their discovery has been serendipitous. For example, *edr* is the only well-characterized eukaryotic cellular gene known to use −1 PRF (Shigemoto et al. 2001). The *edr* −1 PRF signal is very similar to viral signals, containing a heptameric slippery site, spacer, and mRNA pseudoknot, and the frameshift is required for translation of the 3′-region portion of the mRNA (Manktelow et al. 2005). The human *edr* homolog, PEG10, also uses −1 PRF and is highly expressed in placenta (Lux et al. 2005). Edr and PEG10 are members of a large family of functional neogenes called *Mart* (mammalian retro-transposon-derived) that are widely distributed among mammals and appear to be related to the *gag* gene of the Sushi-like long terminal repeat retrotransposons (Brandt et al. 2005).

Computational Approaches

The computational challenges presented to identification of −1 PRF signals are daunting. Nonetheless, this aspect of the field has begun to emerge as computational power and sophistication have increased. Four general bioinformatic strategies for identifying −1 PRF signals have been employed: searches for overlapping reading frames (Moon et al. 2004; Bekaert et al. 2005), queries for known slippery sites (Shah et al. 2002; Wills et al. 2006), neural networks approaches (Bekaert et al. 2003), and programs designed to identify sequence and structure motifs resembling viral −1 PRF signals (Hammell et al. 1999; J.L. Jacobs et al., in prep.). The first method rests on the assumption that −1 PRF events always result in the production of carboxy-terminally extended fusion products and, thus, cannot identify new classes of frameshifted genes. The second method, although computationally fast, only represents a first approximation of potential frameshift sites and does not address the issue of 3′ stimulatory elements. The strength of the third approach is its neutrality with regard to preconceptions as to what may constitute a −1 PRF signal, although, in practice, it has run into computational hurdles that are only beginning to be overcome. The reliance of the fourth method on known stimulatory elements precludes its ability to identify new ones, although as discussed below, it has led to a new paradigm of posttranscriptional regulation of gene expression. On a positive note, these approaches are mutually complementary and, in combination, promise to unveil a new modality of posttranscriptional regulation of cellular gene expression.

At present, the field is progressing through a combination of all approaches. A two-step approach combining a search for overlapping open reading frames (ORFs) with the application of hidden Markov models identified 189 candidate genes in the *Saccharomyces cerevisiae* genome (Bekaert et al. 2005). Follow-up investigations showed that 28 of 58 candidates expressed full-length mRNAs encompassing both ORFs. Furthermore, sequences derived from 11 of these promoted highly efficient −1 PRF. Importantly, this approach revealed that the majority of candidates did not contain typical virus-like −1 PRF signals. Thus, this analysis promises to reveal new classes of −1 PRF promoting *cis*-acting elements. The bioinformatics approach designed to identify sequence and structure motifs resembling viral −1 PRF signals has also yielded surprises. The first such study identified significant numbers of putative −1 PRF signals in multiple genomes, apparently evolutionarily conserved −1 PRF signals in homologous genes from different species, and known disease alleles predicted to abolish frameshifting, and demonstrated that at least two computationally predicted motifs could promote efficient −1 PRF (Hammell et al. 1999). A surprising observation was that, in contrast to viruses, nearly all of the predicted −1 PRF events were predicted to direct elongating ribosomes into premature termination events. This generated the hypothesis that −1 PRF could be used to target mRNAs for rapid degradation via the nonsense-mediated mRNA decay pathway (Chapter 23). Proof-of-principle experiments using a viral −1 PRF signal inserted in the "genomic organization" into a cellular reporter mRNA confirmed this notion (Plant et al. 2004). Additional experiments demonstrating an inverse relationship between −1 PRF efficiency and mRNA half-lives suggested that regulation of −1 PRF could be employed to regulate gene expression posttranscriptionally. More recent investigations along these lines have aimed to boost the speed and power of the computational and predictive methods and have identified a significant number of functional −1 PRF signals that can act as mRNA destabilizing elements (J.L. Jacobs et al., in prep.).

Programmed +1 Ribosomal Frameshifting

Programmed +1 ribosomal frameshifting (+1 PRF) is the result of a net shift of the translational reading frame by one base in the 3′ direction (for review, see Farabaugh 1997). In eukaryotes, the first such example was observed in the Ty1 retrotransposable element. Sequence analysis of Ty1 cDNAs revealed a 38-bp overlap between the *TYA* (*gag*) and *TYB* (*pol*) genes, with the latter shifted into the +1 frame, and immunoblot analyses demonstrated the presence of a TYA-TYB fusion protein. Deletion analysis studies eventually reduced the frameshift signal to a 7-nucleotide sequence in the *TYA-TYB* overlap region. The Ty1 "slippery site" is CUU AGG C, where the 0 frame is indicated by spaces. Frameshifting occurs when the rare AGG codon makes the ribosome pause with the Leu-tRNA^{UAG} in the P site. Slippage of this tRNA one base in the 3′ direction allows decoding of the UUA Leu codon. Overexpression of the rare Arg-tRNA^{CCU} caused a 50-fold decrease in +1 PRF, whereas deleting it caused +1 PRF efficiency to approach 100%. The +1 frameshifts of Ty2 and Ty4, and many other members of the *copia* family of retrotransposable elements, are thought to utilize this mechanism of tRNA slippage.

The Ty3 *gypsy*-like yeast retrotransposon has a similar genome organization (Hansen et al. 1988). The +1 PRF occurs at the sequence GCG AGU U, and the inability of the 0-frame tRNA in the P site to base-pair with the +1 frame codon suggested a completely different mechanism of establishing the frameshift (for review, see Farabaugh 1997). In this case, the frameshift involves skipping the first A of the 0-frame P-site codon and recognition of the +1 frame GUU codon. Stringent analyses of this signal demonstrated that the frameshift depended on a special characteristic of the Ala-tRNA^{UGC} and that four additional tRNAs also were able to promote this. The Ty3 +1 PRF requires a downstream stimulatory element (Farabaugh et al. 1993). It has been suggested that this may base-pair with sequence in the decoding center of 18S rRNA, optimizing the positioning of the slippery site (Li et al. 2001).

Cellular Gene Expression and +1 PRF

Ornithine decarboxylase antizyme (AZ) is a negative regulator of polyamine biosynthesis, and AZ and its substrate (ornithine decarboxylase) coregulate each other in vivo (Heller and Canellakis 1981). Sequence analysis of the rat AZ revealed that translation of the full-length protein required a +1 frameshift (Miyazaki et al. 1992), and the frequency of this event was dependent on polyamine levels (Rom and Kahana 1994). This established an autoregulatory feedback loop between AZ and polyamine levels: Increased levels of polyamines increase +1 frameshifting, increasing AZ production, which in turn decreases the abundance of ornithine decarboxylase, resulting in lower polyamine levels.

Mutagenesis analyses and protein sequencing demonstrated that slippage occurs at the heptameric sequence UCC UGA U in the rat AZ mRNA, and this sequence appears to be conserved in all metazoa (for review, see Ivanov et al. 2000). The sequence is slightly more degenerate when the comparisons are made with fungi and arthropods (BYB UGA U, where B is G, C, or U and Y is a pyrimidine) (Palanimurugan et al. 2004). Changing the rat AZ P-site sequence from UCC to CCC dramatically decreased frameshifting in *S. cerevisiae*, suggesting that tRNA usage is an important difference between systems and that +1 slippage for AZ is primarily dependent on P-site interactions (for review, see Ivanov et al. 2004). AZ frameshifting is also stimulated by sequences both 5′ and 3′ of the slippery site. Their nature is not well understood because they vary extensively between orthologous AZ genes. Although almost all vertebrate AZ genes sequenced to date contain potential pseudoknots, fewer protostome AZ sequences contain predicted pseudoknots, most nematodes lack the structure, and no pseudoknots can be calculated in any yeast/fungi or insect AZ (for review, see Ivanov et al. 2004).

Polyamine levels are critical to AZ-directed frameshifting (Rom and Kahana 1994). Frameshifting efficiency from AZ genes is increased in the presence of polyamines (for review, see Ivanov et al. 2004). Similarly, synthesis of the key polyamine biosynthetic enzyme AdoMetDC is also regulated by intracellular polyamine levels (Hanfrey et al. 2005). In *S. cerevisiae*, increased levels of putrescine consequent to depletion of spermidine synthase (encoded by the yeast *SPE2* gene) were shown to increase Ty*1*-mediated +1 frameshifting from the Ty*1* frameshift (Balasundaram et al. 1994). However, examination of the polyamine biosynthetic pathway suggests that accumulation of putrescine would lead to depletion of arginine, the primary precursor of polyamines. This would decrease the abundance of Arg-tRNA^{CCU}, which in turn would promote increased frequencies of the +1 frameshift. Thus, the reported effect of polyamine depletion on Ty*1* frameshifting is likely an indirect consequence of arginine metabolism.

Other Genomic and Mitochondrial +1 Frameshift Signals

The *S. cerevisiae EST3* gene contains a +1 frameshift signal similar to that of Ty*1* (Lundblad and Morris 1997). The observations that this frameshift signal is not evolutionarily conserved (Singh et al. 2002) and that a mutant which did not require a frameshift event failed to result in any obvious growth defects suggest that frameshifting is not a regulatory feature in this context. The *ABP140* gene of *S. cerevisiae* also contains a +1 frameshift signal (Asakura et al. 1998) that is identical to the Ty*1* frameshift signal, but it has yet to be investigated in detail.

All sequenced mitochondrial *nad3* and *cytb* genes contain single-nucleotide insertions, and the high degree of sequence conservation argues that they are functional (Mindell et al. 1998; Beckenbach et al. 2005). Analyses indicate that the frameshifting mechanism involves a pause at an AGY codon. No other conserved sequences or predicted structures have been identified, and P-site slippage has been suggested based on mitochondrial tRNA content and codon preferences (Beckenbach et al. 2005). In *Euplotes*, partial mass spectrometry analysis was first used to detect a frameshifted telomerase protein (Aigner et al. 2000). Sequence analysis of several nuclear protein kinases and transposons suggested that frameshifting occurs at AAA UAA A (Tan et al. 2001). The presence of insertions and deletions resulting in the requirement for +1 frameshifting has also been described for a number of genes in this genus (Mollenbeck et al. 2004).

Factors Affecting Frameshifting

PRF can be used to investigate how the translational apparatus normally maintains the translational reading frame, to identify targets for new classes of antiviral agents, and to investigate mechanisms of posttranslational regulation of gene expression. PRF has been particularly useful as a probe of ribosome structure and function. This ongoing work has revealed that both −1 and +1 PRF can be influenced by the affinity of ribosomes for peptidyl-tRNAs (Meskauskas and Dinman 2001). Furthermore, −1 PRF can be specifically influenced by changes in (1) ribosome affinities for aa-tRNA, (2) peptidyltransferase activity, and (3) aa-tRNA accommodation rates (Dinman et al. 1997; Peltz et al. 1999; Kinzy et al. 2002; Meskauskas et al. 2003b, 2005; Petrov et al. 2004). Ongoing molecular genetic studies involving mutagenesis of ribosomal proteins and rRNAs coupled with biochemical and rRNA structural probing methods are helping to link function to ribosome structure (see, e.g., Dontsova and Dinman 2005). In particular, studies using mutants of ribosomal protein L3, rRNA methylation enzymes, and 25S rRNA are beginning to suggest that the topology of the A and P sites of the peptidyltransferase center may differentially influence −1 PRF efficiency and virus propagation (Meskauskas et al. 2005; J.D. Dinman, in prep.). *trans*-Acting factors can also influence PRF, including the elongation factors eEF1A and eEF2 (for review, see Harger et al. 2002), the histone deacetylase *RPD3* (Meskauskas et al. 2003b), and the ribosome-associated chaperone complex (Muldoon-Jacobs and Dinman 2006). Small-molecule antibiotics have also been implicated in −1 PRF. Anisomycin, which inhibits accommodation by competing with the acceptor stem of aa-tRNA for the A-site binding pocket (Hansen et al. 2003), specifically inhibits −1 PRF (Dinman et al. 1997; Jacobs and Dinman 2004), lending support to the concept that −1 PRF occurs after this step of the elongation program. In contrast, sparsomycin, which inhibits peptidyltransfer by binding to and displacing the peptidyl-tRNA (Hansen et al. 2003), stimulates −1 PRF (Dinman et al. 1997). Both of these drugs promote loss of the yeast killer virus, supporting the idea of −1 PRF as a target for antiviral therapies (Dinman et al. 1997, 1998). This idea is further explored in Chapter 30.

Termination Suppression

As discussed in Chapter 7, both *cis*-acting sequences and *trans*-acting factors contribute to the efficiency of termination codon recognition by the translational apparatus. Programmed suppression of termination codons represents another common solution that viruses have evolved to solve the problem of regulating the ratios of structural to enzymatic proteins. Studies of the translational readthrough signal of murine leukemia virus (MLV) have identified sequence and structural features, including the requirement for an mRNA pseudoknot (for review, see Farabaugh 1997). The demonstration that MLV reverse transcriptase can interact with the eukaryotic release factor-1 (eRF1) suggests that depletion of this termination factor is used to enhance the frequency of translational readthrough (Goff 2004). Analysis of numerous other examples of programmed translational readthrough (for review, see Maia et al. 1996) suggest that the general rules pertaining to the makeup and context of termination codons that govern termination efficiency (Bonetti et al. 1995) have been adopted by many other RNA viruses. Examples of translational readthrough in cellular mRNAs are also accumulating. A computational analysis of the yeast genome identified eight genes containing "poor context" termination signals (Namy et al. 2002). Follow-up studies demonstrated that termination suppression in one of these, the PDE2 mRNA, is used to modulate cAMP levels. A few examples of translational readthrough have also been documented in *Drosophila: oaf* (Bergstrom et al. 1995), *hdc* (Stenberg et al. 1998), and *keltch* (Xue and Cooley 1993). Regulation of translational readthrough has implicated all three of these genes in *Drosophila* development. In addition, approximately 4% of predicted mRNAs in the mouse genome contain premature termination codons, suggesting that this mechanism is widely used to posttranscriptionally regulate vertebrate gene expression (Xing and Lee 2004). Thus, similar to PRF signals, these emerging findings suggest that termination suppression may also represent a significant mode for the posttranscriptional regulation of gene expression.

Selenocysteine: The 21st Amino Acid

The defining feature of the selenoprotein family is the presence of the 21st amino acid, selenocysteine. Selenoproteins have sparked the interest of researchers in the translation field, due to their unusual mechanisms of biosynthesis, which have yielded new insights into termination of translation and the nonsense-mediated decay pathway. The selenoprotein family has also attracted considerable attention of researchers in other fields, through the elucidation of essential functions of several selenoproteins, and the characterization of selenoproteomes from a wide diversity of organisms. Finally, the functions of selenoproteins in antioxidant protection, cellular redox balance, male fertility, and thyroid hormone metabolism have important implications for human health and disease. In the last few years, researchers in all of these areas have contributed greatly to our understanding of this intriguing protein family, while at the same time unveiling many new questions for future studies.

Selenium and Disease

The earliest information on the effects of selenium in diet pointed to its toxicity when consumed in excess. The first description of what was likely to be selenium poisoning is attributed to Marco Polo's 13th century writings of a necrotic hoof disease of horses in China. Association of the disease with selenium was defined in the 1930s and 1940s, when the symptoms of chronic selenosis in livestock (hair loss and hoof lesions) were attributed to grazing on plants that accumulate selenium from soil. Subsequent studies revealed selenium to be an essential trace element, with its deficiency in animal diets being implicated in liver necrosis and male infertility. Most of the United States and many other regions of the world contain adequate selenium in the soil, with a few notable exceptions. Selenium deficiency has been directly linked to a fatal cardiomyopathy termed Keshan disease, named for the province in China where it was identified; an osteoarthropathy known as Kashin-Beck disease, found in selenium-deficient regions of China and Russia; and myxedematous endemic cretinism in Zaire. Selenium depletion and supplementation studies in animal models, as well as biochemical and molecular biological approaches to elucidate function, including targeted gene disruption in mouse models, have revealed roles for specific selenoproteins in antioxidant defenses, thyroid hormone metabolism, spermatogenesis, neuronal development and function, and many other biological processes. In fact, disruption of the entire repertoire of selenoproteins results in early embryonic lethality, underscoring the essential nature of selenium in life (Bosl et al. 1997). Finally, human epidemiological and supplementation studies provide compelling evidence for the efficacy of selenium in the prevention and treatment of some forms of cancer (Clark et al. 1996 and references therein).

Incorporation of selenium into proteins requires specific signals in the selenoprotein mRNAs, as well as *trans*-acting factors that appear to have evolved specifically for the unique processes of selenocysteine biosynthesis and incorporation. Selenocysteine is found in proteins in the prokaryotic, archaeal, and eukaryotic kingdoms. The three kingdoms employ common features as well as numerous differences in their selenoprotein biosynthesis pathways. Among the common features are (1) the use of UGA, typically read as a stop codon, to specify the position of selenocysteine in the coding region; (2) the utilization of a unique tRNA species, tRNA$^{(Ser)Sec}$, which serves as both the site of selenocysteine biosynthesis and the means of delivering the amino acid to the ribosome; and (3) the requirement for specialized structures in selenoprotein mRNAs, which serve to recruit and deliver the factors that mediate recoding of UGA from "stop" to sense. Distinguishing features among the different kingdoms include the selenocysteine biosynthesis pathways, the specific features and locations of the mRNA structures, and the identity and functions of the factors they recruit. Furthermore, the presence of nuclei in the eukarya introduces additional complexities into the pathway. Finally, although widely employed, selenocysteine incorporation is not universal. Sequencing of the genomes of yeast and higher plants, as well as numerous species of bacteria and archaea, revealed the absence in these organisms of components essential for selenocysteine incorporation.

Selenocysteine Biosynthesis

The factors involved in selenocysteine biosynthesis in prokaryotes were first identified through genetic studies in the Böck laboratory (Leinfelder et al. 1988; Böck 2000). *selC* encodes tRNA$^{(Ser)Sec}$, a unique gene product with an anticodon complementary to UGA. The tRNA is initially charged with serine by seryl-tRNA synthetase and then serves as the substrate for conversion of the seryl moiety into the selenocysteinyl product. This conversion is catalyzed by the *selA* gene product, selenocysteine synthase. *selD* encodes selenophosphate synthetase, an enzyme catalyzing the synthesis of the active selenium donor, monoselenophosphate, from selenide and ATP. Conversion to selenocysteine proceeds through a selenocysteine-synthase-enzyme-linked dehydroalanine intermediate. The pathway in archaea has not been fully elucidated, but the enzymatic components appear to be similar to those in prokaryotes.

In eukaryotes, a different picture has emerged from studies in several laboratories. Although the tRNA is conserved, both the seryl to selenocysteinyl-tRNA conversion and selenophosphate biosynthesis pathways in eukaryotes differ from their prokaryotic counterparts. Following the identification of a human selenophosphate synthase and demonstration of its role in selenoprotein synthesis in transfected mammalian cells, a second such enzyme was identified and shown to be a selenoenzyme. The two enzymes were subsequently designated SPS1 and SPS2. Biochemical studies suggest that SPS2 may be the more crucial form for selenocysteine biosynthesis, with SPS1 possibly functioning in recycling selenite from selenocysteine (Tamura et al. 2004). Recent studies have elucidated a further difference, the presence of a phosphoseryl kinase in eukaryotes that phosphorylates seryl-tRNA$^{(Ser)Sec}$ (Carlson et al. 2004). This form may actually serve as the substrate for a putative selenocysteine synthase. A recent report has identified a similar pathway for cysteinyl-tRNA biosynthesis in archaea, which proceeds through a phosphointermediate (Sauerwald et al. 2005). Finally, tRNA$^{(Ser)Sec}$ in eukaryotes undergoes several nucleoside modifications at positions including the wobble base and the base adjacent to it. Generation of mutant forms of tRNA$^{(Ser)Sec}$ unable to accommodate these modifications, and introduction of these tRNAs into transgenic animals in both a whole-animal and tissue-specific manner, have provided new insights into the importance of these modifications. Strikingly, different patterns of selenoproteins are produced in the presence versus the absence of a specific wobble-base methylation (Carlson et al. 2005; Xu et al. 2005), suggesting differential decoding by an as yet unknown mechanism.

Selenocysteine Incorporation

As indicated above, the sequences in selenoprotein mRNAs that recode UGA codons, termed SECIS elements, differ among the three kingdoms in their secondary and tertiary structures, and their location within the mRNAs. SECIS elements are found immediately 3′ of the UGA codons in bacteria, in the 3′ or 5′ untranslated regions (UTRs) in archaea, and in the 3′UTRs in eukaryotes. In all cases, they function ultimately in delivering sec-tRNA$^{(Ser)Sec}$ to the ribosome, albeit via different mechanisms. In prokaryotes, sec-tRNA$^{(Ser)Sec}$ is bound by the specialized elongation factor, SelB. The carboxy-terminal domain of SelB interacts with the bacterial SECIS element, serving both to recruit sec-tRNA$^{(Ser)Sec}$ to the selenoprotein mRNA and to deliver it to the ribosomal A site, as depicted in Figure 2A. Binding of the charged tRNA increases the affinity of SelB for the SECIS RNA, whereas delivery of the tRNA decreases affinity for the SECIS, leading to dissociation of SelB from the mRNA, a requirement for translation of codons forming this structure.

Specialized elongation factors for sec-tRNA$^{(Ser)Sec}$ were subsequently identified in archaea and eukaryotes (Fagegaltier et al. 2000; Rother et al. 2000; Tujebajeva et al. 2000), but further studies revealed differences in strategies for recruiting these factors, presumably a consequence of the distal location of SECIS elements from the UGA codons. The eukaryotic factor, eEFsec, contains a carboxy-terminal extension analogous to that in bacterial SelB, but this extension does not bind the SECIS element. Independent studies identified a SECIS-binding protein, SBP2 (Copeland et al. 2000), which was subsequently shown to interact with eEFsec. Further experiments in transfected cells showed that the SBP2–eEFsec

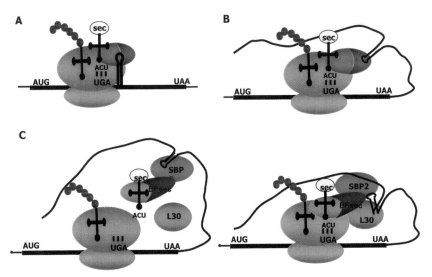

Figure 2. Models for selenocysteine incorporation in prokarya, archaea, and eukarya. (*Aqua*) mRNAs; (*purple*) ribosomes; (*yellow*) tRNAs; (*pink*) nascent peptide; (*black*) codons. Prokaryal (*A*) and archaeal (*B*) SELB are depicted as red and blue ovals, with blue representing the elongation factor domain and red representing the SECIS-binding domain. Eukaryal EFsec is depicted in light blue and dark blue, with light blue representing the elongation factor domain and dark blue representing the SBP2 interaction domain (*C*). (*Red*) SBP2; (*green*) L30. The kink-turn in the SECIS element is depicted in the right panel. (Reprinted, with permission, from Berry 2005 [Nature Publishing Group].)

interaction was greatly enhanced when the tRNA$^{(Ser)Sec}$ was cotransfected, providing a means for ensuring that eEFsec was only recruited when complexed with sec-tRNA$^{(Ser)Sec}$ (Zavacki et al. 2003).

The 3′ or 5′ location of archaeal SECIS elements might have suggested separate factors for SECIS binding and tRNA$^{(Ser)Sec}$ delivery, as in eukaryotes. However, the search for archaeal SBP2 homologs did not yield such a factor. Recent studies elucidating the crystal structure of archaeal SelB have shed new light on this dilemma, with the identification of a potential RNA-binding domain in the carboxy-terminal extension of archaeal SELB, analogous to the bacterial protein (Leibundgut et al. 2005). Modeling this protein on the ribosome, the putative SECIS-binding domain points toward the 3′ mRNA entrance cleft, positioning it to interact with the downstream secondary structure, as depicted in Figure 2B. Thus, archaea appear to exhibit the tethering feature of the prokaryotic selenocysteine incorporation mechanism, while at the same time utilizing distal SECIS elements for recoding, as in eukaryotes.

SECIS-binding Protein 2

Following the initial studies identifying the eukaryotic SECIS element, site-directed mutagenesis and deletion mapping were employed to delineate features crucial for function. Identifying these features provided both a ligand to be used in searches for cognate binding proteins as well as a tool for in silico approaches to identifying selenoproteomes (see below). On the basis of structural probing in solution, the Krol laboratory proposed a three-dimensional model for the SECIS RNA where the phosphodiester backbone is bent at conserved non-Watson-Crick base pairs (Walczak et al. 1996). The proposed folding of the SECIS RNA suggested that it could form a canonical kink turn or kink-turn-related structure.

After several false leads from other laboratories, the efforts of Copeland and Driscoll proved fruitful in identifying SBP2, a protein that specifically interacts with conserved features of the SECIS structure, including the conserved non-Watson-Crick quartet (Lesoon et al. 1997; Copeland and Driscoll 1999). SBP2 was shown to be limiting for Sec incorporation in vitro and in vivo (Copeland et al. 2000). Subsequent studies resulted in mapping of the SECIS RNA-binding domain in the carboxy-terminal region (amino acids 517–777), and delineation of a central functional domain required for Sec incorporation in vitro but not for RNA-binding activity (amino acids 399–517). The amino-terminal domain was found to be dispensable for UGA recoding in vitro and in transfected cells. Interestingly, this region of the protein is found in higher eukaryotes, including primates, rodents, and fish, but is absent in flies and nematodes, organisms with minimal selenoproteome complexity. The amino-terminal region also contains multiple putative localization signals that may be of importance to its function in higher eukaryotes but not lower eukaryotes.

The SECIS-binding domain was found to contain an L7Ae RNA-binding motif, named after the archaeal ribosomal protein L7Ae. This motif is also found in ribosomal proteins L30 and S12, 15.5-kD spliceosomal protein, eukaryotic release factor-1, and other functionally unrelated proteins (Allmang et al. 2002). Cocrystal structures of several of these proteins in complex with their cognate RNAs revealed that the proteins fold into a highly conserved domain that binds specifically to RNAs possessing a kink-turn motif, the motif proposed for the SECIS element by Krol, based on earlier structure studies. The RNA-binding domain of SBP2 is considerably larger than the core L7Ae motif, thus additional sequences in SBP2 may be required for SECIS-binding activity and/or specificity.

Ribosomal Protein L30

The discoveries of SBP2 and eEFsec as two distinct proteins serving the functions of prokaryotic or archaeal SelB, i.e., SECIS binding and sec-tRNA$^{(Ser)Sec}$ binding, shed new light on the eukaryotic mechanism, but the question of how recoding is mediated from a distance remains unsolved. An additional piece of this puzzle has fallen into place with the identification of ribosomal protein L30 as a SECIS-binding protein. L30 was identified in UV cross-linking assays as a SECIS-binding activity and purified by RNA affinity chromatography using a SECIS element as ligand, the same approach that previously resulted in identification of SBP2. The non-Watson-Crick quartet was shown to be essential for L30 binding to the SECIS, indicating overlapping specificity with SBP2. Mass spectrometry analysis of the purified protein identified it as L30 (Chavatte et al. 2005). L30 is an essential component of the large ribosomal subunit in eukaryotes, but little is known about its function in protein synthesis. L30 interacts with the large ribosomal subunit by binding to kink-turn motifs in 28S rRNA (Halic et al. 2005). This invites the question of whether L30 binds simultaneously to the ribosome and to the SECIS element or if the interactions are mutually exclusive. Purified recombinant L30 binds to SECIS elements in vitro, and the endogenous protein is bound to selenoprotein mRNAs in vivo. In addition, overexpression of L30 in transfected cells stimulated UGA recoding activity, implying a role for the protein in selenocysteine incorporation.

Demonstration that L30 and SBP2 have overlapping specificity for SECIS elements suggested that they might interact with the RNA in either a cooperative or competitive manner. The latter was shown to be the case, as the two proteins do not bind the same RNA molecule simultaneously (Chavatte et al. 2005). At low magnesium concentrations, SBP2 is preferentially bound, whereas higher magnesium results in a shift in favor of L30 binding. Magnesium is known to promote formation of kink-turn structures, consistent with the finding that this motif is recognized by L30. On the basis of these findings, Chavatte et al. proposed that the SECIS element acts as a molecular switch, alternating between SBP2 and L30 binding during the recoding process. These investigators have further shown that SBP2 and L30 have the ability to displace each other from preformed protein:SECIS complexes. These findings suggest a mechanism in which L30 and SBP2 may bind and act sequentially during UGA recoding to recruit eEFsec and deliver sec-tRNA$^{(Ser)Sec}$ to the ribosomal A site, as depicted in Figure 2C. Thus, the functions of prokaryotic SelB appear to be carried out by at least three eukaryotic proteins, although the mechanistic details have not been elucidated. Finally, no function has yet been identified for a conserved region in the apical loop of SECIS elements typically composed of an AAR motif. Additional studies will be needed to resolve the role of this motif and any proteins it may recruit.

Circumventing Translation Termination

Although the efforts of a number of laboratories have focused on elucidating the mechanism of UGA recoding, the means by which termination is circumvented at these sites has also remained elusive. Studies to unravel this mystery initially focused primarily on context effects—the ability of sequences in proximity to UGA codons to differentially affect readout. These include the nucleotides immediately downstream from the UGA codon and the codons upstream. In both prokaryotes and eukaryotes, recognition of termination codons by release factors (Chapter 7) is influenced by downstream context (Poole et al. 1995; Liu et al. 1999; Mansell et al. 2001). Likewise, interactions between the tRNAs occupying the A and P sites of the ribosome may explain the differential incorporation versus termination efficiencies observed with different codons immediately upstream of UGA (Grundner-Culemann et al. 2001). Recent studies have unveiled an additional context effect, the potential in some selenoprotein mRNAs to form stem-loop structures downstream from the UGA codon, analogous to the stem-loops or pseudoknots downstream from many frameshift sites. These putative structures have been identified in the mRNAs for several selenoproteins, and comparative sequence analyses and mutagenesis studies support a role for these structures in increasing decoding efficiency (Howard et al. 2005).

Nucleocytoplasmic Shuttling of Selenocysteine Incorporation Factors: A Role in Eluding Nonsense-mediated Decay?

The detrimental effects of termination at UGA selenocysteine codons include not only the production of truncated proteins due to inefficient incorporation, but also the potential for the mRNA to be eliminated early in its life via nonsense-mediated decay (NMD; Chapter 23). Selenoprotein mRNAs have been shown to be sensitive to NMD under conditions of inefficient decoding, as occurs upon selenium deprivation (Moriarty et al. 1998; Weiss and Sunde 1998). However, even when selenium stores are sufficient, these mRNAs must still recruit all of the necessary factors for UGA decoding in order to avoid NMD. Furthermore, they presumably must do so prior to an early proofreading round of translation, at which time premature termination codons are typically recognized as such and the corresponding mRNAs targeted for degradation (Ishigaki et al. 2001).

Recent studies have provided intriguing new insights into how this early recruitment may take place. SBP2 and eEFsec both contain nuclear localization and nuclear export signals, and the minimal functional domain of SBP2 has been shown to undergo nucleocytoplasmic shuttling (de Jesus et al. 2006). Furthermore, coexpression of this protein results in increased nuclear localization of eEFsec, possibly through increased retention upon interaction with SBP2. Thus, these two components of the decoding machinery have the potential to interact with selenoprotein mRNAs within the nucleoplasm.

Additional factors implicated in selenocysteine biosynthesis and incorporation have also been shown to undergo nuclear localization. SECp43, the putative methylase that modifies the wobble base in sec-tRNA$^{(Ser)Sec}$, not only exhibits nuclear localization itself, but its expression results in redistribution of another factor, SLA/LP, to the nucleus (Xu et al. 2005; Small-Howard et al. 2006). SLA/LP has been speculated, based on structural homology, to function in the seryl to the sec-tRNA$^{(Ser)Sec}$ conversion reaction (Sauerwald et al. 2005). Nuclear localization or shuttling of these factors implicated in generating mature sec-tRNA$^{(Ser)Sec}$ may thus also contribute to nuclear assembly of selenocysteine incorporation complexes and circumvention of NMD (Small-Howard et al. 2006).

Evolution of Selenoproteins: Lessons from Bioinformatics and Selenoproteomes

The brief overview provided above gives only a glimpse of the considerable efforts aimed at elucidating the mechanisms of selenoprotein synthesis in the three kingdoms of life. Much of this information has been obtained through bioinformatics approaches to identify new factors based on homology with known factors in different taxa. Use of bioinformatics has also yielded valuable information on the complexity of selenoproteomes in different organisms. This information, in turn, is shedding light on the evolution of selenoproteomes.

SECIS search programs developed using consensus eukaryotic SECIS element features were initially utilized to search several of the lower eukaryotic genomes, including *Caenorhabditis elegans* and *Drosophila melanogaster*. These approaches identified only one and three selenoprotein genes in the nematode and fruit fly genomes, respectively (Castellano et al. 2001; Taskov et al. 2005). More recently, the same approach has identified 25 selenoprotein genes in the mouse and human genomes

(Kryukov et al. 2003). Finally, the widespread application of these search programs to an increasing number of eukaryotic genomes is revealing what has been referred to as a "mosaic-like" evolution, whereby the selenoproteomes of different taxa exhibit wide variation in the number and types of selenoprotein families (Castellano et al. 2005). Examples of this include the presence of a single true selenoprotein but multiple cysteine-containing homologs of selenoproteins in *C. elegans*, the restriction of methionine sulfoxide reductase as a selenoprotein to the alga, *Chlamydomonas reinhardtii*, and the recent identification of two novel selenoproteins, SelJ and SelU, both found only in fish. SelU has a widespread distribution of cysteine homologs among eukaryotes, whereas SelJ has no known homologs, either cysteine or selenocysteine-containing, in any other taxa, making it truly unique to fish. Future studies will undoubtedly unveil further surprises and exciting insights into the evolution and functions of this intriguing family of proteins.

Pyrrolysine: The 22nd Amino Acid

The discovery of pyrrolysine in archaea and bacteria, encoded by UAG, resulted in its designation as the 22nd amino acid (Atkins and Gesteland 2002; Hao et al. 2002; Srinivasan et al. 2002). Although many questions remain regarding the selenocysteine incorporation pathway, studies on the mechanism of pyrrolysine incorporation are in their infancy, but clearly the two new additions to the genetic code exhibit a number of differences. First, in contrast to the wider distribution of selenocysteine, pyrrolysine has only been identified in a small number of methanogenic archaea and a few other microbes. Second, unlike selenocysteine, pyrrolysine appears to be ligated directly to its cognate tRNA, and this tRNA is recognized by the standard elongation factor EF-Tu. A third distinction may lie in how UAG is decoded as pyrrolysine. Three possibilities that have been put forth are (1) redefinition of a subset of UAG codons by *cis*-acting mRNA signals, (2) reassignment of all UAG codons within an organism to pyrrolysine, and (3) competition between readthrough and termination, resulting in either event at any given UAG codon. Recently, evidence has been presented for both of the first two possibilities (Theobald-Dietrich et al. 2005; Zhang et al. 2005). Strikingly, the total usage of UAG codons is under 5% in the archaea species that incorporate pyrrolysine, minimizing the potential deleterious effects of whole organism reassignment. Whether a single or multiple strategies are employed for incorporation of the 22nd amino acid in different organisms will certainly be the focus of considerable future efforts.

ACKNOWLEDGMENTS

The authors acknowledge members of the translational recoding field for their invaluable input. This work was supported by grants from the U.S. Public Health Service and the National Science Foundation to J.D.D. and M.J.B.

REFERENCES

Aigner S., Lingner J., Goodrich K.J., Grosshans C.A., Shevchenko A., Mann M., and Cech T.R. 2000. *Euplotes* telomerase contains an La motif protein produced by apparent translational frameshifting. *EMBO J.* **19:** 6230–6239.

Allmang C., Carbon P., and Krol A. 2002. The SBP2 and 15.5 kD/Snu13p proteins share the same RNA binding domain: Identification of SBP2 amino acids important to SECIS RNA binding. *RNA* **8:** 1308–1318.

Asakura T., Sasaki T., Nagano F., Satoh A., Obaishi H., Nishioka H., Imamura H., Hotta K., Tanaka K., Nakanishi H., and Takai Y. 1998. Isolation and characterization of a novel actin filament-binding protein from *Saccharomyces cerevisiae*. *Oncogene* **16:** 121–130.

Atkins J.F. and Gesteland R. 2002. Biochemistry. The 22nd amino acid. *Science* **296:** 1409–1410.

Balasundaram D., Dinman J.D., Wickner R.B., Tabor C.W., and Tabor H. 1994. Spermidine deficiency increases +1 ribosomal frameshifting efficiency and inhibits Ty1 retrotransposition in *Saccharomyces cerevisiae*. *Proc. Natl. Acad. Sci.* **91:** 172–176.

Baranov P.V., Gesteland R.F., and Atkins J.F. 2002. Recoding: Translational bifurcations in gene expression. *Gene* **286:** 187–201.

Baril M., Dulude D., Steinberg S.V., and Brakier-Gingras L. 2003. The frameshift stimulatory signal of human immunodeficiency virus type 1 group O is a pseudoknot. *J. Mol. Biol.* **331:** 571–583.

Barry J.K. and Miller W.A. 2002. A −1 ribosomal frameshift element that requires base pairing across four kilobases suggests a mechanism of regulating ribosome and replicase traffic on a viral RNA. *Proc. Natl. Acad. Sci.* **99:** 11133–11138.

Beckenbach A.T., Robson S.K., and Crozier R.H. 2005. Single nucleotide +1 frameshifts in an apparently functional mitochondrial cytochrome b gene in ants of the genus *Polyrhachis*. *J. Mol. Evol.* **60:** 141–152.

Bekaert M., Richard H., Prum B., and Rousset J.P. 2005. Identification of programmed translational −1 frameshifting sites in the genome of *Saccharomyces cerevisiae*. *Genome Res.* **15:** 1411–1420.

Bekaert M., Bidou L., Denise A., Duchateau-Nguyen G., Forest J.P., Froidevaux C., Hatin I., Rousset J.P., and Termier M. 2003. Towards a computational model for −1 eukaryotic frameshifting sites. *Bioinformatics* **19:** 327–335.

Bergstrom D.E., Merli C.A., Cygan J.A., Shelby R., and Blackman R.K. 1995. Regulatory autonomy and molecular characterization of the *Drosophila* out at first gene. *Genetics* **139:** 1331–1346.

Böck A. 2000. Biosynthesis of selenoproteins: An overview. *Biofactors* **11:** 77–78.

Bonetti B., Fu L., Moon J., and Bedwell D.M. 1995. The efficiency of translation termination is determined by a synergistic interplay between upstream and downstream sequences in *Saccharomyces cerevisiae*. *J. Mol. Biol.* **251:** 334–345.

Bosl M.R., Takaku K., Oshima M., Nishimura S., and Taketo M.M. 1997. Early embryonic lethality caused by targeted disruption of the mouse selenocysteine tRNA gene (Trsp). *Proc. Natl. Acad. Sci.* **94:** 5531–5534.

Brandt J., Schrauth S., Veith A.M., Froschauer A., Haneke T., Schultheis C., Gessler M., Leimeister C., and Volff J.N. 2005. Transposable elements as a source of genetic innovation: Expression and evolution of a family of retrotransposon-derived neogenes in mammals. *Gene* **345:** 101–111.

Brierley I. 1995. Ribosomal frameshifting on viral RNAs. *J. Gen. Virol.* **76:** 1885–1892.

Brierley I. and Dos Ramos F.J. 2006. Programmed ribosomal frameshifting in HIV-1 and the SARS-CoV. *Virus Res.* **119:** 29–42.

Brierley I.A., Dingard P., and Inglis S.C. 1989. Characterization of an efficient coronavirus ribosomal frameshifting signal: Requirement for an RNA pseudoknot. *Cell* **57:** 537–547.

Carlson B.A, Xu X.M, Kryukov G.V., Rao M., Berry M.J., Gladyshev V.N., and Hatfield D.L. 2004. Identification and characterization of phosphoseryl-tRNA[Ser]Sec kinase. *Proc. Natl. Acad. Sci.* **101:** 12848–12853.

Castellano S., Morozova N., Morey M., Berry M.J., Serras F., Corominas M., and Guigo R. 2001. In silico identification of novel selenoproteins in the *Drosophila melanogaster* genome. *EMBO Rep.* **2:** 697–702.

Castellano S., Lobanov A.V., Chapple C., Novoselov S.V., Albrecht M., Hua D., Lescure A., Lengauer T., Krol A., Gladyshev V.N., and Guigo R. 2005. Diversity and functional plasticity of eukaryotic selenoproteins: Identification and characterization of the SelJ family. *Proc. Natl. Acad. Sci.* **102:** 16188–16193.

Chambers I., Frampton J., Goldfarb P., Affara N., McBain W., and Harrison P.R. 1986. The structure of the mouse glutathione peroxidase gene: The selenocysteine in the active site is encoded by the "termination" codon, TGA. *EMBO J.* **5:** 1221–1227.

Chavatte L., Brown B.A., and Driscoll D.M. 2005. Ribosomal protein L30 is a component of the UGA-selenocysteine recoding machinery in eukaryotes. *Nat. Struct. Mol. Biol.* **12:** 408–416.

Clark L.C., Combs G.F., Jr., Turnbull B.W., Slate E.H., Chalker D.K., Chow J., Davis L.S., Glover R.A., Graham G.F., Gross E.G., et al. 1996. Effects of selenium supplementation for cancer prevention in patients with carcinoma of the skin. A randomized controlled trial. Nutritional Prevention of Cancer Study Group. *J. Am. Med. Assoc.* **276:** 1957–1963.

Cobucci-Ponzano B., Rossi M., and Moracci M. 2005. Recoding in archaea. *Mol. Microbiol.* **55:** 339–348.

Copeland P.R. and Driscoll D.M. 1999. Purification, redox sensitivity, and RNA binding properties of SECIS-binding protein 2, a protein involved in selenoprotein biosynthesis. *J. Biol. Chem.* **274:** 25447–25454.

Copeland P.R., Fletcher J.E., Carlson B.A., Hatfield D.L., and Driscoll D.M. 2000. A novel RNA binding protein, SBP2, is required for the translation of mammalian selenoprotein mRNAs. *EMBO J.* **19:** 306–314.

Cornish P.V., Hennig M., and Giedroc D.P. 2005. A loop 2 cytidine-stem 1 minor groove interaction as a positive determinant for pseudoknot-stimulated −1 ribosomal frameshifting. *Proc. Natl. Acad. Sci.* **102:** 12694–12699.

de Jesus L.A., Hoffmann P.R., Michaud T., Forry E.P., Small-Howard A., Stillwell R.J., Morozova N., Harney J.W., and Berry M.J. 2006. Nuclear assembly of UGA decoding complexes on selenoprotein mRNAs: A mechanism for eluding nonsense mediated decay? *Mol. Cell. Biol.* **26:** 1795–1805.

Dinman J.D. 1995. Ribosomal frameshifting in yeast viruses. *Yeast* **11**: 1115–1127.

Dinman J.D., Ruiz-Echevarria M.J., and Peltz S.W. 1998. Translating old drugs into new treatments: Identifying compounds that modulate programmed −1 ribosomal frameshifting and function as potential antiviral agents. *Trends Biotechnol.* **16**: 190–196.

Dinman J.D., Ruiz-Echevarria M.J., Czaplinski K., and Peltz S.W. 1997. Peptidyl transferase inhibitors have antiviral properties by altering programmed −1 ribosomal frameshifting efficiencies: Development of model systems. *Proc. Natl. Acad. Sci.* **94**: 6606–6611.

Dinman J.D., Richter S., Plant E.P., Taylor R.C., Hammell A.B., and Rana T.M. 2002. The frameshift signal of HIV-1 involves a potential intramolecular triplex RNA structure. *Proc. Natl. Acad. Sci.* **99**: 5331–5336.

Dontsova O.A. and Dinman J.D. 2005. 5S rRNA: Structure and function from head to toe. *Int. J. Biomed. Sci.* **1**: 2–7.

Egli M., Minasov G., Su L., and Rich A. 2002. Metal ions and flexibility in a viral RNA pseudoknot at atomic resolution. *Proc. Natl. Acad. Sci.* **99**: 4302–4307.

Fagegaltier D., Hubert N., Yamada K., Mizutani T., Carbon P., and Krol A. 2000. Characterization of mSelB, a novel mammalian elongation factor for selenoprotein translation. *EMBO J.* **19**: 4796–4805.

Farabaugh P.J. 1997. *Programmed alternative reading of the genetic code.* R.G. Landes Company, Austin, Texas.

Farabaugh P.J., Zhao H., and Vimaladithan A. 1993. A novel programmed frameshift expresses the Pol3 gene of retrotransposon-Ty3 of yeast: Frameshifting without transfer-RNA slippage. *Cell* **74**: 93–103.

Fredrick K. and Noller H.F. 2002. Accurate translocation of mRNA by the ribosome requires a peptidyl group or its analog on the tRNA moving into the 30S P site. *Mol. Cell* **9**: 1125–1131.

Gaudin C., Mazauric M.H., Traikia M., Guittet E., Yoshizawa S., and Fourmy D. 2005. Structure of the RNA signal essential for translational frameshifting in HIV-1. *J. Mol. Biol.* **349**: 1024–1035.

Giedroc D.P., Theimer C.A., and Nixon P.L. 2000. Structure, stability and function of RNA pseudoknots involved in stimulating ribosomal frameshifting. *J. Mol. Biol.* **298**: 167–185.

Goff S.P. 2004. Genetic reprogramming by retroviruses: Enhanced suppression of translational termination. *Cell Cycle* **3**: 123–125.

Gramstat A., Prufer D., and Rohde W. 1994. The nucleic acid-binding zinc-finger protein of potato-virus-M is translated by internal initiation as well as by ribosomal frameshifting involving a shifty stop codon and a novel mechanism of P-site slippage. *Nucleic Acids Res.* **22**: 3911–3917.

Grundner-Culemann E., Martin G.W., III, Tujebajeva R., Harney J.W., and Berry M.J. 2001. Interplay between termination and translation machinery in eukaryotic selenoprotein synthesis. *J. Mol. Biol.* **310**: 699–707.

Halic M., Becker T., Frank J., Spahn C.M., and Beckmann R. 2005. Localization and dynamic behavior of ribosomal protein L30e. *Nat. Struct. Mol. Biol.* **12**: 467–468.

Hammell A.B., Taylor R.L., Peltz S.W., and Dinman J.D. 1999. Identification of putative programmed −1 ribosomal frameshift signals in large DNA databases. *Genome Res.* **9**: 417–427.

Hanfrey C., Elliott K.A., Franceschetti M., Mayer M.J., Illingworth C., and Michael A.J. 2005. A dual upstream open reading frame-based autoregulatory circuit controlling polyamine-responsive translation. *J. Biol. Chem.* **280**: 39229–39237.

Hansen J.L., Moore P.B., and Steitz T.A. 2003. Structures of five antibiotics bound at the peptidyl transferase center of the large ribosomal subunit. *J. Mol. Biol.* **330:** 1061–1075.

Hansen L.J., Chalker D.L., and Sandmeyer S.B. 1988. Ty3, a yeast retrotransposon associated with tRNA genes, has homology to animal retroviruses. *Mol. Cell. Biol.* **8:** 5245–5256.

Hao B., Gong W., Ferguson T.K., James C.M., Krzycki J.A., and Chan M.K. 2002. A new UAG-encoded residue in the structure of a methanogen methyltransferase. *Science* **296:** 1462–1466.

Harger J.W., Meskauskas A., and Dinman J.D. 2002. An "integrated model" of programmed ribosomal frameshifting and post-transcriptional surveillance. *Trends Biochem. Sci.* **27:** 448–454.

Heller J.S. and Canellakis E.S. 1981. Cellular control of ornithine decarboxylase activity by its antizyme. *J. Cell. Physiol.* **107:** 209–217.

Herold J. and Siddell S.G. 1993. An "elaborated" pseudoknot is required for high frequency frameshifting during translation of HCV 229E polymerase mRNA. *Nucleic Acids Res.* **21:** 5838–5842.

Horsfield J.A., Wilson D.N., Mannering S.A., Adamski F.M., and Tate W.P. 1995. Prokaryotic ribosomes recode the HIV-1 gag-pol-1 frameshift sequence by an E/P site posttranslocation simultaneous slippage mechanism. *Nucleic Acids Res.* **23:** 1487–1494.

Howard M.T., Gesteland R.F., and Atkins J.F. 2004. Efficient stimulation of site-specific ribosome frameshifting by antisense oligonucleotides. *RNA* **10:** 1653–1661.

Howard M.T., Aggarwal G., Anderson C.B., Khatri S., Flanigan K.M., and Atkins J.F. 2005. Recoding elements located adjacent to a subset of eukaryal selenocysteine-specifying UGA codons. *EMBO J.* **24:** 1596–1607.

Ishigaki Y., Li X., Serin G., Maquat L.E. 2001. Evidence for a pioneer round of mRNA translation: mRNAs subject to nonsense-mediated decay in mammalian cells are bound by CBP80 and CBP20. *Cell* **106:** 607–617.

Ivanov I.P., Anderson C.B., Gesteland R.F., and Atkins J.F. 2004. Identification of a new antizyme mRNA +1 frameshifting stimulatory pseudoknot in a subset of diverse invertebrates and its apparent absence in intermediate species. *J. Mol. Biol.* **339:** 495–504.

Ivanov I.P., Matsufuji S., Murakami Y., Gesteland R.F., and Atkins J.F. 2000. Conservation of polyamine regulation by translational frameshifting from yeast to mammals. *EMBO J.* **19:** 1907–1917.

Jacks T. 1990. Translational suppression in gene expression in retroviruses and retrotransposons. *Curr. Top. Microbiol. Immunol.* **157:** 93–124.

Jacks T., Power M.D., Masiarz F.R., Luciw P.A., Barr P.J., and Varmus H.E. 1988. Characterization of ribosomal frameshifting in HIV-1 *gag-pol* expression. *Nature* **331:** 280–283.

Jacobs J.L. and Dinman J.D. 2004. Systematic analysis of bicistronic reporter assay data. *Nucleic Acids Res.* **32:** e160–e170.

Kinzy T.G., Harger J.W., Carr-Schmid A., Kwon J., Shastry M., Justice M.C., and Dinman J.D. 2002. New targets for antivirals: The ribosomal A-site and the factors that interact with it. *Virology* **300:** 60–70.

Kolk M.H., van der Graaf M., Wijmenga S.S., Pleij C.W., Heus H.A., and Hilbers C.W. 1998. NMR structure of a classical pseudoknot: Interplay of single- and double-stranded RNA. *Science* **280:** 434–438.

Kollmus H., Hentze M.W., and Hauser H. 1996. Regulated ribosomal frameshifting by an RNA-protein interaction. *RNA* **2:** 316–323.

Kryukov G.V., Castellano S., Novoselov S.V., Lobanov A.V., Zehtab O., Guigo R., and Gladyshev V.N. 2003. Characterization of mammalian selenoproteomes. *Science* **300:** 1439–1443.

Leger M., Sidani S., and Brakier-Gingras L. 2004. A reassessment of the response of the bacterial ribosome to the frameshift stimulatory signal of the human immunodeficiency virus type 1. *RNA* **10:** 1225–1235.

Leibundgut M., Frick C., Thanbichler M., Böck A., and Ban N. 2005. Selenocysteine tRNA-specific elongation factor SelB is a structural chimaera of elongation and initiation factors. *EMBO J.* **24:** 11–22.

Leinfelder W., Forchhammer K., Zinoni F., Sawers G., Mandrand-Berthelot M.A., and Böck A. 1988. *Escherichia coli* genes whose products are involved in selenium metabolism. *J. Bacteriol.* **170:** 540–546.

Lesoon A., Mehta A., Singh R., Chisolm G.M., and Driscoll D.M. 1997. An RNA-binding protein recognizes a mammalian selenocysteine insertion sequence element required for cotranslational incorporation of selenocysteine. *Mol. Cell. Biol.* **17:** 1977–1985.

Li Z., Stahl G., and Farabaugh P.J. 2001. Programmed +1 frameshifting stimulated by complementarity between a downstream mRNA sequence and an error-correcting region of rRNA. *RNA* **7:** 275–284.

Liston P. and Briedis D.J. 1995. Ribosomal frameshifting during translation of measles-virus P protein mRNA is capable of directing synthesis of a unique protein. *J. Virol.* **69:** 6742–6750.

Liu Z., Reches M., and Engelberg-Kulka H. 1999. A sequence in the *Escherichia coli* fdhF "selenocysteine insertion sequence" (SECIS) operates in the absence of selenium. *J. Mol. Biol.* **294:** 1073–1086.

Lundblad V. and Morris D.K. 1997. Programmed translational frameshifting in a gene required for yeast telomere replication. *Curr. Biol.* **7:** 969–976.

Lux A., Beil C., Majety M., Barron S., Gallione C.J., Kuhn H.M., Berg J.N., Kioschis P., Marchuk D.A., and Hafner M. 2005. Human retroviral gag- and gag-pol-like proteins interact with the transforming growth factor-beta receptor activin receptor-like kinase 1. *J. Biol. Chem.* **280:** 8482–8493.

Maia I.G., Seron K., Haenni A.L., and Bernardi F. 1996. Gene expression from viral RNA genomes. *Plant Mol. Biol.* **32:** 367–391.

Manktelow E., Shigemoto K., and Brierley I. 2005. Characterization of the frameshift signal of Edr, a mammalian example of programmed −1 ribosomal frameshifting. *Nucleic Acids Res.* **33:** 1553–1563.

Mansell J.B., Guevremont D., Poole E.S., and Tate W.P. 2001. A dynamic competition between release factor 2 and the tRNA(Sec) decoding UGA at the recoding site of *Escherichia coli* formate dehydrogenase H. *EMBO J.* **20:** 7284–7293.

Meskauskas A. and Dinman J.D. 2001. Ribosomal protein L5 helps anchor peptidyl-tRNA to the P-site in *Saccharomyces cerevisiae*. *RNA* **7:** 1084–1096.

Meskauskas A., Petrov A.N., and Dinman J.D. 2005. Identification of functionally important amino acids of ribosomal protein L3 by saturation mutagenesis. *Mol. Cell. Biol.* **25:** 10863–10874.

Meskauskas A., Harger J.W., Jacobs K.L.M., and Dinman J.D. 2003a. Decreased peptidyl-transferase activity correlates with increased programmed −1 ribosomal frameshifting and viral maintenance defects in the yeast *Saccharomyces cerevisiae*. *RNA* **9:** 982–992.

Meskauskas A., Baxter J.L., Carr E.A., Yasenchak J., Gallagher J.E.G., Baserga S.J., and Dinman J.D. 2003b. Delayed rRNA processing results in significant ribosome biogenesis and functional defects. *Mol. Cell. Biol.* **23:** 1602–1613.

Michiels P.J., Versleijen A.A., Verlaan P.W., Pleij C.W., Hilbers C.W., and Heus H.A. 2001. Solution structure of the pseudoknot of SRV-1 RNA, involved in ribosomal frameshifting. *J. Mol. Biol.* **310:** 1109–1123.

Mindell D.P., Sorenson M.D., and Dimcheff D.E. 1998. Multiple independent origins of mitochondrial gene order in birds. *Proc. Natl. Acad. Sci.* **95:** 10693–10697.

Miyazaki Y., Matsufuji S., and Hayashi S. 1992. Cloning and characterization of a rat gene encoding ornithine decarboxylase antizyme. *Gene* **113:** 191–197.

Mollenbeck M., Gavin M.C., and Klobutcher L.A. 2004. Evolution of programmed ribosomal frameshifting in the TERT genes of *Euplotes*. *J. Mol. Evol.* **58:** 701–711.

Moon S., Byun Y., and Han K. 2004. Computational identification of −1 frameshift signals. In *Proceedings of the 4th International Conference on Computational Science*, Kraków, Poland. Part I (ed. M. Bubaket al.), pp. 334–341. Springer-Verlag, Heidelberg, Germany.

Moosmayer D., Reil H., Ausmeier M., Scharf J.G., Hauser H., Jentsch K.D., and Hunsmann G. 1991. Expression and frameshifting but extremely inefficient proteolytic processing of the HIV-1 gag and pol gene products in stably transfected rodent cell lines. *Virology* **183:** 215–224.

Moriarty P.M., Reddy C.C., and Maquat L.E. 1998. Selenium deficiency reduces the abundance of mRNA for Se-dependent glutathione peroxidase 1 by a UGA-dependent mechanism likely to be nonsense codon-mediated decay of cytoplasmic mRNA. *Mol. Cell. Biol.* **18:** 2932–2939.

Muldoon-Jacobs K.L. and Dinman J.D. 2006. Specific effects of ribosome-tethered molecular chaperones on programmed −1 ribosomal frameshifting. *Eukaryot. Cell* **5:** 762–770.

Namy O., Duchateau-Nguyen G., and Rousset J.P. 2002. Translational readthrough of the PDE2 stop codon modulates cAMP levels in *Saccharomyces cerevisiae*. *Mol. Microbiol.* **43:** 641–652.

Napthine S., Vidakovic M., Girnary R., Namy O., and Brierley I. 2003. Prokaryotic-style frameshifting in a plant translation system: Conservation of an unusual single-tRNA slippage event. *EMBO J.* **22:** 3941–3950.

Nixon P.L. and Giedroc D.P. 2000. Energetics of a strongly pH dependent RNA tertiary structure in a frameshifting pseudoknot. *J. Mol. Biol.* **296:** 659–671.

Noller H.F., Yusupov M.M., Yusupova G.Z., Baucom A., and Cate J.H. 2002. Translocation of tRNA during protein synthesis. *FEBS Lett.* **514:** 11–16.

Olsthoorn R.C., Laurs M., Sohet F., Hilbers C.W., Heus H.A., and Pleij C.W. 2004. Novel application of sRNA: Stimulation of ribosomal frameshifting. *RNA* **10:** 1702–1703.

Palanimurugan R., Scheel H., Hofmann K., and Dohmen R.J. 2004. Polyamines regulate their synthesis by inducing expression and blocking degradation of ODC antizyme. *EMBO J.* **23:** 4857–4867.

Peltz S.W., Hammell A.B., Cui Y., Yasenchak J., Puljanowski L., and Dinman J.D. 1999. Ribosomal protein L3 mutants alter translational fidelity and promote rapid loss of the yeast killer virus. *Mol. Cell. Biol.* **19:** 384–391.

Petrov A., Meskauskas A., and Dinman J.D. 2004. Ribosomal protein L3: Influence on ribosome structure and function. *RNA Biol.* **1:** 59–65.

Plant E.P. and Dinman J.D. 2005. Torsional restraint: A new twist on frameshifting pseudoknots. *Nucleic Acids Res.* **33:** 1825–1833.

———. 2006. Comparative study of the effects of heptameric slippery site composition on −1 frameshifting among different translational assay systems. *RNA* **12:** 666–673.

Plant E.P., Wang P., Jacobs J.L., and Dinman J.D. 2004. A programmed −1 ribosomal frameshift signal can function as a *cis*-acting mRNA destabilizing element. *Nucleic Acids Res.* **32:** 784–790.

Plant E.P., Perez-Alvarado G.C., Jacobs J.L., Mukhopadhyay B., Hennig M., and Dinman J.D. 2005. A three-stemmed mRNA pseudoknot in the SARS coronavirus frameshift signal. *PLoS Biol.* **3:** e172.

Poole E.S., Brown C.M., and Tate W.P. 1995. The identity of the base following the stop codon determines the efficiency of in vivo translational termination in *Escherichia coli. EMBO J.* **14:** 151–158.

Rice N.R., Stephens R.M., Burny A., and Gilden R.V. 1985. The *gag* and *pol* genes of bovine leukemia virus: Nucleotide sequence and analysis. *Virology* **142:** 357–377.

Rom E. and Kahana C. 1994. Polyamines regulate the expression of orinithine decarboxylase antizyme *in vitro* by inducing ribosomal frameshifting. *Proc. Natl. Acad. Sci.* **91:** 3959–3963.

Rother M., Wilting R., Commans S., and Böck A. 2000. Identification and characterisation of the selenocysteine-specific translation factor SelB from the archaeon *Methanococcus jannaschii. J. Mol. Biol.* **299:** 351–358.

Sarkhel S., Rich A., and Egli M. 2003. Water-nucleobase "stacking": H-pi and lone pair-pi interactions in the atomic resolution crystal structure of an RNA pseudoknot. *J. Am. Chem. Soc.* **125:** 8998–8999.

Sauerwald A., Zhu W., Major T.A., Roy H., Palioura S., Jahn D., Whitman W.B., Yates J.R. III, Ibba M., and Soll D. 2005. RNA-dependent cysteine biosynthesis in archaea. *Science* **307:** 1969–1972.

Shah A.A., Giddings M.C., Parvaz J.B., Gesteland R.F., Atkins J.F., and Ivanov I.P. 2002. Computational identification of putative programmed translational frameshift sites. *Bioinformatics* **18:** 1046–1053.

Shigemoto K., Brennan J., Walls E., Watson C.J., Stott D., Rigby P.W.J., and Reith A.D. 2001. Identification and characterisation of a developmentally regulated mammalian gene that utilises −1 programmed ribosomal frameshifting. *Nucleic Acids Res.* **29:** 4079–4088.

Singh S.M., Steinberg-Neifach O., Mian I.S., and Lue N.F. 2002. Analysis of telomerase in *Candida albicans:* Potential role in telomere end protection. *Eukaryot. Cell* **1:** 967–977.

Small-Howard A., Morozova N., Stoytcheva Z., Forry E.P., Mansell J.B., Harney J.W., Carlson B.A., Xu X.M., Hatfield D.L., and Berry M.J. 2006. Supramolecular complexes mediate selenocysteine incorporation *in vivo. Mol. Cell. Biol.* **26:** 2337–2346.

Srinivasan G., James C.M., and Krzycki J.A. 2002. Pyrrolysine encoded by UAG in Archaea: Charging of a UAG-decoding specialized tRNA. *Science* **296:** 1459–1462.

Staple D.W. and Butcher S.E. 2005. Solution structure and thermodynamic investigation of the HIV-1 frameshift inducing element. *J. Mol. Biol.* **349:** 1011–1023.

Stenberg P., Englund C., Kronhamn J., Weaver T.A., and Samakovlis C. 1998. Translational readthrough in the *hdc* mRNA generates a novel branching inhibitor in the *Drosophila* trachea. *Genes Dev.* **12:** 956–967.

Su L., Chen L., Egli M., Berger J.M., and Rich A. 1999. Minor groove RNA triplex in the crystal structure of a ribosomal frameshifting viral pseudoknot. *Nat. Struct. Biol.* **6:** 285–292.

Tamura T., Yamamoto S., Takahata M., Sakaguchi H., Tanaka H., Stadtman T.C., and Inagaki K. 2004. Selenophosphate synthetase genes from lung adenocarcinoma cells: Sps1 for recycling L-selenocysteine and Sps2 for selenite assimilation. *Proc. Natl. Acad. Sci.* **101:** 16162–16167.

Tan M., Liang A., Brunen-Nieweler C., and Heckmann K. 2001. Programmed translational frameshifting is likely required for expressions of genes encoding putative nuclear protein kinases of the ciliate *Euplotes octocarinatus. J. Eukaryot. Microbiol.* **48:** 575–582.

Taskov K., Chapple C., Kryukov G.V., Castellano S., Lobanov A.V., Korotkov K.V., Guigo R., and Gladyshev V.N. 2005. Nematode selenoproteome: The use of the selenocysteine insertion system to decode one codon in an animal genome? *Nucleic Acids Res.* **33:** 2227–2238.

ten Dam E., Brierley I., Inglis S., and Pleij C. 1994. Identification and analysis of the pseudoknot-containing gag-pro ribosomal frameshift signal of simian retrovirus-1. *Nucleic Acids Res.* **22:** 2304–2310.

Theimer C.A. and Giedroc D.P. 2000. Contribution of the intercalated adenosine at the helical junction to the stability of the gag-pro frameshifting pseudoknot from mouse mammary tumor virus. *RNA* **6:** 409–421.

Theobald-Dietrich A., Giege R., and Rudinger-Thirion J. 2005. Evidence for the existence in mRNAs of a hairpin element responsible for ribosome dependent pyrrolysine insertion into proteins. *Biochimie* **87:** 813–817.

Tujebajeva R.M., Copeland P.R., Xu X.M., Carlson B.A., Harney J.W., Driscoll D.M., Hatfield D.L., and Berry M.J. 2000. Decoding apparatus for eukaryotic selenocysteine incorporation. *EMBO Rep.* **2:** 158–163.

Walczak R., Westhof E., Carbon P., and Krol A. 1996. A novel RNA structural motif in the selenocysteine insertion element of eukaryotic selenoprotein mRNAs. *RNA* **2:** 367–379.

Weiss S.L. and Sunde R.A. 1998. Cis-acting elements are required for selenium regulation of glutathione peroxidase-1 mRNA levels. *RNA* **4:** 816–827.

Wills N.M., Moore B., Hammer A., Gesteland R.F., and Atkins J.F. 2006. A functional −1 ribosomal frameshift signal in the human paraneoplastic Ma3 gene. *J. Biol. Chem.* **281:** 7082–7088.

Wilson W., Braddock M., Adams S.E., Rathjen P.D., Kingsman S.M., and Kingsman A.J. 1988. HIV expression strategies: Ribosomal frameshifting is directed by a short sequence in both mammalian and yeast systems. *Cell* **55:** 1159–1169.

XingY. and Lee C.J. 2004. Negative selection pressure against premature protein truncation is reduced by alternative splicing and diploidy. *Trends Genet.* **20:** 472–475.

Xu X.M., Mix H., Carlson B.A., Grabowski P.J., Gladyshev V.N., Berry M.J., and Hatfield D.L. 2005. Evidence for direct roles of two additional factors, SECp43 and SLA, in the selenoprotein synthesis machinery. *J. Biol. Chem.* **280:** 41568–41575.

Xue F. and Cooley L. 1993. *kelch* encodes a component of intercellular bridges in *Drosophila* egg chambers. *Cell* **72:** 681–693.

Yusupova G.Z., Yusupov M.M., Cate J.H., and Noller H.F. 2001. The path of messenger RNA through the ribosome. *Cell* **106:** 233–241.

Zavacki A.M., Mansell J.B., Chung M., Klimovitsky B., Harney J.W., and Berry M.J. 2003. Coupled tRNASec dependent assembly of the selenocysteine decoding apparatus. *Mol. Cell* **11:** 773–781.

Zhang Y., Baranov P.V., Atkins J.F., and Gladyshev V.N. 2005. Pyrrolysine and selenocysteine use dissimilar decoding strategies. *J. Biol. Chem.* **280:** 20740–20751.

23

Nonsense-mediated mRNA Decay: From Yeast to Metazoans

Allan Jacobson
Department of Molecular Genetics and Microbiology
University of Massachusetts Medical School
Worcester, Massachusetts 01655-0122

Elisa Izaurralde
Max-Planck-Institute for Developmental Biology
D-72076 Tübingen, Germany

MUTATIONS RESULTING IN THE LOSS OF PRODUCTION of specific proteins are among the major causes of inherited diseases. When defective, any of several steps in the gene expression pathway can be the underlying cause of such diseases (Kazazian 1990), but one of the more common types of mutation that inactivates gene function does so by promoting premature translational termination. Nonsense mutations result in the occurrence of UAA, UAG, or UGA codons in the protein-coding region of an mRNA template, leading to the termination of polypeptide elongation and, generally, to the triggering of a cellular surveillance mechanism dubbed nonsense-mediated mRNA decay, or NMD. The NMD pathway is operative in all eukaryotic cells and ensures that nonsense-containing mRNAs do not accumulate as substrates for the translation apparatus. In turn, the elimination of these transcripts prevents the accumulation of potentially toxic polypeptide fragments.

NMD has been extensively studied in yeast, worms, flies, and mammals, leading to the identification of a key set of regulatory factors and to the elaboration of models explaining the role of these factors in the discrimination of normal versus premature termination and in the promotion of rapid mRNA decay. These studies have provided insight into cellular quality control mechanisms, elucidated fundamental interrelationships between the pathways of mRNA translation and mRNA decay, and set the stage for a potentially major advance in the treatment of a subset of all genetic disorders.

NORMAL VERSUS PREMATURE TRANSLATION TERMINATION

Unlike initiation and elongation, which involve scores of factors, the events that take place at termination appear to be facilitated by only two classes of proteins, designated release factors (RFs) (Chapter 7). Class I RFs recognize stop codons within the ribosomal A site and trigger the hydrolysis of the ester bond connecting the polypeptide chain and the tRNA in the P site. Class II RFs are GTPases that stimulate class I RF activity and confer GTP dependency on the termination process (Nakamura et al. 2000; Kisselev et al. 2003; Nakamura and Ito 2003). In eukaryotes, translation termination is mediated by eukaryotic release factor 1 (eRF1) and eRF3, two interacting proteins that execute class I and class II functions, respectively (Frolova et al. 1994; Stansfield et al. 1995; Zhouravleva et al. 1995). The GTP hydrolysis activity of eRF3 is dependent on ribosome-bound eRF1 (Frolova et al. 1996) and is required to couple eRF1 recognition of the termination signal in mRNA to efficient polypeptide chain release (Salas-Marco and Bedwell 2004). Hence, discrimination of sense from nonsense codons by eRF1 mimics the recognition of a cognate codon by a tRNA. Normal termination is also stimulated by the interaction of eRF3 with the poly(A)-binding protein (PABP), a highly conserved polypeptide that associates with the $3'$ poly(A) tail present on almost all eukaryotic mRNAs (Frolova et al. 1994; Stansfield et al. 1995; Zhouravleva et al. 1995; Hoshino et al. 1999; Cosson et al. 2002a,b; Uchida et al. 2002). PABP's function in termination is not well understood, but it is thought to have a role in the enhancement of termination efficiency and the *cis*-recycling of ribosomes (i.e., the reuse of ribosomes on the same mRNA that has just been translated) (Uchida et al. 2002; Amrani et al. 2006).

Premature termination has generally been considered to be the mechanistic equivalent of normal termination, but recent experiments indicate that this conclusion is unwarranted. In a series of toeprinting analyses of normal versus premature termination in yeast cell-free extracts, ribosomes at a premature termination codon failed to be released and yielded toeprint signals consistent with A-site occupancy by the nonsense codon (Amrani et al. 2004). Normal termination codons did not yield any toeprint signals unless eRF1 was inactivated by a temperature-sensitive (ts) lesion, suggesting a higher level of termination efficiency than premature terminators (Amrani et al. 2004). Addition of the elongation inhibitor cycloheximide

(CHX) to the in vitro translation reactions also failed to reveal toeprints from normal terminators, but did allow detection of additional toeprints in close proximity to the locations of premature stop codons. The latter toeprints were shown to be attributable to posttermination ribosomes that failed to be released at premature terminators and were able to scan backward and reinitiate translation. When translated in eRF1-defective extracts, all mRNAs harboring premature terminators yielded toeprints corresponding to ribosomes stalled with the stop codon in their A sites, suggesting that prior to any reinitiation event, a premature stop codon must be recognized by eRF1 and peptide hydrolysis must be triggered (Stansfield et al. 1997; Song et al. 2000). Most importantly, the aberrant toeprints associated with reinitiation on nonsense-containing mRNAs were linked to NMD because they were not detected in extracts that lacked one of the key NMD regulatory factors, Upf1 (Amrani et al. 2004; see below).

FACTORS THAT REGULATE NMD

The powerful genetic systems of *Saccharomyces cerevisiae* and *Caenorhabditis elegans* have been particularly fruitful in the identification of genes whose products are involved in the decay of nonsense-containing mRNAs. NMD regulatory factors were originally identified in screens for informational suppressors and subsequent two-hybrid screens employing genes encoding known regulators (Culbertson et al. 1980; Hodgkin et al. 1989; Leeds et al. 1991, 1992; Pinto et al. 1992; Pulak and Anderson 1993; Cui et al. 1995, 1999; He and Jacobson 1995; He et al. 1997; Cali et al. 1999). These screens led to the identification of the *UPF1*, *UPF2* (*NMD2*), and *UPF3* genes (*smg-2*, *smg-3*, and *smg-4* in *C. elegans*) as the principal NMD regulators in eukaryotes. Mutations in these genes, or other means of inactivating their function, generally result in the stabilization and increased accumulation of nonsense-containing mRNAs while having little or no effect on the abundance and stability of most wild-type transcripts. An exception to this rule is the function of *UPF3* in mammalian cells, where the existence of two *UPF3* paralogs (*UPF3a* and *UPF3b*; also known as *UPF3* and *UPF3X*) has made it more difficult to assign a unique *UPF3* function (He et al. 1997; Lykke-Andersen et al. 2000; Serin et al. 2001; Gatfield et al. 2003). In yeast, simultaneous inactivation of more than one of the three principal NMD regulatory genes yields the same mRNA decay phenotype as single inactivation (He et al. 1997), indicating that all three gene products must be functionally related and act in a common pathway. Additional studies, discussed below, indicate that these three proteins either interact sequentially or function as a complex to regulate both NMD and premature translation termination.

In worms and other multicellular organisms, regulators of NMD also include Smg-1, Smg-5, Smg-6, and Smg-7, all of which appear to be regulators of the phosphorylation status of Upf1 (Ohnishi et al. 2003; Grimson et al. 2004). Mammalian NMD is also regulated by numerous protein components of the exon-junction complex (EJC; see below), but the potential role of these factors is complicated by the observation that their homologs in *Drosophila* are not essential for NMD (Gatfield et al. 2003). In yeast, mutations in *PRT1, HRP1, MOF2, MOF5, MOF8,* and *DBP2* also result in selective stabilization of nonsense-containing mRNAs, an observation indicating that RNA-binding proteins and other regulators of translation can modulate NMD activity (Dinman and Wickner 1994; Cui et al. 1995, 1999; Welch and Jacobson 1999; Gonzalez et al. 2000).

The prototypical yeast *UPF1* gene encodes a 109-kD protein with a cysteine- and histidine-rich region near its amino terminus (which may comprise two Zn^{2+} fingers) and seven conserved motifs common to the members of helicase superfamily I (Altamura et al. 1992; Koonin 1992; Leeds et al. 1992). Yeast and human Upf1 are localized to the cytoplasm and exhibit RNA binding, as well as RNA-dependent ATPase and RNA helicase activities (Atkin et al. 1995, 1997; Czaplinski et al. 1995; Weng et al. 1996a,b, 1998; Bhattacharya et al. 2000). The amino-terminal zinc-finger-like domain of yeast Upf1 interacts with a carboxy-terminal domain of Upf2/Nmd2, with the polypeptide release factors eRF3/Sup35 and eRF1/Sup45, and with the catalytic component of the decapping enzyme Dcp2 (He and Jacobson 1995; He et al. 1997; Czaplinski et al. 1998; Mendell et al. 2000; Lykke-Andersen 2002; F. He and A. Jacobson, unpubl.). Upf1–eRF interactions may facilitate ribosome dissociation from premature termination codons, and Upf1–Upf2 interaction may be a precursor to Upf1 recruitment of the decapping enzyme and the initiation of mRNA decay (He and Jacobson 1995; He et al. 1996; Lykke-Andersen 2002; Amrani et al. 2006; see below).

Upf2/Nmd2 is an acidic 127-kD protein with three conserved MIF4G (middle portion of eIF4G) domains (Cui et al. 1995; He and Jacobson 1995; Mendell et al. 2000). One of these domains, in the carboxy-terminal half of Upf2/Nmd2, interacts with Upf3, a basic 45-kD protein with a ribonucleoprotein (RNP)-type RNA-binding domain (RBD) (He et al. 1997; Kadlec et al. 2004). Crystallographic analysis of a complex formed by the respective interaction domains of human Upf2 and Upf3b revealed that the third Upf2 MIF4G domain interacts with the β-sheet surface of the Upf3 RBD (Kadlec et al. 2004). Even though β-sheet surfaces of RBDs are often used for nucleic acid binding by other proteins, additional experiments have shown that Upf3b cannot bind RNA in the absence of Upf2 and that the nonspecific RNA-binding activity of the Upf2/Nmd2–Upf3 complex is attributable to a conserved basic region of Upf2/Nmd2 centered on Arg-796 (Kadlec et al. 2004). Crystallographic analyses also revealed that a putative NES sequence previously thought to be the basis of apparent nucleocytoplasmic shuttling activity of Upf3 (Lee and Culbertson 1995; Shirley et al. 1998) was localized to an internal segment of RNP2 and that mutations thought to demonstrate the dependence of NMD on such shuttling activity simply disrupted Upf2/Nmd2–Upf3 interaction (Kadlec et al. 2004). Although nothing is known of the specific interaction domains, Upf2/Nmd2 and Upf3 have also been shown to interact with eRF3/Sup35 (He et al. 1997; Czaplinski et al. 1998; Mendell et al. 2000), an event that may facilitate Upf1 association with the prematurely terminating ribosome (see below).

MODULATION OF Upf1 AND Upf2/Nmd2 ACTIVITY BY PHOSPHORYLATION AND DEPHOSPHORYLATION

Upf1 activity is regulated by phosphorylation and dephosphorylation of multiple serine residues present in the amino- and carboxy-terminal domains of the protein (Page et al. 1999). Phosphorylation of Upf1 is catalyzed by Smg-1, a protein kinase related to phosphoinositide 3-kinase (Denning et al. 2001; Pal et al. 2001; Yamashita et al. 2001; Grimson et al. 2004). In addition to Smg-1, phosphorylation of Upf1 requires Upf2/Nmd2 and Upf3, suggesting that the formation of a trimeric Upf1–Upf2/Nmd2–Upf3 complex triggers Upf1 phosphorylation (Page et al. 1999). The dephosphorylation of Upf1 is mediated by Smg-5, Smg-6, and Smg-7, three similar but not functionally redundant proteins (Cali et al. 1999; Page et al. 1999; Anders et al. 2003; Ohnishi et al. 2003). Smg5–7 are thought to trigger Upf1 dephosphorylation by recruiting protein phosphatase 2A (PP2A). Indeed, Smg-5 and Smg-7 form a heterodimer that associates with phosphatase PP2A and phosphorylated Upf1 (Anders et al. 2003; Ohnishi et al. 2003). Smg-6 is also part of a protein complex comprising PP2A and phosphorylated Upf1 (Chiu et al. 2003), although it is not clear whether this complex is distinct from the Smg-5/Smg-7 complex.

Structural studies have provided insights into the molecular mechanism by which phosphorylated Upf1 is recognized by the Smg-5–7 proteins. The structure of the amino-terminal domain of Smg-7 resembles that of 14-3-3, a phosphoserine-binding protein involved in signal transduction pathways (Fukuhara et al. 2005). Moreover, residues that line the phosphoserine binding pocket of 14-3-3 are conserved at the corresponding pocket of Smg-7, suggesting that Smg-7 functions as a phosphoserine-binding protein. In agreement with this notion, mutation of residues in the 14-3-3-like phosphoserine-binding site of Smg-7 impairs its binding to Upf1 (Fukuhara et al. 2005). The 14-3-3-like domain of

Smg-7 is conserved in Smg-5 and Smg-6, suggesting that these proteins can also recognize phosphoserine residues, possibly in Upf1 (Fukuhara et al. 2005). In addition to its amino-terminal 14-3-3-like domain, Smg-7 has a carboxy-terminal domain that elicits decay of bound mRNAs (Unterholzner and Izaurralde 2004). Consistent with a role of Smg-7 in recruiting decay enzymes, ectopically expressed Smg-7 accumulates in cytoplasmic foci corresponding to endogenous P bodies and causes the accumulation of Smg-5 or Upf1 in P bodies (Unterholzner and Izaurralde 2004; Fukuhara et al. 2005). Together with the observation that Smg-7 interacts with phosphorylated Upf1 and triggers its dephosphorylation, this suggests that Smg-7 could have a role in coupling changes in the phosphorylation state of Upf1 to mRNA degradation.

Until recently, the phosphorylation/dephosphorylation cycle of Upf1 was thought to be a feature of the NMD pathway restricted to multicellular organisms. However, recent studies have shown that yeast Upf1 and Upf2 are also phosphorylated (Wang et al. 2006). There are no clear Smg5–7 orthologs in yeast, and the mechanism and role of Upf1 phosphorylation in the yeast NMD pathway remain to be established. Phosphorylation of Upf2, in contrast, increases its interaction with the RNA-binding protein, Hrp1, and is required for NMD (Wang et al. 2006). Thus, the regulation of NMD factors through cycles of phosphorylation and dephosphorylation is a conserved event in eukaryotes.

A ROLE FOR THE NMD FACTORS IN TRANSLATION TERMINATION

Mutations in the yeast *UPF/NMD* genes not only lead to the stabilization of nonsense-containing mRNAs, but also promote nonsense suppression (Leeds et al. 1992; Weng et al. 1996a,b; Maderazo et al. 2000; Stahl et al. 2000; Keeling et al. 2004). Suppression occurs when a near-cognate tRNA effectively competes with the RFs for binding to the A site of the ribosome while a nonsense codon is in residence at the decoding site. As a consequence, the nascent peptidyl-tRNA bond is not hydrolyzed and the polypeptide continues to be extended. Normally, a nonsense allele of an essential gene leads to complete loss of gene function and inviability. That the efficiency of termination codon readthrough is sufficient for restoration of normal growth when Upf1, Upf2/Nmd2, or Upf3 is inactivated strongly suggests an important role for these proteins in the regulation of termination, at least at premature nonsense codons. Nonsense suppression in *upf* strains could simply be due to a combination of enhanced mRNA abundance and an inherent rate of nonsense codon readthrough that is sufficient to generate the minimal amount of protein required for function of the respective genes. However, the nonsense-suppressing effects engendered by *upf* mutations in yeast can be separated from changes in mRNA levels by mutation, by independent alterations in the abundance of nonsense-containing mRNAs, and by normalization of suppression data to RNA content (Weng et al. 1996a,b; Maderazo et al. 2000; Keeling et al. 2004); i.e., these are two independent phenomena.

Mutations in eRF1/Sup45 and eRF3/Sup35 also promote nonsense suppression (Stansfield et al. 1997; Keeling et al. 2004), but these effects are additive with those of *upf* mutations (Keeling et al. 2004), indicating distinct functions in termination. The apparent role of the Upf factors in termination fidelity is underscored by their interactions with the RFs. As noted above, Upf1, Upf2/Nmd2, and Upf3 all interact with the eRF3/Sup35, and Upf1 can also interact with eRF1/Sup45 (Czaplinski et al. 1998; Wang et al. 2001). Upf1 protein has a unique binding domain on Sup35 protein, but Upf2/Nmd2, Upf3, and eRF1/Sup45 all compete for binding to eRF3/Sup35 and may thus share a common interaction domain (Wang et al. 2001). Interaction with either RF inhibits the ATPase activity of Upf1, and interaction with Sup35 prevents formation of a Upf1–RNA complex (Czaplinski et al. 1998; Wang et al. 2001). Because eRF3/Sup35 and RNA may compete for binding to Upf1, factors such as ATP, which is capable of decreasing Upf1 affinity for RNA, might promote RF binding (Weng et al. 1996a,b, 1998) or regulate Upf1 interaction with the RFs in the presence of competing RNAs (Czaplinski et al. 1998). The identification of a mutant form of Upf1 that can bind but not hydrolyze ATP, and which is active in translation termination but inactive in mRNA decay, suggests that ATP hydrolysis may promote conformational changes that accompany a Upf1 switch from participation in translation termination to a role in promoting mRNA decay (Weng et al. 1996a,b, 1998).

Yeast strains harboring a *upf1Δ* mutation show higher levels of nonsense suppression than strains harboring either the *upf2Δ/nmd2Δ* or *upf3Δ* mutations, and the *upf1Δ* phenotype predominates in doubly mutant strains (Maderazo et al. 2000). These data, and the demonstration that overexpression of *UPF1* can compensate for *upf2Δ/nmd2Δ* and *upf3Δ* mutations but not vice versa, suggest that Upf2/Nmd2 and Upf3 can regulate Upf1 activity or that an imbalance in the concentration of Upf2/Nmd2 and Upf3 can alter the efficiency of the translation termination factors (Czaplinski et al. 2000; Maderazo et al. 2000).

PATHWAYS FOR DECAY OF NONSENSE-CONTAINING mRNAs

Most wild-type mRNAs are degraded through a decapping-dependent 5′ to 3′ pathway or an exosome-mediated 3′ to 5′ pathway (Muhlrad et al. 1995; Anderson and Parker 1998; Cao and Parker 2001; Mangus et al. 2003;

Parker and Song 2004). The initial decay event in both pathways is the shortening of the poly(A) tail to a length of 10–12 nucleotides in yeast (Decker and Parker 1993; Beelman and Parker 1995; Muhlrad et al. 1995) or 25–30 nucleotides in metazoans (Jacobson 1996). After poly(A) shortening, transcripts can be degraded by either pathway. In 5′ to 3′ decay, transcripts are decapped by the Dcp1/Dcp2 decapping enzyme complex and then digested exonucleolytically by the 5′ to 3′ exoribonuclease, Xrn1 (Hsu and Stevens 1993; Muhlrad and Parker 1994; Muhlrad et al. 1995; Beelman et al. 1996; Dunckley and Parker 1999; Steiger et al. 2003). In 3′ to 5′ decay, transcripts are further deadenylated and then degraded by a ten-subunit 3′ to 5′ exonuclease complex, the exosome (Mitchell and Tollervey 2003).

Poly(A) shortening includes at least two steps: an initial trimming by poly(A) nuclease (PAN) and its more extensive removal by the Ccr4p/Pop2p/Notp complex (Mangus et al. 2003; Parker and Song 2004). Poly(A) shortening minimizes the association of PABP with the mRNP, a step that appears to either stimulate exosome activity or prompt the loss of the cap-binding initiation complex and accelerate the transition of the mRNP to a new state in which the binding of additional proteins enhances the enzymology of decapping and determines the subcellular site of its occurrence (Mangus et al. 2003; Parker and Song 2004). The latter mRNP rearrangement allows binding of the Lsm1-7/Pat1 complex which, in turn, appears to promote interaction of the mRNP with the Dcp1/Dcp2 complex (Tharun and Parker 2001). All steps subsequent to association of the Lsm1-7/Pat1 complex are thought to occur at a limited number of subcellular sites called mRNA-processing bodies or P-bodies (Sheth and Parker 2003).

It was initially thought that degradation of nonsense-containing transcripts utilized an abbreviated normal pathway, proceeding from decapping to 5′ to 3′ decay without prior poly(A) shortening or involvement of the Lsm1-7 complex (Shyu et al. 1991; Muhlrad and Parker 1994; Beelman et al. 1996; Boeck et al. 1998; Dunckley and Parker 1999; Bouveret et al. 2000). More recent studies indicate that the notion of a specific pathway for degrading these RNAs is too simplistic. In an initial step, these mRNAs are recognized as nonsense-containing (see below). Subsequently, "targeted" transcripts appear to be subject to accelerated decapping, accelerated deadenylation, both 5′ to 3′ and 3′ to 5′ decay, or, in *Drosophila*, endonucleolytic cleavage followed by exonucleolytic removal of the products (Cao and Parker 2003; Chen and Shyu 2003; He et al. 2003; Lejeune et al. 2003; Mitchell and Tollervey 2003; Gatfield and Izaurralde 2004).

Microarray analysis of yeast NMD substrates (see below) showed that the transcripts up-regulated in *upf1Δ*, *upf2Δ/nmd2Δ*, or *upf3Δ* cells displayed more than 70% overlap with those in *xrn1Δ* cells, thus indicating that the substrates of the NMD pathway are preferentially degraded by the decapping-dependent 5′ to 3′ pathway (He et al. 2003). Although accelerated decapping is indeed deadenylation-independent, its rate is dependent on the proximity of the nonsense codon to the mRNA 5′ end (Cao and Parker 2003). Moreover, accelerated deadenylation need not be directly related to decay since a premature nonsense codon will also destabilize an mRNA in which 5′/3′ interactions are established independent of a poly(A) tail (Coller et al. 1998). These observations, as well as the diminished translational efficiencies of nonsense-containing mRNAs (Muhlrad and Parker 1999b; Keeling et al. 2004), have given rise to the suggestion that these mRNAs have an altered mRNP structure that changes their accessibility to different nucleases or decay pathways, precludes their transition to a more stable state, or forces their localization to P bodies (Hilleren and Parker 1999; Jacobson and Peltz 2000; Cao and Parker 2003; He et al. 2003; Baker and Parker 2004; Teixeira et al. 2005). Simply put, nonsense codon recognition appears to trigger availability for rapid decay, rather than acceleration of a predetermined decay pathway. This, in turn, indicates that premature termination leads an mRNP to exit the conventional cycle of translation and decay and enter a pool in which RNA degradation is rapid but not necessarily discriminate.

NMD IS DEPENDENT ON mRNA TRANSLATION

The mere presence of a premature nonsense codon within an mRNA is not sufficient to promote its degradation by the NMD pathway. This conclusion follows from experiments showing that the normal rate of decay can be restored to a nonsense-containing mRNA if a nonsense-suppressing tRNA is coexpressed in the same cells as the nonsense-containing transcript (Losson and Lacroute 1979; Gozalbo and Hohmann 1990) or if the initiator AUG is deleted from the transcript (Ruiz-Echevarria et al. 1998). It therefore follows that mRNA destabilization depends on recognition of the nonsense codon as such by the translational apparatus. This conclusion is consistent with several other observations including those demonstrating that NMD factors and decay intermediates are localized to polysomes and interact with the RFs (Peltz et al. 1993b; Atkin et al. 1995, 1997; Zhang et al. 1997; Czaplinski et al. 1998; Shirley et al. 1998; Mangus and Jacobson 1999; Welch and Jacobson 1999; Wang et al. 2001), that NMD can be antagonized by drugs or RNA structures that interfere with protein synthesis (Belgrader et al. 1993; Zhang et al. 1997) or by altered RFs that suppress termination (Keeling et al. 2004), and that polysome-associated nonsense-containing mRNAs are immediately destabilized after release from cycloheximide-promoted stabilization (Zhang et al. 1997).

Additional support for the involvement of the translational apparatus in NMD includes experiments showing that (1) mutations in genes encoding the NMD regulatory factors promote nonsense suppression (Leeds et al. 1992; Weng et al. 1996a,b; Maderazo et al. 2000; Keeling et al. 2004; see above), (2) a dominant-negative form of Upf2/Nmd2 is only active when localized to the cytoplasm (He et al. 1996), and (3) NMD of the yeast *CPA1* mRNA is dependent on the extent of ribosome occupancy of the termination codon that triggers decay (Gaba et al. 2005). The latter example is particularly illustrative. The yeast *CPA1* gene encodes the small subunit of arginine-specific carbamoyl phosphate synthetase, and expression of its mRNA is subject to negative translational regulation by the arginine attenuator peptide (AAP). The AAP is encoded by an upstream open reading frame (uORF) of the *CPA1* mRNA and regulation occurs when the nascent AAP responds to arginine by stalling ribosomes at the uORF termination codon, thereby reducing translation initiation downstream. This stalling promotes NMD of the *CPA1* mRNA. Whereas the wild-type *CPA1* uORF stalls ribosomes efficiently and is an efficient trigger of NMD, a mutated uORF (D13N), which stalls ribosomes poorly or not at all, is an inefficient trigger. However, improving the initiation context of the uORF leads to efficient NMD triggering by both the wild-type and D13N uORFs, and these effects were not attributable to ribosome stalling in the latter case. Rather, the NMD-inducing effects of the context-improved D13N uORF were due to increased ribosomal occupancy of the uORF termination codon, a result indicating that the levels of ribosomes at an early stop codon can control NMD. This conclusion is of particular significance when considering the roles of factors thought to regulate NMD (see below).

MECHANISTIC CONSIDERATIONS: NMD-SPECIFIC *CIS*-ACTING REGULATORY ELEMENTS OR QUALITY CONTROL OF DEFECTIVE TERMINATION?

mRNA Destabilization Requires a Premature Nonsense Codon and Additional Sequences

The destabilizing effects of premature nonsense codons are position-dependent. Generally, nonsense codons located within the first two-thirds to three-quarters of the coding region accelerate mRNA decay up to 20-fold, whereas nonsense codons located within the remaining portions of the coding region, including the normal termination codon, have little or no effect on mRNA decay (Losson and Lacroute 1979; Peltz et al. 1993a; Yun and Sherman 1995). What appears to distinguish a normal nonsense codon from one that promotes mRNA destabilization is its sequence context, or more precisely, the presence of specific sequences 3′ to the nonsense codon.

In yeast, a requirement for such downstream elements, or DSEs, was initially indicated by experiments demonstrating that deletion of most of the *PGK1*-coding protein region downstream from an early nonsense codon reduced mRNA decay rates markedly and that reinsertion of a small segment of the *PGK1*-coding region into the construct harboring the large deletion was sufficient to activate NMD (Peltz et al. 1993a). Characterization of a *PGK1* DSE, and sequences from other mRNAs with comparable activity, has shown that DSE-like sequences can be found in the coding regions of most yeast mRNAs but that they have only a weak consensus sequence (Peltz et al. 1993a; Hagan et al. 1995; Yun and Sherman 1995; Zhang et al. 1995). The presence of DSEs in the coding regions of wild-type mRNAs of diverse decay rates has been taken to indicate that these elements are inactive unless preceded by an upstream termination codon. Likewise, termination codons that do not promote NMD, including the normal termination codon, have been thought to be inactive because they lack a 3′ DSE. Consistent with this interpretation, mRNAs with substantially extended 3′UTRs, generated by mutation of RNA-processing sites, become substrates for NMD because the extended sequence presumably encompasses a DSE (Pulak and Anderson 1993; Muhlrad and Parker 1999a; Das et al. 2000).

Studies of mammalian NMD substrates, although not uncovering a consensus DSE, did indicate that mRNA destabilization was dependent on prior pre-mRNA splicing (Cheng et al. 1994; Carter et al. 1996; Thermann et al. 1998; Zhang et al. 1998a,b), even though nonsense codon recognition took place postsplicing (Zhang and Maquat 1996). A higher-resolution look at this phenomenon, based on the analysis of approximately 1500 mRNA sequences, strongly suggested that mammalian NMD required an exon–exon junction at least 50–55 nucleotides downstream from the premature terminator (Nagy and Maquat 1998; Sun and Maquat 2000). In support of this conclusion, the levels of nonsense-containing mRNAs derived from intronless histone H4, Hsp-70, or melanocortin-4 receptor genes were relatively unaffected when compared to those of the respective wild-type transcripts (Maquat and Li 2001; Brocke et al. 2002). Despite the considerable number of examples supporting the notion of a requirement for an exon–exon junction >50 nucleotides downstream from a premature terminator, the rule is not absolute. Some mRNAs with considerably closer spacing between the terminator and the exon–exon junction, as well as some mRNAs derived from genes in which there is no intron downstream from a premature termination codon, have also been shown to be NMD substrates (Chan et al. 1998; Zhang et al. 1998a; Rajavel and Neufeld 2001; Delpy et al. 2004; LeBlanc and Beemon 2004; Buhler et al. 2006).

Surveillance Complex and Pioneer Round of Translation Models

Mechanistic models for NMD must take into account the selective recognition of premature versus normal termination codons, a requirement for downstream sequences and ongoing translation, the roles and interactions of the principal NMD factors, and the eventual shunting of the nonsense-containing mRNP into one of the mRNA decay pathways. One of two popular models that accommodate these objectives was originally developed to explain NMD in yeast and is known as the surveillance complex model (Czaplinski et al. 1998; Wang et al. 2001). Its major tenet is that all termination events are monitored by a complex of proteins that includes Upf1, Upf2/Nmd2, and Upf3 and that interaction of this complex with a DSE-associated factor(s) is sufficient to trigger mRNA decay. In this model, binding of eRF1 and eRF3 to a ribosome paused at a termination codon is thought to recruit Upf1 and enhance termination factor activity (Czaplinski et al. 1998; Wang et al. 2001, 2006). Subsequent to peptide hydrolysis and dissociation of eRF1, the association of Upf2/Nmd2 and Upf3 and the release of eRF3 are postulated to activate the RNA-binding and ATPase activities of Upf1, allowing the complex to scan 3′ of the termination codon. In turn, an interaction between components of the surveillance complex and a 3′ DSE-bound protein is hypothesized to transform the mRNP into a substrate for rapid decapping and degradation of the body of the transcript (Czaplinski et al. 1998; Gonzalez et al. 2000; Wang et al. 2001, 2006).

In support of this model, Hrp1 has been identified as the putative DSE-binding protein, shown to interact with both Upf1 and Upf2, and to be required for NMD activity (Gonzalez et al. 2000; Kebaara et al. 2003; Wang et al. 2006). Hrp1 shuttles between the nucleus and the cytoplasm (Kessler et al. 1997), and mutations in the RNA-binding domain of Hrp1 lead to both specific stabilization of nonsense-containing mRNAs and loss of DSE and Upf1 interaction (Gonzalez et al. 2000). Furthermore, specific phosphorylated amino acids in Upf2/Nmd2 have been shown to be required for both interaction with Hrp1 and NMD function (Wang et al. 2006). To allow discrimination between normal and premature termination, Hrp1 is postulated to bind to newly synthesized transcripts, but to be displaced when translation extends to the normal end of an ORF. Failure to translate an mRNA completely would thus lead to retention of bound Hrp1 and allow it to interact with the surveillance complex. The weaknesses of this model include the postulated association of the low-abundance NMD factors (Maderazo et al. 2000) with all termination events and the hypothetical immunity from NMD that transcripts must acquire if they are not degraded in an initial round of translation (see below).

The notion of a limited opportunity for decay of nonsense-containing transcripts is also a principal component of the mammalian variation on the surveillance complex model, known as the pioneer round model (Ishigaki et al. 2001). Much like the proposed role of the DSE and Hrp1 in yeast, a central concept of the pioneer round model is that a maturing mRNP is "marked" by proteins deposited on the RNA during splicing and that these proteins are required for the definition of a premature termination codon. A pivotal discovery for the development of this model was the identification of the exon junction complex, a group of proteins that associate with an mRNA postsplicing, 20–24 nucleotides 5′ of exon–exon junctions (Le Hir et al. 2000a,b, 2001). The composition of the EJC is dynamic, but it has been shown to include at least Upf2/Nmd2 and Upf3, several splicing and export factors, Y14, Magoh, and eIF4AIII, the last of which may be the core or "anchoring" factor for the complex (Tange et al. 2004). A role for EJC proteins in NMD is consistent with experiments indicating that depletion of these factors or reductions in splicing efficiencies can antagonize NMD (Maquat 1995, 2000, 2004; Hentze and Kulozik 1999; Le Hir et al. 2001; Palacios et al. 2004; Shibuya et al. 2004; Gudikote et al. 2005) and that tethering of EJC proteins downstream from stop codons can promote NMD (Lykke-Andersen et al. 2001; Fribourg et al. 2003; Gehring et al. 2003; Palacios et al. 2004).

EJC proteins are displaced by translating ribosomes (Dostie and Dreyfuss 2002) and, much like the surveillance complex model, the pioneer round model suggests that such displacement of EJC factors by the initial ribosomes traversing a normal mRNA eliminates the possibility that conventional termination events can destabilize the mRNA. However, when premature termination prevents ribosomes from translating an entire ORF, a set of sequential interactions is thought to activate NMD (Fig. 1). These include the association of Upf1 and SMG1 with the RFs bound to the prematurely terminating ribosome to form a "SURF" complex, subsequent interaction of the SURF with EJC-associated Upf2/Nmd2, Upf3, and other factors, activation of Upf1 phosphorylation, and interaction of phosphorylated Upf1 with SMG7 to trigger mRNA decay (Behm-Ansmant and Izaurralde 2006; Kashima et al. 2006). The precise function of SMG7 in these events is not known, but its localization to P-bodies has been thought to indicate that it may target mRNAs to these subcellular structures. Consistent with this notion, tethering of SMG7 to otherwise normal mRNAs is sufficient to promote their degradation (Unterholzner and Izaurralde 2004; Fukuhara et al. 2005).

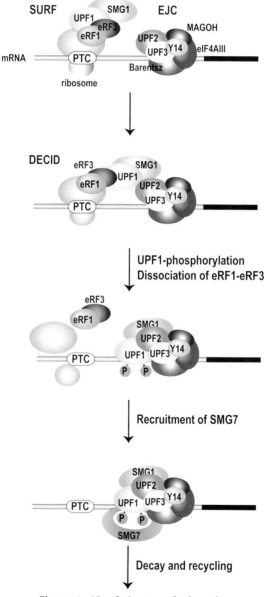

Figure 1. (*See facing page for legend.*)

Can Transcripts Acquire Immunity to NMD?

Both the surveillance complex model and the pioneer round model imply that NMD is a decay process that occurs during an early round of mRNA translation, or not at all. Clearly, if essential NMD factors were removed from an mRNA by translating ribosomes, then the mRNA should acquire immunity to further decay by the NMD pathway. One set of indirect experiments in mammalian cells, demonstrating that nonsense-containing cytoplasmic TPI mRNA remains associated with polysomes while being resistant to accelerated decay, suggests that this may be the case (Stephenson and Maquat 1996). However, the notion has also been tested directly by three different experimental approaches in yeast, all of which demonstrate clearly that nonsense-containing mRNAs never acquire immunity and are always available for NMD regardless of their translational lifetimes. In the first approach, inducible *UPF1*, *UPF2/NMD2*, and *UPF3* alleles were used to test whether nonsense-containing mRNAs stabilized in the absence of an individual NMD factor could be destabilized upon restitution of the factor. Induction of galactose-regulated *UPF1*, *UPF2/NMD2*, or *UPF3* alleles all led to the rapid turnover of previously stabilized nonsense-containing mRNAs (Maderazo et al. 2003). Similarly, in a second set of experiments, stalling of ribosomes at the *CPA1* mRNA uORF by the supplementation of yeast culture media with arginine (see above) led to the immediate induction of NMD of *CPA1* mRNA that had been synthesized and translated well in advance of NMD induction (Gaba et al. 2005). Finally, analyses of the competition between NMD and nonsense codon readthrough, using all three nonsense codons in multiple contexts and a nearly 300-fold variation in the efficiency of termination, revealed that the maximal extent of NMD occurred when readthrough was at or below a threshold of 0.5% (Keeling et al. 2004). Clearly, a reduction in the rate of NMD caused by the readthrough of as

Figure 1. Sequential interactions postulated to promote NMD in mammalian cells. A premature translation termination event leads to the assembly of the SURF complex consisting of Smg-1, Upf1, eRF1, and eRF3. SURF interacts with Upf2, Upf3, and additional EJC proteins bound to a downstream exon–exon boundary. This interaction results in the formation of the DECID complex that triggers Upf1 phosphorylation and the dissociation of eRF1 and eRF3. Phosphorylated Upf1 recruits Smg-7, most likely in association with Smg-5. The interaction of Smg-7 (and probably additional proteins) with decay enzymes targets the bound mRNA for degradation (not shown). Smg-7 also recruits PP2A, resulting in Upf1 dephosphorylation and dissociation from Smg-7 (not shown). This may enable the recycling of these proteins for a new round of NMD. (PTC) Premature translation termination codon. (Modified from Behm-Ansmant and Izaurralde 2006.)

few as 1 ribosome in 100 implies multiple translation events on the mRNA subject to NMD. Collectively, these three sets of results are consistent with the idea that conventional, translatable nonsense-containing transcripts cannot acquire immunity to NMD, at least in yeast, and imply strongly that NMD must be a cytoplasmic pathway.

A Role for Nonsense Codon Context: The *faux* UTR Model

The ability of yeast mRNAs to maintain NMD substrate status throughout their cellular lifetimes implies a need for a mechanistic model applicable to any round of translation. An alternative to the surveillance complex model, the "*faux* UTR" model (Jacobson and Peltz 2000; Maderazo et al. 2003; Amrani et al. 2004, 2006), suggests that proximity to the poly(A) tail (and its bound PABP) has a qualitative and/or quantitative influence on the nature of the termination event. In light of the closed-loop structure of cytoplasmic mRNPs (Jacobson 1996; Mangus et al. 2003), it is not surprising that 3'UTRs can serve as binding sites for numerous factors that regulate mRNA translation or stability (Dominski and Marzluff 1999; Wilson and Brewer 1999; Zhang et al. 2002; Wilkie et al. 2003; Richter and Sonenberg 2005). If subsets of such factors have a role in normal termination, for example, by influencing the activity of the RFs, then there is a possibility that the mRNA sequence distal to a premature terminator, which forms an unconventional 3'UTR, will not be the functional equivalent of a normal 3'UTR. In support of this notion, deletions that eliminate most coding sequences downstream from premature terminators stabilize yeast mRNAs that would otherwise be substrates for NMD (Peltz et al. 1993a; Muhlrad and Parker 1999b) and eliminate the aberrant toeprints characteristic of premature termination in yeast cell-free extracts (Amrani et al. 2004; see above). In addition, mutations in polyadenylation sites that lead to abnormally extended 3'UTRs convert mRNAs into substrates for NMD (Muhlrad and Parker 1999a).

To test whether proximal PABP could influence termination activity and NMD, the in vivo stability of yeast nonsense-containing mRNAs harboring different RNA-binding proteins tethered 3' to premature terminators was evaluated (Amrani et al. 2004). These experiments showed that mimicking a normal 3'UTR by tethering PABP (Pab1 in yeast) selectively stabilizes nonsense-containing mRNAs in vivo, provided the tethering site is sufficiently close to the termination codon so as to resemble the linear distance normally separating Pab1 and the terminator (Amrani et al. 2004). Similar results have been obtained in *Drosophila* (E. Izaurralde, unpubl.). Coimmunoprecipitation experiments, utilized to assess possible indirect effects of tethered Pab1, showed that tethered

Pab1 interacted with eRF3 and that tethered eRF3 could also stabilize nonsense transcripts (Amrani et al. 2004). Because Pab1 interacts with several posttranscriptional regulatory proteins (Mangus et al. 2003), it is possible that other 3′UTR-associated factors, in addition to eRF3, are involved in preventing NMD during normal termination.

A role for 3′UTR-associated factors in the maintenance of normal mRNA decay rates is also suggested by experiments demonstrating that the stability of unspliced Rous sarcoma virus RNA is dependent on a specific sequence element located just downstream from the termination codon for the *gag* ORF (Weil and Beemon 2006). This element is pertinent to NMD because its insertion 3′ to a premature termination codon in the *gag* gene was sufficient to inhibit NMD of the transcript (Weil and Beemon 2006). Similarly, the four uORFs of the yeast *GCN4* mRNA are followed by a sequence element that appears to antagonize NMD (Ruiz-Echevarria et al. 1998; Ruiz-Echevarria and Peltz 2000). This element inactivates the NMD pathway when positioned downstream from a termination codon but still 5′ of a DSE. The *GCN4* stabilizer element is homologous to a comparably positioned sequence within the *YAP1* mRNA that also appears to have NMD-inactivating activity (Ruiz-Echevarria and Peltz 2000). These stabilizing elements bind the Pub1 protein, and Pub1-deficient strains fail to stabilize *GCN4* mRNA (Ruiz-Echevarria and Peltz 2000). These observations suggest that Pub1 bound 3′ to a premature termination codon can interfere directly with activation of the decay apparatus and/or mimic the RNP context of a normal 3′UTR.

Figure 2 presents a more-detailed look at the *faux* UTR model. Normal termination (Fig. 2A) and normal rates of mRNA decay are postulated to be dependent on eRF1 and eRF3, as well as proximal Pab1 that is typically associated with a conventional poly(A) tail. As previously proposed, eRF3–Pab1 interaction may enhance termination efficiency and ensure efficient ribosome release and recycling, possibly on the same mRNA (Hoshino et al. 1999; Cosson et al. 2002b; Uchida et al. 2002; Hosoda et al. 2003). Pab1 is thought to preclude eRF3 interactions with Upf/Nmd (Wang et al. 2001), either because of inherently greater affinity for eRF3 or because it promotes other, competing, interactions. Premature termination (Fig. 2B) is postulated to be characterized by the absence of proximal Pab1 (or another factor dependent on Pab1 for binding) which, in turn, is thought to reduce the efficiency of termination and allow the low-abundance Upf/Nmd factors to associate with this relatively rare form of the termination complex. Although not shown, Upf2/Nmd2 and Upf3 may bind eRF3 first and facilitate subsequent Upf1 binding to the eRFs (Wang et al. 2001).

(A)   (B)

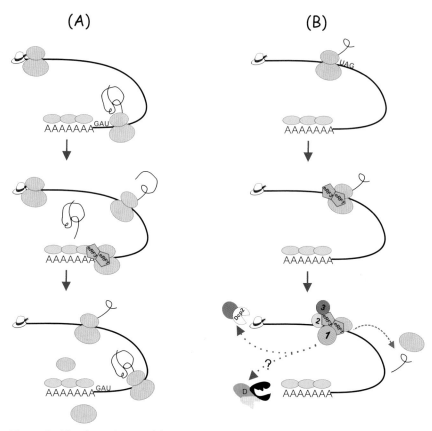

Figure 2. The *faux* UTR model. In normal termination (*A*), the ribosome (*light blue*) harboring a nascent polypeptide approaches the UAG termination codon (*top*) and then engages the UAG codon in its A site, binds eRF1 and eRF3 (*purple pentagons*), and releases the completed polypeptide (*middle*). The 40S and 60S ribosomal sub-units are then released from the mRNA (*bottom*). Normal termination is thought to be highly efficient, in part because interactions between Pab1 (*green ovals*) and ribosome-associated eRF3 enhance the ability of eRF3 to stimulate the activity of eRF1. The spacing of Pab1 from the termination site is minimized to illustrate that the effect of Pab1 depends on its proximity to the terminating ribosome. In premature termina-tion (*B*), a ribosome harboring a nascent polypeptide approaches a premature UAG termination codon (*top*) and then engages the UAG codon in its A site, binds eRF1 and eRF3, and fails to release the incomplete polypeptide (*middle*). Premature termi-nation is thought to be inefficient because the termination site lacks proximal Pab1 and/or other factors associated with a natural 3′UTR. Subsequently, the NMD factors Upf1 (1), Upf2/Nmd2 (2), and Upf3 (3) bind to the eRFs, stimulating peptide hydroly-sis and release of the 60S ribosomal subunit (*bottom*). An interaction between ribosome-associated Upf1 and the Dcp2 subunit of the decapping enzyme is shown promoting removal of the 5′ cap. Also shown is a hypothetical interaction between Upf1 and a deadenylase (D) that may promote poly(A) shortening for some mRNAs.

Three distinct functions are envisaged for the association of the Upf/Nmd factors with the prematurely terminating ribosome. First, the efficiency of peptide hydrolysis is thought to be enhanced, a conclusion that follows from observations that the absence of these factors promotes nonsense suppression (Weng et al. 1996b; Maderazo et al. 2000). Second, the RNA helicase activity of Upf1 (Czaplinski et al. 1995; Bhattacharya et al. 2000) is thought to address the problem of inefficient termination, by promoting the dissociation of the 60S ribosomal subunit and leaving its 40S partner with the opportunity to reinitiate translation. Although this activity has yet to be characterized, such dissociation could conceivably involve Upf1 unwinding of a segment of 28S rRNA. Finally, the amino-terminal zinc finger domain of Upf1, known to interact with Dcp2 (He and Jacobson 1995; Lykke-Andersen 2002), is postulated to complete the targeting of the mRNA for rapid decay by recruiting the decapping enzyme (Dcp1/Dcp2), promoting decapping and thus rendering the transcript susceptible to 5′ to 3′ decay (not shown). These multiple functions for Upf1 are consistent with the existence of mutations and high-copy suppressors that appear to separate its translation and mRNA decay activities (Weng et al. 1996a,b; Maderazo et al. 2000).

Although not shown in Figure 2, it is important to note that other interactions between Upf1, Upf2/Nmd2, or Upf3 and other mRNA decay factors, such as the deadenylase and the exosome, could stimulate additional modes of decay. Certainly, the interaction of mammalian Upf1 with SMG7 (Unterholzner and Izaurralde 2004; Fukuhara et al. 2005) or a hypothetical interaction between *Drosophila* Upf1 and a factor promoting endonucleolytic cleavage (Gatfield and Izaurralde 2004) could be accommodated by the model. Moreover, although the decapping and exonucleolytic events of NMD may occur in P-bodies (Sheth and Parker 2003; Teixeira et al. 2005), localization is not indicated in the model, largely because the precise site of these events does not affect the overall outline of the key steps in NMD.

Are the Different Models Really That Disparate?

The pathways of gene expression in higher and lower eukaryotes are strikingly similar mechanistically, and the respective factors that have key roles in transcription, RNA processing, export, translation, and mRNA decay are highly conserved. Why should the pioneer round model, the prevalent explanation for mammalian NMD, appear to be so different from the *faux* UTR model that accommodates the available data in yeast, *Drosophila*, and *C. elegans*? There may be more in common than is immediately obvious. For example, both models invoke a central role for Upf1 in the triggering of mRNA decay. Whether it is SMG7 or Dcp2, the principal point is that once Upf1 has associated with a prematurely terminating ribosome, one of its key functions is to ensure that the mRNA in question is targeted for rapid decay. A key point of contention that is likely to be tested soon is whether Upf1 associates with all terminating ribosomes or just those that are terminating prematurely.

One of the major differences between the models centers around the proposed activity of the EJC and its associated NMD factors. In the pioneer round model, the EJC is thought to be poised for action in case a premature terminator is encountered by an upstream ribosome. (Most yeast mRNAs lack introns, but the surveillance complex model assigns an EJC-like role to Hrp1.) The *faux* UTR model rejects such anticipatory activities because yeast experiments have shown that nonsense-containing mRNAs are always available for rapid decay, not just in an initial round of translation. Justification for the importance of the EJC to NMD comes largely from experiments showing that inactivation of EJC proteins inhibits NMD of specific mRNAs. However, *Drosophila* NMD can be fully functional without EJC proteins (Gatfield et al. 2003), some mammalian mRNAs are subject to NMD without prior splicing or proper EJC spacing (Chan et al. 1998; Zhang et al. 1998a; Rajavel and Neufeld 2001; Delpy et al. 2004; LeBlanc and Beemon 2004; Buhler et al. 2006), and EJC proteins may have other functions that affect NMD indirectly. More specifically, EJC proteins appear to have important roles in promoting the translatability of a mature mRNP (Wiegand et al. 2003; Nott et al. 2004; Gudikote et al. 2005; Wilkinson 2005). For example, mRNAs derived from intronless β-globin transcripts associate with smaller polysomes than those derived from intron-containing pre-mRNAs, and tethered EJC factors can restore the translational efficiency of mRNAs otherwise lacking EJCs (Nott et al. 2004). Similar experiments have shown that the translational yield of efficiently spliced TCR-β mRNAs is considerably greater than that derived from transcripts in which splicing is impaired (Gudikote et al. 2005). Since NMD in all cells examined is translation-dependent (see above), the apparent NMD requirement for the EJC could be principally a function of its translational stimulatory activity.

In this "compromise" interpretation, the components of the EJC (and possibly Hrp1 in yeast) are not critical NMD factors, but are modulators of the severity of NMD, largely because they are determinants of translational efficiency. If the EJC does not have to be a direct activator of NMD, then mammalian NMD and yeast NMD may indeed be more similar than they currently appear to be. The absolute requirement for a pioneer round of translation would be eliminated, and NMD in all cells could become a problem of aberrant termination that is solved by NMD factors that both undo the termination mess and ensure that the offending mRNA is eliminated.

A PLETHORA OF ENDOGENOUS NMD SUBSTRATES

The existence of a pathway that promotes rapid decay of nonsense-containing mRNAs raised the question of whether the substrates of this pathway are restricted to aberrant mRNAs, i.e., those derived from genes in which a mutation or an error in transcription or processing has given rise to a premature nonsense codon. Additional substrates were sought by identifying mRNAs that were selectively stabilized in strains harboring mutations in one or more of the genes encoding factors essential for NMD. Initial studies showed that endogenous NMD substrates included inefficiently spliced pre-mRNAs that enter the cytoplasm with their introns intact, mRNAs in which a leaky scanning ribosome bypasses the initiator AUG and commences translation further downstream, some mRNAs containing uORFs, and transcripts with extended 3′UTRs (He et al. 1993; Pulak and Anderson 1993; Cui et al. 1995; Vilela et al. 1998, 1999; Muhlrad and Parker 1999a; Welch and Jacobson 1999; Das et al. 2000). Genome-wide expression profiles of strains containing single or multiple deletions of *UPF1*, *UPF2/NMD2*, or *UPF3*, as well as genes encoding other mRNA decay factors, identified 765 "core transcripts" that are regulated by NMD in yeast (He et al. 2003). The similar transcript profiles of *upf1Δ*, *upf2Δ/nmd2Δ*, and *upf3Δ* strains supported the notion that Upf2/Nmd2 and Upf3 are regulators of the activity of Upf1 (He et al. 1997; Maderazo et al. 2000; He and Jacobson 2001), confirmed the original classes of NMD substrates, and identified several additional classes. The latter included mRNAs subject to +1 frameshifting, bicistronic mRNAs, and transcripts of pseudogenes and transposable elements. Some of these substrates, e.g., the standard nonsense-containing mRNAs, the intron-containing pre-mRNAs, and the mRNAs derived from pseudogenes and transposable elements, can all be considered to be targets of a quality control system that eliminates RNAs capable of giving rise to potentially deleterious translation products. Given the prevalence of transposable elements in eukaryotic genomes, these results suggest that at least one important function of the NMD pathway is the degradation of transcripts generated from these genetic elements. However, the existence of other substrate classes, e.g., mRNAs with uORFs, indicates that NMD has additional regulatory capabilities that include the modulation of specific biochemical pathways. Since many of the uORF-containing mRNAs encode transcription factors (Vilela et al. 1998),

NMD-regulated changes in the expression of these transcripts may be the primary source of indirect effects of *nmd∆* mutations, i.e., changes in mRNA levels that are not accompanied by changes in mRNA decay rates.

Similar types of experiments done in metazoan systems have confirmed several of the substrate classes identified in yeast, but they have not identified specific evolutionarily conserved NMD-regulated transcripts (Mitrovich and Anderson 2000, 2005; Mendell et al. 2004; Rehwinkel et al. 2005; Wittmann et al. 2006). In addition, although common targets of the different NMD factors are apparent from transcript-profiling studies in *Drosophila* (Rehwinkel et al. 2005), similar experiments in mammalian cells raise the possibility that some mRNAs are selectively targeted by only a subset of NMD factors (Gehring et al. 2005; Wittmann et al. 2006). Of particular note are studies indicating that a major function of the NMD pathway in higher organisms is the elimination of alternately spliced transcripts that lack complete ORFs (Hillman et al. 2004). Empirically, this class comprises approximately one-third of alternately spliced mRNAs (Hillman et al. 2004).

THERAPEUTIC APPROACHES TO NONSENSE SUPPRESSION

Nonsense mutations have been implicated in hundreds of inherited diseases, including cystic fibrosis (CF), Duchenne muscular dystrophy (DMD), hemophilias, lysosomal storage disorders, skin disorders, and various cancers (Culbertson 1999). Pharmacological manipulation of premature translation termination has been investigated as a potential approach to treating a patient population with a substantial unmet medical need. Aminoglycoside antibiotics, normally used to inhibit bacterial protein synthesis, are also capable of promoting low-level readthrough of premature nonsense codons in mammalian cells when used at high concentrations (Bedwell et al. 1997; Howard et al. 2000; Keeling et al. 2001; Keeling and Bedwell 2002; Lai et al. 2004; Chapter 30). This treatment does elevate some mRNA levels, but it does not appear to promote general inhibition of NMD (E.M. Welch et al., in prep.).

In mouse models of muscular dystrophy and cystic fibrosis caused by nonsense mutations, aminoglycoside treatment has led to the synthesis of full-length dystrophin and CFTR proteins, as well as to the respective restoration of muscle and chloride channel function (Barton-Davis et al. 1999; Du et al. 2002). The results obtained in these mouse models, and others, have led to the testing of aminoglycoside therapy in patients with disease caused by nonsense mutations (Wilschanski et al. 2000, 2003; Clancy et al. 2001; Wagner et al. 2001; Politano et al. 2003; James et al. 2005; Chapter 30). These important proof-of-concept experiments have demonstrated that small molecules which promote readthrough of premature termination codons have the potential to restore gene function. However, the need for parenteral administration and the potential for serious renal and otic toxicities limit the clinical utility of long-term aminoglycoside therapy, and additional compounds that promote readthrough of disease-causing nonsense codons is an important medical objective. In this regard, recent work has reported the identification and characterization of PTC124, a novel orally bioavailable small molecule with excellent pharmaceutical properties that selectively promotes readthrough of premature nonsense codons (E.M. Welch et al., in prep.; Chapter 30). Two key attributes of this drug are that it does not affect NMD and only promotes readthrough of premature termination codons, not normal termination codons (E.M. Welch et al., in prep.). These features are likely to not only minimize undesirable side effects, but also underscore the notions that the termination and mRNA decay functions of NMD are separable and that premature termination is not the same biochemical event as normal termination.

ACKNOWLEDGMENTS

This work was supported by grants from the National Institutes of Health to A.J. (GM27757 and HD048137) and by grants from the Human Frontiers Science Program Organization (HFSPO) to E.I. We are indebted to members of the Jacobson and Izaurralde labs, present and past, for their contributions to many of the experiments and ideas discussed in this review.

REFERENCES

Altamura N., Groudinsky O., Dujardin G., and Slonimski P.P. 1992. *NAM7* nuclear gene encodes a novel member of a family of helicases with a Zn-ligand motif and is involved in mitochondrial functions in *Saccharomyces cerevisiae. J. Mol. Biol.* **224:** 575–587.

Amrani N., Sachs M.S., and Jacobson A. 2006. Early nonsense: mRNA decay solves a translational problem. *Nature Rev. Mol. Cell. Biol.* **7:** 415–425.

Amrani N., Ganesan R., Kervestin S., Mangus D.A., Ghosh S., and Jacobson A. 2004. A *faux* 3′-UTR promotes aberrant termination and triggers nonsense-mediated mRNA decay. *Nature* **432:** 112–118.

Anders K.R., Grimson A., and Anderson P. 2003. SMG-5, required for *C. elegans* nonsense-mediated mRNA decay, associates with SMG-2 and protein phosphatase 2A. *EMBO J.* **22:** 641–650.

Anderson J.S. and Parker R. 1998. The 3′ to 5′ degradation of yeast mRNAs is a general mechanism for mRNA turnover that requires the *SKI2* DEVH box protein and 3′ to 5′ exonucleases of the exosome complex. *EMBO J.* **17:** 1497–1506.

Atkin A.L., Altamura N., Leeds P., and Culbertson M.R. 1995. The majority of yeast UPF1 co-localizes with polyribosomes in the cytoplasm. *Mol. Biol. Cell* **6:** 611–625.

Atkin A.L., Schenkman L.R., Eastham M., Dahlseid J.N., Lelivelt M.J., and Culbertson M.R. 1997. Relationship between yeast polyribosomes and Upf proteins required for nonsense mRNA decay. *J. Biol. Chem.* **272:** 22163–22172.

Baker K.E. and Parker R. 2004. Nonsense-mediated mRNA decay: Terminating erroneous gene expression. *Curr. Opin. Cell Biol.* **16:** 293–299.

Barton-Davis E.R., Cordier L., Shoturma D.I., Leland S.E., and Sweeney H.L. 1999. Aminoglycoside antibiotics restore dystrophin function to skeletal muscles of mdx mice. *J. Clin. Invest.* **104:** 375–381.

Bedwell D.M., Kaenjak A., Benos D.J., Bebok Z., Bubien J.K., Hong J., Tousson A., Clancy J.P., and Sorscher E.J. 1997. Suppression of a CFTR premature stop mutation in a bronchial epithelial cell line. *Nat. Med.* **3:** 1280–1284.

Beelman C.A. and Parker R. 1995. Degradation of mRNA in eukaryotes. *Cell* **81:** 179–183.

Beelman C.A., Stevens A., Caponigro G., LaGrandeur T.E., Hatfield L., Fortner D.M., and Parker R. 1996. An essential component of the decapping enzyme required for normal rates of mRNA turnover. *Nature* **382:** 642–646.

Behm-Ansmant I. and Izaurralde E. 2006. Quality control of gene expression: A stepwise assembly pathway for the surveillance complex that triggers nonsense-mediated mRNA decay. *Genes Dev.* **20:** 391–398.

Belgrader P., Cheng J., and Maquat L.E. 1993. Evidence to implicate translation by ribosomes in the mechanism by which nonsense codons reduce the nuclear level of human triosephosphate isomerase mRNA. *Proc. Natl. Acad. Sci.* **90:** 482–486.

Bhattacharya A., Czaplinski K., Trifillis P., He F., Jacobson A., and Peltz S.W. 2000. Characterization of the biochemical properties of the human Upf1 gene product that is involved in nonsense-mediated mRNA decay. *RNA* **6:** 1226–1235.

Boeck R., Lapeyre B., Brown C.E., and Sachs A.B. 1998. Capped mRNA degradation intermediates accumulate in the yeast *spb8-2* mutant. *Mol. Cell. Biol.* **18:** 5062–5072.

Bouv.eret E., Rigaut G., Shevchenko A., Wilm M., and Seraphin B. 2000. A Sm-like protein complex that participates in mRNA degradation. *EMBO J.* **19:** 1661–1671.

Brocke K.S., Neu-Yilik G., Gehring N.H., Hentze M.W., and Kulozik A.E. 2002. The human intronless melanocortin 4-receptor gene is NMD insensitive. *Hum. Mol. Genet.* **11:** 331–335.

Buhler M., Steiner S., Mohn F., Paillusson A., and Muhlemann O. 2006. Exon junction complex-independent degradation of nonsense immunoglobulin μ mRNA suggests an evolutionary conserved mechanism for nonsense-mediated mRNA decay. *Nat. Struct. Mol. Biol.* **13:** 462–464.

Cali B.M., Kuchma S.L., Latham J., and Anderson P. 1999. smg-7 is required for mRNA surveillance in *Caenorhabditis elegans*. *Genetics* **151:** 605–616.

Cao D. and Parker R. 2001. Computational modeling of eukaryotic mRNA turnover. *RNA* **7:** 1192–1212.

———. 2003. Computational modeling and experimental analysis of nonsense-mediated decay in yeast. *Cell* **113:** 533–545.

Carter M.S., Li S., and Wilkinson M.F. 1996. A splicing-dependent regulatory mechanism that detects translation signals. *EMBO J.* **15:** 5965–5975.

Chan D., Weng Y.M., Graham H.K., Sillence D.O., and Bateman J.F. 1998. A nonsense mutation in the carboxyl-terminal domain of type X collagen causes haploinsufficiency in schmid metaphyseal chondrodysplasia. *J. Clin. Invest.* **101:** 1490–1499.

Chen C.Y. and Shyu A.B. 2003. Rapid deadenylation triggered by a nonsense codon precedes decay of the RNA body in a mammalian cytoplasmic nonsense-mediated decay pathway. *Mol. Cell. Biol.* **23:** 4805–4813.

Cheng J., Belgrader P., Zhou X., and Maquat L.E. 1994. Introns are cis effectors of the nonsense-codon-mediated reduction in nuclear mRNA abundance. *Mol. Cell. Biol.* **14:** 6317–6325.

Chiu S.Y., Serin G., Ohara O., and Maquat L.E. 2003. Characterization of human Smg5/7a: A protein with similarities to *Caenorhabditis elegans* SMG5 and SMG7 that functions in the dephosphorylation of Upf1. *RNA* **9:** 77–87.

Clancy J.P., Bebok Z., Ruiz F., King C., Jones J., Walker L., Greer H., Hong J., Wing L., Macaluso M., et al. 2001. Evidence that systemic gentamicin suppresses premature stop mutations in patients with cystic fibrosis. *Am. J. Respir. Crit. Care Med.* **163:** 1683–1692.

Coller J.M., Gray N.K., and Wickens M.P. 1998. mRNA stabilization by poly(A) binding protein is independent of poly(A) and requires translation. *Genes Dev.* **12:** 3226–3235.

Cosson B., Berkova N., Couturier A., Chabelskaya S., Philippe M., and Zhouravleva G. 2002a. Poly(A)-binding protein and eRF3 are associated in vivo in human and *Xenopus* cells. *Biol. Cell* **94:** 205–216.

Cosson B., Couturier A., Chabelskaya S., Kiktev D., Inge-Vechtomov S., Philippe M., and Zhouravleva G. 2002b. Poly(A)-binding protein acts in translation termination via eukaryotic release factor 3 interaction and does not influence *PSI*(+) propagation. *Mol. Cell. Biol.* **22:** 3301–3315.

Cui Y., Hagan K.W., Zhang S., and Peltz S.W. 1995. Identification and characterization of genes that are required for the accelerated degradation of mRNAs containing a premature translational termination codon. *Genes Dev.* **9:** 423–436.

Cui Y., Gonzalez C.I., Kinzy T.G., Dinman J.D., and Peltz S.W. 1999. Mutations in the *MOF2/SUI1* gene affect both translation and nonsense-mediated mRNA decay. *RNA* **5:** 794–804.

Culbertson M.R. 1999. RNA surveillance. Unforeseen consequences for gene expression, inherited genetic disorders and cancer. *Trends Genet.* **15:** 74–80.

Culbertson M.R., Underbrink K.M., and Fink G.R. 1980. Frameshift suppression *Saccharomyces cerevisiae*. Genetic properties of group II suppressors. *Genetics* **95:** 833–853.

Czaplinski K., Majlesi N., Banerjee T., and Peltz S.W. 2000. Mtt1 is a Upf1-like helicase that interacts with the translation termination factors and whose overexpression can modulate termination efficiency. *RNA* **6:** 730–743.

Czaplinski K., Weng Y., Hagan K.W., and Peltz S.W. 1995. Purification and characterization of the Upf1 protein: A factor involved in translation and mRNA degradation. *RNA* **1:** 610–623.

Czaplinski K., Ruiz-Echevarria M.J., Paushkin S.V., Han X., Weng Y., Perlick H.A., Dietz H.C., Ter-Avanesyan M.D., and Peltz S.W. 1998. The surveillance complex interacts with the translation release factors to enhance termination and degrade aberrant mRNAs. *Genes Dev.* **12:** 1665–1677.

Das B., Guo Z., Russo P., Chartrand P., and Sherman F. 2000. The role of nuclear cap binding protein Cbc1p of yeast in mRNA termination and degradation. *Mol. Cell. Biol.* **20:** 2827–2838.

Decker C.J. and Parker R. 1993. A turnover pathway for both stable and unstable mRNAs in yeast: Evidence for a requirement for deadenylation. *Genes Dev.* **7:** 1632–1643.

Delpy L., Sirac C., Magnoux E., Duchez S., and Cogne M. 2004. RNA surveillance down-regulates expression of nonfunctional kappa alleles and detects premature termination within the last kappa exon. *Proc. Natl. Acad. Sci.* **101:** 7375–7380.

Denning G., Jamieson L., Maquat L.E., Thompson E.A., and Fields A.P. 2001. Cloning of a novel phosphatidylinositol kinase-related kinase: Characterization of the human SMG-1 RNA surveillance protein. *J. Biol. Chem.* **276:** 22709–22714.

Dinman J.D. and Wickner R.B. 1994. Translational maintenance of frame: Mutants of *Saccharomyces cerevisiae* with altered −1 ribosomal frameshifting efficiencies. *Genetics* **136**: 75–86.

Dominski Z. and Marzluff W.F. 1999. Formation of the 3′ end of histone mRNA. *Gene* **239**: 1–14.

Dostie J. and Dreyfuss G. 2002. Translation is required to remove Y14 from mRNAs in the cytoplasm. *Curr. Biol.* **12**: 1060–1067.

Du M., Jones J.R., Lanier J., Keeling K.M., Lindsey J.R., Tousson A., Bebok Z., Whitsett J.A., Dey C.R., Colledge W.H., et al. 2002. Aminoglycoside suppression of a premature stop mutation in a Cftr−/− mouse carrying a human CFTR-G542X transgene. *J. Mol. Med.* **80**: 595–604.

Dunckley T. and Parker R. 1999. The DCP2 protein is required for mRNA decapping in *Saccharomyces cerevisiae* and contains a functional MutT motif. *EMBO J.* **18**: 5411–5422.

Fribourg S., Gatfield D., Izaurralde E., and Conti E. 2003. A novel mode of RBD-protein recognition in the Y14-Mago complex. *Nat. Struct. Biol.* **10**: 433–439.

Frolova L., Le Goff X., Zhouravleva G., Davydova E., Philippe M., and Kisselev L. 1996. Eukaryotic polypeptide chain release factor eRF3 is an eRF1- and ribosome-dependent guanosine triphosphatase. *RNA* **2**: 334–341.

Frolova L., Le Goff X., Rasmussen H.H., Cheperegin S., Drugeon G., Kress M., Arman I., Haenni A.L., Celis J.E., and Philippe M., et al. 1994. A highly conserved eukaryotic protein family possessing properties of polypeptide chain release factor. *Nature* **372**: 701–703.

Fukuhara N., Ebert J., Unterholzner L., Lindner D., Izaurralde E., and Conti E. 2005. SMG7 is a 14-3-3-like adaptor in the nonsense-mediated mRNA decay pathway. *Mol. Cell* **17**: 537–547.

Gaba A., Jacobson A., and Sachs M.S. 2005. Ribosome occupancy of the yeast *CPA1* upstream open reading frame termination codon modulates nonsense-mediated mRNA decay. *Mol. Cell* **20**: 449–460.

Gatfield D. and Izaurralde E. 2004. Nonsense-mediated messenger RNA decay is initiated by endonucleolytic cleavage in *Drosophila*. *Nature* **429**: 575–578.

Gatfield D., Unterholzner L., Ciccarelli F.D., Bork P., and Izaurralde E. 2003. Nonsense-mediated mRNA decay in *Drosophila:* At the intersection of the yeast and mammalian pathways. *EMBO J.* **22**: 3960–3970.

Gehring N.H., Neu-Yilik G., Schell T., Hentze M.W., and Kulozik A.E. 2003. Y14 and hUpf3b form an NMD-activating complex. *Mol. Cell* **11**: 939–949.

Gehring N.H., Kunz J.B., Neu-Yilik G., Breit S., Viegas M.H., Hentze M.W., and Kulozik A.E. 2005. Exon-junction complex components specify distinct routes of nonsense-mediated mRNA decay with differential cofactor requirements. *Mol. Cell* **20**: 65–75.

Gonzalez C.I., Ruiz-Echevarria M.J., Vasudevan S., Henry M.F., and Peltz S.W. 2000. The yeast hnRNP-like protein Hrp1/Nab4 marks a transcript for nonsense-mediated mRNA decay. *Mol. Cell* **5**: 489–499.

Gozalbo D. and Hohmann S. 1990. Nonsense suppressors partially revert the decrease of the mRNA level of a nonsense mutant allele in yeast. *Curr. Genet.* **17**: 77–79.

Grimson A., O'Connor S., Newman C.L., and Anderson P. 2004. SMG-1 is a phosphatidylinositol kinase-related protein kinase required for nonsense-mediated mRNA decay in *Caenorhabditis elegans*. *Mol. Cell. Biol.* **24**: 7483–7490.

Gudikote J.P., Imam J.S., Garcia R.F., and Wilkinson M.F. 2005. RNA splicing promotes translation and RNA surveillance. *Nat. Struct. Mol. Biol.* **12**: 801–819.

Hagan K.W., Ruiz-Echevarria M.J., Quan Y., and Peltz S.W. 1995. Characterization of *cis*-acting sequences and decay intermediates involved in nonsense-mediated mRNA turnover. *Mol. Cell. Biol.* **15:** 809–823.

He F. and Jacobson A. 1995. Identification of a novel component of the nonsense-mediated mRNA decay pathway by use of an interacting protein screen. *Genes Dev.* **9:** 437–454.

———. 2001. Upf1p, Nmd2p, and Upf3p regulate the decapping and exonucleolytic degradation of both nonsense-containing mRNAs and wild-type mRNAs. *Mol. Cell. Biol.* **21:** 1515–1530.

He F., Brown A.H., and Jacobson A. 1996. Interaction between Nmd2p and Upf1p is required for activity but not for dominant-negative inhibition of the nonsense-mediated mRNA decay pathway in yeast. *RNA* **2:** 153–170.

———. 1997. Upf1p, Nmd2p, and Upf3p are interacting components of the yeast nonsense-mediated mRNA decay pathway. *Mol. Cell. Biol.* **17:** 1580–1594.

He F., Peltz S.W., Donahue J.L., Rosbash M., and Jacobson A. 1993. Stabilization and ribosome association of unspliced pre-mRNAs in a yeast *upf1-* mutant. *Proc. Natl. Acad. Sci.* **90:** 7034–7038.

He F., Li X., Spatrick P., Casillo R., Dong S., and Jacobson A. 2003. Genome-wide analysis of mRNAs regulated by the nonsense-mediated and 5′ to 3′ mRNA decay pathways in yeast. *Mol. Cell* **12:** 1439–1452.

Hentze M.W. and Kulozik A.E. 1999. A perfect message: RNA surveillance and nonsense-mediated decay. *Cell* **96:** 307–310.

Hilleren P. and Parker R. 1999. mRNA surveillance in eukaryotes: Kinetic proofreading of proper translation termination as assessed by mRNP domain organization? *RNA* **5:** 711–719.

Hillman R.T., Green R.E., and Brenner S.E. 2004. An unappreciated role for RNA surveillance. *Genome Biol.* **5:** R8.1–R8.16.

Hodgkin J., Papp A., Pulak R., Ambros V., and Anderson P. 1989. A new kind of informational suppression in the nematode *Caenorhabditis elegans. Genetics* **123:** 301–313.

Hoshino S., Imai M., Kobayashi T., Uchida N., and Katada T. 1999. The eukaryotic polypeptide chain releasing factor (eRF3/GSPT) carrying the translation termination signal to the 3′-poly(A) tail of mRNA. Direct association of erf3/GSPT with polyadenylate-binding protein. *J. Biol. Chem.* **274:** 16677–16680.

Hosoda N., Kobayashi T., Uchida N., Funakoshi Y., Kikuchi Y., Hoshino S., and Katada T. 2003. Translation termination factor eRF3 mediates mRNA decay through the regulation of deadenylation. *J. Biol. Chem.* **278:** 38287–38291.

Howard M.T., Shirts B.H., Petros L.M., Flanigan K.M., Gesteland R.F., and Atkins J.F. 2000. Sequence specificity of aminoglycoside-induced stop codon readthrough: Potential implications for treatment of Duchenne muscular dystrophy. *Ann. Neurol.* **48:** 164–169.

Hsu C.L. and Stevens A. 1993. Yeast cells lacking 5′→3′ exoribonuclease 1 contain mRNA species that are poly(A) deficient and partially lack the 5′ cap structure. *Mol. Cell. Biol.* **13:** 4826–4835.

Ishigaki Y., Li X., Serin G., and Maquat L.E. 2001. Evidence for a pioneer round of mRNA translation: mRNAs subject to nonsense-mediated decay in mammalian cells are bound by CBP80 and CBP20. *Cell* **106:** 607–617.

Jacobson A. 1996. Poly(A) metabolism and translation: The closed loop model. In *Translational control* (ed. J.W.B. Hershey et al.), pp. 451–480. Cold Spring Harbor Laboratory Press, Cold Spring Harbor, New York.

Jacobson A. and Peltz S.W. 2000. Destabilization of nonsense-containing transcripts in *Saccharomyces cerevisiae*. In *Translational control of gene expression* (ed. N. Sonenberg et al.), pp. 827–847. Cold Spring Harbor Laboratory Press, Cold Spring Harbor, New York.

James P.D., Raut S., Rivard G.E., Poon M.C., Warner M., McKenna S., Leggo J., and Lillicrap D. 2005. Aminoglycoside suppression of nonsense mutations in severe hemophilia. *Blood* **106:** 3043–3048.

Kadlec J., Izaurralde E., and Cusack S. 2004. The structural basis for the interaction between nonsense-mediated mRNA decay factors UPF2 and UPF3. *Nat. Struct. Mol. Biol.* **11:** 330–337.

Kashima I., Yamashita A., Izumi N., Kataoka N., Morishita R., Hoshino S., Ohno M., Dreyfuss G., and Ohno S. 2006. Binding of a novel SMG-1-Upf1-eRF1-eRF3 complex (SURF) to the exon junction complex triggers Upf1 phosphorylation and nonsense-mediated mRNA decay. *Genes Dev.* **20:** 355–367.

Kazazian H.H., Jr. 1990. The thalassemia syndromes: Molecular basis and prenatal diagnosis in 1990. *Semin. Hematol.* **27:** 209–228.

Kebaara B., Nazarenus T., Taylor R., Forch A., and Atkin A.L. 2003. The Upf-dependent decay of wild-type PPR1 mRNA depends on its 5′-UTR and first 92 ORF nucleotides. *Nucleic Acids Res.* **31:** 3157–3165.

Keeling K.M. and Bedwell D.M. 2002. Clinically relevant aminoglycosides can suppress disease-associated premature stop mutations in the IDUA and P53 cDNAs in a mammalian translation system. *J. Mol. Med.* **80:** 367–376.

Keeling K.M., Brooks D.A., Hopwood J.J., Li P., Thompson J.N., and Bedwell D.M. 2001. Gentamicin-mediated suppression of Hurler syndrome stop mutations restores a low level of alpha-L-iduronidase activity and reduces lysosomal glycosaminoglycan accumulation. *Hum. Mol. Genet.* **10:** 291–299.

Keeling K.M., Lanier J., Du M., Salas-Marco J., Gao L., Kaenjak-Angeletti A., and Bedwell D.M. 2004. Leaky termination at premature stop codons antagonizes nonsense-mediated mRNA decay in *S. cerevisiae*. *RNA* **10:** 691–703.

Kessler M.M., Henry M.F., Shen E., Zhao J., Gross S., Silver P.A., and Moore C.L. 1997. Hrp1, a sequence-specific RNA-binding protein that shuttles between the nucleus and the cytoplasm, is required for mRNA 3′-end formation in yeast. *Genes Dev.* **11:** 2545–2556.

Kisselev L., Ehrenberg M., and Frolova L. 2003. Termination of translation: Interplay of mRNA, rRNAs and release factors? *EMBO J.* **22:** 175–182.

Koonin E. 1992. A new group of putative RNA helicases. *Trends. Biochem. Sci.* **17:** 495–497.

Lai C.H., Chun H.H., Nahas S.A., Mitui M., Gamo K.M., Du L., and Gatti R.A. 2004. Correction of ATM gene function by aminoglycoside-induced readthrough of premature termination codons. *Proc. Natl. Acad. Sci.* **101:** 15676–15681.

LeBlanc J.J. and Beemon K.L. 2004. Unspliced Rous sarcoma virus genomic RNAs are translated and subjected to nonsense-mediated mRNA decay before packaging. *J. Virol.* **78:** 5139–5146.

Lee B.S. and Culbertson M.R. 1995. Identification of an additional gene required for eukaryotic nonsense mRNA turnover. *Proc. Natl. Acad. Sci.* **92:** 10354–10358.

Leeds P., Peltz S.W., Jacobson A., and Culbertson M.R. 1991. The product of the yeast *UPF1* gene is required for rapid turnover of mRNAs containing a premature translational termination codon. *Genes Dev.* **5:** 2303–2314.

Leeds P., Wood J.M., Lee B.S., and Culbertson M.R. 1992. Gene products that promote mRNA turnover in *Saccharomyces cerevisiae*. *Mol. Cell. Biol.* **12:** 2165–2177.

Le Hir H., Moore M.J., and Maquat L.E. 2000a. Pre-mRNA splicing alters mRNP composition: Evidence for stable association of proteins at exon-exon junctions. *Genes Dev.* **14:** 1098–1108.

Le Hir H., Gatfield D., Izaurralde E., and Moore M.J. 2001. The exon-exon junction complex provides a binding platform for factors involved in mRNA export and nonsense-mediated mRNA decay. *EMBO J.* **20:** 4987–4997.

Le Hir H., Izaurralde E., Maquat L.E., and Moore M.J. 2000b. The spliceosome deposits multiple proteins 20–24 nucleotides upstream of mRNA exon-exon junctions. *EMBO J.* **19:** 6860–6869.

Lejeune F., Li X., and Maquat L.E. 2003. Nonsense-mediated mRNA decay in mammalian cells involves decapping, deadenylating, and exonucleolytic activities. *Mol. Cell* **12:** 675–687.

Losson R. and Lacroute F. 1979. Interference of nonsense mutations with eukaryotic messenger RNA stability. *Proc. Natl. Acad. Sci.* **76:** 5134–5137.

Lykke-Andersen J. 2002. Identification of a human decapping complex associated with hUpf proteins in nonsense-mediated decay. *Mol. Cell. Biol.* **22:** 8114–8121.

Lykke-Andersen J., Shu M.D., and Steitz J.A. 2000. Human Upf proteins target an mRNA for nonsense-mediated decay when bound downstream of a termination codon. *Cell* **103:** 1121–1131.

―――. 2001. Communication of the position of exon-exon junctions to the mRNA surveillance machinery by the protein RNPS1. *Science* **293:** 1836–1839.

Maderazo A.B., Belk J.P., He F., and Jacobson A. 2003. Nonsense-containing mRNAs that accumulate in the absence of a functional nonsense-mediated mRNA decay pathway are destabilized rapidly upon its restitution. *Mol. Cell. Biol.* **23:** 842–851.

Maderazo A.B., He F., Mangus D.A., and Jacobson A. 2000. Upf1p control of nonsense mRNA translation is regulated by Nmd2p and Upf3p. *Mol. Cell. Biol.* **20:** 4591–4603.

Mangus D.A. and Jacobson A. 1999. Linking mRNA turnover and translation: Assessing the polyribosomal association of mRNA decay factors and degradative intermediates. *Methods* **17:** 28–37.

Mangus D.A., Evans M.C., and Jacobson A. 2003. Poly(A)-binding proteins: Multifunctional scaffolds for the post-transcriptional control of gene expression. *Genome Biol.* **4:** 223.221–223.214.

Maquat L.E. 1995. When cells stop making sense: Effects of nonsense codons on RNA metabolism in vertebrate cells. *RNA* **1:** 453–465.

―――. 2000. Nonsense-mediated mRNA decay in mammalian cells: A splicing-dependent means to down-regulate the levels of mRNA that prematurely terminate translation. In *Translational control of gene expression* (ed. N. Sonenberg et al.), pp. 849–868. Cold Spring Laboratory Press, Cold Spring Harbor, New York.

―――. 2004. Nonsense-mediated mRNA decay: Splicing, translation, and mRNP dynamics. *Nat. Rev. Mol. Cell Biol.* **5:** 89–99.

Maquat L.E. and Li X. 2001. Mammalian heat shock p70 and histone H4 transcripts, which derive from naturally intronless genes, are immune to nonsense-mediated decay. *RNA* **7:** 445–456.

Mendell J.T., Medghalchi S.M., Lake R.G., Noensie E.N., and Dietz H.C. 2000. Novel Upf2p orthologues suggest a functional link between translation initiation and nonsense surveillance complexes. *Mol. Cell. Biol.* **20:** 8944–8957.

Mendell J.T., Sharifi N.A., Meyers J.L., Martinez-Murillo F., and Dietz H.C. 2004. Nonsense surveillance regulates expression of diverse classes of mammalian transcripts and mutes genomic noise. *Nat. Genet.* **36:** 1073–1078.

Mitchell P. and Tollervey D. 2003. An NMD pathway in yeast involving accelerated dead-enylation and exosome-mediated 3′→5′ degradation. *Mol. Cell* **11:** 1405–1413.

Mitrovich Q.M. and Anderson P. 2000. Unproductively spliced ribosomal protein mRNAs are natural targets of mRNA surveillance in *C. elegans. Genes Dev.* **14:** 2173–2184.

———. 2005. mRNA surveillance of expressed pseudogenes in *C. elegans. Curr. Biol.* **15:** 963–967.

Muhlrad D. and Parker R. 1994. Premature translational termination triggers mRNA decapping. *Nature* **370:** 578–581.

———. 1999a. Aberrant mRNAs with extended 3′UTRs are substrates for rapid degrada-tion by mRNA surveillance. *RNA* **5:** 1299–1307.

———. 1999b. Recognition of yeast mRNAs as "nonsense containing" leads to both inhi-bition of mRNA translation and mRNA degradation: Implications for the control of mRNA decapping. *Mol. Biol. Cell* **10:** 3971–3978.

Muhlrad D., Decker C.J., and Parker R. 1995. Turnover mechanisms of the stable yeast PGK1 mRNA. *Mol. Cell. Biol.* **15:** 2145–2156.

Nagy E. and Maquat L.E. 1998. A rule for termination-codon position within intron-containing genes: When nonsense affects RNA abundance. *Trends Biochem. Sci.* **23:** 198–199.

Nakamura Y. and Ito K. 2003. Making sense of mimic in translation termination. *Trends Biochem. Sci.* **28:** 99–105.

Nakamura Y., Ito K., and Ehrenberg M. 2000. Mimicry grasps reality in translation termi-nation. *Cell* **101:** 349–352.

Nott A., Le Hir H., and Moore M.J. 2004. Splicing enhances translation in mammalian cells: An additional function of the exon junction complex. *Genes Dev.* **18:** 210–222.

Ohnishi T., Yamashita A., Kashima I., Schell T., Anders K.R., Grimson A., Hachiya T., Hentze M.W., Anderson P., and Ohno S. 2003. Phosphorylation of hUPF1 induces formation of mRNA surveillance complexes containing hSMG-5 and hSMG-7. *Mol. Cell* **12:** 1187–1200.

Page M.F., Carr B., Anders K.R., Grimson A., and Anderson P. 1999. *SMG-2* is a phospho-rylated protein required for mRNA surveillance in *Caenorhabditis elegans* and related to Upf1p of yeast. *Mol. Cell. Biol.* **19:** 5943–5951.

Pal M., Ishigaki Y., Nagy E., and Maquat L.E. 2001. Evidence that phosphorylation of human Upf1 protein varies with intracellular location and is mediated by a wortmannin-sensi-tive and rapamycin-sensitive PI 3-kinase-related kinase signaling pathway. *RNA* **7:** 5–15.

Palacios I.M., Gatfield D., St. Johnston D., and Izaurralde E. 2004. An eIF4AIII-containing complex required for mRNA localization and nonsense-mediated mRNA decay. *Nature* **427:** 753–757.

Parker R. and Song H. 2004. The enzymes and control of eukaryotic mRNA turnover. *Nat. Struct. Mol. Biol.* **11:** 121–127.

Peltz S.W., Brown A.H., and Jacobson A. 1993a. mRNA destabilization triggered by prema-ture translational termination depends on at least three *cis*-acting sequence elements and one *trans*-acting factor. *Genes Dev.* **7:** 1737–1754.

Peltz S.W., Trotta C., He F., Brown A., Donahue J.L., Welch E., and Jacobson A. 1993b. Iden-tification of the *cis*-acting sequences and *trans*-acting factors involved in nonsense-me-diated mRNA decay. In *Protein synthesis and targeting in yeast* (ed. A.J.P. Brown et al.), pp. 1–10. Springer-Verlag, New York.

Pinto I., Na J.G., Sherman F., and Hampsey M. 1992. *cis*- and *trans*-acting suppressors of a translation initiation defect at the *cyc1* locus of *Saccharomyces cerevisiae. Genetics* **132:** 97–112.

Politano L., Nigro G., Nigro V., Piluso G., Papparella S., Paciello O., and Comi L.I. 2003. Gentamicin administration in Duchenne patients with premature stop codon. Preliminary results. *Acta Myol.* **22:** 15–21.

Pulak R. and Anderson P. 1993. mRNA surveillance by the *Caenorhabditis elegans smg* genes. *Genes Dev.* **7:** 1885–1897.

Rajavel K.S. and Neufeld E.F. 2001. Nonsense-mediated decay of human HEXA mRNA. *Mol. Cell. Biol.* **21:** 5512–5519.

Rehwinkel J., Letunic I., Raes J., Bork P., and Izaurralde E. 2005. Nonsense-mediated mRNA decay factors act in concert to regulate common mRNA targets. *RNA* **11:** 1530–1544.

Richter J.D. and Sonenberg N. 2005. Regulation of cap-dependent translation by eIF4E inhibitory proteins. *Nature* **433:** 477–480.

Ruiz-Echevarria M.J. and Peltz S.W. 2000. The RNA binding protein Pub1 modulates the stability of transcripts containing upstream open reading frames. *Cell* **101:** 741–751.

Ruiz-Echevarria M.J., Gonzalez C.I., and Peltz S.W. 1998. Identifying the right stop: Determining how the surveillance complex recognizes and degrades an aberrant mRNA. *EMBO J.* **17:** 575–589.

Salas-Marco J. and Bedwell D.M. 2004. GTP hydrolysis by eRF3 facilitates stop codon decoding during eukaryotic translation termination. *Mol. Cell. Biol.* **24:** 7769–7778.

Serin G., Gersappe A., Black J.D., Aronoff R., and Maquat L.E. 2001. Identification and characterization of human orthologues to *Saccharomyces cerevisiae* Upf2 protein and Upf3 protein (*Caenorhabditis elegans SMG-4*). *Mol. Cell. Biol.* **21:** 209–223.

Sheth U. and Parker R. 2003. Decapping and decay of messenger RNA occur in cytoplasmic processing bodies. *Science* **300:** 805–808.

Shibuya T., Tange T.O., Sonenberg N., and Moore M.J. 2004. eIF4AIII binds spliced mRNA in the exon junction complex and is essential for nonsense-mediated decay. *Nat. Struct. Mol. Biol.* **11:** 346–351.

Shirley R.L., Lelivelt M.J., Schenkman L.R., Dahlseid J.N., and Culbertson M.R. 1998. A factor required for nonsense-mediated mRNA decay in yeast is exported from the nucleus to the cytoplasm by a nuclear export signal sequence. *J. Cell Sci.* **111:** 3129–3143.

Shyu A.B., Belasco J.G., and Greenberg M.E. 1991. Two distinct destabilizing elements in the c-fos message trigger deadenylation as a first step in rapid mRNA decay. *Genes Dev.* **5:** 221–231.

Song H., Mugnier P., Das A.K., Webb H.M., Evans D.R., Tuite M.F., Hemmings B.A., and Barford D. 2000. The crystal structure of human eukaryotic release factor eRF1—Mechanism of stop codon recognition and peptidyl-tRNA hydrolysis. *Cell* **100:** 311–321.

Stahl G., Bidou L., Hatin I., Namy O., Rousset J.P., and Farabaugh P. 2000. The case against the involvement of the NMD proteins in programmed frameshifting. *RNA* **6:** 1687–1688.

Stansfield I., Kushnirov V.V., Jones K.M., and Tuite M.F. 1997. A conditional-lethal translation termination defect in a *sup45* mutant of the yeast *Saccharomyces cerevisiae. Eur. J. Biochem.* **245:** 557–563.

Stansfield I., Jones K.M., Kushnirov V.V., Dagkesamanskaya A.R., Poznyakovski A.I., Paushkin S.V., Nierras C.R., Cox B.S., Ter-Avanesyan M.D., and Tuite M.F. 1995. The products of the *SUP45* (eRF1) and *SUP35* genes interact to mediate translation termination in *Saccharomyces cerevisiae. EMBO J.* **14:** 4365–4373.

Steiger M., Carr-Schmid A., Schwartz D.C., Kiledjian M., and Parker R. 2003. Analysis of recombinant yeast decapping enzyme. *RNA* **9:** 231–238.

Stephenson L. and Maquat L. 1996. Cytoplasmic mRNA for human triosephosphate isomerase is immune to nonsense-mediated decay despite forming polysomes. *Biochimie* **78:** 1043–1047.

Sun X. and Maquat L.E. 2000. mRNA surveillance in mammalian cells: The relationship between introns and translation termination. *RNA* **6:** 1–8.

Tange T.O., Nott A., and Moore M.J. 2004. The ever-increasing complexities of the exon junction complex. *Curr. Opin. Cell Biol.* **16:** 279–284.

Teixeira D., Sheth U., Valencia-Sanchez M.A., Brengues M., and Parker R. 2005. Processing bodies require RNA for assembly and contain nontranslating mRNAs. *RNA* **11:** 371–382.

Tharun S. and Parker R. 2001. Targeting an mRNA for decapping: Displacement of translation factors and association of the Lsm1p-7p complex on deadenylated yeast mRNAs. *Mol. Cell* **8:** 1075–1083.

Thermann R., Neu-Yilik G., Deters A., Frede U., Wehr K., Hagemeier C., Hentze M.W., and Kulozik A.E. 1998. Binary specification of nonsense codons by splicing and cytoplasmic translation. *EMBO J.* **17:** 3484–3494.

Uchida N., Hoshino S., Imataka H., Sonenberg N., and Katada T. 2002. A novel role of the mammalian GSPT/eRF3 associating with poly(A)-binding protein in cap/poly(A)-dependent translation. *J. Biol. Chem.* **277:** 50286–50292.

Unterholzner L. and Izaurralde E. 2004. SMG7 acts as a molecular link between mRNA surveillance and mRNA decay. *Mol. Cell* **16:** 587–596.

Vilela C., Linz B., Rodrigues-Pousada C., and McCarthy J.E. 1998. The yeast transcription factor genes *YAP1* and *YAP2* are subject to differential control at the levels of both translation and mRNA stability. *Nucleic Acids Res.* **26:** 1150–1159.

Vilela C., Ramirez C.V., Linz B., Rodrigues-Pousada C., and McCarthy J.E. 1999. Post-termination ribosome interactions with the 5′-UTR modulate yeast mRNA stability. *EMBO J.* **18:** 3139–3152.

Wagner K.R., Hamed S., Hadley D.W., Gropman A.L., Burstein A.H., Escolar D.M., Hoffman E.P., and Fischbeck K.H. 2001. Gentamicin treatment of Duchenne and Becker muscular dystrophy due to nonsense mutations. *Ann. Neurol.* **49:** 706–711.

Wang W., Czaplinski K., Rao Y., and Peltz S.W. 2001. The role of Upf proteins in modulating the translation read-through of nonsense-containing transcripts. *EMBO J.* **20:** 880–890.

Wang W., Cajigas I.J., Peltz S.W., Wilkinson M.F., and Gonzalez C.I. 2006. A role for Upf2p phosphorylation in *Saccharomyces cerevisiae* nonsense-mediated mRNA decay. *Mol. Cell. Biol.* **26:** 3390–4000.

Weil J.E. and Beemon K.L. 2006. A 3′UTR sequence stabilizes termination codons in the unspliced RNA of Rous sarcoma virus. RNA **12:** 102–110.

Welch E.M. and Jacobson A. 1999. An internal open reading frame triggers nonsense-mediated decay of the yeast SPT10 mRNA. *EMBO J.* **18:** 6134–6145.

Weng Y., Czaplinski K., and Peltz S.W. 1996a. Genetic and biochemical characterization of mutations in the ATPase and helicase regions of the Upf1 protein. *Mol. Cell. Biol.* **16:** 5477–5490.

———. 1996b. Identification and characterization of mutations in the *UPF1* gene that affect nonsense suppression and the formation of the Upf protein complex but not mRNA turnover. *Mol. Cell. Biol.* **16:** 5491–5506.

———. 1998. ATP is a cofactor of the Upf1 protein that modulates its translation termination and RNA binding activities. *RNA* **4:** 205–214.

Wiegand H.L., Lu S., and Cullen B.R. 2003. Exon junction complexes mediate the enhancing effect of splicing on mRNA expression. *Proc. Natl. Acad. Sci.* **100:** 11327–11332.

Wilkie G.S., Dickson K.S., and Gray N.K. 2003. Regulation of mRNA translation by 5′- and 3′-UTR-binding factors. *Trends Biochem. Sci.* **28:** 182–188.

Wilkinson M.F. 2005. A new function for nonsense-mediated mRNA-decay factors. *Trends Genet.* **21:** 143–148.

Wilschanski M., Famini C., Blau H., Rivlin J., Augarten A., Avital A., Kerem B., and Kerem E. 2000. A pilot study of the effect of gentamicin on nasal potential difference measurements in cystic fibrosis patients carrying stop mutations. *Am. J. Respir. Crit. Care Med.* **161:** 860–865.

Wilschanski M., Yahav Y., Yaacov Y., Blau H., Bentur L., Rivlin J., Aviram M., Bdolah-Abram T., Bebok Z., Shushi L., et al. 2003. Gentamicin-induced correction of CFTR function in patients with cystic fibrosis and CFTR stop mutations. *N. Engl. J. Med.* **349:** 1433–1441.

Wilson G.M. and Brewer G. 1999. The search for *trans*-acting factors controlling messenger RNA decay. *Prog. Nucleic Acid Res. Mol. Biol.* **62:** 257–291.

Wittmann J., Hol E.M., and Jack H.M. 2006. hUPF2 silencing identifies physiologic substrates of mammalian nonsense-mediated mRNA decay. *Mol. Cell. Biol.* **26:** 1272–1287.

Yamashita A., Ohnishi T., Kashima I., Taya Y., and Ohno S. 2001. Human SMG-1, a novel phosphatidylinositol 3-kinase-related protein kinase, associates with components of the mRNA surveillance complex and is involved in the regulation of nonsense-mediated mRNA decay. *Genes Dev.* **15:** 2215–2228.

Yun D.F. and Sherman F. 1995. Initiation of translation can occur only in a restricted region of the CYC1 mRNA of *Saccharomyces cerevisiae*. *Mol. Cell. Biol.* **15:** 1021–1033.

Zhang J. and Maquat L.E. 1996. Evidence that the decay of nucleus-associated nonsense mRNA for human triosephosphate isomerase involves nonsense codon recognition after splicing. *RNA* **2:** 235–243.

Zhang J., Sun X., Qian Y., and Maquat L.E. 1998a. Intron function in the nonsense-mediated decay of beta-globin mRNA: Indications that pre-mRNA splicing in the nucleus can influence mRNA translation in the cytoplasm. *RNA* **4:** 801–815.

Zhang J., Sun X., Qian Y., LaDuca J.P., and Maquat L.E. 1998b. At least one intron is required for the nonsense-mediated decay of triosephosphate isomerase mRNA: A possible link between nuclear splicing and cytoplasmic translation. *Mol. Cell. Biol.* **18:** 5272–5283.

Zhang S., Ruiz-Echevarria M.J., Quan Y., and Peltz S.W. 1995. Identification and characterization of a sequence motif involved in nonsense-mediated mRNA decay. *Mol. Cell. Biol.* **15:** 2231–2244.

Zhang S., Welch E.M., Hogan K., Brown A.H., Peltz S.W., and Jacobson A. 1997. Polysome-associated mRNAs are substrates for the nonsense-mediated mRNA decay pathway in *Saccharomyces cerevisiae*. *RNA* **3:** 234–244.

Zhang T., Kruys V., Huez G., and Gueydan C. 2002. AU-rich element-mediated translational control: Complexity and multiple activities of *trans*-activating factors. *Biochem. Soc. Trans.* **30:** 952–958.

Zhouravleva G., Frolova L., Le Goff X., Le Guellec R., Inge-Vechtomov S., Kisselev L., and Philippe M. 1995. Termination of translation in eukaryotes is governed by two interacting polypeptide chain release factors, eRF1 and eRF3. *EMBO J.* **14:** 4065–4072.

24

Localized Translation through Messenger RNA Localization

Elizabeth R. Gavis

Department of Molecular Biology
Princeton University
Princeton, New Jersey 08544

Robert H. Singer

Department of Anatomy and Structural Biology
Albert Einstein College of Medicine
Bronx, New York 10461

Stefan Hüttelmaier

ZAMED
Medical Faculty
University of Halle-Wittenberg
06907 Halle(Saale), Germany

THE ESTABLISHMENT OF CELLULAR ASYMMETRIES and the functions of polarized cells depend on the asymmetric distribution of cytoplasmic proteins. Messenger RNA (mRNA) localization provides an important mechanism for generating asymmetric protein distributions by targeting the synthesis of cytoplasmic proteins to specific subcellular domains. Polarized cellular functions like motility in fibroblasts and synaptic plasticity in dendrites, asymmetric division of budding yeast, and embryonic axis formation in *Xenopus* and *Drosophila* all require proteins that are synthesized from localized mRNAs. Furthermore, mRNA localization and translation are often tightly coupled, such that protein synthesis is restricted to the target destination. This spatial control of protein synthesis afforded by mRNA localization provides advantages over direct protein targeting, allowing rapid responses to local requirements. Moreover, localized mRNA translation prevents untoward effects due to premature or ectopic protein function and ensures spatial control of protein concentrations needed for the coordinated assembly and maintenance of multifunctional protein complexes.

mRNA transport is facilitated by the packaging of mRNAs into large ribonucleoprotein (RNP) particles or granules, which may also function to silence translation during transport. Recent evidence indicates that localized mRNAs are first "marked" in the nucleus, prior to their export, and in some cases this nuclear marking appears to be necessary for translational repression during transport. Factors that regulate localization and translation are frequently shared among mRNAs within a particular cell type, and an increasing number appear to have evolutionarily conserved functions. Here we review the relationship between mRNA localization and localized translation in representative experimental systems, focusing on specific examples of mRNAs whose local translation underlies functional cellular and developmental asymmetries.

mRNA LOCALIZATION IN YEAST: *ASH1* mRNA

The budding yeast *Saccharomyces cerevisiae* exists in both diploid and, under conditions of nutrient deprivation, haploid states. Diploidy results from the

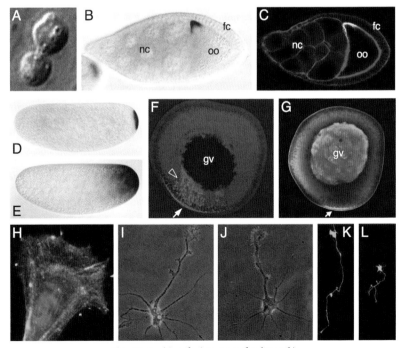

Figure 1. (*See facing page for legend.*)

mating of two haploid cells of the opposite mating type. Mating-type switching, which occurs uniquely in the mother cell, allows mating between recently budded cells. Because the daughter cell retains the original mating type of the mother, a single spore can produce diploid cells through mating between its descendants. Mating-type switching occurs through a genomic rearrangement of the MAT locus initiated by the HO endonuclease, which is expressed specifically in the mother cell. This asymmetry in HO transcription arises from the restriction of the transcriptional repressor Ash1p to the daughter cell by asymmetric localization of *ASH1* mRNA. *ASH1* RNA localizes specifically to the bud tip where it is translated and subsequently imported into the daughter cell nucleus (Fig. 1A) (Long et al. 1997; Takizawa et al. 1997). Although nearly two dozen RNAs have now been shown to be localized to the bud tip (for review, see Gonsalvez et al. 2005), *ASH1* is the only one whose localization is essential for protein targeting.

Figure 1. Localized RNAs and their regulators. (*A*) Localization of *ASH1* mRNA (*red*) in budding yeast. Ash1 protein (*green*) is translated following localization. The nucleus is stained in blue. (*B*) *Grk* mRNA localized to the future dorsal anterior corner of the *Drosophila* oocyte. Nurse cells (nc), oocyte (oo), and follicle cells (fc) are indicated. (*C*) Grk protein (*green*) translated from localized *grk* RNA. The actin cytoskeleton of the nurse cells (nc), oocyte (oo), and follicle cells (fc) is visualized in red. (*D*) Maternal *nos* mRNA localized to the germ plasm at the posterior pole of a *Drosophila* embryo. The small localized pool of *nos* RNA is readily visible by in situ hybridization due to its high concentration at the posterior. The large unlocalized pool produces a diffuse signal throughout the embryo. (*E*) The Nos protein gradient produced by translation of posteriorly localized *nos* RNA and translational repression of unlocalized *nos* RNA. (*F*) RNA encoding the Vg1 localization element is transported to the vegetal cortex after injection into a *Xenopus* oocyte. Arrowhead indicates RNA at an intermediate stage in the localization process, arrow indicates RNA that has reached the vegetal cortex. The oocyte nucleus or germinal vesicle (gv) is indicated. (*G*) Vg1RBP60 protein at the vegetal pole of a *Xenopus* oocyte (*arrow*). Vg1RBP60 is also detected in the oocyte nucleus (gv). (*H*) β-Actin mRNA (*red*) localized to the leading edge of a moving fibroblast. The actin cytoskeleton is shown in green and the nucleus in blue. (*I*) β-Actin mRNA (*red*) localization to growth cone and neurite extensions. (*J*) β-actin protein (*red*) concentrated in the growth cone and filopodia of a neurite. (*K*) ZBP1 (*green*) localized to the growth cone of a primary hippocampal neuron. (*L*) Mutation of ZBP1 (*green*) at Tyr-396 prevents ZBP1 phosphorylation and translational activation of β-actin mRNA, resulting in retardation of neurite outgrowth. (*A,* Adapted from Long et al. [1997]; *B,* courtesy of R. Ray and T. Schüpbach; *C,* courtesy of C. Van Buskirk and T. Schüpbach; *D,* reprinted, with permission, from Bergsten and Gavis [1999]; *E,* courtesy of E. Gavis; *F,* courtesy of Y. Yoon and K. Mowry; *G,* courtesy of T. Kress and K. Mowry; *H,* courtesy of A. Wells; *I, J,* reprinted, with permission, from Zhang et al. [1999]; *K, L,* reprinted, with permission, from Hüttelmaier et al. [2005] © Nature Publishing Group.)

ASH1 mRNA is packaged in an RNP particle or "locasome" that is actively transported to the bud in a repressed state, ensuring the exclusive translation of Ash1p in the daughter cell (Bertrand et al. 1998). Formation and localization of this particle is mediated by four unique *cis*-acting localization elements or "zipcodes" that lie within or overlap the *ASH1* coding region. Although each is able to direct the mRNA to the bud independently, the four act synergistically to ensure efficient transport (Chartrand et al. 1999; Gonzalez et al. 1999). Each of the four localization elements is recognized by She2p, a novel protein that forms a charged RNA-binding face upon dimerization (Bohl et al. 2000; Long et al. 2000; Niessing et al. 2004). She2p shuttles from the nucleus to the cytoplasm in an RNA-dependent manner, indicating that newly transcribed *ASH1* mRNA is likely to be marked for cytoplasmic regulation before its nuclear export (Kruse et al. 2002). She2p forms the core of a complex that assembles when the *ASH1* mRNA arrives in the cytoplasm. This complex includes She3p, which binds to She2p and bridges its connection to myosin (She1p/Myo4p) (Bohl et al. 2000; Takizawa and Vale 2000). Since each of the four zipcode elements is recognized by this complex, *ASH1* can be attached simultaneously to four myosins, ensuring that the mRNA will translocate processively on actin filaments polarized toward the bud tip (Takizawa et al. 1997; Munchow et al. 1999).

In addition to their role in movement of *ASH1* mRNA to the bud tip, the *ASH1* zipcodes also promote the asymmetric distribution of Ash1p by regulating *ASH1* mRNA translation. By forming stem-loop structures within the *ASH1* coding region, these elements, and potentially the She2p/She3p/Myo4p complex bound to them, are thought to impede ribosome transit, thereby allowing *ASH1* mRNA to reach the bud tip before the first protein is fully synthesized (Chartrand et al. 2002). Two additional factors have been identified that bind to *ASH1* mRNA and participate in both its localization and translational repression, providing further evidence that these processes are closely coordinated. Puf6p is a member of the PUF protein family (Chapter 19) that most likely becomes associated with *ASH1* mRNA in the nucleus (Gu et al. 2004). Although PUF proteins have not been previously implicated in mRNA localization, the founding member of this family, *Drosophila* Pumilio (Pum) protein, is involved in spatial control of translation in the *Drosophila* embryo (Chapter 19; see below). Khd1p contains a KH domain and is proposed to join the *ASH1* RNP following nuclear export (Irie et al. 2002). The mechanism by which translational repressors like Khd1p and Puf6p are inactivated at the bud tip to allow translation of Ash1p is currently not known. Interestingly, translation of *ASH1* is required for the correct localization of the mRNA; if translation is inhibited, the mRNA apparently never reaches or does not remain anchored at the bud tip (Gonzalez et al. 1999; Irie et al. 2002). Thus, Ash1p may participate in anchoring its own mRNA. Alternatively, anchoring may depend on association of *ASH1* with the translational apparatus.

mRNA LOCALIZATION AND LOCALIZED TRANSLATION IN OOCYTES AND EMBRYOS

Spatial regulation of gene expression by posttranscriptional mechanisms like mRNA localization and localized translation is particularly important during development in organisms like *Drosophila* and *Xenopus*, where zygotic gene expression is delayed after fertilization and the early period of embryogenesis depends on maternally synthesized mRNAs. In both *Drosophila* and *Xenopus*, asymmetries in localization of mRNAs within the oocyte ultimately give rise to axial asymmetries in the embryo. In many cases, tight coupling between localization and translation is essential to ensure that protein synthesis is appropriately targeted.

Drosophila

A progression of mRNA localization events during oogenesis in *Drosophila* results in the local synthesis of Gurken (Grk), Oskar (Osk), Nanos (Nos), and Bicoid (Bcd) proteins necessary to determine the anterior–posterior and dorsal–ventral body axes. Oogenesis proceeds in a series of 14 morphologically defined stages whereby the oocyte grows at the expense of 15 germ-line-derived nurse cells connected to its anterior end (for review, see Spradling 1993). These nurse cells synthesize maternal mRNAs, including *grk*, *osk*, *nos*, and *bcd*, that are transferred to the oocyte through cytoplasmic channels. Similarly to maternal mRNAs in *Xenopus*, *grk* also appears to be synthesized by the oocyte nucleus (Saunders and Cohen 1999). The entire oocyte–nurse cell complex is surrounded by a layer of somatic follicle cells. Eventually, the nurse cells extrude or "dump" their contents into the oocyte and are eliminated through apoptosis, leading to formation of the mature oocyte (Spradling 1993).

gurken mRNA

Localization of *grk* mRNA to the posterior of the early oocyte leads to local synthesis of the Grk TGF-α ligand, which then signals to the *Drosophila* EGF receptor on a special set of posterior follicle cells (González-Reyes et al. 1995; Roth et al. 1995). This signaling event results in reorientation of the oocyte microtubule cytoskeleton, setting up an overall microtubule polarity that is thought to provide the scaffold for mRNA transport along the anterior–posterior axis of the oocyte (Theurkauf et al. 1992; González-Reyes et al. 1995). Translation of *grk* at these early stages of oogenesis requires Vasa (Vas), an ATP-dependent RNA helicase (Styhler et al. 1998; Tomancak et al. 1998; Ghabrial and Schüpbach 1999). Vas interacts with eIF5B, and *vas* mutations that block this interaction prevent *grk* translation, suggesting that Vas regulates 60S subunit joining on *grk* mRNA (Johnstone and Lasko 2004). Intriguingly, Vas is a target of a meiotic checkpoint that is thought to coordinate establishment of oocyte polarity by Grk with progression through meiosis. Activation of the checkpoint by unrepaired double-stranded breaks results in posttranslational modification of Vas and failure to accumulate Grk (Ghabrial and Schüpbach 1999).

As a consequence of the cytoskeletal reorganization induced by Grk signaling, *grk* mRNA is transported to the anterior margin of the oocyte and then to the future anterodorsal corner (Fig. 1B) by dynein motors (MacDougall et al. 2003). There it is translated once more to produce a local concentration of TGF-α ligand (Fig. 1C) that signals to overlying follicle cells (Neuman-Silberberg and Schüpbach 1993; MacDougall et al. 2003), this time inducing dorsal fates and setting in motion a series of events that establishes the dorsal–ventral axis of the embryo (Nilson and Schüpbach 1999; van Eeden and St. Johnston 1999). Two classes of mutants have been isolated that disrupt the movement of *grk* to the anterodorsal corner, leaving *grk* mislocalized around the entire anterior cortex of the oocyte. In oocytes mutant for *fs(1)K10, squid (sqd), ovarian tumor (otu), half pint (hfp)*, and *hrb27C/hrp48*, mislocalized *grk* is translated, producing dorsalized embryos (Karlin-Mcginness et al. 1996; Norvell et al. 1999; Saunders and Cohen 1999; Van Buskirk and Schüpbach 2002; Goodrich et al. 2004). In contrast, in *encore (enc)* mutant oocytes, mislocalized *grk* is not translated, producing ventralized embryos (Hawkins et al. 1997). Biochemical and genetic evidence supports a model in which K10, Squid, Otu, Hfp, and Hrb27C/Hrp48 (referred to hereafter as Hrp48) participate in the assembly of a translationally repressed *grk* transport RNP. Hrp48 and Sqd, both hnRNP A/B homologs, bind to the *grk* 3'UTR, and Hrp48 interacts with Otu, suggesting that these three proteins are components of the RNP (Norvell et al. 1999; Goodrich et al. 2004). The repressed state of this RNP may be conferred by the translational repressor Bruno (Bru), which interacts with *grk* genetically and binds to the *grk* 3'UTR in vitro and to Sqd in vitro and in vivo (Kim-Ha et al. 1995; Norvell et al. 1999; Filardo and Ephrussi 2003). By an as-yet-unknown mechanism, Enc may facilitate anchoring of *grk* at the anterodorsal corner and, together with Vas, activate *grk* translation (Styhler et al. 1998; Tomancak et al. 1998; Van Buskirk and Schüpbach 2002).

oskar mRNA

Concurrent with transport of *grk* to the anterodorsal corner of the oocyte, *osk* mRNA accumulates at the oocyte posterior. Osk protein synthesized from localized *osk* mRNA maintains *osk* at the posterior and initiates assembly of the germ plasm, a specialized cytoplasm rich in RNA–protein complexes that contains the germ-cell determinant(s) (Ephrussi et al. 1991; Kim-Ha et al. 1991). Osk-dependent assembly of germ plasm is also essential for the posterior localization of *nos* mRNA and synthesis of Nos protein, which determines abdominal fates in the embryo (Ephrussi et al. 1991; Kim-Ha et al. 1991; Gavis and Lehmann 1994). The ease with which Osk can assemble germ plasm and recruit *nos* mRNA when ectopically produced, leading to ectopic germ cells and patterning defects, necessitates tight coupling between *osk* mRNA localization and translation (Ephrussi and Lehmann 1992; Smith et al. 1992). Genetic and biochemical studies have identified a bevy of proteins that participate in localization and/or translational regulation of *osk* mRNA and suggest that the complex regulatory needs of this mRNA are met by its incorporation into RNPs that are remodeled throughout its life cycle.

Like *grk*, *osk* mRNA accumulates first in the early oocyte, but unlike *grk*, its translation is repressed during this period. Mutations in *armitage (armi), aubergine (aub), spindleE (spnE),* and *maelstrom (mael)*, which encode components of the RNAi pathway, cause precocious accumulation of Osk protein in the early oocyte (Cook et al. 2004). Potentially relevant in this regard are predicted binding sites for several microRNAs within the *osk* 3′UTR, although the validity of these sites in vivo has not been tested (Enright et al. 2003; Stark et al. 2003; Nakahara et al. 2005). Thus, *osk* may be a target of microRNA-mediated repression in the early oocyte. Mutation of *me31B*, which encodes a DEAD-box family member, causes ectopic synthesis of Osk in nurse cells at these stages, suggesting that a distinct complex regulates translation of *osk* as it moves from the nurse cells to oocyte (Nakamura et al. 2001).

Transport of *osk* to the oocyte posterior during mid-oogenesis depends on microtubules and kinesin, although whether kinesin plays a direct role in this movement has been debated (Glotzer et al. 1997; Brendza et al. 2000; Cha et al. 2002). Transport of *osk* is thus far unique in its requirement not only for 3′UTR sequences, but also for a splicing event that removes its first intron (Kim-Ha et al. 1993; Hachet and Ephrussi 2004). The additional requirement for exon junction complex

(EJC) components Mago nashi (Mago), Y14/Tsunagi, and eIF4AIII in posterior transport of *osk* (Newmark and Boswell 1994; Hachet and Ephrussi 2001; Mohr et al. 2001; Palacios et al. 2004) suggests that splicing of the first intron positions the EJC so that it can initiate the assembly of an *osk* transport RNP prior to nuclear export.

In the cytoplasm, additional proteins are recruited to the *osk* RNP, either through interactions with *osk* mRNA itself, or through protein–protein interactions. All of these proteins, including the EJC components, are present in the oocyte in distributions similar to *osk* mRNA, consistent with their inclusion in an *osk* transport RNP. Barentsz (Btz), recruited through an interaction with eIF4AIII and Cup (see below), is the only protein identified that appears to function solely for posterior transport of *osk* (van Eeden et al. 2001; Wilhelm et al. 2003; Palacios et al. 2004). The double-stranded RNA-binding protein Staufen (Stau) also plays a distinct role, being required not only for transport, but also for anchoring and translation of *osk* at the posterior (St. Johnston et al. 1991; Rongo et al. 1995; Micklem et al. 2000). Paralleling their roles in regulation of *grk*, Hrp48, Sqd, and Bru interact with the *osk* 3'UTR, and mutations in *hrp48* and *sqd* cause defects in *osk* localization and ectopic production of Osk protein, suggesting that Hrp48 and Sqd participate in coupling of *osk* localization and translation (Huynh et al. 2004; Yano et al. 2004; Norvell et al. 2005). Some alleles of *hrp48* affect localization without affecting translation, suggesting that different domains of Hrp48 affect these different functions (Huynh et al. 2004).

For *osk* as well as for *grk*, Bru's role appears to be limited to translational repression. Bru interacts with Cup, an eIF4E-binding protein that is thought to block translation initiation by preventing the interaction of eIF4E with eIF4G and thus recruitment of the 43S preinitiation complex (Nakamura et al. 2004; Nelson et al. 2004; Zappavigna et al. 2004; Chapter 19). Analysis of Bru-dependent repression in an ovarian in vitro translation system has shown that Bru does indeed inhibit 43S recruitment to a reporter mRNA containing Bru-binding sites (Chekulaeva et al. 2006). Surprisingly, however, this inhibition can occur in both a Cup-dependent and a Cup-independent manner. In this second, Cup-independent mechanism, Bru mediates oligomerization of *osk* mRNA and assembly of large RNPs that sequester *osk* from the translation machinery (Chekulaeva et al. 2006). Repression of *grk* by Bru does not require Cup (Nakamura et al. 2004), suggesting that Bru may repress *grk* by this mechanism as well. Whether Hrp48 or Sqd interfaces with either of these mechanisms is not yet known.

Because Osk protein is required to maintain *osk* mRNA at the oocyte posterior, translation of *osk* must be activated effectively upon localization (Ephrussi et al. 1991; Kim-Ha et al. 1991). Whereas some components of the *osk* transport complex, like the EJC proteins and Btz, are not retained at the oocyte posterior, repressors including Bru, Cup, and Hrp48 persist along with *osk* RNA. Thus, the repressive activities of these proteins must be overcome for *osk* translation to ensue. The mechanism by which *osk* translation is activated is not yet known, but it requires a derepressor element in the *osk* 5′UTR and Stau protein (Rongo et al. 1995; Gunkel et al. 1998; Micklem et al. 2000).

bicoid and *nanos* mRNAs

Localization of *bcd* and *nos* mRNAs to the anterior and posterior poles of the oocyte, respectively, provides the embryo with sources for opposing gradients of Bcd and Nos proteins that pattern the anterior–posterior body axis (Berleth et al. 1988; Gavis and Lehmann 1992). It is interesting to note that Bcd and Nos proteins are themselves translational regulators, such that their local synthesis confers spatial regulation on the synthesis of proteins from target mRNAs that are not themselves localized (Chapter 19).

Whereas nearly all *bcd* mRNA in the embryo is localized at the anterior pole, inefficient localization leaves the vast majority of *nos* mRNA distributed throughout the embryo (Bergsten and Gavis 1999). This large pool of unlocalized *nos* mRNA creates a problem for the embryo, because production of Nos ectopically in the anterior region blocks head and thorax development (Wharton and Struhl 1989; Gavis and Lehmann 1992). Proper development of the embryo is ensured, therefore, by obligate linkage of translation and posterior localization of *nos*. Translational repression of unlocalized *nos* mRNA prevents production of Nos in the anterior of the embryo, whereas localization of *nos* to the germ plasm at the posterior activates translation to produce the Nos protein gradient (Fig. 1D, E) (Gavis and Lehmann 1994). Posterior localization is indeed essential for *nos* translation, as mutations in components of the germ plasm prevent *nos* localization and production of Nos protein. Consequently, these embryos fail to develop abdominal segments (Gavis and Lehmann 1994; Wang et al. 1994).

Unlike *grk* and *osk*, *nos* mRNA is not localized until late stages of oogenesis, after the nurse cells initiate apoptosis and extrude their contents into the oocyte (Wang et al. 1994; Forrest and Gavis 2003). A bolus of *nos* enters the oocyte at this time and diffuses to the posterior, where it is trapped by the localized germ plasm. Microtubule-dependent streaming movements that mix the incoming nurse contents with the oocyte cytoplasm facilitate this passive accumulation of *nos* at the posterior (Forrest and Gavis 2003). Targeting of *nos* to the posterior is mediated by a complex *cis*-acting localization signal within the *nos* 3'UTR; interaction of localization factors with this signal is thought to package *nos* into an RNP competent for germ plasm association (Gavis et al. 1996a; Bergsten and Gavis 1999; Bergsten et al. 2001). Translational repression of *nos* is mediated by a translational control element (TCE) that overlaps but is functionally distinct from the *nos* localization signal (Dahanukar and Wharton 1996; Gavis et al. 1996b; Smibert et al. 1996; Crucs et al. 2000). The TCE functions through the formation of two stem-loops, designated II and III, that are required differentially in the embryo and ovary (Crucs et al. 2000; Forrest et al. 2004). Stem-loop II mediates translational repression of unlocalized *nos* RNA during early embryogenesis through its interaction with Smaug (Smg), a conserved SAM domain protein (Dahanukar et al. 1999; Smibert et al. 1999; Crucs et al. 2000; Forrest et al. 2004) whereas stem-loop III functions primarily in the ovary, through its interaction with Glo, an hnRNP F/H homolog (Forrest et al. 2004; Kalifa et al. 2006).

Whether Glo is also a component of the *nos* transport RNP, and how Glo represses translation of *nos* that fails to become localized, is not yet known. Smg interacts with Cup and is therefore thought to repress *nos* translation by blocking the interaction of eIF4E with eIF4G (Nelson et al. 2004). Contrary to this model, *nos* mRNA remains associated with polysomes under conditions where *nos* mRNA localization is blocked and Nos protein cannot be detected (Clark et al. 2000), suggesting a repression mechanism that acts downstream of initiation. Thus, Smg, like Bru, may act at two different steps to block *nos* translation. Alternatively, Glo could be the effector of a postinitiation block. The repressive activity of Glo and Smg must be blocked at the posterior pole of the oocyte and embryo to allow *nos* translation. In a model suggested by overlap between the TCE and the *nos* RNA localization signal, localized RNA is not subject to repression because binding of repressors and binding of localization factors are mutually exclusive (Bergsten and Gavis 1999; Crucs et al. 2000). Other scenarios include inactivation of Glo and Smg by factors localized at the posterior pole.

In contrast to *grk*, *osk*, and *nos*, translation of *bcd* is temporally rather than spatially regulated. Translation of *bcd*, regardless of its location, is repressed throughout oogenesis, and is activated only upon fertilization, through cytoplasmic polyadenylation (Driever and Nüsslein-Volhard 1988; Sallés et al. 1994). The particular requirement for temporal control of *bcd* translation is not entirely clear, as there are no known mutations that disrupt repression of *bcd* without blocking cytoplasmic polyadenylation in general. However, temporal control effectively prevents premature translation of *bcd* mRNA during transport. The efficiency of *bcd* localization, which suffices to target Bcd synthesis to the anterior (Bergsten and Gavis 1999), further obviates the need for spatial control of *bcd* translation. Moreover, the ability of Nos to repress translation of *bcd* RNA provides a fail-safe mechanism to ensure that *bcd* is translated only at the anterior of the embryo (Wharton and Struhl 1989; Gavis and Lehmann 1992).

Xenopus: Vg1 mRNA

At least sixteen different RNAs are localized to the vegetal pole of the *Xenopus* oocyte by one of two temporally and mechanistically distinct pathways (King et al. 2005). The best characterized of these, Vg1, has provided a model for dissecting the microtubule-dependent late pathway. Localization of Vg1 in the oocyte results in segregation of Vg1 RNA, and consequently Vg1 protein, to the vegetal cells of the early-cleavage-stage embryo (Weeks and Melton 1987; Tannahill and Melton 1989). Presumably, restriction of Vg1, a TGF-β family member, to these cells is required for its functions in mesoderm induction and patterning (Birsoy et al. 2006). Localization of Vg1 is directed by repeated sequence motifs within the Vg1 3'UTR (Fig. 1F) (Deshler et al. 1997; Gautreau et al. 1997; Havin et al. 1998). In lieu of genetics, biochemical purification of proteins that interact with these motifs identified three candidate Vg1 localization factors: Vg1RPB/Vera, the *Xenopus* homolog of ZBP1 (Deshler et al. 1998; Havin et al. 1998; see below); Vg1RBP60, an hnRNP I family member (Cote et al. 1999); and 40LoVe, a *Xenopus* hnRNP D homolog (Czaplinski et al. 2005). Although all three accumulate with Vg1 at the vegetal cortex (see Fig. 1G for Vg1RBP60), they first become associated with Vg1 in the oocyte nucleus, suggesting that assembly of a Vg1 transport RNP is initiated while the mRNA is still nuclear (Deshler et al. 1998; Havin et al. 1998; Cote et al. 1999; Kress et al. 2004; Czaplinski et al. 2005; Czaplinski and Mattaj 2006). This complex is then modified in the cytoplasm, through phosphorylation of Vg1RBP60 and the recruitment of additional proteins, including the *Xenopus* Stau homolog (Xstau) (Xie et al. 2003; Kress et al. 2004; Yoon and Mowry 2004). The Vg1 RNP also contains kinesin I (Yoon and Mowry 2004) and perturbation of kinesin II significantly inhibits Vg1 localization (Betley et al. 2004), suggesting that the Vg1 RNP is transported to the vegetal cortex by one or more kinesin motors.

Prior to localization, Vg1 RNA is translationally repressed via a translational control element located downstream of the mRNA localization element within the Vg1 3'UTR (Mowry and Melton 1992; Wilhelm et al. 2000; Otero et al. 2001). An ELAV family member, ElrB, binds to a motif within the translational control element and has been implicated in mediating repression of Vg1 (Campagnoni et al. 1980). It is not yet known whether ElrB is a component of the Vg1 transport RNP, although its cytoplasmic location indicates that it would likely join the complex following export from the nucleus. Upon localization of Vg1 mRNA to the vegetal cortex, translational repression is alleviated through the action of Vg1RBP71, a ZBP2 homolog (see below) (Kolev and Huber 2003). Vg1RBP71 binds to the Vg1 mRNA localization element and stimulates cleavage of Vg1 RNA at a polyadenylation signal that lies upstream of the translational control element. Cleavage and polyadenylation at this site remove the translational control element, activating translation. Since the Vg1RPB71-binding site overlaps the motif recognized by Vg1RBP60, Vg1RBP71 may displace Vg1RBP60 to remodel the Vg1 RNP and allow cleavage of the RNA (Kolev and Huber 2003). Whether Vg1RBP71 associates with Vg1 RNA only upon localization, and how its function is coordinated with localization, is not clear, however. Vg1RBP71 also binds to other localized mRNAs, including mRNAs localized by the microtubule-independent early pathway and mRNAs localized to the animal hemisphere, suggesting that these mRNAs could be similarly regulated (Kroll et al. 2002).

LOCALIZATION AND TRANSLATION OF mRNAs THAT MEDIATE POLARITY AND POLARIZED FUNCTION IN SOMATIC CELLS

β-Actin mRNA

Localization of β-actin mRNA to the leading edge of motile fibroblasts, growth cones of developing neurites, and dendritic spines of neuroblastoma cells provides a mechanism to enrich β-actin protein at sites characterized by high actin dynamics (Fig. 1H–J) (Sundell and Singer 1990; Bassell et al. 1998; Shestakova et al. 2001; Zhang et al. 2001). Because the rate of actin polymerization depends on the availability of monomeric actin, localized synthesis of β-actin facilitates rapid de novo polymerization of actin monomers in a spatially restricted fashion. Perturbations that impair localized synthesis of β-actin mRNA affect fibroblast motility, growth of dendritic filopodia and synapse formation, and axonal growth cone extension (Kislauskis et al. 1997; Shestakova et al. 1999; Eom et al. 2003; Hüttelmaier et al. 2005), all of which require controlled actin dynamics.

In all cellular systems for which asymmetric sorting of the β-actin mRNA has been described, transport is dependent on a zipcode located in the β-actin 3′UTR (Kislauskis et al. 1993, 1994; Zhang et al. 2001). The β-actin zipcode consists of a core element of 54 nucleotides that includes a highly conserved ACACC motif. This element is absolutely essential to target β-actin mRNA and mediates localization when fused to heterologous mRNAs (Kislauskis et al. 1994). The zipcode is recognized by Zip-code Binding Protein (ZBP1) (Ross et al. 1997), a member of the conserved VICKZ protein family that also includes *Xenopus* Vg1 RBP/Vera. Members of this family are characterized by the presence of two amino-terminal RNA recognition motifs (RRMs) and four carboxy-terminal hnRNP K homology (KH) domains (for review, see Yisraeli 2005). Binding is mediated by the two most carboxy-terminal KH domains, and over-expression of these domains alone results in delocalization of β-actin mRNA and inhibition of fibroblast motility, presumably by hindering the assembly of transport complexes (Farina et al. 2003). Direct involvement of ZBP1-mediated mRNA transport in regulating actin dynamics is further supported by the observations that knock-down of ZBP1 in primary hippocampal neurons by morpholino oligonucleotides interferes with F-actin density in dendritic spines (Eom et al. 2003) and siRNA knock-down of ZBP1 reduces peripheral β-actin protein concentrations in neuroblastoma cells and impairs neurite outgrowth (Hüttelmaier et al. 2005).

Although ZBP1 is primarily a cytoplasmic protein, it probably associates with β-actin mRNA in the nucleus near the site of β-actin mRNA transcription (Oleynikov and Singer 2003; Hüttelmaier et al. 2005). ZBP2, a predominantly nuclear protein, also associates with β-actin mRNA and is implicated in its localization (Gu et al. 2002), but factors that join the RNP in the cytoplasm have yet to be identified. Granules containing ZBP1 and β-actin mRNA can be detected at the periphery of chick embryonic fibroblasts and in the dendrites of cultured hippocampal neurons (Farina et al. 2003; Tiruchinapalli et al. 2003). Transport of these granules to the leading edge of motile fibroblasts is mediated by actin filaments, possibly via myosin motors (Sundell and Singer 1991; Latham et al. 2001). In contrast, in neurites and growth cones of cultured neurons, β-actin mRNA granules appear to be associated with microtubules (Bassell et al. 1998), and live imaging of ZBP1 granules in dendrites of cultured hippocampal neurons reveals velocities consistent with microtubule-dependent transport (Fusco et al. 2003; Tiruchinapalli et al. 2003). Following transport, β-actin mRNA is thought to be anchored at the cell periphery by the actin cytoskeleton, probably through interactions with the F-actin-binding protein EF1-α (eEF1A) (Liu et al. 2002).

ZBP1 also plays a key role in linking translational activity with localization of β-actin mRNA. ZBP1 bound to the β-actin 3′UTR represses translation by preventing joining of the large ribosomal subunit to the 48S initiation complex positioned at the AUG (Ross et al. 1997; Hüttelmaier et al. 2005), a mechanism first described for repression of 15-lipoxygenase (lox) mRNA translation during erythroid differentiation by hnRNPs K and E1 (Ostareck et al. 1997, 2001). Repression of lox mRNA is relieved by tyrosine-phosphorylation of hnRNP K by c-Src, which inhibits binding of hnRNP K to the lox 3'UTR (Ostareck-Lederer et al. 2002). Similarly, phosphorylation of a specific tyrosine residue near the ZBP1 KH-3 domain that is essential for binding to the β-actin 3'UTR abolishes ZBP1 binding, allowing β-actin mRNA translation (Fig. 1K, L) (Farina et al. 2003; Hüttelmaier et al. 2005). In this case, the restriction of Src activity to the cell periphery results in derepression of β-actin mRNA only after it has reached its target destination, thereby conferring spatial control on β-actin protein synthesis (Hüttelmaier et al. 2005). Additional studies will be required to determine whether spatial control of β-actin translational activation is accomplished solely by the subcellular restriction of Src kinase or, as in the case of the *Drosophila* mRNAs, multiple mechanisms are required to prevent premature activation of translation.

A Cancer Relevance for ZBP1-mediated Targeting of β-Actin mRNA

Members of the VICKZ protein family are highly expressed in a variety of human tumors (for review, see Yaniv and Yisraeli 2002). In metastatic rat mammary tumor cells, however, ZBP1 is down-regulated relative to its expression in noninvasive cells from the same primary tumor. Moreover, increasing expression of ZBP1 in metastatic cells to levels found in noninvasive tumor cells suppresses invasion in vivo (Wang et al. 2004). A possible connection between ZBP1-mediated mRNA transport and metastasis is suggested by differences in β-actin mRNA localization in MTLn3 (metastatic) versus MTC (nonmetastatic) cell lines (Shestakova et al. 1999). Localization of β-actin mRNA at the leading edge in MTC cells correlates with higher levels of F-actin in the leading edge and reduced metastatic potential relative to MTLn3 cells, where β-actin mRNA is delocalized. The nonmetastatic cells also show directional movement consistent with their morphological polarization, whereas cells with greater metastatic potential and delocalized β-actin mRNA exhibit unpolarized, amoeboid movement (Shestakova et al. 1999). Furthermore, reduction of β-actin mRNA localization by treatment with antisense oligonucleotides to the zipcode, which disrupts the interaction between ZBP1 and β-actin mRNA, results in loss of both cell polarity and directional motility (Shestakova et al. 2001). These studies implicate ZBP1 as a "metastasis suppressor" that limits the migratory behavior of tumor cells in response to chemotactic factors or other signals by promoting stable cytoskeletal polarization and polarized motility through β-actin mRNA localization.

Myelin Basic Protein mRNA

In the central nervous system, axons are insulated by myelin sheaths produced by oligodendrocytes. One of the major components of myelin, myelin basic protein (MBP), is found on free polyribosomes recovered from the myelin fraction after subcellular fractionation, suggesting that it is synthesized on site in the myelinating sheaths (Campagnoni et al. 1980; Colman et al. 1982). This finding is supported by the observation that both endogenous MBP mRNA and synthetic MBP mRNA injected into the soma of oligodendrocytes are localized to the distal oligodendrocyte processes, near sites of myelin assembly (Ainger et al. 1993). Because MBP is involved in compacting membranes to form myelin, on-site synthesis via localization of MBP mRNA presumably prevents insertion of MBP into the cell's internal membranes where it would have deleterious effects.

Transport of MBP mRNA has been studied largely through the analysis of fluorescently labeled mRNA injected into the soma of cultured oligodendrocytes. Injected MBP mRNA, like endogenous MBP mRNA, assembles into RNP granules (Ainger et al. 1993). Movement of these granules to the distal oligodendrocyte processes requires an intact microtubule cytoskeleton and kinesin motors (Carson et al. 1997), and velocity measurements are consistent with microtubule-dependent transport (Ainger et al. 1993). Analysis of injected MBP mRNA identified a 21-nucleotide RNA transport signal (RTS, also termed 2ARE) that is both necessary and sufficient for transport to processes (Ainger et al. 1997). This motif is different from the β-actin zipcode and is found in a variety of mRNAs, some of which are localized to distal processes in oligodendrocytes or to axons in primary hippocampal neurons (LoPresti et al. 1995; Aronov et al. 2001). As revealed by in vitro binding studies, the RTS/2ARE of MBP mRNA associates with the multifunctional RNA-binding protein hnRNP A2, which resides primarily in the nucleus of steady-state cells (Munro et al. 1999). As myelination occurs during oligodendrocyte maturation, hnRNP A2 protein becomes enriched in the cytoplasm where it assembles into granules with MBP mRNA (Munro et al. 1999; Brumwell et al. 2002; Maggipinto et al. 2004). Depletion of both cytoplasmic and nuclear hnRNP A2 following treatment of oligodendrocytes with antisense oligonucleotides results in decreased translocation of injected MBP mRNA to processes (Kwon et al. 1999), supporting a role for hnRNP A2 in mRNA transport. Analysis of the myelination phenotype observed in quaking viable (qk(v)) mice suggests that the QKI RNA-binding protein may also participate in the nuclear export and translocation of the MBP mRNA in oligodendrocytes (Larocque et al. 2002).

To date, little is known about regulatory factors and mechanisms involved in coordinating the localized translation of MBP mRNA, and repression of MBP mRNA translation during the transport process has not been explicitly demonstrated. However, MBP mRNA granules contain components of the translational machinery including arginyl-tRNA synthetase, elongation factor 1A (eEF1A), and 18S ribosomal RNA in addition to hnRNP A2, suggesting that they carry along some of the factors they need for translation in the myelin compartment (Barbarese et al. 1995; Ainger et al. 1997; Carson and Barbarese 2005). The MBP mRNA RTS enhances cap-dependent translation of a GFP reporter in vitro and in vivo, and this enhancement is abolished by depletion of hnRNP A2 (Kwon et al. 1999). Although the significance of this RTS enhancer function is not known, it may help to activate synthesis of MBP following transport.

mRNAs Localized in Neuronal Cells

Localized translation plays an important role in the highly polarized functions of neurons. First suggested by the discovery of polyribosomes beneath postsynaptic sites on dendrites (Steward and Levy 1982), local synthesis of proteins in dendrites enables synapses to modulate their strength in response to stimulation, a phenomenon known as synaptic plasticity. Emerging roles for localized translation within dendrites in regulating specific forms of synaptic plasticity linked to memory consolidation and learning have garnered much interest in recent years (Chapter 18). In situ hybridization experiments together with isolation and amplification of mRNAs from neuronal processes have identified a large number of dendritically localized mRNAs encoding a variety of proteins (Eberwine et al. 2001; Steward and Schuman 2001), implicating dendritic mRNA localization in dendritically localized translation. Several mRNAs have been detected within different axonal domains in mature vertebrate neurons, although evidence for ribosomes in the distal portions of mature axons is lacking (Mohr and Richter 2000). Thus, the functional significance of mRNA localization in mature axons is unclear. In contrast, mRNAs localized to exploratory growth cones in developing neurons are likely to play a major role in directing protein synthesis to distal regions for the regulation of axon development and guidance (for review, see Martin 2004; van Horck et al. 2004). The importance of β-actin mRNA localization in local synthesis of actin in dendrites and growth cones is described above. Below, we discuss several additional mRNAs whose local translation in neuronal processes appears to depend on selective mRNA transport.

Axonal mRNA Localization—Tau mRNA

In cultured primary neurons, in differentiated embryonic carcinoma (P19) cells, as well as in the rat cortex, mRNA encoding the microtubule-associated protein tau is localized along developing axons and in the initial region of mature axons that are characterized by the presence of polysomes (Steward and Ribak 1986; Litman et al. 1993). Whereas both tau mRNA and protein are present primarily in the cell body and axon, a second microtubule-associated protein, Map2, and its mRNA are segregated to dendrites (Kindler et al. 1996; Aronov et al. 2002). This partitioning of microtubule binding and organizing proteins is presumed to modulate microtubule architecture differentially in axonal versus dendritic processes of neuronal cells, and therefore may contribute to the polarization of neurons.

Transport of tau mRNA from the cell body to the axon is microtubule-dependent and is facilitated by an axonal localization sequence (ALS) located in the tau 3′UTR (Aronov et al. 2001). Removal of this signal prevents axonal localization of both tau mRNA and protein. Moreover, swapping the tau ALS with the dendritic localization signal from MAP2 mRNA targets MAP2 mRNA and protein to axons, and tau mRNA and protein to dendrites (Aronov et al. 2001). Together, these experiments provide evidence that local synthesis of tau in axons requires axonal targeting of tau mRNA. The ALS also contains an AU-rich region that is required for the stabilization of tau mRNA via association with HuD, a protein involved in stabilization of several axonal mRNAs (Good 1995; Aranda-Abreu et al. 1999). Both HuD and KIF3, a kinesin family member, colocalize with tau RNA granules in differentiated P19 neuronal cells, and partial depletion of KIF3 with antisense oligonucleotides results in a decrease in tau mRNA and protein levels as well as axonal targeting (Aronov et al. 2002). Thus, tau mRNA stability may be linked to transport via HuD. Biochemical studies have also identified KIF3, HuD, the ZBP1 homolog IMP-1, and a RAS-Gap-SH3 domain binding protein G3BP-1 as components of the tau RNP granule (Aronov et al. 2001, 2002; Atlas et al. 2004). Complexes immunoprecipitated with an antibody to HuD contain β-actin mRNA as well as tau, suggesting that both mRNAs can be part of the same transport granule in P19 axons (Atlas et al. 2004). Given the role of ZBP1 in transport of β-actin mRNA granules to growth cones of primary chick forebrain neurons (Zhang et al. 2001), IMP-1 may mediate inclusion of β-actin mRNA in the HuD-associated granule. Antibodies to 60S ribosomal proteins show colocalization with tau RNA granules in growth cones (Aronov et al. 2001), although it is not clear whether ribosomal proteins are actual components of the transport granule in the same manner that components of the translational apparatus travel with MBP mRNA.

Dendritic mRNA Localization—CamKIIα

Among dendritically localized mRNAs, the mRNA encoding the α isoform of the calcium/calmodulin-dependent protein kinase II (CaMKIIα) is one of the best characterized. CaMKIIα protein is present in both presynaptic nerve terminals and postsynaptic specializations. Presynaptically, CaMKIIα binds to vesicles and regulates their movement to the site of release through phosphorylation of the vesicle protein synapsin I (for review, see Bayer and Schulman 2001). Synaptic activity promotes translocation of CaMKIIα to postsynaptic densities, where it associates with several protein targets, including the glutamate (NMDA) receptor. As an essential regulatory component of the NMDA receptor complex, CaMKIIα kinase activity is required in hippocampal neurons for long-term potentiation (LTP), an NMDA-dependent form of synaptic plasticity that contributes to learning and memory. In addition, roles have been identified for CaMKIIα in regulating formation of synapses and dendritic spines and in dendrite morphogenesis (for review, see Bayer and Schulman 2001; Colbran and Brown 2004).

CaMKIIα and a number of other neuronal mRNAs have been identified as components of granules isolated from neurons biochemically (Krichevsky and Kosik 2001; Mallardo et al. 2003; Kanai et al. 2004), although it is not known how many distinct populations of granules are represented. At a population level, these granules also contain various translation factors, ribosomes, motor proteins, and RNA-binding proteins. Among the RNA-binding proteins are homologs of several proteins required for RNA localization and translational control in *Drosophila*, including Staufen, Vasa, Me31B, and hnRNP A/B (Kanai et al. 2004). Thus, these factors may have evolutionarily conserved roles in assembly or function of RNA transport particles. Sucrose gradient sedimentation analysis indicates that RNAs are released from granules upon depolarization and become polysome-associated, suggesting that these granules represent translationally silenced transport particles (Krichevsky and Kosik 2001).

Accumulation of the CaMKIIα subunit at synapses occurs following NMDA receptor activation and is accomplished at least in part by transport of CaMKIIα mRNA into dendrites and local translation of this mRNA at synapses. Dendritic localization and translation of CaMKIIα mRNA are mediated by its 3′UTR (Mayford et al. 1996; Mori et al. 2000). The functional significance of this dendritic localization is illustrated by mice in which the CaMKIIα 3′UTR is selectively replaced by a heterologous 3′UTR. In the hippocampus of these mutant mice, dendritic localization of CaMKIIα RNA is abolished and CaMKIIα protein at postsynaptic densities is greatly reduced, consistent with a role for mRNA localization in local synthesis of CaMKIIα protein. Moreover, the mutant mice are impaired in LTP and long-term memory (Miller et al. 2002). Use of a membrane-bound GFP reporter gene fused to the CaMKIIα 5′ and 3′UTRs has enabled visualization of dendritic protein synthesis in cultured hippocampal neurons in real time, revealing GFP "hot spots" that are positioned near ribosomes or synapses. These results further implicate a role for local protein synthesis in synaptic function (Aakalu et al. 2001).

Two cytoplasmic polyadenylation elements (CPEs) within the CaMKIIα 3′UTR appear to link transport and translational control of CaMKIIα mRNA through their interaction with the cytoplasmic polyadenylation element binding protein (CPEB). The CPEs and CPEB facilitate dendritic transport of CPEB, and GFP-tagged CPEB forms particles containing RNA and motor proteins that exhibit microtubule-dependent transport in cultured hippocampal neurons (Huang et al. 2003). These particles also contain maskin, which interacts with CPEB to promote translational repression of CPE-containing mRNAs (for details, see Chapter 18), suggesting that CPEB and maskin may participate both in maintaining RNAs like CaMKIIα in a translationally repressed state during transport to dendrites and in microtubule association of transport particles containing these RNAs. Activation of Aurora kinase at synapses following NMDA receptor activation results in CPEB phosphorylation and CaMKIIα polyadenylation (Huang et al. 2002), providing a mechanism for localization-dependent translational activation (for details, see Chapter 18).

SUMMARY AND PERSPECTIVES

The mRNAs described here represent only a small subset of the transcripts whose translation is thought to be spatially regulated by mRNA localization. Eberwine et al. (2001) have provided a compendium of over 200 dendritically localized mRNAs, and it is likely that all asymmetric cells harbor localized mRNAs. Although much remains to be determined about how individual mRNAs are specifically recognized and transported and how localization and translation are coordinately regulated, current models incorporate the following features. One or more zipcodes present within an mRNA provide the signal to create the core "locasome." The assembly of this complex initiates in the nucleus, as the RNA is being transcribed, and some of the factors assembled at this point "lock up" the mRNA in a non-translatable form. This configuration prevents ribosomes from initiating when the RNA enters the cytoplasm. Additional components required for localization and/or translational repression are added to the RNP following nuclear export; their interaction with the complex may be indirect or may require distinct *cis*-acting regulatory elements within the mRNA. For the majority of RNAs analyzed, cytoplasmic locasomes contain motors responsible for translocating the RNA particle along microtubules or actin filaments. When the particle reaches its destination, it comes in contact with cytoskeleton or membrane-associated factors that anchor the mRNA in place. These or additional factors then inactivate translational repressors; for example, by phosphorylation or dephosphorylation, allowing translation of the localized mRNA.

Analysis of single mRNA movements revealed that mRNAs move within the cytoplasm primarily by diffusion, but that a zipcode can increase the probability that an mRNA will associate with motors and then with cytoskeletal elements (Fusco et al. 2003). Moreover, this analysis suggests that mRNAs can be translated intermittently and are not in motion during periods of translation. This hypothesis is further supported by peripherin mRNA translation. Peripherin is a neurofilament subunit protein that forms insoluble filaments at the site of its translation. Simultaneous imaging of fluorescently labeled peripherin protein and peripherin mRNA shows translationally silent peripherin RNPs moving away from newly synthesized peripherin protein to other regions of the cell, where they presumably resume translation (Chang et al. 2006). Therefore, mRNA localization may be seen as a stochastic phenomenon wherein a probabilistic series of events incrementally increase the chances of an RNA molecule ending up at its target destination, be it the posterior or vegetal pole of an oocyte, or the distal reaches of the cytoplasm of the oligodendrocyte, neuron, or fibroblast.

ACKNOWLEDGMENTS

We regret the recent passing of our friend and colleague in the field of mRNA localization, Larry Etkin. We thank Agata Becalska for helpful comments on the manuscript. The authors' work is supported by grants from the National Institutes of Health (to E.R.G. and R.H.S.).

REFERENCES

Aakalu G., Smith W.B., Nguyen N., Jiang C., and Schuman E.M. 2001. Dynamic visualization of local protein synthesis in hippocampal neurons. *Neuron* **30:** 489–502.

Ainger K., Avossa D., Diana A.S., Barry C., Barbarese E., and Carson J.H. 1997. Transport and localization elements in myelin basic protein mRNA. *J. Cell Biol.* **138:** 1077–1087.

Ainger K., Avossa D., Morgan F., Hill S.J., Barry C., Barbarese E., and Carson J.H. 1993. Transport and localization of exogenous myelin basic protein mRNA microinjected into oligodendrocytes. *J. Cell Biol.* **123:** 431–441.

Aranda-Abreu G.E., Behar L., Chung S., Furneaux H., and Ginzburg I. 1999. Embryonic lethal abnormal vision-like RNA-binding proteins regulate neurite outgrowth and tau expression in PC12 cells. *J. Neurosci.* **19:** 6907–6917.

Aronov S., Aranda G., Behar L., and Ginzburg I. 2001. Axonal tau mRNA localization coincides with tau protein in living neuronal cells and depends on axonal targeting signal. *J. Neurosci.* **21:** 6577–6587.

———. 2002. Visualization of translated tau protein in the axons of neuronal P19 cells and characterization of tau RNP granules. *J. Cell Sci.* **115:** 3817–3827.

Atlas R., Behar L., Elliott E., and Ginzburg I. 2004. The insulin-like growth factor mRNA binding-protein IMP-1 and the Ras-regulatory protein G3BP associate with tau mRNA and HuD protein in differentiated P19 neuronal cells. *J. Neurochem.* **89:** 613–626.

Barbarese E., Koppel D.E., Deutscher M.P., Smith C.L., Ainger K., Morgan F., and Carson J.H. 1995. Protein translation components are colocalized in granules in oligodendrocytes. *J. Cell Sci.* **108:** 2781–2790.

Bassell G.J., Zhang H., Byrd A.L., Femino A.M., Singer R.H., Taneja K.L., Lifshitz L.M., Herman I.M., and Kosik K.S. 1998. Sorting of beta-actin mRNA and protein to neurites and growth cones in culture. *J. Neurosci.* **18:** 251–265.

Bayer K.U. and Schulman H. 2001. Regulation of signal transduction by protein targeting: The case for CaMKII. *Biochem. Biophys. Res. Commun.* **289:** 917–923.

Bergsten S.E. and Gavis E.R. 1999. Role for mRNA localization in translational activation but not spatial restriction of *nanos* RNA. *Development* **126:** 659–669.

Bergsten S.E., Huang T., Chatterjee S., and Gavis E.R. 2001. Recognition and long-range interactions of a minimal *nanos* RNA localization signal element. *Development* **128:** 427–435.

Berleth T., Burri M., Thoma G., Bopp D., Richstein S., Frigerio G., Noll M., and Nüsslein-Volhard C. 1988. The role of localization of *bicoid* RNA in organizing the anterior pattern of the *Drosophila* embryo. *EMBO J.* **7:** 1749–1756.

Bertrand E., Chartrand P., Schaefer M., Shenoy S.M., Singer R.H., and Long R.M. 1998. Localization of ASH1 mRNA particles in living yeast. *Mol. Cell* **2:** 437–445.

Betley J.N., Heinrich B., Vernos I., Sardet C., Prodon F., and Deshler J.O. 2004. Kinesin II mediates Vg1 mRNA transport in *Xenopus* oocytes. *Curr. Biol.* **14:** 219–224.

Birsoy B., Kofron M., Schaible K., Wylie C., and Heasman J. 2006. Vg 1 is an essential signaling molecule in *Xenopus* development. *Development* **133:** 15–20.

Bohl F., Kruse C., Frank A., Ferring D., and Jansen R.P. 2000. She2p, a novel RNA-binding protein tethers ASH1 mRNA to the Myo4p myosin motor via She3p. *EMBO J.* **19:** 5514–5524.

Brendza R.P., Serbus L.R., Duffy J.B., and Saxton W.M. 2000. A function for kinesin I in the posterior transport of *oskar* mRNA and Staufen protein. *Science* **289:** 2120–2122.

Brumwell C., Antolik C., Carson J.H., and Barbarese E. 2002. Intracellular trafficking of hnRNP A2 in oligodendrocytes. *Exp. Cell Res.* **279:** 310–320.

Campagnoni A.T., Carey G.D., and Yu Y.T. 1980. In vitro synthesis of the myelin basic proteins: Subcellular site of synthesis. *J. Neurochem.* **34:** 677–686.

Carson J.H. and Barbarese E. 2005. Systems analysis of RNA trafficking in neural cells. *Biol. Cell* **97:** 51–62.

Carson J.H., Worboys K., Ainger K., and Barbarese E. 1997. Translocation of myelin basic protein mRNA in oligodendrocytes requires microtubules and kinesin. *Cell Motil. Cytoskel.* **38:** 318–328.

Cha B.-J., Serbus L.R., Koppetsch B.S., and Theurkauf W.E. 2002. Kinesin I-dependent cortical exclusion restricts pole plasm to the oocyte posterior. *Nat. Cell Biol.* **4:** 592–598.

Chang L., Shav-Tal Y., Trcek T., Singer R.H., and Goldman R.D. 2006. Assembling an intermediate filament network by dynamic cotranslation. *J. Cell Biol.* **172:** 747–758.

Chartrand P., Meng X.H., Singer R.H., and Long R.M. 1999. Structural elements required for the localization of *ASH1* mRNA and of a green fluorescent protein reporter particle in vivo. *Curr. Biol.* **9:** 333–336.

Chartrand P., Meng X.H., Hüttelmaier S., Donato D., and Singer R.H. 2002. Asymmetric sorting of ash1p in yeast results from inhibition of translation by localization elements in the mRNA. *Mol. Cell* **10:** 1319–1330.

Chekulaeva M., Hentze M.W., and Ephrussi A. 2006. Bruno acts as a dual repressor of *oskar* translation, promoting mRNA oligomerization and formation of silencing particles. *Cell* **124:** 521–533.

Clark I.E., Wyckoff D., and Gavis E.R. 2000. Synthesis of the posterior determinant Nanos is spatially restricted by a novel cotranslational mechanism. *Curr. Biol.* **10:** 1311–1314.

Colbran R.J. and Brown A.M. 2004. Calcium/calmodulin-dependent protein kinase II and synaptic plasticity. *Curr. Opin. Neurobiol.* **14:** 318–327.

Colman D.R., Kreibich G., Frey A.B., and Sabatini D.D. 1982. Synthesis and incorporation of myelin polypeptides into CNS myelin. *J. Cell Biol.* **95:** 598–608.

Cook H.A., Koppetsch B.S., Wu J., and Theurkauf W.E. 2004. The *Drosophila* SDE3 homolog *armitage* is required for *oskar* mRNA silencing and embryonic axis specification. *Cell* **116:** 817–829.

Cote C.A., Gautreau D., Denegre J.M., Kress T., Terry N.A., and Mowry K.L. 1999. A *Xenopus* protein related to hnRNP I has a role in cytoplasmic RNA localization. *Mol. Cell* **4:** 431–437.

Crucs S., Chatterjee S., and Gavis E.R. 2000. Overlapping but distinct RNA elements control repression and activation of *nanos* translation. *Mol. Cell* **5:** 457–467.

Czaplinski K. and Mattaj I.W. 2006. 40LoVe interacts with Vg1RBP/Vera and hnRNP I in binding the Vg1-localization element. *RNA* **12:** 213–222.

Czaplinski K., Kocher T., Schelder M., Segref A., Wilm M., and Mattaj I.W. 2005. Identification of 40LoVe, a *Xenopus* hnRNP D family protein involved in localizing a TGF-beta-related mRNA during oogenesis. *Dev. Cell* **8:** 505–515.

Dahanukar A. and Wharton R.P. 1996. The Nanos gradient in *Drosophila* embryos is generated by translational regulation. *Genes Dev.* **10:** 2610–2620.

Dahanukar A., Walker J.A., and Wharton R.P. 1999. Smaug, a novel RNA-binding protein that operates a translational switch in *Drosophila*. *Mol. Cell* **4:** 209–218.

Deshler J.O., Highett M.I., and Schnapp B.J. 1997. Localization of *Xenopus* Vg1 mRNA by Vera protein and the endoplasmic reticulum. *Science* **276:** 1128–1131.

Deshler J.O., Highett M.I., Abramson T., and Schnapp B.J. 1998. A highly conserved RNA-binding protein for cytoplasmic mRNA localization in vertebrates. *Curr. Biol.* **8:** 489–496.

Driever W. and Nüsslein-Volhard C. 1988. The *bicoid* protein determines position in the *Drosophila* embryo in a concentration-dependent manner. *Cell* **54:** 95–104.

Eberwine J., Miyashiro K., Kacharmina J.E., and Job C. 2001. Local translation of classes of mRNAs that are targeted to neuronal dendrites. *Proc. Natl. Acad. Sci.* **98:** 7080–7085.

Enright A.J., John B., Gaul U., Tuschl T., Sander C., and Marks D.S. 2003. MicroRNA targets in *Drosophila*. *Genome Biol.* **5:** R1.

Eom T., Antar L.N., Singer R.H., and Bassell G.J. 2003. Localization of a beta-actin messenger ribonucleoprotein complex with zipcode-binding protein modulates the density of dendritic filopodia and filopodial synapses. *J. Neurosci.* **23:** 10433–10444.

Ephrussi A. and Lehmann R. 1992. Induction of germ cell formation by *oskar*. *Nature* **358:** 387–392.

Ephrussi A., Dickinson L.K., and Lehmann R. 1991. *oskar* organizes the germ plasm and directs localization of the posterior determinant *nanos*. *Cell* **66:** 37–50.

Farina K.L., Hüttelmaier S., Musunuru K., Darnell R., and Singer R.H. 2003. Two ZBP1 KH domains facilitate beta-actin mRNA localization, granule formation, and cytoskeletal attachment. *J. Cell Biol.* **160:** 77–87.

Filardo P. and Ephrussi A. 2003. Bruno regulates *gurken* during *Drosophila* oogenesis. *Mech. Dev.* **120:** 289–297.

Forrest K.M. and Gavis E.R. 2003. Live imaging of endogenous mRNA reveals a diffusion and entrapment mechanism for *nanos* mRNA localization in *Drosophila*. *Curr. Biol.* **13:** 1159–1168.

Forrest K.M., Clark I.E., Jain R.A., and Gavis E.R. 2004. Temporal complexity within a translational control element in the *nanos* mRNA. *Development* **131:** 5849–5857.

Fusco D., Accornero N., Lavoie B., Shenoy S.M., Blanchard J.M., Singer R.H., and Bertrand E. 2003. Single mRNA molecules demonstrate probabilistic movement in living mammalian cells. *Curr. Biol.* **13:** 161–167.

Gautreau D., Cote C.A., and Mowry K.L. 1997. Two copies of a subelement from the Vg1 RNA localization sequence are sufficient to direct vegetal localization in *Xenopus* oocytes. *Development* **124:** 5013–5020.

Gavis E.R. and Lehmann R. 1992. Localization of *nanos* RNA controls embryonic polarity. *Cell.* **71:** 301–313.

———. 1994. Translational regulation of *nanos* by RNA localization. *Nature* **369:** 315–318.

Gavis E.R., Curtis D., and Lehmann R. 1996a. Identification of *cis*-acting sequences that control nanos RNA localization. *Dev. Biol.* **176:** 36–50.

Gavis E.R., Lunsford L., Bergsten S.E., and Lehmann R. 1996b. A conserved 90 nucleotide element mediates translational repression of *nanos* RNA. *Development* **122:** 2791–2800.

Ghabrial A. and Schüpbach T. 1999. Activation of a meiotic checkpoint regulates translation of Gurken during *Drosophila* oogenesis. *Nat. Cell Biol.* **1:** 354–357.

Glotzer J.B., Saffrich R., Glotzer M., and Ephrussi A. 1997. Cytoplasmic flows localize injected *oskar* RNA in *Drosophila* oocytes. *Curr. Biol.* **7:** 326–337.

Gonzalez I., Buonomo S.B., Nasmyth K., and von Ahsen U. 1999. ASH1 mRNA localization in yeast involves multiple secondary structural elements and Ash1 protein translation. *Curr. Biol.* **9:** 337–340.

González-Reyes A., Elliot H., and St. Johnston D. 1995. Polarization of both major body axes in *Drosophila* by *gurken-torpedo* signalling. *Nature* **375:** 654–658.

Gonsalvez G.B., Urbinati C.R., and Long R.M. 2005. RNA localization in yeast: Moving towards a mechanism. *Biol. Cell* **97:** 75–86.

Good P.J. 1995. A conserved family of elav-like genes in vertebrates. *Proc. Natl. Acad. Sci.* **92:** 4557–4561.

Goodrich J.S., Clouse K.N., and Schüpbach T. 2004. Hrb27C, Sqd and Otu cooperatively regulate *gurken* RNA localization and mediate nurse cell chromosome dispersion in *Drosophila* oogenesis. *Development* **131:** 1949–1958.

Gu W., Deng Y., Zenklusen D., and Singer R.H. 2004. A new yeast PUF family protein, Puf6p, represses ASH1 mRNA translation and is required for its localization. *Genes Dev.* **18:** 1452–1465.

Gu W., Pan F., Zhang H., Bassell G.J., and Singer R.H. 2002. A predominantly nuclear protein affecting cytoplasmic localization of beta-actin mRNA in fibroblasts and neurons. *J. Cell Biol.* **156:** 41–51.

Gunkel N., Yano T., Markussen F.H., Olsen L.C., and Ephrussi A. 1998. Localization-dependent translation requires a functional interaction between the 5′ and 3′ ends of *oskar* mRNA. *Genes Dev.* **12:** 1652–1664.

Hachet O. and Ephrussi A. 2001. *Drosophila* Y14 shuttles to the posterior of the oocyte and is required for *oskar* mRNA transport. *Curr. Biol.* **11:** 1666–1674.

———. 2004. Splicing of *oskar* RNA in the nucleus is coupled to its cytoplasmic localization. *Nature* **428:** 959–963.

Havin L., Git A., Elisha Z., Oberman F., Yaniv K., Schwartz S.P., Standart N., and Yisraeli J.K. 1998. RNA-binding protein conserved in both microtubule- and microfilament-based RNA localization. *Genes Dev.* **12:** 1593–1598.

Hawkins N.C., Van Buskirk C., Grossniklaus U., and Schüpbach T. 1997. Post-transcriptional regulation of *gurken* by encore is required for axis determination in *Drosophila*. *Development* **124:** 4801–4810.

Huang Y.S., Carson J.H., Barbarese E., and Richter J.D. 2003. Facilitation of dendritic mRNA transport by CPEB. *Genes Dev.* **17:** 638–653.

Huang Y.S., Jung M.Y., Sarkissian M., and Richter J.D. 2002. N-methyl-D-aspartate receptor signaling results in Aurora kinase-catalyzed CPEB phosphorylation and alpha CaMKII mRNA polyadenylation at synapses. *EMBO J.* **21:** 2139–2148.

Hüttelmaier S., Zenklusen D., Lederer M., Dictenberg J., Lorenz M., Meng X., Bassell G.J., Condeelis J., and Singer R.H. 2005. Spatial regulation of beta-actin translation by Src-dependent phosphorylation of ZBP1. *Nature* **438:** 512–515.

Huynh J.R., Munro T.P., Smith-Litiere K., Lepesant J.A., and St. Johnston D.S. 2004. The *Drosophila* hnRNPA/B homolog, Hrp48, is specifically required for a distinct step in *osk* mRNA localization. *Dev. Cell* **6:** 625–635.

Irie K., Tadauchi T., Takizawa P.A., Vale R.D., Matsumoto K., and Herskowitz I. 2002. The Khd1 protein, which has three KH RNA-binding motifs, is required for proper localization of ASH1 mRNA in yeast. *EMBO J.* **21:** 1158–1167.

Johnstone O. and Lasko P. 2004. Interaction with eIF5B is essential for Vasa function during development. *Development* **131:** 4167–4178.

Kalifa Y., Huang T., Rosen L.N., Chatterjee S., and Gavis E.R. 2006. Glorund, a *Drosophila* hnRNP F/H homolog, is an ovarian repressor of *nanos* translation. *Dev. Cell* **10:** 291–301.

Kanai Y., Dohmae N., and Hirokawa N. 2004. Kinesin transports RNA: Isolation and characterization of an RNA-transporting granule. *Neuron* **43:** 513–525.

Karlin-Mcginness M., Serano T.L., and Cohen R.S. 1996. Comparative analysis of the kinetics and dynamics of K10, bicoid, and oskar mRNA localization in the *Drosophila* oocyte. *Dev. Genet.* **19:** 238–248.

Kim-Ha J., Kerr K., and Macdonald P.M. 1995. Translational regulation of *oskar* mRNA by Bruno, an ovarian RNA-binding protein, is essential. *Cell* **81:** 403–412.

Kim-Ha J., Smith J.L., and Macdonald P.M. 1991. *oskar* mRNA is localized to the posterior pole of the *Drosophila* oocyte. *Cell* **66:** 23–35.

Kim-Ha J., Webster P.J., Smith J.L., and Macdonald P.M. 1993. Multiple RNA regulatory elements mediate distinct steps in localization of *oskar* mRNA. *Development* **119:** 169–178.

Kindler S., Muller R., Chung W.J., and Garner C.C. 1996. Molecular characterization of dendritically localized transcripts encoding MAP2. *Brain Res. Mol. Brain Res.* **36:** 63–69.

King M.L., Messitt T.J., and Mowry K.L. 2005. Putting RNAs in the right place at the right time: RNA localization in the frog oocyte. *Biol. Cell* **97:** 19–33.

Kislauskis E.H., Zhu X., and Singer R.H. 1994. Sequences responsible for intracellular localization of beta-actin messenger RNA also affect cell phenotype. *J. Cell Biol.* **127:** 441–451.

————. 1997. β-Actin messenger RNA localization and protein synthesis augment cell motility. *J. Cell Biol.* **136:** 1263–1270.

Kislauskis E.H., Li Z., Singer R.H., and Taneja K.L. 1993. Isoform-specific 3'-untranslated sequences sort alpha-cardiac and beta-cytoplasmic actin messenger RNAs to different cytoplasmic compartments. *J. Cell Biol.* **123:** 165–172.

Kolev N.G. and Huber P.W. 2003. VgRBP71 stimulates cleavage at a polyadenylation signal in Vg1 mRNA, resulting in the removal of a *cis*-acting element that represses translation. *Mol. Cell* **11:** 745–755.

Kress T.L., Yoon Y.J., and Mowry K.L. 2004. Nuclear RNP complex assembly initiates cytoplasmic RNA localization. *J. Cell Biol.* **165:** 203–211.

Krichevsky A.M. and Kosik K.S. 2001. Neuronal RNA granules: A link between RNA localization and stimulation-dependent translation. *Neuron* **32:** 683–696.

Kroll T.T., Zhao W.M., Jiang C., and Huber P.W. 2002. A homolog of FBP2/KSRP binds to localized mRNAs in *Xenopus* oocytes. *Development* **129:** 5609–5619.

Kruse C., Jaedicke A., Beaudouin J., Bohl F., Ferring D., Guttler T., Ellenberg J., and Jansen R.P. 2002. Ribonucleoprotein-dependent localization of the yeast class V myosin Myo4p. *J. Cell Biol.* **159:** 971–982.

Kwon S., Barbarese E., and Carson J.H. 1999. The *cis*-acting RNA trafficking signal from myelin basic protein mRNA and its cognate *trans*-acting ligand hnRNP A2 enhance cap-dependent translation. *J. Cell Biol.* **147:** 247–256.

Larocque D., Pilotte J., Chen T., Cloutier F., Massie B., Pedraza L., Couture R., Lasko P., Almazan G., and Richard S. 2002. Nuclear retention of MBP mRNAs in the quaking viable mice. *Neuron* **36:** 815–829.

Latham V.M., Yu E.H., Tullio A.N., Adelstein R.S., and Singer R.H. 2001. A Rho-dependent signaling pathway operating through myosin localizes beta-actin mRNA in fibroblasts. *Curr. Biol.* **11:** 1010–1016.

Litman P., Barg J., Rindzoonski L., and Ginzburg I. 1993. Subcellular localization of tau mRNA in differentiating neuronal cell culture: Implications for neuronal polarity. *Neuron* **10:** 627–638.

Liu G., Grant W.M., Persky D., Latham V.M. Jr., Singer R.H., and Condeelis J. 2002. Interactions of elongation factor 1alpha with F-actin and beta-actin mRNA: Implications for anchoring mRNA in cell protrusions. *Mol. Biol. Cell* **13:** 579–592.

Long R.M., Gu W., Lorimer E., Singer R.H., and Chartrand P. 2000. She2p is a novel RNA-binding protein that recruits the Myo4p-She3p complex to ASH1 mRNA. *EMBO J.* **19:** 6592–6601.

Long R.M., Singer R.H., Meng X., Gonzalez I., Nasmyth K., and Jansen R.P. 1997. Mating type switching in yeast controlled by asymmetric localization of ASH1 mRNA. *Science* **277:** 383–387.

LoPresti P., Szuchet S., Papasozomenos S.C., Zinkowski R.P., and Binder L.I. 1995. Functional implications for the microtubule-associated protein tau: Localization in oligodendrocytes. *Proc. Natl. Acad. Sci.* **92:** 10369–10373.

MacDougall N., Clark A., MacDougall E., and Davis I. 2003. *Drosophila gurken* (TGFalpha) mRNA localizes as particles that move within the oocyte in two dynein-dependent steps. *Dev. Cell* **4:** 307–319.

Maggipinto M., Rabiner C., Kidd G.J., Hawkins A.J., Smith R., and Barbarese E. 2004. Increased expression of the MBP mRNA binding protein HnRNP A2 during oligodendrocyte differentiation. *J. Neurosci. Res.* **75:** 614–623.

Mallardo M., Deitinghoff A., Muller J., Goetze B., Macchi P., Peters C., and Kiebler M.A. 2003. Isolation and characterization of Staufen-containing ribonucleoprotein particles from rat brain. *Proc. Natl. Acad. Sci.* **100:** 2100–2105.

Martin K.C. 2004. Local protein synthesis during axon guidance and synaptic plasticity. *Curr. Opin. Neurobiol.* **14:** 305–310.

Mayford M., Baranes D., Podsypanina K., and Kandel E.R. 1996. The 3′-untranslated region of CaMKII alpha is a *cis*-acting signal for the localization and translation of mRNA in dendrites. *Proc. Natl. Acad. Sci.* **93:** 13250–13255.

Micklem D.R., Adams J., Grunert S., and St. Johnston D. 2000. Distinct roles of two conserved Staufen domains in oskar mRNA localization and translation. *EMBO J.* **19:** 1366–1377.

Miller S., Yasuda M., Coats J.K., Jones Y., Martone M.E., and Mayford M. 2002. Disruption of dendritic translation of CaMKIIalpha impairs stabilization of synaptic plasticity and memory consolidation. *Neuron* **36:** 507–519.

Mohr E. and Richter D. 2000. Axonal mRNAs: Functional significance in vertebrates and invertebrates. *J. Neurocytol.* **29:** 783–791.

Mohr S.E., Dillon S.T., and Boswell R.E. 2001. The RNA-binding protein Tsunagi interacts with Mago Nashi to establish polarity and localize *oskar* mRNA during *Drosophila* oogenesis. *Genes Dev.* **15:** 2886–2899.

Mori Y., Imaizumi K., Katayama T., Yoneda T., and Tohyama M. 2000. Two *cis*-acting elements in the 3′-untranslated region of alpha-CaMKII regulate its dendritic targeting. *Nat. Neurosci.* **3:** 1079–1084.

Mowry K.L. and Melton D.A. 1992. Vegetal messenger RNA localization directed by a 340-nt RNA sequence element in *Xenopus* oocytes. *Science* **255:** 991–994.

Munchow S., Sauter C., and Jansen R.P. 1999. Association of the class V myosin Myo4p with a localised messenger RNA in budding yeast depends on She proteins. *J. Cell Sci.* **112:** 1511–1518.

Munro T.P., Magee R.J., Kidd G.J., Carson J.H., Barbarese E., Smith L.M., and Smith R. 1999. Mutational analysis of a heterogeneous nuclear ribonucleoprotein A2 response element for RNA trafficking. *J. Biol. Chem.* **274:** 34389–34395.

Nakahara K., Kim K., Sciulli C., Dowd S.R., Minden J.S., and Carthew R.W. 2005. Targets of microRNA regulation in the *Drosophila* oocyte proteome. *Proc. Natl. Acad. Sci.* **102:** 12023–12028.

Nakamura A., Sato K., and Hanyu-Nakamura K. 2004. *Drosophila* cup is an eIF4E binding protein that associates with Bruno and regulates *oskar* mRNA translation in oogenesis. *Dev. Cell* **6:** 69–78.

Nakamura A., Amikura R., Hanyu K., and Kobayashi S. 2001. Me31B silences translation of oocyte-localizing RNAs through the formation of cytoplasmic RNP complex during *Drosophila* oogenesis. *Development* **128:** 3233–3242.

Nelson M.R., Leidal A.M., and Smibert C.A. 2004. *Drosophila* Cup is an eIF4E-binding protein that functions in Smaug-mediated translational repression. *EMBO J.* **23:** 150–159.

Neuman-Silberberg F.S. and Schüpbach T. 1993. The *Drosophila* dorsoventral patterning gene *gurken* produces a dorsally localized RNA and encodes a TGFα-like protein. *Cell* **75:** 165–174.

Newmark P.A. and Boswell R.E. 1994. The *mago nashi* locus encodes an essential product required for germ plasm assembly in *Drosophila*. *Development* **120:** 1303–1313.

Niessing D., Hüttelmaier S., Zenklusen D., Singer R.H., and Burley S.K. 2004. She2p is a novel RNA binding protein with a basic helical hairpin motif. *Cell* **119:** 491–502.

Nilson L.A. and Schüpbach T. 1999. EGF receptor signaling in *Drosophila* oogenesis. *Curr. Top. Dev. Biol.* **44:** 203–243.

Norvell A., Kelley R.L., Wehr K., and Schüpbach T. 1999. Specific isoforms of *squid*, a *Drosophila* hnRNP, perform distinct roles in Gurken localization during oogenesis. *Genes Dev.* **13:** 864–876.

Norvell A., Debec A., Finch D., Gibson L., and Thoma B. 2005. Squid is required for efficient posterior localization of *oskar* mRNA during *Drosophila* oogenesis. *Dev. Genes Evol.* **215:** 340–349.

Oleynikov Y. and Singer R.H. 2003. Real-time visualization of ZBP1 association with beta-actin mRNA during transcription and localization. *Curr. Biol.* **13:** 199–207.

Ostareck D.H., Ostareck-Lederer A., Shatsky I.N., and Hentze M.W. 2001. Lipoxygenase mRNA silencing in erythroid differentiation: The 3′UTR regulatory complex controls 60S ribosomal subunit joining. *Cell* **104:** 281–902.

Ostareck D.H., Ostareck-Lederer A., Wilm M., Thiele B.J., Mann M., and Hentze M.W. 1997. mRNA silencing in erythroid differentiation: hnRNP K and hnRNP E1 regulate 15-lipoxygenase translation from the 3′ end. *Cell* **89:** 597–606.

Ostareck-Lederer A., Ostareck D.H., Cans C., Neubauer G., Bomsztyk K., Superti-Furga G., and Hentze M.W. 2002. c-Src-mediated phosphorylation of hnRNP K drives translational activation of specifically silenced mRNAs. *Mol. Cell. Biol.* **22:** 4535–4543.

Otero L.J., Devaux A., and Standart N. 2001. A 250-nucleotide UA-rich element in the 3′-untranslated region of *Xenopus laevis* Vg1 mRNA represses translation both in vivo and in vitro. *RNA* **7:** 1753–1767.

Palacios I.M., Gatfield D., St. Johnston D., and Izaurralde E. 2004. An eIF4AIII-containing complex required for mRNA localization and nonsense-mediated mRNA decay. *Nature* **427:** 753–757.

Rongo C., Gavis E.R., and Lehmann R. 1995. Localization of *oskar* RNA regulates *oskar* translation and requires Oskar protein. *Development* **121:** 2737–2746.

Ross A.F., Oleynikov Y., Kislauskis E.H., Taneja K.L., and Singer R.H. 1997. Characterization of a beta-actin mRNA zipcode-binding protein. *Mol. Cell. Biol.* **17:** 2158–2165.

Roth S., Neuman-Silberberg F.S., Barcelo G., and Schüpbach T. 1995. *cornichon* and the EGF receptor signaling process are necessary for both anterior-posterior and dorsal-ventral pattern formation in *Drosophila*. *Cell* **81:** 967–978.

Sallés F.J., Lieberfarb M.E., Wreden C., Gergen J.P., and Strickland S. 1994. Coordinate initiation of *Drosophila* development by regulated polyadenylation of maternal mRNAs. *Science* **266:** 1996–1999.

Saunders C. and Cohen R.S. 1999. The role of oocyte transcription, the 5′UTR, and translation repression and derepression in *Drosophila gurken* mRNA and protein localization. *Mol. Cell* **3:** 43–54.

Shestakova E.A., Singer R.H., and Condeelis J. 2001. The physiological significance of beta-actin mRNA localization in determining cell polarity and directional motility. *Proc. Natl. Acad. Sci.* **98:** 7045–7050.

Shestakova E.A., Wyckoff J., Jones J., Singer R.H., and Condeelis J. 1999. Correlation of beta-actin messenger RNA localization with metastatic potential in rat adenocarcinoma cell lines. *Cancer Res.* **59:** 1202–1205.

Smibert C.A., Wilson J.E., Kerr K., and Macdonald P.M. 1996. Smaug protein represses translation of unlocalized *nanos* mRNA in the *Drosophila* embryo. *Genes Dev.* **10:** 2600–2609.

Smibert C.A., Lie Y.S., Shillingaw W., Henzel W.J., and Macdonald P.M. 1999. Smaug, a novel and conserved protein contributes to repression of *nanos* mRNA translation in vitro. *RNA* **5:** 1535–1547.

Smith J.L., Wilson J.E., and Macdonald P.M. 1992. Overexpression of *oskar* directs ectopic activation of *nanos* and presumptive pole cell formation in *Drosophila* embryos. *Cell* 70: 849–859.

Spradling A.C. 1993. Developmental genetics of oogenesis. In *The development of Drosophila melanogaster* (ed. M. Bate and A. Martinez Arias), vol. 1, pp. 1–70. Cold Spring Harbor Laboratory Press, Cold Spring Harbor, New York.

Stark A., Brennecke J., Russell R.B., and Cohen S.M. 2003. Identification of *Drosophila* microRNA targets. *PLoS Biol.* 1: 397–409.

Steward O. and Levy W.B. 1982. Preferential localization of polyribosomes under the base of dendritic spines in granule cells of the dentate gyrus. *J. Neurosci.* 2: 284–291.

Steward O. and Ribak C.E. 1986. Polyribosomes associated with synaptic specializations on axon initial segments: Localization of protein-synthetic machinery at inhibitory synapses. *J. Neurosci.* 6: 3079–3085.

Steward O. and Schuman E.M. 2001. Protein synthesis at synaptic sites on dendrites. *Annu. Rev. Neurosci.* 24: 299–325.

St. Johnston D., Beuchle D., and Nüsslein-Volhard C. 1991. *staufen*, a gene required to localize maternal RNAs in the *Drosophila* egg. *Cell* 66: 51–63.

Styhler S., Nakamura A., Swan A., Suter B., and Lasko P. 1998. *vasa* is required for GURKEN accumulation in the oocyte, and is involved in oocyte differentiation and germline cyst development. *Development* 125: 1569–1578.

Sundell C.L. and Singer R.H. 1990. Actin mRNA localizes in the absence of protein synthesis. *J. Cell Biol.* 111: 2397–2403.

———. 1991. Requirement of microfilaments in sorting of actin messenger RNA. *Science* 253: 1275–1277.

Takizawa P.A. and Vale R.D. 2000. The myosin motor, Myo4p, binds Ash1 mRNA via the adapter protein, She3p. *Proc. Natl. Acad. Sci.* 97: 5273–5278.

Takizawa P.A., Sil A., Swedlow J.R., Herskowitz I., and Vale R.D. 1997. Actin-dependent localization of an RNA encoding a cell-fate determinant in yeast. *Nature* 389: 90–93.

Tannahill D. and Melton D.A. 1989. Localized synthesis of the Vg1 protein during early *Xenopus* development. *Development* 106: 775–785.

Theurkauf W.E., Smiley S., Wong M.L., and Alberts B.M. 1992. Reorganization of the cytoskeleton during *Drosophila* oogenesis: Implications for axis specification and intercellular transport. *Development* 115: 923–936.

Tiruchinapalli D.M., Oleynikov Y., Kelic S., Shenoy S.M., Hartley A., Stanton P.K., Singer R.H., and Bassell G.J. 2003. Activity-dependent trafficking and dynamic localization of zipcode binding protein 1 and beta-actin mRNA in dendrites and spines of hippocampal neurons. *J. Neurosci.* 23: 3251–3261.

Tomancak P., Guichet A., Zavorszky P., and Ephrussi A. 1998. Oocyte polarity depends on regulation of gurken by Vasa. *Development* 125: 1723–1732.

Van Buskirk C. and Schüpbach T. 2002. *Half pint* regulates alternative splice site selection in *Drosophila*. *Dev. Cell* 2: 343–353.

van Eeden F. and St. Johnston D. 1999. The polarisation of the anterior-posterior and dorsal-ventral axes during *Drosophila* oogenesis. *Curr. Opin. Genet. Dev.* 9: 396–404.

van Eeden F.J., Palacios I.M., Petronczki M., Weston M.J., and St. Johnston D. 2001. Barentsz is essential for the posterior localization of *oskar* mRNA and colocalizes with it to the posterior pole. *J. Cell Biol.* 154: 511–523.

van Horck F.P., Weinl C., and Holt C.E. 2004. Retinal axon guidance: Novel mechanisms for steering. *Curr. Opin. Neurobiol.* 14: 61–66.

Wang C., Dickinson L.K., and Lehmann R. 1994. Genetics of *nanos* localization in *Drosophila. Dev. Dynam.* **199:** 103–115.

Wang W., Goswami S., Lapidus K., Wyckoff J., Sahai E., Singer R.H., Segall J., and Condeelis J. 2004. Identification and testing of a gene expression signature of invasive carcinoma cells within primary mammary tumors. *Cancer Res.* **64:** 8585–8594.

Weeks D.L. and Melton D.A. 1987. A maternal mRNA localized to the vegetal hemisphere in *Xenopus* codes for a growth factor related to TGF-β. *Cell* **51:** 861–867.

Wharton R.P. and Struhl G. 1989. Structure of the *Drosophila* BicaudalD protein and its role in localizing the posterior determinant *nanos. Cell* **59:** 881–892.

Wilhelm J.E., Vale R.D., and Hegde R.S. 2000. Coordinate control of translation and localization of Vg1 mRNA in *Xenopus* oocytes. *Proc. Natl. Acad. Sci.* **97:** 13132–13137.

Wilhelm J.E., Hilton M., Amos Q., and Henzel W.J. 2003. Cup is an eIF4E binding protein required for both the translational repression of *oskar* and the recruitment of Barentsz. *J. Cell Biol.* **163:** 1197–1204.

Xie J., Lee J.A., Kress T.L., Mowry K.L., and Black D.L. 2003. Protein kinase A phosphorylation modulates transport of the polypyrimidine tract-binding protein. *Proc. Natl. Acad. Sci.* **100:** 8776–8781.

Yaniv K. and Yisraeli J.K. 2002. The involvement of a conserved family of RNA binding proteins in embryonic development and carcinogenesis. *Gene* **287:** 49–54.

Yano T., de Quinto S.L., Matsui Y., Shevchenko A., and Ephrussi A. 2004. Hrp48, a *Drosophila* hnRNPA/B homolog, binds and regulates translation of *oskar* mRNA. *Dev. Cell* **6:** 637–648.

Yisraeli J.K. 2005. VICKZ proteins: A multi-talented family of regulatory RNA-binding proteins. *Biol. Cell* **97:** 87–96.

Yoon Y.J. and Mowry K.L. 2004. *Xenopus* Staufen is a component of a ribonucleoprotein complex containing Vg1 RNA and kinesin. *Development* **131:** 3035–3045.

Zappavigna V., Piccioni F., Villaescusa J.C., and Verrotti A.C. 2004. Cup is a nucleocytoplasmic shuttling protein that interacts with the eukaryotic translation initiation factor 4E to modulate *Drosophila* ovary development. *Proc. Natl. Acad. Sci.* **101:** 14800–14805.

Zhang H.L., Singer R.H., and Bassell G.J. 1999. Neurotrophin regulation of beta-actin mRNA and protein localization within growth cones. *J. Cell Biol.* **147:** 59–70.

Zhang H.L., Eom T., Oleynikov Y., Shenoy S.M., Liebelt D.A., Dictenberg J.B., Singer R.H., and Bassell G.J. 2001. Neurotrophin-induced transport of a beta-actin mRNP complex increases beta-actin levels and stimulates growth cone motility. *Neuron* **31:** 261–275.

25

The Interface between mRNA Turnover and Translational Control

Carlos I. Gonzalez
Department of Biology
University of Puerto Rico-Rio Piedras
San Juan, Puerto Rico 00931

Carol J. Wilusz and Jeffrey Wilusz
Department of Microbiology, Immunology,
and Pathology, Colorado State University
Fort Collins, Colorado 80523

THE ABUNDANCE OF AN MRNA IS DETERMINED by the balance between transcription and decay. Thus, not surprisingly, a large percentage of the change in gene expression in response to stimuli can be attributed to regulated mRNA turnover (Garcia-Martinez et al. 2004). However, the relationship between mRNA stability and translation is complex: Some mRNAs can be very stable but produce virtually no protein (Mukhopadhyay et al. 2003), whereas other unstable mRNAs are efficiently translated (Kontoyiannis et al. 1999). This reflects, in part, the fact that an mRNA can assume three states: active translation, translationally silent, and targeted for decay. The issue is further complicated when we take into account that progress from one state to another is not necessarily linear and that there are multiple mRNA turnover pathways. Moreover, the mechanisms that regulate translation and turnover are so intricately interwoven that we are only just beginning to unravel the details. For example, some experiments suggest that translation is absolutely necessary for ongoing mRNA decay; inhibition of translation elongation with cycloheximide stabilizes the vast majority of cellular transcripts (Herrick et al. 1990). In contrast, other data argue that translation and turnover are in direct competition; a reduction in translation initiation efficiency can lead to destabilization of mRNAs (Schwartz and Parker 1999).

To help the reader understand the complex nature of the translation/turnover paradox, our goal is to describe the various pathways and enzymes of mRNA decay, highlighting interactions with the translation process. We go on to discuss the recent finding that mRNAs can be segregated away from the translation machinery into cytoplasmic foci. These foci appear to represent sites of transcript sorting where mRNAs are directed either for storage or for turnover. Finally, we focus on examples of *cis*-acting sequences and *trans*-acting factors that are linked with both translation and mRNA decay.

PATHWAYS OF DECAY

In the predominant cytoplasmic mRNA turnover pathways, degradation is initiated by poly(A) tail shortening. This deadenylation event targets the transcript for decay either by a 3′ to 5′ route mediated by the exosome (Raijmakers et al. 2004) or by decapping and 5′ to 3′ decay mediated by Dcp1/2 and Xrn1 (Coller and Parker 2004). The details of these general mRNA turnover pathways are discussed in detail below, but we should note that alternative mRNA-specific pathways exist as well, including deadenylation-independent decapping (Badis et al. 2004) and endonucleolytic decay (Yang and Schoenberg 2004). In addition, powerful surveillance mechanisms exist in cells to rapidly annihilate aberrant mRNAs. The best understood of these is nonsense-mediated mRNA degradation (NMD; Chapter 23), but there is also a pathway directed at mRNAs that lack a stop codon (Frischmeyer et al. 2002; van Hoof et al. 2002). Very recently, a mechanism known as no-go decay which is targeted at mRNAs bearing stalled ribosomes has been uncovered in yeast (Doma and Parker 2006).

DEADENYLATION

The poly(A) tail, with its associated poly(A)-binding protein (PABP), is a major determinant of both translation efficiency and stability. Thus, deadenylation simultaneously modulates both translation (de Moor and Richter 2001) and decay (Chen et al. 1995). Moreover, poly(A) tail removal is the only step in the decay pathway shown to be reversible. There are multiple examples of mRNAs, particularly those involved in development and neuronal functions, which undergo regulated deadenylation and subsequent polyadenylation (de Moor and Richter 2001; Du and Richter 2005). In these instances, poly(A) tail shortening is used as a means of translational silencing, but it is not clear how the cell differentiates between mRNAs that should be degraded following deadenylation and those that should be stored. To date, there are three main enzymes that are thought to mediate deadenylation in eukaryotic cells: CCR4/CAF1, PAN2/3, and PARN. In addition to their role in poly(A) shortening, the activity of these enzymes also appears to be networked to translation through a variety of mechanisms. These enzymes may, therefore, be targets for integrating aspects of mRNA metabolism.

The CCR4–CAF1/POP2–NOT complex is responsible for the majority of cytoplasmic deadenylation in many organisms (Tucker et al. 2001, 2002; Temme et al. 2004; Yamashita et al. 2005). CCR4 is thought to be the major catalytic subunit of the complex, and CCR4 mutations cause very slow growth in yeast (Tucker et al. 2002). CAF1/POP2, a member of the RNase D family of exonucleases, also has poly(A)-shortening activity in many organisms (Daugeron et al. 2001; Viswanathan et al. 2004; Bianchin et al. 2005) and is essential for early development in *Caenorhabditis elegans* (Molin and Puisieux 2005). Multiple isoforms of CAF1/POP2 and CCR4 are present in humans (Dupressoir et al. 2001), suggesting that different versions of the enzyme complex exist that may be differentially regulated. In addition to being involved in mRNA decay, several lines of evidence suggest that the CCR4–CAF1/POP2–NOT complex may have a role in transcriptional regulation (Denis and Chen 2003). Moreover, the human Caf1 protein has been found to change its localization during the cell cycle, being predominantly cytoplasmic during S phase and predominantly nuclear during G_0 and G_1. This would be consistent with the role of the CCR4–CAF1–NOT complex switching from modulating transcription to mediating deadenylation in a cell-cycle-dependent manner (Morel et al. 2003). Recent findings suggest that the activity of the CCR4–CAF1–NOT complex is also associated with translation. First, PABP, which has a positive role in translation initiation, is a negative regulator of this deadenylase in yeast (Tucker et al. 2002). Second, the *Drosophila* CCR4 homolog, Twin, regulates poly(A) tail lengths of several mRNAs such as *cyclin A* and *bag of marbles* and thereby influences their translation (Morris et al. 2005). Finally, the known translation regulatory proteins Smaug and the yeast pumilio-like Puf3p recruit this deadenylase complex to specific mRNA targets (Jackson et al. 2004; Semotok et al. 2005). Therefore, the activity of this important enzyme in mRNA decay appears to be linked to translational control at multiple levels.

The PAN2/PAN3 deadenylase is thought to be responsible for the posttranscriptional maturation of poly(A) tails in yeast (Brown et al. 1996; Mangus et al. 2004) and has recently been suggested to have a role in the initial slow/synchronous poly(A) shortening of mammalian mRNAs (Yamashita et al. 2005). In contrast to other known deadenylases, this enzyme is stimulated by PABP through an interaction with the PAN3 regulatory subunit (Brown et al. 1996). Therefore, the association of PAN2/PAN3 with translational control may be exclusively related to its influence on poly(A) tail length. It is important, however, to emphasize that additional networking may be revealed in future studies.

The PARN (poly[A] ribonuclease) deadenylase is a member of the RNase D family of nucleases and likely acts as a homodimer (Wu et al. 2005). PARN is highly active and regulated by *cis*-acting sequences in mammalian cell extracts (Gao et al. 2000). Functional homologs of PARN have been found in both vertebrate (*Xenopus*; Copeland and Wormington 2001) and invertebrate organisms (*Aedes* mosquitoes; Opyrchal et al. 2005), as well as in plants (Chiba et al. 2004), making it somewhat surprising that the protein appears to be absent from the *Saccharomyces cerevisiae* and *Drosophila* genomes.

Several studies indicate that PARN has important biological roles. The *Arabidopsis* PARN is essential to development (Chiba et al. 2004; Reverdatto et al. 2004) and appears to be very important in the stress response (Nishimura et al. 2005). In *Xenopus* oocytes, PARN functions in default deadenylation, a process that silences the translation of maternal mRNAs (Korner et al. 1998). PARN may be targeted to only a subset of mRNAs (Reverdatto et al. 2004), and in vertebrate proteins such as KSRP, RHAU, and EDEN-BP/CUG-BP may have a role in this targeting (Gherzi et al. 2004; Tran et al. 2004; Osborne et al. 2005; Moraes et al. 2006).

Like the CCR4 complex, PARN activity is inhibited by PABP interactions with the poly(A) tail (Gao et al. 2000). However, PARN is apparently unique in that it can interact with the 5′ cap structure, and the presence of a 5′ cap on an RNA substrate stimulates the processivity of the enzyme (Dehlin et al. 2000; Gao et al. 2000; Martinez et al. 2001). This highlights the importance of communication between the 5′ and 3′ ends for both mRNA translation and mRNA degradation. Both eIF4E (eukaryotic translation initiation factor 4E) and the nuclear cap-binding complex protein CBP80 have been shown to inhibit PARN deadenylation activity on RNA substrates (Gao et al. 2000). Importantly, competition between the eIF4E and PARN proteins for cap binding appears to have consequences in cells as changes in phosphorylation of both eIF4E and PARN during serum starvation are associated with an increase in deadenylation (Seal et al. 2005; Balatsos et al. 2006). PARN is, therefore, unique among the known deadenylases in that its ability to interface with the 5′ cap clearly implies a tight coupling of enzyme activity with translational status of the mRNA.

3′ TO 5′ EXONUCLEOLYTIC DECAY

Following deadenylation, if an mRNA is destined for decay, it then enters one of two irreversible pathways. The exosome is an evolutionarily conserved complex of at least 11 proteins in yeast and humans (Raijmakers et al. 2004). The core exosome comprises six RNase PH domain-containing exonucleases (PM-Scl75/Rrp45, Rrp41, Rrp42, Rrp46, OIP2 and Mtr3) along with three proteins containing S1 and/or KH-type RNA-binding domains (Csl4, Rrp4, Rrp40). The nuclear exosome contains an additional subunit, PM-Scl100/Rrp6, that bears similarity to RNase D. The association of component 11, Dis3/Rrp44, is not well-characterized. The exosome is responsible for the 3′ to 5′ degradation of mRNAs, as well as the maturation of many small RNAs including 5.8S and 7SL RNAs (Allmang et al. 1999; Grosshans et al. 2001). The activity of the cytoplasmic exosome is influenced by a complex of Ski proteins. Deletion of *ski* genes in yeast increases the translation of poly(A)$^-$ mRNAs by preventing 3′ to 5′ decay (Brown and Johnson 2001). The Ski7 protein of this complex has been proposed to directly interface with translation in two processes. First, the 3′ to 5′ decay experienced by mRNAs that lack a stop codon (i.e., nonstop decay) is thought to be mediated by the recognition of the empty A site of the ribosome by the elongation factor 1A (eEF1A)-like GTPase domain of Ski7 (Frischmeyer et al. 2002; van Hoof et al. 2002). Second, Upf1p–Ski7 interactions can be used to recruit the exosome to mRNAs that contain premature termination codons (Mitchell and Tollervey 2003; Takahashi et al. 2003). Following exosome-mediated degradation of the body of the mRNA, a scavenger mRNA decapping activity, DcpS, removes the cap from short RNA oligomers (Liu et al. 2002). This might alleviate the accumulation of unproductive complexes between eIF4E and cap structures and also prevent the by-products of this mRNA decay pathway from inhibiting translation or being erroneously incorporated into nucleic acids.

DECAPPING AND 5′ TO 3′ DECAY

The alternative pathway following deadenylation consists of decapping followed by 5′ to 3′ decay (Coller and Parker 2004). This pathway seems to be more elaborate than the 3′ to 5′ decay mechanism, as decapping requires multiple auxiliary factors in addition to the enzymes themselves. In yeast, the decapping enzyme is a complex of two proteins, Dcp1p and Dcp2p, with Dcp2p being the catalytic subunit (Steiger et al. 2003). Dcp1p and Dcp2p are also present in mammalian cells (van Dijk et al. 2002; Wang et al. 2002; Piccirillo et al. 2003); however, a third subunit, Hedls, is essential for decapping activity and appears to act as a bridge between Dcp1p and Dcp2p (Fenger-Gron et al. 2005). Hedls is also known as Ge-1 (Yu et al. 2005), and although it is conserved in *Drosophila* and plants, it is not present in yeast and may therefore be specific for higher eukaryotes. Some sort of checkpoint seems to be present before an mRNA is decapped, and this may be because deadenylated mRNAs are not always destined for degradation; they can be diverted for storage and subsequently reenter the translating pool (Brengues et al. 2005). The Lsm proteins are essential for decapping and appear to associate with the 3′ end of yeast mRNAs following deadenylation (Tharun and Parker 2001; Tharun et al. 2005). Other auxiliary factors include Edc proteins (Dunckley et al. 2001; Schwartz et al. 2003; Kshirsagar and Parker 2004; Fenger-Gron et al. 2005), the Dhh1 helicase (Coller et al. 2001; Fischer and Weis 2002; Coller and Parker 2005), Pat1 (Bonnerot et al. 2000; He and Parker 2001; Coller and Parker 2005), and GW182 (Rehwinkel et al. 2005). Several of these proteins also have roles in translation regulation, and their functions are discussed further below. Once the mRNA is decapped, the remainder is rapidly degraded by the 5′ to 3′ exonuclease Xrn1p (Muhlrad et al. 1994).

Because translation initiation relies heavily on the cap structure, connections between protein synthesis and decapping-dependent decay are apparent. The eIF4E cap-binding protein has been shown to inhibit mRNA decapping (Schwartz and Parker 2000; Gao et al. 2001), and PABP interactions with both the cap and poly(A) tail may directly stabilize RNAs to decapping (Khanna and Kiledjian 2004).

NO-GO mRNA DECAY: AN ENDONUCLEOLYTIC MECHANISM

In addition to the standard pathways followed by normal mRNAs, there are specific pathways directed at clearing aberrant mRNAs from the pool. These include nonsense-mediated decay (for detailed discussion, see Chapter 23), nonstop decay (Frischmeyer et al. 2002; van Hoof et al. 2002), and no-go decay (Doma and Parker 2006). Intriguingly, both of these latter decay mechanisms involve proteins that mimic the translation release factors eRF3 and/or eRF1. Nonstop decay is directed specifically at mRNAs lacking a stop codon and represents a means for the cell to remove damaged or mutated RNAs (Vasudevan et al. 2002).

In contrast, no-go decay is targeted to transcripts bearing a stalled ribosome (Doma and Parker 2006; Tollervey 2006). Stalling could be induced by factors such as a strong secondary structure or chemical damage to the mRNA. In yeast, two factors resembling the release factors, Hbs1p (homologous to eRF1) and Dom34p (homologous to eRF3) are required for activation of the no-go decay pathway, likely through interaction with the ribosome. No-go decay is unique in that it appears to be initiated by an endonucleolytic cleavage event close to the site of the stalled ribosome (Doma and Parker 2006). The enzyme responsible for cleavage has not yet been identified, although it has been postulated that the ribosome itself may perform this step. Following the cleavage event, the resulting 3′ and 5′ fragments of the mRNA are degraded by Xrn1p and the exosome, respectively (Doma and Parker 2006). Importantly, the presence of a stalled ribosome is essential for no-go decay: mRNAs that do not undergo translation are not susceptible (Doma and Parker 2006).

Although homologs of Hbs1p and Dom34p are present in mammals, the existence of the no-go pathway in higher eukaryotes has not yet been evaluated. It also remains to be seen whether no-go decay can be used as a means to regulate gene expression, in addition to acting as a surveillance pathway.

P-BODIES, STRESS GRANULES, AND POLYSOMES

Although one might have expected the decay enzymes to be concentrated near sites of translation where their mRNA targets are found, it seems that a large part of the posttranscriptional regulation of gene expression is dependent on compartmentalization of mRNAs within the cytoplasm. Actively translating mRNAs are found on polysomes, but those that are either translationally silent or destined for decay are localized to two types of cytoplasmic bodies (see Fig. 1): stress granules (SGs) and processing bodies (P-bodies).

P-bodies were first visualized by a number of groups by immunofluorescence of various factors involved in mRNA decay and particularly decapping factors. Instead of being evenly distributed throughout the cytoplasm, multiple mRNA decay enzymes and accessory factors, including Dcp1p, Dcp2p, Xrn1p, and Lsm protein are concentrated in small foci in both yeast and mammalian cells (Bashkirov et al. 1997; Ingelfinger et al. 2002; van Dijk et al. 2002; Eystathioy et al. 2003; Sheth and Parker 2003; Cougot et al. 2004). These cytoplasmic foci are now thought to be sites of mRNA decay and have been shown to contain proteins involved in a variety of processes from nonsense-mediated decay (Unterholzner and Izaurralde 2004) to translation initiation (Andrei et al. 2005; Ferraiuolo et al. 2005), to microRNA (miRNA) metabolism (Liu et al. 2005; Sen and Blau 2005). P-body constituents are summarized in Table 1. Several interesting observations give insight into the nature of P-bodies. First, the number and/or size of P-bodies was found to increase when 5′ to 3′ mRNA decay was inhibited or translation initiation was impaired (Sheth and Parker 2003; Cougot et al. 2004). This is likely due to accumulation of untranslated mRNAs awaiting processing in both cases. In contrast, P-bodies virtually disappeared when translation elongation or transcription was blocked with drugs (Cougot et al. 2004; Teixeira et al. 2005). In both these instances, the amount of mRNA available for

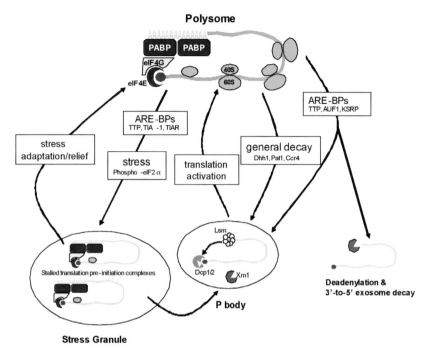

Figure 1. Pathways for travel from polysomes to P-bodies and stress granules.

turnover would be reduced. Thus, it seems that P-body assembly requires mRNAs. This conclusion is also supported by the finding that RNase treatment results in dissolution of P-bodies (Sen and Blau 2005; Teixeira et al. 2005).

The hypothesis that P-bodies are a site of mRNA decay is strongly supported by the fact that intermediates in the 5′ to 3′ degradation pathway actually accumulate in yeast P-bodies (Sheth and Parker 2003). However, it is not clear whether P-bodies are the only site of 5′ to 3′ decay because the enzymes required for this process are also found elsewhere in the cytoplasm in both yeast (Heyer et al. 1995) and mammalian cells (Bashkirov et al. 1997). Moreover, other mRNA turnover pathways converge at P-bodies (see Table 1): Components of the NMD pathway (Smg7, Smg5, and Upf1) (Unterholzner and Izaurralde 2004) as well as of the miRNA pathway (Argonaute proteins; Liu et al. 2005; Sen and Blau 2005) are found in these foci in mammalian cells. In fact, to date, only 3′ to 5′ decay factors have been found excluded from P-bodies (Sheth and Parker 2003; Cougot et al. 2004), although because these proteins are generally enriched in the nucleus, their cytoplasmic distribution has not been extensively investigated.

Table 1. Translation and mRNA turnover factors located in cytoplasmic P-bodies and stress granules

Factor	Present in P-bodies mammals	yeast	Present in stress granules (mammals)
mRNA turnover factors			
Dcp1A/Dcp1B	√	√	x
Dcp2	√	√	x
Xrn1	√	√	√
Lsm proteins	√	√	–
Pat1	–	√	–
Dhh1/rck/p54/Me31B	√	√	–
GW182	√	n.a.	x?
Edc3	√	–	–
Hedls/Ge-1	√	n.a.	x
Ccr4	√	√?	–
Ski7p	–	x	–
DcpS	x	–	–
Nonsense-mediated decay factors			
Upf1	√	–	–
Smg5	√	n.a.	–
Smg7	√	n.a.	–
Translation initiation factors			
eIF4E	√	x	√
4E-T	√	n.a.	–
4E-BP	x	–	x
eIF4G	x	–	√
eIF4A	x	–	–
PABP1	x	–	√
eIF2 subunits	–	–	x
eIF3	x	x	√
eIF5	–	–	–
Translation elongation/termination factors			
Tef4p	–	x	–
Sup45p	–	x	–
Ribosomal proteins			
RpS3, RpS19	–	–	√
RpL5	–	x	x
RNA-binding proteins			
TIA-1/TIAR	x	n.a.	√
HuR	–	n.a.	√
TTP	√	n.a.	√/x
AUF1/hnRNP D	–	n.a.	x
CPEB1	√	n.a.	√
Staufen	–	n.a.	√
Miscellaneous factors			
G3BP endonuclease	x	–	√
FAST	√	n.a.	√
Argonautes	√	n.a.	–

Key: (√) Present; (x) absent; (n.a.) not applicable; (–) not tested; (?) questionable.

Because components of both deadenylation-dependent and deadeny-lation-independent decay pathways are found in P-bodies, it is pertinent to ask whether mRNAs are deadenylated as a prerequisite to entering P-bodies. The answer is not obvious, as the Ccr4p deadenylase in yeast does not visibly localize to P-bodies (Sheth and Parker 2003), but the mam-malian Ccr4 protein clearly does (Cougot et al. 2004). In addition, poly(A)$^+$ RNAs have been detected in P-bodies when the 5′ to 3′ decay pathway is blocked, although it is not clear what length of poly(A) is present (Cougot et al. 2004). In support of deadenylation prior to P-body entry, depletion of Ccr4, in mammalian cells and yeast, results in a re-duction in P-bodies (Sheth and Parker 2003; Andrei et al. 2005). Inter-estingly, the mammalian CPEB1 protein, which can act as a polyadeny-lation factor, has been localized to P-bodies (Wilczynska et al. 2005), giving rise to the intriguing possibility that both deadenylation and polyadenylation occur in these foci.

In yeast, mRNAs are able to exit P-bodies intact, suggesting that mRNA decay is not their only function. When yeast cells are put under certain types of stress, such as glucose deprivation, general translation is inhibited at the level of initiation and mRNAs accumulate in enlarged P-bodies (Coller and Parker 2005). Upon removal of the stress, a signifi-cant portion of these transcripts is permitted to leave the P-bodies and return to polysomes (Brengues et al. 2005). In this respect, the yeast P-bodies seem to act as mRNA storage depots. This is highly reminiscent of the role of stress granules, distinct cytoplasmic entities found in higher eukaryotic cells, but not, as yet, in *S. cerevisiae*.

Stress granules (SGs) are observed only during stress and are thought to form as a result of the global reprogramming of the trans-lational machinery during such times (Kedersha and Anderson 2002). Large quantities of "housekeeping" mRNAs are released from poly-somes to SGs where they undergo triage, being sorted for storage, rerouting to polysomes, or decay. When the stress is removed, or the cell adapts, mRNAs can be shuttled back to polysomes to facilitate rapid resumption of normal gene expression. SGs are composed largely of translationally silent mRNAs bound with stalled translation preini-tiation complexes lacking the eIF2.GTP.Met-tRNA$_i$ ternary complex (Kedersha and Anderson 2002). These stalled complexes are induced upon phosphorylation of eIF2α, which results in a reduction in levels of the ternary complex. The characterized components of SGs are shown in Table 1. They contain small ribosomal subunits and transla-tion initiation factors (eIF3, eIF4G, eIF4E, PABP1), as well as various RNA-binding proteins and other factors (Kedersha et al. 2005). Two related RNA-binding proteins, TIA-1 and TIAR, are integral to the assembly of SGs (Kedersha et al. 1999). They have "prion-like" domains that seem to promote aggregation of nontranslating mRNAs under stress conditions (Gilks et al. 2004). In addition, TIA-1 and TIAR facilitate AU-rich element (ARE)-mediated translational silencing of tumor necrosis factor-α (TNF-α) mRNA in immune cells (Piecyk et al. 2000). Indeed, two other ARE-binding proteins, HuR and TTP, have also been localized to SGs (Kedersha et al. 2002; Stoecklin et al. 2004), suggesting that ARE-mediated effects on translation and mRNA turnover may be initiated in SGs.

The relationship between P-bodies and SGs in mammalian cells is enigmatic. Although P-bodies and SGs share some components (e.g., eIF4E, CPEB1, TTP, Dhh1/rck and FAST; see Table 1), they are distinct entities (Cougot et al. 2004; Kedersha et al. 2005; Wilczynska et al. 2005). As P-bodies often assemble adjacent to SGs and can even be engulfed by them, one hypothesis that has been put forward is that mRNAs are sorted in SGs, and those targeted for turnover are then transferred to P-bodies where they undergo decay (Kedersha et al. 2005). In support of this idea, inhibition of translation elongation, which traps mRNAs on polysomes, results in disassembly of the SGs before disassembly of the adjacent P-bodies, suggesting that the mRNAs move from SGs into the P-bodies (Kedersha et al. 2005). Importantly, however, P-bodies and SGs are apparently induced by different mechanisms and are capable of functioning independently. As mentioned above, SGs are normally initiated by stress-induced phosphorylation of eIF2α, which results in reduced translation initiation (Kedersha and Anderson 2002; Chapters 12 and 13). In the presence of a dominant-negative eIF2α that cannot be phosphorylated due to mutation of serine residues, SGs do not form during oxidative stress. However, P-bodies still assemble normally under these conditions, showing that they are not dependent on SGs, and mRNAs can presumably enter P-bodies without first passing through SGs. In addition, there are several stress conditions that induce SGs but leave P-bodies unaffected, demonstrating that diversion of mRNAs to P-bodies is not an integral part of all stress responses (Kedersha et al. 2005).

So how do mRNAs dissociate from polysomes and travel to SGs and/or P-bodies? There may be multiple answers to this question as three proteins (Dhh1p, Pat1p, and eIF4E-T) have already been implicated in remodeling the mRNP to favor the untranslated state, and all of them have been localized to P-bodies (Coller and Parker 2005; Ferraiuolo et al. 2005). Yeast Dhh1p and Pat1p have been connected to decapping; both single and double mutants exhibit mRNA decapping defects (Bonnerot et al. 2000; Coller et al. 2001; Coller and Parker 2005). In addition, both proteins show interactions with factors involved in translational regulation, and their overexpression promotes dissociation of mRNAs from polysomes (Coller and Parker 2005). However, it appears that Dhh1p and Pat1p act independently for three reasons. First, the double mutant exhibits more severe phenotypes than either single mutant, especially during stress. Normal cells undergo a severe drop in translation during glucose deprivation, and a concomitant increase in P-bodies is observed. Both $dhh1\Delta$ and $pat1\Delta$ mutant strains have a near wild-type response under these conditions, whereas a $dhh1\Delta pat1\Delta$ double mutant exhibits virtually no change in translation or P-body accumulation. Second, overexpression of Pat1p represses translation just as effectively in a $dhh1\Delta$ strain as in a wild-type strain, and vice versa, again suggesting that the two proteins work through independent mechanisms (Coller and Parker 2005). Finally, Dhh1p appears to affect mRNA decay only indirectly as a result of its effects on translation, whereas Pat1p stabilizes even untranslated mRNAs, indicating that it has separable roles in decay and translation (Coller and Parker 2005).

Results of in vitro translation assays performed in the presence of recombinant yeast Dhh1p and its mammalian homolog rck/p54 suggest that the protein functions by inhibiting translation initiation at a step prior to formation of the 48S preinitiation complex. Importantly, Dhh1p is able to repress translation independent of the 5′ cap and 3′ poly(A) tail (Coller and Parker 2005). Thus, it may act directly on the 40S ribosomal subunit and/or simply facilitate the sequestration of mRNAs into P-bodies and away from polysomes.

The eIF4E transporter (4E-T) appears to function to inhibit translation and promote P-body recruitment through a third unrelated pathway unique to mammalian cells. 4E-T binds to the cap-binding factor eIF4E through the same site on eIF4E that interacts with eIF4G and 4E-BP (Dostie et al. 2000). Overexpression of 4E-T results in inhibition of cap-dependent translation, but has no effect on cap-independent translation (Ferraiuolo et al. 2005). At least two mechanisms may be involved here. First, binding of 4E-T prevents eIF4E from interacting with eIF4G, which would negatively affect translation initiation (Dostie et al. 2000). Second, binding of eIF4E to 4E-T results in accumulation of eIF4E in P-bodies (Ferraiuolo et al. 2005), which would obviously segregate the associated mRNA from the translation machinery. Significantly, depletion of 4E-T by RNA interference (RNAi) results in loss of P-bodies and stabilization of at least some mRNAs (Ferraiuolo et al. 2005). It seems possible that binding of 4E-T to eIF4E results in a remodeling of the translating mRNP structure with consequent dissociation from polysomes and transport to P-bodies. Interestingly, 4E-T interacts with rck/p54 within P-bodies by fluorescence resonance energy transfer (FRET), suggesting that these two proteins may facilitate dissociation from polysomes through a shared mechanism (Andrei et al. 2005). This is intriguing because 4E-T clearly acts only on cap-dependent mRNAs, whereas Dhh1/rck/p54 seems to function independent of the cap structure.

The *Drosophila* homolog of 4E-T, Cup, is involved in translational control of Oskar mRNA during oogenesis. Similar to 4E-T, Cup binds to eIF4E and is thought to be recruited to the mRNA through interaction with Bruno, a specific RNA-binding protein (Nakamura et al. 2004). It is interesting to note that there is no identifiable homolog of 4E-T in yeast. Because mammalian eIF4E is excluded from P-bodies in the absence of 4E-T, this observation is consistent with the finding that eIF4E has not been found in P-bodies in yeast (Ferraiuolo et al. 2005).

Depletion or mutation of 4E-T, Dhh1/rck/p54, or Pat1 results in a reduction in P-bodies, comparable to the effects of blocking translation elongation (Coller and Parker 2005; Ferraiuolo et al. 2005). Similarly, RNAi-mediated knockdowns of Lsm1 and Ccr4 cause decreased P-body accumulation (Sheth and Parker 2003; Andrei et al. 2005). Thus, these proteins appear to act early in the pathway that diverts mRNAs to P-bodies. In contrast, inhibition of the actual decapping pathway or 5' to 3' decay by mutation or depletion of Dcp1, Dcp2, or Xrn1 causes an increase in P-bodies, consistent with these proteins acting after mRNAs have been targeted to the cytoplasmic foci (Sheth and Parker 2003; Cougot et al. 2004). Therefore, there are distinct steps in the pathway, each requiring a distinct cohort of factors, to chaperone an mRNA from translation on polysomes to SGs to P-bodies. Moreover, there is likely to be a whole set of factors dedicated to shuttling mRNAs from P-bodies and SGs back to polysomes when they are required.

It is becoming clear that translation and mRNA decay are not completely opposite fates for an mRNA. The factors involved in storing mRNAs in an untranslated state overlap partially with those required for decapping and turnover. Our goal now must be to uncover the switches that trigger decapping and, conversely, the factors that protect an mRNA from this irreversible step in the decay pathway.

SEQUENCE-SPECIFIC REGULATION OF mRNA DECAY

So far, we have described the general pathways of mRNA decay and how they interface with the translation process. However, there are several examples of mRNAs whose turnover is uniquely linked to translation through interaction with sequence-specific RNA-binding factors. Here, we focus on four sequence elements and their associated RNA-binding factors: the major coding region determinant (mCRD) in c-*fos* mRNA, the α-complex-binding site in the α-globin transcript, the Puf-binding site found in several translationally regulated mRNAs, and the AU-rich element.

c-*fos* mCRD: Destabilizing mRNA by Translation

The sequence elements that determine mRNA half-life are primarily located in the 3'UTR (untranslated region), and to a lesser extent within the open reading frame (ORF) or 5'UTR of the mRNA. Elements found in the 5'UTR, such as internal ribosome entry sites (IRESs) and uORFs, often have somewhat indirect effects on decay by modulating translation initiation, although some appear to be bona fide stability elements. Those found in the coding region are unique in that they are influenced strongly by the passage of translating ribosomes that disrupt association with RNA-binding proteins. The best studied of these is the mammalian c-*fos* mCRD that mediates instability of the c-*fos* mRNA only when it is undergoing translation (Grosset et al. 2000). The c-*fos* mCRD binds a complex of at least five proteins (including Unr, PABP, PAIP-1, hnRNP D/AUF1, and NSAP1) which has been suggested to form a bridge between the mCRD and the poly(A) tail, thus inhibiting deadenylation (Grosset et al. 2000). When ribosome transit disrupts this complex, rapid deadenylation ensues, facilitated by an interaction between Unr and the CCR4 deadenylase (Chang et al. 2004).

Stabilization by α-Complex through Two Independent Mechanisms

Another stability element that has been shown to be affected by ribosome transit is found in the α-globin 3'UTR. In erythroid cells, this stabilizing sequence element normally associates with a protein complex (α complex) that protects it from decay (Waggoner and Liebhaber 2003). However, in one naturally occurring variant, Constant Spring, the termination codon is mutated and ribosomes erroneously enter the 3'UTR (Weiss and Liebhaber 1994). The passage of ribosomes through the α-complex-binding site results in destabilization of the mRNA. The α-complex contains two KH-type RNA-binding proteins (αCP1 and αCP2) that recognize cytosine-rich sequences (Kiledjian et al. 1995). Interestingly, the αCP proteins appear to interact with PABP, and this may form the basis of their ability to prevent mRNA deadenylation (Wang et al. 1999). In addition, the αCP-binding site has been shown to overlap with the cleavage site of an erythroid-enriched endonuclease (ErEN); thus, binding of αCP blocks decay by preventing ErEN from reaching its target (Rodgers et al. 2002). The relative importance of these two stabilizing mechanisms in erythroid cells is not clear. However, the fact that tethering of αCP to an mRNA in the absence of its natural binding site allows stabilization indicates that the endonuclease pathway may be less significant (Kong et al. 2003).

Pumilio Family Proteins as Global Controllers of mRNA Metabolism

The Pumilio family (PUF) of mRNA-binding proteins is conserved from yeast to mammals and is characterized by the presence of a unique RNA-binding domain consisting of eight tandem PUF repeats (Spassov and Jurecic 2003). The Puf repeat domain recognizes specific sequence elements within the 3'UTR of target mRNAs. Different Puf proteins have related target sequences based on a UGUR tetramer. Puf proteins were initially characterized as translational repressors, but recent studies in yeast show that they are also capable of modulating mRNA decay rates and intracellular localization of mRNAs (Gerber et al. 2004; Gu et al. 2004). Interestingly, Puf proteins in yeast, and probably in higher eukaryotes, are used as a means to coordinate expression of genes with related functions. Microarray analysis has shown that yeast Puf3p is associated preferentially with mRNAs encoding mitochondrial proteins, whereas Puf1p and Puf2p recognize mRNAs encoding membrane-associated proteins. Puf4p associates with mRNAs encoding nucleolar ribosomal RNA-processing factors and Puf5p with transcripts coding for chromatin modification factors (Gerber et al. 2004).

Several mechanistic models for the way in which Puf proteins regulate gene expression have been proposed (Wickens et al. 2002). An important question raised by these models is whether the Puf proteins directly promote deadenylation, thereby inhibiting translation, or perhaps inhibit translation by an alternate means that then leads to an increase in deadenylation rates. Binding of Puf3p accelerates the deadenylation of the COX17 mRNA, without affecting polysome distribution of this transcript (Olivas and Parker 2000). Puf3p also stimulates decapping of the deadenylated transcript. Similarly, *Drosophila* Pumilio is involved in the repression of *hunchback* mRNA translation by accelerating its deadenylation rate (Wreden et al. 1997), but it also inhibits translation by a poly(A)-independent mechanism (Chagnovich and Lehmann 2001).

So far, the partners of Puf proteins have given few clues to how these proteins function. In *Drosophila*, *pumilio* recruits *nanos* and *brat* to its target mRNAs, but it remains unclear how formation of this complex inhibits translation and/or recruits a deadenylase (Wreden et al. 1997; Sonoda and Wharton 2001). However, it is interesting to note that in neurons, *nanos-*

encoded protein has been localized to cytoplasmic RNA granules reminiscent of SGs (Ye et al. 2004). All together, these studies suggest that the Puf proteins might be involved in the transition of a translationally competent mRNP to a translationally silent structure that can be accessible to the deadenylase and subsequently degraded by the mRNA turnover machinery or maybe readenylated and routed back to polysomes.

AU-rich Elements: Coordinating Posttranscriptional Control

AU-rich elements are sequences of 50–150 nucleotides located in the 3'UTR of a wide variety of short-lived mammalian mRNAs (Bakheet et al. 2003; Khabar 2005) and also in yeast transcripts (Vasudevan and Peltz 2001; Vasudevan et al. 2005). Generally, AREs are typified by the presence of either single or tandem repeats of the AUUUA pentamer (Chen and Shyu 1995; Bakheet et al. 2003), although it is now clear that these core sequences are not an absolute requirement among AREs. For example, the destabilizing element present in the 3'UTR of c-*jun* mRNA, although rich in uridine residues, does not contain any AUUUA motifs (Peng et al. 1996). AREs can have a role at multiple steps in the posttranscriptional life of an mRNA, including regulation of mRNA stability, export, and translation (Chen et al. 1995; Brennan et al. 2000; Piecyk et al. 2000). Moreover, recent studies have shown that as many as 8% of human mRNAs possess AREs within their 3'UTRs (Khabar 2005).

To date, many vertebrate ARE-BPs have been identified, and their regulatory role is widely supported by experimental evidence. The most studied of these are AUF1/hnRNP D, tristetraprolin (TTP), KSRP (KH-type splicing regulatory protein), TIA-1/TIAR, PABP, and HuR (Zhang et al. 1993; Myer et al. 1997; Carballo et al. 1998; Piecyk et al. 2000; Gherzi et al. 2004; Sladic et al. 2004). There is evidence for competition between different ARE-binding proteins for the same site on an mRNA, and in some instances, two different ARE-BPs can bind nonoverlapping sites simultaneously (Lal et al. 2004). Each ARE-BP interacts specifically with the ARE, and they appear to alter gene expression by different mechanisms.

For example, TTP and KSRP appear to directly recruit enzymes such as the decapping enzyme (Fenger-Gron et al. 2005), PARN (Lai et al. 2003; Gherzi et al. 2004), and/or the exosome (Gherzi et al. 2004) to the ARE-containing mRNA, thereby destabilizing it. In contrast, AUF1 may act by changing the RNP structure in response to phosphorylation and consequently facilitate binding of other *trans*-acting factors (Wilson et al. 2003). In support of this idea, AUF1 has been reported to have both stabilizing and destabilizing influences on ARE-containing mRNAs (Loflin et al. 1999; Xu et al. 2001; Sarkar et al. 2003). A recent study suggests that phosphorylation of AUF1 can also result in isomerization of the protein by the peptidyl prolyl isomerase, Pin1, causing dissociation and decay of the AUF1 and allowing other factors to bind the ARE (Shen et al. 2005).

The mammalian ELAV-like protein, HuR, stabilizes ARE-containing mRNAs in vitro by blocking exosome-mediated decay (Ford et al. 1999) and also influences stability and/or translation of its substrates in vivo (Myer et al. 1997; Fan and Steitz 1998; Katsanou et al. 2005). PABP is normally associated with the poly(A) tail, but it has also been identified as an ARE-binding factor in both yeast and mammalian cells (Voeltz et al. 2001; Sladic et al. 2004; Vasudevan et al. 2005). It is not clear how interaction of PABP with the ARE affects gene expression, although in yeast, it has been suggested to mediate translation repression (Vasudevan et al. 2005). Finally, TIA-1 and TIAR, two related RRM-type RNA-binding proteins, are thought to inhibit translation of ARE-containing mRNAs perhaps by a mechanism similar to that used to silence general mRNA translation during stress (Piecyk et al. 2000).

The role of SGs and P-bodies in mediating ARE function has only recently come to light. The ARE-binding proteins, TTP, TIA-1, TIAR, and HuR, have all been localized to SGs, and TTP is also found in P-bodies (Kedersha et al. 2005). Moreover, the recent finding that miRNAs may be involved in ARE-mediated mRNA turnover strengthens this link, because Ago proteins essential for miRNA function are found in P-bodies (Jing et al. 2005; Liu et al. 2005; Sen and Blau 2005). Bearing in mind that SGs and P-bodies are sites of mRNA turnover and of storage for translationally silent mRNAs, the localization of ARE-binding factors to these foci is fully consistent with the dual role of AREs as translational regulators and mRNA stability elements.

The ability of AREs and their associated *trans*-acting factors to rapidly modulate both mRNA decay and translation makes them an efficient means to alter gene expression in response to extracellular cues. Different cellular conditions have been shown to modulate the ARE-dependent mRNP complexes via specific signaling transduction pathways. These include the p38 mitogen-activated protein kinase (MAPK), the Jun amino-terminal kinase pathways, heat shock, and calcium signals (Koeffler et al. 1988; Winzen et al. 1999; Gallouzi et al. 2000; Stoecklin et al. 2001; Espel 2005). More recent studies have demonstrated that the p38 MAPK pathway also influences ARE-mediated control in yeast (Vasudevan et al. 2005). These signal transduction pathways can affect both mRNA turnover and translation (Chapter 14).

One of the most elegant examples of the downstream effects of the p38 MAPK pathway on ARE modulation is the modification and relocalization of TTP during stress (Stoecklin et al. 2004). In the presence of arsenite, the p38 MAPK pathway is activated and as a result TTP becomes phosphorylated. This phosphorylation event leads to the association of TTP with the phospho-dependent chaperones 14-3-3 proteins, which in turn results in exclusion of TTP from cytoplasmic SGs. Under these conditions, HuR binds to ARE-containing mRNAs present in SGs and consequently stabilizes such transcripts. In contrast, under mitochondrial stress, TTP is not phosphorylated and can enter the SGs where it binds ARE-containing mRNAs and promotes their degradation.

Stimulating macrophages with bacterial endotoxin (LPS) activates numerous intracellular signaling pathways including four MAP kinase pathways (ERK, JNK, p38, and BMK; Zhu et al. 2000). Both the JNK and p38 MAP kinase pathways are involved in relieving translational silencing of TNF-α (Swantek et al. 1997; Neininger et al. 2002). Inhibition of p38 MAP kinase with specific drugs results in reduced TNF-α translation (Kontoyiannis et al. 1999), whereas inactivation of a downstream target of p38, MK-2, completely prevents LPS-induced TNF-α expression (Neininger et al. 2002). It seems that the p38 pathway acts on repressive factors bound to the ARE, as deletion of the ARE bypasses the requirement for p38/MK-2 for translational activation (Kontoyiannis et al. 1999). The JNK pathway can be inhibited by glucocorticoids such as dexamethasone, and these drugs reduce TNF-α production at the level of translation (Swantek et al. 1997). In addition, the ERK pathway also regulates expression of TNF-α in an ARE-dependent manner by modulating export of the TNF-α mRNA from the nucleus (Dumitru et al. 2000). The ultimate targets of these pathways are likely to be ARE-binding proteins, although the specific factors have not yet been identified.

A related pathway appears to act in yeast to control translational efficiency of the ARE-containing, constitutively unstable MFA2 transcript in response to carbon source. Interestingly, the carbon-source-regulated translational control of the MFA2 transcript is dependent on the ELAV-like protein Pub1 and binding of Pab1 to the ARE (Vasudevan et al. 2005).

As outlined above, AREs are able to coordinately modulate both translation and turnover of mRNAs. However, although a large number of ARE-BPs have been identified, we are only just beginning to understand the complexities of this regulation.

CONCLUSIONS AND FUTURE DIRECTIONS

Since the last edition of this book in 2000, we have made substantial advances in our understanding of mRNA decay and its complex relationships with the translation process. Multiple new enzymes and factors have been identified and characterized, including the CCR4 and CAF1 deadenylases

(Tucker et al. 2001, 2002), the mammalian decapping enzyme component Hedls (Fenger-Gron et al. 2005), and the decapping accessory Edc proteins (Dunckley et al. 2001; Kshirsagar and Parker 2004). Structural data have been generated for PARN (Wu et al. 2005), Dcp1 (She et al. 2004), and Dcp2 (She et al. 2006), providing clues as to their mechanisms of action. Perhaps the most significant advance has been the finding that the turnover machinery is concentrated in cytoplasmic foci and that these foci are also sites of accumulation of untranslated mRNAs (Bashkirov et al. 1997; van Dijk et al. 2002; Eystathioy et al. 2003; Sheth and Parker 2003; Cougot et al. 2004). In future editions of *Translational Control*, perhaps we will be discussing how mRNAs are diverted from the translational machinery to P-bodies and stress granules and how the cell distinguishes between transcripts destined for decay and those that are to be stored.

ACKNOWLEDGMENTS

We are grateful to many of our colleagues who shared their work prior to publication. In addition, we thank Nancy Kedersha for her insight into the complexities of P-bodies and stress granules. Research in the Wilusz laboratory is supported by the National Institutes of Health (grants GM072481, GM063832, and AI063434 to J.W.). C.I.G. is supported by the National Institutes of Health (HL-04355, GM008102, and U54 CA96297).

REFERENCES

Allmang C., Kufel J., Chanfreau G., Mitchell P., Petfalski E., and Tollervey D. 1999. Functions of the exosome in rRNA, snoRNA and snRNA synthesis. *EMBO J.* **18:** 5399–5410.

Andrei M.A., Ingelfinger D., Heintzmann R., Achsel T., Rivera-Pomar R., and Luhrmann R. 2005. A role for eIF4E and eIF4E-transporter in targeting mRNPs to mammalian processing bodies. *RNA* **11:** 717–727.

Badis G., Saveanu C., Fromont-Racine M., and Jacquier A. 2004. Targeted mRNA degradation by deadenylation-independent decapping. *Mol. Cell* **15:** 5–15.

Bakheet T., Williams B.R., and Khabar K.S. 2003. ARED 2.0: An update of AU-rich element mRNA database. *Nucleic Acids Res.* **31:** 421–423.

Balatsos N.A., Nilsson P., Mazza C., Cusack S., and Virtanen A. 2006. Inhibition of mRNA deadenylation by the nuclear cap binding complex (CBC). *J. Biol. Chem.* **281:** 4517–4522.

Bashkirov V.I., Scherthan H., Solinger J.A., Buerstedde J.M., and Heyer W.D. 1997. A mouse cytoplasmic exoribonuclease (mXRN1p) with preference for G4 tetraplex substrates. *J. Cell Biol.* **136:** 761–773.

Bianchin C., Mauxion F., Sentis S., Seraphin B., and Corbo L. 2005. Conservation of the deadenylase activity of proteins of the Caf1 family in human. *RNA* **11:** 487–494.

Bonnerot C., Boeck R., and Lapeyre B. 2000. The two proteins Pat1p (Mrt1p) and Spb8p interact in vivo, are required for mRNA decay, and are functionally linked to Pab1p. *Mol. Cell. Biol.* **20:** 5939–5946.

Brengues M., Teixeira D., and Parker R. 2005. Movement of eukaryotic mRNAs between polysomes and cytoplasmic processing bodies. *Science* **310:** 486–489.

Brennan C.M., Gallouzi I.E., and Steitz J.A. 2000. Protein ligands to HuR modulate its interaction with target mRNAs in vivo. *J. Cell Biol.* **151:** 1–14.

Brown C.E., Tarun S.Z., Jr., Boeck R., and Sachs A.B. 1996. PAN3 encodes a subunit of the Pab1p-dependent poly(A) nuclease in *Saccharomyces cerevisiae*. *Mol. Cell. Biol.* **16:** 5744–5753.

Brown J.T. and Johnson A.W. 2001. A cis-acting element known to block 3′ mRNA degradation enhances expression of polyA-minus mRNA in wild-type yeast cells and phenocopies a ski mutant. *RNA* **7:** 1566–1577.

Carballo E., Lai W.S., and Blackshear P.J. 1998. Feedback inhibition of macrophage tumor necrosis factor-alpha production by tristetraprolin. *Science* **281:** 1001–1005.

Chagnovich D. and Lehmann R. 2001. Poly(A)-independent regulation of maternal hunchback translation in the *Drosophila* embryo. *Proc. Natl. Acad. Sci.* **98:** 11359–11364.

Chang T.C., Yamashita A., Chen C.Y., Yamashita Y., Zhu W., Durdan S., Kahvejian A., Sonenberg N., and Shyu A.B. 2004. UNR, a new partner of poly(A)-binding protein, plays a key role in translationally coupled mRNA turnover mediated by the c-*fos* major coding-region determinant. *Genes Dev.* **18:** 2010–2023.

Chen C.Y. and Shyu A.B. 1995. AU-rich elements: Characterization and importance in mRNA degradation. *Trends Biochem. Sci.* **20:** 465–470.

Chen C.Y., Xu N., and Shyu A.B. 1995. mRNA decay mediated by two distinct AU-rich elements from c-*fos* and granulocyte-macrophage colony-stimulating factor transcripts: Different deadenylation kinetics and uncoupling from translation. *Mol. Cell. Biol.* **15:** 5777–5788.

Chiba Y., Johnson M.A., Lidder P., Vogel J.T., van Erp H., and Green P.J. 2004. AtPARN is an essential poly(A) ribonuclease in *Arabidopsis*. *Gene* **328:** 95–102.

Coller J. and Parker R. 2004. Eukaryotic mRNA decapping. *Annu. Rev. Biochem.* **73:** 861–890.

———. 2005. General translational repression by activators of mRNA decapping. *Cell* **122:** 875–886.

Coller J.M., Tucker M., Sheth U., Valencia-Sanchez M.A., and Parker R. 2001. The DEAD box helicase, Dhh1p, functions in mRNA decapping and interacts with both the decapping and deadenylase complexes. *RNA* **7:** 1717–1727.

Copeland P.R. and Wormington M. 2001. The mechanism and regulation of deadenylation: Identification and characterization of *Xenopus* PARN. *RNA* **7:** 875–886.

Cougot N., Babajko S., and Seraphin B. 2004. Cytoplasmic foci are sites of mRNA decay in human cells. *J. Cell Biol.* **165:** 31–40.

Daugeron M.C., Mauxion F., and Seraphin B. 2001. The yeast POP2 gene encodes a nuclease involved in mRNA deadenylation. *Nucleic Acids Res.* **29:** 2448–2455.

Dehlin E., Wormington M., Korner C.G., and Wahle E. 2000. Cap-dependent deadenylation of mRNA. *EMBO J.* **19:** 1079–1086.

de Moor C.H. and Richter J.D. 2001. Translational control in vertebrate development. *Int. Rev. Cytol.* **203:** 567–608.

Denis C.L. and Chen J. 2003. The CCR4-NOT complex plays diverse roles in mRNA metabolism. *Prog. Nucleic Acid Res. Mol. Biol.* **73:** 221–250.

Doma M.K. and Parker R. 2006. Endonucleolytic cleavage of eukaryotic mRNAs with stalls in translation elongation. *Nature* **440:** 561–564.

Dostie J., Ferraiuolo M., Pause A., Adam S.A., and Sonenberg N. 2000. A novel shuttling protein, 4E-T, mediates the nuclear import of the mRNA 5′ cap-binding protein, eIF4E. *EMBO J.* **19:** 3142–3156.

Du L. and Richter J.D. 2005. Activity-dependent polyadenylation in neurons. *RNA* **11:** 1340–1347.

Dumitru C.D., Ceci J.D., Tsatsanis C., Kontoyiannis D., Stamatakis K., Lin J.H., Patriotis C., Jenkins N.A., Copeland N.G., Kollias G., and Tsichlis P.N. 2000. TNF-alpha induction by LPS is regulated posttranscriptionally via a Tpl2/ERK-dependent pathway. *Cell* **103:** 1071–1083.

Dunckley T., Tucker M., and Parker R. 2001. Two related proteins, Edc1p and Edc2p, stimulate mRNA decapping in *Saccharomyces cerevisiae*. *Genetics* **157:** 27–37.

Dupressoir A., Morel A.P., Barbot W., Loireau M.P., Corbo L., and Heidmann T. 2001. Identification of four families of yCCR4- and Mg2+-dependent endonuclease-related proteins in higher eukaryotes, and characterization of orthologs of yCCR4 with a conserved leucine-rich repeat essential for hCAF1/hPOP2 binding. *BMC Genomics* **2:** 9.

Espel E. 2005. The role of the AU-rich elements of mRNAs in controlling translation. *Semin. Cell Dev. Biol.* **16:** 59–67.

Eystathioy T., Jakymiw A., Chan E.K., Seraphin B., Cougot N., and Fritzler M.J. 2003. The GW182 protein colocalizes with mRNA degradation associated proteins hDcp1 and hLSm4 in cytoplasmic GW bodies. *RNA* **9:** 1171–1173.

Fan X.C. and Steitz J.A. 1998. Overexpression of HuR, a nuclear-cytoplasmic shuttling protein, increases the in vivo stability of ARE-containing mRNAs. *EMBO J.* **17:** 3448–3460.

Fenger-Gron M., Fillman C., Norrild B., and Lykke-Andersen J. 2005. Multiple processing body factors and the ARE-binding protein TTP activate mRNA decapping. *Mol. Cell* **20:** 905–915.

Ferraiuolo M.A., Basak S., Dostie J., Murray E.L., Schoenberg D.R., and Sonenberg N. 2005. A role for the eIF4E-binding protein 4E-T in P-body formation and mRNA decay. *J. Cell Biol.* **170:** 913–924.

Fischer N. and Weis K. 2002. The DEAD box protein Dhh1 stimulates the decapping enzyme Dcp1. *EMBO J.* **21:** 2788–2797.

Ford L.P., Watson J., Keene J.D., and Wilusz J. 1999. ELAV proteins stabilize deadenylated intermediates in a novel in vitro mRNA deadenylation/degradation system. *Genes Dev.* **13:** 188–201.

Frischmeyer P.A., van Hoof A., O'Donnell K., Guerrerio A.L., Parker R., and Dietz H.C. 2002. An mRNA surveillance mechanism that eliminates transcripts lacking termination codons. *Science* **295:** 2258–2261.

Gallouzi I.E., Brennan C.M., Stenberg M.G., Swanson M.S., Eversole A., Maizels N., and Steitz J.A. 2000. HuR binding to cytoplasmic mRNA is perturbed by heat shock. *Proc. Natl. Acad. Sci.* **97:** 3073–3078.

Gao M., Fritz D.T., Ford L.P., and Wilusz J. 2000. Interaction between a poly(A)-specific ribonuclease and the 5′ cap influences mRNA deadenylation rates in vitro. *Mol. Cell* **5:** 479–488.

Gao M., Wilusz C.J., Peltz S.W., and Wilusz J. 2001. A novel mRNA-decapping activity in HeLa cytoplasmic extracts is regulated by AU-rich elements. *EMBO J.* **20:** 1134–1143.

Garcia-Martinez J., Aranda A., and Perez-Ortin J.E. 2004. Genomic run-on evaluates transcription rates for all yeast genes and identifies gene regulatory mechanisms. *Mol. Cell* **15:** 303–313.

Gerber A.P., Herschlag D., and Brown P.O. 2004. Extensive association of functionally and cytotopically related mRNAs with Puf family RNA-binding proteins in yeast. *PLoS Biol.* **2:** E79.

Gherzi R., Lee K.Y., Briata P., Wegmuller D., Moroni C., Karin M., and Chen C.Y. 2004. A KH domain RNA binding protein, KSRP, promotes ARE-directed mRNA turnover by recruiting the degradation machinery. *Mol. Cell* **14:** 571–583.

Gilks N., Kedersha N., Ayodele M., Shen L., Stoecklin G., Dember L.M., and Anderson P. 2004. Stress granule assembly is mediated by prion-like aggregation of TIA-1. *Mol. Biol. Cell* **15**: 5383–5398.

Grosset C., Chen C.Y., Xu N., Sonenberg N., Jacquemin-Sablon H., and Shyu A.B. 2000. A mechanism for translationally coupled mRNA turnover: Interaction between the poly(A) tail and a c-*fos* RNA coding determinant via a protein complex. *Cell* **103**: 29–40.

Grosshans H., Deinert K., Hurt E., and Simos G. 2001. Biogenesis of the signal recognition particle (SRP) involves import of SRP proteins into the nucleolus, assembly with the SRP-RNA, and Xpo1p-mediated export. *J. Cell Biol.* **153**: 745–762.

Gu W., Deng Y., Zenklusen D., and Singer R.H. 2004. A new yeast PUF family protein, Puf6p, represses ASH1 mRNA translation and is required for its localization. *Genes Dev.* **18**: 1452–1465.

He W. and Parker R. 2001. The yeast cytoplasmic LsmI/Pat1p complex protects mRNA 3′ termini from partial degradation. *Genetics* **158**: 1445–1455.

Herrick D., Parker R., and Jacobson A. 1990. Identification and comparison of stable and unstable mRNAs in *Saccharomyces cerevisiae*. *Mol. Cell. Biol.* **10**: 2269–2284.

Heyer W.D., Johnson A.W., Reinhart U., and Kolodner R.D. 1995. Regulation and intracellular localization of *Saccharomyces cerevisiae* strand exchange protein 1 (Sep1/Xrn1/Kem1), a multifunctional exonuclease. *Mol. Cell. Biol.* **15**: 2728–2736.

Ingelfinger D., Arndt-Jovin D.J., Luhrmann R., and Achsel T. 2002. The human LSm1-7 proteins colocalize with the mRNA-degrading enzymes Dcp1/2 and Xrnl in distinct cytoplasmic foci. *RNA* **8**: 1489–1501.

Jackson J.S., Houshmandi S.S., Lopez L.F., and Olivas W.M. 2004. Recruitment of the Puf3 protein to its mRNA target for regulation of mRNA decay in yeast. *RNA* **10**: 1625–1636.

Jing Q., Huang S., Guth S., Zarubin T., Motoyama A., Chen J., Di Padova F., Lin S.C., Gram H., and Han J. 2005. Involvement of microRNA in AU-rich element-mediated mRNA instability. *Cell* **120**: 623–634.

Katsanou V., Papadaki O., Milatos S., Blackshear P.J., Anderson P., Kollias G., and Kontoyiannis D.L. 2005. HuR as a negative posttranscriptional modulator in inflammation. *Mol. Cell* **19**: 777–789.

Kedersha N. and Anderson P. 2002. Stress granules: Sites of mRNA triage that regulate mRNA stability and translatability. *Biochem. Soc. Trans.* **30**: 963–969.

Kedersha N.L., Gupta M., Li W., Miller I., and Anderson P. 1999. RNA-binding proteins TIA-1 and TIAR link the phosphorylation of eIF-2 alpha to the assembly of mammalian stress granules. *J. Cell Biol.* **147**: 1431–1442.

Kedersha N., Chen S., Gilks N., Li W., Miller I.J., Stahl J., and Anderson P. 2002. Evidence that ternary complex (eIF2-GTP-tRNA(i)(Met))-deficient preinitiation complexes are core constituents of mammalian stress granules. *Mol. Biol. Cell* **13**: 195–210.

Kedersha N., Stoecklin G., Ayodele M., Yacono P., Lykke-Andersen J., Fitzler M.J., Scheuner D., Kaufman R.J., Golan D.E., and Anderson P. 2005. Stress granules and processing bodies are dynamically linked sites of mRNP remodeling. *J. Cell Biol.* **169**: 871–884.

Khabar K.S. 2005. The AU-rich transcriptome: More than interferons and cytokines, and its role in disease. *J. Interferon Cytokine Res.* **25**: 1–10.

Khanna R. and Kiledjian M. 2004. Poly(A)-binding-protein-mediated regulation of hDcp2 decapping in vitro. *EMBO J.* **23**: 1968–1976.

Kiledjian M., Wang X., and Liebhaber S.A. 1995. Identification of two KH domain proteins in the alpha-globin mRNP stability complex. *EMBO J.* **14**: 4357–4364.

Koeffler H.P., Gasson J., and Tobler A. 1988. Transcriptional and posttranscriptional modulation of myeloid colony-stimulating factor expression by tumor necrosis factor and other agents. *Mol. Cell. Biol.* **8**: 3432–3438.

Kong J., Ji X., and Liebhaber S.A. 2003. The KH-domain protein alpha CP has a direct role in mRNA stabilization independent of its cognate binding site. *Mol. Cell. Biol.* **23:** 1125–1134.

Kontoyiannis D., Pasparakis M., Pizarro T.T., Cominelli F., and Kollias G. 1999. Impaired on/off regulation of TNF biosynthesis in mice lacking TNF AU-rich elements: Implications for joint and gut-associated immunopathologies. *Immunity* **10:** 387–398.

Korner C.G., Wormington M., Muckenthaler M., Schneider S., Dehlin E., and Wahle E. 1998. The deadenylating nuclease (DAN) is involved in poly(A) tail removal during the meiotic maturation of *Xenopus* oocytes. *EMBO J.* **17:** 5427–5437.

Kshirsagar M. and Parker R. 2004. Identification of Edc3p as an enhancer of mRNA decapping in *Saccharomyces cerevisiae*. *Genetics* **166:** 729–739.

Lai W.S., Kennington E.A., and Blackshear P.J. 2003. Tristetraprolin and its family members can promote the cell-free deadenylation of AU-rich element-containing mRNAs by poly(A) ribonuclease. *Mol. Cell. Biol.* **23:** 3798–3812.

Lal A., Mazan-Mamczarz K., Kawai T., Yang X., Martindale J.L., and Gorospe M. 2004. Concurrent versus individual binding of HuR and AUF1 to common labile target mRNAs. *EMBO J.* **23:** 3092–3102.

Liu H., Rodgers N.D., Jiao X., and Kiledjian M. 2002. The scavenger mRNA decapping enzyme DcpS is a member of the HIT family of pyrophosphatases. *EMBO J.* **21:** 4699–4708.

Liu J., Valencia-Sanchez M.A., Hannon G.J., and Parker R. 2005. MicroRNA-dependent localization of targeted mRNAs to mammalian P-bodies. *Nat. Cell Biol.* **7:** 719–723.

Loflin P., Chen C.Y., and Shyu A.B. 1999. Unraveling a cytoplasmic role for hnRNP D in the in vivo mRNA destabilization directed by the AU-rich element. *Genes Dev.* **13:** 1884–1897.

Mangus D.A., Evans M.C., Agrin N.S., Smith M., Gongidi P., and Jacobson A. 2004. Positive and negative regulation of poly(A) nuclease. *Mol. Cell. Biol.* **24:** 5521–5533.

Martinez J., Ren Y.G., Nilsson P., Ehrenberg M., and Virtanen A. 2001. The mRNA cap structure stimulates rate of poly(A) removal and amplifies processivity of degradation. *J. Biol. Chem.* **276:** 27923–27929.

Mitchell P. and Tollervey D. 2003. An NMD pathway in yeast involving accelerated deadenylation and exosome-mediated 3′→5′ degradation. *Mol. Cell* **11:** 1405–1413.

Molin L. and Puisieux A. 2005. *C. elegans* homologue of the Caf1 gene, which encodes a subunit of the CCR4-NOT complex, is essential for embryonic and larval development and for meiotic progression. *Gene* **358:** 73–81.

Moraes K.C.M., Wilusz C.J., and Wilusz J. 2006. CUG-BP binds to RNA substrates and recruits PARN deadenylase. *RNA* **12:** 1084–1091.

Morel A.P., Sentis S., Bianchin C., Le Romancer M., Jonard L., Rostan M.C., Rimokh R., and Corbo L. 2003. BTG2 antiproliferative protein interacts with the human CCR4 complex existing in vivo in three cell-cycle-regulated forms. *J. Cell Sci.* **116:** 2929–2936.

Morris J.Z., Hong A., Lilly M.A., and Lehmann R. 2005. twin, a CCR4 homolog, regulates cyclin poly(A) tail length to permit *Drosophila* oogenesis. *Development* **132:** 1165–1174.

Muhlrad D., Decker C.J., and Parker R. 1994. Deadenylation of the unstable mRNA encoded by the yeast MFA2 gene leads to decapping followed by 5′→3′ digestion of the transcript. *Genes Dev.* **8:** 855–866.

Mukhopadhyay D., Houchen C., Kennedy S., Dieckgraefe B.K., and Anant S. 2003. Coupled mRNA stabilization and translational silencing of cyclooxygenase-2 by a novel RNA binding protein, CUGBP2. *Mol. Cell* **11:** 113–126.

Myer V.E., Fan X.C., and Steitz J.A. 1997. Identification of HuR as a protein implicated in AUUUA-mediated mRNA decay. *EMBO J.* **16:** 2130–2139.

Nakamura A., Sato K., and Hanyu-Nakamura K. 2004. *Drosophila* cup is an eIF4E binding protein that associates with Bruno and regulates oskar mRNA translation in oogenesis. *Dev. Cell* **6:** 69–78.

Neininger A., Kontoyiannis D., Kotlyarov A., Winzen R., Eckert R., Volk H.D., Holtmann H., Kollias G., and Gaestel M. 2002. MK2 targets AU-rich elements and regulates biosynthesis of tumor necrosis factor and interleukin-6 independently at different post-transcriptional levels. *J. Biol. Chem.* **277:** 3065–3068.

Nishimura N., Kitahata N., Seki M., Narusaka Y., Narusaka M., Kuromori T., Asami T., Shinozaki K., and Hirayama T. 2005. Analysis of ABA hypersensitive germination2 revealed the pivotal functions of PARN in stress response in *Arabidopsis*. *Plant J.* **44:** 972–984.

Olivas W. and Parker R. 2000. The Puf3 protein is a transcript-specific regulator of mRNA degradation in yeast. *EMBO J.* **19:** 6602–6611.

Opyrchal M., Anderson J.R., Sokoloski K.J., Wilusz C.J., and Wilusz J. 2005. A cell-free mRNA stability assay reveals conservation of the enzymes and mechanisms of mRNA decay between mosquito and mammalian cell lines. *Insect Biochem. Mol. Biol.* **35:** 1321–1334.

Osborne H.B., Gautier-Courteille C., Graindorge A., Barreau C., Audic Y., Thuret R., Pollet N., and Paillard L. 2005. Posttranscriptional regulation in *Xenopus* embryos: Role and targets of EDEN-BP. *Biochem. Soc. Trans.* **33:** 1541–1543.

Peng S.S., Chen C.Y., and Shyu A.B. 1996. Functional characterization of a non-AUUUA AU-rich element from the c-jun proto-oncogene mRNA: Evidence for a novel class of AU-rich elements. *Mol. Cell. Biol.* **16:** 1490–1499.

Piccirillo C., Khanna R., and Kiledjian M. 2003. Functional characterization of the mammalian mRNA decapping enzyme hDcp2. *RNA* **9:** 1138–1147.

Piecyk M., Wax S., Beck A.R., Kedersha N., Gupta M., Maritim B., Chen S., Gueydan C., Kruys V., Streuli M., and Anderson P. 2000. TIA-1 is a translational silencer that selectively regulates the expression of TNF-alpha. *EMBO J.* **19:** 4154–4163.

Raijmakers R., Schilders G., and Pruijn G.J. 2004. The exosome, a molecular machine for controlled RNA degradation in both nucleus and cytoplasm. *Eur. J. Cell Biol.* **83:** 175–183.

Rehwinkel J., Behm-Ansmant I., Gatfield D., and Izaurralde E. 2005. A crucial role for GW182 and the DCP1:DCP2 decapping complex in miRNA-mediated gene silencing. *RNA* **11:** 1640–1647.

Reverdatto S.V., Dutko J.A., Chekanova J.A., Hamilton D.A., and Belostotsky D.A. 2004. mRNA deadenylation by PARN is essential for embryogenesis in higher plants. *RNA* **10:** 1200–1214.

Rodgers N.D., Wang Z., and Kiledjian M. 2002. Characterization and purification of a mammalian endoribonuclease specific for the alpha-globin mRNA. *J. Biol. Chem.* **277:** 2597–2604.

Sarkar B., Xi Q., He C., and Schneider R.J. 2003. Selective degradation of AU-rich mRNAs promoted by the p37 AUF1 protein isoform. *Mol. Cell. Biol.* **23:** 6685–6693.

Schwartz D.C. and Parker R. 1999. Mutations in translation initiation factors lead to increased rates of deadenylation and decapping of mRNAs in *Saccharomyces cerevisiae*. *Mol. Cell. Biol.* **19:** 5247–5256.

———. 2000. mRNA decapping in yeast requires dissociation of the cap binding protein, eukaryotic translation initiation factor 4E. *Mol. Cell. Biol.* **20:** 7933–7942.

Schwartz D., Decker C.J., and Parker R. 2003. The enhancer of decapping proteins, Edc1p and Edc2p, bind RNA and stimulate the activity of the decapping enzyme. *RNA* **9:** 239–251.

Seal R., Temperley R., Wilusz J., Lightowlers R.N., and Chrzanowska-Lightowlers Z.M. 2005. Serum-deprivation stimulates cap-binding by PARN at the expense of eIF4E, consistent with the observed decrease in mRNA stability. *Nucleic Acids Res.* **33:** 376–387.

Semotok J.L., Cooperstock R.L., Pinder B.D., Vari H.K., Lipshitz H.D., and Smibert C.A. 2005. Smaug recruits the CCR4/POP2/NOT deadenylase complex to trigger maternal transcript localization in the early *Drosophila* embryo. *Curr. Biol.* **15:** 284–294.

Sen G.L. and Blau H.M. 2005. Argonaute 2/RISC resides in sites of mammalian mRNA decay known as cytoplasmic bodies. *Nat. Cell Biol.* **7:** 633–636.

She M., Decker C.J., Chen N., Tumati S., Parker R., and Song H. 2006. Crystal structure and functional analysis of Dcp2p from *Schizosaccharomyces pombe. Nat. Struct. Mol. Biol.* **13:** 63–70.

She M., Decker C.J., Sundramurthy K., Liu Y., Chen N., Parker R., and Song H. 2004. Crystal structure of Dcp1p and its functional implications in mRNA decapping. *Nat. Struct. Mol. Biol.* **11:** 249–256.

Shen Z.J., Esnault S., and Malter J.S. 2005. The peptidyl-prolyl isomerase Pin1 regulates the stability of granulocyte-macrophage colony-stimulating factor mRNA in activated eosinophils. *Nat. Immunol.* **6:** 1280–1287.

Sheth U. and Parker R. 2003. Decapping and decay of messenger RNA occur in cytoplasmic processing bodies. *Science* **300:** 805–808.

Sladic R.T., Lagnado C.A., Bagley C.J., and Goodall G.J. 2004. Human PABP binds AU-rich RNA via RNA-binding domains 3 and 4. *Eur. J. Biochem.* **271:** 450–457.

Sonoda J. and Wharton R.P. 2001. *Drosophila* Brain Tumor is a translational repressor. *Genes Dev.* **15:** 762–773.

Spassov D.S. and Jurecic R. 2003. The PUF family of RNA-binding proteins: Does evolutionarily conserved structure equal conserved function? *IUBMB. Life* **55:** 359–366.

Steiger M., Carr-Schmid A., Schwartz D.C., Kiledjian M., and Parker R. 2003. Analysis of recombinant yeast decapping enzyme. *RNA* **9:** 231–238.

Stoecklin G., Stoeckle P., Lu M., Muehlemann O., and Moroni C. 2001. Cellular mutants define a common mRNA degradation pathway targeting cytokine AU-rich elements. *RNA* **7:** 1578–1588.

Stoecklin G., Stubbs T., Kedersha N., Wax S., Rigby W.F., Blackwell T.K., and Anderson P. 2004. MK2-induced tristetraprolin:14-3-3 complexes prevent stress granule association and ARE-mRNA decay. *EMBO J.* **23:** 1313–1324.

Swantek J.L., Cobb M.H., and Geppert T.D. 1997. Jun N-terminal kinase/stress-activated protein kinase (JNK/SAPK) is required for lipopolysaccharide stimulation of tumor necrosis factor alpha (TNF-alpha) translation: Glucocorticoids inhibit TNF-alpha translation by blocking JNK/SAPK. *Mol. Cell. Biol.* **17:** 6274–6282.

Takahashi S., Araki Y., Sakuno T., and Katada K. 2003. Interaction between Ski7p and Upf1p is required for nonsense-mediated 3′-to-5′ mRNA decay in yeast. *EMBO J.* **22:** 3951–3959.

Teixeira D., Sheth U., Valencia-Sanchez M.A., Brengues M., and Parker P. 2005. Processing bodies require RNA for assembly and contain nontranslating mRNAs. *RNA* **11:** 371–382.

Temme C., Zaessinger S., Meyer S., Simonelig M., and Wahle E. 2004. A complex containing the CCR4 and CAF1 proteins is involved in mRNA deadenylation in *Drosophila. EMBO J.* **23:** 2862–2871.

Tharun S. and Parker R. 2001. Targeting an mRNA for decapping: Displacement of translation factors and association of the Lsm1p-7p complex on deadenylated yeast mRNAs. *Mol. Cell* **8:** 1075–1083.

Tharun S., Muhlrad D., Chowdhury A., and Parker R. 2005. Mutations in the *Saccharomyces cerevisiae* LSM1 gene that affect mRNA decapping and 3′ end protection. *Genetics* **170:** 33–46.

Tollervey D. 2006. Molecular biology: RNA lost in translation. *Nature* **440:** 425–426.

Tran H., Schilling M., Wirbelauer C., Hess D., and Nagamine Y. 2004. Facilitation of mRNA deadenylation and decay by the exosome-bound, DExH protein RHAU. *Mol. Cell* **13:** 101–111.

Tucker M., Staples R.R., Valencia-Sanchez M.A., Muhlrad D., and Parker R. 2002. Ccr4p is the catalytic subunit of a Ccr4p/Pop2p/Notp mRNA deadenylase complex in *Saccharomyces cerevisiae*. *EMBO J.* **21:** 1427–1436.

Tucker M., Valencia-Sanchez M.A., Staples R.R., Chen J., Denis C.L., and Parker R. 2001. The transcription factor associated Ccr4 and Caf1 proteins are components of the major cytoplasmic mRNA deadenylase in *Saccharomyces cerevisiae*. *Cell* **104:** 377–386.

Unterholzner L. and Izaurralde E. 2004. SMG7 acts as a molecular link between mRNA surveillance and mRNA decay. *Mol. Cell* **16:** 587–596.

van Dijk E., Cougot N., Meyer S., Babajko S., Wahle E., and Seraphin B. 2002. Human Dcp2: A catalytically active mRNA decapping enzyme located in specific cytoplasmic structures. *EMBO J.* **21:** 6915–6924.

van Hoof A., Frischmeyer P.A., Dietz H.C., and Parker R. 2002. Exosome-mediated recognition and degradation of mRNAs lacking a termination codon. *Science* **295:** 2262–2264.

Vasudevan S. and Peltz S.W. 2001. Regulated ARE-mediated mRNA decay in *Saccharomyces cerevisiae*. *Mol. Cell* **7:** 1191–1200.

Vasudevan S., Peltz S.W., and Wilusz C.J. 2002. Non-stop decay: A new mRNA surveillance pathway. *Bioessays* **24:** 785–788.

Vasudevan S., Garneau N., Tu K.D., and Peltz S.W. 2005. p38 mitogen-activated protein kinase/Hog1p regulates translation of the AU-rich-element-bearing MFA2 transcript. *Mol. Cell. Biol.* **25:** 9753–9763.

Viswanathan P., Ohn T., Chiang Y.C., Chen J., and Denis C.L. 2004. Mouse CAF1 can function as a processive deadenylase/3′-5′-exonuclease in vitro but in yeast the deadenylase function of CAF1 is not required for mRNA poly(A) removal. *J. Biol. Chem.* **279:** 23988–23995.

Voeltz G.K., Ongkasuwan J., Standart N., and Steitz J.A. 2001. A novel embryonic poly(A) binding protein, epab, regulates mRNA deadenylation in *Xenopus* egg extracts. *Genes Dev.* **15:** 774–788.

Waggoner S.A. and Liebhaber S.A. 2003. Regulation of alpha-globin mRNA stability. *Exp. Biol. Med.* **228:** 387–395.

Wang Z., Day N., Trifillis P., and Kiledjian M. 1999. An mRNA stability complex functions with poly(A)-binding protein to stabilize mRNA in vitro. *Mol. Cell. Biol.* **19:** 4552–4560.

Wang Z., Jiao X., Carr-Schmid A., and Kiledjian M. 2002. The hDcp2 protein is a mammalian mRNA decapping enzyme. *Proc. Natl. Acad. Sci.* **99:** 12663–12668.

Weiss I.M. and Liebhaber S.A. 1994. Erythroid cell-specific determinants of alpha-globin mRNA stability. *Mol. Cell. Biol.* **14:** 8123–8132.

Wickens M., Bernstein D.S., Kimble J., and Parker R. 2002. A PUF family portrait: 3′UTR regulation as a way of life. *Trends Genet.* **18:** 150–157.

Wilczynska A., Aigueperse C., Kress M., Dautry F., and Weil D. 2005. The translational regulator CPEB1 provides a link between dcp1 bodies and stress granules. *J. Cell Sci.* **118:** 981–992.

Wilson G.M., Lu J., Sutphen K., Suarez Y., Sinha S., Brewer B., Villanueva-Feliciano E.C., Ysla R.M., Charles S., and Brewer G. 2003. Phosphorylation of p40AUF1 regulates binding to A + U-rich mRNA-destabilizing elements and protein-induced changes in ribonucleoprotein structure. *J. Biol. Chem.* **278:** 33039–33048.

Winzen R., Kracht M., Ritter B., Wilhelm A., Chen C.Y., Shyu A.B., Muller M., Gaestel M., Resch K., and Holtmann H. 1999. The p38 MAP kinase pathway signals for cytokine-induced mRNA stabilization via MAP kinase-activated protein kinase 2 and an AU-rich region-targeted mechanism. *EMBO J.* **18:** 4969–4980.

Wreden C., Verrotti A.C., Schisa J.A., Lieberfarb M.E., and Strickland S. 1997. Nanos and pumilio establish embryonic polarity in *Drosophila* by promoting posterior deadenylation of hunchback mRNA. *Development* **124:** 3015–3023.

Wu M., Reuter M., Lilie H., Liu Y., Wahle E., and Song H. 2005. Structural insight into poly(A) binding and catalytic mechanism of human PARN. *EMBO J.* **24:** 4082–4093.

Xu N., Chen C.Y., and Shyu A.B. 2001. Versatile role for hnRNP D isoforms in the differential regulation of cytoplasmic mRNA turnover. *Mol. Cell. Biol.* **21:** 6960–6971.

Yamashita A., Chang T.C., Yamashita Y., Zhu W., Zhong Z., Chen C.Y., and Shyu A.B. 2005. Concerted action of poly(A) nucleases and decapping enzyme in mammalian mRNA turnover. *Nat. Struct. Mol. Biol.* **12:** 1054–1063.

Yang F. and Schoenberg D.R. 2004. Endonuclease-mediated mRNA decay involves the selective targeting of PMR1 to polyribosome-bound substrate mRNA. *Mol. Cell* **14:** 435–445.

Ye B., Petritsch C., Clark I.E., Gavis E.R., Jan L.Y., and Jan Y.N. 2004. nanos and pumilio are essential for dendrite morphogenesis in *Drosophila* peripheral neurons. *Curr. Biol.* **14:** 314–321.

Yu J.H., Yang W.H., Gulick T., Bloch K.D., and Bloch D.B. 2005. Ge-1 is a central component of the mammalian cytoplasmic mRNA processing body. *RNA* **11:** 1795–1802.

Zhang W., Wagner B.J., Ehrenman K., Schaefer A.W., DeMaria C.T., Crater D., DeHaven K., Long L., and Brewer G. 1993. Purification, characterization, and cDNA cloning of an AU-rich element RNA-binding protein, AUF1. *Mol. Cell. Biol.* **13:** 7652–7665.

Zhu W., Downey J.S., Gu J., Di Padova F., Gram H., and Han J. 2000. Regulation of TNF expression by multiple mitogen-activated protein kinase pathways. *J. Immunol.* **164:** 6349–6358.

26

Translational Control in Plants and Chloroplasts

Daniel R. Gallie
Department of Biochemistry
University of California
Riverside, California 92521-0129

THE USE OF PLANTS AS A HIGHER EUKARYOTIC MODEL for translation has distinct advantages. The translation lysate from wheat germ has long been used to study protein synthesis. Moreover, because plants exhibit complex development, possess signaling pathways that are conserved in many instances with those of other eukaryotes, and have evolved complex pathogen resistance or stress responses, they are useful for the analysis of translation in differentiated cells types and in responses to internal cues and external conditions. Plants produce many seeds, are inexpensive and easy to grow, and provide a ready source of material for analysis. The genomes of *Arabidopsis thaliana* and rice have been sequenced and those of several other plant species are nearing completion, enabling cross-species comparisons. Mutations can be easily generated, and if the species is self-pollinated (e.g., *Arabidopsis*), recessive mutations quickly become homozygous. Insertion mutants resulting in loss of function are available for most genes in *Arabidopsis*. *Arabidopsis* and other plant species can be readily transformed for purposes of complementation, introduction of mutant genes, or for RNA interference (RNAi)-mediated repression of gene expression. Because plants are essential to mankind in agriculture, understanding the regulation of translation has direct application to crop improvement. Finally, the study of translation in plants enables cross-kingdom comparisons to determine what features of the machinery and its regulation have been conserved during the evolution of higher eukaryotes and how these have been adapted during the evolution of the unique aspects of plants.

TRANSLATION OF NUCLEUS-ENCODED mRNAs

One significant change in the translation system that occurred during the evolution of eukaryotes was the adoption of 5′-capped, monocistronic mRNAs as the predominant templates for protein synthesis. In the absence of a Shine-Dalgarno (SD)-mediated mechanism to recruit and direct ribosomes to the correct start site, eukaryotic translation became largely 5′-end-dependent, requiring 40S subunit scanning to identify the correct initiation codon. The numerous initiation factors that evolved to assist in initiation reflect this fundamental difference in how ribosomal subunits are recruited to an mRNA and represent the most significant increase in complexity of the eukaryotic translation machinery. Because the initiation factors are conserved from higher to lower eukaryotes, they arose prior to the divergence of the animal, plant, and fungal kingdoms. Despite this conservation, significant differences exist in the regulation of the plant translation machinery, which may be related to the sessile nature of plants, their autotrophic growth, and the development of unique organs such as leaves, flowers, and roots. Thus, those aspects that are unique to the plant machinery or have not yet been reported in other species are highlighted in this chapter.

TRANSLATION FACTORS

Factors Involved in Binding the 5′ Cap of the mRNA

Early in initiation, the 5′-cap structure is bound by the eukaryotic initiation factor 4E (eIF4E), the small subunit of eIF4F, a complex that contains eIF4G, the large subunit, and eIF4A, an RNA helicase. In addition to eIF4F, plants contain an isoform, eIFiso4F, composed of eIFiso4E and eIFiso4G (Browning 1996). eIF4A copurifies as part of eIF4F or eIFiso4F from crude extracts using m^7GTP-Sepharose chromatography (Gallie 2001), but its association is weak and eIF4A is not normally retained during extensive purification of eIF4F or eIFiso4F (Browning 1996). Thus, purified eIF4F or eIFiso4F used in most studies lacks eIF4A. Although eIF4E and eIFiso4G can form a heterodimeric complex in vitro, as can eIFiso4E and eIF4G (K. Browning, pers. comm.), these complexes are not observed following purification of eIF4F and eIFiso4F, suggesting that the specificity of their interaction is maintained in vivo in a manner that is not yet understood.

eIF4E and eIFiso4E are similar in size but share only about 50% sequence similarity to each other and 30–40% similarity to yeast or animal eIF4E. Plant eIF4E exhibits slightly more conservation with animal eIF4E than does eIFiso4E (Browning 1996). eIF4E is encoded by a three-member gene family in *Arabidopsis*, whereas eIFiso4E is encoded by a single gene. eIFiso4G is considerably smaller than eIF4G (86 kD vs.

162 kD in wheat) (Browning et al. 1992). eIFiso4G is encoded by a two-member gene family in *Arabidopsis*, whereas eIF4G is encoded by a single gene. Although two forms of eIF4G are also observed in other eukaryotes, they do not differ substantially in molecular weight: They share 46% identity in mammals (Gradi et al. 1998) and 53% identity in yeast (Goyer et al. 1993). In contrast, plant eIF4G and eIFiso4G share only 30% identity (K. Browning, unpubl.).

eIFiso4E binding to m^7GTP and mRNA analogs occurs in a two-step process with the first step determined by diffusion (Sha et al. 1995). eIFiso4F undergoes a concentration-independent conformational change following its binding to m^7GpppG (Sha et al. 1995). eIF4B accelerates eIFiso4F binding, and the rate of its dissociation from m^7GpppG is reduced by eIF4B and poly(A)-binding protein (PABP) (Wei et al. 1998; Khan and Goss 2005). Addition of eIF4B to eIFiso4F or the eIFiso4F/PABP complex lowers the activation energy of cap binding (Luo and Goss 2001; Khan and Goss 2005), suggesting that eIF4B and/or PABP enhances cap binding by providing a lower energy barrier, perhaps by inducing a conformational change that is propagated to the cap-binding site. The lower activation energy associated with eIF4B suggests an intermediate of the complex that more easily achieves a stable conformation. Thus, eIF4B and PABP promote stable binding of eIFiso4F (and presumably eIF4F) at the cap (Fig. 1) (Wei et al. 1998; Khan and Goss 2005). Together, eIF4F/eIF4A/eIF4B/PABP have been suggested to function as an initiation surveillance complex that evaluates the structural integrity of an mRNA prior to the initiation of protein synthesis (Gallie 2004). eIF4B and PABP have a larger effect on binding to m^7GpppG than to a mononucleotide cap analog, indicating that the second base is involved in the binding (Khan and Goss 2005). For eIFiso4F, the second base in m^7GpppG slows binding relative to m^7GTP, whereas eIFiso4E and the eIFiso4F/eIF4B/PABP complex bind m^7GpppG faster. Thus, when present in the eIFiso4F/eIF4B/PABP complex, eIFiso4E may be present in a conformation that enables it to form a stable complex with the 5′ cap more rapidly than eIFiso4F alone.

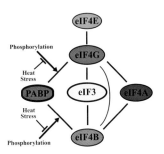

Figure 1. Interaction map of the initiation surveillance complex. (*Solid lines*) Interactions between components of the complex; (*arrows*) regulation of an interaction by phosphorylation. Interaction between PABP and eIF4B is promoted by their phosphorylation, whereas interaction between PABP and eIF4G is promoted by phosphorylation of eIF4G. Heat stress disrupts each interaction through the dephosphorylation of eIF4B and eIF4G.

Although eIFiso4E and eIF4E lack Ser-209, the residue phosphorylated in mammalian eIF4E, Ser-207 of eIFiso4E can be phosphorylated in vitro, which reduces its cap-binding affinity by 1.2–2.6-fold (Khan and Goss 2004). Two phosphorylated isoforms of eIF4E and two of eIFiso4E are observed in metabolically active tissues, and a third isoform for each is present in young wheat shoots and roots in which protein synthesis is particularly active (Gallie et al. 1997). Maize eIF4E is constitutively phosphorylated and undergoes additional phosphorylation following oxygen deprivation (Manjunath et al. 1999). Modification of eIF4E could be mimicked by caffeine treatment under aerobic conditions, and phosphorylation was blocked by ruthenium red under O_2 deprivation (Manjunath et al. 1999), implicating Ca^{2+} as a second messenger in eIF4E modification.

A novel plant cap-binding protein (nCBP) exhibits sequence similarity to eIF4E and eIFiso4E but differs from them by the substitution of two of the eight conserved tryptophan residues with other aromatic amino acids, and it binds m^7GTP 5–20-fold more tightly than does eIFiso4E. nCBP interacts with eIFiso4G in a yeast two-hybrid assay and can support translation of capped mRNA in vitro (Ruud et al. 1998), representing a significant difference from the mammalian eIF4E homologous protein (4EHP) that fails to interact with eIF4G (Cho et al. 2005).

No plant eIF4E-binding protein (4E-BP) ortholog has been identified, suggesting that plants differ from other eukaryotes (Chapter 14) in this respect. The ability of *Arabidopsis* lipoxygenase and a homolog of the mammalian BTF3 factor to bind eIF4E has been observed in vitro, but whether this interaction occurs in vivo is not known (Freire et al. 2000; Freire 2005). The absence of a 4E-BP-like activity in plants would represent a substantial difference from other eukaryotes in the regulation of eIF4F activity and, consequently, of protein synthesis.

Plant eIF4G differs most from eIFiso4G in that it contains an additional 700 amino acids at its amino terminus and, in this respect, is more similar to human eIF4G than is eIFiso4G (Chapter 4). Plant eIF4G and eIFiso4G, like the yeast homologs, contain just one conserved eIF4A-binding site corresponding to the central domain (which also contains the binding site for eIF3) of mammalian eIF4G. The central domain of plant eIF4G and eIFiso4G exhibits the highest degree of conservation with eIF4G from other eukaryotes (Metz and Browning 1996). Plant eIF4G and eIFiso4G lack the carboxy-terminal domain responsible for binding eIF4A and Mnk1, the mitogen-activated protein kinase (MAPk)-activated kinase that phosphorylates eIF4G-bound eIF4E (Metz and Browning 1996), a significant difference from animal eIF4G. Although proteins identified as calcium-dependent protein kinases (CDPKs) exhibit limited similarity to

Mnk1, no formal demonstration of a Mnk-1 homolog has been reported in plants, suggesting that another kinase may phosphorylate eIF4E.

eIF4G and eIFiso4G exhibit some functional differences. eIF4F is present in plants at 11% the level of eIFiso4F yet is at least 20–30-fold more effective in stimulating translation (Browning et al. 1990; Gallie and Browning 2001). The affinity of wheat eIF4F for hypermethylated cap structures is lower than that of eIFiso4F, and the RNA-dependent ATPase activity of eIF4F is greater than that of eIFiso4F (Lax et al. 1986; Carberry et al. 1991). eIF4F binding is more sensitive to the presence of mRNA secondary structure than is eIFiso4F, although the latter exhibits a binding preference for unstructured nucleic acids (Carberry and Goss 1991). eIF4F and eIFiso4F engage different mRNAs for translation (see below), although a systematic identification of genes that exhibit a preference has yet to be performed.

eIF4A stimulates the RNA-dependent ATP hydrolysis and ATP-dependent RNA helicase activities of eIF4F and eIFiso4F, which eIF4B stimulates further by increasing the ATP-binding affinity of eIF4A (Browning 1996). The ATPase activity of wheat eIF4A is stimulated by RNA and eIF4F (or eIFiso4F) together but not by RNA alone (Lax et al. 1986). Wheat eIF4A does not bind mRNA tightly and ATP increases the association of eIF4A with eIFiso4F likely through enhanced protein–protein interactions instead of enhanced binding to mRNA (Balasta et al. 1993). eIF4A interacts directly with eIFiso4G and functionally with eIF4B (Fig. 1) (Lax et al. 1986; Metz and Browning 1996). A cyclin-dependent protein-kinase-containing complex interacts with eIF4A in proliferating cells but not in cells that have ceased dividing (Hutchins et al. 2004), suggesting a molecular mechanism that links cell proliferation with translational control. However, the substrate(s) for this kinase has yet to be identified. Phosphorylation of eIF4A from wheat and maize is induced by hypoxia or heat shock, correlating with translational repression (Webster et al. 1991; Gallie et al. 1997). Whether phosphorylation of eIF4A modifies its interaction with partner proteins such as eIFiso4G or eIF4B or regulates its activity remains to be determined. eIF4A is encoded by two genes in *Arabidopsis* that are widely expressed with the exception of pollen.

Wheat eIF4B stimulates but is not required for the ATP hydrolysis activity of eIF4A and eIF4F (Browning et al. 1989). The nonessential contribution of eIF4B to ATP hydrolysis and RNA helicase activities may be similar to *Saccharomyces cerevisiae* (Altmann et al. 1995) but differs from mammals in which eIF4B and eIF4A are sufficient for ATP-dependent double-stranded unwinding in the absence of eIF4F (Jaramillo et al. 1990). eIF4B is encoded by two genes in *Arabidopsis* that are widely expressed. The two eIF4B proteins in *Arabidopsis* are 84% identical but share only

52–54% identity with wheat eIF4B. Only limited conservation is observed between plant eIF4B and its orthologs from *S. cerevisiae* and animals, ranging from 24% to 31% identity (Metz et al. 1999), which may account for the degree to which eIF4B contributes to the ATP hydrolysis and RNA helicase activities of eIF4A and eIF4F among eukaryotes. Little is known about plant eIF4H, which, in animals, is functionally similar to eIF4B (Richter-Cook et al. 1998). eIF4H is encoded by a single gene in *Arabidopsis* whose expression appears to be limited to seeds.

Six to eight phosphorylated isoforms of eIF4B are present in plants as in mammals, and its phosphorylation state is regulated by developmental cues and external signals that correlate with the level of translational activity (Gallie et al. 1995a, 1997; Le et al. 1998). Phosphorylated

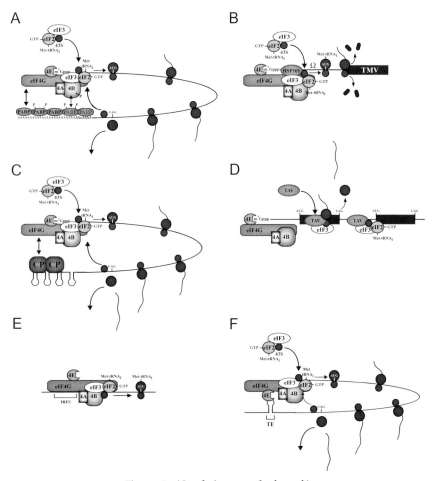

Figure 2. (*See facing page for legend.*)

eIF4B enhanced the RNA-binding affinity of phosphorylated PABP (Le et al. 2000). Although hypophosphorylated eIF4B promoted RNA binding of hypophosphorylated PABP, substantially higher levels of eIF4B were required, and, in growing tissues, hypophosphorylated eIF4B is not present at significant concentrations (Le et al. 1998, 2000). These observations suggest that the phosphorylation state of PABP determines whether it interacts with eIF4B or with eIF4G, whereas the phosphorylation state of eIF4B and eIF4G may regulate the formation of a productive interaction between the 5′ and 3′ termini (Figs. 1 and 2A).

Factors Involved in Ternary Complex Formation

The three subunits of wheat eIF2 are α (42 kD), β (38 kD), and γ (50 kD) (Metz and Browning 1997). eIF2α is encoded by a two-member gene family

Figure 2. Translation of cellular and plant viral mRNAs. (*A*) Phosphorylation regulates the assembly of the PABP/poly(A) tail complex and its interaction with eIF4G and eIF4B. Multiple molecules of PABP of differing phosphorylation states are shown bound to the poly(A) tail. eIF4G preferentially interacts with hypophosphorylated PABP, whereas phosphorylated eIF4B preferentially interacts with phosphorylated PABP as indicated by double-headed arrows. Their interaction likely promotes 40S subunit recycling to the 5′ terminus promoted by the interactions indicated. (P) Phosphorylation. (*B*) The 5′ leader of TMV binds HSP101, which, in turn, recruits eIF4G and eIF3 to promote cotranslational disassembly of the virion particle. HSP101 is shown bound to Ω, the TMV 5′ leader, and recruits eIF4G and eIF3 to the viral mRNA. Following translation initiation, the 80S ribosome cotranslationally strips the viral coat protein from the genomic mRNA. (*C*) Translation of AMV mRNAs is promoted by coat protein bound to the 3′UTR. An AMV mRNA is shown with its 3′UTR in the hairpin conformation to which AMV coat protein binds. The coat protein interacts directly with eIF4G, an association required for efficient translation and viral infection. (CP) Coat protein. (*D*) *trans*-Activation of the polycistronic CaMV 35S RNA is promoted by TAV through its interaction with eIF3 and L24. Following translation initiation, eIF4B is released from its interaction with eIF3g to be replaced by TAV. The interaction of TAV with L24 of 60S subunit maintains the association of eIF3 with the translocating ribosome until termination, at which point the TAV/eIF3 complex associates with the 40S subunit and promotes reinitiation at the next ORF. (*E*) The 5′ leader of TEV functions as an IRES that recruits eIF4G to promote cap-independent translation. The naturally uncapped TEV 5′ leader is shown. A 5′-proximal pseudoknot and adjacent sequence function to promote cap-independent translation and exhibit IRES activity. eIF4G is required for IRES activity, which is enhanced by eIF4A and eIF4B. eIF4G is shown associated with the IRES, but whether the interaction is direct or mediated by other *trans*-acting factors is unknown. (*F*) Translation of STNV satellite viral mRNA requires a TE element in the 3′UTR that recruits eIF4F. The translational enhancer (TE) in the 3′UTR of the naturally uncapped and poly(A)⁻ satellite viral mRNA recruits eIF4E. This binding is enhanced by eIF4G. The interaction promotes efficient translation initiation.

in *Arabidopsis*, whereas eIF2β and eIF2γ are encoded by three-member gene families. Wheat eIF2 differs significantly from mammalian eIF2 in that the latter binds GDP much more tightly than GTP (K_d 100-fold lower for GDP than GTP), whereas the K_d for GDP of the wheat factor is only 10–20-fold lower than for GTP and decreases further in the presence of Met-tRNA$_i$ (Browning 1996). Although the similarity in the relative binding affinities of eIF2 for GDP and GTP might suggest that eIF2B, responsible for promoting the GDP:GTP exchange in eIF2, may not be required in plants, orthologs of all five eIF2B subunits are present in the *Arabidopsis* genome.

The activity of wheat eIF2 is regulated by phosphorylation on Ser-51 of the α subunit (Gil et al. 2000). Up to approximately ten different phosphorylated eIF2α isoforms can be observed, in striking contrast to the single phosphorylated species observed in animals and yeast (Chapters 9, 12, and 13). With the exception of Ser-51, the residues subject to phosphorylation have not been identified (Gallie et al. 1997; Le et al. 1998). It remains to be determined whether phosphorylation of plant eIF2α regulates selective translation of transcripts as occurs in yeast, e.g., *GCN4* (Hinnebusch 2000). In addition to the phosphorylation of plant eIF2α, up to four eIF2β phosphorylated isoforms are observed in wheat leaves and seeds, correlating with a high level of protein synthetic activity (Gallie et al. 1997).

Wild-type eIF2α (but not the S51A mutant) is phosphorylated by a plant protein kinase R (PKR)-like activity (Langland et al. 1996; Chang et al. 1999), suggesting the existence of plant eIF2α kinase activity. P58[IPK], a plant ortholog of the PKR inhibitor, shares 50% similarity with bovine P58[IPK] (Bilgin et al. 2003). Although eIF2α in *Nicotiana benthamiana* does not undergo phosphorylation following viral infection, it is phosphorylated in virus-infected P58[IPK] mutants, resulting in death in contrast to the survival of infected wild-type plants (Bilgin et al. 2003). Expression of bovine P58[IPK] complemented the P58[IPK] mutants, suggesting that the plant P58[IPK] is a functional homolog. Expression of the eIF2α S51A mutant blocked virus-induced cell death in the absence of P58[IPK], data supporting the regulation of eIF2 activity through phosphorylation of the α subunit at Ser-51. Therefore, P58[IPK] is essential to prevent virus-induced cell death. Although these data implicate the existence of PKR activity in plants, no plant homolog has been reported. Unambiguous evidence for a plant PKR will require its isolation and demonstration of function.

Factors Involved in 48S and 80S Complex Formation

eIF1A and eIF3 serve several functions including preventing premature 60S ribosomal subunit joining to 40S subunits, stabilizing ternary complex (eIF2:GTP:Met-tRNA$_i$) formation and binding to 40S subunits, and promoting 40S subunit binding to mRNA (stimulated by eIF1A), functions that are conserved in wheat (Checkley et al. 1981; Lauer et al. 1985). Wheat eIF3 exhibits mRNA binding independent of the presence of a cap or ATP hydrolysis (Carberry and Goss 1992), but its binding is enhanced by eIF4F or eIFiso4F (Carberry and Goss 1991), supporting the notion that recruitment of plant eIF3 to mRNA involves protein interaction with eIF4G and direct binding to the RNA.

The subunit composition of plant eIF3 is largely similar to that of mammalian eIF3, but eIF3j appears to be lacking (Burks et al. 2001). eIF3l, present in wheat and *Arabidopsis* as in animal eIF3, is not present in *Schizosaccharomyces pombe* (Burks et al. 2001). The a, e, f, h, and k subunits of eIF3 are each encoded by a single gene in *Arabidopsis*, whereas the b, c, d, g, i, and l subunits are each encoded by two genes. Most of the eIF3 subunit genes are expressed widely in *Arabidopsis* with the exception of pollen. A direct interaction between eIF4B and eIF3g was demonstrated in plants as in *S. cerevisiae* (Vornlocher et al. 1999; Park et al. 2004), whereas eIF4B interacts with eIF3a in mammals (Methot et al. 1996).

Wheat eIF1A (14 kD) is a heat-stable single polypeptide that is 68% conserved with mammalian eIF1A and 56% conserved with *S. cerevisiae* eIF1A (Browning 1996). Wheat eIF1A can substitute for animal eIF1A in a rabbit reticulocyte lysate and can partially complement *tif11* yeast (Timmer et al. 1993a; Rausell et al. 2003). eIF1A is encoded by a two-member gene family in *Arabidopsis*, and both appear to be expressed in most tissues. The concentration of eIF1A in wheat-germ lysate is sufficiently high (1–2 µM) that it is not a rate-limiting factor for translation initiation, although it is required (Timmer et al. 1993a).

Little is known about eIF1 in plants, but a gene encoding this factor has been isolated in rice (De Pater et al. 1992). In *Arabidopsis*, eIF1 is encoded by a four-member gene family, and all gene members appear to be expressed in most organs. Little has been reported about the role of eIF5, eIF5B, or eIF6 in plant translation.

Poly(A)-binding Protein

The observation that the cap and the poly(A) tail synergistically increase translation (Gallie 1991) was an early indication that PABP functions in a step during translation initiation involving one or more cap-associated initiation factors. PABP interacts directly with eIF4G in plants as in other eukaryotes (Le et al. 1997; Gallie 2002a). An interaction between PABP and eIF4B originally observed in plants (Fig. 1) (Le et al. 1997) was also observed in animals (Bushell et al. 2001). The effect of eIF4B on the RNA-binding activity of PABP is greater than the effect of eIF4G or eIFiso4G, suggesting that interaction with eIF4B may be of greater importance in plants. eIF4F (or eIFiso4F) and eIF4B synergistically increase the binding affinity of PABP for poly(A) RNA (Le et al. 1997). These data, in addition to those discussed above, suggest that PABP increases the cap-binding affinity of eIF4E, increases 48S complex formation, and increases the removal of secondary structure present in a 5′ leader that results in more efficient translation.

The phosphorylation state of PABP determines its affinity for, and type of binding to, poly(A) RNA. Phosphorylated PABP binds poly(A) RNA cooperatively but with a tenfold lower affinity than hypophosphorylated PABP, which binds noncooperatively (Le et al. 2000). The combination of hypophosphorylated and phosphorylated PABP, however, exhibited multimeric binding with the highest affinity for RNA (Le et al. 2000). The interaction of eIF4G or eIFiso4G with hypophosphorylated PABP promotes the multimeric binding of PABP to poly(A) RNA and decreases the dissociation rate of PABP (Le et al. 2000). eIF4B increases the RNA-binding affinity of phosphorylated, but not hypophosphorylated, PABP by decreasing its rate of dissociation (Le et al. 2000). The PABP/poly(A) tail complex would be expected to contain PABP of differing phosphorylated states (Fig. 2A), and the interaction of eIF4G and eIF4B with separate molecules of PABP would result in a heterotrimeric complex that is more stable than the heterodimeric complexes alone, consistent with the observation that eIF4G and eIF4B synergistically increase PABP-binding activity to poly(A) RNA (Le et al. 1997).

Arabidopsis PABP is encoded by a divergent eight-member gene family that is divided into at least three distinct classes based on phylogenetic analysis, gene structure, and expression patterns that predate the evolutionary split of monocots and dicots (Belostotsky 2003). Expression of PAB5 and PAB3 is limited to floral organs, with the highest level of expression in pollen, whereas PAB2 is expressed widely. *Arabidopsis* PAB2 restored viability, participated in poly(A) tail shortening, and partially restored the linkage between deadenylation, decapping, and mRNA decay when expressed in *pab1* yeast, suggesting that it functions in translation (Palanivelu et al. 2000). *Arabidopsis* PAB5 partially complemented the poly(A) shortening and translation defects of *pab1* yeast but did not restore the linkage between deadenylation and decapping (Belostotsky and Meagher 1996). *Arabidopsis* PAB3 restored viability to *pab1* yeast but did not prevent decapping, support poly(A)-dependent translation, or mediate a synergistic interaction between a cap and a poly(A) tail during translation (Chekanova et al. 2001). PAB3 was localized to the nucleus when expressed in *pab1* yeast and partially restored the control of poly(A) tail length during polyadenylation in a PAN-dependent in vitro assay. These data suggest a role of PAB3 in mRNA biogenesis that is supported by the synthetic lethality of PAB3 with PAN3 mutants (Chekanova et al. 2001; Chekanova and Belostotsky 2003). Expression of PAB3 alleviated the mRNA export defect of an *nab2-1* in yeast, supporting a possible role in mRNA export (Chekanova and Belostotsky 2003; Chapter 23).

Factors Involved in Elongation and Termination

Eukaryotic elongation factor 1A (eEF1A) is one of the most abundant proteins in the cell, present at up to 5% of the soluble protein (Browning et al. 1990). It is associated with microfilaments and microtubules in plants as in animals (Condeelis 1995). Expression of eEF1A is highest in actively growing tissues and is stimulated by light and auxin (Browning 1996). eEF1A is phosphorylated (Yang and Boss 1994), although whether this regulates its activity during translation is unknown. In *Arabidopsis*, eEF1A is encoded by a four-member gene family, all of which are actively transcribed (Axelos et al. 1989).

Rice and wheat eEF1Bα lack Ser-89, the residue that is phosphorylated in *Artemia salina* by casein kinase II with consequent inhibition of the guanine nucleotide exchange activity of eEF1A (Janssen et al. 1988). eEF1Bα, eEF1Bβ, and eEF1Bγ are each encoded by a two-member gene family in *Arabidopsis*. eEF1B-A1, one of the members encoding eEF1Bα, is poorly expressed throughout most tissues, whereas eEF1B-A2 is widely expressed.

Like eEF2 from other species, plant eEF2 is posttranslationally modified at His-715 to generate diphthamide (Lauer et al. 1984). eEF2 activity is inhibited following phosphorylation by a rabbit eEF2 kinase (Smailov et al. 1993), although no such kinase activity has yet been identified in plants. eEF2 is encoded by a two-member gene family in *Arabidopsis* in which eEF2-1 is widely expressed, whereas eEF2-2 is poorly expressed throughout most tissues.

Although most organisms have a single gene encoding eukaryotic release factor 1 (eRF1), three genes encoding eRF1 are present in *Arabidopsis*. Each eRF1 isoform is able to rescue a *S. cerevisiae* eRF1 temperature-sensitive mutant (Chapman and Brown 2004), suggesting functional redundancy. *Arabidopsis* eRF1 proteins show a high degree of conservation with the *Homo sapiens* orthologs (up to 75% identity), but a greater degree of divergence exists among *Arabidopsis* eRF1 proteins than between eRF1 from *H. sapiens* and *Xenopus laevis* (Chapman and Brown 2004; Chapter 7).

EFFECT OF HEAT STRESS

Heat stress results in a rapid loss of cap- and poly(A)-dependent translation in wheat (Gallie et al. 1995a). Loss of cap-dependent translation was observed only for polyadenylated transcripts and was accompanied by a loss in synergy between these regulatory elements, suggesting that heat stress disrupts their functional interaction. A substantial increase in cap-independent translation accompanied the loss in cap-dependent translation (Gallie et al. 1995a), which may have been a result of increased availability of unengaged translational machinery released following the heat-mediated disassembly of polysomes (Key et al. 1981). The 5′ leader from heat shock protein (HSP) 70 mRNA was sufficient to confer translational competence in heat-stressed plant cells, suggesting that the leader is required to overcome the loss in cap-independent translation (Pitto et al. 1992).

An increase in mRNA stability is observed following a heat stress regardless of whether the mRNA is capped or polyadenylated, and the degree of the increase correlates with the severity of the stress (Gallie et al. 1995a). These data indicate that, unlike the specific repression of cap-dependent translation, the heat-stress-mediated changes in the RNA degradatory machinery do not discriminate between capped or uncapped mRNAs (Gallie et al. 1995a). This observation suggests that heat may repress the activity of RNases responsible for mRNA turnover or that mRNAs are sequestered from the degradation apparatus. The observation that non-HSP mRNAs are sequestered in heat-stress granules (HSGs) (Nover et al. 1989; Apuya and Zimmerman 1992) supports sequestration. During recovery, the mRNA may be released from the HSGs at a constant rate for subsequent recruitment onto polysomes.

Heat stress results in the rapid dephosphorylation eIF4B (and likely eIF4G and eIFiso4G), temporally correlating with the thermal-induced repression of translation (Gallie et al. 1995a, 1997; Le et al. 2000). Phosphorylation of eIF4A occurs only after a prolonged exposure to stress (Webster et al. 1991; Gallie et al. 1997), suggesting that its phosphorylation might be an adaptive response to stress. No change in the phosphorylation state of the eIF4E isoforms, eIF2α, or PABP, was observed following exposure to heat stress (Gallie et al. 1997). Thus, plants are similar to animals in their response to thermal stress in that eIF4B undergoes dephosphorylation following heat stress, but they differ in two key points: in the regulation of eIF4E/eIFiso4E and eIF2α.

Heat stress substantially reduced the association of PABP and eIF4B with the cap-binding complex and reduced the RNA-binding activity of PABP (Le et al. 2000). The correlation among the heat-stress-mediated dephosphorylation of eIF4B, eIF4G, and eIFiso4G, dissociation of PABP from the cap-binding complex, and the reduction in PABP RNA-binding activity is consistent with the loss of cap- and poly(A)-dependent translation (Gallie et al. 1995a).

These observations suggest that the interaction among PABP, eIF4F, and eIF4B may be regulated by heat stress through the regulation of the phosphorylation state of eIF4G and eIF4B, rather than through phosphorylation of PABP (see Fig. 1). Thus, the effect of the phosphorylation state on the interactions between PABP and eIF4B, eIFiso4G, or eIF4G may be germane to translation initiation in two ways: (1) by providing the specificity for the interaction between the cap-associated factors and PABP and (2) by regulating the strength of the interaction in response to internal or external cues that globally affect protein synthesis.

TRANSLATIONAL REGULATION OF CELLULAR AND VIRAL MRNAS

uORF-mediated Regulation of Cellular mRNA Translation

Expression of S-adenosylmethionine decarboxylase (AdoMetDC) is regulated by two conserved upstream open reading frames (uORFs) that are present in the 5′ leader of AdoMetDC mRNA (Franceschetti et al. 2001). A tiny uORF (which is 3–4 codons in length depending on the plant species) lies upstream and overlaps a small uORF (50–54 codons) by one nucleotide, an arrangement that predates the origin of flowering plants (Franceschetti et al. 2001). The sequence context of the tiny uORF initiation codon is poor, whereas that of the small uORF is good, suggesting that the small uORF is translated through leaky scanning. The small uORF represses downstream translation under normal growth conditions (Hanfrey et al. 2002). Truncation or elimination of the small uORF increases translation from the AdoMetDC main ORF, suggesting a repressive function through a sequence-dependent mechanism (Hanfrey et al. 2005). The tiny uORF is required for polyamine-responsive translation of the AdoMetDC main ORF, likely by repressing the translation of the small uORF (Hanfrey et al. 2005).

Although disruption of eIF3h function in an *Arabidopsis eif3h* mutant does not result in lethality or disrupt the eIF3 complex or polysomes, the translation of specific mRNAs is compromised. One such mRNA encodes the AtbZip11/ATB2 transcription factor, translation of which is regulated by sucrose (Kim et al. 2004; Wiese et al. 2004). Four uORFs present in the leader of the AtbZip11/ATB2 mRNA are responsible for the sensitivity to eIF3h, suggesting that eIF3 is involved in scanning, recognition of AUG codons, and/or reinitiation (Kim et al. 2004). Abolishing one of the four uORFs improved AtbZip11/ATB2 translation only marginally in the *eif3h* mutant, but the elimination of all but one of them restored wild-type translation, suggesting the importance of multiple uORFs in eIF3h-dependent translation (Kim et al. 2004).

Role of the TMV 5′ Leader as a Translational Enhancer

The 68-nucleotide 5′ leader (called Ω) of tobacco mosaic virus (TMV) was one of the first examples of a translational enhancer, followed by reports of other enhancers in several viral and cellular mRNAs (Gallie et al. 1987a, b; Gallie and Walbot 1992; Turner and Foster 1995; Satoh et al. 2004). The TMV genome is a capped, single-stranded, positive-sense RNA that serves as mRNA for the 5′-proximal cistron encoding the replicase. A poly(CAA) region within Ω is responsible for the translational enhancement (Gallie and Walbot 1992). Ω requires no viral proteins or other viral RNA sequences, but its function is stimulated by heat shock (Ling et al. 2000). HSP101, a heat shock protein isolated for its ability to bind Ω (Tanguay and Gallie 1996), is necessary and sufficient to mediate the translational enhancement and genetically requires eIF4G and eIF3 to do so (Fig. 2B) (Wells et al. 1998). Ω preferentially employs eIF4F over eIFiso4F whether the mRNA is capped or not (Gallie 2002b). Because eIF4F is substantially more effective in stimulating translation (Browning et al. 1990; Gallie and Browning 2001), this selective recruitment by Ω may be responsible for its translational enhancement activity. These observations indicate that the genomic mRNA of TMV evolved to access a cellular regulatory mechanism in which HSP101 binds to those mRNAs containing a high-affinity binding site for the heat shock protein. The *Ferredoxin 1* transcript was recently identified as a client cellular mRNA whose translation is enhanced by HSP101 (Ling et al. 2000). HSP101 expression is developmentally regulated and is translated via a cap-independent mechanism that involves an internal ribosome entry site (IRES)-like element in its 5′ leader (Ling et al. 2000; Young et al. 2001; Dinkova et al. 2005).

TMV genomic mRNA, like several other positive-sense RNA viruses that infect plants, is unusual in that it is not polyadenylated but instead terminates in a 204-nucleotide 3′-untranslated region (3′UTR) that functions as a translational enhancer (Gallie and Walbot 1990). Upstream of a 3′-terminal 105-nucleotide tRNA-like structure (TLS) is a 72-nucleotide region, referred to as the upstream pseudoknot domain (UPD), that folds into three coaxially stacked, phylogenetically conserved RNA pseudoknots (van Belkum et al. 1985). Both the conserved primary sequence and the higher-ordered structure of the third pseudoknot are required for the translational enhancement (Leathers et al. 1993). The UPD requires a 5′ cap in order to enhance translation, functionally similar to the requirement of a 5′ cap for a poly(A) tail to stimulate translation (Gallie 1991, 2002b; Leathers et al. 1993).

The presence of coat protein in the virion particle would be expected to inhibit viral translation by preventing binding of initiation factors. However, the weak association of coat protein to Ω (Mundry et al. 1991) allows 40S subunits to be recruited as a consequence of HSP101-mediated recruitment of eIF4F. Coat protein bound to the region of the first cistron is then displaced during translation (see Fig. 2B). Because the 3′ end of the viral RNA remains encapsidated (Wu et al. 1994), the UPD is not able to function during the initial round of translation. Thus, Ω enhances the first round of translation and facilitates virion particle disassembly while enhancing translation of the genomic RNA following disassembly.

Role of Viral 3′UTRs in Enhancing Cap-dependent Translation

Although the cell-cycle-regulated histone mRNAs of metazoans and some algae naturally lack a poly(A) tail, those in higher plants are polyadenylated (Lindauer et al. 1993), and the only known mRNAs in higher plants that naturally lack a poly(A) tail are viral in origin. The TMV UPD binds eEF1A, and mutations affecting its binding correlate with reduced translation, although how eEF1A might function to enhance translation has not been determined (Zeenko et al. 2002). The 3′UTR of brome mosaic virus also folds into a pseudoknot-containing, 3′-terminal TLS, enhances translation, and requires a 5′ cap in order to do so (Gallie and Kobayashi 1994), suggesting that the evolution of translational enhancers within the 3′UTR of nonpolyadenylated viral mRNAs may be widespread.

The 3′UTR of turnip yellow mosaic virus (TYMV) also adopts a TLS conformation and binds eEF1A (Zeenko et al. 2002; Matsuda and Dreher 2004). As observed with the TMV UPD, mutations that reduce binding of eEF1A to the TLS reduce translational enhancement (Matsuda and Dreher 2004). The TLS serves as the origin of viral minus-strand replication where synthesis initiates opposite the penultimate C of the 3′-terminal sequence of the viral RNA that controls replication (Singh and Dreher 1997, 1998; Deiman et al. 1998). Because minus-strand RNA synthesis and translation are presumed to be competing processes for such viral RNAs, the observation that eEF1A inhibits TYMV minus-strand RNA synthesis suggests that eEF1A may indirectly affect translation by maintaining the 3′-terminal sequence in a TLS conformation that negatively regulates viral replication (Matsuda et al. 2004).

Alfalfa mosaic virus (AMV) requires viral coat protein to initiate infection (Bol 1999, 2003; Jaspars 1999). The 3′UTR of AMV RNA can adopt one of two conformations: either a series of hairpins with a high affinity for coat protein or a pseudoknot-containing TLS structure that is not aminoacylated (Olsthoorn et al. 1999). Binding of coat protein to the hairpin conformation is required to enhance translation (Gallie and Kobayashi 1994; Neeleman et al. 2001; Krab et al. 2005), whereas the TLS conformation is required for minus-strand RNA synthesis (Olsthoorn et al. 1999, 2004). The effect of coat protein on the translation of mRNAs terminating in the AMV 3′UTR could be recapitulated in yeast (Krab et al. 2005), demonstrating that no plant-specific protein is required for the coat-protein-mediated enhancement of translation. The coat protein specifically interacts with eIF4F (see Fig. 2C) (Krab et al. 2005), similar to the interaction reported between the rotavirus NSP3 and eIF4G (Piron et al. 1998). Although NSP3 represses cellular mRNA translation by evicting PABP from eIF4F (Piron et al. 1998), AMV does not inhibit cellular protein synthesis (Hooft van Huijsduijnen et al. 1986). Thus, AMV coat protein functionally resembles the histone stem-loop-binding protein (SLBP) that binds to the 3′ end of nonpolyadenylated animal histone mRNAs, interacts with eIF4G, and promotes translation initiation, but does not evict PABP from eIF4F (Ling et al. 2002).

Shunting and *trans*-Activation as Mechanisms Promoting Reinitiation

The 5′ leader of cauliflower mosaic virus (CaMV) 35S mRNA is 612 nucleotides long and contains extensive secondary structure and several short upstream ORFs (uORFs). Translation of the first main ORF (ORF VII) occurs through a shunting mechanism that requires 40S subunits to scan from the 5′ end of the capped mRNA, bypass a significant portion of the structured leader, and resume scanning to the ORF VII initiation codon (Hohn et al. 1998). Shunting requires a 5′ cap, a 5′-proximal uORF (i.e., uORF A), and a stable stem at a correct distance downstream from the uORF (Hohn et al. 1998). The stable secondary structure in the leader facilitates shunting by bringing the "shunt landing site" that lies upstream of ORF VII close to the shunt "take-off site" downstream from uORF A (Futterer et al. 1993; Dominguez et al. 1998). Translation initiation and termination at uORF A is essential for both shunting and linear ribosome migration to occur, suggesting that shunting is a special case of reinitiation that may be used by plant pararetroviruses in general (Pooggin et al. 1999; Ryabova and Hohn 2000; Chapter 8).

Although shunting from the CaMV uORF A to ORF VII is a basal activity, the translation of ORF VII following shunting is promoted by the *trans*-activator (TAV) protein that also facilitates reinitiation or *trans*-activation at five distal ORFs (Bonneville et al. 1989; Pooggin et al. 2001). TAV associates with eIF3 through a direct interaction with eIF3g and interacts with 60S subunits via the ribosomal L24 protein such that TAV/eIF3/40S and TAV/eIF3/60S ternary complexes can assemble in vitro (Fig. 2D) (Park et al. 2001). TAV may disrupt the eIF4B/eIF3 interaction during the first initiation event as it competes with eIF4B for binding to eIF3g through overlapping interaction domains on eIF3g (Park et al. 2004). Following dissociation of eIF3 from the 40S subunit, the TAV/eIF3 complex may associate with the 60S subunit of the translocating ribosome where it would be available for reinitiation of downstream ORFs (Park et al. 2004).

Role of the Potyviral VPg and 5′ Leader in Promoting Cap-independent Translation

The genomic RNA of potyviruses, members of the picornaviral super-family, is naturally uncapped, but a virus-encoded protein, VPg, is cova-lently linked to the 5′ terminus. eIFiso4E null *Arabidopsis* grow normally and are resistant to potyviruses but not to other viruses (Robaglia and Caranta 2006). Several loci (e.g., *pvr1* and *pvr2* in pepper, *pot-1* in tomato, *mo1* in lettuce, and *sbm1* in pea) encode mutant forms of either eIF4E or eIFiso4E that confer potyviral resistance as a result of the mutations (Robaglia and Caranta 2006). The VPg interacts with eIF4E or eIFiso4E which m^7GTP (but not GTP) can inhibit, suggesting that VPg and cellu-lar mRNAs may compete for binding to eIFiso4E (Robaglia and Caranta 2006). Supporting this idea, the resistance provided by some *mo1* alleles results from point mutations that map close to the cap-binding pocket (Nicaise et al. 2003). However, among three pepper resistance alleles, only one failed to bind m^7GTP, suggesting that disruption of cap binding is not required for potyvirus resistance (Kang et al. 2005). Although these data support the possibility that the VPg–eIF4E interaction is required for potyviral translation, direct evidence demonstrating this has yet to be reported. Moreover, the observation that eIF4E is required for cell-to-cell movement of the potyviral pea-seed-borne mosaic virus (Gao et al. 2004) suggests a nontranslational function for the VPg–eIF4E interaction. The demonstration that potyviral leaders can confer cap-independent transla-tion (Carrington and Freed 1990; Gallie et al. 1995b), thus obviating the need for eIF4E following viral RNA release from the virion particle, sug-gests that any role of the VPg–eIF4E interaction during translation may be limited to the initial round in order to promote virion disassembly.

The 143-nucleotide 5′ leader of tobacco etch virus (TEV) RNA is sufficient to confer cap-independent translation in the absence of the VPg and is analogous to a cap in that it functionally interacts with the poly(A) tail to promote efficient translation (Gallie et al. 1995b; Niepel and Gallie

1999). The TEV leader also exhibits IRES activity (Chapters 5 and 6) but differs from picornaviral leaders in that it is considerably shorter (143 nucleotides), less structured, and contains no upstream AUG triplets. TEV IRES activity requires eIF4G and is further stimulated by eIF4A, eIF4B, or PABP but not eIF4E (Fig. 2E) (Gallie 2001), suggesting that the requirement for eIF4E in potyviral infection is not to enable TEV IRES activity.

Within the TEV leader, a 5′-proximal RNA pseudoknot in which one loop exhibits complementarity to nucleotides 1117–1123 of 18S rRNA is required for mediating cap-independent translation together with sequences flanking the pseudoknot (Zeenko and Gallie 2005). Mutations disrupting this potential base-pairing with 18S rRNA reduce cap-independent translation, whereas mutations maintaining base-pairing have the least effect. Nucleotides 1115–1124 of plant 18S rRNA are accessible for intermolecular base-pairing, and mRNAs able to base-pair with this region exhibit a high affinity for 40S subunits and an enhanced rate of translation (Akbergenov et al. 2004). The ability of mRNA–rRNA base-pairing to facilitate translation initiation has been shown with a 9-nucleotide element found in the mouse Gtx homeodomain mRNA that can base-pair to 18S rRNA (Chappell et al. 2000, 2004). mRNAs containing this sequence are more efficiently translated in a yeast system that uses ribosomes containing a mouse–yeast hybrid 18S rRNA able to base-pair with the Gtx element when there is an exact complementary match between the element and the hybrid 18S rRNA (Dresios et al. 2006), supporting the notion that translation from some mRNAs may depend on base-pairing to 18S rRNA.

Role of the 5′ Leader and 3′UTR of Uncapped, Nonpolyadenylated Viral RNAs in Promoting Cap-independent Translation

Satellite tobacco necrosis virus (STNV) genomic RNA is naturally uncapped and lacks a poly(A) tail. Despite these apparent handicaps, the RNA is efficiently translated with a requirement for the 5′ leader and 3′UTR (Timmer et al. 1993b). A 105-nucleotide translational enhancer (TE) located in the STNV 3′UTR confers cap-independent translation and requires eIF4F to do so (Fig. 2F) (Browning et al. 1988; Gazo et al. 2004). eIF4F and eIFiso4F can be cross-linked to the STNV TE (Gazo et al. 2004). eIF4E or eIFiso4E also binds the TE (Gazo et al. 2004), and their affinity is increased by eIF4G or eIFiso4G, respectively. These data suggest that eIF4F or eIFiso4F binds the TE through direct contacts between their eIF4E or eIFiso4E subunits, respectively, to recruit the factor to the 3′UTR; the TE may promote ribosome assembly at the 5′ end of the viral RNA through RNA–RNA and/or RNA–protein interactions with the 5′ leader (Danthinne et al. 1993; Timmer et al. 1993b).

Barley yellow dwarf virus (BYDV) genomic RNA also lacks both a 5′ cap and a poly(A) tail and contains a 105-nucleotide TE element located in the 3′UTR that confers cap-independent translation at the 5′-most proximal AUG (Wang et al. 1997). A 7-base loop sequence within the cruciform structure of the TE enables the viral mRNA to form a closed circle by base-pairing with a stem-loop in the 5′ leader, thereby directing translation initiation to the 5′-proximal AUG (Guo et al. 2000, 2001). TE activity is inhibited by TE RNA *in trans*, and the inhibition can be reversed by the addition of eIF4F (Wang et al. 1997), suggesting that, like STNV, eIF4F may be recruited to the BYDV TE to promote translation initiation.

TRANSLATION IN CHLOROPLASTS

Half of the approximately 100 genes in chloroplast genomes encode components involved in translation, including 16S, 23S, and 5S RNAs and about 30 tRNAs (Sugiura et al. 1998). The chloroplast translational machinery is similar, although not identical, to that in *Escherichia coli*, consistent with its origins from an ancestral photosynthetic prokaryote (Margulis 1970). In addition to orthologs of all *E. coli* 30S ribosomal proteins, four nucleus-encoded, chloroplast/plastid-specific ribosomal proteins (PSRPs) have been identified (Yamaguchi et al. 2000). Although orthologs of *E. coli* L25 and L30 are absent from the chloroplast 50S subunit, two nucleus-encoded PSRPs are present (Yamaguchi and Subramanian 2000). An additional protein, the plastid ribosome recycling factor (pRRF), is associated with the chloroplast ribosome in a stoichiometric amount (Yamaguchi and Subramanian 2000). In spinach, 12 proteins of the 30S subunit and 8 proteins of the 50S subunit are chloroplast-encoded (Yamaguchi and Subramanian 2000; Yamaguchi et al. 2000). Chloroplast orthologs of the bacterial initiation factors IF1, IF2, and IF3; elongation factors EF-Tu, EF-Ts, EF-G, and EF-P; release factors RF1, RF2, and RF3; and the ribosome recycling factor RRF, have been identified as encoded mostly by nuclear genes in *Chlamydomonas* and higher plants (Manuell et al. 2004).

Chloroplast mRNAs, initially transcribed as polycistronic pre-mRNAs, are endonucleolytically processed into mostly monocistronic mRNAs. This processing, at least in some cases, has been shown to affect the translational efficiency of the processed mRNAs (Zerges 2000). As in *E. coli*, chloroplast mRNAs have no 5′ cap or poly(A) tail, except short tracts that target the mRNAs for degradation (Kudla et al. 1996). A canonical anti-Shine-Dalgarno (anti-SD) sequence is present in all chloroplast 16S rRNAs, but only about 30% of chloroplast mRNAs have ribosome-binding sites and, of those that do, their positions are highly variable (Sugiura et al. 1998). Moreover, deletion of the SD-like sequence from some chloroplast mRNAs has little effect on translation (Sugiura et al. 1998). There-fore, other *cis*-acting elements, including secondary and tertiary structural features of the element, may be required to promote translation initiation (Higgs et al. 1999; Zerges 2000). The plastid-specific ribosomal proteins, together with *trans*-acting factors, may be involved in translation initia-tion of certain chloroplast RNAs and may have a role in the positioning of the start site for translation initiation (Manuell et al. 2004).

CONCLUSIONS

Following completion of the genome sequence of *Arabidopsis*, the iden-tification of every gene encoding a translation factor and their expression profilings provides the opportunity to investigate aspects of translational control during the development of differentiated cell types and in response to disease or adverse environmental conditions. Some transla-tion gene family members in *Arabidopsis* are ubiquitously expressed, whereas others are limited to specific cells or organs. The availability of insertion mutants resulting in knockdown or knockout of expression of individual members of a gene family provides the genetic resources to uncover possible tissue-specific roles for each gene member. Moreover, microarray analysis of polysome-associated mRNAs in knockdown or knockout mutants will identify which mRNAs require specific translation factors to regulate their expression. The recent identification of microR-NAs encoded by the *Arabidopsis* genome (Lu et al. 2005) also makes pos-sible a systematic analysis of their role in regulating translation during plant growth and development.

REFERENCES

Akbergenov R.Z., Zhanybekova S.S., Kryldakov R.V., Zhigailov A., Polimbetova N.S., Hohn T., and Iskakov B.K. 2004. ARC-1, a sequence element complementary to an internal 18S rRNA segment, enhances translation efficiency in plants when present in the leader or intercistronic region of mRNAs. *Nucleic Acids Res.* **32:** 239–247.

Altmann M., Wittmer B., Methot N., Sonenberg N., and Trachsel H. 1995. The *Saccha-romyces cerevisiae* translation initiation factor Tif3 and its mammalian homologue, eIF-4B, have RNA annealing activity. *EMBO J.* **14:** 3820–3827.

Apuya N.R. and Zimmerman J.L. 1992. Heat shock gene expression is controlled primarily at the translational level in carrot cells and somatic embryos. *Plant Cell* **4:** 657–665.

Axelos M., Bardet C., Liboz T., Le Van Thai A., Curie C., and Lescure B. 1989. The gene family encoding the *Arabidopsis thaliana* translation elongation factor EF-1α: Molecu-lar cloning, characterization and expression. *Mol. Gen. Genet.* **219:** 106–112.

Balasta M.L., Carberry S.E., Friedland D.E., Perez R.A., and Goss D.J. 1993. Characteriza-tion of the ATP-dependent binding of wheat germ protein synthesis initiation factors eIF-(iso)4F and eIF-4A to mRNA. *J. Biol. Chem.* **268:** 18599–18603.

Belostotsky D.A. 2003. Unexpected complexity of poly(A)-binding protein gene families in flowering plants: Three conserved lineages that are at least 200 million years old and possible auto- and cross-regulation. *Genetics* **163:** 311–319.

Belostotsky D.A. and Meagher R.B. 1996. A pollen-, ovule-, and early embryo-specific poly(A) binding protein from *Arabidopsis* complements essential functions in yeast. *Plant Cell* **8:** 1261–1275.

Bilgin D.D., Liu Y., Schiff M., and Dinesh-Kumar S.P. 2003. P58^{IPK}, a plant ortholog of double-stranded RNA-dependent protein kinase PKR inhibitor, functions in viral pathogenesis. *Dev. Cell.* **4:** 651–661.

Bol J.F. 1999. Alfalfa mosaic virus and ilarviruses: Involvement of coat protein in multiple steps of the replication cycle. *J. Gen. Virol.* **80:** 1089–1102.

———. 2003. Alfalfa mosaic virus: Coat protein-dependent initiation of infection. *Mol. Plant Pathol.* **4:** 1–8.

Bonneville J.-M., Sanfacon H., Futterer J., and Hohn T. 1989. Posttranscriptional trans-activation in cauliflower mosaic virus. *Cell* **59:** 1135–1143.

Browning K.S. 1996. The plant translational apparatus. *Plant Mol. Biol.* **32:** 107–144.

Browning K.S., Fletcher L., and Ravel J.M. 1988. Evidence that the requirements for ATP and wheat germ initiation factors 4A and 4F are affected by a region of satellite tobacco necrosis virus RNA that is 3′ to the ribosomal binding site. *J. Biol. Chem.* **263:** 8380–8383.

Browning K.S., Fletcher L., Lax S.R., and Ravel J.M. 1989. Evidence that the 59-kDa protein synthesis initiation factor from wheat germ is functionally similar to the 80-kDa initiation factor 4B from mammalian cells. *J. Biol. Chem.* **264:** 8491–8494.

Browning K.S., Webster C., Roberts J.K., and Ravel J.M. 1992. Identification of an isozyme form of protein synthesis initiation factor 4F in plants. *J. Biol. Chem.* **267:** 10096–10100.

Browning K.S., Humphreys J., Hobbs W., Smith G.B., and Ravel J.M. 1990. Determination of the amounts of the protein synthesis initiation and elongation factors in wheat germ. *J. Biol. Chem.* **265:** 17967–17973.

Burks E.A., Bezerra P.P., Le H., Gallie D.R., and Browning K.S. 2001. Plant initiation factor 3 subunit composition resembles mammalian initiation factor 3 and has a novel subuit. *J. Biol. Chem.* **276:** 2122–2131.

Bushell M., Wood W., Carpenter G., Pain V.M., Morley S.J., and Clemens M.J. 2001. Disruption of the interaction of mammalian protein synthesis initiation factor 4B with the poly(A) binding protein by caspase- and viral protease-mediated cleavages. *J. Biol. Chem.* **276:** 23922–23928.

Carberry S.E. and Goss D.J. 1991. Interaction of wheat germ protein synthesis initiation factors eIF-3, eIF-(iso)4F, and eIF-4F with mRNA analogues. *Biochemistry* **30:** 6977–6982.

———. 1992. Characterization of the interaction of wheat germ protein synthesis initiation factor eIF-3 with mRNA oligonucleotide and cap analogues. *Biochemistry* **31:** 296–299.

Carberry S.E., Darzynkiewicz E., and Goss D.J. 1991. A comparison of the binding of methylated cap analogues to wheat germ protein synthesis initiation factors 4F and (iso)4F. *Biochemistry* **30:** 1624–1627.

Carrington J.C. and Freed D.D. 1990. Cap-independent enhancement of translation by a plant potyvirus 5′ nontranslated region. *J. Virol.* **64:** 1590–1597.

Chang L.Y., Yang W.Y., Browning K., and Roth D. 1999. Specific in vitro phosphorylation of plant eIF2α by eukaryotic eIF2α kinases. *Plant Mol. Biol.* **41:** 363–370.

Chapman B. and Brown C. 2004. Translation termination in *Arabidopsis thaliana*: Characterisation of three versions of release factor 1. *Gene* **341:** 219–225.

Chappell S.A., Edelman G.M., and Mauro V.P. 2000. A 9-nt segment of a cellular mRNA can function as an internal ribosome entry site (IRES) and when present in linked multiple copies greatly enhances IRES activity. *Proc. Natl. Acad. Sci.* **97:** 1536–1541.

———. 2004. Biochemical and functional analysis of a 9-nt RNA sequence that affects translation efficiency in eukaryotic cells. *Proc. Natl. Acad. Sci.* **101:** 9590–9594.

Checkley J.W., Cooley L., and Ravel J.M. 1981. Characterization of initiation factor eIF-3 from wheat germ. *J. Biol. Chem.* **256:** 1582–1586.

Chekanova J.A. and Belostotsky D.A. 2003. Evidence that poly(A) binding protein has an evolutionarily conserved function in facilitating mRNA biogenesis and export. *RNA* **9:** 1476–1490.

Chekanova J.A., Shaw R.J., and Belostotsky D.A. 2001. Analysis of an essential requirement for the poly(A) binding protein function using cross-species complementation. *Curr. Biol.* **11:** 1207–1214.

Cho P.F., Poulin F., Cho-Park Y.A., Cho-Park I.B., Chicoine J.D., Lasko P., and Sonenberg N. 2005. A new paradigm for translational control: Inhibition via 5′-3′ mRNA tethering by Bicoid and the eIF4E cognate 4EHP. *Cell* **121:** 411–423.

Condeelis J. 1995. Elongation factor 1α, translation and the cytoskeleton. *Trends Biochem. Sci.* **20:** 169–170.

Danthinne X., Seurinck J., Meulewaeter F., Van Montagu M., and Cornelissen M. 1993. The 3′ untranslated region of satellite tobacco necrosis virus RNA stimulates translation in vitro. *Mol. Cell. Biol.* **13:** 3340–3349.

Deiman B.A., Koenen A.K., Verlaan P.W., and Pleij C.W. 1998. Minimal template requirements for initiation of minus-strand synthesis in vitro by the RNA-dependent RNA polymerase of turnip yellow mosaic virus. *J. Virol.* **72:** 3965–3972.

De Pater B.S., van der Mark F., Rueb S., Katagiri F., Chua N.H., Schilperoort R.A., and Hensgens L.A. 1992. The promoter of the rice gene GOS2 is active in various different monocot tissues and binds rice nuclear factor ASF-1. *Plant J.* **2:** 837–844.

Dinkova T.D., Zepeda H., Martinez-Salas E., Martinez L.M., Nieto-Sotelo J., and de Jimenez E.S. 2005. Cap-independent translation of maize Hsp101. *Plant J.* **41:** 722–731.

Dominguez D.I., Ryabova L.A., Pooggin M.M., Schmidt-Puchta W., Futterer J., and Hohn T. 1998. Ribosome shunting in cauliflower mosaic virus. Identification of an essential and sufficient structural element. *J. Biol. Chem.* **273:** 3669–3678.

Dresios J., Chappell S.A., Zhou W., and Mauro V.P. 2006. An mRNA-rRNA base-pairing mechanism for translation initiation in eukaryotes. *Nat. Struct. Mol. Biol.* **13:** 30–34.

Franceschetti M., Hanfrey C., Scaramagli S., Torrigiani P., Bagni N., Burtin D., and Michael A.J. 2001. Characterization of monocot and dicot plant S-adenosyl-l-methionine decarboxylase gene families including identification in the mRNA of a highly conserved pair of upstream overlapping open reading frames. *Biochem. J.* **353:** 403–409.

Freire M.A. 2005. Translation initiation factor (iso) 4E interacts with BTF3, the beta subunit of the nascent polypeptide-associated complex. *Gene* **345:** 271–277.

Freire M.A., Tourneur C., Granier F., Camonis J., El Amrani A., Browning K.S., and Robaglia C. 2000. Plant lipoxygenase 2 is a translation initiation factor-4E-binding protein. *Plant Mol. Biol.* **44:** 129–140.

Futterer J., Kiss-Laszlo Z., and Hohn T. 1993. Nonlinear ribosome migration on cauliflower mosaic virus 35S RNA. *Cell* **73:** 789–802.

Gallie D.R. 1991. The cap and poly(A) tail function synergistically to regulate mRNA translational efficiency. *Genes Dev.* **5:** 2108–2116.

———. 2001. Cap-independent translation conferred by the 5′ leader of tobacco etch virus is eukaryotic initiation factor 4G dependent. *J. Virol.* **75:** 12141–12152.

————. 2002a. Protein-protein interactions required during translation. *Plant Mol. Biol.* **50:** 949–970.

————. 2002b. The 5′-leader of tobacco mosaic virus promotes translation through enhanced recruitment of eIF4F. *Nucleic Acids Res.* **30:** 3401–3411.

————. 2004. The role of the initiation surveillance complex in promoting efficient protein synthesis. *Biochem. Soc. Trans.* **32:** 585–588.

Gallie D.R. and Browning K.S. 2001. eIF4G functionally differs from eIFiso4G in promoting internal initiation, cap-independent translation, and translation of structured mRNAs. *J. Biol. Chem.* **276:** 36951–36960.

Gallie D.R. and Kobayashi M. 1994. The role of the 3′-untranslated region of non-polyadenylated plant viral mRNAs in regulating translational efficiency. *Gene* **142:** 159–165.

Gallie D.R. and Walbot V. 1990. RNA pseudoknot domain of tobacco mosaic virus can functionally substitute for a poly(A) tail in plant and animal cells. *Genes Dev.* **4:** 1149–1157.

————. 1992. Identification of the motifs within the tobacco mosaic virus 5′-leader responsible for enhancing translation. *Nucleic Acids Res.* **20:** 4631–4638.

Gallie D.R., Caldwell C., and Pitto L. 1995a. Heat shock disrupts cap and poly(A) tail function during translation and increases mRNA stability of introduced reporter mRNA. *Plant Physiol.* **108:** 1703–1713.

Gallie D.R., Tanguay R.L., and Leathers V. 1995b. The tobacco etch viral 5′ leader and poly(A) tail are functionally synergistic regulators of translation. *Gene* **165:** 233–238.

Gallie D.R., Sleat D.E., Watts J.W., Turner P.C., and Wilson T.M. 1987a. The 5′-leader sequence of tobacco mosaic virus RNA enhances the expression of foreign gene transcripts in vitro and in vivo. *Nucleic Acids Res.* **15:** 3257–3273.

————. 1987b. In vivo uncoating and efficient transient expression of recombinant RNA packaged into pseudovirus particles. *Science* **236:** 1122–1124.

Gallie D.R., Le H., Caldwell C., Tanguay R.L., Hoang N.X., and Browning, K.S. 1997. The phosphorylation state of translation initiation factors is regulated developmentally and following heat shock in wheat. *J. Biol. Chem.* **272:** 1046–1053.

Gao Z., Johansen E., Eyers S., Thomas C.L., Noel Ellis T.H., and Maule A.J. 2004. The potyvirus recessive resistance gene, *sbm1*, identifies a novel role for translation initiation factor eIF4E in cell-to-cell trafficking. *Plant J.* **40:** 376–385.

Gazo B.M., Murphy P., Gatchel J.R., and Browning K.S. 2004. A novel interaction of cap-binding protein complexes eukaryotic initiation factor (eIF) 4F and eIF(iso)4F with a region in the 3′-untranslated region of satellite tobacco necrosis virus. *J. Biol. Chem.* **279:** 13584–13592.

Gil J., Esteban M., and Roth D. 2000. In vivo regulation of protein synthesis by phosphorylation of the alpha subunit of wheat eukaryotic initiation factor 2. *Biochemistry* **39:** 7521–7530.

Goyer C., Altmann M., Lee H.S., Blanc A., Deshmukh M., Woolford J.L., Trachsel H., and Sonenberg N. 1993. *TIF4631* and *TIF4632*—Two yeast genes encoding the high-molecular-weight subunits of the cap-binding protein complex (eukaryotic initiation factor 4F) contain an RNA recognition motif-like sequence and carry out an essential function. *Mol. Cell. Biol.* **13:** 4860–4874.

Gradi A., Imataka H., Svitkin Y.V., Rom E., Raught B., Morino S., and Sonenberg N. 1998. A novel functional human eukaryotic translation initiation factor 4G. *Mol. Cell. Biol.* **18:** 334–342.

Guo L., Allen E., and Miller W.A. 2000. Structure and function of a cap-independent translation element that functions in either the 3′ or the 5′ untranslated region. *RNA* **6:** 1808–1820.

————. 2001. Base-pairing between untranslated regions facilitates translation of uncapped, nonpolyadenylated viral RNA. *Mol. Cell* **7**: 1103–1109.

Hanfrey C., Franceschetti M., Mayer M.J., Illingworth C., and Michael A.J. 2002. Abrogation of upstream open reading frame-mediated translational control of a plant S-adenosylmethionine decarboxylase results in polyamine disruption and growth perturbations. *J. Biol. Chem.* **277**: 44131–44139.

Hanfrey C., Elliott K.A., Franceschetti M., Mayer M.J., Illingworth C., and Michael A.J. 2005. A dual upstream open reading frame-based autoregulatory circuit controlling polyamine-responsive translation. *J. Biol. Chem.* **280**: 39229–39237.

Higgs D.C., Shapiro R.S., Kindle K.L., and Stern D.B. 1999. Small cis-acting sequences that specify secondary structures in a chloroplast mRNA are essential for RNA stability and translation. *Mol. Cell. Biol.* **19**: 8479–8491.

Hinnebusch A.G. 2000. Mechanism and regulation of initiator methionyl-tRNA binding to ribosomes. In *Translational control of gene expression* (ed. N. Sonenberg et al.), pp. 185–243. Cold Spring Harbor Laboratory Press, Cold Spring Harbor, New York.

Hohn T., Dominguez D.I., Scharer-Hernandez N., Pooggin M.M., Schmidt-Puchta W., Hemmings-Mieszczak M., and Futterer J. 1998. Ribosome shunting in eukaryotes: What the viruses tell me. In *A look beyond transcription: Mechanisms determining mRNA stability and translation in plants* (ed. J. Bailey-Serres and D.R. Gallie), pp. 84–95. American Society of Plant Physiologists, Rockville, Maryland.

Hooft van Huijsduijnen R.A.M., Alblas S.W., de Rijk R.H., and Bol J.F. 1986. Induction by salicylic acid of pathogenesis-related proteins and resistance to alfalfa mosaic virus infection in various plant species. *J. Gen. Virol.* **67**: 2135–2143.

Hutchins A.P., Roberts G.R., Lloyd C.W., and Doonan J.H. 2004. In vivo interaction between CDKA and eIF4A: A possible mechanism linking translation and cell proliferation. *FEBS Lett.* **556**: 91–94.

Janssen G.M., Maessen G.D., Amons R., and Moller W. 1988. Phosphorylation of elongation factor 1β by an endogenous kinase affects its catalytic nucleotide exchange activity. *J. Biol. Chem.* **263**: 11063–11066.

Jaramillo M., Browning K., Dever T.E., Blum S., Trachsel H., Merrick W.C., Ravel J.M., and Sonenberg N. 1990. Translation initiation factors that function as RNA helicases from mammals, plants and yeast. *Biochim. Biophys. Acta* **1050**: 134–139.

Jaspars E.M.J. 1999. Genome activation in alfamo- and ilarviruses. *Arch. Virol.* **144**: 843–863.

Kang B.C., Yeam I., Frantz J.D., Murphy J.F., and Jahn M.M. 2005. The pvr1 locus in *Capsicum* encodes a translation initiation factor eIF4E that interacts with tobacco etch virus Vpg. *Plant J.* **42**: 392–405.

Key J.L., Lin C.Y., and Chen Y.M. 1981. Heat shock proteins of higher plants. *Proc. Natl. Acad. Sci.* **78**: 3526–3530.

Khan M.A. and Goss D.J. 2004. Phosphorylation states of translational initiation factors affect mRNA cap binding in wheat. *Biochemistry* **43**: 9092–9097.

————. 2005. Translation initiation factor (eIF) 4B affects the rates of binding of the mRNA m^7G cap analogue to wheat germ eIFiso4F and eIFiso4F·PABP. *Biochemistry* **44**: 4510–4516.

Kim T.H., Kim B.H., Yahalom A., Chamovitz D.A., and von Arnim A.G. 2004. Translational regulation via 5′ mRNA leader sequences revealed by mutational analysis of the *Arabidopsis* translation initiation factor subunit eIF3h. *Plant Cell* **16**: 3341–3356.

Krab I.M., Caldwell C., Gallie D.R., and Bol J.F. 2005. Coat protein enhances translational efficiency of alfalfa mosaic virus RNAs and interacts with the eIF4G component of initiation factor eIF4F. *J. Gen. Virol.* **86**: 1841–1849.

Kudla J., Hayes R., and Gruissem W. 1996. Polyadenylation accelerates degradation of chloroplast mRNA. *EMBO J.* **15:** 7137–7146.

Langland J.O., Langland L.A., Browning K.S., and Roth D.A. 1996. Phosphorylation of plant eukaryotic initiation factor-2 by the plant-encoded double-stranded RNA-dependent protein kinase, pPKR, and inhibition of protein synthesis in vitro. *J. Biol. Chem.* **271:** 4539–4544.

Lauer S.J., Browning K.S., and Ravel J.M. 1985. Characterization of initiation factor 3 from wheat germ. 2. Effects of polyclonal and monoclonal antibodies on activity. *Biochemistry* **24:** 2928–2931.

Lauer S.J., Burks E., Irvin J.D., and Ravel J.M. 1984. Purification and characterization of three elongation factors, EF-1α, EF-1βγ, and EF-2, from wheat germ. *J. Biol. Chem.* **259:** 1644–1648.

Lax S., Browning K.S., Maia D.M., and Ravel J.M. 1986. ATPase activities of wheat germ initiation factors 4A, 4B, and 4F. *J. Biol. Chem* **261:** 15632–15636.

Le H., Browning K.S., and Gallie D.R. 1998. The phosphorylation state of the wheat translation initiation factors eIF4B, eIF4A, and eIF2 is differentially regulated during seed development and germination. *J. Biol. Chem.* **273:** 20084–20089.

———. 2000. The phosphorylation state of poly(A)-binding protein specifies its binding to poly(A) RNA and its interaction with eukaryotic initiation factor (eIF) 4F, eIFiso4F, and eIF4B. *J. Biol. Chem.* **275:** 17452–17462.

Le H., Tanguay R.L., Balasta M.L., Wei C.-C., Browning K.S., Metz A.M., Goss D.J., and Gallie D.R. 1997. Translation initiation factors eIF-iso4G and eIF-4B interact with the poly(A)-binding protein and increase its RNA binding activity. *J. Biol. Chem.* **272:** 16247–16255.

Leathers V., Tanguay R., Kobayashi M., and Gallie D.R. 1993. A phylogenetically conserved sequence within viral 3′ untranslated RNA pseudoknots regulates translation. *Mol. Cell. Biol.* **13:** 5331–5347.

Lindauer A., Muller K., and Schmitt R. 1993. Two histone H1-encoding genes of the green alga *Volvox carteri* with features intermediate between plant and animal genes. *Gene* **129:** 59–68.

Ling J., Morley S.J., Pain V.M., Marzluff W.F., and Gallie D.R. 2002. The histone 3′-terminal stem-loop-binding protein enhances translation through a functional and physical interaction with eukaryotic initiation factor 4G (eIF4G) and eIF3. *Mol. Cell. Biol.* **22:** 7853–7867.

Ling J., Wells D.R., Tanguay R.L., Dickey L.F., Thompson W.F., and Gallie D.R. 2000. Heat shock protein HSP101 binds to the *Fed-1* internal light regulatory element and mediates its high translational activity. *Plant Cell* **12:** 1213–1228.

Lu C., Tej S.S., Luo S., Haudenschild C.D., Meyers B.C., and Green P.J. 2005. Elucidation of the small RNA component of the transcriptome. *Science* **309:** 1567–1569.

Luo Y. and Goss D.J. 2001. Homeostasis in mRNA initiation: Wheat germ poly(A)-binding protein lowers the activation energy barrier to initiation complex formation. *J. Biol. Chem.* **276:** 43083–43086.

Manjunath S., Williams A.J., and Bailey-Serres J. 1999. Oxygen deprivation stimulates Ca^{2+}-mediated phosphorylation of mRNA cap-binding protein eIF4E in maize roots. *Plant J.* **19:** 21–30.

Manuell A., Beligni M.V., Yamaguchi K., and Mayfield S.P. 2004. Regulation of chloroplast translation: Interactions of RNA elements, RNA-binding proteins and the plastid ribosome. *Biochem. Soc. Trans.* **32:** 601–605.

Margulis L. 1970. *Origin of eukaryotic cells.* Yale University Press, New Haven, Connecticut.

Matsuda D. and Dreher T.W. 2004. The tRNA-like structure of turnip yellow mosaic virus RNA is a 3′-translational enhancer. *Virology* **321:** 36–46.

Matsuda D., Yoshinari S., and Dreher T.W. 2004. eEF1A binding to aminoacylated viral RNA represses minus strand synthesis by TYMV RNA-dependent RNA polymerase. *Virology* **321:** 47–56.

Methot N., Song M.S., and Sonenberg N. 1996. A region rich in aspartic acid, arginine, tyrosine, and glycine (DRYG) mediates eukaryotic initiation factor 4B (eIF4B) self-association and interaction with eIF3. *Mol. Cell. Biol.* **16:** 5328–5334.

Metz A.M. and Browning K.S. 1996. Mutational analysis of the functional domains of the large subunit of the isozyme form of wheat initiation factor eIF4F. *J. Biol. Chem.* **271:** 31033–31036.

————. 1997. Assignment of the β-subunit of wheat eIF2 by protein and DNA sequence analysis and immunoanalysis. *Arch. Biochem. Biophys.* **342:** 187–189.

Metz A.M., Wong K.C., Malmstrom S.A., and Browning K.S. 1999. Eukaryotic initiation factor 4B from wheat and *Arabidopsis thaliana* is a member of a multigene family. *Biochem. Biophys. Res. Commun.* **266:** 314–321.

Mundry K.W., Watkins P.A., Ashfield T., Plaskitt K.A., Eisele-Walter S., and Wilson T.M. 1991. Complete uncoating of the 5′ leader sequence of tobacco mosaic virus RNA occurs rapidly and is required to initiate cotranslational virus disassembly in vitro. *J. Gen. Virol.* **72:** 769–777.

Neeleman L., Olsthoorn R.C.L., Linthorst H.J.M., and Bol J.F. 2001. Translation of a non-polyadenylated viral RNA is enhanced by binding of viral coat protein or polyadenylation of the RNA. *Proc. Natl. Acad. Sci.* **98:** 14286–14291.

Nicaise V., German-Retana S., Sanjuan R., Dubrana M.P., Mazier M., Maisonneuve B., Candresse T., Caranta C., and LeGall O. 2003. The eukaryotic translation initiation factor 4E controls lettuce susceptibility to the potyvirus lettuce mosaic virus. *Plant Physiol.* **132:** 1272–1282.

Niepel M. and Gallie D.R. 1999. Identification and characterization of the functional elements within the tobacco etch virus 5′ leader required for cap-independent translation. *J. Virol.* **73:** 9080–9088.

Nover L., Scharf K.D., and Neumann D. 1989. Cytoplasmic heat shock granules are formed from precursor particles and are associated with a specific set of mRNAs. *Mol. Cell. Biol.* **9:** 1298–1308.

Olsthoorn R.C.L., Haasnoot P.C.J., and Bol J.F. 2004. Similarities and differences between the subgenomic and minus-strand promoters of an RNA plant virus. *J. Virol.* **78:** 4048–4053.

Olsthoorn R.C.L., Mertens S., Brederode F.T., and Bol J.F. 1999. A conformational switch at the 3′ end of a plant virus RNA regulates viral replication. *EMBO J.* **18:** 4856–4864.

Palanivelu R., Belostotsky D.A., and Meagher R.B. 2000. *Arabidopsis thaliana* poly (A) binding protein 2 (PAB2) functions in yeast translational and mRNA decay processes. *Plant J.* **22:** 187–198.

Park H.S., Browning K.S., Hohn T., and Ryabova L.A. 2004. Eucaryotic initiation factor 4B controls eIF3-mediated ribosomal entry of viral reinitiation factor. *EMBO J.* **23:** 1381–1391.

Park H.S., Himmelbach A., Browning K.S., Hohn T., and Ryabova L.A. 2001. A plant viral "reinitiation" factor interacts with the host translational machinery. *Cell* **106:** 723–733.

Piron M., Vende P., Cohen J., and Poncet D. 1998. Rotavirus RNA-binding protein NSP3 interacts with eIF4GI and evicts the poly(A) binding protein from eIF4F. *EMBO J.* **17:** 5811–5821.

Pitto L., Gallie D.R., and Walbot V. 1992. Functional analysis of sequence required for post-transcriptional regulation of the maize HSP70 gene in monocots and dicots. *Plant Physiol.* **100:** 1827–1833.

Pooggin M.M., Futterer J., Skryabin K.G., and Hohn T. 1999. A short open reading frame terminating in front of a stable hairpin is the conserved feature in pregenomic RNA leaders of plant pararetroviruses. *J. Gen. Virol.* **80:** 2217–2228.

———. 2001. Ribosome shunt is essential for infectivity of cauliflower mosaic virus. *Proc. Natl. Acad. Sci.* **98:** 886–891.

Rausell A., Kanhonou R., Yenush L., Serrano R., and Ros R. 2003. The translation initiation factor eIF1A is an important determinant in the tolerance to NaCl stress in yeast and plants. *Plant J.* **34:** 257–267.

Richter-Cook N.J., Dever T.E., Hensold J.O., and Merrick W.C. 1998. Purification and characterization of a new eukaryotic protein translation factor. Eukaryotic initiation factor 4H. *J. Biol. Chem.* **273:** 7579–7587.

Robaglia C. and Caranta C. 2006. Translation initiation factors: A weak link in plant RNA virus infection. *Trends Plant Sci.* **11:** 40–45.

Ruud K.A., Kuhlow C., Goss D.J., and Browning K.S. 1998. Identification and characterization of a novel cap-binding protein from *Arabidopsis thaliana. J. Biol. Chem.* **273:** 10325–10330.

Ryabova L.A. and Hohn T. 2000. Ribosome shunting in the cauliflower mosaic virus 35S RNA leader is a special case of reinitiation of translation functioning in plant and animal systems. *Genes Dev.* **14:** 817–829.

Satoh J., Kato K., and Shinmyo A. 2004. The 5′-untranslated region of the tobacco alcohol dehydrogenase gene functions as an effective translational enhancer in plant. *J. Biosci. Bioeng.* **98:** 1–8.

Sha M., Wang Y., Xiang T., van Heerden A., Browning K.S., and Goss D.J. 1995. Interaction of wheat germ protein synthesis initiation factor eIFiso4F and its subunits p28 and p86 with m7GTP and mRNA analogues. *J. Biol. Chem.* **270:** 29904–29909.

Singh R.N. and Dreher T.W. 1997. Turnip yellow mosaic virus RNA-dependent RNA polymerase: Initiation of minus strand synthesis in vitro. *Virology* **233:** 430–439.

———. 1998. Specific site selection in RNA resulting from a combination of nonspecific secondary structure and -CCR-boxes: Initiation of minus strand synthesis by turnip yellow mosaic virus RNA-dependent RNA polymerase. *RNA* **4:** 1083–1095.

Smailov S.K., Lee A.V., and Iskakov B.K. 1993. Study of phosphorylation of translation elongation factor 2 (EF-2) from wheat germ. *FEBS Lett.* **321:** 219–223.

Sugiura M., Hirose T., and Sugita M. 1998. Evolution and mechanism of translation in chloroplasts. *Annu. Rev. Genet.* **32:** 437–459.

Tanguay R.L. and Gallie D.R. 1996. Isolation and characterization of the 102-kilodalton RNA-binding protein that binds to the 5′ and 3′ translational enhancers of tobacco mosaic virus RNA. *J. Biol. Chem.* **271:** 14316–14322.

Timmer R.T., Lax S.R., Hughes D.L., Merrick W.C., Ravel J.M., and Browning K.S. 1993a. Characterization of wheat germ protein synthesis initiation factor eIF-4C and comparison of eIF-4C from wheat germ and rabbit reticulocytes. *J. Biol. Chem.* **268:** 24863–24867.

Timmer R.T., Benkowski L.A., Schodin D., Lax S.R., Metz A.M., Ravel J.M., and Browning K.S. 1993b. The 5′ and 3′ untranslated regions of satellite tobacco necrosis virus RNA affect translational efficiency and dependence on a 5′ cap structure. *J. Biol. Chem.* **268:** 9504–9510.

Turner R. and Foster G.D. 1995. The potential exploitation of plant viral translational enhancers in biotechnology for increased gene expression. *Mol. Biotechnol.* **3:** 225–236.

van Belkum A., Abrahams J.P., Pleij C.W., and Bosch L. 1985. Five pseudoknots are present at the 204 nucleotides long 3′ noncoding region of tobacco mosaic virus RNA. *Nucleic Acids Res.* **13:** 7673–7686.

Vornlocher H.P., Hanachi P., Ribeiro S., and Hershey J.W. 1999. A 110-kilodalton subunit of translation initiation factor eIF3 and an associated 135-kilodalton protein are encoded by the *Saccharomyces cerevisiae TIF32* and *TIF31* genes. *J. Biol. Chem.* **274:** 16802–16812.

Wang S., Browning K.S., and Miller W.A. 1997. A viral sequence in the 3′-untranslated region mimics a 5′ cap in facilitating translation of uncapped mRNA. *EMBO J.* **16:** 4107–4116.

Webster C., Gaut R.L., Browning K.S., Ravel J.M., and Roberts J.K. 1991. Hypoxia enhances phosphorylation of eukaryotic initiation factor 4A in maize root tips. *J. Biol. Chem.* **266:** 23341–23346.

Wei C.-C., Balasta M.L., Ren J., and Goss D.J. 1998. Wheat germ poly(A) binding protein enhances the binding affinity of eukaryotic initiation factor 4F and (iso)4F for cap analogues. *Biochemistry* **37:** 1910–1916.

Wells D.R., Tanguay R.L., Le H., and Gallie D.R. 1998. HSP101 functions as a specific translational regulatory protein whose activity is regulated by nutrient status. *Genes Dev.* **12:** 3236–3251.

Wiese A., Elzinga N., Wobbes B., and Smeekens S. 2004. A conserved upstream open reading frame mediates sucrose-induced repression of translation. *Plant Cell* **16:** 1717–1729.

Wu X., Xu Z., and Shaw J.G. 1994. Uncoating of tobacco mosaic virus RNA in protoplasts. *Virology* **200:** 256–262.

Yamaguchi K. and Subramanian A.R. 2000. The plastid ribosomal proteins. Identification of all the proteins in the 50 S subunit of an organelle ribosome (chloroplast). *J. Biol. Chem.* **275:** 28466–28482.

Yamaguchi K., von Knoblauch, K., and Subramanian A.R. 2000. The plastid ribosomal proteins. Identification of all the proteins in the 30 S subunit of an organelle ribosome (chloroplast). *J. Biol. Chem.* **275:** 28455–28465.

Yang W. and Boss W.F. 1994. Regulation of phosphatidylinositol 4-kinase by the protein activator PIK-A49. Activation requires phosphorylation of PIK-A49. *J. Biol. Chem.* **269:** 3852–3857.

Young T.E., Ling J., Geisler-Lee C.J., Tanguay R.L., Caldwell C., and Gallie D.R. 2001. Developmental and thermal regulation of the maize heat shock protein, HSP101. *Plant Physiol.* **127:** 777–791.

Zeenko V. and Gallie D.R. 2005. Cap-independent translation of tobacco etch virus is conferred by an RNA pseudoknot in the 5′-leader. *J. Biol. Chem.* **280:** 26813–26824.

Zeenko V.V., Ryabova L.A., Spirin A.S., Rothnie H.M., Hess D., Browning K.S., and Hohn T. 2002. Eukaryotic elongation factor 1A interacts with the upstream pseudoknot domain in the 3′ untranslated region of tobacco mosaic virus RNA. *J. Virol.* **76:** 5678–5691.

Zerges W. 2000. Translation in chloroplasts. *Biochimie* **82:** 583–601.

27

Mitochondrial Translation and Human Disease

Eric A. Shoubridge and Florin Sasarman
Montreal Neurological Institute and Department
of Human Genetics, McGill University
Montreal, Quebec, Canada H3A 2B4

Most eukaryotic cells rely on oxidative phosphorylation for cellular ATP production. The machinery for oxidative phosphorylation consists of five large hetero-oligomeric enzyme complexes, located in the inner mitochondrial membrane. The majority of the approximately 85 structural components of this system are encoded in the nuclear genome, but a small number of essential protein subunits—13 in mammals—have been retained on the mitochondrial genome (mtDNA), and these are synthesized on a dedicated protein translation apparatus in the mitochondrial matrix. All of the proteins necessary for the replication, transcription, and translation of the genes encoded in mtDNA are encoded in the nuclear genome. This genetic investment is far out of proportion to the number of proteins involved, and it is likely that a small, semiautonomous mitochondrial genome has persisted because the proteins it encodes are hydrophobic proteins that need to be cotranslationally inserted into the inner mitochondrial membrane during assembly of the oxidative phosphorylation complexes. As might be expected from the α-proteobacterial origins of mitochondria, many of the features of mitochondrial translation are similar to those found in prokaryotes. In this chapter, we review the organization and control of mitochondrial translation, with a particular emphasis on the system in mammals and on mechanisms of disease.

ORGANIZATION OF THE MAMMALIAN MITOCHONDRIAL TRANSLATION SYSTEM

Mammalian mtDNA is a small (~16.5 kb) double-stranded circular genome that codes for 13 proteins, 22 tRNAs, and 2 rRNAs. It contains no introns, and the genetic code is different from the universal code: Nuclear arginine (AGA, AGG) is a stop codon, nuclear isoleucine (AUA) is methionine, nuclear stop (UGA) is tryptophan, and there is only one methionine tRNA. No tRNAs are known to be imported into mammalian mitochondria. Many of the tRNAs show marked structural modifications as compared to the conventional cloverleaf structure (Ling et al. 1997). Mammalian mitochondrial mRNAs are not capped, and they lack significant 5′ untranslated regions (5′UTRs). Yeast mitochondrial mRNAs are also uncapped, but have 5′UTRs ranging from 50 to more than 900 bases that, however, lack typical Shine-Dalgarno sequences.

Mitochondrial ribosomes (mitoribosomes) are distinct from both prokaryotic and eukaryotic cytosolic ribosomes. Although antibiotic sensitivity of mitoribosomes is generally similar to that of prokaryotic ribosomes, mitochondrial translation is more resistant to several antibiotics as compared to that in *Escherichia coli* (Zhang et al. 2005). Mitoribosomes are 55S particles composed of a small (28S) and a large (39S) subunit, containing a much higher protein–RNA ratio than the bacterial 70S ribosome (O'Brien 2002). The mitochondrial 12S and 16S rRNAs are considerably reduced in size as compared to their 16S and 23S prokaryotic counterparts, and this is attributable to the deletion of specific domains rather than to overall shortening. Although a 5S rRNA is reportedly present in highly purified mammalian mitochondria (Magalhaes et al. 1998), it does not form part of the mature ribosome, and its role in translation remains uncertain. The protein constituents of both mammalian ribosomal subunits have been identified in their entirety: The small subunit contains 29 proteins, 14 of which have homologs in *E. coli* (Koc et al. 2000; Suzuki et al. 2001a), and the large subunit has 48 distinct proteins, 28 of which have homologs in *E. coli* (Koc et al. 2001; Suzuki et al. 2001b). Thus, almost half the proteins in mitoribosomes have apparently evolved independently in eukaryotes. The fact that most of these have orthologs in yeast and in model metazoan organisms (*Caenorhabditis elegans*, *Drosophila*) argues that this was an early feature of mitochondrial evolution. Some of these new proteins may have taken over roles played by sequences that have been deleted in the shortened mitochondrial rRNAs (Suzuki et al. 2001b), and there are suggestions that others are bifunctional proteins, having assumed additional roles outside of translation, such as apoptosis (Cavdar Koc et al. 2001b).

A cryo-electron microscopic (cryo-EM) structure of the bovine mito-chondrial ribosome reveals that these new proteins by and large occupy new positions, such as protein–protein bridges between the subunits, and peripheral shielding of the rRNA components (Sharma et al. 2003). The mRNA entry site is dominated by a unique gate-like structure that has been suggested to be involved in the recruitment and translation initiation of the leaderless mammalian mRNAs (Sharma et al. 2003). The structure of the polypeptide exit tunnel also suggests two sites for the emergence of nascent chains: a polypeptide-accessible site and a conventional exit site. It has been speculated that the polypeptide-accessible site may allow chaperones such as Oxa1, which has a key role in membrane insertion of the nascent polypeptides (Stuart 2002), an opportunity to tether the nascent chain while the transmembrane domains insert into the inner membrane in the correct topology (Szyrach et al. 2003). The carboxy-terminal domain of yeast Oxa1 can be cross-linked to Mrp20, which is homologous to the L23 protein, that in bacteria surrounds the exit site tunnel (Jia et al. 2003; Szyrach et al. 2003). This might facilitate both recruitment of the ribosomes to the inner mitochondrial membrane and cotranslational insertion of nascent polypeptides into the membrane.

The pathways of assembly of mammalian mitoribosomes have not been well studied. Impaired mitochondrial translation and respiratory deficiency were shown to result from mutations in two yeast genes, *MTG1* and *MTG2*, encoding putative assembly factors of the large ribosomal subunit (Barrientos et al. 2003; Datta et al. 2005). The human homolog of the *MTG1* factor was identified and shown to partially rescue the mitochondrial translation defect and respiratory phenotype of the corresponding mutant yeast strain (Barrientos et al. 2003). Recently, a mitochondrial AAA protease that is responsible for quality control of inner-membrane proteins was shown to process the yeast mitochondrial ribosomal protein MrpL32, allowing it to associate with preassembled ribosomal particles close to the inner membrane (Nolden et al. 2005).

MECHANISMS OF MITOCHONDRIAL TRANSLATION

The translation process requires a number of initiation, elongation, and termination (or release) factors. The cDNAs for two initiation factors, $IF2_{mt}$ (Ma and Spremulli 1995) and $IF3_{mt}$ (Koc and Spremulli 2002); four elongation factors, $EF-Tu_{mt}$ (EF1A) (Woriax et al. 1995; Ling et al. 1997), $EF-Ts_{mt}$ (EF1B) (Xin et al. 1995), $EF-G1_{mt}$ (EF2A) (Gao et al. 2001), and $EF-G2_{mt}$ (EF2B) (Hammarsund et al. 2001); one release factor, $RF1_{mt}$; and one ribosomal recycling factor, RRR_{mt} (Zhang and Spremulli 1998), have been cloned and sequenced in several mammalian species, including human. An exhaustive review of the biochemical characteristics of the initiation and elongation factors involved in mammalian mitochondrial translation was published recently (Spremulli et al. 2004). A schematic diagram of mitochondrial translation as it is thought to occur in mammals is shown in Figure 1.

Translation Initiation

Translation initiation is the least well understood aspect of mitochondrial translation, largely because of the structural peculiarities of mitochondrial mRNAs, in particular, the lack of significant 5′UTRs in mammals.

Initiation of translation of yeast mitochondrial mRNAs requires membrane-bound, mRNA-specific activator proteins (Fox 1996; Green-Willms et al. 1998). These activators recognize specific sequences in the 5′UTRs of individual mRNAs and mediate the interactions among the mitochondrial mRNA, the small subunit of the mitochondrial ribosome, and the inner mitochondrial membrane (Fox 1996). Several yeast mitochondrial mRNAs (e.g., COX1-3, coding for subunits of complex IV; ATP9, coding for a subunit of complex V; cytochrome *b*, coding for a subunit of complex III) require specific activators for their translation. Yeast mitochondrial ribosomes appear to recognize a common feature of all mRNA 5′UTRs, which, in conjunction with specific mRNA translational activation, is required for the initiation of mitochondrial translation (Green-Willms et al. 1998). Balanced levels of expression of translational activators appear to be crucial for the cooperative regulation of translation of the individual subunits required for the biogenesis of the oxidative phosphorylation complexes. Overexpression of the COX2 mRNA-specific translational activator Pet111 leads to a decrease in the synthesis of the COX1 mRNA, reduced levels of cytochrome *c* oxidase, and impaired respiratory growth (Fiori et al. 2005), a phenotype that is partially corrected by the overproduction of either Pet309 or MSS51, the two COX1 mRNA-specific activators. A 900-kD complex containing several proteins including the cytochrome *b* transcript-specific activator Cbp1 and the COX1 message-specific activator Pet309 (Krause et al. 2004) has been identified, and protein–protein interactions have been demonstrated among the translational activators of the COX1, COX2, and COX3 mRNAs by immunoprecipitation and two-hybrid experiments (Naithani et al. 2003). These observations suggest that translation of all mitochondrial mRNAs is coordinated in the vicinity of the inner membrane.

It is not clear how selection of the initiation codon takes place in yeast mitochondria. It has been proposed that the mRNA-specific

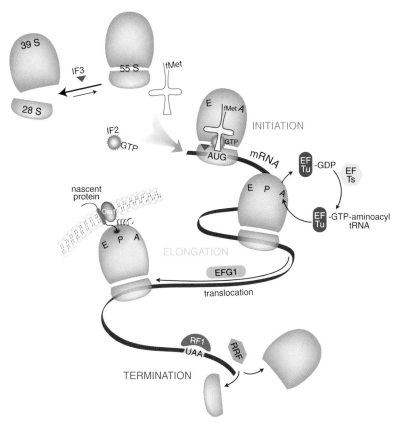

Figure 1. Mechanism of protein synthesis in mammalian mitochondria. The initiation factor IF2$_{mt}$ promotes the binding of fMet-tRNA to mitochondrial ribosomes in the presence of GTP and an RNA template containing a 5′ terminal AUG. In vitro, the initiation factor IF3$_{mt}$ promotes the dissociation of the 55S mitoribosomes and may promote positioning of the 5′ start codon in the P (peptidyl) site of the ribosome. Although it has not been firmly established, translation is likely to initiate on 55S ribosomes. The elongation factor EF-Tu$_{mt}$ participates in the formation of the -EF-Tu$_{mt}$:GTP:aminoacyl-tRNA ternary complex, which delivers the aminoacyl-tRNA to the acceptor (A) site of the ribosome. EF-Tu$_{mt}$ is released from the ribosome as EF-Tu:GDP, which is converted back to EF-Tu:GTP by the exchange factor EF-Ts$_{mt}$. The elongation factor EF-G$_{mt}$ catalyzes the translocation of peptidyl-tRNA from the ribosomal A site to the P site following peptide bond formation. The concomitant movement of the mRNA exposes the next codon in the A site. Two mitochondrial translocases have been described, EF-G1$_{mt}$ and EF-G2$_{mt}$. The association of the mitoribosomes with the Oxa1 translocase through its matrix-oriented, carboxy-terminal domain promotes cotranslational insertion of nascent polypeptides in the inner mitochondrial membrane. The release factor RF$_{mt}$ and the ribosomal recycling factor RRF$_{mt}$ are expected to have equivalent functions in mitochondria and prokaryotes, namely, recognition of the stop codon leading to release of the nascent protein and discharge of ribosomes from the mRNA.

translational activators might position ribosomes over the correct AUGs (Fox 1996). Simple scanning of the 5′UTRs for the initiation codon by mitochondrial ribosomes was ruled out, based on the identification of different numbers of AUG triplets upstream of the initiation codon in various yeast mitochondrial mRNAs. Site-directed mutagenesis of the AUG initiation codons of both COX2 and COX3 mRNAs to AUA decreased translation severalfold, but the residual translation initiated at the altered initiation codon, suggesting that sites outside the initiation codon dictate the correct positioning of the ribosome (Fox 1996).

Translation initiation and selection of the initiation codon in mammalian mitochondria remain poorly understood. As the start codon is generally located within three nucleotides from the 5′ end of the mammalian mitochondrial mRNAs (Montoya et al. 1981), translational activators, if they exist, cannot operate through 5′UTRs. Except for LRPPRC (leucine-rich pentatricopeptide repeat cassette), a weak homolog of Pet309 that has an as yet ill-defined role in COX1 translation (Xu et al. 2004), there are no other obvious human homologs of the yeast translational activators. This does not necessarily rule them out in mammals, because these proteins and their 5′UTR targets are highly divergent even among the budding yeasts (Costanzo et al. 2000). A search for RNA-binding proteins in bovine mitochondria uncovered several matrix and inner-membrane polypeptides with RNA-binding activity; however, none were specific for binding mitochondrial mRNAs (Koc and Spremulli 2003).

Leaderless mRNAs are also found in certain types of bacteria, where they encode regulatory proteins, proteins conferring antibiotic resistance, and proteins involved in light-mediated ATP synthesis (Moll et al. 2002). These mRNAs bind the small ribosomal subunit to a significant extent only in the presence of fMet-tRNA$_\mathrm{f}^\mathrm{Met}$. Specific molar ratios between the two initiation factors IF2 and IF3 appear to be required for efficient translation of these mRNAs (Grill et al. 2001), as binding and translation initiation are promoted by IF2 (Grill et al. 2000) and antagonized by IF3 (Tedin et al. 1999). The ribosomal protein S1 mediates the discriminatory action of IF3 against translation initiation of leaderless mRNAs (Moll et al. 1998). No homolog of the S1 protein has been found in the small subunit of mammalian mitoribosomes (Cavdar Koc et al. 2001a), suggesting that its absence could be important for efficient translation of the leaderless mitochondrial mRNAs.

Efficient translation of leaderless mRNAs has been observed in vitro with 70S ribosomes that had been chemically cross-linked to prevent dissociation of the individual subunits, and in vivo, in a temperature-sensitive *E. coli* mutant strain containing stable 70S ribosomes (Moll et al. 2004). The latter experiments demonstrated continued translation of leaderless mRNAs at the nonpermissive temperature, when translation of canonical mRNAs was drastically reduced. The aminoglycoside antibiotic kasugamycin is known to inhibit initiation of prokaryotic protein synthesis by removing fMet-tRNA from 30S ribosomes, but not from 70S ribosomes (Poldermans et al. 1979), and formation of a translation initiation complex on a leaderless mRNA was demonstrated even at elevated concentration of the antibiotic in the presence of 70S ribosomes (Moll and Blasi 2002). In contrast, even modest amounts of the antibiotic interfere with the formation of an initiation complex on a leaderless mRNA with 30S ribosomes (Moll and Blasi 2002). Sensitivity to kasugamycin is conferred by two N^6,N^6-dimethyladenosines present at conserved positions in the *E. coli* 16S rRNA, and mutations in the methyltransferase responsible for this modification are associated with resistance to the antibiotic (Helser et al. 1972). Studies with antibodies raised against dimethyladenosine showed that the dimethylated adenosines in 30S ribosomal subunits are accessible to the antibody, but are inaccessible in 70S ribosomes (Politz and Glitz 1977). Consequently, this site is likely located between the two ribosomal subunits.

Several observations suggest that translation initiation in mammalian mitochondria occurs on 55S ribosomes. First, given the prokaryotic antibiotic sensitivity of mitochondria, kasugamycin would be expected to inhibit organellar translation. However, mitochondrial translation in human cultured cells is not inhibited by kasugamycin (E.A. Shoubridge and F. Sasarman, unpubl.), suggesting that most mitoribosomes are present in a 55S configuration. Although it is not known for certain whether kasugamycin enters the mitochondrion, other aminoglycosides apparently do access the mitochondrial matrix, as they are known to affect mitochondrial translation (see below). Second, the tandem adenine residues in the human 12S mitochondrial ribosomal rRNA, which are homologous to the methylated adenosines in the bacterial 16S rRNA, appear to be mostly unmethylated (Seidel-Rogol et al. 2003). This could occur in 55S ribosomes, where the residues of the small ribosomal subunit facing the large subunit would be inaccessible for methylation. Finally, no mammalian mitochondrial initiation factor equivalent to bacterial IF1 has been identified (Koc and Spremulli 2002). Bacterial IF1 binds to the A site of the 30S ribosomal subunit (Carter et al. 2001), promoting correct positioning of the fMet-tRNA at the P site. If most mammalian mitoribosomes exist as 55S particles, the large ribosomal subunit might prevent accidental binding of the fMet-tRNA to the A site, thus circumventing the need for IF1.

Initiation Factors

The initiation factor IF2 belongs to the family of GTPases that are molecular switches, alternating between an active (GTP-bound) form and an inactive (GDP-bound) form without the assistance of an exchange factor. The human gene is ubiquitously expressed, although at vastly different levels across tissues (Ma and Spremulli 1995). Mammalian $IF2_{mt}$ promotes the binding of fMet-tRNA$_i^{Met}$ to both 28S and 55S ribosomes in the presence of a poly(A,U,G) template and GTP, although under these conditions, two to three times more fMet-tRNA was found to be bound to 55S ribosomes than to 28S subunits (Liao and Spremulli 1990, 1991). In fact, $IF2_{mt}$ can bind to the 55S bovine mitochondrial ribosome in the absence of GTP, initiator tRNA, or a mRNA template (Ma and Spremulli 1996), suggesting that unlike the sequence of events in eukaryotic cytoplasmic translation, the ternary $IF2_{mt}$:fMet-tRNA:GTP complex does not seem to have a role in translation initiation in mitochondria. In vitro binding experiments show that $IF2_{mt}$ exhibits a 50-fold preference for fMet-tRNA over Met-tRNA in promoting initiator tRNA binding to mitochondrial ribosomes (Spencer and Spremulli 2004), suggesting that formylation of the initiatior tRNA is important for initiation.

Much less information is currently available about $IF3_{mt}$. Mammalian $IF3_{mt}$ was shown to promote initiation complex formation on mitochondrial 55S ribosomes in the presence of $IF2_{mt}$, fMet-tRNA, and a poly(A,U,G) mRNA template, an activity that was attributed to its ability to promote dissociation of the 55S ribosome into its subunits (Koc and Spremulli 2002). Recent investigations demonstrate that $IF3_{mt}$ can dissociate fMet-tRNA bound to the 28S subunit in the absence of mRNA, leading to the suggestion that the role of $IF3_{mt}$ may be to position the start codon in the P site prior to $IF2_{mt}$-dependent binding of fMet-tRNA (Bhargava and Spremulli 2005).

Elongation Factors EF-Tu and EF-Ts

Elongation factor Tu (EF-Tu) has a central role in protein synthesis by delivering aminoacyl-tRNAs to the acceptor (A) site of the ribosome during the elongation stage of translation. Reminiscent of its prokaryotic counterpart, mammalian $EF-Tu_{mt}$ participates in the formation of the (EF-Tu$_{mt}$:GTP:aminoacyl-tRNA) ternary complex, which promotes binding of the aminoacyl-tRNA to the ribosomal A site (Woriax et al. 1997; Cai et al. 2000).

Like IF2, EF-Tu is a GTPase; however, the switch between an active (GTP-bound) form and an inactive (GDP-bound) form generally requires the elongation factor Ts (EF-Ts). Both $EF-Tu_{mt}$ and $EF-Ts_{mt}$ are ubiquitously expressed, with highest levels present in skeletal muscle, kidney, and liver (Woriax et al. 1995; Xin et al. 1995). Analysis of the steady-state levels of EF-Tu$_{mt}$ and EF-Ts$_{mt}$ indicated a 1:1 ratio between the two factors in bovine liver mitochondria and in cultured cells (Woriax et al. 1997). We have obtained similar results in human liver, muscle, fibroblasts, and cultured myoblasts; however, the relative ratio of EF-Tu$_{mt}$ to EF-Ts$_{mt}$ in human heart is 1:6 (Antonicka et al. 2006). In contrast, the relative ratio between EF-Tu$_{mt}$ and EF-Ts$_{mt}$ in prokaryotic organisms is 8:1, and *Saccharomyces cerevisiae* (but not *Schizosaccharomyces pombe*) completely lacks EF-Ts$_{mt}$ (Chiron et al. 2005). Thus, the organization of this part of the translational machinery appears to be different in prokaryotes and in mitochondria from lower and higher eukaryotes. Bovine EF-Tu$_{mt}$ was shown to be active without its exchange factor in in vitro experiments, albeit at a ten times lower specific activity than when EF-Ts$_{mt}$ was also present (Woriax et al. 1995). Exchange-factor-independent EF-Tu$_{mt}$ variants have been identified in *S. pombe* that contain mutations in their G-nucleotide-binding site, presumably leading to a change in their affinities for GTP and GDP (Chiron et al. 2005). Overexpression of EF-Tu$_{mt}$ in cultured human fibroblasts and myoblasts, which changed the EF-Tu$_{mt}$:EF-Ts$_{mt}$ ratio from 1:1 to 4:1, resulted in significant decreases in mitochondrial translation, assembly of the OXPHOS complexes, and COX activity (Antonicka et al. 2006).

Mechanistic and kinetic studies have demonstrated considerable differences in the elongation stage of protein synthesis between prokaryotes and mitochondria. The mitochondrial EF-Tu:EF-Ts complex is very stable and, unlike the prokaryotic complex, cannot be dissociated even by elevated concentrations of guanine nucleotides (Schwartzbach and Spremulli 1989). Mitochondrial EF-Tu has a much higher affinity for EF-Ts than for either GDP or GTP; thus, after GTP hydrolysis by EF-Tu$_{mt}$, the EF-Tu:EF-Ts complex forms readily (Cai et al. 2000). In *E. coli*, the affinity of EF-Tu for GDP is lower, but comparable to that for EF-Ts, resulting in a weaker EF-Tu:EF-Ts complex. However, the EF-Tu:GTP:tRNA ternary complex is still the major form of EF-Tu in both mammalian mitochondria and *E. coli* (Cai et al. 2000). This is due to the significantly higher ratio of tRNAs to EF-Tu in mitochondria, which results in shifting of the equilibrium between the EF-Tu:EF-Ts complex and EF-Tu:GTP:tRNA toward formation of the ternary complex.

Elongation Factors EF-G1 and EF-G2

The elongation factor G (EF-G) catalyzes the translocation of peptidyl-tRNA from the ribosomal A site to the P site following peptide bond formation, with the concomitant movement of the mRNA and exposure of the next codon in the A site. Similar to IF2, EF-G is a GTPase that functions without an auxiliary exchange factor. Hydrolysis of GTP confers high translational efficiency to EF-G, as both translocation and turnover of EF-G are 50- to 100-fold slower when GTP is replaced with nonhydrolyzable GTP analogs or GDP (Katunin et al. 2002). Human EF-G1$_{mt}$ is active on both mitochondrial and bacterial ribosomes (Bhargava et al. 2004). Two isoforms, EF-G1 and EF-G2, have been identified in mitochondria (Hammarsund et al. 2001), but the significance of the two functional homologs is not understood. EF-G1$_{mt}$ and EF-G2$_{mt}$ are ubiquitously expressed, with highest levels in the skeletal muscle, heart, liver, and kidney (Gao et al. 2001; Hammarsund et al. 2001).

Deletion mutants of *MEF1*, encoding the EF-G1$_{mt}$ yeast homolog, are viable, but they display a clear respiratory phenotype, with impaired mitochondrial protein synthesis (Vambutas et al. 1991). Deletion of *MEF2* also results in a growth defect on a nonfermentable carbon source (Steinmetz et al. 2002), suggesting that both factors are important in the translocation step of mitochondrial protein synthesis. Overexpression studies in fibroblasts from patients with mutations in EF-G1$_{mt}$ demonstrate nonoverlapping functions for the two isoforms of EF-G (Coenen et al. 2004).

Release Factors

Two additional translation factors have been identified in mitochondria: the translation release factor RF1, which is responsible for the recognition of stop codons, and the ribosome recycling factor RRF, which is critical for the release of ribosomes from the mRNA at the stop codon (Zhang and Spremulli 1998). Although these have not been directly studied in mammalian cells, it is likely that they perform functions similar to those of their prokaryotic homologs.

MITOCHONDRIAL TRANSLATION AND DISEASE

Defects in the mitochondrial translation system are the most frequently observed causes of mitochondrial disease in humans. Most of the mutations described so far occur in mtDNA, reflecting the relative ease of study of this small genome and the unique characteristics of mitochondrial genetics. During the last few years, however, several mutations have been reported in the nucleus-encoded components of the translation system. Below, we review the current knowledge of these mutations and the mechanisms of pathogenesis.

Mitochondrial tRNAs

Mutations in mtDNA-encoded tRNA genes are the most common cause of oxidative phosphorylation deficiencies in humans. Since the first mutation in tRNA^{lys} (A8344G) was identified in 1990 (Shoffner et al. 1990), more than 90 pathogenic tRNA mutations have been reported (see MITOMAP, http://www.mitomap.org). An extremely broad range of clinical phenotypes has been associated with these mutations (Fig. 2, top). Cells of the nervous system, skeletal or cardiac muscle, and the endocrine system (specifically, the insulin-secreting cells of the pancreas) are the most commonly affected cell types. Because all of these cell types rely heavily on oxidative production of ATP, this is perhaps not too surprising.

What is unexpected, and remains largely unexplained, is the striking association of different mutations with particular clinical phenotypes. On the other hand, a specific mutation can produce a number of distinct phenotypes. A few examples will serve to illustrate these points. The tRNA$^{leu(UUR)}$ appears to be a hot spot for mutations (Fig. 2, bottom). At least 19 different point mutations have been reported in this tRNA, but by far the most common is the A3243G mutation in the D arm. This mutation is most often associated with a neurological syndrome called MELAS (mitochondrial encephalopathy, lactic acidosis, and stroke-like episodes), characterized by metabolic strokes at an early age. About 80% of patients with this phenotype have the A3243G mutation; another 10% have the T3271C mutation in the anticodon stem. The A3243G mutation is also found in patients with CPEO (chronic progressive external opthalmoplegia—paralysis of the extraocular muscles) or in families with sensorineural deafness and diabetes. Thus, the same mutation can cause at least three different phenotypes. In contrast, most patients with the A8344G mutation in the TψC loop of tRNA^{lys} have a distinct clinical phenotype called MERRF (myoclonus epilepsy and ragged red fibers). Ragged-red fibers are a pathological feature of muscle, reflecting the abnormal proliferation of structurally and biochemically abnormal mitochondria. Finally, almost all patients with mutations in tRNA$^{ser(UCN)}$ (most of which occur in the aminoacyl stem) suffer from sensorineural deafness.

How can we account for these patterns? At least part of this variability of expression in clinical phenotype can be attributed to the unique features of mitochondrial genetics. mtDNA is a multicopy genome that is present in hundreds to thousands of copies in most cells. Most patients with tRNA mutations are segregating mutant and wild-type versions of the genome, the proportions of which can vary from cell to cell and tissue to tissue, a situation referred to as mtDNA heteroplasmy. Mitochondria and mtDNA turn over throughout life, even in

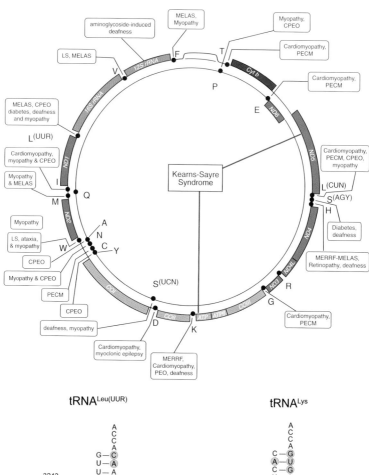

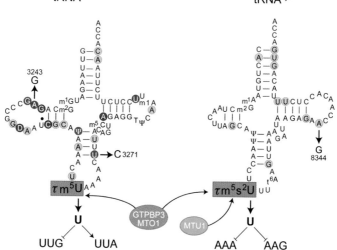

Figure 2. (*See facing page for legend.*)

postmitotic cells, so the relative proportions of mutant and wild-type genomes can vary both spatially and temporally. Even in dividing cells, replication of mtDNA is not tightly linked to the cell cycle. Thus, some templates may be replicated more than once, others not at all, during the cell cycle, and these will be randomly apportioned to daughter cells. Once the proportion of mutant mtDNAs exceeds a threshold (typically 85–90%), a translation defect results in cultured cells (Boulet et al. 1992). This has been shown to account for the fact that some patients with A3243G have CPEO (high proportion of mutant mtDNAs in individual muscle fibers), whereas others have MELAS (low proportion in individual muscle fibers) (Petruzzella et al. 1994). The mechanisms underlying different patterns of segregation of pathogenic mtDNA

Figure 2. Mitochondrial disease phenotypes associated with mutations in mtDNA-encoded tRNA genes. (*Top*) Structure of the human mitochondrial genome. The tRNA genes are depicted as solid circles using the single-letter amino acid code, and the major clinical phenotypes associated with mutations in the individual genes are shown in boxes next to the position of the gene. The extent of the common deletion, which is associated with Kearns-Sayre syndrome and removes five tRNA genes, is indicated by lines inside the genome. The structural genes are shown as shaded bars: (COI-III) subunits of cytochrome *c* oxidase (complex IV); (ND1-5) subunits of NADH CoQ Reductase (complex I); (ATP6,8) subunits of ATP synthase (complex V); (Cyt *b*) cytochrome *b* subunit of CoQ-cytochrome *c* reductase (complex III); (CPEO) chronic progressive external ophthalmoplegia; (LS) Leigh syndrome; (PECM) progressive encephalomyopathy; (MELAS) mitochondrial encephalomyopathy, lactic acidosis and stroke-like episodes; (MERRF) myoclonus epilepsy and ragged-red fibers. (*Bottom*) Secondary structures of tRNA$^{\text{leu (UUR)}}$ and tRNA^{lys}. The most common mutations at positions 3243 and 3271 in tRNA$^{\text{leu (UUR)}}$ and at position 8344 in tRNA^{lys} are shown. Colored bases are those in which pathogenic mutations have been reported. The wobble base at position 34 is modified in both tRNAs as shown in a shaded box. Loss of the wobble base modification in tRNA$^{\text{leu (UUR)}}$ results in a severe decrease in the ability to decode UUG codons. Mutations at five different positions in tRNA$^{\text{leu (UUR)}}$ (*red circles*) abrogate this modification; mutations in four other bases (*blue circles*) do not significantly impair wobble base modification. Wobble base modification has not been investigated in the other reported mutations (*gold circles*). Loss of wobble base modification in the tRNA^{lys} carrying the A8344G mutation prevents efficient decoding of both AAA and AAG codons. The enzymes thought to be involved in wobble base modification are shown in ovals next to the wobble bases: GTPBP3 and MTO1 for the taurine modification at position 5, and MTU1 for the thio modification at position 2. ($\tau m^5 U$) 5-Taurinomethyluridine; ($\tau m^5 U s^2 U$) 5-taurinomethyl-2-thiouridine; (ψ) pseudouridine; (D) dihydrouridine; ($m^1 A$) 1-methyladenosine; ($m^1 G$) 1-methylguanosine; ($t^6 A$) N^6-threoninocarbonyladenosine; ($m^2 G$) 2-methylguanosine; ($m^5 C$) 5-methylcytidine; (T) ribothymidine.

mutations, however, remain unknown. Nuclear loci have been mapped that control segregation of even neutral mtDNA sequence variants (Battersby et al. 2003), but the genes involved have not yet been identified.

Although this goes some way to explaining why only some tissues are affected, it cannot explain why mutations in different tRNAs are generally associated with different clinical phenotypes. This fact suggests that the details of exactly how the mutations influence translation, and which and how many of the respiratory chain complexes are affected, are important determinants of the nature of the biochemical, and thus, the clinical phenotype.

Mechanisms of Pathogenesis of Mitochondrial tRNA Mutations

There are a number of places where tRNA mutations could affect tRNA function: aminoacylation, codon:anticodon binding and recognition, tRNA base modification, tRNA stability, and tRNA processing. All of these have been shown to have a role in pathogenesis (for review, see Jacobs 2003). The most informative studies on molecular pathogenesis have been performed by Watanabe, Suzuki, and coworkers on the alterations in wobble base modification that are associated with pathogenic tRNA point mutations (see below).

Modification of the wobble base U is important in codon recognition and codon–anticodon binding at the ribosomal A site. An investigation of the nature of the translation defect in cybrid (cytoplasmic hybrid) cells derived from patients with the A3243G tRNA$^{leu(UUR)}$ mutation showed no correlation between the severity of the translation defect in individual polypeptides and the proportion of UUR codons in their mRNAs in cells (Chomyn et al. 2000), prompting Watanabe and coworkers to investigate modification of the wobble base U in tRNA$^{leu(UUR)}$. They established that the wobble base remained unmodified in tRNAs carrying both the A3243G and T3271C mutations (Yasukawa et al. 2000) and subsequently identified the modified base in the wild-type tRNA as 5-taurinomethyluridine (Suzuki et al. 2002). Parallel studies on tRNA^{lys} showed that the wobble base is additionally modified at position 2 with a thio group (Suzuki et al. 2002). These modifications are completely abrogated by the common MERRF (A8344G) and MELAS (A3243G) mutations in cybrid cells and tissues from patients (Fig. 2, bottom) (Yasukawa et al. 2001, 2005). Using a very elegant molecular surgery technique on tRNA$^{leu(UUR)}$, these authors were able to dissect the independent effects of the lack of wobble base modification from the effects of the A3243G mutation per se on tRNA function by creating tRNAs with a wild-type sequence, but lacking the wobble base modification (Kirino et al. 2004). Using both an A-site-binding assay and an in vitro translation assay employing synthetic mRNAs on *E. coli* ribosomes, they demonstrated that lack of the wobble modification in tRNA$^{leu(UUR)}$ produced a severe effect on decoding the UUG codon and a moderate effect on the UUA codon. UUG is a relatively underutilized leucine codon, except in the gene coding for ND6, a subunit of complex I, where more than 40% of the leucine codons are UUG. Consistent with the predictions of the above experiment, translation of ND6 is severely depressed in cybrid cells established from MELAS patients, and there is a marked reduction in the activity of complex I (Dunbar et al. 1996). Lack of the wobble base modification in tRNA^{lys} prevents translation of both AAA and AAG codons (Kirino et al. 2004), explaining why a severe translation defect is seen in almost all mitochondrial polypeptides in cells from MERRF patients (Boulet et al. 1992), the magnitude of which correlates with the proportion of AAR codons (Enriquez et al. 1995).

Suzuki and coworkers went on to develop a sensitive primer extension assay that allowed them to measure the extent of wobble modification on minute quantities of tRNAs (Kirino et al. 2005). Remarkably, they demonstrated that any of five different point mutations in tRNA$^{leu(UUR)}$, all of which had been associated with the MELAS phenotype in patients (red circles, Fig. 2, bottom), lacked wobble base modification, whereas four other mutations in tRNA$^{leu(UUR)}$ (blue circles, Fig. 2, bottom), all of which were associated with myopathy (but not MELAS), maintained the modification. This is the first real molecular insight into the molecular mechanisms that might differentiate at least some patients who develop MELAS from those who do not, as a result of mutations in the same tRNA. All other base modifications in tRNA$^{leu(UUR)}$ were largely unaffected by any of the mutations tested, so it seems that the structural change introduced by a subset of the mutations is sufficient to prevent wobble base modification, while allowing recognition of the other ten bases that are modified in tRNA$^{leu(UUR)}$ (Fig. 2, bottom).

A phenotypic revertant, in which the respiratory defect caused by the A3243G mutation was suppressed by a mutation (G12300A) in the anticodon of the leucine isoacceptor tRNA$^{leu(CUN)}$, has been isolated in cybrid cells (El Meziane et al. 1998). Although the mutation converts the anticodon from UAG to UAA, the mutant tRNA would still not be able to decode UUG without an unmodified wobble U. Recently, it was shown that the wobble base in the tRNA carrying the mutation acquires the $\tau m^5 U$ modification de novo, permitting it to function as a suppressor of the defective tRNA$^{leu(UUR)}$ associated with MELAS (Kirino et al. 2006).

Ribosomal RNA and Ribosomal Proteins

A homoplasmic mutation (A1555G) in the 12S rRNA was first reported in several pedigrees with aminoglycoside-induced deafness or nonsyndromic deafness (Prezant et al. 1993). The clinical phenotype in families not exposed to the antibiotic ranged from congenital deafness to completely normal hearing, suggesting that other environmental or genetic influences modified expression of the mutant gene. Several other mutations in the 12S gene have since been reported in many different ethnic groups, and all are associated with hearing loss (see MITOMAP, http://www.mitomap.org).

The A1555G mutation is located in a conserved region of the rRNA that forms part of the A site, and which binds aminoglycoside antibiotics (Moazed and Noller 1987). The mutation facilitates antibiotic binding (Hamasaki and Rando 1997), and this presumably accounts for the hearing loss in patients exposed to the antibiotic. Studies on cell lines from patients and from cybrid cells carrying patient mitochondria harboring the A1555G mutation have demonstrated translation defects, the severity of which depends on nuclear background, suggesting that modifier genes influence the clinical phenotype (Guan et al. 2001).

Two candidate modifiers (MSS1, MTO1) have been identified in paromomycin-resistant mutants of *S. cerevisiae* (Decoster et al. 1993; Colby et al. 1998). Paromomycin binds to a site on the small rRNA (15S) in yeast that is homologous to the 1555 site on human 12S rRNA. MSS1 (human GTPBP3; Li and Guan 2002) and MTO1 are homologous to two *E. coli* genes that are required for the modification of the uridine wobble base at the C5 position, suggesting that they may also be involved in the C5 taurinomethyl modification in mitochondrial tRNAs (Fig. 2, bottom) (Umeda et al. 2005), although this has not yet been directly demonstrated. They apparently exist in a heterodimeric complex in yeast mitochondria and, when mutated, express a respiratory defect that seems to be rather specific for the COX1 subunit, but only in a paromomycin-resistant background (Colby et al. 1998). The molecular basis for these effects remains unknown. Although these observations suggest tRNA-modifying enzymes as candidate genes for modifiers of the deafness phenotype associated with mutations at the 1555 site in human 12S rRNA, there is as yet no evidence for variants in these genes that confer such effects, and genetic studies are only suggestive (Bykhovskaya et al. 2004b). A candidate locus on human chromosome 8 has been identified as a modifier of the A1555G mutation, but the gene involved has not been identified (Bykhovskaya et al. 2000). Recently, MTO2/MTU1/TRMU was shown to be necessary for the 2-thio modification of uridine (Umeda et al. 2005; Yan et al. 2005), and this could be considered as another potential nuclear modifier (Yan et al. 2006).

Mutations have so far been identified in only a single mitochondrial ribosomal protein (MRPS16) in humans (Miller et al. 2004). The patient, from a consanguineous family, presented with intractable lactic acidosis, agenesis of the corpus callosum, dysmorphism, and marked decreases in the activities of complexes I and IV. Mitochondrial translation in patient fibroblasts was impaired, and there was a marked reduction in the steady-state level of the 12S rRNA. DNA sequence analysis of the genes coding for small ribosomal subunit proteins that are conserved between *E. coli* and mammals revealed a homozygous premature stop codon in MRPS16. This is one of the most highly conserved proteins between mammals and yeast, and it is 40% identical to the bacterial homolog. S16 has been shown to have a role in assembly of the small ribosomal subunit in *Thermus thermophilus* (Allard et al. 2000), but its role in the mitoribosome is unknown.

The *Drosophila melanogaster* mutant *"technical knockout"* (*tko*) contains a missense mutation in the gene encoding the mitochondrial ribosomal protein S12, which leads to an amino acid substitution at a phylogenetically conserved position (Toivonen et al. 2001; Jacobs et al. 2004). The resulting phenotype includes hypersensitivity to doxycyclin, severely reduced activities of the respiratory chain enzymes in the mutant larvae, decreased levels of 12S rRNA, and an array of features reminiscent of mitochondrial disease in humans, of which the most prominent is deafness (Toivonen et al. 2001).

The unexpected involvement of an AAA protease in mitochondrial ribosome assembly has recently been reported by Langer and colleagues (Nolden et al. 2005). Mitochondria contain two AAA protease activities, directed to opposite surfaces of the inner membrane, that are important in quality control of inner mitochondrial membrane proteins (Leonhard et al. 1996). The matrix-directed protease is composed of two subunits, which in humans are called AFG3l2 and paraplegin (Atorino et al. 2003). Loss-of-function mutations in paraplegin are associated with a dominant form of hereditary spastic paraplegia (Casari et al. 1998), a neurodegenerative disease caused by axonal degeneration of motor neurons of the corticospinal tracts. The ribosomal protein MrpL32 was shown to be a substrate for the matrix-directed AAA protease in yeast and in mouse (Nolden et al. 2005), and it was demonstrated that proteolytic processing of this protein was essential for the recruitment of preassembled ribosomal particles and completion of ribosomal assembly. When the protease is defective, maturation of MrpL32 is prevented and a translation deficiency results (Nolden et al. 2005).

tRNA-modifying Enzymes

A missense mutation in the *PUS1* gene, coding for pseudouridine synthase 1, has been reported in families with mitochondrial myopathy and sideroblastic anemia (Bykhovskaya et al. 2004a). Cell lines from these patients lack Pus1 activity, and both cytoplasmic and mitochondrial tRNAs lack pseudouridine at sites known to be modified by Pus1 (Patton et al. 2005). Pus1 is found in the nucleus, cytoplasm, and mitochondria, therefore the phenotype caused by mutations in this gene could be much broader than just an oxidative phosphorylation deficiency; however, the fact that patient muscles show combined respiratory chain defects (Zeharia et al. 2005) demonstrates that mitochondrial dysfunction is a major part of the disease.

Putative Translational Activators

Although translational activators similar to those found in yeast have not been found in mammals, the biochemical challenge of recruiting mRNAs to the matrix face of the inner mitochondrial membrane where the protein synthetic machinery is located is similar in yeast and mammals. Recent studies of a form of Leigh syndrome (subacute necrotizing encephalopathy) associated with complex IV deficiency, and so far found only in the French Canadian population, have uncovered mutations in LRPPRC (Mootha et al. 2003). This form of Leigh syndrome is associated with severe complex IV deficiency in brain and liver, and modest deficiencies in heart, skeletal muscle, and kidney (Merante et al. 1993). As mentioned earlier, LRPPRC is a weak homolog of Pet309, the specific translational activator of the COX1 subunit in *S. cerevisiae*, and is part of a family of proteins that are involved in RNA–protein interactions in mitochondria. Consistent with this, LRPPRC has been shown to bind both nuclear and mitochondrial RNA in vivo (Mili and Pinol-Roma 2003). Fibroblasts carrying the common missense mutation found in these patients display reduced amounts of LRPPRC protein, translational defects in COXI (and possibly COXIII), and the presence of an abnormal translation product that has not yet been identified (Xu et al. 2004). These data suggest that this protein has some role in the translation of COX subunits in humans, although the molecular mechanism of action remains unknown.

Translation Elongation Factors

Mutations in EF-G1 have been found in two pedigrees in patients who presented with combined defects in the activities of respiratory chain enzymes associated with a hepatoencephalopathy (Coenen et al. 2004; Antonicka et al. 2006). The effects of the mutation were polypeptide and tissue-specific: Translation of two complex I subunits (ND5, ND6) was about 10% of control levels, whereas that of other complex I subunits was 30–60% of control, and translation of all three COX subunits was reduced to 15% of control levels. The reason for these differences is not known. Analysis of all five oxidative phosphorylation complexes by blue-native gel electrophoresis (a technique that allows one to quantify the amount of the individual complexes in their native state) in a patient who was a compound heterozygote for a missense mutation and a nonsense mutation showed striking differences in the nature and severity of the biochemical defect among tissues. Whereas heart muscle showed only a mild deficiency in complex IV assembly, both complex IV and complex V were severely reduced in skeletal muscle, and there was a small decrease in complex I. Complexes I and IV were severely reduced in fibroblasts and liver, and there was a more modest reduction in complex V (Anton-icka et al. 2006). Consistent with these results, immunoblot analysis demonstrated that EF-G1 was reduced in heart tissue, but it was unde-tectable in skeletal muscle and liver. These results suggest a different or-ganization and control of mitochondrial translation in different tissues.

SUMMARY AND PERSPECTIVE

The mitochondrial translation system has evolved as a specialized system to translate a handful of hydrophobic inner mitochondrial membrane proteins, all of which are essential components of the oxidative phosphorylation system. Translation of these polypeptides requires recruitment of the mRNAs, ribosomes, and the rest of the translational machinery to the inner mitochondrial membrane where the nascent polypeptides are most probably cotranslationally inserted into the inner membrane with the aid of molecular chaperones. The mechanism of translation is broadly similar to that found in prokaryotes, but there are striking differences in the composition and structure of mitoribosomes, and in the structure of the mRNAs. It is not clear why it became advantageous to employ leader-less mRNAs in mammalian mitochondria. The fact that mtDNA itself is an extremely compact molecule with almost no extraneous genetic material suggests that there was some evolutionary pressure to reduce nucleic acid content. One highly speculative possibility is that this was an important adjunct to the acquisition of innate immunity, a system that involved the elaboration of toll-like receptors to recognize relatively unmodified bacterial RNA and DNA species, while ignoring highly modified nuclear DNA and RNA molecules (Kariko et al. 2005). As in bacteria, mitochondrial nucleic acids remain relatively unmodified compared to nuclear DNA, so minimizing their concentration may have been important to avoid inappropriate immune responses.

One of the most intriguing and largely unexplained aspects of mitochondrial translation disease remains the tissue specificity associated with mutations in the RNAs and proteins of the translation machinery. In the case of mtDNA-encoded tRNAs and rRNAs, progress in understanding the mechanisms of pathogenesis has been hampered by the lack of a system to transform mammalian mtDNA. Thus far, it has not been possible to make animal models of these disorders without first isolating the appropriate mutant species of mtDNA in a murine cell line. Some progress has been made in understanding the molecular mechanisms underlying MELAS, but even here we do not know why this particular phenotype results from lack of wobble base modification in a particular tRNA. It seems likely that the relative deficiencies in the enzymatic activities of different respiratory chain complexes will be important determinants of whether or not a particular cell type will be affected, but our understanding of which cells might, for instance, rely more heavily on complex I versus complex IV activity for their normal function remains rudimentary. One other possibility is that defects in translation result in partial assemblies of different complexes and that these might have other effects, such as the production of reactive oxygen species that might act as gain-of-function signals. It is surprising to find that nucleus-encoded components of the translation apparatus, such as EF-G1, produce such striking tissue-specific defects in translation and assembly of the respiratory chain components. It appears that the organization of the mitochondrial translation apparatus and its regulation might vary considerably among different tissues in mammals. An understanding of this diversity at the molecular level will be a major focus of future research.

Finally, although it has long been thought that translation on mito-ribosomes was restricted to those mRNAs encoded in the organellar genome, it has recently been shown that translation of a subset of mRNAs transcribed from nuclear genes occurs during capacitation in mammalian sperm (a series of physiological changes that renders sperm cells competent for fertilization) and that this new protein synthesis is sensitive to a mitochondrial translation inhibitor (chloramphenicol), but insensitive to an inhibitor of cytoplasmic translation (cycloheximide) (Gur and Breitbart 2006). This is a very surprising and unexpected result. If the mRNAs were being translated on bona fide mitoribosomes in the mitochondrial matrix, then there would have to be a mechanism to import the mRNAs, to circumvent the different mitochondrial genetic codes, and to export the proteins produced. If the mitoribosomes were located in the cytoplasm, then there would have to be a mechanism to allow them to escape the matrix space, and to use the cytosolic initiation, elongation factors, tRNAs, etc. Neither of these scenarios seems very probable, and a satisfactory explanation for the above results will have to await further investigation. Nonetheless, these observations alert us to the possibility that atypical translational control might be important in the regulation of specific physiological processes.

ACKNOWLEDGMENTS

E.A.S. is an International Scholar of the Howard Hughes Medical Institute and a Senior Investigator of the Canadian Institutes of Health Research (CIHR). Research in his laboratory is supported by grants from the CIHR, National Institutes of Health, Muscular Dystrophy Association, Muscular Dystrophy Canada, and the March of Dimes.

REFERENCES

Allard P., Rak A.V., Wimberly B.T., Clemons W.M., Jr., Kalinin A., Helgstrand M., Garber M.B., Ramakrishnan V., and Hard T. 2000. Another piece of the ribosome: Solution structure of S16 and its location in the 30S subunit. *Structure* **8:** 875–882.

Antonicka H., Sasarman F., Kennaway N.G., and Shoubridge E.A. 2006. The molecular basis for tissue specificity of the oxidative phosphorylation deficiencies in patients with mutations in the mitochondrial translation factor EFG1. *Hum. Mol. Genet.* **15:** 1835–1846.

Atorino L., Silvestri L., Koppen M., Cassina L., Ballabio A., Marconi R., Langer T., and Casari G. 2003. Loss of m-AAA protease in mitochondria causes complex I deficiency and increased sensitivity to oxidative stress in hereditary spastic paraplegia. *J. Cell Biol.* **163:** 777–787.

Barrientos A., Korr D., Barwell K.J., Sjulsen C., Gajewski C.D., Manfredi G., Ackerman S., and Tzagoloff A. 2003. MTG1 codes for a conserved protein required for mitochondrial translation. *Mol. Biol. Cell* **14:** 2292–2302.

Battersby B.J., Loredo-Osti J.C., and Shoubridge E.A. 2003. Nuclear genetic control of mitochondrial DNA segregation. *Nat. Genet.* **33:** 183–186.

Bhargava K. and Spremulli L.L. 2005. Role of the N- and C-terminal extensions on the activity of mammalian mitochondrial translational initiation factor 3. *Nucleic Acids Res.* **33:** 7011–7018.

Bhargava K., Templeton P., and Spremulli L.L. 2004. Expression and characterization of isoform 1 of human mitochondrial elongation factor G. *Protein Expr. Purif.* **37:** 368–376.

Boulet L., Karpati G., and Shoubridge E.A. 1992. Distribution and threshold expression of the tRNA(Lys) mutation in skeletal muscle of patients with myoclonic epilepsy and ragged-red fibers (MERRF). *Am. J. Hum. Genet.* **51:** 1187–1200.

Bykhovskaya Y., Casas K., Mengesha E., Inbal A., and Fischel-Ghodsian N. 2004a. Missense mutation in pseudouridine synthase 1 (PUS1) causes mitochondrial myopathy and sideroblastic anemia (MLASA). *Am. J. Hum. Genet.* **74:** 1303–1308.

Bykhovskaya Y., Mengesha E., Wang D., Yang H., Estivill X., Shohat M., and Fischel-Ghodsian N. 2004b. Human mitochondrial transcription factor B1 as a modifier gene for hearing loss associated with the mitochondrial A1555G mutation. *Mol. Genet. Metab.* **82:** 27–32.

Bykhovskaya Y., Estivill X., Taylor K., Hang T., Hamon M., Casano R.A., Yang H., Rotter J.I., Shohat M., and Fischel-Ghodsian N. 2000. Candidate locus for a nuclear modifier gene for maternally inherited deafness. *Am. J. Hum. Genet.* **66:** 1905–1910.

Cai Y.C., Bullard J.M., Thompson N.L., and Spremulli L.L. 2000. Interaction of mammalian mitochondrial elongation factor EF-Tu with guanine nucleotides. *Protein Sci.* **9:** 1791–1800.

Carter A.P., Clemons W.M., Jr., Brodersen D.E., Morgan-Warren R.J., Hartsch T., Wimberly B.T., and Ramakrishnan V. 2001. Crystal structure of an initiation factor bound to the 30S ribosomal subunit. *Science* **291:** 498–501.

Casari G., De Fusco M., Ciarmatori S., Zeviani M., Mora M., Fernandez P., De Michele G., Filla A., Cocozza S., Marconi R., et al. 1998. Spastic paraplegia and OXPHOS impairment caused by mutations in paraplegin, a nuclear-encoded mitochondrial metalloprotease. *Cell* **93:** 973–983.

Cavdar Koc E., Burkhart W., Blackburn K., Moseley A., and Spremulli L.L. 2001a. The small subunit of the mammalian mitochondrial ribosome. Identification of the full comple-ment of ribosomal proteins present. *J. Biol. Chem.* **276:** 19363–19374.

Cavdar Koc E., Ranasinghe A., Burkhart W., Blackburn K., Koc H., Moseley A., and Spremulli L.L. 2001b. A new face on apoptosis: Death-associated protein 3 and PDCD9 are mito-chondrial ribosomal proteins. *FEBS Lett.* **492:** 166–170.

Chiron S., Suleau A., and Bonnefoy N. 2005. Mitochondrial translation: Elongation factor tu is essential in fission yeast and depends on an exchange factor conserved in hu-mans but not in budding yeast. *Genetics* **169:** 1891–1901.

Chomyn A., Enriquez J.A., Micol V., Fernandez-Silva P., and Attardi G. 2000. The mito-chondrial myopathy, encephalopathy, lactic acidosis, and stroke-like episode syndrome-associated human mitochondrial tRNALeu(UUR) mutation causes aminoacylation defi-ciency and concomitant reduced association of mRNA with ribosomes. *J. Biol. Chem.* **275:** 19198–19209.

Coenen M.J., Antonicka H., Ugalde C., Sasarman F., Rossi R., Heister J.G., Newbold R.F., Tri-jbels F.J., van den Heuvel L.P., Shoubridge E.A., and Smeitink J.A. 2004. Mutant mitochondrial elongation factor G1 and combined oxidative phosphorylation deficiency. *N. Engl. J. Med.* **351:** 2080–2086.

Colby G., Wu M., and Tzagoloff A. 1998. MTO1 codes for a mitochondrial protein required for respiration in paromomycin-resistant mutants of *Saccharomyces cerevisiae*. *J. Biol. Chem.* **273:** 27945–27952.

Costanzo M.C., Bonnefoy N., Williams E.H., Clark-Walker G.D., and Fox T.D. 2000. Highly diverged homologs of *Saccharomyces cerevisiae* mitochondrial mRNA-specific transla-tional activators have orthologous functions in other budding yeasts. *Genetics* **154:** 999–1012.

Datta K., Fuentes J.L., and Maddock J.R. 2005. The yeast GTPase Mtg2p is required for mito-chondrial translation and partially suppresses an rRNA methyltransferase mutant, mrm2. *Mol. Biol. Cell* **16:** 954–963.

Decoster E., Vassal A., and Faye G. 1993. MSS1, a nuclear-encoded mitochondrial GTPase involved in the expression of COX1 subunit of cytochrome c oxidase. *J. Mol. Biol.* **232:** 79–88.

Dunbar D.R., Moonie P.A., Zeviani M., and Holt I.J. 1996. Complex I deficiency is associ-ated with 3243G:C mitochondrial DNA in osteosarcoma cell cybrids. *Hum. Mol. Genet.* **5:** 123–129.

El Meziane A., Lehtinen S.K., Hance N., Nijtmans L.G., Dunbar D., Holt I.J., and Jacobs H.T. 1998. A tRNA suppressor mutation in human mitochondria. *Nat. Genet.* **18:** 350–353.

Enriquez J.A., Chomyn A., and Attardi G. 1995. MtDNA mutation in MERRF syndrome causes defective aminoacylation of tRNA(Lys) and premature translation termination. *Nat. Genet.* **10:** 47–55.

Fiori A., Perez-Martinez X., and Fox T.D. 2005. Overexpression of the COX2 translational activator, Pet111p, prevents translation of COX1 mRNA and cytochrome c oxidase assembly in mitochondria of *Saccharomyces cerevisiae*. *Mol. Microbiol.* **56:** 1689–1704.

Fox T.D. 1996. Genetics of mitochondrial translation. In *Translational control* (ed. J.W.B. Hershey et al.), pp. 733–758. Cold Spring Harbor Laboratory Press, Cold Spring Harbor, New York.

Gao J., Yu L., Zhang P., Jiang J., Chen J., Peng J., Wei Y., and Zhao S. 2001. Cloning and characterization of human and mouse mitochondrial elongation factor G, GFM and Gfm, and mapping of GFM to human chromosome 3q25.1-q26.2. *Genomics* **74:** 109–114.

Green-Willms N.S., Fox T.D., and Costanzo M.C. 1998. Functional interactions between yeast mitochondrial ribosomes and mRNA 5′ untranslated leaders. *Mol. Cell. Biol.* **18:** 1826–1834.

Grill S., Gualerzi C.O., Londei P., and Blasi U. 2000. Selective stimulation of translation of leaderless mRNA by initiation factor 2: Evolutionary implications for translation. *EMBO J.* **19:** 4101–4110.

Grill S., Moll I., Hasenohrl D., Gualerzi C.O., and Blasi U. 2001. Modulation of ribosomal recruitment to 5′-terminal start codons by translation initiation factors IF2 and IF3. *FEBS Lett.* **495:** 167–171.

Guan M.X., Fischel-Ghodsian N., and Attardi G. 2001. Nuclear background determines biochemical phenotype in the deafness-associated mitochondrial 12S rRNA mutation. *Hum. Mol. Genet.* **10:** 573–580.

Gur Y. and Breitbart H. 2006. Mammalian sperm translate nuclear-encoded proteins by mitochondrial-type ribosomes. *Genes Dev.* **20:** 411–416.

Hamasaki K. and Rando R.R. 1997. Specific binding of aminoglycosides to a human rRNA construct based on a DNA polymorphism which causes aminoglycoside-induced deafness. *Biochemistry* **36:** 12323–12328.

Hammarsund M., Wilson W., Corcoran M., Merup M., Einhorn S., Grander D., and Sangfelt O. 2001. Identification and characterization of two novel human mitochondrial elongation factor genes, hEFG2 and hEFG1, phylogenetically conserved through evolution. *Hum. Genet.* **109:** 542–550.

Helser T.L., Davies J.E., and Dahlberg J.E. 1972. Mechanism of kasugamycin resistance in *Escherichia coli*. *Nat. New Biol.* **235:** 6–9.

Jacobs H.T. 2003. Disorders of mitochondrial protein synthesis. *Hum. Mol. Genet.* (spec. no. 2) **12:** R293–R301.

Jacobs H.T., Fernandez-Ayala D.J., Manjiry S., Kemppainen E., Toivonen J.M., and O'Dell K.M. 2004. Mitochondrial disease in flies. *Biochim. Biophys. Acta* **1659:** 190–196.

Jia L., Dienhart M., Schramp M., McCauley M., Hell K., and Stuart R.A. 2003. Yeast Oxa1 interacts with mitochondrial ribosomes: The importance of the C-terminal region of Oxa1. *EMBO J.* **22:** 6438–6447.

Kariko K., Buckstein M., Ni H., and Weissman D. 2005. Suppression of RNA recognition by Toll-like receptors: The impact of nucleoside modification and the evolutionary origin of RNA. *Immunity* **23:** 165–175.

Katunin V.I., Savelsbergh A., Rodnina M.V., and Wintermeyer W. 2002. Coupling of GTP hydrolysis by elongation factor G to translocation and factor recycling on the ribosome. *Biochemistry* **41:** 12806–12812.

Kirino Y., Goto Y., Campos Y., Arenas J., and Suzuki T. 2005. Specific correlation between the wobble modification deficiency in mutant tRNAs and the clinical features of a human mitochondrial disease. *Proc. Natl. Acad. Sci.* **102:** 7127–7132.

Kirino Y., Yasukawa T., Marjavaara S.K., Jacobs H.T., Holt I.J., Watanabe K., and Suzuki T. 2006. Acquisition of the wobble modification in mitochondrial tRNALeu(CUN) bearing the G12300A mutation suppresses the MELAS molecular defect. *Hum. Mol. Genet.* **15:** 897–904.

Kirino Y., Yasukawa T., Ohta S., Akira S., Ishihara K., Watanabe K., and Suzuki T. 2004. Codon-specific translational defect caused by a wobble modification deficiency in mutant tRNA from a human mitochondrial disease. *Proc. Natl. Acad. Sci.* **101:** 15070–15075.

Koc E.C. and Spremulli L.L. 2002. Identification of mammalian mitochondrial translational initiation factor 3 and examination of its role in initiation complex formation with natural mRNAs. *J. Biol. Chem.* **277:** 35541–35549.

———. 2003. RNA-binding proteins of mammalian mitochondria. *Mitochondrion* **2:** 277–291.

Koc E.C., Burkhart W., Blackburn K., Moseley A., Koc H., and Spremulli L.L. 2000. A proteomics approach to the identification of mammalian mitochondrial small subunit ribosomal proteins. *J. Biol. Chem.* **275:** 32585–32591.

Koc E.C., Burkhart W., Blackburn K., Moyer M.B., Schlatzer D.M., Moseley A., and Spremulli L.L. 2001. The large subunit of the mammalian mitochondrial ribosome. Analysis of the complement of ribosomal proteins present. *J. Biol. Chem.* **276:** 43958–43969.

Krause K., Lopes de Souza R., Roberts D.G., and Dieckmann C.L. 2004. The mitochondrial message-specific mRNA protectors Cbp1 and Pet309 are associated in a high-molecular weight complex. *Mol. Biol. Cell* **15:** 2674–2683.

Leonhard K., Herrmann J.M., Stuart R.A., Mannhaupt G., Neupert W., and Langer T. 1996. AAA proteases with catalytic sites on opposite membrane surfaces comprise a proteolytic system for the ATP-dependent degradation of inner membrane proteins in mitochondria. *EMBO J.* **15:** 4218–4229.

Li X. and Guan M.X. 2002. A human mitochondrial GTP binding protein related to tRNA modification may modulate phenotypic expression of the deafness-associated mitochondrial 12S rRNA mutation. *Mol. Cell. Biol.* **22:** 7701–7711.

Liao H.X. and Spremulli L.L. 1990. Identification and initial characterization of translational initiation factor 2 from bovine mitochondria. *J. Biol. Chem.* **265:** 13618–13622.

———. 1991. Initiation of protein synthesis in animal mitochondria. Purification and characterization of translational initiation factor 2. *J. Biol. Chem.* **266:** 20714–20719.

Ling M., Merante F., Chen H.S., Duff C., Duncan A.M., and Robinson B.H. 1997. The human mitochondrial elongation factor tu (EF-Tu) gene: cDNA sequence, genomic localization, genomic structure, and identification of a pseudogene. *Gene* **197:** 325–336.

Ma J. and Spremulli L.L. 1996. Expression, purification, and mechanistic studies of bovine mitochondrial translational initiation factor 2. *J. Biol. Chem.* **271:** 5805–5811.

Ma L. and Spremulli L.L. 1995. Cloning and sequence analysis of the human mitochondrial translational initiation factor 2 cDNA. *J. Biol. Chem.* **270:** 1859–1865.

Magalhaes P.J., Andreu A.L., and Schon E.A. 1998. Evidence for the presence of 5S rRNA in mammalian mitochondria. *Mol. Biol. Cell* **9:** 2375–2382.

Merante F., Petrova-Benedict R., MacKay N., Mitchell G., Lambert M., Morin C., De Braekeleer M., Laframboise R., Gagne R., and Robinson B.H. 1993. A biochemically distinct form of cytochrome oxidase (COX) deficiency in the Saguenay-Lac-Saint-Jean region of Quebec. *Am. J. Hum. Genet.* **53:** 481–487.

Mili S. and Pinol-Roma S. 2003. LRP130, a pentatricopeptide motif protein with a non-canonical RNA-binding domain, is bound in vivo to mitochondrial and nuclear RNAs. *Mol. Cell. Biol.* **23:** 4972–4982.

Miller C., Saada A., Shaul N., Shabtai N., Ben-Shalom E., Shaag A., Hershkovitz E., and Elpeleg O. 2004. Defective mitochondrial translation caused by a ribosomal protein (MRPS16) mutation. *Ann. Neurol.* **56:** 734–738.

Moazed D. and Noller H.F. 1987. Interaction of antibiotics with functional sites in 16S ribosomal RNA. *Nature* **327:** 389–394.

Moll I. and Blasi U. 2002. Differential inhibition of 30S and 70S translation initiation complexes on leaderless mRNA by kasugamycin. *Biochem. Biophys. Res. Commun.* **297:** 1021–1026.

Moll I., Resch A., and Blasi U. 1998. Discrimination of 5′-terminal start codons by translation initiation factor 3 is mediated by ribosomal protein S1. *FEBS Lett.* **436:** 213–217.

Moll I., Grill S., Gualerzi C.O., and Blasi U. 2002. Leaderless mRNAs in bacteria: Surprises in ribosomal recruitment and translational control. *Mol. Microbiol.* **43:** 239–246.

Moll I., Hirokawa G., Kiel M.C., Kaji A., and Blasi U. 2004. Translation initiation with 70S ribosomes: An alternative pathway for leaderless mRNAs. *Nucleic Acids Res.* **32:** 3354–3363.

Montoya J., Ojala D., and Attardi G. 1981. Distinctive features of the 5′-terminal sequences of the human mitochondrial mRNAs. *Nature* **290:** 465–470.

Mootha V.K., Lepage P., Miller K., Bunkenborg J., Reich M., Hjerrild M., Delmonte T., Villeneuve A., Sladek R., Xu F., et al. 2003. Identification of a gene causing human cytochrome c oxidase deficiency by integrative genomics. *Proc. Natl. Acad. Sci.* **100:** 605–610.

Naithani S., Saracco S.A., Butler C.A., and Fox T.D. 2003. Interactions among COX1, COX2, and COX3 mRNA-specific translational activator proteins on the inner surface of the mitochondrial inner membrane of *Saccharomyces cerevisiae. Mol. Biol. Cell* **14:** 324–333.

Nolden M., Ehses S., Koppen M., Bernacchia A., Rugarli E.I., and Langer T. 2005. The -m-AAA protease defective in hereditary spastic paraplegia controls ribosome assembly in mitochondria. *Cell* **123:** 277–289.

O'Brien T.W. 2002. Evolution of a protein-rich mitochondrial ribosome: Implications for human genetic disease. *Gene* **286:** 73–79.

Patton J.R., Bykhovskaya Y., Mengesha E., Bertolotto C., and Fischel-Ghodsian N. 2005. Mitochondrial myopathy and sideroblastic anemia (MLASA): Missense mutation in the pseudouridine synthase 1 (PUS1) gene is associated with the loss of tRNA pseudouridylation. *J. Biol. Chem.* **280:** 19823–19828.

Petruzzella V., Moraes C.T., Sano M.C., Bonilla E., DiMauro S., and Schon E.A. 1994. Extremely high levels of mutant mtDNAs co-localize with cytochrome c oxidase-negative ragged-red fibers in patients harboring a point mutation at nt 3243. *Hum. Mol. Genet.* **3:** 449–454.

Poldermans B., Roza L., and Van Knippenberg P.H. 1979. Studies on the function of two adjacent N6,N6-dimethyladenosines near the 3′ end of 16 S ribosomal RNA of *Escherichia coli.* III. Purification and properties of the methylating enzyme and methylase-30 S interactions. *J. Biol. Chem.* **254:** 9094–9100.

Politz S.M. and Glitz D.G. 1977. Ribosome structure: Localization of N6,N6-dimethyladenosine by electron microscopy of a ribosome-antibody complex. *Proc. Natl. Acad. Sci.* **74:** 1468–1472.

Prezant T.R., Agapian J.V., Bohlman M.C., Bu X., Oztas S., Qiu W.Q., Arnos K.S., Cortopassi G.A., Jaber L., and Rotter J.I., et al. 1993. Mitochondrial ribosomal RNA mutation associated with both antibiotic-induced and non-syndromic deafness. *Nat. Genet.* **4:** 289–294.

Schwartzbach C.J. and Spremulli L.L. 1989. Bovine mitochondrial protein synthesis elongation factors. Identification and initial characterization of an elongation factor Tu-elongation factor Ts complex. *J. Biol. Chem.* **264:** 19125–19131.

Seidel-Rogol B.L., McCulloch V., and Shadel G.S. 2003. Human mitochondrial transcription factor B1 methylates ribosomal RNA at a conserved stem-loop. *Nat. Genet.* **33:** 23–24.

Sharma M.R., Koc E.C., Datta P.P., Booth T.M., Spremulli L.L., and Agrawal R.K. 2003. Structure of the mammalian mitochondrial ribosome reveals an expanded functional role for its component proteins. *Cell* **115:** 97–108.

Shoffner J.M., Lott M.T., Lezza A.M., Seibel P., Ballinger S.W., and Wallace D.C. 1990. Myoclonic epilepsy and ragged-red fiber disease (MERRF) is associated with a mitochondrial DNA tRNA(Lys) mutation. *Cell* **61:** 931–937.

Spencer A.C. and Spremulli L.L. 2004. Interaction of mitochondrial initiation factor 2 with mitochondrial fMet-tRNA. *Nucleic Acids Res.* **32:** 5464–5470.

Spremulli L.L., Coursey A., Navratil T., and Hunter S.E. 2004. Initiation and elongation factors in mammalian mitochondrial protein biosynthesis. *Prog. Nucleic Acid Res. Mol. Biol.* **77:** 211–261.

Stuart R. 2002. Insertion of proteins into the inner membrane of mitochondria: The role of the Oxa1 complex. *Biochim. Biophys. Acta* **1592:** 79–87.

Suzuki T., Wada T., Saigo K., and Watanabe K. 2002. Taurine as a constituent of mitochondrial tRNAs: New insights into the functions of taurine and human mitochondrial diseases. *EMBO J.* **21:** 6581–6589.

Suzuki T., Terasaki M., Takemoto-Hori C., Hanada T., Ueda T., Wada A., and Watanabe K. 2001a. Proteomic analysis of the mammalian mitochondrial ribosome. Identification of protein components in the 28 S small subunit. *J. Biol. Chem.* **276:** 33181–33195.

———. 2001b. Structural compensation for the deficit of rRNA with proteins in the mammalian mitochondrial ribosome. Systematic analysis of protein components of the large ribosomal subunit from mammalian mitochondria. *J. Biol. Chem.* **276:** 21724–21736.

Szyrach G., Ott M., Bonnefoy N., Neupert W., and Herrmann J.M. 2003. Ribosome binding to the Oxa1 complex facilitates co-translational protein insertion in mitochondria. *EMBO J.* **22:** 6448–6457.

Tedin K., Moll I., Grill S., Resch A., Graschopf A., Gualerzi C.O., and Blasi U. 1999. Translation initiation factor 3 antagonizes authentic start codon selection on leaderless mRNAs. *Mol. Microbiol.* **31:** 67–77.

Toivonen J.M., O'Dell K.M., Petit N., Irvine S.C., Knight G.K., Lehtonen M., Longmuir M., Luoto K., Touraille S., Wang Z., et al. 2001. Technical knockout, a *Drosophila* model of mitochondrial deafness. *Genetics* **159:** 241–254.

Umeda N., Suzuki T., Yukawa M., Ohya Y., Shindo H., and Watanabe K. 2005. Mitochondria-specific RNA-modifying enzymes responsible for the biosynthesis of the wobble base in mitochondrial tRNAs. Implications for the molecular pathogenesis of human mitochondrial diseases. *J. Biol. Chem.* **280:** 1613–1624.

Vambutas A., Ackerman S.H., and Tzagoloff A. 1991. Mitochondrial translational-initiation and elongation factors in *Saccharomyces cerevisiae*. *Eur. J. Biochem.* **201:** 643–652.

Woriax V.L., Burkhart W., and Spremulli L.L. 1995. Cloning, sequence analysis and expression of mammalian mitochondrial protein synthesis elongation factor Tu. *Biochim. Biophys. Acta* **1264:** 347–356.

Woriax V.L., Bullard J.M., Ma L., Yokogawa T., and Spremulli L.L. 1997. Mechanistic studies of the translational elongation cycle in mammalian mitochondria. *Biochim. Biophys. Acta* **1352:** 91–101.

Xin H., Woriax V., Burkhart W., and Spremulli L.L. 1995. Cloning and expression of mitochondrial translational elongation factor Ts from bovine and human liver. *J. Biol. Chem.* **270:** 17243–17249.

Xu F., Morin C., Mitchell G., Ackerley C., and Robinson B.H. 2004. The role of the LRPPRC (leucine-rich pentatricopeptide repeat cassette) gene in cytochrome oxidase assembly: Mutation causes lowered levels of COX (cytochrome c oxidase) I and COX III mRNA. *Biochem. J.* **382:** 331–336.

Yan Q., Li X, Faye G., and Guan M.X. 2005. Mutations in MTO2 related to tRNA modification impair mitochondrial gene expression and protein synthesis in the presence of a paromomycin resistance mutation in mitochondrial 15 S rRNA. *J. Biol. Chem.* **280:** 29151–29157.

Yan Q., Bykhovskaya Y., Li R., Mengesha E., Shohat M., Estivill X., Fischel-Ghodsian N., and Guan M.X. 2006. Human TRMU encoding the mitochondrial 5-methylamino-methyl-2-thiouridylate-methyltransferase is a putative nuclear modifier gene for the phenotypic expression of the deafness-associated 12S rRNA mutations. *Biochem. Biophys. Res. Commun.* **342:** 1130–1136.

Yasukawa T., Suzuki T., Ishii N., Ohta S., and Watanabe K. 2001. Wobble modification defect in tRNA disturbs codon-anticodon interaction in a mitochondrial disease. *EMBO J.* **20:** 4794–4802.

Yasukawa T., Suzuki T., Ueda T., Ohta S., and Watanabe K. 2000. Modification defect at anticodon wobble nucleotide of mitochondrial tRNAs(Leu)(UUR) with pathogenic mutations of mitochondrial myopathy, encephalopathy, lactic acidosis, and stroke-like episodes. *J. Biol. Chem.* **275:** 4251–4257.

Yasukawa T., Kirino Y., Ishii N., Holt I.J., Jacobs H.T., Makifuchi T., Fukuhara N., Ohta S., Suzuki T., and Watanabe K. 2005. Wobble modification deficiency in mutant tRNAs in patients with mitochondrial diseases. *FEBS Lett.* **579:** 2948–2952.

Zeharia A., Fischel-Ghodsian N., Casas K., Bykhocskaya Y., Tamari H., Lev D., Mimouni M., and Lerman-Sagie T. 2005. Mitochondrial myopathy, sideroblastic anemia, and lactic acidosis: An autosomal recessive syndrome in Persian Jews caused by a mutation in the PUS1 gene. *J. Child Neurol.* **20:** 449–452.

Zhang L., Ging N.C., Komoda T., Hanada T., Suzuki T., and Watanabe K. 2005. Antibiotic susceptibility of mammalian mitochondrial translation. *FEBS Lett.* **579:** 6423–6427.

Zhang Y. and Spremulli L.L. 1998. Identification and cloning of human mitochondrial translational release factor 1 and the ribosome recycling factor. *Biochim. Biophys. Acta* **1443:** 245–250.

28

Translational Control in Prokaryotes

Pascale Romby
UPR 9002 CNRS, Université Louis Pasteur
Institut de Biologie Moléculaire et Cellulaire
67084 Strasbourg, France

Mathias Springer
UPR 9073 CNRS, Université de Paris VII-Denis Diderot
Institut de Biologie Physico-Chimique
75005 Paris, France

BACTERIA AND PHAGES HAVE EXTREMELY SOPHISTICATED WAYS to adapt the level of gene expression to growth condition changes or to specific needs at a given stage of their life cycle. A general model for gene regulation was proposed by Jacob and Monod (1961) that adequately explained many aspects of phage and bacterial regulation at the level of transcription but paid almost no attention to translation as a level at which gene expression could be regulated. In prokaryotes, protein-mediated translational regulation was discovered first in RNA bacteriophages (Lodish et al. 1964) and was later found to occur in DNA phages and in *Escherichia coli*. The first *trans*-acting RNA regulators were found in plasmids (Light and Molin 1981) and later in bacteria (Mizuno et al. 1984) and phages (Wu et al. 1987). The most significant recent finding is that translation (and transcription) can be regulated without the help of a regulatory RNA or protein, abandoning this cornerstone of the Jacob and Monod model. It is now accepted that mRNAs can act as direct sensors of the physical and/or metabolic state of the cell and modulate transcription or translation via switches in mRNA conformation.

There are three different ways to temporally regulate gene expression at the translational level: through *trans*-acting proteins, through *trans*-acting RNAs, and through *cis*-acting mRNA elements acting as sensors.

The three mechanisms are reviewed with a limited number of examples that illustrate the amazing diversity of these controls. Regulation of translation at the elongation/termination steps and at the level of mRNA maturation/ degradation will be excluded. For a description of how translation is initiated in bacteria, see Chapter 2.

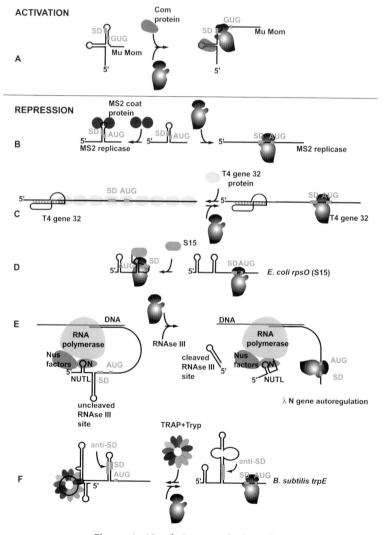

Figure 1. (*See facing page for legend.*)

PROTEIN-MEDIATED TRANSLATIONAL CONTROL

The vast majority of proteins regulating translation in bacteria are repressors, although phage studies have provided counterexamples (Fig. 1A) (Oppenheim et al. 1993; Hattman 1999). A characteristic of prokaryotic repressors is their frequent involvement in biochemical activities distinct from regulation. We begin by describing the class of translational regulators that exhibit dual roles and end with a few examples of dedicated regulators.

Direct Regulation by Major Biochemical Activity

In several cases, the major biochemical activity of a protein is directly utilized for regulation. For example, the coat protein of RNA phages in the MS2 and R17 family binds to a single site, which serves both to nucleate phage assembly and to repress the translation of the replicase cistron (Fig. 1B) (for review, see Witherell et al. 1991). The interaction between the coat protein and a stem-loop structure containing the replicase translation initiation site has been extensively studied (Witherell et al. 1991). The three-dimensional structure of this stem-loop structure either free

Figure 1. Different examples of protein-mediated translational regulation. (A) Positive regulation of bacteriophage Mu Mom gene by Com. Com binds to a hairpin located a few nucleotides upstream of the Shine-Dalgarno (SD) sequence and destabilizes a second hairpin containing the SD and the initiation codon, allowing the 30S subunit to bind and initiate translation (Hattman 1999). (B) Repression of the MS2/R17 replicase translation by the coat protein. (C) T4 gene 32 negative feedback regulation. The pseudoknot at the 5′ end of the operator (40 nucleotides upstream of the initiation codon) serves as a nucleation site for gp32 binding. Cooperative binding allows the protein to cover the mRNA from the pseudoknot to the RBS, inhibiting ribosome binding. (D) Negative feedback control of E. coli rpsO (S15). The S15 traps the translation initiation complex in an inactive form (see text). (E) Negative feedback control of bacteriophage λ N protein. The transcription complex (RNA polymerase, Nus factors, and N) blocks translation of N when the RNase III site is not cleaved. (F) TRAP-mediated translational regulation of B. subtilis trpE. Tryptophan-activated TRAP binds to the nascent RNA and causes the readthrough transcript to fold into a conformation occluding the SD sequence. In the absence of TRAP, the transcript adopts an alternative structure where the translation initiation site can bind the ribosome. In each of the panels, the heavy black line represents the mRNA and the gray object represents the 30S ribosomal subunit. Green boxes, the Shine-Dalgarno and AUG regions of the mRNA.

(Borer et al. 1995) or complexed to the coat protein dimer (Valegard et al. 1997) shows that the RNA–protein recognition relies on a few interactions and that the protein binds to the RNA by an induced-fit mechanism involving conformational changes of the RNA.

Another example of control by direct exploitation of the primary activity of the protein is the translation initiation factor IF3, which negatively controls its own synthesis (Butler et al. 1987). IF3 binds to the 30S subunit and inhibits translation initiation at codons other than AUG, GUG, or UUG (Sacerdot et al. 1996; Sussman et al. 1996). The initiation codon of IF3 is noncanonical, generally AUU. When the IF3 concentration increases, the number of 30S subunits bound to IF3 increases, causing a decrease in the translation of its own mRNA. One unique aspect of this regulation is that translational feedback is not due to IF3 binding to its own mRNA but rather to the ribosome.

A related example is ribosomal protein S1, which is involved in translation initiation in *E. coli* and whose gene is under translational feedback regulation (Boni et al. 2001). S1 recognizes AU-rich regions of the ribosomal binding site (RBS) of cellular mRNAs through its amino-terminal domain, whereas the carboxy-terminal domain of S1 binds the ribosome via protein–protein interactions (Boni et al. 1982). When S1 is in excess over ribosomes, free S1 preferentially binds to the AU-rich regions of its own RBS, most probably through the cooperative binding of several S1 proteins (Boni et al. 2001). Therefore, the AU-rich regions of this mRNA are no longer available to bind the ribosome-bound S1, inhibiting its translation initiation. Thus, the ability of S1 to bind AU-rich regions through its amino-terminal domain appears to be responsible for both its activity and regulation.

Regulation Only Partially Related to Primary Activity

In other cases, regulators are DNA- or RNA-binding proteins which recognize their primary and regulatory targets similarly. For example, the bacteriophage T4 gene 32 protein (gp32), which cooperatively binds to single-stranded DNA during replication, binds its own mRNA in a similar way, causing inhibition of translation (Fig. 1C) (for review, see Miller et al. 1994).

Many similar examples can be found concerning gene expression of the translational machinery of *E. coli*. The ribosomal protein (r-protein) genes are often organized as long operons (for review, see Zengel and Lindahl 1994). The expression of most of these operons is under negative translational feedback regulation. In general, one cistron of the operon encodes a primary rRNA-binding protein that also binds to its own mRNA, thereby inhibiting its translation. The binding of this regulatory r-protein takes place at the translational operator, generally located close to the RBS of the first cistron. The inhibition of translation of the first cistron is transmitted downstream by translational coupling (Chapter 8). Although the mechanism of coupling has been investigated only in a very limited number of cases, it seems that it is explained by the ability of translating ribosomes to open up the secondary structure that forms between the end of a cistron and the start of the next one (Lesage et al. 1992). Once opened, ribosome binding to the start of the next cistron is allowed. When the distance between the end of a cistron and the next cistron is short, or if the cistrons are overlapping, ribosomes might scan the intergenic region in either direction to reinitiate (Adhin and van Duin 1990). Nomura and coworkers proposed that r-protein synthesis is coupled to that of rRNA: If rRNA synthesis increases, it titrates all free regulatory r-proteins, causing a general derepression of r-protein synthesis (for review, see Zengel and Lindahl 1994). Nomura and coworkers also proposed that the regulatory r-proteins might recognize their rRNA and mRNA binding sites similarly. One such example is found in the *spc* operon, which is regulated by r-protein S8 (Cerretti et al. 1988; Wu et al. 1994). The three-dimensional structure of the protein within the ribosome and associated with its operator clearly indicates that S8 recognizes its mRNA- and rRNA-binding sites similarly (Merianos et al. 2004).

In other cases, mimicry appears to be rather limited. The translational operator of the gene encoding r-protein S15 folds into a pseudoknot (Philippe et al. 1990) that straddles the RBS of its own gene (Fig. 1D). S15 binds to the lower and upper stem of the pseudoknot, the latter carrying a C-G/G-U motif (Benard et al. 1998; Serganov et al. 2002). The S15-binding site of the 16S rRNA is also bipartite, composed of a three-way junction, with no equivalent on the mRNA, and a C-G/G-U motif. Although this motif is essential for regulation, it has only a limited role in S15 binding to 16S rRNA. At the present time, it is difficult to conclude that mimicry is a general feature in r-protein regulation. In cases with no obvious similarity, only detailed studies and three-dimensional structures of the regulatory complexes will give a definitive answer. Fortunately, the structure of the *E. coli* ribosome (Schuwirth et al. 2005) will facilitate further studies on this topic.

The regulatory scheme described above is not restricted to r-proteins. *E. coli* threonyl-tRNA synthetase (ThrRS) is also feedback-regulated (for review, see Romby and Springer 2003). The operator consists of four domains, two of which are recognized by the synthetase in a manner similar to that of the anticodon stem and loop of tRNA[Thr]. The crystal structure of ThrRS has been solved in complex with the main operator domain and with tRNA[Thr] (for review, see Romby and Springer 2003).

These structures confirm previous genetic and biochemical data showing that the recognition of the tRNA anticodon loop and the anticodon-like loop of the operator are identical.

In most of the cases described in this section, the repressor and 30S subunit compete for binding to the mRNA target. However, both the repressor and the 30S subunit can simultaneously bind to the mRNA in the case of the S15 and the *alpha* operons. In these cases, the 30S subunit is "trapped" in a dead-end preinitiation complex. Interestingly, the translation initiation regions of both mRNAs fold into a pseudoknot that is essential for control, suggesting that this type of mechanism requires specific mRNA structures capable of blocking translation. Repression by entrapment is more efficient than competition and can work when the repressor affinity or concentration would not suffice for competition, because the repressor only needs to increase the stability of an existing mRNA–ribosomal complex for entrapment, whereas competition requires inhibition of the de novo formation of an mRNA–ribosomal complex (Draper et al. 1998).

Translational Regulation by Complexes

In the cases reviewed above, repressors bind to their regulatory targets as either homopolymers (T4 gp32), preformed homodimers (R17 coat protein and *E. coli* ThrRS), or monomers (r-proteins). Regulation can also be performed by large complexes. Bacteriophage N protein is a transcriptional regulator that prevents transcription termination in the early λ operons. The antitermination complex consists of N and four host-encoded Nus factors that assemble with RNA polymerase on specific sites of the nascent RNA called NutL or NutR (for review, see Wilson et al. 2004). The complex consisting of the transcribing RNA polymerase, N, and Nus factors bound at the NutL site acts as a very efficient translational repressor of N expression. It has been proposed that the formation of a long hairpin, an RNase III recognition site, brings the RBS of N close to the antitermination complex assembled at NutL, thereby inhibiting translation (Fig. 1E). The Nun protein of phage HK022 also represses the translation of N by a similar mechanism (Kim et al. 2003). Both N- and Nun-mediated translational controls are effective only under slow growth conditions where RNase III is limiting in the cell (Wilson et al. 2002).

Translational Attenuation

The genes conferring resistance to chloramphenicol in gram-positive and -negative bacteria and to the macrolide, lincosamide, and streptogramin type B (MLS) antibiotics in gram-positive bacteria are induced at sublethal antibiotic concentrations by a mechanism called translational attenuation (Weisblum 1995; Lovett 1996). A similar regulatory mechanism has recently been characterized for the *secA* gene, which encodes an ATPase involved in the translocation of a major class of extracytoplasmic proteins across the inner membrane (for review, see Nakatogawa et al. 2004). The expression of *secA* is induced when the translocation activity is lowered. The *secA* gene is preceded by an open reading frame called *secM*. During *secM* translation, ribosomes stall five codons upstream of the *secM* termination codon, allowing the nascent membrane-inserted SecM protein to assess its own export status. If translocation is normal, the SecM protein is fully translated and the *secM-secA* intergenic region forms a stem-loop structure that occludes the RBS of *secA*. If, on the other hand, translocation is slowed down, ribosomes stall for a longer time at the end of *secM*, which opens the occluding secondary structure at the start of *secA* and allows its translation. Second-site mutations that partially alleviate pausing were found close to the constriction of the nascent peptide exit tunnel of the 50S subunit, either in the 23S rRNA or in r-protein L22. These experiments implicate a specific interaction between the nascent SecM protein and the exit tunnel. Interestingly, ribosome stalling at the end of *secM* has been shown to induce mRNA cleavage in the *secM-secA* message or in an artificial mRNA carrying the *secM* sequences (Collier et al. 2004; Sunohara et al. 2004). Thus, *secA* translation may be the consequence of mRNA cleavage and release of a translation-competent *secA* mRNA, rather than the opening of the inhibitory secondary structure. The stalling of ribosomes in the leader peptide of the MLS resistance genes *ermA* and *ermC* in *Bacillus subtilis* also results in mRNA cleavage (Drider et al. 2002).

Dedicated Translational Regulators

The first characterized protein whose sole function in bacteria is translational regulation was RegA of bacteriophage T4, which negatively regulates the translation of about a dozen early T4 mRNAs (including its own) to extremely variable extents (for review, see Miller et al. 1994). The RegA target sites are single-stranded AU-rich regions of varying length, always straddling the initiation codon. Competition between the RegA protein and the 30S ribosomal subunit for mRNA binding causes translational repression.

A second example of a dedicated translational regulator is the TRAP protein, originally characterized as a regulator of transcription termination in the *trp* operon of *B. subtilis* (Gollnick et al. 2005). TRAP protein is activated in the presence of tryptophan and binds to the nascent mRNA of the *trp* operon (Fig. 1F). This binding causes a rho-independent terminator to form. However, some RNA polymerase molecules escape termination and proceed into *trpE*. TRAP-bound transcripts that escape termination fold into a structure that sequesters the *trpE* translation initiation site (Fig. 1F). This translational repression causes a further 15-fold decrease in *trpE* expression on top of transcriptional regulation. The TRAP protein represses translation of three other genes by different mechanisms. The TRAP-binding site overlaps the RBS of two regulated mRNAs, namely, *trpP*, encoding a putative tryptophan transporter, and *trpG*, encoding a glutamine amidotransferase, and directly inhibits ribosome binding. For *ycbK* mRNA encoding a protein of unknown function, the TRAP-binding site is located within the coding sequence and inhibits translation by an unknown mechanism. The TRAP-binding site consists of multiple NAG repeats, where N is preferably G or U. The three-dimensional structure of TRAP has been solved in the presence of an RNA-binding site with 11 (G/U)AG repeats. The RNA wraps around the protein in such a way that each repeat interacts with a subunit of TRAP, forming a wheel-like structure (Fig. 1F).

In *E. coli*, CsrA regulates the utilization of carbon sources, glycogen synthesis, biofilm formation, and motility. In *Erwinia carotovora,* the homolog of CsrA, RsmA, controls the expression of extracellular enzymes and type III secretion. In *Pseudomonas fluorescens,* the two CsrA homologs, RsmA and RsmE, regulate the synthesis of extracellular secondary metabolites (Romeo 1998; Kay et al. 2005) and in *Vibrio cholerae,* CsrA regulates quorum sensing (Lenz et al. 2005). Regulation by CsrA can either be positive or negative; in *E. coli,* the expression of more than 400 genes is either directly or indirectly affected, indicating that CsrA is a global regulator (Wang et al. 2005). *E. coli* CsrA binds to the translation initiation regions of the target mRNAs at two to six sites (Wang et al. 2005). Although the sequence of CsrA-binding sites varies considerably, they all contain a highly conserved GGA triplet. In all cases studied, the binding of CsrA affects translation by competing with ribosome binding. Interestingly, the regulatory effects of CsrA are modulated by noncoding RNAs called CsrB/CsrC in *E. coli* (see below).

TRANS-ACTING RNA-MEDIATED TRANSLATIONAL CONTROL

It has been known for a long time that antisense RNAs (asRNAs) regulate essential functions of extrachromosomal genetic elements in bacteria, mainly plasmids, phages, and transposons (Wagner et al. 2002). In all cases, the asRNA and the target mRNA are transcribed in opposite orientations from the same DNA sequence and, hence, are fully complementary. More recently, diverse experimental approaches (Wagner and Vogel 2005) have led to the identification of more than 80 small noncoding RNAs (srnas, not to be confused with the old nomenclature for tRNAs) that originate from the bacterial chromosome and are likely to be bona fide regulators. Several of these novel sRNAs control translation by base-pairing at or near the RBS of target mRNAs (Storz et al. 2005). Other sRNAs specifically titrate regulatory proteins, modifying the level of expression of their target genes. Examples representing the different categories of RNA regulators are described below. Aspects of the bacterial asRNAs and sRNAs are discussed in other reviews (Wagner et al. 2002; Gottesman 2005; Storz et al. 2005).

Fully Complementary Antisense RNAs

Different groups of low-copy-number bacterial plasmids use asRNAs to regulate their copy number through inhibition of the synthesis of the replication initiator protein (Wagner et al. 2002 and references therein). In many cases, binding to the target mRNA curtails translational initiation and/or allows the formation of a target for RNase III, causing degradation. Plasmid R1 (IncFI), for example, uses an asRNA called CopA as the main element of replication control. Its target, CopT, is located in the leader region of the *repA* mRNA, encoding the replication initiator protein RepA (Light and Molin 1981). The synthesis of RepA is translationally coupled to that of a short leader peptide (*tap*) located upstream of *repA*. The coupling is mediated by a stable stem-loop structure, which sequesters the *repA* translation initiation site when *tap* is not translated (Blomberg et al. 1994). CopA binds to CopT just two nucleotides upstream of the *tap* Shine-Dalgarno sequence. This complex forms an unusual structure sufficient to directly inhibit translation initiation of the *tap* leader peptide and indirectly that of RepA (Kolb et al. 2000). The CopA–CopT complex is also cleaved by RNase III, although this second event has only a minor effect on repression efficiency (Blomberg et al. 1990).

Repression by plasmid-encoded asRNAs is extremely efficient and mainly depends on the association rate constant for asRNA–mRNA complex formation. In addition, many of these systems cause translation inhibition without the formation of a fully extended duplex (Wagner et al. 2002). Instead, rapidly formed and stable binding intermediates suffice to repress translation initiation. The recognition pathway differs considerably from system to system and appears to be dependent on the RNA structure. For the most efficient asRNAs, initial recognition involves a restricted number of intermolecular base-pairings between two complementary hairpin loops (kissing complex). These initial contacts are then rapidly followed by a stepwise progression of intermolecular interactions that lead to the formation of stable complexes (Wagner et al. 2002).

srNAs and Regulatory Cascades

Chromosomally encoded srNAs can interact with numerous target mRNAs, forming imperfect duplexes. In addition, many of them regulate the expression of mRNAs encoding global regulatory proteins such as transcription factors, thereby indirectly affecting the expression of many genes via regulatory cascades. Furthermore, a single mRNA can be regulated by multiple srNAs, thus subjecting the target to different environmental signals. The transcription of many srNAs is growth-phase-dependent and their promoters are highly regulated (Gottesman 2005). In *E. coli*, srNAs mediate responses to environmental signals such as iron starvation, temperature changes, oxidative and osmotic stress, and cell density. More recently, srNAs have been found as key regulators of genes involved in pathogenicity (Romby et al. 2006).

srNAs as Negative Regulators

In *E. coli*, the regulatory protein Fur (Ferric uptake regulator) senses iron concentration and represses the transcription of many genes involved in iron acquisition. The RyhB srNA whose transcription is Fur-dependent affects the expression of at least 18 operons in response to iron limitation, thus achieving coordinated regulation of a class of target genes involved in iron homeostasis (Masse et al. 2005). The binding site overlaps the RBS of its target mRNAs (Geissmann and Touati 2004). Binding blocks translation initiation and, in the case of the *sodB* mRNA, triggers its rapid RNase-E-dependent degradation (Fig. 2A) (Masse et al. 2003). Binding of the RyhB srNA requires the Sm-like Hfq protein (Geissmann and Touati 2004). Interestingly, Hfq has different roles for different srNAs (Valentin-Hansen et al. 2004). It stabilizes some srNAs against RNase-E-dependent degradation when they are not in interaction with their target mRNAs, and it also mediates srNA interaction with target mRNAs by changing the conformation of either partner for efficient binding. Hfq also forms different ribonucleoprotein complexes with RNase E and *E. coli* srNAs such as RyhB and SgrS (see below), providing mechanistic clues as to how mRNA destabilization is mediated by srNAs (Morita et al. 2005). Rapid degradation of mRNA is accompanied by concomitant degradation of the srNA following hybridization (Masse et al. 2003). Thus, rapid depletion of the srNA pool occurs if synthesis has stopped, allowing instantaneous readaptation of the target mRNAs to the absence of the srNA.

    E. coli SgrS srNA, which is induced by the accumulation of glucose-6-phosphate, regulates glucose transport by binding to the RBS of *ptsG*

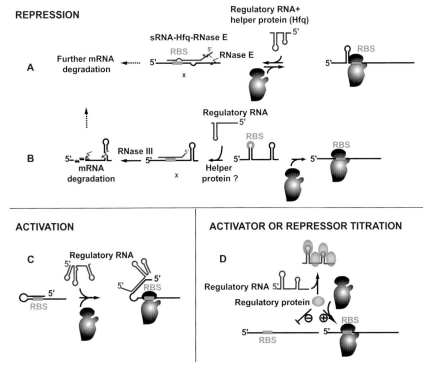

REPRESSION

ACTIVATION

ACTIVATOR OR REPRESSOR TITRATION

Figure 2. Examples of regulatory RNA-mediated controls. (*A,B*) Two cases of sRNA-mediated repression where both the sRNA and the target mRNA are directly cleaved by RNase E (*A*) or by RNase III, followed by other RNases (*B*). (*C*) Activation of translation as for DsrA/RprA and *E. coli rpoS*, or *S. aureus* RNAIII and *hla* gene encoding hemolysin α (see text). (*D*) Regulation by titration of a translational regulator such as *E. coli* CsrA by CsrB/C RNAs (see text). The objects or symbols are either labeled or defined in the legend to Fig. 1.

mRNA encoding the major glucose transporter. Like RyhB, SgrS sRNA binding causes translational repression and rapid RNase-E-dependent degradation of the target mRNA (Vanderpool and Gottesman 2004; Kawamoto et al. 2005). Interestingly, efficient repression requires localization of the mRNA to the inner membrane, coupled with the insertion of the nascent peptide into the membrane. This targeting is thought to reduce the access of ribosomes but not SgrS to the translation initiation region of *ptsG* mRNA (Kawamoto et al. 2005).

The three major outer-membrane protein genes (*ompA/C/F*) are also regulated by sRNAs that are induced under different stress conditions. MicA, MicC, and MicF bind functionally equivalent regions in their respective target mRNAs, and all of them require Hfq to repress translation in vivo and to induce rapid mRNA degradation (Aiba et al. 1987;

Chen et al. 2004; Rasmussen et al. 2005; Udekwu et al. 2005). An additional srna, RseX, was recently shown to repress the expression of both *ompA* and *ompC* mRNAs (Douchin et al. 2006). The *E. coli* srnas, OmrA and OmrB, also inhibit the expression of several genes encoding outer-membrane proteins, e.g., *ompT* (Guillier and Gottesman 2006). These data suggest that numerous srnas modulate and fine-tune the cell surface composition in *E. coli* in response to changes in the environment.

Some srnas that form more extended base pairings do not require Hfq protein, and, in several cases, RNase III initiates the degradation of the duplex formed (Fig. 2B). The short srna LstR inhibits the synthesis of an SOS-induced toxic peptide called TisB in *E. coli* (Vogel et al. 2004). Under normal conditions, LstR is present in large excess over its target *tisAB* mRNA and binds to an internal region of the mRNA. The duplex formed is cleaved by RNase III, producing an untranslatable form of the *tisB* mRNA. Under SOS conditions, the *tisAB* mRNA accumulates to levels above that of LstR, leading to the synthesis of the toxic peptide. Another example is *Staphylococcus aureus* RNAIII, which inhibits the synthesis of the main surface adhesin, protein A, encoded by the *spa* gene (Huntzinger et al. 2005). Binding of RNAIII sequesters the RBS of the *spa* mRNA and induces rapid mRNA degradation initiated by RNase III (Fig. 2B).

For many of these systems, the relative contributions of ribosome occlusion and mRNA degradation in the efficiency of control are still not well known. However, since the half-lives of bacterial mRNAs are strongly affected by their association with ribosomes, it is possible that acceleration of Hfq/RNase-E- or RNase-III-mediated degradation of target mRNAs is a consequence of srna-mediated translation inhibition, rather than being the primary control mechanism (Deana and Belasco 2005). Recent work supports this hypothesis, since srna-dependent translational repression occurs in the absence of RNase-E-dependent mRNA destabilization in the case of SgrS and RyhB srnas acting on *ptsG* and *sodB* (the latter encoding superoxide dismutase), respectively (Morita et al. 2006). Degradation of target mRNAs may then serve to render the arrest of translation irreversible.

srNAs as Positive Regulators

In some cases, srnas promote the conversion of a translationally inert mRNA to an active conformation (Fig. 2C). Two srnas, DsrA and RprA, induced by cold and osmotic stress, respectively, activate the translation of the gene encoding the stationary phase factor σ^S (*rpoS*) in *E. coli* (Repoila et al. 2003). Binding of DsrA to the long 5′-untranslated region (5′UTR) of the *rpoS* mRNA prevents the formation of an inhibitory structure that sequesters the RBS (Fig. 2C). Recognition of *rpoS* mRNA is stimulated by Hfq by a mechanism that is not well understood (Sledjeski et al. 2001). DsrA might also exert its action during *rpoS* transcription to kinetically trap the translationally active form of the mRNA.

In addition to acting as a negative regulator, *S. aureus* RNAIII is a positive regulator of hemolysin α synthesis (Novick et al. 1993). Like DsrA, the binding of the 5′ end of RNAIII to the hemolysin α (*hla*) mRNA prevents the formation of an untranslatable form of the mRNA (Novick et al. 1993; Morfeldt et al. 1995).

srNA Regulation by Titration of Regulatory Proteins

Several srNAs regulate gene expression indirectly by titrating regulatory proteins (Fig. 2D) (Romeo 1998). Two noncoding RNAs of *E. coli*, CsrB and CsrC, bind to the regulatory protein CsrA, which controls the expression of many genes at the translational level (see above). Although the lengths of CsrB and CsrC RNAs are different, they each carry several purine-rich repeats (5′-CAGGA(U/A/C)G-3′). These repeats mimic the CsrA-binding sites present in the 5′UTRs of target mRNAs (Baker et al. 2002). *E. coli* CsrB RNA contains 18 repeats (CsrC RNA has only 9) and binds 18 CsrA molecules cooperatively, thus sequestering the protein within a compact ribonucleoprotein complex. Orthologs of CrsA and CsrB RNA have been found in several human pathogens such as *Pseudomonas aeruginosa*, *Salmonella enterica*, and *V. cholerae* (Romby et al. 2006 and references therein).

CIS-MRNA ELEMENTS AS DIRECT SENSORS

One of the most fascinating recent discoveries in the field of bacterial gene expression is that mRNAs can function as direct sensors of both the physical and metabolic environment of the cell in the absence of any macromolecular *trans*-acting factors such as proteins or srNAs.

Thermosensor mRNAs

The first evidence for temperature-mediated gene regulation via alternative mRNA structures was provided by the bacteriophage λ *c*III gene (Altuvia et al. 1989). Later, the synthesis of several cellular heat-shock or cold-shock chaperones was shown to be regulated by temperature-induced conformational changes of their mRNAs (Fig. 3A) (Nagai et al. 1991; Narberhaus 2002). Several pathogenic bacteria also exploit mRNA thermosensors to turn on virulence functions under favorable conditions. Thermoregulation of virulence functions was first shown in *Yersinia pestis* (Hoe et al. 1992) and later in *Listeria monocytogenes* (Johansson et al. 2002). In both cases, the gene for a transcriptional activator (LcrF in *Y. pestis* and PrfA in *L. monocytogenes*) of the synthesis of several virulence factors is transcribed at equal rates at 26°C and 37°C, whereas translation is thermally controlled. A secondary structure in the 5′ leader of the *prfA* mRNA of *L. monocytogenes* that inhibits translation at low temperature by sequestering the RBS melts at temperatures above 37°C, allowing virulence genes to be expressed (Johansson et al. 2002).

Metabolite-sensing mRNAs, the So-called Riboswitches

Regulatory regions of some mRNAs can also directly sense the intracellular concentration of small metabolites, such as vitamins (thiamine pyrophosphate [TPP], flavin mononucleotide [FMN], adenosylcobalamin), amino acids and their derivatives (*S*-adenosylmethionine [SAM], glycine, lysine, etc.), purine precursors (Lim et al. 2005; Winkler and Breaker 2005), and even divalent ions such as Mg^{2+} (Cromie et al. 2006). Riboswitches are characterized by highly structured domains that reside within the 5′UTRs of mRNAs and can adopt mutually exclusive alternative conformations (Fig. 3B,C). One of the two conformers is stabilized upon metabolite binding, thus regulating the expression of the downstream gene at either the transcriptional level or the translational level (Vitreschak et al. 2004; Winkler and Breaker 2005). Riboswitches consist of two functional domains, an "aptamer" domain, which carries the recognition determinants for binding the effector metabolite, and a so-called expression platform, which triggers changes in gene expression (Winkler and Breaker 2005). Repression of translation occurs via a conformational switch in which binding of the metabolite induces the formation of a hairpin structure that sequesters the RBS (Fig. 3B). An example of such a riboswitch is the conserved aptamer domain (*thi* box) found in the leader regions of the *Rhizobium etli thiCOGE* operon and the *E. coli thiM/C* genes. Binding of the *thi* box by TPP causes occlusion of the downstream RBS (Rodionov et al. 2002; Winkler et al. 2002).

Several types of riboswitches have been discovered recently. Binding of glucosamine-6-phosphate (GlcN6P) induces a site-specific metal-ion-dependent self-cleavage in the 5′UTR of the GlcN6P synthase (*glmS*) mRNA in gram-positive bacteria. This cleavage event leads to repression of *glmS* mRNA translation through an as yet unknown mechanism (Winkler et al. 2004). A glycine-dependent riboswitch activates the translation of mRNAs encoding proteins involved in glycine efflux and catab-

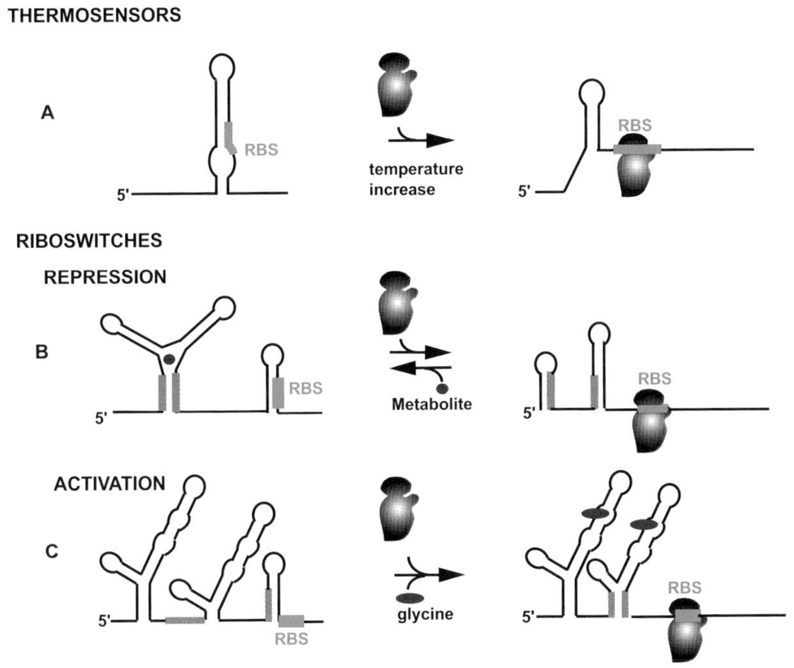

THERMOSENSORS

RIBOSWITCHES

REPRESSION

ACTIVATION

Figure 3. Thermosensors and riboswitches. In all cases, mRNA undergoes conformational switches, dependent on changes in temperature (*A*) or metabolite concentration (*B,C*). (*A*) At low temperatures, the mRNA (e.g., *L. monocytogenes prfA* gene) is not translated due to a hairpin structure which sequesters the RBS. At higher temperature, the structure melts, allowing translation initiation. (*B*) The *V. cholerae* VC1422 mRNA contains two nearly identical domains in the 5' leader region which bind glycine in a cooperative manner (see text for details). (*C*) Binding of metabolites such as thiamine pyrophosphate or adenosyl-cobalamin to mRNA occluding the downstream RBS by promoting the formation of inhibitory structures. The objects or symbols are either labeled or defined in the legend to Fig. 1.

olism in *V. cholerae* (Fig. 3C) (Mandal et al. 2004). The 5' leader of this mRNA contains two adjacent and nearly identical structural domains. In the absence of glycine, the RBS is embedded in a secondary structure that inhibits translation. In the presence of an excess of glycine, one glycine molecule binds to the upstream domain and increases the affinity of the downstream binding site for a second glycine through cooperative interactions. This leads to the disruption of the inhibitory helix, allowing ribosome access and translation. Finally, a novel SAM riboswitch in lactic acid bacteria regulates the translation of SAM synthetase. Binding of SAM is dependent on the formation of a pseudoknot structure sequestering the Shine-Dalgarno sequence within the core element (Fuchs et al.

2006). In this case, the aptamer domain and the expression platform are structurally related.

Three-dimensional crystal structures have recently been solved for the *B. subtilis xpt* aptamer bound to either hypoxanthine (Batey et al. 2004) or guanine, as well as for the *Vibrio vulnificus add* 5'UTR bound to adenine (Serganov et al. 2004). Despite their sequence differences, the three RNAs fold into a similar structure characterized by a loop–loop interaction and a side-by-side helical alignment. Binding of the purine base induces conformational changes in the binding pocket, and specificity is provided by interactions with the Watson-Crick edge of the purine. These structures reveal that part of the antiterminator structure becomes embedded in the aptamer fold, explaining why terminator formation is favored upon metabolite binding. Although these riboswitches control transcription attenuation, the same concepts can be applied to translational control.

Regulation by riboswitches is kinetically driven in that the metabolite-dependent genetic switch depends in part on the speed of RNA transcription (Wickiser et al. 2005). Both the association rate constant of FMN binding to the *B. subtilis ribD* riboswitch and RNA polymerase processivity are determining factors for efficient transcription attenuation in vitro. Rapid binding of the metabolite is also expected to be essential for translational control.

PHYLOGENIC CONSERVATION OF REGULATION

The two model bacterial organisms, *E. coli* and *B. subtilis,* diverged about 10^9 years ago and offer a good opportunity to evaluate the conservation of regulation in distantly related organisms. Orthologous sets of genes, such as those involved in r-protein synthesis and in tryptophan biosynthesis, for example, are not only well-conserved between these two bacteria, but also respond to similar stimuli; tryptophan biosynthesis genes are repressed by tryptophan (Gollnick et al. 2005) and r-protein synthesis is negatively autoregulated, at least in some cases (Grundy and Henkin 1992). However, the regulatory mechanisms are widely divergent between *E. coli* and *B. subtilis* and are only conserved among closely related species. Tryptophan biosynthesis is regulated at the transcriptional level in *E. coli* and at both the transcriptional and translational levels in *B. subtilis* by very different mechanisms. Similarly, the mechanism of autogenous control for r-proteins characterized in *E. coli* seems to be found only in γ-proteobacteria for the S10, *spc* (Allen et al. 2004), L35/L20 (Guillier et al. 2005) operons and the *rpsA* gene (Tchufistova et al. 2003). Although restricted conservation of regulatory mechanisms seems to be the rule, *Thermus thermophilus* S15 (Serganov et al. 2003), *Bacillus stearothermophilus* S15 (Scott and Williamson 2005), and more surprisingly, the archaeal *Methanococcus vannielii* L1 (Kraft et al. 1999) have been shown to bind their own mRNA close to their RBS, suggesting that translational autoregulation might operate in distant eubacterial species and even in archaea.

Numerous srNAs involved in stress responses in *E. coli* are also conserved in closely related bacteria, a criterion that has been used to select novel srNAs by computational screens (Wagner and Vogel 2005). However, this conservation is limited even in closely related species, because srNAs involved in virulence gene expression are usually species-specific.

The situation with riboswitches is exceptional in that the aptamer domain that binds a specific metabolite is often conserved in bacteria as distant as *E. coli* and *B. subtilis*. Interestingly enough, a thiamine-pyrophosphate-binding aptamer domain was found adjacent to putative splice sites in plants and fungal species, indicating that such domains might have a biological role in eukaryotes (Sudarsan et al. 2005). However, the expression platform associated with a specific aptamer domain varies in some cases from translation inhibition to transcription termination platforms in *E. coli* and *B. subtilis*, respectively. The fact that aptamer-binding domains are conserved between very distant species indicates that they existed more than 10^9 years ago and might constitute relics from primordial organisms (Vitreschak et al. 2004).

PARALLELS BETWEEN PROKARYOTIC AND EUKARYOTIC REGULATION

Although there are some indications that riboswitches exist in plants (see above), it remains to be seen whether "direct" regulatory mechanisms are widespread in higher organisms. Regarding protein- and RNA-mediated regulation, it appears that there are both similarities and major differences between bacteria and eukaryotes. One of the major differences is that regulation through initiation factor modification is widespread in eukaryotes, but does not seem to have a major role in bacteria. Most of the other key differences between prokaryotic and eukaryotic translational control mechanisms are due to differences in their initiation mechanisms, e.g., ribosome scanning and the part played by eukaryotic 3'UTRs. However, some common features exist. For instance, short upstream open reading frames have an important role in translational control, both in prokaryotes (translational attenuation) and in many eukaryotic genes, from viruses to mammals (Geballe and Sachs 2000; Sachs and Geballe 2006; Chapter 9). Most of the prokaryotic protein-mediated regulatory mechanisms are based on the recognition of sequences in the 5'UTR of the target mRNA by a specific regulator (see above). An example in eukaryotes concerns the ferritin gene, whose mRNA contains a hairpin in its 5'UTR that is recognized by the iron regulatory protein, allowing translation to be repressed in iron-depleted cells (Gebauer and Hentze 2004), although this mechanism of regulation is very rare in eukaryotes.

In eukaryotes, the translation of mRNAs encoding functionally related proteins can be regulated in a coordinated manner through the binding of specific RNA-binding proteins. By analogy with bacterial regulons, the genes subject to this kind of regulation have been described as forming posttranscriptional regulons (Keene and Lager 2005). Interestingly, small noncoding eukaryotic RNAs, the so-called microRNAs, also regulate multiple mRNAs; one mRNA can be targeted by several microRNAs and one microRNA can regulate several mRNAs (Chapter 11). In prokaryotes, global protein regulators such as CsrA and many sRNAs control the translation of several different mRNAs, and some mRNAs, such as those encoding the stationary phase sigma factor σ^S, are under the control of multiple sRNAs. Thus, mRNAs subject to multiple translational controls exist in both eukaryotes and prokaryotes.

CONCLUDING REMARKS AND FUTURE PROSPECTS

Despite the large number of genes under translational control, transcription remains the main target of regulation in phage and bacteria, where genes of related functions are often expressed in operons. However, in transcriptionally controlled operons, translational control of individual cistrons allows differential regulation within the operon and may increase the efficiency of control of a subset of cistrons. Translational control is also used for adaptive responses since it has the advantage of providing a fast response. The most striking feature of prokaryotic translational regulation is its mechanistic diversity. This feature is probably essential for the prokaryotic cell to adapt to the extremely large range of growth conditions it may face. The diversity comes from the varying nature of the regulators, which can range from small monomeric proteins or sRNAs to large multiprotein and ribonucleoprotein complexes. The most extreme case is that of an mRNA sensor, which has the obvious evolutionary advantage of functioning directly without any *trans*-acting regulator.

Recent progress in studying translation regulation concerns the discovery of numerous sRNAs and riboswitches. Are new classes of regulatory RNAs still to be discovered in prokaryotes? Possibly, since many stable 5'UTRs of bacterial mRNAs accumulate in bacteria, raising the question of their potential regulatory role *in trans* on other unidentified genes (Vogel et al. 2003, Kawano et al. 2005). In addition, some sRNA candidates carry short open reading frames that are suspected to produce a protein of as yet unknown function. Although the regulation of whole operons by a protein regulator is rather common in prokaryotes, it is surprising that no such example has been described for sRNAs to date.

One feature common to all translational regulatory networks is that structural switches in mRNA are critical for the recognition of proteins, sRNAs, or metabolites in preference to (or in addition to, in the case of trapping) the ribosome. Thus, translational regulation, perhaps more than any other RNA-mediated biological process, underlines the importance of the plasticity of mRNA. An important aspect, often neglected, is the fact that sequential formation of RNA interactions during transcription ultimately determines the functional state of the nascent mRNA (Heilman-Miller and Woodson 2003). This is particularly important in prokaryotes where transcription and translation are coupled. The rate of RNA elongation and the extent of pausing may also alter the distribution of the folding intermediates by limiting the conversion of one structure to another. Thus, kinetically trapped intermediates can create a time window for gene regulation and can be influenced by *trans*-acting factors. Future studies of the kinetics and tertiary structures of mRNA and regulatory complexes will provide an urgently needed basis for addressing fundamental questions in translational control.

ACKNOWLEDGMENTS

We thank members of our research groups for discussions, and Ciarán Condon and Marc Dreyfus for critical reading of the manuscript. M.S. and P.R. are grateful for financial support from the Centre National de la Recherche Scientifique (CNRS) and from the Agence Nationale de la Recherche (ANR), M.S. from the Ministère de la Recherche (AC DRAB), and P.R. from the European Community (FOSRAK EC005120, BacRNA EC).

REFERENCES

Adhin M.R. and van Duin J. 1990. Scanning model for translational reinitiation in eubacteria. *J. Mol. Biol.* **213:** 811–818.

Aiba H., Matsuyama S., Mizuno T., and Mizushima S. 1987. Function of micF as an antisense RNA in osmoregulatory expression of the ompF gene in *Escherichia coli*. *J. Bacteriol.* **169:** 3007–3012.

Allen T.D., Watkins T., Lindahl L., and Zengel J.M. 2004. Regulation of ribosomal protein synthesis in *Vibrio cholerae*. *J. Bacteriol.* **186:** 5933–5937.

Altuvia S., Kornitzer D., Teff D., and Oppenheim A.B. 1989. Alternative mRNA structures of the cIII gene of bacteriophage l determine the rate of its translation initiation. *J. Mol. Biol.* **210:** 265–280.

Baker C.S., Morozov I., Suzuki K., Romeo T., and Babitzke P. 2002. CsrA regulates glycogen biosynthesis by preventing translation of *glgC* in *Escherichia coli. Mol. Microbiol.* **44:** 1599–1610.

Batey R.T., Gilbert S.D., and Montange R.K. 2004. Structure of a natural guanine-responsive riboswitch complexed with the metabolite hypoxanthine. *Nature* **432:** 411–415.

Benard L., Mathy N., Grunberg-Manago M., Ehresmann B., Ehresmann C., and Portier C. 1998. Identification in a pseudoknot of a U·G motif essential for the regulation of the expression of ribosomal protein S15. *Proc. Natl. Acad. Sci.* **95:** 2564–2567.

Blomberg P., Wagner E.G., and Nordstrom K. 1990. Control of replication of plasmid R1: The duplex between the antisense RNA, CopA, and its target, CopT, is processed specifically in vivo and in vitro by RNase III. *EMBO J.* **9:** 2331–2340.

Blomberg P., Engdahl H.M., Malmgren C., Romby P., and Wagner E.G. 1994. Replication control of plasmid R1: Disruption of an inhibitory RNA structure that sequesters the repA ribosome-binding site permits tap-independent RepA synthesis. *Mol. Microbiol.* **12:** 49–60.

Boni I.V., Zlatkin I.V., and Budowsky E.I. 1982. Ribosomal protein S1 associates with *Escherichia coli* ribosomal 30-S subunit by means of protein-protein interactions. *Eur. J. Biochem.* **121:** 371–376.

Boni I.V., Artamonova V.S., Tzareva N.V., and Dreyfus M. 2001. Non-canonical mechanism for translational control in bacteria: Synthesis of ribosomal protein S1. *EMBO J.* **20:** 4222–4232.

Borer P.N., Lin Y., Wang S., Roggenbuck M.W., Gott J.M., Uhlenbeck O.C., and Pelczer I. 1995. Proton NMR and structural features of a 24-nucleotide RNA hairpin. *Biochemistry* **34:** 6488–6503.

Butler J.S., Springer M., and Grunberg-Manago M. 1987. AUU-to-AUG mutation in the initiator codon of the translation initiation factor IF3 abolishes translational autocontrol of its own gene (infC) in vivo. *Proc. Natl. Acad. Sci.* **84:** 4022–4025.

Cerretti D.P., Mattheakis L.C., Kearney K.R., Vu L., and Nomura M. 1988. Translational regulation of the spc operon in *Escherichia coli*. Identification and structural analysis of the target site for S8 repressor protein. *J. Mol. Biol.* **204:** 309–329.

Chen S., Zhang A., Blyn L.B., and Storz G. 2004. MicC, a second small-RNA regulator of Omp protein expression in *Escherichia coli. J. Bacteriol.* **186:** 6689–6697.

Collier J., Bohn C., and Bouloc P. 2004. SsrA tagging of *Escherichia coli* SecM at its translation arrest sequence. *J. Biol. Chem.* **279:** 54193–54201.

Cromie M.J., Shi Y., Latifi T., and Groisman E.A. 2006. An RNA sensor for intracellular mg(2+). *Cell* **125:** 71–84.

Deana A. and Belasco J.G. 2005. Lost in translation: The influence of ribosomes on bacterial mRNA decay. *Genes Dev.* **19:** 2526–2533.

Douchin V., Bohn C., and Bouloc P. 2006. Down-regulation of porins by a small RNA bypasses the essentiality of the RIP protease RseP in *E. coli. J. Biol. Chem.* **281:** 12253–12259.

Draper D.E., Gluick T.C., and Schlax P.J. 1998. Pseudoknots, RNA folding, and translational regulation. In *RNA structure and function* (ed. R.W. Simons and M. Grunberg-Manago), pp. 415–436. Cold Spring Harbor Laboratory Press, Cold Spring Harbor, New York.

Drider D., DiChiara J.M., Wei J., Sharp J.S., and Bechhofer D.H. 2002. Endonuclease cleavage of messenger RNA in *Bacillus subtilis*. *Mol. Microbiol.* **43:** 1319–1329.

Fuchs R.T., Grundy F.J., and Henkin T.M. 2006. The S(MK) box is a new SAM-binding RNA for translational regulation of SAM synthetase. *Nat. Struct. Mol. Biol.* **13:** 226–233.

Geballe A.P. and Sachs M.S. 2000. Translational control by upstream open reading frames. In *Translational control of gene expression* (ed. N. Sonenberg et al.), pp. 595–614. Cold Spring Harbor Laboratory Press, Cold Spring Harbor, New York.

Gebauer F. and Hentze M.W. 2004. Molecular mechanisms of translational control. *Nat. Rev. Mol. Cell Biol.* **5:** 827–835.

Geissmann T.A. and Touati D. 2004. Hfq, a new chaperoning role: Binding to messenger RNA determines access for small RNA regulator. *EMBO J.* **23:** 396–405.

Gollnick P., Babitzke P., Antson A., and Yanofsky C. 2005. Complexity in regulation of tryptophan biosynthesis in *Bacillus subtilis*. *Annu. Rev. Genet.* **39:** 47–68.

Gottesman S. 2005. Micros for microbes: Non-coding regulatory RNAs in bacteria. *Trends Genet.* **21:** 399–404.

Grundy F.J. and Henkin T.M. 1992. Characterization of the *Bacillus subtilis* rpsD regulatory target site. *J. Bacteriol.* **174:** 6763–6770.

Guillier M. and Gottesman S. 2006. Remodelling of the *Escherichia coli* outer membrane by two small regulatory RNAs. *Mol. Microbiol.* **59:** 231–247.

Guillier M., Allemand F., Dardel F., Royer C.A., Springer M., and Chiaruttini C. 2005. Double molecular mimicry in *Escherichia coli*: Binding of ribosomal protein L20 to its two sites in mRNA is similar to its binding to 23S rRNA. *Mol. Microbiol.* **56:** 1441–1456.

Hattman S. 1999. Unusual transcriptional and translational regulation of the bacteriophage Mu mom operon. *Pharmacol. Ther.* **84:** 367–388.

Heilman-Miller S.L. and Woodson S.A. 2003. Effect of transcription on folding of the *Tetrahymena* ribozyme. *RNA* **9:** 722–733.

Hoe N.P., Minion F.C., and Goguen J.D. 1992. Temperature sensing in *Yersinia pestis*: Regulation of *yopE* transcription by *lcrF*. *J. Bacteriol.* **174:** 4275–4286.

Huntzinger E., Boisset S., Saveanu C., Benito Y., Geissmann T., Namane A., Lina G., Etienne J., Ehresmann B., Ehresmann C., et al. 2005. *Staphylococcus aureus* RNAIII and the endoribonuclease III coordinately regulate spa gene expression. *EMBO J.* **24:** 824–835.

Jacob F. and Monod J. 1961. Genetic regulatory mechanisms in the synthesis of proteins. *J. Mol. Biol.* **3:** 318–356.

Johansson J., Mandin P., Renzoni A., Chiaruttini C., Springer M., and Cossart P. 2002. An RNA thermosensor controls expression of virulence genes in *Listeria monocytogenes*. *Cell* **110:** 551–561.

Kawamoto H., Morita T., Shimizu A., Inada T., and Aiba H. 2005. Implication of membrane localization of target mRNA in the action of a small RNA: Mechanism of post-transcriptional regulation of glucose transporter in *Escherichia coli*. *Genes Dev.* **19:** 328–338.

Kawano M., Reynolds A.A., Miranda-Rios J., and Storz G. 2005. Detection of 5′- and 3′-UTR-derived small RNAs and cis-encoded antisense RNAs in *Escherichia coli*. *Nucleic Acids Res.* **33:** 1040–1050.

Kay E., Dubuis C., and Haas D. 2005. Three small RNAs jointly ensure secondary metabolism and biocontrol in *Pseudomonas fluorescens* CHA0. *Proc. Natl. Acad. Sci.* **102:** 17136–17141.

Keene J.D. and Lager P.J. 2005. Post-transcriptional operons and regulons co-ordinating gene expression. *Chromosome Res.* **13:** 327–337.

Kim H.C., Zhou J.G., Wilson H.R., Mogilnitskiy G., Court D.L., and Gottesman M.E. 2003. Phage HK022 Nun protein represses translation of phage lambda N (transcription termination/translation repression). *Proc. Natl. Acad. Sci.* **100:** 5308–5312.

Kolb F.A., Malmgren C., Westhof E., Ehresmann C., Ehresmann B., Wagner E.G., and Romby P. 2000. An unusual structure formed by antisense-target RNA binding involves an extended kissing complex with a four-way junction and a side-by-side helical alignment. *RNA* **6:** 311–324.

Kraft A., Lutz C., Lingenhel A., Grobner P., and Piendl W. 1999. Control of ribosomal protein L1 synthesis in mesophilic and thermophilic archaea. *Genetics* **152:** 1363–1372.

Lenz D.H., Miller M.B., Zhu J., Kulkarni R.V., and Bassler B.L. 2005. CsrA and three redundant small RNAs regulate quorum sensing in *Vibrio cholerae*. *Mol. Microbiol.* **58:** 1186–1202.

Lesage P., Chiaruttini C., Graffe M., Dondon J., Milet M., and Springer M. 1992. Messenger RNA secondary structure and translational coupling in the *Escherichia coli* operon encoding translation initiation factor IF3 and the ribosomal proteins, L35 and L20. *J. Mol. Biol.* **228:** 366–386.

Light J. and Molin S. 1981. Replication control functions of plasmid R1 act as inhibitors of expression of a gene required for replication. *Mol. Gen. Genet.* **184:** 56–61.

Lim J., Winkler W.C., Nakamura S., Scott V., and Breaker R.R. 2005. Molecular-recognition characteristics of SAM-binding riboswitches. *Angew. Chem. Int. Ed. Engl.* **45:** 964–968.

Lodish H.F., Cooper S., and Zinder N.D. 1964. Host-dependent mutants of the bacteriophage f2. IV. On the biosynthesis of viral polymerase. *Virology* **24:** 60–70.

Lovett P.S. 1996. Translation attenuation regulation of chloramphenicol resistance in bacteria: A review. *Gene* **179:** 157–162.

Mandal M., Lee M., Barrick J.E., Weinberg Z., Emilsson G.M., Ruzzo W.L., and Breaker R.R. 2004. A glycine-dependent riboswitch that uses cooperative binding to control gene expression. *Science* **306:** 275–279.

Masse E., Escorcia F.E., and Gottesman S. 2003. Coupled degradation of a small regulatory RNA and its mRNA targets in *Escherichia coli*. *Genes Dev.* **17:** 2374–2383.

Masse E., Vanderpool C.K., and Gottesman S. 2005. Effect of RyhB small RNA on global iron use in *Escherichia coli*. *J. Bacteriol.* **187:** 6962–6971.

Merianos H.J., Wang J., and Moore P.B. 2004. The structure of a ribosomal protein S8/spc operon mRNA complex. *RNA* **10:** 954–964.

Miller E.S., Karam J.D. and Spicer E. 1994. Control of translation initiation: mRNA structure and protein repressors. In *Molecular biology of bacteriophage T4* (ed. J.D. Karam et al.), pp. 193–205. American Society for Microbiology, Washington, D.C.

Mizuno T., Chou M.Y., and Inouye M. 1984. A unique mechanism regulating gene expression: Translational inhibition by a complementary RNA transcript (micRNA). *Proc. Natl. Acad. Sci.* **81:** 1966–1970.

Morfeldt E., Taylor D., von Gabain A., and Arvidson S. 1995. Activation of alpha-toxin translation in *Staphylococcus aureus* by the trans-encoded antisense RNA, RNAIII. *EMBO J.* **14:** 4569–4577.

Morita T., Maki K., and Aiba H. 2005. RNase E-based ribonucleoprotein complexes: Mechanical basis of mRNA destabilization mediated by bacterial noncoding RNAs. *Genes Dev.* **19:** 2176–2186.

Morita T., Mochizuki Y., and Aiba H. 2006. Translational repression is sufficient for gene silencing by bacterial small noncoding RNAs in the absence of mRNA destruction. *Proc. Natl. Acad. Sci.* **103:** 4858–4863.

Nagai H., Yuzawa H., and Yura T. 1991. Interplay of two cis-acting mRNA regions in translational control of s32 synthesis during the heat shock response of *Escherichia coli. Proc. Natl. Acad. Sci.* **88:** 10515–10519.

Nakatogawa H., Murakami A., and Ito K. 2004. Control of SecA and SecM translation by protein secretion. *Curr. Opin. Microbiol.* **7:** 145–150.

Narberhaus F. 2002. mRNA-mediated detection of environmental conditions. *Arch. Microbiol.* **178:** 404–410.

Novick R.P., Ross H.F., Projan S.J., Kornblum J., Kreiswirth B., and Moghazeh S. 1993. Synthesis of staphylococcal virulence factors is controlled by a regulatory RNA molecule. *EMBO J.* **12:** 3967–3975.

Oppenheim A.B., Kornitzer D., Altuvia S., and Court D.L. 1993. Posttranscriptional control of the lysogenic pathway in bacteriophage lambda. *Prog. Nucleic Acid Res. Mol. Biol.* **46:** 37–49.

Philippe C., Portier C., Mougel M., Grunberg-Manago M., Ebel J.P., Ehresmann B., and Ehresmann C. 1990. Target site of *Escherichia coli* ribosomal protein S15 on its messenger RNA. Conformation and interaction with the protein. *J. Mol. Biol.* **211:** 415–426.

Rasmussen A.A., Eriksen M., Gilany K., Udesen C., Thomas F., Petersen C., and Valentin-Hansen P. 2005. Regulation of ompA mRNA stability: The role of a small regulatory RNA in growth phase-dependent control. *Mol. Microbiol.* **58:** 1421–1429.

Repoila F., Majdalani N., and Gottesman S. 2003. Small non-coding RNAs, co-ordinators of adaptation processes in *Escherichia coli:* The RpoS paradigm. *Mol. Microbiol.* **48:** 855–861.

Rodionov D.A., Vitreschak A.G., Mironov A.A., and Gelfand M.S. 2002. Comparative genomics of thiamin biosynthesis in procaryotes. New genes and regulatory mechanisms. *J. Biol. Chem.* **277:** 48949–48959.

Romby P. and Springer M. 2003. Bacterial translational control at atomic resolution. *Trends Genet.* **19:** 155–161.

Romby P., Vandenesch F., and Wagner E.G. 2006. The role of RNAs in the regulation of virulence-gene expression. *Curr. Opin. Microbiol.* **9:** 229–236.

Romeo T. 1998. Global regulation by the small RNA-binding protein CsrA and the noncoding RNA molecule CsrB. *Mol. Microbiol.* **29:** 1321–1330.

Sacerdot C., Chiaruttini C., Engst K., Graffe M., Milet M., Mathy N., Dondon J., and Springer M. 1996. The role of the AUU initiation codon in the negative feedback regulation of the gene for translation initiation factor IF3 in *Escherichia coli. Mol. Microbiol.* **21:** 331–346.

Sachs M. and Geballe A. 2006. Downsream control of upstream open reading frames. *Genes Dev.* **20:** 915–921

Schuwirth B.S., Borovinskaya M.A., Hau C.W., Zhang W., Vila-Sanjurjo A., Holton J.M., and Cate J.H. 2005. Structures of the bacterial ribosome at 3.5 Å resolution. *Science* **310:** 827–834.

Scott L.G. and Williamson J.R. 2005. The binding interface between *Bacillus stearothermophilus* ribosomal protein S15 and its 5′-translational operator mRNA. *J. Mol. Biol.* **351:** 280–290.

Serganov A., Ennifar E., Portier C., Ehresmann B., and Ehresmann C. 2002. Do mRNA and rRNA binding sites of *E. coli* ribosomal protein S15 share common structural determinants? *J. Mol. Biol.* **320:** 963–978.

Serganov A., Polonskaia A., Ehresmann B., Ehresmann C., and Patel D.J. 2003. Ribosomal protein S15 represses its own translation via adaptation of an rRNA-like fold within its mRNA. *EMBO J.* **22:** 1898–1908.

Serganov A., Yuan Y.R., Pikovskaya O., Polonskaia A., Malinina L., Phan A.T., Hobartner C., Micura R., Breaker R.R., and Patel D.J. 2004. Structural basis for discriminative regulation of gene expression by adenine- and guanine-sensing mRNAs. *Chem. Biol.* **11:** 1729–1741.

Sledjeski D.D., Whitman C., and Zhang A. 2001. Hfq is necessary for regulation by the untranslated RNA DsrA. *J. Bacteriol.* **183:** 1997–2005.

Storz G., Altuvia S., and Wassarman K.M. 2005. An abundance of RNA regulators. *Annu. Rev. Biochem.* **74:** 199–217.

Sudarsan N., Cohen-Chalamish S., Nakamura S., Emilsson G.M., and Breaker R.R. 2005. Thiamine pyrophosphate riboswitches are targets for the antimicrobial compound pyrithiamine. *Chem. Biol.* **12:** 1325–1335.

Sunohara T., Jojima K., Yamamoto Y., Inada T., and Aiba H. 2004. Nascent-peptide-mediated ribosome stalling at a stop codon induces mRNA cleavage resulting in nonstop mRNA that is recognized by tmRNA. *RNA* **10:** 378–386.

Sussman J.K., Simons E.L., and Simons R.W. 1996. *Escherichia coli* translation initiation factor 3 discriminates the initiation codon in vivo. *Mol. Microbiol.* **21:** 347–360.

Tchufistova L.S., Komarova A.V., and Boni I.V. 2003. A key role for the mRNA leader structure in translational control of ribosomal protein S1 synthesis in gammaproteobacteria. *Nucleic Acids Res.* **31:** 6996–7002.

Udekwu K.I., Darfeuille F., Vogel J., Reimegård J., Holmqvist E., and Wagner E.G.H. 2005. Hfq-dependent regulation of OmpA synthesis is mediated by an antisense RNA. *Genes Dev.* **19:** 2355–2366.

Valegard K., Murray J.B., Stonehouse N.J., van den Worm S., Stockley P.G., and Liljas L. 1997. The three-dimensional structures of two complexes between recombinant MS2 capsids and RNA operator fragments reveal sequence-specific protein-RNA interactions. *J. Mol. Biol.* **270:** 724–738.

Valentin-Hansen P., Eriksen M., and Udesen C. 2004. The bacterial Sm-like protein Hfq: A key player in RNA transactions. *Mol. Microbiol.* **51:** 1525–1533.

Vanderpool C.K. and Gottesman S. 2004. Involvement of a novel transcriptional activator and small RNA in post-transcriptional regulation of the glucose phosphoenolpyruvate phosphotransferase system. *Mol. Microbiol.* **54:** 1076–1089.

Vitreschak A.G., Rodionov D.A., Mironov A.A., and Gelfand M.S. 2004. Riboswitches: The oldest mechanism for the regulation of gene expression? *Trends Genet.* **20:** 44–50.

Vogel J., Argaman L., Wagner E.G.H., and Altuvia S. 2004. The small RNA IstR inhibits synthesis of an SOS-induced toxic peptide. *Curr. Biol.* **14:** 2271–2276.

Vogel J., Bartels V., Tang T.H., Churakov G., Slagter-Jager J.G., Huttenhofer A., and Wagner E.G. 2003. RNomics in *Escherichia coli* detects new sRNA species and indicates parallel transcriptional output in bacteria. *Nucleic Acids Res.* **31:** 6435–6443.

Wagner E.G.H. and Vogel J. 2005. Approaches to identify novel non-messenger RNAs in bacteria and to investigate their biological functions: Functional analysis of identified non-mRNAs. In *Handbook of RNA biochemistry* (ed. R.K. Hartmann et al.), pp. 614–654. Wiley-VCH, Weinheim, Germany.

Wagner E.G.H., Altuvia S., and Romby P. 2002. Antisense RNAs in bacteria and their genetic elements. *Adv. Genet.* **46:** 361–398.

Wang X., Dubey A.K., Suzuki K., Baker C.S., Babitzke P., and Romeo T. 2005. CsrA post-transcriptionally represses pgaABCD, responsible for synthesis of a biofilm polysaccharide adhesin of *Escherichia coli*. *Mol. Microbiol.* **56:** 1648–1663.

Weisblum B. 1995. Erythromycin resistance by ribosome modification. *Antimicrob. Agents Chemother.* **39:** 577–585.

Wickiser J.K., Winkler W.C., Breaker R.R., and Crothers D.M. 2005. The speed of RNA transcription and metabolite binding kinetics operate an FMN riboswitch. *Mol. Cell* **18:** 49–60.

Wilson H.R., Zhou J.G., Yu D., and Court D.L. 2004. Translation repression by an RNA polymerase elongation complex. *Mol. Microbiol.* **53:** 821–828.

Wilson H.R., Yu D., Peters H.K., III, Zhou J.G., and Court D.L. 2002. The global regulator RNase III modulates translation repression by the transcription elongation factor N. *EMBO J.* **21:** 4154–4161.

Winkler W.C. and Breaker R.R. 2005. Regulation of bacterial gene expression by riboswitches. *Annu. Rev. Microbiol.* **59:** 487–517.

Winkler W.C., Cohen-Chalamish S., and Breaker R.R. 2002. An mRNA structure that controls gene expression by binding FMN. *Proc. Natl. Acad. Sci.* **99:** 15908–15913.

Winkler W.C., Nahvi A., Roth A., Collins J.A., and Breaker R.R. 2004. Control of gene expression by a natural metabolite-responsive ribozyme. *Nature* **428:** 281–286.

Witherell G.W., Gott J.M., and Uhlenbeck O.C. 1991. Specific interaction between RNA phage coat proteins and RNA. *Prog. Nucleic Acid Res. Mol. Biol.* **40:** 185–220.

Wu H., Jiang L., and Zimmermann R.A. 1994. The binding site for ribosomal protein S8 in 16S rRNA and spc mRNA from *Escherichia coli:* Minimum structural requirements and the effects of single bulged bases on S8-RNA interaction. *Nucleic Acids Res.* **22:** 1687–1695.

Wu T.H., Liao S.M., McClure W.R., and Susskind M.M. 1987. Control of gene expression in bacteriophage P22 by a small antisense RNA. II. Characterization of mutants defective in repression. *Genes Dev.* **1:** 204–212.

Zengel J.M. and Lindahl L. 1994. Diverse mechanisms for regulating ribosomal protein synthesis in *Escherichia coli*. *Prog. Nucleic Acid Res. Mol. Biol.* **47:** 331–370.

29

Noncanonical Functions of Aminoacyl-tRNA Synthetases in Translational Control

Paul L. Fox, Partho S. Ray, Abul Arif, and Jie Jia
Department of Cell Biology
Lerner Research Institute, Cleveland Clinic
Cleveland, Ohio 44195

AMINOACYL-TRNA SYNTHETASES (AARSs) ARE ANCIENT ENZYMES, ubiquitous in the three domains of life, that catalyze the ligation of amino acids to cognate tRNAs (Ibba and Söll 2000; Ribas de Pouplana and Schimmel 2001). They are uniquely responsible for deciphering the genetic code, reading the genetic information in the tRNA anticodon, and ligating the appropriate amino acid to the terminal ribose of the conserved CCA sequence at the 3′ end of the tRNA. In most prokaryotes, there are 20 AARSs, one for each major amino acid. Lower eukaryotes have separate cytoplasmic and nuclear-encoded mitochondrial (as well as chloroplastic) AARSs (Sissler et al. 2005). In all vertebrates, and in some invertebrates, the 20 cytoplasmic AARS activities are contained in 19 proteins; the bifunctional GluProRS expresses two enzyme activities in a single polypeptide chain. All synthetases contain catalytic and tRNA anticodon recognition sites in separate domains, and belong to one of two structurally distinct classes (Ibba and Söll 2000). The 10 Class I enzymes have a Rossman fold in the active site, bind the minor groove of the tRNA acceptor stem, and aminoacylate ribose at the 2′-OH position. In contrast, the 10 Class II enzymes have an antiparallel β-sheet in the active site, bind the major groove of the acceptor stem, and aminoacylate ribose at 3′-OH. Class I and II enzymes can be further grouped into subclasses that exhibit additional structural similarities and that recognize related amino acid substrates. In vertebrate cells, 9 AARS activities in 8 enzymes (including the bifunctional GluProRS, also known as EPRS), and 3 nonsynthetase proteins, reside in a 1.5-mD, cytosolic tRNA multisynthetase complex (Rho et al. 1999; Robinson et al. 2000). The function of this complex is unclear, but its association with elongation factors and ribosomes suggests it may improve efficiency of protein synthesis by channeling charged tRNAs; i.e., vectorially transferring them from AARSs to ribosomes as a ternary complex with elongation factor and GTP, thereby reducing their diffusion into the cellular fluid (Negrutskii and Deutscher 1991).

Unique functions of AARSs in translational control may have evolved due to their intracellular localization near the translation machinery and translating mRNA. In this chapter, we provide an overview of the non-canonical activities of AARSs, and discuss in detail the role of GluProRS, ValRS, and AlaRS in the regulation of translation at the levels of initiation, elongation, and ribosome recycling. Autoinhibition of *Escherichia coli thrS* mRNA translation by ThrRS (Romby et al. 1996) is described in Chapter 28. In addition, consideration is given to the role of AARSs in genetic and autoimmune diseases, and their potential therapeutic uses as pharmacologic agents or targets.

NONCANONICAL FUNCTIONS OF AMINOACYL-tRNA SYNTHETASES

Translation machinery components, including ribosomal proteins, AARSs, tRNAs, and initiation and elongation factors, often exhibit noncanonical functions unrelated to their primary function in protein synthesis. These activities include regulation of gene expression, cell signaling, protein folding and trafficking, and biosynthetic pathways (Kinzy and Goldman 2000). Indeed, members of the translation apparatus may be considered as "hub proteins," establishing connectivity between nodes in protein networks that control key cell functions (Barabasi and Oltvai 2004). AARS function is particularly significant, because in addition to their essential role as catalysts of the first committed reaction in protein synthesis, several AARSs exhibit important, noncanonical regulatory functions. In eukaryotes, most of these functions are performed by AARSs carrying an appended domain that is neither a part of the enzymatic core nor present in bacterial homologs. These domains are usually appended to the amino or carboxyl terminus, and include glutathione-*S*-transferase (GST)-like domains, an endothelial monocyte-activating polypeptide II (EMAP II)-like domain, and WHEP-TRS domains (Shiba 2002). The WHEP-TRS domain is a 50-amino acid, helix-turn-helix structure (Cahuzac et al. 2000; Jeong et al. 2000) named after three AARSs containing them, i.e., Trp(*W*)RS, His(*H*)RS, and GluPro(*EP*)RS. The domain is also present in GlyRS and MetRS, but not in any other non-AARS proteins. Appended domains in several eukaryotic AARSs may contribute to assembly or stability of the tRNA multisynthetase complex (Rho et al. 1999); however, the presence of appendages in some AARSs not in the complex suggests a specific role in noncanonical functions of these enzymes (Shiba et al. 1998; Francklyn et al. 2002; Ko et al. 2002).

In a well-studied example of an AARS with an appended domain, human TyrRS contains a carboxy-terminal domain with 49% sequence identity to the cytokine, EMAP II. TyrRS is secreted from cultured cells under apoptotic conditions and cleaved into two fragments by leukocyte elastase (Wakasugi and Schimmel 1999b). The carboxy-terminal domain has cytokine-like activity and induces migration of leukocytes and stimulates production of tumor necrosis factor (TNF)-α, tissue factor, and myeloperoxidase (Wakasugi and Schimmel 1999b). The cleaved, amino-terminal catalytic domain of TyrRS has sequence features common to CXC chemokines such as IL-8, and becomes an angiogenic factor (Wakasugi et al. 2002a). Angiogenic activity depends on a Glu-Leu-Arg motif, also present in CXC chemokines, within the Rossman-fold domain in the catalytic site (Wakasugi and Schimmel 1999a). The closely related TrpRS has a WHEP-TRS domain appended to its amino terminus. A TrpRS fragment lacking the amino terminus is generated by IFN-γ-induced alternative splicing or by proteolytic cleavage (Wakasugi et al. 2002b). This fragment, referred to as "mini-TrpRS," is an efficient angiostatic agent that induces endothelial cell apoptosis and inhibits blood vessel growth (Otani et al. 2002). An 8-amino acid peptide in the tRNA-binding domain is required for angiostatic activity (Kise et al. 2004). Recently, several other noncanonical activities of AARSs have been described. LysRS is secreted from TNF-α-activated cells and triggers a pro-inflammatory response in target macrophages, increasing TNF-α secretion and cell migration (Park et al. 2005b). LysRS also translocates to the nucleus and activates MITF and USF2 transcription factors in activated mast cells (Lee et al. 2004; Lee and Razin 2005). GlnRS interacts with apoptosis signal-regulating kinase 1 (ASK1) in a glutamine-dependent manner and inhibits its activity, thereby playing an antiapoptotic role (Ko et al. 2001). MetRS translocates to the nucleolus in response to growth factors and enhances rRNA synthesis (Ko et al. 2000). HisRS and AsnRS activate chemokine receptors on T lymphocytes and immature dendritic cells, and act as autoantigens in myositis (Howard et al. 2002).

Several AARSs directly bind DNA or RNA, either by the tRNA-binding domain or via appended nonenzymatic domains, to regulate gene expression at the level of transcription, mRNA splicing, or translation. For example, *E. coli* AlaRS represses transcription of its own gene by binding to a palindromic sequence flanking the transcription start site (Putney and Schimmel 1981). The self-splicing reactions of group I introns in two fungi are facilitated by mitochondrial AARSs (Lambowitz and Perlman 1990). *Neurospora crassa* mitochondrial TyrRS (CYT-18) binds a conserved tRNA-like structure in the group I intron catalytic core via its tRNA-binding domain and a unique amino-terminal domain (Cherniack et al. 1990; Caprara et al. 1996). *Saccharomyces cerevisiae* LysRS, in cooperation with b14 maturase, facilitates splicing of group I introns in two closely related genes encoding essential respiratory proteins (Herbert et al. 1988). LysRS splicing activity requires the connective peptide 1 (CP1) domain, which is also involved in aminoacyl-tRNA proofreading and is conserved in several AARSs (Rho et al. 2002). *S. cerevisiae* AspRS specifically binds a region near the 5′ end of its own transcript (Frugier and Giegé 2003; Frugier et al. 2005) via a 70-amino acid, α-helical appendage at the amino terminus (Sellami et al. 1986). The interaction regulates AspRS expression by a mechanism that has not been elucidated, but depends on relative concentrations of AspRS and its cognate tRNA.

INHIBITION OF CERULOPLASMIN mRNA TRANSLATION BY GluProRS IN MACROPHAGES

The bifunctional GluProRS (Cerini et al. 1991; Kaiser et al. 1994) is a 172-kD protein consisting of a short GST-like domain at the amino terminus followed by the GluRS enzymatic domain, a linker consisting of three tandem WHEP-TRS domains, and finally, the carboxy-terminal ProRS enzymatic domain (Cahuzac et al. 2000; Jeong et al. 2000). All five AARSs bearing WHEP-TRS domains have noncanonical functions unrelated to tRNA synthetase activity (Shiba 2002; Sampath et al. 2004). Vertebrate GluProRS is localized almost entirely in the cytosolic multisynthetase complex (Han et al. 2003). Recent experiments indicate GluProRS has a noncanonical function in transcript-selective regulation of translation in human macrophages (Sampath et al. 2004).

Function of GluProRS in GAIT-mediated, Transcript-selective Translational Silencing

Ceruloplasmin (Cp) is an acute-phase plasma protein made by liver hepatocytes and monocyte/macrophages. The protein serves important roles in iron homeostasis and inflammation. The pro-inflammatory function of macrophage-derived Cp is uncertain, but roles in bactericidal activity, defense against oxidant stress, and lipoprotein oxidation have been reported (Musci 2001). Cp mRNA is undetectable in human monocytic U937 cells, but treatment with interferon (IFN)-γ causes a robust induction of Cp mRNA and protein within 2–4 hours of treatment (Mazumder et al. 1997). Despite the continued presence of abundant Cp mRNA for at least 24 hours, synthesis of Cp protein stops abruptly, and almost completely, about 12–16 hours after IFN-γ treatment. Silencing of Cp translation is highly selective, since metabolic labeling studies show that global protein synthesis is unaffected by IFN-γ.

Translational silencing of Cp mRNA requires a structural element (termed IFN-Gamma-Activated Inhibitor of Translation, or GAIT element) in its 3′UTR (Mazumder and Fox 1999; Sampath et al. 2003). Deletion and mutation analysis indicates that the GAIT element is a 29-nucleotide, bipartite stem-loop with sequence and structural features that distinguish it from other defined translational control elements; for example, the iron-responsive element (Klausner et al. 1993) and the selenocysteine insertion sequence element (Copeland and Driscoll 2001). Application of a pattern-matching algorithm against a 3′UTR database has identified similar structural elements in the 3′UTRs of about 30 human transcripts (Sampath et al. 2004). The Cp GAIT element is specifically recognized by an IFN-γ-activated, multimeric assembly referred to as the GAIT complex. By proteomic and genetic methods, four GAIT complex constituents have been identified: GluProRS, NS1-associated protein-1 (NSAP1), glyceraldehyde 3-phosphate dehydrogenase (GAPDH), and ribosomal protein L13a (Mazumder et al. 2003; Sampath et al. 2004). GAIT complex assembly takes place in two distinct stages (Fig. 1). During the first 2–4 hours, an inactive pre-GAIT complex, which does not bind the GAIT element, is formed from GluProRS and NSAP1. During the second stage, approximately 12 hours later, L13a joins GAPDH and the pre-GAIT complex to form the functional, quaternary GAIT complex that binds the 3′UTR GAIT element of Cp mRNA and silences its translation (Sampath et al. 2004). The molecular mechanism of inhibition is not known; however, translationally silenced Cp mRNA is not bound to polysomes, indicating a block at the initiation step (Mazumder and Fox 1999). Silencing may involve interaction of a GAIT complex component with a ribosomal subunit or an initiation complex component (Fig. 2A). The requirement for poly(A) tail, poly(A)-binding protein, and eIF4G, the essential elements of mRNA circularization, suggests that an end-to-end interaction of the mRNA is required for regulation. Possibly, these interactions bring the 3′UTR-localized GAIT complex into the vicinity of the translation-initiation site to facilitate the transcript-specific inhibition.

GluProRS directly binds the GAIT element as shown by UV-crosslinking of cell lysates and by RNA electrophoretic mobility shift

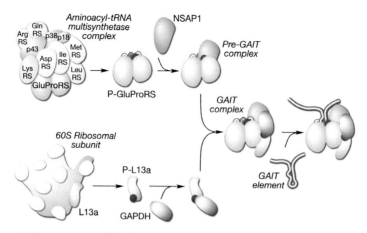

Figure 1. Two-stage model of GAIT complex activation. The inactive, pre-GAIT complex contains phosphorylated GluProRS (P-GluProRS) and NSAP1. They are subsequently joined by phosphorylated ribosomal protein L13a (P-L13a) and GAPDH to form the functional GAIT complex that binds the 3'UTR GAIT element of Cp mRNA and silences its translation.

assay using purified or recombinant proteins. A dilemma raised by finding that GluProRS is the GAIT element-binding protein is that the pre-GAIT complex (which contains GluProRS and NSAP1) does not bind the element. Possibly, NSAP1 binds GluProRS at or near the GAIT element-binding site and prevents that interaction. A parallel inhibitory mechanism was shown previously for GRY-RBP, a larger NSAP1 isoform. GRY-RBP binds APOBEC-1 complementation factor and inhibits its binding to apoB RNA, preventing apolipoprotein B RNA editing (Blanc et al. 2001). Thus, NSAP1 and its isoforms may generally function as negative regulators of RNA binding. Binding of L13a or GAPDH (or both) to the pre-GAIT complex may cause a conformational change that permits the interaction of the holocomplex with the GAIT element, despite the presence of NSAP1. The GluProRS domains involved in formation of the GAIT complex and binding to the GAIT element have not been identified. The association of WHEP-TRS domains with noncanonical AARS functions (Shiba 2002), and the binding of WHEP-TRS domains to RNA (although with low affinity and specificity) (Cahuzac et al. 2000; Jeong et al. 2000), suggest the three WHEP-TRS domains in the GluProRS linker as GAIT-element-binding site candidates. Likewise, NSAP1 may bind at this site, since WHEP-TRS domains and viral protein NS1, the defining NSAP1 target (Harris et al. 1999), have almost identical helix-turn-helix structures.

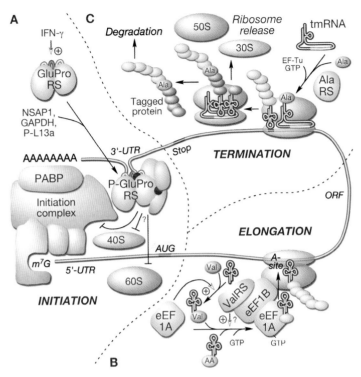

Figure 2. Regulation of translation by AARSs. (*A*) Inhibition of translation initiation by GluProRS. P-GluProRS and three other proteins bind the 3′UTR of Cp mRNA (*green tube*) and block translation initiation, possibly by interfering with an initiation complex protein or a ribosomal subunit. (*B*) Interaction of ValRS with eEF1 and its regulation of translation elongation. ValRS interacts with eEF1B in the eEF1AB complex. ValRS may facilitate formation of the ternary eEF1A·GTP·AA-t-NA^{AA} complex and loading into the A site of ribosome. Alternatively, free eEF1A may increase the aminoacylation rate of ValRS. (*C*) Aminoacylation of tmRNA by AlaRS and its role in *trans*-translation. AlaRS aminoacylates a tmRNA to form a ternary complex with EF-Tu and GTP. The complex enters the A site of a ribosome stalled on an mRNA (*red tube segment*), tmRNA-ligated Ala is transferred to the nascent peptide, and the tmRNA translocates to the ribosome P site. The original mRNA is released from the ribosome, and the tmRNA becomes the new template for translation (*violet tube segment*). The short ORF is translated to produce a protein tagged at the carboxyl terminus by the tmRNA-encoded peptide. The translation complex dissociates to release the ribosomal subunits and the tagged peptide that is targeted for protease degradation.

Mobilization and Activation of GluProRS

Under unstimulated conditions, essentially all GluProRS is present in the cytosolic, tRNA multisynthetase complex. The multisynthetase complex is the likely source of GluProRS in the GAIT complex, since its loss from the multisynthetase complex upon IFN-γ treatment coincides temporally and quantitatively with its appearance in the pre-GAIT complex (Sampath et al. 2004). In addition, translocation of GluProRS to the pre-GAIT complex occurs even in the presence of cycloheximide, thereby eliminating de novo synthesis as a source. Although the mechanism of GluProRS release from the tRNA multisynthetase complex is not understood in detail, phosphorylation may be a critical event. IFN-γ-induced serine phosphorylation of GluProRS just precedes its release from the tRNA multisynthetase complex, and Ser/Thr kinase inhibitors block release (Sampath et al. 2004). The responsible kinase has not been identified. Although other multisynthetase complex residents have noncanonical functions outside the complex, e.g., MetRS, LysRS, and several nonsynthetase components, GluProRS is the only component shown to be released from the complex. Remarkably, a second component of the GAIT complex, ribosomal protein L13a, is also released from its residence in a large macromolecular complex, i.e., the 60S ribosomal subunit (Mazumder et al. 2003). Despite the phosphorylation and release of about half of the GluProRS, and all of the L13a, from their parent complexes, global protein synthesis is unaltered. These findings have been generalized in a "depot hypothesis" which contends that macromolecular assemblies, in addition to serving their primary purpose coordinating complex tasks, also function as depots for releasable regulatory proteins (Mazumder et al. 2003; Sampath et al. 2004).

The physiological function of GAIT complex-mediated translational silencing is not known. The abundance of the four GAIT complex components, and the near-stoichiometric release of two proteins from their parent complexes, indicate that the GAIT complex is likely to be in marked excess of the Cp transcript. This, together with the identification of putative GAIT elements in multiple human mRNAs, suggests that the GAIT complex interacts with and silences the translation of additional transcripts (Sampath et al. 2004). Identification of other members of the "posttranscriptional operon," i.e., the family of co-regulated transcripts containing similar structural elements recognized by the same *trans*-acting factor(s), may provide clues to the function of the GAIT pathway (Keene and Tenenbaum 2002). Defects in the GAIT pathway may prolong or increase macrophage-mediated inflammatory processes, a hypothesis consistent with reports of disease-causing defects in noncoding regions of mRNA, and the concept that the 3'UTR is a pathological "hot spot" (Conne et al. 2000).

INTERACTION OF VALRS WITH ELONGATION
FACTOR EEF1AB COMPLEX

Several AARSs bind eukaryotic elongation factors (eEF), either stably or transiently (Lee et al. 2002). Mammalian ValRS has a special relationship with eEFs since it forms a stable, defined 700-kD complex with eEF1A (a G-protein, formerly called eEF1α) and eEF1B (a trimeric guanine nucleotide exchange complex, formerly called eEF1$\beta\gamma\delta$,) (Motorin et al. 1988; Bec et al. 1989; Negrutskii et al. 1999; Jiang et al. 2005). Essentially all cellular ValRS is present in the complex, to the exclusion of all other AARSs. The amino-terminal, GST-like domain of ValRS is homologous to the amino terminus of eEF1Bγ (Hsieh and Campbell 1991; Bec et al. 1994). In the holocomplex, eEF1Bβ interacts with eEF1A and eEF1Bγ subunits, whereas the eEF1Bδ subunit interacts with ValRS to form ValRS/eEF1AB (Bec et al. 1994; Negrutskii et al. 1999; Jiang et al. 2005). According to one model, the ValRS/eEF1AB complex facilitates two parallel reactions: ValRS-mediated aminoacylation of valine, and eEF1B-mediated exchange of GTP for GDP on eEF1A·GDP (Negrutskii et al. 1999). Finally, the products of both reactions form the ternary eEF1A·GTP·Val-tRNA^{Val} that directly transfers the charged tRNA to the ribosome.

The function of the ValRS/eEF1AB complex in translation is unclear. Two possibilities have been considered; namely, that ValRS affects protein elongation rate, or conversely, that the elongation factor complex influences ValRS activity (Fig. 2B). Experimental evidence against the first case has been reported: ValRS activates neither nucleotide exchange activity of elongation factors (Peters et al. 1995) nor poly(U)-directed binding of Phe-tRNA^{Phe} to ribosomes (Bec et al. 1994). The possibility of ValRS-mediated regulation of elongation cannot be completely excluded, as regulation may require specific posttranslational modifications of ValRS or the elongation factors, or regulation may be condition-dependent and require specific cofactors, for example, charged or uncharged tRNA^{Val}. In contrast, regulation of ValRS activity by elongation factors has been reported. The catalytic constant of ValRS is nearly the same when measured as free enzyme or as part of the elongation factor complex (Bec et al. 1989). However, aminoacylation activity of ValRS in the ValRS/eEF1AB complex is stimulated up to 3-fold by an excess of free eEF1A·GTP (Negrutskii et al. 1999). It is tempting to speculate that this mechanism has evolved as a "rheostat" to tune the rate of protein synthesis as a function of the cellular level of eEF1A·GTP. In fact, previous evidence is consistent with eEF1A having a critical, rate-controlling role in protein synthesis (Vargas and Castaneda 1981; Thomas 1986).

The physiological importance of the ValRS/eEF1AB complex is supported by evidence of stimulus-dependent regulation of elongation by posttranslational modification of the components (Mulner-Lorillon et al. 1992; Minella et al. 1998). Activation of protein kinase C in rabbit reticulocytes induces phosphorylation of all five protein components of ValRS/eEF1AB and increases the rate of chain elongation by up to 3-fold measured as poly(U)-directed phenylalanine synthesis (Venema et al. 1991a,b). Phosphorylation of the complex enhances GDP/GTP exchange, formation of eEF1A•GTP•Phe-tRNA^{Phe}, and binding of Phe-tRNA^{Phe} to ribosomes (Peters et al. 1995). Concurrently, ValRS synthetase activity is increased nearly 2-fold. These data suggest that, in addition to its physical role in shuttling Val-tRNA to ribosomes, the ValRS-elongation factor complex may be a regulatable sensor of the cell metabolic state, which in turn regulates the global rate of protein synthesis.

ROLE OF AlaRS IN *TRANS*-TRANSLATION AND RIBOSOME RESCUE

AlaRS is unique among AARSs in its noncanonical roles in regulation of both transcription and translation. AlaRS represses transcription of its own gene by binding to a palindromic sequence flanking the gene's start site, providing one of the earliest examples of a noncanonical function of an AARS (Putney and Schimmel 1981). AlaRS regulates bacterial translation by aminoacylation of transfer-messenger RNA (tmRNA), also known as SsrA or 10Sa RNA, a small bacterial RNA that mediates ribosome rescue and *trans*-translation (Keiler et al. 1996). In bacteria, particularly under stress conditions, aberrant mRNAs lacking a termination codon may be generated by mRNA degradation or premature transcription termination. Ribosomes tend to stall and collect on these messages, leading to depletion of free ribosome levels and production of truncated polypeptides, and consequent cell injury. The tmRNA rescues the ribosome and facilitates the degradation of incompletely synthesized peptides (Fig. 2C). tmRNA acts both as a tRNA and an mRNA to mediate the *trans*-translation reaction (Withey and Friedman 2003). In this highly unusual case, the canonical aminoacylation function of an AARS (AlaRS) is applied to a noncanonical target (tmRNA) to perform a noncanonical activity (*trans*-translation).

tmRNA was first identified as a small RNA of unknown function in *E. coli* (Lee et al. 1978). A role in translation regulation was suggested by observations that the 5′ and 3′ ends of tmRNAs from multiple bacteria resembled those of tRNA^{Ala} (Komine et al. 1994; Ushida et al. 1994). AlaRS aminoacylates tmRNAs in vitro, although more slowly than acylation of tRNA^{Ala} (Komine et al. 1994). tmRNA interacts with 70S ribosomes, but not with 50S or 30S subunits, and charging with alanine is required for the interaction (Ushida et al. 1994; Tadaki et al. 1996). These observations, coupled with the mRNA-like qualities of tmRNA, led to the recognition of its role in translation and the development of the *trans*-translation model (Keiler et al. 1996).

The mechanism of *trans*-translation can be described in four steps: First, a ribosome stalls on an mRNA, either because of a missing stop codon or at a cluster of rare codons. A tmRNA, which has been amino-acylated by AlaRS, enters the stalled ribosome much like a tRNA, in a ternary complex with EF-Tu and GTP (Rudinger-Thirion et al. 1999). In the second step, tmRNA-ligated alanine is transferred to the nascent peptide and the tmRNA translocates from the ribosome A site to the P site. Third, the original mRNA is released from the ribosome and the tmRNA becomes the new template for translation, having a 10-amino acid open reading frame (ORF) starting with alanine as the "resume" codon. The short ORF is translated to produce a protein tagged at the carboxyl terminus by the tmRNA-encoded peptide. Finally, the translation complex dissociates, releasing the ribosomal subunits and the tagged peptide that is degraded rapidly by specific proteases (Keiler et al. 1996; Gottesman et al. 1998).

Aminoacylation of tmRNA is the first and essential step of the *trans*-translation process. Despite being about five times larger than $tRNA^{Ala}$, tmRNAs are specifically recognized and charged by AlaRS. The presence of a single G3•U70 base pair in the acceptor stem of $tRNA^{Ala}$ is responsible for AlaRS recognition and is conserved in all tmRNA sequences (Hou and Schimmel 1988; McClain and Foss 1988a). Replacement of the G•U pair by either G•A or G•C abrogates alanine aminoacylation of tmRNA (Komine et al. 1994). The 5′ and 3′ termini of tmRNA sequences base-pair to form a tRNA-acceptor stem with a 3′ CCA overhang (Felden et al. 1997). The tRNA-like structure extends to form a TψC stem-loop and a D-like loop but lacks an anticodon stem-loop (Felden et al. 1996). The conserved G3•U357 base pair is present in the acceptor stem of tmRNAs and permits aminoacylation by AlaRS. Mutation of these bases causes failure of aminoacylation and abolishes tmRNA association with ribosomes (Gillet and Felden 2001). Although the structural basis of the interaction between AlaRS and tmRNA is not known, it may resemble the interaction between AlaRS and $tRNA^{Ala}$, where the enzyme interacts with both the acceptor stem and the T-loop of the tRNA (Pleiss et al. 2000).

Besides AlaRS, elongation factor EF-Tu also interacts with tmRNA and is responsible for transport to the ribosome (Rudinger-Thirion et al. 1999). tRNA access to the ribosomal A site requires an interaction between the tRNA anticodon and the mRNA codon. Because tmRNA lacks an anticodon, it must access the ribosome by an alternate mechanism. Cross-linking studies show that EF-Tu binds tmRNA in the elongated helix between the D-like stem and pseudoknot 1, whereas it binds tRNA near the acceptor stem (Zvereva et al. 2001). This unique binding mechanism may permit tmRNA access to the ribosome despite the absence of an anticodon. Although EF-Tu binds equally well to both charged and uncharged tmRNA, only the former can access the ribosome. Two additional proteins, SmpB and the ribosomal protein S1, interact with tmRNAs (Karzai et al. 1999; Wower et al. 2000). SmpB enhances the aminoacylation of tmRNA and is required for *trans*-translation (Shimizu and Ueda 2002). S1 binding causes a large conformational change in tmRNA and may facilitate the transition from tRNA to mRNA activity (Bordeau and Felden 2002).

trans-Translation initially was suggested to be a physiological mechanism to tag aberrant peptides for proteolysis (Keiler et al. 1996). However, subsequent studies indicate ribosome rescue as the principal function of tmRNA. Certain phages, e.g., λ*imm*P22 hybrid phage, require functional tmRNA for growth in *E. coli* (Retallack et al. 1994). A mutant tmRNA that encodes a defective peptide tag not recognized by proteases supports phage growth; in contrast, a mutant that fails to be aminoacylated does not. Thus, charging of tmRNA with alanine, but not proteolysis of tagged peptides, is essential for phage growth (Withey and Friedman 1999). A similar role of tmRNA[Ala] for *Neisseria gonorrhoeae* viability has been reported (Huang et al. 2000). These experiments suggest that the principal role of *trans*-translation by tmRNAs may be the rescue of stalled ribosomes from mRNAs to maintain high efficiency of translation (Withey and Friedman 2002). The presence of tmRNA-encoded peptide tags after the normal carboxyl terminus in several *E. coli* and bacteriophage proteins, e.g., YbeL and λ cI repressor, shows that *trans*-translation also may occur at bona fide stop codons in intact mRNAs (Roche and Sauer 2001; Fujihara et al. 2002). The specific codons just upstream of the stop codon may determine whether *trans*-translation occurs at a stop codon (Collier et al. 2002). For example, a pair of rare arginine codons just before a UGA stop codon in the ribokinase gene results in *trans*-translation at the arginine codons and at the stop codon (Hayes et al. 2002a). Proline residues at the carboxyl termini of proteins often result in *trans*-translation at the stop codon, as observed for the *E. coli* YbeL protein (Hayes et al. 2002b). *trans*-Translation at stop codons is likely due to competition between tmRNAs and release factors, particularly when the stop codon context causes inefficient translation termination (Hayes et al. 2002b). tmRNA-tagged, full-length proteins undergo rapid proteolysis, which may be used as a mechanism to regulate expression of the *trans*-translated protein. For example, tmRNA-tagging and degradation of λ repressor may ensure that intracellular repressor level is maintained within narrow limits, because operator binding by λ repressor is highly cooperative, and small changes in repressor significantly affect activity (Roche and Sauer 2001).

The selection of AlaRS, among all AARSs, for tmRNA aminoacylation is intriguing. AlaRS is atypical in its ability to aminoacylate cog-nate tRNAs independent of anticodon recognition. tmRNAs lack the anticodon loop and thus cannot be a substrate for a synthetase utilizing anticodon recognition. The major determinant for charging tRNA^{Ala} is the G3•U70 base pair in the acceptor stem, which is also present in tm-RNAs (Hou and Schimmel 1988). Substitution of the anticodon of a tR-NA^{Ala} isoacceptor from a GGC to an amber suppressor CUA does not cause mischarging with other amino acids (McClain and Foss 1988b). This feature of AlaRS, which makes it dependent on the "second genetic code" (de Duve 1988) embedded in the nucleotides of the tRNA accep-tor stem for aminoacylation of its cognate tRNAs, possibly made AlaRS the synthetase of choice for tmRNAs and for this unique regulatory role in translation.

tRNA SYNTHETASES IN MEDICINE AND DISEASE

Inherited mutations in AARSs that inactivate synthetase enzymatic activity will certainly be lethal when both alleles are affected. Recent dis-coveries of human diseases involving defective AARSs argue for the phys-iological significance of the noncanonical functions of these enzymes. In the following sections, diseases featuring a mutated AARS, or a mutated AARS-interacting protein, are described. The special importance of AARS in autoimmune disease is also discussed. In no case has the defective function of the AARS been elucidated, and no effects on translational control have been reported to date.

Neurological Diseases Related to tRNA Synthetases

In the first report of an AARS associated with a human genetic disease, several mutations and deletions in GlyRS have been implicated in Charcot-Marie-Tooth (CMT) disease (Antonellis et al. 2003). CMT dis-ease is a group of related, autosomal-dominant neurological disorders expressed in the extremities. The type 2D form of the disease (CMT2D) is characterized by impaired motor and sensory neuron response, despite normal axon myelination, and maps to four mutations in the *GlyRS* cod-ing region. The dominant nature of the mutation(s) may be explained by the known dimeric state of the protein, and by inactivation of wild-type protein by heterodimerization with mutant monomers. The specific relationship between the mutant protein and the phenotype is unknown. Low synthetase activity is a possible cause, but it is difficult to reconcile this defect with restricted localization in peripheral neurons. The fact that two of the four mutations are far from the catalytic or anticodon binding sites suggests that a noncanonical function of GlyRS may be affected. Interestingly, the E71G mutation is adjacent to the amino-terminal WHEP-TRS domain that may convey noncanonical activities to AARSs bearing this domain (Shiba 2002). The RNA-binding activity of the WHEP-TRS domain in *Bombyx mori* GlyRS is not required for catalysis, consistent with a noncanonical activity (Wu et al. 1995).

Additional insight into the role of AARS mutants in peripheral neuropathies is provided by the discovery of *TyrRS* mutations in three families with CMT type C disease, a phenotype similar to CMT2D, but with demyelination in addition to axonal defects (Jordanova et al. 2006). TyrRS, like GlyRS, is a dimeric protein, and mutant forms have a dominant-negative effect on function of the wild-type protein. Immunolocalization experiments show a polarized localization of TyrRS in differentiating neuronal cells; it concentrates in the growth cone and in the most distal regions of projecting neurites. Importantly, the distinctive "teardrop" distribution in neurites is TyrRS-specific and is not observed for two other AARSs, ArgRS and TrpRS. The causal relationship between mutant TyrRS and the resulting neuropathology is unclear. TyrRS may be required for protein synthesis in nerve endings since the region harbors its own cell-body-independent translational machinery, including mRNAs, ribosomes, translation initiation factors, and AARSs (Zheng et al. 2001). However, the requirement for the full complement of AARSs for protein synthesis, and the unique polarization of TyrRS, suggests a noncanonical, neurite-extension-specific function. An amino-terminal TyrRS fragment (which contains the sites of the CMT type C mutations) exhibits chemokine-like and angiogenic activity (Wakasugi and Schimmel 1999b; Wakasugi et al. 2002b); however, there is no evidence that neuronal cells produce or secrete this fragment.

An AARS may be involved in another inherited neuropathy, amyotrophic lateral sclerosis (ALS, Lou Gehrig's disease), a common, degenerative motor neuron disease. A subset of ALS (which incidentally, like CMT, was discovered by Jean-Martin Charcot, in 1867) is caused by mutations in the gene encoding Cu,Zn superoxide dismutase (*SOD1*). Several mechanisms have been proposed to link the SOD1 defect with selective neuron death, e.g., defective antioxidant activity or exacerbated oxidant stress due to surface-bound copper (Bruijn et al. 2004). More recently, inappropriate protein–protein interactions have received attention based on the ubiquitous finding of SOD1-containing aggregates in neurons from *SOD1*-mediated familial ALS patients, and from *SOD1* ALS mouse models. The potential pathological consequences of neuronal aggregates are unclear, but they may bind and deplete a critical cellular component(s), or possibly saturate the ubiquitin-proteasome system. The yeast 2-hybrid method has shown (and coimmunoprecipitation studies confirm) that LysRS selectively and specifically binds SOD1 with ALS-causing mutations, but not wild-type SOD1 (Kunst et al. 1997). The ensuing toxicity may result from alteration of a noncanonical LysRS function; e.g., activation of transcription factors. Alternatively, interaction of SOD1 with LysRS could remove it from its residence in the tRNA multisynthetase complex and disturb the function of this complex in global protein synthesis. Given the ubiquitous presence of both SOD1 and LysRS in all tissues, the specificity of the target cell may reflect elevated sensitivity of motor neurons, or the additional involvement of neuron-specific, SOD1-binding proteins.

Linkage analysis and other population studies have suggested single-nucleotide polymorphisms in AARS as risk factors for several non-neurodegenerative diseases. Association between the AARS mutation and disease has been excluded in two studies (Shah et al. 2001; Zee et al. 2005). However, an association between the H324Q variant of mitochondrial LeuRS and type 2 diabetes has been reported ('t Hart et al. 2005). Functional studies of the mutant protein have not revealed defective aminoacylation, and a noncanonical function may be involved.

Aminoacyl-tRNA Synthetases in Autoimmune Disease

Idiopathic inflammatory myopathy (IIM) comprises a group of autoimmune diseases characterized by inflammation of skeletal muscle, lung interstitium, and bone joints (Targoff et al. 1997). The two most common IIM forms are polymyositis, an inflammatory muscle condition, and dermatomyositis, which affects skin and muscle. Unexpectedly, highly specific anti-AARS antibodies are often observed in patient sera (Targoff 2000) in conditions referred to as "anti-synthetase syndrome" (Kron and Härtlein 2005). In every case, an autoantibody against a specific AARS inhibits aminoacylation activity, and likely inhibits protein synthesis (Mathews and Bernstein 1983). Anti-HisRS antibodies (Jo-1) are the most common, and are observed in 15–25% of patients with polymyositis and in about 70% of patients with concurrent myositis and interstitial lung disease (Targoff 2000). Autoantibodies against five other AARSs—AlaRS, GlyRS, IleRS, ThrRS, and AsnRS—are seen with lesser frequency in IIM patients (Targoff et al. 1988; Targoff 1990; García-Lozano et al. 1998; Hirakata et al. 1999). The mechanism by which AARSs are converted into autoantigens has not been elucidated.

The recognition of noncanonical functions of AARSs provides clues on the mechanism of initiation of myositis, and how an AARS contributes to the generation of an autoimmune response. The major HisRS epitopes recognized by Jo-1 anti-HisRS antibodies from IIM patients differ significantly from those recognized by antibodies produced by immunization with purified HisRS (Miller et al. 1990b). Fine epitope mapping by Jo-1 antibodies shows that the major autoantigenic epitope in HisRS is in the amino-terminal 60 amino acids constituting the WHEP-TRS domain (Raben et al. 1994). Anti-HisRS autoantibodies specifically recognize mammalian HisRS, which contains the WHEP-TRS domain, but not bacterial or yeast HisRS, which lack this domain, or recombinant HisRS missing the first 60 amino acids (Miller et al. 1990a; Raben et al. 1994). Interestingly, GlyRS, another autoantigenic AARS in IIM, also contains an amino-terminal WHEP-TRS domain. Charge-rich α-helical motifs, like those in WHEP-TRS domains, may be highly antigenic, as similar structures are found in approximately 37% of autoantigens targeted in autoimmune diseases (Dohlman et al. 1993). A similar peptide appendage in AsnRS may be the antigenic epitope for the KS autoantibody found in some IIM cases (Beaulande et al. 2001). In summary, the nonenzymatic appended peptides in AARSs, some of which are implicated in noncanonical functions, may be critical for their autoantigenicity.

Several AARSs express cytokine-like activities and elicit inflammatory responses (Park et al. 2005a). Secretion and cleavage of TyrRS from apoptotic cells yield a product that induces migration of mononuclear phagocytes and neutrophils, possibly as a mechanism to clear dead cells (Wakasugi and Schimmel 1999b). Interestingly, HisRS and AsnRS, AARSs commonly targeted by the autoimmune response in IIM, also have cytokine-like activities (Howard et al. 2002). Both induce in vitro migration of CD4+ and CD8+ T cells, monocytes, and immature dendritic cells. The cytokine activity of HisRS and AsnRS is mediated by CCR5 and CCR3 chemokine receptors, respectively, expressed on immature dendritic cells that infiltrate muscle tissue from myositis patients (Howard et al. 2002). Furthermore, three AARSs targeted in IIM—HisRS, IleRS, and AlaRS—are cleaved by granzyme B, a serine protease critical in the T-cell granule exocytosis pathway involved in apoptosis (Casciola-Rosen et al. 1999). Granzyme B cleaves HisRS in the amino terminus, to release the antigenic fragment (Levine et al. 2003).

These studies provide evidence for a link between the noncanonical functions of AARSs and their role as autoantigens in IIM. A general model has been proposed in which injury in muscle or lung tissue, caused, for example, by viral infection or chemical agents, induces an initial immune response (Levine et al. 2003). Exposure to immune mediators, such as cytotoxic T cells or NK cells, activates an apoptotic pathway that causes release of AARSs and cleavage by proteases. The newly generated cytokine-like fragments recruit macrophages, neutrophils, and dendritic cells to the site of injury. AARS-derived peptides are processed by dendritic cells and presented as novel antigens, giving rise to the autoimmune response targeted against cellular AARSs. Neutralizing antibodies against AARSs inhibit aminoacylation of tRNAs, which in turn inhibits translation. Because translation inhibition may be accompanied by apoptosis (Clemens et al. 2000), a feed-forward cycle could be established in which cell death generates additional antigenic AARS peptides and amplification of the autoimmune response. Autoimmune myositis may represent a scenario in which the noncanonical, cytokine-like function of AARSs, when taking place in an unregulated manner, impedes their canonical role in global protein synthesis.

tRNA Synthetases as Therapeutic Targets and Agents

Antibiotic resistance of pathogenic bacteria is a major threat to public health and provides a rationale for seeking new therapeutic bactericidal agents (Chapter 30). Screening of natural products for antibiotics has revealed translation system components as common targets (Kim et al. 2003). Clinical effectiveness of an AARS inhibitor demands high selectivity for bacterial versus human AARS targets. AARSs possess several potential advantages as targets; e.g., the evolutionary divergence of the prokaryotic and eukaryotic AARSs, the full complement of bacterial AARSs provides 20 distinct targets, the abundance of natural inhibitors that can be used for starting points for drug development, and the wealth of structural information on bacterial and eukaryotic synthetases. The structural similarity of AARS domains within a subclass provides an additional unique advantage; namely, multiple AARSs can be targeted by a single agent (Hurdle et al. 2005). This characteristic may be valuable in development of drugs that reduce the acquisition of drug resistance, since bacteria would require multiple, simultaneous mutations to overcome the agent. Currently, muciprocin (pseudomonic acid) is the only AARS inhibitor commercially used as an antibiotic. It is a natural product isolated from *Pseudomonas fluorescens* that inhibits bacterial IleRS catalytic activity and is topically effective against gram-positive pathogens, e.g., *Streptococcus aureus* (Hurdle et al. 2005). Other AARS inhibitors are being sought by screening chemical libraries and by chemical optimization of aminoacylation reaction intermediates and natural product inhibitors (Pohlmann and Brötz-Oesterhelt 2004).

PERSPECTIVES

We are at an early stage of understanding the noncanonical activities of AARSs, particularly those involved in translational control, and central questions need to be addressed. For example, besides the translational control activities already shown for one bacterial and two vertebrate AARSs, do these or other AARSs have as-yet-undiscovered functions in translation? A systematic search for these functions is hampered by the absolute essentiality of the canonical activity of these enzymes that will cause certain lethality in a knockout model. A related question is whether other AARSs in the tRNA multisynthetase complex participate in translational control. This possibility is indicated by the residence of both vertebrate AARSs involved in translational control, i.e., GluProRS and ValRS, in multicomponent complexes (i.e., the multisynthetase and ValRS/eEF1AB complexes) spatially near the ribosomes and translating mRNA. What are the cellular conditions and posttranslational mechanisms that induce an AARS to switch to its noncanonical function? In the case of GluProRS, activation requires phosphorylation and release from the multisynthetase complex (Sampath et al. 2004); however, the kinase, signaling pathway that activates the kinase, mechanism of release from the parent complex, and molecular mechanism of translational silencing are unknown. Finally, what is the physiological significance of translational control by AARSs? Virtually nothing is known about the in vivo role of noncanonical AARS activities related to translational control, in either normal or pathological settings.

Further elucidation of the noncanonical functions of eukaryotic AARSs will provide unique opportunities for therapeutic intervention. The pro-angiogenic activity of the TyrRS fragment may be useful in situations calling for new blood vessel formation, e.g., revascularization of injured cardiac tissue or reduced collateral blood flow in peripheral artery disease (Wakasugi et al. 2002a). The anti-angiogenic properties of TrpRS fragments may be useful to arrest neovascularization of tumors or to inhibit abnormal angiogenesis in ocular diseases (Tzima and Schimmel 2006). Looking farther into the future, stimulation of GluProRS phosphorylation and GAIT complex formation may provide a therapy for reduction of macrophage inflammation (Sampath et al. 2004). Because many therapeutic agents with high potency also exhibit adverse side effects, e.g., elevated risk of coronary heart disease by nonsteroidal antiinflammatory drugs, the application or stimulation of endogenous agents, such as AARSs, may provide unique health benefits.

ACKNOWLEDGMENTS

This work was supported by National Institutes of Health grants HL29582, HL67725, and HL76491 (to P.L.F.), and by an AHA Postdoctoral Fellowship, Ohio Valley Affiliate (to J.J.).

REFERENCES

Antonellis A., Ellsworth R.E., Sambuughin N., Puls I., Abel A., Lee-Lin S.Q., Jordanova A., Kremensky I., Christodoulou K., Middleton L.T., et al. 2003. Glycyl tRNA synthetase mutations in Charcot-Marie-Tooth disease type 2D and distal spinal muscular atrophy type V. *Am. J. Hum. Genet.* **72:** 1293–1299.

Barabasi A.L. and Oltvai Z.N. 2004. Network biology: Understanding the cell's functional organization. *Nat. Rev. Genet.* **5:** 101–113.

Beaulande M., Kron M., Hirakata M., and M. Hartlein M. 2001. Human anti-asparaginyl-tRNA synthetase autoantibodies (anti-KS) increase the affinity of the enzyme for its tRNA substrate. *FEBS Lett.* **494:** 170–174.

Bec G., Kerjan P., and Waller J.P. 1994. Reconstitution *in vitro* of the valyl-tRNA synthetase-elongation factor (EF) 1βγδ complex. Essential roles of the NH$_2$-terminal extension of valyl-tRNA synthetase and of the EF-1δ subunit in complex formation. *J. Biol. Chem.* **269:** 2086–2092.

Bec G., Kerjan P., Zha X.D., and Waller J.P. 1989. Valyl-tRNA synthetase from rabbit liver. I. Purification as a heterotypic complex in association with elongation factor 1. *J. Biol. Chem.* **264:** 21131–21137.

Blanc V., N. Navaratnam, Henderson J.O., Anant S., Kennedy S., Jarmuz A., Scott J., and Davidson N.O. 2001. Identification of GRY-RBP as an apolipoprotein B RNA-binding protein that interacts with both apobec-1 and apobec-1 complementation factor to modulate C to U editing. *J. Biol. Chem.* **276:** 10272–10283.

Bordeau V. and Felden B. 2002. Ribosomal protein S1 induces a conformational change of tmRNA; more than one protein S1 per molecule of tmRNA. *Biochimie* **84:** 723–729.

Bruijn L.I., Miller T.M., and Cleveland D.W. 2004. Unraveling the mechanisms involved in motor neuron degeneration in ALS. *Annu. Rev. Neurosci.* **27:** 723–749.

Cahuzac B., Berthonneau E., Birlirakis N., Guittet E., and Mirande M. 2000. A recurrent RNA-binding domain is appended to eukaryotic aminoacyl-tRNA synthetases. *EMBO J.* **19:** 445–452.

Caprara M.G., Lehnert V., Lambowitz A.M., and Westhof E. 1996. A tyrosyl-tRNA synthetase recognizes a conserved tRNA-like structural motif in the group I intron catalytic core. *Cell* **87:** 1135–1145.

Casciola-Rosen L., Andrade F., Ulanet D., Wong W.B., and Rosen A. 1999. Cleavage by granzyme B is strongly predictive of autoantigen status: Implications for initiation of autoimmunity. *J. Exp. Med.* **190:** 815–826.

Cerini C., Kerjan P., Astier M., Gratecos D., Mirande M., and Semeriva M. 1991. A component of the multisynthetase complex is a multifunctional aminoacyl-tRNA synthetase. *EMBO J.* **10:** 4267–4277.

Cherniack A.D., Garriga G., Kittle J.D., Jr., Akins R.A., and Lambowitz A.M. 1990. Function of *Neurospora* mitochondrial tyrosyl-tRNA synthetase in RNA splicing requires an idiosyncratic domain not found in other synthetases. *Cell* **62:** 745–755.

Clemens M.J., Bushell M., Jeffrey I.W., Pain V.M., and Morley S.J. 2000. Translation initiation factor modifications and the regulation of protein synthesis in apoptotic cells. *Cell Death Differ.* **7:** 603–615.

Collier J., Binet E., and Bouloc P. 2002. Competition between SsrA tagging and translational termination at weak stop codons in *Escherichia coli. Mol. Microbiol.* **45:** 745–754.

Conne B., Stutz A., and Vassalli J.D. 2000. The 3′ untranslated region of messenger RNA: A molecular 'hotspot' for pathology? *Nat. Med.* **6:** 637–641.

Copeland P.R. and Driscoll D.M. 2001. RNA binding proteins and selenocysteine. *BioFactors* **14:** 11–16.

de Duve C. 1988. Transfer Rnas: The second genetic code. *Nature* **333:** 117–118.

Dohlman J.G., Lupas A., and Carson M. 1993. Long charge-rich α-helices in systemic autoantigens. *Biochem. Biophys. Res. Commun.* **195:** 686–696.

Felden B., Himeno H., Muto A., Atkins J.F., and Gesteland R.F. 1996. Structural organization of *Escherichia coli* tmRNA. *Biochimie* **78:** 979–983.

Felden B., Himeno H., Muto A., McCutcheon J.P., Atkins J.F., and Gesteland R.F. 1997. Probing the structure of the *Escherichia coli* 10Sa RNA (tmRNA). *RNA* **3:** 89–103.

Francklyn C., Perona J.J., Puetz J., and Hou Y.M. 2002. Aminoacyl-tRNA synthetases: Versatile players in the changing theater of translation. *RNA* **8:** 1363–1372.

Frugier M. and Giegé R. 2003. Yeast aspartyl-tRNA synthetase binds specifically its own mRNA. *J. Mol. Biol.* **331:** 375–383.

Frugier M., Ryckelynck M., and Giegé R. 2005. tRNA-balanced expression of a eukaryal aminoacyl-tRNA synthetase by an mRNA-mediated pathway. *EMBO Rep.* **6:** 860–865.

Fujihara A., Tomatsu H., Inagaki S., Tadaki T., Ushida C., Himeno H., and Muto A. 2002. Detection of tmRNA-mediated *trans*-translation products in *Bacillus subtilis. Genes Cells* **7:** 343–350.

García-Lozano J.R., González-Escribano M.F., Rodríguez R., Rodriguez-Sanchez J.L., Targoff I.N., Wichmann I., and Núñez-Roldán A. 1998. Detection of anti-PL-12 autoantibodies by ELISA using a recombinant antigen; study of the immunoreactive region. *Clin. Exp. Immunol.* **114:** 161–165.

Gillet R. and Felden B. 2001. Transfer RNA[Ala] recognizes transfer-messenger RNA with specificity; a functional complex prior to entering the ribosome? *EMBO J.* **20:** 2966–2976.

Gottesman S., Roche E., Zhou Y., and Sauer R.T. 1998. The ClpXP and ClpAP proteases degrade proteins with carboxy-terminal peptide tails added by the SsrA-tagging system. *Genes Dev.* **12:** 1338–1347.

Han J.M., Kim J.Y., and Kim S. 2003. Molecular network and functional implications of macromolecular tRNA synthetase complex. *Biochem. Biophys. Res. Commun.* **303:** 985–993.

Harris C.E., Boden R.A., and Astell C.R. 1999. A novel heterogeneous nuclear ribonucleoprotein-like protein interacts with NS1 of the minute virus of mice. *J. Virol.* **73:** 72–80.

Hayes C.S., Bose B., and Sauer R.T. 2002a. Stop codons preceded by rare arginine codons are efficient determinants of SsrA tagging in *Escherichia coli. Proc. Natl. Acad. Sci.* **99:** 3440–3445.

———. 2002b. Proline residues at the C terminus of nascent chains induce SsrA tagging during translation termination. *J. Biol. Chem.* **277:** 33825–33832.

Herbert C.J., Labouesse M., Dujardin G., and Slonimski P.P. 1988. The NAM2 proteins from *S. cerevisiae* and *S. douglasii* are mitochondrial leucyl-tRNA synthetases, and are involved in mRNA splicing. *EMBO J.* **7:** 473–483.

Hirakata M., Suwa A., Nagai S., Kron M.A., Trieu E.P., Mimori T., Akizuki M., and Targoff I.N. 1999. Anti-KS: Identification of autoantibodies to asparaginyl-transfer RNA synthetase associated with interstitial lung disease. *J. Immunol.* **162:** 2315–2320.

Hou Y.M. and Schimmel P. 1988. A simple structural feature is a major determinant of the identity of a transfer RNA. *Nature* **333:** 140–145.

Howard O.M., Dong H.F., Yang D., Raben N., Nagaraju K., Rosen A., Casciola-Rosen L., Härtlein M., Kron M., Yiadom K., et al. 2002. Histidyl-tRNA synthetase and asparaginyl-tRNA synthetase, autoantigens in myositis, activate chemokine receptors on T lymphocytes and immature dendritic cells. *J. Exp. Med.* **196:** 781–791.

Hsieh S.L. and Campbell R.D. 1991. Evidence that gene G7a in the human major histocompatibilty complex encodes valyl-tRNA synthetase. *Biochem. J.* **278:** 809–816.

Huang C., Wolfgang M.C., Withey J., Koomey M., and Friedman D.I. 2000. Charged tmRNA but not tmRNA-mediated proteolysis is essential for *Neisseria gonorrhoeae* viability. *EMBO J.* **19:** 1098–1107.

Hurdle J.G., O'Neill A.J., and Chopra I. 2005. Prospects for aminoacyl-tRNA synthetase inhibitors as new antimicrobial agents. *Antimicrob. Agents Chemother.* **49:** 4821–4833.

Ibba M. and Söll D. 2000. Aminoacyl-tRNA synthesis. *Annu. Rev. Biochem.* **69:** 617–650.

Jeong E.J., Hwang G.S., Kim K.H., Kim M.J., Kim S., and Kim K.S. 2000. Structural analysis of multifunctional peptide motifs in human bifunctional tRNA synthetase: Identification of RNA-binding residues and functional implications for tandem repeats. *Biochemistry* **39:** 15775–15782.

Jiang S., Wolfe C.L., Warrington J.A., and Norcum M.T. 2005. Three-dimensional reconstruction of the valyl-tRNA synthetase/elongation factor-1H complex and localization of the δ subunit. *FEBS Lett.* **579:** 6049–6054.

Jordanova A., Irobi J., Thomas F.P., Van Dijck P., Meerschaert K., Dewil M., Dierick I., Jacobs A., De Vriendt E., Guergueltcheva V., et al. 2006. Disrupted function and axonal distribution of mutant tyrosyl-tRNA synthetase in dominant intermediate Charcot-Marie-Tooth neuropathy. *Nat. Genet.* **38:** 197–202.

Kaiser E., Hu B., Becher S., Eberhard D., Schray B., Baack M., Hameister H., and Knippers R. 1994. The human EPRS locus (formerly the QARS locus): A gene encoding a class I and a class II aminoacyl-tRNA synthetase. *Genomics* **19:** 280–290.

Karzai A.W., Susskind M.M., and Sauer R.T. 1999. SmpB, a unique RNA-binding protein essential for the peptide-tagging activity of SsrA (tmRNA). *EMBO J.* **18:** 3793–3799.

Keene J.D. and Tenenbaum S.A. 2002. Eukaryotic mRNPs may represent posttranscriptional operons. *Mol. Cell* **9:** 1161–1167.

Keiler K.C., Waller P.R., and Sauer R.T. 1996. Role of a peptide tagging system in degradation of proteins synthesized from damaged messenger RNA. *Science* **271:** 990–993.

Kim S., Lee S.W., Choi E.C., and Choi S.Y. 2003. Aminoacyl-tRNA synthetases and their inhibitors as a novel family of antibiotics. *Appl. Microbiol. Biotechnol.* **61:** 278–288.

Kinzy T.G. and Goldman E. 2000. Nontranslational functions of components of the translational apparatus. In *Translational control of gene expression* (ed. N. Sonenberg et al.), pp. 973–997. Cold Spring Harbor Laboratory Press, Cold Spring Harbor, New York.

Kise Y., Lee S.W., Park S.G., Fukai S., Sengoku T., Ishii R., Yokoyama S., Kim S., and Nureki O. 2004. A short peptide insertion crucial for angiostatic activity of human tryptophanyl-tRNA synthetase. *Nat. Struct. Mol. Biol.* **11:** 149–156.

Klausner R.D., Rouault T.A., and Harford J.B. 1993. Regulating the fate of mRNA: The control of cellular iron metabolism. *Cell* **72:** 19–28.

Ko Y.G., Park H., and Kim S. 2002. Novel regulatory interactions and activities of mammalian tRNA synthetases. *Proteomics* 2: 1304–1310.

Ko Y.G., Kang Y.S., Kim E.K., Park S.G., and Kim S. 2000. Nucleolar localization of human methionyl-tRNA synthetase and its role in ribosomal RNA synthesis. *J. Cell Biol.* 149: 567–574.

Ko Y.G., Kim E.Y., Kim T., Park H., Park H.S., Choi E.J., and Kim S. 2001. Glutamine-dependent antiapoptotic interaction of human glutaminyl-tRNA synthetase with apoptosis signal-regulating kinase 1. *J. Biol. Chem.* 276: 6030–6036.

Komine Y., Kitabatake M., Yokogawa T., Nishikawa K., and Inokuchi H. 1994. A tRNA-like structure is present in 10Sa RNA, a small stable RNA from *Escherichia coli*. *Proc. Natl. Acad. Sci.* 91: 9223–9227.

Kron M. and Härtlein M. 2005. Aminoacyl-tRNA synthetases and disease. In *The aminoacyl-tRNA synthetases* (ed. M. Ibba et al.), pp. 397–404. Landes Bioscience, Georgetown, Texas.

Kunst C.B., Mezey E., Brownstein M.J., and Patterson D. 1997. Mutations in *SOD1* associated with amyotrophic lateral sclerosis cause novel protein interactions. *Nat. Genet.* 15: 91–94.

Lambowitz A.M. and Perlman P.S. 1990. Involvement of aminoacyl-tRNA synthetases and other proteins in group I and group II intron splicing. *Trends Biochem. Sci.* 15: 440–444.

Lee J.S., Gyu Park S., Park H., Seol W., Lee S., and Kim S. 2002. Interaction network of human aminoacyl-tRNA synthetases and subunits of elongation factor 1 complex. *Biochem. Biophys. Res. Commun.* 291: 158–164.

Lee S.Y., Bailey S.C., and Apirion D. 1978. Small stable RNAs from *Escherichia coli*: Evidence for the existence of new molecules and for a new ribonucleoprotein particle containing 6S RNA. *J. Bacteriol.* 133: 1015–1023.

Lee Y.N. and Razin E. 2005. Nonconventional involvement of LysRS in the molecular mechanism of USF2 transcriptional activity in FcεRI-activated mast cells. *Mol. Cell. Biol.* 25: 8904–8912.

Lee Y.N., Nechushtan H., Figov N., and Razin E. 2004. The function of lysyl-tRNA synthetase and Ap4A as signaling regulators of MITF activity in FcεRI-activated mast cells. *Immunity* 20: 145–151.

Levine S.M., Rosen A., and Casciola-Rosen L.A. 2003. Anti-aminoacyl tRNA synthetase immune responses: Insights into the pathogenesis of the idiopathic inflammatory myopathies. *Curr. Opin. Rheumatol.* 15: 708–713.

Mathews M.B. and Bernstein R.M. 1983. Myositis autoantibody inhibits histidyl-tRNA synthetase: A model for autoimmunity. *Nature* 304: 177–179.

Mazumder B. and Fox P.L. 1999. Delayed translational silencing of ceruloplasmin transcript in gamma interferon-activated U937 monocytic cells: Role of the 3' untranslated region. *Mol. Cell. Biol.* 19: 6898–6905.

Mazumder B., Mukhopadhyay C.K., Prok A., Cathcart M.K., and Fox P.L. 1997. Induction of ceruloplasmin synthesis by IFN-γ in human monocytic cells. *J. Immunol.* 159: 1938–1944.

Mazumder B., Sampath B., Seshadri V., Maitra R.K., DiCorleto P., and Fox P.L. 2003. Regulated release of L13a from the 60S ribosomal subunit as a mechanism of transcript-specific translational control. *Cell* 115: 187–198.

McClain W.H. and Foss K. 1988a. Changing the acceptor identity of a transfer RNA by altering nucleotides in a "variable pocket." *Science* 241: 1804–1807.

———. 1988b. Changing the identity of a tRNA by introducing a G-U wobble pair near the 3' acceptor end. *Science* 240: 793–796.

Miller F.W., Twitty S.A., Biswas T., and Plotz P.H. 1990a. Origin and regulation of a disease-specific autoantibody response. Antigenic epitopes, spectrotype stability, and isotype restriction of anti-Jo-1 autoantibodies. *J. Clin. Invest.* **85:** 468–475.

Miller F.W., Waite K.A., Biswas T., and Plotz P.H. 1990b. The role of an autoantigen, histidyl-tRNA synthetase, in the induction and maintenance of autoimmunity. *Proc. Natl. Acad. Sci.* **87:** 9933–9937.

Minella O., Mulner-Lorillon O., Bec G., Cormier P., and Belle R. 1998. Multiple phosphorylation sites and quaternary organization of guanine-nucleotide exchange complex of elongation factor-1 (EF-1βγδ/ValRS) control the various functions of EF-1α. *Biosci. Rep.* **18:** 119–127.

Motorin Y.A., Wolfson A.D., Orlovsky A.F., and Gladilin K.L. 1988. Mammalian valyl-tRNA synthetase forms a complex with the first elongation factor. *FEBS Lett.* **238:** 262–264.

Mulner-Lorillon O., Cormier P., Cavadore J.C., Morales J., Poulhe R., and Belle R. 1992. Phosphorylation of *Xenopus* elongation factor-1γ by cdc2 protein kinase: Identification of the phosphorylation site. *Exp. Cell Res.* **202:** 549–551.

Musci G. 2001. Ceruloplasmin, the unique multi-copper oxidase of vertebrates. *Protein Pept. Lett.* **8:** 159–169.

Negrutskii B.S. and Deutscher M.P. 1991. Channeling of aminoacyl-tRNA for protein synthesis *in vivo*. *Proc. Natl. Acad. Sci.* **88:** 4991–4995.

Negrutskii B.S., Shalak V.F., Kerjan P., El'skaya A.V., and Mirande M. 1999. Functional interaction of mammalian valyl-tRNA synthetase with elongation factor EF-1α in the complex with EF-1H. *J. Biol. Chem.* **274:** 4545–4550.

Otani A., Slike B.M., Dorrell M.I., Hood J., Kinder K., Ewalt K.L., Cheresh D., Schimmel P., and Friedlander M. 2002. A fragment of human TrpRS as a potent antagonist of ocular angiogenesis. *Proc. Natl. Acad. Sci.* **99:** 178–183.

Park S.G., Ewalt K.L., and Kim S. 2005a. Functional expansion of aminoacyl-tRNA synthetases and their interacting factors: New perspectives on housekeepers. *Trends Biochem. Sci.* **30:** 569–574.

Park S.G., Kim H.J., Min Y.H., Choi E.C., Shin Y.K., Park B.J., Lee S.W., and Kim S. 2005b. Human lysyl-tRNA synthetase is secreted to trigger proinflammatory response. *Proc. Natl. Acad. Sci.* **102:** 6356–6361.

Peters H.I., Chang Y.W., and Traugh J.A. 1995. Phosphorylation of elongation factor 1 (EF-1) by protein kinase C stimulates GDP/GTP-exchange activity. *Eur. J. Biochem.* **234:** 550–556.

Pleiss J.A., Wolfson A.D., and Uhlenbeck O.C. 2000. Mapping contacts between *Escherichia coli* alanyl tRNA synthetase and 2′ hydroxyls using a complete tRNA molecule. *Biochemistry* **39:** 8250–8258.

Pohlmann J. and Brötz-Oesterhelt H. 2004. New aminoacyl-tRNA synthetase inhibitors as antibacterial agents. *Curr. Drug Targets Infect. Disord.* **4:** 261–272.

Putney S.D. and Schimmel P. 1981. An aminoacyl tRNA synthetase binds to a specific DNA sequence and regulates its gene transcription. *Nature* **291:** 632–635.

Raben N., Nichols R., Dohlman J., McPhie P., Sridhar V., Hyde C., Leff R., and Plotz P. 1994. A motif in human histidyl-tRNA synthetase which is shared among several aminoacyl-tRNA synthetases is a coiled-coil that is essential for enzymatic activity and contains the major autoantigenic epitope. *J. Biol. Chem.* **269:** 24277–24283.

Retallack D.M., Johnson L.L., and Friedman D.I. 1994. Role for 10Sa RNA in the growth of λ-P22 hybrid phage. *J. Bacteriol.* **176:** 2082–2089.

Rho S.B., Lincecum T.L., Jr., and Martinis S.A. 2002. An inserted region of leucyl-tRNA synthetase plays a critical role in group I intron splicing. *EMBO J.* **21:** 6874–6881.

Rho S.B., Kim M.J., Lee J.S., Seol W., Motegi H., Kim S., and Shiba K. 1999. Genetic dissection of protein-protein interactions in multi-tRNA synthetase complex. *Proc. Natl. Acad. Sci.* **96:** 4488–4493.

Ribas de Pouplana L. and Schimmel P. 2001. Aminoacyl-tRNA synthetases: Potential markers of genetic code development. *Trends Biochem. Sci.* **26:** 591–596.

Robinson J.C., Kerjan P., and Mirande M. 2000. Macromolecular assemblage of aminoacyl-tRNA synthetases: Quantitative analysis of protein-protein interactions and mechanism of complex assembly. *J. Mol. Biol.* **304:** 983–994.

Roche E.D. and Sauer R.T. 2001. Identification of endogenous SsrA-tagged proteins reveals tagging at positions corresponding to stop codons. *J. Biol. Chem.* **276:** 28509–28515.

Romby P., Caillet J., Ebel C., Sacerdot C., Graffe M., Eyermann F., Brunel C., Moine H., Ehresmann C., Ehresmann B., and Springer M. 1996. The expression of *E. coli* threonyl-tRNA synthetase is regulated at the translational level by symmetrical operator-repressor interactions. *EMBO J.* **15:** 5976–5987.

Rudinger-Thirion J., Giegé R., and Felden B. 1999. Aminoacylated tmRNA from *Escherichia coli* interacts with prokaryotic elongation factor Tu. *RNA* **5:** 989–992.

Sampath P., Mazumder B., Seshadri V., and Fox P.L. 2003. Transcript-selective translational silencing by gamma interferon is directed by a novel structural element in the ceruloplasmin mRNA 3′ untranslated region. *Mol. Cell. Biol.* **23:** 1509–1519.

Sampath P., Mazumder B., Seshadri V., Gerber C.A., Chavatte L., Kinter M., Ting S.M., Dignam J.D., Kim S., Driscoll D.M., and Fox P.L. 2004. Noncanonical function of glutamyl-prolyl-tRNA synthetase: Gene-specific silencing of translation. *Cell* **119:** 195–208.

Sellami, M., F. Fasiolo, G. Dirheimer, J.P. Ebel, and J. Gangloff. 1986. Nucleotide sequence of the gene coding for yeast cytoplasmic aspartyl-tRNA synthetase (APS); mapping of the 5′ and 3′ termini of AspRS mRNA. *Nucleic Acids Res.* **14:** 1657–1666.

Shah Z.H., Toompuu M., Hakkinen T., Rovio A.T., van Ravenswaay C., De Leenheer E.M., Smith R.J., Cremers F.P., Cremers C.W., and Jacobs H.T. 2001. Novel coding-region polymorphisms in mitochondrial seryl-tRNA synthetase (*SARSM*) and mitoribosomal protein S12 (*RPMS12*) genes in *DFNA4* autosomal dominant deafness families. *Hum. Mutat.* **17:** 433–434.

Shiba K. 2002. Intron positions delineate the evolutionary path of a pervasively appended peptide in five human aminoacyl-tRNA synthetases. *J. Mol. Evol.* **55:** 727–733.

Shiba K., Motegi H., Yoshida M., and Noda T. 1998. Human asparaginyl-tRNA synthetase: Molecular cloning and the inference of the evolutionary history of Asx-tRNA synthetase family. *Nucleic Acids Res.* **26:** 5045–5051.

Shimizu Y. and Ueda T. 2002. The role of SmpB protein in *trans*-translation. *FEBS Lett.* **514:** 74–77.

Sissler M., Pütz J., Fasiolo F., and Florentz C. 2005. Mitochondrial aminoacyl-tRNA synthetases. In *The aminoacyl-tRNA synthetases* (ed. M. Ibba et al.), pp. 271–284. Landes Bioscience, Georgetown, Texas.

Tadaki T., Fukushima M., Ushida C., Himeno H., and Muto A. 1996. Interaction of 10Sa RNA with ribosomes in *Escherichia coli*. *FEBS Lett.* **399:** 223–226.

Targoff I.N. 1990. Autoantibodies to aminoacyl-transfer RNA synthetases for isoleucine and glycine. Two additional synthetases are antigenic in myositis. *J. Immunol.* **144:** 1737–1743.

————. 2000. Update on myositis-specific and myositis-associated autoantibodies. *Curr. Opin. Rheumatol.* **12:** 475–481.

Targoff I.N., Arnett F.C., and Reichlin M. 1988. Antibody to threonyl-transfer RNA synthetase in myositis sera. *Arthritis Rheum.* **31:** 515–524.

Targoff I.N., Miller F.W., Medsger T.A., Jr., and Oddis C.V. 1997. Classification criteria for the idiopathic inflammatory myopathies. *Curr. Opin. Rheumatol.* **9:** 527–535.

't Hart L.M., Hansen T., Rietveld I., Dekker J.M., Nijpels G., Janssen G.M., Arp P.A., Uitterlinden A.G., Jørgensen T., Borch-Johnsen K., et al. 2005. Evidence that the mitochondrial leucyl tRNA synthetase (*LARS2*) gene represents a novel type 2 diabetes susceptibility gene. *Diabetes* **54:** 1892–1895.

Thomas G. 1986. Translational control of mRNA expression during the early mitogenic response in Swiss mouse 3T3 cells: Identification of specific proteins. *J. Cell Biol.* **103:** 2137–2144.

Tzima E. and Schimmel P. 2006. Inhibition of tumor angiogenesis by a natural fragment of a tRNA synthetase. *Trends Biochem. Sci.* **31:** 7–10.

Ushida C., Himeno H., Watanabe T., and Muto A. 1994. tRNA-like structures in 10Sa RNAs of *Mycoplasma capricolum* and *Bacillus subtilis*. *Nucleic Acids Res.* **22:** 3392–3396.

Vargas R. and Castaneda M. 1981. Role of elongation factor 1 in the translational control of rodent brain protein synthesis. *J. Neurochem.* **37:** 687–694.

Venema R.C., Peters H.I., and Traugh J.A. 1991a. Phosphorylation of elongation factor 1 (EF-1) and valyl-tRNA synthetase by protein kinase C and stimulation of EF-1 activity. *J. Biol. Chem.* **266:** 12574–12580.

————. 1991b. Phosphorylation of valyl-tRNA synthetase and elongation factor 1 in response to phorbol esters is associated with stimulation of both activities. *J. Biol. Chem.* **266:** 11993–11998.

Wakasugi K. and P. Schimmel. 1999a. Highly differentiated motifs responsible for two cytokine activities of a split human tRNA synthetase. *J. Biol. Chem.* **274:** 23155–23159.

————. 1999b. Two distinct cytokines released from a human aminoacyl-tRNA synthetase. *Science* **284:** 147–151.

Wakasugi K., Slike B.M., Hood J., Ewalt K.L., Cheresh D.A., and Schimmel P. 2002a. Induction of angiogenesis by a fragment of human tyrosyl-tRNA synthetase. *J. Biol. Chem.* **277:** 20124–20126.

Wakasugi K., Slike B.M., Hood J., Otani A., Ewalt K.L., Friedlander M., Cheresh D.A., and Schimmel P. 2002b. A human aminoacyl-tRNA synthetase as a regulator of angiogenesis. *Proc. Natl. Acad. Sci.* **99:** 173–177.

Withey J. and Friedman D. 1999. Analysis of the role of *trans*-translation in the requirement of tmRNA for λ*imm*[P22] growth in *Escherichia coli*. *J. Bacteriol.* **181:** 2148–2157.

————. 2002. The biological roles of trans-translation. *Curr. Opin. Microbiol.* **5:** 154–159.

————. 2003. A salvage pathway for protein structures: tmRNA and trans-translation. *Annu. Rev. Microbiol.* **57:** 101–123.

Wower I.K., Zwieb C.W., Guven S.A., and Wower J. 2000. Binding and cross-linking of tmRNA to ribosomal protein S1, on and off the *Escherichia coli* ribosome. *EMBO J.* **19:** 6612–6621.

Wu H., Nada S., and Dignam J.D. 1995. Analysis of truncated forms of *Bombyx mori* glycyl-tRNA synthetase: Function of an N-terminal structure in RNA binding. *Biochemistry* **34:** 16327–16336.

Zee R.Y., Hegener H.H., Chasman D.I., and Ridker P.M. 2005. Tryptophanyl-tRNA synthetase gene polymorphisms and risk of incident myocardial infarction. *Atherosclerosis* **181:** 137–141.

Zheng J.Q., Kelly T.K., Chang B., Ryazantsev S., Rajasekaran A.K., Martin K.C., and Twiss J.L. 2001. A functional role for intra-axonal protein synthesis during axonal regeneration from adult sensory neurons. *J. Neurosci.* **21:** 9291–9303.

Zvereva M.I., Ivanov P.V., Teraoka Y., Topilina N.I., Dontsova O.A., Bogdanov A.A., Kalkum M., Nierhaus K.H., and Shpanchenko O.V. 2001. Complex of transfer-messenger RNA and elongation factor Tu. Unexpected modes of interaction. *J. Biol. Chem.* **276:** 47702–47708.

30

Therapeutic Opportunities in Translation

Jerry Pelletier

Department of Biochemistry and McGill Cancer Center
McIntyre Medical Sciences Building, McGill University
Montreal, Quebec, Canada H3G 1Y6

Stuart W. Peltz

PTC Therapeutics
South Plainfield, New Jersey 07080

THE PROTEIN SYNTHESIS APPARATUS AND SIGNALING PATHWAYS that regulate its activity represent excellent, largely unexploited targets for small-molecule discovery. Approaches that disrupt this process can cause either qualitative or quantitative changes in mRNA expression. Interference with the function of rRNA, tRNA, or general protein factors is likely to exert effects on global protein synthesis. On the other hand, compounds that target the ribosome recruitment phase of translation have the potential to selectively inhibit gene expression. A significant portion of our current understanding of the translation process is a consequence of utilizing small molecules to chemically dissect this complex process (Pestka 1977; Vazquez 1979). Such probes have been used to perturb the translation process in vitro and in vivo, freeze short-lived intermediates that otherwise could not be studied, identify new initiation factors, and therapeutically target this process in pathogenic organisms. At a time when novel approaches for discovering new drugs to treat a range of microbial, viral, and metabolic diseases are sought, it would seem opportune to review our understanding of small molecules that target translation. Herein, we discuss various aspects of the translation process that have recently been explored as targets for small-molecule discovery. The potential for targeting this process as an anticancer approach is also addressed. Finally, we review examples of small-molecule inhibitors of translation that are clinically used as anti-infective agents.

SMALL-MOLECULE APPROACHES THAT QUALITATIVELY ALTER mRNA TRANSLATION

Treating Genetic Disorders by Promoting Readthrough of Nonsense Mutations

Genetic disorders often arise as a consequence of mutations that abolish production of specific proteins. Although this type of impairment could result from defects at any of several steps in the gene expression pathway, a common type of mutation causing genetic disorders inactivates gene function by promoting premature translation termination. Nonsense mutations occurring in the protein-coding region of the mRNA template lead to the production of a truncated polypeptide product and can also be associated with mRNA destabilization. Such mutations have been implicated in many inherited diseases, including cystic fibrosis (CF), Duchenne muscular dystrophy (DMD), hemophilias, lysosomal storage disorders, skin disorders, and various cancers.

One therapeutic approach to treat genetic disorders caused by nonsense mutations is to develop drugs that promote their readthrough, thus allowing the production of full-length active proteins. The decoding site of eukaryotic 18S rRNA is similar to, but not identical to, the 16S rRNA-decoding site. One major difference is that A1408 in the prokaryotic 16S rRNA is a guanosine in 18S rRNA, a change associated with reduced aminoglycoside binding (Recht et al. 1999). In addition, the aminoglycoside-binding site in eukaryotes appears to be a more shallow pocket (Lynch and Puglisi 2001). Although these differences have enabled aminoglycosides to be used therapeutically as antimicrobial agents, this class of compounds is known to induce low levels of translational misreading in eukaryotes (Palmer and Wilhelm 1978; Howard et al. 2000; Manuvakhova et al. 2000). The sequence context of the termination codon and the nature of the aminoglycoside influence the efficiency with which nonsense mutations are suppressed (Howard et al. 2000; Manuvakhova et al. 2000).

On the basis of these observations, a pharmacological approach using aminoglycosides to treat genetic disorders caused by nonsense mutations has been investigated. In cell-based assays, gentamicin can promote readthrough of nonsense mutations resulting in production of full-length CFTR (cystic fibrosis transmembrane conductance regulator) (Howard et al. 1996; Bedwell et al. 1997), dystrophin (Barton-Davis et al. 1999), α-ʟ-iduronidase (Keeling et al. 2001), tripeptidyl-peptidase 1 (Sleat et al. 2001), cystinosis (Helip-Wooley et al. 2002), and ataxia-telangiectasia mutated (Lai et al. 2004). The aminoglycoside amikacin and the dipeptide antibiotic, negamycin, have also shown efficacy in suppressing disease- associated premature stop mutations (Keeling and Bedwell 2002; Arakawa et al. 2003). In preclinical mouse models of DMD, CF, and nephrogenic diabetes insipidus, aminoglycosides have been shown to suppress nonsense mutations and ameliorate the clinical symptoms of these disease-causing mutations (Barton-Davis et al. 1999; Du et al. 2002; Sangkuhl et al. 2004).

The preclinical results obtained with gentamicin therapy in mouse models led to the testing of similar approaches in patients with CF and DMD. A comparison of CF patients harboring nonsense alleles with those carrying the ΔF508 mutation demonstrated that only the former group responded to two 2-week periods of gentamicin application to the nasal mucosa, manifesting increased local CFTR protein with corresponding functional improvements in chloride channel activity (Wilschanski et al. 2003). A study with patients harboring nonsense mutations in the DMD gene demonstrated that intravenous administration of gentamicin increased production of full-length dystrophin (Politano et al. 2003), although another study did not observe gentamicin-promoted readthrough (Wagner et al. 2001). The reason for the discrepancy in results is not known, but gentamicin consists of a mixture of different isomers with varying activities in promoting readthrough, and different batches of gentamicin have varying levels of these isomers (H.L. Sweeney and E.R. Barton-Davis, unpubl.). The identification of non-aminoglycoside compounds that suppress nonsense mutations circumvents the need for parenteral administration, avoids the renal toxicity and ototoxicity associated with long-term gentamicin treatment, and opens the avenue for treating genetic disorders that are the consequence of nonsense mutations, with compounds having a safer profile. To this end, PTC124, an orally bioavailable small-molecule heterocycle, is currently in clinical trials for the treatment of CF and DMD arising from nonsense mutations (S. Peltz, unpubl.).

Targeting Frameshifting as an Antiviral Approach

A number of viral mRNAs frameshift during translation, an event that generally requires two components: a "slippery" sequence followed by

RNA secondary structure that is either a hairpin or pseudoknot (Chapter 22). The likely function of the slippery site and secondary structure is to impede the progressing ribosomes such that the P- and A-site tRNAs slip backward in the 5′ direction simultaneously from their initial positions in the zero frame. The importance of the slippery sequence is highlighted by sequence comparison of 1000 human immunodeficiency virus type 1 (HIV-1) sequences that found this site to be invariant, a somewhat surprising result for a virus known for its variability (Biswas et al. 2004). Changes to the HIV-1 slippery site that reduce frameshifting levels dramatically inhibit infectivity (Biswas et al. 2004). This process is an attractive drug target since it is essential for HIV-1 replication, resistant to mutation, and is not commonly found during translation of cellular mRNAs. A screen for compounds that affect this process utilized a split luciferase reporter in which the HIV-1 frameshift signal interrupted the firefly-luciferase-coding region, placing the carboxy-terminal sequences in the −1 reading frame with respect to the luciferase start codon. One compound was identified that stimulated frameshifting in vitro and in vivo and doubled the production of the Gag-Pol product. This compound inhibits viral replication and appears to exert its effect by binding to and stabilizing the RNA stem-loop within the frameshift signal (Hung et al. 1998).

Anisomycin and sparsomycin, two compounds known to affect A-site function, also alter the efficiency of −1 ribosomal frameshifting, whereas other general inhibitors such as cycloheximide, paromomycin, and hygromycin have no effect on this process (Dinman et al. 1997). Ribosomal frameshifting decreases in the presence of anisomycin, an effect rationalized by the ability of anisomycin to destabilize A-site-specific tRNA–ribosome interaction (see Appendix I at end of chapter), which together with the unstable codon/anticodon interaction of the A-site aminoacylated tRNA (aa-tRNA) during decoding of the slippery sequence, leads to aborted translation and reduced frameshifting (Dinman et al. 1997). On the other hand, sparsomycin stimulates −1 frameshifting, possibly by increasing binding of the peptidyl-tRNA stem to the donor site of the peptidyltransferase center (PTC) and slowing the rate of the peptidyltransferase reaction (Dinman et al. 1997). Mutations in both eukaryotic elongation factor 1A (eEF1A) and eEF2 have been described that alter frameshifting efficiency, suggesting that small molecules targeting these *trans*-acting factors may be able to alter this process and thus provide a novel antiviral strategy (Goss Kinzy et al. 2002).

SMALL-MOLECULE APPROACHES THAT QUANTITATIVELY ALTER mRNA TRANSLATION

mRNA as a Drug Target

Most current drug discovery efforts are aimed at identifying small molecules that inhibit protein function. However, the discovery of riboswitches (Chapter 28) and that rRNA is the primary target for many antibiotics has underscored the possibility of regulating gene expression by targeting RNA with small molecules. Engineered riboswitches were first described by Werstuck and Green (1998), who inserted an aptamer directed against Hoechst 33258 into the 5′-untranslated region (5′UTR) of a reporter mRNA and showed that this could be used to regulate gene expression. This report demonstrated the feasibility of directly modulating mRNA expression utilizing small molecules. Subsequently, Harvey et al. (2002b) demonstrated that small-molecule ligands targeting engineered riboswitches could inhibit the ribosome recruitment phase of eukaryotic translation initiation. Compounds with known RNA-binding activity have also been used to selectively inhibit expression of thymidylate synthase (TS), apparently as a consequence of a fortuitous binding site(s) present near the TS initiation codon (Tok et al. 1999; Cho and Rando 2000).

Unfortunately, the promiscuous binding nature of many RNA-binding compounds (e.g., aminoglycosides and intercalators), along with the conformational flexibility of RNA, has made direct targeting of RNA with small molecules difficult. One approach to overcome these problems is to restrict RNA conformation by utilizing chemical inducers of dimerization (CID). Small molecules that bind two proteins simultaneously have been extensively used to modulate the activity of different cellular proteins to evoke physiological changes (Crabtree and Schreiber 1996). Two features define CIDs: One is their ability to induce proximity and the other is that the stability of the trimeric complex formed is greater than that of the individual bimolecular components (Crabtree and Schreiber 1996). A CID approach to recruit a protein to a specific mRNA target was first demonstrated by positioning tobramycin aptamers in the 5′UTR of a reporter mRNA. Tobramycin–biotin conjugates were designed to recruit streptavidin to the tobramycin-binding site (Harvey et al. 2002a). Ribosome recruitment to, and translation of, the mRNA reporter was inhibited only when translation reactions were supplemented with the CID (Harvey et al. 2002a). This targeting method takes advantage of the fact that RNA undergoes induced structural reorganizations when binding to specific protein surfaces.

The antibiotic thiostrepton may function as a CID when exerting its inhibitory effects on translation. During elongation, contacts are made between the amino-terminal domain of rpL11 and EF-G (Agrawal et al. 2001), as well as with release factors (Tate et al. 1986). Thiostrepton is thought to interfere with these interactions by binding to the 50S ribosomal subunit at the rpL11-binding site (Rodnina et al. 1999; Cameron et al. 2002) and making contacts with both rRNA and rpL11 (Lentzen et al. 2003). Binding of thiostrepton to the ribosome is cooperative with rpL11 (Xing and Draper 1996).

Targeting RNA–Protein Interactions as Therapeutic Intervention Strategies

Iron regulation in a cell is controlled by RNA-binding proteins, called iron-regulatory proteins (IRPs), that interact with mRNAs containing iron-responsive elements (IREs) to suppress translation or promote mRNA stabilization (Rouault and Harford 2000). When the iron concentration is low, IRP binds to the 5′UTR of ferritin mRNA to inhibit ferritin production, whereas binding to the 3′UTR of the transferrin receptor stabilizes the mRNA and produces more of the iron receptor. IREs have been found in a number of other transcripts, including the erythroid 5-aminolevulinate synthase (Cox et al. 1991), mammalian mitochondrial aconitase (Gray et al. 1996), *Drosophila melanogaster* succinate dehydrogenase subunit b (Kohler et al. 1995), and the Alzheimer's β-amyloid precursor protein transcript (amyloid-β peptide) (Rogers et al. 2002). Recently, the natural product yohimbine was identified as being capable of binding to the ferritin IRE (Tibodeau et al. 2006). Yohimbine prevents IRP binding to the IRE, causing an increase in translation of the ferritin mRNA. In contrast, the drug phenserine reduces the production of amyloid-β peptide by decreasing translation of the β-amyloid precursor protein transcript in tissue culture (Shaw et al. 2001). The effect is mediated through the 5′UTR of mRNA and may involve regulation of the IRP–IRE interaction (Shaw et al. 2001). These examples highlight the potential to regulate mRNA expression by interfering with specific mRNA–protein interactions.

Inhibitors of Eukaryotic Translation as Antitumor Agents

Deregulation of global translation and selective mRNA expression have emerged as important components of cancer etiology and chemoresponsiveness (Chapter 15). The treatment of acute lymphoblastic leukemia (ALL) with L-asparaginase highlights how inhibiting protein synthesis can be used as a chemotherapeutic approach in the clinic. Transformed hematopoietic cells are often unable to synthesize sufficient asparagine for their own metabolism, requiring them to import asparagine from the serum. Depletion of the systemic asparagine pool from the serum by administration of L-asparaginase leads to inhibition of protein synthesis in leukemic cells, followed by cell death (Muller and Boos 1998). This approach is currently the standard of care in ALL.

Several inhibitors of eukaryotic protein synthesis (see Appendix I) have been tested as potential chemotherapeutic agents in preclinical murine models (for a description of results with individual compounds, see http://dtp.nci.nih.gov/docs/dtp_search.html). In addition, a synthetic-lethal screen for compounds capable of killing engineered tumor cells, but not their isogenic nontransformed counterparts, identified the translation inhibitor bouvardin as a genotype-selective agent (Dolma et al. 2003). Bouvardin was also independently identified in a screen for antileukemic agents (Valeriote et al. 1996). Recently, a screen for compounds with the ability to overcome human papillomavirus (HPV) E6 oncogene-induced drug resistance identified the translation inhibitors emetine, dihydrolycorine, and cycloheximide (see Appendix I) as potentiators of doxorubicin cytotoxicity (Smukste et al. 2006). Translation elongation inhibitors have also been shown to sensitize cells to tumor necrosis factor (TNF) family members (Choi et al. 2004) and to cisplatin (Budihardjo et al. 2000). However, these effects may be genotype-selective or concentration-dependent since in some systems, cycloheximide antagonizes doxorubicin-induced apoptosis (Bonner and Lawrence 1989; Furusawa et al. 1995).

Two inhibitors of protein synthesis have been tested in clinical trials, didemnin B and homoharringtonine. Results from phase II clinical trials indicate that didemnin B has little or no activity against advanced human cancers, possibly due to its rapid conversion to an inactive form (Rinehart 2000; Vera and Joullie 2002). A closely related derivative, dehydrodidemnin B (aplidine) is ten times more active than didemnin B against murine leukemia cells and human tumor xenografts while not showing the cardiotoxicity of didemnin B. Results from a phase I study of dehydrodidemnin B on 67 patients with a variety of advanced malignancies indicated no objective response in antitumor activity, although six patients with endocrine tumors (four with thyroid carcinoma, one with bronchial carcinoma, and one with malignant pheochromocytoma) showed prolonged disease stabilization (>6 months) (Faivre et al. 2005).

Clinical trials with homoharringtonine have reported encouraging results in patients with acute myeloid leukemia, myelodysplastic syndrome, acute promyelocytic leukemia, and chronic myeloid leukemia (for review, see Kantarjian et al. 2001). Currently, structural derivatives are being generated to improve on some of the pharmacological properties of this compound (i.e., cardiotoxicity). Phase II clinical trials are in progress to assess the efficacy of homoharringtonine in chronic myeloid leukemia or as salvage therapy in patients with refractory acute promyelocytic leukemia (see http://clinicaltrials.gov/ct).

Structure-activity-relationship (SAR) studies performed with sparsomycin and didemnin B demonstrated equivalent rank order potency for inhibition of translation and for their antiproliferative effects (Ottenheijm and van den Broek 1988; van den Broek et al. 1989; Ahuja et al. 2000). These results correlate the antitumor activity of these compounds with their ability to inhibit protein synthesis. More controlled studies are required in which anticancer efficacy is monitored in tumors harboring genetic lesions known to affect translation rates (e.g., *Akt*/mTOR pathway). The identification of reliable surrogate markers to monitor translation in a tumor undergoing drug exposure is also required to allow assessment of the efficacy of compound treatment and help define dosing strategies. Targeting the initiation phase of translation to inhibit selective mRNA species may also be a more effective avenue to explore.

TRANSLATION INITIATION AS AN ANTICANCER TARGET

Targeting the mTOR Pathway: The Rapamycin Example

The recruitment of ribosomes to mRNAs is highly regulated by extra- and intracellular environmental cues that impinge on the mTOR (mammalian target of rapamycin) kinase pathway (Chapter 14). Although mTOR has been shown to regulate a large number of cellular processes, its best-characterized function in mammals is regulation of translation initiation (Hay and Sonenberg 2004). It achieves this by influencing the shuttling of eukaryotic initiation factor 4E (eIF4E) between the eIF4F complex and the inhibitory 4E-BP/eIF4E heterodimer (Fig. 1). Phosphorylation of 4E-BPs upon mTOR activation allows eIF4E to assemble into the eIF4F complex, resulting in stimulation of protein synthesis. Conversely, dephosphorylation of 4E-BP1 leads to increased sequestration of eIF4E from eIF4G and decreases translation initiation rates (Fig. 1). mTOR activity is elevated in many human cancers as a consequence of mutations in the upstream negative regulator PTEN, amplification of the p110 catalytic subunit of phosphoinositide 3-kinase (PI3K), or amplification of the *Akt* gene (Bjornsti and Houghton 2004). Moreover, constitutive growth factor activation can also lead to aberrant activation of this pathway. This includes increased signaling from the HER (ErbB) family of growth factor receptors, the insulin receptor, the insulin-like growth factor receptor (Pollak et al. 2004),

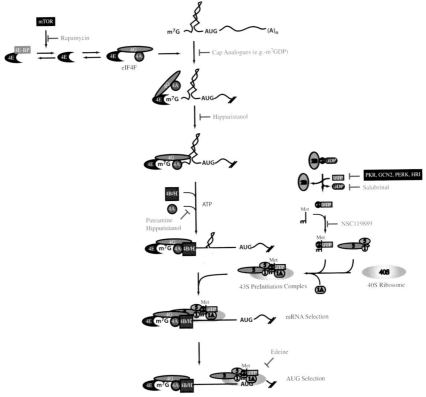

Figure 1. Schematic representation of the ribosome recruitment phase of translation initiation, illustrating the steps targeted by various translation inhibitors. (*Black rectangles*) Kinases that affect the activity of different steps. See text for additional details.

the estrogen receptor (Dancey and Sausville 2003), and oncogenic Ras (Li et al. 2004). A large number of studies have assessed the therapeutic potential of inhibiting PI3K/*Akt* prosurvival signaling in cancers (Hennessy et al. 2005; Morgensztern and McLeod 2005).

Rapamycin (Sirolimus, Rapamune, Wyeth Ayerst Laboratories) is a CID that forms a complex with FK506-binding protein, FKBP12, which in turn binds to a region in the carboxyl terminus of TOR to inhibit its kinase activity. Although initially discovered as a fungicide and later established as an immunosuppressant, it was soon realized that rapamycin exhibits potent antitumor activity (Douros and Suffness 1981). Rapamycin treatment induces G_1 phase arrest in lymphocytes, although in most other cell types, it delays cell cycle progression rather than inducing an absolute block (Abraham and Wiederrecht 1996). In some cells, rapamycin induces apoptosis (Hosoi et al. 1999; Huang et al. 2001) and also exerts prominent antiangiogenic properties (Guba et al. 2002). Because of formulation and chemical stability problems with rapamycin, analogs with improved pharmaceutical properties and comparable efficacy have been developed. These include CCI-779 (Wyeth Ayerst), RAD001 (everolimus, Novartis Pharmaceuticals), and AP23573 (Ariad Pharmaceuticals). Inhibition of TOR activity by rapamycin is thought to affect translation of specific classes of mRNAs, including transcripts that encode cell cycle and apoptotic regulators (for review, see Chapter 15). Compounds that alter the ribosome recruitment phase of translation initiation may be more specific than rapamycin since these would avoid disrupting other cellular processes under TOR control.

Inhibitors of Translation Initiation

Of all the initiation factors, eIF4E is the one that is most implicated in cancer (Chapter 15). eIF4E is the subunit of eIF4F responsible for binding of the complex to the mRNA cap structure (m^7GpppN; where N is any nucleotide) (Fig. 1). The eIF4F complex also consists of eIF4A, a DEAD-box RNA helicase thought to unwind local mRNA secondary structure to facilitate access of the 43S ribosomal complex to the mRNA template and a scaffolding protein (eIF4GI or eIF4GII, encoded by two different genes) that mediates mRNA binding to the 43S preinitiation complex via interactions with eIF3 (present on the 40S ribosome) (Fig. 1). Unfortunately, few small-molecule inhibitors of translation initiation have been described that could be tested in preclinical models (Fig. 1). Cap analogs (e.g., m^7GDP) are effective at inhibiting cap-dependent ribosome recruitment but cannot be used in vivo since they are not cell-permeable. Edeine is not expected to selectively inhibit protein synthesis since it acts downstream from the ribosome recruitment phase (see Appendix I and Fig. 1).

Recently, two inhibitors of translation that affect activity of the eIF4A subunit of eIF4F have been identified and characterized. Pateamine is a marine natural product that stimulates several eIF4A-associated activities, including RNA-dependent ATP hydrolysis, ATP and RNA binding, and helicase activity, and appears to "set" eIF4A in a high-affinity RNA-binding conformation (Bordeleau et al. 2005; Low et al. 2005). How stimulation of eIF4A activity results in inhibition of protein synthesis is not clear. Low et al. (2005) have suggested that pateamine acts as a CID to induce dimer formation between eIF4A and eIF4B. However, this apparent association may be an indirect consequence of eIF4B's nonsequence-specific RNA-binding activity accompanied by the induced binding of eIF4A to RNA by pateamine (M.E. Bordeleau and J. Pelletier, unpubl.). Hippuristanol, another marine natural product that inhibits translation initiation, blocks the RNA-binding properties of free eIF4A and of eIF4A present in the eIF4F complex (Bordeleau et al. 2006). Although hippuristanol and pateamine have been previously recognized as being cytotoxic (Higa et al. 1981; Hood et al. 2001), it remains to be determined whether they (or any other translation initiation inhibitor) can exert a selective inhibitory growth effect on transformed cells relative to nontransformed counterparts.

Modulating Ternary Complex Formation as a Therapeutic Approach

Phosphorylation of eIF2α causes a decrease in general translation by inhibiting exchange of GDP for GTP by eIF2B, thereby preventing reconstitution of a functional ternary complex for a new round of translation initiation (Fig. 1). Decreased phosphorylation of eIF2α (and hence increased protein synthetic rates) is associated with transformation. Introduction of a dominant-negative form of the eIF2α kinase, PKR, into NIH-3T3 cells leads to cellular transformation (Koromilas et al. 1992). Furthermore, overexpression of the non-phosphorylatable eIF2α mutant (S51A), but not wild-type eIF2α, causes cells to become malignant (Donze et al. 1995). The antiproliferative effects of small molecules that induce phosphorylation of eIF2α and limit the availability of the eIF2·Met-tRNA$_i^{Met}$·GTP ternary complex have been assessed. Clotrimazole (Benzaquen et al. 1995; Aktas et al. 1998), troglitazone (Palakurthi et al. 2001), and eicosapentaenoic acid (Palakurthi et al. 2000) have all been shown to cause partial depletion of intracellular Ca^{2+} stores, activation of PKR, and phosphorylation of eIF2α. Although these compounds show antiproliferative activities, it remains to be established whether this is a direct consequence of altering eIF2α phosphorylation status (and hence ternary complex availability) or due to secondary effects arising from increased intracellular Ca^{2+} concentrations.

Recently, Boyce et al. (2005) described a chemical screen for compounds capable of protecting PC12 cells from tunicamycin-induced endoplasmic reticulum stress. These authors identified a small molecule they named salubrinal, which prevents dephosphorylation of eIF2α, causing up-regulation of proteins (e.g., GADD34 and CHOP) whose expressions are induced by eIF2α phosphorylation (Boyce et al. 2005). Although the mechanism of action of salubrinal remains to be defined, it reduced the activity of PP1–GADD34, a complex required for dephosphorylation of eIF2α. Salubrinal, and compounds with similar modes of action, might also be useful for the treatment of protein-misfolding disorders by reducing the load on the endoplasmic reticulum.

ANTIMICROBIAL DRUGS THAT TARGET TRANSLATION

Antimicrobial agents are the best examples of small-molecule drug classes shown to inhibit translation. These include tetracyclines, spectinomycin, aminoglycosides, chloramphenicol, macrolides, clindamycin (lincosamides), streptogramins, and oxazolidinones. The antimicrobial actions, pharmacological properties, therapeutic uses, and toxicology of these compounds have been described in detail (Chambers 2001a,b). The high-resolution crystal structure of antibiotics bound to prokaryotic ribosomes, along with decades of biochemical characterization, have yielded unprecedented insight into the molecular details of how these compounds inhibit translation. These examples also provide an appreciation of the extent to which small molecule–RNA interactions can perturb molecular interactions responsible for ribosome function. Below we describe clinically used antimicrobial drugs that target the different ribosomal subunits.

Drugs That Target the 30S Ribosomal Subunit

The 30S ribosomal subunit harbors the decoding center for translation and controls translational fidelity. Drugs with demonstrated clinical benefit that target this subunit either affect A-site tRNA binding (tetracyclines, aminoglycosides) and ribosomal mobility (spectinomycin, hygromycin B) or cause miscoding (streptomycin, paromomycin).

The tetracyclines and aminoglycosides affect A-site occupancy in very different ways. Crystal structures of ribosome-bound tetracycline have been obtained with 30S *Thermus thermophilus* ribosomal subunits showing two (Brodersen et al. 2000) or six (Pioletti et al. 2001) binding sites. In both reports, the site with the highest occupancy (called Tet-1) is situated above the A-site tRNA and involves interactions between tetracycline and RNA components of the 16S rRNA (Brodersen et al. 2000; Carter et al. 2000; Pioletti et al. 2001). This is consistent with tetracycline's mode of action in preventing binding of the aa-tRNA to the A site (Maxwell 1967). These studies also allow rationalization of SAR studies demonstrating that modification of functional groups in the tetracyclines participating in rRNA binding destroys activity, whereas modifications to regions not implicated in binding have little impact (Chopra and Roberts

2001; Zhanel et al. 2004). Since the Tet-1-binding site is not conserved in eukaryotic ribosomes, structural information also provides an explanation for the selectivity of tetracyclines (Brodersen et al. 2000).

Aminoglycosides bind to the 30S ribosomal subunit and target the decoding RNA loop. Crystallization studies of streptomycin (Carter et al. 2000), paromomycin (Carter et al. 2000), and hygromycin B (Brodersen et al. 2000) bound to the 30S ribosomal subunit have provided molecular insight into the mode of action of this class of antibiotics. Streptomycin appears to preferentially stabilize the ribosome ambiguity (ram) state, allowing increased binding of noncognate aa-tRNAs to the A site and perturbing proofreading (Ruusala and Kurland 1984). On the other hand, paromomycin binds near the decoding center and causes a reorientation of two bases, A1492 and A1493, that form interactions with the A-site tRNA (Carter et al. 2000). This increases the affinity of cognate and noncognate tRNAs for the A site, resulting in increased error rates during translation. Hygromycin B also interacts with RNA bases of H44 near the decoding center and appears to restrict the movement of H44 during translocation, resulting in accumulation of tRNA in the A site (Cabanas et al. 1978; Brodersen et al. 2000; Frank and Agrawal 2000). This high-resolution structural information has enabled the design of novel ligands that target the 30S-decoding center. An example of this is the generation of novel compounds in which an RNA specificity determinant scaffold shared by the aminoglycoside, 2-deoxystreptamine, was replaced by an azepane heterocyclic scaffold (Barluenga et al. 2004). These azepane-glycosides inhibit translation, prevent growth of *S. aureus* (including aminoglycoside-resistant strains), and may not be compromised by some of the resistance mechanisms that hamper aminoglycosides (Barluenga et al. 2004).

Spectinomycin inhibits EF-G-catalyzed translocation of the peptidyl-tRNA from the A site to the P site (Bilgin et al. 1990) by binding to a minor groove at one end of H34 and making several contacts with nucleotides within this region. Inhibition of translation is thought to be a consequence of preventing H34 conformational changes required for translocation (Carter et al. 2000).

Drugs That Target the 50S Ribosomal Subunit

The 50S ribosomal subunit harbors the peptidyl transferase center (PTC) and the peptide exit tunnel, regions that are targeted by several antibiotics. Chloramphenicol has been used to treat typhoid fever, bacterial meningitis, and rickettsial disease. One significant side effect of this compound is that it can also inhibit mammalian mitochondrial ribosomes (Pestka 1977). For this reason, it is only used in the clinic when less toxic alternatives are not available. It binds to 50S ribosomal subunits (Vazquez 1966) and blocks peptidyl transferase activity (Moazed and Noller 1987). Structural studies with eubacteria *Deinococcus radiodurans* 50S ribosomal subunits are consistent with this and demonstrate that chloramphenicol interacts with only nucleotides in the PTC (Schlunzen et al. 2001). However, crystals obtained with archaea *Haloarcula marismortui* 50S ribosomal subunits reveal that chloramphenicol is located at the hydrophobic crevice at the entrance of the peptide exit tunnel, placing the binding site further away from the PTC (Hansen et al. 2003). Both sites are likely physiologically relevant, and the discrepancies between the two studies have been addressed by Hansen et al. (2003).

Macrolides are the most widely used translation inhibitors in the clinic to treat mycoplasma, legionnaires' disease, chlamydia infections, diphtheria, pertussis, tetanus, and infections of streptococcus, staphylococcus, campylobacter, helicobacter, and mycobacteria (Chambers 2001b). Three-dimensional structures are available for several macrolides bound to *H. marismortui* or *D. radiodurans* 50S ribosomes (Schlunzen et al. 2001, 2003; Hansen et al. 2002; Berisio et al. 2003a,b; Tu et al. 2005). Structural analysis revealed that these drugs prevent access to the peptide exit tunnel through which the nascent polypeptide is threaded. Six to eight peptide bonds are formed on the nascent polypeptide before it reaches the bound macrolide (Mao and Robishaw 1971). This is consistent with the fact that macrolides are not effective at inhibiting activity of already elongating ribosomes (Odom et al. 1991).

Clindamycin is a semisynthetic derivative of lincomycin used to treat streptococcal and staphylococcal infections (Chambers 2001b). Structural studies using *H. marismortui* 50S ribosomes have shown that this compound binds to the floor of the peptide exit tunnel (Tu et al. 2005), whereas in *D. radiodurans*, it is positioned somewhat higher, overlapping with both the A and P sites and interacting with nucleotides in the PTC (Schlunzen et al. 2001).

Synercid is a streptogramin combination consisting of a Streptogramin A (dalfopristin) and a Streptogramin B (quinpristin) in a 70:30 ratio. This combination is effective against gram-positive cocci and indicated for treatment of infections caused by multidrug-resistant bacteria. Dalfopristin and quinpristin act synergistically, with each component exhibiting weak bacteriostatic activity, but together being bacteriocidal. Streptogramin-A members bind within the PTC, impair A- and P-site occupancy, and cause conformational changes near the PTC (Hansen et al. 2003; Harms et al. 2004; Tu et al. 2005). Streptogramin-B compounds bind to the entrance of the peptide exit tunnel at a location similar to that of macrolides (Harms et al. 2004; Tu et al. 2005). The synergistic activity of Streptogramins A and B appears to be the consequence of interactions between the two classes of compounds and of shared interactions with a common nucleotide on the 23S rRNA (A2062Dr) (Harms et al. 2004).

Linezolid is an oxazolidinone that is active against gram-positive organisms and effective against VRE, MRSA, and penicillin-susceptible strains of *Streptococcus pneumoniae* (Chambers 2001b). Although no structural information has been published on the binding of linezolid to ribosomes, linezolid interacts with the 50S ribosomal subunit (Zhou et al. 2002). It competes with chloramphenicol for binding, does not affect peptidyl transferase activity, and inhibits P-site function during initiation and elongation (Lin et al. 1997; Swaney et al. 1998; Aoki et al. 2002; Bobkova et al. 2003). Oxazolidinones also promote frameshifting and nonsense suppression (Thompson et al. 2002).

Aminoacyl-tRNA Synthetases as Drug Targets

The emergence of multidrug-resistant bacterial strains has renewed interest in compounds known to selectively inhibit prokaryotic translation. One set of targets that have been extensively explored as potential antimicrobial targets are the aa-tRNA synthetases (Finn and Tao 2005; Chapter 29). A number of natural products have been identified that specifically target aa-tRNA synthetases (RS). These include pseudomonic acid (IleRS), borrelidin (ThrRS), furanomycin (IleRS), ochratoxin A (PheRS), granaticin (LeuRS), and indolmycin (TrpRS) (Finn and Tao 2005). Pseudomonic acid (mupirocin) is used clinically as a topical agent for skin infections and against nasal colonization by methicillin-resistant *S. aureus* (Henkel and Finlay 1999). The rational design of aa-tRNA synthetase inhibitors, based on mimics of aminoacyl adenylate intermediates, has yielded compounds with nanomolar potencies, but which are not active systemically in animal models due to high serum protein binding and poor selectivity for the bacterial enzyme (Lee et al. 2001; Finn and Tao 2005). On the other hand, high-throughput screening campaigns for inhibitors of synthetases identified compounds showing selective antibacterial activity (Gallant et al. 2000). Chemical optimization of inhibitors to *S. aureus* MetRS has recently led to the development of a series with very good selectivity and potency (Critchley et al. 2005).

CONCLUSIONS AND FUTURE PROSPECTS

At the research level, small-molecule inhibitors of translation have provided excellent tools to assign functional roles to components of the translation apparatus. They have been invaluable to help dissect the translation process, as well as evaluate the contribution of translation to a number of physiological and pathological processes. With both academic and industrial efforts aiming to identify and characterize small molecules that affect the translation process, we anticipate that novel compounds will help us understand how regulation of translation is integrated to cellular signaling events, as well as explore the potential of targeting the translation process in various diseases. Finally, an ever-increasing number of available genomic sequences from bacteria, fungi, and parasites are enabling structural and functional differences in translation factor activity and mechanistic differences among organisms in the protein synthetic process to be identified. These represent interesting starting points for drug screening and designing new therapies for currently unmet medical diseases.

ACKNOWLEDGMENTS

We sincerely apologize to those authors whose work is not cited herein due to space constraints. We are grateful to Marie-Eve Bordeleau, Dr. Francis Robert, and Dr. Regina Cencic for their helpful comments on the manuscript. Work in the laboratory of J.P. is supported by grants from the Canadian Institutes of Health Research, National Cancer Institute of Canada, and the National Institutes of Health (USA).

REFERENCES

Abraham R.T. and Wiederrecht G.J. 1996. Immunopharmacology of rapamycin. *Annu. Rev. Immunol.* **14:** 483–510.

Agrawal R.K., Linde J., Sengupta J., Nierhaus K.H., and Frank J. 2001. Localization of L11 protein on the ribosome and elucidation of its involvement in EF-G-dependent translocation. *J. Mol. Biol.* **311:** 777–787.

Ahuja D., Geiger A., Ramanjulu J.M., Vera M.D., SirDeshpande B., Pfizenmayer A., Abazeed M., Krosky D.J., Beidler D., Joullie M.M., and Toogood P.L. 2000. Inhibition of protein synthesis by didemnins: Cell potency and SAR. *J. Med. Chem.* **43:** 4212–4218.

Aktas H., Fluckiger R., Acosta J.A., Savage J.M., Palakurthi S.S., and Halperin J.A. 1998. Depletion of intracellular Ca^{2+} stores, phosphorylation of eIF2alpha, and sustained inhibition of translation initiation mediate the anticancer effects of clotrimazole. *Proc. Natl. Acad. Sci.* **95:** 8280–8285.

Aoki H., Ke L., Poppe S.M., Poel T.J., Weaver E.A., Gadwood R.C., Thomas R.C., Shinabarger D.L., and Ganoza M.C. 2002. Oxazolidinone antibiotics target the P site on *Escherichia coli* ribosomes. *Antimicrob. Agents Chemother.* **46:** 1080–1085.

Arakawa M., Shiozuka M., Nakayama Y., Hara T., Hamada M., Kondo S., Ikeda D., Takahashi Y., Sawa R., Nonomura Y., et al. 2003. Negamycin restores dystrophin expression in skeletal and cardiac muscles of mdx mice. *J. Biochem.* **134:** 751–758.

Barluenga S., Simonsen K.B., Littlefield E.S., Ayida B.K., Vourloumis D., Winters G.C., Takahashi M., Shandrick S., Zhao Q., Han Q., and Hermann T. 2004. Rational design of azepane-glycoside antibiotics targeting the bacterial ribosome. *Bioorg. Med. Chem. Lett.* **14:** 713–718.

Barton-Davis E.R., Cordier L., Shoturma D.I., Leland S.E., and Sweeney H.L. 1999. Aminoglycoside antibiotics restore dystrophin function to skeletal muscles of mdx mice. *J. Clin. Invest.* **104:** 375–381.

Bedwell D.M., Kaenjak A., Benos D.J., Bebok Z., Bubien J.K., Hong J., Tousson A., Clancy J.P., and Sorscher E.J. 1997. Suppression of a CFTR premature stop mutation in a bronchial epithelial cell line. *Nat. Med.* **3:** 1280–1284.

Benzaquen L.R., Brugnara C., Byers H.R., Gatton-Celli S., and Halperin J.A. 1995. Clotrimazole inhibits cell proliferation in vitro and in vivo. *Nat. Med.* **1:** 534–540.

Berisio R., Harms J., Schluenzen F., Zarivach R., Hansen H.A., Fucini P., and Yonath A. 2003a. Structural insight into the antibiotic action of telithromycin against resistant mutants. *J. Bacteriol.* **185:** 4276–4279.

Berisio R., Schluenzen F., Harms J., Bashan A., Auerbach T., Baram D., and Yonath A. 2003b. Structural insight into the role of the ribosomal tunnel in cellular regulation. *Nat. Struct. Biol.* **10:** 366–370.

Bilgin N., Richter A.A., Ehrenberg M., Dahlberg A.E., and Kurland C.G. 1990. Ribosomal RNA and protein mutants resistant to spectinomycin. *EMBO J.* **9:** 735–739.

Biswas P., Jiang X., Pacchia A.L., Dougherty J.P., and Peltz S.W. 2004. The human immunodeficiency virus type 1 ribosomal frameshifting site is an invariant sequence determinant and an important target for antiviral therapy. *J. Virol.* **78:** 2082–2087.

Bjornsti M.-A. and Houghton P.J. 2004. The TOR pathway: A target for cancer therapy. *Nat. Rev. Cancer* **4:** 335–348.

Bobkova E.V., Yan Y.P., Jordan D.B., Kurilla M.G., and Pompliano D.L. 2003. Catalytic properties of mutant 23 S ribosomes resistant to oxazolidinones. *J. Biol. Chem.* **278:** 9802–9807.

Bonner J.A. and Lawrence T.S. 1989. Protection of doxorubicin cytotoxicity by cycloheximide. *Int. J. Radiat. Oncol. Biol. Phys.* **16:** 1209–1212.

Bordeleau M.-E., Matthews J., Wojnar J.M., Lindqvist J.M., Novac O., Jankowsky E., Sonenberg N., Northcote P.T., Teesdale-Spittle P., and Pelletier J. 2005. Stimulation of mammalian translation initiation factor eIF4A activity a small molecule inhibitor of eukaryotic translation. *Proc. Natl. Acad. Sci.* **102:** 10460–10465.

Bordeleau M.-E., Mori A., Oberer M., Lindqvist L., Chard L.S., Higa T., Belsham G.J., Wagner G., Tanaka J., and Pelletier J. 2006. Functional characterization of internal ribosome entry sites by a novel inhibitor of the DEAD box RNA helicase, eIF4A. *Nat. Chem. Biol.* **2:** 213–220.

Boyce M., Bryant K.F., Jousse C., Long K., Harding H.P., Scheuner D., Kaufman R.J., Ma D., Coen D.M., Ron D., and Yuan J. 2005. A selective inhibitor of eIF2alpha dephosphorylation protects cells from ER stress. *Science* **307:** 935–939.

Brodersen D.E., Clemons W.M., Jr., Carter A.P., Morgan-Warren R.J., Wimberly B.T., and Ramakrishnan V. 2000. The structural basis for the action of the antibiotics tetracycline, pactamycin, and hygromycin B on the 30S ribosomal subunit. *Cell* **103:** 1143–1154.

Budihardjo I.I., Boerner S.A., Eckdahl S., Svingen P.A., Rios R., Ames M.M., and Kaufmann S.H. 2000. Effect of 6-aminonicotinamide and other protein synthesis inhibitors on formation of platinum-DNA adducts and cisplatin sensitivity. *Mol. Pharmacol.* **57:** 529–538.

Cabanas M.J., Vazquez D., and Modolell J. 1978. Dual interference of hygromycin B with ribosomal translocation and with aminoacyl-tRNA recognition. *Eur. J. Biochem.* **87:** 21–27.

Cameron D.M., Thompson J., March P.E., and Dahlberg A.E. 2002. Initiation factor IF2, thiostrepton and micrococcin prevent the binding of elongation factor G to the *Escherichia coli* ribosome. *J. Mol. Biol.* **319:** 27–35.

Carter A.P., Clemons W.M., Brodersen D.E., Morgan-Warren R.J., Wimberly B.T., and Ramakrishnan V. 2000. Functional insights from the structure of the 30S ribosomal subunit and its interactions with antibiotics. *Nature* **407:** 340–348.

Chambers H.F. 2001a. Antimicrobial agents: The aminoglycosides. In *Goodman and Gilman's the pharmacological basis of therapeutics*, 10th edition (ed. J.G. Hardman et al.), pp. 1219–1238. McGraw-Hill, New York.

———. 2001b. Antimicrobial agents: Protein synthesis inhibitors and miscellaneous antibacterial agents. In *Goodman and Gilman's the pharmacological basis of therapeutics*, 10th edition (ed. J.G. Hardman et al.), pp. 1239–1272. McGraw-Hill, New York.

Cho J. and Rando R.R. 2000. Specific binding of Hoechst 33258 to site 1 thymidylate synthase mRNA. *Nucleic Acids Res.* **28:** 2158–2163.

Choi K.H., Choi H.Y., Ko J.K., Park S.S., Kim Y.N., and Kim C.W. 2004. Transcriptional regulation of TNF family receptors and Bcl-2 family by chemotherapeutic agents in murine CT26 cells. *J. Cell. Biochem.* **91:** 410–422.

Chopra I. and Roberts M. 2001. Tetracycline antibiotics: Mode of action, applications, molecular biology, and epidemiology of bacterial resistance. *Microbiol. Mol. Biol. Rev.* **65:** 232–260.

Cox T.C., Bawden M.J., Martin A., and May B.K. 1991. Human erythroid 5-aminolevulinate synthase: Promoter analysis and identification of an iron-responsive element in the mRNA. *EMBO J.* **10:** 1891–1902.

Crabtree G.R. and Schreiber S.L. 1996. Three-part inventions: Intracellular signaling and induced proximity. *Trends Biochem. Sci.* **21:** 418–422.

Critchley I.A., Young C.L., Stone K.C., Ochsner U.A., Guiles J., Tarasow T., and Janjic N. 2005. Antibacterial activity of REP8839, a new antibiotic for topical use. *Antimicrob. Agents Chemother.* **49:** 4247–4252.

Dancey J. and Sausville E.A. 2003. Issues and progress with protein kinase inhibitors for cancer treatment. *Nat. Rev. Drug Discov.* **2:** 296–313.

Dinman J.D., Ruiz-Echevarria M.J., Czaplinski K., and Peltz S.W. 1997. Peptidyl-transferase inhibitors have antiviral properties by altering programmed −1 ribosomal frameshifting efficiencies: Development of model systems. *Proc. Natl. Acad. Sci.* **94:** 6606–6611.

Dolma S., Lessnick S.L., Hahn W.C., and Stockwell B.R. 2003. Identification of genotype-selective antitumor agents using synthetic lethal chemical screening in engineered human tumor cells. *Cancer Cell* **3:** 285–296.

Donze O., Jagus R., Koromilas A.E., Hershey J.W., and Sonenberg N. 1995. Abrogation of translation initiation factor eIF-2 phosphorylation causes malignant transformation of NIH 3T3 cells. *EMBO J.* **14:** 3828–3834.

Douros J. and Suffness M. 1981. New antitumor substances of natural origin. *Cancer Treat. Rev.* **8:** 63–87.

Du M., Jones J.R., Lanier J., Keeling K.M., Lindsey J.R., Tousson A., Bebok Z., Whitsett J.A., Dey C.R., Colledge W.H., et al. 2002. Aminoglycoside suppression of a premature stop mutation in a Cftr−/− mouse carrying a human CFTR-G542X transgene. *J. Mol. Med.* **80:** 595–604.

Faivre S., Chieze S., Delbaldo C., Ady-Vago N., Guzman C., Lopez-Lazaro L., Lozahic S., Jimeno J., Pico F., Armand J.P., et al. 2005. Phase I and pharmacokinetic study of aplidine, a new marine cyclodepsipeptide in patients with advanced malignancies. *J. Clin. Oncol.* **23:** 7871–7880.

Finn J. and Tao J. 2005. Aminoacyl-tRNA synthetases as anti-infective drug targets. In *The aminoacyl-tRNA synthetases* (ed. M. Ibba et al.), pp. 405–413. Landes Bioscience, Georgetown, Texas.

Frank J. and Agrawal R.K. 2000. A ratchet-like inter-subunit reorganization of the ribosome during translocation. *Nature* **406:** 318–322.

Furusawa S., Nakano S., Kosaka K., Takayanagi M., Takayanagi Y., and Sasaki K. 1995. Inhibition of doxorubicin-induced cell death in vitro and in vivo by cycloheximide. *Biol. Pharm. Bull.* **18:** 1367–1372.

Gallant P., Finn J., Keith D., and Wendler P. 2000. The identification of quality antibacterial drug discovery targets: A case study with aminoacyl-tRNA synthetases. *Emer. Ther. Targets* **4:** 1–9.

Goss Kinzy T., Harger J.W., Carr-Schmid A., Kwon J., Shastry M., Justice M., and Dinman J.D. 2002. New targets for antivirals: The ribosomal A-site and the factors that interact with it. *Virology* **300:** 60–70.

Gray N.K., Pantopoulos K., Dandekar T., Ackrell B.A., and Hentze M.W. 1996. Translational regulation of mammalian and *Drosophila* citric acid cycle enzymes via iron-responsive elements. *Proc. Natl. Acad. Sci.* **93:** 4925–4930.

Guba M., von Breitenbuch P., Steinbauer M., Koehl G., Flegel S., Hornung M., Bruns C.J., Zuelke C., Farkas S., Anthuber M., et al. 2002. Rapamycin inhibits primary and metastatic tumor growth by antiangiogenesis: Involvement of vascular endothelial growth factor. *Nat. Med.* **8:** 128–135.

Hansen J.L., Moore P.B., and Steitz T.A. 2003. Structures of five antibiotics bound at the peptidyl transferase center of the large ribosomal subunit. *J. Mol. Biol.* **330:** 1061–1075.

Hansen J.L., Ippolito J.A., Ban N., Nissen P., Moore P.B., and Steitz T.A. 2002. The structures of four macrolide antibiotics bound to the large ribosomal subunit. *Mol. Cell* **10:** 117–128.

Harms J.M., Schlunzen F., Fucini P., Bartels H., and Yonath A. 2004. Alterations at the peptidyl transferase centre of the ribosome induced by the synergistic action of the streptogramins dalfopristin and quinupristin. *BMC Biol.* **2:** 4.

Harvey I., Garneau P., and Pelletier J. 2002a. Forced engagement of a RNA/protein complex by a chemical inducer of dimerization to modulate gene expression. *Proc. Natl. Acad. Sci.* **99:** 1882–1887.

———. 2002b. Inhibition of translation by RNA-small molecule interactions. *RNA* **8:** 452–463.

Hay N. and Sonenberg N. 2004. Upstream and downstream of mTOR. *Genes Dev.* **18:** 1926–1945.

Helip-Wooley A., Park M.A., Lemons R.M., and Thoene J.G. 2002. Expression of CTNS alleles: Subcellular localization and aminoglycoside correction in vitro. *Mol. Genet. Metab.* **75:** 128–133.

Henkel T. and Finlay J. 1999. Emergence of resistance during mupirocin treatment: Is it a problem in clinical practice? *J. Chemother.* **11:** 331–337.

Hennessy B.T., Smith D.L., Ram P.T., Lu Y., and Mills G.B. 2005. Exploiting the PI3K/AKT pathway for cancer drug discovery. *Nat. Rev. Drug Discov.* **4:** 988–1004.

Higa T., Tanaka J., Tsukitani Y., and Kikuchi H. 1981. Hippuristanols, cytotoxic polyoxygenated steroids from the gorgonian *Isis hippuris*. *Chem. Lett.* **11:** 1647–1650.

Hood K.A., West L.M., Northcote P.T., Berridge M.V., and Miller J.H. 2001. Induction of apoptosis by the marine sponge (Mycale) metabolites, mycalamide A and pateamine. *Apoptosis* **6:** 207–219.

Hosoi H., Dilling M.B., Shikata T., Liu L.N., Shu L., Ashmun R.A., Germain G.S., Abraham R.T., and Houghton P.J. 1999. Rapamycin causes poorly reversible inhibition of mTOR and induces p53-independent apoptosis in human rhabdomyosarcoma cells. *Cancer Res.* **59:** 886–894.

Howard M., Frizzell R.A., and Bedwell D.M. 1996. Aminoglycoside antibiotics restore CFTR function by overcoming premature stop mutations. *Nat. Med.* **2:** 467–469.

Howard M.T., Shirts B.H., Petros L.M., Flanigan K.M., Gesteland R.F., and Atkins J.F. 2000. Sequence specificity of aminoglycoside-induced stop codon readthrough: Potential implications for treatment of Duchenne muscular dystrophy. *Ann. Neurol.* **48:** 164–169.

Huang S., Liu L.N., Hosoi H., Dilling M.B., Shikata T., and Houghton P.J. 2001. p53/p21(CIP1) cooperate in enforcing rapamycin-induced G(1) arrest and determine the cellular response to rapamycin. *Cancer Res.* **61:** 3373–3381.

Hung M., Patel P., Davis S., and Green S.R. 1998. Importance of ribosomal frameshifting for human immunodeficiency virus type 1 particle assembly and replication. *J. Virol.* **72:** 4819–4824.

Kantarjian H.M., Talpaz M., Santini V., Murgo A., Cheson B., and O'Brien S.M. 2001. Homoharringtonine: History, current research, and future direction. *Cancer* **92:** 1591–1605.

Keeling K.M. and Bedwell D.M. 2002. Clinically relevant aminoglycosides can suppress disease-associated premature stop mutations in the IDUA and P53 cDNAs in a mammalian translation system. *J. Mol. Med.* **80:** 367–376.

Keeling K.M., Brooks D.A., Hopwood J.J., Li P., Thompson J.N., and Bedwell D.M. 2001. Gentamicin-mediated suppression of Hurler syndrome stop mutations restores a low level of alpha-L-iduronidase activity and reduces lysosomal glycosaminoglycan accumulation. *Hum. Mol. Genet.* **10:** 291–299.

Kohler S.A., Henderson B.R., and Kuhn L.C. 1995. Succinate dehydrogenase b mRNA of *Drosophila melanogaster* has a functional iron-responsive element in its 5'-untranslated region. *J. Biol. Chem.* **270:** 30781–30786.

Koromilas A.E., Roy S., Barber G.N., Katze M.G., and Sonenberg N. 1992. Malignant transformation by a mutant of the IFN-inducible dsRNA-dependent protein kinase. *Science* **257:** 1685–1689.

Lai C.H., Chun H.H., Nahas S.A., Mitui M., Gamo K.M., Du L., and Gatti R.A. 2004. Correction of ATM gene function by aminoglycoside-induced read-through of premature termination codons. *Proc. Natl. Acad. Sci.* **101:** 15676–15681.

Lee J., Kang S.U., Kim S.Y., Kim S.E., Kang M.K., Jo Y.J., and Kim S. 2001. Ester and hydroxamate analogues of methionyl and isoleucyl adenylates as inhibitors of methionyl-tRNA and isoleucyl-tRNA synthetases. *Bioorg. Med. Chem. Lett.* **11:** 961–964.

Lentzen G., Klinck R., Matassova N., Aboul-ela F., and Murchie A.I. 2003. Structural basis for contrasting activities of ribosome binding thiazole antibiotics. *Chem. Biol.* **10:** 769–778.

Li W., Zhu T., and Guan K.L. 2004. Transformation potential of Ras isoforms correlates with activation of phosphatidylinositol 3-kinase but not ERK. *J. Biol. Chem.* **279:** 37398–37406.

Lin A.H., Murray R.W., Vidmar T.J., and Marotti K.R. 1997. The oxazolidinone eperezolid binds to the 50S ribosomal subunit and competes with binding of chloramphenicol and lincomycin. *Antimicrob. Agents Chemother.* **41:** 2127–2131.

Low W.K., Dang Y., Schneider-Poetsch T., Shi Z., Choi N.S., Merrick W.C., Romo D., and Liu J.O. 2005. Inhibition of eukaryotic translation initiation by the marine natural product pateamine A. *Mol. Cell* **20:** 709–722.

Lynch S.R. and Puglisi J.D. 2001. Structural origins of aminoglycoside specificity for prokaryotic ribosomes. *J. Mol. Biol.* **306:** 1037–1058.

Manuvakhova M., Keeling K., and Bedwell D.M. 2000. Aminoglycoside antibiotics mediate context-dependent suppression of termination codons in a mammalian translation system. *RNA* **6:** 1044–1055.

Mao J.C. and Robishaw E.E. 1971. Effects of macrolides on peptide-bond formation and translocation. *Biochemistry* **10:** 2054–2061.

Maxwell I.H. 1967. Partial removal of bound transfer RNA from polysomes engaged in protein synthesis in vitro after addition of tetracycline. *Biochim. Biophys. Acta* **138:** 337–346.

Moazed D. and Noller H.F. 1987. Chloramphenicol, erythromycin, carbomycin and vernamycin B protect overlapping sites in the peptidyl transferase region of 23S ribosomal RNA. *Biochimie* **69:** 879–884.

Morgensztern D. and McLeod H.L. 2005. PI3K/Akt/mTOR pathway as a target for cancer therapy. *Anticancer Drugs* **16:** 797–803.

Muller H.J. and Boos J. 1998. Use of L-asparaginase in childhood ALL. *Crit. Rev. Oncol. Hematol.* **28:** 97–113.

Odom O.W., Picking W.D., Tsalkova T., and Hardesty B. 1991. The synthesis of polyphenylalanine on ribosomes to which erythromycin is bound. *Eur. J. Biochem.* **198:** 713–722.

Ottenheijm H.C. and van den Broek L.A. 1988. The development of sparsomycin as an anti-tumour drug. *Anticancer Drug Des.* **2:** 333–337.

Palakurthi S.S., Aktas H., Grubissich L.M., Mortensen R.M., and Halperin J.A. 2001. Anticancer effects of thiazolidinediones are independent of peroxisome proliferator-activated receptor gamma and mediated by inhibition of translation initiation. *Cancer Res.* **61:** 6213–6218.

Palakurthi S.S., Fluckiger R., Aktas H., Changolkar A.K., Shahsafaei A., Harneit S., Kilic E., and Halperin J.A. 2000. Inhibition of translation initiation mediates the anticancer effect of the n-3 polyunsaturated fatty acid eicosapentaenoic acid. *Cancer Res.* **60:** 2919–2925.

Palmer E. and Wilhelm J.M. 1978. Mistranslation in a eucaryotic organism. *Cell* **13:** 329–334.

Pestka S. 1977. Inhibitors of protein synthesis. In *Molecular mechanisms of protein biosynthesis* (ed. H. Weissbach and S. Pestka), pp. 467–553. Academic Press, New York.

Pioletti M., Schlunzen F., Harms J., Zarivach R., Gluhmann M., Avila H., Bashan A., Bartels H., Auerbach T., Jacobi C., et al. 2001. Crystal structures of complexes of the small ribosomal subunit with tetracycline, edeine and IF3. *EMBO J.* **20:** 1829–1839.

Politano L., Nigro G., Nigro V., Piluso G., Papparella S., Paciello O., and Comi L.I. 2003. Gentamicin administration in Duchenne patients with premature stop codon. Preliminary results. *Acta Myol.* **22:** 15–21.

Pollak M.N., Schernhammer E.S., and Hankinson S.E. 2004. Insulin-like growth factors and neoplasia. *Nat. Rev. Cancer* **4:** 505–518.

Recht M.I., Douthwaite S., and Puglisi J.D. 1999. Basis for prokaryotic specificity of action of aminoglycoside antibiotics. *EMBO J.* **18:** 3133–3138.

Rinehart K.L. 2000. Antitumor compounds from tunicates. *Med. Res. Rev.* **20:** 1–27.

Rodnina M.V., Savelsbergh A., Matassova N.B., Katunin V.I., Semenkov Y.P., and Wintermeyer W. 1999. Thiostrepton inhibits the turnover but not the GTPase of elongation factor G on the ribosome. *Proc. Natl. Acad. Sci.* **96:** 9586–9590.

Rogers J.T., Randall J.D., Cahill C.M., Eder P.S., Huang X., Gunshin H., Leiter L., McPhee J., Sarang S.S., Utsuki T., et al. 2002. An iron-responsive element type II in the 5′-untranslated region of the Alzheimer's amyloid precursor protein transcript. *J. Biol. Chem.* **277:** 45518–45528.

Rouault T.A. and Harford J.B. 2000. Translational control of ferritin synthesis. In *Translational control of gene expression* (ed. N. Sonenberg et al.), pp. 655–670. Cold Spring Harbor Laboratory Press, Cold Spring Harbor, New York.

Ruusala T. and Kurland C.G. 1984. Streptomycin preferentially perturbs ribosomal proofreading. *Mol. Gen. Genet.* **198:** 100–104.

Sangkuhl K., Schulz A., Rompler H., Yun J., Wess J., and Schoneberg T. 2004. Aminoglycoside-mediated rescue of a disease-causing mutation in the V2 vasopressin receptor gene in vitro and in vivo. *Hum. Mol. Genet.* **13:** 893–903.

Schlunzen F., Harms J.M., Franceschi F., Hansen H.A., Bartels H., Zarivach R., and Yonath A. 2003. Structural basis for the antibiotic activity of ketolides and azalides. *Structure* **11:** 329–338.

Schlunzen F., Zarivach R., Harms J., Bashan A., Tocilj A., Albrecht R., Yonath A., and Franceschi F. 2001. Structural basis for the interaction of antibiotics with the peptidyl transferase centre in eubacteria. *Nature* **413:** 814–821.

Shaw K.T., Utsuki T., Rogers J., Yu Q.S., Sambamurti K., Brossi A., Ge Y.W., Lahiri D.K., and Greig N.H. 2001. Phenserine regulates translation of beta-amyloid precursor protein mRNA by a putative interleukin-1 responsive element, a target for drug development. *Proc. Natl. Acad. Sci.* **98:** 7605–7610.

Sleat D.E., Sohar I., Gin R.M., and Lobel P. 2001. Aminoglycoside-mediated suppression of nonsense mutations in late infantile neuronal ceroid lipofuscinosis. *Eur. J. Paediatr. Neurol.* (suppl A) **5:** 57–62.

Smukste I., Bhalala O., Persico M., and Stockwell B.R. 2006. Using small molecules to overcome drug resistance induced by a viral oncogene. *Cancer Cell* **9:** 133–146.

Swaney S.M., Aoki H., Ganoza M.C., and Shinabarger D.L. 1998. The oxazolidinone linezolid inhibits initiation of protein synthesis in bacteria. *Antimicrob. Agents Chemother.* **42:** 3251–3255.

Tate W.P., McCaughan K.K., Ward C.D., Sumpter V.G., Trotman C.N., Stoffler-Meilicke M., Maly P., and Brimacombe R. 1986. The ribosomal binding domain of the *Escherichia coli* release factors. Modification of tyrosine in the N-terminal domain of ribosomal protein L11 affects release factors 1 and 2 differentially. *J. Biol. Chem.* **261:** 2289–2293.

Thompson J., O'Connor M., Mills J.A., and Dahlberg A.E. 2002. The protein synthesis inhibitors, oxazolidinones and chloramphenicol, cause extensive translational inaccuracy in vivo. *J. Mol. Biol.* **322:** 273–279.

Tibodeau J.D., Fox P.M., Ropp P.A., Theil E.C., and Thorp H.H. 2006. The up-regulation of ferritin expression using a small-molecule ligand to the native mRNA. *Proc. Natl. Acad. Sci.* **103:** 253–257.

Tok J.B., Cho J., and Rando R.R. 1999. Aminoglycoside antibiotics are able to specifically bind the 5′-untranslated region of thymidylate synthase messenger RNA. *Biochemistry* **38:** 199–206.

Tu D., Blaha G., Moore P.B., and Steitz T.A. 2005. Structures of MLSBK antibiotics bound to mutated large ribosomal subunits provide a structural explanation for resistance. *Cell* **121:** 257–270.

Valeriote F., Corbett T., Edelstein M., and Baker L. 1996. New in vitro screening model for the discovery of antileukemic anticancer agents. *Cancer Invest.* **14:** 124–141.

van den Broek L.A., Lazaro E., Zylicz Z., Fennis P.J., Missler F.A., Lelieveld P., Garzotto M., Wagener D.J., Ballesta J.P., and Ottenheijm H.C. 1989. Lipophilic analogues of sparsomycin as strong inhibitors of protein synthesis and tumor growth: A structure-activity relationship study. *J. Med. Chem.* **32:** 2002–2015.

Vazquez D. 1966. Binding of chloramphenicol to ribosomes. The effect of a number of antibiotics. *Biochim. Biophys. Acta* **114:** 277–288.

———. 1979. Inhibitors of protein biosynthesis. *Mol. Biol. Biochem. Biophys.* **30:** 1–312.

Vera M.D. and Joullie M.M. 2002. Natural products as probes of cell biology: 20 years of didemnin research. *Med. Res. Rev.* **22:** 102–145.

Wagner K.R., Hamed S., Hadley D.W., Gropman A.L., Burstein A.H., Escolar D.M., Hoffman E.P., and Fischbeck K.H. 2001. Gentamicin treatment of Duchenne and Becker muscular dystrophy due to nonsense mutations. *Ann. Neurol.* **49:** 706–711.

Werstuck G. and Green M.R. 1998. Controlling gene expression in living cells through small molecule-RNA interactions. *Science* **282:** 296–298.

Wilschanski M., Yahav Y., Yaacov Y., Blau H., Bentur L., Rivlin J., Aviram M., Bdolah-Abram T., Bebok Z., Shushi L., et al. 2003. Gentamicin-induced correction of CFTR function in patients with cystic fibrosis and CFTR stop mutations. *N. Engl. J. Med.* **349:** 1433–1441.

Xing Y. and Draper D.E. 1996. Cooperative interactions of RNA and thiostrepton antibiotic with two domains of ribosomal protein L11. *Biochemistry* **35:** 1581–1588.

Zhanel G.G., Homenuik K., Nichol K., Noreddin A., Vercaigne L., Embil J., Gin A., Karlowsky J.A., and Hoban D.J. 2004. The glycylcyclines: A comparative review with the tetracyclines. *Drugs* **64:** 63–88.

Zhou C.C., Swaney S.M., Shinabarger D.L., and Stockman B.J. 2002. 1H nuclear magnetic resonance study of oxazolidinone binding to bacterial ribosomes. *Antimicrob. Agents Chemother.* **46:** 625–629.

Appendix 1. Small-molecule inhibitors of translation[a]

Phase[b] (I, E, T)	Inhibitor	Site and Mode of Inhibition[c]	Inhibition[d] in vivo	in vitro	PDB ID[e]
E	Actinobolin/Bactobolin	Targets the large ribosomal subunit and inhibits binding of aa-tRNA to the A site [1].	Prok, Euk	Prok, Euk	
E	Althiomycin	Inhibits peptidyl transferase activity [2].	Prok	Prok	
E	Amaryllidaceae alkaloids (dihydrolycorine, haemanthamine, lycorine, narciclasine, pretazzetine, pseudolycorine)	Bind to the 60S ribosomal subunit [3]. Narciclasine inhibits peptidyl transferase activity [4]. Lycorine has been reported to prevent aa-tRNA binding to the A site [5].[f]	Euk	Euk	
E	Aminoacylaminonucleosides (e.g., puromycin, gougerotin, blasticidin S, amicetin, SAN-Gly)	Binds to large ribosomal subunit and inhibits peptidyl transferase activity.	Prok, Euk	Prok, Euk	Blasticidin: 1KC8 Puromycin derivative: 1FFZ
E	Aminoglycosides	Bind to decoding center of 40S subunit and affect A-site tRNA binding, cause miscoding and reducing ribosome mobility [6, 7].	Prok, Euk[g]	Prok, Euk[g]	Streptomycin: 1FJG Paromomycin: 1FJG, 1IBK Hygromycin B: 1HNZ
E	Anisomycin	Binds to the large ribosomal subunit and inhibits peptidyl transferase activity [8] by blocking the accommodation step of translation (i.e., the active insertion of the 3' end of the aa-tRNA	Euk	Euk	1K73

	Name	Mechanism			PDB
E	Arginine-aminoglycoside conjugates	into the A site of the ribosomal peptidyl transferase center by eEF1). Also activates mitogen- and stress-activated pathways [9]. Inhibits peptidyl transferase activity [10].	—	Prok[h], Euk	
I, (E)[j]	Avilamycin,[i] evernimicin	Evernimicin interacts with the 23S of the 50S subunit [11–14] and prevents 70S initiation complex formation by interfering with IF-2 activity [15]. Evernimicin also inhibits synthesis of 50S ribosomal subunits [16].	Prok	Prok	
—	Baciphelacin	NR[j]	Prok[k], Euk	Euk[l]	
E	Bisamidine I	Inhibits translocation by eEF2 [17].	NR	Euk	
E	Bottromycin A$_2$	Binds to the 50S r-bosomal subunit and interferes with binding of aa-tRNA to the A site [18].	Prok	Prok	
E	Bouvardin, RA-VII	Binds to 80S ribosomes and inhibits EF1-dependent binding of aa-tRNA and EF2-dependent translocation [19]. It has been proposed that RA-VII, a structurally related compound, inhibits protein synthesis by preventing peptidyl transferase [20].	Euk	Euk	
I	Cap analogs (e.g., m^{7}GDP, m^{7}GTP, m^{7}GpppG, 7-benzyl-GMP)	Block cap-dependent ribosome recruitment to mRNAs by eIF4E [21, 22].	—	Euk	1WKW, 1RF8, 1L8B, 1EJ1, 1AP8
E	Cephalotaxus alkaloids (homoharringtonine, harringtonine, isoharringtonine)	Inhibit peptidyl transferase activity [23, 24].[m] These alkaloids were initially classified as inhibitors of irritiation because they caused polysome breakdown in intact cells and cell-free	Euk	Euk	

(Continued)

Appendix I. (*Continued*)

Phase[b] (I, E, T)	Inhibitor	Site and Mode of Inhibition[c]	Inhibition[d] in vivo	Inhibition[d] in vitro	PDB ID[e]
		systems. This effect appears to be due to low affinity of the harringtonine alkaloids for polysomes, allowing these to run off [23]. The available data suggest that they block translation in the early rounds of elongation, prior to polysome formation.			
E	Chloramphenicol	Binds to the PTC of *D. radiodurans* 50S ribosomes and blocks peptidyl transferase activity [25, 26].	Prok, Euk[n]	Prok	1K01, 1NJI
E	Chlorolissoclimides	Inhibit translocation [27].	Euk	Euk	
E	13-Deoxytedanolide	Bind to the 60S ribosomal subunit [28].	Euk	Euk	
E	Didemnins/Tamandarins	Bind to pretranslocative ribosome·EF1α complexes, preventing EF2 binding to 60S ribosome [29, 30]. Also binds to palmitoyl thioesterase [31].	Euk	Euk	
I, E	Edeine A	In prokaryotes, edeine inhibits binding of fMet-tRNA to 30S subunits [32]. In eukaryotes, edeine destabilizes GTP-dependent binding of Met-tRNA to ribosomes [33], resulting in 40S ribosomes that contain Met-tRNA but do not recognize AUG codons [34].	Euk [35]	Prok, Euk	1I95
E	Enacyloxin IIa	Prevents the release of EF-Tu·GDP from the ribosome [36]. The binding site overlaps with that of kirromycin [36].	Prok	Prok	

E	Fusidic acid	Stabilizes EF-G·GDP complexes on the ribosome preventing its dissociation after GTP hydrolysis [37–39].	Prok	Prok, Euk	
E	GE2270	Interferes with the interaction between EF-Tu·GTP and aa-tRNA [40].	Prok	Prok	
I	GE81112	Targets the 30S ribosomal subunit and inhibits fMet-tRNA binding [41].	Prok	Prok	
–	Gephyronic acid	NR [42]	Euk	Euk	
–	Girodazole	Interferes with release of newly synthesized peptides, but does not inhibit peptidyl transferase activity or poly(U)-dependent translation [43].	Euk	Euk	
E	Glutarimides[o] (cycloheximide, streptimidone, acetoxycycloheximide)	Binds to 60S ribosomal subunit and interferes with E-site function when it contains deacylated tRNA [44, 45].	Euk	Euk	
I	Hippuristanol	Inhibits eIF4A RNA-binding activity [46].	Euk	Euk	
E	Ipecac alkaloids (emetine, tubulosine, cephaeline)	Binds to 40S ribosomal subunits, but the mode of action is not defined.[P]	Euk	Euk	
E	Kirromycin	Prevents release of EF-Tu·GDP from the ribosome by forming a complex with EF-Tu·GDP, aa-tRNA, and the ribosome [47, 48].	Prok	Prok	
E	Lincosamides (e.g., clindamycin)	In *H. marismortui* 50S ribosomes, clindamycin is positioned at the floor of the peptide exit tunnel [49]. In *D. radiodurans* 50S ribosomes, clindamycin overlaps both the A and P sites and interacts with nucleotides at the PTC [25].	Prok	Prok	1YJN, 1JZX

(Continued)

Appendix I. (*Continued*)

Phase[b] (I, E, T)	Inhibitor	Site and Mode of Inhibition[c]	Inhibition[d] in vivo	Inhibition[d] in vitro	PDB ID[e]
E	Macrolides	Bind to the 50S ribosome and block access to the peptide exit tunnel [25, 49–53]. Macrolides also inhibit formation of the 50S ribosomal subunit [54, 55].	Prok	Prok	Azithromycin: 1NWY, 1M1K, 1YHQ Erythromycin: 1JZY, 1Y12 Clarithromycin: 1J5A Roxithromycin: 1JZZ Troleandomycin: 1P9X Telithromycin: 1P9X, 1YIJ Carbomycin: 1K8A Spiramycin: 1KD1 Tylosin: 1K9M
I	2-(4-methyl-2, 6-Dinitroanilino)-N-methylpropionamide (MDMP)	Blocks 60S subunit joining to the 40S subunit [56].	Euk	Euk	

E	Nagilactone	Inhibits EF1α-dependent aa-tRNA loading and peptidyl transferase [57].	Euk	Euk	
–	Naphthyridine-type compounds [58]	Bind to tRNA and perturb the tRNA/30S complex at the decoding site [58].	Prok	Prok	
I,E,T	Negamycin	In prokaryotes, negamycin has been documented to inhibit binding of fMet-tRNA to ribosomes [59], induce miscoding [60], and inhibit termination [61]. In eukaryotes, negamycin binds to the 18S decoding A site.[q]	Prok, Euk	Prok, Euk	
–	Nicotinamide, 6-amino- [62]	NR. Probably needs to be metabolized to 6-amino NAD– to inhibit protein synthesis [62]. Also inhibits 6-phosphogluconate dehydrogenase [63].	Euk	NR	
I	NSC119889	Inhibits ternary complex formation by targeting eIF2 [64].	Euk	Prok, Euk	
I/E[r]	Nucleic acid intercalators (chartreusin, doxorubicin, quinoline, Hoechst 33258, acriflavine, ethidium bromide)	Exert pleiotropic effects on ribosome recruitment and elongation [65].	Prok, Euk	Prok, Euk	
I, E	Oxazolidinones (e.g., linezolid)	Bind to 50S ribosome. Inhibit fMet-tRNA binding to the P site, preventing 70S formation. Inhibit translocation during the elongation phase. Also promote frameshifting and nonsense suppression [66].	Prok	Prok	
E[s]	Pactamycin	Binds to small ribosomal subunit and inhibits translocation.[s]	Prok, Euk	Prok, Euk	1HNX

(Continued)

Appendix 1. (*Continued*)

Phase[b] (I, E, T)	Inhibitor	Site and Mode of Inhibition[c]	Inhibition[d] in vivo	Inhibition[d] in vitro	PDB ID[e]
I	Pateamine	Stimulates RNA binding, ATP binding, ATP hydrolysis, and helicase activity of eIF4A [67, 68].	Euk	Euk	
E	Pederins (e.g., mycalamides, theopederin, onnamide A)	Inhibit translocation [38].[t]	Euk	Euk	
E	Phenanthroindolizidine alkaloids (e.g., tylophorine, tylocrebrine, and cryptopleurine)	Bind to 40S ribosome and inhibit translocation [69–72]. Inhibition is irreversible [73–75].	Euk	Euk	
I	Phenserine	Reduces the production of amyloid-β peptide by decreasing translation of the β-amyloid precursor protein transcript in tissue culture [76].	Euk	–	
E	Phyllanthoside	Inhibits EF1α-dependent aa-tRNA loading and EF2-mediated translocation [57].	Euk	Euk	
E	Pleuromutilin, tiamulin, valnemulin	Bind to the peptidyl transferase center and interfere with aa-tRNA binding to A and P sites [77].	Prok	Prok	1ZBP
E	Pulvomycin	Prevents ternary complex formation between EF-Tu·GTP and aa-tRNA.[u]	Prok	Prok	
E	Quassinoids (e.g., bruceantin)	Inhibit peptidyl transferase in the first round of peptide bond formation prior to polysome formation [78, 79].[v]	Euk	Euk	
E	RG501 {1,4-bis-[N-(3-N,N-dimethylpropyl)amidino] benzene tetrahydrochloride	Stimulates frameshifting by binding to and stabilizing the RNA stem-loop within the HIV frameshift signal [80].	Euk	Euk	
E	Sordarin	Binds to eEF2 preventing dissociation from the ribosome [81–83]. Inhibits the transition from the	Euk (fungus-	Euk (fungus-	

			specific)	specific)	
		pretranslocated ribosome·eEF2·GTP state to the posttranslocated ribosome·eEF2·GTP state [84].			
E	Sparsomycin	Binds above, and interacts with, the CCA end of P-site bound substrate on the large ribosomal subunit [26].	Prok, Euk	Prok, Euk	1M90
E	Spectinomycin	Binds to the 30S and prevents EF-G-catalyzed translocation [7].	Prok	Prok	1FJG
E	Streptogramin A (e.g., virginiamycin M, dalfopristin)	Streptogramin A inhibits peptide bond formation. The compounds bind within the PTC and prevent A- and P-site occupancy [26, 49, 85].	Prok	Prok	Dalfopristin: 1SM1 Virginiamycin M: 1N8R, 1YIT
E	Streptogramin B (e.g., quinpristin)	Streptogramin B compounds bind in the peptide exit tunnel at a location similar to that of macrolides.	Prok	Prok	Quinpristin: 1SM1, 1YJW
E	Streptothricins	Inhibit EF-Tu-dependent aa-tRNA binding to the ribosome and EF-G-dependent translocation [86]. Also reported to cause misreading [87].	Prok	Prok	
E	TAN1057 (GS7128)	Inhibits peptidyl transferase [88].	Prok, Euk[w]	Prok, Euk	
E	Tenuazonic acid	Inhibits peptidyl transferase [89–91].	Euk	Euk	
E	Tetracycline	Binds to the 30S ribosome and prevents access of aa-tRNA to the A site of the mRNA–ribosome complex [6, 92]. Tetracyclines inhibit mitochondrial ribosomes [93].	Prok	Prok	1HNW, 1I97

(Continued)

Appendix I. (*Continued*)

Phase[b] (I, E, T)	Inhibitor	Site and Mode of Inhibition[c]	Inhibition[d] in vivo	Inhibition[d] in vitro	PDB ID[e]
E	Thiopeptides (micrococcin, thiostrepton, nosiheptide, siomycin)	EF-G-dependent GTP hydrolysis is inhibited by thiostrepton but enhanced by micrococcin [94]. The available data suggest that these interact with both rRNA and rpL11. Thiostrepton inhibits mitochondrial ribosomes [93].	Prok	Prok	
E[x]	Trichothecenes (e.g., trichothecin, diacetoxyscirpenol, nivalenol, T-2 toxin, verrucarin A)	Bind to the 60S ribosome and inhibit peptidyl transferase. Also activate mitogen and stress [95].	Euk	Euk	
I	Triphenylmethane dyes (e.g., aurintricarboxylic acid [ATA])	In mammals, prevent mRNA recruitment to the ribosome by inhibiting $eIF2 \cdot Met\text{-}tRNA_i^{Met} \cdot GTP$ ternary complex formation [96, 97]. ATA exerts nonspecific binding to many proteins, including interfering with protein–nucleic acid interactions. Hence, data from the use of these compounds must be carefully interpreted [98].	–	Prok, Euk	
E	Tuberactinomycins (viomycin/capreomycin)	Bind to 30S and 50S ribosomal subunits. Increase the affinity of the aa-tRNA for the A site and inhibit translocation [99].	Bact[y]	Bact	

| I | Yohimbine | Prevents binding of the iron-regulatory protein to the iron-responsive element, thus increasing expression from the ferritin mRNA [100]. | NR | Euk |

[a]Summary of literature data on the properties of small-molecule translation inhibitors. Inhibitors of protein synthesis are listed alphabetically, either as single compounds or as major chemical groups when appropriate. When listed as primarily chemical families, the site of inhibition given is for the majority of compounds in that particular family. In providing citations for some of the information in the table, we focused on more recent reports characterizing the mode of action of these compounds. Inhibitors of aa-tRNA synthetases and of kinases that affect translation are not listed in this table.

[b]Denotes whether the compound affects initiation (I), elongation (E), or termination (T) of translation. NR indicates not reported.

[c]Refers to the specific partial reaction(s) affected. NR indicates not reported.

[d]Denotes whether the compounds are active in vivo and/or in vitro in prokaryotes (Prok) or eukaryotic (Euk) systems. NR indicates not reported. – means inactive. If some compounds are active on some bacterial species, but not others, they were considered active.

[e]When available, the PDB ID is provided for the structure of inhibitors and their biological targets. Structures of compounds bound to RNA aptamers are not provided.

[f]It has been suggested that lycorine may also affect termination [101], but this remains to be confirmed.

[g]Some aminoglycosides can bind to eukaryotic 40S ribosomes and cause miscoding of eukaryotic translation. See text for details.

[h]Bacterial S10 extracts are ~20-fold less sensitive than eukaryotic extracts [10].

[i]Avilamycin interacts with 23S rRNA proximal to the tRNA in the A site, and this may account for its ability to interfere with IF2 activity (which is positioned close to or in the A site), as well as to inhibit poly(U)-directed poly-phe synthesis [102]. This compound is structurally related to evernimicin and can compete for binding to 50S ribosomes with [¹⁴C]evernimicin [11].

[j]Had no activity in ribosome/mRNA binding, peptide bond formation, and polyphenylalanine synthesis assays. Hence, may affect initiation or tRNA charging [103].

[k]Was effective in vivo against S. aureus, B. subtilis, S. lutea but not against E. coli or P. aeruginosa. Hence, penetration of compound in gram-negative organisms may be an issue [104].

[l]Effective in retic lysate, but not in wheat-germ extracts [103].

[m]They have also been reported to displace prebound substrate in the A site and not to affect substrate bound to the P site [23], although this has been disputed [105].

[n]Inhibits mitochondrial ribosomes [37].

[o]The conclusions presented are for cycloheximide. This mechanism of inhibition explains why one complete elongation cycle is observed before elongation is blocked [45, 106].

[p]Emetine-resistant CHO cells display cross-resistance to cryptopleurine, tubulosine, and tylocrebrine, but not to cycloheximide, trichodermin, anisomycin, pactamycin, and sparsomycin, suggesting overlapping binding sites with the former set of compounds [107].

(Continued)

(Notes continued from previous page.)

[q]Based on binding of compound to an RNA oligonucleotide encompassing the 18S rRNA A site [108].

[r]Although generally nonspecific, some intercalators show concentration-dependent selectivity. For example, at concentrations less than 10 μM, aurintricarboxylic acid blocks initiation without affecting elongation in mammalian in vitro translation extracts [109]. At 70 μM aurintricarboxylic acid, initiation is inhibited in *E. coli* S-30 extracts, whereas elongation remains unaffected [110]. Although ethidium bromide and acriflavine inhibit initiation and elongation, at concentrations below the IC_{50} for inhibition of peptidyl transferase, they preferentially block internal initiation from the HCV IRES relative to cap-dependent initiation [65].

[s]Pactamycin has been previously reported to exert effects on initiation and elongation [37, 111, 112]. However, recent experiments have shown that pactamycin affects translocation [32]. A model consistent with the bulk of the data on pactamycin is that monosomes are formed in the presence of pactamycin, but these are not competent for elongation due to perturbation of E-site function. The presence of dipeptides in cell-free systems in the presence of pactamycin is consistent with this [113, 114]. The compound appears to bind to the 30S E-site, in the path of the mRNA, with two of the pactamycin rings mimicking the last two nucleotides of the E-site codon. This displaces the mRNA and precludes interaction with an E-site-bound tRNA [6] and may account for the poor binding of pactamycin to prokaryotic or eukaryotic ribosomes already engaged in protein synthesis [115].

[t]Pederins can compete with radiolabeled 13-deoxydihydrotedanolide (radiolabel derivative of 13-deoxytedanolide) for binding to the 60S ribosome, suggesting that pederins bind to 60S ribosomes [28].

[u]Pulvomycin increases the binding affinity of EF-Tu for GTP and decreases the affinity of EF-Tu for GDP [116].

[v]Bruceantin binds to the peptidyl transferase center on ribosomes with a $K_d = 0.34$ μM, whereas the affinity for the compound on polysomes is much weaker ($K_d = 557$ μM) [79]. Hence, translating ribosomes are resistant to inhibition by quassinoids and must release their completed polypeptide chains before inhibition is effected, explaining why a 2–4-minute lag in inhibition is observed when they are added to extracts actively engaged in synthesis [78, 117].

[w]Although not directly shown, the compound is also probably active in vivo since initial studies in mice showed toxic effects [88].

[x]Some trichothecenes were initially thought to be inhibitors of initiation since they caused polysome breakdown (verrucarin A, nivalenol, T-2 toxin, diacetoxyscirpenol), whereas others froze polysomes on the mRNA template (trichodermin, trichodermol, crotocin, trichothecin) [38]. However, they all inhibit peptidyl transferase and are classified here as inhibitors of elongation [38].

[y]Mainly effective against mycobacteria and to a lesser degree on gram-positive bacteria.

REFERENCES

1. Smithers D., Bennett L.L., Jr., and Struck F. 1969. Inhibition of protein synthesis in mammalian cells by actinobolin. *Mol. Pharmacol.* **5:** 433–445.

2. Fujimoto H., Kinoshita T., Suzuki H., and Umezawa H. 1970. Studies on the mode of action of althiomycin. *J. Antibiot.* **23:** 271–275.

3. Baez A. and Vazquez D. 1978. Binding of [3H]narciclasine to eukaryotic ribosomes. A study on a structure-activity relationship. *Biochim. Biophys. Acta* **518:** 95–103.

4. Carrasco L., Fresno M., and Vazquez D. 1975. Narciclasine: An antitumour alkaloid which blocks peptide bond formation by eukaryotic ribosomes. *FEBS Lett.* **52:** 236–239.

5. Kukhanova M., Victorova L., and Krayevsky A. 1983. Peptidyltransferase center of ribosomes. On the mechanism of action of alkaloid lycorine. *FEBS Lett.* **160:** 129–133.

6. Brodersen D.E., Clemons W.M., Jr., Carter A.P., Morgan-Warren R.J., Wimberly B.T., and Ramakrishnan V. 2000. The structural basis for the action of the antibiotics tetracycline, pactamycin, and hygromycin B on the 30S ribosomal subunit. *Cell* **103:** 1143–1154.

7. Carter A.P., Clemons W.M., Brodersen D.E., Morgan-Warren R.J., Wimberly B.T., and Ramakrishnan V. 2000. Functional insights from the structure of the 30S ribosomal subunit and its interactions with antibiotics. *Nature* **407:** 340–348.

8. Grollman A.P. 1967. Inhibitors of protein biosynthesis. II. Mode of action of anisomycin. *J. Biol. Chem.* **242:** 3226–3233.

9. Bébien M., Salinas S., Becamel C., Richard V., Linares L., and Hipskind R.A. 2003. Immediate-early gene induction by the stresses anisomycin and arsenite in human osteosarcoma cells involves MAPK cascade signalling to Elk-1, CREB and SRF. *Oncogene* **22:** 1836–1847.

10. Carriere M., Vijayabaskar V., Applefield D., Harvey I., Garneau P., Lorsch J., Lapidot A., and Pelletier J. 2002. Inhibition of protein synthesis by aminoglycoside-arginine conjugates. *RNA* **8:** 1267–1279.

11. McNicholas P.M., Najarian D.J., Mann P.A., Hesk D., Hare R.S., Shaw K.J., and Black T.A. 2000. Evernimicin binds exclusively to the 50S ribosomal subunit and inhibits translation in cell-free systems derived from both gram-positive and gram-negative bacteria. *Antimicrob. Agents Chemother.* **44:** 1121–1126.

12. Adrian P.V., Mendrick C., Loebenberg D., McNicholas P., Shaw K.J., Klugman K.P., Hare R.S., and Black T.A. 2000. Evernimicin (SCH27899) inhibits a novel ribosome target site: Analysis of 23S ribosomal DNA mutants. *Antimicrob. Agents Chemother.* **44:** 3101–3106.

13. Adrian P.V., Zhao W., Black T.A., Shaw K.J., Hare R.S., and Klugman K.P. 2000. Mutations in ribosomal protein L16 conferring reduced susceptibility to evernimicin (SCH27899): Implications for mechanism of action. Antimicrob. *Agents Chemother.* **44:** 732–738.

14. McNicholas P.M., Mann P.A., Najarian D.J., Miesel L., Hare R.S., and Black T.A. 2001. Effects of mutations in ribosomal protein L16 on susceptibility and accumulation of evernimicin. *Antimicrob. Agents Chemother.* **45:** 79–83.

15. Belova L., Tenson T., Xiong L., McNicholas P.M., and Mankin A.S. 2001. A novel site of antibiotic action in the ribosome: Interaction of evernimicin with the large ribosomal subunit. *Proc. Natl. Acad. Sci.* **98:** 3726–3731.

16. Champney W.S. and Tober C.L. 2000. Evernimicin (SCH27899) inhibits both translation and 50S ribosomal subunit formation in *Staphylococcus aureus* cells. *Antimicrob. Agents Chemother.* **44:** 1413–1417.

17. Gajko-Galicka A., Bielawski K., Sredzinska K., Bielawska A., and Gindzienski A. 2002. Elongation factor 2 as a target for selective inhibition of protein synthesis in vitro by the novel aromatic bisamidine. *Mol. Cell. Biochem.* **233:** 159–164.

18. Otaka T. and Kaji A. 1983. Mode of action of bottromycin A2: Effect of bottromycin A2 on polysomes. *FEBS Lett.* **153:** 53–59.

19. Zalacain M., Zaera E., Vázquez D., and Jiménez A. 1982. The mode of action of the antitumor drug bouvardin, an inhibitor of protein synthesis in eukaryotic cells. *FEBS Lett.* **148:** 95–97.

20. Sirdeshpande B. and Toogood P.L. 1995. Inhibition of protein synthesis by RA–VII. *Bioorg. Chem.* **23:** 460–470.

21. Sonenberg N. and Shatkin A.J. 1977. Reovirus mRNA can be covalently crosslinked via the 5′ cap to proteins in initiation complexes. *Proc. Natl. Acad. Sci.* **74:** 4288–4292.

22. Marcotrigiano J., Gingras A.C., Sonenberg N., and Burley S.K. 1997. Cocrystal structure of the messenger RNA 5′ cap-binding protein (eIF4E) bound to 7-methyl-GDP. *Cell* **89:** 951–961.

23. Fresno M., Jimenez A., and Vazquez D. 1977. Inhibition of translation in eukaryotic systems by harringtonine. *Eur. J. Biochem.* **72:** 323–330.

24. Tujebajeva R.M., Graifer D.M., Karpova G.G., and Ajtkhozhina N.A. 1989. Alkaloid homoharringtonine inhibits polypeptide chain elongation on human ribosomes on the step of peptide bond formation. *FEBS Lett.* **257:** 254–256.

25. Schlunzen F., Zarivach R., Harms J., Bashan A., Tocilj A., Albrecht R., Yonath A., and Franceschi F. 2001. Structural basis for the interaction of antibiotics with the peptidyl transferase centre in eubacteria. *Nature* **413:** 814–821.

26. Hansen J.L., Moore P.B., and Steitz T.A. 2003. Structures of five antibiotics bound at the peptidyl transferase center of the large ribosomal subunit. *J. Mol. Biol.* **330:** 1061–1075.

27. Robert F., Gao H.Q., Marwa D., Merrick W.C., Hamann M.T., and Pelletier J. 2006. Chlorolissoclimides—New inhibitors of eukaryotic protein synthesis. *RNA* **12:** 717–725.

28. Nishimura S., Matsunaga S., Yoshida M., Hirota H., Yokoyama S., and Fusetani N. 2005. 13-Deoxytedanolide, a marine sponge-derived antitumor macrolide, binds to the 60S large ribosomal subunit. *Bioorg. Med. Chem.* **13:** 449–454.

29. Crews C.M., Collins J.L., Lane W.S., Snapper M.L., and Schreiber S.L. 1994. GTP-dependent binding of the antiproliferative agent didemnin to elongation factor 1 alpha. *J. Biol. Chem.* **269:** 15411–15414.

30. Ahuja D., Vera M.D., SirDeshpande B.V., Morimoto H., Williams P.G., Joullie M.M., and Toogood P.L. 2000. Inhibition of protein synthesis by didemnin B: How EF-1alpha mediates inhibition of translocation. *Biochemistry* **39:** 4339–4346.

31. Crews C.M., Lane W.S., and Schreiber S.L. 1996. Didemnin binds to the protein palmitoyl thioesterase responsible for infantile neuronal ceroid lipofuscinosis. *Proc. Natl. Acad. Sci.* **93:** 4316–4319.

32. Dinos G., Wilson D.N., Teraoka Y., Szaflarski W., Fucini P., Kalpaxis D., and Nierhaus K.H. 2004. Dissecting the ribosomal inhibition mechanisms of edeine and pactamycin: The universally conserved residues G693 and C795 regulate P-site RNA binding. *Mol. Cell* **13:** 113–124.

33. Odon O.W., Kramer G., Henderson A.B., Pinphanichakarn P., and Hardesty B. 1978. GTP hydrolysis during methionyl-tRNAf binding to 40 S ribosomal subunits and the site of edeine inhibition. *J. Biol. Chem.* **253:** 1807–1816.

34. Kozak M. and Shatkin A.J. 1978. Migration of 40 S ribosomal subunits on messenger RNA in the presence of edeine. *J. Biol. Chem.* **253:** 6568–6577.

35. Kurylo-Borowska Z. and Heaney-Kieras J. 1979. The uptake and subcellular localization of the peptide antibiotic edeine A in HeLa cells in suspension culture. *Exp. Cell Res.* **124:** 371–379.

36. Parmeggiani A., Krab I.M., Watanabe T., Nielsen R.C., Dahlberg C., Nyborg J., and Nissen P. 2005. Enacyloxin IIa pinpoints a binding pocket of elongation factor Tu for development of novel antibiotics. *J. Biol. Chem.* **281:** 2893–2900.

37. Pestka S. 1977. Inhibitors of protein synthesis. In *Molecular mechanisms of protein biosynthesis* (ed. H. Weissbach and S. Pestka), pp. 467–553. Academic Press, New York.

38. Vazquez D. 1979. Inhibitors of protein biosynthesis. *Mol. Biol. Biochem. Biophys.* **30:** 1–312.

39. Agrawal R.K., Penczek P., Grassucci R.A., and Frank J. 1998. Visualization of elongation factor G on the *Escherichia coli* 70S ribosome: The mechanism of translocation. *Proc. Natl. Acad. Sci.* **95:** 6134–6138.

40. Heffron S.E. and Jurnak F. 2000. Structure of an EF-Tu complex with a thiazolyl peptide antibiotic determined at 2.35 Å resolution: Atomic basis for GE2270A inhibition of EF-Tu. *Biochemistry.* **39:** 37–45.

41. Brandi L., Fabbretti A., La Teana A., Abbondi M., Losi D., Donadio S., and Gualerzi C.O. 2005. Specific, efficient, and selective inhibition of prokaryotic translation initiation by a novel peptide antibiotic. *Proc. Natl. Acad. Sci.* **103:** 39–44.

42. Sasse F., Steinmetz H., Hofle G., and Reichenbach H. 1995. Gephyronic acid, a novel inhibitor of eukaryotic protein synthesis from *Archangium gephyra* (myxobacteria). Production, isolation, physico-chemical and biological properties, and mechanism of action. *J. Antibiot.* **48:** 21–25.

43. Colson G., Rabault B., Lavelle F., and Zerial A. 1992. Mode of action of the antitumor compound girodazole (RP 49532A, NSC 627434). *Biochem. Pharmacol.* **43:** 1717–1723.

44. Obrig T.G., Culp W.J., McKeehan W.L., and Hardesty B. 1971. The mechanism by which cycloheximide and related glutarimide antibiotics inhibit peptide synthesis on reticulocyte ribosomes. *J. Biol. Chem.* **246:** 174–181.

45. Pestova T.V. and Hellen C.U. 2003. Translation elongation after assembly of ribosomes on the Cricket paralysis virus internal ribosomal entry site without initiation factors or initiator tRNA. *Genes Dev.* **17:** 181–186.

46. Bordeleau M.-E., Mori A., Oberer M., Lindqvist L., Chard L.S., Higa T., Belsham G.J., Wagner G., Tanaka J., and Pelletier J. 2006. Functional characterization of internal ribosome entry sites by a novel inhibitor of the DEAD box RNA helicase, eIF4A. *Nat. Chem. Biol.* **2:** 213–220.

47. Kristensen O., Reshetnikova L., Nissen P., Siboska G., Thirup S., and Nyborg J. 1996. Isolation, crystallization and X-ray analysis of the quaternary complex of Phe-tRNA(Phe), EF-Tu, a GTP analog and kirromycin. *FEBS Lett.* **399:** 59–62.

48. Vogeley L., Palm G.J., Mesters J.R., and Hilgenfeld R. 2001. Conformational change of elongation factor Tu (EF-Tu) induced by antibiotic binding. Crystal structure of the complex between EF-Tu.GDP and aurodox. *J. Biol. Chem.* **276:** 17149–17155.

49. Tu D., Blaha G., Moore P.B., and Steitz T.A. 2005. Structures of MLSBK antibiotics bound to mutated large ribosomal subunits provide a structural explanation for resistance. *Cell* **121:** 257–270.

50. Hansen J.L., Ippolito J.A., Ban N., Nissen P., Moore P.B., and Steitz T.A. 2002. The structures of four macrolide antibiotics bound to the large ribosomal subunit. *Mol. Cell* **10:** 117–128.

51. Schlunzen F., Harms J.M., Franceschi F., Hansen H.A., Bartels H., Zarivach R., and Yonath A. 2003. Structural basis for the antibiotic activity of ketolides and azalides. *Structure* **11:** 329–338.

52. Berisio R., Schluenzen F., Harms J., Bashan A., Auerbach T., Baram D., and Yonath A. 2003. Structural insight into the role of the ribosomal tunnel in cellular regulation. *Nat. Struct. Biol.* **10:** 366–370.

53. Berisio R., Harms J., Schluenzen F., Zarivach R., Hansen H.A., Fucini P., and Yonath A. 2003. Structural insight into the antibiotic action of telithromycin against resistant mutants. *J. Bacteriol.* **185:** 4276–4279.

54. Champney W.S. and Tober C.L. 1998. Inhibition of translation and 50S ribosomal subunit formation in *Staphylococcus aureus* cells by 11 different ketolide antibiotics. *Curr. Microbiol.* **37:** 418–425.

55. Champney W.S., Tober C.L., and Burdine R. 1998. A comparison of the inhibition of translation and 50S ribosomal subunit formation in *Staphylococcus aureus* cells by nine different macrolide antibiotics. *Curr. Microbiol.* **37:** 412–417.

56. Weeks D.P. and Baxter R. 1972. Specific inhibition of peptide-chain initiation by 2-(4-methyl-2,6dinitro anilino)-N-methyl propionamide. *Biochemistry* **11:** 3060–3064.

57. Chan J., Khan S.N., Harvey I., Merrick W., and Pelletier J. 2004. Eukaryotic protein synthesis inhibitors identified by comparison of cytotoxicity profiles. *RNA* **10:** 528–543.

58. Shen L.L., Black-Schaefer C., Cai Y., Dandliker P.J., and Beutel B.A. 2005. Mechanism of action of a novel series of naphthyridine-type ribosome inhibitors: Enhancement of tRNA footprinting at the decoding site of 16S rRNA. *Antimicrob. Agents Chemother.* **49:** 1890–1897.

59. Mizuno S., Nitta K., and Umezawa H. 1970. Mechanism of action of negamycin in *Escherichia coli* K12. I. Inhibition of initiation of protein synthesis. *J. Antibiot.* **23:** 581–588.

60. Mizuno S., Nitta K., and Umezawa H. 1970. Mechanism of action of negamycin in *Escherichia coli* K12. II. Miscoding activity in polypeptide synthesis directed by synthetic polynucleotide. *J. Antibiot.* **23:** 589–594.

61. Uehara Y., Hori M., and Umezawa H. 1974. Negamycin inhibits termination of protein synthesis directed by phage f2 RNA in vitro. *Biochim. Biophys. Acta* **374:** 82–95.

62. Budihardjo I.I., Boerner S.A., Eckdahl S., Svingen P.A., Rios R., Ames M.M., and Kaufmann S.H. 2000. Effect of 6-aminonicotinamide and other protein synthesis inhibitors on formation of platinum-DNA adducts and cisplatin sensitivity. *Mol. Pharmacol.* **57:** 529–538.

63. Herken H. and Lange K. 1969. Blocking of pentose phosphate pathway in the brain of rats by 6-aminonicotinamide. *Naunyn-Schmiedebergs Arch. Exp. Pathol. Pharmakol.* **263:** 496–499.

64. Robert F., Kapp L.D., Khan S.N., Acker M., Kazemi S., Kaufman R.J., Merrick W.C., Koromilas A.E., Lorsch J.R., and Pelletier J. 2006. Translation initiation by the HCV IRES is refractory to reduced eIF2·GTP·Met-tRNA$_i^{met}$ ternary complex availability. (in prep.)

65. Malina A., Khan S., Carlson C.B., Svitkin Y., Harvey I., Sonenberg N., Beal P.A., and Pelletier J. 2005. Inhibitory properties of nucleic acid-binding ligands on protein synthesis. *FEBS Lett.* **579:** 79–89.

66. Thompson J., O'Connor M., Mills J.A., and Dahlberg A.E. 2002. The protein synthesis inhibitors, oxazolidinones and chloramphenicol, cause extensive translational inaccuracy in vivo. *J. Mol. Biol.* **322:** 273–279.

67. Bordeleau M.-E., Matthews J., Wojnar J.M., Lindgvist J.M., Novac O., Jankowsky E., Sonenberg N., Northcote P.T., Teesdale-Spittle P., and Pelletier J. 2005. Stimulation of mammlian translation initiation factor eIF4A activity a small molecule inihibitor of eukaryotic translation. *Proc. Natl. Acad. Sci.* **102:** 10460–10465.

68. Low W.K., Dang Y., Schneider-Poetsch T., Shi Z., Choi N.S., Merrick W.C., Romo D., and Liu J.O. 2005. Inhibition of eukaryotic translation initiation by the marine natural product pateamine A. *Mol. Cell* **20:** 709–722.

69. Sanchez L., Vasquez D., and Jimenez A. 1977. Genetics and biochemistry of cryptopleurine resistance in the yeast *Saccharomyces cerevisiae. Mol. Gen. Genet.* **156:** 319–326.

70. Paulovich A.G., Thompson J.R., Larkin J.C., Li Z., and Woolford J.L., Jr. 1993. Molecular genetics of cryptopleurine resistance in *Saccharomyces cerevisiae:* Expression of a ribosomal protein gene family. *Genetics* **135:** 719–730.

71. Carrasco L., Jimenez A., and Vazquez D. 1976. Specific inhibition of translocation by tubulosine in eukaryotic polysomes. *Eur. J. Biochem.* **64:** 1–5.

72. Bucher K. and Skogerson L. 1976. Cryptopleurine—An inhibitor of translocation. *Biochemistry* **15:** 4755–4759.

73. Donaldson G.R., Atkinson M.R., and Murray A.W. 1968. Inhibition of protein synthesis in Ehrlich ascites-tumour cells by the phenanthrene alkaloids tylophorine, tylocrebrine and cryptopleurine. *Biochem. Biophys. Res. Commun.* **31:** 104–109.

74. Huang M.T. and Grollman A.P. 1972. Mode of action of tylocrebrine: Effects on protein and nucleic acid synthesis. *Mol. Pharmacol.* **8:** 538–550.

75. Haslam J.M., Davey P.J., Linnane A.W., and Atkinson M.R. 1968. Differentiation in vitro by phenanthrene alkaloids of yeast mitochondrial protein synthesis from ribosomal systems of both yeast and bacteria. *Biochem. Biophys. Res. Commun.* **33:** 368–373.

76. Shaw K.T., Utsuki T., Rogers J., Yu Q.S., Sambamurti K., Brossi A., Ge Y.W., Lahiri D.K., and Greig N.H. 2001. Phenserine regulates translation of beta-amyloid precursor protein mRNA by a putative interleukin-1 responsive element, a target for drug development. *Proc. Natl. Acad. Sci.* **98:** 7605–7610.

77. Schlunzen F., Pyetan E., Fucini P., Yonath A., and Harms J.M. 2004. Inhibition of peptide bond formation by pleuromutilins: The structure of the 50S ribosomal subunit from *Deinococcus radiodurans* in complex with tiamulin. *Mol. Microbiol.* **54:** 1287–1294.

78. Liao L.-L., Kupchan S.M., and Horwitz S.B. 1976. Mode of action of the antitumor compound bruceantin, an inhibitor of protein synthesis. *Mol. Pharmacol.* **12:** 167–176.

79. Fresno M., Gonzales A., Vazquez D., and Jimenez A. 1978. Bruceantin, a novel inhibitor of peptide bond formation. *Biochim. Biophys. Acta* **518:** 104–112.

80. Hung M., Patel P., Davis S., and Green S.R. 1998. Importance of ribosomal frameshifting for human immunodeficiency virus type 1 particle assembly and replication. *J. Virol.* **72:** 4819–4824.

81. Justice M.C., Hsu M.J., Tse B., Ku T., Balkovec J., Schmatz D., and Nielsen J. 1998. Elongation factor 2 as a novel target for selective inhibition of fungal protein synthesis. *J. Biol. Chem.* **273:** 3148–3151.

82. Dominguez J.M. and Martin J.J. 1998. Identification of elongation factor 2 as the essential protein targeted by sordarins in *Candida albicans. Antimicrob. Agents Chemother.* **42:** 2279–2283.

83. Spahn C.M., Gomez-Lorenzo M.G., Grassucci R.A., Jorgensen R., Andersen G.R., Beckmann R., Penczek P.A., Ballesta J.P., and Frank J. 2004. Domain movements of elongation factor eEF2 and the eukaryotic 80S ribosome facilitate tRNA translocation. *EMBO J.* **23:** 1008–1019.

84. Dominguez J.M., Gomez-Lorenzo M.G., and Martin J.J. 1999. Sordarin inhibits fungal protein synthesis by blocking translocation differently to fusidic acid. *J. Biol. Chem.* **274:** 22423–22427.

85. Harms J.M., Schlunzen F., Fucini P., Bartels H., and Yonath A. 2004. Alterations at the peptidyl transferase centre of the ribosome induced by the synergistic action of the streptogramins dalfopristin and quinupristin. *BMC Biol.* **2:** 4.

86. Haupt I., Jonak J., Rychlik I., and Thrum H. 1980. Action of streptothricin F on ribosomal functions. *J. Antibiot.* **33:** 636–641.

87. Haupt I., Hubener R., and Thrum H. 1978. Streptothricin F, an inhibitor of protein synthesis with miscoding activity. *J. Antibiot.* **31:** 1137–1142.

88. Boddeker N., Bahador G., Gibbs C., Mabery E., Wolf J., Xu L., and Watson J. 2002. Characterization of a novel antibacterial agent that inhibits bacterial translation. *RNA* **8:** 1120–1128.

89. Carrasco L. and Vazquez D. 1972. Survey of inhibitors in different steps of protein synthesis by mammalian ribosomes. *J. Antibiot.* **25:** 732–737.

90. Carrasco L. and Vazquez D. 1973. Differences in eukaryotic ribosomes detected by the selective action of an antibiotic. *Biochim. Biophys. Acta* **319:** 209–215.

91. Neth R., Monro R.E., Heller C., Battaner E., and Vazquez D. 1970. Catalysis of peptidyl transfer by human ribosomes and effects of some antibiotics. *FEBS Lett.* **6:** 198–202.

92. Pioletti M., Schlunzen F., Harms J., Zarivach R., Gluhmann M., Avila H., Bashan A., Bartels H., Auerbach T., Jacobi C., et al. 2001. Crystal structures of complexes of the small ribosomal subunit with tetracycline, edeine and IF3. *EMBO J.* **20:** 1829–1839.

93. Zhang L., Ging N.C., Komoda T., Hanada T., Suzuki T., and Watanabe K. 2005. Antibiotic susceptibility of mammalian mitochondrial translation. *FEBS Lett.* **579:** 6423–6427.

94. Cundliffe E. and Thompson J. 1981. Concerning the mode of action of micrococcin upon bacterial protein synthesis. *Eur. J. Biochem.* **118:** 47–52.

95. Shifrin V.I. and Anderson P. 1999. Trichothecene mycotoxins trigger a ribotoxic stress response that activates c-Jun N-terminal kinase and p38 mitogen-activated protein kinase and induces apoptosis. *J. Biol. Chem.* **274:** 13985–13992.

96. Dettman G.L. and Stanley W.M., Jr. 1973. The ternary complex of initiation factor IF-I, MET-tRNA Met f and GTTP. An aurintricarboxylate-sensitive intermediate in the initiation of eukaryotic protein synthesis. *Biochim. Biophys. Acta* **299:** 142–147.

97. Fresno M., Carrasco L., and Vazquez D. 1976. Initiation of the polypeptide chain by reticulocyte cell-free systems. Survey of different inhibitors of translation. *Eur. J. Biochem.* **68:** 355–364.

98. Gonzalez R.G., Haxo R.S., and Schleich T. 1980. Mechanism of action of polymeric aurintricarboxylic acid, a potent inhibitor of protein–nucleic acid interactions. *Biochemistry* **19:** 4299–4303.

99. Peske F., Savelsbergh A., Katunin V.I., Rodnina M.V., and Wintermeyer W. 2004. Conformational changes of the small ribosomal subunit during elongation factor G-dependent tRNA-mRNA translocation. *J. Mol. Biol.* **343:** 1183–1194.

100. Tibodeau J.D., Fox P.M., Ropp P.A., Theil E.C., and Thorp H.H. 2006. The up-regulation of ferritin expression using a small-molecule ligand to the native mRNA. *Proc. Natl. Acad. Sci.* **103:** 253–257.

101. Vrijsen R., Vanden Berghe D.A., Vlietinck A.J., and Boeye A. 1986. Lycorine: A eukaryotic termination inhibitor? *J. Biol. Chem.* **261:** 505–507.

102. Wolf H. 1973. Avilamycin, an inhibitor of the 30 S ribosomal subunits function. *FEBS Lett.* **36:** 181–186.

103. Carrasco L. 1987. Baciphelacin: A new eukaryotic translation inhibitor. *Biochimie* **69:** 797–802.

104. Okazaki H., Kishi T., Beppu T., and Arima K. 1975. A new antibiotic, baciphelacin. *J. Antibiot.* **28:** 717–719.

105. Tujebajeva R.M., Graifer D.M., Matasova N.B., Fedorova O.S., Odintsov V.B., Ajtkhozhina N.A., and Karpova G.G. 1992. Selective inhibition of the polypeptide chain elongation in eukaryotic cells. *Biochim. Biophys. Acta* **1129:** 177–182.

106. Dmitriev S.E., Pisarev A.V., Rubtsova M.P., Dunaevsky Y.E., and Shatsky I.N. 2003. Conversion of 48S translation preinitiation complexes into 80S initiation complexes as revealed by toeprinting. *FEBS Lett.* **533:** 99–104.

107. Gupta R.S. and Siminovitch L. 1977. Mutants of CHO cells resistant to the protein synthesis inhibitors, cryptopleurine and tylocrebrine: Genetic and biochemical evidence for common site of action of emetine, cryptopleurine, tylocrebine, and tubulosine. *Biochemistry* **16:** 3209–3214.

108. Arakawa M., Shiozuka M., Nakayama Y., Hara T., Hamada M., Kondo S., Ikeda D., Takahashi Y., Sawa R., Nonomura Y., et al. 2003. Negamycin restores dystrophin expression in skeletal and cardiac muscles of mdx mice. *J. Biochem.* **134:** 751–758.

109. Mathews M.B. 1971. Mammalian chain initiation: The effect of aurintricarboxylic acid. *FEBS Lett.* **15:** 201–204.

110. Siegelman F. and Apirion D. 1971. Aurintricarboxylic acid, a preferential inhibitor of initiation of protein synthesis. *J. Bacteriol.* **105:** 902–907.

111. Tai P.C., Wallace B.J., and Davis B.D. 1973. Actions of aurintricarboxylate, kasugamycin, and pactamycin on *Escherichia coli* polysomes. *Biochemistry* **12:** 616–620.

112. Vazquez D. 1966. Binding of chloramphenicol to ribosomes. The effect of a number of antibiotics. *Biochim. Biophys. Acta* **114:** 277–288.

113. Kappen L.S. and Goldberg I.H. 1973. Inhibition of globin chain initiation in reticulocyte lysates by pactamycin: Accumulation of methionyl-valine. *Biochem. Biophys. Res. Commun.* **54:** 1083–1091.

114. Cheung C.P., Stewart M.L., and Gupta N.K. 1973. Protein synthesis in rabbit reticulocytes: Evidence for the synthesis of initial dipeptides in the presence of pactamycin. *Biochem. Biophys. Res. Commun.* **54:** 1092–1101.

115. Macdonald J.S. and Goldberg I.H. 1970. An effect of pactamycin on the initiation of protein synthesis in reticulocytes. *Biochem. Biophys. Res. Commun.* **41:** 1–8.

116. Anborgh P.H., Okamura S., and Parmeggiani A. 2004. Effects of the antibiotic pulvomycin on the elongation factor Tu-dependent reactions. Comparison with other antibiotics. *Biochemistry* **43:** 15550–15556.

117. Willingham W., Jr., Stafford E.A., Reynolds S.H., Chaney S.G., Lee K.-H., Okano M., and Hall I.H. 1981. Mechanism of eukaryotic protein synthesis inhibition by brusatol. *Biochem. Biophys. Res. Commun.* **654:** 169–174.

Index

G

Q